aplia™

STUDENTS: Accessing Your Aplia Course Through CengageBrain

CENGAGEbrain.com

Creating Your Aplia Account On CengageBrain By Visiting: login.cengagebrain.com

What is login.cengagebrain.com? Imagine that you are using Aplia along with another online study tool from Cengage Learning. Rather than having separate logins and passwords for both applications, Cengage offers a single access point through login.cengagebrain.com.

When you visit login.cengagebrain.com, you will see the following:

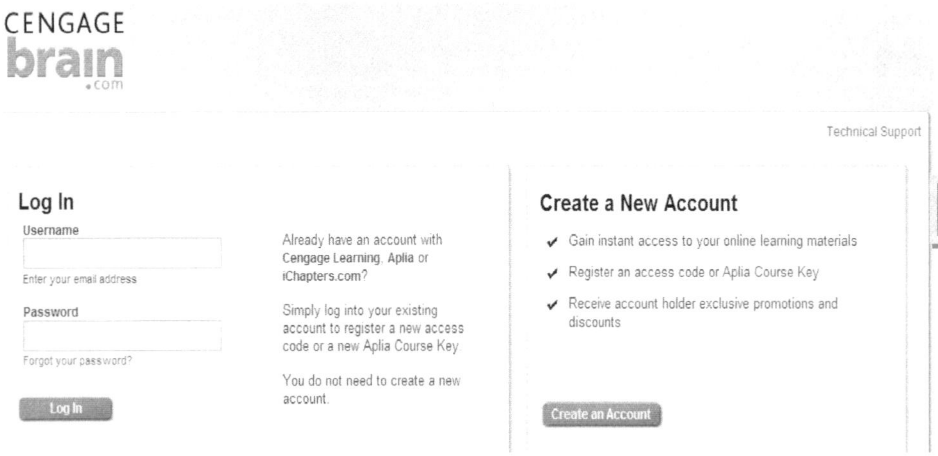

A Single Sign-On (SSO) account provides for one login to access Cengage Learning digital resources.

For the purpose of this document, we will focus on the right side of this screen or "**Create a New Account**".

Note: If you *do* have an existing account, CengageBrain accepts all prior login credentials, including Aplia and iChapters.

Step 1 | Begin by Creating an Account

1. Click the "Create an Account" button
2. The following page will load. Enter in the Aplia Course Key

YOUR COURSE KEY IS PROVIDED BY YOUR INSTRUCTOR TO ACCESS APLIA.

Instructors often provide the Aplia Course Key on their **syllabus**, or the course web site. If you need help locating the **Aplia Course Key**, ask your instructor.

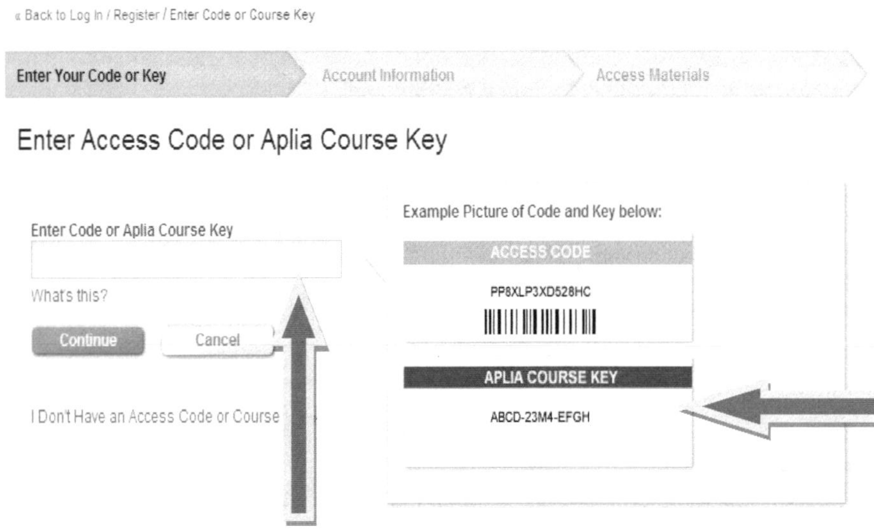

Your Aplia Course Key will look like this.

Step 2 | Confirm Your Course Information

The graphic below serves as an illustration for what you will see. Please confirm the displayed screen represents your enrolled course.

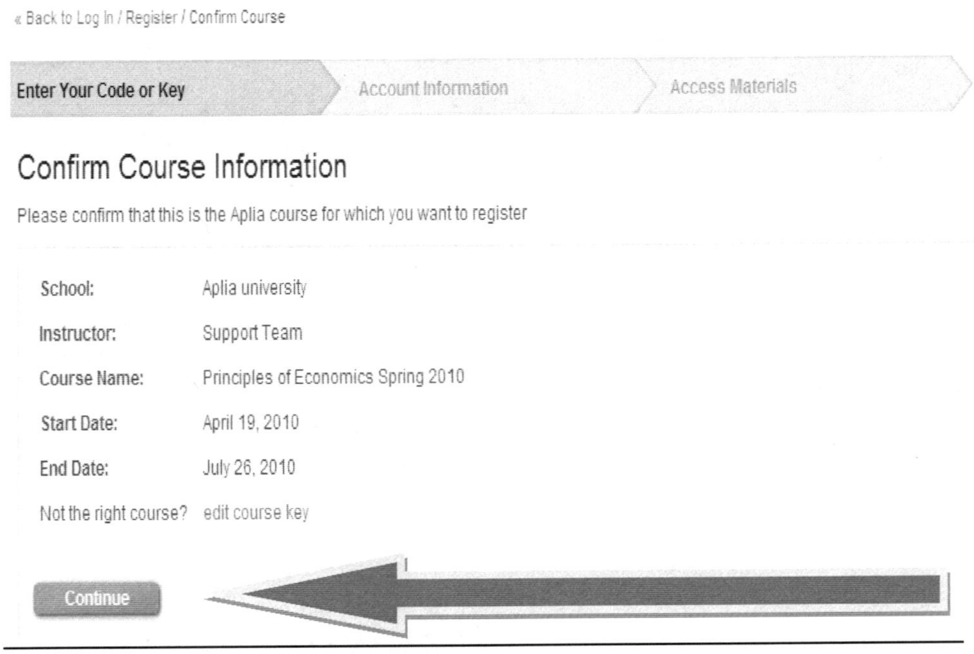

Confirm that the course information shown matches the course you are enrolled in. If yes, click the "**Continue**" button. If not, click on the "edit course key" link to re-enter the correct Course Key.

Step 3 | Complete Registration Form

As a new user, you must complete all fields in the registration form and click "Continue"

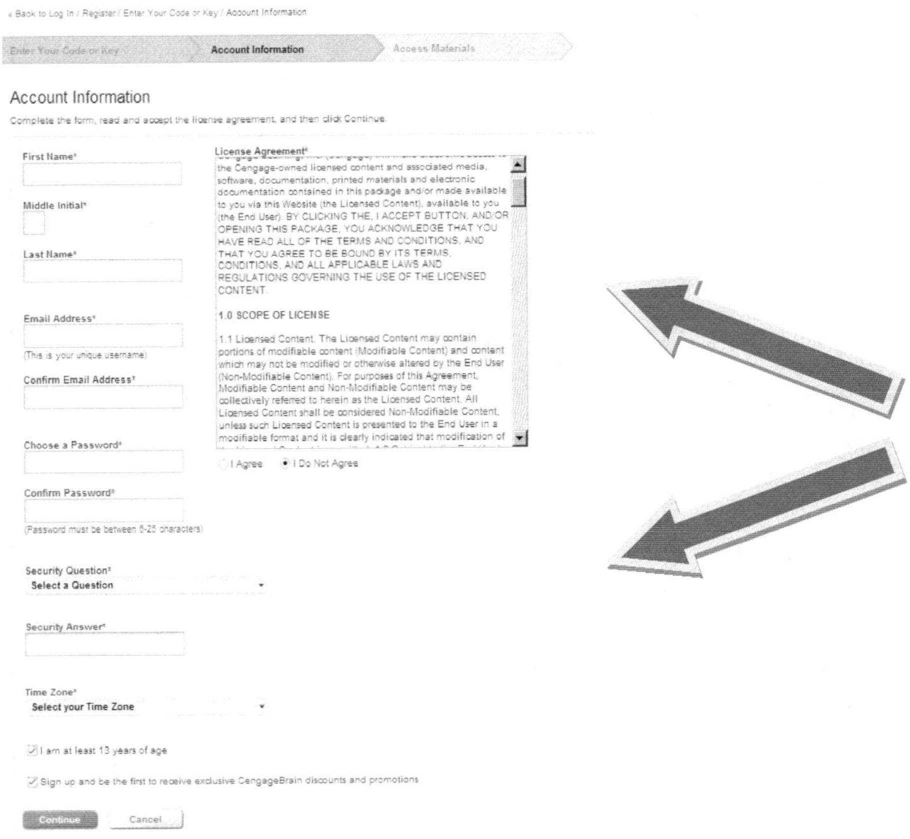

Please complete all fields to establish your account and select a security question. Select "Continue" to proceed with your Aplia registration.

Step 4 | My Home Dashboard Review

You have arrived at your "**My Home**" page. Here you can verify your e-Mail address with CengageBrain and pay for your course **(See Step 5).** Note that any prior (and future) Cengage Learning purchases will appear on this page. Additionally, the page provides support resources and a transaction history of your Cengage purchases – digital products, textbooks or rentals.

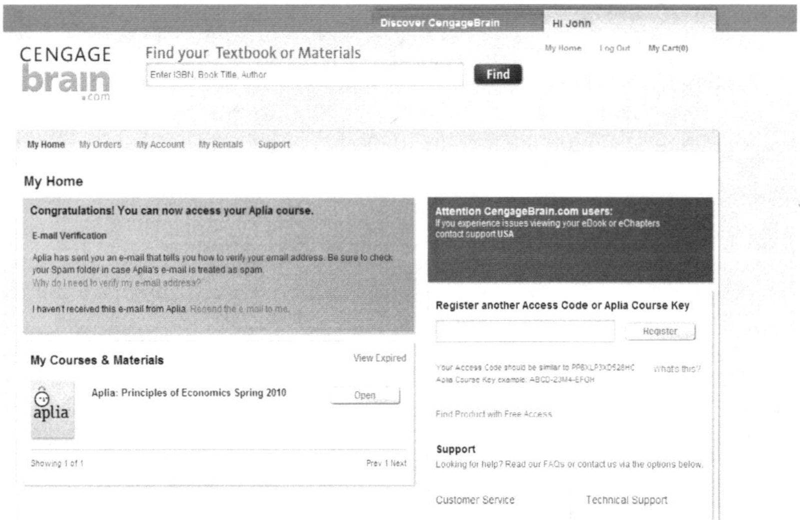

IMPORTANT: CengageBrain requires you to verify your e-mail address. Please go to your e-mail and check for a message from CengageBrain.com with the Subject "**Aplia Verification E-Mail**". If you don't see the message, check your SPAM folder.

Click the link to verify your e-mail. **Once verified, you will be prompted to log back in again with your e-mail and password from Step 3 above. Once you do, you will be brought back to your "My Home" Dashboard as shown below.**

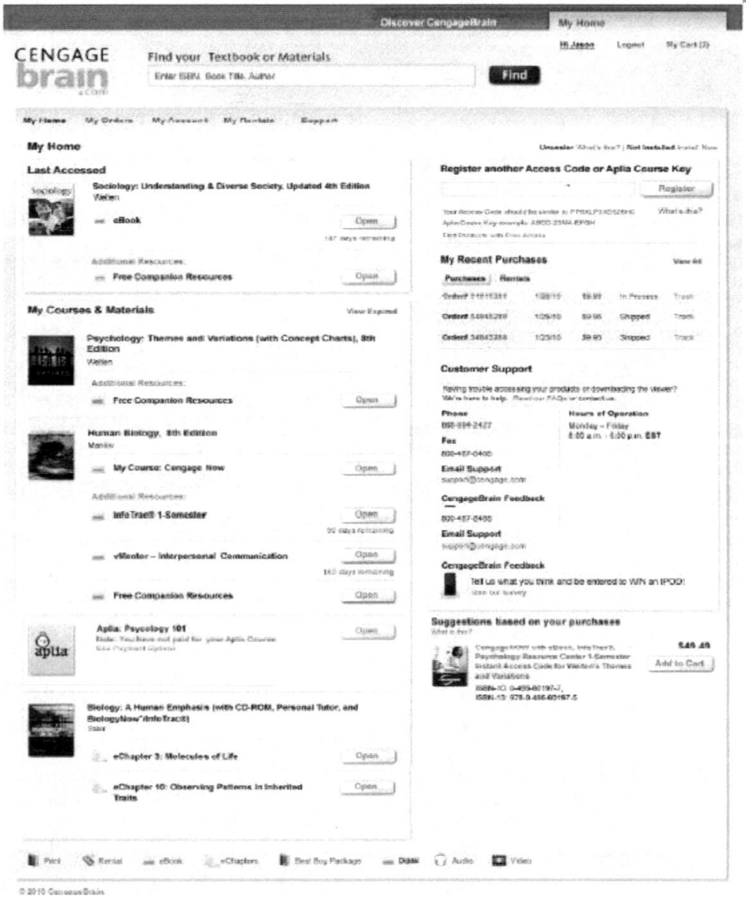

Here is a complete view of the Dashboard showing all Cengage Learning resources instantly available in one spot. Also note transaction history and support options in the right hand field.

Step 5 | Completing Payment For Your Aplia Course

NOTE: YOUR APLIA PAYMENT CODES ARE BOUND WITHIN THIS BOOK. THERE ARE <u>NO ADDITIONAL CHARGES TO ACCESS APLIA</u>. SIMPLY COMPLETE THE PAYMENT PROCESS BY CLICKING ON THE "SEE PAYMENT OPTIONS LINK" CIRCLED IN GREEN.

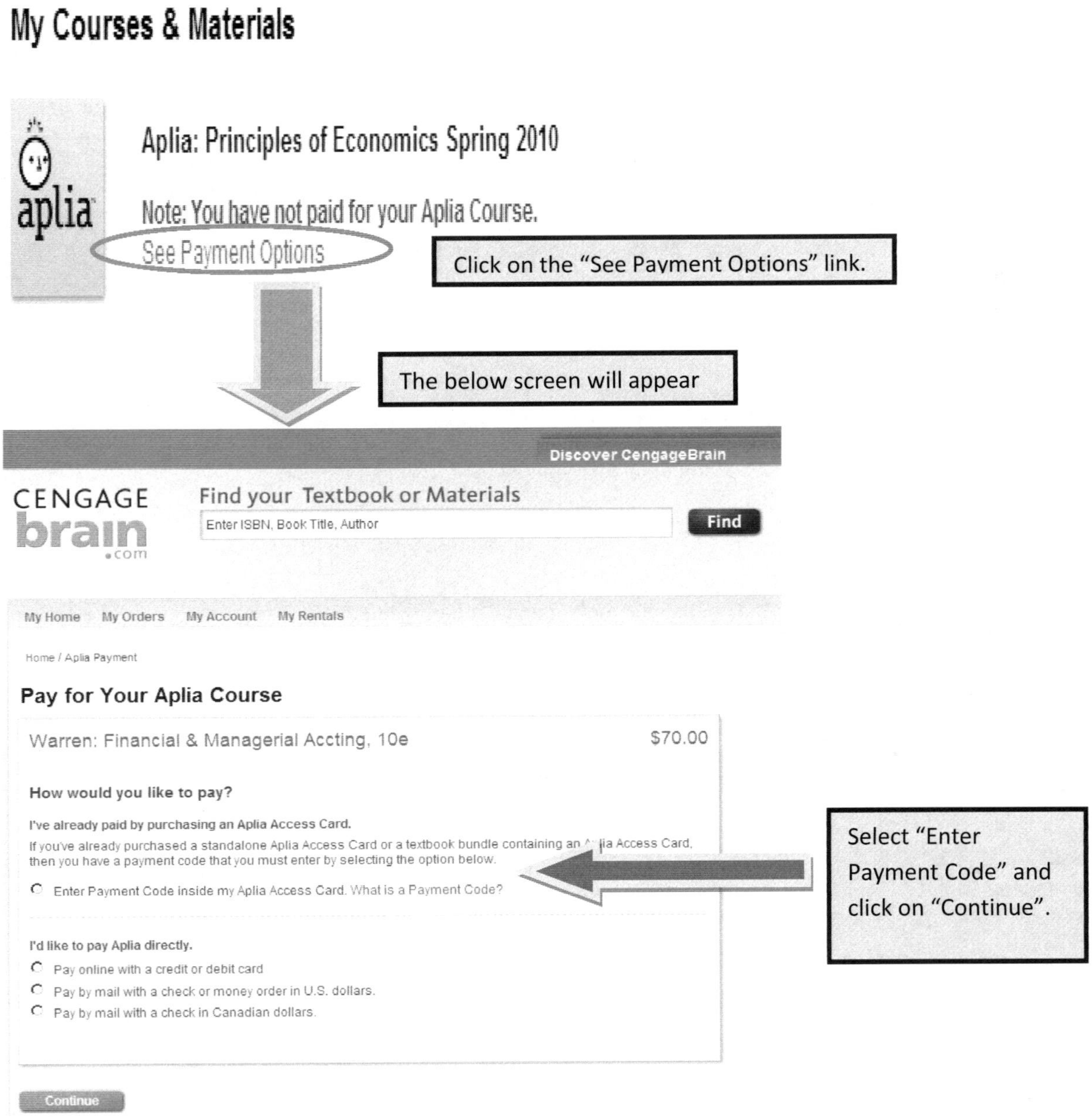

Step 5 | Completing Payment For Your Aplia Course (cont.)

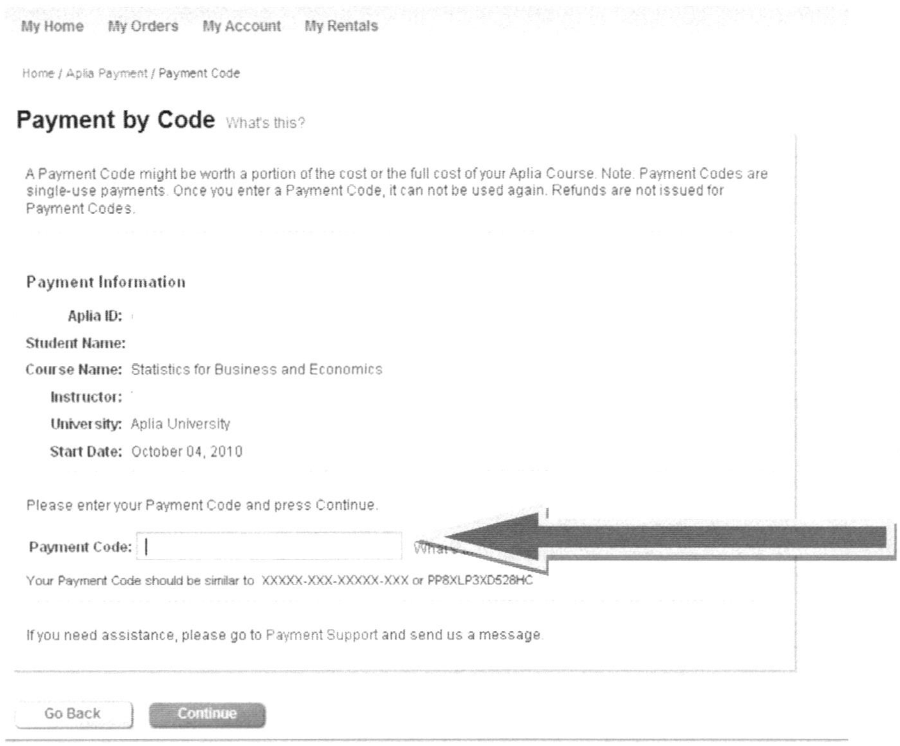

Enter the payment code that is **bound within your book**. Your payment code will look like the one shown below.

Payment Code
BQQFF-FNB-JPCPQ-PXH

NOTE: Your payment code ***is not*** the same as your Course Key provided by your Instructor.

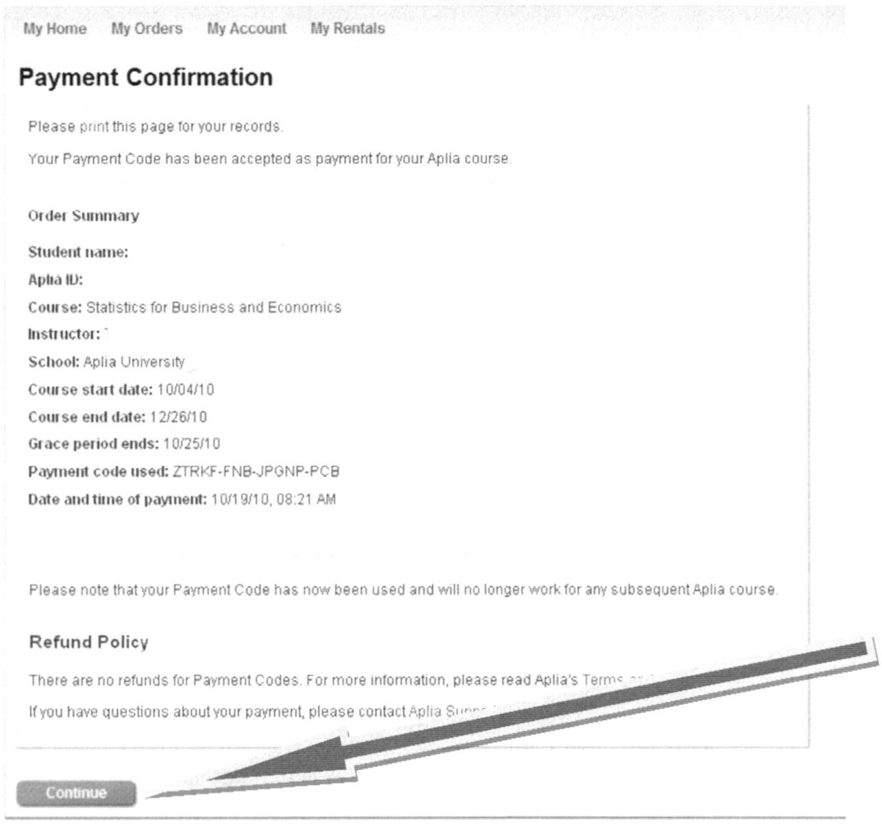

This is a **Payment Confirmation** screen. Feel free to print the screen by clicking on the link that says "print" in blue at the top of the page.

Click "Continue" to return to your My Home Dashboard and click on the "OPEN" button next to your Aplia course.

Step 6 | Accessing Your Course

Once click on the "Open" button, you will be directed to your Aplia course as shown below.

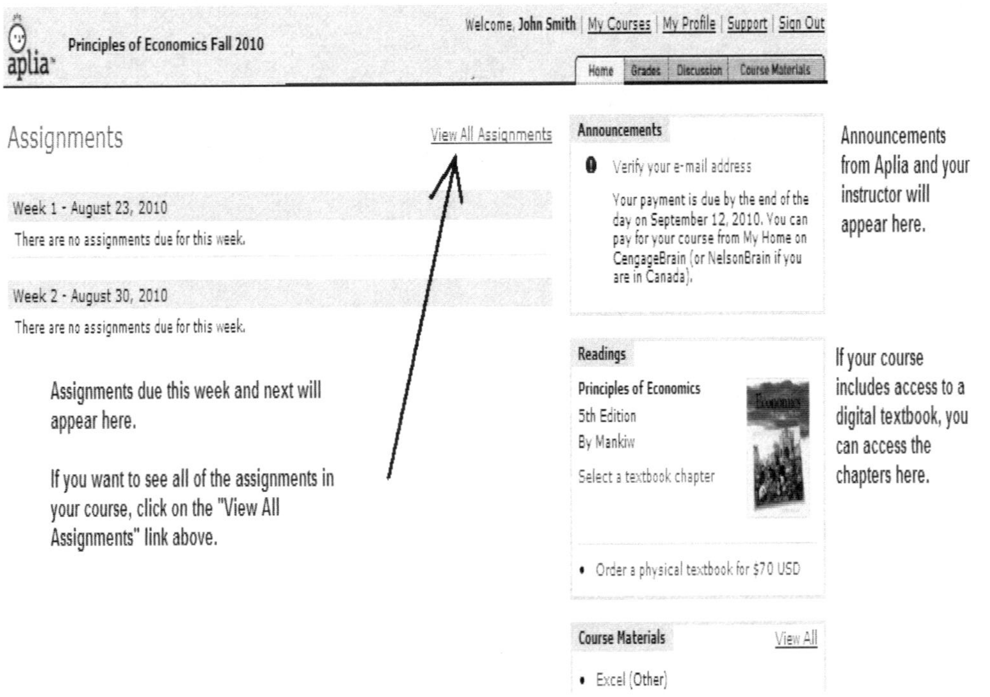

Congratulations!

You have successfully accessed your Aplia course! This page is your Aplia Home Page. Make note of Assignments, Instructor Announcements, access your eBook, etc.

Step 7 | Re-Entering Your Course After Logging Out

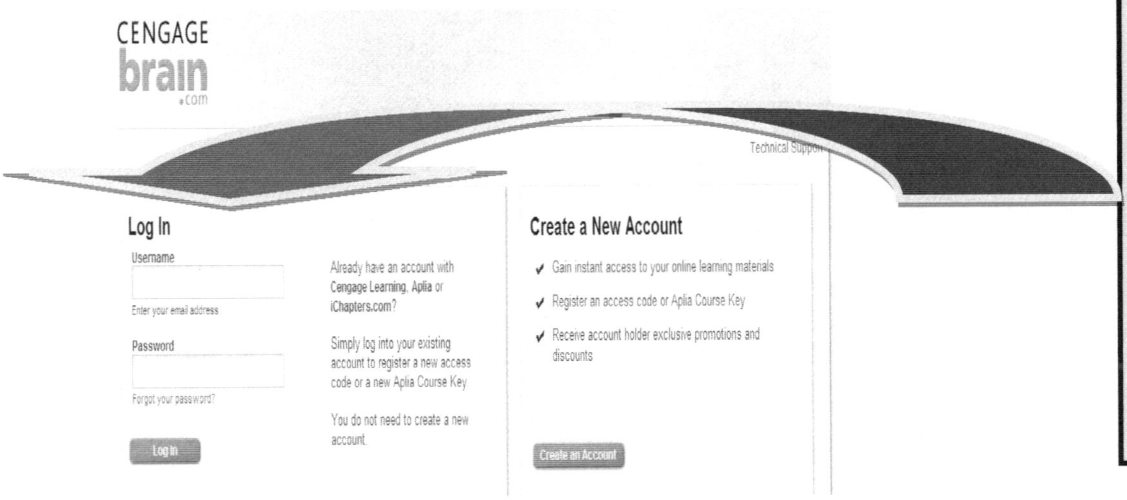

Now that you have created an account at login.cengagebrain.com, you can gain immediate access to your "My Home" page, simply by logging in at login.cengagebrain.com. Fill in your username and password to go directly to your "My Home" page and access your course!

Still Need Help?

Problems with your CengageBrain account?

- Check the FAQs in the Support area of your CengageBrain home.

  OR

- Write to **cengagebrain.support@cengage.com**

  OR

- Call 866-994-2427 Monday through Friday from 8 AM to 6 PM EST

Problems with your Aplia course?

- Click on the Support link in your Aplia course

  OR

- Write to **support@aplia.com**

  OR

- Start a chat with a Support representative Monday through Friday from 8 AM to 5 PM PST.

Biology

University of New Hampshire

Custom Edition

Peter J. Russell | Paul E. Hertz | Eldra Solomon
Charles Martin | Diana W. Martin | Linda R. Berg

CENGAGE Learning®

Australia • Brazil • Japan • Korea • Mexico • Singapore • Spain • United Kingdom • United States

CENGAGE Learning

Biology, University of New Hampshire, Custom Edition

Senior Manager, Student Engagement:
Linda deStefano

Manager, Student Engagement:
Julie Dierig

Marketing Manager:
Rachael Kloos

Manager, Premedia:
Kim Fry

Manager, Intellectual Property Project Manager:
Brian Methe

Senior Manager, Production:
Donna M. Brown

Manager, Production:
Terri Daley

Biology: The Dynamic Science, 3rd Edition
Peter J. Russell, Paul E. Hertz

Biology, 10th Edition
Eldra Solomon, Charles Martin, Diana W. Martin, Linda R. Berg

ISBN: 978-1-305-31101-5

WCN: 01-100-101

Cengage Learning
20 Channel Center Street
Boston, MA 02210
USA

Cengage Learning is a leading provider of customized learning solutions with office locations around the globe, including Singapore, the United Kingdom, Australia, Mexico, Brazil, and Japan. Locate your local office at: **international.cengage.com/region.**

Cengage Learning products are represented in Canada by Nelson Education, Ltd.
For your lifelong learning solutions, visit **www.cengage.com/custom.**
Visit our corporate website at **www.cengage.com.**

Brief Contents

STUDENTS Accessing Your Aplia Course Through CengageBrain

From Biology: The Dynamic Science
Peter J. Russell, Paul E. Hertz

Chapter 1 Introduction to Biological Concepts and Research 1
Chapter 20 Development of Evolutionary Thought 439
Chapter 21 Microevolution Genetic Changes within Populations 456
Chapter 22 Speciation 480
Chapter 23 Paleobiology and Macroevolution 499
Chapter 24 Systematics and Phylogenetics Revealing the Tree of Life 528
Chapter 25 The Origin of Life 553

From Biology
Eldra Solomon, Charles Martin, Diana W. Martin, Linda R. Berg

Chapter 24 Viruses and Subviral Agents 495

From Biology: The Dynamic Science
Peter J. Russell, Paul E. Hertz

Chapter 26 Prokaryotes Bacteria and Archaea 567
Chapter 27 Protists 587
Chapter 28 Seedless Plants 611
Chapter 29 Seed Plants 630
Chapter 30 Fungi 649
Chapter 31 Animal Phylogeny, Acoelomates, and Protostomes 670
Chapter 32 Deuterostomes Vertebrates and Their Closest Relatives 706
Chapter 36 Reproduction and Development in Flowering Plants 811
Chapter 38 Introduction to Animal Organization and Physiology 867
Chapter 42 The Endocrine System 943
Chapter 48 Regulating the Internal Environment 1069

Chapter 49 Animal Reproduction 1096
Chapter 50 Animal Development 1118
Appendix A Answers A-1
Appendix B Classification System A-36
Glossary G-1
Index I-1

Earth, a planet teeming with life, is seen here in a satellite photograph.

1

Introduction to Biological Concepts and Research

STUDY OUTLINE

1.1 What Is Life? Characteristics of Living Organisms

1.2 Biological Evolution

1.3 Biodiversity and the Tree of Life

1.4 Biological Research

Why it matters . . . Life abounds in almost every nook and cranny on our planet Earth. A lion creeps across an African plain, ready to spring at a zebra. The leaves of a sunflower in Kansas turn slowly through the day, keeping their surfaces fully exposed to rays of sunlight. Fungi and bacteria in the soil of a Canadian forest obtain nutrients by decomposing dead organisms. A child plays in a park in Madrid, laughing happily as his dog chases a tennis ball. In one room of a nearby hospital, a mother hears the first cry of her newborn baby; in another room, an elderly man sighs away his last breath. All over the world, countless organisms are born, live, and die every moment of every day. How did life originate, how does it persist, and how is it changing? Biology, the science of life, provides scientific answers to these questions.

What *is* life? Offhandedly, you might say that although you cannot define it, you know it when you see it. The question has no simple answer, because life has been unfolding for billions of years, ever since nonliving materials assembled into the first organized, living cells. Clearly, any list of criteria for the living state only hints at the meaning of "life." Deeper scientific insight requires a wide-ranging examination of the characteristics of life, which is what this book is all about.

Over the next semester or two, you will encounter examples of how organisms are constructed, how they function, where they live, and what they do. The examples provide evidence in support of concepts that will greatly enhance your appreciation and understanding of the living world, including its fundamental unity and striking diversity. This chapter provides a brief overview of these basic concepts. It also describes some of the ways in which biologists conduct research, the process by which they observe nature, formulate explanations of their observations, and test their ideas.

FIGURE 1.1 Living organisms and inanimate objects. Living organisms, such as this lizard *(Iguana iguana)*, have characteristics that are fundamentally different from those of inanimate objects, like the rock on which it is sitting.

Image copyright Mishella, 2010. Used under license from Shutterstock.com

1.1 What Is Life? Characteristics of Living Organisms

Picture a lizard on a rock, slowly turning its head to follow the movements of another lizard nearby **(Figure 1.1)**. You know that the lizard is alive and that the rock is not. At the atomic and molecular levels, however, the differences between them blur. Lizards, rocks, and all other matter are composed of atoms and molecules, which behave according to the same physical laws. Nevertheless, living organisms share a set of characteristics that collectively set them apart from nonliving matter.

The differences between a lizard and a rock depend not only on the kinds of atoms and molecules present, but also on their organization and their interactions. Individual organisms are at the middle of a hierarchy that ranges from the atoms and molecules within their bodies to the assemblages of organisms that occupy Earth's environments. Within every individual, certain biological molecules contain instructions for building other molecules, which, in turn, are assembled into complex structures. Living organisms must gather energy and materials from their surroundings to build new biological molecules, grow in size, maintain and repair their parts, and produce offspring. They must also respond to environmental changes by altering their chemistry and activity in ways that allow them to survive. Finally, the structure and function of living organisms often change from one generation to the next.

Life on Earth Exists at Several Levels of Organization, Each with Its Own Emergent Properties

The organization of life extends through several levels of a hierarchy **(Figure 1.2)**. Complex biological molecules exist at the lowest level of organization, but by themselves, these molecules are not alive. The properties of life do not appear until they are arranged into cells. A **cell** is an organized chemical system that includes many spe-

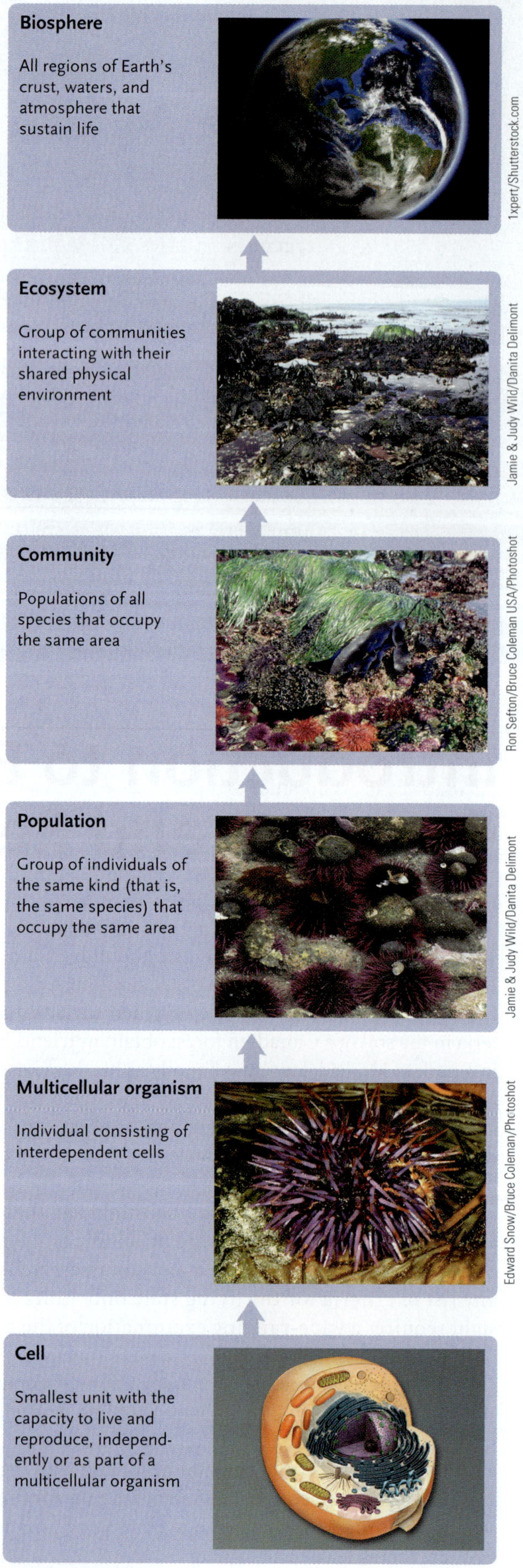

FIGURE 1.2 The hierarchy of life. Each level in the hierarchy of life exhibits emergent properties that do not exist at lower levels. The middle four photos depict a rocky intertidal zone on the coast of Washington State.

© Cengage Learning 2014

cialized molecules surrounded by a membrane. A cell is the lowest level of biological organization that can survive and reproduce—as long as it has access to a usable energy source, the necessary raw materials, and appropriate environmental conditions. However, a cell is alive only as long as it is organized as a cell; if broken into its component parts, a cell is no longer alive even if the parts themselves are unchanged. Characteristics that depend on the level of organization of matter, but do not exist at lower levels of organization, are called **emergent properties.** Life is thus an emergent property of the organization of matter into cells.

Many single cells, such as bacteria and protozoans, exist as **unicellular organisms.** By contrast, plants and animals are **multicellular organisms.** Their cells live in tightly coordinated groups and are so interdependent that they cannot survive on their own. For example, human cells cannot live by themselves in nature because they must be bathed in body fluids and supported by the activities of other cells. Like individual cells, multicellular organisms have emergent properties that their individual components lack; for example, humans can learn biology.

The next, more inclusive level of organization is the **population,** a group of organisms of the same kind that live together in the same place. The humans who occupy the island of Tahiti and a group of sea urchins living together on the coast of Washington State are examples of populations. Like multicellular organisms, populations have emergent properties that do not exist at lower levels of organization. For example, a population has characteristics such as its birth or death rate—that is, the number of individual organisms who are born or die over a period of time—that do not exist for single cells or individual organisms.

Working our way up the biological hierarchy, all the populations of different organisms that live in the same place form a **community.** The algae, snails, sea urchins, and other organisms that live along the coast of Washington State, taken together, make up a community. The next higher level, the **ecosystem,** includes the community *and* the nonliving environmental factors with which it interacts. For example, a coastal ecosystem comprises a community of living organisms, as well as rocks, air, seawater, minerals, and sunlight. The highest level, the **biosphere,** encompasses all the ecosystems of Earth's waters, crust, and atmosphere. Communities, ecosystems, and the biosphere also have emergent properties. For example, communities can be described in terms of their *diversity*—the number and types of different populations they contain—and their *stability*—the degree to which the populations within the community remain the same through time.

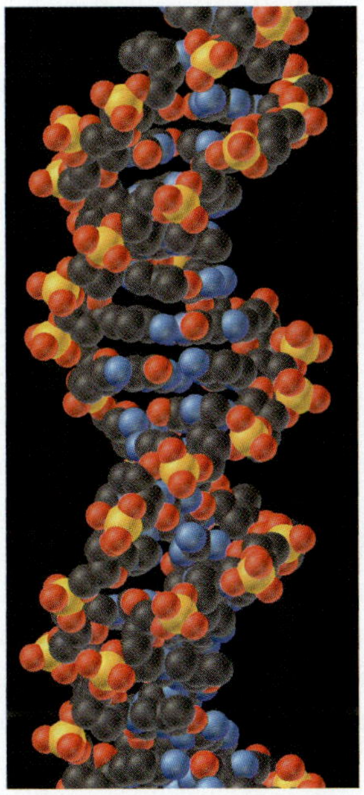

FIGURE 1.3 Deoxyribonucleic acid (DNA). A computer-generated model of DNA illustrates that it is made up of two strands twisted into a double helix.

Living Organisms Contain Chemical Instructions That Govern Their Structure and Function

The most fundamental and important molecule that distinguishes living organisms from nonliving matter is **deoxyribonucleic acid (DNA; Figure 1.3).** DNA is a large, double-stranded, helical molecule that contains instructions for assembling a living organism from simpler molecules. We recognize bacteria, trees, fishes, and humans as different because differences in their DNA produce differences in their appearance and function. (Some nonliving systems, notably certain viruses, also contain DNA, but biologists do not consider viruses to be alive because they cannot reproduce independently of the organisms they infect.)

DNA functions similarly in all living organisms. As you will discover in Chapters 14 and 15, the instructions in DNA are copied into molecules of a related substance, **ribonucleic acid (RNA),** which then directs the synthesis (production) of different protein molecules **(Figure 1.4). Proteins** carry out most of the activities of life, including the synthesis of all other biological

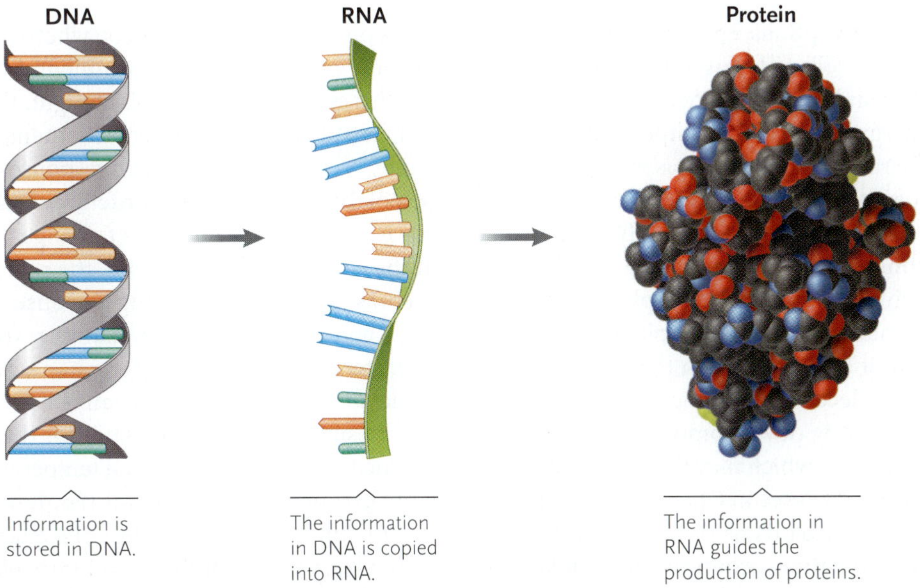

FIGURE 1.4 The pathway of information flow in living organisms. Information stored in DNA is copied into RNA, which then directs the construction of protein molecules. The protein shown here is lysozyme.

molecules. This pathway is preserved from generation to generation by the ability of DNA to copy itself so that offspring receive the same basic molecular instructions as their parents.

Living Organisms Engage in Metabolic Activities

Metabolism, described in Chapters 8 and 9, is another key property of living cells and organisms. **Metabolism** describes the ability of a cell or organism to extract energy from its surroundings and use that energy to maintain itself, grow, and reproduce. As a part of metabolism, cells carry out chemical reactions that assemble, alter, and disassemble molecules **(Figure 1.5).** For example, a growing sunflower plant carries out **photosynthesis,** in which the electromagnetic energy in sunlight is absorbed and converted into chemical energy. The cells of the plant store some chemical energy in sugar and starch molecules, and they use the rest to manufacture other biological molecules from simple raw materials obtained from the environment.

Sunflowers concentrate some of their energy reserves in seeds from which more sunflower plants may grow. The chemical energy stored in the seeds also supports other organisms, such as insects, birds, and humans, that eat them. Most organisms, including sunflower plants, tap stored chemical energy through another metabolic process, **cellular respiration.** In cellular respiration complex biological molecules are broken down with oxygen, releasing some of their energy content for cellular activities.

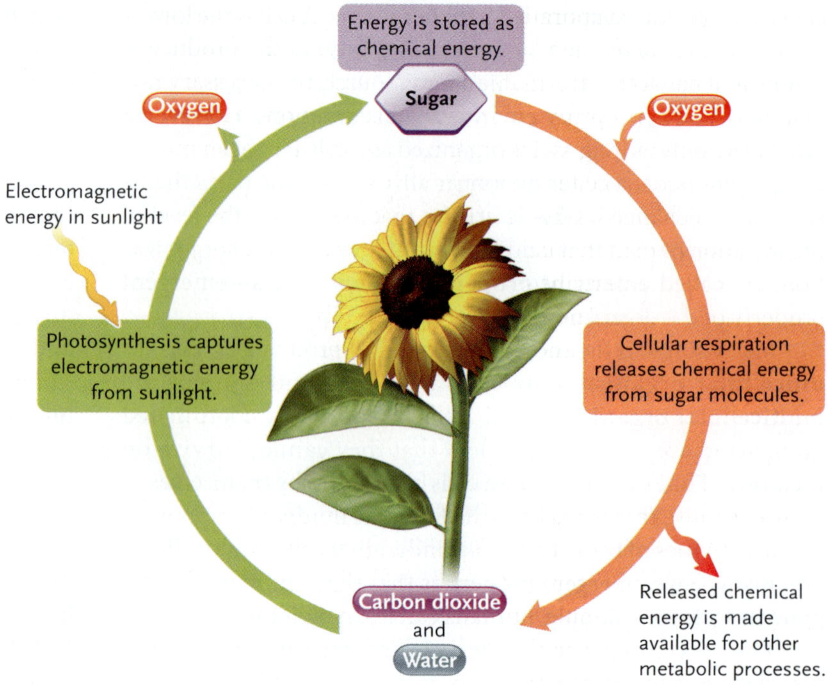

FIGURE 1.5 Metabolic activities. Photosynthesis converts the electromagnetic energy in sunlight into chemical energy, which is stored in sugars and starches built from carbon dioxide and water; oxygen is released as a by-product of the reaction. Cellular respiration uses oxygen to break down sugar molecules, releasing their chemical energy and making it available for other metabolic processes.

Energy Flows and Matter Cycles through Living Organisms

With few exceptions, energy from sunlight supports life on Earth. Plants and other photosynthetic organisms absorb energy from sunlight and convert it into chemical energy. They use this chemical energy to assemble complex molecules, such as sugars, from simple raw materials, such as water and carbon dioxide. As such, photosynthetic organisms are the **primary producers** of the food on which all other organisms rely **(Figure 1.6).** By contrast, animals are **consumers:** directly or indirectly, they feed on the complex molecules manufactured by plants. For example, zebras tap directly into the molecules of plants when they eat grass, and lions tap into it indirectly when they eat zebras. Certain bacteria and fungi are **decomposers:** they feed on the remains of dead organisms, breaking down complex biological molecules into simpler raw materials, which may then be recycled by the producers.

As you will see in Chapter 54, much of the energy that photosynthetic organisms trap from sunlight *flows* within and between populations, communities, and ecosystems. But because the transfer of energy from one organism to another is not 100% efficient, a portion of that energy is lost as heat. Although some animals can use this form of energy to maintain body temperature, it cannot sustain other life processes. By contrast, matter—nutrients such as carbon and nitrogen—*cycles* between living organisms and the nonliving components of the biosphere, to be used again and again (see Figure 1.6).

Living Organisms Compensate for Changes in the External Environment

All objects, whether living or nonliving, respond to changes in the environment; for example, a rock warms up on a sunny day and cools at night. But only living organisms have the capacity to detect environmental changes and *compensate* for them through controlled responses. Diverse and varied *receptors*—molecules or larger structures located on individual cells and body surfaces—can detect changes in external and internal conditions. When stimulated, the receptors trigger reactions that produce a compensating response.

For example, your internal body temperature remains reasonably constant, even though the environment in which you live is usually either cooler or warmer than you are. Your body compensates for these environmental variations and maintains its internal temperature at about 37° Celsius (C). When the environmental temperature drops significantly, receptors in your skin detect the change and transmit that information to your brain. Your brain may send a signal to your muscles, causing you to shiver, thereby releasing heat that keeps your body temperature from dropping below its optimal level. When the environmental temperature rises significantly, glands in your skin secrete sweat, which evaporates, cooling the skin and its underlying blood supply. The cooled blood circulates internally and keeps your body temperature from rising above 37°C. People also compensate behaviorally by dressing warmly on a cold winter day or jumping into a swimming pool in the heat of summer. Keeping your internal temperature within a narrow range is one example of **homeostasis**—a steady internal condition maintained by responses that compensate for changes in the external environment. As described in Units 5 and 6, all organisms have mechanisms that maintain homeostasis in relation to temperature, blood chemistry, and other important factors.

Living Organisms Reproduce and Many Undergo Development

Humans and all other organisms are part of an unbroken chain of life that began at least 3.5 billion years ago. This chain continues today through **reproduction,** the process by which parents produce offspring. Offspring generally resemble their parents because the parents pass copies of their DNA—with all the accompanying instructions for virtually every life process—to their offspring. The transmission of DNA (that is, genetic information) from one generation to the next is called **inheritance.** For example, the eggs produced by storks hatch into little storks, not into pelicans, because they inherited stork DNA, which is different from pelican DNA.

Multicellular organisms also undergo a process of **development,** a series of programmed changes encoded in DNA, through which a fertilized egg divides into many cells that ultimately are transformed into an adult, which is itself capable of reproduction. As an example, consider the development of a moth **(Figure 1.7).** This insect begins its life as a tiny egg that contains all the instructions necessary for its development into an adult moth. Following these instructions, the egg first hatches into a caterpillar, a larval form adapted for feeding and rapid growth. The caterpillar increases in size until internal chemical signals indicate that it is time to spin a cocoon and become a pupa. Inside its cocoon, the pupa undergoes profound developmental changes that remodel its body completely. Some cells die; others multiply and become organized in different patterns. When these transformations are complete, the adult moth emerges from the cocoon. It is equipped with structures and behaviors, quite different from those of the caterpillar, that enable it to reproduce.

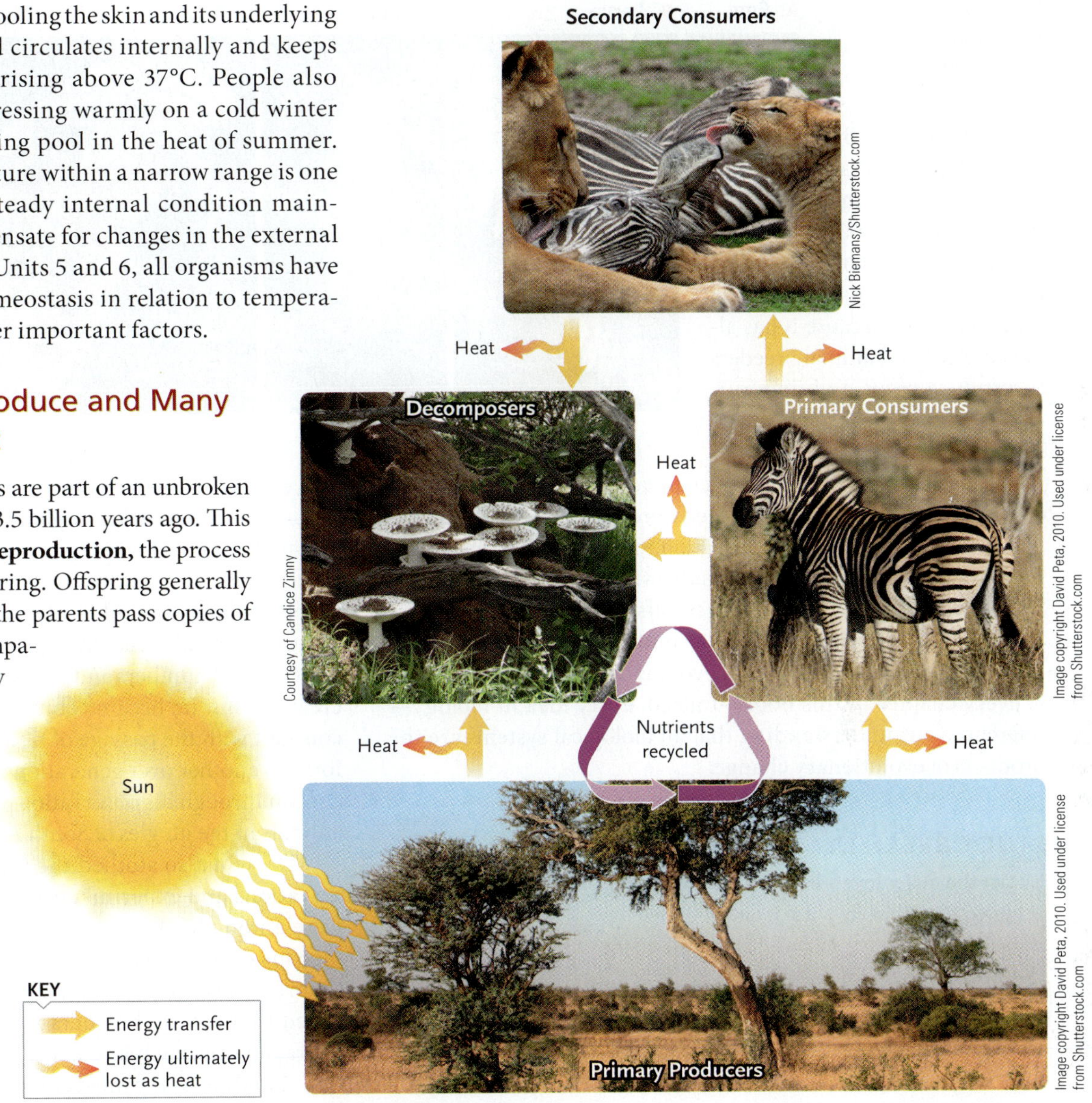

FIGURE 1.6 Energy flow and nutrient recycling. In most ecosystems, energy flows from the sun to producers to consumers to decomposers. On the African savanna, the sun provides energy to grasses (producers); zebras (primary consumers) then feed on the grasses before being eaten by lions (secondary consumers); fungi (decomposers) absorb nutrients and energy from the digestive wastes of animals and from the remains of dead animals and plants. All of the energy that enters an ecosystem is ultimately lost from the system as heat. Nutrients move through the same pathways, but they are conserved and recycled.

The sequential stages through which individuals develop, grow, maintain themselves, and reproduce are known collectively as the **life cycle** of an organism. The moth's life cycle includes egg, larva, pupa, and adult stages. Through reproduction, adult moths continue the cycle by producing the sperm and eggs that unite to form the fertilized egg, which starts the next generation.

Populations of Living Organisms Change from One Generation to the Next

Although offspring generally resemble their parents, individuals with unusual characteristics sometimes suddenly appear in a population. Moreover, the features that distinguish these oddballs are

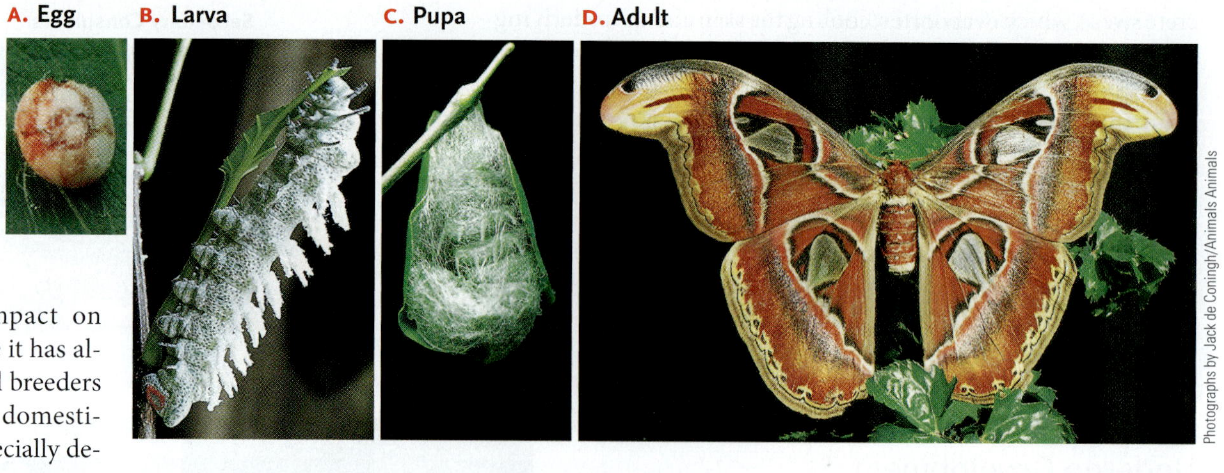

FIGURE 1.7 Life cycle of an atlas moth *(Attacus atlas).*

often inherited by their offspring. Our awareness of the inheritance of unusual characteristics has had an enormous impact on human history because it has allowed plant and animal breeders to produce crops and domesticated animals with especially desirable characteristics.

Biologists have observed that similar changes also take place under natural conditions. In other words, populations of all organisms change from one generation to the next, because some individuals experience changes in their DNA and they pass those modified instructions along to their offspring. We introduce this fundamental process, **biological evolution,** in the next section. Although we explore biological evolution in great detail in Unit 3, every chapter in this book—indeed, every idea in biology—references our understanding that all biological systems are the products of evolutionary change.

STUDY BREAK 1.1

1. List the major levels in the hierarchy of life, and identify one emergent property of each level.
2. What do living organisms do with the energy they collect from the external environment?
3. What is a life cycle?

1.2 Biological Evolution

All research in biology—ranging from analyses of the precise structure of biological molecules to energy flow through the biosphere—is undertaken with the knowledge that biological evolution has shaped life on Earth. Our understanding of the evolutionary process reveals several truths about the living world: (1) all populations change through time, (2) all organisms are descended from a common ancestor that lived in the distant past, and (3) evolution has produced the spectacular diversity of life that we see around us. Evolution is the unifying theme that links all the subfields of the biological sciences, and it provides cohesion to our treatment of the many topics discussed in this book.

Darwin and Wallace Explained How Organisms Change through Time

How do evolutionary changes take place? One important mechanism was first explained in the mid-nineteenth century by two British naturalists, Charles Darwin and Alfred Russel Wallace. On a five-year voyage around the world, Darwin observed many "strange and wondrous" organisms. He also found fossils of species that are now extinct (that is, all members of the species are dead). The extinct forms often resembled living species in some traits but differed in others. Darwin originally believed in special creation—the idea that living organisms were placed on Earth in their present numbers and kinds and have not changed since their creation. But he became convinced that species do not remain constant with the passage of time: instead, they change from one form to another over generations. Wallace came to the same conclusion through his observations of the great variety of plants and animals in the jungles of South America and Southeast Asia.

Darwin also studied the process of evolution through observations and experiments on domesticated animals. Pigeons were among his favorite experimental subjects. Domesticated pigeons exist in a variety of sizes, colors, and shapes, but all of them are descended from the wild rock dove **(Figure 1.8).** Darwin noted that pigeon breeders who wished to promote a certain characteristic, such as elaborately curled tail feathers, selected individuals with the most curl in their feathers as parents for the next generation. By permitting only these birds to mate, the breeders fostered the desired characteristic and gradually eliminated or reduced other traits. The same practice is still used today to increase the frequency of desirable traits in tomatoes, dogs, and other domesticated plants and animals. Darwin called this practice **artificial selection.** He termed the equivalent process that occurs in nature **natural selection.**

In 1858, Darwin and Wallace formally summarized their observations and conclusions explaining biological evolution. (1) Most organisms can produce numerous offspring, but environmental factors limit the number that actually survive and reproduce. (2) Heritable variations allow some individuals to compete more successfully for space, food, and mates. (3) These successful individuals somehow pass the favorable characteristics to their offspring. (4) As a result, the favorable traits become more common in the next generation, and less successful traits become less common. This process of natural selection results in evolutionary change. Today, evolutionary biologists recognize that natural selection is just one of several potent evolutionary processes, as described in Chapter 21.

Over many generations, the evolutionary changes in a population may become extensive enough to produce a population of organisms that is distinct from its ancestors. Nevertheless, parental and descendant species often share many characteristics, allowing researchers to understand their relationships and reconstruct their shared evolutionary history, as described below and in Chapter 24. Starting with the first organized cells, this aspect of evolutionary change has contributed to the diversity of life that exists today.

Darwin and Wallace described evolutionary change largely in terms of how natural selection changes the commonness or rarity of particular variations over time. Their intellectual achievement was remarkable for its time. Although Darwin and Wallace understood the central importance of variability among organisms to the process of evolution, they could not explain how new variations arose or how they were passed to the next generation.

Wild rock dove

Photographs courtesy Derrell Fowler, Tecumseh, Oklahoma

FIGURE 1.8 Artificial selection. Using artificial selection, pigeon breeders have produced more than 300 varieties of domesticated pigeons from ancestral wild rock doves *(Columba livia)*.

Mutations in DNA Are the Raw Materials That Allow Evolutionary Change

Today, we know that both the origin and the inheritance of new variations arise from the structure and variability of DNA, which is organized into functional units called **genes.** Each gene contains the code (that is, the instructions for building) for a protein molecule or one of its parts. Proteins are the molecules that establish the structures and perform important biological functions within organisms.

Variability among individuals—the raw material molded by evolutionary processes—arises ultimately through **mutations,** random changes in the structure, number, or arrangement of DNA molecules. Mutations in the DNA of reproductive cells (that is, sperm and eggs) may change the instructions for the development of offspring that the reproductive cells produce. Many mutations are of no particular value to individuals bearing them, and some turn out to be harmful. On rare occasions, however, a mutation is beneficial under the prevailing environmental conditions. Beneficial mutations increase the likelihood that individuals carrying the mutation will survive and reproduce. Thus, through the persistence and spread of beneficial mutations among individuals and their descendants, the genetic makeup of a population will change from one generation to the next.

Adaptations Enable Organisms to Survive and Reproduce in the Environments Where They Live

Favorable mutations may produce **adaptations,** characteristics that help an organism survive longer or reproduce more under a particular set of environmental conditions. To understand how organisms benefit from adaptations, consider an example from the recent literature on *cryptic coloration* (camouflage) in animals. Many animals have skin, scales, feathers, or fur that matches the color and appearance of the background in their environment, enabling them to blend into their surroundings. Camouflage makes it harder for predators to identify and then catch them—an obvious advantage to survival. Animals that are not camouflaged are often just sitting ducks.

The rock pocket mouse *(Chaetodipus intermedius),* which lives in the deserts of the southwestern United States, is mostly nocturnal (that is, active at night). At most desert localities, the rocks are pale brown, and rock pocket mice have sandy-colored fur on their backs. However, at several sites, the rocks—remnants of lava flows from now-extinct volcanoes—are black; here, the rock pocket mice have black fur on their backs. Thus, like the sandy-colored mice in other areas, they are camouflaged in their habitats, the types of areas in which they live **(Figure 1.9A).** Camouflage appears to be important to these mice because owls, which locate prey using their exceptionally keen eyesight, frequently eat nocturnal desert mice.

Examples of cryptic coloration are well documented in scientific literature, and biologists generally interpret them as adaptations that reduce the likelihood of being captured by a predator. Michael W. Nachman, Hopi E. Hoekstra, and their colleagues at the University of Arizona explored the genetic and evolutionary basis for the color difference between rock pocket mice that live on light and dark backgrounds. In an article published in 2003, they reported the results of an analysis of mice sampled at six sites in southern Arizona and New Mexico. In two regions (Pinacate, AZ, and Armendaris, NM), both light and dark rocks were present, allowing the researchers to compare mice that lived on differently colored backgrounds. Two other sites had only light rocks and sandy-colored mice.

A. Camouflage in rock pocket mice *(Chaetodipus intermedius)*

Sandy-colored mice are well camouflaged on pale rocks, and black mice are well camouflaged on dark rocks (top); but mice with fur that does not match their backgrounds (bottom) are easy to see.

Photographs courtesy of Hopi Hoekstra, Harvard University

B. Distributions of rock pocket mice with light and dark fur

At sites in Arizona and New Mexico, mouse fur color closely matched the color of the rocks where they lived. The pie charts show the proportion of mice with sandy-colored or black fur, N = the number of mice sampled at each site. The bars beneath the pie charts indicate the rock color.

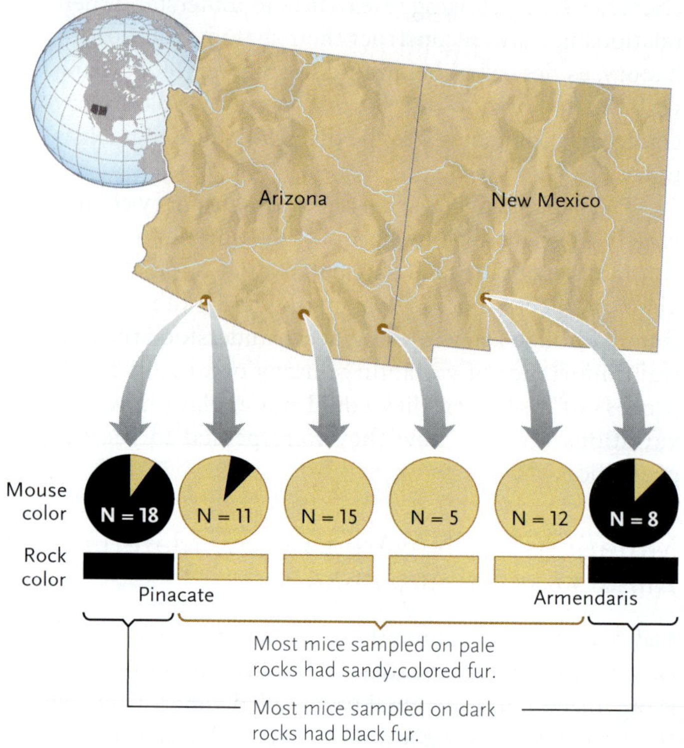

FIGURE 1.9 **Adaptive coloration in rock pocket mice *(Chaetodipus intermedius)*.**

Nachman and his colleagues found that nearly all of the mice they captured on dark rocks had dark fur and that nearly all of the mice they captured on light rocks had light fur **(Figure 1.9B)**. The researchers then studied the structure of *Mc1r*, a gene known to influence fur color in laboratory mice; random mutations in this gene can produce fur colors ranging from light to dark in any population of mice, regardless of the habitat it occupies. The 17 black mice from Pinacate all shared certain mutations in their *Mc1r* gene, which established four specific changes in the structure of the Mc1r protein. However, none of the 12 sandy-colored mice from Pinacate carried these mutations. The exact match between the presence of the mutations and the color of the mouse strongly suggests that these mutations in the *Mc1r* gene are responsible for the dark fur in the mice from Pinacate. These data on the distributions of light and dark mice coupled with analyses of their DNA suggest that the color difference is the product of specific mutations that were favored by natural selection. In other words, natural selection *conserved* random mutations that produced black fur in mice that live on black rocks.

Nachman's team then analyzed the *Mc1r* gene in the dark and light mice from Armendaris and in the light mice at two intermediate sites. Because the mice in these regions also closely matched the color of their environments, the researchers expected to find the *Mc1r* mutations in the dark mice but not in the light mice. However, none of the mice from Armendaris shared any of the mutations that apparently contribute to the dark color of mice from Pinacate. Thus, mutations in some other gene or genes, which the researchers have not yet identified, must be responsible for the camouflaging black coloration of mice that live on black rocks in Armendaris.

The example of an adaptation provided by the rock pocket mice illustrates the observation that genetic differences often develop between populations. Sometimes these differences become so great that the organisms develop different appearances and adopt different ways of life. If they become different enough, biologists may regard them as distinct types, as described in Chapter 22. Over immense spans of time, evolutionary processes have produced many types of organisms, which constitute the diversity of life on Earth. In the next section, we survey this diversity and consider how it is studied.

STUDY BREAK 1.2

1. **What is the difference between artificial selection and natural selection?**
2. **How do random changes in the structure of DNA affect the characteristics of organisms?**
3. **What is the usefulness of being camouflaged in natural environments?**

FIGURE 1.10 Traditional hierarchical classification. The classification of the American black bear *(Ursus americanus)* illustrates how each species fits into a nested hierarchy of ever-more inclusive categories. The following sentence can help you remember the order of categories in a classification, from *Domain* to *Species:* Diligent Kindly Professors Cannot Often Fail Good Students.

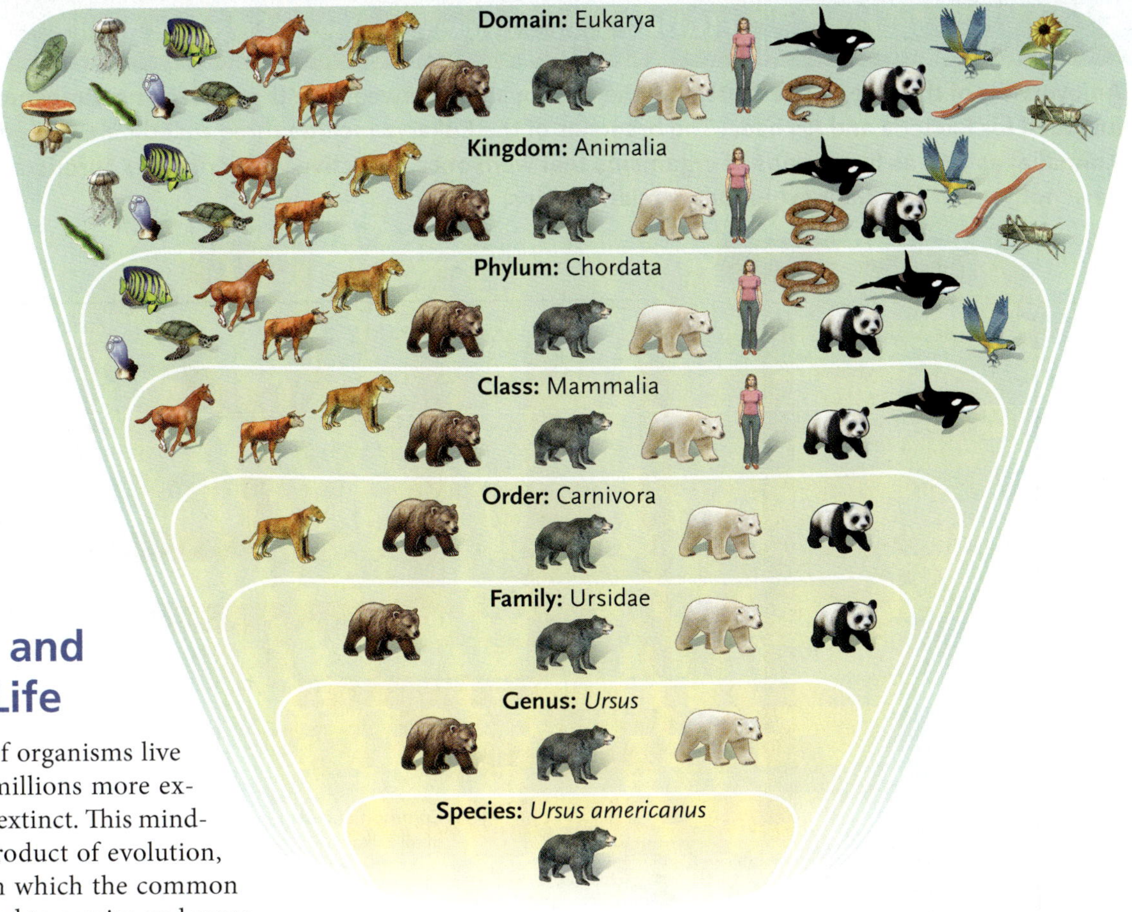

1.3 Biodiversity and the Tree of Life

Millions of different kinds of organisms live on Earth today, and many millions more existed in the past and became extinct. This mind-boggling biodiversity, the product of evolution, represents the many ways in which the common elements of life have combined to survive and reproduce. To make sense of the past and present diversity of life on Earth, biologists analyze the evolutionary relationships of these organisms and use classification systems to keep track of them. As described in Chapter 24, the task is daunting, and there is no clear consensus on the numbers and kinds of divisions and categories to use. Moreover, our understanding of evolutionary relationships is constantly changing as researchers develop new analytical techniques and learn more about extinct and living organisms.

Researchers Traditionally Defined Species and Grouped Them into Successively More Inclusive Hierarchical Categories

Biologists generally consider the species to be the most fundamental grouping in the diversity of life. As described in Chapter 22, a **species** is a group of populations in which the individuals are so similar in structure, biochemistry, and behavior that they can successfully interbreed. Biologists recognize a **genus** (plural, *genera*) as a group of similar species that share recent common ancestry. Species in the same genus usually also share many characteristics. For example, a group of closely related animals that have large bodies, four stocky legs, long snouts, shaggy hair, non-retractable claws, and short tails are classified together in the genus *Ursus,* commonly known as bears.

Each species is assigned a two-part **scientific name:** the first part identifies the genus to which it belongs, and the second part designates a particular species within that genus. In the genus *Ursus,* for example, *Ursus americanus* is the scientific name of the American black bear; *Ursus maritimus,* the polar bear, and *Ursus arctos,* the brown bear, are two other species in the same genus. Scientific names are always written in italics, and only the genus name is capitalized. After its first mention in a discussion, the genus name is frequently abbreviated to its first letter, as in *U. americanus.*

In a traditional classification, biologists first identified species and then grouped them into successively more inclusive categories **(Figure 1.10):** related genera are placed in the same **family,** related families in the same **order,** and related orders in the same **class.** Related classes are grouped into a **phylum** (plural, *phyla*), and related phyla are assigned to a **kingdom.** In recent years, biologists have added the **domain** as the most inclusive group.

Today Biologists Identify the Trunks, Branches, and Twigs on the Tree of Life

For hundreds of years, biologists classified biodiversity within the hierarchical scheme described above, mostly using structural similarities and differences as clues to evolutionary relationships. With the development of new techniques late in the twentieth century, biologists began to use the precise structure of DNA and other biological molecules to trace the evolutionary pathways

CLOSER LOOK **FIGURE 1.11** Phylogenetic View of the Tree of Life

An overview of the Tree of Life illustrates the relationships between the three domains. Branches and twigs are not included for Bacteria and Archaea. The branches of the Eukarya include three well-defined kingdoms (Plantae, Fungi, and Animalia) as well as five groups of organisms that were once collectively described as "protists" (marked with *); biologists have not yet clarified their evolutionary relationships.

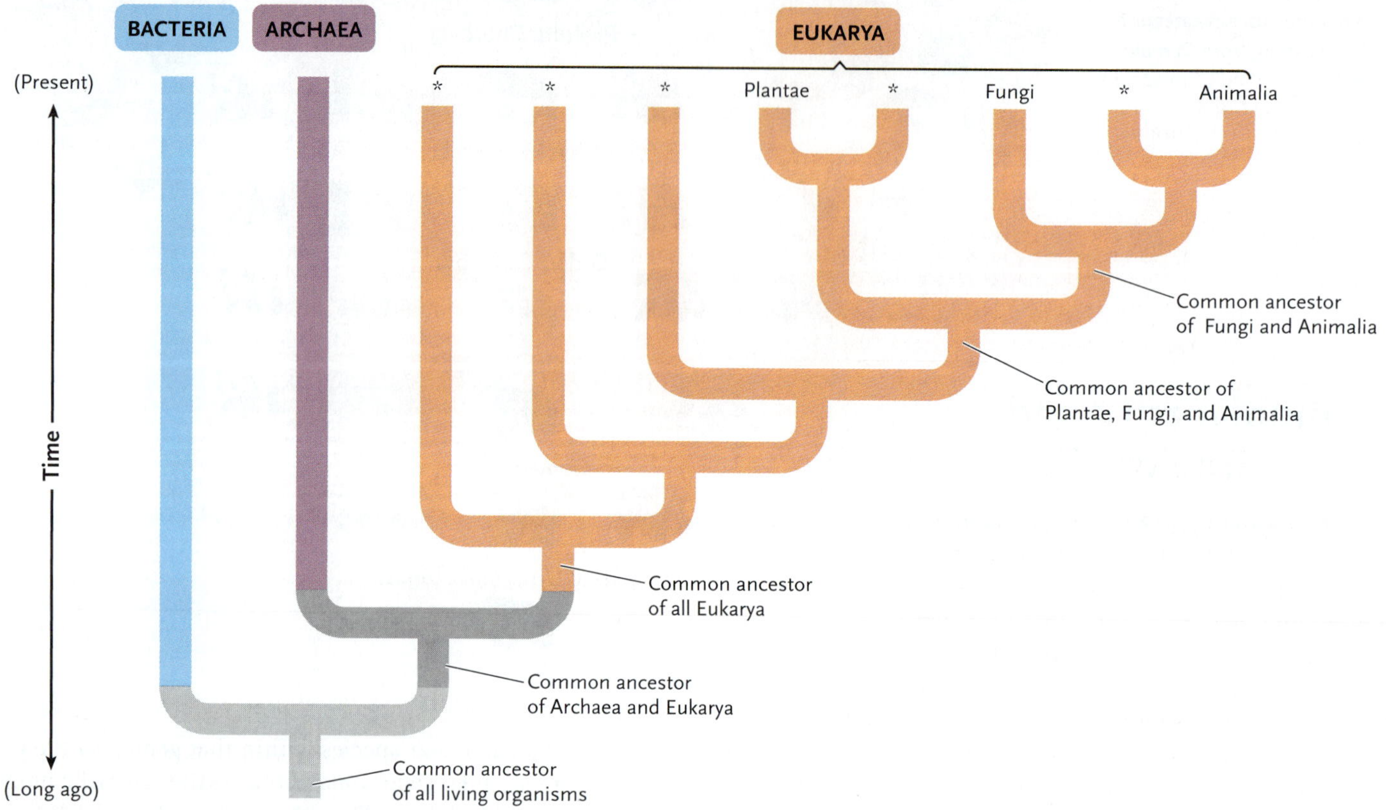

SUMMARY Phylogenetic trees contain more information than simple hierarchical classifications do because the trees illustrate which ancestors gave rise to which descendants, as well as when those evolutionary events occurred. Each fork between trunks, branches, and twigs on the phylogenetic tree represents an evolutionary event in which one ancestral species gave rise to two descendant species. Detailed phylogenetic trees illustrate how, over time, descendant species gave rise to their own descendants, producing the great diversity of life.

THINK LIKE A SCIENTIST Given the structure of the Tree of Life represented in this figure, do you think that animals are more closely related to fungi or to plants?

through which biodiversity evolved. This new approach allows the comparison of species as different as bread molds and humans because all living organisms share the same genetic code. It also provides so much data that biologists are now able to construct very detailed **phylogenetic trees** (*phylon* = race; *genetikos* = origin)—illustrations of the evolutionary pathways through which species and more inclusive groups appeared—for all organisms **(Figure 1.11)**. Phylogenetic trees, described further in Chapter 24, are like family genealogies spanning the many millions of years that evolution has been occurring. In the phylogenetic trees you will encounter in Units 3 and 4 of this book, time is usually represented vertically, with forks closer to the base of the tree representing evolutionary events in the distant past and those near the top representing more recent evolutionary events.

In many ways, information in a phylogenetic tree parallels the traditional hierarchical classification, because organisms on the same branch share the common ancestor that is represented at the base of their branch. If the base of a branch that includes two species is near the bottom of the tree, biologists would judge the species to be only distant relatives because their ancestries separated very long ago. By contrast, if the base of the branch containing two species is close to the top of the tree, we would describe the species as being close relatives. Major branches on the tree are therefore roughly equivalent to kingdoms and phyla;

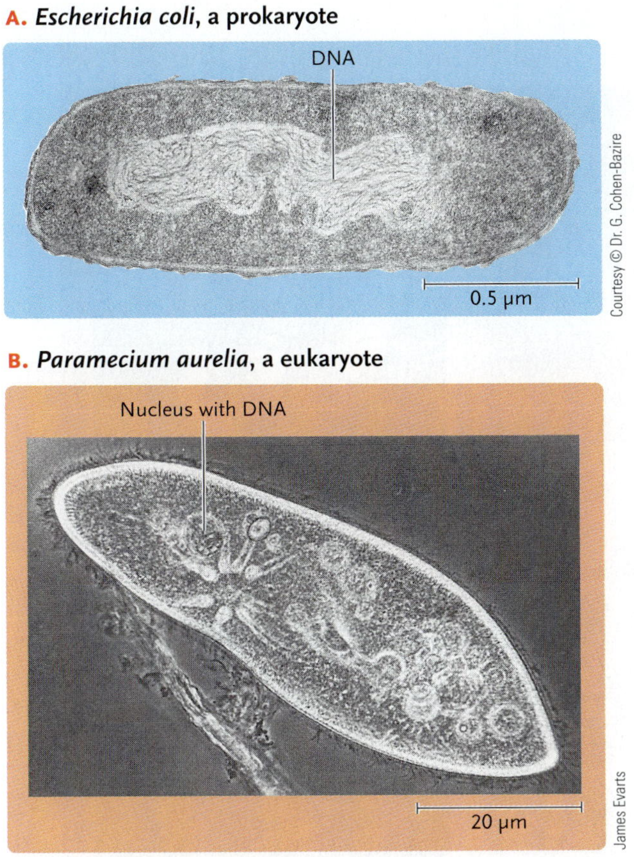

FIGURE 1.12 **Prokaryotic and eukaryotic cells. (A)** *Escherichia coli*, a prokaryote, lacks the complex internal structures apparent in **(B)** *Paramecium aurelia*, a eukaryote.

progressively smaller branches represent classes, orders, families, and genera. The twigs represent species or the individual populations they comprise.

Since 1994, with substantial support from the National Science Foundation, biologists have collaborated on the Tree of Life web project to share and disseminate their discoveries about how all organisms on Earth are related. The "Tree of Life" has been reconstructed from data on the genetics, structure, metabolic processes, and behavior of living organisms, as well as data gathered from the fossils of extinct species. It is constantly updated and revised as scientists accumulate new data.

Three Domains and Several Kingdoms Form the Major Trunks and Branches on the Tree of Life

Biologists distinguish three domains—Bacteria, Archaea, and Eukarya—each of which is a group of organisms with characteristics that set it apart as a major trunk on the Tree of Life. Species in two of the three domains, Bacteria and Archaea, are described as **prokaryotes** (*pro* = before; *karyon* = nucleus). Their DNA is suspended inside the cell without being separated from other cellular components **(Figure 1.12A)**. By contrast, the domain Eukarya comprises organisms that are described as **eukaryotes** (*eu* = typical) because their DNA is enclosed in a nucleus, a separate structure within the cells **(Figure 1.12B)**. The nucleus and other specialized internal compartments of eukaryotic cells are called **organelles** ("little organs").

THE DOMAIN BACTERIA The Domain Bacteria **(Figure 1.13A)** comprises unicellular organisms (bacteria) that are generally visible only under the microscope. These prokaryotes live as producers, consumers, or decomposers almost everywhere on Earth, utilizing metabolic processes that are the most varied of any group of organisms. They share with the archaeans a relatively simple cellular organization of internal structures and DNA, but bacteria have some unique structural molecules and mechanisms of photosynthesis.

THE DOMAIN ARCHAEA Similar to bacteria, species in the Domain Archaea (*arkhaios* = ancient) **(Figure 1.13B)**, known as archaeans, are unicellular, microscopic organisms that live as producers or decomposers. Many archaeans inhabit extreme environments—hot springs, extremely salty ponds, or habitats with little or no oxygen—that other organisms cannot tolerate. They have some distinctive structural molecules and a primitive form of photosynthesis that is unique to their domain. Although archaeans are prokaryotic, they have some molecular and biochemical characteristics that are typical of eukaryotes, including features of DNA and RNA organization and processes of protein synthesis.

THE DOMAIN EUKARYA All the remaining organisms on Earth, including the familiar plants and animals, are members of the Domain Eukarya **(Figure 1.13C)**. Organisms with eukaryotic cell structure are currently described as "protists" or classified as members of one of three well-defined kingdoms: Plantae, Fungi, and Animalia.

The "Protists". The term "protists" describes a diverse set of single-celled and multicellular eukaryotic species. They do not constitute a kingdom, because they do not share a unique common ancestry (see asterisks in Figure 1.11). The most familiar "protists" are "protozoans," which are primarily unicellular, and algae, which range from single-celled, microscopic species to large, multicellular seaweeds. "Protozoans" are consumers and decomposers, but almost all algae are photosynthetic producers.

The Kingdom Plantae. Members of the Kingdom Plantae are multicellular organisms that, with few exceptions, carry out photosynthesis; they, therefore, function as producers in ecosystems. Except for the reproductive cells (pollen) and seeds of some species, plants do not move from place to place. The kingdom includes the familiar flowering plants, conifers, and mosses.

The Kingdom Fungi. The Kingdom Fungi includes a highly varied group of unicellular and multicellular species, among them the yeasts and molds. Most fungi live as decomposers by breaking down and then absorbing biological molecules from dead organisms. Fungi do not carry out photosynthesis.

FIGURE 1.13 Three domains of life. (A) This member of the Domain Bacteria *(Helicobacter pylori)* causes ulcers in the digestive systems of humans. **(B)** These organisms from the Domain Archaea *(Pyrococcus furiosus)* live in hot ocean sediments near an active volcano. **(C)** The Domain Eukarya includes the "protists" and three kingdoms in this book. The "protists" are represented by a trichomonad (*Trichonympha* species) that lives in the gut of a termite. Coast redwoods *(Sequoia sempervirens)* are among the largest members of the Kingdom Plantae; the picture shows the large trunk of an older tree in the foreground. The Kingdom Fungi includes the big laughing mushroom (*Gymnopilus* species), which lives on the forest floor. Members of the Kingdom Animalia are consumers, as illustrated by the fishing spider (*Dolomedes* species), which is feasting on a minnow it has captured.

The Kingdom Animalia. Members of the Kingdom Animalia are multicellular organisms that live as consumers by ingesting "protists" and organisms from all three domains. The ability to move actively from one place to another during some stage of their life cycles is a distinguishing feature of animals. The kingdom encompasses a great range of organisms, including groups as varied as sponges, worms, insects, fishes, amphibians, reptiles, birds, and mammals.

Now that we have introduced the characteristics of living organisms, basic concepts of evolution, and biological diversity, we turn our attention to the ways in which biologists examine the living world to make new discoveries and gain new insights about life on Earth.

STUDY BREAK 1.3

1. What is a major difference between prokaryotic and eukaryotic organisms?
2. In which domain and kingdom are humans classified?

THINK OUTSIDE THE BOOK

Learn more about the Tree of Life web project by visiting this web site: http://tolweb.org/tree/.

1.4 Biological Research

The entire content of this book—every observation, experimental result, and generality—is the product of **biological research,** the collective effort of countless individuals who have worked to understand every aspect of the living world. This section describes how biologists working today pose questions and find answers to them.

Biologists Confront the Unknown by Conducting Basic and Applied Research

As you read this book, you may at first be uncomfortable discovering how many fundamental questions have not yet been answered. How and where did life begin? How exactly do genes govern the growth and development of an organism? What triggers the signs of aging? Scientists embrace these "unknowns" as opportunities to apply creative thinking to important problems. To show you how exciting it can be to venture into unknown territory, most chapters close with a discussion of Unanswered Questions. Although the concepts and facts that you will learn are profoundly interesting, you will discover that unanswered questions are even more exciting. In many cases, we do not even know *how* you and other scientists of your generation will answer these questions.

Research science is often broken down into two complementary activities—basic research and applied research—that constantly inform one another. Biologists who conduct **basic research** often seek explanations about natural phenomena to satisfy their own curiosity and to advance our collective knowledge. Sometimes, they may not have a specific practical goal in mind. For example, some biologists study how lizards control their body temperatures in different environments. At other times, basic research is inspired by specific practical concerns. For example, understanding how certain bacteria attack the cells of larger organisms might someday prove useful for the development of a new antibiotic (that is, a bacteria-killing agent). Many chapters in this book include a *Focus on Basic Research,* which describes particularly elegant or insightful basic research that advanced our knowledge.

Other scientists conduct **applied research,** with the goal of solving specific practical problems. For example, biomedical scientists conduct applied research to develop new drugs and to learn how illnesses spread from animals to humans or through human populations. Similarly, agricultural scientists try to develop varieties of important crop plants that are more productive and more pest-resistant than the varieties currently in use. Examples of applied research are presented throughout this book, some of them described in detail as a *Focus on Applied Research.*

The Scientific Method Helps Researchers Crystallize and Test Their Ideas

People have been adding to our knowledge of biology ever since our distant ancestors first thought about gathering food or hunting game. However, beginning about 500 years ago in Europe, inquisitive people began to understand that direct observation is the most reliable and productive way to study natural phenomena. By the nineteenth century, researchers were using the **scientific method,** an investigative approach to acquiring knowledge in which scientists make observations about the natural world, develop working explanations about what they observe, and then test those explanations by collecting more information.

Grade school teachers often describe the scientific method as a stepwise, linear procedure for observing and explaining the world around us **(Figure 1.14).** But because scientists usually think faster than they can work, they often undertake the different steps simultaneously. Application of the scientific method requires both curiosity and skepticism: successful scientists question the current state of our knowledge and challenge old concepts with new ideas and new observations. Scientists like to be shown *why* an idea is correct, rather than simply being told that it is: explanations of natural phenomena must be backed up by objective evidence rooted in observation and measurement.

Most important, scientists share their ideas and results through the publication of their work. Publications typically include careful descriptions of the methods employed and details of the results obtained so that other researchers can repeat and verify their findings at a later time.

Biologists Conduct Research by Collecting Observational and Experimental Data

Biologists generally use one of two complementary approaches—*descriptive science* and *experimental science*—or a combination of the two to advance our knowledge. In many cases, they collect **observational data,** basic information on biological structures or the details of biological processes. This approach, which is sometimes called descriptive science, provides information about systems that have not yet been well studied. For example, biologists are now collecting observational data about the precise chemical structure of the DNA in different species of organisms for the purpose of understanding their evolutionary relationships.

In other cases, researchers collect **experimental data,** information that describes the result of a careful manipulation of the system under study. This approach, which is known as experimental science, often answers questions about why or how systems work as they do. For example, a biologist who wonders whether a particular snail species influences the distribution of algae on a rocky shoreline might remove the snail from some enclosed patches of shoreline and examine whether the distribution of algae changes as a result. Similarly, a geneticist who wants to understand the role of a particular gene in the functioning of an organism might make mutations in the gene and examine the consequences.

Researchers Often Test Hypotheses with Controlled Experiments

Research on a previously unexplored system usually starts with basic observations (step 1 in Figure 1.14). Once the facts have been carefully observed and described, scientists may develop a

FIGURE 1.14 Research Method

The Scientific Method

Purpose: The scientific method is a method of inquiry that allows researchers to crystallize their thoughts about a topic and devise a formal way to test their ideas by making observations and collecting measurable data.

Protocol:

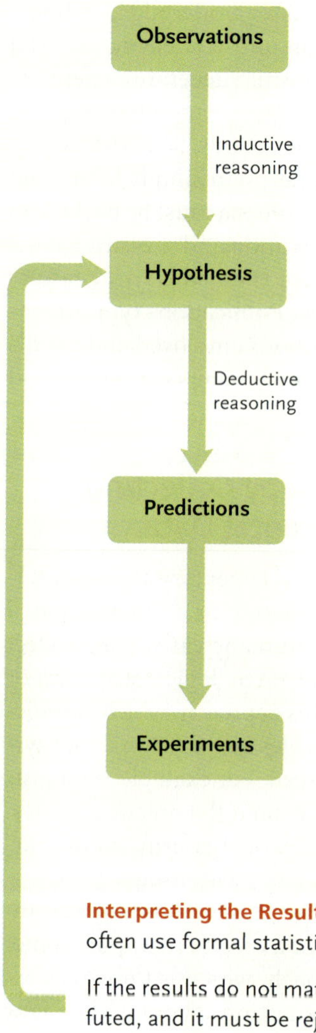

1. Make detailed observations about a phenomenon of interest.

2. Use inductive reasoning to create a testable hypothesis that provides a working explanation of the observations. Hypotheses may be expressed in words or in mathematical equations. Many scientists also formulate alternative hypotheses (that is, alternative explanations) at the same time.

3. Use deductive reasoning to make predictions about what you would observe if the hypothesis were applied to a novel situation.

4. Design and conduct a controlled experiment (or new observational study) to test the predictions of the hypothesis. The experiment must be clearly defined so that it can be repeated in future studies. It must also lead to the collection of measurable data that other researchers can evaluate and reproduce if they choose to repeat the experiment themselves.

Interpreting the Results: Compare the results of the experiment or new observations with those predicted by the hypothesis. Scientists often use formal statistical tests to determine whether the results match the predictions of the hypothesis.

If the results do not match the predictions, the hypothesis is refuted, and it must be rejected or revised.

If the statistical tests suggest that the prediction was correct, the hypothesis is confirmed—until new data refute it in the future.

hypothesis to explain them (step 2). In a 2008 report, the National Academy of Sciences defined a *hypothesis* as "a tentative explanation for an observation, phenomenon, or scientific problem that can be tested by further investigation." And whenever scientists create a hypothesis, they simultaneously define—either explicitly or implicitly—a **null hypothesis,** a statement of what they would see if the hypothesis being tested is not correct.

Many scientists structure their hypotheses with one crucial requirement: it must be *falsifiable* by experimentation or further observation. In other words, scientists must describe an idea in such a way that, if it is wrong, they will be able to demonstrate that it is wrong. The principle of falsifiability helps scientists define testable, focused hypotheses. Hypotheses that are testable and falsifiable fall within the realm of science, whereas those that cannot be falsified—although possibly valid and true—do not fall within the realm of science.

Hypotheses generally explain the relationship between **variables,** environmental factors that may differ among places or organismal characteristics that may differ among individuals. Thus, hypotheses yield testable **predictions** (step 3), statements about what the researcher expects to happen to one variable if another variable changes. Scientists then test their hypotheses and prediction with experimental or observational tests that generate relevant data (step 4). And if data from just one study refute a scientific hypothesis (that is, demonstrate that its predictions are incorrect), the scientist must modify the hypothesis and test it again or abandon it altogether. However, no amount of data can *prove* beyond a doubt that a hypothesis is correct; there may always be a contradictory example somewhere on Earth, and it is impossible to test every imaginable example. That is why scientists say that positive results *are consistent with, support,* or *confirm* a hypothesis.

To make these ideas more concrete, consider a simple example of hypothesis development and testing. Say that a friend gives you a plant that she grew on her windowsill. Under her loving care, the plant always flowered. You place the plant on your windowsill and water it regularly, but the plant never blooms. You know that your friend always gave fertilizer to the plant—your observation—and you wonder whether fertilizing the plant would make it flower. In other words, you create a hypothesis with a specific prediction: "This type of plant will flower if it receives fertilizer." This is a good scientific hypothesis because it is falsifiable. To test the hypothesis, you would simply give the plant fertilizer. If it blooms, the data—the fact that it flowers—confirm your hypothesis. If it does not bloom, the data force you to reject or revise your hypothesis.

One problem with this experiment is that the hypothesis does not address other possible reasons that the plant did not flower. Maybe it received too little water. Maybe it did not get enough sunlight. Maybe your windowsill was too cold. All of these explanations could be the basis of **alternative hypotheses,** which a conscientious scientist always considers when designing experiments. You could easily test any of these hypotheses by providing more water, more hours of sunlight, or warmer temperatures to the plant.

But even if you provide each of these necessities in turn, your efforts will not definitively confirm or refute your hypothesis unless you introduce a control treatment. The **control,** as it is often called, represents a null hypothesis; it tells us what we would see in the absence of the experimental manipulation. For example, your experiment would need to compare plants that received fertilizer (the experimental treatment) with plants grown without fertilizer (the control treatment). The presence or absence of fertilizer is the **experimental variable,** and in a controlled experiment, everything except the experimental variable—the flower pots, the soil, the amount of water, and exposure to sunlight—is exactly the same, or as close to exactly the same as possible. Thus, if your experiment is well controlled **(Figure 1.15),** any difference in flowering pattern observed between plants that receive the experimental treatment (fertilizer) and those that receive the control treatment (no fertilizer) can be attributed to the experimental variable. If the plants that receive fertilizer did not flower more than the control plants, you would reject your initial hypothesis. The elements of a typical experimental approach, as well as our hypothetical experiment, are summarized in Figure 1.15. Figures that present observational and experimental research using this basic format are provided throughout this book.

Notice that in the preceding discussion we discussed plants (plural) that received fertilizer and plants that did not. Nearly all experiments in biology include **replicates,** multiple subjects that receive either the same experimental treatment or the same control treatment. Scientists use replicates in experiments because individuals typically vary in genetic makeup, size, health, or other characteristics—and because accidents may disrupt a few replicates. By exposing multiple subjects to both treatments, we can use a statistical test to compare the average result of the experimental treatment with the average result of the control treatment, giving us more confidence in the overall findings. Thus, in the fertilizer experiment we described, we might expose six or more individual plants to each treatment and compare the results obtained for the experimental group with those obtained for the control group. We would also try to ensure that the individuals included in the experiment were as similar as possible. For example, we might specify that they all must be the same age or size.

When Controlled Experiments Are Unfeasible, Researchers Employ Null Hypotheses to Evaluate Observational Data

In some fields of biology, especially ecology and evolution, the systems under study may be too large or complex for experimental manipulation. In such cases, biologists can use a null hypothesis to evaluate observational data. For example, Paul E. Hertz of Barnard College studies temperature regulation in lizards. As in many other animals, a lizard's body temperature can vary substantially as environmental temperatures change. Research on many lizard species has demonstrated that they often compensate for fluctuations in environmental temperature—that is, maintain thermal homeostasis—by perching in the sun to warm up or in the shade when they feel hot.

Hertz hypothesized that the crested anole, *Anolis cristatellus,* a lizard species in Puerto Rico, regulates its body temperature by perching in patches of sun when environmental temperatures are low. To test this hypothesis, Hertz needed to determine what he would see if lizards were *not* trying to control their body temperatures. In other words, he needed to know the predictions of a null hypothesis that states: "Lizards do not regulate their body temperature, and they select perching sites at random with respect to factors that influence body temperature" **(Figure 1.16).** Of course, it would be impossible to force a natural population of lizards to perch in places that define the null hypothesis. Instead, he and his students created a population of artificial lizards, copper models that served as lizard-sized, lizard-shaped, and lizard-colored thermometers. Each hollow copper model was equipped with a built-in temperature-sensing wire that can be plugged into an electronic thermometer. After constructing

FIGURE 1.15 Experimental Research

Hypothetical Experiment Illustrating the Use of Control Treatment and Replicates

Question: Your friend fertilizes a plant that she grows on her windowsill, and it flowers. After she gives you the plant, you put it on your windowsill, but you do not give it any fertilizer and it does not flower. Will giving the plant fertilizer induce it to flower?

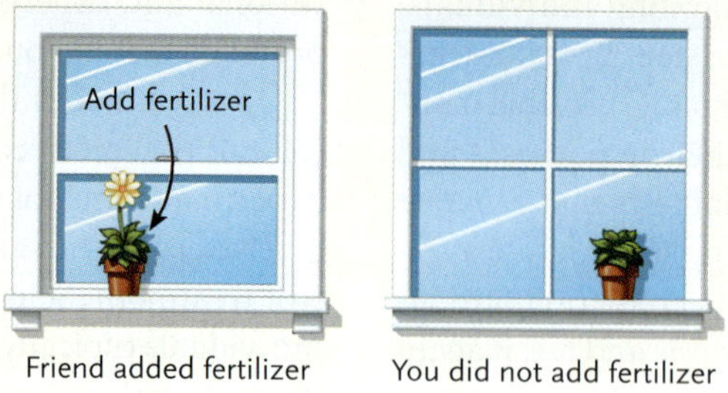

Experiment: Establish six replicates of an experimental treatment (identical plants grown with fertilizer) and six replicates of a control treatment (identical plants grown without fertilizer).

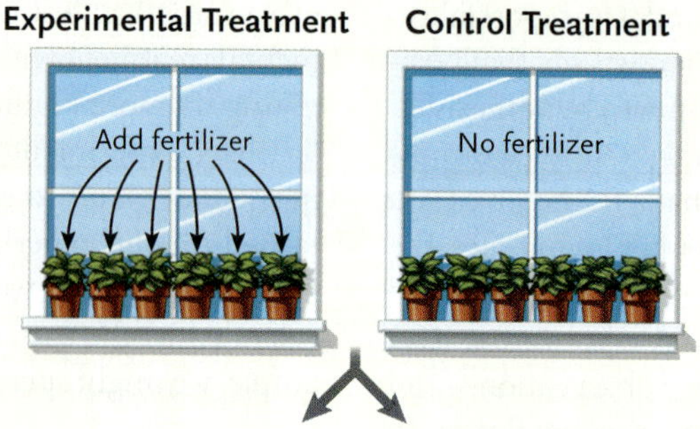

Possible Result 1: Neither experimental nor control plants flower.

Conclusion: Fertilizer alone does not cause the plants to flower. Consider alternative hypotheses and conduct additional experiments, each testing a different experimental treatment, such as the amount of water or sunlight the plant receives or the temperature to which it is exposed.

Possible Result 2: Plants in the experimental group flower, but plants in the control group do not.

Conclusion: The application of fertilizer induces flowering in this type of plant, confirming your original hypothesis. Pat yourself on the back and apply to graduate school in plant biology.

THINK LIKE A SCIENTIST Suppose you were interested in studying the effects of both fertilizer and water on the flowering of your plants. How would you design an experiment that addressed the two experimental variables at the same time?

the copper models, Hertz and his students verified that the models reached the same internal temperatures as live lizards under various laboratory conditions. They then traveled to Puerto Rico and hung 60 models at randomly selected positions in each of the habitats where this lizard species lives.

How did the copper models allow Hertz and his students to interpret their data? Because the researchers placed these inanimate objects at random positions in the lizards' habitats, the percentages of models observed in sun and in shade provided a measure of how sunny or shady a particular habitat was. In other words, the copper models established the null hypothesis about the percentage of lizards that would perch in sunlit spots just by chance. Similarly, the temperatures of the models provided a null hypothesis about what the temperatures of lizards would be if they perched at random in their habitats. Hertz and his students gathered data on the use of sunny perching places and temperatures from both the copper models and live lizards. By comparing the behavior and temperatures of live lizards with

FIGURE 1.16 Observational Research

A Field Study Using a Null Hypothesis

Anolis cristatellus

Alejandro Sanchez

Hypothesis: *Anolis cristatellus,* the crested anole, uses patches of sun and shade to regulate its body temperature.

Null Hypothesis: If these lizards do not regulate their body temperature, individuals would select perching sites at random with respect to environmental factors that influence body temperature.

Method: The researchers created a set of hollow, copper lizard models, each equipped with a temperature-sensing wire. At study sites where the lizards live in Puerto Rico, the researchers hung 60 models at random positions in trees. They observed how often live lizards and the randomly positioned copper models were perched in patches of sun or shade, and they measured the temperatures of live lizards and the copper models. Data from the randomly positioned copper models define the predictions of the null hypothesis.

Results: The researchers compared the frequency with which live lizards and the copper models perched in sun or shade as well as the temperatures of live lizards and the copper models. The data revealed that the behavior and temperatures of *A. cristatellus* were different from those of the randomly positioned models, therefore confirming the original hypothesis.

Copper *Anolis* model

Kevin de Queiroz, National Museum of Natural History, Smithsonian Institution

Percentage of models and lizards perched in sun or shade

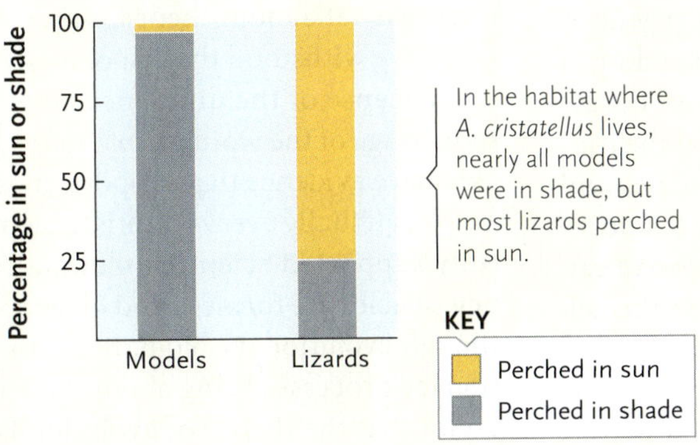

Temperatures of models and lizards

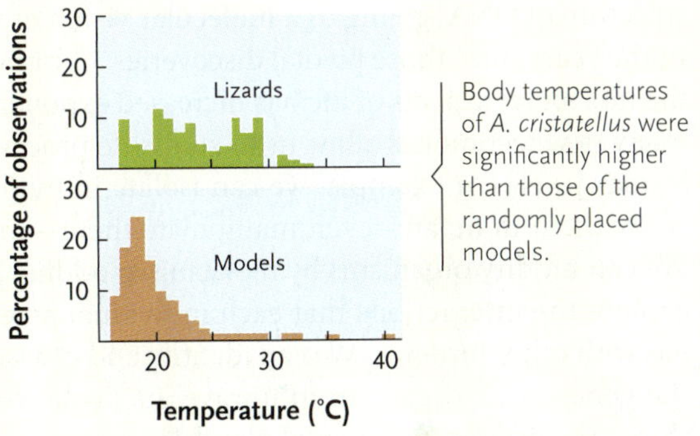

Conclusion: *A. cristatellus* uses patches of sun and shade to regulate its body temperature.

THINK LIKE A SCIENTIST Based on the data presented in the two graphs, how would you describe the variability in temperature among shaded sites in this environment?

Source: P.E. Hertz. 1992. Temperature regulation in Puerto Rican *Anolis* lizards: A field test using null hypotheses. *Ecology* 73:1405–1417.

the random "behavior" and random temperatures of the copper models, they demonstrated that *A. cristatellus* did, in fact, regulate its body temperature (see Figure 1.16).

Biologists Often Use Model Organisms to Study Fundamental Biological Processes

Certain species or groups of organisms have become favorite subjects for laboratory and field studies because their characteristics make them relatively easy subjects of research. In most cases, biologists began working with these **model organisms** because they have rapid development, short life cycles, and small adult size. Thus, researchers can rear and house large numbers of them in the laboratory. Also, as fuller portraits of their genetics and other aspects of their biology emerge, their appeal as research subjects grows because biologists have a better understanding of the biological context within which specific processes occur.

Because many forms of life share similar molecules, structures, and processes, research on these small and often simple organisms provides insights into biological processes that operate in larger and more complex organisms. For example, early analyses of inheritance in a fruit fly *(Drosophila melanogaster)* established our basic understanding of genetics in all eukaryotic organisms. Research in the mid-twentieth century with the bacterium *Escherichia coli* demonstrated the mechanisms that control whether the information in any particular gene is used to manufacture a protein molecule. In fact, the body of research with *E. coli* formed the foundation that now allows scientists to make and clone (that is, produce multiples copies of) DNA molecules. Similarly, research on a tiny mustard plant *(Arabidopsis thaliana)* is providing information about the genetic and molecular control of development in all plants, including important agricultural crops. Other model organisms facilitate research in ecology and evolution. For example, *Anolis cristatellus* is just 1 of more than 400 *Anolis* species. The geographic distribution of these species allows researchers to study general processes and interactions that affect the ecology and evolution of all forms of life. You will read about eight of the organisms most frequently used in research in *Focus on Model Organisms* boxes distributed throughout this book.

Molecular Techniques Have Revolutionized Biological Research

In 1941, George Beadle and Edward Tatum used a simple bread mold *(Neurospora crassa)* as a model organism to demonstrate that genes provide the instructions for constructing certain proteins. Their work represents the beginning of the molecular revolution. In 1953, James Watson and Francis Crick determined the structure of DNA, giving us a molecular vision of what a gene is. In the years since those pivotal discoveries, our understanding of the molecular aspects of life has increased exponentially because many new techniques allow us to study life processes at the molecular level. For example, we can isolate individual genes and study them in detail—even manipulate them—in the test tube. We can modify organisms by replacing or adding genes. We can explore the interactions that each individual protein in the cell has with other proteins. We can identify and characterize each of the genes in an organism to unravel its evolutionary relationships. The list of experimental possibilities is nearly endless.

This molecular revolution has made it possible to answer questions about biological systems that we could not even ask just a few years ago. For example: What specific DNA changes are responsible for genetic diseases? How is development controlled at the molecular level? What genes do humans and chimpanzees share? In particular, the continuing analysis of the structure of DNA in many organisms is fueling a new intensity of scientific enquiry focused on the role of whole genomes (all of the DNA of an organism) in directing biological processes. To give you a sense of the exciting impact of molecular research on all areas of biology, most chapters in this book include a box that focuses on *Insights from the Molecular Revolution*.

Advances in molecular biology have also revolutionized applied research. DNA "fingerprinting" allows forensic scientists to identify individuals who left molecular traces at crime scenes. **Biotechnology,** the manipulation of living organisms to produce useful products, has also revolutionized the pharmaceutical industry. For example, insulin—a protein used to treat the metabolic disorder diabetes—is now routinely produced by bacteria into which the gene coding for this protein has been inserted. Current research on gene therapy and the cloning of stem cells also promises great medical advances in the future.

Scientific Theories Are Grand Ideas That Have Withstood the Test of Time

When a hypothesis stands up to repeated experimental tests, it is gradually accepted as an accurate explanation of natural events. This acceptance may take many years, and it usually involves repeated experimental confirmations. When many different tests have consistently confirmed a hypothesis that addresses many broad questions, it may become regarded as a **scientific theory.** In a 2008 report, the National Academy of Sciences defined a *scientific theory* as "a plausible or scientifically acceptable, well-substantiated explanation of some aspect of the natural world ... that applies in a variety of circumstances to explain a specific set of phenomena and predict the characteristics of as yet unobserved phenomena."

Most scientific theories are supported by exhaustive experimentation; thus, scientists usually regard them as established truths that are unlikely to be contradicted by future research. Note that this use of the word *theory* is quite different from its informal meaning in everyday life. In common usage, the word *theory* most often labels an idea as either speculative or downright suspect, as in the expression "It's only a theory." But when scientists talk about theories, they do so with respect for ideas that have withstood the test of many experiments.

Because of the difference between the scientific and common usage of the word *theory,* many people fail to appreciate the extensive evidence that supports most scientific theories. For example, virtually every scientist accepts the theory of evolution as fully supported scientific truth: all species change with time, new species are formed, and older species eventually die off. Although evolutionary biologists debate the details of how evolutionary processes bring about these changes, very few scientists doubt that the theory of evolution is essentially correct. Moreover, *no scientist who has tried to cast doubt on the theory of evolution has ever devised or conducted a study that disproves any part of it.* Unfortunately, the confusion between the scientific and common usage of the word *theory* has led, in part, to endless public debate about supposed faults and inadequacies in the theory of evolution.

Curiosity and the Joy of Discovery Motivate Scientific Research

What drives scientists in their quest for knowledge? The motivations of scientists are as complex as those driving people toward any goal. Intense curiosity about ourselves, our fellow creatures, and the chemical and physical objects of the world and their interactions is a basic ingredient of scientific research. The discovery of information that no one knew before is as exciting to a scientist as finding buried treasure. There is also an element of play in science, a joy in the manipulation of scientific ideas and apparatus, and the chase toward a scientific goal. Biological research also has practical motivations—for example, to cure disease or improve agricultural productivity. In all of this research, one strict requirement of science is honesty—without honesty in the gathering and reporting of results, the work of science is meaningless. Dishonesty is actually rare in science, not least because repetition of experiments by others soon exposes any funny business.

Whatever the level of investigation or the motivation, the work of every scientist adds to the fund of knowledge about us and our world. For better or worse, the scientific method—that inquiring and skeptical approach—has provided knowledge and technology that have revolutionized the world and improved the quality of human life immeasurably. This book presents the fruits of the biologists' labors in the most important and fundamental areas of biological science—cell and molecular biology, genetics, evolution, systematics, physiology, developmental biology, ecology, and behavioral science.

STUDY BREAK 1.4

1. **In your own words, explain the most important requirement of a scientific hypothesis.**
2. **What information did the copper lizard models provide in the study of temperature regulation described earlier?**
3. **Why do biologists often use model organisms in their research?**
4. **How would you respond to a nonscientist who told you that Darwin's ideas about evolution were "just a theory"?**

THINK OUTSIDE THE BOOK

Learn more about the scientific method and the richness of the scientific enterprise by visiting this web site, maintained by researchers at the University of California at Berkeley: http://undsci.berkeley.edu/.

REVIEW KEY CONCEPTS

To access the course materials and companion resources for this text, please visit www.cengagebrain.com.

1.1 What Is Life? Characteristics of Living Organisms

- Living systems are organized in a hierarchy, each level having its own emergent properties (Figure 1.2): cells, the lowest level of organization that is alive, are organized into unicellular or multicellular organisms; populations are groups of organisms of the same kind that live together in the same area; an ecological community comprises all the populations living in an area, and ecosystems include communities that interact through their shared physical environment; the biosphere includes all of Earth's ecosystems.
- Living organisms have complex structures established by instructions coded in their DNA (Figure 1.3). The information in DNA is copied into RNA, which guides the production of protein molecules (Figure 1.4). Proteins carry out most of the activities of life.
- Living cells and organisms engage in metabolism, obtaining energy and using it to maintain themselves, grow, and reproduce. The two primary metabolic processes are photosynthesis and cellular respiration (Figure 1.5).
- Energy that flows through the hierarchy of life is eventually released as heat. By contrast, matter is recycled within the biosphere (Figure 1.6).
- Cells and organisms use receptors to detect environmental changes and trigger a compensating reaction that allows the organism to survive.
- Organisms reproduce, and their offspring develop into mature, reproductive adults (Figure 1.7).
- Populations of living organisms undergo evolutionary change as generations replace one another over time.

Animation: Insect development

Animation: One-way energy flow and materials cycling

Animation: Levels of Organization of Life

1.2 Biological Evolution

- The structure, function, and types of organisms in populations change with time. According to the theory of evolution by natural selection, certain characteristics allow some organisms to survive better and reproduce more than others in their population. If the instructions that produce those characteristics are coded in DNA, successful characteristics will become more common in later generations. As a result, the characteristics of the offspring generation will differ from those of the parent generation (Figure 1.8).
- The instructions for many characteristics are coded by segments of DNA called *genes,* which are passed from parents to offspring in reproduction.
- Mutations—changes in the structure, number, or arrangement of DNA molecules—create variability among individuals. Variability is the raw material of natural selection and other processes that cause biological evolution.
- Over many generations, the accumulation of favorable characteristics may produce adaptations, which enable individuals to survive longer or reproduce more (Figure 1.9).
- Over long spans of time, the accumulation of different adaptations and other genetic differences between populations has produced the diversity of life on Earth.

Animation: Adaptive coloration in Moths

1.3 Biodiversity and the Tree of Life

- Scientists classify organisms in a hierarchy of categories. The species is the most fundamental category, followed by genus, family, order, class, phylum, and kingdom as increasingly inclusive categories (Figure 1.10). Analyses of DNA structure now allow biologists to construct the Tree of Life, a model of the evolutionary relationships among all known living organisms (Figure 1.11).
- Most biologists recognize three domains—Bacteria, Archaea, and Eukarya—based on fundamental characteristics of cell structure and molecular analysis. The Bacteria and Archaea each include one kingdom; the Eukarya is divided into the "protists" and three kingdoms: Plantae, Fungi, and Animalia (Figures 1.12 and 1.13).

Animation: Life's diversity

1.4 Biological Research

- Biologists conduct basic research to advance our knowledge of living organisms and applied research to solve practical problems.
- The scientific method allows researchers to crystallize and test their ideas. Scientists develop hypotheses—working explanations about the relationships between variables. Scientific hypotheses must be falsifiable (Figure 1.14).
- Scientists may collect observational data, which describe particular organisms or the details of biological processes, or experimental data, which describe the results of an experimental manipulation.
- A well-designed experiment considers alternative hypotheses and includes control treatments and replicates (Figure 1.15). When experiments are unfeasible, biologists often use null hypotheses, explanations of what they would see if their hypothesis was wrong, to evaluate data (Figure 1.16).
- Model organisms, which are easy to maintain in the laboratory, have been the subject of much research.
- Molecular techniques allow detailed analysis of the DNA of many species and the manipulation of specific genes in the laboratory.
- A scientific theory is a set of broadly applicable hypotheses that have been supported by repeated tests under many conditions and in many different situations. The theory of evolution by natural selection is of central importance to biology because it explains how life evolved through natural processes.

Animation: Bacteriophage mice experiment

UNDERSTAND & APPLY

Test Your Knowledge

1. What is the lowest level of biological organization that biologists consider to be alive?
   a. a protein
   b. DNA
   c. a cell
   d. a multicellular organism
   e. a population of organisms
2. Which category falls immediately below "class" in the systematic hierarchy?
   a. species
   b. order
   c. family
   d. genus
   e. phylum
3. Which of the following represents the application of the "scientific method"?
   a. comparing one experimental subject to one control subject
   b. believing an explanation that is too complex to be tested
   c. using controlled experiments to test falsifiable hypotheses
   d. developing one testable hypothesis to explain a natural phenomenon
   e. observing a once-in-a-lifetime event under natural conditions
4. Houseflies develop through a series of programmed stages from egg, to larva, to pupa, to flying adult. This series of stages is called:
   a. artificial selection.
   b. respiration.
   c. homeostasis.
   d. a life cycle.
   e. metabolism.
5. Which structure allows living organisms to detect changes in the environment?
   a. a protein
   b. a receptor
   c. a gene
   d. RNA
   e. a nucleus
6. Which of the following is *not* a component of Darwin's theory as he understood it?
   a. Some individuals in a population survive longer than others.
   b. Some individuals in a population reproduce more than others.
   c. Heritable variations allow some individuals to compete more successfully for resources.
   d. Mutations in genes produce new variations in a population.
   e. Some new variations are passed to the next generation.
7. What role did the copper lizard models play in the field study of temperature regulation?
   a. They attracted live lizards to the study site.
   b. They measured the temperatures of live lizards.
   c. They established null hypotheses about basking behavior and temperatures.
   d. They scared predators away from the study site.
   e. They allowed researchers to practice taking lizard temperatures.
8. Which of the following questions best exemplifies basic research?
   a. How did life begin?
   b. How does alcohol intake affect aging?
   c. How fast does H1N1 flu spread among humans?
   d. How can we reduce hereditary problems in purebred dogs?
   e. How does the consumption of soft drinks promote obesity?
9. When researchers say that a scientific hypothesis must be falsifiable, they mean that:
   a. the hypothesis must be proved correct before it is accepted as truth.
   b. the hypothesis has already withstood many experimental tests.
   c. they have an idea about what will happen to one variable if another variable changes.
   d. appropriate data can prove without question that the hypothesis is correct.
   e. if the hypothesis is wrong, scientists must be able to demonstrate that it is wrong.
10. Which of the following characteristics would *not* qualify an animal as a model organism?
    a. It has rapid development.
    b. It has a small adult size.
    c. It has a rapid life cycle.
    d. It has unique genes and unusual cells.
    e. It is easy to raise in the laboratory.

Discuss the Concepts

1. Viruses are infectious agents that contain either DNA or RNA surrounded by a protein coat. They cannot reproduce on their own, but they can take over the cells of the organisms they infect and force those cells to produce more virus particles. Based on the characteristics of living organisms described in this chapter, should viruses be considered living organisms?
2. While walking in the woods, you discover a large rock covered with a gelatinous, sticky substance. What tests could you perform to determine whether the substance is inanimate, alive, or the product of a living organism?
3. Explain why control treatments are a necessary component of well-designed experiments.

Design an Experiment

Design an experiment to test the hypothesis that the color of farmed salmon is produced by pigments in their food.

Interpret the Data

While working in Puerto Rico, Paul E. Hertz and his students studied a second species of lizard, the yellow-chinned anole, *Anolis gundlachi,* using the procedures described in Figure 1.16. Their results for copper models and living lizards are in the **Figure**. Based on these data, do you think that *A. gundlachi* regulates its body temperature? Why or why not?

Apply Evolutionary Thinking

When a biologist first tested a new pesticide on a population of insects, she found that only 1% of the insects survived their exposure to the poison. She allowed the survivors to reproduce and discovered that 10% of the offspring survived exposure to the same concentration of pesticide. One generation later, 50% of the insects survived this experimental treatment. What is a likely explanation for the increasing survival rate of these insects over time?

Percentage of models and lizards perched in sun or shade

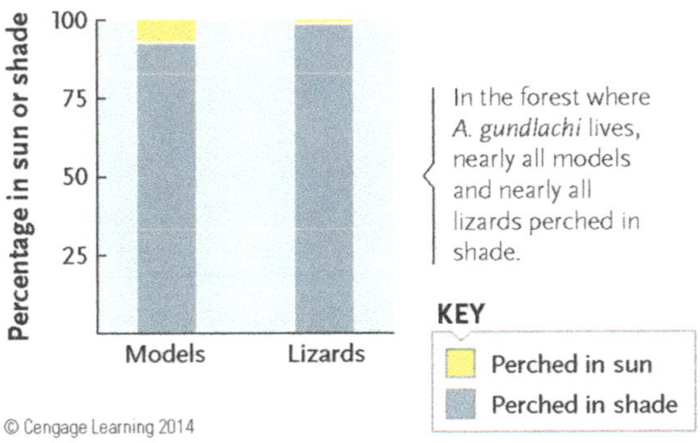

Temperatures of models and lizards

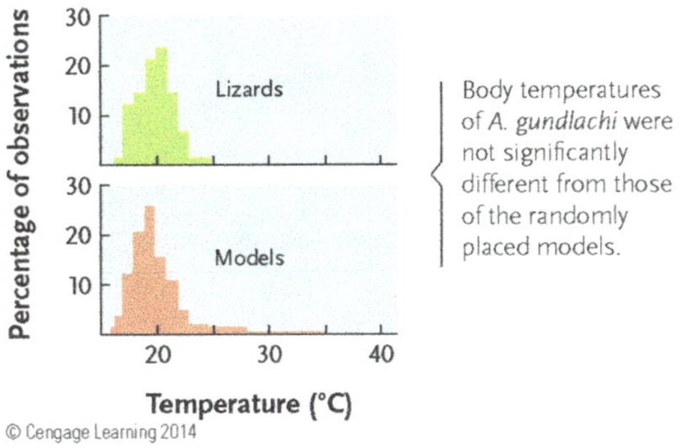

Anolis gundlachi

A rendering of H.M.S. *Beagle*, the ship that carried Charles Darwin on his round-the-world journey of discovery.

20

STUDY OUTLINE

20.1 Recognition of Evolutionary Change

20.2 Darwin's Journeys

20.3 Evolutionary Biology since Darwin

Development of Evolutionary Thought

Why it matters . . . On June 18, 1858, Charles Darwin received the shock of his life. Alfred Russel Wallace, a young naturalist working in the Asian tropics, solicited Darwin's opinion of a manuscript he had written about how species change through time. Darwin quickly realized that Wallace had independently described a mechanism for biological evolution that was nearly identical to the one he had been studying for more than 20 years, but had not yet described in print.

Like researchers today, scientists in the nineteenth century had to publish their work quickly to establish the "priority" on which their scientific reputations were made. Darwin's friend and colleague, the geologist Charles Lyell, had encouraged him to publish a preliminary essay on evolution two years before Wallace's letter arrived. But Darwin procrastinated, and because Wallace was the first to prepare his work for publication, Darwin feared that history would credit the younger man with these new ideas. Despite his anxiety, Darwin forwarded Wallace's manuscript to Lyell, who passed it along to the botanist Joseph Hooker. Lyell and Hooker engineered a solution that gave credit to both men **(Figure 20.1)**. On July 1, 1858, papers by Darwin and Wallace were presented to the Linnaean Society of London, a prestigious scientific organization.

Darwin worked feverishly after this harrowing experience, and his now-famous book, *On the Origin of Species by Means of Natural Selection,* was published on November 24, 1859. The first printing of 1,250 copies sold out in one day. Today, we honor

FIGURE 20.1 Pioneers of evolutionary theory. Charles Darwin (1809–1882) and Alfred Russel Wallace (1823–1913) independently discovered the mechanism of natural selection.

Darwin for developing the seminal idea about how biological evolution occurs and for accumulating and documenting vast quantities of evidence over decades of study.

In *The Origin,* Darwin proposed that natural mechanisms produce and transform the diversity of life on Earth. His concept of evolution still forms the unifying intellectual paradigm within which all biological research is undertaken. Even when researchers do not address explicitly evolutionary questions, their observations, theories, hypotheses, and experiments are formulated with the implicit knowledge that all forms of life are related and have evolved from ancestral forms.

Biological evolution occurs in populations when specific *processes* cause the genomes of organisms to differ from those of their ancestors. These genetic changes, and the phenotypic modifications they cause, are the *products* of evolution. By studying the products of evolution, biologists strive to understand the processes that cause evolutionary change.

The theory of evolution is so widely accepted that most people cannot think about the biological world in any other way. But the biological changes implied by Darwin's ideas and by modern evolutionary theory had not been included in earlier European worldviews.

20.1 Recognition of Evolutionary Change

The historical development of evolutionary theory is a fascinating tale of scientists struggling to reconcile evidence of change with a prevailing philosophy that change was impossible in a perfectly created universe.

Europeans Integrated Ideas from Ancient Greek Philosophy into Christian Doctrine

The Greek philosopher Aristotle (384–322 B.C.) was a keen observer of nature; he is generally considered to have been the first student of **natural history,** the branch of biology that examines the form and variety of organisms in their natural environments. Aristotle believed that both inanimate objects and living species had fixed characteristics. Careful study of their differences and similarities enabled him to create a ladder-like classification of nature, from the simplest to the most complex forms: minerals ranked below plants, plants below animals, animals below humans, and humans below the gods of the spiritual realm.

By the fourteenth century, Europeans had merged Aristotle's classification system with the biblical account of creation: each organism had been specially created by God; species could never change or become extinct; and new species could never arise. Biological research became dominated by **natural theology,** which sought to name and catalog all of God's creation. Careful study of each species would identify its position and purpose in the *Scala Naturae,* or Great Chain of Being, as Aristotle's ladder of life was called. In the eighteenth century, the Swedish botanist Carolus Linnaeus (1707–1778), who developed the science of **taxonomy,** the branch of biology that classifies organisms (see Chapter 24), undertook this important work *ad majorem Dei gloriam* ("for the greater glory of God").

Scholars also used a literal interpretation of scripture to date the time of creation precisely. By tabulating the human generations described in the Bible, they determined that the creation had occurred around 4000 B.C., making Earth a bit less than 6,000 years old—hardly old enough for much change to have taken place.

Scientists Slowly Became Aware of Change in the Natural World

Modern science came of age in the fifteenth through eighteenth centuries. The English philosopher and statesman Sir Francis Bacon (1561–1626) established the importance of observation, experimentation, and inductive reasoning. Other scientists, notably Nicolaus Copernicus (1473–1543), Galileo Galilei (1564–1642), René Descartes (1596–1650), and Sir Isaac Newton (1643–1727), proposed mechanistic theories to explain physical events. In addition, three new disciplines—biogeography, comparative morphology, and geology—promoted a growing awareness of change.

QUESTIONS ABOUT BIOGEOGRAPHY As long as naturalists encountered organisms only from Europe and surrounding lands, the task of understanding the *Scala Naturae* was manageable. But global explorations in the fifteenth through seventeenth centuries provided naturalists with thousands of unknown plants and animals from Asia, Africa, the Pacific Islands, and the Americas. Although some were similar to European species, others were new and very strange.

Studies of the world distribution of plants and animals, now called **biogeography,** raised puzzling questions. Was there no limit to the number of species created by God? Where did all these species fit in the *Scala Naturae*? If they had all been created in the Garden of Eden, why did some species have limited geographical distributions, whereas others were widespread? And why were some species found in Africa or Asia different from those found in Europe, whereas other species from far-flung locations were similar to each other **(Figure 20.2)**?

QUESTIONS ABOUT COMPARATIVE MORPHOLOGY When biologists began to compare the **morphology** (anatomical structure) of organisms, they discovered interesting similarities and differences. For example, the front legs of pigs, the flippers of dolphins, and the wings of bats differ markedly in size, shape, and function **(Figure 20.3).** But these appendages have similar locations in the animals' bodies; all are constructed of bones, muscles, and skin; and all develop similarly in the animals' embryos. If these limbs were specially created for different means of locomotion, naturalists wondered, why didn't the Creator use entirely different materials and structures for walking, swimming, and flying?

FIGURE 20.2 Large, flightless birds. Three large bird species with greatly reduced wings occupy similar habitats in geographically separated regions.

Natural theologians countered this argument by stating that the body plans were perfect, and there was no need to invent a new plan for every species. But a French scientist, George-Louis Leclerc (1707–1788), le Comte (Count) de Buffon, was still puzzled by the existence of body parts with no apparent function. For example, he noted that the feet of pigs and some other mammals have two toes that never touch the ground (pig digits 2 and 5 in Figure 20.3). If each species was anatomically perfect for its particular way of life, Buffon asked, why did useless structures exist?

Buffon proposed that some animals must have *changed* since their creation; he suggested that **vestigial structures,** the useless parts we observe today, must have functioned in ancestral organisms. Buffon offered no explanation of how functional structures became vestigial, but he clearly recognized that some species were "conceived by Nature and produced by Time."

QUESTIONS ABOUT FOSSILS By the mid-eighteenth century, geologists were mapping the stratification, or horizontal layering, of sedimentary rocks beneath the soil surface (see Figure 23.2). Different layers held different kinds of **fossils** (*fossilis* = dug up). Relatively small and simple fossils appeared in the deepest layers. Fossils in the layers above them were more complex. Those in the uppermost layers often resembled living organisms. Moreover, fossils found in any particular layer were often similar, even if they were collected from geographically separated sites. What were these fossils, and why did they vary more from one layer of rock to another than from one geographical region to another?

Some scientists suggested that fossils were the remains of extinct organisms—but natural theology did not allow extinction. Thomas Jefferson, the third president of the United States and an amateur fossil hunter, thought that fossils were the remains of species that were now extremely rare; he believed that nature could not have "permitted any one race of her animals to become extinct" or "formed any link in her great works so weak [as] to be broken." He even asked Lewis and Clark to keep an eye out for giant ground sloths, now known to be extinct, during their exploration of the Pacific Northwest.

Georges Cuvier (1769–1832), a French zoologist and a founder of comparative morphology, as well as **paleobiology** (the study of ancient organisms), realized that the layers of fossils represented organisms that had lived at successive times in the past. He suggested that the abrupt changes between geological strata marked dramatic shifts in ancient environments. Cuvier and his followers developed the theory of **catastrophism,** reasoning that each layer of fossils represented the remains of organisms that had died in a local catastrophe, such as a flood. Somewhat different species then recolonized the area, and when another catastrophe struck, they formed a different set of fossils in the next, higher layer of rock.

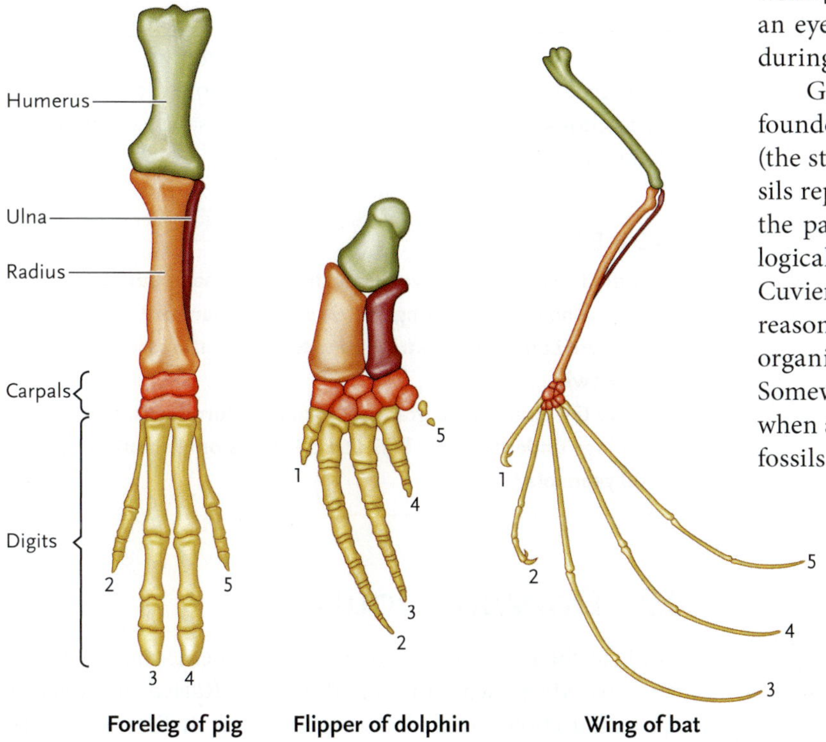

FIGURE 20.3 Mammalian forelimbs and locomotion. Pigs use their legs to walk or run, dolphins use their flippers to swim, and bats use their wings to fly. Homologous (equivalent) bones are pictured in the same color, and digits (fingers) are numbered; pigs have lost the first digit over evolutionary time. (Limbs are not drawn to scale.)

Lamarck Developed an Early Theory of Biological Evolution

A contemporary of Cuvier and a student of Buffon, Jean Baptiste de Lamarck (1744–1829) proposed the first comprehensive theory of biological evolution that was based on specific mechanisms. He proposed that a metaphysical "perfecting principle" caused organisms to become better suited to their environments. Simple organisms evolved into more complex ones, moving up the ladder of life, and microscopic organisms were replaced at the bottom of the ladder by spontaneous generation.

Lamarck theorized that two mechanisms fostered evolutionary change. According to his *principle of use and disuse,* body parts grow in proportion to how much they are used, as anyone who "pumps iron" well knows. Conversely, structures that are not often used get weaker and shrink, as do the muscles of an arm immobilized in a cast. According to his second principle, the *inheritance of acquired characteristics,* changes that an animal acquires during its lifetime are inherited by its offspring. Thus, Lamarck argued that long-legged wading birds, such as herons **(Figure 20.4),** are descended from short-legged ancestors that stretched their legs to stay dry while feeding in shallow water. Their offspring inherited slightly longer legs, and after many generations, their legs became extremely long.

Today, we know that Lamarck's proposed mechanisms do not cause evolutionary change. Although muscles do grow larger through continued use, most structures do not respond in the way Lamarck predicted. Moreover, structural changes acquired during an organism's lifetime are not inherited by the next generation. Even in his own day, Lamarck's ideas were not widely accepted.

© McPHOTO/AgeFotostock

FIGURE 20.4 A great blue heron *(Ardea herodias).* Like many other wading birds, herons have long, stilt-like legs. Lamarck hypothesized that as wading birds stretched their legs to keep their bodies dry while feeding, successive generations of their offspring would have progressively longer legs.

Despite the shortcomings of his theory, Lamarck made four tremendously important contributions to the development of an evolutionary worldview. First, he proposed that all species change through time. Second, he recognized that new characteristics are passed from one generation to the next. Third, he suggested that organisms change in response to their environments. And fourth, he hypothesized the existence of specific mechanisms that fostered evolutionary change. All four of these ideas became cornerstones of Darwin's evolutionary theory, although he proposed a different mechanism, natural selection, as the cause of evolutionary change. Perhaps Lamarck's most important contribution was that he fostered discussion. By the mid-nineteenth century, most educated Europeans were talking about evolutionary change, whether they believed in it or not.

Geologists Recognized That Earth Had Changed over Time

In 1795, the Scottish geologist James Hutton (1726–1797) argued that slow and continuous physical processes, *acting over long periods of time,* produced Earth's major geological features; for example, the movement of water in a river slowly erodes the land and deposits sediments near the mouth of the river. Given enough time, erosion creates deep canyons, and sedimentation creates thick topsoil on flood plains. Hutton's **gradualism,** the view that Earth changed *slowly* over its history, contrasted sharply with Cuvier's catastrophism.

The English geologist Charles Lyell (1797–1875) championed and extended Hutton's ideas in an influential series of books, *Principles of Geology.* Lyell argued that the geological processes that sculpted Earth's surface over long periods of time—such as volcanic eruptions, earthquakes, erosion, and the formation and movement of glaciers—are exactly the same as the processes observed today. This concept, called **uniformitarianism,** undermined any remaining notions of an unchanging Earth. Also, because geological processes proceed very slowly, it must have taken millions of years, not just a few thousand, to mold the landscape into its current configuration.

STUDY BREAK 20.1

1. **Why did the existence of vestigial structures make Buffon question the idea that living systems never changed?**
2. **What were Lamarck's contributions to an evolutionary worldview?**
3. **How do the concepts of gradualism and uniformitarianism in geology undermine the belief that Earth is only about 6,000 years old?**

20.2 Darwin's Journeys

In 1831, in the midst of this intellectual ferment, young Charles Darwin wondered what to do with his life. Raised in a wealthy English household, he had always collected shells and studied the habits of insects and birds; he preferred hunting and fishing to classical studies. Despite his lackluster performance as a student, Darwin was expected to continue the family tradition of practicing medicine. But after two years, he abandoned his medical studies at the University of Edinburgh. Instead, he followed his interest in natural history despite the objections of his father, who reputedly told him, "You care for nothing but shooting, dogs, and rat-catching and you will be a disgrace to yourself and all of your family."

At the suggestion of his father, Darwin studied for a career as a clergyman, earning a degree from Cambridge University. There, he found a mentor in the Reverend John Henslow, a leading botanist, who arranged for Darwin to travel as the captain's dining companion aboard H.M.S. *Beagle,* a naval surveying ship. Darwin thus embarked on a sea voyage and an intellectual journey that altered the foundations of modern thought.

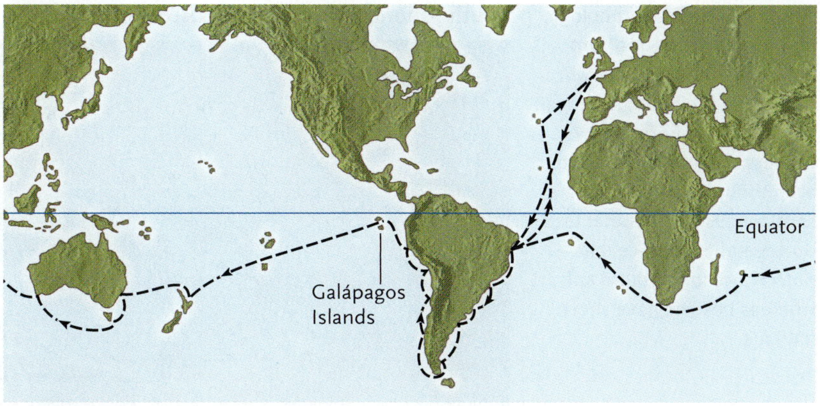

FIGURE 20.5 Darwin's voyage. H.M.S. *Beagle* circumnavigated the globe between 1831 and 1836.

Darwin Saw the World on the Voyage of the *Beagle*

The *Beagle* sailed southwest to map the coastline of South America and then circumnavigated the globe **(Figure 20.5).** When the ship's naturalist quit his post mid-journey, Darwin replaced him in an unofficial capacity. For nearly five years Darwin toured the world, and because he suffered from seasickness, he seized every chance to go ashore. He collected plants and animals in Brazilian rain forests and fossils in Patagonia. He hiked the grasslands of the Pampas and climbed the Andes in Chile. Armed with Henslow's parting gift, the first volume of Lyell's *Principles of Geology,* Darwin was primed to apply gradualism and uniformitarianism to the living world.

WHAT DARWIN SAW When he began his travels, Darwin had no clue that biological evolution had produced the mind-boggling variety of species that he would encounter. Three broad sets of observations later helped him unravel the mystery of evolutionary change.

First, while exploring along the coast of Argentina, Darwin discovered fossils that often resembled organisms inhabiting the same region today. For example, despite an enormous size difference, living armadillos and fossilized glyptodonts had similar body armor, but they were unlike any other species known to science **(Figure 20.6).** If both species had been created at the same time and both were found in South America, why didn't glyptodonts live alongside armadillos? Darwin later wondered whether armadillos might be living relatives of the now-extinct glyptodonts.

Second, Darwin observed that the animals he encountered in different South American habitats clearly resembled each other, but they differed from species that occupied similar habitats in Europe. For example, he noted that nutria *(Myocastor coypus),* a semiaquatic rodent native to South America, bore a closer resemblance to other rodents from the mountains or grasslands of that continent than it did to the beaver *(Castor fiber),* a semiaquatic rodent native to Europe **(Figure 20.7).** Why did animals from markedly different South American environments resemble each other so closely? Why did animals that lived in similar environments on separate continents look so different? Darwin later understood that animals in South America resembled each other because they had inherited their similarities from a common ancestor.

Third, Darwin observed fascinating patterns in the distributions of species on the Galápagos Islands **(Figure 20.8).** There he found strange and wonderful creatures, including giant tortoises and lizards that dove into the sea to feed on algae. Darwin quickly noted that the animals on different islands varied slightly in form. Indeed, experienced sailors could easily identify a tortoise's island of origin by the shape of its shell. More-

Charles R. Knight painting (negative CK21T), Field Museum of Natural History.

Ron Kacmarcik/Shutterstock.com

FIGURE 20.6 Ancestors and descendants. Darwin hypothesized that even though an extinct glyptodont (top) probably weighed 300 to 400 times as much as a living nine-banded armadillo *(Dasypus novemcinctus),* their obvious resemblance suggested that they are related.

FIGURE 20.7 Morphologic differences in species from different continents. Darwin noted that the South American nutria and the European beavers differ in appearance, even though both species are semiaquatic rodents that feed on vegetation. Notice that nutria have long, round tails, whereas beavers have short, flat tails.

over, many species resembled those on the distant South American mainland. Why did so many different organisms occupy one small island cluster, and why did these species resemble others from the nearest continent? Darwin later hypothesized that the plants and animals of the Galápagos Islands were descended from South American ancestors, and that each species had changed after being isolated on a particular island.

DARWIN'S REFLECTIONS AFTER HIS VOYAGE The *Beagle* returned to England in 1836, and Darwin began his first notebook on the "transmutation of species" the following year. He realized that changes in species over time provided the only plausible explanation for his observations.

A diverse group of finches from the Galápagos Islands **(Figure 20.9)** provided the single greatest spark for Darwin's work. He had noticed great variability in the shapes of their bills, but he had incorrectly assumed that birds on different islands belonged to the same species. Thus, he had not recorded the island where he had captured each specimen. Luckily, the *Beagle's* captain, Robert Fitzroy, had more thoroughly documented his own collection, allowing Darwin to study the relationships and geographical distributions of a dozen species. As Darwin re-

FIGURE 20.8 The Galápagos Islands and some of their unusual animal inhabitants. (A) Volcanic eruptions created the Galápagos archipelago (located 1,000 km west of Ecuador) between 3 and 5 million years ago. **(B)** The islands were named for the giant tortoises found there (in Spanish, *galápa* means tortoise); this tortoise is native to Isla Santa Cruz. **(C)** Marine iguanas dive into the Pacific Ocean to feed on algae.

A. Warbler finch (*Certhidea olivacea*)

B. Common cactus-finch (*Geospiza scandens*)

C. Large ground-finch (*Geospiza magnirostris*)

D. Woodpecker finch (*Camarhynchus pallidus*)

FIGURE 20.9 Bill shape and food habits. The 13 finch species that inhabit the Galápagos Islands are descended from a common ancestor, a seed-eating ground finch that migrated to the islands from South America. **(A)** The warbler finch uses its slender bill to probe for insects in vegetation. **(B)** The common cactus-finch has a medium-sized bill suitable for eating cactus flowers and fruit. **(C)** The large ground-finch uses its thick, strong bill to crush cactus seeds. **(D)** The woodpecker finch uses its bill to hammer at bark and to hold cactus spines, with which it probes for wood-boring insects, such as termites.

viewed the data, he began to focus on two aspects of a general problem. Why were the finches on a particular island slightly different from those on nearby islands, and how did all these different species arise?

Darwin Used Common Knowledge and Several Inferences to Develop His Theory

With a substantial inheritance, and burdened by chronic illness, Darwin led a reclusive life as he embarked on an intellectual journey every bit as exciting as his voyage on the *Beagle* (see *Focus on Basic Research*). His lifetime goal was to accumulate evidence of evolutionary change and identify the mechanism that caused it.

SELECTIVE BREEDING AND HEREDITY Having grown up in the country, Darwin was well aware that "like begets like"; that is, offspring frequently resemble their parents. Plant breeders and animal breeders had applied this basic truth of inheritance for thousands of years. By selectively breeding individuals with desired characteristics, they enhanced those traits in future generations.

Farmers use selective breeding to improve domesticated plants and animals. If one cow produces more milk than any other, the farmer selectively breeds her (rather than others), hoping that her offspring will also be good milk producers. Although the mechanism of heredity was not yet understood, this principle had been applied countless times to produce bigger beets, plumper pigs, and prize-winning pigeons (see Figure 1.8). Darwin was well aware of this process, which he called **artificial selection,** but he puzzled over how it could operate in nature.

THE STRUGGLE FOR EXISTENCE Darwin had a revelation about how selective breeding could occur naturally when he read the famous publication by Thomas Malthus, *Essay on the Principles of Population.* Malthus, an English clergyman and economist, was worried about the fate of the nation's poor. England's population was growing much faster than its agricultural capacity, and with individuals competing for limited food resources, some would inevitably starve.

Darwin applied Malthus's argument to organisms in nature. Species typically produce many more offspring than are needed to replace the parent generation, yet the world is not overrun with sunflowers, tortoises, or bears. Darwin even calculated that, if its reproduction went unchecked, a single pair of elephants, the slowest breeding animal known, would leave roughly 19 million descendants after only 750 years. Happily for us (and all other species that might get underfoot), the world is not so crowded with elephants. Instead, some members of every population survive and reproduce, whereas others die without reproducing.

DARWIN'S INFERENCES Darwin's discovery of a mechanism for evolutionary change required him to infer the nature of a process that no one had envisioned, much less documented **(Figure 20.10).** First, individuals within populations vary in size, form, color, behavior, and other characteristics. Second, many of these variations are hereditary. What if variations in hereditary traits enabled some individuals to survive and reproduce more than others? Organisms with such advantageous traits would leave many offspring, whereas those that lacked such traits would die leaving few, if any, descendants. Thus, advantageous hereditary traits would become more common in the next generation. If the next generation was subjected to the same process of selection, the traits would be even more common in the third generation. Because this process is analogous to artificial selection, Darwin called it **natural selection.**

As an evolutionary mechanism, natural selection favors **adaptive traits,** genetically based characteristics that make or-

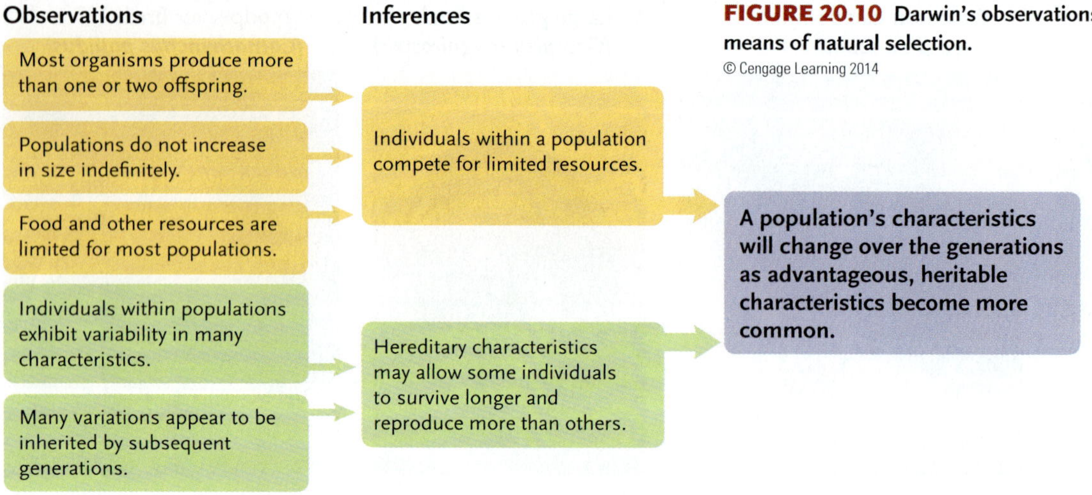

FIGURE 20.10 **Darwin's observations and inferences about evolution by means of natural selection.**

ganisms more likely to survive and reproduce. And by favoring individuals that are well adapted to the environments in which they live, natural selection causes species to change through time. As shown in Figure 20.9, each species of Galápagos finch has a distinctive bill. Variations in bill size and shape make some birds better adapted for crushing seeds and others for capturing insects. Imagine an island where large seeds were the only food available; individuals with a stout bill would be more likely to survive and reproduce than would birds with slender bills. These favored individuals would pass the genes that produce stout bills to their descendants, and after many generations, their bills might resemble those of *Geospiza magnirostris* (see Figure 20.9C). Natural selection also changes nonmorphologic characteristics of populations; for example, insect populations that are exposed to insecticides develop resistance to these toxic chemicals over time (see Figure 20.12).

Darwin realized that natural selection could also account for striking differences between populations and, given enough time, for the production of new species. For example, suppose that small insects were the only food available to finches on a different island. Birds with long, thin bills might be favored by natural selection, and the population of finches might eventually possess a bill shaped like that of *Certhidea olivacea* (see Figure 20.9A). If we apply parallel reasoning to the many characteristics that affect survival and reproduction, natural selection would cause the populations to become more different over time, a process called **evolutionary divergence.**

Darwin's Theory Revolutionized the Way We Think about the Living World

It would be hard to overestimate the impact of Darwin's theory on Western thought. In *The Origin*, Darwin proposed a logical mechanism for evolutionary change and provided enough supporting evidence to convince the educated public.

Darwin argued that all the organisms that have ever lived arose through **descent with modification,** the evolutionary alteration and diversification of ancestral species. He envisioned this pattern of descent as a tree growing through time **(Figure 20.11).** The base of the trunk represents the ancestor of all organisms. Branching points above it represent the evolutionary divergence of ancestors into their descendants. Each limb represents a body plan suitable for a particular way of life; smaller branches represent more narrowly defined groups of organisms; and the uppermost twigs represent living species. Biologists still apply this analogy today when analyzing the Tree of Life (see Figure 1.11 and Chapter 24).

Darwin proposed natural selection as the mechanism that drives evolutionary change. In fact, most of *The Origin* was an explanation of how natural selection acted on the variability within groups of organisms, preserving advantageous traits and eliminating disadvantageous ones.

FIGURE 20.11 **The Tree of Life.** Darwin envisioned the history of life as a tree. Branching points represent the origins of new lineages; branches that do not reach the top represent extinct groups.

FOCUS ON **Basic Research**

Charles Darwin's Life as a Scientist

Darwin's observations during the voyage of H.M.S. *Beagle* convinced him that species change through time, and that natural processes produced Earth's biodiversity. He spent the rest of his life gathering data to test his ideas and unravel the workings of natural selection.

Shortly after the *Beagle* returned to England in 1836, Darwin started to publish the geological and biological research that he had undertaken during the voyage. This task took him 10 years to complete—twice as long as the journey itself. The results of these efforts were numerous articles and several books, including the now famous *Journal of the Voyage of the Beagle,* published in 1839.

After preparing a sketch of his ideas about evolution in 1844, Darwin was puzzled by one species of barnacle, a small marine invertebrate, which he had collected in Chile. Over the next eight years he studied more than 10,000 barnacle specimens and revised their entire classification. Although colleagues viewed this study as a strange diversion from his work on evolution, the project honed Darwin's observational skills and provided a test case in which he could apply his ideas about descent with modification to a large and diverse group of organisms. He published four volumes about barnacles in 1854.

While studying barnacles, Darwin continued to think about "the species question." He kept notebooks about variation in plants and animals, focusing on variation that was amplified by selective breeding. He was a tireless collector of facts, which he sought from every possible source. He badgered dog breeders, horse farmers, and horticulturists with long lists of questions about their work. His enthusiasm was infectious, and workers throughout the world supplied him with data and specimens. Darwin was also an eager and skilled experimentalist, and he took up pigeon breeding, marveling at the huge variety of morphological traits that he and other breeders could produce. In the late 1850s, a communication from Alfred Russel Wallace (see *Why It Matters*) forced him to complete *The Origin,* which revolutionized the study of biology.

Even after *The Origin* was published, Darwin continued to gather facts and write about evolution, working almost to the day he died in 1882 at age 74. He published a detailed analysis of how earthworms improve the soil *(The Formation of Vegetable Mould through the Action of Worms)* and wrote books on several botanical topics, among them plants that eat animals *(Insectivorous Plants)*, and the tendency of plants to grow toward sunlight *(The Power of Movement in Plants)*. Darwin's work always had an evolutionary focus, however, and he produced several revisions of *The Origin,* as well as books on artificial selection *(Variation of Animals and Plants under Domestication)*, human ancestry *(The Descent of Man)*, and animal behavior *(The Expression of the Emotions in Men and Animals)*.

William Perlman/Star Ledger/Corbis

Darwin's study. Darwin undertook most of his life's work in this room at Down House. He hesitated to discard old papers and specimens, believing that he would find a use for them as soon as they were carried away in the trash.

Four characteristics distinguish Darwin's theory from earlier explanations of biological diversity and adaptive traits:

1. Darwin provided purely physical, rather than spiritual, explanations for the origins of biological diversity.
2. Darwin recognized that evolutionary change occurs in groups of organisms, rather than in individuals: some members of a group survive and reproduce more successfully than others.
3. Darwin described evolution as a multistage process: variations arise within groups, natural selection eliminates unsuccessful variations, and the next generation inherits successful variations.
4. Like Lamarck, Darwin understood that evolution occurs because some organisms function better than others *in a particular environment.*

What is most amazing about Darwin's intellectual achievement is that he knew nothing about genetics (see Chapter 12). Thus, he had no clear idea of how variation arose or how it was passed from one generation to the next.

Evolution was a popular topic in Victorian England, and Darwin's theory was both praised and ridiculed. Although he had not speculated about the evolution of humans in *The Origin,* many readers were quick to extrapolate Darwin's ideas to our own species. Needless to say, certain influential Victorians were not amused by the suggestion that humans and apes share a common ancestry.

Nevertheless, Darwin's painstaking logic and careful documentation convinced most readers that evolution really does take place. Thomas Huxley, so staunch an advocate that he was known as "Darwin's bulldog," summed up the reaction of many when he quipped that the theory was so obvious, once articulated, that he was surprised he had not thought of it himself.

Darwin's vision of common ancestry quickly became the intellectual framework for nearly all biological research. Many readers, however, did not readily accept the mechanism of natural selection. The major stumbling block was that Darwin had not provided any plausible theory of heredity.

STUDY BREAK 20.2

1. What observations that Darwin made on his round-the-world voyage influenced his later thoughts about evolution?
2. How did Darwin's understanding of artificial selection enable him to envision the process of natural selection?
3. What were the four great intellectual triumphs of Darwin's theory?

THINK OUTSIDE THE BOOK

Find a copy of *On the Origin of Species by Means of Natural Selection* in a library or bookstore or online, and read a chapter or two. How did Darwin's careful observation of organisms allow him to draw conclusions? Do you think that straightforward observation still has a role in twenty-first–century science?

20.3 Evolutionary Biology since Darwin

Although Gregor Mendel published his work on genetics in 1866, it was not well known in England until 1900. At that time, scientists perceived a fundamental conflict between Darwin's and Mendel's theories. One problem was that Darwin had used complex characteristics, such as the structure of bird bills, to illustrate how natural selection worked. We now know that several genes often control such traits (see Section 12.2). By contrast, Mendel had studied simpler characteristics, such as the height of pea plants. A single gene often controls simple traits, which is one reason Mendel could interpret his experimental results so clearly. Biologists had a hard time applying Mendel's straightforward experimental results to Darwin's complex examples.

A second problem arose because Darwin believed that biological evolution occurred gradually over many generations. However, early twentieth-century geneticists, focusing on simple traits such as those Mendel had studied, sometimes observed very rapid and dramatic changes in certain characteristics. A widely accepted theory, *mutationism,* suggested that evolution occurred in spurts, induced by the chance appearance of "hopeful monsters," rather than by gradual change.

The Modern Synthesis Created a Unified Theory of Evolution

In the early twentieth century, Thomas Hunt Morgan of Columbia University discerned that genes are carried on chromosomes. His experiments, described in Chapter 13, enabled geneticists and mathematicians to forge a critical link between Darwin's and Mendel's ideas. The new discipline, **population genetics,** recognized the importance of genetic variation as the raw material of evolution. Population geneticists constructed mathematical models, which applied equally well to simple and complex traits, to predict how natural selection and other processes influence a population's genetics.

In the 1930s and 1940s, a unified theory of evolution, the **modern synthesis,** integrated data from biogeography, comparative morphology, comparative embryology, paleontology, and taxonomy within an evolutionary framework. The authors of the modern synthesis focused on evolutionary change within populations, and although they considered natural selection the primary mechanism of evolution, they acknowledged the importance of other processes. Proponents of the modern synthesis also embraced Darwin's idea of gradualism and de-emphasized the significance of mutations that changed traits suddenly and dramatically.

The modern synthesis also tried to link the two levels of evolutionary change that Darwin had identified: microevolution and macroevolution. **Microevolution** describes the small-scale genetic changes that populations undergo, often in response to shifting environmental circumstances; a small evolutionary shift in the size of the bill of a finch species is an example of microevolution (see Chapter 21). **Macroevolution** describes larger-scale evolutionary changes observed in species and more inclusive groups. According to the modern synthesis, macroevolution results from the gradual accumulation of microevolutionary changes; researchers have recently begun to unravel the genetic mechanisms that establish a relationship between these two levels of evolutionary change (see Chapter 23).

Research in Many Fields Has Provided Evidence of Evolutionary Change

Since the emergence of the modern synthesis, scientists have assembled a huge and compelling body of evidence from many biological disciplines indicating that biological evolution is a fact of life on Earth.

ADAPTATION BY NATURAL SELECTION Biologists interpret the products of natural selection as evolutionary adaptations. For example, the wings of birds, which have been modified by evolutionary processes over millions of years, have an obvious function that helps these animals survive and reproduce. In Units 5 and 6 of this book, you will encounter many examples of adaptive structures in plants and animals that have been modeled by natural selection. Sometimes, however, natural selection operates on a short time scale, as illustrated by the development of pesticide resistance in insects. When we first use a new pesticide, a low concentration often kills a large percentage of the pests. However, just by chance, a few insects may have genetic characteristics that confer resistance to the poison. These indi-

FIGURE 20.12 Experimental Research

How Exposure to Insecticide Fosters the Evolution of Insecticide Resistance

Question: Does exposure to insecticide foster the evolution of insecticide resistance in insect populations?

Experiment: Researchers studied samples of wild mosquitoes *(Anopheles culicifacies)* captured at a small village in India, where public health officials frequently sprayed the insecticide dichloro-diphenyl-trichloroethane (DDT) to control these pests. For each test, the researchers exposed samples of mosquitoes to a 4% concentration of DDT for 1 hour and then measured the percentage that died during the next 24 hours. Tests were repeated 12 months and 16 months after the first experiment.

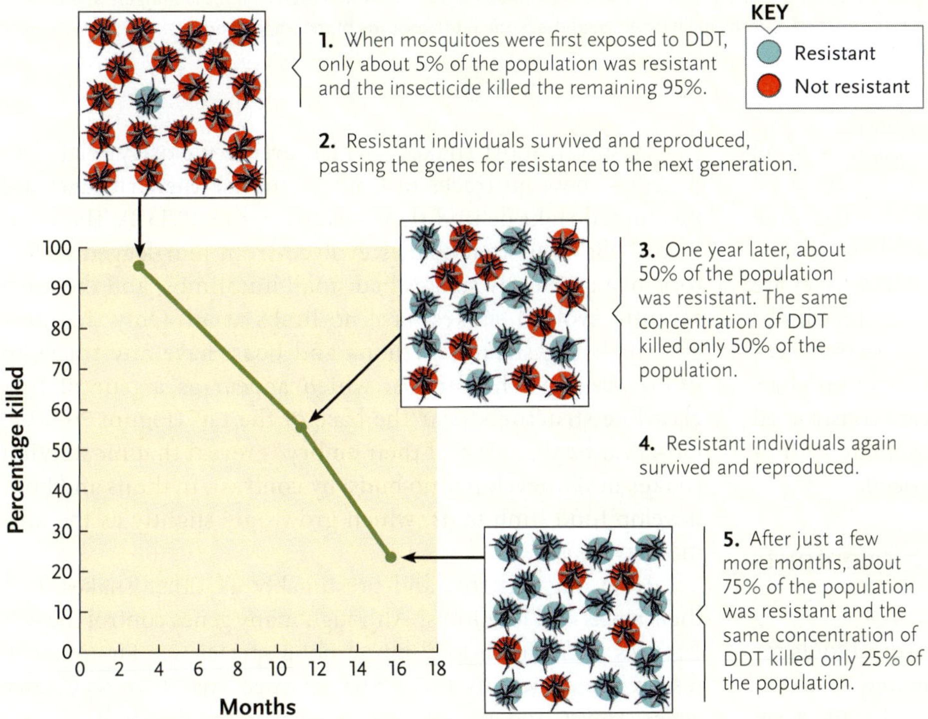

Results: Over the course of the experiment, smaller and smaller percentages of the mosquitoes died after their exposure to the test concentration of the insecticide.

Conclusion: When public health officials used DDT indiscriminately, they created an environment in which natural selection favored DDT-resistant individuals. Exposure to DDT therefore fostered the evolution of an adaptive resistance to DDT in the mosquito population.

THINK LIKE A SCIENTIST What effect would the continued use of DDT be likely to have on the mosquitoes' susceptibility to that chemical?

Source: A.M. Shalaby. 1968. Susceptibility studies on *Anopheles culicifacies* with DDT and Dieldrin in Gujarat state, India. *Journal of Economic Entomology* 61:533–541.

viduals survive and produce offspring, many of which inherit the resistance. As a result, a given concentration of the poison kills a smaller percentage of insects in the next generation, and over time, the entire population may become highly resistant **(Figure 20.12)**.

THE FOSSIL RECORD Because evolution results from the modification of existing species, Darwin's theory proposes that all species that have ever lived are genetically related. The fossil record documents such continuity in morphological characteristics, providing clear evidence of ongoing change in **biological lineages,** evolutionary sequences of ancestral organisms and their descendants (see Chapter 23). For example, the evolution of modern birds can be traced from a dinosaur ancestor through fossils such as *Archaeopteryx lithographica* **(Figure 20.13).** This species, discovered only two years after *The Origin* was published, resembled both dinosaurs and birds. Like the small carnivorous dinosaur *Dromaeosaurus, Archaeopteryx* walked on its hind legs and had teeth, long fingers with claws on its forelimbs, and a long, bony tail. Like modern birds, it had enlarged flight feathers on its forelimbs. Recently discovered fossils reveal that many bird ancestors were feathered (discussed further in Chapter 23).

HISTORICAL BIOGEOGRAPHY Analyses of **historical biogeography,** the study of the geographical distributions of plants and animals in relation to their evolutionary history, are generally consistent with Darwin's theory of evolution. Species on oceanic islands often closely resemble species on the nearest mainland, suggesting that the island and mainland species share a common ancestry. Moreover, species on a continental land mass are clearly related to one another and are often distinct from those on other continents. For example, monkeys in South America have long, prehensile tails and broad noses, traits that they inherited from a shared South American ancestor. By contrast, monkeys in Africa and Asia evolved from a different common ancestor in the Old World, and their shorter tails and narrower noses distinguish them from their American cousins.

COMPARATIVE MORPHOLOGY Other evidence of evolution comes from **comparative morphology,** analyses of the structure of living and extinct organisms. Such analyses are based on the comparison of **homologous traits,** characteristics that are similar in two species because they inherited the genetic basis of the trait

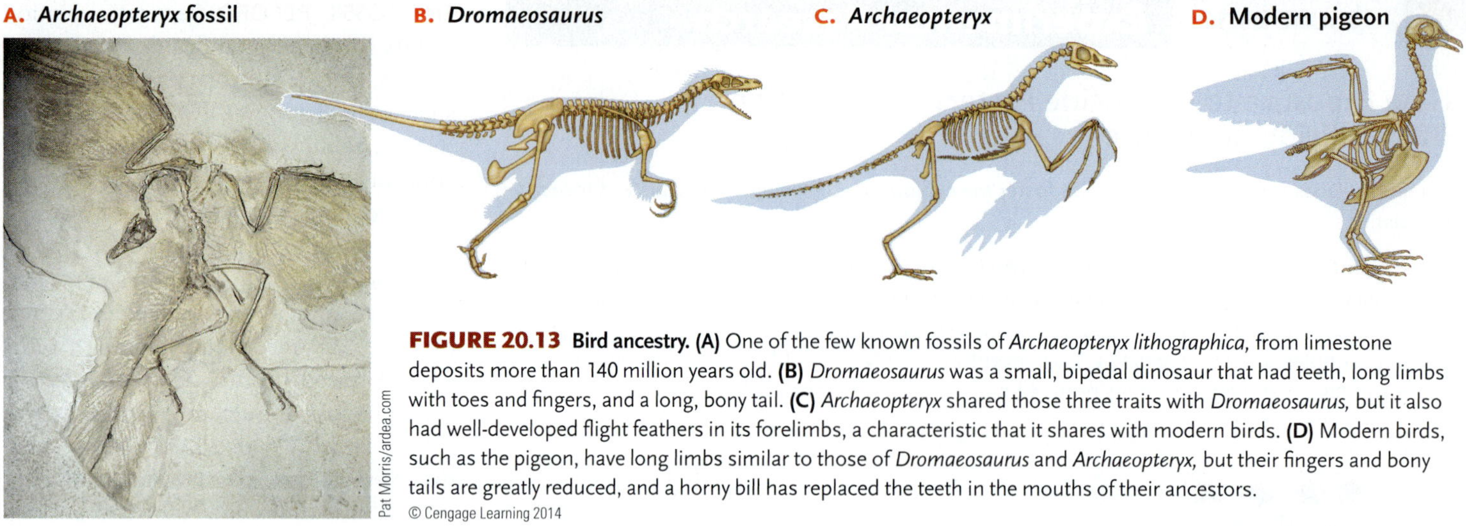

FIGURE 20.13 Bird ancestry. (A) One of the few known fossils of *Archaeopteryx lithographica*, from limestone deposits more than 140 million years old. **(B)** *Dromaeosaurus* was a small, bipedal dinosaur that had teeth, long limbs with toes and fingers, and a long, bony tail. **(C)** *Archaeopteryx* shared those three traits with *Dromaeosaurus*, but it also had well-developed flight feathers in its forelimbs, a characteristic that it shares with modern birds. **(D)** Modern birds, such as the pigeon, have long limbs similar to those of *Dromaeosaurus* and *Archaeopteryx*, but their fingers and bony tails are greatly reduced, and a horny bill has replaced the teeth in the mouths of their ancestors.

from their common ancestor. For example, the forelimbs of all four-legged vertebrates are homologous because they evolved from a common ancestor with a forelimb composed of the same component parts (see Figure 20.3, which shows homologous bones in the same color). Even though the shapes of the bones are different in pigs, dolphins, and bats, similarities in the three limbs are apparent. The differences in structural details arose over evolutionary time, allowing pigs to walk, dolphins to swim, and bats to fly. The arms of humans and the wings of birds are constructed of comparable elements, suggesting that they, too, share a common ancestor with the three species that are illustrated.

Molecular Techniques Extend the Achievements of the Modern Synthesis

Molecular techniques provide biologists with powerful tools for exploring all aspects of life—and evolutionary biology is no exception. In this section, we describe two examples that illustrate how molecular techniques have furthered our understanding of evolution since the modern synthesis was articulated.

HOW SNAKES LOST THEIR LEGS From Darwin's time until the mid-twentieth century, biologists tried to discern the evolutionary history of animals by comparing their embryos and patterns of development **(Figure 20.14).** The early embryos of related species are often strikingly similar, but morphological differences appear as the embryos grow and develop their adult forms (see Chapter 50). For example, the early embryos of most four-legged vertebrates (such as lizards, mammals, and birds) develop "limb buds" from which the legs or wings typical of each species later grow. Forelimbs and their supporting structures grow at the base of the neck, just in front of the ribcage, and hindlimbs grow right behind the ribcage (Figure 20.14A). Similarities in the limb buds and the positions of the limbs in these animals provide evidence of their descent from a shared ancestor. Differences in their adult structures are caused by additional genetic instructions that have evolved over time.

However, most snakes—which are very closely related to lizards—show no traces of limbs or necks; their ribcages are positioned right behind their heads (Figure 20.14B). The fossil record shows us that snakes evolved from four-legged ancestors in stages: early snakes had small hindlimbs, and the most recently evolved snakes have no limbs at all. Only the most ancient living snakes, pythons and boas, have any traces of limbs—vestigial hindlimbs, which appear as a pair of tiny claw-like structures near the base of the tail (Figure 20.14C). Observational studies of their embryos reveal that most living snakes never develop limb buds; by contrast, pythons and boas develop hind limb buds, which grow only slightly as the animal develops.

How did pythons, and presumably all other snakes, lose their necks and forelimbs? Although many genes control the differences in the adult forelimbs among species (see Figure 20.3), research reported in the 1990s revealed that two regulatory genes, *Hoxc6* and *Hoxc8*, determine whether forelimbs or ribs grow at a particular site along an animal's backbone. The *Hox* genes either activate or suppress other genes that direct the development of these structures. Forelimbs—but not ribs—grow just in front of the tissues where only *Hoxc6* is expressed. By contrast, ribs—but not forelimbs—grow where both *Hoxc6* and *Hoxc8* are expressed. In chickens and other vertebrates with four limbs, only *Hoxc6* is expressed just in front of the ribcage, and forelimbs grow nearby (Figure 20.14D).

In 1999, Martin J. Cohn of the University of Reading and Cheryll Tickle of University College London reported that in pythons (primitive snakes), both *Hoxc6* and *Hoxc8* are expressed all along the backbone, beginning at the base of the skull; as a result, a python's ribcage develops right behind its head, and no limb buds or forelimbs develop (Figure 20.14E). Thus, snakes have no forelimbs or necks because a mutation causes the expression of *Hoxc8* to extend into a more forward region of the animal's body. All descendants of the ancestor with that original mutation now lack necks and forelimbs. Cohn and Tickle's research also suggests that the second stage in snake

CLOSER LOOK **FIGURE 20.14** Genetics of limb loss in snakes.

A. Most lizards, like this monitor lizard (*Varanus* species), have four limbs attached to their backbones.

NHPA/Photoshot

B. Most snakes, like this grass snake (*Natrix* species), lack limbs altogether.

Simon D. Pollard/Photo Researchers, Inc.

C. Some primitive snakes, like this ball python (*Python regius*), have vestigial hindlimbs, visible as a pair of claw-like mating spurs near the base of the tail.

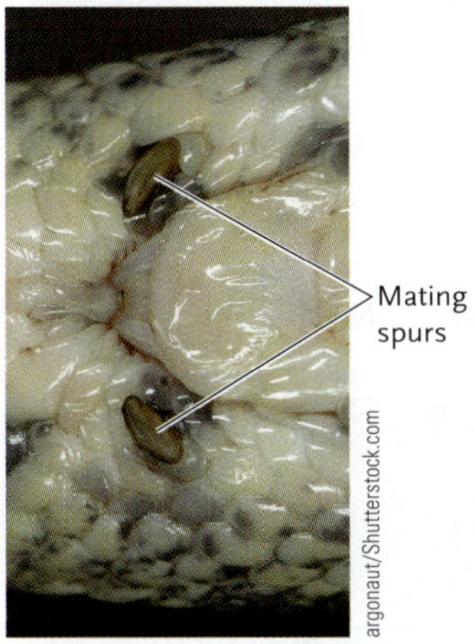

argonaut/Shutterstock.com

D. Molecular analyses reveal that the expression of *Hoxc6*, but not *Hoxc8*, at the base of the neck in a chick embryo (see arrow) causes limb buds and forelimbs to develop nearby.

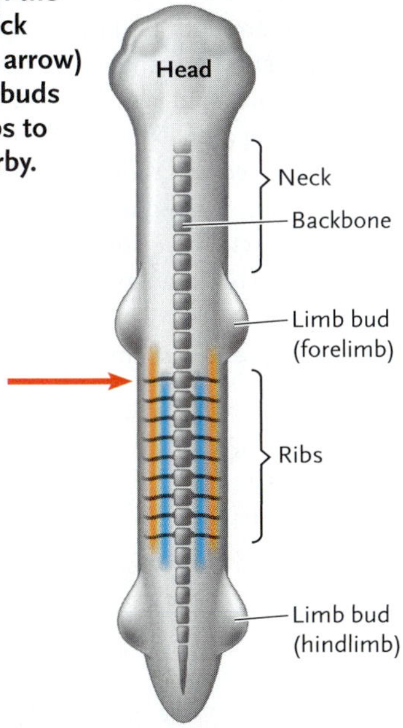

E. In pythons, both *Hoxc6* and *Hoxc8* are expressed alongside the backbone, beginning just behind the head (see arrow). The expression of these two genes appears to suppress the development of limb buds and forelimbs and promote the development of ribs at that location.

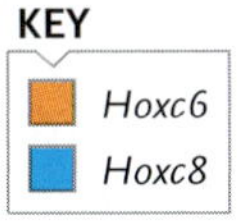

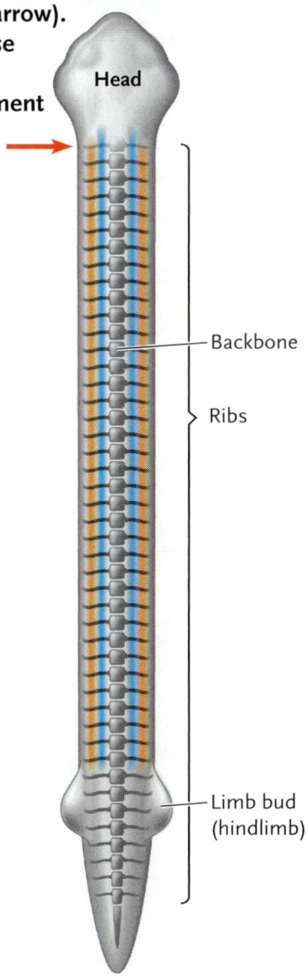

SUMMARY New molecular techniques from evolutionary developmental biology, coupled with traditional embryological observations, solved a long-standing evolutionary mystery. Evolutionary changes in the spatial pattern of gene expression caused the loss of forelimbs in snakes.

THINK LIKE A SCIENTIST This research identifies the genetic mechanism underlying the loss of forelimbs in snakes, but why might natural selection have favored this anatomical change?

evolution, the reduction or complete absence of hindlimbs, is caused by other genetic variations that appeared some time after the altered expression pattern of *Hoxc8*. Thus, molecular research has identified the genetic changes that caused snakes to lose their forelimbs and necks before losing their hindlimbs.

THE WOOLLY MAMMOTH'S CLOSEST LIVING RELATIVE Evolutionary biologists now routinely compare the genetic sequences of different species to determine their evolutionary relationships (see Chapter 24). The process is straightforward as long as researchers can collect small tissue samples from which they extract DNA. For

Asian Elephant or African Elephant: Which is the woolly mammoth's closest living relative?

Recent technical innovations now allow biologists to extract DNA from some fossils of extinct organisms, providing answers to long-puzzling questions. Based on morphological evidence, paleobiologists suspected that woolly mammoths *(Mammuthus primigenius)* were more closely related to living Asian elephants *(Elephas maximus)* than to living African elephants *(Loxodonta africana)*. When they recovered samples of mammoth DNA, researchers knew they could settle the matter.

Research Question

Were woolly mammoths genetically more similar to Asian elephants than to African elephants?

Experiment

Svante Pääbo and his colleagues at the Max Planck Institute for Evolutionary Anthropology, Leipzig, Germany, and collaborating researchers at three other institutions in England, Germany, and the United States, used molecular techniques—specifically, PCR amplification, cloning, and sequencing of mitochondrial DNA (see Chapter 18)—to answer the question **(Figure).**

1. They extracted DNA from a mammoth bone, estimated to be between 11,900 and 13,400 years old, from Berelek, Yakutia (in Siberia).
2. Based on known mitochondrial DNA sequences from the African and Asian elephant, they designed and synthesized 46 pairs of PCR primers. Assuming homology with the elephant sequences, these primer pairs were predicted to amplify the entire circular mitochondrial genome of the mammoth (yellow circle) as overlapping DNA fragments. All of the primer pairs were used in a single PCR, and the resulting amplified DNA fragments were cloned and sequenced. Based on their overlaps, the sequences could be arranged in a circle (blue and red), which showed that the researchers had succeeded in amplifying the entire mitochondrial genome.
3. Using computer algorithms, this mitochondrial DNA was compared with the mitochondrial genomes of Asian and African elephants.

Conclusion

The mammoth DNA sequences were more similar to those from the Asian elephant than those from the African elephant. Thus, Asian elephants are the woolly mammoth's closest living relatives. This conclusion makes biogeographic sense because woolly mammoths and Asian elephants occupied the same land mass (that is, Asia), whereas African elephants live on a different continent.

THINK LIKE A SCIENTIST How does the conclusion from this experiment match some of the observations Darwin made while traveling the globe on H.M.S *Beagle*?

Source: J. Krause et al. 2006. Multiplex amplification of the mammoth mitochondrial genome and the evolution of the Elephantidae. *Nature* 439:724–727.

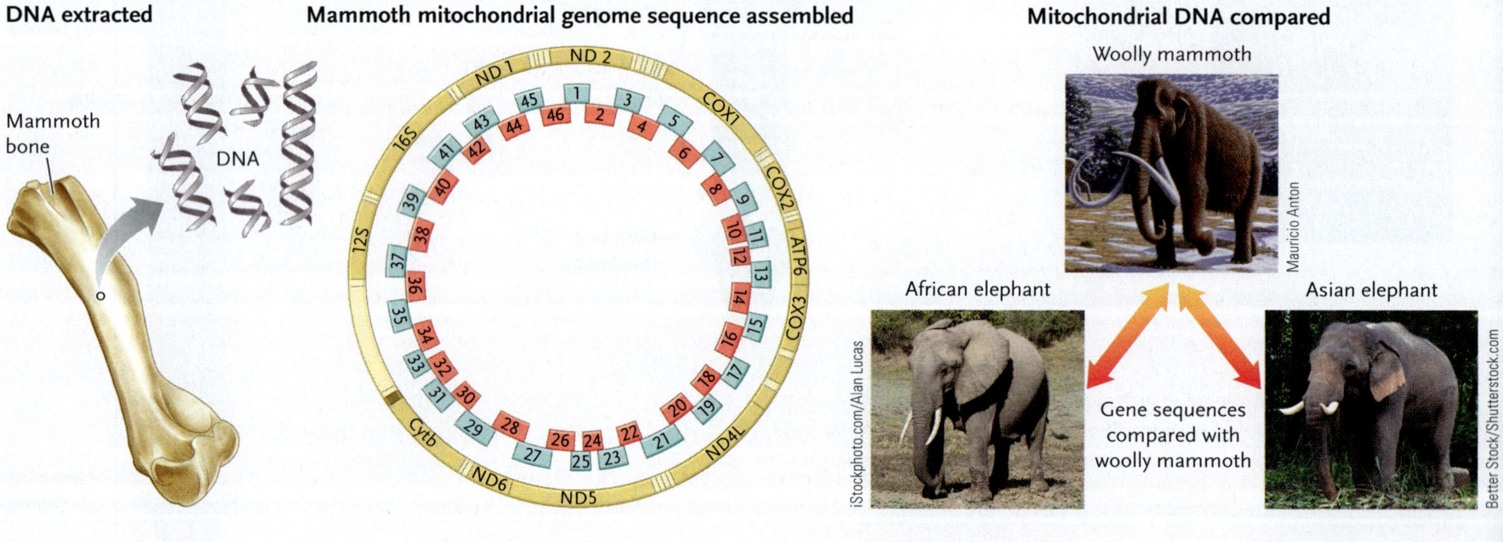

many years, paleobiologists could not use these techniques to determine the evolutionary relationships between extinct organisms and those living today because they could not obtain the necessary samples from fossilized material. Sequencing the DNA of extinct organisms was the "holy grail" of paleobiology.

Recent advances in molecular techniques now allow researchers to sequence DNA preserved in some fossils of extinct organisms. In a study published in 2006, Hendrick Poinar of McMaster University and colleagues at other institutions sequenced 13 million base pairs of mitochondrial and nuclear DNA extracted from the jawbone of a woolly mammoth *(Mammuthus primigenius)*, a species that has been extinct for at least 4,000 years **(Figure 20.15)**. The mammoth, which died 27,000 years ago, had been preserved in a frigid Siberian ice cave. When the researchers compared its DNA sequences to those from a living African elephant *(Loxodonta africana)*, they discovered that more than 98% of the sequence was identical in the two species, confirming their close evolutionary relationship. *Insights from the Molecular Revolution* describes an-

FIGURE 20.15 Woolly mammoths.

other study, published the same year, in which several European researchers analyzed the complete mitochondrial genome sequence of a woolly mammoth and compared it to the sequences from two living elephant species.

Some People Misinterpret the Theory of Evolution

The theory of evolution has always been a contentious subject because it challenges deeply held traditional views of how living organisms originated. Many of Darwin's contemporaries were dismayed by the suggestion that all organisms share a common ancestry. Some people even misinterpreted this assertion as "humans evolved from chimpanzees or gorillas." But the theory of evolution makes no such claims. Instead, it suggests that humans and apes are descended from an apelike common ancestor (see Section 32.13). In other words, an ancient population of organisms left descendants, which now include the living species of apes, as well as our own species. Moreover, the theory recognizes that evolution is an ongoing process: humans and apes have been evolving up until this very moment and will continue to evolve for as long as their descendants persist.

Early in the twentieth century, some scientists embraced the notion of **orthogenesis,** or progressive, goal-oriented evolution. This idea, derived from the *Scala Naturae,* suggests that evolution produces new species with the goal of improvement. We now know that evolution proceeds as an ongoing process of dynamic adjustment, not toward any fixed goal. Natural selection preserves the genes of organisms that function well in particular environments, but it cannot predict future environmental change. Imagine a population of plants with genes that affect how well they function under wet versus dry conditions. After a five-year drought, the population would include mostly dry-adapted plants. If a series of wet years follows the drought, these plants will be poorly adapted to the altered conditions. The process that favored drought-adapted plants operated under the prevailing dry conditions, not in anticipation of how conditions might change in the future.

Evolution is the core theory of modern biology because its explanatory power touches on every aspect of the living world. The application of molecular techniques to the study of evolutionary biology has greatly enhanced our knowledge. In the remaining chapters of this unit, you will discover how contemporary evolutionary theory explains changes at every level of biological organization, from adaptive modifications within populations (see Chapter 21), to the evolution of new species (see Chapter 22), major patterns in the history of life (see Chapter 23), and unraveling the evolutionary history of all organisms on Earth (see Chapter 24). Moreover, we will revisit many concepts from evolutionary biology in the discussions of biodiversity, the structure and function of plants and animals, and ecology and behavior in subsequent units.

STUDY BREAK 20.3

1. What two problems slowed the acceptance of Darwin's theory among scientists?
2. What is the difference between microevolution and macroevolution?
3. What types of data provide evidence that evolution has adapted organisms to their environments and promoted the diversification of species?

THINK OUTSIDE THE BOOK

Find examples from popular publications or advertisements for consumer products that misrepresent the theory of biological evolution. Explain how the theory is misrepresented.

UNANSWERED QUESTIONS

What determines whether a species adapts to a changing environment or becomes extinct?

Natural selection has produced marvelous adaptations in every species on Earth, and we know that evolutionary adaptation to certain environmental changes has allowed many species to persist. But we also know that more than 99% of the species that have ever lived became extinct, evidently because they failed to adapt to changes in climate, natural competitors or enemies, or other environmental factors. But what kinds of genetic variation are required for adaptation, and what kinds of characteristics must evolve to allow survival? This is a critical question today, because human activities are changing environments so rapidly and drastically that many species face the threat of extinction. Can aquatic species adapt to various kinds of water pollution? Can animals and plants that lived in prairies adapt to different habitats now that most prairies have been destroyed?

Is adaptation by natural selection responsible for most of the genetic differences between species?

New genetic variations sometimes become more common within populations or species because the proteins for which they code are advantageous and preserved by natural selection. But biologists who study molecular evolution have discovered that a large part of the genome in most organisms (about 98% of the human genome, for example) does not code for proteins and therefore appears to have no function. If this observation is generally correct, why do the noncoding parts of genomes exist? Are evolutionary changes in noncoding regions and the differences in noncoding sequences among species adaptive? For example, only about 1% of the DNA base pairs differ between human and chimpanzee genomes—but this amounts to about 34 million base-pair differences altogether, at least 60,000 of which alter the amino acid sequences of proteins. How can we determine which of these differences are adaptive and which differences underlie the unique characteristics of humans?

How do pathways of embryonic development evolve?

The characteristics of adult organisms are the product of developmental events, starting with the fertilized egg, that include growth in size, changes in the shape of various body parts, and the differentiation of cell types. These processes are largely controlled by genes, with input from the environment. Although biologists are beginning to learn how the genetic foundations of developmental processes evolve, many questions remain. For example, how do genetic changes induce differences in the branching patterns of antlers among species of deer, or differences in the length of the tails of monkeys and apes (including humans), or differences in the number and size of scales among species of lizards? We know that the proteins forming the lens of the eye are actually enzymes that play different roles in other cells, and that they have been "recruited" to form the lens, but what mechanisms induce them to assume this new role? And why do different enzymes form the lenses in eyes of birds and mammals? Evolutionary developmental biology, which is discussed in Chapter 23, is one of the most active, exciting fields in biology at this time.

THINK LIKE A SCIENTIST As our production of carbon dioxide increases Earth's average temperature faster than ever before, will Arctic species be able to adapt to changes in climate?

Courtesy of Douglas Futuyama

Douglas J. Futuyma is Distinguished Professor of evolution and ecology at Stony Brook University. His research interests focus on speciation and the evolution of ecological interactions among species, and in particular on insect–plant interactions. Learn more about Dr. Futuyma's work at http://life.bio.sunysb.edu/ee/people.htm.

REVIEW KEY CONCEPTS

To access the course materials and companion resources for this text, please visit www.cengagebrain.com.

20.1 Recognition of Evolutionary Change

- Ancient Greek philosophers classified the natural world, ranking inanimate objects and living organisms from simple to complex.
- Natural theologians believed that all species were specially created and perfectly adapted; that existing species do not change or become extinct; and that new species do not arise. Studies in biogeography, comparative morphology, and paleontology led scientists to question whether species change through time (Figures 20.2 and 20.3).
- Lamarck proposed that species evolved into more complex forms that functioned better in their environments. He hypothesized that structures in an organism changed when they were used, and that those changes were inherited by the organism's offspring (Figure 20.4). Experiments have refuted Lamarck's proposed mechanisms.
- Two geologists, Hutton and Lyell, recognized that major features on Earth were created by the long-term action of the very slow geological processes that scientists observe today. Their insights suggested that Earth was much older than natural theologians had supposed.

Animation: Evolution of limb bones

20.2 Darwin's Journeys

- Darwin's observations during his voyage on the *Beagle* provided much of the data and inspiration for the development of his theory of evolution (Figures 20.5–20.8).
- Darwin based the theory of evolution by means of natural selection on three inferences: (1) individuals within a population compete for limited resources, (2) hereditary characteristics allow some individuals to survive longer and reproduce more than others, and (3) a population's characteristics change over time as advantageous heritable characteristics become more common (Figure 20.10).
- Darwin also proposed that the accumulation of differences fostered by natural selection could cause populations to diverge over time. Such evolutionary divergence can lead to the production of new species, which can, in turn, give rise to new evolutionary lineages (Figures 20.9 and 20.11).

Animation: Finches of the Galápagos

Animation: The Galápagos Islands

20.3 Evolutionary Biology since Darwin

- Scientists working in population genetics developed theories of evolutionary change by integrating Darwin's ideas with Mendel's research on genetics.
- In the 1930s and 1940s, the modern synthesis provided a unified view of evolution that drew on studies from many biological disciplines. It emphasized evolution within populations, the central role of variation in the evolutionary process, and the gradualism of evolutionary change.
- Studies of adaptation, the fossil record, historical biogeography, and comparative morphology provide compelling evidence of evolutionary change (Figures 20.12 and 20.13).
- Molecular techniques have extended the achievements of the modern synthesis, allowing precise analysis of the genetic basis of evolutionary change (Figure 20.14) and the genetic relatedness of living and extinct organisms (Figure 20.15).

Animation: Comparative pelvic anatomy

Animation: Antibiotic resistance

UNDERSTAND & APPLY

Test Your Knowledge

1. Which of the following statements about evolutionary studies is *not* true?
   a. Biologists study the products of evolution to understand the processes causing it.
   b. Biologists design molecular experiments to examine evolutionary processes operating over short time periods.
   c. Biologists study the inheritance of characteristics that a parent acquired during its lifetime.
   d. Biologists study variation in homologous structures among related organisms.
   e. Biologists examine why a huge variety of species may inhabit a small island cluster.
2. Which of the following ideas is *not* included in Darwin's theory?
   a. All organisms that have ever existed arose through evolutionary modifications of ancestral species.
   b. The great variety of species alive today resulted from the diversification of ancestral species.
   c. Natural selection drives some evolutionary change.
   d. Natural selection preserves advantageous traits.
   e. Natural selection eliminates adaptive traits.
3. The "father" of taxonomy is:
   a. Charles Darwin.
   b. Charles Lyell.
   c. Alfred Wallace.
   d. Carolus Linnaeus.
   e. Jean Baptiste de Lamarck.
4. The wings of birds, the legs of pigs, and the flippers of dolphins provide examples of:
   a. vestigial structures.
   b. homologous structures.
   c. acquired characteristics.
   d. artificial selection.
   e. uniformitarianism.
5. Which of the following statements is *not* compatible with Darwin's theory?
   a. All organisms have arisen by descent with modification.
   b. Evolution has altered and diversified ancestral species.
   c. Evolution occurs in individuals rather than in groups.
   d. Natural selection eliminates unsuccessful variations.
   e. Evolution occurs because some individuals function better than others in a particular environment.
6. Which of the following does *not* contribute to the study of evolution?
   a. population genetics
   b. inheritance of acquired characteristics
   c. the fossil record
   d. DNA sequencing
   e. comparative morphology
7. Which of the following could be an example of microevolution?
   a. a slight change in a bird population's color due to a small genetic change in the population
   b. large differences between fossils found near the ground surface and those found in deep rock layers
   c. the sudden disappearance of an entire genus
   d. the direct evolutionary link between living primates and humans
   e. a flood that drowns all members of a population
8. Which of the following ideas proposed by Lamarck was *not* included in Darwin's theory?
   a. Organisms change in response to their environments.
   b. Changes that an organism acquires during its lifetime are passed to its offspring.
   c. All species change with time.
   d. New variations may be passed from one generation to the next.
   e. Specific mechanisms cause evolutionary change.
9. Medical advances now allow many people who suffer from genetic diseases to survive and reproduce. These advances:
   a. refute Darwin's theory.
   b. support Lamarck's theory.
   c. disprove descent with modification.
   d. reduce the effects of natural selection.
   e. eliminate adaptive traits.
10. The belief that evolution is progressive or goal-oriented is called:
   a. gradualism.
   b. uniformitarianism.
   c. taxonomy.
   d. orthogenesis.
   e. the modern synthesis.

Discuss the Concepts

1. Explain why the characteristics we see in living organisms adapt them to the environments in which their ancestors lived rather than to the environments in which they live today.
2. Imagine a population of mice that includes both brown and black individuals. They live in a habitat with brown soil, where predatory hawks can see black mice more easily than they can see brown ones. Design a study that would allow you to determine whether the brown mice are better adapted to this environment than the black mice.

Design an Experiment

Design an experiment to test Lamarck's hypothesis that characteristics acquired during an organism's lifetime are inherited by their offspring. (You may wish to review the components of a well-designed experiment in Chapter 1 before formulating your answer.) Can you think of examples of acquired characteristics that are *not* inherited by offspring?

Interpret the Data

For centuries, animal breeders have used artificial selection to increase the speed of racehorses. The results of such efforts are illustrated in a graph that identifies how quickly winning horses ran the 1 1/4 mile track at the Kentucky Derby, a famous race held annually at Churchill Downs, Kentucky. (The time it takes a horse to run a track of fixed length is inversely proportional to the horse's speed.) Based on the data in the graph, has artificial selection increased the speed of horses consistently between 1900 and 2000? Do you think that artificial selection can increase the speed of racehorses in the future?

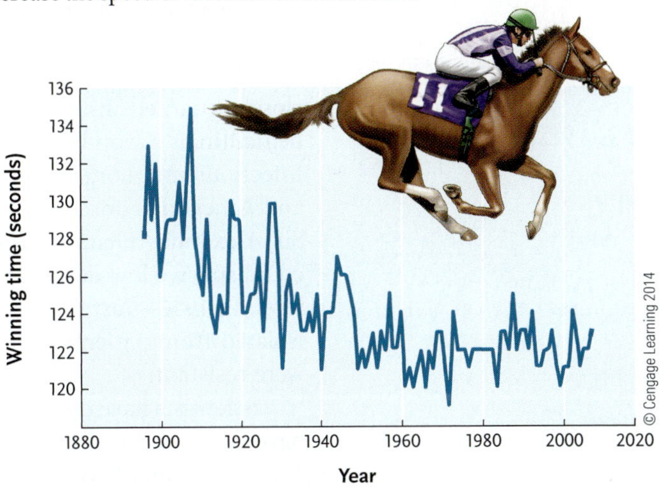

Apply Evolutionary Thinking

Identify three discoveries or inventions that have changed how humans are affected by natural selection. Describe in detail how each discovery influences survival or reproduction in our species.

21

Phenotypic variation. The frog *Dendrobates pumilio* exhibits dramatic color variation in populations that inhabit the Bocas del Toro Islands, Panama.

STUDY OUTLINE

21.1 Variation in Natural Populations

21.2 Population Genetics

21.3 The Agents of Microevolution

21.4 Maintaining Genetic and Phenotypic Variation

21.5 Adaptation and Evolutionary Constraints

Microevolution: Genetic Changes within Populations

Why it matters . . . On November 28, 1942, at the height of American involvement in World War II, a disastrous fire killed more than 400 people in Boston's Cocoanut Grove nightclub. Many more would have died later but for a new experimental drug, penicillin. A product of *Penicillium* mold, penicillin fought the usually fatal infections of *Staphylococcus aureus,* a bacterium that enters the body through damaged skin. Penicillin was the first antibiotic drug based on a naturally occurring substance that kills bacteria.

Until the disaster at the Cocoanut Grove, the production and use of penicillin had been a closely guarded military secret. But after its public debut, the pharmaceutical industry hailed penicillin as a wonder drug, promoting its use for the treatment of the many diseases caused by infectious microorganisms. Penicillin became widely available as an over-the-counter remedy, and Americans dosed themselves with it, hoping to cure all sorts of ills **(Figure 21.1).** But in 1945, Sir Alexander Fleming, the scientist who discovered penicillin, predicted that some bacteria could survive low doses, and that the offspring of those germs would be more resistant to its effects. In 1946—just four years after penicillin's use in Boston—14% of the *Staphylococcus* strains isolated from patients in a London hospital were resistant. By 1950, more than half the strains were resistant.

Scientists have discovered numerous antibiotics since the 1940s, and many strains of bacteria have developed resistance to these drugs. In fact, according to the Centers for Disease Control and Prevention, between 30,000 and 40,000 Americans die each year from infections caused by antibiotic-resistant bacteria.

How do bacteria become resistant to antibiotics? The genomes of bacteria—like those of all other organisms—vary among individuals, and some bacteria have genetic traits that allow them to withstand attack by antibiotics. When we administer antibiotics to an infected patient,

Image Courtesy of The Advertising Archives

FIGURE 21.1 Selling penicillin. This ad, from a 1944 issue of *Life* magazine, credits penicillin with saving the lives of wounded soldiers.

we create an environment favoring bacteria that are even slightly resistant to the drug. The surviving bacteria reproduce, and resistant microorganisms—along with the genes that confer antibiotic resistance—become more common in later generations. In other words, bacterial strains adapt to the presence of antibiotics in their environment through the evolutionary process of selection. Our use of antibiotics is comparable to artificial selection by plant and animal breeders (see Chapter 20), but when we use antibiotics, we inadvertently select for the success of the organisms we are trying to eradicate.

The evolution of antibiotic resistance in bacteria is an example of **microevolution,** which is a heritable change in the genetics of a population. A **population** of organisms includes all the individuals of a single species that live together in the same place and time. Today, when scientists study microevolution, they analyze variations—the differences between individuals—in natural populations and determine how and why these variations are inherited. Darwin recognized the importance of heritable variation within populations; he also realized that natural selection can change the pattern of variation in a population from one generation to the next. Scientists have since learned that microevolutionary change results from several processes, not just natural selection, and that sometimes these processes counteract each other.

In this chapter, we first examine the extensive variation that exists within natural populations. We then take a detailed look at the most important processes that alter genetic variation within populations, causing microevolutionary change. Finally, we consider how microevolution can fine-tune the functioning of populations within their environments.

21.1 Variation in Natural Populations

In some species, individuals vary dramatically in appearance, but in most species, the members of a population look pretty much alike **(Figure 21.2).** However, even those that look alike, such as the *Cerion* snails on the right in Figure 21.2, are not identical. With a scale and ruler, you could detect differences in their weight as well as in the length and diameter of their shells. With suitable techniques, you could also document variations in their individual biochemistry, physiology, internal anatomy, and behavior. All of these are examples of **phenotypic variation,** differences in appearance or function that are passed from generation to generation.

Evolutionary Biologists Describe and Quantify Phenotypic Variation

Darwin's theory recognized the importance of heritable phenotypic variation, and today, microevolutionary studies often begin by assessing phenotypic variation within populations. Most characters exhibit **quantitative variation:** individuals differ in small, incremental ways. If you weighed everyone in your biology class, for example, you would see that weight varies almost continuously from your lightest to your heaviest classmate. Humans also exhibit quantitative variation in the length of their toes, the number of hairs on their heads, and their height, as discussed in Section 12.2.

We usually display data on quantitative variation in a bar graph or, if the sample is large enough, as a curve **(Figure 21.3).** The width of the curve is proportional to the variability—the

A. European garden snails ***(Cepaea nemoralis)***

B. Bahaman land snails ***(Cerion christophei)***

George Bernard/Photo Researchers, Inc.

Timothy A. Pearce, Ph.D./Section of Mollusks/Carnegie Museum of Natural History. Photograph by Mindy McNaugher

FIGURE 21.2 Phenotypic variation. (A) Shells of European garden snails from a population in Scotland vary considerably in appearance. **(B)** By contrast, shells of land snails from a population in the Bahamas look very similar.

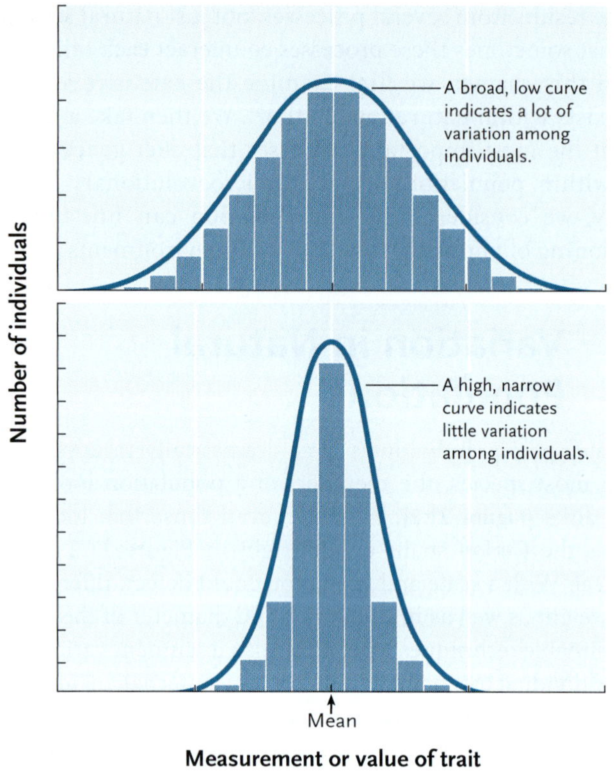

FIGURE 21.3 Quantitative variation. Many traits vary continuously among members of a population, and a bar graph of the data often approximates a bell-shaped curve. The mean defines the average value of the trait in the population, and the width of the curve is proportional to the variability among individuals.

amount of variation—among individuals, and the *mean* describes the average value of the character. As you will see shortly, natural selection often changes the mean value of a character or its variability within populations.

Other characters, like those Mendel studied (see Section 12.1), exhibit **qualitative variation:** they exist in two or more discrete states, and intermediate forms are often absent. Snow geese, for example, have *either* blue *or* white feathers **(Figure 21.4).** The existence of discrete variants of a character is called a **polymorphism** (*poly* = many, *morphos* = form); we describe such traits as *polymorphic.* The *Cepaea nemoralis* snail shells in Figure 21.2A are polymorphic in background color, number of stripes, and color of stripes. Biochemical polymorphisms, like the human A, B, AB, and O blood types (described in Section 12.2), are also common.

FIGURE 21.4 Qualitative variation. Individual snow geese *(Chen caerulescens)* are either blue or white. Although both colors are present in many populations, geese tend to associate with others of the same color.

We describe phenotypic polymorphisms quantitatively by calculating the percentage or *frequency* of each trait. For example, if you counted 123 blue snow geese and 369 white ones in a population of 492 geese, the frequency of the blue phenotype would be 123/492 or 0.25, and the frequency of the white phenotype would be 369/492 or 0.75.

Phenotypic Variation Can Have Genetic and Environmental Causes

Phenotypic variation within populations may be caused by genetic differences between individuals, by differences in the environmental factors that individuals experience, or by an interaction between genetics and the environment. As a result, genetic and phenotypic variations may not be perfectly correlated. Under some circumstances, organisms with different genotypes exhibit the same phenotype. For example, the black coloration of some rock pocket mice from Arizona is caused by certain mutations in the *Mc1r* gene (see Section 1.2); but black rock pocket mice from New Mexico do not share those mutations—that is, they have different genotypes—even though they exhibit the same phenotype. Conversely, organisms with the same genotype sometimes exhibit different phenotypes. For example, the acidity of soil influences flower color in some plants **(Figure 21.5).**

Knowing whether phenotypic variation is caused by genetic differences, environmental factors, or an interaction of the two is important because *only genetically based variation is subject to evolutionary change.* Moreover, knowing the causes of phenotypic variation has important practical applications. Suppose, for

FIGURE 21.5 Environmental effects on phenotype. Soil acidity affects the expression of the gene controlling flower color in the common garden plant *Hydrangea macrophylla.* When grown in acidic soil, it produces deep blue flowers. In neutral or alkaline soil, its flowers are bright pink.

FIGURE 21.6 Experimental Research

Using Artificial Selection to Demonstrate That Activity Level in Mice Has a Genetic Basis

Question: Do observed differences in activity level among house mice have a genetic basis?

Experiment: Swallow, Carter, and Garland knew that a phenotypic character responds to artificial selection only if it has a genetic, rather than an environmental, basis. In an experiment with house mice *(Mus domesticus)*, they selected for the phenotypic character of increased wheel-running activity. In four experimental lines, they bred those mice that ran the most. Four other lines, in which breeders were selected at random with respect to activity level, served as controls.

Results: After 10 generations of artificial selection, mice in the experimental lines ran longer distances and ran faster than mice in the control lines. Thus, artificial selection on wheel-running activity in house mice increased **(A)** the distance that mice run per day and **(B)** their average speed. The data illustrate responses of females in four experimental lines and four control lines. Males showed similar responses.

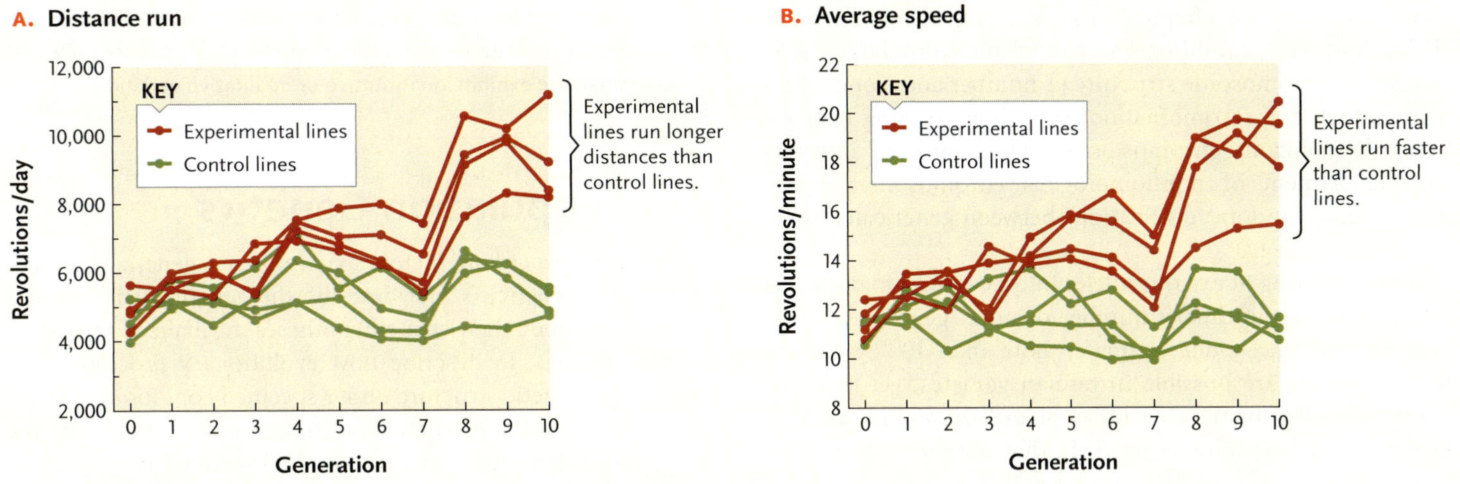

Conclusion: Because the two measures of activity level responded to artificial selection, researchers concluded that variation in this behavioral character has a genetic basis.

THINK LIKE A SCIENTIST Do the data presented in the graphs suggest that continued artificial selection would further increase the distance that the experimental lines run or their average speed?

Source: J. G. Swallow, P. A. Carter, and T. Garland, Jr. 1998. Artificial selection for increased wheel-running behavior in house mice. *Behavior Genetics* 28:227–237.

example, that one field of wheat produced more grain than another. If a difference in the availability of nutrients or water caused the difference in yield, a farmer might choose to fertilize or irrigate the less productive field. But if the difference in productivity resulted from genetic differences between plants in the two fields, a farmer might plant only the more productive genotype. Because environmental factors can influence the expression of genes, an organism's phenotype is frequently the product of an interaction between its genotype and its environment. In our hypothetical example, the farmer may maximize yield by fertilizing and irrigating the more productive genotype of wheat.

How can we determine whether phenotypic variation is caused by environmental factors or by genetic differences? We can test for an environmental cause experimentally by changing one environmental variable and measuring the effects on genetically similar subjects. You can try this yourself by growing some cuttings from an ivy plant in shade and other cuttings from the same plant in full sun. Although they all have the same genotype, the cuttings grown in sun will produce smaller leaves and shorter stems.

Breeding experiments can demonstrate the genetic basis of phenotypic variation. For example, Mendel inferred the genetic basis of qualitative traits, such as flower color in peas, by crossing plants with different phenotypes. Although simple crosses will not reveal the genetic basis of variations in quantitative traits, these characteristics will respond to artificial selection if the variation has some genetic basis. For example, researchers observed that individual house mice *(Mus domesticus)* differ in activity levels, as measured by how much they use an exercise wheel and how fast they run. John G. Swallow and his colleagues at the University of Wisconsin, Madison, used artificial selection to produce lines of mice that exhibit increased wheel-running behavior, demonstrating that the observed differences in these two aspects of activity level have a genetic basis **(Figure 21.6).**

Breeding experiments are not always practical, however, particularly for organisms with long generation times. Ethical concerns also render these techniques unthinkable for humans. Instead, researchers sometimes study the inheritance of particular traits by analyzing genealogical pedigrees, as discussed in Section 13.2, but this approach often provides poor results for analyses of complex traits.

Several Processes Generate Genetic Variation

Genetic variation, the raw material molded by microevolutionary processes, has two potential sources: the production of new alleles and the rearrangement of existing alleles. Most new alleles probably arise from small-scale mutations in DNA (described later in this chapter). The rearrangement of existing alleles into new combinations can result from larger scale changes in chromosome structure or number and from several forms of genetic recombination, including crossing over between homologous chromosomes during meiosis, independent assortment of nonhomologous chromosomes during meiosis, and random fertilizations between genetically different sperm and eggs.

The shuffling of *existing* alleles into new combinations can produce an extraordinary number of novel genotypes in the next generation. By one estimate, more than 10^{600} combinations of alleles are possible in human gametes, yet there are fewer than 10^{10} humans alive today. So unless you have an identical twin, it is extremely unlikely that another person with your genotype has ever lived or ever will.

Populations Often Contain Substantial Genetic Variation

How much genetic variation actually exists within populations? In the 1960s, evolutionary biologists began to use gel electrophoresis (see Figure 18.7) to identify biochemical polymorphisms in diverse organisms. This technique separates two or more forms of a given protein if they differ significantly in shape, mass, or net electrical charge. The identification of a protein polymorphism allowed researchers to infer genetic variation at the locus coding for that protein. This approach revealed much more genetic variation in natural populations than anyone had imagined. For example, nearly half the loci surveyed in many populations of plants and invertebrates are polymorphic. Advances in molecular biology now allow scientists to survey genetic variation directly, and researchers have accumulated an astounding knowledge of the structure of DNA and its nucleotide sequences. In general, studies of chromosomal and mitochondrial DNA suggest that every locus exhibits some variability in its nucleotide sequence among individuals from a single population, between populations of the same species, and between related species. However, some variations detected in the protein-coding regions of DNA may not affect phenotypes because, as explained in Section 15.1, they do not change the amino acid sequences of the proteins for which the genes code.

STUDY BREAK 21.1

1. If a population of skunks includes some individuals with stripes and others with spots, would you describe the variation as quantitative or qualitative?
2. In the experiment on house mice described in Figure 21.6, how did researchers demonstrate that variations in activity level had a genetic basis?
3. What factors contribute to phenotypic variation in a population?

THINK OUTSIDE THE BOOK

Go to a local park, field, or woodland and choose a population of plants or animals to observe. Make a list of the characteristics that show phenotypic variation in this population, describe their patterns of variation, and identify whether each characteristic exhibits quantitative or qualitative variation.

21.2 Population Genetics

To predict how certain factors may influence genetic variation, population geneticists first describe the genetic structure of a population. They then create hypotheses, formalized in mathematical models, to describe how evolutionary processes may change the genetic structure under specified conditions. Finally, researchers test the predictions of these models to evaluate the ideas about evolution that are embodied within them.

All Populations Have a Genetic Structure

Populations are made up of individuals, each with its own genotype. In diploid organisms, which have pairs of homologous chromosomes, an individual's genotype includes two copies of every gene. The sum of all gene copies at all gene loci in all individuals is called the population's **gene pool.**

To describe the structure of a gene pool, scientists first identify the genotypes in a representative sample and calculate **genotype frequencies,** the percentages of individuals possessing each genotype. Knowing that each diploid organism has two copies of each gene (either two copies of the same allele or two different alleles), a scientist can then calculate **allele frequencies,** the relative abundances of the different alleles. For a gene with two alleles, scientists use the symbol p to identify the frequency of one allele, and q the frequency of the other.

The calculation of genotype and allele frequencies for the two alleles at the gene locus governing flower color in snapdragons (genus *Antirrhinum*) is straightforward **(Table 21.1).** This locus is easy to study because it exhibits incomplete dominance (see Section 12.2). Individuals that are homozygous for the C^R allele (C^RC^R) have red flowers; those homozygous for the C^W allele (C^WC^W) have white flowers; and heterozygotes (C^RC^W) have pink flowers. Genotype frequencies represent how the C^R and C^W alleles are distributed among individuals. In this example, examination of the plants reveals that 45% of individuals have the C^RC^R genotype, 50% have the heterozygous C^RC^W genotype, and

TABLE 21.1 Calculation of Genotype Frequencies and Allele Frequencies for the Snapdragon Flower Color Locus

Because each diploid individual has two alleles at each gene locus, the entire sample of 1,000 individuals has a total of 2,000 alleles at the *C* locus.

Flower Color Phenotype	Genotype	Number of Individuals	Genotype Frequency[1]	Total Number of C^R Alleles[2]	Total Number of C^W Alleles[2]
Red	C^RC^R	450	450/1,000 = 0.45	2 × 450 = 900	0 × 450 = 0
Pink	C^RC^W	500	500/1,000 = 0.50	1 × 500 = 500	1 × 500 = 500
White	C^WC^W	50	50/1,000 = 0.05	0 × 50 = 0	2 × 50 = 100
	Total	1,000	0.45 + 0.50 + 0.05 = 1.0	1,400	600

Calculate allele frequencies using the total of 1,400 + 600 = 2,000 alleles in the sample:

p = frequency of C^R allele = 1,400/2,000 = 0.7

q = frequency of C^W allele = 600/2,000 = 0.3

$p + q = 0.7 + 0.3 = 1.0$

[1]Genotype frequency = the number of individuals possessing a particular genotype divided by the total number of individuals in the sample.

[2]Total number of C^R or C^W alleles = the number of C^R or C^W alleles present in one individual with a particular genotype multiplied by the number of individuals with that genotype.

the remaining 5% have the C^WC^W genotype. Allele frequencies represent the commonness or rarity of each allele in the gene pool. As calculated in the table, 70% of the alleles in the population are C^R and 30% are C^W. Remember that for a gene locus with two alleles, there are three genotype frequencies, but only two allele frequencies (p and q). The sum of the three genotype frequencies must equal 1; so must the sum of the two allele frequencies.

The Hardy–Weinberg Principle Is a Null Model That Defines How Evolution Does Not Occur

When designing experiments, scientists often use control treatments to evaluate the effect of a particular factor: the control tells us what we would see if the experimental treatment had no effect. As you may recall from the hypothetical example presented in Chapter 1 (see Figure 1.15), to determine whether fertilizer has an effect on plant growth, you must compare the growth of fertilized plants (the experimental treatment) to the growth of plants that received no fertilizer (the control treatment). However, in studies that use observational rather than experimental data, there is often no suitable control. In such cases, investigators develop conceptual models, called **null models,** which predict what they would see if a particular factor had no effect. Null models serve as theoretical reference points against which observations can be evaluated.

Early in the twentieth century, geneticists were puzzled by the persistence of recessive traits because they assumed that natural selection replaced recessive or rare alleles with dominant or common ones. An English mathematician, G. H. Hardy, and a German physician, Wilhelm Weinberg, tackled this problem independently in 1908. Their analysis, now known as the **Hardy–Weinberg principle,** specifies the conditions under which a population of diploid organisms achieves **genetic equilibrium,** the point at which neither allele frequencies nor genotype frequencies change in succeeding generations. Their work also showed that dominant alleles need not replace recessive ones, and that the shuffling of genes in sexual reproduction does not in itself cause the gene pool to change.

The Hardy–Weinberg principle describes how genotype frequencies are established in sexually reproducing organisms. According to this mathematical model, genetic equilibrium is possible only if *all* of the following conditions are met:

1. No mutations are occurring.
2. The population is closed to migration from other populations.
3. The population is infinitely large.
4. All genotypes in the population survive and reproduce equally well.
5. Individuals in the population mate randomly with respect to genotypes.

If the conditions of the model are met, the allele frequencies of the population will never change, and the genotype frequencies will stop changing after one generation. In short, under these restrictive conditions, microevolution will *not* occur (see *Focus on Basic Research* on pp. 462–463). The Hardy–Weinberg principle is thus a null model that serves as a reference point for evaluating the circumstances under which evolution *may* occur.

If a population's genotype frequencies do not match the predictions of this model or if its allele frequencies change over time, microevolution may be occurring. Determining which of the model's conditions are *not* met is a first step in understanding how and why the gene pool is changing.

STUDY BREAK 21.2

1. **What is the difference between the genotype frequencies and the allele frequencies in a population?**
2. **Why is the Hardy–Weinberg principle considered a null model of evolution?**
3. **If the conditions of the Hardy–Weinberg principle are met, when will genotype frequencies stop changing?**

Using the Hardy–Weinberg Principle

To see how the Hardy–Weinberg principle can be applied, we will analyze the snapdragon flower color locus, using the hypothetical population of 1,000 plants described in Table 21.1. This locus includes two alleles—C^R (with its frequency designated as p) and C^W (with its frequency designated as q)—and three genotypes—homozygous C^RC^R, heterozygous C^RC^W, and homozygous C^WC^W. Table 21.1 lists the number of plants with each genotype: 450 have red flowers (C^RC^R), 500 have pink flowers (C^RC^W), and 50 have white flowers (C^WC^W). It also shows the calculation of both the genotype frequencies (C^RC^R = 0.45, C^RC^W = 0.50, and C^WC^W = 0.05) and the allele frequencies (p = 0.7 and q = 0.3) for the population.

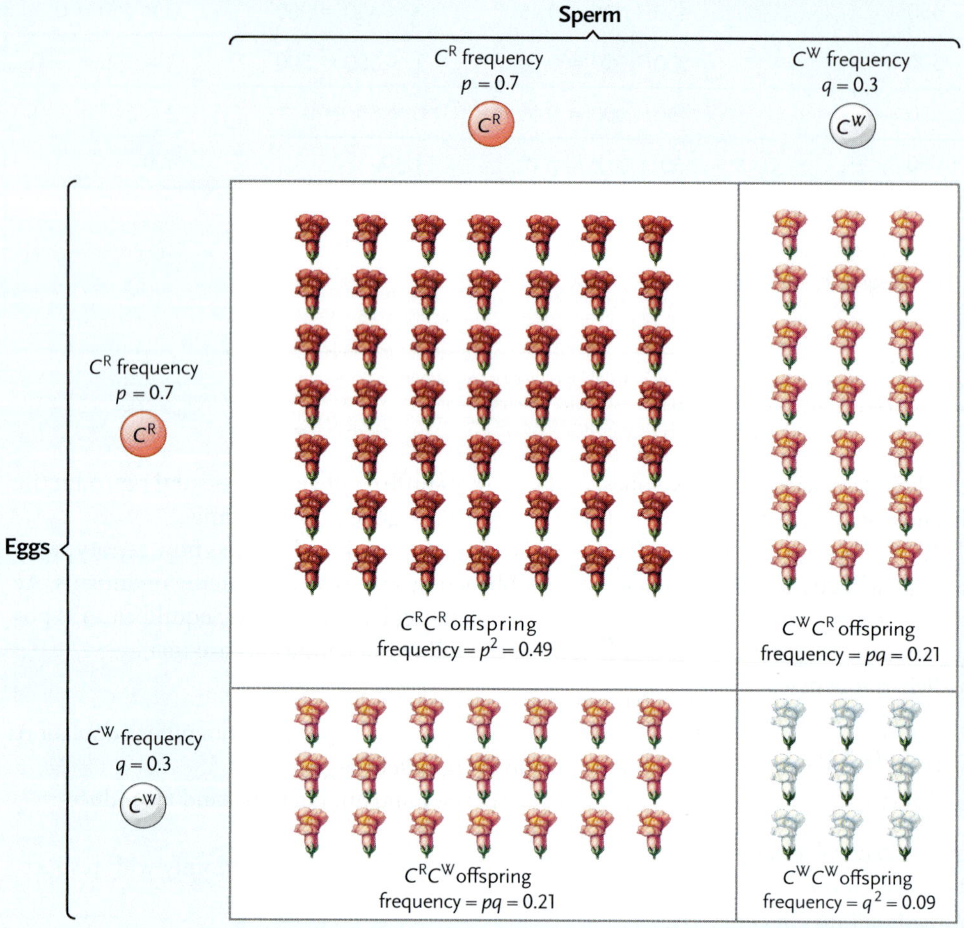

Let's assume for simplicity that each individual produces only two gametes and that both gametes contribute to the production of offspring. This assumption is unrealistic, of course, but it meets the Hardy–Weinberg requirement that all individuals in the population contribute equally to the next generation. In each parent, the two alleles segregate and end up in different gametes:

450 C^RC^R individuals produce	→	900 C^R gametes		
500 C^RC^W individuals produce	→	500 C^R gametes	+	500 C^W gametes
50 C^WC^W individuals produce	→			100 C^W gametes

21.3 The Agents of Microevolution

A population's allele frequencies will change over time if one or more conditions of the Hardy–Weinberg model are violated. In this section, we describe the processes that foster microevolutionary change—mutation, gene flow, genetic drift, natural selection, and nonrandom mating—which are summarized in **Table 21.2**.

Mutations Create New Genetic Variations

A **mutation** is a spontaneous and heritable change in DNA. Mutations are rare events: roughly one gamete in 100,000 to one in 1 million will include a new mutation at a particular gene locus. New mutations are so infrequent, in fact, that they exert little or no immediate effect on allele frequencies in most populations. But over evolutionary time scales, their numbers are significant—mutations have been accumulating in biological lineages for billions of years. And because it creates entirely new genetic variations, *mutation is a major source of heritable variation.*

For most animals, only mutations in the germ line (the cell lineage that produces gametes) are heritable; mutations in other cell lineages have no direct effect on the next generation. In plants, however, mutations may occur in meristem cells, which eventually produce flowers as well as nonreproductive structures (see Chapter 33); in such cases, a mutation may be passed to the next generation and ultimately influence the gene pool.

Deleterious mutations alter an individual's structure, function, or behavior in harmful ways. In mammals, for example, a protein called collagen is an essential component of most extracellular structures. Several simple mutations in humans cause forms of Ehlers–Danlos syndrome, a disruption of collagen synthesis that may result in loose skin, weak joints, or sudden death from the rupture of major blood vessels, the colon, or the uterus.

Lethal mutations can cause great harm to organisms carrying them. If a lethal allele is dominant, both homozygous and

You can readily see that 1,400 of the 2,000 total gametes carry the C^R allele and 600 carry the C^W allele. The frequency of C^R gametes is 1,400/2,000 or 0.7, which is equal to *p;* the frequency of C^W gametes is 600/2,000 or 0.3, which is equal to *q*. Thus, the allele frequencies in the gametes are exactly the same as the allele frequencies in the parent generation—it could not be otherwise because each gamete carries one allele at each locus.

Now assume that these gametes, both sperm and eggs, encounter each other at random. In other words, individuals reproduce without regard to the genotype of a potential mate. We can visualize the process of random mating in the mating table on the left.

We can also describe the consequences of random mating—$(p + q)$ sperm fertilizing $(p + q)$ eggs—with an equation that predicts the genotype frequencies in the offspring generation:

$$(p + q) \times (p + q) = p^2 + 2pq + q^2$$

If the population is at genetic equilibrium for this locus, p^2 is the predicted frequency of the C^RC^R genotype; $2pq$, the predicted frequency of the C^RC^W genotype; and q^2, the predicted frequency of the C^WC^W genotype. Using the gamete frequencies determined above, we can calculate the predicted genotype frequencies in the next generation:

$$\text{frequency of } C^RC^R = p^2 = (0.7 \times 0.7) = 0.49$$

$$\text{frequency of } C^RC^W = 2pq = 2(0.7 \times 0.3) = 0.42$$

$$\text{frequency of } C^WC^W = q^2 = (0.3 \times 0.3) = 0.09$$

Notice that the predicted genotype frequencies in the offspring generation have changed from the genotype frequencies in the parent generation: the frequency of heterozygous individuals has decreased, and the frequencies of both types of homozygous individuals have increased. This result occurred because the starting population was *not in equilibrium* at this gene locus. In other words, the distribution of parent genotypes did not conform to the predicted $p^2 + 2pq + q^2$ distribution.

The 2,000 gametes in our hypothetical population produced 1,000 offspring. Using the genotype frequencies we just calculated, we can predict how many offspring will carry each genotype:

490 red (C^RC^R)

420 pink (C^RC^W)

90 white (C^WC^W)

In a real study, we would examine the offspring to see how well their numbers match these predictions.

What about the allele frequencies in the offspring? The Hardy–Weinberg principle predicts that they did not change. Let's calculate them and see. Using the method shown in Table 21.1 and the prime symbol (′) to indicate offspring allele frequencies:

$$p' = ([2 \times 490] + 420)/2{,}000 = 1{,}400/2{,}000 = 0.7$$

$$q' = ([2 \times 90] + 420)/2{,}000 = 600/2{,}000 = 0.3$$

You can see from this calculation that the allele frequencies did not change from one generation to the next, even though the alleles were rearranged to produce different proportions of the three genotypes. Thus, the population is now at genetic equilibrium for the flower color locus; neither the genotype frequencies nor the allele frequencies will change in succeeding generations as long as the population meets the conditions specified in the Hardy–Weinberg model.

To verify this, you can calculate the allele frequencies of the gametes for this offspring generation and predict the genotype frequencies and allele frequencies for a third generation. You could continue calculating until you ran out of either paper or patience, but these frequencies will not change.

Researchers use calculations like these to determine whether an actual population is near its predicted genetic equilibrium for one or more gene loci. When they discover that a population is not at equilibrium, they infer that microevolution is occurring and can investigate the factors that might be responsible.

heterozygous carriers will, by definition, die from its effects; if recessive, it kills only homozygous recessive individuals. Any lethal mutation that causes an individual to die before it reproduces is eliminated from the population.

Neutral mutations are neither harmful nor helpful. Because of the degeneracy of the genetic code, several codons with different nucleotides in the third position may specify the same amino acid in the construction of a polypeptide chain (see Section 15.1). As a result, some DNA sequence changes—especially those in the third nucleotide of the codon—do not alter the amino acid sequence of the protein under construction. Not surprisingly, mutations at the third position appear to persist longer in populations than those at the first two positions. In other instances, mutations that change the amino acid sequence in a protein or even an organism's phenotype may have no influence on its survival and reproduction. A neutral mutation might even prove to be beneficial later if the environment changes.

Sometimes a change in DNA produces an *advantageous mutation,* which confers some benefit on an individual that carries it. However slight the advantage, natural selection may preserve the new allele and even increase its frequency over time. Once the mutation has been passed to a new generation, other agents of microevolution determine its long-term fate.

Gene Flow Introduces Novel Genetic Variants into Populations

Organisms or their gametes (for example, pollen) sometimes move from one population to another. If the immigrants reproduce, they may introduce novel alleles into a population, shifting its allele and genotype frequencies away from the values predicted by the Hardy–Weinberg model. This phenomenon, called **gene flow,** violates the Hardy–Weinberg requirement that populations must be closed to migration.

TABLE 21.2	Agents of Microevolutionary Change	
Agent	**Definition**	**Effect on Genetic Variation**
Mutation	Heritable change in DNA	Introduces new genetic variation into population; does not change allele frequencies quickly
Gene flow	Change in allele frequencies as individuals join a population and reproduce	May introduce genetic variation from another population
Genetic drift	Random changes in allele frequencies caused by chance events	Reduces genetic variation, especially in small populations; can eliminate rare alleles
Natural selection	Differential survivorship or reproduction of individuals with different genotypes	One allele can replace another or allelic variation can be preserved
Nonrandom mating	Choice of mates based on their phenotypes and genotypes	Does not directly affect allele frequencies, but usually prevents genetic equilibrium

FIGURE 21.7 Gene flow. Blue jays *(Cyanocitta cristata)* serve as agents of gene flow for oaks (genus *Quercus*) when the birds carry acorns from one oak population to another. An uneaten acorn may germinate and contribute to the gene pool of the population into which it was carried.

Gene flow is common in some animal species. For example, young male baboons typically move from one local population to another after experiencing aggressive behavior by older males. And many marine invertebrates disperse long distances as larvae carried by ocean currents.

Dispersal agents, such as pollen-carrying wind or seed-carrying animals, are responsible for gene flow in most plant populations. For example, blue jays foster gene flow among populations of oaks by carrying acorns from nut-bearing trees to their winter caches, which may be as much as a mile away **(Figure 21.7)**. Transported acorns that go uneaten may germinate and contribute to the gene pool of a neighboring oak population.

The movement of individuals from one population to another is not sufficient to foster gene flow between two populations. The immigrants must also reproduce in the population they join, thereby contributing to its gene pool. In the San Francisco Bay area, for example, Bay checkerspot butterflies *(Euphydryas editha bayensis)* rarely move from one population to another because they are poor fliers (see Figure 52.19). But when adult females do change populations, it is often late in the breeding season, and their offspring have virtually no chance of finding enough food to mature. Thus, many immigrant females do not foster gene flow because they do not contribute to the gene pool of the population they join.

The evolutionary importance of gene flow depends on the degree of genetic differentiation between populations and the rate of gene flow between them. If two gene pools are very different, a little gene flow may increase genetic variability within the population that receives immigrants, and it will make the two populations more similar. But if populations are already genetically similar, even lots of gene flow will have little effect.

Genetic Drift Reduces Genetic Variability within Populations

Chance events sometimes cause the allele frequencies in a population to change unpredictably. This phenomenon, known as **genetic drift,** has especially dramatic effects on small populations, which clearly violate the Hardy–Weinberg assumption of infinitely large population size.

A simple analogy clarifies why genetic drift is more pronounced in small populations than in large ones. When individuals reproduce, male and female gametes often pair up randomly, as though the allele in any particular sperm or ovum was determined by a coin toss. Imagine that "heads" specifies the *R* allele and that "tails" specifies the *r* allele. If the two alleles are equally common (that is, their frequencies, *p* and *q*, are both equal to 0.5), heads should be as likely an outcome as tails. But if you toss the coin 20 or 30 times to simulate random mating in a small population, you won't often see a 50:50 ratio of heads and tails. Sometimes heads will predominate and sometimes tails will—just by chance. Tossing the coin 500 times to simulate random mating in a somewhat larger population is more likely to produce a 50:50 ratio of heads and tails. And if you tossed the coin 5,000 times, you would get even closer to a 50:50 ratio.

Chance deviations from expected results—which cause genetic drift—occur whenever organisms engage in sexual reproduction, simply because their population sizes are not infinitely large. But genetic drift is particularly common in small populations because only a few individuals contribute to the gene pool and because any given allele may be present in very few individuals.

Genetic drift generally leads to the loss of alleles and reduced genetic variability; it therefore causes allele and genotype frequencies to differ from those predicted by the Hardy–Weinberg model. Two general circumstances, population bottlenecks and founder effects, often foster genetic drift.

POPULATION BOTTLENECKS On occasion, a stressful factor such as disease, starvation, or drought kills a large proportion of the individuals in a population, producing a **population bottleneck,** a dramatic reduction in population size. This cause of genetic drift greatly reduces genetic variation even if the population numbers later rebound **(Figure 21.8).**

In the late nineteenth century, for example, hunters nearly wiped out northern elephant seals *(Mirounga angustirostris)* along the Pacific coast of North America. Since the 1880s, when the species received protected status, the population has increased to more than 30,000, all descended from a group of about 20 survivors. Today the population exhibits no variation in 24 proteins studied by gel electrophoresis. This low level of genetic variation, which is unique among seal species, is consistent with the hypothesis that genetic drift eliminated many alleles when the population experienced the bottleneck.

FOUNDER EFFECT When a few individuals colonize a distant locality and start a new population, they carry only a small sample of the parent population's genetic variation. By chance, some alleles may be totally missing from the new population, whereas other alleles that were rare "back home" might occur at relatively high frequencies. This change in the gene pool is called the **founder effect.**

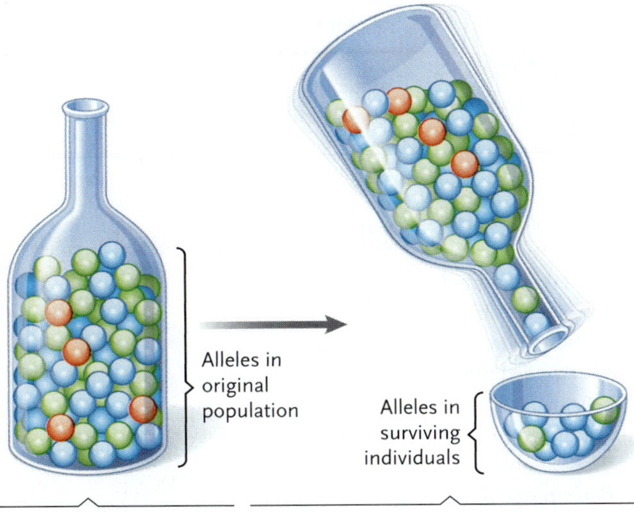

The gene pool of the original population, represented by a bottle filled with colored marbles, includes a locus with three alleles. Two of the alleles, represented by blue and green marbles, occur at high frequency; the third allele, represented by red marbles, occurs at low frequency.

If an environmental event randomly kills a large number of individuals in the population, the drastic reduction in population size is described as a population bottleneck. The process is analogous to shaking only a few of the marbles—the survivors—through the neck of the bottle. As a consequence of chance events associated with population bottlenecks, surviving individuals may not have the same allele frequencies as the original population. Rare alleles are inevitably lost.

FIGURE 21.8
Population bottlenecks, genetic drift, and the loss of genetic variability.

The human medical literature provides some of the best-documented examples of the founder effect. The Old Order Amish, an essentially closed religious community in Lancaster County, Pennsylvania, have an exceptionally high incidence of Ellis–van Creveld syndrome, a genetic disorder caused by a recessive allele. In the homozygous state, the allele produces dwarfism, shortened limbs, and polydactyly (extra fingers). Genetic analysis suggests that, although this syndrome affects less than 1% of the Amish in Lancaster County, as many as 13% may be heterozygous carriers of the allele. All of the individuals exhibiting the syndrome are descended from one couple who helped found the community in the mid-1700s.

CONSERVATION IMPLICATIONS Genetic drift has important implications for conservation biology. By definition, endangered species experience severe population bottlenecks, which result in the loss of genetic variability. Moreover, the small number of individuals available for captive breeding programs may not fully represent a species' genetic diversity. Without such variation, no matter how large a population may become in the future, it will be less resistant to diseases or less able to cope with environmental change.

For example, scientists believe that an environmental catastrophe produced a bottleneck in the African cheetah *(Acinonyx jubatus)* population 10,000 years ago. Cheetahs today are remarkably uniform in genetic make-up. Their populations are highly susceptible to diseases; males also have a high proportion of sperm cell abnormalities and a reduced reproductive capacity. Thus, limited genetic variation, as well as small numbers, threatens populations of endangered species. *Insights from the Molecular Revolution* describes techniques used to determine whether hunting has had the same effect on humpback whales.

Natural Selection Shapes Genetic Variability by Favoring Some Traits over Others

The Hardy–Weinberg model requires all genotypes in a population to survive and reproduce equally well. But as you know from reading Section 20.2, heritable traits enable some individuals to survive better and reproduce more than others. **Natural selection** is the process by which such traits become more common in subsequent generations. Thus, natural selection violates a requirement of the Hardy–Weinberg equilibrium and causes allele and genotype frequencies to differ from those predicted by the model.

Although natural selection can change allele frequencies in a population, *it is the phenotype of an individual organism, rather than any particular allele, that is successful or not.* When individuals survive and reproduce, their alleles—both favorable and unfavorable—are passed to the next generation. Of course, an organism with harmful or lethal dominant al-

INSIGHTS FROM THE Molecular Revolution

Back from the Brink: Did humpback whales lose genetic variability when they squeezed through a population bottleneck?

As a result of more than 250 years of commercial whaling, humpback whales *(Megaptera novaeangliae)* experienced a disastrous population decline—from about 125,000 to 5,000 individuals. After an international agreement limited whaling in 1966, humpbacks have rebounded strongly; about 80,000 individuals now form three distinct populations in the North Atlantic, North Pacific, and Southern oceans. Yet, conservation biologists wondered if the population bottleneck may have reduced their genetic variability. Such a loss could have adverse effects on the population's reproductive capacity, resistance to disease, and ability to survive unfavorable environmental changes.

Research Question

Do populations of humpback whales exhibit reduced genetic variability as a result of the population bottleneck they experienced before 1966?

Experiment

In the early 1990s, about 25 years after the ban on whaling took effect, a large group of researchers working in Hawaii, the continental United States, Australia, South Africa, Canada, Mexico, and the Dominican Republic measured genetic variability in surviving humpback populations. They studied mitochondrial DNA (mtDNA) because it is small, easily extracted and identified, and almost all of its variability comes from chance mutations that occur at a steady rate (rather than from genetic recombination; see Section 13.5). Except for variations produced by mutation that occurred after the population bottleneck—which can be estimated from the mutation rate and subtracted from the total variation detected—the variability in mtDNA should reflect the level present in the population during the bottleneck.

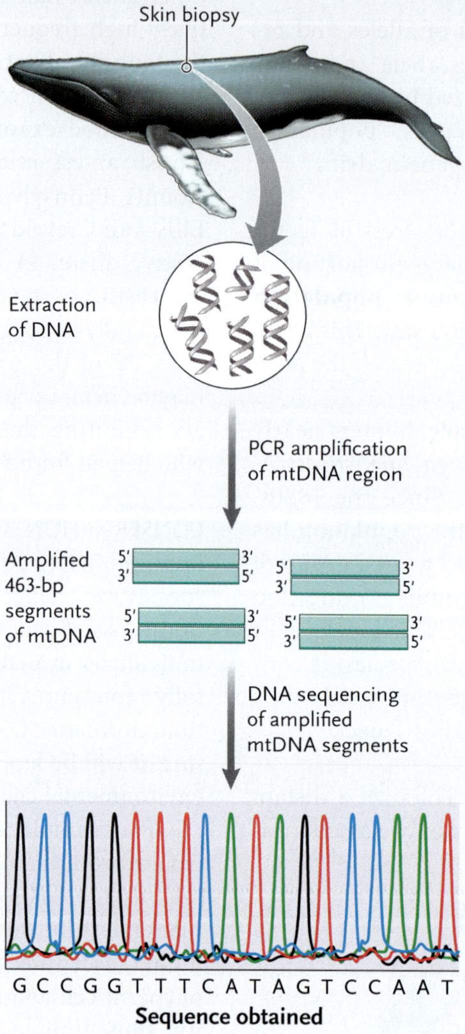

The researchers used biopsy darts to obtain skin samples from 90 humpback whales distributed among the three populations. They extracted DNA from each sample and used the polymerase chain reaction (PCR; see Figure 18.6) to amplify a 463-base pair segment of mtDNA that includes most of the variable nucleotide positions. They then determined the DNA sequences of the PCR products. (DNA sequencing is described in Section 19.2.)

Results

The researchers were surprised to find that the mtDNA sequence variation was relatively high in most of their sample. However, a subpopulation of the North Pacific population near Hawaii showed no variability at all in the mtDNA segment examined. One possible hypothesis for this result is that the Hawaiian subpopulation originated recently, perhaps during the twentieth century. There is indirect support for that idea: whaling records from the nineteenth century list no humpback sightings around Hawaii, and the native Hawaiian people have no legends or words describing whales of the humpback type (baleen whales). Perhaps the Hawaiian subpopulation was started by a few whales with limited genetic variability, an example of the founder effect.

Conclusion

With the exception of this Hawaiian subpopulation, humpback whales appear to have retained genetic variability comparable to that seen in other animals. Humpbacks have a potential life span of about 50 years. Thus, some individuals still alive at the time of the study had been born before the most intense period of commercial hunting in the mid-twentieth century; those individuals provided a reservoir of genetic variability from the old populations. These results suggest that the hunting ban came just in time to prevent a significant loss of genetic variability in humpback whales.

THINK LIKE A SCIENTIST If the ban on whaling had taken effect in 2000 instead of 1966, how much genetic variability do you think we would find in populations of humpback whales today?

Source: C. S. Baker et al. 1993. Abundant mitochondrial DNA variation and world-wide population structure in humpback whales. *Proceedings of the National Academy of Sciences USA* 90:8239–8243.

leles will probably die before reproducing, and all the alleles it carries will share that unhappy fate, even those that are advantageous.

To evaluate reproductive success, evolutionary biologists consider **relative fitness,** the number of surviving offspring that an individual produces compared with the number left by others in the population. Thus, a particular allele will increase in frequency in the next generation if individuals carrying that allele leave *more* offspring than individuals carrying other alleles. Differences in the *relative* success of individuals are the essence of natural selection.

Natural selection tests fitness differences at nearly every stage of an organism's life cycle. One plant may be fitter than others in the population because its seeds survive colder conditions, be-

CLOSER LOOK | **FIGURE 21.9 Three modes of natural selection.**

A hypothetical example using tail length of birds as the quantitative trait subject to selection.

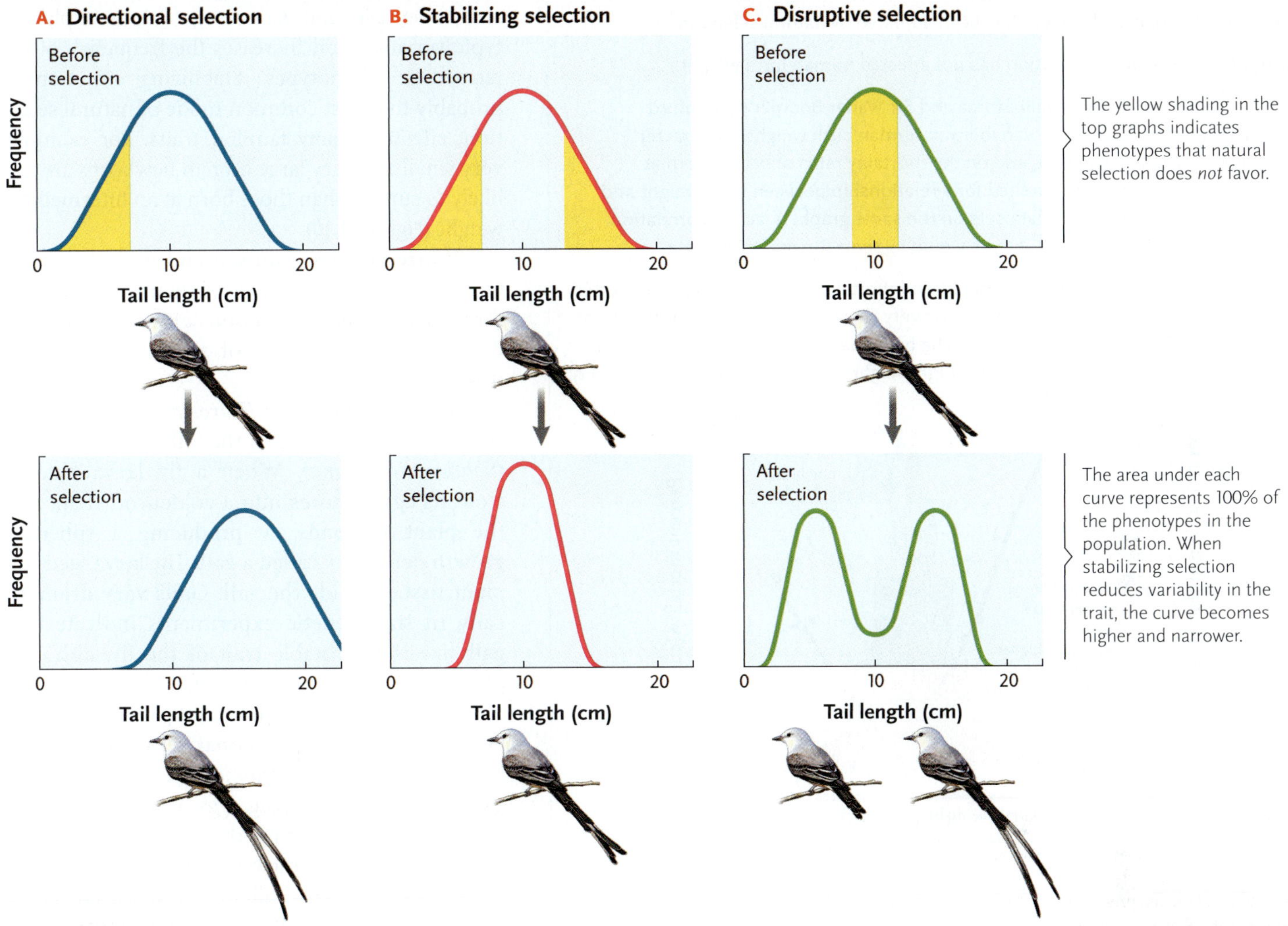

SUMMARY Directional selection shifts the mean phenotype toward the end of the distribution favored by natural selection. After selection has operated, the average tail length in the population has increased.

Stabilizing selection favors the mean phenotype over extreme phenotypes. After selection has operated, the mean tail length has not changed, but the variability in tail length among individuals in the population has decreased.

Disruptive selection decreases the frequency of the mean phenotype, but increases the frequencies of extreme phenotypes. After selection, the mean tail length may be unchanged, but the variability in tail length among individuals has increased.

THINK LIKE A SCIENTIST When people used artificial selection to produce the long body and short legs of a dachshund, which mode of selection did they use?

cause the arrangement of its leaves captures sunlight more efficiently, or because its flowers are more attractive to pollinators. However, natural selection exerts little or no effect on traits that appear during an individual's postreproductive life. For example, Huntington disease, a dominant-allele disorder that first strikes humans after the age of 40, is not subject to strong selection. Carriers of the disease-causing allele can reproduce before the onset of the condition, passing it to the next generation.

Biologists measure the effects of natural selection on phenotypic variation by recording changes in the mean and variability of characters over time (see Figure 21.3). Three modes of natural selection have been identified: directional selection, stabilizing selection, and disruptive selection **(Figure 21.9).**

DIRECTIONAL SELECTION Traits undergo **directional selection** when individuals near one end of the phenotypic spectrum have the highest relative fitness. Directional selection shifts a trait away from the existing mean and toward the favored extreme (see Figure 21.9A). After selection, the trait's mean value is higher or lower than before, and variability in the trait may be reduced.

FIGURE 21.10 Observational Research

Do Humans Experience Stabilizing Selection?

Hypothesis: Human birth weight has been adjusted by natural selection.

Null Hypothesis: Natural selection has not affected human birth weight.

Method: Geneticists Luigi Cavalli-Sforza and Sir Walter Bodmer of Stanford University collected data on the variability in human birth weight, a character exhibiting quantitative variation, and on the mortality rates of babies born at different weights. They then searched for a relationship between birth weight and mortality rate by plotting both data sets on the same graph. A lack of correlation between birth weight and mortality rate would support the null hypothesis.

Results: When birth weight (the bar graph) and mortality rate (the curve) are plotted together, the mean birth weight is seen to be very close to the optimum birth weight (the weight at which mortality is lowest). The two data sets also show that few babies are born at the very low weights and very high weights associated with high mortality.

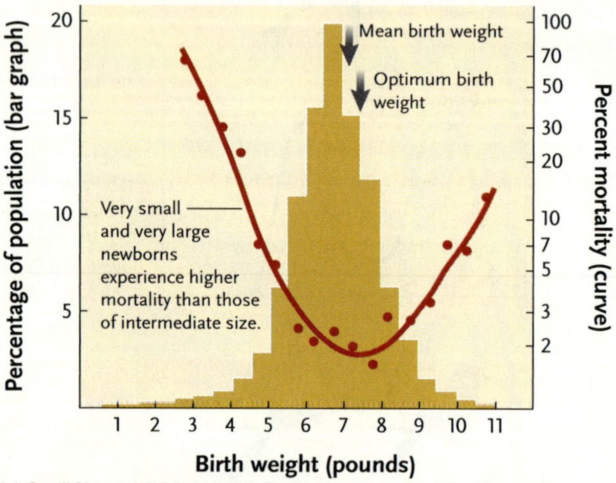

L. L. Cavalli-Sforza and W. F. Bodmer. 1971. *The Genetics of Human Populations.* Freeman, San Francisco. Reprinted by permission of John Wiley and Sons.

Conclusion: The shapes and positions of the birth weight bar graph and the mortality rate curve suggest that stabilizing selection has adjusted human birth weight to an average of 7 to 8 pounds.

THINK LIKE A SCIENTIST Having multiple births (that is, carrying three or more children in the same pregnancy) is historically very rare in humans. Why might natural selection have selected against a genetic predisposition for multiple births?

Source: L. L. Cavalli-Sforza and W. F. Bodmer. 1971. *The Genetics of Human Populations.* Freeman, San Francisco.

Directional selection is extremely common. For example, predatory fish promote directional selection for larger body size in guppies when they selectively feed on the smallest individuals in a guppy population (see *Focus on Basic Research* in Chapter 52). And most cases of artificial selection, including the experiment on the activity levels of house mice described above, are directional, aimed at increasing or decreasing specific phenotypic traits. Humans routinely use directional selection to produce domestic animals and crops with desired characteristics, such as the small size of Chihuahuas and the intense "bite" of chili peppers.

STABILIZING SELECTION Traits undergo **stabilizing selection** when individuals expressing intermediate phenotypes have the highest relative fitness (see Figure 21.9B). By eliminating phenotypic extremes, stabilizing selection reduces genetic and phenotypic variation and increases the frequency of intermediate phenotypes. Stabilizing selection is probably the most common mode of natural selection, affecting many familiar traits. For example, very small and very large human newborns are less likely to survive than those born at an intermediate weight **(Figure 21.10).**

Warren G. Abrahamson and Arthur E. Weis of Bucknell University have shown that opposing forces of directional selection can sometimes produce an overall pattern of stabilizing selection **(Figure 21.11).**

The gallmaking fly *(Eurosta solidaginis)* is a small insect that feeds on the tall goldenrod plant *(Solidago altissima).* When a fly larva hatches from its egg, it bores into a goldenrod stem, and the plant responds by producing a spherical growth deformity called a gall. The larva feeds on plant tissues inside the gall. Galls vary dramatically in size; genetic experiments indicate that gall size is a heritable trait of the fly, although plant genotype also has an effect.

Fly larvae inside galls are subjected to two opposing patterns of directional selection. On one hand, a tiny wasp *(Eurytoma gigantea)* parasitizes gallmaking flies by laying eggs in fly larvae inside their galls. After hatching, the young wasps feed on the fly larvae, killing them in the process. However, adult wasps are so small that they cannot easily penetrate the thick walls of a large gall; they generally lay eggs in fly larvae occupying small galls. Thus, wasps establish directional selection favoring flies that produce large galls, which are less likely to be parasitized. On the other hand, several bird species open galls to feed on mature fly larvae; these predators preferentially open large galls, fostering directional selection in favor of small galls.

In about one-third of the populations surveyed in central Pennsylvania, wasps and birds attacked galls with equal frequency, and flies producing galls of intermediate size had the highest survival rate. The smallest and largest galls—as well as the genetic predisposition to make very small or very large galls—were eliminated from the population.

DISRUPTIVE SELECTION Traits undergo **disruptive selection** when extreme phenotypes have higher relative fitness than intermediate phenotypes (see Figure 21.9C). Thus, alleles producing extreme phenotypes become more common, promoting polymorphism. Under natural conditions, disruptive selection is much less common than directional selection and stabilizing selection.

FIGURE 21.11 Observational Research

How Opposing Forces of Directional Selection Produce Stabilizing Selection

Hypothesis: The size of galls made by larvae of the gallmaking fly *(Eurosta solidaginis)* is governed by conflicting selection pressures established by parasitic wasps and predatory birds.

Prediction: If parasitic wasps and predatory birds exert selection for different sizes of galls made by gallmaking fly larvae, fly populations that suffer from wasp parasitism and bird predation will make galls of intermediate size.

Method: Abrahamson and his colleagues surveyed galls made by the larvae of the gallmaking fly in Pennsylvania. They measured the diameters of the galls they encountered, and, for those galls in which the larvae had died, they determined whether they had been killed by a parasitic wasp or a predatory bird, such as the downy woodpecker.

A. **Goldenrod (*Solidago altissima*) galls**

B. **Gallmaking fly (*Eurosta solidaginis*) inside a gall**

C. **Fly larvae parasitized by wasps (*Eurytoma gigantea*, inset) versus gall size**

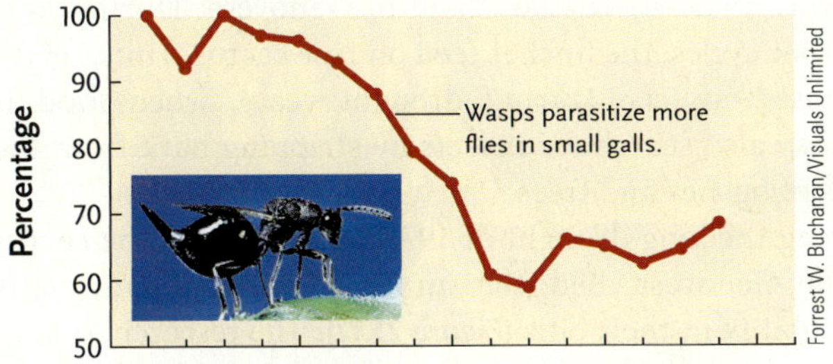

D. **Fly larvae eaten by birds (*Picoides pubescens*, inset) versus gall size**

E. **Fly larvae alive or killed versus gall size**

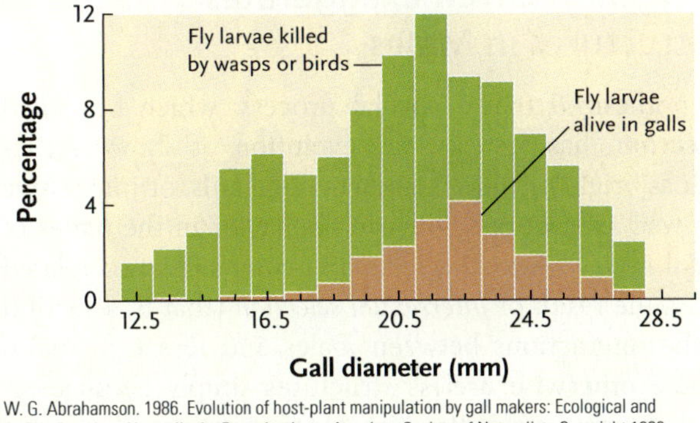

Results: Tiny wasps are more likely to parasitize gallmaking fly larvae inside small galls **(C),** fostering directional selection in favor of large galls. By contrast, birds usually feed on fly larvae inside large galls **(D),** fostering directional selection in favor of small galls. These opposing patterns of directional selection create stabilizing selection for the size of galls that the fly larvae make **(E).**

Conclusion: Because wasps preferentially parasitize fly larvae in small galls and birds preferentially eat fly larvae in large galls, the opposing forces of directional selection establish an overall pattern of stabilizing selection in favor of medium-sized galls.

THINK LIKE A SCIENTIST As you will learn in Chapter 55, human activities are causing many North American bird populations to decline in numbers. What effect would declining bird populations be likely to have on the sizes of galls made by gallmaking flies?

Source: A. E. Weis and W. G. Abrahamson. 1986. Evolution of host-plant manipulation by gall makers: Ecological and genetic factors in the *Solidago–Eurosta* system. *The American Naturalist* 127:681–695.

FIGURE 21.12
Disruptive selection. Cactus finches *(Geospiza conirostris)* on Genovesa exhibit extreme variability in the size and shape of their bills.

Peter and Rosemary Grant of Princeton University, the world's experts on the ecology and evolution of the Galápagos finches, have analyzed a likely case of disruptive selection on the size and shape of the bill in a population of cactus finches *(Geospiza conirostris)* on the island of Genovesa. During normal weather cycles, the finches feed on ripe cactus fruits, seeds, and exposed insects. During drought years, when food is scarce, they also search for insects by stripping bark from the branches of bushes and trees.

During the long drought of 1977, about 70% of the cactus finches on Genovesa died; the survivors exhibited unusually high variability in their bills **(Figure 21.12).** The researchers suggested that this morphological variability allowed birds to specialize on particular foods. Birds that stripped bark from branches to look for insects had particularly deep bills, and birds that opened cactus fruits to feed on the fleshy interior had especially long bills. Thus, birds with extreme bill phenotypes appeared to feed more efficiently on specific resources, establishing disruptive selection on the size and shape of their bills. The selection may be particularly strong when drought limits the variety and overall availability of food. However, intermediate bill morphologies may be favored during nondrought years when insects and small seeds are abundant.

Sexual Selection Often Exaggerates Showy Structures in Males

Darwin hypothesized that a special process, which he called **sexual selection,** has fostered the evolution of showy structures—such as brightly colored feathers, long tails, or impressive antlers—as well as elaborate courtship behavior in the males of many animal species. Sexual selection encompasses two related processes. As the result of *intersexual selection* (that is, selection based on the interactions between males and females), males produce these otherwise useless structures simply because females find them irresistibly attractive. Under *intrasexual selection* (that is, selection based on the interactions between members of the same sex), males use their large body size, antlers, or tusks to intimidate, injure, or kill rival males. In many species, sexual selection is the most probable cause of **sexual dimorphism,** differences in the size or appearance of males and females.

Like directional selection, sexual selection pushes phenotypes toward one extreme. But the products of sexual selection are sometimes bizarre—such as the ridiculously long tail feathers of male African widowbirds. How could evolutionary processes favor such elaborate structures, which are costly to produce? Research by Malte Andersson of the University of Gothenburg, Sweden, suggests that the males' long tail feathers are a product of intersexual selection because females are more strongly attracted to males with long tails than to males with short tails, while tail length had no effect on a male's ability to compete with other males for space in the habitat **(Figure 21.13).** Behavioral aspects of sexual selection are described further in Chapter 56.

Nonrandom Mating Can Influence Genotype Frequencies

The Hardy–Weinberg model requires individuals to select mates randomly with respect to their genotypes. This requirement is, in fact, often met; humans, for example, generally marry one another in total ignorance of their genotypes for digestive enzymes or blood types.

Nevertheless, many organisms mate nonrandomly, selecting a mate with a particular phenotype and underlying genotype. Snow geese, for example, usually select mates of their own color, and a tall woman is more likely to marry a tall man than a short man. If no one phenotype is preferred by all potential mates, nonrandom mating does not change allele frequencies by establishing selection for one phenotype over another. But because individuals with similar genetically based phenotypes mate with each other, the next generation will contain fewer heterozygous offspring—and more homozygous offspring—than the Hardy–Weinberg model predicts.

Inbreeding is a special form of nonrandom mating in which genetically related individuals mate with each other. Self-fertilization in plants (see Chapter 36) and a few animals (see Chapter 49) is an extreme example of inbreeding because offspring are produced from the gametes of a single parent. However, other organisms that live in small, relatively closed populations often mate with related individuals. Because relatives often carry the same alleles, inbreeding generally increases the frequency of homozygous genotypes and decreases the frequency of heterozygotes. Thus, recessive phenotypes are often expressed.

For example, the high incidence of Ellis–van Creveld syndrome among the Old Order Amish population, mentioned earlier, is caused by inbreeding. Although the founder effect originally established the disease-causing allele in this population, inbreeding increases the likelihood that it will be expressed. Most human societies discourage matings between genetically

FIGURE 21.13 Experimental Research

Sexual Selection in Action

Question: Is the long tail of the male long-tailed widowbird *(Euplectes progne)* the product of intrasexual selection, intersexual selection, or both?

Experiment: Andersson counted the number of females that associated with individual male widowbirds in the grasslands of Kenya. He then shortened the tails of some males by cutting the feathers, lengthened the tails of others by gluing feather extensions to their tails, and left a third group essentially unaltered as a control. One month later, he again counted the number of females associating with each male and compared the results from the three groups.

Results: Males with experimentally lengthened tails attracted more than twice as many mates as males in the control group, and males with experimentally shortened tails attracted fewer. Andersson observed no differences in the ability of altered males and control group males to maintain their display areas.

Francois Loubser/Shutterstock.com

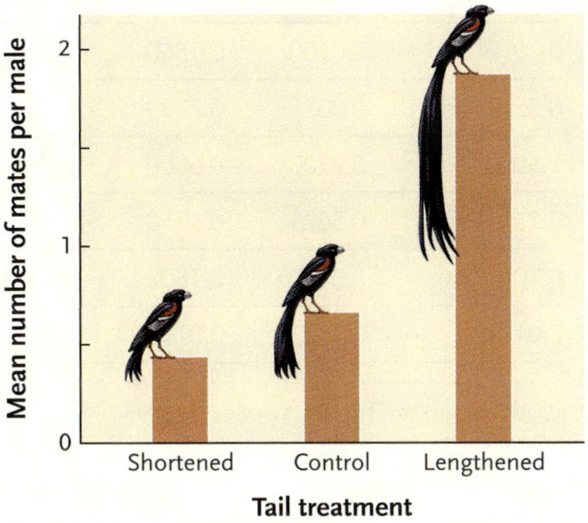

Conclusion: Female widowbirds clearly prefer males with experimentally lengthened tails to those with normal tails or experimentally shortened tails. Tail length had no obvious effect on the interactions between males. Thus, the long tail of male widowbirds is the product of intersexual selection.

THINK LIKE A SCIENTIST Which experimental treatment had the greater effect on the number of females with which male widowbirds mated?

Source: M. Andersson. 1982. Female choice selects for extreme tail length in a widowbird. *Nature* 299:818–820.

close relatives, thereby reducing inbreeding and the inevitable production of recessive homozygotes that inbreeding causes.

STUDY BREAK 21.3

1. **Which agents of microevolution tend to increase genetic variation within populations, and which ones tend to decrease it?**
2. **Which mode of natural selection increases the representation of the average phenotype in a population?**
3. **In what way is sexual selection like directional selection?**

THINK OUTSIDE THE BOOK

Search the Internet for examples of diseases or unusual phenotypes caused by inbreeding in specific human populations. Have human geneticists discovered the molecular basis of these conditions?

21.4 Maintaining Genetic and Phenotypic Variation

Evolutionary biologists continue to discover extraordinary amounts of genetic and phenotypic variation in most natural populations. How can so much variation persist in the face of stabilizing selection and genetic drift?

Diploidy Can Hide Recessive Alleles from the Action of Natural Selection

The diploid condition reduces the effectiveness of natural selection in eliminating harmful recessive alleles from a population. Although such alleles are disadvantageous in the homozygous state, they may have little or no effect on heterozygotes. Thus, recessive alleles can be protected from natural selection by the phenotypic expression of the dominant allele.

TABLE 21.3 **Masking of Recessive Alleles in Diploid Organisms**

When a recessive allele is common in a population (top of table), most copies of the allele are present in homozygotes. But when the allele is rare (bottom of table), most copies of it exist in heterozygotes. Thus, rare alleles that are completely recessive are protected from the action of natural selection because they are masked by dominant alleles in heterozygous individuals.

Frequency of Allele *a*	Genotype Frequencies*			% of Allele *a* Copies in	
	AA	*Aa*	*aa*	*Aa*	*aa*
0.99	0.0001	0.0198	0.9801	1	99
0.90	0.0100	0.1800	0.8100	10	90
0.75	0.0625	0.3750	0.5625	25	75
0.50	0.2500	0.5000	0.2500	50	50
0.25	0.5625	0.3750	0.0625	75	25
0.10	0.8100	0.1800	0.0100	90	10
0.01	0.9801	0.0198	0.0001	99	1

*Population is assumed to be in genetic equilibrium.

In most cases, the masking of recessive alleles in heterozygotes makes it almost impossible to eliminate them completely through selective breeding. Experimentally, we can prevent homozygous recessive organisms from mating. But, as the frequency of a recessive allele decreases, an increasing proportion of its remaining copies is "hidden" in heterozygotes **(Table 21.3).** Thus, the diploid state preserves recessive alleles at low frequencies, at least in large populations. In small populations, a combination of natural selection and genetic drift can eliminate harmful recessive alleles.

Natural Selection Can Maintain Balanced Polymorphisms

A **balanced polymorphism** is one in which two or more phenotypes are maintained in fairly stable proportions over many generations. Natural selection preserves balanced polymorphisms when heterozygotes have higher relative fitness, when different alleles are favored in different environments, and when the rarity of a phenotype provides an advantage.

HETEROZYGOTE ADVANTAGE A balanced polymorphism can be maintained by **heterozygote advantage,** when heterozygotes have higher relative fitness than either homozygote. As Darwin first discovered in his experiments on corn, the offspring of crosses between two homozygous strains of the same species often exhibit a robustness described as "hybrid vigor." Apparently, being heterozygous at many gene loci provides some advantage, perhaps by allowing organisms to respond effectively to environmental variation.

The best-documented example of heterozygote advantage with reference to a specific gene locus is the maintenance of the *HbS* (sickle) allele, which codes for a defective form of hemoglobin in humans. As you learned in Chapter 12, hemoglobin is an oxygen-transporting molecule in red blood cells. The hemoglobin produced by the *HbS* allele differs from normal hemoglobin (coded by the *HbA* allele) by just one amino acid. In *HbS*/*HbS* homozygotes, the faulty hemoglobin forms long, fibrous chains under low oxygen conditions, causing red blood cells to assume a sickle shape (as shown in Figure 12.1). Homozygous *HbS*/*HbS* individuals often die of sickle-cell disease before reproducing, yet in tropical and subtropical Africa, *HbS*/*HbA* heterozygotes make up nearly 25% of many populations.

Why is the harmful allele maintained at such high frequency in some populations? It turns out that sickle-cell disease is common in regions where malarial parasites infect red blood cells in humans **(Figure 21.14).** When heterozygous *HbA*/*HbS* individuals contract malaria, their in-

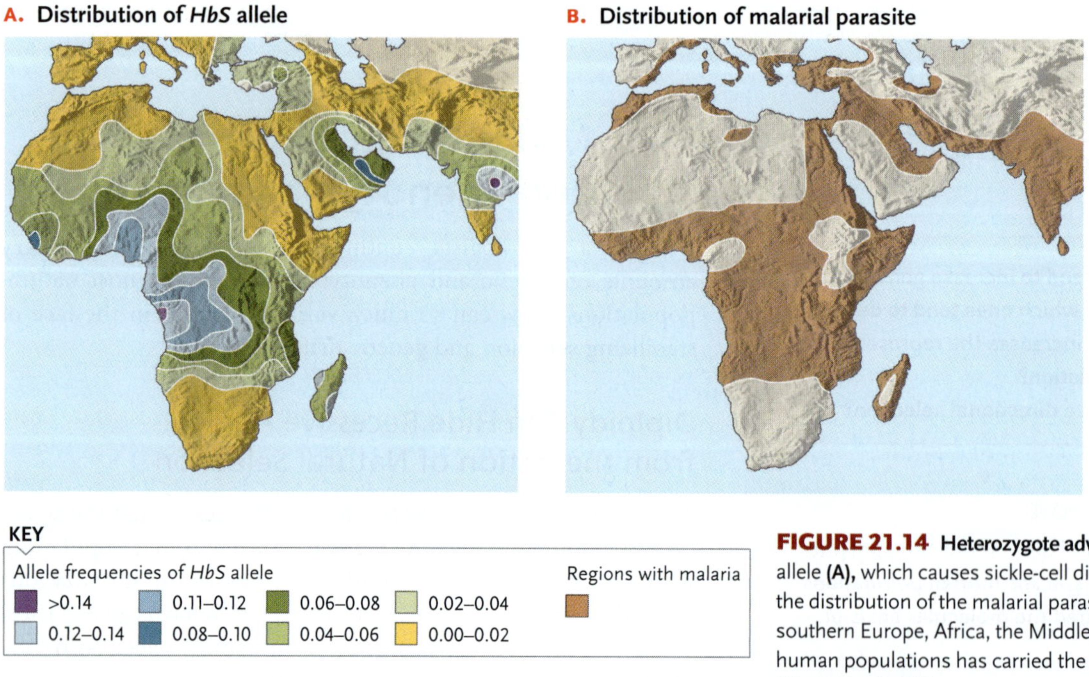

FIGURE 21.14 **Heterozygote advantage.** The distribution of the *HbS* allele **(A),** which causes sickle-cell disease in homozygotes, roughly matches the distribution of the malarial parasite *Plasmodium falciparum* **(B)** in southern Europe, Africa, the Middle East, and India. Gene flow among human populations has carried the *HbS* allele to some malaria-free regions.

FIGURE 21.15 **Observational Research**

Habitat Variation in Color and Striping Patterns of European Garden Snails

Hypothesis: Birds and other visual predators are more likely to find snails that are not well camouflaged in their habitats than snails that blend in with their habitats.

Prediction: Genetically based variations in the shell color and striping patterns of the European garden snail *(Cepaea nemoralis)* will differ substantially between local populations of snails that live in different types of vegetation.

Method: British researchers A. J. Cain and P. M. Shepard surveyed the distribution of color and striping patterns of snails in many local populations. They plotted the data on a graph showing the percentage of snails with yellow shells versus the percentage of snails with striped shells, noting the vegetation type where each local population lived.

Results: The shell color and striping patterns of snails living in a particular vegetation type tend to be clustered on the graph, reflecting phenotypic differences that enable the snails to be camouflaged in different habitats. Thus, the alleles that control these characters vary from one local population to another.

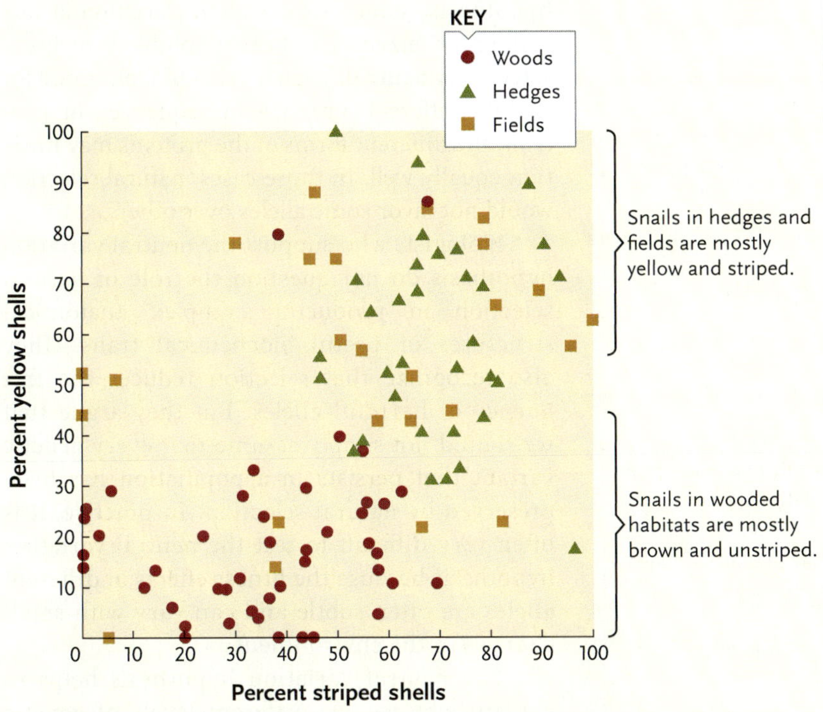

Conclusion: Variations in the color and striping patterns on the shells of European garden snails allow most snails to be camouflaged in whatever habitat they occupy. Because these traits are genetically based, the frequencies of the alleles that control them also differ among snails living in different vegetation types. Natural selection therefore favors different alleles in different local populations, maintaining genetic variability in populations that span several vegetation types.

THINK LIKE A SCIENTIST Do the data in the graph suggest that populations of snails in wooded habitats include *only* snails with brown, unstriped shells and that populations of snails in hedges and fields include *only* snails with yellow striped shells?

Source: A. J. Cain and P. M. Shepard. 1954. Natural selection in *Cepaea*. *Genetics* 39:89–116.

fected red blood cells assume the same sickle shape as those of homozygous *HbS*/*HbS* individuals. The sickled cells lose potassium, killing the parasites, which limits their spread within the infected individual. Heterozygous individuals often survive malaria because the parasites do not multiply quickly inside them, their immune systems can effectively fight the infection, and they retain a large population of uninfected red blood cells. Homozygous *HbA*/*HbA* individuals are also subject to malarial infection, but because their infected cells do not sickle, the parasites multiply rapidly, causing a severe infection with a high mortality rate.

Therefore, *HbA*/*HbS* heterozygotes have greater resistance to malaria and are more likely to survive severe infections in areas where malaria is prevalent. Natural selection preserves the *HbS* allele in these populations because heterozygotes in malaria-prone areas have higher relative fitness than homozygotes for the normal *HbA* allele.

SELECTION IN VARYING ENVIRONMENTS Genetic variability can also be maintained within a population when different alleles are favored in different places or at different times. For example, the shells of European garden snails range in color from nearly white to pink, yellow, or brown, and may be patterned by one to five stripes of varying color (see Figure 21.2A). This polymorphism, which is relatively stable through time, is controlled by several gene loci. The variability in color and striping pattern can be partially explained by selection for camouflage in different habitats.

Predation by song thrushes *(Turdus ericetorum)* is a major agent of selection on the color and pattern of these snails in England. When a thrush finds a snail, it smacks it against a rock to break the shell. The bird eats the snail, but leaves the shell near its "anvil." Researchers used the broken shells near an anvil to compare the phenotypes of captured snails to a random sample of the entire snail population. Their analyses indicated that thrushes are visual predators, usually capturing snails that are easy to find. Thus, well-camouflaged snails survive, and the alleles that specify their phenotypes increase in frequency.

The success of camouflage varies with habitat, however; local subpopulations of the snail, which occupy different habitats, often differ markedly in shell color and pattern. The predators eliminate the most conspicuous individuals in each habitat; thus, natural selection differs from place to place **(Figure 21.15).** In woods where the ground is covered with dead leaves, snails with unstriped pink or brown shells predominate. In hedges and fields, where the vegetation includes thin stems and grass, snails with striped yellow shells are the most common. In populations that span

FIGURE 21.16 Experimental Research

Demonstration of Frequency-Dependent Selection

Question: How does the frequency of a prey type influence the likelihood that it will be captured by predators?

Experiment: Water boatmen *(Sigara distincta)* occur in three color forms, which vary in the effectiveness of their camouflage. Researchers offered different proportions of the three color forms to predatory fishes in the laboratory and recorded how many of each form were eaten.

Result: When all three phenotypes were available, predatory fishes consumed a disproportionately large number of the most common form, thereby reducing its frequency in the population.

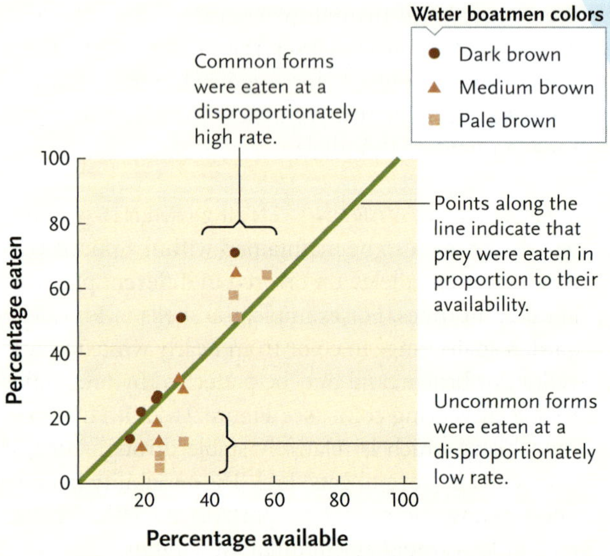

B. Clarke. 1962. Balanced polymorphism and the diversity of sympatric species. *Systematics Association Publications* 4:47–70. Reprinted by permission of The Systematics Association.

Conclusion: Predators tend to feed disproportionately on whatever form of their prey is most abundant, thereby reducing its frequency in the prey population.

THINK LIKE A SCIENTIST Do the data in the graph indicate that predatory fishes had a preference for one color form of water boatman over the other color forms?

Source: B. Clarke. 1962. Balanced polymorphism and the diversity of sympatric species. *Systematics Association Publications* 4:47–70.

several habitats, selection preserves different alleles in different places, thus maintaining variability in the population as a whole.

FREQUENCY-DEPENDENT SELECTION Sometimes, genetic variability is maintained in a population simply because rare phenotypes—whatever they happen to be—have higher relative fitness than more common phenotypes. The rare phenotype will increase in frequency until it becomes so common that it loses its advantage. Such phenomena are examples of **frequency-dependent selection** because the selective advantage enjoyed by a particular phenotype depends on its frequency in the population.

Predator–prey interactions can establish frequency-dependent selection because predators often focus their attention on the most common types of prey (see Chapter 53). For example, the aquatic insects called water boatmen *(Sigara distincta)* occur in three different shades of brown. When all three shades are available at moderate frequencies, fish preferentially feed on the darkest individuals, which are the least camouflaged. But if any one phenotype is very common, fish will learn to focus their attention on that phenotype, consuming it in disproportionately large numbers **(Figure 21.16).**

Some Genetic Variations May Be Selectively Neutral

Many biologists believe that some genetic variations are neither preserved nor eliminated by natural selection. According to the **neutral variation hypothesis,** some of the genetic variation at loci coding for enzymes and other soluble proteins is **selectively neutral.** Even if various alleles code for slightly different amino acid sequences in proteins, the different forms of the proteins may function equally well. In those cases, natural selection would not favor some alleles over others.

Biologists who support the neutral variation hypothesis do not question the role of natural selection in producing complex anatomical structures or useful biochemical traits. They also recognize that selection reduces the frequency of harmful alleles. But they argue that we should not simply assume that every genetic variant that persists in a population has been preserved by natural selection. In practice, it is often very difficult to test the neutral variation hypothesis because the fitness effects of different alleles are often subtle and can vary with small changes in the environment.

The neutral variation hypothesis helps to explain why we see different levels of genetic variation in different populations. It proposes that genetic variation is directly proportional to a population's size and the length of time over which variations have accumulated. Small populations experience fewer mutations than large populations simply because they include fewer replicating genomes. Small populations also lose rare alleles more readily through genetic drift. Thus, small populations should exhibit less genetic variation than large ones, and a population, like the northern elephant seals, that has experienced a recent population bottleneck should exhibit an exceptionally low level of genetic variation. These predictions of the neutral variation hypothesis are generally supported by empirical data.

STUDY BREAK 21.4

1. How does the diploid condition protect harmful recessive alleles from natural selection?
2. What is a balanced polymorphism?
3. Why is the allele that causes sickle-cell disease very rare in human populations that are native to northern Europe?

21.5 Adaptation and Evolutionary Constraints

Although natural selection preserves alleles that confer high relative fitness on the individuals that carry them, researchers are cautious about interpreting the benefits that particular traits may provide.

Scientists Construct Hypotheses about the Evolution of Adaptive Traits

An **adaptive trait** is any product of natural selection that increases the relative fitness of an organism in its environment. **Adaptation** is the accumulation of adaptive traits over time, and most chapters of this book describe examples that range across all levels of biological organization, from the molecular to the ecological. For example, the change in the oxygen-binding capacity of hemoglobin in response to carbon dioxide concentration, the water-retaining structures and special photosynthetic pathways of desert plants, and the warning coloration of poisonous animals can all be interpreted as adaptive traits.

In fact, we can concoct an adaptive explanation for almost any characteristic we observe in nature. But such explanations are just fanciful stories unless they are framed as testable hypotheses about the relative fitness of different phenotypes and genotypes. Unfortunately, evolutionary biologists cannot always conduct straightforward experiments because they sometimes study traits that do not vary much within a population or species. In such cases, they may compare variations of a trait in closely related species living in different environments. For example, one can test how the traits of desert plants are adaptive by comparing them to traits in related species from moister habitats.

When biologists try to unravel how and why a particular characteristic evolved, they must also remember that a trait they observe today may have had a different function in the past. For example, the structure of the shoulder joint in birds allows them to move their wings first upward and backward and then downward and forward during flapping flight. But analyses of the fossil record reveal that this adaptation, which is essential for flight, did not originate in birds: some predatory nonflying dinosaurs, including the ancestors of birds, had a similarly constructed shoulder joint. Researchers hypothesize that these fast-running predators may have struck at prey with a flapping motion similar to that used by modern birds. Thus, the structure of the shoulder may have evolved first as an adaptation for capturing prey, and only later proved useful for flapping flight. This hypothesis—however plausible it may be—cannot be tested by direct experimentation because the nonflying ancestors of birds have been extinct for millions of years. Instead, evolutionary biologists must use anatomical studies of birds and their ancestors as well as theoretical models about the mechanics of movement to challenge and refine the hypothesis.

Finally, although evolution has produced all the characteristics of organisms, not all are necessarily adaptive. Some traits may be the products of chance events and genetic drift. Others are produced by alleles that were selected for unrelated reasons (see Section 12.2). And still other characteristics result from the action of basic physical laws. For example, the seeds of many plants fall to the ground when they mature, reflecting the inevitable effect of gravity.

Several Factors Constrain Adaptive Evolution

When we analyze the structure and functions of an organism, we often marvel at how well adapted it is to its environment and mode of life. However, the adaptive traits of most organisms are compromises produced by competing selection pressures. Sea turtles, for example, must lay their eggs on beaches because their embryos cannot acquire oxygen under water. Although flippers allow females to crawl to nesting sites on beaches, they are not ideally suited for terrestrial locomotion. Their structure reflects their primary function, swimming.

Moreover, no organism can be perfectly adapted to its environment because environments change over time. When selection occurs in a population, it preserves alleles that are successful under the prevailing environmental conditions. Thus, each generation is adapted to the environmental conditions under which its parents lived. If the environment changes from one generation to the next, adaptation will always lag behind.

Another constraint on the evolution of adaptive traits is historical. Natural selection is not an engineer that designs new organisms from scratch. Instead, it acts on new mutations and existing genetic variation. Because new mutations are fairly rare, natural selection works primarily with alleles that have been present for many generations. Thus, adaptive changes in the morphology of an organism are often based on small modifications of existing structures. The bipedal (two-footed) posture of humans, for example, evolved from the quadrupedal (four-footed) posture of our ancestors. Natural selection did not produce an entirely new skeletal design to accompany this radical behavioral shift. Instead, existing characteristics of the spinal column and the musculature of the legs and back were modified, albeit imperfectly, for an upright stance.

The agents of evolution cause microevolutionary changes in the gene pools of populations. In the next chapter, we examine how microevolution in different populations can cause their gene pools to diverge. The extent of genetic divergence is sometimes sufficient to cause the populations to evolve into different species.

STUDY BREAK 21.5

1. **How can a biologist test whether a trait is adaptive?**
2. **Why are most organisms adapted to the environments in which their parents lived?**

UNANSWERED QUESTIONS

What are the evolutionary forces affecting molecular variation within populations?

This question may sound like a simple restatement of the entire chapter you have just read, but it is one of the *fundamental* questions in population genetics today—and we have only begun to scratch its surface. The Hardy–Weinberg principle provides a useful null hypothesis, but since we know that evolution happens routinely, that null hypothesis is very frequently rejected. Recent studies have attempted to address this question using theoretical models, extensive DNA sequence data, and detailed measurements of recombination rate.

Recombination generates new variation, and, most important, it causes the evolutionary forces acting on some genes to become independent of forces acting on other genes. Let's imagine that genes *A* and *B* are on the same chromosome, as shown in this depiction of chromosomes sampled from different individuals within a population:

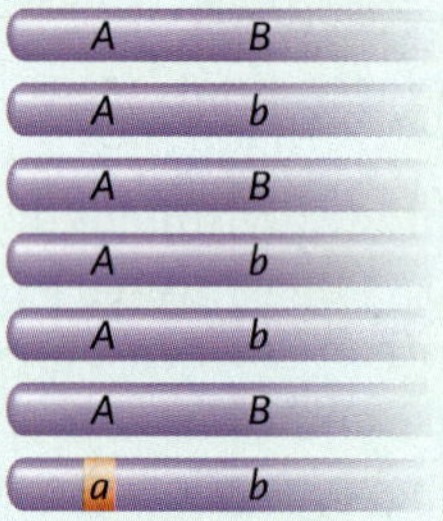

© Cengage Learning 2014

Gene *B* has two alleles (*B* and *b*), but they have no phenotypic effect, and natural selection does not act on them. Suppose that a new advantageous allele at gene *A* (designated *a*) arises in one chromosome. If there is no recombination between genes *A* and *B*, then as allele *a* spreads in the population by selection, so too will allele *b*, even though there was no selection directly favoring the *b* allele. This effect of selection on nearby genes is called a *selective sweep.* By contrast, if genes *A* and *B* frequently recombine, then allele *a* may not remain associated with allele *b*. Under frequent recombination, the spread of allele *a* may have little or no effect on gene *B*: sometimes *a* will be associated with *b*, but at other times *a* will be associated with *B*.

In the 1990s, evolutionary geneticists were greatly excited by several studies that identified a strong and positive relationship between the recombination rate between particular genes and the amount of genetic variation within those genes. In other words, genes that experienced a lot of recombination also exhibited a great deal of variability. This relationship is consistent with the hypothesis that natural selection often occurs throughout the genome—new advantageous alleles arise frequently, and the impact of their "sweeps" is proportional to their recombination rates. This relationship between recombination and genetic variation was first documented in *Drosophila* (fruit flies) by Chip Aquadro and his team at Cornell University, but it has since been demonstrated in humans and various plants. Hence, this pattern appears to be very general.

However, our initial interpretation may be too simplistic. Brian Charlesworth, then at the University of Chicago, suggested that the observed pattern may result from the frequent appearance of detrimental mutations that eliminate variation in regions of low recombination—called *background selection*—rather than from sweeps associated with the spread of advantageous alleles. Given that detrimental mutations arise far more frequently than advantageous ones, background selection surely explains some of this general pattern, and perhaps much of it.

An alternative hypothesis that may explain the relationship between recombination rate and genetic variation suggests that recombination rate and the level of genetic variation may be mechanistically connected. The connection may be direct if recombination itself induces mutations, resulting in higher mutation rates in regions of high recombination. Or the connection may be indirect: recombination rate is related to the base composition in specific regions of the genome, and base composition influences mutation rates. In 2006, Chris Spencer and his colleagues at Oxford University examined the impact of recombination rates on patterns of nucleotide variation at a very fine scale across the human genome. They found that recombination rates had very local effects on variation, an observation that is consistent with the hypothesis of a mechanistic connection between recombination and mutation rate; their results are not consistent with explanations involving natural selection.

Although biologists first thought that the observed relationship between recombination rate and genetic variation had solved questions about the evolutionary forces that affect molecular variation, this observation has become a puzzle in and of itself. Many researchers continue to address this question, now using whole-genome sequences and theoretical and empirical tools for estimating recombination rates. Evolutionary biologists know that the "final answer" will be that all of the processes described above contribute to this relationship, but knowing their specific contributions will help us understand how, how much, and what kinds of natural selection shape variation within genomes.

THINK LIKE A SCIENTIST If recombination itself induces mutations, should genes in regions of high recombination have many differences or few differences when the sequences of these genes are compared among closely related species? Does this test suggest a means for evaluating whether recombination rate and genetic variation are connected mechanistically or connected by the action of natural selection?

Mohammed Noor

Mohamed Noor is professor of biology at Duke University. His research interests include speciation and evolutionary genetics, and recombination. To learn more about Dr. Noor's research, go to http://www.biology.duke.edu/noorlab/Noorlab.html.

Dr. Noor was a PhD student with Dr. Jerry Coyne, who contributed the Unanswered Questions for Chapter 22.

REVIEW KEY CONCEPTS

To access the course materials and companion resources for this text, please visit www.cengagebrain.com.

21.1 Variation in Natural Populations

- Phenotypic traits exhibit either quantitative or qualitative variation within populations (Figures 21.2 and 21.3).
- Genetic variation, environmental factors, or an interaction between the two cause phenotypic variation within populations (Figures 21.4 and 21.5). Only genetically based phenotypic variation is heritable and subject to evolutionary change.
- Genetic variation arises within populations largely through mutation and genetic recombination. Artificial selection experiments and analyses of protein and DNA sequences reveal that most populations include significant genetic variation (Figure 21.6).

21.2 Population Genetics

- All the gene copies in a population comprise its gene pool, which can be described in terms of allele frequencies and genotype frequencies (Table 21.1).
- The Hardy–Weinberg principle of genetic equilibrium is a null model that describes the conditions under which microevolution, a change in allele frequencies through time, will not take place. Microevolution occurs in populations when the restrictive requirements of the model are not met.

Animation: How to find out if a population is evolving

21.3 The Agents of Microevolution

- Several processes cause microevolution in populations (Table 21.2). Mutation introduces completely new genetic variation. Gene flow carries novel genetic variation into a population through the arrival and reproduction of immigrants (Figure 21.7). Genetic drift causes random changes in allele frequencies, especially in small populations (Figure 21.8). Natural selection occurs when the genotypes of some individuals enable them to survive and reproduce more than others.
- Natural selection alters phenotypic variation in three ways (Figure 21.9). Directional selection increases or decreases the mean value of a trait, shifting it toward a phenotypic extreme. Stabilizing selection increases the frequency of the mean phenotype and reduces variability in the trait (Figures 21.10 and 21.11). Disruptive selection increases the frequencies of extreme phenotypes and decreases the frequency of intermediate phenotypes (Figure 21.12).
- Sexual selection promotes the evolution of exaggerated structures and behaviors (Figure 21.13).
- Although nonrandom mating does not change allele frequencies, it can produce more homozygotic and fewer heterozygotic genotypes than the Hardy–Weinberg model predicts.

Animation: Disruptive selection among African finches

Animation: Simulation of genetic drift

Animation: Directional selection

Animation: Disruptive selection

Animation: Change in moth population

Animation: Stabilizing selection

21.4 Maintaining Genetic and Phenotypic Variation

- Diploidy can maintain genetic variation in a population if recessive alleles are not expressed in heterozygotes and are thus hidden from natural selection (Table 21.3).
- Polymorphisms are maintained in populations when heterozygotes have higher relative fitness than both homozygotes (Figure 21.14), when natural selection occurs in variable environments (Figure 21.15), or when the relative fitness of a phenotype varies with its frequency in the population (Figure 21.16).
- The neutral variation hypothesis proposes that many genetic variations are selectively neutral, conferring neither advantages nor disadvantages on the individuals that carry them. It explains why large populations and those that have not experienced a recent population bottleneck exhibit the highest levels of genetic variation.

21.5 Adaptation and Evolutionary Constraints

- Adaptive traits increase the relative fitness of individuals carrying them. Adaptive explanations of traits must be framed as testable hypotheses.
- Natural selection cannot result in perfectly adapted organisms because most adaptive traits represent compromises among conflicting needs, because most environments change constantly, and because natural selection can affect only existing genetic variation.

Animation: Adaptation to what?

UNDERSTAND & APPLY

Test Your Knowledge

1. Which of the following represents an example of qualitative phenotypic variation?
   a. the lengths of people's toes
   b. the body sizes of pigeons
   c. human ABO blood types
   d. the birth weights of humans
   e. the number of leaves on oak trees
2. A population of mice is at Hardy–Weinberg equilibrium at a gene locus that controls fur color. The locus has two alleles, *M* and *m*. A genetic analysis of one population reveals that 60% of its gametes carry the *M* allele. What percentage of mice contains both the *M* and *m* alleles?
   a. 60%
   b. 48%
   c. 40%
   d. 36%
   e. 16%

3. If the genotype frequencies in a population are 0.60 *AA*, 0.20 *Aa*, and 0.20 *aa*, and if the requirements of the Hardy–Weinberg principle apply, the genotype frequencies in the offspring generation will be:
   a. 0.60 *AA*, 0.20 *Aa*, 0.20 *aa*.
   b. 0.36 *AA*, 0.60 *Aa*, 0.04 *aa*.
   c. 0.49 *AA*, 0.42 *Aa*, 0.09 *aa*.
   d. 0.70 *AA*, 0.00 *Aa*, 0.30 *aa*.
   e. 0.64 *AA*, 0.32 *Aa*, 0.04 *aa*.
4. The reason spontaneous mutations do not have an immediate effect on allele frequencies in a large population is that:
   a. mutations are random events, and mutations may be either beneficial or harmful.
   b. mutations usually occur in males and have little effect on eggs.
   c. many mutations exert their effects after an organism has stopped reproducing.
   d. mutations are so rare that mutated alleles are greatly outnumbered by nonmutated alleles.
   e. most mutations do not change the amino acid sequence of a protein.
5. The phenomenon in which chance events cause unpredictable changes in allele frequencies is called:
   a. gene flow.
   b. genetic drift.
   c. inbreeding.
   d. balanced polymorphism.
   e. stabilizing selection.
6. An Eastern European immigrant carrying the allele for Tay-Sachs disease settled in a small village on the St. Lawrence River. Many generations later, the frequency of the allele in that village is statistically higher than it is in the immigrant's homeland. The high frequency of the allele in the village probably provides an example of:
   a. natural selection.
   b. the concept of relative fitness.
   c. the Hardy–Weinberg genetic equilibrium.
   d. phenotypic variation.
   e. the founder effect.
7. If a storm kills many small sparrows in a population, but only a few medium-sized and large ones, which type of selection is probably operating?
   a. directional selection
   b. stabilizing selection
   c. disruptive selection
   d. intersexual selection
   e. intrasexual selection
8. Which of the following phenomena explains why the allele for sickle-cell hemoglobin is common in some tropical and subtropical areas where the malaria parasite is prevalent?
   a. balanced polymorphism
   b. heterozygote advantage
   c. sexual dimorphism
   d. neutral selection
   e. stabilizing selection
9. The neutral variation hypothesis proposes that:
   a. complex structures in most organisms have not been fostered by natural selection.
   b. most mutations have a strongly harmful effect.
   c. some mutations are not affected by natural selection.
   d. natural selection cannot counteract the action of gene flow.
   e. large populations are subject to stronger natural selection than small populations.
10. Phenotypic characteristics that increase the fitness of individuals are called:
   a. mutations.
   b. founder effects.
   c. heterozygote advantages.
   d. adaptive traits.
   e. polymorphisms.

Discuss the Concepts

1. Most large commercial farms routinely administer antibiotics to farm animals to prevent the rapid spread of diseases through a flock or herd. Explain why you think that this practice is either wise or unwise.
2. Many human diseases are caused by recessive alleles that are not expressed in heterozygotes. Some people think that eugenics—the selective breeding of humans to eliminate undesirable genetic traits—provides a way for us to rid our populations of such harmful alleles. Explain why eugenics cannot eliminate such genetic traits from human populations.
3. Using two types of beans to represent two alleles at the same gene locus, design an exercise to illustrate how population size affects genetic drift.
4. In what ways are the effects of sexual selection, disruptive selection, and nonrandom mating different? How are they similar?

Design an Experiment

Design an experiment to test the hypothesis that the differences in size among adult guppies are determined by the amount of food they eat rather than by genetic factors.

Interpret the Data

Peter and Rosemary Grant of Princeton University have studied the ecology and evolution of finches on the Galápagos Islands since the early 1970s. They have shown that finches with large bills (as measured by bill depth; see **Figure**) can eat both small seeds and large seeds, but finches with small bills can only eat small seeds. In 1977, a severe drought on the island of Daphne Major reduced seed production by plants. After the birds consumed whatever small seeds they found, only large seeds were still available. The resulting food shortage killed a majority of the medium ground finches *(Geospiza fortis)* on Daphne Major; their population plummeted from 751 in 1976 to just 90 in 1978. The Grants' research also documented a change in the distributions of bill depths in the birds from 1976 to 1978, as illustrated in the graphs to the right. In light of what you now know about the relationship between bill size and food size for these birds, interpret the change illustrated in the graph. What type of natural selection does this example illustrate?

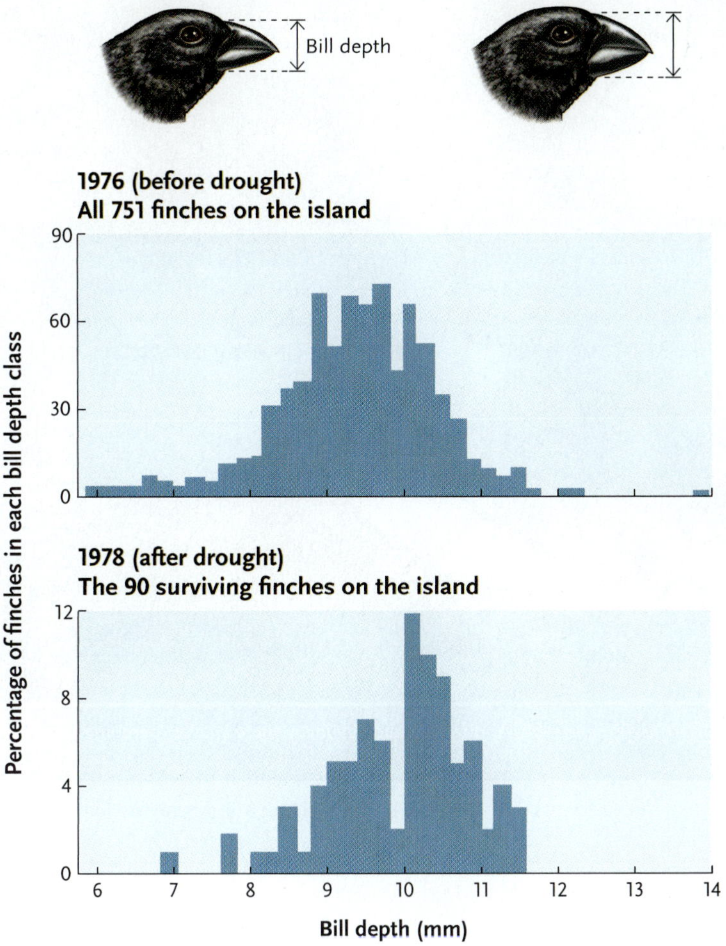

Source: P. R. Grant. 1986. *Ecology and Evolution of Darwin's Finches.* Princeton University Press.

Apply Evolutionary Thinking

Captive breeding programs for endangered species often have access to a limited supply of animals for a breeding stock. As a result, their offspring are at risk of being highly inbred. Why and how might zoological gardens and conservation organizations avoid or minimize inbreeding?

22

Two closely related bird species, purple martins *(Progne subis)* and tree swallows *(Tachycineta bicolor)* perching together on a branch in Crane Creek, Ohio.

STUDY OUTLINE

22.1 What Is a Species?

22.2 Maintaining Reproductive Isolation

22.3 The Geography of Speciation

22.4 Genetic Mechanisms of Speciation

Speciation

Bruce Beehler

FIGURE 22.1 Birds of paradise. A male Count Raggi's bird of paradise *(Paradisaea raggiana)* tries to attract the attention of a female (not pictured) with his showy plumage and conspicuous display. There are 43 known bird of paradise species, 35 of them found only on the island of New Guinea.

Why it matters . . . In 1927, nearly 100 years after Darwin boarded the *Beagle,* a young German naturalist named Ernst Mayr embarked on his own journey, to the highlands of New Guinea. He was searching for rare "birds of paradise" **(Figure 22.1).** These birds were known in Europe only through their ornate and colorful feathers, which were used to decorate ladies' hats. On his trek through the remote Arfak Mountains, Mayr identified 137 bird species (including many birds of paradise) based on differences in their size, plumage, color, and other external characteristics.

To Mayr's surprise, the native Papuans—who were untrained in the ways of Western science, but who hunted these birds for food and feathers—had their own names for 136 of the 137 species he had identified. The close match between the two lists confirmed Mayr's belief that the *species* is a fundamental level of organization in nature. Each species has a unique combination of genes underlying its distinctive appearance and habits. Thus, people who observe them closely—whether indigenous hunters or Western scientists—can often distinguish one species from another.

Mayr also discovered some remarkable patterns in the geographical distributions of the bird species in New Guinea. For example, each mountain range he explored was home to some species that lived nowhere else. Closely related species often lived on different mountaintops, separated by deep valleys of unsuitable habitat. In 1942, Mayr published the book *Systematics and the Origin of Species,* in which he described the role of geography in the evolution of new species; the book quickly became a cornerstone of the modern synthesis (which was outlined in Section 20.3).

What mechanisms produce distinct species? As you discovered in Chapter 21, microevolutionary processes alter the pattern and extent of genetic and phenotypic variation within populations. When these processes differ between populations, the populations will diverge genetically, and they may eventually become so different that we recognize them as distinct species. Although Darwin's famous book was titled *On the Origin of Species,* he didn't dwell on the question of *how* new species arise. But the concept of **speciation**—the process of species formation—was implicit in his insight that similar species often share inherited characteristics and a common ancestry. Darwin also recognized that "descent with modification" had generated the amazing diversity of organisms on Earth.

Today evolutionary biologists view speciation as a *process,* a series of events that occur through time. However, they usually study the *products* of speciation, species that are alive today. Because they can rarely witness the process of speciation from start to finish, scientists make inferences about it by studying organisms in various stages of species formation. In this chapter, we consider four major topics: how biologists define and recognize species; how species maintain their genetic identity; how the geographical distributions of organisms influence speciation; and how different genetic mechanisms produce new species.

22.1 What Is a Species?

Like the hunters of the Arfak Mountains, most of us recognize the different species that we encounter every day. We can distinguish a cat from a dog and sunflowers from roses. The concept of species is based on our perception that Earth's biological diversity is packaged in discrete, recognizable units, and not as a continuum of forms grading into one another. As biologists have learned more about evolutionary processes—and the dazzling biodiversity those processes have produced—they have developed a variety of complementary species concepts.

The Morphological Species Concept Is a Practical Way to Identify Species

Biologists often describe new species on the basis of visible anatomical characteristics, a process that dates back to Linnaeus' classification of organisms in the eighteenth century (described in Chapter 24). This approach is based on the **morphological species concept,** the idea that all individuals of a species share measurable traits that distinguish them from individuals of other species.

The morphological species concept has many practical applications. For example, paleobiologists use morphological criteria to identify the species of fossilized organisms (see Chapter 23). And because we can observe the external traits of organisms in nature, field guides to plants and animals list diagnostic (that is, distinguishing) physical characters that allow us to recognize them **(Figure 22.2).**

FIGURE 22.2 Diagnostic characters. Yellow-throated warblers and myrtle warblers can be distinguished by the color of feathers on the throat and rump.

Nevertheless, relying exclusively on morphology to identify species can present problems. Consider the variation in the shells of the European garden snail (*Cepaea nemoralis;* shown earlier, in Figure 21.2). How could anyone imagine that so variable a collection of shells represents just one species of snail? Conversely, morphology does not help us distinguish some closely related species that are nearly identical in appearance. Finally, morphological species definitions tell us little about the evolutionary processes that produce new species.

The Biological Species Concept Is Based on Reproductive Characteristics

The **biological species concept** emphasizes the dynamic nature of species. Ernst Mayr defined biological species as "groups of . . . interbreeding natural populations that are reproductively isolated from [do not produce fertile offspring with] other such groups." The concept is based on reproductive criteria and is easy to apply, at least in principle: if the members of two populations interbreed and produce fertile offspring *under natural conditions,* they belong to the same species; their fertile offspring will, in turn, produce the next generation of that species. If two populations do not interbreed in nature, or fail to produce fertile offspring when they do, they belong to different species.

The biological species concept defines species in terms of population genetics and evolutionary theory. The first half of Mayr's definition notes the genetic *cohesiveness* of species: populations of the same species experience gene flow, which mixes their genetic material. Thus, we can think of a species as one large gene pool, which may be subdivided into local populations.

The second part of the biological species concept emphasizes the genetic *distinctness* of each species. Because populations of different species are reproductively isolated, they cannot exchange genetic information. In fact, the process of speciation is frequently defined as the evolution of reproductive isolation between populations.

The biological species concept also explains why individuals of a species generally look alike: members of the same gene pool share genetic traits that determine their appearance. Individuals of different species generally do not resemble one another as closely because they share fewer genetic characteristics. In practice, biologists often still use similarities or differences in morphological traits as convenient markers of genetic similarity or reproductive isolation.

However, the biological species concept does not apply to the many forms of life that reproduce asexually, including most bacteria; some protists, fungi, and plants; and a few animals. In these species, individuals don't interbreed, so it is pointless to ask whether different populations do. Similarly, we cannot use the biological species concept to study extinct organisms, because we have little or no data on their specific reproductive habits. These species must all be defined using morphological or biochemical criteria. Yet, despite its limitations, the biological species concept currently provides the best evolutionary definition of a sexually reproducing species.

The Phylogenetic Species Concept Focuses on Evolutionary History

Recognizing the limitations of the biological species concept, biologists have developed dozens of other ways to define a species. A widely accepted alternative is the **phylogenetic species concept.** Using both morphological and genetic sequence data, scientists first reconstruct the evolutionary tree for the organisms of interest. They then define a phylogenetic species as a cluster of populations—the tiniest twigs on this part of the Tree of Life—that emerge from the same small branch. Thus, a phylogenetic species comprises populations that share a recent evolutionary history. We will consider this approach for understanding the evolutionary relationships of organisms in Chapter 24.

One advantage of the phylogenetic species concept is that biologists can apply it to any group of organisms, including species that have long been extinct, as well as living organisms that reproduce asexually. Proponents of this approach also argue that the morphological and genetic distinctions between organisms on different branches of the Tree of Life reflect the absence of gene flow between them—one of the key requirements of the biological species definition. Nevertheless, because detailed evolutionary histories have been described for relatively few groups of organisms, biologists are not yet able to apply the phylogenetic species concept to all forms of life. As noted in *Unanswered Questions* for Chapter 24, continued research on the details of evolutionary relationships will increase this concept's usefulness in the future.

Many Species Exhibit Substantial Geographical Variation

Just as individuals within populations exhibit genotypic and phenotypic variation (see Section 21.1, Figures 21.2 and 21.15), populations within species also differ both genetically and phenotypically. Neighboring populations often have shared characteristics because they live in similar environments, exchange individuals, and experience comparable patterns of natural selection. Widely separated populations, by contrast, may live under different conditions and experience different patterns of selection; because gene flow is less likely to occur between distant populations, their gene pools and phenotypes often differ.

When geographically separated populations of a species exhibit dramatic, easily recognized phenotypic variation, biologists may identify them as different **subspecies (Figure 22.3),** which are local variants of a species. Individuals from different subspecies usually interbreed where their geographical distributions meet, and their offspring often exhibit intermediate phenotypes. Biologists sometimes use the word "race" as shorthand for the term "subspecies."

Various patterns of geographical variation—as well as analyses of how the variation may relate to climatic or habitat variation—have provided great insight into the speciation process. Two of the best-studied patterns are *ring species* and *clinal variation.*

RING SPECIES Some plant and animal species have a ring-shaped geographical distribution that surrounds uninhabitable terrain. Adjacent populations of these **ring species** can exchange genetic material directly, but gene flow between distant populations occurs only through the intermediary populations.

The lungless salamander *Ensatina eschscholtzii,* an example of a ring species, is widely distributed in the coastal mountains and the Sierra Nevada of California, but it cannot survive in the hot, dry Central Valley **(Figure 22.4).** Seven subspecies differ in biochemical traits, color, size, and ecology. Individuals from adjacent subspecies often interbreed where their geographical distributions overlap, and intermediate phenotypes are fairly common.

FIGURE 22.3 Subspecies. Five subspecies of rat snake *(Elaphe obsoleta)* in eastern North America differ in color and in the presence or absence of stripes or blotches.

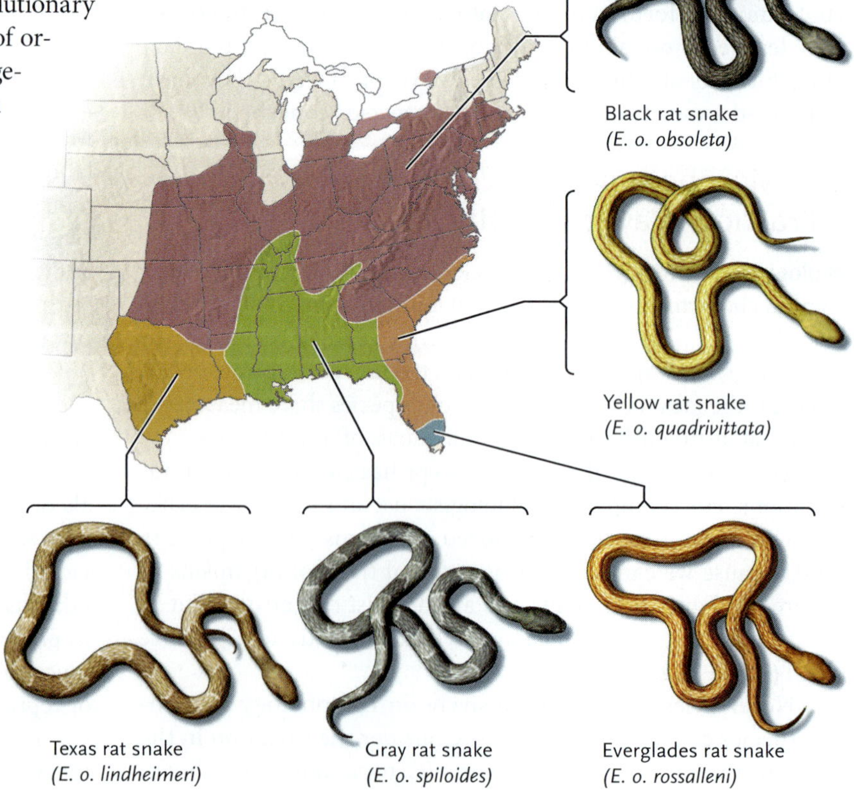

But at the southern end of the Central Valley, adjacent subspecies rarely interbreed. Apparently, they have differentiated to such an extent that they can no longer exchange genetic material directly.

Are the southernmost populations of this salamander subspecies or different species? A biologist who saw *only* the southern populations, which coexist without interbreeding, might define them as separate species. However, they still have the potential to exchange genetic material through the intervening populations that form the ring. Hence, biologists recognize these populations as belonging to the same species. Most likely, the southern subspecies are in an intermediate stage of species formation.

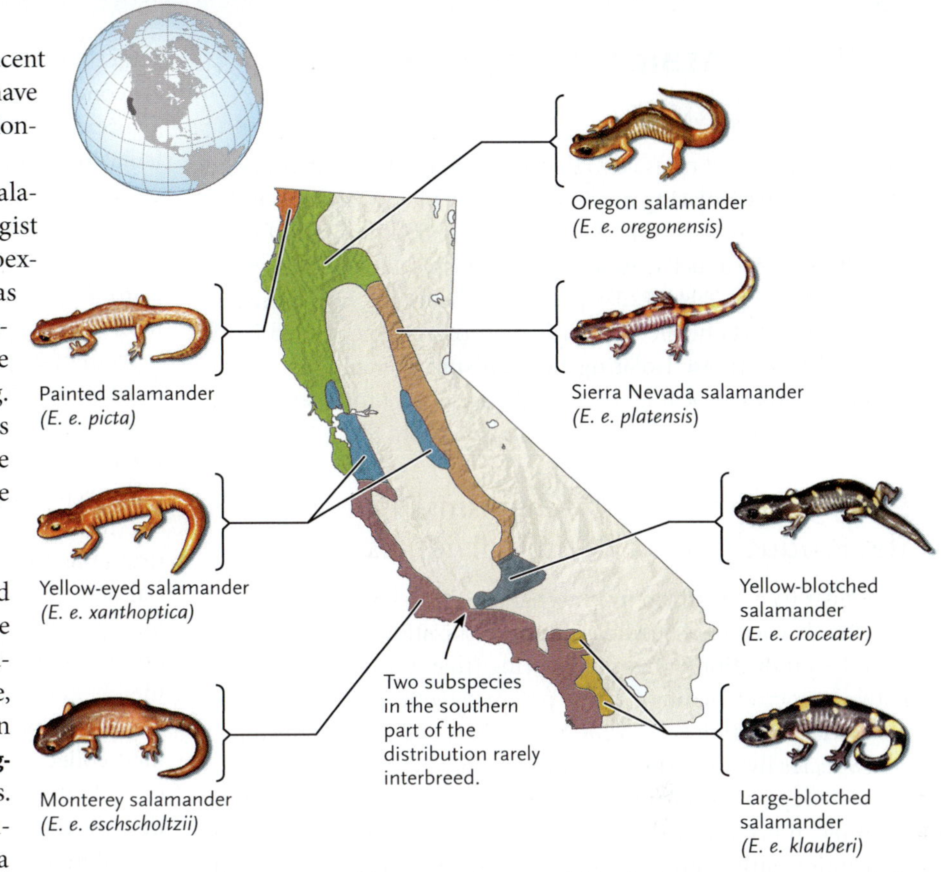

FIGURE 22.4 Ring species. Six of the seven subspecies of the salamander *Ensatina eschscholtzii* are distributed in a ring around California's Central Valley. Subspecies often interbreed where their geographical distributions overlap. However, the two subspecies that nearly close the ring in the south (marked with an arrow), the Monterey salamander and the yellow-blotched salamander, rarely interbreed.

CLINAL VARIATION When a species is distributed over a large, environmentally diverse area, some traits may exhibit a **cline,** a pattern of smooth variation along a geographical gradient. For example, many birds and mammals in the northern hemisphere show clinal variation in body size **(Figure 22.5)** and the relative length of their appendages. In general, populations living in colder environments have larger bodies and shorter appendages, a pattern that is usually interpreted as a mechanism to conserve body heat (see Chapter 48).

Clinal variation usually results from gene flow between adjacent populations that are each adapting to slightly different conditions. However, if populations at opposite ends of a cline are separated by great distances, they may exchange very little genetic material through reproduction. Thus, when a cline extends over a large geographical gradient, distant populations may be genetically and morphologically distinct.

Despite the geographical variation that many species exhibit, even closely related species are genetically and morphologically different from each other. In the next section, we consider the mechanisms that maintain the genetic distinctness of closely related species by preventing their gene pools from mixing.

STUDY BREAK 22.1

1. **How do the morphological, biological, and phylogenetic species concepts differ?**
2. **What is clinal variation?**

THINK OUTSIDE THE BOOK

Search the biological literature and the Internet for a definition and application of the "ecological species concept." How does the ecological species concept differ from the three species concepts described above? Under what circumstances or for what purpose would the ecological species concept be useful?

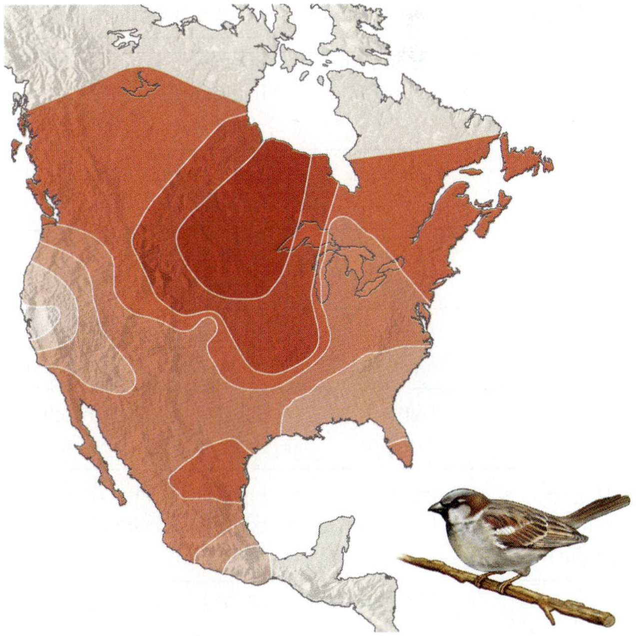

FIGURE 22.5 Clinal variation. House sparrows *(Passer domesticus)* exhibit clinal variation in overall body size, which was summarized from measurements of 16 skeletal features. Darker shading in the map indicates larger size.

22.2 Maintaining Reproductive Isolation

Reproductive isolation is central to the biological species concept. A **reproductive isolating mechanism** is any biological characteristic that prevents the gene pools of two species from mixing. Biologists classify reproductive isolating mechanisms into two categories (summarized in **Table 22.1**): **prezygotic isolating mechanisms** exert their effects before the production of a zygote, or fertilized egg, and **postzygotic isolating mechanisms** operate after zygote formation. These isolating mechanisms are not mutually exclusive; two or more of them may operate simultaneously.

Prezygotic Isolating Mechanisms Prevent the Production of Hybrid Individuals

Biologists have identified five mechanisms that can prevent interspecific (between species) matings or fertilizations, and thus prevent the production of hybrid offspring (that is, the offspring of a between-species mating). These five prezygotic mechanisms are *ecological, temporal, behavioral, mechanical,* and *gametic isolation.*

Species living in the same geographical region may experience **ecological isolation** if they live in different habitats. For example, lions and tigers were both common in India until the mid-nineteenth century, when hunters virtually exterminated the Asian lions. However, because lions live in open grasslands and tigers in dense forests, the two species did not encounter one another and did not interbreed. Lion-tiger hybrids are sometimes born in captivity, but they do not occur under natural conditions.

TABLE 22.1 Isolating Mechanisms

Timing Relative to Fertilization	Mechanism	Mode of Action
Prezygotic ("premating") mechanisms	Ecological isolation	Species live in different habitats
	Temporal isolation	Species breed at different times
	Behavioral isolation	Species cannot communicate
	Mechanical isolation	Species cannot physically mate
	Gametic isolation	Species have nonmatching receptors on gametes
Postzygotic ("postmating") mechanisms	Hybrid inviability	Hybrid offspring do not complete development
	Hybrid sterility	Hybrid offspring cannot produce gametes
	Hybrid breakdown	Hybrid offspring have reduced survival or fertility

Species living in the same habitat can experience **temporal isolation** if they mate at different times of day or different times of year. For example, the fruit flies *Drosophila persimilis* and *Drosophila pseudoobscura* overlap extensively in their geographical distributions, but they do not interbreed, in part because *D. persimilis* mates in the morning and *D. pseudoobscura* in the afternoon. Similarly, two species of pine in California are reproductively isolated where their geographical distributions overlap: even though both rely on the wind to carry male gametes (pollen grains) to female gametes (ova) in other cones, *Pinus radiata* releases pollen in February and *Pinus muricata* releases pollen in April.

Many animals rely on specific signals, which may differ dramatically between species, to identify the species of a potential mate. **Behavioral isolation** results when the signals used by one species are not recognized by another. For example, female birds rely on the song, color, and displays of males to identify members of their own species. Similarly, female fireflies identify males by their flashing patterns **(Figure 22.6).** These behaviors (collectively called *courtship displays*) are often so complicated that signals sent by one species are like a foreign language that another species simply does not understand.

Mate choice by females and sexual selection (discussed in Section 21.3) generally drive the evolution of mate recognition signals. Females often spend substantial energy in reproduction, and choosing an appropriate mate—that is, a male of her own species—is critically important for the production of successful young. By contrast, a female that mates with a male from a different species is unlikely to leave any surviving offspring at all. Over time, the number of males with recognizable traits, as well as the number of females able to recognize the traits, increases in the population.

Differences in the structure of reproductive organs or other body parts—**mechanical isolation**—may prevent individuals of different species from interbreeding. In particular, many plants have anatomical features that allow only certain pollinators, usually particular bird or insect species, to collect and distribute pollen (see Chapter 29). For example, the flowers and nectar of two native California plants, the purple monkey-flower *(Mimulus lewisii)* and the scarlet monkey-flower *(Mimulus cardinalis),* attract different animal pollinators **(Figure 22.7).** *Mimulus lewisii* is pollinated by bumblebees. It has shallow purple flowers with broad petals that provide a landing platform for the bees. Bright yellow streaks on the petals serve as "nectar guides," directing bumblebees to the short nectar tube and reproductive parts, which are located among the petals. Bees enter the flowers to drink their concentrated nectar, and they pick up and deliver pollen as their legs and bodies brush against the reproductive parts of the flowers. *Mimulus cardinalis,* by contrast, is pollinated by hummingbirds. It has long red flowers with no yellow streaks, and the reproductive parts extend above the petals. The red color attracts hummingbirds but lies outside the color range detected by bumblebees. The nectar of *M. cardinalis* is more dilute than that of *M. lewisii* but is produced in much greater quantity, making it easier for hummingbirds to ingest. When a hummingbird visits *M. cardinalis* flowers, it pushes its long bill down the nectar tube, and its forehead touches the reproductive parts, picking up and delivering pollen. Recent research has demonstrated that where the two monkey-flower species grow side-by-side, animal pollinators restrict their visits to either one species or the other 98% of the time, providing nearly complete reproductive isolation.

Even when individuals of different species mate, **gametic isolation,** an incompatibility between the sperm of one species and the eggs of another, may prevent fertilization. Many marine invertebrates release gametes into the environment for external fertilization. The sperm and eggs of each species recognize one another's complementary surface proteins (see Chapter 49), but the surface proteins on the gametes of different species don't match. In animals with internal fertilization, sperm of one species may not survive and function within the reproductive tract of another. Interspecific matings between some *Drosophila* species, for example, induce a reaction in the female's reproductive tract that blocks "foreign" sperm from reaching eggs. Parallel physiological incompatibilities between a pollen tube and a stigma prevent interspecific fertilization in some plants.

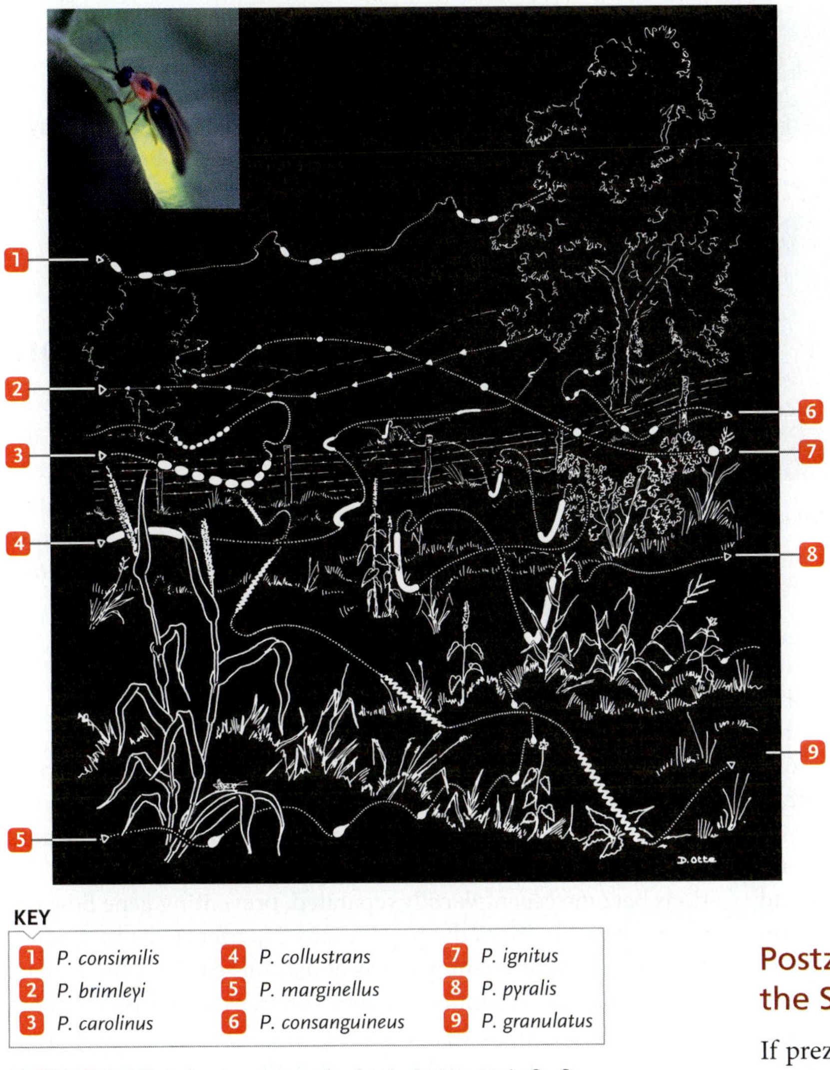

FIGURE 22.6 Behavioral reproductive isolation. Male fireflies use bioluminescent signals to attract potential mates. The different flight paths and flashing patterns of males in nine North American *Photinus* species are represented here. Females respond only to the display given by males of their own species. The inset photo shows *P. pyralis.* (Illustration courtesy of James E. Lloyd. Miscellaneous Publications of the Museum of Zoology of the University of Michigan, 130:1–195, 1966. Photo by Gail Shumway/Photographer's Choice/Getty Images)

FIGURE 22.7 Mechanical reproductive isolation. Because of differences in floral structure, two species of monkey-flower attract different animal pollinators. *Mimulus lewisii* attracts bumblebees, and *Mimulus cardinalis* attracts hummingbirds.

Postzygotic Isolating Mechanisms Reduce the Success of Hybrid Individuals

If prezygotic isolating mechanisms between two closely related species are incomplete or ineffective, sperm from one species sometimes fertilizes an egg of the other species. In such cases the two species will be reproductively isolated if their offspring, called interspecific (between species) hybrids, have lower fitness than those produced by intraspecific (within species) matings. Three postzygotic isolating mechanisms—*hybrid inviability, hybrid sterility,* and *hybrid breakdown*—can reduce the fitness of hybrid individuals.

Many genes govern the complex processes that transform a zygote into a mature organism. Hybrid individuals have two sets of developmental instructions, one from each parent species, which may not interact properly for the successful completion of embryonic development. As a result, hybrid organisms frequently die as embryos or at an early age, a phenomenon called **hybrid inviability.** For example, domestic sheep and goats can mate and fertilize one another's ova, but the hybrid embryos always die before coming to term, presumably because the developmental programs of the two parent species are incompatible.

Although some hybrids between closely related species develop into healthy and vigorous adults, they may not produce functional gametes. This **hybrid sterility** often results when the parent species differ in the number or structure of their chro-

Jen and Des Bartlett/Bruce Coleman USA

FIGURE 22.8 Interspecific hybrids. Horses and zebroids (hybrid offspring of horses and zebras) run in a mixed herd. Zebroids are usually sterile.

mosomes, which cannot pair properly during meiosis. Such hybrids have zero fitness because they leave no descendants. The most familiar example is a mule, the product of mating between a female horse ($2n = 64$) and a male donkey ($2n = 62$). Zebroids, the offspring of matings between horses and zebras, are also usually sterile **(Figure 22.8).**

Some first-generation hybrids (F_1; see Section 12.1) are healthy and fully fertile. They can breed with other hybrids and with both parental species. However, the second generation (F_2), produced by matings between F_1 hybrids, or between F_1 hybrids and either parental species, may exhibit reduced survival or fertility, a phenomenon known as **hybrid breakdown.** For example, experimental crosses between *Drosophila* species may produce functional hybrids, but their offspring experience a high rate of chromosomal abnormalities and harmful types of genetic recombination. Thus, reproductive isolation is maintained between the species because there is little long-term mixing of their gene pools.

STUDY BREAK 22.2

1. What is the difference between prezygotic and postzygotic isolating mechanisms?
2. When a male duck of one species performed a courtship display to a female of another species, she interpreted his behavior as aggressive rather than amorous. What type of reproductive isolating mechanism does this scenario illustrate?

22.3 The Geography of Speciation

As Ernst Mayr recognized, geography has a huge impact on whether gene pools have the opportunity to mix. Biologists define three modes of speciation, based on the geographical relationship of populations as they become reproductively isolated: *allopatric speciation* (*allo* = different, *patria* = homeland), *parapatric speciation* (*para* = beside), and *sympatric speciation* (*sym* = together).

Allopatric Speciation Occurs between Geographically Separated Populations

Allopatric speciation may take place when a physical barrier subdivides a large population or when a small population becomes separated from a species' main geographical distribution. Allopatric speciation occurs in two stages. First, two populations become *geographically* separated, preventing gene flow between them. Then, as the populations experience distinct mutations as well as different patterns of natural selection and genetic drift, they may accumulate genetic differences that isolate them *reproductively.* Allopatric speciation is probably the most common mode of speciation in large animals.

Geographical separation sometimes occurs when a barrier divides a large population into two or more units **(Figure 22.9).** For example, hurricanes may create new channels that divide low coastal islands and the populations inhabiting them. Uplifting mountains or landmasses as well as rivers or advancing glaciers can also produce barriers that subdivide populations. The uplift of the Isthmus of Panama, caused by movements of Earth's crust about 5 million years ago, separated a once-

FIGURE 22.9 The model of allopatric speciation and secondary contact.

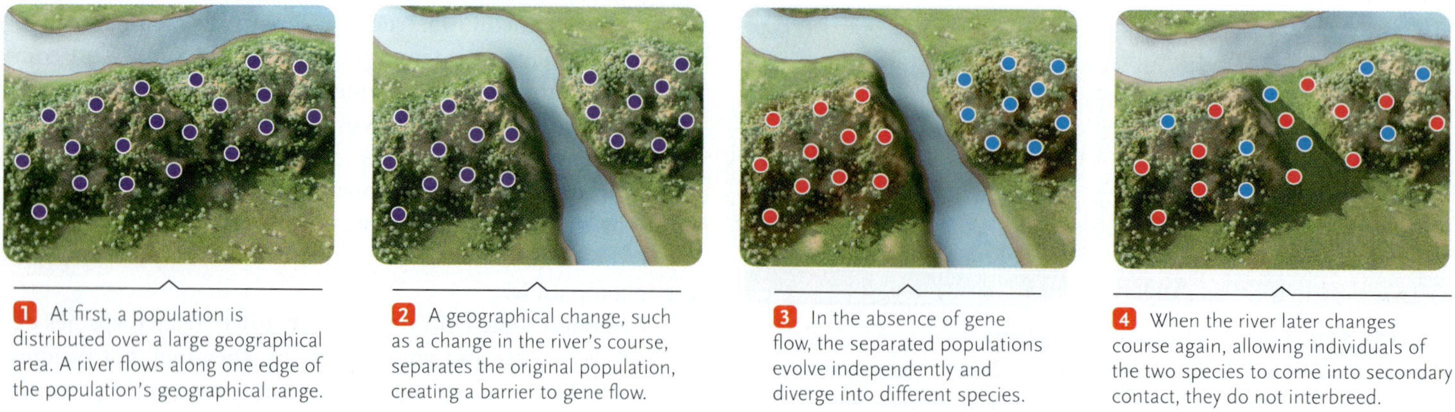

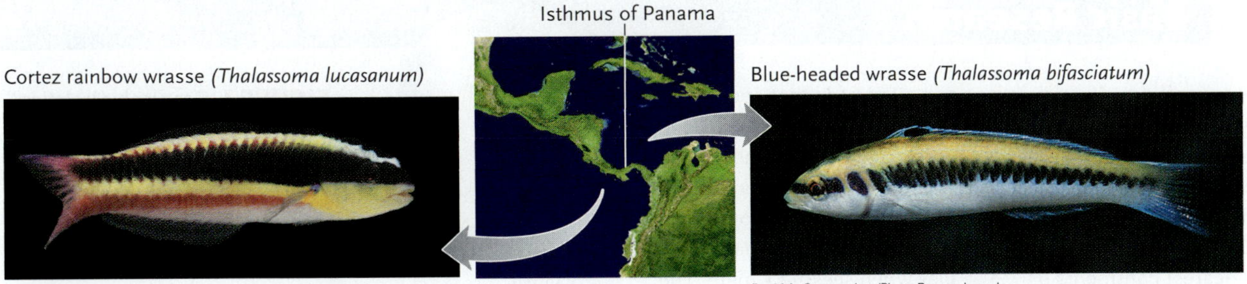

Patrice Ceisel/Visuals Unlimited NASA Fred Mc Connaughey/Photo Researchers, Inc.

FIGURE 22.10 Geographical separation. The uplift of the Isthmus of Panama divided an ancestral wrasse population. The Cortez rainbow wrasse now occupies the eastern Pacific Ocean, and the blue-headed wrasse now occupies the western Atlantic Ocean.

FIGURE 22.11 Evolution of a species cluster on an archipelago. Letters identify four islands in a hypothetical archipelago, and colored dots represent different species. The ancestor of all the species is represented by black dots on the mainland. At the end of the process, islands A and B are each occupied by two species, and islands C and D are each occupied by one species, all of which evolved on the islands.

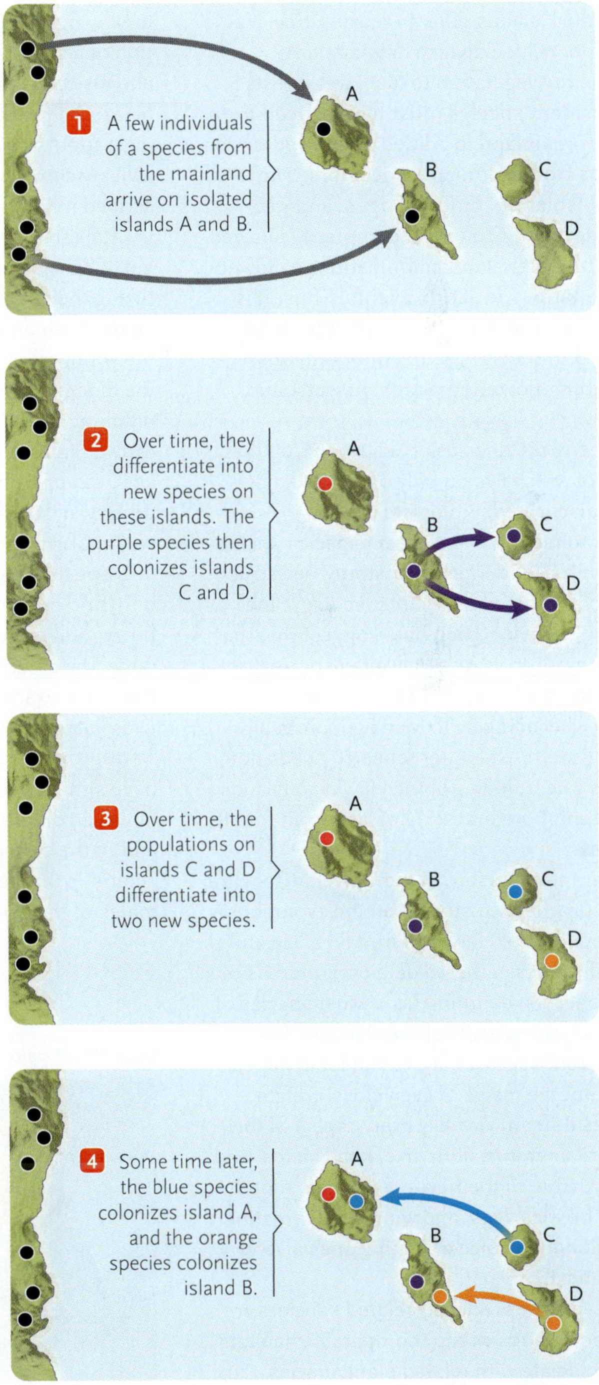

continuous shallow sea into the eastern tropical Pacific Ocean and the western tropical Atlantic Ocean. Populations of marine organisms were subdivided by this event. In the tropical Atlantic Ocean, populations experienced patterns of mutation, natural selection, and genetic drift that were different from those experienced by populations in the tropical Pacific Ocean. As a result, the populations diverged genetically, and pairs of closely related species now live on either side of this divide **(Figure 22.10).**

In other cases, small populations may become isolated at the edge of a species' geographical distribution. Such peripheral populations often differ genetically from the central population because they are adapted to somewhat different environments. Once a small population is isolated, founder effects and small population size may promote genetic drift (see page 464), and natural selection may favor the evolution of distinctive traits. If the isolated population experiences limited gene flow from the parent population, these agents of evolution will foster genetic differentiation between them. In time, the accumulated genetic differences may lead to reproductive isolation.

Populations established by colonization of oceanic islands represent extreme examples of this phenomenon. The founder effect makes the populations genetically distinct. And on oceanic archipelagos, such as the Galápagos and Hawaiian islands, individuals from one island may colonize nearby islands, founding populations that differentiate into distinct species. Each island may experience multiple invasions, and the process may be repeated many times within the archipelago, leading to the evolution of a **species cluster,** a group of closely related species recently descended from a common ancestor **(Figure 22.11).** Sometimes a species cluster can evolve relatively quickly; for example, the nearly 800 species of fruit flies now living on the Hawaiian Islands (described in *Focus on Basic Research*) evolved in less than 5 million years, an average of just over 6,000 years per species.

FOCUS ON **Basic Research**

Speciation in Hawaiian Fruit Flies

The islands of the Hawaiian archipelago have been geographically isolated throughout their history, lying at least 3,200 km (approximately 2,000 miles) from the nearest continents or other islands **(Figure 1).** Built by undersea volcanic eruptions over millions of years, they emerged from northwest to southeast: Kauai is at least 5 million years old, and Hawaii, the "Big Island," is less than 1 million years old. Individual islands differ in maximum elevation and include diverse habitats, from sparse, dry vegetation to lush, wet forests.

Resident species must have arrived from distant mainland localities or evolved on the islands from colonizing ancestors. The islands' isolation, different ages, and geographical and ecological complexity allowed repeated interisland colonizations followed by allopatric speciation events. Thus, it is not surprising that species clusters have evolved in several groups of organisms (including flowering plants, insects, and birds).

Nearly 800 species of Hawaiian fruit flies, most of which live on only one island, have been discovered. Biologists used many characters to identify the different species, including external and internal anatomy, cell structure, chromosome structure, ecology, and mating behavior. Their data suggest that the vast majority of native Hawaiian species arose from a single ancestral species that colonized the archipelago long ago, probably from eastern Asia. After repeated speciation events, the fruit flies of the Hawaiian Islands represent more than 25% of all known fruit fly species.

Hampton Carson, of the University of Hawaii, spearheaded studies on the evolutionary relationships of Hawaiian fruit flies. He and his colleagues gathered data on hundreds of fly species—a daunting task. Most species of fruit flies are sexually dimorphic. The females of different species may be similar in appearance, but the males of even closely related species differ in virtually every aspect of their external anatomy: body size, head shape, and the structure of their eyes, antennae, mouthparts, bristles, legs, and wings. Their mating behavior and choice of mating sites also vary dramatically.

Nevertheless, closely related species on different islands occupy comparable habitats and associate with related plant species. Carson suggested that speciation in these flies resulted from the evolution of different genetically determined *mating systems,* the behaviors and morphological characteristics that males display when seeking a mate. The mating systems serve as prezygotic isolating mechanisms.

The 100 or more species of "picture-wing" *Drosophila,* relatively large flies with patterns on their wings, illustrate the evolution of a species cluster. Carson and his colleagues used similarities and differences in the banding patterns on the flies' giant salivary chromosomes (described in *Focus on Model Research Organisms* in Chapter 13), to trace the evolutionary origin of species on the younger islands by identifying their closest relatives on the older islands. Their analysis of 26 species on Hawaii, the youngest island, suggested that flies from the older islands colonized Hawaii at least 19 different times, and each founder population evolved into a new species there. Additional species apparently evolved when lava flows on Hawaii subdivided existing populations.

Among the picture-wing fruit flies, some interspecies matings result in hybrid sterility or hybrid breakdown. But for most species, prezygotic reproductive isolation is maintained by differences in their mating systems. For example, *Drosophila silvestris* and *Drosophila heteroneura,* which produce healthy and fertile hybrids in the laboratory, have similar geographical distributions; however, differences in courtship behavior and in the shape of the males' heads, a characteristic that females use to recognize males of their own species **(Figure 2),** keep these two species reproductively isolated. In nature, they hybridize only in one small geographical area.

The work of Carson and his colleagues suggests that most speciation in Hawaiian *Drosophila* has resulted from founder effects and genetic drift. When a fertile female—or a small group of males and females—moves to a new island, this founding population responds to novel selection pressures in its new environment. Sexual selection then exaggerates distinctive morphological and behavioral characteristics, maintaining the population's reproductive isolation from its new neighbors. The tremendous variety of Hawaiian fruit flies has undoubtedly been produced by repeated colonizations of newer islands by flies from older islands and by the back-colonization of older islands by newly evolved species. Thus, they represent what evolutionary biologists describe as an *adaptive radiation,* a cluster of closely related species that are ecologically different (as described further in Chapter 23).

FIGURE 1 Geographic isolation of the Hawaiian Islands.

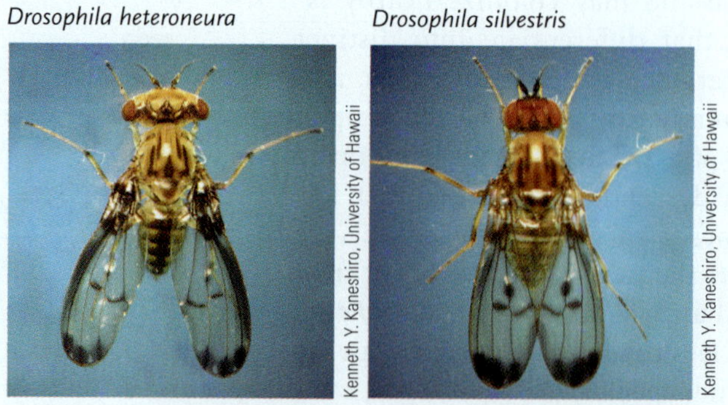

FIGURE 2 Two *Drosophila* species in which the males' head shapes differ.

Bullock's oriole *(Icterus bullockii)*

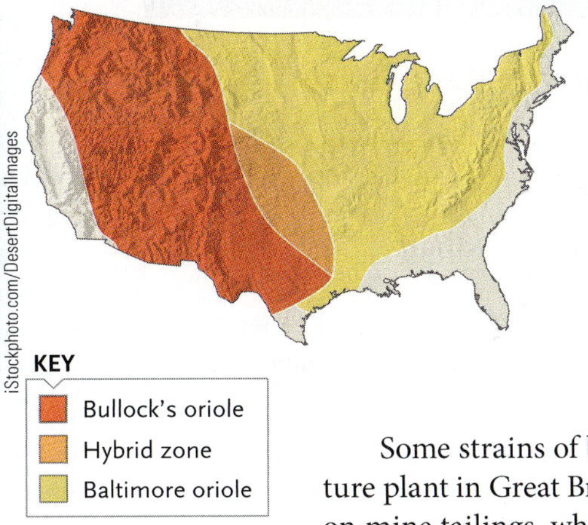

Baltimore oriole *(Icterus galbula)*

FIGURE 22.12 Hybrid zones. Males of the Bullock's oriole and Baltimore oriole differ in color and courtship song. The populations have maintained a hybrid zone for hundreds of years, and once were considered subspecies of the same species. The American Ornithologists' Union recognized them as separate species in 1997. They now hybridize less frequently than they once did, leading some researchers to suggest that their reproductive isolation evolved recently.

Sometimes, allopatric populations reestablish contact when a geographical barrier is eliminated or breached (see Figure 22.9, step 4). This *secondary contact* provides a test of whether or not the populations have diverged into separate species. If their gene pools did not differentiate much during geographical separation, the populations will interbreed and merge. But if the populations have differentiated enough to be reproductively isolated, they have become separate species.

During the early stages of secondary contact, prezygotic reproductive isolation may be incomplete. Some members of each population may mate with individuals from the other, producing viable, fertile offspring in areas called **hybrid zones.** Although some hybrid zones have persisted for hundreds or thousands of years **(Figure 22.12),** they are generally narrow, and ecological or geographical factors maintain the separation of the gene pools for the majority of individuals in both species.

If postzygotic isolating mechanisms cause hybrid offspring to have lower fitness than those produced within each population, natural selection will promote the evolution of prezygotic isolating mechanisms, favoring individuals that mate only with members of their own population. Recent studies of *Drosophila* suggest that this phenomenon, called **reinforcement,** enhances reproductive isolation that had begun to develop while the populations were geographically separated.

Parapatric Speciation May Occur between Adjacent Populations

Sometimes a single species is distributed across a discontinuity in environmental conditions, such as a major change in soil type. Although organisms on one side of the discontinuity may interbreed freely with those on the other side, natural selection may favor different alleles on either side, limiting gene flow. In such cases, **parapatric speciation**—speciation arising between adjacent populations—may occur if hybrid offspring have low relative fitness.

Some strains of bent grass *(Agrostis tenuis),* a common pasture plant in Great Britain, have the physiological ability to grow on mine tailings, where the soil is heavily polluted by copper or other metals. Plants of the copper-tolerant strains grow well on polluted soils, but plants of the pasture strain do not. Conversely, copper-tolerant plants don't survive as well as pasture plants on unpolluted soils. These strains often grow within a few meters of each other where polluted and unpolluted soils form an intricate environmental mosaic. Because bent grass is wind-pollinated, pollen is readily transferred from one strain to another. Laboratory tests showed that the strains are fully interfertile. However, copper-tolerant plants flower about one week earlier than nearby pasture plants, which promotes prezygotic (temporal) isolation of the two strains **(Figure 22.13).** If the flowering times become further separated, the two strains may attain complete reproductive isolation and become separate species.

Some biologists argue that the places where parapatric populations of bent grass interbreed are really hybrid zones where previously allopatric populations have established secondary contact. Unfortunately, there is no way to determine whether the hybridizing populations were parapatric or allopatric in the past. Thus, a thorough evaluation of the parapatric speciation hypothesis must await the development of techniques that enable biologists to distinguish clearly between the products of allopatric and parapatric speciation.

Sympatric Speciation Occurs within One Continuously Distributed Population

In **sympatric speciation,** reproductive isolation evolves between distinct subgroups that arise within one population. Models of sympatric speciation do not require that the populations be either geographically or environmentally separated as their gene pools diverge. We examine below general models of sympatric speciation in animals and plants; the genetic basis of sympatric speciation is one of the topics we consider in the next section.

Insects that feed on just one or two plant species are among the animals most likely to evolve by sympatric speciation. These insects generally carry out most important life cycle activities on or near their "host" plants. Adults mate on the host plant; females lay their eggs on it; and larvae feed on the host plant's tissues, eventually developing into adults, which initiate another round of the life cycle. Host plant choice is genetically determined in many

FIGURE 22.13 Observational Research

Evidence for Reproductive Isolation in Bent Grass

Question: Do adjacent populations of bent grass living on different soil types exhibit signs of reproductive isolation?

Hypothesis: Adjacent populations of bent grass *(Agrostis tenuis)* flower at slightly different times, which could foster prezygotic reproductive isolation between them.

Prediction: On any given summer day, the flowers in one population will be more mature (thus, being pollinated earlier) than flowers in the adjacent population.

Method: On a late summer day in 1965, Thomas McNeilly and Janis Antonovics of University College of North Wales compared the flowers of bent grass growing on polluted soil at a copper mine with the flowers of plants growing on unpolluted soil in a nearby pasture. A meter-wide stretch of polluted pasture (indicated by cross-hatching in the graph) formed a boundary between the two populations. Researchers assigned a score to every flower, with immature flowers scored as 3 and mature flowers as 4.

Results: On the day that they were surveyed, flowers of the copper-tolerant plants had higher scores, indicating that they were more mature—and thus would complete pollination earlier—than the flowers of the pasture plants.

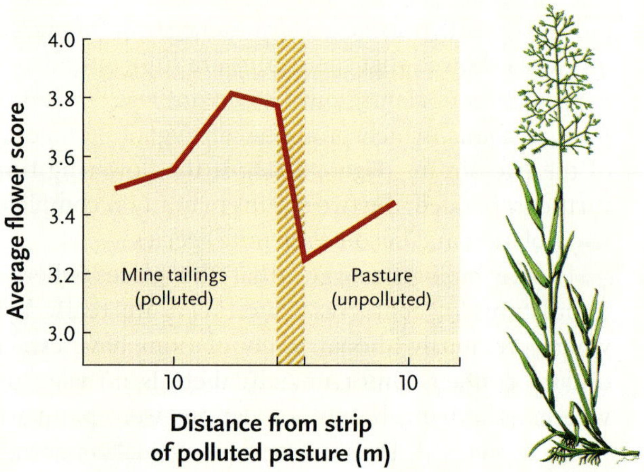

Conclusion: Because adjacent populations of bent grass flower at slightly different times, temporal reproductive isolation may be developing between them.

THINK LIKE A SCIENTIST If bent grass were insect pollinated instead of being wind pollinated, would the observed difference in the timing of flower maturation still establish temporal reproductive isolation between the two strains?

Source: T. McNeilly and J. Antonovics. 1968. Evolution in closely adjacent plant populations. IV. Barriers to gene flow. *Heredity* 23:205–218.

FIGURE 22.14 Sympatric speciation in animals. Male and female apple maggots *(Rhagoletis pomonella)* court on a hawthorn leaf. The female will later lay her eggs on the fruit, and the offspring will feed, mate, and lay their eggs on hawthorns as well.

insect species. In others, individuals associate with the host plant species they ate as larvae.

Theoretically, a genetic mutation could suddenly change some insects' choice of host plant. Mutant individuals would shift their life cycle activities to the new host, and then interact primarily with others preferring the same new host, an example of ecological isolation. These individuals would collectively form a separate subpopulation, called a **host race.** Reproductive isolation could evolve between different host races if the individuals of each host race are more likely to mate with members of their own host race than with members of another. Some biologists criticize this model, however, because it assumes that the genes controlling two traits, the insects' host plant choice and their mating preferences, change simultaneously. Moreover, host plant choice is controlled by multiple gene loci in some insect species, and it is clearly influenced by prior experience in others.

The apple maggot *(Rhagoletis pomonella)* is the most thoroughly studied example of possible sympatric speciation in animals **(Figure 22.14).** This fly's natural host plant in eastern North America is the hawthorn (*Crataegus* species), but at least two host races have appeared in the past 150 years. The larvae of a new host race were first discovered feeding on apples in New York State in the 1860s. In the 1960s, a cherry-feeding host race appeared in Wisconsin.

Genetic analyses have shown that variations at just a few gene loci underlie differences in the feeding preferences of *Rhagoletis* host races; other genetic differences cause the host races to develop at different rates. Moreover, adults of the three races mate during different summer months. Nevertheless, individuals show no particular preference for mates of their own host race, at least under simplified laboratory conditions. Thus, although behavioral isolation has not developed between races, ecological and temporal isolation may separate adults in nature. Researchers are still not certain that the different host races are reproductively isolated under natural conditions.

In 2010, Andrew P. Michel of the University of Notre Dame and colleagues elsewhere in the United States and Germany published a genomic analysis of the apple- and hawthorn-feeding races of *Rhagoletis.* Their results suggest that over the past 150 years, the two races have diverged at many loci in their genomes—not just at the loci that influence food choice and developmental rate—and that the divergence has largely been driven by natural selection. Ongoing genetic divergence may prevent them from interbreeding in the future.

Sympatric speciation often occurs in plants through a genetic phenomenon, **polyploidy,** in which an individual has one or more *extra* copies of the entire haploid complement of chromosomes (see Section 13.3). Polyploidy can lead to speciation because these large-scale genetic changes may prevent polyploid individuals from breeding with individuals of the parent species. Nearly half of all flowering plant species are polyploid, including many important crops and ornamental species. The genetic mechanisms that produce polyploid individuals in plant populations are well understood; we describe them in the next section as part of a larger discussion of the genetics of speciation.

STUDY BREAK 22.3

1. What are the two stages required for allopatric speciation?
2. What factor appears to promote parapatric speciation in bent grass?
3. Why might insects from different host races be unlikely to mate with each other?

THINK OUTSIDE THE BOOK

Figure 22.12 presents an example of a hybrid zone between two closely related bird species in North America. Search the biological literature and the Internet for three other examples of hybrid zones in animals. How large are the areas of hybridization? What prezygotic isolating mechanisms prevent the species from interbreeding outside the hybrid zones? Is the frequency of hybrid matings increasing or decreasing through time?

22.4 Genetic Mechanisms of Speciation

What genetic changes lead to reproductive isolation between populations? In this section we examine three genetic mechanisms that can lead to reproductive isolation: *genetic divergence* between allopatric populations, *polyploidy* in sympatric populations, and *chromosome alterations,* which occur independently of the geographical distributions of populations.

Genetic Divergence in Allopatric Populations Can Lead to Speciation

In the absence of gene flow, geographically separated populations inevitably accumulate genetic differences through the action of mutation, genetic drift, and natural selection.

How much genetic divergence is necessary for speciation to occur? To understand the genetic basis of speciation in closely related species, researchers first identify the specific causes of reproductive isolation. They then use standard techniques of genetic analysis along with new molecular approaches such as gene mapping and sequencing to analyze the genetic mechanisms that establish reproductive isolation. As explained in *Insights from the Molecular Revolution,* these techniques now allow researchers to determine the minimum number of genes responsible for reproductive isolation in particular pairs of species.

In cases of postzygotic reproductive isolation, mutations in just a few gene loci can establish reproductive isolation. For example, if two common aquarium fishes, swordtails *(Xiphophorus helleri)* and platys *(Xiphophorus maculatus),* mate, two genes induce the development of lethal tumors in their hybrid offspring. When hybrid sterility is the primary cause of reproductive isolation between *Drosophila* species, at least 5 gene loci are responsible. About 55 gene loci contribute to postzygotic reproductive isolation between the European fire-bellied toad *(Bombina bombina)* and the yellow-bellied toad *(Bombina variegata).*

In cases of prezygotic reproductive isolation, some mechanisms have a surprisingly simple genetic basis. For example, a single mutation reverses the direction of coiling in the shells of some snails (*Bradybaena* species). Snails with shells that coil in opposite directions cannot approach each other closely enough to mate, making reproduction between them mechanically impossible.

Many traits that now function as prezygotic isolating mechanisms may originally have evolved in response to sexual selection (described on page 470). In sexually dimorphic species, this evolutionary process exaggerates showy structures and courtship behaviors in males, the traits that females use to identify appropriate mates. When two populations encounter one another on secondary contact, these traits may also prevent interspecific mating. For example, many closely related duck species exhibit dramatic variation in the appearance of males, but not females **(Figure 22.15),** an almost certain sign of sexual selection. Yet these species hybridize readily in captivity, producing offspring that are both viable and fertile.

Reproductive isolation and speciation in ducks and other sexually dimorphic birds probably results from geographical

Mallard ducks *(Anas platyrhynchos)*

MUMUISM/Shutterstock.com

Pintail ducks *(Anas acuta)*

David Chapman/Alamy

FIGURE 22.15 Sexual selection and prezygotic isolation. In closely related species, such as mallard and pintail ducks, males have much more distinctive coloration than females, a sure sign of sexual selection at work.

Monkey-Flower Speciation: How many genes contribute to reproductive isolation?

Reproductive isolation is the primary criterion that biologists use to distinguish species. As noted earlier, the monkey-flower species *Mimulus lewisii* and *Mimulus cardinalis* experience mechanical reproductive isolation in nature because differences in flower structure keep bumblebees or hummingbirds from carrying pollen from one species to the other (see Figure 22.7). However, the two species are easily crossed in the laboratory and produce fertile F_1 hybrids. The F_2 offspring have flowers with various forms intermediate between the two parental types (see bottom row of **Figure**), suggesting that several gene loci control the traits separating the species.

Research Question

How many gene loci contribute to the differences in flower traits that foster prezygotic reproductive isolation between the two *Mimulus* species?

Experiment

Because little was known about the genetics of the two monkey-flower species, it was not possible to use a direct genetic analysis to identify and map the flower trait genes to the chromosomes. Instead, H. D. ("Toby") Bradshaw and other researchers at the University of Washington used an indirect molecular approach that identified DNA sequence variations (analogous to those used in DNA fingerprinting; see Chapter 18) at various loci in the genome. Just as morphological and biochemical traits vary within a population, DNA sequences vary at particular sites in the genome; and the different DNA sequence alleles can be distinguished using PCR.

The researchers used a random set of 153 DNA sequence variations. They correlated the segregation of those variations with the segregation of flower traits in 93 plants of the F_2 generation. Some of the DNA sequences segregated closely with a particular flower trait. This result indicated that the particular DNA sequence variation locus was very near the gene for that flower trait on the chromosome. In other words, the flower trait locus was identified indirectly through the close linkage between the DNA variation locus and the flower trait locus. But where are the genes? To answer this question, the researchers used the DNA sequences linked to flower trait loci as probes to find the sites on the chromosomes where they originated. (The use of DNA probes to identify complementary DNA in the genome is described in Figure 18.9.) For any given DNA variation locus, once its position on the chromosomes was determined, the investigators knew that the flower trait locus correlated with it must be nearby.

Purple monkey-flower
(Mimulus lewisii)

F_1 hybrid

Scarlet monkey-flower
(Mimulus cardinalis)

F_2 hybrids

H. D. Bradshaw Jr., S. M. Wilbert, K. G. Otto, and D. W. Schemske. 1995. Genetic mappings of floral traits associated with reproductive isolation in monkey flowers (Mimulus). Nature 376: 762–765.

Results

Their results indicate that reproductive isolation of *M. lewisii* and *M. cardinalis* results from differences in eight floral traits: the amount of (1) anthocyanin pigments and (2) carotenoid pigments in petals; (3) flower width; (4) petal width; (5) nectar volume; (6) nectar concentration; and the lengths of the stalks supporting the (7) male and (8) female reproductive parts. Although the investigators could not directly determine the number of genes controlling each trait, the characteristics of the traits, their locations at eight sites on six of the eight chromosomes, and their pattern of inheritance make it most likely that each trait is controlled by a single gene, giving a likely minimum of eight genes.

Conclusion

Mutations in as few as eight genes may have established reproductive isolation between these two species of monkey-flower. Thus, it appears that in some cases surprisingly little genetic change is required for the evolution of a new species. This research was the first in which random differences in DNA sequences were used to answer the fundamental evolutionary question of how much genetic change is needed to produce a new species.

THINK LIKE A SCIENTIST Given what you learned about the two species of monkey-flowers and their pollinators in Section 22.2, explain how differences in the eight floral traits might contribute to reproductive isolation between these two species.

Source: H. D. Bradshaw Jr., S. M. Wilbert, K. G. Otto, and D. W. Schemske. 1995. Genetic mapping of floral traits associated with reproductive isolation in monkey-flowers *(Mimulus)*. *Nature* 376:762–765.

isolation and sexual selection on just a few morphological and behavioral characteristics that influence their mating behavior Thus, sometimes the evolution of reproductive isolation may not require much genetic change at all. Indeed, sexual selection appears to increase the rate at which new species arise: bird lineages that are sexually dimorphic generally include more species than do related lineages in which males and females have a similar appearance.

CLOSER LOOK | **FIGURE 22.16 Polyploidy in plants.**

A. Speciation by autopolyploidy in plants
A spontaneous doubling of chromosomes during meiosis produces diploid gametes. If the plant fertilizes itself, a tetraploid zygote will be produced.

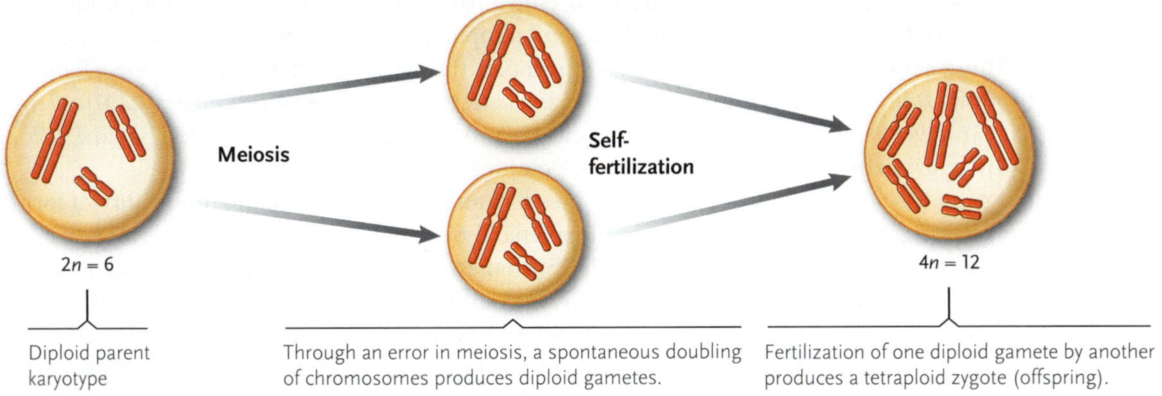

B. Speciation by hybridization and allopolyploidy in plants
A hybrid mating between two species followed by a doubling of chromosomes during mitosis in gametes of the hybrid can instantly create sets of homologous chromosomes. Self-fertilization can then generate polyploid individuals that are reproductively isolated from both parent species.

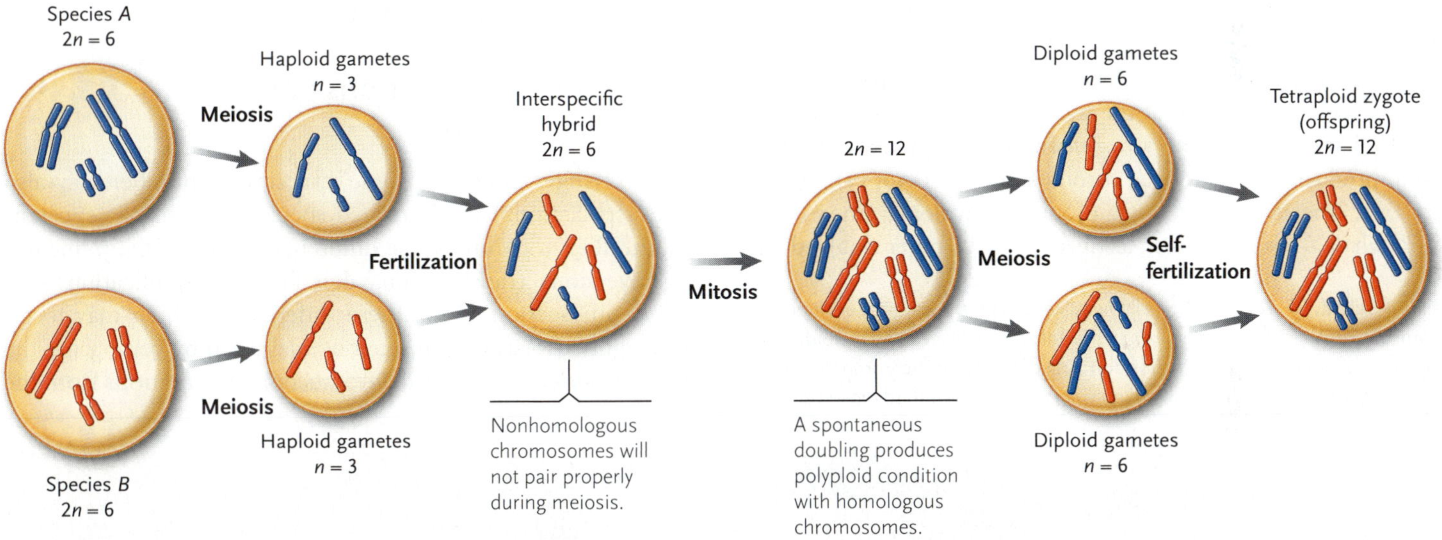

SUMMARY In both autopolyploidy and allopolyploidy, a spontaneous doubling of chromosome number produces gametes with twice the original number of chromosomes, but the timing of doubling is different. In autopolyploidy, the doubling occurs during a meiotic cell division that produces $2n$ gametes in the parent. In allopolyploidy, the doubling occurs after a hybrid offspring is produced, when some of its cells are undergoing mitosis; meiosis in the polyploid hybrid then produces polyploid gametes.

THINK LIKE A SCIENTIST How many chromosomes would you be likely to find in the allopolyploid hybrid offspring of parents that had $2n = 8$ and $2n = 12$, respectively?

Polyploidy Is a Common Mechanism of Sympatric Speciation in Plants

Polyploidy is common among plants, and it may be an important factor in the evolution of some fish, amphibian, and reptile species. Polyploid individuals can arise from chromosome duplications within a single species (autopolyploidy) or through hybridization of different species (allopolyploidy).

AUTOPOLYPLOIDY In **autopolyploidy (Figure 22.16A),** a diploid ($2n$) individual may produce, for example, tetraploid ($4n$) offspring, each of which has four complete chromosome sets. Autopolyploidy often results when gametes, through an error in either mitosis or meiosis, spontaneously receive the same number of chromosomes as a somatic cell. Such gametes are called **unreduced gametes** because their chromosome number has not been reduced compared with that of somatic cells.

Diploid pollen can fertilize the diploid ovules of a self-fertilizing individual, or it may fertilize diploid ovules on another plant with unreduced gametes. The resulting tetraploid offspring can reproduce either by self-pollination or by breeding with other tetraploid individuals. However, a tetraploid plant cannot produce fertile offspring by hybridizing with its diploid parents. The fusion of a diploid gamete with a normal haploid gamete produces a triploid ($3n$) offspring, which is usually sterile because its odd number of chromosomes cannot segregate properly during meiosis. Thus, the tetraploid is reproductively isolated from the original diploid population. Many species of grasses, shrubs, and ornamental plants, including violets, chrysanthemums, and nasturtiums, are autopolyploids, having anywhere from 4 to 20 complete chromosome sets.

ALLOPOLYPLOIDY In **allopolyploidy (Figure 22.16B),** two closely related species hybridize and subsequently form polyploid offspring. Hybrid offspring are sterile if the two parent species have diverged enough that their chromosomes do not pair properly during meiosis. However, if the hybrid's chromosome number is doubled, the chromosome complement of the gametes is also doubled, producing homologous chromosomes that *can* pair during meiosis. The hybrid can then produce polyploid gametes and, through self-fertilization or fertilization with other doubled hybrids, establish a population of a new polyploid species. Compared with speciation by genetic divergence, speciation by allopolyploidy is extremely rapid, causing a new species to arise in one generation without geographical isolation.

Even when sterile, polyploids are often robust, growing larger than either parent species. For that reason, both autopolyploids and allopolyploids have been important to agriculture. For example, the wheat used to make flour *(Triticum aestivum)* has six sets of chromosomes **(Figure 22.17).** Other polyploid crop plants include plantains (cooking bananas), coffee, cotton, potatoes, sugarcane, and tobacco.

Plant breeders often try to increase the probability of forming an allopolyploid by using chemicals that foster nondisjunction of chromosomes during mitosis. In the first such experiment, undertaken in the 1920s, scientists crossed a radish and a cabbage, hoping to develop a plant with both edible roots and leaves. Instead, the new species, *Raphanobrassica,* combined the least desirable characteristics of each parent, growing a cabbage-like root and radishlike leaves. Recent experiments have been more successful. For example, plant scientists have produced an allopolyploid grain, triticale, that has the disease-resistance of its rye parent and the high productivity of its wheat parent.

Chromosome Alterations Can Foster Speciation

Other changes in chromosome structure or number may also foster speciation. Closely related species often have a substantial number of chromosome differences between them, including inversions, translocations, deletions, and duplications (described in Section 13.3). These differences, which may foster postzygotic isolation, can often be identified by comparing the *banding patterns* in stained chromosome preparations from the

FIGURE 22.17 The evolution of wheat. Researchers believe the evolution of common bread wheat resulted from a hybridization between two diploid species, followed by a spontaneous doubling of the hybrid's chromosomes, and a second hybridization between the polyploid hybrid and a third diploid species.

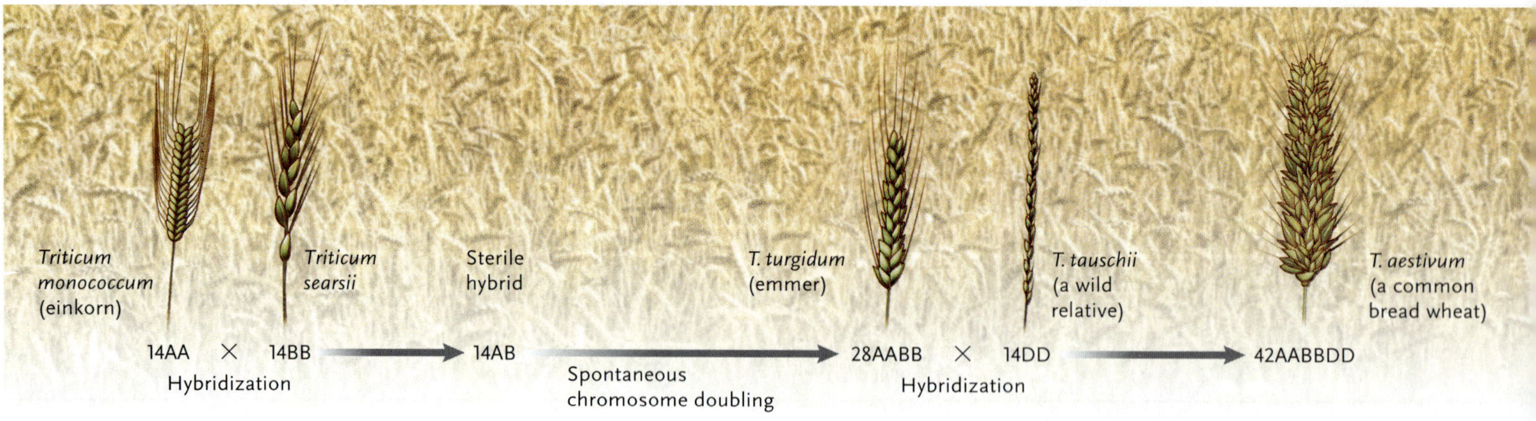

1 Wild diploid wheat, *Triticum monococcum* (einkorn), has two sets of 7 chromosomes (shown above as 14AA). Long ago, einkorn hybridized with *T. searsii*, another wild diploid species that also had two sets of 7 chromosomes (14BB). The resulting hybrid, which had 14 chromosomes (14AB), was sterile because its A chromosomes could not pair with its B chromosomes during meiosis.

2 Between 10,000 and 8,000 years ago, the sterile hybrid experienced a spontaneous doubling of its chromosomes, producing wild emmer (*T. turgidum*). These tetraploid plants had four sets of 7 chromosomes (28AABB) and were fertile. At meiosis, the A chromosomes pair with each other, and the B chromosomes pair with each other.

3 Later, emmer, the tetraploid hybrid (AABB), hybridized with *T. tauschii*, a wild diploid species with two sets of 7 chromosomes (14DD). Their hybrid descendants include common bread wheats, such as *T. aestivum*, a hexaploid that has six sets of 7 chromosomes (42AABBDD).

FIGURE 22.18 Observational Research

Chromosomal Similarities and Differences among the Great Apes

Question: Does chromosome structure differ between humans and their closest relatives among the apes?

Hypothesis: Large scale chromosome rearrangements contributed to the development of reproductive isolation between species within the evolutionary lineage that includes humans and apes.

Prediction: Chromosome structure differs markedly between humans and their close relatives among the great apes: chimpanzees, gorillas, and orangutans.

Method: Jorge J. Yunis and Om Prakash of the University of Minnesota Medical School used Giemsa stain to visualize the banding patterns on metaphase chromosome preparations from humans, chimpanzees, gorillas, and orangutans. They identified about 1,000 bands that are present in humans and in the three ape species. By matching the banding patterns on the chromosomes, the researchers verified that they were comparing the same segments of the genomes in the four species. They then searched for similarities and differences in the structure of the chromosomes.

Results: Analysis of human chromosome 2 reveals that it was produced by the fusion of two smaller chromosomes that are still present in the other three species. Although the position of the centromere in human chromosome 2 matches that of the centromere in one of the chimpanzee chromosomes, in gorillas and orangutans it falls within an inverted segment of the chromosome.

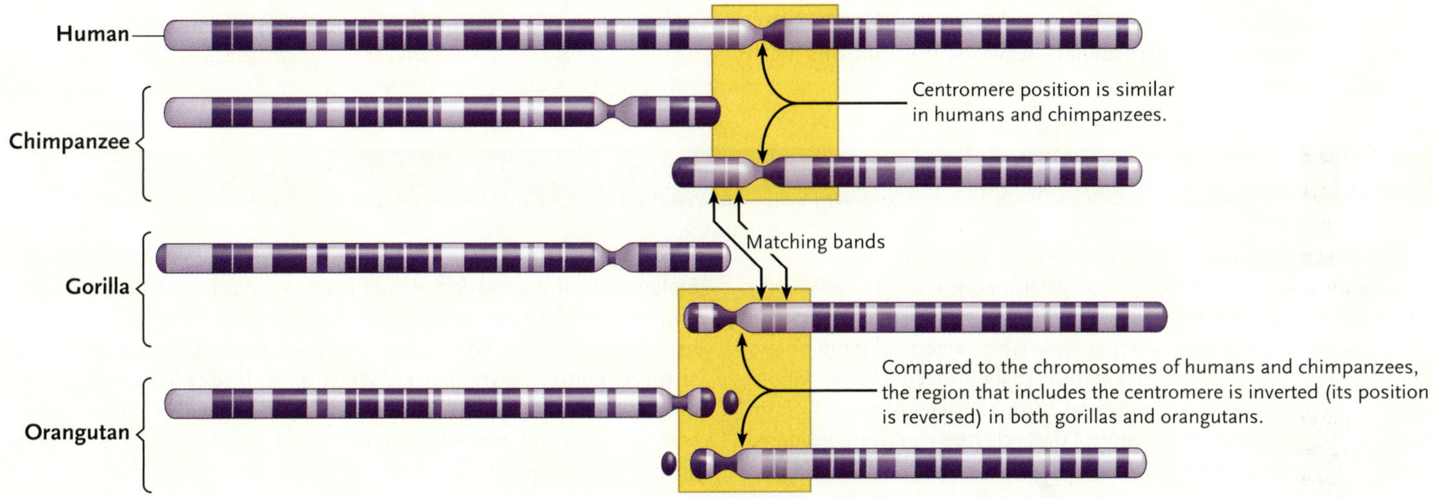

Conclusion: Differences in chromosome structure between humans and both gorillas and orangutans are more pronounced than they are between humans and chimpanzees. Structural differences in the chromosomes of these four species may contribute to their reproductive isolation.

THINK LIKE A SCIENTIST Do you think that the differences in chromosome structure between humans and their closest relatives among the apes represent a prezygotic or postzygotic reproductive isolating mechanism?

Source: J. J. Yunis and O. Prakash. 1982. The origin of man: A chromosomal pictorial legacy. *Science* 215:1525–1530.

different species. In all species, banding patterns vary from one chromosome segment to another. When researchers find identical banding patterns in chromosome segments from two or more related species, they know that they are examining comparable portions of the species' genomes. Thus, the banding patterns allow scientists to identify specific chromosome segments and compare their positions in the chromosomes of different species.

The banding patterns of humans and their closest relatives among the apes—chimpanzees, gorillas, and orangutans—reveal that whole sections of chromosomes have been rearranged over evolutionary time **(Figure 22.18)**. For example, humans have a diploid complement of 46 chromosomes, whereas chimpanzees, gorillas, and orangutans have 48 chromosomes. The difference can be traced to the fusion (that is, the joining together) of two ancestral chromosomes into chromosome 2 of humans; the ancestral chromosomes are separate in the other three species.

Moreover, banding patterns suggest that the position of the centromere in human chromosome 2 closely matches that of a centromere in one of the chimpanzee chromosomes, reflecting their close evolutionary relationship. But this centromere falls within an inverted region of the chromosome in gorillas and orangutans, reflecting their evolutionary divergence from chimpanzees and humans. (Recall from Section 13.3 that an inverted chromosome segment has a reversed orientation, so the order of genes on it is reversed relative to the order in a segment that is not inverted.) Nevertheless, humans and chimps differ from each other in centromeric inversions in six other chromosomes.

How might such chromosome rearrangements promote speciation? In a paper published in 2003, Arcadi Navarro of the Universitat Pompeu Fabra in Spain and Nick H. Barton of the University of Edinburgh in Scotland compared the rates of evolution in protein-coding genes that lie within rearranged chromosome segments of humans and chimpanzees to those in genes outside the rearranged segments. They discovered that proteins evolved more than twice as quickly in the rearranged chromosome segments. Navarro and Barton reasoned that because chromosome rearrangements inhibit chromosome pairing and recombination during meiosis, new genetic variations favored by natural selection would be conserved within the rearranged segments. These variations accumulate over time, contributing to genetic divergence between populations with the rearrangement and those without it. Thus, chromosome rearrangements can be a trigger for speciation: once a chromosome rearrangement becomes established within a population, that population will diverge more rapidly from populations lacking the rearrangement. The genetic divergence eventually causes reproductive isolation.

Because speciation results from microevolutionary processes that divide one gene pool into two, it is sometimes the first step in the evolution of new biological lineages. In the next chapter we consider the effects of speciation over vast spans of time as we examine paleobiology and patterns of macroevolution. In Unit 4, we further explore the evolution of each of the major lineages in the Tree of Life.

STUDY BREAK 22.4

1. **How can natural selection promote reproductive isolation in allopatric populations?**
2. **How does polyploidy promote speciation in plants?**

UNANSWERED QUESTIONS

Do asexual organisms form species?

As you learned in this chapter, the biological species concept applies only to sexually reproducing organisms because only those organisms can evolve barriers to gene flow (asexual organisms reproduce more or less clonally). Nevertheless, research is starting to show that organisms whose reproduction is almost entirely asexual, such as bacteria, seem to form distinct and discrete clusters in nature. (These clusters could be considered "species.") That is, bacteria and other asexual forms may be as distinct as the species of birds described by Ernst Mayr in New Guinea. Workers are now studying the many species of bacteria in nature (only a small number of which have been discovered) to see if they indeed fall into distinct groups. If they do, then scientists will need a special theory, independent of reproductive isolation, to explain this distinctness. Scientists are now working on theories of whether the existence of discrete ecological niches in nature might explain the possible discreteness of asexual "species."

How often does speciation occur allopatrically versus sympatrically or parapatrically?

Scientists do not know how often speciation occurs between populations that are completely isolated geographically (allopatric speciation), compared to how often it occurs in populations that exchange genes (parapatric or sympatric speciation). The relative frequency of these modes of speciation in nature is an active area of research. The ongoing work includes studies on small, isolated islands: if an invading species divides into two or more species in this situation, it probably did so sympatrically or parapatrically, since geographical isolation of populations in small islands is unlikely. In addition, biologists are reconstructing the evolutionary history of speciation using molecular tools and correlating this history with the species' geographical distributions. If this research were to show, for example, that the most closely related pairs of species always had geographically isolated distributions, it would imply that speciation was usually allopatric. These lines of research should eventually answer the controversial question of the relative frequency of various forms of speciation.

What are the genetic changes underlying speciation?

Biologists know a great deal about the types of reproductive isolation that prevent gene flow between species, but almost nothing about their genetic underpinnings. Which genes control the difference between flower shape in monkey-flower species? Which genes lead to inviability and sterility of *Drosophila* hybrids? Which genes cause species of ducks to preferentially mate with members of their own species over members of other species? Do the genetic changes that lead to reproductive isolation tend to occur repeatedly at the same gene loci in a group of organisms, or at different gene loci? Do the changes occur mostly in protein-coding regions of genes, or in the noncoding regions that control the production of proteins? Were the changes produced by natural selection or by genetic drift? Biologists are now isolating "speciation genes" and sequencing their DNA. With only a handful of such genes known, and all of these causing hybrid sterility or inviability, there will undoubtedly be a lot to learn about the genetics of speciation in the next decade.

THINK LIKE A SCIENTIST Professor Coyne suggested that biologists will need to devise a special theory—or an additional type of species definition—to explain the distinctness of clusters of asexual organisms that are observed in nature. Do you think that this new explanation would be more similar to the morphological species concept or the phylogenetic species concept as described in this chapter?

Jerry Coyne

Jerry Coyne conducts research on speciation and teaches at the University of Chicago. To learn more about his research, go to http://pondside.uchicago.edu/ecol-evol/faculty/coyne_j.html.

REVIEW KEY CONCEPTS

To access the course materials and companion resources for this text, please visit www.cengagebrain.com.

22.1 What Is a Species?

- In practice, most biologists describe, identify, and recognize species on the basis of morphological characteristics that serve as indicators of their genetic similarity to or divergence from other species (Figure 22.2).
- The biological species concept defines species as groups of interbreeding populations that are reproductively isolated from populations of other species in nature. A biological species thus represents a gene pool within which genetic material is potentially shared among populations. The biological species concept cannot be applied to organisms that reproduce only asexually, to those that are extinct, or to geographically separated populations.
- The phylogenetic species concept defines a species as a group of populations with a recently shared evolutionary history.
- Most species exhibit geographical variation of phenotypic and genetic traits. When marked geographical variation in phenotypes is discontinuous, biologists sometimes name subspecies (Figure 22.3). In ring species, populations are distributed in a ring around unsuitable habitat (Figure 22.4). Many species exhibit clinal variation of characteristics, which change smoothly over a geographical gradient (Figure 22.5).

22.2 Maintaining Reproductive Isolation

- Reproductive isolating mechanisms are characteristics that prevent two species from interbreeding (Table 22.1).
- Prezygotic isolating mechanisms either prevent individuals of different species from mating or prevent fertilization between their gametes. Prezygotic isolation occurs because species live in different habitats, breed at different times, use different courtship behavior (Figure 22.6), or differ anatomically (Figure 22.7). Prezygotic isolation can also result from genetic and physiological incompatibilities between male and female gametes.
- Postzygotic isolating mechanisms reduce the fitness of interspecific hybrids through hybrid inviability, hybrid sterility (Figure 22.8), or hybrid breakdown.

Animation: Reproductive isolating mechanisms

Animation: Temporal isolation among cicadas

22.3 The Geography of Speciation

- The model of allopatric speciation proposes that speciation results from divergent evolution in geographically separated populations (Figures 22.9–22.11). If allopatric populations accumulate enough genetic differences, they will be reproductively isolated upon secondary contact. Nevertheless, some species hybridize over small areas of secondary contact (Figure 22.12).
- The model of parapatric speciation suggests that reproductive isolation can evolve between parts of a population that occupy opposite sides of an environmental discontinuity (Figure 22.13).
- A model of sympatric speciation in insects suggests that reproductive isolation may evolve between host races that rarely contact one another under natural conditions (Figure 22.14). Sympatric speciation commonly occurs in flowering plants by allopolyploidy.

Animation: Allopatric speciation on an archipelago

Animation: Models of speciation

22.4 Genetic Mechanisms of Speciation

- Allopatric populations inevitably accumulate genetic differences, some of which contribute to their reproductive isolation. Reproductive isolating mechanisms evolve as by-products of genetic changes that occur during divergence. Prezygotic isolating mechanisms may evolve in populations experiencing secondary contact (Figure 22.15).
- We cannot yet generalize about how many gene loci participate in the process of speciation, but at least several gene loci are usually involved.
- Speciation by polyploidy in flowering plants involves the duplication of an entire chromosome complement through nondisjunction of chromosomes during meiosis or mitosis. Polyploids can arise among the offspring of a single species (autopolyploidy; Figure 22.16A) or, more commonly, after hybridization between closely related species (allopolyploidy; Figures 22.16B and 22.17).
- Chromosome alterations can promote speciation by fostering the genetic divergence of, and reproductive isolation between, populations with different numbers of chromosomes or different chromosome structure (Figure 22.18).

Animation: Sympatric speciation in wheat

UNDERSTAND & APPLY

Test Your Knowledge

1. The biological species concept defines species on the basis of:
   a. reproductive characteristics.
   b. biochemical characteristics.
   c. morphological characteristics.
   d. behavioral characteristics.
   e. all of the above.
2. Biologists can apply the biological species concept *only* to species that:
   a. reproduce asexually.
   b. lived in the past.
   c. are allopatric to each other.
   d. hybridize in captivity.
   e. reproduce sexually.
3. A characteristic that exhibits smooth changes in populations distributed along a geographical gradient is called a:
   a. ring species.
   b. hybrid.
   c. cline.
   d. hybrid breakdown.
   e. subspecies.
4. If two species of holly (genus *Ilex*) flower during different months, their gene pools may be kept separate by:
   a. mechanical isolation.
   b. ecological isolation.
   c. gametic isolation.
   d. temporal isolation.
   e. behavioral isolation.

5. Prezygotic isolating mechanisms:
   a. reduce the fitness of hybrid offspring.
   b. generally prevent individuals of different species from producing zygotes.
   c. are found only in animals.
   d. are found only in plants.
   e. are observed only in organisms that reproduce asexually.
6. In the model of allopatric speciation, the geographical separation of two populations:
   a. is sufficient for speciation to occur.
   b. occurs only after speciation is complete.
   c. allows gene flow between them.
   d. reduces the relative fitness of hybrid offspring.
   e. inhibits gene flow between them.
7. Adjacent populations that produce hybrid offspring with low relative fitness may be undergoing:
   a. clinal isolation.
   b. parapatric speciation.
   c. allopatric speciation.
   d. sympatric speciation.
   e. geographical isolation.
8. An animal breeder, attempting to cross a llama with an alpaca for finer wool, found that the hybrid offspring rarely lived more than a few weeks. This outcome probably resulted from:
   a. genetic drift.
   b. prezygotic reproductive isolation.
   c. postzygotic reproductive isolation.
   d. sympatric speciation.
   e. polyploidy.
9. Which of the following could be an example of allopolyploidy?
   a. One parent has 8 chromosomes, the other has 10, and their offspring have 36.
   b. Gametes and somatic cells have the same number of chromosomes.
   c. Chromosome number increases by one in a gamete and in the offspring it produces.
   d. Chromosome number decreases by one in a gamete and in the offspring it produces.
   e. Chromosome number in the offspring is exactly half of what it is in the parents.
10. Which of the following genetic characteristics is shared by humans and chimpanzees?
    a. They have the same number of chromosomes.
    b. The position of the centromere on human chromosome 2 matches the position of a centromere on a chimpanzee chromosome.
    c. A fusion of ancestral chromosomes formed chromosome 2.
    d. Centromeres on all of their chromosomes fall within inverted chromosome segments.
    e. all of the above

Discuss the Concepts

1. All domestic dogs are classified as members of the species *Canis familiaris*. But it is hard to imagine how a tiny Chihuahua could breed with a gigantic Great Dane. Do you think that artificial selection for different breeds of dogs will eventually create different dog species?
2. Human populations often differ dramatically in external morphological characteristics. On what basis are all human populations classified as a single species?
3. If intermediate populations in a ring species go extinct, eliminating the possibility of gene flow between populations at the two ends of the ring, would you now identify those remaining populations as full species? Explain your answer.

Design an Experiment

Design an experiment to test whether populations of birds on different islands belong to the same species.

Interpret the Data

David Hillis of Baylor University noted that three closely related species of leopard frog (genus *Rana*) exhibit substantial—but not complete—postzygotic reproductive isolation when crossed in the laboratory. Field surveys of numerous populations in Texas and surrounding states revealed that populations of the three species breed at various times during the year. Data on the breeding schedule of both allopatric and sympatric populations of these species are presented in the **Figure** below. Interpret the data in the figure and explain how they may demonstrate that these frogs experience prezygotic reproductive isolation in nature. What type of prezygotic reproductive isolation do the data suggest?

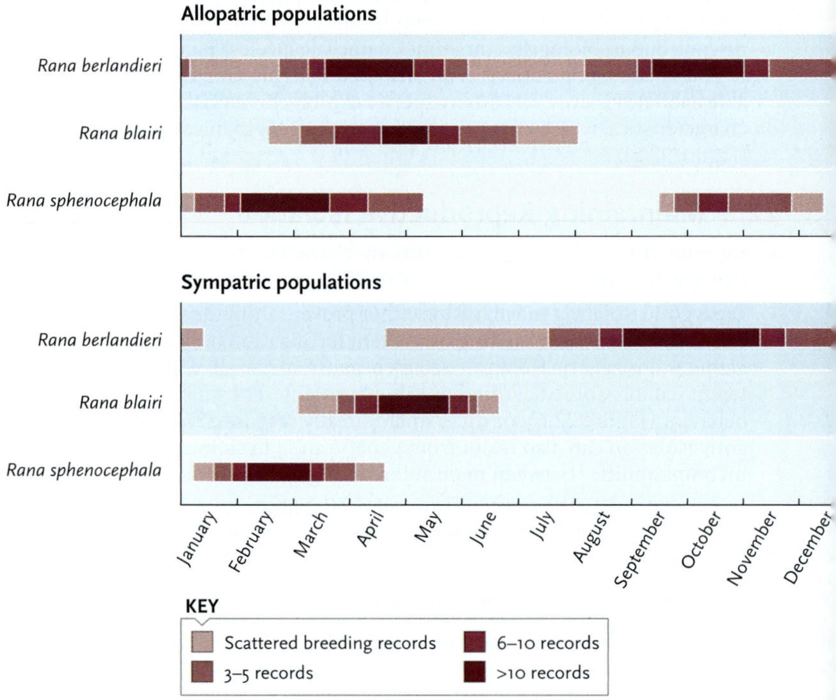

D. M. Hillis. 1981. Premating isolating mechanisms among three species of the *Rana pipiens* complex in Texas and southern Oklahoma. *Copeia* 1981:312–319.

Source: D. M. Hillis. 1981. Premating isolating mechanisms among three species of the *Rana pipiens* complex in Texas and southern Oklahoma. *Copeia* 1981:312–319.

Apply Evolutionary Thinking

How do human activities (such as destruction of natural habitats, diversion of rivers, and the construction of buildings) influence the chances that new species of plants and animals will evolve in the future? Frame your answer in terms of the geographical and genetic factors that foster speciation.

Fossil of a trilobite *(Asaphus kowalewskii)* that lived 490 million years ago.

23

STUDY OUTLINE

23.1 The Fossil Record

23.2 Earth History

23.3 Historical Biogeography and Convergent Biotas

23.4 The History of Biodiversity

23.5 Interpreting Evolutionary Lineages

23.6 The Evolution of Morphological Novelties

Paleobiology and Macroevolution

Why it matters . . . In a report on the fossil reptiles of Britain, published in April 1842, Sir Richard Owen described fragmentary fossils of three species that another researcher had unearthed in 1818. Recognizing that these gigantic creatures were distinct from lizards and snakes, Owen assigned them to a new group, which he named Dinosauria (*deinos* = terrible; *sauros* = lizards). In 1851, he brought the dinosaurs to the attention of a broader public in a display of the "inhabitants of the ancient world" at the Crystal Palace exhibition in London.

Owen's published accounts and Crystal Palace exhibition models led to a dinosaur craze that still grips scientists and the public today. Dinosaur discoveries continued apace, especially in North America, where the most complete fossils are found. In 1858, a nearly intact skeleton of the duck-billed dinosaur *Hadrosaurus foulkii* was uncovered in New Jersey. Analyses of its leg bones showed clearly that it was bipedal (that is, it stood on its hind legs rather than on all fours), a shocking hypothesis at the time. Ten years later, the specimen was exhibited at the Philadelphia Academy of Science, where it is still on display. The *Hadrosaurus* exhibit tripled attendance at the Philadelphia Academy, setting off a race to uncover other complete dinosaur skeletons. In the late 1870s, abundant fossils turned up in Colorado, Wyoming, and Alberta; and scientists and entrepreneurs scoured the region for fossils that would bring them fame and fortune. Their quest was not unlike the California gold rush of the mid-nineteenth century, with rivals reputedly hijacking one another's fossils and sometimes coming to blows over their ownership.

As natural history museums exhibited these discoveries, a major question plagued preparators (professionals who clean and reassemble fossils) as they mounted the specimens. What did dinosaurs really look like? How did these animals stand when they were alive? How did they hold their heads and tails?

FIGURE 23.1 Dinosaurs in living color. By examining the fine structure of pigment-containing organelles inside fossilized feathers, scientists were able to reconstruct the colors of *Anchiornis huxleyi*, a dinosaur that lived about 150 million years ago.

The continued study of dinosaur biology has answered many of these questions. In the 1960s, biologists hypothesized that dinosaurs were very active creatures, like mammals and birds are today. According to this hypothesis, many dinosaurs held their heads up and their tails off the ground; others adopted a bipedal stance. As our knowledge of dinosaurs increased, natural history museums have remounted their specimens and models in more active poses. But one question seemed impossible to answer. Did dinosaurs blend in with their surroundings, or were they brightly colored?

As you will discover later in this chapter, many dinosaurs were feathered, just as birds are today. In 2010, two groups of researchers published analyses of the colors of fossilized feathers from several species of small dinosaurs. Using scanning electron microscopes, the researchers compared the fine structure of melanosomes (pigment-bearing organelles) in the fossilized feathers to those of living birds, allowing them to reconstruct the feather colors of the dinosaurs. One analysis suggests that *Anchiornis huxleyi*, a dinosaur that lived about 150 million years ago in what is now northern China, had gray or black feathers over most of its body, black and white feathers on its limbs, and a crown of red feathers on its head **(Figure 23.1)**. These animals must surely have used their colorful feathers in social displays.

This description of a truly remarkable discovery opens our chapter on paleobiology and macroevolution, the large-scale changes in morphology and diversity observed over life's 3.8-billion-year history. Macroevolution has occurred over so vast a span of time and space that the evidence for it is fundamentally different from that for microevolution and speciation. In this chapter we consider what paleobiology and evolutionary developmental biology tell us about macroevolutionary patterns.

23.1 The Fossil Record

Paleobiologists discover, describe, and name new fossil species and analyze the morphology and ecology of extinct organisms. Because fossils provide physical evidence of life in the past, they are a primary source of data about the evolutionary history of many organisms.

Fossils Form When Organisms Are Buried by Sediments or Preserved in Oxygen-Poor Environments

Most fossils form in sedimentary rocks. Rain and runoff constantly erode the land, carrying fine particles of rock and soil downstream to a swamp, a lake, or the sea. Particles settle to the bottom as sediments, forming successive layers, called *strata*, over millions of years **(Figure 23.2)**. The weight of newer sediments compresses the older layers beneath them into a solid matrix: sand into sandstone and silt or mud into shale. Fossils form within the layers when the remains of organisms are buried in the accumulating sediments. Because sedimentation superimposes new layers over old ones, the lowest strata in a sedimentary rock formation are usually the oldest and the highest layers are the newest.

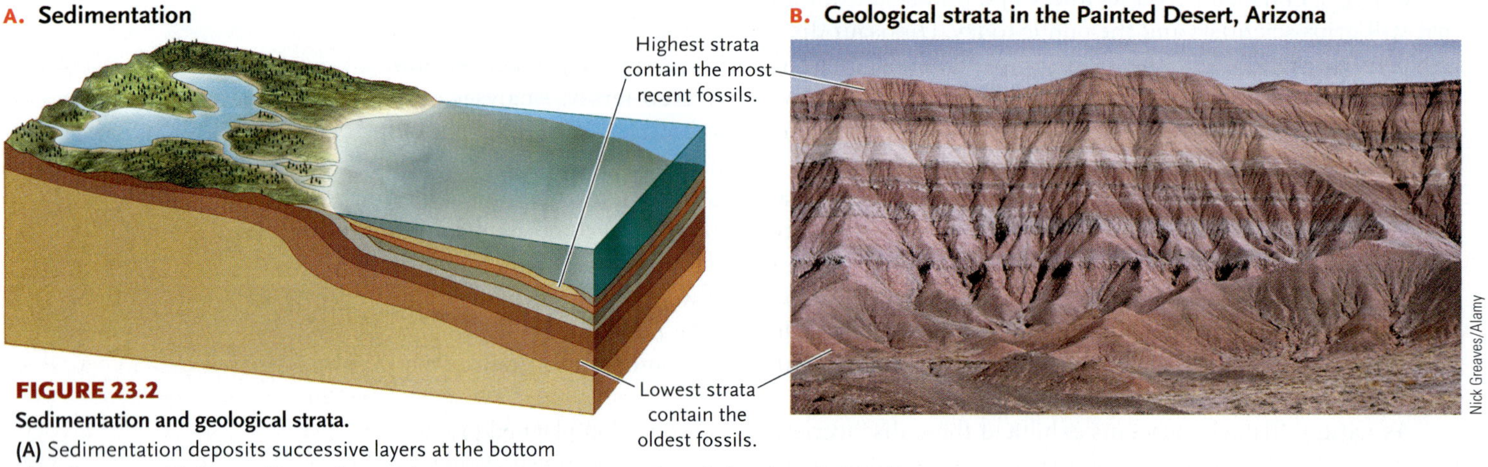

FIGURE 23.2
Sedimentation and geological strata.
(A) Sedimentation deposits successive layers at the bottom of a lake or sea. **(B)** Over millions of years, the upper layers compress those below them into rock. When the rocks are later exposed by uplifting or erosion, the different layers are evident as geological strata.

The process of fossilization is a race against time because the soft remains of organisms are quickly consumed by scavengers or decomposed by microorganisms. Thus, fossils usually preserve the details of hard structures, such as the bones, teeth, and shells of animals and the wood, leaves, and pollen of plants. During fossilization, dissolved minerals replace some parts molecule by molecule, leaving a fossil made of stone **(Figure 23.3A)**; other fossils form as molds, casts, or impressions in material that is later transformed into solid rock **(Figure 23.3B)**.

In some environments, the near absence of oxygen prevents decomposition, and even soft-bodied organisms are preserved. Some insects, plants, and tiny lizards and frogs are embedded in amber, the fossilized resin of coniferous trees **(Figure 23.3C)**. Other organisms are preserved in glacial ice, coal, tar pits, or the highly acidic water of peat bogs **(Figure 23.3D)**. Sometimes organisms are so well preserved that researchers can examine their internal anatomy, cell structure, and even food in their digestive tracts.

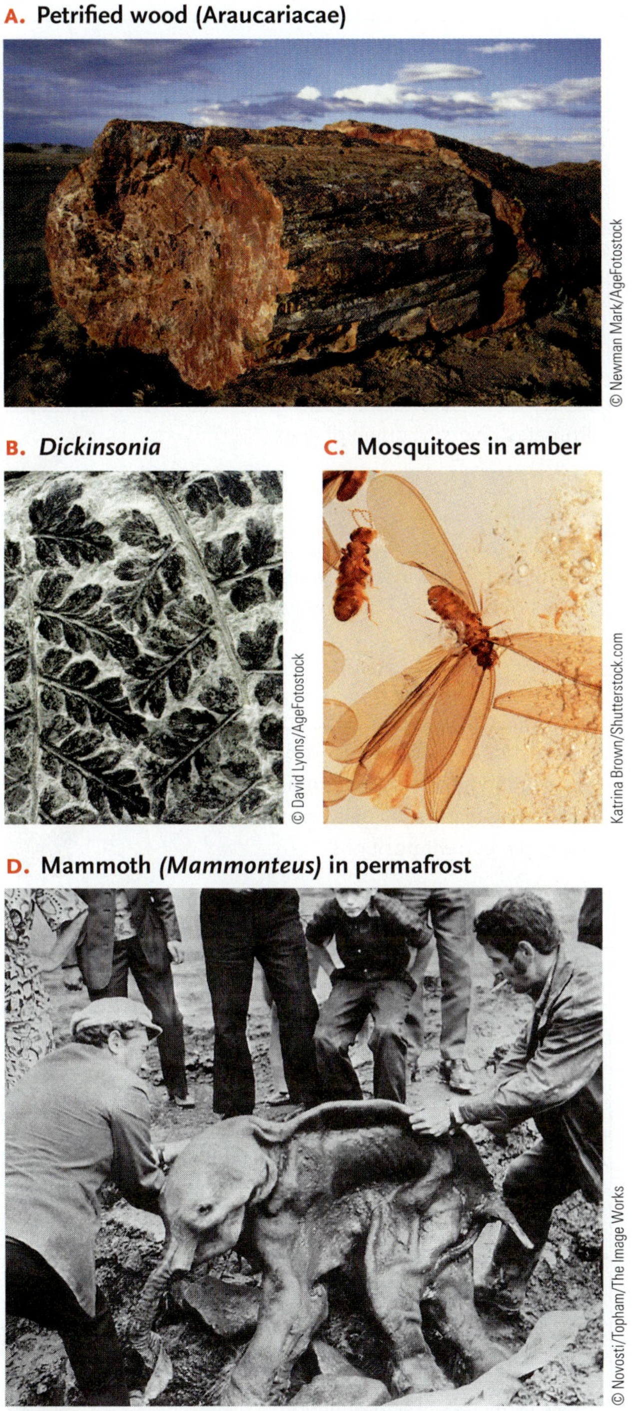

FIGURE 23.3 Fossils. **(A)** Petrified wood in Arizona formed when minerals replaced the wood of dead trees, molecule by molecule. These forests lived during the late Triassic period, about 225 million years ago. **(B)** The remains of a fern *(Sphenopteris)* from the Carboniferous period, 300 million years ago, were preserved in coal. **(C)** These 10-million-year-old mosquitoes were trapped in the oozing resin of a coniferous tree and are now encased in amber. **(D)** A frozen baby mammoth that lived about 40,000 years ago was discovered embedded in Siberian permafrost in 1977.

The Fossil Record Provides an Incomplete Portrait of Life in the Past

The 300,000 described fossil species represent less than 1% of all the species that have ever lived. Several factors make the fossil record incomplete. First, soft-bodied organisms do not fossilize as readily as species with hard body parts. Moreover, we are unlikely to find fossilized remains of species that were rare or locally distributed. Finally, fossils rarely form in habitats where sediments do not accumulate, such as mountain forests. The most common fossils are those of hard-bodied, widespread, and abundant organisms that lived in swamps or shallow seas, where sedimentation is ongoing.

Although most fossils are composed of stone, they don't last forever. Many are deformed by pressure from overlying rocks or destroyed by geological disturbances like volcanic eruptions and earthquakes. Once they are exposed on Earth's surface, where scientists are most likely to find them, rain and wind cause them to erode. Because the effects of these destructive processes are additive, old fossils are much less common than those formed more recently.

Scientists Assign Relative and Absolute Dates to Geological Strata and the Fossils They Contain

The sediments found in any one place form distinctive strata (layers) that differ in color, mineral composition, particle size, and thickness (see Figure 23.2B). If they have not been disturbed, the strata are arranged in the order in which they formed, with the youngest layers on top. However, strata are sometimes uplifted, warped, or even inverted by geological processes.

Geologists of the early nineteenth century deduced that the fossils discovered in a particular sedimentary stratum, no matter where on Earth it is found, represent organisms that lived and died at roughly the same time in the past. Because each stratum formed at a specific time, the sequence of fossils in the lowest (oldest) to the highest (newest) strata reveals their *relative ages*. Geologists used the sequence of strata and their distinctive fossil assemblages to establish the geological time scale **(Table 23.1)**.

TABLE 23.1 The Geological Time Scale and Major Evolutionary Events

Eon	Era	Period	Epoch	Millions of Years Ago	Major Evolutionary Events	
Phanerozoic	Cenozoic	Quaternary	Holocene	0.01		
			Pleistocene	2.6	Origin of humans; major glaciations	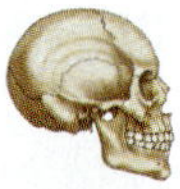
		Neogene	Pliocene	5.3	Origin of apelike human ancestors	
			Miocene	23.0	Angiosperms and mammals further diversify and dominate terrestrial habitats	 
		Paleogene	Oligocene	33.9	Primates diversify; origin of apes	
			Eocene	55.8	Angiosperms and insects diversify; modern orders of mammals differentiate	 
			Paleocene	65.5	Grasslands and deciduous woodlands spread; modern birds, mammals, snakes, pollinating insects diversify; continents approach current positions	 
	Mesozoic	Cretaceous		145.5	Angiosperms, insects, marine invertebrates, fishes, dinosaurs diversify; asteroid impact causes mass extinction at end of period, eliminating dinosaurs and many other groups	 
		Jurassic		201.6	Gymnosperms abundant in terrestrial habitats; modern fishes diversify; dinosaurs diversify and dominate terrestrial habitats; frogs, salamanders, lizards, and birds appear; continents continue to separate	 
		Triassic		251.0	Predatory fishes and reptiles dominate oceans; gymnosperms dominate terrestrial habitats; radiation of dinosaurs; early mammals; Pangaea starts to break up; mass extinction at end of period	 

Eon	Era	Period	Epoch	Millions of Years Ago	Major Evolutionary Events	
Phanerozoic	Paleozoic	Permian		299.0	Insects and reptiles abundant and diverse in swamp forests; some reptiles colonize oceans; fishes colonize freshwater habitats; continents coalesce into Pangaea, causing glaciation and decline in sea level; mass extinction at end of period eliminates 85% of species	
		Carboniferous		359.0	Vascular plants form large swamp forests; first flying insects; amphibians diversify; first reptiles appear	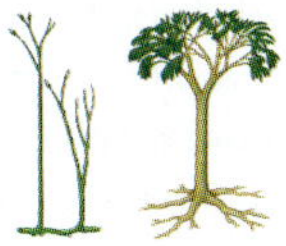
		Devonian		416.0	Terrestrial vascular plants diversify; fungi, invertebrates, amphibians colonize land; first insects and seed plants; major glaciation at end of period; mass extinction, mostly of marine life	
		Silurian		444.0	Jawless fishes diversify; first jawed fishes, arthropods, terrestrial vascular plants	
		Ordovician		488.0	Major radiations of marine invertebrates and jawless fishes; major glaciation at end of period causes mass extinction of marine life	
		Cambrian		542.0	Appearance of modern animal phyla, including earliest vertebrates (Cambrian explosion); simple marine communities	
Proterozoic				2,500	High concentration of oxygen in atmosphere; origin of eukaryotic cells; evolution and diversification of "protists," fungi, soft-bodied animals	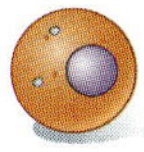
Archean				3,850	Evolution of prokaryotes, including anaerobic and photosynthetic bacteria; oxygen starts to accumulate in atmosphere; origin of aerobic respiration	
Hadean				4,600	Formation of Earth, including crust, atmosphere, and oceans; origin of life	

FIGURE 23.4 Research Method

Radiometric Dating

Purpose: Radiometric dating allows researchers to estimate the absolute age of a rock sample or fossil.

Protocol:

1. Knowing the approximate age of a rock or fossil, select a radioisotope that has an appropriate half-life. Because different radioisotopes have half-lives ranging from seconds to billions of years, it is usually possible to choose one that brackets the estimated age of the sample under study. For example, if you think that your fossil is more than 10 million years old, you might use uranium-235. The half life of ^{235}U, which decays into the lead isotope ^{207}Pb, is about 700 million years. Or if you think that your fossil is less than 70,000 years old, you might select carbon-14. The half-life of ^{14}C, which decays into the nitrogen isotope ^{14}N, is 5,730 years.

Radioisotopes Commonly Used in Radiometric Dating

Radioisotope (Unstable)		More Stable Breakdown Product	Half-Life (Years)	Useful Range (Years)
Samarium-147	⟶	Neodymium-143	106 billion	>100 million
Rubidium-87	⟶	Strontium-87	48 billion	>10 million
Thorium-232	⟶	Lead-208	14 billion	>10 million
Uranium-238	⟶	Lead-206	4.5 billion	>10 million
Uranium-235	⟶	Lead-207	700 million	>10 million
Potassium-40	⟶	Argon-40	1.25 billion	>100,000
Carbon-14	⟶	Nitrogen-14	5,730	<70,000

2. Prepare a sample of the material and measure the quantities of the parent radioisotope and its more stable breakdown product.

Interpreting the Results: Compare the relative quantities of the parent radioisotope and its breakdown product (or some other stable isotope) to determine what percentage of the original parent radioisotope remains in the sample. Then use a graph of radioactive decay for that isotope to determine how many half-lives have passed since the sample formed.

Theory of radiometric dating

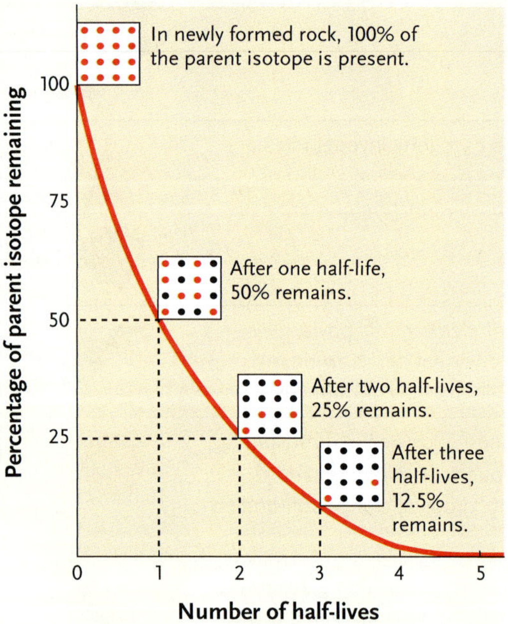

Knowing the number of half-lives that have passed allows you to estimate the age of the sample.

Radiometric dating of a fossilized mollusk

A living mollusk absorbed trace amounts of ^{14}C, a rare radioisotope of carbon, and large amounts of ^{12}C, which is the more stable and common isotope of carbon.

When the mollusk died, it was buried in silt and fossilized. From the moment of its death, the ratio of ^{14}C to ^{12}C began to decline through radioactive decay. Because the half-life of ^{14}C is 5,730 years, half of the original ^{14}C was eliminated from the fossil in 5,730 years and half of what remained was eliminated in another 5,730 years.

After the fossil was discovered, a scientist determined that its ^{14}C to ^{12}C ratio was one-eighth (12.5%) of the ^{14}C to ^{12}C ratio in living organisms. Thus, radioactive decay had proceeded for three half-lives—about 17,000 years—since the mollusk's death.

Although the geological time scale provides a relative dating system for sedimentary strata, it does not tell us how old the rocks and fossils actually are. But many rocks contain radioisotopes, which, from the moment they form, begin to break down into other, more stable elements. The breakdown proceeds at a steady rate that is unaffected by chemical reactions or environmental conditions such as temperature or pressure. Using a technique called **radiometric dating,** scientists can estimate the age of a rock by noting how much of an unstable "parent" isotope has decayed to another form. By measuring the relative amounts of the parent radioisotope and its breakdown products and comparing this ratio with the isotope's **half-life**—the time it takes for half of a given amount of radioisotope to decay—researchers can estimate the *absolute age* of the rock **(Figure 23.4).** Table 23.1

presents these age estimates along with the major geological and evolutionary events of each period.

Radiometric dating works best with volcanic rocks, which form when lava cools and solidifies. But most fossils are found in sedimentary rocks. To date a sedimentary fossil, scientists determine the age of volcanic rocks from the same stratum. Using this method, investigators have linked fossils to deposits that are hundreds of millions of years old.

Fossils that still contain organic matter, such as the remains of bones or wood, can be dated directly by measuring their content of the radioactive carbon isotope ^{14}C, which decays to ^{14}N. Living organisms absorb traces of ^{14}C and large quantities of ^{12}C, a stable carbon isotope, from the environment and incorporate them into biological molecules. As long as an organism is alive, its ^{14}C content remains constant because the ratio of ^{14}C to ^{12}C in its tissues is at equilibrium with the ratio in the environment. But as soon as the organism dies, ^{14}C begins its steady radioactive decay. Scientists use the ratio of ^{14}C to ^{12}C in a fossil to determine its age, as explained in Figure 23.4.

To develop a feeling for geological time, imagine the 4.5-billion-year history of Earth scaled onto a calendar year; each day represents a little over 12 million years **(Figure 23.5).** The planet was formed on January 1. Animal life originated in mid-November, dinosaurs lived between December 14 and December 26, and the genus that includes modern humans appeared during the last 4 hours of December 31.

FIGURE 23.5 A calendar of Earth's history. Geological time can be scaled onto a calendar year in which each day is a little more than 12 million years.

Fossils Provide Abundant Information about Life in the Past

Although imperfect, the fossil record provides our only direct information about life in the past. Fossilized skeletons, shells, stems, leaves, and flowers tell us about the size and appearance of ancient animals and plants. The fossil record also allows scientists to see how structures were modified as they became adapted for specialized uses (see Figure 21.3). Moreover, fossils chronicle the proliferation and extinction of evolutionary lineages and provide data on their past geographical distributions.

Recent applications of new techniques allow paleobiologists to make truly remarkable discoveries about extinct organisms. For example, researchers appear to have solved a long-standing puzzle about lambeosaurines, a group of huge duck-billed dinosaurs that lived in the swampy habitats of western North America in the late Cretaceous period. The heads of lambeosaurines sported bony crests with gigantic nasal passages. Paleobiologists had proposed that the crests served as weapons in male combat, adornments that attracted mates, snorkels that facilitated breathing underwater, radiators that cooled the dinosaurs' bodies, structures that enhanced the sense of smell, or resonating chambers that produced honking vocalizations. Based on anatomical analyses, researchers accepted the vocalization hypothesis as the most probable: computerized acoustic models predicted that air flowing through the nasal passages would have produced low frequency sounds (30–375 hertz).

But could lambeosaurines hear sounds in that frequency range? In 2009, David Evans of the Royal Ontario Museum in Toronto, Canada, and colleagues in Ohio used computed tomography and 3-D visualization software to scan and reconstruct the interior anatomy of the skulls and brains of several lambeosaurine species **(Figure 23.6).** Their findings indicate that

FIGURE 23.6 Honking dinosaurs. Analysis of the sinuses and braincase of lambeosaurines (*Corythosaurus* species) revealed that the nasal passages in their crests served as resonating chambers for the production of low-frequency sounds that their inner ears could detect.

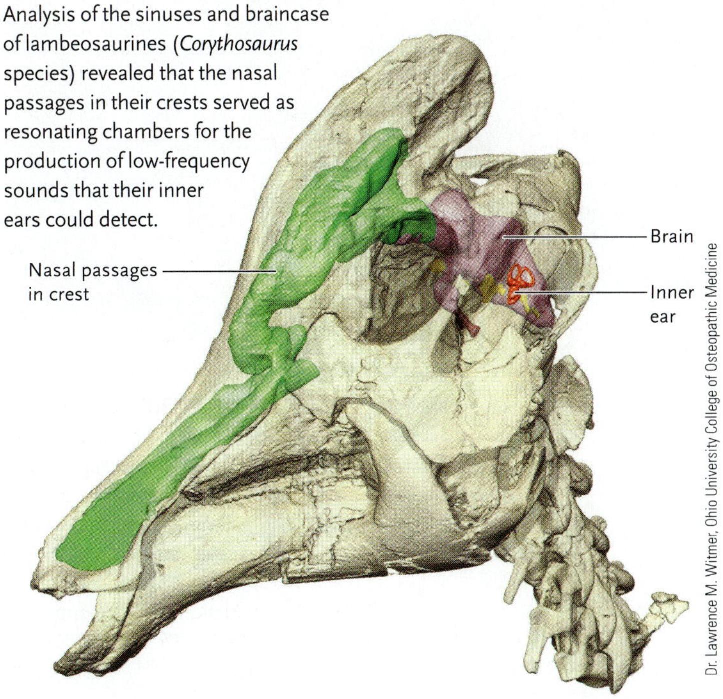

Dr. Lawrence M. Witmer, Ohio University College of Osteopathic Medicine

the inner ears of lambeosaurines were attuned to hear low-frequency sounds that matched those predicted by the earlier research. The authors concluded that the elaborate nasal passages in lambeosaurine crests as well as the structure of their inner ears facilitated vocal communication, perhaps between parents and their offspring.

Fossils also provide indirect data about behavior, physiology, and ecology. For example, the fossilized footprints of some dinosaurs suggest that adults surrounded their young when the group moved, possibly to protect them from predators. Complex scrolls of bone in the nasal passages of early mammals suggest that they had a well-developed sense of smell, and fossilized teeth and dung provide data about the diets of extinct animals. The study of fossilized pollen allows paleobiologists to reconstruct the vegetation and climate of ancient sites. The changing arrays of fossils that document biological evolution partly reflect large-scale shifts in Earth's physical environments, a topic that we explore in the next section.

STUDY BREAK 23.1

1. What biological materials are the most likely to fossilize?
2. Why does the fossil record provide an incomplete portrait of life in the past?
3. What sorts of information can paleobiologists discern from the fossil record?

23.2 Earth History

Organisms have interacted constantly with their environments over geological time. These interactions have triggered fundamental changes in Earth's physical environment as well as in the organisms that occupy them. Perhaps the most striking effect of living systems on the physical environment was the development of an oxidizing atmosphere, discussed extensively in Chapter 25. Photosynthesis by prokaryotic organisms increased the concentration of oxygen in the atmosphere, facilitating the evolution of eukaryotic cells and later allowing animals to invade terrestrial habitats. In this section we focus on long-term shifts in geography and climate—as well as some brief but catastrophic events—that have significantly altered the environments where organisms live.

Continental Drift Has Altered the Configuration of Landmasses and Oceans

Major geological and climatic shifts occur because the planet's crust is constantly in motion. According to the theory of **plate tectonics,** Earth's crust is broken into irregularly shaped plates of rock—roughly 40 km thick—that float on its semisolid mantle **(Figure 23.7).** Currents in the mantle cause the plates—and the continents embedded in them—to move, a phenomenon called **continental drift.**

Continent-sized landmasses have coalesced and broken apart several times in Earth's history. About 250 million years

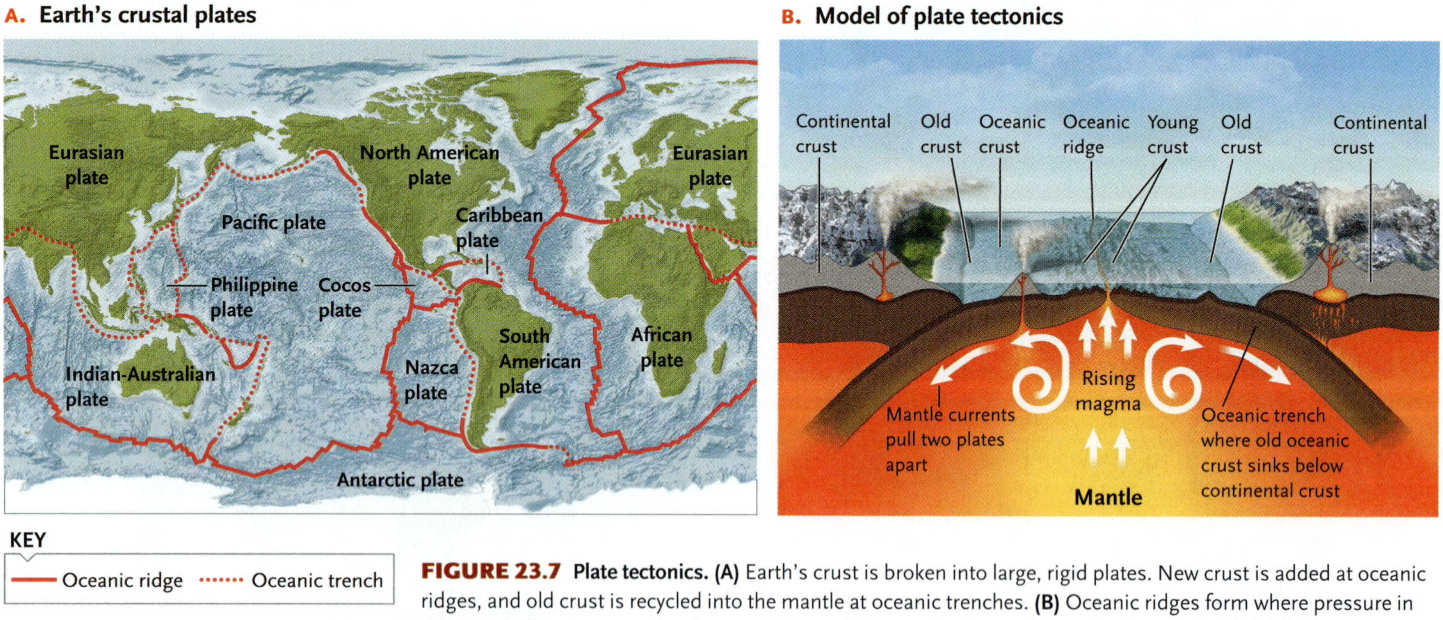

FIGURE 23.7 Plate tectonics. (A) Earth's crust is broken into large, rigid plates. New crust is added at oceanic ridges, and old crust is recycled into the mantle at oceanic trenches. **(B)** Oceanic ridges form where pressure in the mantle forces magma (molten rock) through fissures in the sea floor. Mantle currents pull the plates apart on either side of the ridge, forcing the sea floor to move laterally away from the ridge. This phenomenon, seafloor spreading, is widening the Atlantic Ocean about 3 cm per year. Oceanic trenches form where plates collide. The heavier oceanic crust sinks below the lighter continental crust, and it is recycled into the mantle, a process called subduction. The highest mountain ranges (including the Rockies, Himalayas, Alps, and Andes) formed where subduction uplifted continental crust. Earthquakes and volcanoes are also common near trenches.

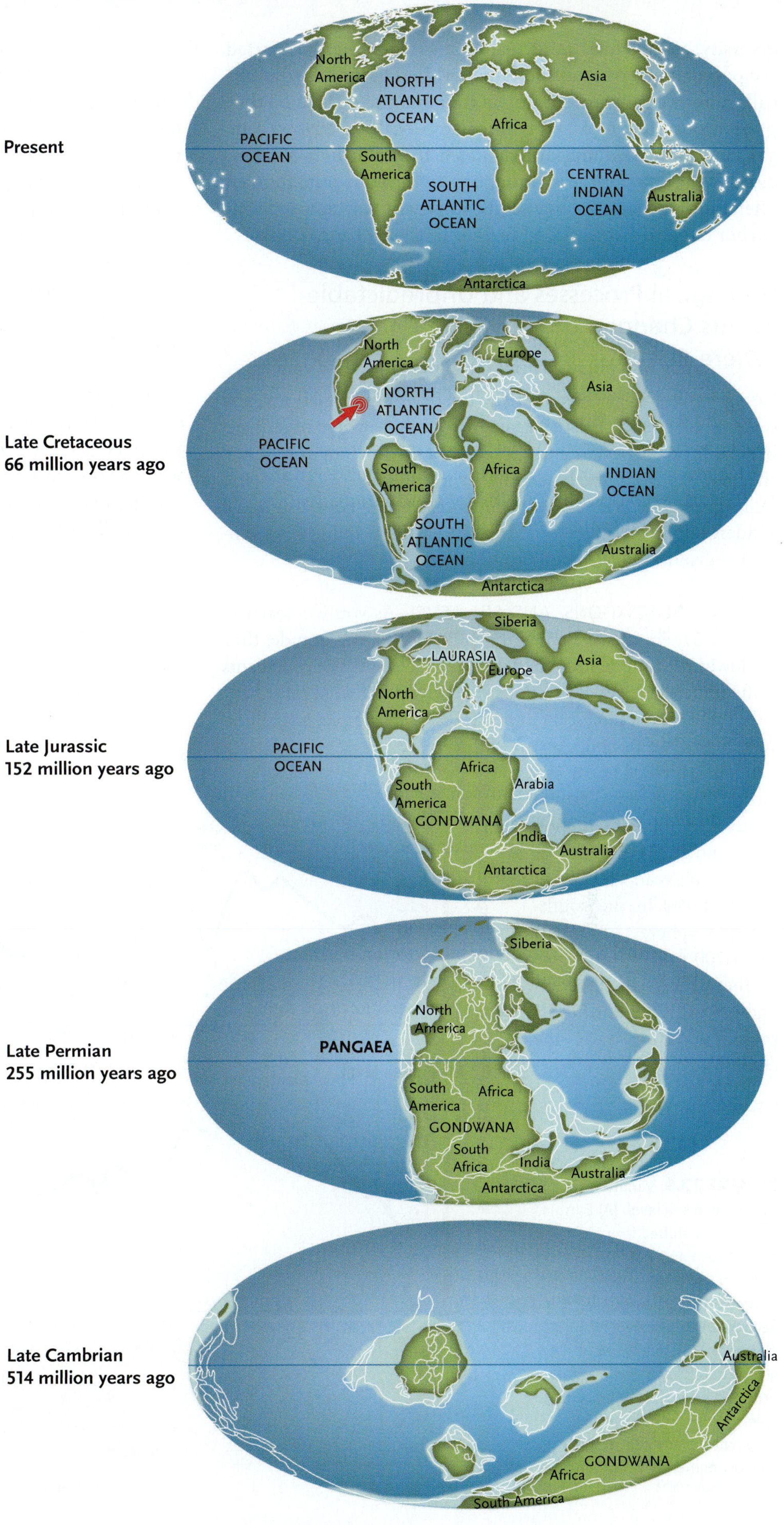

FIGURE 23.8 Continental drift. Earth's many landmasses coalesced during the Permian period, forming the supercontinent Pangaea. About 160 million years ago, Pangaea began to separate into a southern continent, Gondwana, and a northern continent, Laurasia. Then Gondwana began to break apart. The landmasses that would form Africa and India pulled away first, opening the South Atlantic and Indian Oceans. Australia separated from Antarctica about 55 million years ago and slowly drifted northward. South America separated from Antarctica shortly thereafter. Laurasia remained nearly intact until 43 million years ago when North America and Greenland both separated from Europe and Asia. Movement of the continents also changed the shapes and the sizes of the oceans.

ago, they merged into a single supercontinent called Pangaea; continental drift later separated Pangaea into a northern continent, Laurasia, and a southern continent, Gondwana. Laurasia and Gondwana subsequently broke into the continents we know today **(Figure 23.8)**.

The breakup and separation of the continents had a huge impact on the evolution of terrestrial and aquatic organisms. For example, the widespread distribution of many terrestrial groups on Pangaea was a defining characteristic of the Paleozoic era. But continental drift later established large-scale patterns of geographical isolation. Each continent became a separate arena in which organisms evolved independently of those on other landmasses. Thus, the organisms that live on southern continents, the remnants of Gondwana, are very different from those that live on northern continents, which are the remnants of Laurasia. For example, the large flightless birds known as ratites (pictured in Figure 20.2) occur only in Africa, Australia, South America and on nearby islands; no comparable or closely related species live in Eurasia or North America.

Moreover, the smaller landmasses produced by plate tectonics are surrounded by more extensive coastlines and shallow marine habitats than were the larger continents. Coastal waters have always harbored tremendous biodiversity, and as the continents became increasingly separated, the populations living in these shallow seas became geographically isolated from each other. As the continents assumed their present positions at different latitudes, locally distributed organisms diversified and differentiated from those in other regions. Today, shallow marine habitats in the temperate zone are often occupied by giant kelp beds, whereas those in the tropics harbor coral reefs (described further in Chapter 51).

Geological Processes and Unpredictable Events Changed the Environments Where Organisms Lived

Continental drift affected climate, the extent of glaciations, and sea levels, changing the physical environments where organisms live on local, regional, and global scales. Most environmental changes occurred incrementally over millions of years—and millions of generations—but some catastrophes have also had a sudden impact. The combined effects of these changes on biological evolution have been profound.

CLIMATE, GLACIATIONS, AND SEA LEVEL As you will learn in Chapter 51, environmental temperatures vary with latitude: they are lower near the poles than near the equator. As the continents drifted, their latitudinal positions shifted, causing what must have been huge changes in local temperatures. For example, one billion years ago Queensland, Australia, was situated near the North Pole; it then drifted southward for 800 million years, moving to the equator 440 million years ago and to mid-latitudes in the southern hemisphere (40° S latitude) 200 million years ago. It then reversed direction and moved northward to its present position near the equator (12° S).

On a regional level, the sizes of landmasses also have an effect on their climates. Coastal regions generally experience smaller daily and seasonal fluctuations in temperature than interior regions because their proximity to the sea moderates their climate. Whenever landmasses were joined into large continents, vast expanses of the interior landscape must have experienced frigid winters and hot summers. As the landmasses broke up into smaller continents, a larger fraction of the land was close to the sea and its moderating influence.

On a global scale, the overall climate has shifted from warm and wet to cool and dry several times in Earth's history **(Figure 23.9).** The changing positions of continents altered the flow of ocean currents, which, along with small changes in Earth's orbit around the sun, contributed to these climatic shifts. At times when Earth's climate was cooler than it is today, the polar ice caps grew larger. And because much of Earth's water was incorporated into these enormous glaciers, rainfall was greatly reduced and sea level fell dramatically. Between the cold spells, Earth's climate was warmer than it is today. Under those conditions, glaciers retreated, rainfall was luxuriant, temperature differences between equatorial and polar regions were reduced, and sea levels rose.

Changing climate patterns and the accompanying rise or fall of sea level influenced the evolution of living systems. Organisms adapted to the altered conditions, shifted their geographical ranges into suitable environments, or became ex-

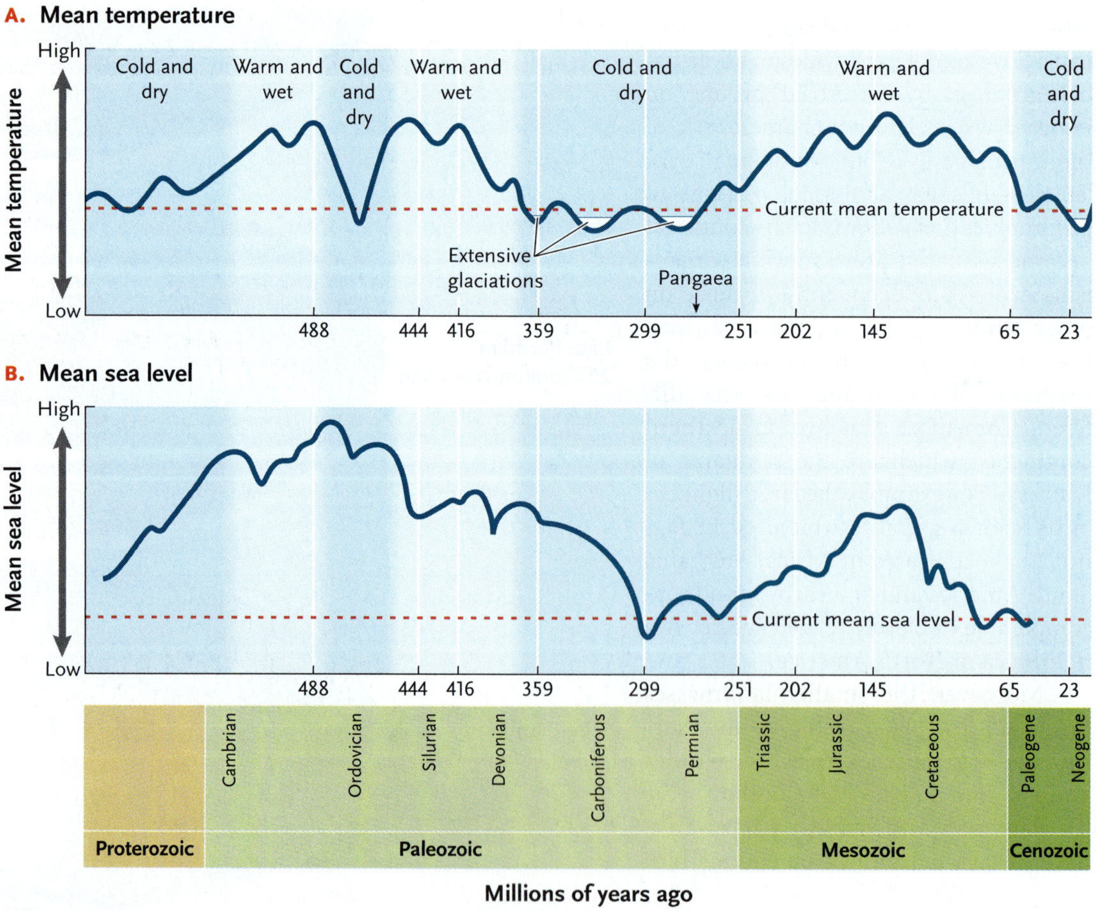

FIGURE 23.9 Variations in climate and sea level. (A) Earth's climate has shifted from cold and dry to warm and wet multiple times in Earth's history. Particularly cold periods were accompanied by extensive glaciation. **(B)** As a result of climatic shifts and continental drift, sea level has often risen and fallen. Large drops in sea level eliminated many near-shore, shallow marine environments.

tinct. Declining sea levels took an especially severe toll on organisms living in shallow marine environments. As their habitats disappeared, many marine species ceased to exist. Nevertheless, some organisms always survived, and their descendants flourished and later diversified in the altered environments.

Today we are living in one of the cooler periods of Earth's history. But human activities, including the combustion of fossil fuels and industrial-scale agriculture, are causing an extraordinarily rapid warming of the atmosphere and oceans. Although Earth's climate has experienced repeated temperature shifts, these changes occurred slowly relative to the lifetimes of individual organisms, and many populations, species, and evolutionary lineages adapted to them. But global climate change is occurring much faster today, and scientists have documented its effects: melting glaciers, rising temperatures, and rising sea level. They have also documented changes in the geographical ranges and breeding schedules of many species. We explore the effects of global climate change more fully in Unit Seven.

VOLCANIC ERUPTIONS AND ASTEROID IMPACTS Sudden, unpredictable events have also changed physical environments on Earth. For example, volcanoes are especially active near oceanic trenches where old crust is recycled into Earth's mantle. Volcanic eruptions often spew enormous quantities of ash and gas into the atmosphere, blocking incoming sunlight. Massive eruptions may cause Earth's surface temperature to decrease several degrees for as long as a year. Other types of tectonic activity also release molten rock on Earth's surface, sometimes with the opposite effect on climate, as you will discover in Section 23.4.

Asteroid strikes have also had devastating effects on living systems. In addition to obliterating a gigantic area around the site of impact, the collisions threw massive amounts of material into the atmosphere, blocking sunlight and potentially altering the planet's climate. As we describe in Section 23.4, these cataclysmic events have sometimes caused many forms of life to disappear over relatively short periods of geological time.

STUDY BREAK 23.2

1. How did continental drift affect the geographical distributions of organisms?
2. What effect did glaciations have on sea level?

THINK OUTSIDE THE BOOK

Search Internet and library resources for information about how continental drift and other major geological phenomena affected the area where you live. At what latitude was your home or university 65 million years ago, 100 million years ago, and 250 million years ago? Was it ever submerged below a shallow sea?

23.3 Historical Biogeography and Convergent Biotas

Continental drift facilitated the diversification of distinct evolutionary lineages in different regions on Earth. In this section we consider the large-scale geographical distributions of organisms.

Historical Biogeography Explains the Broad Geographical Distributions of Organisms

More than a century after Darwin published his observations, the theory of plate tectonics refocused attention on biogeography. Historical biogeographers try to explain how organisms acquired their geographical distributions over evolutionary time.

CONTINUOUS AND DISJUNCT DISTRIBUTIONS Many species have a **continuous distribution:** they live in suitable habitats throughout a geographical area. For example, herring gulls *(Larus argentatus)* live along the coastlines of all northern continents. Continuous distributions, especially in mobile organisms like birds, usually require no special historical explanation.

Other groups exhibit **disjunct distributions,** in which closely related species live in widely separated locations. Two phenomena—dispersal and vicariance—can create disjunct distributions. **Dispersal** is the movement of organisms away from their place of origin; it can produce a disjunct distribution if a new population becomes established on the far side of a geographical barrier. **Vicariance** is the fragmentation of a once-continuous geographical distribution by external factors.

How can researchers determine whether a disjunct distribution is the product of dispersal or vicariance? Analysis of the fossil record for a group of organisms, interpreted in the context of their evolutionary relationships as well as continental drift, can provide an answer. For example, southern beech trees (genus *Nothofagus*) are currently found in Australasia and South America **(Figure 23.10).** (Australasia is the region that includes Australia, New Guinea, New Zealand, and other islands of the South Pacific.) For many years, the genus' distribution only on southern continents and islands has been the classic example of vicariance: the group presumably originated on Gondwana, and its subsequent fragmentation apparently isolated ancestral *Nothofagus* populations on different continents and islands where their descendants live today. The fossil record generally supports that interpretation: researchers have identified fossilized *Nothofagus* pollen, dating from 55 to 34 million years ago, in locations that still formed a nearly contiguous landmass at that time (Figure 23.10C).

However, Lyn G. Cook and Michael D. Crisp of The Australian National University recently undertook a molecular analysis of the relationships and geographical distributions of living *Nothofagus* species. Their research reveals that the genus had a much more complicated biogeographic history. It confirms that the distributions of some *Nothofagus* species in Australia and South America are the products of vicariance: their common ancestor originated on Gondwana *before* the two con-

A. ***Nothofagus***

Geoff Bryant/Photo Researchers© Cengage Learning 2014

FIGURE 23.10 Distribution of southern beech trees *(Nothofagus)*. The genus *Nothofagus* **(A)** exhibits a disjunct distribution **(B)**; it includes species living in South America and in Australasia. **(C)** A reconstruction of the positions of Australia and South America in the late Cretaceous reveals that they were parts of a nearly continuous landmass at that time.

B. Current distribution of *Nothofagus* species

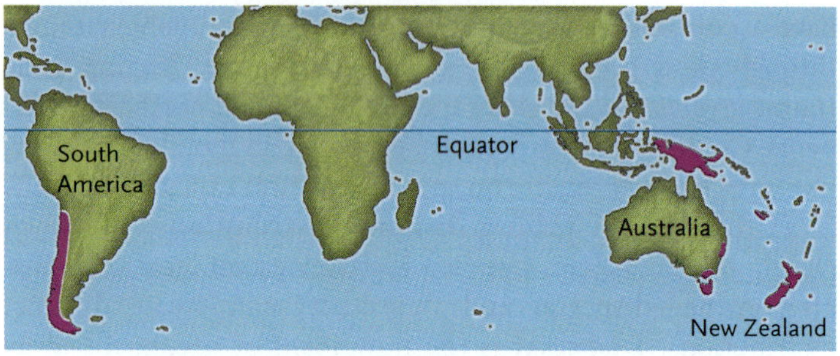

C. Position of southern continents in the late Cretaceous

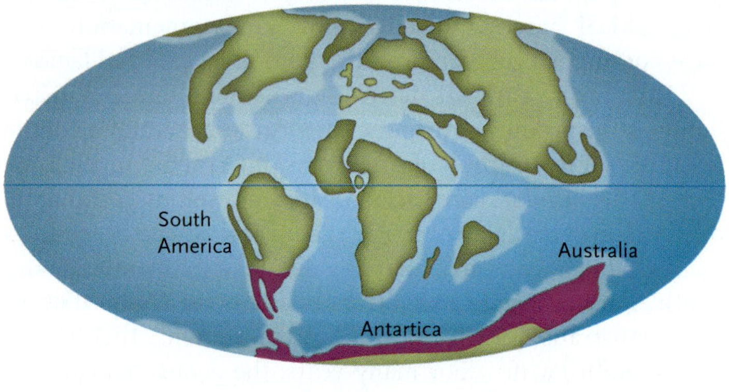

FIGURE 23.11 Wallace's biogeographical realms. Each realm contains a distinctive biota.

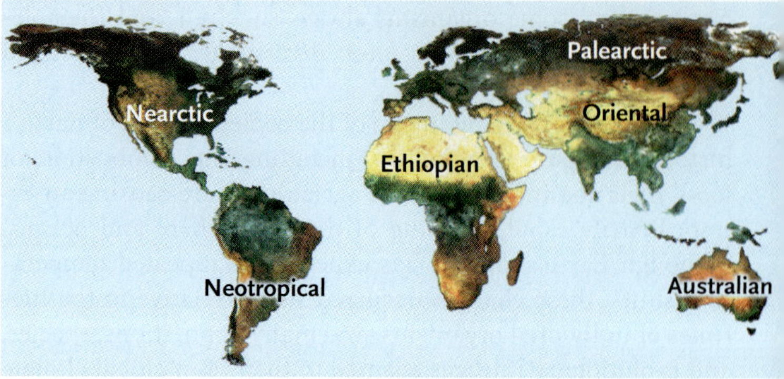

tinents separated from each other, and their descendants now live in both places. However, the analysis also suggests that the common ancestor of certain closely related *Nothofagus* species in New Zealand and Australia originated more than 30 million years *after* New Zealand had separated from Australia. Thus, vicariance cannot explain their current distribution. Instead, seeds of these trees must have dispersed across the open ocean from one place to the other. Therefore, both vicariance and dispersal have played important roles in establishing the present geographical distribution of *Nothofagus*.

BIOGEOGRAPHICAL REALMS For species that were widespread in the Mesozoic era, Pangaea's breakup was a powerful vicariant experience. The subsequent geographical isolation of continents fostered the evolution of distinctive regional **biotas.** (The word *biota* refers to all of the organisms living in a region; *fauna* to all of the animals; and *flora* to all of the plants.) Alfred Russel Wallace used the biotas to define six **biogeographical realms,** which we still recognize today **(Figure 23.11).**

The Australian and Neotropical realms, which have been geographically isolated since the Mesozoic, contain many **endemic species** (those that occur nowhere else on Earth). The Australian realm, in particular, has had no complete land connection to any other continent for approximately 55 million years. As a result, Australia's mammalian fauna is unique, made up almost entirely of endemic marsupials (mammals that give birth after a short gestation period and then carry their young in a pouch). Although Eocene fossils of placental mammals (mammals that do not rear their young in a pouch) have been found in Australia, all of them became extinct. The only native placental mammals that live there now are bats and rodents, which apparently dispersed to Australia from either Asia or New Guinea after the marsupial fauna had already diversified there.

The biotas of the Nearctic and Palearctic realms are, by contrast, fairly similar. North America and Eurasia were frequently connected by land bridges: eastern North America was attached to Western Europe until the breakup of Laurasia 43 million years ago, and northwestern North America had periodic contact with northeastern Asia over the Bering land bridge during much of the past 60 million years. As a result, many plant and animal species on these continents are closely related.

Evolution Has Produced Convergent Biotas in Widely Separated Regions

Distantly related species living in different biogeographical realms are sometimes very similar in appearance. For example, the overall form of cactuses in the Americas is extraordinarily similar to that of spurges in Africa **(Figure 23.12).** But these lineages arose independently long after those continents had separated; thus, cactuses and spurges did not inherit their similarities from a shared ancestor. Their overall resemblance is the product of **convergent evolution,** the evolution of similar adaptations in distantly related organisms that occupy similar environments.

Convergent evolution has also fostered morphological similarities in distantly related animals that feed on similar foods and occupy similar habitats in widely separated geographic ranges. Sometimes, large portions of entire faunas develop convergent morphologies. For example, the marsupial mammals of Australia and the placental mammals of North America—groups that arose long after the breakup of Pangaea—include many pairs of morphologically convergent species **(Figure 23.13).** Although the species in each pair differ in most of their anatomical details, their overall appearance and structure are remarkably similar. Biologists interpret these convergent morphologies as evolutionary responses to similar patterns of natural selection in the habitats where these animals live.

A. Cactus (*Echinocereus* species)

B. Spurge (*Euphorbia* species)

FIGURE 23.12 Convergent evolution in plants. (A) North American cactuses (family Cactaceae) are strikingly similar to **(B)** African spurges (family Euphorbiaceae). Convergent evolution adapted both groups to desert environments with thick, water-storing stems, spiny structures that discourage animals from feeding on them, CAM photosynthesis (see Section 9.4), and stomata that open only at night.

STUDY BREAK 23.3

1. **Which type of geographical distribution requires no special explanation?**
2. **Why do distantly related species that live in different biogeographical realms sometimes resemble each other?**

A. Mammals and continental drift

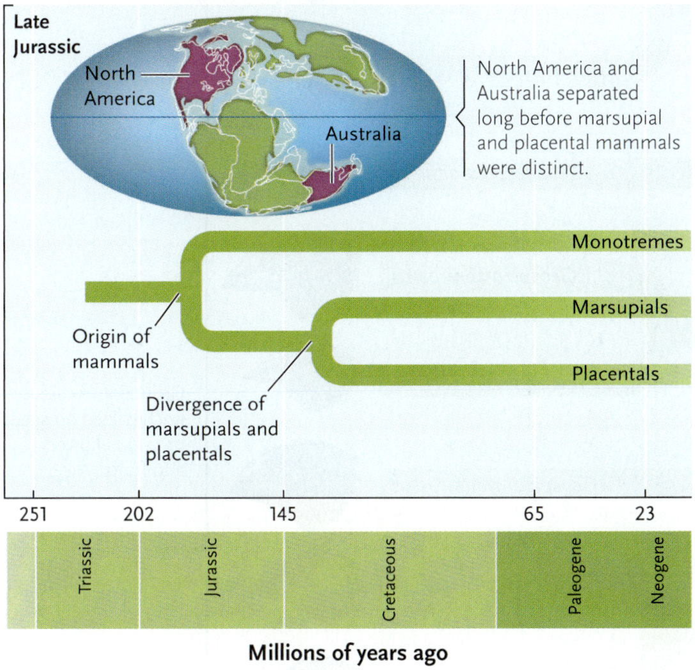

B. Convergence of placental and marsupial mammals

FIGURE 23.13 Convergent evolution in mammalian faunas. (A) Two distinctive lineages of mammals, the placentals and the marsupials, diverged after Pangaea had already split into northern and southern continents. **(B)** Thus, biologists conclude that convergent evolution, rather than shared ancestry, produced similar body forms in North American placental mammals (on left) and Australian marsupial mammals (on right). The members of each pair differ in their anatomical details, but their overall forms are similar even though they are only distantly related. Biologists interpret these similarities as convergent adaptations in animals that fulfill similar ecological roles on the two continents.

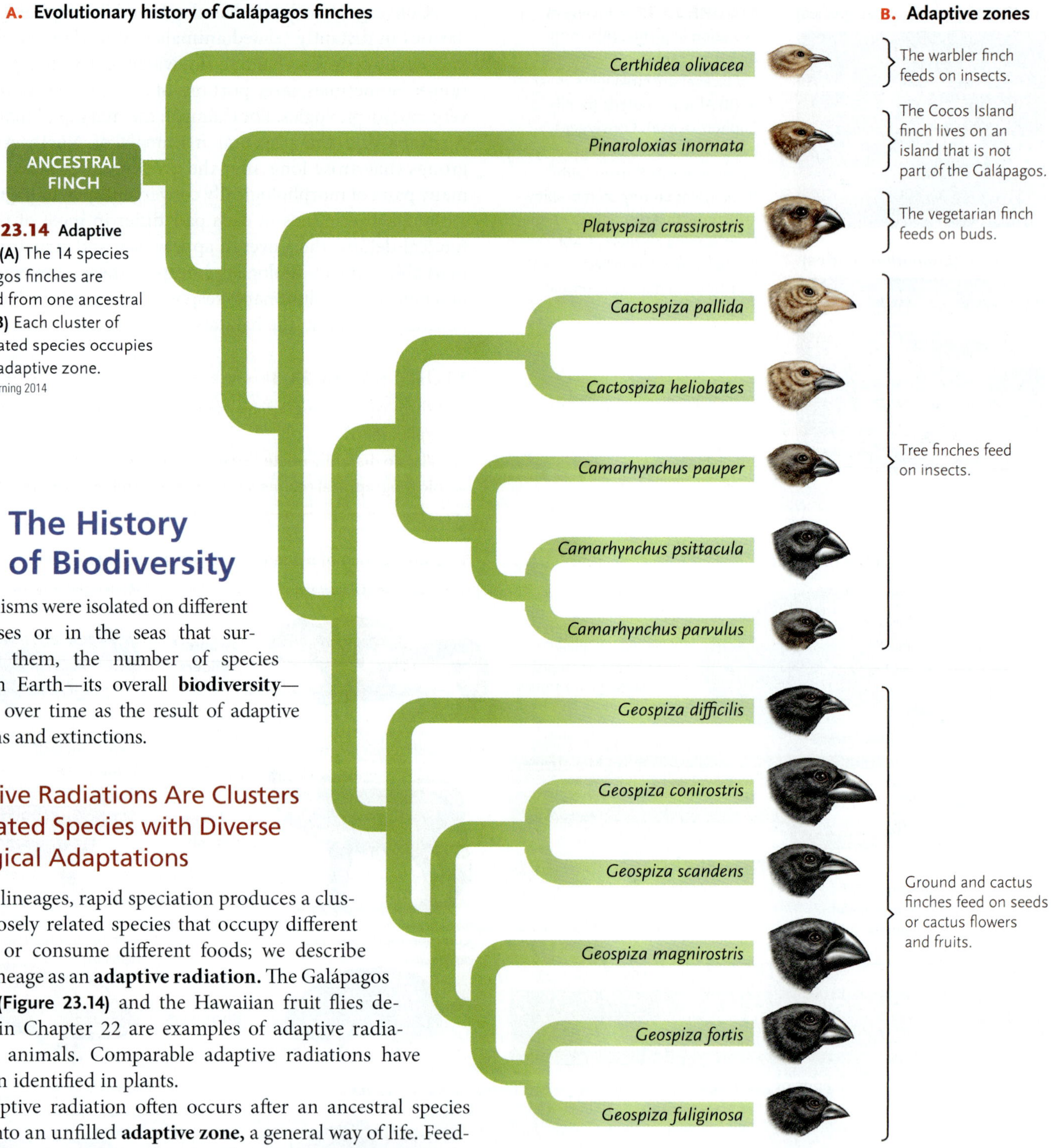

FIGURE 23.14 Adaptive radiation. (A) The 14 species of Galápagos finches are descended from one ancestral species. **(B)** Each cluster of closely related species occupies a specific adaptive zone.

23.4 The History of Biodiversity

As organisms were isolated on different landmasses or in the seas that surrounded them, the number of species living on Earth—its overall **biodiversity**—changed over time as the result of adaptive radiations and extinctions.

Adaptive Radiations Are Clusters of Related Species with Diverse Ecological Adaptations

In some lineages, rapid speciation produces a cluster of closely related species that occupy different habitats or consume different foods; we describe such a lineage as an **adaptive radiation.** The Galápagos finches **(Figure 23.14)** and the Hawaiian fruit flies described in Chapter 22 are examples of adaptive radiations in animals. Comparable adaptive radiations have also been identified in plants.

Adaptive radiation often occurs after an ancestral species moves into an unfilled **adaptive zone,** a general way of life. Feeding on other marine animals is the adaptive zone that most sharks occupy, and grazing on grass in open habitats is the adaptive zone that living horses exploit. Feeding on plastic in landfills might become an adaptive zone in the future if some organism develops the ability to digest that now-abundant resource.

An organism's move into a new adaptive zone is sometimes triggered by the chance evolution of a key morphological innovation that allows it to use the environment in a unique way. For example, the dehydration-resistant eggs of early reptiles enabled them to complete their life cycle on land, opening terrestrial habitats to them. Similarly, the evolution of flowers that attract insect pollinators was a key innovation in the history of flowering plants.

An adaptive radiation may also be triggered after the demise of a successful group. Mammals, for example, were relatively inconspicuous during their first 150 million years on Earth, presumably because dinosaurs dominated terrestrial habitats. But after dinosaurs declined in the late Mesozoic era, mammals underwent an explosive adaptive radiation. Today they are the dominant vertebrates in many terrestrial habitats.

Extinctions Have Been Common in the History of Life

Biodiversity has not always increased through the history of life. The additions to biodiversity caused by adaptive radiations and other speciation events are counteracted by **extinction,** the death of the last individual in a species or the last species in a lineage. Paleobiologists recognize two distinct patterns of extinction in the fossil record, background extinction and mass extinction.

Species and lineages have been going extinct since life first appeared. We should expect species to disappear at some low rate, the **background extinction rate;** as environments change, poorly adapted organisms will not survive and reproduce. In all likelihood, more than 99.9% of the species that have ever lived are now extinct. David Raup of the University of Chicago has suggested that, on average, as many as 10% of species go extinct every million years and that more than 50% go extinct every 100 million years. Thus, the history of life has been characterized by an ongoing turnover of species.

On at least five occasions, however, extinction rates rose well above the background rate. During these **mass extinctions,** large numbers of species and lineages died out over relatively short periods of geological time **(Figure 23.15).** The Permian extinction was the most severe: more than 85% of the species alive at that time—including all trilobites, many insects and amphibians, and the trees of the coal swamp forests—disappeared forever. During the last mass extinction, at the end of the Cretaceous, half the species on Earth, including most dinosaurs, became extinct. A sixth mass extinction, potentially the largest of all, may be occurring now as a result of global climate change and human degradation of the environment (see Unit Seven).

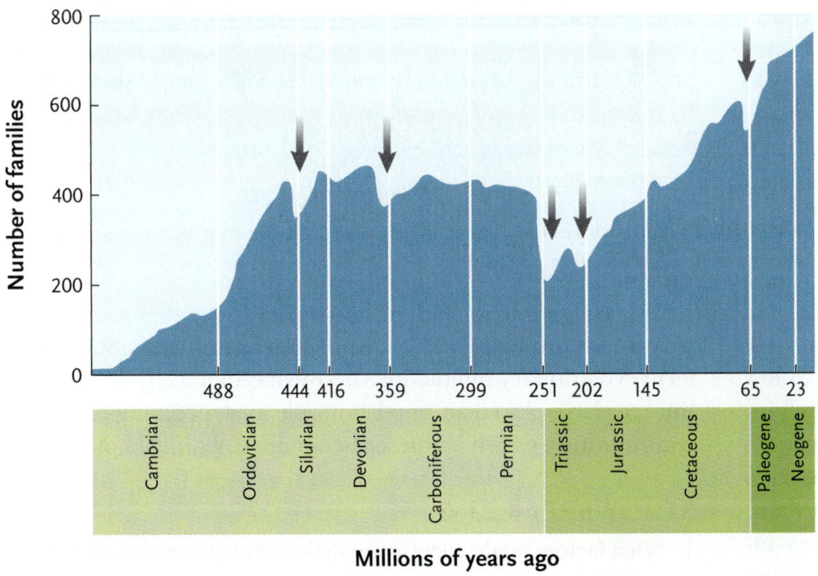

FIGURE 23.15 Mass extinctions. Biodiversity, indicated by the height of the dark blue area in the graph, was temporarily reduced by at least five mass extinctions (arrows) during the history of life. The data presented in this graph record the family-level diversity of marine animals. A family is a group of genera descended from a common ancestor.

Different factors were responsible for the five mass extinctions. For example, the Ordovician extinction occurred after Gondwana moved toward the South Pole, triggering a glaciation that cooled the world's climate and lowered sea levels, eliminating many shallow marine environments.

Scientists hypothesize that the massive Permian extinction was caused by a chain of events that resulted in severe climate warming. The trigger was apparently a series of ongoing eruptions that released a flood of molten lava hundreds—or even thousands—of meters thick. Once solidified, the lava formed the "Siberian traps," a rock formation that covers 1.6 million km^2 of northeastern Asia. Scientists estimate that the lava flood released so much carbon dioxide, a greenhouse gas (see Section 54.4), into the atmosphere that it may have increased the average global surface temperature by 6°C. The resulting higher ocean temperature melted frozen submarine deposits of methane, which then bubbled into the atmosphere, causing even more climate warming. Michael J. Benton and Richard J. Twitchett of the University of Bristol, United Kingdom, describe the situation as a "runaway greenhouse" in which global temperature increased so much that organisms simply could not adapt to the change. One hundred million years elapsed before global diversity returned to pre-extinction levels.

Many researchers agree that an asteroid impact was a major cause of the Cretaceous mass extinction. The resulting dust cloud may have blocked the sunlight necessary for photosynthesis, setting up a chain reaction of extinctions that began with microscopic marine organisms. Geological evidence supports this hypothesis. Rocks dating to the end of the Cretaceous period (65 million years ago) contain a highly concentrated layer of iridium, a metal that is rare on Earth but common in asteroids. The impact from an iridium-laden asteroid only 10 km in diameter could have caused an explosion equivalent to that of 100 million megatons of TNT, scattering iridium dust around the world. Geologists have identified the Chicxulub crater, 180 km in diameter, on the edge of Mexico's Yucatán peninsula (see the red arrow in Figure 23.8) as the likely site of the impact.

Biodiversity Has Increased Repeatedly over Evolutionary History

Although mass extinctions temporarily reduce biodiversity, they also create evolutionary opportunities. Some species survive because they have highly adaptive traits, large population sizes, or widespread distributions. And some of the surviving species undergo adaptive radiation, filling adaptive zones that mass extinctions made available.

Sometimes, the success of one lineage comes at the expense of another. Although the diversity of terrestrial vascular plants has increased almost continuously since the Devonian period, this trend includes booms and busts in several lineages **(Figure 23.16).**

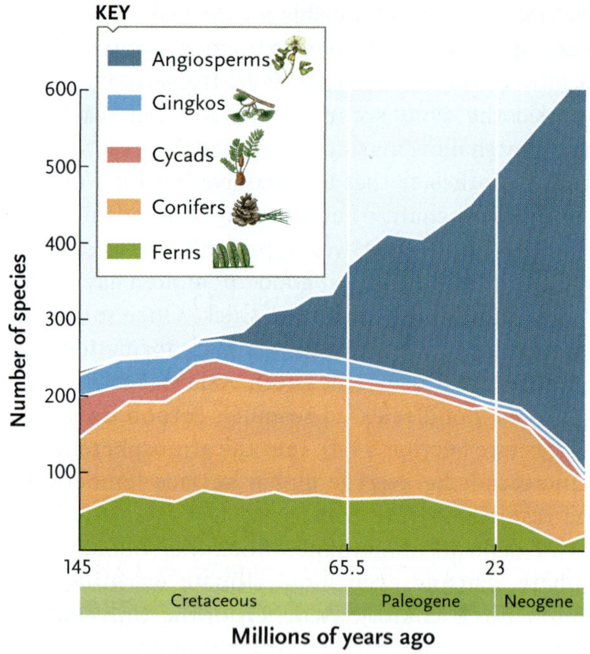

FIGURE 23.16 History of vascular plant diversity. The diversity of angiosperms increased during the Cretaceous period as the diversity of other groups declined. Changes in the diversity of each group through time are indicated by changes in the height of the group's colored band in the graph.

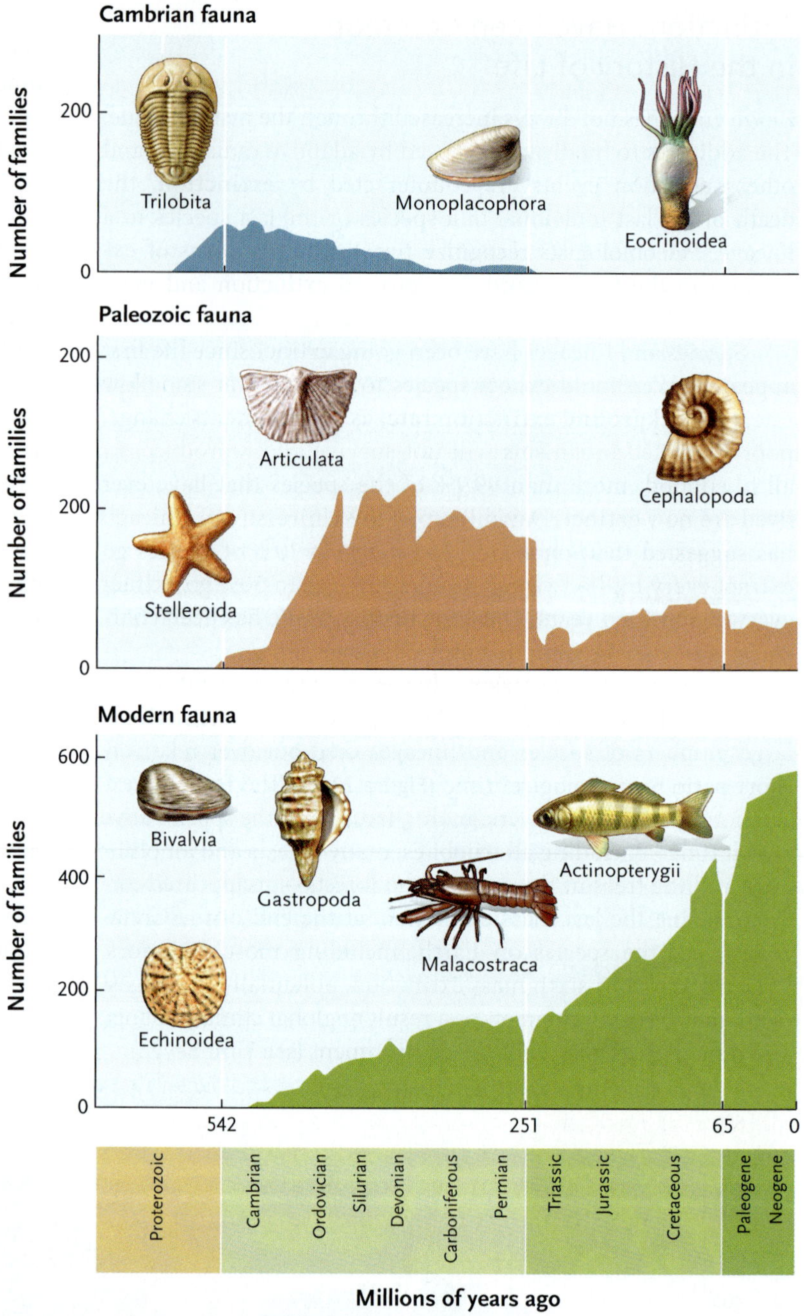

FIGURE 23.17 History of marine animal diversity. Marine animals have undergone three major radiations. Few remnants of the Cambrian fauna remain alive today.

Ferns and conifers recovered rapidly after the Permian extinction, maintaining their diversity into the early Cenozoic era. However, angiosperms, which arose and began to diversify in the early Cretaceous period, may have hastened the decline of these groups by replacing them in many environments.

The superb fossil record left by certain marine animals reveals three major periods of adaptive radiation **(Figure 23.17).** The first occurred during the Cambrian, more than 500 million years ago, when many animal phyla, the major categories of animal life, first appeared. Most of these phyla became extinct, and a second wave of radiations established the dominant Paleozoic fauna during the Ordovician period. A third evolutionary fauna diversified in the Triassic period, right after the great Permian extinction; it produced the immediate ancestors of modern marine animals. The diversity of marine animals has increased consistently since the early Triassic, in large measure because of continental drift. As continents and shallow seas became increasingly isolated, regional biotas diversified independently of one another, increasing worldwide biodiversity.

Historical increases in biodiversity can also be attributed to the evolution of ecological interactions. For example, the number of plant species found *within* fossil assemblages has increased over time, suggesting the evolution of mechanisms that allow more species to coexist. In addition, insects continued to diversify dramatically in the Cretaceous period, possibly because the angiosperms created a new adaptive zone for them. New insect species then provided a novel set of pollinators that may have stimulated the radiation of angiosperms. Such long-term evolutionary interactions between ecologically intertwined lineages have played an important role in structuring ecological communities, which are described more fully in Chapter 53.

STUDY BREAK 23.4

1. **What factors might allow a population of organisms to occupy a new adaptive zone?**
2. **What events apparently triggered the mass extinction at the end of the Permian period?**
3. **When did the first major adaptive radiation of animals occur?**

THINK OUTSIDE THE BOOK

You have learned how the geography of the Galápagos and the Hawaiian Islands fostered adaptive radiations in finches and fruit flies. Search Internet or library sources for examples of adaptive radiations in other types of organisms on these archipelagos. Also look for information about the adaptive radiation of mammals on continental land masses. Do adaptive radiations on archipelagos and on continental landmasses differ in substantive ways?

23.5 Interpreting Evolutionary Lineages

As newly discovered fossils demand the reinterpretation of old hypotheses, biologists constantly refine their ideas about the history of life. In this section we describe how our interpretation of evolutionary lineages is constantly changing.

Modern Horses Are Living Representatives of a Once-Diverse Lineage

The earliest known ancestors of modern horses were first identified by Othniel C. Marsh of Yale University just a year after Darwin published *On the Origin of Species.* These early horses, *Hyracotherium,* stood 25 to 50 cm high and weighed no more than 20 kg. Their toes (four on the front feet and three on the hind) were each capped with a tiny hoof, but the animals walked on soft pads as dogs do today. Their faces were short, their teeth were small, and they browsed on soft leaves in woodland habitats.

In 1879, Marsh published his analysis of 55 million years of horse family history. He described the evolution of this group of mammals as a sequence of stages from the tiny *Hyracotherium* through intermediates represented by *Mesohippus, Merychippus,* and *Pliohippus* to the modern *Equus* **(Figure 23.18A).** (Each of these names refers to a genus, a group of closely related species.) Marsh inferred a pattern of descent characterized by gradual, directional evolution in several skeletal features. Changes in the legs and feet allowed horses to run more quickly, and changes in the face and teeth accompanied a switch in diet from soft leaves to tough grasses.

The fossil record for horses is superb, and we now have fossils of more than 100 extinct species from five continents. These data reveal a macroevolutionary history very different from Marsh's interpretation. *Hyracotherium* was not gradually transformed into *Equus* along a linear track. Instead, the evolutionary tree for horses was highly branched **(Figure 23.18B),** and *Hyracotherium's* descendants differed in size, number of toes, tooth structure, and other traits. Although many branches of this lineage lived in the Miocene and Pliocene epochs, all but one are now extinct. The species of the genus *Equus* living today (horses, donkeys, and zebras) are the surviving tips of that one branch.

When we study extinct organisms, we tend to focus on traits that characterize modern species. Marsh, for example, assumed that the differences between *Hyracotherium* and *Equus* were typical of the changes that characterized the group's evolutionary history. But not all fossil horses were larger **(Figure 23.18C),** had fewer toes, or were better adapted to feed on grass than their ancestors.

The Tempo of Morphological Change Varies among Lineages

Analyses of the fossil record indicate that the tempo, or timing, of morphological change varies among lineages. The discovery of transitional fossils provides evidence in support of the **phyletic gradualism hypothesis** (*phylon* = race), which proposes that most morphological change occurs gradually over long periods of time. For example, a study of Ordovician trilobites revealed that the number of "ribs" in their tail region changed continuously over 3 million years. The change was so gradual that a sample from any given stratum was almost always intermediate between samples from the strata just above and below it. The changes in rib number probably evolved without the evolution of new, reproductively isolated species **(Figure 23.19).**

But for most groups of organisms, the fossil record is not nearly as good as it is for horses or trilobites. In fact, the discovery of intermediate or transitional fossils is fairly rare. Instead, most species appear suddenly in a particular layer, persist for some time with little change, and then disappear from the fossil record. Then another species with different traits suddenly appears in the next higher stratum.

In the early 1970s, Niles Eldredge of the American Museum of Natural History and Stephen Jay Gould of Harvard University published an explanation for the absence of transitional forms, or "missing links." Their **punctuated equilibrium hypothesis** suggested that speciation usually occurs in isolated populations at the edge of a species' geographical distribution. Such populations experience substantial genetic drift and distinctive patterns of natural selection (as described in Section 23.3). According to this hypothesis, morphological variations appear rapidly as new species arise; then, most species exhibit long periods of morphological equilibrium (that is, little change in form), punctuated by brief periods of speciation and rapid morphological evolution. Eldridge and Gould further argued that transitional forms live only for short periods of geological time in small, localized populations—the very conditions that discourage broad representation in the fossil record. Darwin himself used this line of reasoning to explain puzzling gaps in the fossil record: new species appear as fossils only after they become abundant and widespread and begin a period of morphological stasis.

Some evolutionists challenged the hypothesis' definition of rapid morphological change, particularly given our inability to resolve time precisely in the fossil record. To a paleobiologist with a geological perspective, "instantaneous" events

CLOSER LOOK

FIGURE 23.18 Evolving views of horse ancestry.

A. Marsh's reconstruction of horse evolution

Marsh's description of horse evolution focused on increased body size, reduced number of toes, and larger, grinding teeth at the back of the mouth.

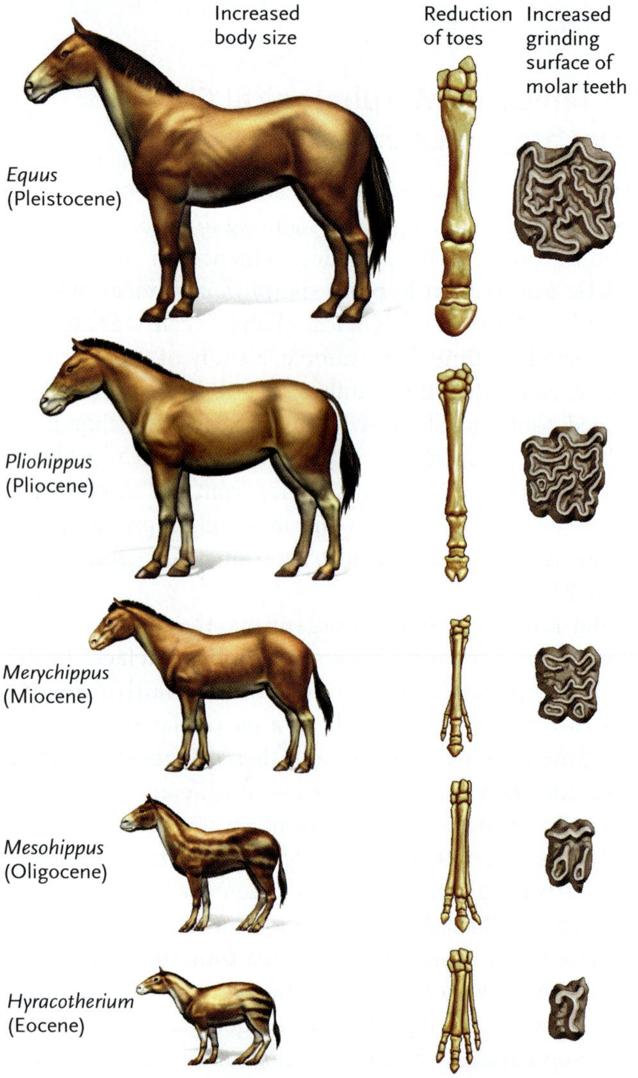

B. Modern reconstruction of horse evolution

Recent studies revealed that the horse family includes numerous evolutionary branches with variable morphology. The horses in Marsh's analysis are highlighted in green.

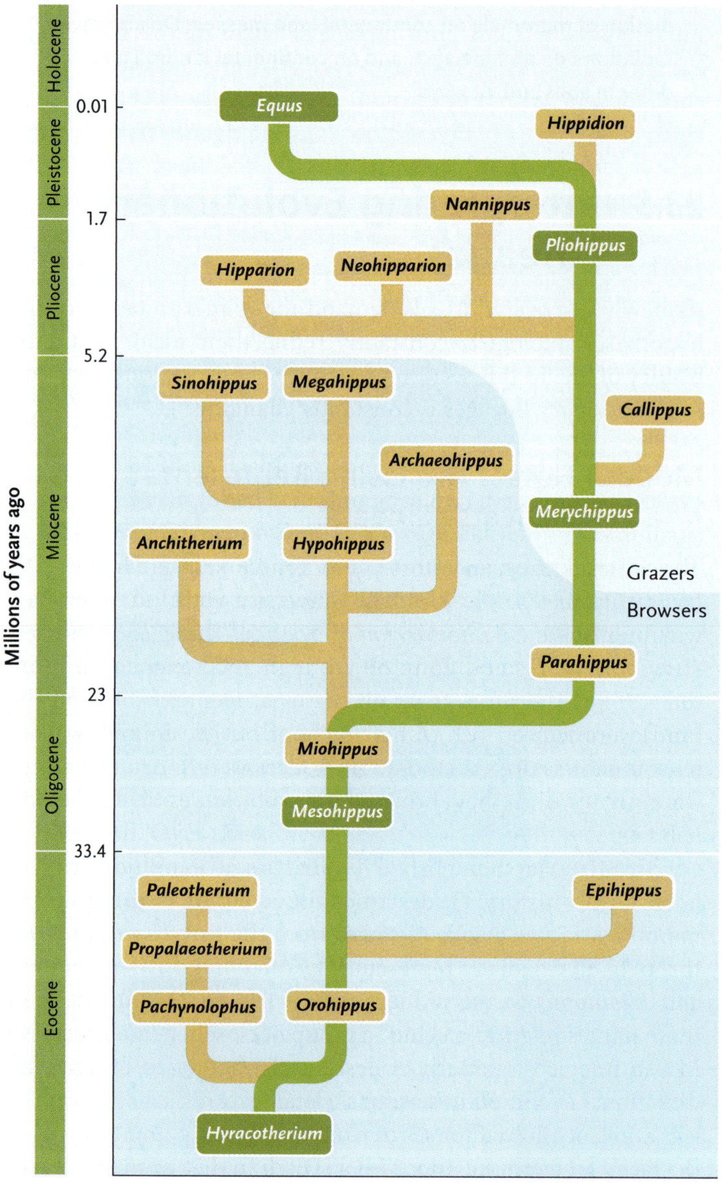

C. Changes in body size of horse species over time

Some branches of the lineage remained as small as the first horses.

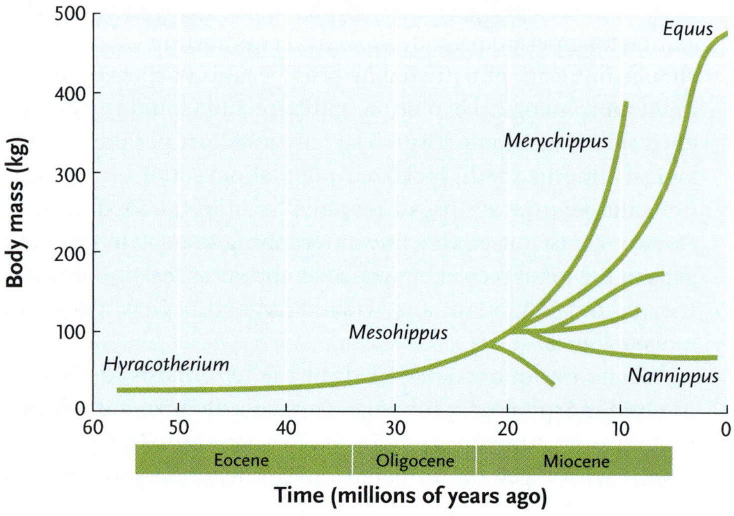

SUMMARY The lineage that includes modern horses includes many branches that did not leave any living descendants. If a branch other than *Equus* had survived, Marsh's description of trends in horse evolution would have been very different. All evolutionary lineages have extinct branches, and any attempt to trace a linear evolutionary path—as Marsh did for horses and many people do for humans—imposes artificial order on an inherently disorderly history.

THINK LIKE A SCIENTIST Modern horse breeders have used artificial selection to produce miniature horses that resemble their full-sized ancestors in all respects except body size. Does artificial selection for miniature horses constitute a reversal of the evolutionary trends observed in the horse lineage?

FIGURE 23.19 **Observational Research**

Evidence of Phyletic Gradualism

Hypothesis: The phyletic gradualism hypothesis states that most morphological change within evolutionary lineages results from the accumulation of small, incremental changes over long periods of time.

Prediction: The morphology of fossils from a given stratum will be intermediate between those of fossils from the strata immediately below and above it.

Method: Peter R. Sheldon of Trinity College, Dublin, Ireland, counted the number of "ribs" in the tail region of the exoskeletons of approximately 15,000 trilobite fossils from central Wales, United Kingdom. The fossils had formed over a span of about 3 million years during the Ordovician period. Shelton plotted the mean number of ribs found in successive samples of eight lineages.

Results: Sheldon's data reveal gradual changes in the mean number of "ribs" in these animals with no evidence of speciation.

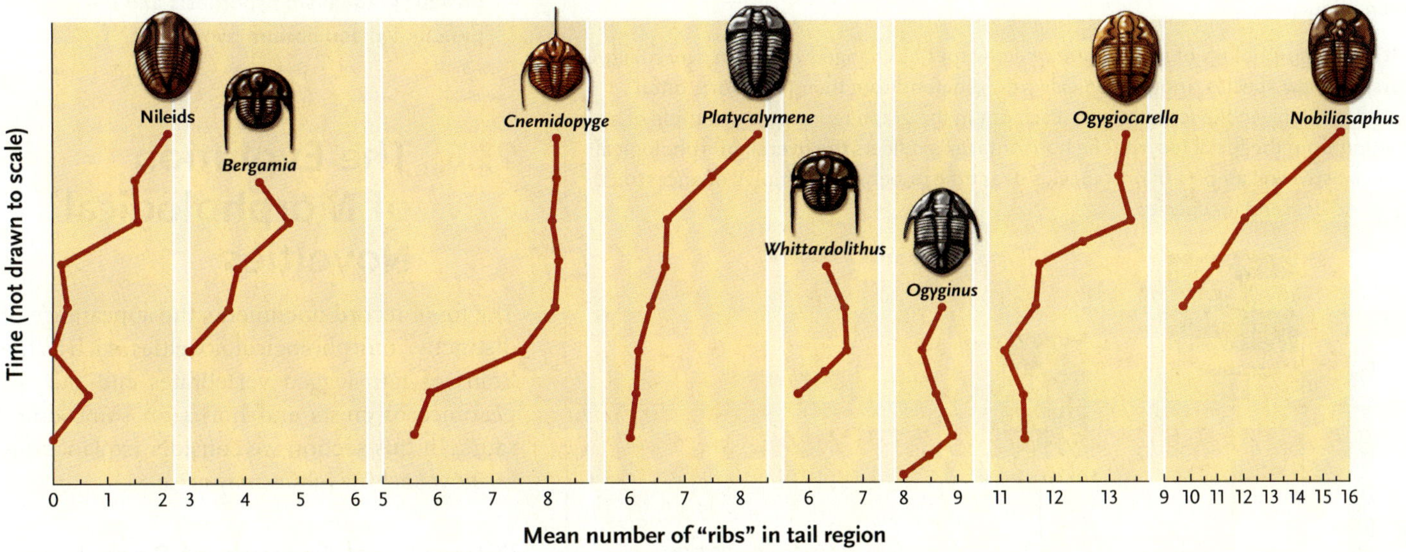

Reprinted by permission from Macmillan Publishers Ltd: P. R. Sheldon. 1987. Parallel gradualistic evolution of Ordovician trilobites. *Nature* 330:561–563.

Conclusion: Morphological changes in Ordovician trilobites from central Wales are consistent with the predictions of the phyletic gradualism hypothesis.

THINK LIKE A SCIENTIST Do you think that the changes in rib number over time in *Bergamia* and *Ogyginus* are the products of balancing selection, directional selection, or some other agent of evolution?

Source: Grant, Peter R., Ecology and Evolution of Darwin's Finches. © Princeton University Press. Reprinted by permission of Princeton University Press.

occur over tens or hundreds of thousands of years. But to a population geneticist, those time scales may encompass thousands of generations, ample time for gradual microevolutionary change.

Moreover, examples of evolutionary stasis may not be as static as they appear. Alternating periods of directional selection that favor opposite patterns of change could produce the appearance of stasis. For example, if natural selection favored slight increases in body size for 2,000 years and then favored slight decreases for the next 2,000 years, paleobiologists would probably detect no change in body size at all.

Nevertheless, punctuated patterns of evolution have been observed in some groups, such as *Metrarabdotos*, a genus of ectoprocts from the Caribbean Sea. Ectoprocts are small colonial animals that build hard skeletons (see Figure 31.15A), the details of which are well preserved in fossils. Alan Cheetham of the Smithsonian Institution measured the morphological differences between populations of individual *Metrarabdotos* species over time and between ancestral species and their descendants. His results indicate that most species did not change much for millions of years, but new species, which were morphologically different from their ancestors, often appeared suddenly **(Figure 23.20)**.

Researchers have found evidence of both gradual and punctuated morphological changes as well as some intermediate patterns. The punctuated equilibrium hypothesis caused a stir when it was first proposed because it challenged long-held assumptions about the tempo of evolution. But its publication rekindled interest in paleobiology and macroevolution, and

FIGURE 23.20 Observational Research

Evidence of a Punctuated Pattern of Morphological Change

Hypothesis: The punctuated equilibrium hypothesis states that most morphological change within evolutionary lineages appears during speciation.

Prediction: The fossil record will reveal that most species experienced relatively little morphological change for long periods of time, but that new, morphologically distinctive species arose suddenly.

Method: Cheetham examined numerous fossilized samples of populations of a small marine invertebrate, the ectoproct *Metrarabdotos,* from the Dominican Republic. He measured 46 morphological characters in populations representing 18 species and then used a complex statistical analysis to summarize the extent of morphological difference between populations.

Results: The morphology of most *Metrarabdotos* species changed very little over millions of years, but new species, morphologically very different from their ancestors, often appeared suddenly in the fossil record. Each dot in the graph represents a sample of a population from the fossil record. The horizontal axis reflects the overall morphological difference between samples of one species over time or between samples of ancestral species and descendant species.

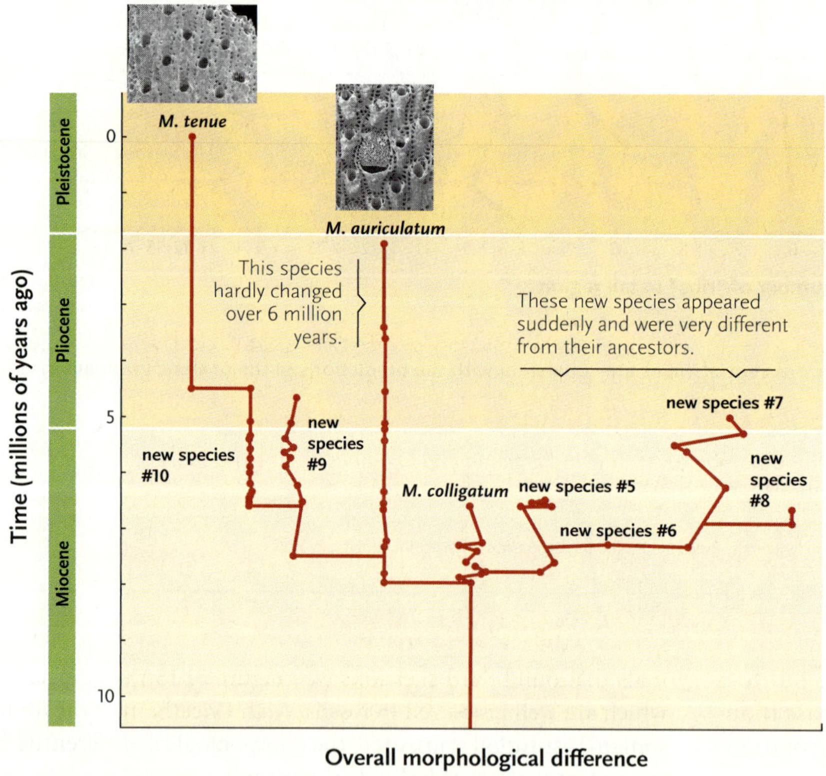

A. H. Cheetham. 1987. Tempo of evolution in a Neogene bryozoan: Are trends in single morphologic characters misleading? *Paleobiology* 13:286–296. Used with permission of the Paleontological Society.

Conclusion: The *Metrarabdotos* lineage exhibits a pattern of morphological evolution that is consistent with the predictions of the punctuated equilibrium hypothesis.

THINK LIKE A SCIENTIST Which of the *Metrarabdotos* species in the illustration shows some evidence of gradual morphological change?

Source: A. H. Cheetham. 1987. Tempo of evolution in a Neogene bryozoan: Are trends in single morphologic characters misleading? *Paleobiology* 13:286–296.

like any good idea in science, it inspired much new research. Some of the most interesting results have focused on the evolution of morphological novelties, which biologists now analyze with studies in paleontology, embryology, and genetics.

STUDY BREAK 23.5

1. **Did the horse lineage undergo a steady increase in body size over its evolutionary history? Explain your answer.**
2. **What is the difference between the morphological change predicted by the phyletic gradualism hypothesis and the punctuated equilibrium hypothesis?**

23.6 The Evolution of Morphological Novelties

The fossil record documents the appearance of distinctive morphological novelties, such as the limbs of four-legged vertebrates and the appearance of crests and horns on some dinosaurs. In this section we consider explanations for how such novelties may arise.

Differential Growth of Body Parts and Changes in the Timing of Developmental Events Produce Distinctive Structures

The morphology of individuals sometimes changes over time because of *allometric growth* (*allos* = other; *metron* = measure), the differential growth of body parts. In humans, for example, the relative sizes of different body parts change because human heads, torsos, and limbs grow at different rates **(Figure 23.21A).**

Allometric growth can also establish morphological differences in closely related species. For example, the skulls of chimpanzees and humans are similar in newborns of both species, but markedly different in adults **(Figure 23.21B).** The jaw region of the chimp skull grows much faster than other regions, while the different parts of the human skull all grow at more similar rates. Differences in the adult skulls may therefore simply reflect changes in one or a few genes that regulate the pattern of growth.

Changes in the timing of developmental events, called *heterochrony* (*heteros* = different;

FIGURE 23.21 Examples of allometric growth.

A. Allometric growth in humans

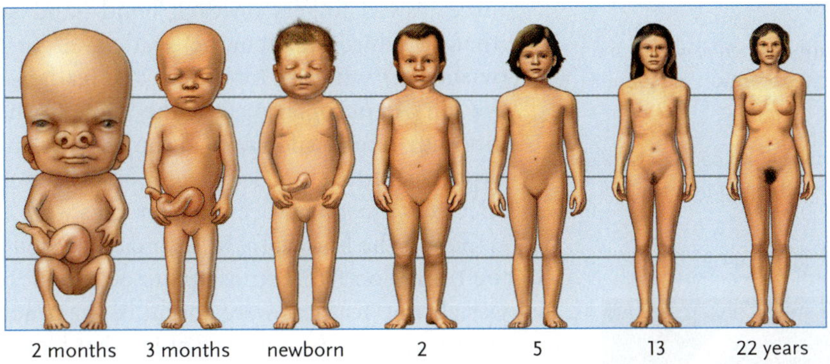

Humans exhibit allometric growth from prenatal development until adulthood. Our heads grow more slowly than other body parts; our legs grow faster.

B. Differential growth in the skulls of chimpanzees and humans

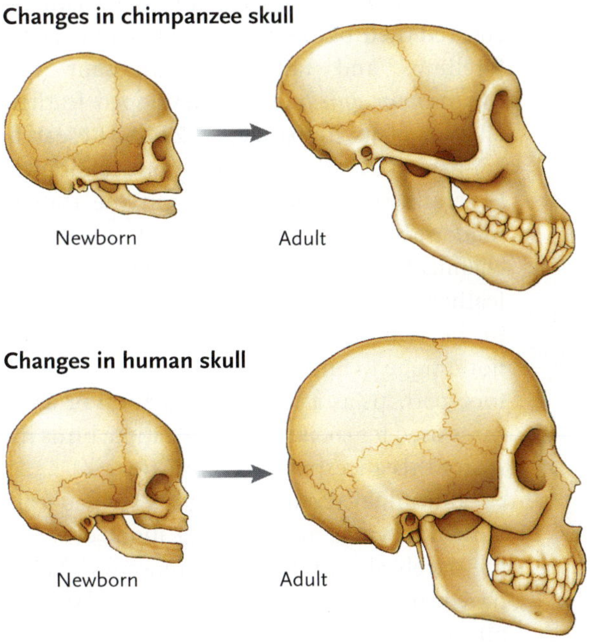

Although the skulls of newborn humans and chimpanzees are remarkably similar, differential patterns of growth make them diverge during development. The jaw of a chimpanzee grows much faster than other regions of the skull. In humans, the different parts of the skull grow at more similar rates.

khronos = time), also cause the morphology of closely related species to differ. **Paedomorphosis** (*paedo-* = child; *morphos* = form or shape), the development of reproductive capability in an organism with juvenile characteristics, is a common form of heterochrony.

Many salamanders, for example, undergo metamorphosis from an aquatic juvenile into a morphologically distinct terrestrial adult. However, populations of several species are paedomorphic—they grow to adult size and become reproductively mature without changing to the adult form **(Figure 23.22).** The evolutionary change causing these differences may be surprisingly simple. In amphibians, including salamanders, the hormone thyroxine induces metamorphosis (see Chapter 42). Paedomorphosis could result from a mutation that either reduces thyroxine production or limits the responsiveness of some developmental processes to thyroxine concentration.

FIGURE 23.22 **Paedomorphosis in salamanders.** Some small-mouthed salamanders *(Ambystoma talpoideum)* undergo metamorphosis, losing their gills and developing lungs (left). Others are paedomorphic: they retain juvenile morphological characteristics, such as gills, after attaining sexual maturity (right).

Changes in developmental rates also influence the morphology of plants **(Figure 23.23).** The flower of a larkspur species, *Delphinium decorum,* includes a ring of petals that guide bees to its nectar tube and structures on which bees can perch. By contrast, *Delphinium nudicaule,* a more recently evolved species, has tight flowers that attract hummingbird pollinators, which can hover in front of the flowers. Slower development in *D. nudicaule* flowers causes the structural difference: a mature flower in the descendant species resembles an unopened (juvenile) flower of the ancestral species.

Morphological Novelties Often Arise as Modifications of Existing Structures

Sometimes a trait that is adaptive in one context is also advantageous under different circumstances. Biologists describe the original version of the trait as an **exaptation** (formerly called a *preadaptation*). Natural selection may then exaggerate the trait or modify it altogether to enhance its new function. Exaptations are just lucky accidents; they never evolve *in anticipation* of future evolutionary needs or benefits.

The forelimbs and feathers of certain dinosaurs provide a good example of an exaptation. Shortly after the first discovery of an *Archaeopteryx* skeleton in 1861, Thomas Henry Huxley, a protégé of Darwin, developed the controversial hypothesis that birds are the living representatives of the theropods, a lineage of bipedal predatory dinosaurs (see Figure 20.13). More than a century passed before John Ostrom of Yale University provided support for Huxley's hypothesis; in 1964, he described a previously unknown theropod dinosaur, *Deinonychus antirrhopus,* with a very birdlike skeleton. But *Archaeopteryx*

FIGURE 23.23 **Observational Research**

Paedomorphosis in *Delphinium* Flowers

Hypothesis: The narrow tubular shape of the flowers of *Delphinium nudicaule,* which are pollinated by hummingbirds, is the product of paedomorphosis, the retention of juvenile characteristics in a reproductive adult.

Prediction: The flowers of *D. nudicaule* grow more slowly and mature at an earlier stage of development than those of *Delphinium decorum,* a species with broad, open flowers that are pollinated by bees.

D. decorum

Gary Head

D. nudicaule

Gary Head

Method: Edward O. Guerrant of the University of California, Berkeley, measured 42 bud and flower characteristics in *D. nudicaule* and *D. decorum* as their flowers developed and used the number of days since the completion of meiosis in pollen grains as a measure of flower maturity. He then used a complex statistical analysis to compare the characteristics of the buds and flowers of both species.

Results: The mature flowers of *D. nudicaule* resemble the buds of both species more closely than they resemble the flowers of *D. decorum.* Although the time required for maturation of the reproductive structures is similar in the two species, the rate of petal growth (measured as petal blade length) is slower in *D. nudicaule.* As a result, the mature flowers of *D. nudicaule* do not open as widely as those of *D. decorum.* Because of these morphological differences, bees can pollinate flowers of *D. decorum,* but they cannot land on the flowers of *D. nudicaule,* which are instead pollinated by hummingbirds.

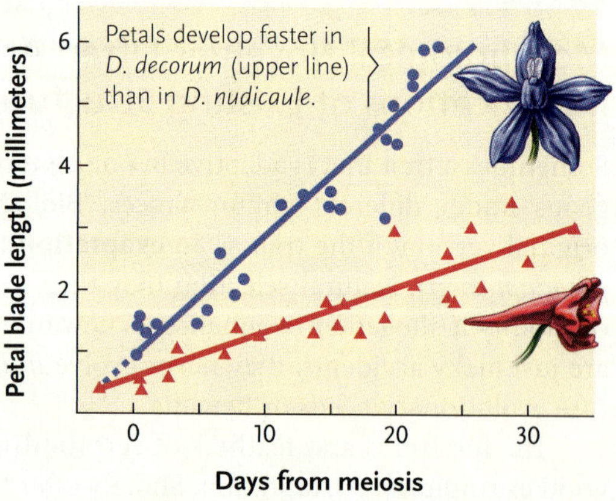

E. O. Guerrant. 1982. Neotenic evolution of *Delphinium nudicaule* (Ranunculaceae): A hummingbird-pollinated larkspur. *Evolution* 36:699–712.

Conclusion: The narrower and more tubular shape of *D. nudicaule* flowers, which mature at an earlier stage of development than *D. decorum* flowers, is the product of paedomorphosis.

THINK LIKE A SCIENTIST How might the structural difference in the flowers of *D. decorum* and *D. nudicaule* maintain reproductive isolation between the two species?

Source: E. O. Guerrant. 1982. Neotenic evolution of *Delphinium nudicaule* (Ranunculaceae): A hummingbird-pollinated larkspur. *Evolution* 36:699–712.

remained a puzzling creature because of its mix of dinosaurian characteristics—teeth, a flat breastbone, a long tail with many vertebrae, and three clawed fingers on each hand—and avian (that is, birdlike) characteristics—a furcula (wishbone), winglike forelimbs, and feathers. Although *Archaeopteryx* was the only feathered dinosaur-like fossil known at the time, scientists hypothesized that at least some of its dinosaur ancestors must have been feathered as well. Nevertheless, the known fossil record suggested that the transition from dinosaurs to feathered birds was abrupt: feathers were absent in the dinosaur ancestors of birds, but present in what biologists consider the earliest known bird.

Since the early 1990s, however, paleontologists have uncovered a wealth of feathered dinosaur fossils; most have been found in early Cretaceous deposits, dated at 124 million years ago, in northern China. The fossilized dinosaur feathers range from filamentous or downy structures that covered their bodies to long stiff feathers on their forelimbs and tails. One dinosaur species, *Microraptor gui,* had well-developed feathers on all four of its limbs **(Figure 23.24).** Research has confirmed that the feathers in one fossil included β-keratin, the protein found in the feathers of modern birds. Based on this anatomical and biochemical evidence, biologists hypothesize that feathers arose as modifications of ancestral scales in the skin of early dinosaurs. Thus, feathers are not unique to birds. They appeared in bird ancestors, perhaps as a form of insulation or as structures used for social communication; birds merely retain this ancestral characteristic.

But flight requires wings as well as feathers. How did wings and flapping flight evolve from the forelimbs of bipedal dinosaurs? Scientists have proposed two hypotheses to explain the origin of flight in birds, both of which view feathered forelimbs as exaptations. According to the "from the ground up" hypothesis, the dinosaur ancestors of birds were bipedal predators that chased prey on the ground. If they jumped at flying insect prey and used their feathered forelimbs as "fly-swatters," natural selection may have enlarged the forelimbs—along with their muscles and feathers—in stages, first to improve jumping ability and later to power flapping flight. The alternative "from the trees down" hypothesis argues that bird ancestors were at least partially arboreal. If they leapt from limb to limb or from tree to tree to chase prey or escape from predators, feathered forelimbs may have allowed them to glide and extend their jumps. Natural selec-

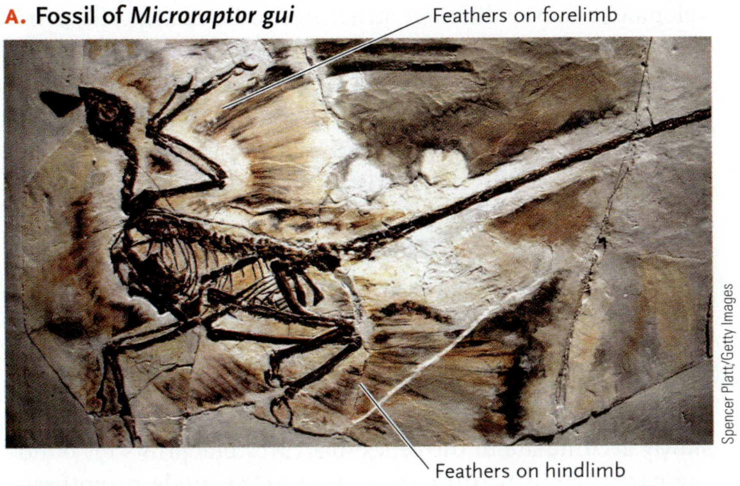

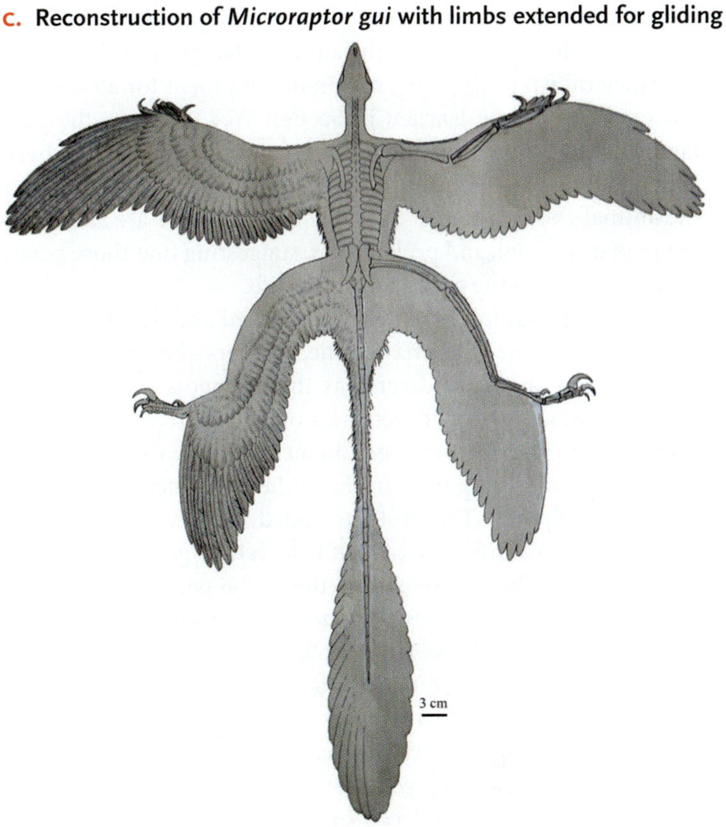

FIGURE 23.24 Feathered dinosaurs. *Microraptor gui* is one of many fossils of feathered dinosaurs that researchers have found in China since the mid-1990s. The reconstruction shows how this species may have looked with its four limbs extended for gliding flight. The scale bar in part A is 6 cm.

tion might then have modified these structures and behaviors to improve gliding ability and eventually allow flapping flight.

Evolutionary Developmental Biology May Explain the Sudden Appearance of Some Morphological Novelties

During much of the twentieth century, the evolutionary biologists who studied morphological novelties in extinct and living organisms worked independently of the scientists studying embryonic development. As a result, evolutionary biologists were unable to construct a coherent picture of the specific developmental mechanisms that contributed to morphological innovations. Since the late 1980s, however, advances in molecular genetics have allowed scientists to explore the genomes of organisms in great detail, fostering a new approach to these studies. **Evolutionary developmental biology**—evo-devo, for short—asks how evolutionary changes in the genes regulating embryonic development can lead to the morphological changes that foster adaptive radiations, increasing biodiversity over geological time.

In the life cycle of a multicellular organism, the many different body parts of the adult develop in a highly controlled sequence of steps that is specified by genetic instructions in the single cell of a fertilized egg. Developmental biologists study how regulatory genes control the development of phenotypes and their variations. As described in Section 16.4, **homeotic genes** (described further for plants in Chapter 36) code for transcription factors that bind regulatory sites on DNA, either activating or repressing the expression of other genes that contribute to an organism's form. In the remainder of this chapter, we describe a few intriguing discoveries about the genetic mechanisms that underlie some macroevolutionary changes in animals. You have already encountered one example when you read about the evolution of limblessness in snakes (Section 20.3).

Most Animals Share the Genetic Tool Kit That Regulates Their Development

Comparisons of genome sequence data reveal that most animals, regardless of their complexity or position in the Tree of Life, share a set of several hundred homeotic genes that control their development. Collectively, these genes have been dubbed the "genetic tool kit," because they govern the basic design of the body plan by controlling the activity of thousands of other genes. Some of the tool-kit genes must be at least 500 million years old, because all living animals inherited them from a common ancestor alive at that time. Not surprisingly, their structure does not differ much among the animals that possess them, and they generally play the same role in development for all species. For example, as you learned in Section 16.4, genes in the *Hox* family control the overall body plan of animals, including where appendages—wings in flies and legs in mice—will develop on the animal's body. Some of the same tool-kit genes are also present in plants, fungi, and prokaryotes, suggesting that those genes may date back to the earliest forms of life.

Another example of a highly conserved and widely distributed tool-kit gene, the *Pax-6* gene, triggers the formation of light-sensing organs as diverse as the eye spots in flatworms, the compound eyes of insects and other arthropods, and the camera eyes of vertebrates (see Chapter 41). Like the *Hox* genes, *Pax-6* also codes for a protein that either activates or represses gene transcription. The proteins coded by *Pax-6* in different animals are so similar that when researchers genetically engineered fruit fly larvae to express the *Pax-6* gene taken from a squid or a mouse, the flies responded by developing eyes. The induced eyes were, however, fruit fly eyes—not squid eyes or mouse eyes. Thus, *Pax-6* triggers activity in the fruit fly genes that carry the specific instructions for making an eye typical of that species. Apparently, the ancient genetic sequence for *Pax-6*, the master regulatory gene for eye development, has been conserved over the hundreds of millions of years since the common ancestor of squids, fruit flies, and mice lived.

Evolutionary Changes in Developmental Switches May Account for Much Evolutionary Change

If most animals share the same tool-kit genes, how has evolution produced different body plans among species? What makes a squid, a fruit fly, and a mouse different? Researchers in evo-devo have proposed that morphological differences among species arise when mutations alter the effects of developmental regulatory genes. As described in Chapter 16, the developmental programs of animals involve complex networks of many interacting genes. Varying combinations of tool-kit genes may be expressed at different times and in different body regions. According to this hypothesis, the several hundred tool-kit genes encode proteins that work as either activators or repressors in a multitude of possible combinations. Thus, they can generate an unimaginably large number of different gene expression patterns, each with the potential to alter morphology.

Sean B. Carroll of the Howard Hughes Medical Institute has described the regulatory sites that transcription factors can bind as *switches*, like those we use to turn lights on or off. When a combination of transcription factors turns on a regulatory switch, a gene further downstream (that is, one that is expressed later in time) is activated. When transcription factors turn off a regulatory switch, a downstream gene is inactivated.

Although all the cells in an animal contain exactly the same set of genes, the differential expression of genes in different body regions and at different times during embryonic development causes different structures to be made. Allometric growth can result from evolutionary changes in developmental switches that cause certain body parts to grow faster or larger than others. Similarly, heterochrony can be explained as an evolutionary change in the switches that either delays the development of adult characteristics or speeds up the development of reproductive maturity.

If Carroll's hypothesis is correct, morphological novelties arise when evolutionary changes in developmental switches alter the expression patterns of *existing* genes. This view contrasts markedly with the explanation proposed in the modern synthesis (the unified theory of evolution described in Chapter 20), that most morphological novelties arise as mutations that slowly accumulate in the genes that carry blueprints for building particular structures. According to the modern synthesis, the accumulated mutations eventually create *new* genes that specify the creation of new structures.

Although Carroll's hypothesis proposes a different genetic mechanisms as the basis for evolutionary changes in morphology, mutations in developmental regulatory genes and their effects on morphology are subject to the action of the same microevolutionary processes—natural selection, genetic drift, and gene flow—that influence the frequencies of genotypes and phenotypes in populations. Thus, every morphological change induced by a mutation in a homeotic gene or developmental switch is tested by the success or failure of the individual that carries it.

Numerous studies have shown that changes in the expression of homeotic genes can have dramatic effects on morphology. *Insights from the Molecular Revolution* explains how a change in the timing of the expression of *Hox* genes may have changed the fins of fishes into the limbs of four-legged vertebrates.

In another example, researchers have determined how an adaptive morphological change in a small fish, the three-spined stickleback *(Gasterosteus aculeatus),* results from the deactivation of a homeotic gene. The freshwater stickleback populations in North American lakes are the descendants of marine ancestors that colonized the lakes after the retreat of glaciers between 10,000 and 20,000 years ago. Marine sticklebacks have bony armor along their sides and prominent spines; lake-dwelling sticklebacks have greatly reduced armor and, in many populations, lack spines on their pelvic fins **(Figure 23.25).**

Natural selection has apparently fostered these morphological differences in response to the dominant predators in each habitat. In marine environments, long spines prevent some predatory fishes from swallowing sticklebacks. But long spines are a liability in lakes, where voracious dragonfly larvae grab sticklebacks by their spines and devour them; freshwater sticklebacks that lack spines are more likely to escape from their clutches.

In a paper published in 2004, Michael D. Shapiro of the Stanford University School of Medicine and the Howard Hughes Medical Institute and colleagues at several other institutions described how the presence or absence of spines on the pelvic fins of these fishes is governed by the expression of the

INSIGHTS FROM THE Molecular Revolution

Fancy Footwork from Fins to Fingers

Both fishes and tetrapods (four-footed animals) have two pairs of appendages, one anterior and one posterior. In both groups the appendages develop from buds of mesoderm—the middle of the three primary embryonic tissue layers—that thicken by increased cell division. As the buds elongate, cartilage is deposited at localized centers, where bones later form.

In fishes, the bones that support the fin develop along a central axis from the base to the tip of the fin. Like fins, tetrapod limbs have a central axis, but they also have distal structures, the digits (that is, fingers and toes). In tetrapods, centers of cartilage formation generate both the long bones of the limb and the five digits of the foot.

Groups of homeobox *(Hox)* genes control limb development in all animals with paired anterior and posterior appendages. Researchers have studied the expression of the *Hoxd* genes by using a nucleic acid probe that can pair with mRNA products of the genes, resulting in the generation of a color. The color is visualized under the microscope in embryos at different stages of development to determine where and when gene expression occurs. For tetrapods, genes at the 5′ end of the *Hoxd* cluster—designated 5′ *Hoxd* genes—are expressed in two distinct phases (**Figure,** part A). First, gene activity is restricted to the posterior half of the limb; this period of activity corresponds to development of the long limb bones. Second, the 5′ *Hoxd* genes became active in a band of cells perpendicular to the central axis; the cartilage centers that form the bones of the digits develop in this anterior-posterior band.

In 1995 Paolo Sordino, Frank van der Hoeven, and Denis Duboule at the University of Geneva in Switzerland published a study on *Hoxd* expression in the zebrafish *Danio rerio* (a common aquarium fish and a model research organism). They found that only a single phase of 5′ *Hoxd* gene activity occurs during zebrafish fin development (**Figure,** part B). Because the second phase of 5′ *Hoxd* gene activity is apparently absent in zebrafishes, they concluded that tetrapod digits must be an entirely novel evolutionary structure fostered by the addition of the second phase of 5′ *Hoxd* gene activity.

Research Question

Is the pattern of 5′ *Hoxd* gene expression seen in zebrafish typical of all fishes? That is, are tetrapods the only vertebrates that have a second phase of 5′ *Hoxd* gene expression?

Experiment

A number of research groups have carried out experiments to answer the question. One study, published in 2007 by Renata Freitas, GuangJun Zhang, and Martin Cohn at the University of Florida, Gainesville, examined the pattern of 5′ *Hoxd* gene expression in developing catshark *(Scyliorhinus canicula)* fins. Catsharks are included in an evolutionary lineage (Chondrichthyes, fishes with cartilaginous skeletons) that predates the lineage that includes zebrafishes (Actinopterygii, ray-finned fishes). Thus, data on fin development in catsharks may portray a more ancient developmental pattern than the one observed in zebrafishes.

Results

Their results showed that in the early stages of catshark fin development, the 5′ *Hoxd* genes are expressed in a phase like that seen during fin development in zebrafish and in the first phase of limb development in tetrapods. Unexpectedly, the researchers observed a *second* phase of expression of two of the 5′ *Hoxd* genes along the distal margin of the catshark fin buds (**Figure,** part C). Strikingly, that second phase of expression is similar to the second phase of 5′ *Hoxd* gene expression in tetrapod limb development.

Conclusion

The second phase of 5′ *Hoxd* gene expression is not uniquely associated with tetrapod limb development, as the earlier zebrafish data had suggested. Indeed, the more recent research suggests that two phases of 5′ *Hoxd* gene expression were present in the common ancestor of sharks, ray-finned fishes, and tetrapods. The authors of the later article hypothesized that digits may have arisen if the *Hoxd* genes in the distal parts of the limb were expressed for a longer period of development in tetrapods than in catsharks. The loss of the second phase of 5′ *Hoxd* gene expression in the zebrafish and their relatives may explain why most ray-finned fishes have a fin skeleton that is much smaller than that of catsharks and their relatives.

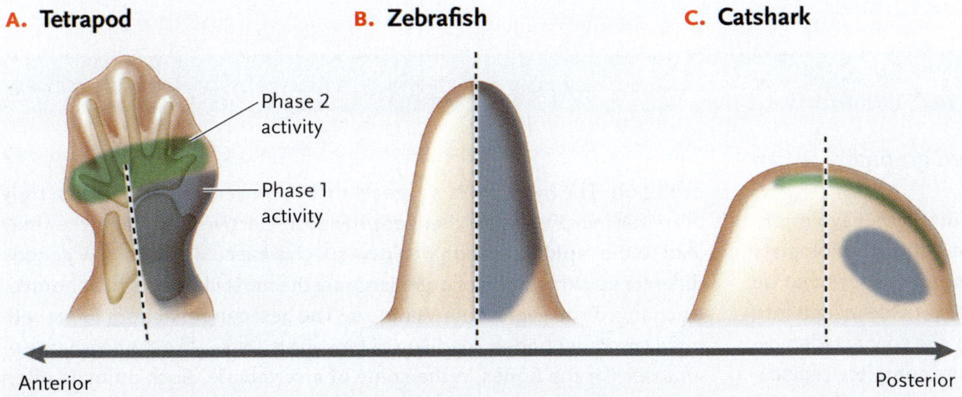

During development of the limb and digits in tetrapods, *Hoxd* genes first become active in cells on the posterior side of the limb (blue). Later, these genes are active in a broad band of cells perpendicular to the central axis of the limb (green).

During development of the fin in zebrafish, *Hoxd* genes become active in cells only on the posterior side of the developing fin (blue). Zebrafish exhibit no second phase of *Hoxd* activity in the developing fin.

During development of the fin in catsharks, *Hoxd* genes first become active in cells on the posterior side of the developing fin (blue). Later these genes become active in a very narrow band of cells near the edge of the developing fin (green).

THINK LIKE A SCIENTIST The research described above suggests that *Hoxd* genes are expressed for a longer period of time during the development of the limbs in tetrapods than during the development of fins in ray-finned fishes. What general phenomenon does this example at the molecular level illustrate?

Sources: P. Sordino, F. van der Hoeven, and D. DuBoule. 1995. *Hox* gene expression in teleost fins and the origin of vertebrate digits. *Nature* 375:678–681; R. Freitas, G. Zhang, and M. J. Cohn. 2007. Biphasic *Hoxd* gene expression in shark paired fins reveals an ancient origin of the distal limb domain. *PLoS ONE* 2:e754.

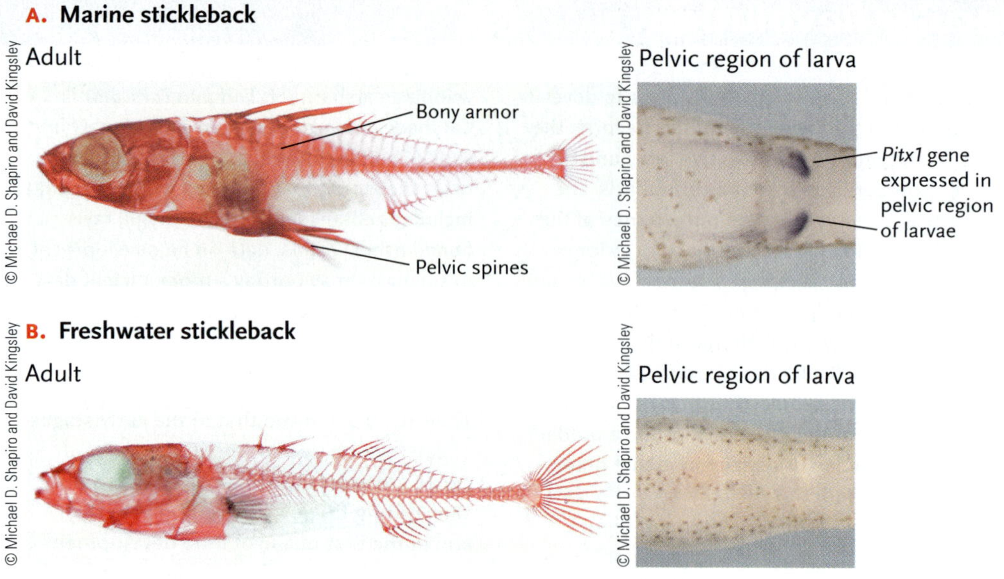

FIGURE 23.25 Sticklebacks with and without pelvic spines. **(A)** Marine populations of three-spined sticklebacks *(Gasterosteus aculeatus)* have prominent bony plates along their sides and large spines on their dorsal and pelvic fins. The growth of pelvic spines is induced by the expression of the *Pitx1* gene (the purple crescents in the photo on the right) in the pelvic region during embryonic development. **(B)** Many freshwater populations of the same species lack the bony plates and spines. Pelvic spines do not develop in the freshwater sticklebacks because the *Pitx1* gene is not expressed in the pelvic region. Natural selection has apparently fostered these morphological differences in response to the dominant predators in each environment. The skeletons of these specimens, each about 8 cm long, were dyed bright red.

gene *Pitx1*. Pelvic spines are part of the pelvic fin skeleton, the fishes' equivalent of a hind limb. In fact, *Pitx1* also contributes to the development of hind limbs in four-legged vertebrates and to certain glands and sensory organs in the head.

In long-spined marine sticklebacks, *Pitx1* is expressed in the embryonic buds from which pelvic fins develop, promoting the development of spines. But *Pitx1* is not expressed in the fin buds of freshwater sticklebacks; hence, pelvic spines do not develop. However, freshwater sticklebacks have not *lost* the *Pitx1* gene; it is still expressed elsewhere in the fishes' bodies. In a follow-up paper, published in 2009, the Shapiro team reported that the failure of freshwater sticklebacks to express the *Pitx1* gene in the developing pelvic region results from the deletion of a tissue-specific enhancer of hind limb expression; without this enhancer, the *Pitx1* gene is not expressed, and the fishes do not grow pelvic spines.

In the next chapter, we examine how biologists explore the evolutionary relationships among species and how they organize that information into a useful framework for researchers in every biological discipline. In the following unit, we will revisit evo-devo and examine some recent discoveries about how changes in homeotic genes have diversified body plans in the major evolutionary groups of organisms.

STUDY BREAK 23.6

1. **What is the difference between allometry and heterochrony?**
2. **What evidence suggests that many developmental control genes have been conserved by evolution?**
3. **What genetic factor is apparently responsible for the presence or absence of spines in stickleback fish?**

UNANSWERED QUESTIONS

Does morphological evolution always proceed gradually or can it occur in great leaps and bounds?

As you read in this chapter, biologists disagree about whether evolutionary changes in morphology can occur very rapidly. Although biologists have proposed various hypotheses to explain the abrupt changes that we sometimes find in the fossil record, evo-devo studies provide insight into one mechanism for how dramatic changes can arise: the spatial redeployment of homeotic genes. *Homeosis* is defined as the complete replacement of one type of organ with another. In one famous example, a *Hox* gene mutation in *Drosophila* replaces the antennae with legs. If such a mutation were to occur in nature, the organism would probably not have a selective advantage. But what if it did? These kinds of mutant phenotypes first inspired Richard Goldschmidt to develop his idea of the "hopeful monster" early in the twentieth century.

Stephen Jay Gould later revised and updated this idea in the context of his punctuated equilibrium hypothesis about the tempo and mode of evolution. The hypothesis suggests that if, on very rare occasions, truly dramatic morphological changes provide a selective advantage, they may lead to the rapid formation of a new species based on only a few genetic differences. What types of organisms are the most likely to exhibit homeotic change in an evolutionary context? The best candidates are those with highly modular bodies made up of repeating units—like the segments of an insect or the bones in the spine of a vertebrate. Such animals often express different organ identities in different modular units—such as the antennae, claws, and legs of a lobster—and these identities may be redeployed to different positions along the body axis. Plants are among the most modular organisms on Earth. They produce serially repeated structures—a leaf, a bud at the base of the leaf, and a stem—to generate their bodies. Many exciting and promising questions in plant evo-devo relate to how evolutionary homeosis may have generated rapid change in plant morphology.

Has homeosis contributed to the appearance and diversification of the angiosperms?

The sudden appearance of flowering plants, the angiosperms, in the fossil record so puzzled Charles Darwin that he dubbed their evolution an "abominable mystery." How did the gymnosperms, which always bear their male and female reproductive structures separately, give rise to the hermaphroditic (that is, bearing both male and female structures) flower? Our current understanding of the genetics of floral developmental provides a simple solution to this puzzle: the genetic program controlling floral organ identity is homeotic. Thus, it is possible for very simple genetic changes to transform an entirely male set of reproductive organs into a combination of male and female parts. Such models have been outlined by Günter Theissen at the Friedrich-Schiller-Universität Jena in Germany and by David Baum and Lena Hileman at the University of Wisconsin–Madison and the University of Kansas, respectively. In addition to fostering the origin of the angiosperms, homeosis may have played a role in the group's diversification. Commonly observed shifts in the morphology of sepals and petals or in the number of stamens (male reproductive structures) are suggestive of homeotic changes. These examples are more suitable for experimental verification than is the question on the origin of the angiosperms because they are much more recent occurrences. Although these hypotheses are very attractive, they remain to be confirmed through molecular genetic analyses.

THINK LIKE A SCIENTIST One point of disagreement among evolutionary biologists is whether evolutionary changes primarily occur in gene regulatory regions or in the coding regions. Which of these two types of changes do you think is more likely with homeotic evolution?

Courtesy of Elena M. Kramer

Elena M. Kramer is an Associate Professor of Organismic and Evolutionary Biology at Harvard University, where she studies the evolution of the genetic mechanisms controlling floral development. To learn more about Dr. Kramer's research, go to http://www.oeb.harvard.edu/faculty/kramer/Site/Home.html.

REVIEW KEY CONCEPTS

To access the course materials and companion resources for this text, please visit www.cengagebrain.com.

23.1 The Fossil Record

- Fossils are the parts of organisms preserved in sedimentary rocks or in oxygen-poor environments (Figures 23.2 and 23.3).
- The fossil record is incomplete because few organisms fossilize completely, because some organisms are more likely to fossilize than others, and because natural processes destroy many fossils.
- Fossils provide a relative dating system, the geological time scale, for the strata in which they occur. Radiometric dating techniques establish the absolute age of rocks and fossils (Figures 23.4 and 23.5, Table 23.1).
- The fossil record provides data on changes in morphology, biogeography, ecology, and the behavior of organisms (Figure 23.6).

Animation: Radiometric dating

Animation: Geologic time scale

Animation: Radioisotope decay

23.2 Earth History

- Earth's crust is composed of plates of solid rock that float on a semisolid mantle (Figure 23.7). New crust is constantly generated and old crust is recycled. Currents in the mantle cause the continents to move over geological time (Figure 23.8). The breakup of the continents profoundly influenced biological evolution by separating populations and creating more shallow marine habitats.
- Continental movements caused variations in climate, the extent of glaciation, and sea levels (Figure 23.9). Asteroid impacts and volcanic eruptions have also influenced Earth's environment, sometimes triggering large-scale extinctions.

Animation: Plate margins

23.3 Historical Biogeography and Convergent Biotas

- Disjunct distributions of species can be produced by dispersal and/or vicariance (Figure 23.10). Dispersal results in a disjunct distribution when a new population is established on the far side of a barrier. Vicariance results in a disjunct distribution when external factors such as continental drift fragment the landscape.
- Continental drift has created six major biogeographical realms, each with a characteristic biota (Figure 23.11).
- Convergent evolution produces similar adaptations in distantly related species that live in similar environments (Figures 23.12 and 23.13).

23.4 The History of Biodiversity

- Adaptive radiation produces morphologically diverse species within lineages (Figure 23.14), increasing biodiversity.
- Extinction decreases species diversity. Mass extinctions have occurred at least five times in the history of life (Figure 23.15). Tectonic activity, climate change, and asteroid strikes are probable causes of mass extinctions.
- Biodiversity has increased repeatedly since life first evolved, partly in response to increased geographical separation of the continents and partly because complex interactions evolve among existing species (Figures 23.16 and 23.17).

Animation: Five major extinctions

Animation: Animal evolution in Phyla

23.5 Interpreting Evolutionary Lineages

- The horse lineage is complex and highly branched. It includes species of various sizes and diverse morphological adaptations (Figure 23.18).
- The tempo of morphological change varies among lineages. Some exhibit a pattern of gradual change through time (Figure 23.19). Other lineages confirm the predictions of the punctuated equilibrium hypothesis (Figure 23.20).

Animation: Evolution of Horses

23.6 The Evolution of Morphological Novelties

- Morphological novelties can arise from evolutionary changes in the relative growth rates of body parts (Figure 23.21) or in the timing of developmental events (Figures 23.22 and 23.23).
- Morphological novelties are often modifications of existing structures. An exaptation is a trait that turns out to be useful in a new environmental context even before natural selection refines its form. For example, the feathers and wings of birds evolved from the ancestral feathers and elongated forelimbs of some bipedal dinosaurs (Figure 23.24).
- Evolutionary developmental biology—evo-devo—examines how evolutionary changes in genes that regulate embryonic development can foster changes in body shape and form.
- Most organisms share an ancient tool kit of several hundred genes that regulate the expression of the thousands of genes involved in development. Evolutionary changes in developmental switches may account for many morphological changes. The differential expression of the *Pitx1* gene in sticklebacks determines whether or not a fish grows pelvic spines (Figure 23.25).

Animation: Mutation and proportional changes

UNDERSTAND & APPLY

Test Your Knowledge

1. The fossil record:
   a. provides direct and indirect evidence about life in the past.
   b. shows that all morphological novelties arise rapidly.
   c. provides abundant data about rare species with local distributions.
   d. is equally good for all organisms that ever lived.
   e. provides no evidence about the physiology or behavior of ancient organisms.
2. The absolute age of a geological stratum is determined by:
   a. the thickness of its rocks.
   b. the particle size in its rocks.
   c. the types of fossils found within it.
   d. its position relative to other layers.
   e. radiometric dating techniques.
3. The observation that fossils of *Premedosaurus* are found only in Argentina and Northern Europe provides an example of:
   a. a continuous distribution.
   b. a disjunct distribution.
   c. punctuated equilibrium.
   d. heterochrony.
   e. gradualism.
4. The evolutionary history of horses demonstrates that:
   a. modern horses are the direct, lineal descendants of the earliest horses.
   b. the leg bones of modern horses are more complex than those of the earliest horses.
   c. horses have always had specialized teeth that allow them to feed on tough grasses.
   d. horses diversified greatly, but only a few types survived to the present.
   e. the first horses lived in open, grassy habitats.
5. The punctuated equilibrium hypothesis:
   a. recognizes that morphological evolution may occur slowly or quickly.
   b. suggests that major morphological novelties arise gradually.
   c. may help explain why there are so many "missing links" in the fossil record.
   d. suggests that the fossil record is usually complete.
   e. links mass extinctions to the impact of asteroids striking Earth.
6. Biologists believe that the overall similarities between marsupial mammals in Australia and placental mammals in North America were caused by:
   a. plate tectonics.
   b. dispersal.
   c. convergent evolution.
   d. homeotic genes.
   e. punctuated equilibria.
7. The differential growth of body parts is called:
   a. allometry.
   b. paedomorphosis.
   c. heterochrony.
   d. phyletic gradualism.
   e. evo-devo.
8. Exaptations are traits that:
   a. prepare some organisms for future environmental changes.
   b. appear in lineages as a result of an adaptive radiation.
   c. evolve in anticipation of a species' future needs.
   d. are useful in new situations before natural selection alters them for a new function.
   e. occur in animals, but not in plants.
9. Adaptive radiations often follow mass extinctions because:
   a. mass extinctions limit the impact of paedomorphosis.
   b. mass extinctions foster allometry and heterochrony.
   c. mass extinctions decimate all forms of life on Earth.
   d. species that form transitional fossils often survive mass extinctions.
   e. extinctions open adaptive zones that had been previously occupied.
10. Homeotic genes are defined as genes that:
    a. bind directly to a regulatory site on DNA.
    b. code for transcription factors activating or repressing genes that influence an organism's form.
    c. determine whether or not a morphological innovation leads to an adaptive radiation.
    d. have been inherited from an ancient ancestor by nearly all forms of life.
    e. help biologists differentiate between plants and animals.

Discuss the Concepts

1. Many millions of years from now, continental drift may obliterate the Pacific Ocean, pushing North America into physical contact with Asia. What effects might these events have on the organisms living at that time?
2. One possible measure of a species' evolutionary success is the number of descendant species it produces. Should our species, *Homo sapiens,* be considered successful under this definition?
3. Extinctions are common in the history of life. Why are biologists alarmed by the current wave of extinctions caused by human activity?

Design an Experiment

Design a study to determine whether the wings of birds, bats, and insects and their ability to fly are the products of convergent evolution.

Interpret the Data

The graphs presented below provide data on the body sizes (above) and brain sizes (below) of modern humans and three extinct species in the evolutionary lineage that includes humans. How would you describe the relative changes of body size and brain size in this lineage through time?

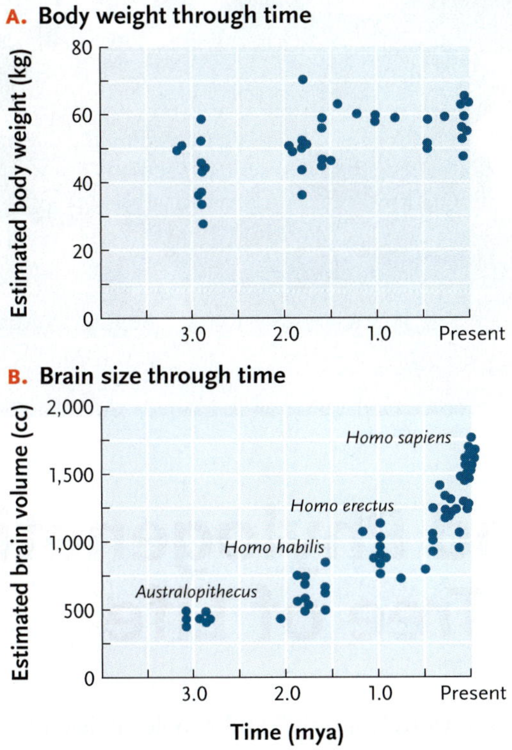

D. J. Futuyma. 2009. *Evolution*, 2e. Sinauer Associates, Sunderland, MA (after S. Jones, R. Martin, and D. Pilbeam, eds. 1992). *The Cambridge Encyclopedia of Human Evolution*. Cambridge University Press, Cambridge. Reprinted with the permission of Cambridge University Press.

Source: D. J. Futuyma. 2009. *Evolution,* 2e. Sinauer Associates, Sunderland, MA (after S. Jones, R. Martin, and D. Pilbeam, eds. 1992. *The Cambridge Encyclopedia of Human Evolution.* Cambridge University Press, Cambridge).

Apply Evolutionary Thinking

The geological evolution of Earth has had an obvious effect on biological evolution. You have read about how the release of oxygen by photosynthetic organisms increased atmospheric oxygen concentration. How are human activities changing the physical environment on Earth? What new selection pressures do these environmental alterations establish?

24

Courtesy of Dr. Marc Ducousso

Barbie pagoda fungus *(Podoserpula miranda)* discovered in 2011 in New Caledonia.

STUDY OUTLINE

24.1 Nomenclature and Classification

24.2 Phylogenetic Trees

24.3 Sources of Data for Phylogenetic Analyses

24.4 Traditional Classification and Paraphyletic Groups

24.5 The Cladistic Revolution

24.6 Phylogenetic Trees as Research Tools

24.7 Molecular Phylogenetic Analyses

Systematics and Phylogenetics: Revealing the Tree of Life

Why it matters . . . Mention the word "malaria," and people envision the tropics: explorers wander through the jungle in pith helmets and sleep under mosquito netting; clouds of insects hover nearby, ready to infect them with *Plasmodium,* the protistan parasite that causes this disease. You may be surprised to learn, however, that less than 100 years ago, malaria was also a serious threat in the southeastern United States and much of Western Europe.

Scientists puzzled over the cause of malaria for thousands of years. Hippocrates, a Greek physician who worked in the fifth century B.C., knew that people who lived near malodorous marshes often suffered from fevers and swollen spleens. Indeed, the name malaria is derived from the Latin for "bad air." By 1900, scientists had established that mosquitoes, *Plasmodium's* intermediate hosts, transmit the parasite to humans. Mosquitoes breed in standing water, and anyone living nearby is likely to suffer their bites.

Until the 1920s, scientists thought that the mosquito species *Anopheles maculipennis* carried malaria in Europe. But some areas with huge mosquito populations had little human malaria, whereas other areas had relatively few mosquitoes and a high incidence of the disease.

Then, a French researcher reported variation in the mosquitoes, and Dutch scientists identified two forms of the "species," only one of which seemed to carry malaria. The breakthrough came in 1924, when a retired public health inspector in Italy discovered that individual mosquitoes—all thought to be the same species—produced eggs with one of six distinctive surface patterns **(Figure 24.1).**

Further research revealed that the name *Anopheles maculipennis* had been applied to six separate mosquito species. Although the adults of these species are almost indistinguishable, their eggs are clearly different. The species are reproductively isolated from each other, and they differ ecologically: some breed in brackish coastal marshes, others in freshwater inland marshes, and still others in slow-moving streams. Only some of these species have a preference for human blood, and researchers eventually determined that only three of them routinely transmit malaria to humans.

These discoveries explained why the geographical distributions of mosquitoes and malaria did not always match. And government agencies could finally fight malaria by eradicating the disease-carrying species. Health workers drained marshes to prevent mosquitoes from breeding. They used insecticides to kill mosquito larvae or introduced fish of the genus *Gambusia,* the mosquitofish, which eats them. These targeted control programs were very successful in the early and middle decades of the twentieth century.

Today, with the increased mobility of humans, agricultural products, and other goods, some mosquito species—as well as many other organisms—are invading habitats made more hospitable by global climate change (discussed further in Unit Seven). Some have expanded their geographical ranges substantially: introduced mosquito species have been discovered on all continents except Antarctica. Mosquitoes carry numerous agents of disease, and biologists are now devising new ways to recognize—and eradicate—the species that pose the greatest threats to human welfare.

The historic eradication of malaria in Europe owed a debt to **systematics,** the branch of biology that studies the diversity of life and its evolutionary relationships. Systematic biologists—systematists for short—identify, describe, name, and classify organisms, organizing their observations within a framework that reflects the organisms' evolutionary relationships. In this chapter we briefly describe the traditional approach to classification. We then focus attention on how systematists working today develop hypotheses about the evolutionary relationships of all the branches, twigs, and leaves on the Tree of Life.

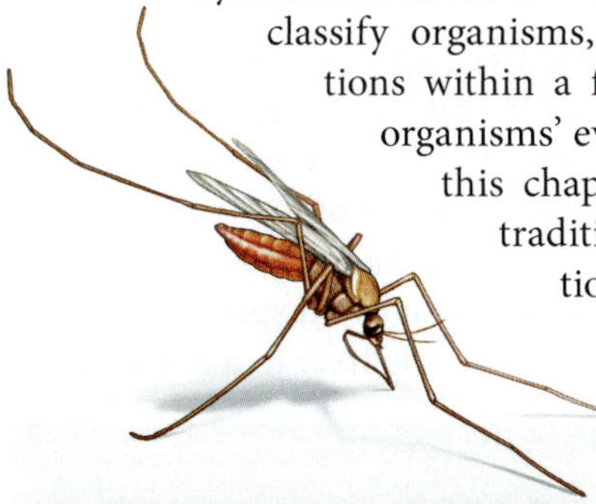

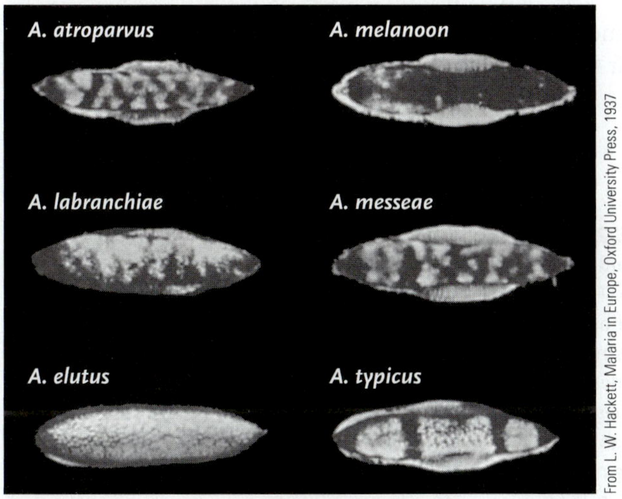

FIGURE 24.1 Eggs of European mosquitoes. Differences in surface patterns on the eggs of *Anopheles* mosquitoes in Europe helped researchers identify six separate species. The adults of all six species look remarkably alike. An adult *Anopheles atroparvus* is illustrated.

24.1 Nomenclature and Classification

The Swedish naturalist Carl von Linné (1707–1778), better known by his Latinized name, Carolus Linnaeus, was the first modern practitioner of **taxonomy,** the science that identifies, names, and classifies new species. A professor at the University of Uppsala, Linnaeus sent ill-prepared students around the world to gather specimens, losing perhaps a third of his followers to the rigors of their expeditions. Although he may not have been a commendable student adviser, Linnaeus developed the basic system of naming and classifying organisms that biologists embraced for two centuries.

Linnaeus Developed the System of Binomial Nomenclature

Linnaeus invented the system of **binomial nomenclature,** in which species are assigned a Latinized two-part name, or **binomial.** The first part of the name identifies a **genus** (plural, *genera*), a group of species with similar characteristics. The second part is the **specific epithet,** or species name. When identifying and naming a new species, Linnaeus used the morphological species concept (described in Section 22.1), assigning the same scientific name to individuals that shared anatomical characteristics.

A combination of the generic name and the specific epithet provides a unique name for every species. For example, *Ursus maritimus* is the polar bear and *Ursus arctos* is the brown bear. By convention, the first letter of a generic name is always capitalized; the specific epithet is never capitalized; and the entire binomial is italicized. In addition, the specific epithet is never used without the full or abbreviated generic name preceding it because the same specific epithet is often given to species in different genera. For instance, *Ursus americanus* is the American black bear, *Homarus americanus* is the Atlantic lobster, and *Bufo americanus* is the American toad. If you were to order just "*americanus*" for dinner, you might be dismayed when your plate arrived—unless you have an adventurous palate!

Nonscientists often use different common names to identify a species. For example, *Bothrops asper,* a poisonous snake native to Central and South America, is called *barba amarilla* (meaning "yellow beard") in some places and *cola blanca* (meaning "white tail") in others; biologists have recorded about 50 local names for this species. Adding to the confusion, the same common name is sometimes used for several different species. Binomials, however, allow people everywhere to discuss organisms unambiguously.

Many binomials are descriptive of the organism or its habitat. *Asparagus horridus,* for example, is a spiny plant. Other species, such as the South American bird *Rhea darwinii,* are named for notable biologists. The naming of newly discovered species follows a formal process of publishing a description of the species in a scientific journal. International commissions meet periodically to settle disputes about scientific names.

Linnaeus Devised the Taxonomic Hierarchy to Organize Information about Species

Linnaeus described and named thousands of species on the basis of their morphological similarities and differences. Keeping track of so many species was no easy task, so he devised a **classification,** a conceptual filing system that arranges organisms into ever more inclusive categories. Linnaeus' classification, called the **taxonomic hierarchy,** includes a nested series of formal categories: domain, kingdom, phylum, class, order, family, genus, species, and subspecies. (You may find it useful to review the traditional taxonomic categories, which were described in Section 1.3 and illustrated in Figure 1.10.) The organisms included within any category of the taxonomic hierarchy compose a **taxon** (plural, *taxa*). Woodpeckers, for example, are a taxon (Picidae) at the family level, and pine trees are a taxon *(Pinus)* at the genus level.

Species that are included in the same taxon at the bottom of the hierarchy (that is, in the same genus or family) generally share many characteristics. By contrast, species that are included in the same taxon only near the top of the hierarchy (that is, the same kingdom or phylum) generally share many fewer characteristics. The hierarchy has been a great convenience for biologists because every taxon is defined by a set of shared characteristics. Thus, when a biologist refers to a member of the family Picidae, all of his or her colleagues understand that the biologist is talking about a medium-sized bird that uses its stout bill to drill holes in tree trunks.

STUDY BREAK 24.1

1. **How does the system of binomial nomenclature minimize ambiguity in the naming and identification of species?**
2. **How does the taxonomic hierarchy help biologists organize information about different species?**

24.2 Phylogenetic Trees

Linnaeus devised the taxonomic hierarchy long before Darwin published his theory of evolution. His goals were to illuminate the details of God's creation and to devise a practical way for naturalists to keep track of their discoveries. But the science of systematics changed in response to Darwin's idea that all organisms in the Tree of Life are descended from a distant common ancestor: systematists began to focus on discovering the evolutionary relationships between groups of organisms.

Systematists Adapted Linnaeus' Approach to a Darwinian Worldview

The taxonomic hierarchy that Linnaeus defined was easily adapted to Darwin's concept of branching evolution, which is itself a hierarchical phenomenon **(Figure 24.2).** As we discussed in Chapters 22 and 23, ancestral species give rise to descendant species through repeated branching of a lineage. Organisms in the same genus generally share a fairly recent common ancestor, whereas those assigned to the same higher taxonomic category, such as a class or phylum, share a common ancestor that lived in the more distant past.

In the second half of the nineteenth century, systematists began to reconstruct the **phylogeny** (that is, the evolutionary history) of organisms. Phylogenies are illustrated in **phylogenetic trees,** which are formal hypotheses that identify likely relationships among species and higher taxonomic groups. And like all hypotheses, they are constantly revised as scientists gather new data.

A Phylogenetic Tree Depicts the Evolutionary History of a Group of Organisms

Contemporary evolutionary biologists construct phylogenetic trees to illustrate the hypothesized evolutionary history of organisms. Researchers tailor the breadth of their analyses to match specific research questions. Thus, some trees might include the evolutionary history of all known organisms, others a small cluster of closely related populations within a species, and still others a group somewhere between those extremes. Regardless of how wide a range of organisms is included, all phylogenetic trees share a specific structure and depict key relationships in similar ways **(Figure 24.3).**

For example, phylogenetic trees are usually drawn along an implicit or explicit time line. In this book, phylogenetic trees are generally depicted vertically; the most ancient organisms and evolutionary events are at the bottom of the tree (often labeled "long ago"), and the most recent are at the top (often labeled "present"). The common ancestor of all species included in the tree is described as the **root** of the tree. In a few cases in this book, trees are presented with the root on the left, with time passing from left to right.

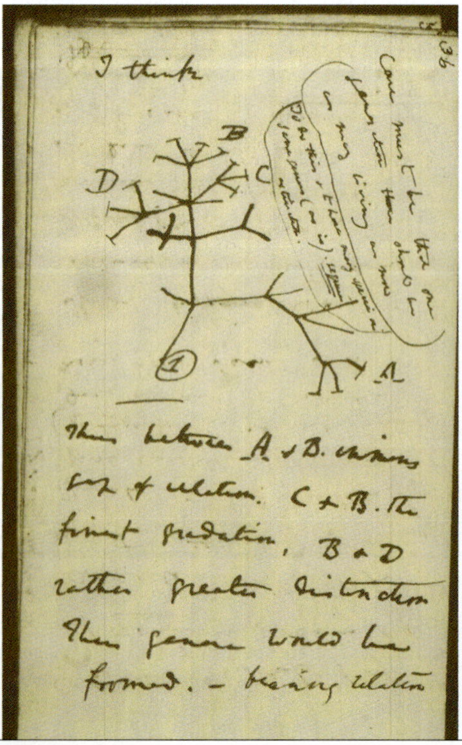

FIGURE 24.2 Darwin and branching evolution. This entry from his notebook on the "transmutation of species" demonstrates that Charles Darwin first thought about the branching pattern of evolution in 1837, more than 20 years before he published *On the Origin of Species.*

As you know from the discussion in Section 23.5, the tempo of evolution varies within and among lineages. In some cases, evolutionary changes may accumulate slowly in a lineage as the environment shifts over time. This pattern of gradual phyletic change is often described as *anagenesis*. If the changes through time are substantial and the fossil record is incomplete, paleontologists who discover morphologically distinct fossils in different strata may assign them different species names and say that "the ancestral species A evolved into the descendant species B." But the production of such "new" species by anagenesis does not increase biodiversity; rather, it is simply the gradual transformation of one "species" into another as its characteristics shifted over time. Anagenesis is often illustrated by a straight line in a phylogenetic tree.

In other circumstances, an ancestral species undergoes speciation, producing two descendant species, both of which are distinct from their common ancestor. This pattern of evolution is described as *cladogenesis*, a process that *does* increase biodiversity. Cladogenesis is depicted in a phylogenetic tree by a branching pattern, with two descendants arising from their common ancestor. When they first emerge, the two branches may represent new species. But as cladogenesis continues repeatedly through evolutionary time—with branches giving rise to branchlets and branchlets to twigs—each of those new species may become the common ancestor of its own many descendants. Thus, each new species produced by cladogenesis has the potential to become the "root" of its own evolutionary lineage.

When describing phylogenetic trees, each branching point is called a **node,** and each evolutionary lineage—a node with all of the branches, branchlets, and twigs that emerge from it—is called a **clade** (*klados* = branch). You can identify a clade on any phylogenetic tree by following a lineage from a node

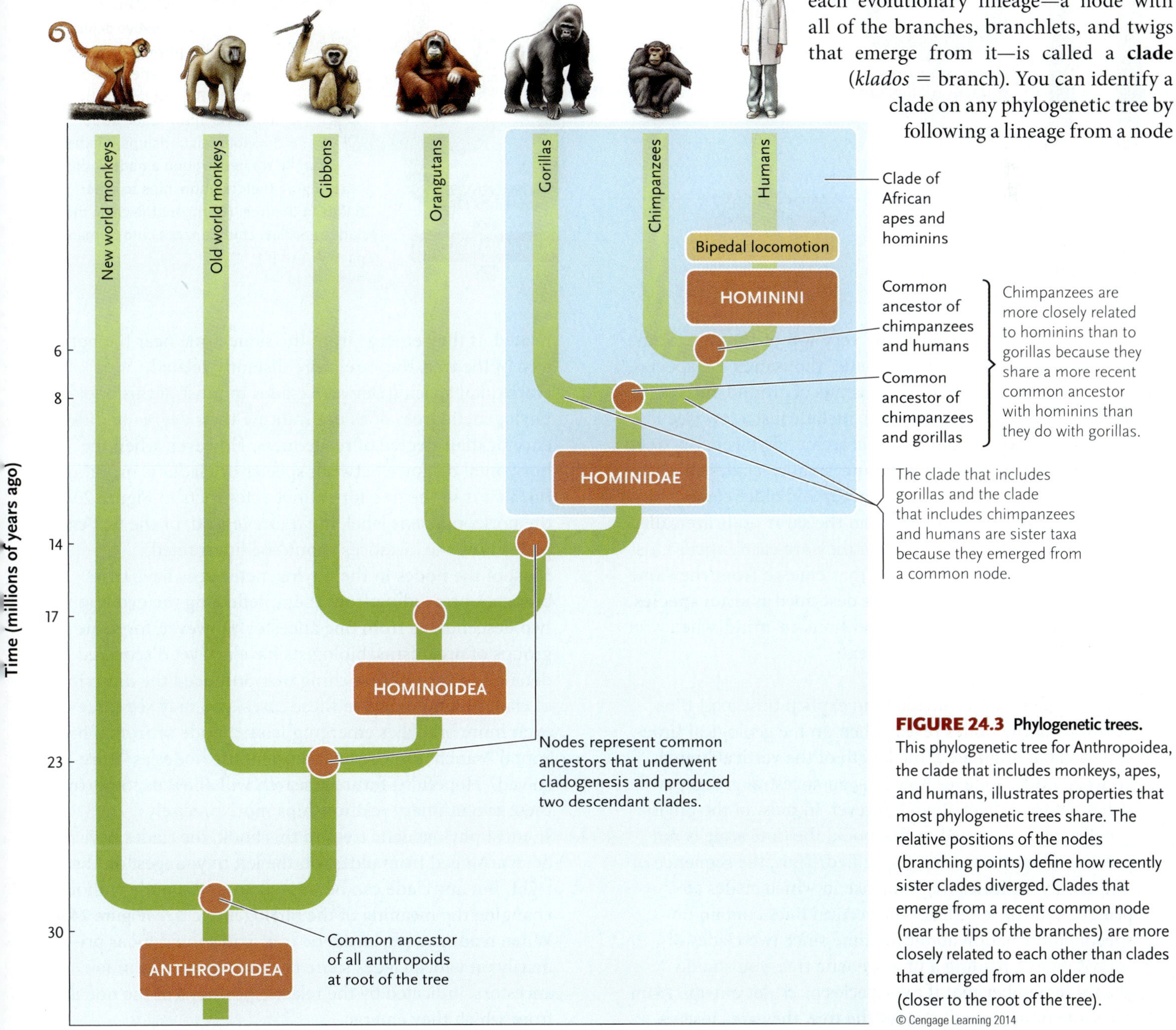

FIGURE 24.3 Phylogenetic trees. This phylogenetic tree for Anthropoidea, the clade that includes monkeys, apes, and humans, illustrates properties that most phylogenetic trees share. The relative positions of the nodes (branching points) define how recently sister clades diverged. Clades that emerge from a recent common node (near the tips of the branches) are more closely related to each other than clades that emerged from an older node (closer to the root of the tree).

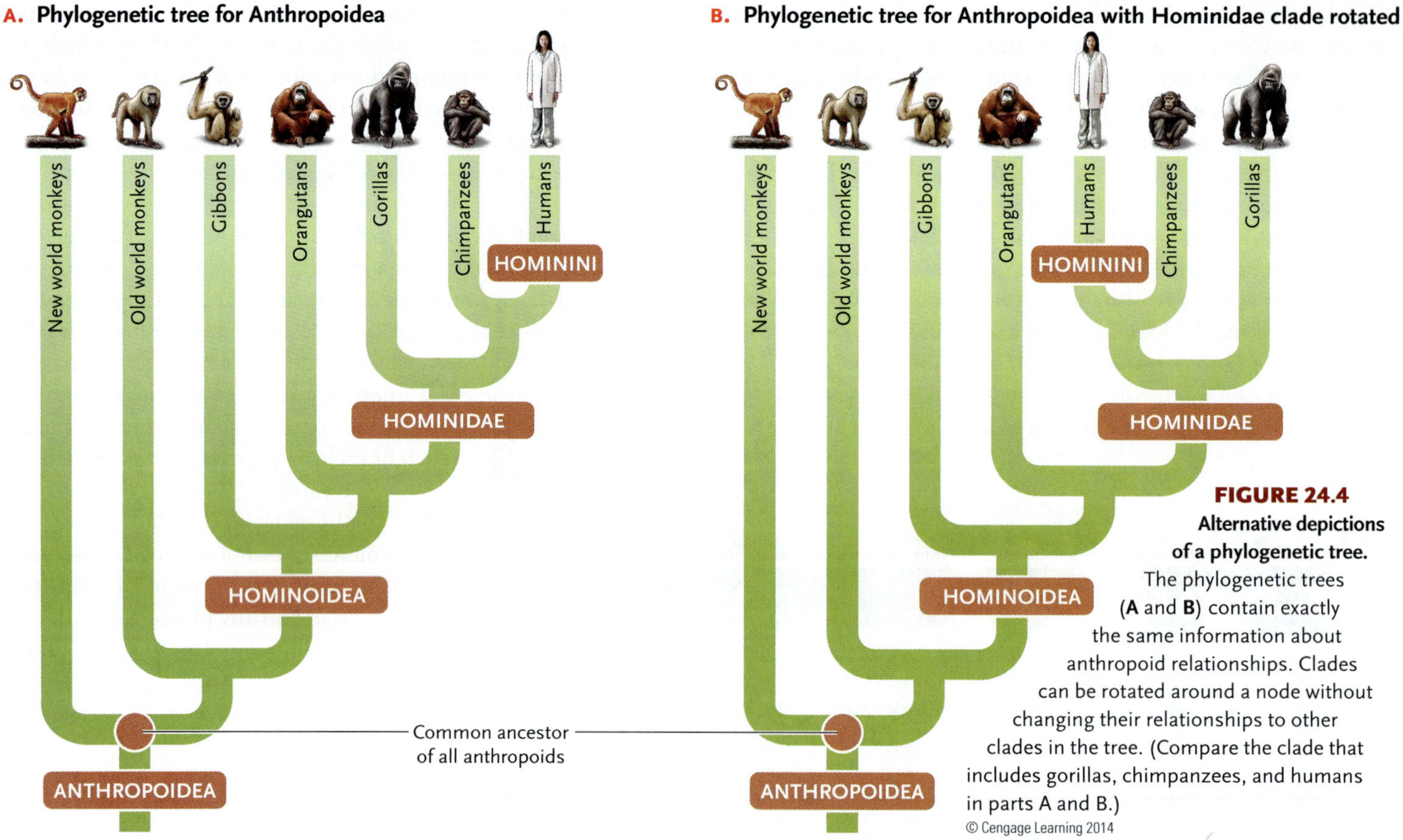

FIGURE 24.4 **Alternative depictions of a phylogenetic tree.** The phylogenetic trees (**A** and **B**) contain exactly the same information about anthropoid relationships. Clades can be rotated around a node without changing their relationships to other clades in the tree. (Compare the clade that includes gorillas, chimpanzees, and humans in parts A and B.)

to the tips of all of its twigs at the very top of the tree. Some clades, such as Aves (birds), include thousands of species, whereas others, such as *Geospiza* (a genus of ground and cactus finches from the Galápagos Islands), include just a few (see Figure 23.14). Like the taxonomic hierarchy, phylogenetic trees have a nested structure: younger and smaller clades, like the genus *Geospiza,* are nested within larger and older clades, such as Aves. Two clades that emerge from the same node are called **sister clades** (or sister taxa) because they are each other's closest relatives; similarly, two species that emerge from the same node near the very top of the tree are described as **sister species.**

You should keep certain conventions in mind when you read the phylogenetic trees in this text:

1. If a phylogenetic tree includes an explicit time axis, the positions of the nodes reveal when on the geological time scale a clade originated; the length of the vertical branch between nodes indexes how long an ancestral group persisted before it diversified. However, in most of the phylogenetic trees presented in this book, the time scale is not precise; often it is not even specified. Thus, the sequence of nodes only indicates the sequence in which clades appeared, and the lengths of the vertical lines contain no specific information about the time since two clades diverged. When reading a phylogenetic tree, you should always remember that if two species or clades emerge from the same node near the top of the tree, they are closely related. If they emerge from the same node near the bottom of the tree, they are more distantly related.
2. Horizontal spacing between clades in most of this book's phylogenetic trees does not indicate their degree of difference or their degree of relatedness. However, when the horizontal distance between species or clades is meaningful, as it is in the tree for extinct ectoprocts in Figure 23.20, the horizontal axis label, the figure legend, or the text explains how the distances should be interpreted.
3. Most of the nodes in the phylogenetic trees have two branches protruding from them, reflecting the evolution of two descendants from one ancestor. However, for some groups of organisms, biologists have not yet discovered the detailed pattern of branching that produced the diversity of clades living today. In these cases, you may see three or even more branches emerging from a node or from a horizontal branch. Biologists describe these nodes as "unresolved." Hopefully, future research will allow us to portray these evolutionary relationships more precisely.
4. In most phylogenetic trees in this book, the clades have been arranged from oldest on the left to youngest on the right. But any clade can be rotated around a node without changing the meaning of the phylogenetic tree **(Figure 24.4).** When reading a phylogenetic tree, you should focus primarily on which clades share the more recent common ancestors, indicated by the relative positions of the nodes from which they emerge.

Phylogenetic Trees Allow Biologists to Define Evolutionary Classifications

Evolutionary biologists working today want a classification that mirrors the branching patterns of a group's phylogenetic history. When converting a phylogenetic tree into a classification, they try to identify only **monophyletic taxa** or lineages. A monophyletic taxon comprises one clade—an ancestral species (represented by a node) and *all* of its descendants, but no other species **(Figure 24.5)**. Monophyletic taxa are defined at every level of the taxonomic hierarchy. For example, biologists consider the Felidae (the traditional family-level taxon that includes all cat species) to be monophyletic: all cat species living on Earth today, from housecats to tigers, are the descendants—and the only living descendants—of a common ancestor that lived about 25 million years ago. Thus, the Felidae is one small, but complete, branch on the Tree of Life. At a much broader scale, the Animalia is a monophyletic taxon (at the traditional kingdom level) that comprises all animals, which are the descendants of one common ancestor. Even if biologists have not yet identified the very first animal in the fossil record, they can infer its existence at the root of the phylogenetic tree for Animalia.

Biologists have not always defined strictly monophyletic taxa. Because of missing data, or as a matter of convenience, they sometimes named taxa that included species from different clades or taxa that included some, but not all, of an ancestral species' descendants. A **polyphyletic taxon** is one that includes organisms from different clades, but not their common ancestor. For example, a taxon that included only birds and bats, two clades of vertebrates that are capable of flight, would be considered polyphyletic because it would not include their last common ancestor, which lived many millions of years before birds and bats first appeared. A **paraphyletic taxon** is one that includes an ancestor and some, *but not all,* of its descendants. For example, people commonly used to define terrestrial dinosaurs and birds as distinct groups. But "terrestrial dinosaurs" was a paraphyletic taxon: as you know from Section 23.6, birds are the descendants of one group of terrestrial dinosaurs. Thus, the monophyletic taxon Dinosauria must include birds as well as their nonflying relatives.

Many of the phylogenetic trees included in this book identify monophyletic taxa with labels on their branches. The names of these clades are identified in capital letters in a bronze-colored plaque; the major characteristics that define major clades are identified in lowercase letters in a tan-colored plaque. As a convenience, some commonly used names for groups that biologists now recognize as paraphyletic are presented in quotation marks in bronze plaques with dashed, rather than solid, borders.

FIGURE 24.5 Defining taxa in a classification. Although systematists can identify different groups of species in a phylogenetic tree as a taxon, the taxon highlighted in blue on the left is the only monophyletic group.

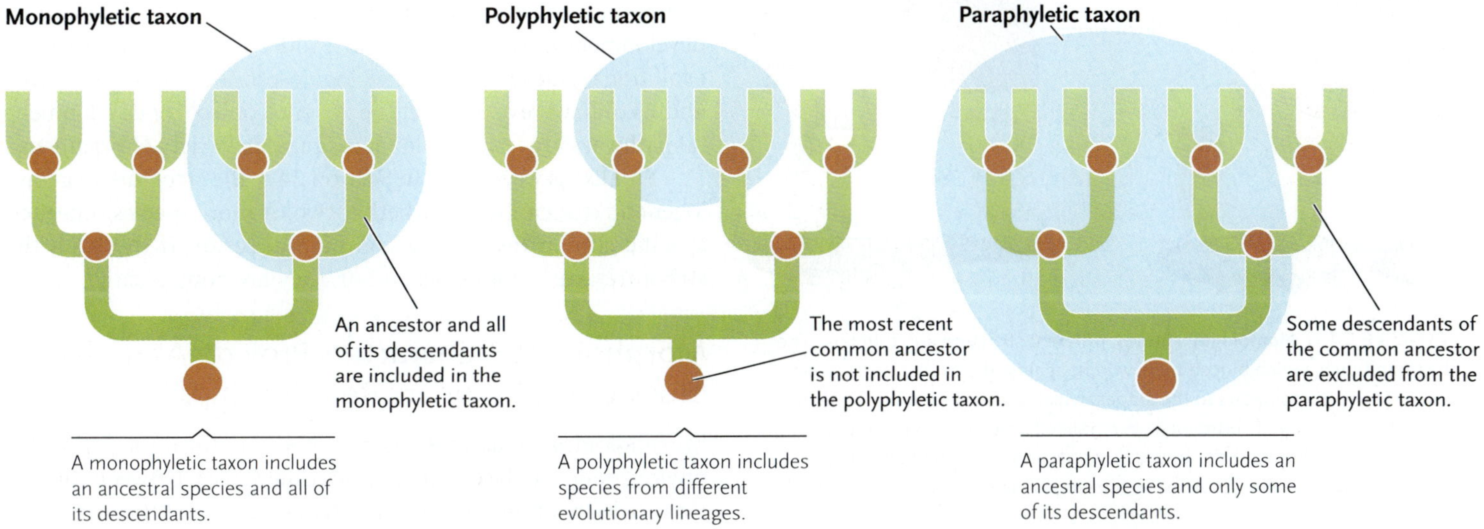

STUDY BREAK 24.2

1. **What is the difference between a phylogenetic tree and a classification?**
2. **What is the difference between a monophyletic taxon, a polyphyletic taxon, and a paraphyletic taxon?**

24.3 Sources of Data for Phylogenetic Analyses

Linnaeus classified organisms on the basis of their morphological similarities and differences—even though he did not understand how those characteristics arose. For example, he defined birds as a class of oviparous ("egg-laying") animals with feathered bodies, two wings, two feet, and a bony beak. No other animals possess all these characteristics, which distinguish birds from "quadrupeds" (his term for mammals), "amphibians" (among which he included reptiles), fishes, insects, and "worms."

Mendel's subsequent work on the heritable basis of morphological variations provided the scientific rationale for this endeavor: modern systematists infer that morphological differences serve as indicators of underlying genetic differences between species and lineages. Today, with our much deeper understanding of the genetic basis of variation, systematists undertake phylogenetic analyses using a variety of organismal and molecular characters. Indeed, any heritable trait (that is, any trait with a genetic basis) that is intrinsic to the organism can be used in a phylogenetic analysis; phenotypic differences caused by environmental variation (see Section 20.1) are excluded. In this section we first consider a general criterion for evaluating characters and then examine examples of how a few specific types of characters are useful in this effort.

The Analysis of Homologous Characters Sheds Light on Evolutionary Relationships

A basic premise of phylogenetic analyses is that phenotypic similarities between organisms reflect their underlying genetic similarities. As you may recall from Figure 20.3, species that are morphologically similar have often inherited the genetic basis of their resemblance from a common ancestor. Similarities that result from shared ancestry, such as the four limbs of all tetrapod vertebrates, are called **homologies** (or homologous characters). Any trait, from genetic sequences to anatomical structures to mating behaviors, can be described as homologous in two or more species as long as they inherited the trait from their common ancestor.

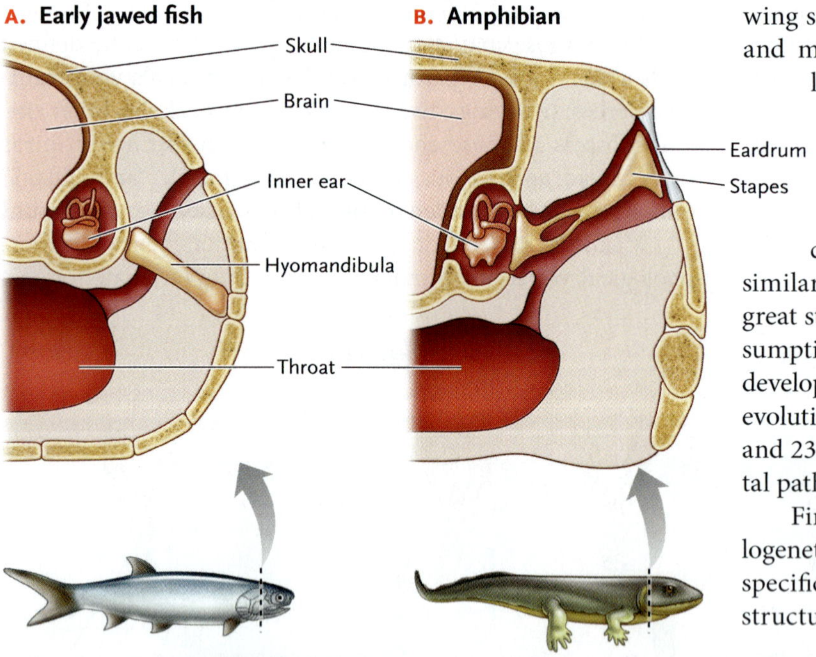

FIGURE 24.6 Homologous bones, different structure and function. The hyomandibula, which braced the jaw joint against the skull in early jawed fishes **(A)**, is homologous to the stapes, which transmits sound to the inner ear in the four-legged vertebrates, exemplified here by an early amphibian **(B)**. Both large illustrations show a cross section through the head just behind the jaw joint, as depicted in the small illustrations.

Even though characters are homologous, they may differ greatly among species, especially if their function has changed over time. For example, the stapes, a bone in the middle ear of tetrapod vertebrates, evolved from—and is therefore homologous to—the hyomandibula, a bone that supported the jaw joint of early fishes. The ancestral function of the bone is retained in some modern fishes, but its structure, position, and function are different in tetrapods **(Figure 24.6)**.

As you know from the discussion of convergent evolution in Section 23.3, organisms that are not closely related sometimes bear a striking resemblance to one another, especially when they are exposed to similar patterns of natural selection. Phenotypic similarities that evolved independently in different lineages are called **homoplasies** (or homoplastic characters). Some biologists use the terms *analogies* or *analogous characters* for homoplastic characters that serve a similar function in different species.

When systematists encounter similar morphological traits, how can they determine whether they are homologous or homoplastic? First, homologous structures are similar in anatomical detail and in their relationship to surrounding structures. For example, the bones in the wings of birds and bats are considered to be homologous **(Figure 24.7)**. They include the same structural elements and have similar connections to the rest of the skeleton. Moreover, the fossil record documents that birds and bats inherited the basic skeletal structure of the forelimb from their most recent common ancestor, a tetrapod vertebrate that lived more than 300 million years ago. However, the large flat surfaces of their wings, as well as flying behavior itself, are homoplastic. The wing surfaces are made of different materials—feathers in birds and membranous skin in bats—and their common ancestor, lacking any hint of such structures on its forelimbs, was confined to life on the ground. Thus, for birds and bats, flight and the anatomical structures that produce it are the products of convergent evolution.

Second, in multicellular organisms, homologous characters grow from the same embryonic tissues and in similar ways during development. Systematists have always put great stock in embryological indications of homology on the assumption that evolution has conserved the pattern of embryonic development in related organisms. Indeed, recent discoveries in evolutionary developmental biology (described in Sections 16.4 and 23.6) have revealed that the genetic controls of developmental pathways are very similar across a wide variety of organisms.

Finally, as we will see in Section 24.4, the structure of a phylogenetic tree can reveal whether two or more species inherited specific similarities from a common ancestor. If they did, the structures are homologous; if not, they are homoplastic.

Morphological Characters Provide Abundant Clues to Evolutionary Relationships

Morphological structures often provide useful information for phylogenetic analyses. Structural differences between organisms, which often reflect underlying genetic differences (see

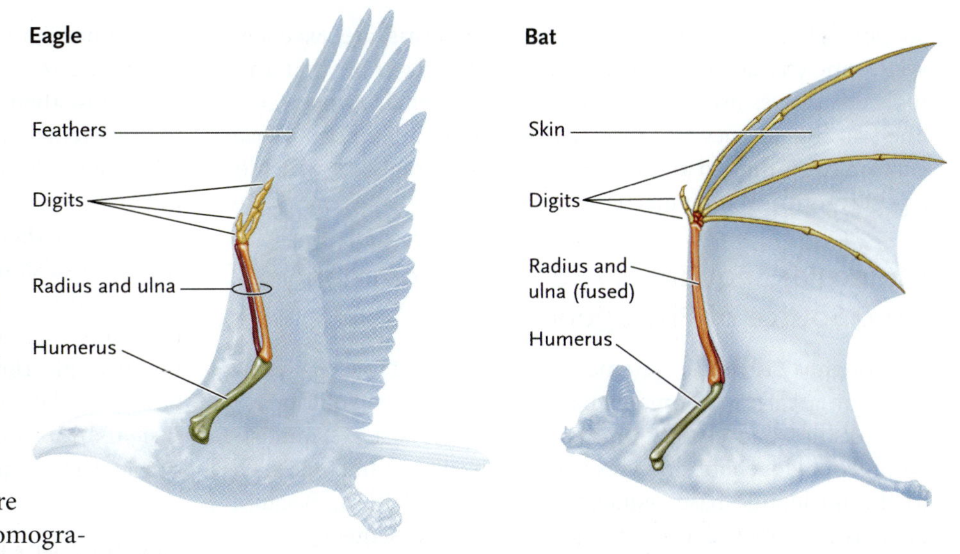

FIGURE 24.7 Assessing homology. The wing skeletons of birds and bats are homologous structures with the same basic elements. However, similarities in the flat wing surfaces are homoplastic because the surfaces are composed of different tissues.

Section 21.1), are easy to measure in preserved or living specimens. Moreover, morphological characteristics are often clearly preserved in the fossil record, allowing the comparison of living species with their extinct relatives. As you discovered in the discussions of dinosaur feathers and the inner ear structure of lambeosaurine dinosaurs in Chapter 23, researchers are applying new techniques, such as computed tomography and electron microscopy, to expand our knowledge of the anatomy of extinct organisms.

The morphological traits that are useful in phylogenetic analyses vary from group to group. In flowering plants, the details of flower anatomy may reveal common ancestry. Among vertebrates, the presence or absence of scales, feathers, and fur as well as the structure of the skull help scientists to reconstruct the evolutionary history of major groups. Sometimes systematists use obscure characters of unknown function. But differences in the number of scales on the backs of lizards or in the curvature of a vein in the wings of bees may be good indicators of the genetic differentiation that accompanied or followed speciation—even if we do not know *why* these differences evolved.

Sometimes characteristics found only in the earliest stages of an organism's life cycle can provide evidence of evolutionary relationships. As described in Chapter 31, analyses of the embryos of vertebrates revealed that they are rather closely related to sea cucumbers, sea stars, and sea urchins and even more closely related to a group of nearly shapeless marine invertebrates called sea squirts or tunicates.

Despite their usefulness, morphological characters alone cannot reveal the details of all evolutionary relationships. For example, some salamander species in North America differ in relatively few morphological features, even when they are genetically, physiologically, and behaviorally distinct. Moreover, researchers cannot easily compare the structures of organisms, such as flatworms and dogs, that share very few morphological traits.

Behavioral Characters Are Useful When Animal Species Are Not Morphologically Distinct

When external morphology cannot be used to differentiate animal species, systematists often examine their behaviors for clues about their relationships. For example, two species of treefrog (*Hyla versicolor* and *H. chrysoscelis*) commonly occur together in forests of the central and eastern United States. Both species have bumpy skin and adhesive pads on their toes that enable them to climb vegetation. They also have gray backs, white bellies, yellowish-orange coloration on their thighs, and large white spots below their eyes. The frogs are so similar that even experts cannot easily tell them apart.

How do we know that these frogs represent two species? During the breeding season, males of each species use a distinctive mating call to attract females **(Figure 24.8).** The difference in calls is a prezygotic reproductive isolating mechanism that prevents females from mating with males of a different species (see

FIGURE 24.8 Morphologically similar frog species. The frogs *Hyla versicolor* and *Hyla chrysoscelis* are so similar in appearance that one photo can depict both species. Male mating calls, visualized in sound spectrograms for the two species, are very different. The spectrograms, which depict call frequency on the vertical axis and time on the horizontal axis, show that *H. chrysoscelis* has a faster trill rate.

Section 22.2). Prezygotic isolating mechanisms are excellent systematic characters because they are often the traits that animals themselves use to recognize members of their own species. The two frog species also differ in chromosome number—*Hyla chrysoscelis* is diploid and *Hyla versicolor* is tetraploid—which is a postzygotic isolating mechanism.

Molecular Sequences Are Now a Commonly Used Source of Phylogenetic Data

For roughly 200 years, systematists building on Linnaeus' work relied on a variety of organismal traits to analyze evolutionary relationships and classify organisms. Today, most systematists conduct phylogenetic analyses using molecular characters, such as the nucleotide base sequences of DNA and RNA. Because DNA is inherited, shared changes in molecular sequences—insertions, deletions, or substitutions—provide clues to the evolutionary relationships of organisms.

The polymerase chain reaction (PCR) makes it easy for researchers to produce numerous copies of specific segments of DNA for analysis (see Section 18.1); the technique is so effective that it allows scientists to sequence minute quantities of DNA taken from dried or preserved specimens in museums and even from some fossils. Technological advances have automated the sequencing process, and researchers use analytical software to compare new data to known sequences filed in online data banks. Nuclear DNA is frequently used in phylogenetic analyses, and the publication of complete genome sequences for an ever-expanding list of organisms allows researchers to undertake broad comparative studies.

Molecular sequences have certain practical advantages over organismal characters. First, they provide abundant data: every base in a nucleic acid can serve as a separate, independent character for analysis. Moreover, because many genes have been conserved by evolution, molecular sequences can be compared between distantly related organisms that share no organismal characteristics. They can also be used to study closely related species with only minor morphological differences. Finally, nucleic acids are not directly affected by the developmental or environmental factors that cause the nongenetic morphological variations described in Section 21.1.

Molecular characters have certain drawbacks, however. For example, only four alternative character states (the four nucleotide bases) exist at each position in a DNA or RNA sequence and only 20 alternative character states (the 20 amino acids) at each position in a protein. (You may want to review Sections 14.2 and 15.1 on the structure of these molecules.)

Because of the limited number of character states, researchers often find it difficult to assess the homology of a nucleotide base substitution that appears at the same position in the DNA of two or more species. For organismal characters, biologists can determine homology by analyzing the characters' embryonic development, details of their function, or their presence in the fossil record. But molecular characters have no embryonic development; biologists still do not understand the functional significance of many molecular differences they discover; and researchers have only recently developed techniques that allow them to sequence DNA found in fossils. Nevertheless, systematists have devised complex statistical tools that allow them to discern whether molecular similarities are likely to be homologous or homoplastic.

Despite their disadvantages, molecular sequences allow researchers to sample the genome directly, and systematists have successfully used sequence data to analyze phylogenetic relationships that organismal characters were unable to resolve. For example, the phylogenetic tree for animals in Figure 31.6 is based on data from several different nucleic acid molecules. *Insights from the Molecular Revolution* describes another example using sequences called transposable elements.

STUDY BREAK 24.3

1. **Why do systematists use homologous characters in their phylogenetic analyses?**
2. **Why are morphological traits often helpful in tracing the long-term evolutionary relationships within a group of organisms?**
3. **What are three advantages of using molecular characters in phylogenetic analyses?**

24.4 Traditional Classification and Paraphyletic Groups

For a century after Darwin published his theory of evolution, systematists followed an approach called **traditional systematics.** Researchers constructed phylogenetic trees and classified organisms by assessing the amount of phenotypic divergence between lineages as well as the patterns of branching evolution that had produced them. In other words, they focused on the products of anagenesis (that is, evolutionary change through the accumulation of new or modified characteristics) as well as the products of cladogenesis (that is, the new species and lineages produced through branching evolution). Thus, their classifications did not always strictly reflect the patterns of branching evolution **(Figure 24.9A).**

For example, the fossil record for tetrapod vertebrates reveals that the amphibian and mammalian lineages each diverged early, followed shortly thereafter by the turtle lineage. (The phylogenetic position of turtles is controversial, as discussed in Chapter 32; here we follow the traditional view based on paleontological evidence.) The remaining organisms then diverged into two groups: one includes living lizards and snakes and another includes living crocodilians and birds. Thus, although crocodilians, with their scaly skin and sprawling posture outwardly resemble lizards, evolutionary biologists have long recognized that crocodilians share a more recent common ancestor with birds.

Even though the phylogenetic tree shows six living clades, the traditional classification recognizes four classes of tetrapod

INSIGHTS FROM THE Molecular Revolution

Evolution of Whales: SINEs yield signs of the LINEs of evolution

The fossil record reveals that whales evolved from terrestrial mammals into streamlined creatures adapted to life in the sea more than 50 million years ago. Using morphological, fossil, and molecular data, researchers have determined that modern cetaceans—whales, dolphins, and porpoises—are closely related to artiodactyls (even-toed ungulates). Systematists now group cetaceans and artiodactyls in the clade Cetartiodactyla, which, in addition to whales and their relatives, includes camels, pigs, ruminants, and hippopotamuses.

Research Question

Which living cetartiodactyls are the closest relatives of whales?

Experiment

A number of research groups have attempted to answer the question using molecular approaches. Norihiro Okada and his research group at the Tokyo Institute of Technology, Yokohama, Japan, and collaborators in the United States used the patterns of locations of transposable elements (TEs) in genomes to determine the evolutionary relationship between living members of the various cetartiodactyl groups. TEs are sequences that can move to new locations in DNA. The type of TEs the researchers studied are retrotransposons, which move by making RNA copies of themselves; the RNA copies then act as templates for making DNA copies, which insert in new locations (see Section 15.5).

Okada and his collaborators analyzed specific representatives of two classes of retrotransposons, SINEs (for *S*hort *IN*terspersed *E*lements) and LINEs (for *L*ong *IN*terspersed *E*lements), which are known to be inserted into particular chromosomal loci. SINEs and LINEs are valuable tools for phylogenetic studies; their pattern of duplication and movement to new locations represent a novel character that is unique to each clade. Once they have inserted themselves into new locations, they are rarely lost. Thus, if SINEs and LINEs occur at the same sites in the nuclear DNA of several species, those species are likely to be members of the same lineage.

To determine whether a chromosomal locus contained the SINE or LINE associated with it, the researchers isolated DNA, amplified the locus using PCR (see Section 18.1), and then used probes—labeled nucleic acid molecules—that could pair specifically with the SINE or LINE of the locus. A signal from the probing indicated that the SINE or LINE was present; no signal indicated the SINE or LINE was absent.

Results

The investigators analyzed 16 retrotransposons in 10 loci in the DNA from living members of all cetartiodactyl groups. Using the patterns of SINEs and LINEs identified in their phylogenetic analysis, the researchers identified the best hypothesis about the evolutionary relationships of the Cetartiodactyla **(Figure)**.

Conclusion

As the phylogenetic tree indicates, hippopotamuses are the closest relatives of whales; ruminants (such as cattle and bison) are more distantly related, and pigs and camels even more distantly related. Other phylogenetic analyses based on combined data sets that include morphological characters from extinct and living species as well other molecular characters generally support this conclusion.

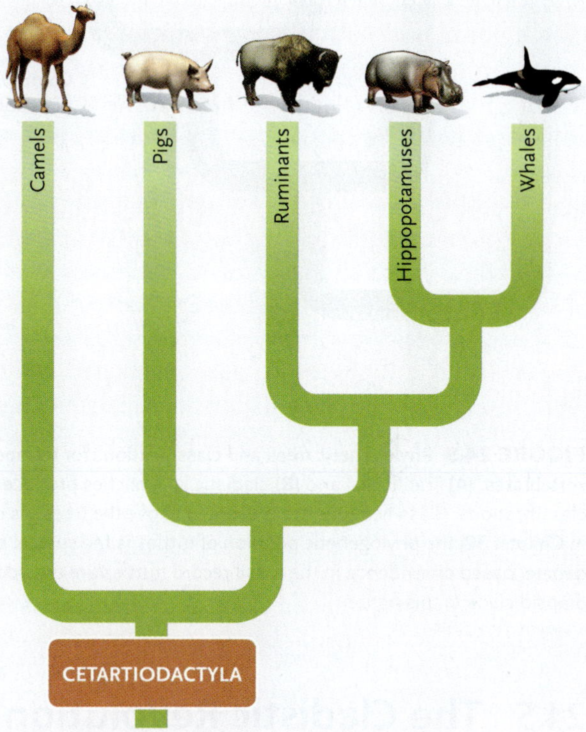

THINK LIKE A SCIENTIST Research by Nikaido and colleagues reveals that whales are closely related to hippos and other artiodactyls. But what sorts of morphological changes must the ancestors of whales have undergone as they changed from being wholly terrestrial to wholly aquatic animals?

Source: M. Nikaido et al. 1999. Phylogenetic relationships among cetartiodactyls based on insertions of short and long interspersed elements: Hippopotamuses are the closest extant relatives of whales. *Proceedings of the National Academy of Sciences USA* 96:10261–10266.

vertebrates: Amphibia, Mammalia, Reptilia, and Aves. These groups are given equal ranking because each represents a distinctive body plan and way of life. The traditionally defined taxon Reptilia is clearly paraphyletic because, even though birds share a common ancestor with reptiles, they are placed in a separate taxon. Traditional systematists justify this definition of Reptilia because it includes morphologically similar animals with close evolutionary relationships. Crocodilians are classified with lizards, snakes, and turtles because they share a distant common ancestry and are covered with dry, scaly skin. Traditional systematists also argued that the key innovations initiating the adaptive radiation of birds—a high metabolic rate, wings, and flight—represent such extreme divergence from the ancestral morphology that birds merit recognition as a separate class.

STUDY BREAK 24.4

Why does a classification produced by traditional systematics sometimes include paraphyletic groups?

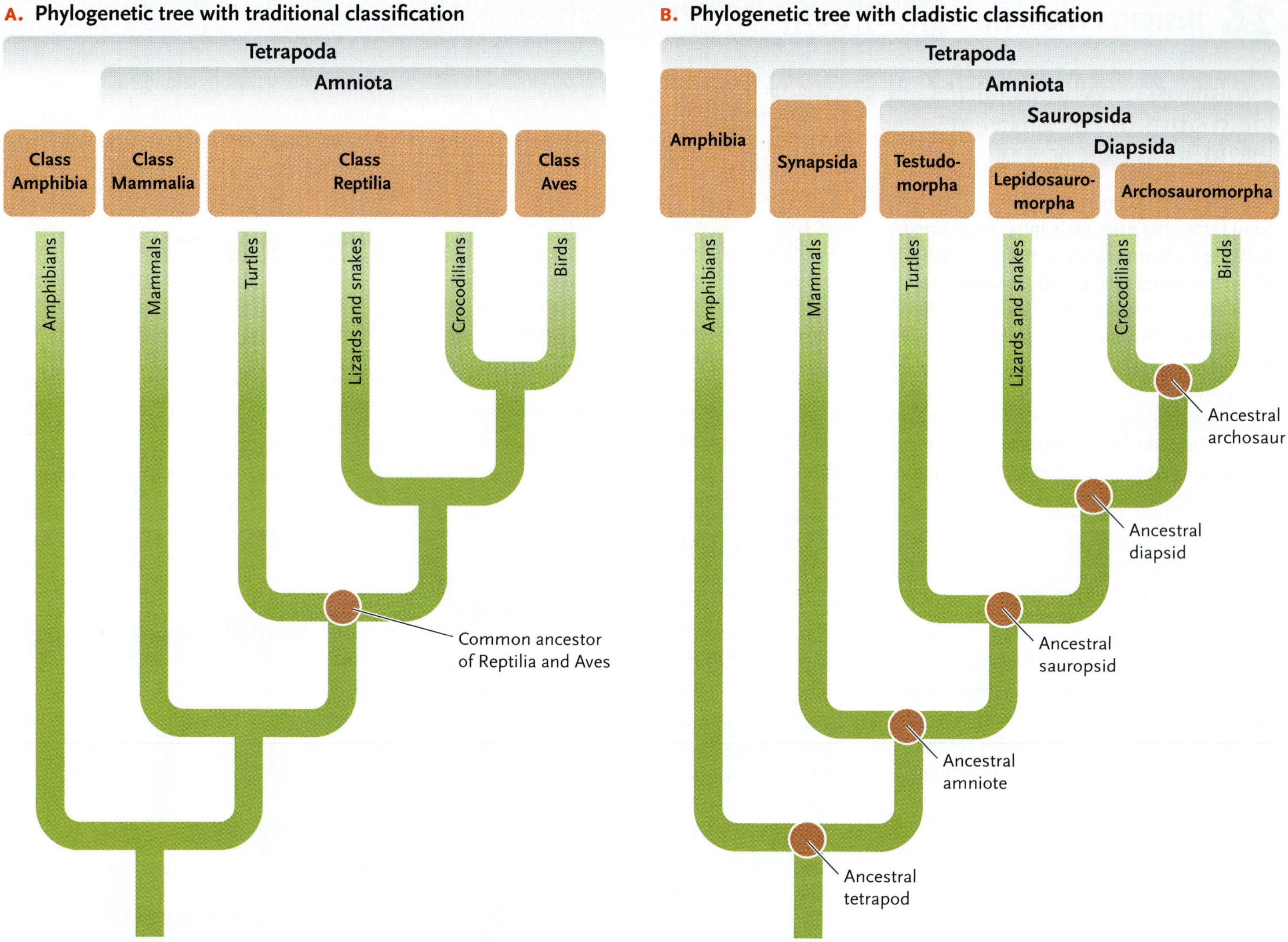

FIGURE 24.9 Phylogenetic trees and classifications for tetrapod vertebrates. (A) Traditional and **(B)** cladistic approaches produce different classifications. Classifications are presented above the trees. As discussed in Chapter 32, the phylogenetic position of turtles is the subject of much debate; based on evidence in the fossil record, turtles are excluded from the diapsid clade in this figure.

24.5 The Cladistic Revolution

In the 1950s and 1960s, some researchers criticized the traditional classifications based on two distinct phenomena—branching evolution and morphological divergence—as inherently unclear. How can we tell *why* two groups are classified in the same higher taxon? They may have shared a recent common ancestor, as did lizards and snakes. Alternatively, they may have retained some ancestral characteristics after being separated on different branches of a phylogenetic tree, as is the case for lizards and crocodilians.

To avoid such confusion, many systematists followed the philosophical and analytical lead of Willi Hennig, a German entomologist (that is, a scientist who studies insects), who published an influential book, *Phylogenetic Systematics,* in 1950; its English translation appeared in 1966. Hennig and his followers argued that classifications should be based solely on evolutionary relationships. This approach, which is called **cladistics,** produces phylogenetic hypotheses and classifications that reflect only the branching pattern of evolution; it ignores morphological divergence altogether.

Cladistic Analyses Focus on Recently Evolved Character States

Like traditional systematics, cladistics analyzes the evolutionary relationships among organisms by comparing their organismal and, more recently, genetic characteristics. Each **character** can exist in two or more forms, described as character states. Evolutionary processes change characters over time from an original, **ancestral character state** to a newer, **derived character state.** Characters that were present in the ancestors of a clade are considered ancestral; those that are new in a descendant are considered derived. For example, ancient fishes, which represent the ancestral vertebrates, had fins; but some of their descendants, the tetrapods, which appeared much later in the fossil record,

FIGURE 24.10 Outgroup comparison. Most adult insects, like **(A)** the caddis fly and **(B)** the orange palm dart butterfly, have six walking legs. This comparison of butterflies with other insects suggests that the four walking legs of **(C)** the monarch butterfly represents the derived character state.

have limbs. In this example, fins are the ancestral character state, and limbs are the derived character state.

In the jargon of cladistics, a derived character state is called an **apomorphy** (*apo* = away from, *morphe* = form), and a derived character state found in two or more species is called a **synapomorphy** (*syn* = together). The presence of a synapomorphy (that is, a *shared derived character state*) among species provides a clue that they may be members of the same clade. As you learned in Section 22.4, morphological innovations often trigger an adaptive radiation. And once a derived character state becomes established in a species, it is likely to be present in that species' descendants. Thus, unless they are lost or replaced by newer characters over evolutionary time, *synapomorphies can serve as markers for monophyletic lineages.*

Systematists define synapomorphies only when comparing character states among species. Thus, any particular character state is derived *only in relation to* an ancestral character state observed in other organisms—either an older version of the character or its absence. For example, most species of animals lack a vertebral column. However, one animal clade—the vertebrates, including fishes, amphibians, reptiles, birds, and mammals—has that structure. Thus, when systematists compare vertebrates to all other animals, the absence of a vertebral column is the ancestral character state and the presence of a vertebral column is derived.

How can systematists distinguish between ancestral and derived character states? In other words, how can they determine the direction in which a character has evolved? The fossil record, if it is detailed enough, can provide unambiguous information. For example, biologists are confident that the presence of a vertebral column is a derived character state because fossils of animals that lived before vertebrates lack that structure.

In the absence of evidence from fossils, systematists frequently use a technique called **outgroup comparison** to identify ancestral and derived character states. Using this approach, systematists compare characters in the *ingroup,* the clade under study, to those in an *outgroup,* a closely related species that is not a member of the clade. Character states observed in the outgroup are considered ancestral, and those observed *only* in the ingroup are considered derived. And because the outgroup and the ingroup are phylogenetically related, outgroup comparison allows researchers to hypothesize the root (that is, the common ancestor shared by the outgroup and the ingroup) of the phylogenetic tree. Most modern butterflies, for example, have six walking legs, but species in two families have four walking legs and two small, nonwalking legs. Which is the ancestral character state, and which is derived? Outgroup comparison with other insects, which are not included in the butterfly clade, demonstrates that most insects have six walking legs as adults; this result suggests that six walking legs is ancestral and four is derived **(Figure 24.10).**

Cladistics Uses Synapomorphies to Reconstruct Evolutionary History and Classify Organisms

Following the cladistic method, biologists construct phylogenetic trees and classifications by grouping together species that *share derived character states.* Ancestral character states, because they are shared by the ingroup and the outgroup, do not help to define the ingroup. For example, mammals are a monophyletic lineage—a clade—because they possess a unique set of synapomorphies: hair, mammary glands, and a reduced number of bones in the lower jaw. The ancestral character states found in mammals, such as a vertebral column and four legs, do not distinguish them from other tetrapod vertebrates. Thus, these shared ancestral character states are not useful in defining the mammal clade.

The results of a cladistic analysis are presented in a phylogenetic tree that illustrates the hypothesized sequence of evolutionary branching that produced the organisms under study **(Figure 24.9B):** a common ancestor is hypothesized at each node, and every branch portrays a strictly monophyletic group. Once a researcher identifies a suitable outgroup and the ancestral and derived character states, a cladistic analysis is straightforward **(Figure 24.11).** The synapomorphies that define each clade are sometimes listed on the branches.

The use of molecular sequence data in phylogenetic analyses relies on the same logic that underlies analyses based on organismal characters, like those considered above: species included within a clade are expected to exhibit more molecular

FIGURE 24.11 **Research Method**

Using Cladistics to Construct a Phylogenetic Tree

Purpose: Systematists construct phylogenetic trees to visualize hypothesized evolutionary relationships among a group of organisms. The cladistic method requires a researcher to group together organisms that share derived characters states. The derived character states identified in the tan plaques are the synapomorphies that define each clade.

Protocol:

1. *Select the organisms to study.* To demonstrate the method, we develop a phylogenetic tree for the nine groups of living vertebrates: lampreys, sharks (and their relatives), bony fishes (and their relatives), amphibians (frogs and salamanders), turtles, lizards (including snakes), crocodilians (including alligators), birds, and mammals. We also include marine animals called lancelets (Chordata, Cephalochordata) as the outgroup. Lancelets are closely related to, but not included within, the vertebrates (see Chapter 32). The inclusion of an outgroup allows biologists to identify ancestral versus derived character states and root the tree.

2. *Choose the characters on which the phylogenetic tree will be based.* Our simplified example is based on the presence or absence of 10 characters: (1) vertebral column, (2) jaws, (3) swim bladder or lungs, (4) paired limbs (with one bone connecting each limb to the body), (5) extraembryonic membranes (such as the amnion), (6) mammary glands, (7) dry, scaly skin somewhere on the body, (8) two openings on each side near the back of the skull, (9) one opening on each side of the skull in front of the eye, and (10) feathers.

3. *Score the character states in each group.* Because lancelets serve as the outgroup in this analysis, we consider character states observed in lancelets as ancestral; any deviation from the lancelet pattern is considered derived. Because lancelets lack all of the characters in our analysis, the presence of each character is the derived condition. We tabulate data on the distribution of ancestral (−) and derived (+) characters in all species included in the analysis.

	Vertebrae	Jaws	Swim bladder or lungs	Paired limbs	Extraembryonic membranes	Mammary glands	Dry, scaly skin	Two openings at back of skull	One opening in front of eye	Feathers
Lancelets	−	−	−	−	−	−	−	−	−	−
Lampreys	+	−	−	−	−	−	−	−	−	−
Sharks	+	+	−	−	−	−	−	−	−	−
Bony fishes	+	+	+	−	−	−	−	−	−	−
Amphibians	+	+	+	+	−	−	−	−	−	−
Mammals	+	+	+	+	+	+	−	−	−	−
Turtles	+	+	+	+	+	−	+	−	−	−
Lizards	+	+	+	+	+	−	+	+	−	−
Crocodilians	+	+	+	+	+	−	+	+	+	−
Birds	+	+	+	+	+	−	+	+	+	+

4. *Construct the phylogenetic tree from information in the table, grouping organisms that share derived character states.* All groups except lancelets have vertebrae. Thus, we group organisms that share this derived character state on the right-hand branch, identifying them as a monophyletic lineage. Lancelets are on their own branch to the left, indicating that they lack vertebrae.

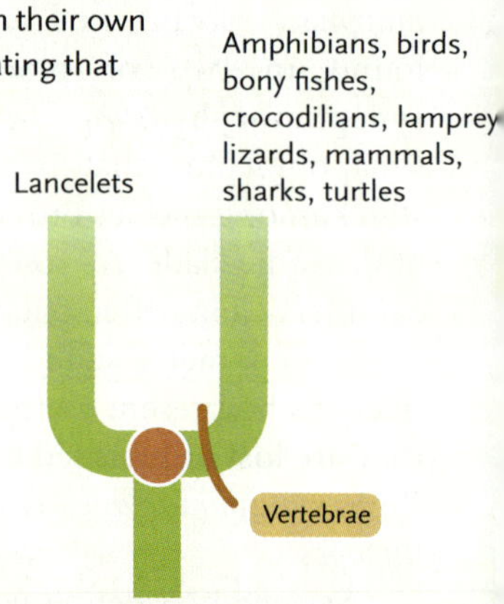

All of the remaining organisms except lampreys have jaws. (Lancelets also lack jaws, but we have already separated them out, and do not consider them further.) Place all groups with jaws, a derived character state, on the right-hand branch. Lampreys are separated out to the left, because they lack jaws. Again, the branch on the right represents a monophyletic lineage.

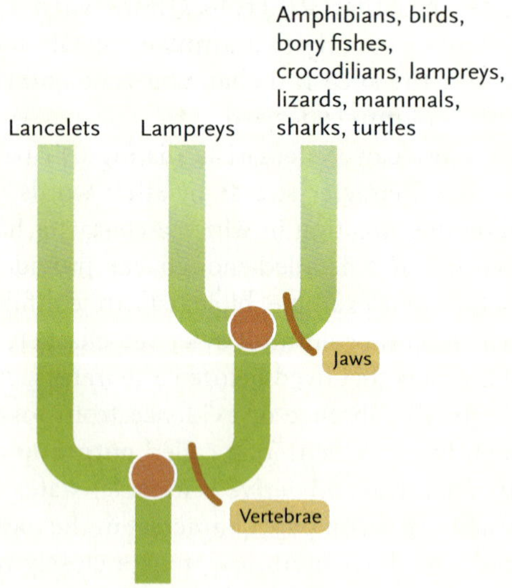

synapomorphies than do species from different clades. The comparison of sequences in the ingroup to those in the outgroup may allow a researcher to define ancestral and derived character states.

The classifications produced by cladistic analysis often differ radically from those of traditional systematics (compare Figure 24.9A with Figure 24.9B). Pairs of nested taxa are defined directly from the two-way branching pattern of the phy-

5. *Construct the rest of the phylogenetic tree using the same step-by-step procedure to separate the remaining groups.* In our completed tree, seven groups share a swim bladder or lungs; six share paired limbs; and five have extraembryonic membranes during development. Some groups are distinguished by the unique presence of a derived character state, such as feathers in birds.

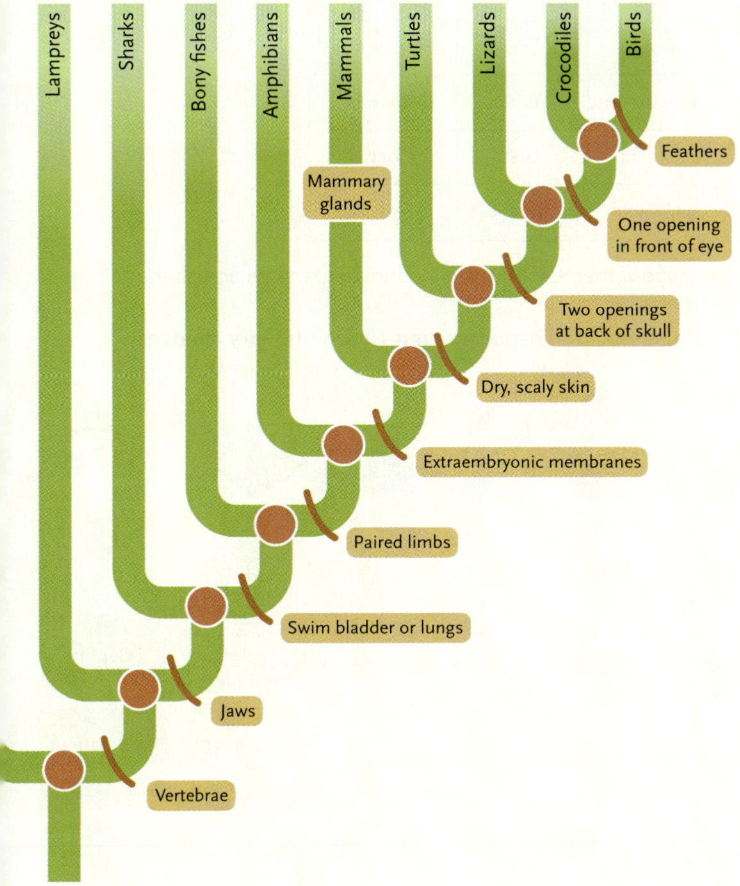

Interpreting the Results: Although phylogenetic trees provide information about evolutionary relationships, the common ancestors represented by the branch points are often hypothetical. You can tell from the tree, however, that birds are more closely related to lizards than they are to mammals. Follow the branches of the tree from birds and lizards back to their node. Next, trace the branches of birds and mammals to their node. You can see that the bird–mammal node is closer to the root of the tree than the bird–lizard node is. Nodes that are closer to the bottom of the tree indicate a more distant common ancestry than those closer to the top. Note also that this simplified example produces a phylogenetic tree that is easy to interpret because the data set includes no homoplastic similarities and no conflicting evidence. Most phylogenetic analyses include many such complications.

logenetic tree. Thus, the clade Tetrapoda (the traditional amphibians, reptiles, birds, and mammals) is divided into two taxa, the Amphibia (tetrapods that do not have an amnion, as discussed in Section 32.3) and the Amniota (tetrapods that have an amnion). The Amniota is subdivided into two taxa on the basis of skull morphology and other characteristics: Synapsida (mammals) and Sauropsida (turtles, lizards, snakes, crocodilians, and birds). The Sauropsida is further divided into the Testudomorpha (turtles) and the Diapsida (lizards and snakes, crocodilians, and birds). Finally, the Diapsida is subdivided into two more recently evolved taxa—the Lepidosauromorpha (lizards and snakes) and the Archosauromorpha (crocodilians and birds). Thus, a strictly cladistic classification exactly parallels the pattern of branching evolution that produced the organisms included in the classification. These parallels are the essence and strength of the cladistic method.

Most biologists value the evolutionary focus, clear goals, and precise methods of the cladistic approach. In fact, some systematists advocate abandoning the Linnaean hierarchy for classifying and naming organisms. They propose using a strictly phylogenetic system, called **PhyloCode,** that identifies and names clades instead of pigeonholing organisms into traditional taxonomic categories. This approach is further described in *Unanswered Questions* at the end of this chapter. Although traditional systematics has guided many people's understanding of biological diversity, we use cladistic analyses to describe evolutionary lineages and taxa in the Biodiversity unit that follows this chapter.

Systematists Use Several Techniques to Identify an Optimal Phylogenetic Tree

In practice, most phylogenetic studies are far more complicated than the examples discussed above and in Figure 24.11. Researchers may collect data on hundreds of characters in dozens of species. After scoring each character state as ancestral or derived in every species, a systematist uses one or more computer programs to generate a set of alternative phylogenetic trees. The output of these analyses is often substantial: an analysis of five species can produce 15 possible phylogenetic trees; an analysis of 50 species can produce 3×10^{76}.

Faced with such an unimaginably large number of alternative hypotheses, how does a systematist decide which phylogenetic tree is the "best" representation of a clade's evolutionary history? This problem is complex, because, when evaluating large data sets, we expect to see some similarities that arise when convergent evolution causes distantly related organisms to evolve similar traits independently; because such traits are not synapomorphies, they are false indicators of relatedness that confound the analysis. We also expect to find some differences between closely related organisms if natural selection or some other microevolutionary process caused a derived character state to be reversed or lost. How can we tell which similarities are synapomorphies and which are homoplasies? Researchers use several approaches to sort through the alternatives, two of which we describe below.

PARSIMONY APPROACH Many systematists adopt a philosophical concept, the **principle of parsimony,** to identify the optimal phylogenetic tree. This principle states that the simplest

CLOSER LOOK **FIGURE 24.12** The principle of parsimony.

A. Distribution of character states in six clades of vascular plants

Ferns represent the outgroup: all of its character states are considered ancestral.

Characters (possible states)	Ferns (outgroup)	Gnetophytes	Cycads	Ginkgophytes	Conifers	Angiosperms
Archegonium (present or lost)	Present	Lost	Present	Present	Present	Lost
Double fertilization (absent or present)	Absent	Present	Absent	Absent	Absent	Present
Pollen tube growth (haustorium or tube)	Absent	Tube	Haustorium	Haustorium	Tube	Tube
Sperm flagella (present or lost)	Present	Lost	Present	Present	Lost	Lost
Vessels (absent or present)	Absent	Present	Absent	Absent	Absent	Present

Two alternative phylogenetic trees for the six clades illustrate different hypotheses about their evolutionary relationships. Bars across the branches mark the hypothesized evolution of derived character states in both trees.

B. Phylogenetic tree hypothesizing five evolutionary changes

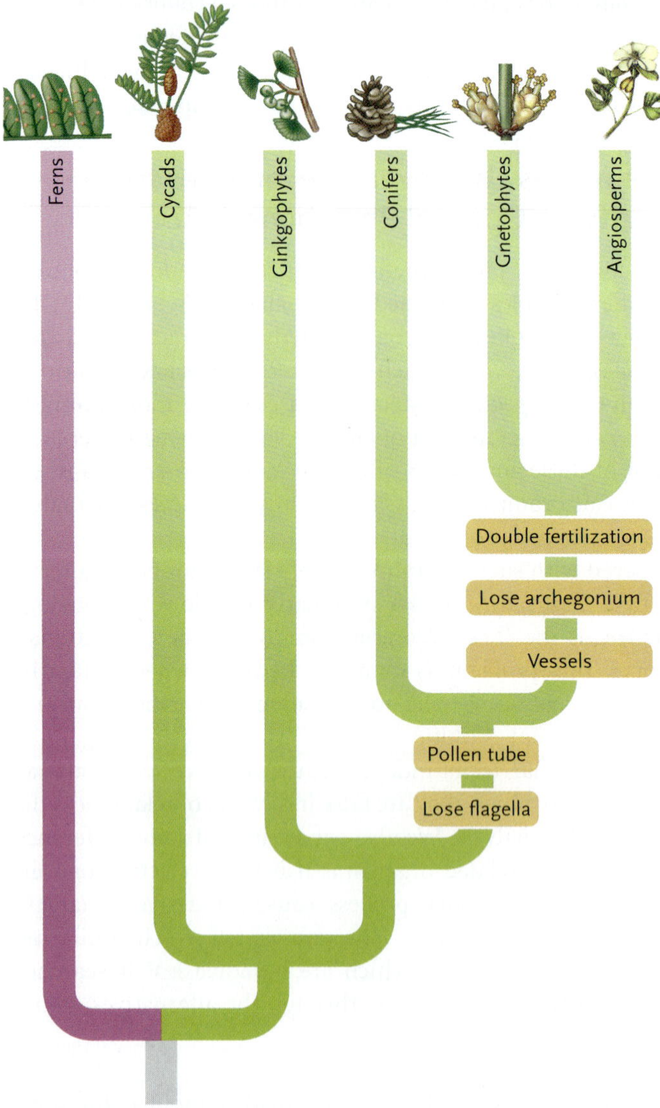

C. Phylogenetic tree hypothesizing 10 evolutionary changes

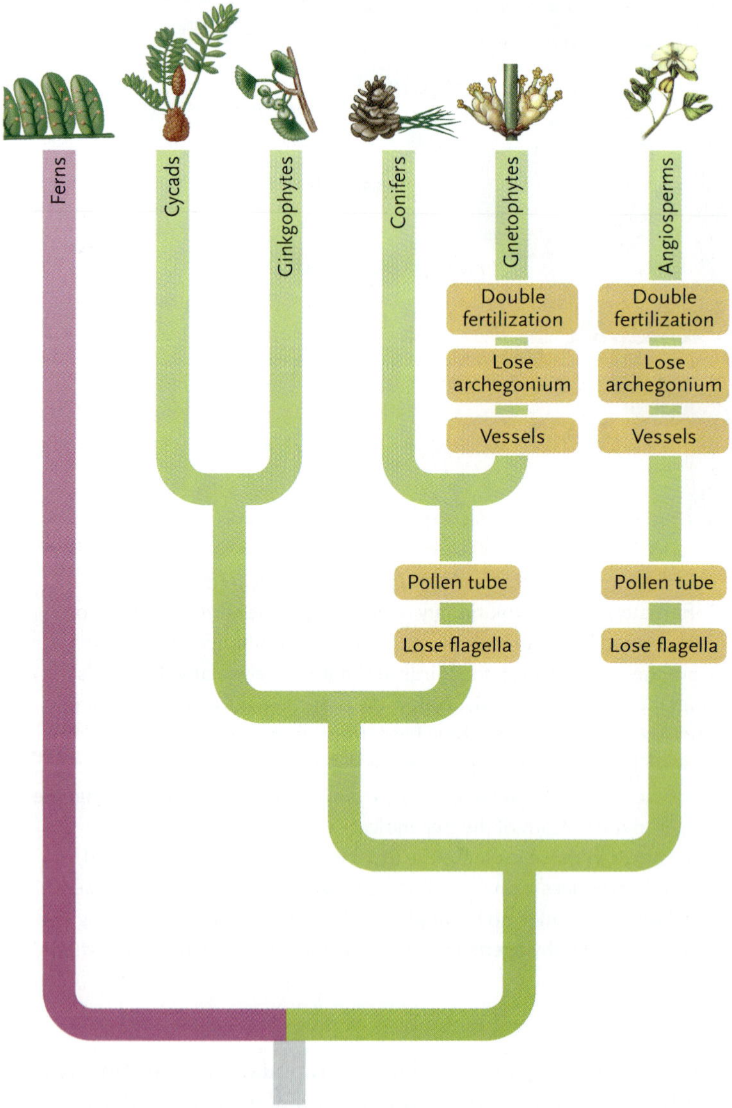

SUMMARY Based on this set of five characters, the phylogenetic tree with five hypothesized evolutionary changes **(B)** is more parsimonious than the tree that hypothesizes 10 evolutionary changes **(C).** In the absence of additional data, a systematist would accept the more parsimonious tree as the best working hypothesis. (The characters and clades of vascular plants are discussed in detail in Chapter 28.)

THINK LIKE A SCIENTIST Do the five characters in the table provide sufficient information to develop an hypothesis about the evolutionary relationship of cycads and ginkgophytes?

plausible explanation of any phenomenon is the best. If we assume that any complex evolutionary change is an unlikely event, then it is extremely unlikely that the same complex change evolved twice in one lineage. Thus, when the principle is applied to phylogenetic analyses, it suggests that the "best" phylogenetic tree is the one that hypothesizes the smallest number of evolutionary changes needed to account for the distribution of character states within a clade; in effect, this approach minimizes the number of homoplasies (that is, the independent evolution of similar traits) in the tree **(Figure 24.12).** To apply the principle, computer programs evaluate the number of evolutionary changes hypothesized by each phylogenetic tree they generate, and the researcher identifies the one with the fewest hypothesized changes as the most plausible.

The principle of parsimony also allows researchers to identify homologous characters and infer their ancestral and derived states. Once the most parsimonious phylogenetic tree is identified, a researcher can visualize the distribution of derived character states and pinpoint when each derived state evolved.

STATISTICAL APPROACHES When comparing two genome sequences, each base in a strand of DNA can be treated as a character with four possible states (A, G, T, and C). One could perform a parsimony analysis on molecular sequence data, like the one illustrated for morphological data in Figure 24.12, to identify the phylogenetic tree that assumes the fewest mutations. But the application of the parsimony approach to molecular data is complicated by several factors. First, given that there are only four possible character states at each position in a nucleic acid, identical changes in nucleotides often arise independently. Second, segments of DNA that do not code for proteins are less likely than coding regions to be affected by natural selection. As a result, mutations accumulate faster in non-coding regions, causing them to evolve rapidly. Third, because of the degeneracy of the genetic code (described in Section 15.1), mutations in the third codon position do not often influence the amino acid composition of the protein for which a gene codes. As a result, third codon mutations are often selectively neutral (see Section 21.4), and they accumulate more rapidly than do mutations in the first or second positions. Finally, certain nucleotide substitutions are more common than others: transitions (the substitution of a purine for another purine or a pyrimidine for another pyrimidine) occur more frequently than transversions (substitutions between purines and pyrimidines).

To avoid this problem, systematists develop statistical models of evolutionary change that take into account variations in the evolutionary rates at different nucleotide positions or in different genes or species as well as changes in evolutionary rates over time. In one statistical approach, the **maximum likelihood method,** systematists compare alternative trees with specific models about the rates of evolutionary change. The tree that is most likely to have produced the observed distribution of molecular character states is identified as the best hypothesis.

To illustrate how phylogenetic trees are constructed from DNA sequence data, we cite an example using the **genetic distance method,** which calculates the overall proportion of bases that differ between two species. The genetic distance between closely related species is smaller than the genetic distance between distantly related species, because the gene pools of closely related species have accumulated distinctive mutations for a shorter period of time. Systematists can construct a phylogenetic tree from these data by making multiple comparisons of genetic distance between pairs of species and then between groups of species **(Figure 24.13).** Although not as powerful as the maximum likelihood method, the genetic distance method does not depend on assumptions about the evolutionary likelihood of different types of mutations. It also requires much less computing power, which is useful when comparing billions of bases of homologous DNA sequences.

STUDY BREAK 24.5

1. **How does outgroup comparison facilitate the identification of ancestral and derived character states?**
2. **What characteristics are used to group organisms in a cladistic analysis?**
3. **How is the principle of parsimony applied in phylogenetic analyses?**

THINK OUTSIDE THE BOOK

Access the web page for the Tree of Life project at http://www.tolweb.org/tree/. Select a group of animals or plants that is of interest to you, and study the structure of its phylogenetic tree. How many major clades does it include? On the basis of what shared derived characters are those clades defined?

FIGURE 24.13 **Research Method**

Using Genetic Distances to Construct a Phylogenetic Tree

Purpose: Systematists use data on genetic distances (the overall differences in DNA sequences) among species to reconstruct their phylogenetic tree.

Protocol:

1. Calculate the genetic distance between each pair of species. The genetic distances between humans and three species of great apes are shown in the first table.

	Chimpanzee	Gorilla	Orangutan
Human	1.37	1.75	3.40
Chimpanzee		1.81	3.44
Gorilla			3.50

2. Identify the pair of species with the smallest genetic distance in the first table; in this example, the smallest distance is between chimpanzee and human (genetic distance = 1.37). These two species therefore form a cluster of two closely related species.

3. Calculate the average genetic distances between the chimpanzee-human cluster and each of the other species in the analysis, gorilla and orangutan. For example, the genetic distance between the chimpanzee-human cluster and gorilla is the average of the human-gorilla genetic distance and the chimpanzee-gorilla genetic distance [(1.75 + 1.81)/2 = 1.78]. The newly calculated genetic distances are shown in the second table.

	Gorilla	Orangutan
Chimpanzee–human cluster	1.78	3.42
Gorilla		3.50

4. Identify the two groups (individual species or clusters) with the smallest genetic distance in the second table. In this example, the next smallest genetic distance is between the chimpanzee-human cluster and gorilla (genetic distance = 1.78). Thus, chimpanzee-human-gorilla forms the next cluster, leaving orangutan as the outgroup.

5. Because there are only four species in our example, these genetic distance calculations define the phylogenetic tree shown below. If the analysis included additional species, we would repeat the calculation described in steps 2 to 4 as many times as necessary to complete the phylogenetic tree.

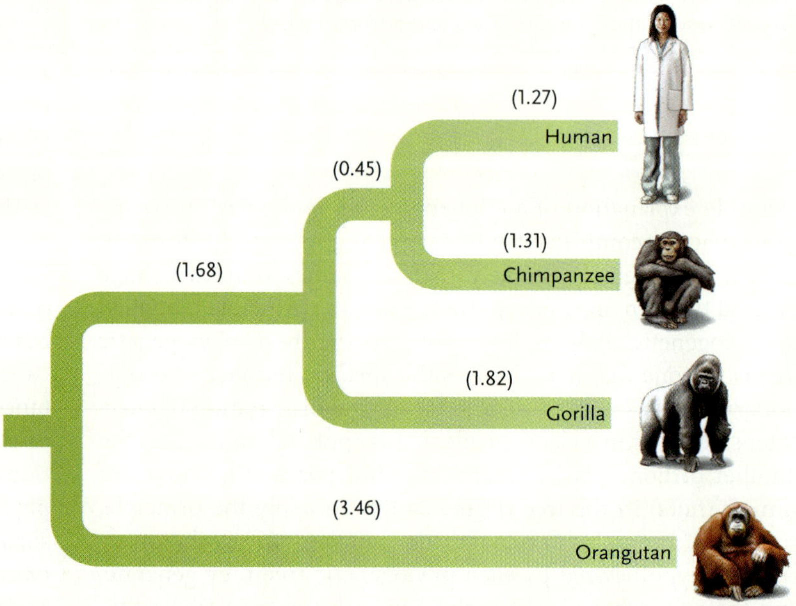

Interpreting the Results: In our phylogenetic tree, the length of each branch reflects the amount of genetic change that has occurred in that branch. A longer branch length indicates that the genome of that species has evolved at a correspondingly higher rate (that is, undergone more evolutionary change since the two sister branches emerged from their common ancestor).

Source: A. Scally et al. 2012. Insights into hominid evolution from the gorilla genome sequence. *Nature* 483:169–175.

24.6 Phylogenetic Trees as Research Tools

In addition to providing a wealth of information about the patterns of branching evolution across the entire spectrum of living organisms, phylogenetic trees are useful tools that facilitate research in all areas of biology.

Molecular Clocks Estimate the Time of Evolutionary Divergences

Although many biological molecules have been conserved by evolution, different adaptive changes and neutral mutations accumulate in separate lineages from the moment they first diverge. Because mutations that arise in noncoding regions of DNA do not affect protein structure, they are probably not often eliminated by natural selection. If mutations accumulate in these segments at a reasonably constant rate, differences in their DNA sequences can serve as a **molecular clock,** indexing the time at which two species diverged. Large differences imply divergence in the distant past, whereas small differences suggest a more recent common ancestor.

Because different molecules exhibit individual rates of evolutionary change, every molecule is an independent clock, ticking at its own rate. Researchers study different molecules to track evolutionary divergences that occurred at different times in the past. For example, mitochondrial DNA (mtDNA) evolves relatively quickly; it is useful for dating evolutionary diver-

FIGURE 24.14 Observational Research

Do Molecular Clocks Tick at a Constant Rate?

Hypothesis: Some DNA segments accumulate mutations at a constant rate, allowing researchers to use genetic differences between species as molecular clocks.

Prediction: The genetic differences observed between pairs of species will be proportional to the time since they diverged from their common ancestor.

Method: Robert C. Fleischer of the Smithsonian Institution and colleagues at the Smithsonian and Pennsylvania State University compared gene sequences among pairs of related bird species (honeycreepers) and fruit fly species *(Drosophila)* on different Hawaiian Islands. They used radiometric dating of volcanic rocks to estimate the age of each island and used the estimates of island age to measure the divergence time between pairs of species. This approach assumes that birds and flies founded populations on each new island shortly after the island's emergence from the Pacific Ocean and that newly founded populations began to diverge from ancestral populations immediately. Thus, if a population of *Drosophila* now lives on Oahu, the researchers estimate that it diverged from its ancestor 1.6 million years ago, which is the estimated time of Oahu's origin. The researchers plotted a summary statistic of genetic differences between species pairs versus estimated island age to test their hypothesis about constant rates of evolution in the DNA segments they studied.

Results: The extent of genetic difference in the cytochrome *b* sequences of Hawaiian honeycreepers and in the *Yp1* gene sequences of Hawaiian fruit flies is strongly correlated with the time since the members of each species pair were separated. The results indicate that these DNA sequences have evolved at constant rates.

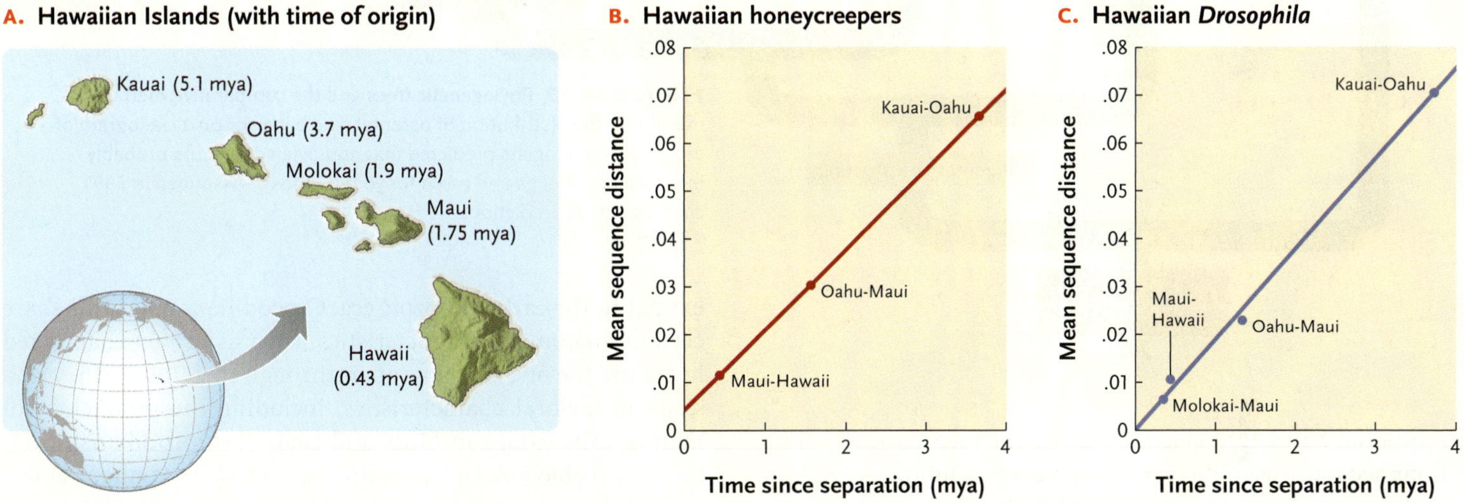

Conclusion: Because the gene sequences evolve at a constant rate, they can be used as molecular clocks. Each 0.016 units of genetic difference in the cytochrome *b* sequence in Hawaiian honeycreepers and each 0.019 units of genetic difference in the *Yp1* sequence in Hawaiian *Drosophila* indexes 1 million years of independent evolution.

THINK LIKE A SCIENTIST Which of the two gene sequences surveyed—cytochrome *b* honeycreepers or *Yp1* sequence in *Drosophila*—appears to have evolved more quickly?

Source: R. C. Fleischer et al. 1998. Evolution on a volcanic conveyor belt: using phylogeographic reconstructions and K–Ar-based ages of the Hawaiian Islands to estimate molecular evolutionary rates. *Molecular Ecology* 7:533–545.

gences that occurred within the last few million years. Studies of mtDNA have illuminated aspects of the evolutionary history of humans, as described in Section 32.13. By contrast, chloroplast DNA (cpDNA) and genes that encode ribosomal RNA evolve much more slowly, providing information about divergences that date back hundreds of millions of years.

To calibrate molecular clocks, researchers examine the degree of genetic difference between species in relation to their time of divergence estimated from the fossil record. Alternatively, the clock can be calibrated biogeographically with independent data on when volcanic islands first emerged from the sea or when landmasses separated **(Figure 24.14).**

The reliability of molecular clocks depends on the constancy of evolutionary change in the DNA segment analyzed. Some researchers have noted that even DNA segments that are thought to be selectively neutral may show variable rates of

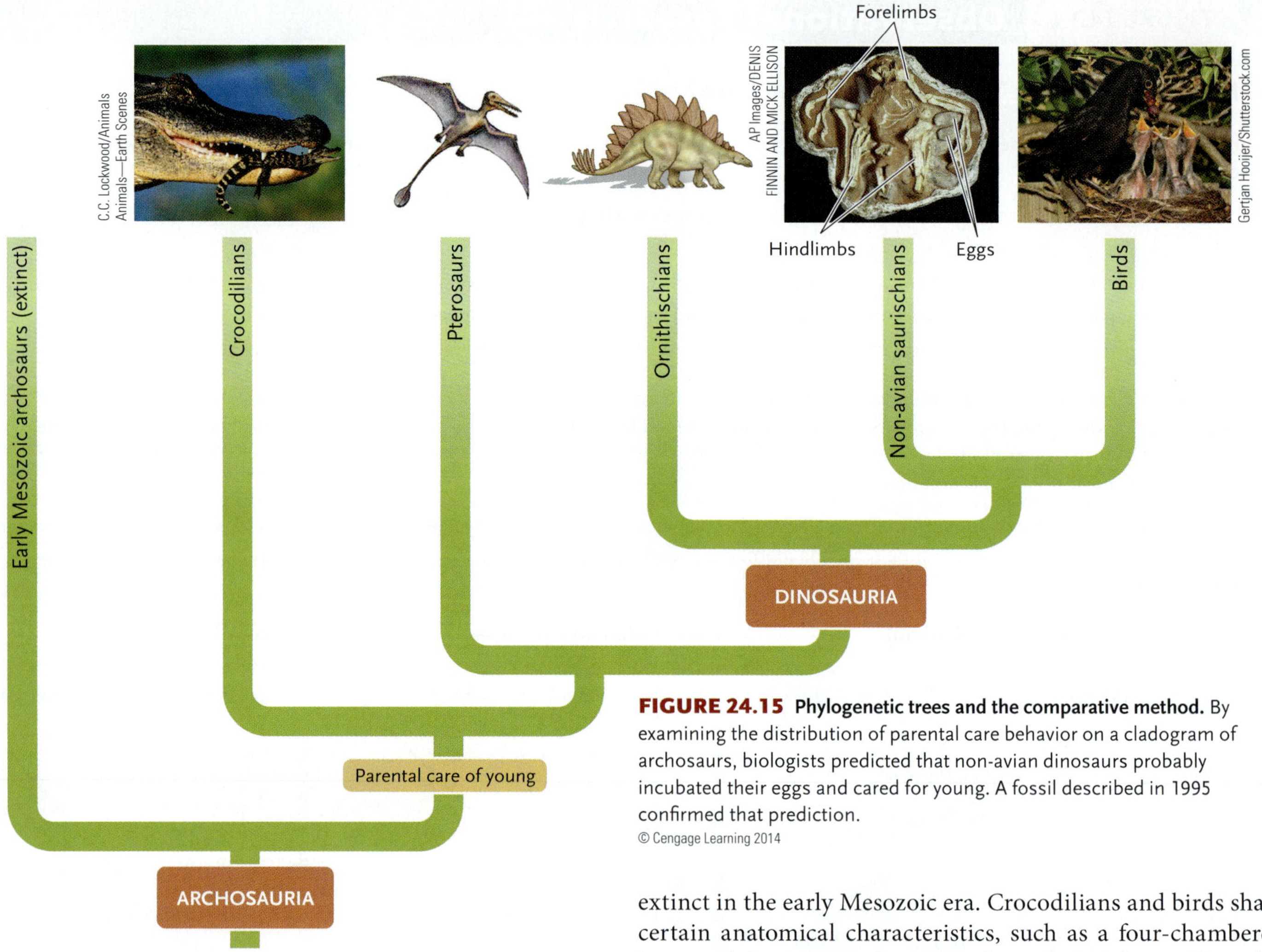

FIGURE 24.15 Phylogenetic trees and the comparative method. By examining the distribution of parental care behavior on a cladogram of archosaurs, biologists predicted that non-avian dinosaurs probably incubated their eggs and cared for young. A fossil described in 1995 confirmed that prediction.

evolution. Many factors can influence the rates at which mutations accumulate, and researchers must be cautious when evaluating divergence times estimated with this technique, especially if there are no independent data to corroborate the estimates.

Phylogenetic Trees Allow Biologists to Propose and Test Hypotheses

Accurate phylogenetic trees are essential tools for analyses that biologists describe as the "comparative method." With this approach, researchers compare the characteristics of different species to assess the homology of their similarities and infer where on the phylogenetic tree a particular trait appeared. The comparative method is used to study almost any sort of organismal trait, but in this section we will focus on parental care behavior.

As noted above, birds and crocodilians are included within the Archosauria, a clade that also includes non-avian dinosaurs (that is, extinct terrestrial dinosaurs that are not included within the bird clade), pterosaurs (an extinct group of flying vertebrates, not closely related to bats, with wing surfaces formed by skin), and a number of other groups that became extinct in the early Mesozoic era. Crocodilians and birds share certain anatomical characteristics, such as a four-chambered heart and the one-way flow of air through their lungs. They also share behavioral characteristics, including the production of mating calls (songs in birds and roars in crocodilians), nest-building behavior, and parental care of their young. Female crocodilians guard their nests and keep them moist with urine. They also excavate the nest as the young hatch, and then carry them to standing water. Young stay with their mother for about a year, feeding on scraps of food that fall from her mouth.

Did similar parental care behavior evolve independently in birds and crocodilians, or is it truly a synapomorphy? Did most Mesozoic archosaurs care for their young as birds and crocodilians do today? The comparative method seeks answers to these questions by examining the phylogenetic tree for archosaurs **(Figure 24.15).** As you can readily see, crocodilians and birds lie on widely separated branches of the archosaur tree, with pterosaurs and non-avian dinosaurs positioned between them. The most parsimonious inference about the evolution of parental care behavior is that it evolved once in the common ancestor of crocodilians and birds. If that inference is correct, then non-avian dinosaurs and pterosaurs probably also cared for their young in the nest. Indeed, that prediction was confirmed in 1995, when Mark A. Norell of George Washington University and his colleagues discovered a fossil of a non-avian dinosaur *(Oviraptor)* sitting on a nest full of eggs.

STUDY BREAK 24.6

1. What assumption underlies the use of genetic sequence differences between species as a molecular clock?
2. Are birds more closely related to non-avian dinosaurs or to crocodilians?

24.7 Molecular Phylogenetic Analyses

The application of molecular techniques to phylogenetic analyses has allowed systematic biologists to resolve some evolutionary puzzles that could not be addressed with older techniques. In this section we describe how molecular analyses have enabled public health researchers to identify sources of HIV infection in humans and elucidate the relationships among the three domains of the Tree of Life.

Molecular Phylogenetic Analyses Pinpoint the Origins of Infectious Diseases

Phylogenetic analyses also allow physicians and public health workers to identify the origin of infectious agents and follow their spread through a population. Many pathogenic organisms and viruses mutate as they proliferate, establishing derived character states that are ripe for phylogenetic analysis.

The human immunodeficiency virus (HIV), the agent that causes acquired immunodeficiency syndrome (AIDS) in humans, began to infect large numbers of people in the 1980s. As its devastating effects on humans became apparent, scientists scrambled to discover its origin. Genetic analyses linked it to the lentiviruses, specifically simian immunodeficiency virus (SIV), which infects dozens of monkey species as well as chimpanzees in Africa. Surprisingly, SIV does not cause illness in those animals, perhaps because their populations developed immunity to it after a long period of exposure.

Two distinct strains of HIV infect humans: HIV-1 is common in central Africa, and HIV-2 is common is West Africa. Did these strains evolve within human hosts, or did they exist before the virus was first transmitted to humans? An analysis by Beatrice H. Hahn of the University of Alabama at Birmingham and the Howard Hughes Medical Institute and colleagues at other institutions identified three major clades of SIV. The clade that infects chimpanzees includes HIV-1, and one of the clades that infect monkeys includes HIV-2 **(Figure 23.16).** Thus, the two strains of HIV apparently originated in nonhuman hosts. Scientists suspect that the transmission to humans occurred multiple times when hunters who were butchering bush meat—chimpanzees in central Africa and sooty mangabey monkeys in West Africa—acquired the virus through cuts on their hands.

Analyses of Gene Sequences Have Revealed the Branching Pattern of the Entire Tree of Life

On a very grand scale, molecular phylogenetics has revolutionized our view of the entire Tree of Life. The first efforts to create a phylogenetic tree for all forms of life were based on morphological analyses. However, these analyses did not resolve the branches of the Tree containing prokaryotes, which lack significant structural variability for analysis, or the relationships of those branches to eukaryotes.

In the 1960s and early 1970s, biologists organized living systems into five kingdoms. All prokaryotes were grouped into the kingdom Monera. The eukaryotic organisms were grouped into four kingdoms: Fungi, Plantae, Animalia, and "Protista." The kingdom "Protista" was always recognized as a polyphy-

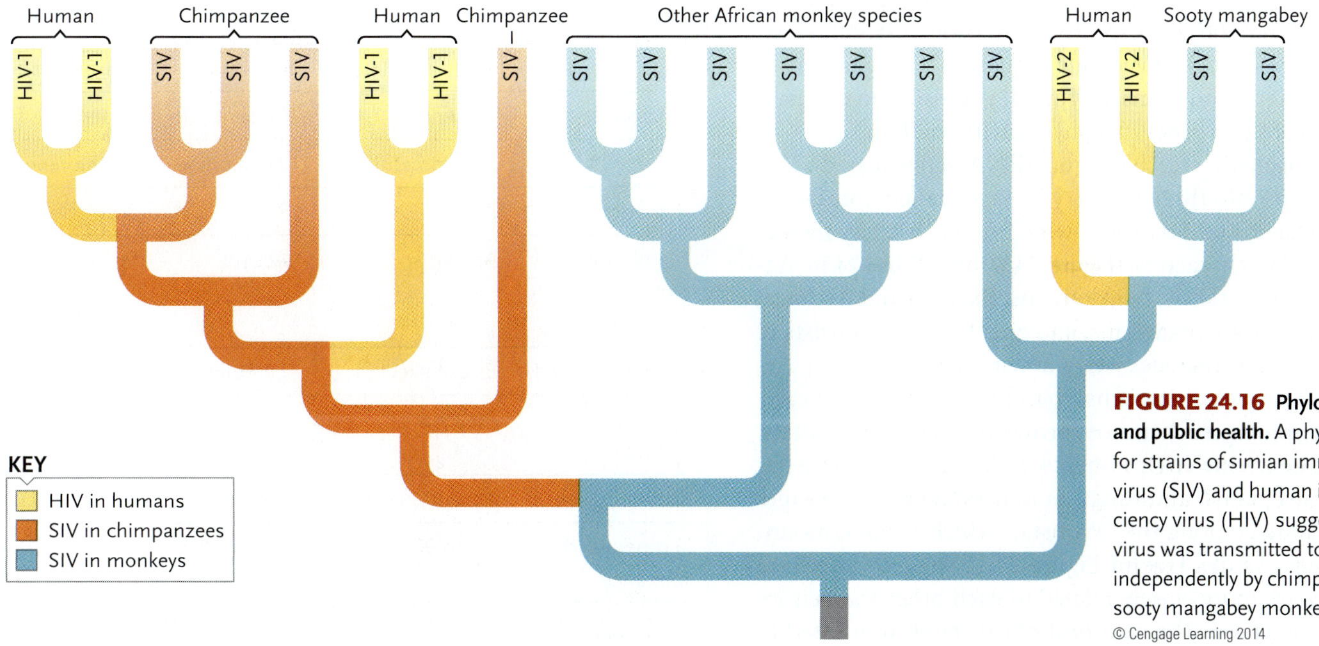

FIGURE 24.16 Phylogenetic trees and public health. A phylogenetic tree for strains of simian immunodeficiency virus (SIV) and human immunodeficiency virus (HIV) suggests that the virus was transmitted to humans independently by chimpanzees and sooty mangabey monkeys.

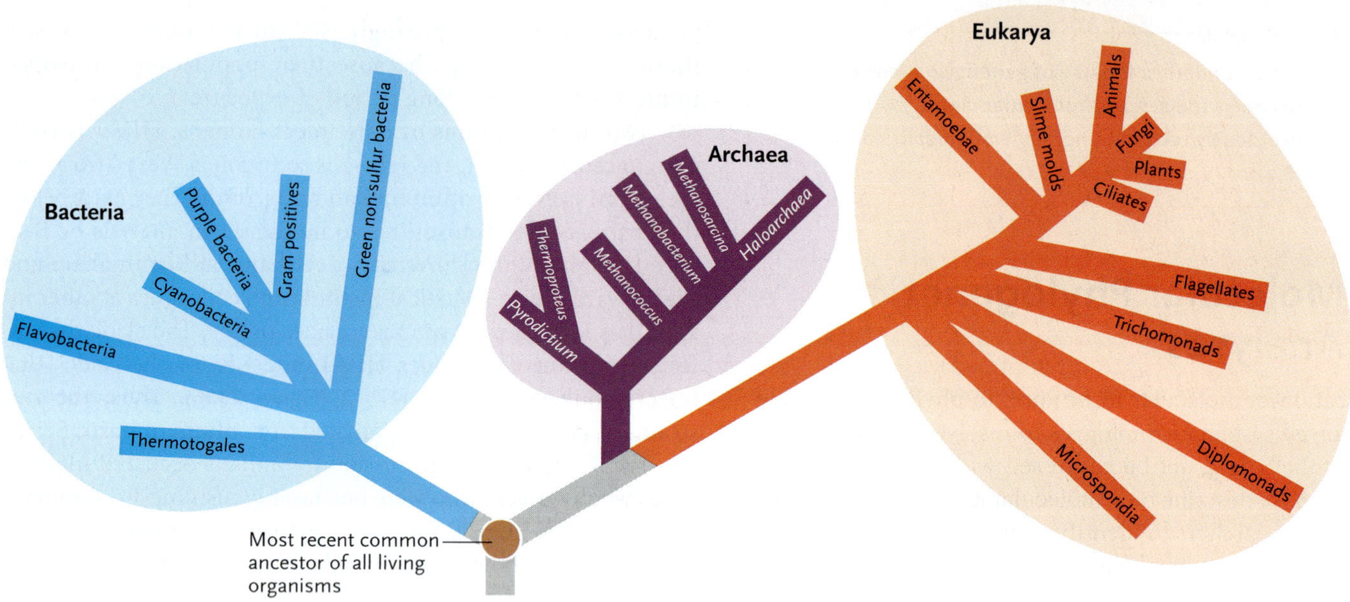

FIGURE 24.17 Three domains in the Tree of Life. Carl R. Woese's analysis of rRNA sequences suggests that all living organisms can be classified into one of three domains: Bacteria, Archaea, and Eukarya.

letic "grab bag" of unicellular or acellular organisms (discussed further in Chapter 27). Unfortunately, phylogenetic analyses based on morphology were unable to sort them into distinct evolutionary lineages.

In the 1970s, biologists realized that molecular phylogenetics provides an alternative approach. They simply needed to identify and analyze molecules that have been conserved by evolution over billions of years. Carl R. Woese, a microbiologist at the University of Illinois at Urbana–Champaign, identified the small subunit of ribosomal RNA (rRNA) as a suitable molecule for analysis. Ribosomes, the structures that translate messenger RNA molecules into proteins (see Section 15.1), are remarkably similar in all forms of life. They are apparently so essential to cellular processes that the genes specifying ribosomal structure exhibit similarities in their nucleotide sequences in organisms as different as bacteria and humans. Thus, it is possible to compare the sequences of these genes in a phylogenetic analysis.

The phylogenetic tree based on rRNA sequences divides living organisms into three primary lineages called domains—Bacteria, Archaea, and Eukarya—which differ in many molecular and cellular characters (**Figure 24.17** and **Table 24.1**). According to this hypothesis, two domains, Bacteria and Archaea, consist of prokaryotic organisms, and one, Eukarya, consists of eukaryotes. Bacteria includes well-known microorganisms. Archaea includes microorganisms that live in physiologically harsh environments, such as hot springs or very salty habitats, as well as less extreme environments. Eukarya includes the familiar animals, plants, and fungi, as well as the many lineages formerly included among the "Protista," which is not a monophyletic group. As the Tree in Figure 24.17 suggests, Archaea and Eukarya are more closely related to each other than either domain is to Bacteria. The next unit of this book is devoted to detailed analyses of the biology and evolutionary relationships between and within these three domains.

Our discussion of phylogenetic analysis has emphasized the importance of direct descent: the transmission of derived traits from ancestors to descendants. But as scientists analyze the complete genome sequences of an ever-growing list of organisms, they are discovering that the three domains in the Tree of Life have not had entirely independent evolutionary

TABLE 24.1 Some Differences among the Three Domains

Character	Bacteria	Archaea	Eukarya
Chromosome structure	Circular	Circular	Linear
DNA location	Nucleoid	Nucleoid	Nucleus
Chromosome segregation	Binary fission	Binary fission	Meiosis/mitosis
Introns in genes	Rare	Common	Common
Operons	Present	Present	Absent
Initiator tRNA	Formylmethionine	Methionine	Methionine
Ribosomes	70S	70S	80S
Membrane-enclosed organelles	Absent	Absent	Present
Membrane lipids	Ester-linked	Ether-linked	Ester-linked
Peptidoglycan in cell wall	Present	Absent	Absent
Methanogenesis	Absent	Present	Absent
Temperature tolerance	Up to 90°C	Up to 120°C	Up to 70°C

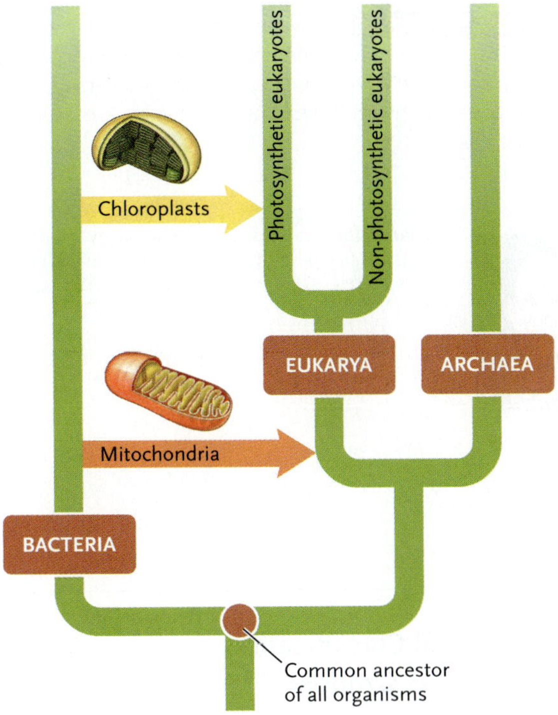

FIGURE 24.18 Horizontal gene transfer. Molecular analyses suggest that horizontal gene transfer from bacteria brought mitochondria to an ancestral eukaryote. A later transfer brought chloroplasts into the cells of some eukaryotes.

histories. Although inheritance from one generation to the next has produced the clades we recognize today, **horizontal gene transfer,** the movement of genetic material from one clade to another, has also been important (discussed further in Chapters 26 and 27). The mechanisms that facilitate horizontal gene transfer include transformation and transduction (discussed in Section 17.1), viral infection, and the incorporation of one organism into another. For example, the genes that regulate metabolism in eukaryotes are more similar to those in Bacteria than to those in Archaea, leading researchers to hypothesize that the mitochondria in eukaryotic cells may have originated when bacterial genes, or entire bacteria, were transferred horizontally into an ancestral eukaryote. The chloroplasts of plants may well have originated in a similar way **(Figure 24.18).** Thus, a more accurate portrait of the evolutionary history of life on Earth may require networklike connections between the three major trunks on the Tree of Life. Researchers are hard at work analyzing these connections.

STUDY BREAK 24.7

1. **Did HIV originate in humans, or was the infection acquired from other animals?**
2. **Why was a phylogenetic analysis of prokaryotes based on molecular sequence data more successful than the analysis based on morphological data?**

UNANSWERED QUESTIONS

Should we abandon the traditional Linnaean hierarchy in favor of a more evolutionary classification?

Diligent Kindly Professors Cannot Often Fail Good Students—or some equally silly mnemonic device for remembering the Linnaean taxonomic hierarchy—is all that many students recall about systematics. Even if they remember the underlying rank names—is G for "group" or "genus"—they often forget that Linnaeus conceived his system of classification more than a century before Darwin articulated his theory of evolution, which revolutionized our understanding of biological diversity. In the more than 150 years since Darwin published *On the Origin of Species,* systematists have sought not only to categorize life's diversity but, more important, to understand its origins. The broad relevance of studies in systematics has become increasingly clear as biologists have discovered that systematic principles are as important in tracing the emergence and spread of HIV as they are in distinguishing a duck from a dove.

More than a century and a half after it was first published, Darwin's theory has an increasingly revolutionary impact. Perhaps the most striking recent example is a call for the complete abandonment of the Linnaean taxonomic hierarchy. Although biologists thought they had reconciled the perspectives of Darwin and Linnaeus, a growing minority of systematists now believe that any effort to catalog and categorize life's diversity must be explicitly phylogenetic and free of the arbitrary ranks that Linnaeus invented. This movement, which has been codified in the PhyloCode initiative, is fueled largely by newly available molecular data, vastly improved phylogenetic methods, and increasingly fast computers. These advances offer the potential to reconstruct accurate and fully resolved phylogenetic trees at a scale never before possible. For the first time, biologists see real progress in accurately reconstructing the entire Tree of Life. Although we are still far from achieving this goal, every day millions of new, phylogenetically informative DNA fragments are being sequenced and analyzed by thousands of computers running around the clock.

Although PhyloCode's synthesis of taxonomy and evolutionary systematics may be long overdue, this attempted coup is not without controversy. For example, some systematists contend that such a radical revision of our taxonomic system will introduce confusion and instability in the naming of species. Even the revolution's adherents recognize that we still face many challenging limitations to the synthesis between taxonomic practice and Darwinian principles. Nowhere is this more evident than in the definition of species.

During Linnaeus' time, species were viewed as immutable natural types created by God. Darwin, however, formulated his theory on the principle that species change over time. Although the truth of this basic hypothesis is no longer a subject of debate, its practical implications for delimiting species boundaries and understanding how new species form are among the most exciting areas of study in modern systematics. Most practicing systematists view species as real (that is, biologically meaningful) categories, but the criteria for recognizing species vary dramatically among systematists working on different types of organisms (plants versus animals, or organisms that reproduce asexually versus those that reproduce sexually). Using new molecular tools and sophisticated genetic experiments, evolutionary biologists are beginning to probe the precise genetic basis of species. Over the past decade, a small number of "speciation genes" have been identified; more such discoveries are sure to follow in the coming years. Although many of these studies have been restricted to model research organisms, such as fruit flies, the new tools offered by the fields of genomics and bioinformatics offer the potential to address similar questions in an increasingly broad array of organisms.

Simply put, the systematics of today is not that of your grandparents. Given the enormous challenge involved in categorizing and understanding the origin and evolutionary relationships of millions of species, many additional changes are on the horizon. For the next generation of systematists, however, a better mnemonic to remember may be "Do Keep Probing Charles' *Origin* For Good Systematics."

THINK LIKE A SCIENTIST How easily does the Linnaean system of classification accommodate the phenomenon of horizontal gene transfer? Is a completely phylogenetic view of biodiversity, like PhyloCode, more consistent with the idea that genes have moved horizontally between clades?

Richard Glor

Richard Glor conducts research on the evolution of *Anolis* lizards at the University of Rochester. To learn more about Dr. Glor's research, go to http://www.lacertilia.com.

REVIEW KEY CONCEPTS

To access the course materials and companion resources for this text, please visit www.cengagebrain.com.

24.1 Nomenclature and Classification

- Linnaeus invented a system of binomial nomenclature in which each species receives a unique two-part name.
- Species are organized into a taxonomic hierarchy, which largely reflects the pattern of branching evolution (Figure 24.2). Species classified in the same lower-level taxon have a more recent common ancestor than species classified in the same higher-level taxon.

24.2 Phylogenetic Trees

- Phylogenetic trees are hypotheses that portray the branching pattern of evolution (Figure 24.3). Most phylogenetic trees have an implicit or explicit time line that indicates the relative times for cladogenesis. Branch points are described as nodes, and monophyletic lineages are called clades. Two lineages that share a node are called sister clades. Clades can be rotated at nodes without changing the meaning of the tree (Figure 24.4).
- Evolutionary biologists define monophyletic taxa directly from the structure of a phylogenetic tree (Figure 24.5). Biologists never intentionally define paraphyletic or polyphyletic taxa.

Animation: Evolutionary tree diagrams

Animation: Primate evolutionary tree

24.3 Sources of Data for Phylogenetic Analyses

- Homologies are characters that two or more descendant species inherited from their common ancestor (Figure 24.6). Homologous structures are similar in anatomical detail, and they often show a similar pattern of embryonic development. Characters that are similar for any other reason are described as homoplasies (Figure 24.7).
- Systematists often use morphological characters in phylogenetic analyses.
- If morphological characters do not distinguish species, behavioral characters may be useful (Figures 24.8).
- Systematists today use genetic sequence data to reconstruct phylogenetic history.

Animation: Evolution of jaws

24.4 Traditional Classification and Paraphyletic Groups

- In traditional systematics, biologists used the similarities and differences between organisms to construct a phylogenetic tree and a classification.
- Traditional classifications sometimes included paraphyletic groups (Figure 24.9A).

24.5 The Cladistic Revolution

- Derived character states can serve as markers of clades.
- Systematists use evidence from the fossil record as well as outgroup comparison to identify which character states are derived and which are ancestral (Figure 24.10).
- Cladistic analyses use synapomorphies (derived character states) to construct phylogenetic hypotheses (Figure 24.9B).
- The construction of a simple using cladistic methods is straightforward (Figure 24.11).
- Systematists use genetic distance data (Figure 24.13), the principle of parsimony (Figure 24.12), and maximum likelihood techniques to evaluate alternate phylogenetic hypotheses.

Animation: Interpreting a cladogram

24.6 Phylogenetic Trees as Research Tools

- If a nucleic acid accumulates mutations at a reasonably constant rate, it can be used as a molecular clock (Figure 24.14).
- Phylogenetic trees allow biologists to propose and test hypotheses about when certain characters evolved (Figure 24.15).

24.7 Molecular Phylogenetic Analyses

- Molecular techniques allow systematists to address questions that they could not have posed in the mid-twentieth century. Phylogenetic analyses are useful tools in medicine because they allow researchers to track the origin and spread of infectious diseases (Figure 24.16).
- Sequence data from the small subunit ribosomal RNA allowed researchers to recognize the three domains of life (Figure 24.17). Horizontal gene transfer complicates our understanding of the history of biodiversity (Figure 24.18).

Animation: Current evolutionary tree

UNDERSTAND & APPLY

Test Your Knowledge

1. The evolutionary history of a group of organisms is called its:
   a. classification.
   b. taxonomy.
   c. phylogeny.
   d. domain.
   e. outgroup.
2. In the Linnaean hierarchy, the organisms classified within the same taxonomic category are called:
   a. a phylum.
   b. a taxon.
   c. a genus.
   d. a binomial.
   e. an epithet.
3. When systematists study morphological or behavioral traits to reconstruct the evolutionary history of a group of animals, they assume that:
   a. similarities and differences in phenotypic characters reflect underlying genetic similarities and differences.
   b. the animals use exactly the same traits to identify appropriate mates.
   c. differences in these traits caused speciation in the past.
   d. the adaptive value of these traits can be explained.
   e. variations in these traits are produced by environmental effects during development.
4. Which of the following pairs of structures are homoplastic?
   a. the wing skeleton of a bird and the wing skeleton of a bat
   b. the wing of a bird and the wing of a fly
   c. the eye of a fish and the eye of a human
   d. the bones in the foot of a duck and the bones in the foot of a chicken
   e. the toes on the foot of a lizard and the toes on the foot of a human
5. Which of the following does *not* help systematists determine whether a morphological character state is ancestral or derived?
   a. outgroup comparison
   b. patterns of embryonic development
   c. studies of the fossil record
   d. studies of the character in more related species
   e. dating of the character by molecular clocks
6. In a cladistic analysis, a systematist groups together organisms that share:
   a. derived homologous traits.
   b. derived homoplastic traits.
   c. ancestral homologous traits.
   d. ancestral homoplastic traits.
   e. all of the above.
7. A monophyletic taxon is one that includes:
   a. an ancestor and all of its descendants.
   b. an ancestor and some of its descendants.
   c. organisms from different evolutionary lineages.
   d. an ancestor and those descendants that still resemble it.
   e. organisms that resemble each other because they live in similar environments.
8. Which of the following is *not* an advantage of using molecular characters in a systematic analysis?
   a. Molecular characters provide abundant data.
   b. Systematists can compare molecules among species that are morphologically very similar.
   c. Systematists can compare molecules among species that are morphologically very different.
   d. Nucleotide sequences in DNA are generally not influenced by environmental factors.
   e. Systematists can easily determine whether base substitutions in the DNA of two species are synapomorphies.
9. To construct a phylogenetic tree by applying the principles of parsimony to molecular sequence data, one would:
   a. start by making assumptions about variations in the rates at which different DNA segments evolve.
   b. group together organisms that share the largest number of ancestral sequences.
   c. group together organisms that share derived sequences, matching the groups to those defined by morphological characters.
   d. group together organisms that share derived sequences, minimizing the number of hypothesized evolutionary changes.
   e. identify derived sequences by studying the embryology of the organisms.
10. Which of the following underlying assumptions allows differences in a particular molecular sequence to be used as a molecular clock?
   a. The sequence never experiences any mutations.
   b. The sequence codes for a protein.
   c. The sequence accumulates mutations at a reasonably constant rate.
   d. The sequence is part of a mitochondrial gene.
   e. The sequence codes for small subunit ribosomal RNA.

Discuss the Concepts

1. In the past, systematists commonly used the amino acid sequences of proteins and DNA sequences in phylogenetic analyses. Think about the genetic code (Section 15.1), and explain why phylogenetic hypotheses based on DNA sequences may be more accurate than those based on amino acid sequences.
2. Traditional evolutionary systematists identify the Reptilia as one class of vertebrates, even though we know that this taxon is paraphyletic. Describe disadvantages of defining paraphyletic taxa in a classification.
3. The following table provides information about the distribution of ancestral and derived character states for six systematic characters (labeled 1 through 6) in five species (labeled A through E). A "d" means that the species has the derived state, and an "a" means that it has the ancestral state. Construct a phylogenetic tree for the five species using the principle of parsimony; in other words, assume that each derived character state evolved only once in this group of organisms. Mark the branches of the phylogenetic tree to show where each character changed from the ancestral to the derived state.

	Character					
Species	1	2	3	4	5	6
A	a	a	a	a	a	a
B	d	a	a	a	a	d
C	d	d	d	a	a	a
D	d	d	d	a	d	a
E	d	d	a	d	a	a

4. Imagine that you are a systematist studying a group of little-known flowering plants. You discover that the phylogenetic tree based on flower morphology differs dramatically from the phylogenetic tree based on DNA sequences. How would you try to resolve the discrepancy? Which tree would you believe is more accurate?
5. Create an imaginary phylogenetic tree for an ancestral species and its 10 descendants. Circle a monophyletic group, a polyphyletic group, and a paraphyletic group on the tree. Explain why the groups you identify match the definitions of the three types of groups.

Design an Experiment

Imagine that you are trying to determine the evolutionary relationships among six groups of animals that look very much alike because they have few measurable morphological characters. What data would you collect to reconstruct their phylogenetic history?

Interpret the Data

The phylogenetic tree for 12 cat species (Felidae) reproduced at right was assembled from molecular sequence data. Which species is the domestic cat's closest relative? Which clade is the sister taxon to tigers? Are bobcats more closely related to cougars or to ocelots?

Apply Evolutionary Thinking

Is the construction of parsimonious phylogenetic trees more consistent with the predictions of the phyletic gradualism hypothesis or the punctuated equilibrium hypothesis of evolutionary change? You may want to review material in Section 23.5 before answering this question.

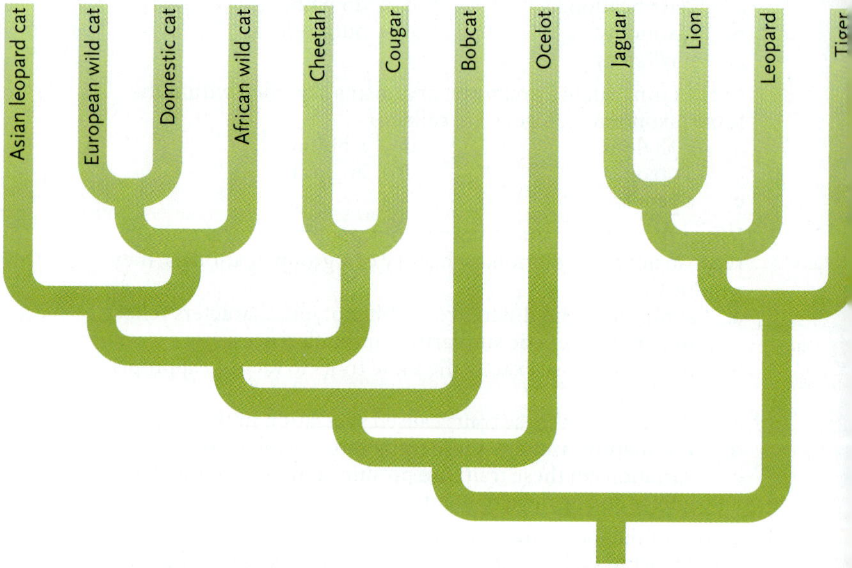

Source: From Warren E. Johnson et al. 2006. The late Miocene radiation of modern Felidae: A genetic assessment. *Science* 311:73–77.

Black smoker hydrothermal vents on the ocean floor. Many scientists support the theory that life developed near hydrothermal vents, where superheated, mineral-rich water is found.

25

The Origin of Life

STUDY OUTLINE

25.1 The Formation of Molecules Necessary for Life

25.2 The Origin of Cells

25.3 The Origins of Eukaryotic Cells

Why it matters . . . In 1927, Belgian priest and astronomer George Lemaître proposed the Big Bang Theory, which is now the dominant scientific theory about the origin of the universe. According to this theory, an incomprehensibly vast explosion about 14 billion years ago produced the matter and energy of our universe. As the universe expanded, slight variations in temperature and density led to an uneven distribution of matter. Gravitational attraction caused gas to form clouds and then collapse into more concentrated collections of matter. Stars, in which nuclear reactions generated the major elements we find in present-day biological molecules, were formed. Stars that were aging before the birth of our solar system expanded to produce red supergiants, which eventually exploded as supernovae. Such an explosion ejects the contents of a star at near light speed, in effect distributing all the elements of the periodic table across the universe. When our Sun formed, some of the elements coalesced with it and also formed the planets of our solar system. In other words, stardust, the remnants of long-ago stars, is the elemental origin of life on Earth. The origin of cellular life required two other fundamental conditions: a continual source of energy, and a temperature range in which liquid water can form. The continual energy source is, of course, the Sun. The relatively narrow temperature range suitable for life relates to the nature of the Sun, the distance of Earth from the Sun, and the properties of the atmosphere that cause trapping of heat.

Earth is estimated to have formed approximately 4.6 billion years ago. When and how did living cells form from inert materials? Answering these questions is difficult because we must rely on indirect geological evidence, and on experimental data that are open to debate and multiple interpretations.

Figure 25.1 outlines the key events in the origin and early evolution of life, which we will examine in this chapter. The earliest events are uncertain, but probably include the development of

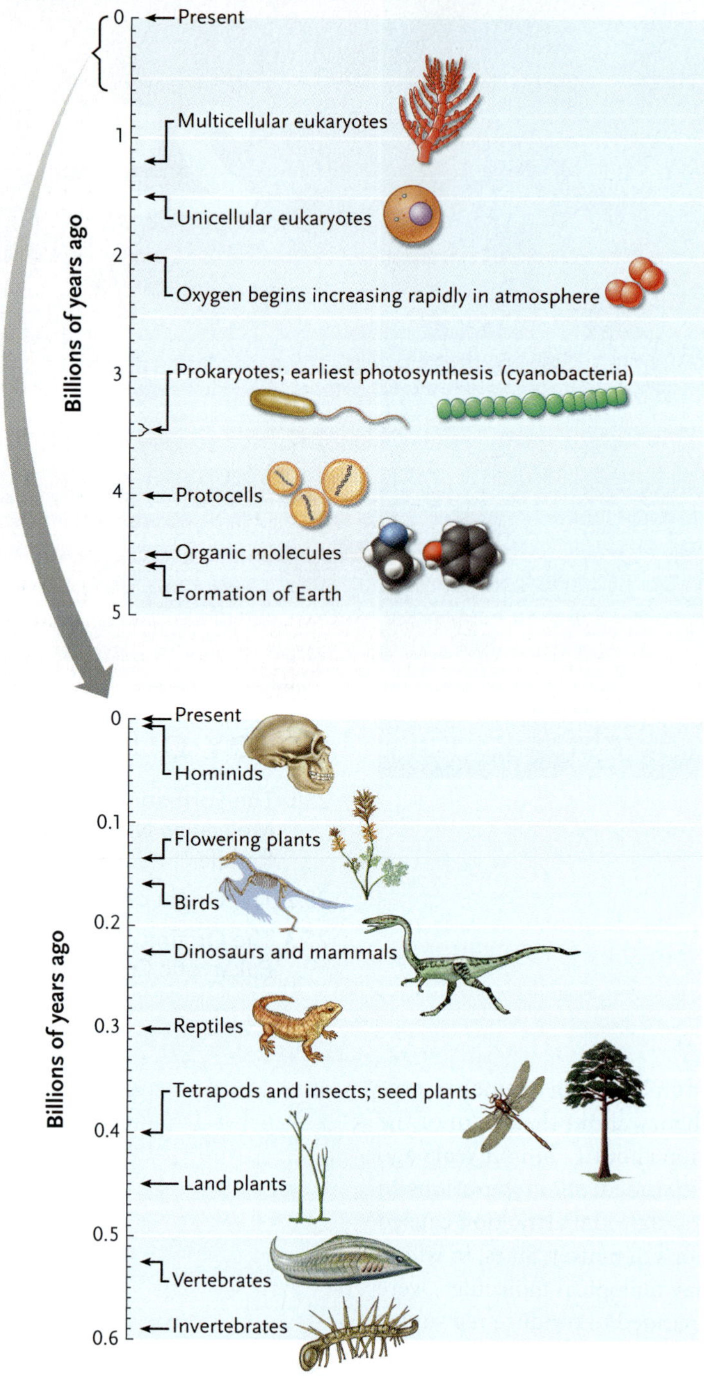

FIGURE 25.1 A timeline for the evolution of cellular life, in the context of the evolution of major present-day organisms. The dates for the origin of cellular life derive from geological evidence. Each date is the subject of debate, as discussed in the text. A more complete presentation of the timeline for the evolution of organismal life is in Table 23.1.

protocells, primitive cell-like structures that have some of the properties of life and that might have been the precursors of cells. Some geological evidence shows that prokaryotic cells were present in the Archaean eon by 3.5 billion years ago (see Table 23.1 for the geological time scale). The activity of photosynthetic prokaryotes generated an oxygen-enriched atmosphere, a condition probably necessary for the development of the first eukaryotic cells. Fossils of cells that resemble modern eukaryotic cells appeared about 1.5 billion years ago.

25.1 The Formation of Molecules Necessary for Life

All present-day living cells have the following characteristics: (1) a boundary membrane separating the cell interior from the exterior; (2) one or more nucleic acid coding molecules located in a nuclear region (a nucleus in eukaryotes and a nucleoid region in prokaryotes); (3) a system using the coded information in the nucleic acids to make proteins and, through them, other biological molecules; and (4) a metabolic system providing energy for these activities. The complexity of these systems makes it highly unlikely that living cells appeared suddenly from nonliving matter. Rather, there must have been a transition from nonliving to living matter, although we lack direct evidence for how it occurred.

Most scientists study the origin of life by assuming that it originated from nonliving matter on Earth, through chemical and physical processes no different from those operating today. Hypotheses made under these assumptions are testable to the extent that the chemical and physical processes can be duplicated in the laboratory.

But some scientists consider a possible extraterrestrial origin of life. Analysis of meteorites has shown that they contain some organic molecules that are characteristic of living organisms. Could a living cell or organism, or perhaps a spore, have arrived in such a way? Most scientists believe it is unlikely that a cell or an organism could have survived a long journey in space, even if protected from radiation, or that it could have survived intense heating while traveling through Earth's atmosphere and the actual impact with Earth. At this point the hypothesis that life arrived on Earth by interplanetary transport is not testable, so it cannot be ruled out. Nonetheless, even if a living organism arrived from space and spawned a population on this planet, life would still have had to arise from nonliving matter in a similar way on the organism's home planet.

Conditions on Primordial Earth Led to the Formation of Organic Molecules

The formation of Earth led to a planet with three layers. In its center is the core, which is surrounded by the mantle. Both the core and the mantle are rich in iron. Surrounding the mantle is the relatively thin crust, which consists mainly of silicates. Volcanic activity on early Earth released gases from the crust and mantle such as CO_2, H_2, and N_2, which helped form the first atmosphere. Water vapor emitted from volcanoes formed the first oceans, lakes, and rivers.

Primordial Earth met several basic conditions necessary for life to begin. Although Earth's gravitational pull was strong enough to retain an atmosphere, it was not strong enough to compress the atmospheric gases into liquid form. Earth's distance from the Sun was such that, on average, sunlight warmed the surface enough to keep much of the liquid water from freezing, but not enough to boil the water. *Liquid water is essential for the chemistry of biological systems* (see Section 2.4).

In the primordial atmosphere, any molecular oxygen would have reacted with elements of the crust and atmosphere to form oxides. Spontaneous reactions of hydrogen, nitrogen, and carbon would have produced ammonia (NH_3) and methane (CH_4).

As Earth's surface cooled, natural sources of energy caused chemical bonds to break and reform, leading to the formation of organic molecules at about 4.5 billion years ago. In addition to sunlight and electrical discharges during storms and volcanic activity, radioactivity from atomic decay and heat from volcanoes, geysers, and hydrothermal (hot water) vents in the sea floor all acted on the primordial atmosphere and crust—as they still do today. As many as a half-billion years may have passed before the concentrations of organic molecules reached levels where their interactions formed more complex organic substances. We now consider the current thinking about how simple molecules were converted into the key molecules of life.

The Oparin–Haldane Hypothesis Initiated Scientific Investigations into the Origin of Life

Scientific efforts to explain the origin of life began with a major hypothesis proposed independently in the 1920s by two investigators, Aleksandr I. Oparin, a Russian plant biochemist at Moscow State University in Russia, and J. B. S. Haldane, a British geneticist and evolutionary biologist at Cambridge University in England. Their hypothesis rested on the critical assumption that Earth's primordial atmosphere was radically different from today's atmosphere. They proposed that, rather than being an oxygen-rich (oxidizing) atmosphere as it is now, the early atmosphere was a strongly reducing atmosphere, due to the concentration of substances such as hydrogen (H_2), methane (CH_4), ammonia (NH_3), and water, which are *fully reduced*—they contain the maximum possible number of electrons and hydrogens (see Section 8.1). They reasoned that the primordial atmosphere would have had an abundance of electrons and hydrogens for reduction reactions, which could create organic molecules from inorganic elements and compounds. Energy to drive the reductions, according to the hypothesis, came from solar energy and other natural sources such as the electrical energy of lightning in atmospheric storms and from volcanic activity.

The absence of oxygen in the primitive atmosphere is essential to the Oparin–Haldane hypothesis. Oxygen can reverse reductions by removing electrons and hydrogens from organic molecules (see Section 8.1). In other words, if oxygen was present, the newly formed molecules would have been broken down quickly by oxidation.

Oparin and Haldane proposed that reductions occurring on the primordial Earth produced great quantities of organic molecules. The molecules accumulated because the two main routes by which such substances break down today, chemical attack by oxygen and decay by microorganisms, could not take place. According to Oparin and Haldane's hypothesis, the organic substances would have become so concentrated that the oceans and other bodies of water resembled a prebiotic soup. Recent geochemical evidence, however, suggests an alternate mechanism for the formation of organic molecules. According to this evidence, early Earth's oceans actually contained some amount of oxidized forms of nitrogen, sulfur, and iron that were generated by UV radiation that penetrated the atmosphere in the absence of an ozone layer. Those oxidized materials could have reacted with reduced molecules of the crust, releasing energy that could have been used to produce some more complex biomolecules.

In their model, Oparin and Haldane assumed that the highly concentrated organic molecules would tend to aggregate in random combinations and that, by chance, some of the combinations were able to carry out one or more primitive reactions characteristic of life, such as increasing in mass by adding new materials. Later, scientists reasoned that these combinations were able to compete successfully against less efficient combinations for space and materials in an organic soup. As a result, they persisted and became more numerous.

Chemistry Simulation Experiments Support the Oparin–Haldane Hypothesis

In 1953, Stanley L. Miller, a graduate student in Harold Urey's laboratory at the University of Chicago, tested the Oparin–Haldane hypothesis. They created a laboratory simulation of conditions Oparin and Haldane believed existed on early Earth; namely involving a reducing atmosphere consisting of H_2, CH_4, and NH_3, as well as H_2O **(Figure 25.2)**. After one week of running the apparatus, as much as 15% of the carbon was now in the form of organic compounds, including amino acids. The significance of the finding at the time was enormous; amino acids, which are essential to cellular life, could be made under the conditions scientists believed existed on early Earth.

The synthesis of complex biological molecules in a reducing atmosphere in the Miller–Urey experiment supported the Oparin–Haldane hypothesis. However, many scientists believe that the atmosphere at the time key organic molecules were formed on early Earth was not reducing. Rather, the atmosphere at that time is thought to have been neutral (neither reducing nor oxidizing), consisting mostly of gases released by volcanic activity, notably CO_2, N_2, CO, and H_2S. Indeed, experiments using these gases in the Miller–Urey apparatus, in addition to the original gases used, resulted in the formation of many other molecules of biological significance. Produced, for example, were all the building blocks of complex biological molecules, including amino acids; fatty acids; the purine and pyrimidine building blocks of nucleic acids; sugars such as glyceraldehyde, ribose, deoxyribose, glucose, and fructose; and phospholipids, which form the lipid bilayers of biological membranes.

FIGURE 25.2 Experimental Research

The Miller–Urey Apparatus Demonstrating That Organic Molecules Can Be Synthesized Spontaneously under Conditions Simulating Primordial Earth

Question: Do chemistry simulation experiments support the Oparin–Haldane hypothesis that reduction reactions on primordial Earth produced organic molecules?

Experiment: Stanley Miller set up an experiment to simulate chemical conditions of early Earth **(Figure).** He placed components of a reducing atmosphere—hydrogen, methane, and ammonia—in a closed apparatus. Water vapor was added to the "atmosphere" by boiling water in one part of the apparatus, and it was removed by cooling and condensation in another part. This simulated the cycle of release of water vapor into the atmosphere, and condensation into water. Miller exposed the gases to an energy source in the form of continuously sparking electrodes to simulate lightning.

~4.5 billion years ago

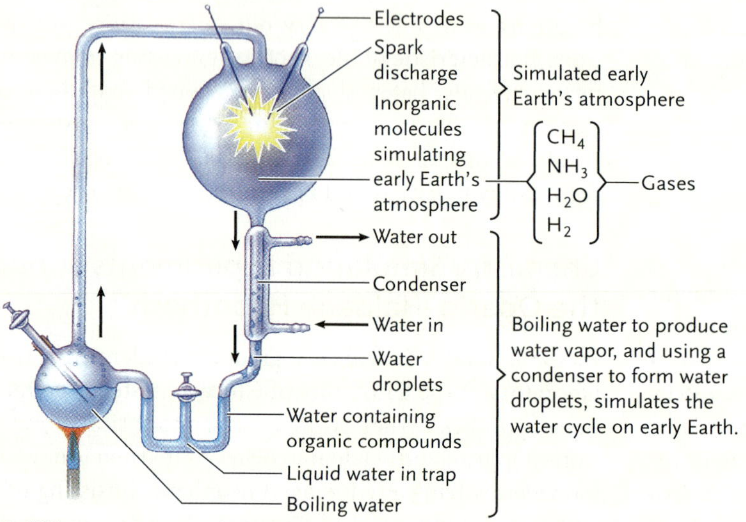

Results: After running the apparatus for only a week, Miller found a large assortment of organic compounds in the water, including urea, amino acids, and lactic, formic, and acetic acids—as much as 15% of the carbon was now in the form of organic compounds. Two percent of the carbon was in the form of amino acids, which form easily under sufficiently reducing conditions.

Conclusion: The synthesis of organic molecules, including biological molecules, in a reducing atmosphere supported the Oparin–Haldane hypothesis.

THINK LIKE A SCIENTIST What aspect of the Miller–Urey experimental design ruled out the possibility that amino acids were synthesized by living organisms such as bacteria?

Source: S. Miller. 1953. A production of amino acids under possible primitive Earth conditions. *Science* 117:528–529. (Figure redrawn from an original courtesy of S. L. Miller, Copyright 1955 by the American Chemical Society.)

Scientists Have New Theories about the Sites for the Origin of Organic Molecules

Scientists have developed a number of other theories for how organic compounds may have arisen on early Earth. All of the theories have the reasonable assumption of the presence of liquid water. Remember that water is essential for the chemistry of biological systems (see Section 2.4). All of the theories have their advocates and critics. It is fair to say that there are controversial issues in each of them. Two of the theories are outlined here.

One current theory for the origin of life, which has significant support among scientists, is that life developed near hydrothermal vents in the sea floor (see photo on first page of the chapter). Many such vents exist in today's oceans, emitting bursts of mineral-rich water superheated to up to 400°C by submarine volcanoes. Scientists exploring hydrothermal vents find complex ecosystems associated with them.

Life might have originated near oceanic hydrothermal vents because reducing conditions existed there along with an abundance of the chemicals that are essential for life. Even now, there are high levels of hydrogen gas, methane, and ammonia around the vents. Indeed, based on simulation experiments, scientists hypothesize that hydrothermal vents could have produced a lot more organic material than that generated in the Miller–Urey experiment. However, if life did evolve near hydrothermal vents, we would expect many present-day hydrothermal-vent life forms to be ancient. This is not the case; in most cases, these organisms are closely related to modern organisms that do not live in hydrothermal vent environments. Critics of the hydrothermal-vent origin of life theory also argue that the temperature at the vents is too high to permit the origin of life. The critics argue that at the high temperature found at vents, the organic molecules are too unstable and would be destroyed as soon as they form. Supporters of the theory counter that the necessary organic molecules for life form not at the vent itself, but somewhere in the gradient between the hot water at the vent and the near-freezing water surrounding the vent.

Another theory is that some organic compounds had an extraterrestrial origin. (As mentioned earlier, scientists do not believe living cells or organisms of extraterrestrial origin could have survived a journey to Earth.) Indeed, many of the compounds made in the Miller–Urey experiment exist in outer space. For example, a meteorite that fell on Murchison, Australia, in 1969 contained more than 90 amino acids, only 19 of which are found on Earth. Since amino acids appear to be able to survive in outer space, they could potentially have been present when Earth was formed. DNA precursors have also been identified in some meteorites. It is possible that other organic compounds arrived by meteor impact, too. Repeated bombardment by meteors occurred during the first period of Earth's existence, the Hadean eon, from 4.6 to 3.85 billion years ago (see Figure 25.1 and Table 23.1), causing a cycle of vaporization of water in the oceans, which then cooled and recondensed.

STUDY BREAK 25.1

1. **Why is the reducing nature of early Earth's atmosphere key to the Oparin–Haldane theory of the origin of molecules necessary for life?**
2. **How do the theories about the sites for the origin of life differ?**

25.2 The Origin of Cells

No matter where organic molecules originated, they do not qualify as life. In this section, we discuss the key stage in the origin of life—the formation of the first cells.

Protocells Formed with Some of the Properties of Life

How did organic building blocks such as amino acids assemble into macromolecules such as proteins and nucleic acids? To answer this question, researchers have proposed and tested several processes. One process is the concentration of subunits by the evaporation of water. Another is *dehydration synthesis (condensation),* in which subunits assemble into larger molecules through removal of the elements of a molecule of water (see Section 3.1). Experiments with these processes under simulated conditions showed that both evaporation and condensation reactions can produce polypeptide chains from amino acids, polysaccharides from glucose and other monosaccharides, and nucleotides and nucleic acids from nitrogenous bases, ribose, and phosphates.

Scientists reason that spontaneous condensations and other reactions produced significant quantities of all the major biological molecules over the hundreds of millions of years following the initial formation of Earth. They hypothesize that the accumulation of organic matter set up the conditions necessary for the next stage, the chance assembly of molecules into aggregations that became membrane-bound to form primitive protocells, perhaps by 4.0 billion years ago **(Figure 25.3A)**. Protocells are key to the origin of life, because life depends on reactions occurring in a controlled and sequestered environment, the cell. Researchers have proposed several mechanisms for the assembly of organic molecules into aggregates, each of which has been successfully duplicated in laboratory experiments simulating primordial conditions. Two of those mechanisms are adsorption into clays and lipid bilayer assembly.

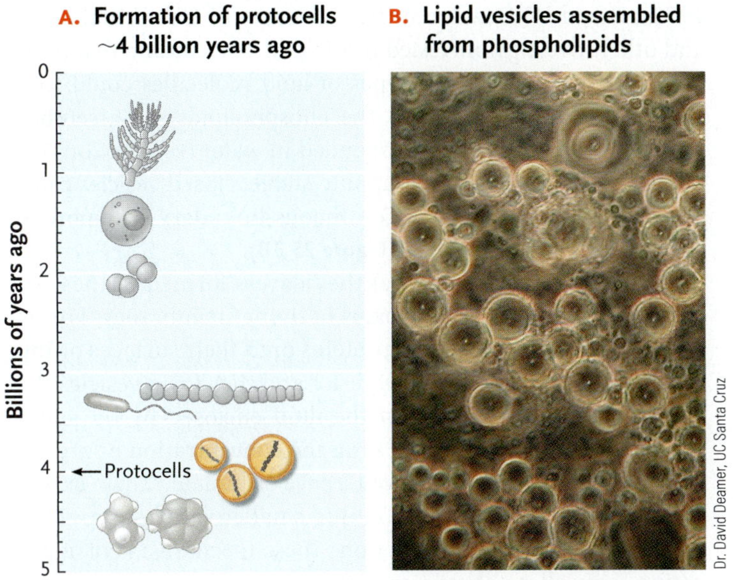

FIGURE 25.3 Protocells. (A) As a key event in the origin of life, protocells are thought to have formed ~4 billion years ago. **(B)** Phase micrograph of lipid vesicles assembled from phospholipids. Structurally, primitive protocells were vesicle-like, with a lipid bilayer membrane surrounding aggregations of major biological molecules.

EXPERIMENTAL SUPPORT FOR ADSORPTION INTO CLAYS Could clays have provided an ideal environment for molecular aggregation and interaction on the primitive Earth? Clays consist of very thin layers of minerals separated by layers of water only a few nanometers thick. The layered structure readily adsorbs ions and organic molecules and promotes their interactions, including condensations and other assembly reactions. Clays can also store potential energy, and therefore could have channeled some of the energy into reactions taking place between molecules adsorbed to them.

Several experiments have supported these proposals. For example, in the 1970s Noam Lahav at the Hebrew University of Jerusalem and Sherwood Chang of NASA's Ames Research Center added amino acids to clays and exposed the mixtures to water-content changes and fluctuating temperatures that are likely to occur in a tidal flat. After several cycles of the fluctuating conditions, polypeptides were detected in the clays. Other researchers found that RNA nucleotides linked to phosphate chains could combine into RNA-like molecules in clays. Accumulation of these and other macromolecules in the clays could have provided an environment in which they could react to carry out the first reactions of life.

However, even if molecules became organized in clay and some of the reactions of life commenced, it is not clear how a lipid bilayer membrane could have formed around them. Such a membrane is necessary to organize the molecules into protocells, the presumed precursors of cells. (The biological importance of lipid bilayers and membranes are discussed in Sections 2.4, 3.3, and 6.1.)

EXPERIMENTAL SUPPORT FOR LIPID BILAYER ASSEMBLY In the 1950s, R. J. Goldacre at Chester Beatty Research Institute, London, hypothesized that protocells could have formed starting with lipid bilayers that had assembled spontaneously. In the

1970s, David W. Deamer at the University of California, Davis, and other investigators tested this hypothesis, finding that phospholipids and some other types of lipid molecules could form under simulated conditions. The phospholipids self-assembled readily into bilayers when suspended in water (see Section 6.1). Often, the bilayers rounded up into stable, closed vesicles called liposomes, consisting of a continuous-boundary "membrane" surrounding an inner space **(Figure 25.3B)**.

Further tests showed that the bilayers formed in these experiments have many properties of living membranes. For example, they can incorporate proteins onto their surfaces or into the hydrophobic membrane interior and they form vesicles that can trap other substances in the fluid enclosed by the membrane. Potentially, on early Earth, the concentration of organic molecules in such vesicles could have stimulated their expansion in size and fragmentation into smaller vesicles, providing a primitive form of reproduction. These mechanisms of aggregation, as well as others, may have worked separately or together to form protocells.

Researchers continue to experiment on producing protocells in the laboratory as a step toward understanding the transition from the protocell to the living cell. Present-day thinking is that a boundary membrane system must have evolved simultaneously with a genetic information system, rather than one being formed before the other, as some earlier models have proposed. Experimental approaches being used are therefore directed toward generating self-replicating nucleic acid and membrane vesicle systems.

Living Cells May Have Developed from Protocells

Eventually the chemical reactions taking place in the primitive protocells became organized enough to make the transition to living cells. Exactly how this occurred is open to speculation. Two of the critical events necessary for the transition are the development of pathways that captured and harnessed the energy required to drive molecular synthesis, and the development of a system for the storage, replication, and translation of information for protein synthesis. Remember that proteins are the catalysts for most cellular reactions. How the information system developed is crucial to our understanding of the origin of life. The development of energy-harnessing reaction pathways is not well understood. Here we focus on the development of the information system.

In contemporary organisms, information flows from DNA to RNA to protein. This nucleic acid-based information system depends mostly on enzymatic proteins for replication, transcription, and translation of the nucleic acids. However, the specificity of enzymatic proteins depends on their amino acid sequences, which are determined by the sequences of nucleotides in nucleic acids. Thus, proteins depend on nucleic acids for their structure, and nucleic acids depend on proteins to catalyze their activities. How could one have appeared before the other? Scientists hypothesize the information system developed in stages.

The prevalent model for the development of the information system is the **RNA world model.** The RNA world model states that the first genes *and* enzymes were RNA molecules. That is, *ribozymes*—RNA molecules capable of catalyzing biochemical reactions—may have functioned both as informational molecules and as catalysts in protocells, without requiring protein enzymes for catalytic reactions (ribozymes are discussed in Section 4.6). Thus, a self-catalyzed RNA world may have been the first step in the development of an information system. (See *Unanswered Questions* in Chapter 4 for more about the RNA world model.)

Ribozymes may have originally developed by the chance assembly of RNA nucleotides taking part in oxidative and other metabolic reactions in protocells (RNA nucleotides such as ATP, NAD, and coenzyme A form important parts of many metabolic pathways, including glycolysis, respiration, and photosynthesis; see Chapters 8 and 9). The RNA molecules then developed the capacity to replicate themselves and other RNA molecules. That is, these RNA molecules acted both as templates—like mRNA—and as catalysts—like ribosomal RNA (see Section 15.4). Then ribozymes could replicate ribozymes, with no need for protein enzymes. Such self-replicating systems may have provided the basis of an RNA-based informational system, and founded the RNA world. *Insights from the Molecular Revolution* describes an experiment in which ribozymes that can replicate RNA were generated in a test tube.

In the RNA world, DNA would have developed as a subsequent step. At first, DNA nucleotides may have been produced by random removal of an oxygen atom from the ribose subunits of the RNA nucleotides. At some point, the DNA nucleotides paired with the RNA informational molecules, and were assembled into complementary copies of the RNA sequences. Some modern day viruses carry out this RNA-to-DNA reaction using the enzyme reverse transcriptase (see Section 18.1). Once the DNA copies were made, selection may have favored DNA as the informational storage molecule because it has greater chemical stability and can be assembled into much longer coding sequences than RNA can. RNA was left to function at intermediate steps between the stored information in DNA and protein synthesis, as it still does today.

As the RNA-based information system evolved, some RNAs may have acted as tRNA-like molecules, linking to amino acids and pairing with the RNA informational molecules. These associations could have led to the assembly of polypeptides with an ordered sequence of amino acids—the development of an RNA genetic code. When DNA took over information storage from RNA, the DNA would have picked up the code information that could then be transferred back to RNA by transcription.

Modern analysis of the ribosome, the organelle responsible for translation of mRNA (see Section 15.4) has shown that the enzymatic activity that catalyzes the formation of a peptide bond between amino acids is a function of one of the RNA molecules of the ribosome. This finding supports the proposal that, in addition to replicating themselves, RNA molecules also generated the first proteins.

INSIGHTS FROM THE Molecular Revolution

Toward the Evolution of Life in the Lab: The construction of protocell-like vesicles containing an active ribozyme

Life is characterized by membrane-bound cells that grow and divide, and in which enzymatic reactions take place in a controlled and sequestered environment. We have learned in this chapter so far that living cells may have developed from protocells. And the discovery of ribozymes led to the proposal that an RNA world was the first step in the development of a molecular information system that could store, replicate, and translate the information required for protein synthesis. Constructing a simple protocell in the lab that contains an active ribozyme would provide a system with which to explore the possible steps for the origin and early evolution of life. That is, the simplest cellular system capable of evolution requires a self-replicating informational molecule, and a means to compartmentalize that molecule.

Research Goal

Construct stable protocell-like vesicles capable of growth, and that contain an active ribozyme.

Experiments

Jack Szostak and his research team at the Howard Hughes Medical Institute and Harvard Medical School set out to construct model protocell vesicles. Many chemicals are capable of forming simple vesicles, but the key requirement here is that the vesicle membrane must be stable and capable of growth under conditions that favor ribozyme activity. For example, fatty acids form vesicles that can grow and divide, but ribozymes cannot function in them because they need magnesium ions (Mg^{2+}) for activity, and divalent cations such as Mg^{2+} cause the fatty acids to precipitate.

Results

The researchers found that a mixture of particular, chemically simple amphiphiles (molecules with both hydrophilic and lipophilic properties) form vesicles with the desired protocell-like properties. That is, these vesicles are stable and grow, but, importantly, remain intact in the presence of Mg^{2+}. The vesicles are also permeable to Mg^{2+}, as shown by an experiment in which a ribozyme included in the vesicles became active and self-cleaved when Mg^{2+} was added to the solution.

Conclusion

The researchers successfully constructed simple vesicles containing active ribozymes. They concluded that the stability of the vesicles, and their ability to grow and allow the diffusion into them of Mg^{2+}, are "critical for building model protocells in the laboratory and may have been important for early cellular evolution."

THINK LIKE A SCIENTIST Why was the observation that a ribozyme could become activated within the protocell-like vesicles important?

Source: I. A. Chen, K. Salehi-Ashtiani, and J. W. Szostak. 2005. RNA catalysis in model protocell vesicles. *Journal of the American Chemical Society* 127:13213–13219.

Of course, we have no way of knowing exactly how life originated. Sifting through the various models and theories, we can perhaps agree that there were some basic steps: (1) the abiotic (nonliving) synthesis of organic molecules such as amino acids; (2) the assembly of complex organic molecules from simple molecules, including protein or RNA or both; and (3) the aggregation of complex organic molecules (likely including those of the information system) inside membrane-bound protocells. Once the information system fully developed in the protocells, and it could replicate, and the protocells could divide, they became true living cells. The advent of living cells marked the beginning of biological evolution, which depends on cells that can reproduce and pass on information to their descendants.

Prokaryotic Cells Were the First Living Cells

The change to biological evolution set the stage for the appearance of all the features of cellular life. One of these features was a nuclear region that contained the DNA of the coding system and the mechanisms replicating the DNA and transcribing it into RNA. Another feature was a cytoplasmic region containing ribosomes and the enzymes required to translate RNA information into sequences of amino acids in proteins. The cytoplasm also contained an oxidative system supplying chemical energy for protein synthesis and assembly of other required molecules. A mechanism of cell division also evolved, allowing replicated DNA to be distributed equally between daughter cells. All these systems were enclosed by a membrane controlling the flow of molecules and ions in and out of the cell. The stages leading to this level may have taken more than a billion years, occupying the period from Earth's formation 4.6 billion years ago to about 3.5 billion years ago.

There is geological evidence of different kinds for early life:

- **Isotope ratios determined by radiometric dating.** Radiometric dating is a technique used to date a rock and is based on the decay of isotopes in the rock (see Section 23.1). Each isotope of an element has a specific decay rate. However, if rock is subjected to biological activity that uses one isotope preferentially over another, then the ratio of the two isotopes will differ from that in inert rock from the same time. Thus, an altered isotope ratio in rock or mineral samples can be interpreted as evidence of biological activity. The earliest rock samples showing a significant alteration of an isotope ratio, interpreted to involve preferential use of $^{12}CO_2$ rather than $^{13}CO_2$ (as is seen in present-day bacteria), date to about 3.7 billion years ago. This may mark the earliest evidence of biochemical reactions by living cells which, by any present theory, would be prokaryotes.
- **Stromatolites.** Fossil stromatolites are layers of carbonate or silicate rock that resemble present-day stromatolites **(Figure 25.4A)**. Geologists think that the fossils formed as layers of phototrophic prokaryotes grew and died, and their remaining forms were filled in by calcium carbonate or silica. ("Phototrophic" refers to an organism that obtains energy from light.) The earliest fossil stromatolites that are accepted by ge-

ologists date to about 3.4 billion years ago in the oldest rocks of the Archaean eon. However, the fossil rock is too deformed to allow details of cell structure to be analyzed, and some researchers question whether they are of biological origin.

- **Biosignatures. Biosignatures** are particular organic molecules in sedimentary rocks that could only have been formed by cellular activity. There are limited examples of biosignature molecules. One is a steroidlike molecule that is found only in the membrane lipids of cyanobacteria, and in no other organism known today. That particular biosignature molecule has been found in rock dated to 2.5 billion years ago, providing evidence for the presence of cyanobacteria by the end of the Archaean eon. Cyanobacteria are photosynthetic organisms. Since photosynthesis is a complex metabolic process, more primitive prokaryotes must have developed at an earlier time.
- **Microfossils.** A **microfossil** is the remains of a cell that has decayed and been filled in by calcium carbonate or silica (**Figure 25.4B**). (Section 23.1 discusses the fossil record and its use in providing information about life in the past.) The earliest microfossils are of filamentous and single-celled prokaryotes, and they date to about 2.0 billion years ago. Those microfossils are considered the most convincing geological evidence of early life. Microfossils from around 1.5 billion years ago include some with larger cells that resemble cells of present-day eukaryotic protists (see Chapter 27).

In sum, prokaryotes probably appeared first about 3.5 billion years ago and eukaryotes first appeared about 1.5 billion years ago, or perhaps earlier.

Subsequent Events Increased the Oxidizing Nature of the Atmosphere

According to Richard E. Dickerson of University of California, Los Angeles, and others, the earliest form of photosynthesis evolved about 3.5 billion years ago in the early prokaryotes. This form of photosynthesis probably used electron donors such as hydrogen sulfide (H_2S) that do not release oxygen. However, at some point, an enzymatic system evolved that could use the most abundant molecule of the environment, water (H_2O), as the electron donor for photosynthesis. This reaction split water into protons, electrons, and oxygen, which was released into the atmosphere.

The oxygen released by the water-splitting reaction accumulated in the atmosphere and set the stage for the development of electron transfer systems using oxygen as the final electron acceptor. These transfer systems arose when some cells developed cytochromes that could deliver low-energy electrons to oxygen (see Section 8.4). These cells were able to tap the greatest possible amount of energy from the electrons before releasing them from electron transfer, making the cells highly successful in their environment.

The first water-splitting photosynthesizers confirmed by fossil evidence are the cyanobacteria. As you learned earlier, convincing evidence from biosignatures indicates cyanobacteria were present 2.5 billion years ago. Present-day stromatolites predominantly contain cyanobacteria but, as stated earlier, the types of prokaryotes in fossil stromatolites cannot be identified with any certainty. If the stromatolites do contain cyanobacteria, that would date early water-splitting photosynthesis to about 3.4 billion years ago.

The photosynthesizing activity of cyanobacteria produced a gradual rise in the amount of oxygen in early Earth's atmosphere. That rising oxygen concentration likely killed off many prokaryotic groups to which oxygen is lethal. Around 2 billion years ago, the concentration of oxygen began to increase more rapidly. This abrupt change may correlate with the evolution of eukaryotic cells containing chloroplasts.

These major events established the preconditions for the evolution of eukaryotic cells. The next section traces this evolution, which was pivotal to the later evolution of large-scale multicellularity and the plants, animals, and the other organisms of the domain Eukarya.

FIGURE 25.4 Geological evidence for early life.
(A) Stromatolites exposed at low tide in Western Australia's Shark Bay. These mounds, which consist of mineral deposits made by prokaryotic communities (predominantly cyanobacteria), are about 2,000 years old; they are highly similar in structure to fossil stromatolites that formed 3.4 billion years ago. **(B)** Microfossils—fossil algae from the Gunflint Formation, Ontario, Canada. The earliest microfossils date to about 2.0 billion years ago.

STUDY BREAK 25.2

Several mechanisms have been proposed for the assembly of organic molecules into protocells. Why is the model involving a lipid bilayer membrane a particularly attractive one?

THINK OUTSIDE THE BOOK

Individually or collaboratively investigate banded iron formations and outline the following in your investigation: (1) what they are; and (2) how researchers consider that they provide evidence for the presence of particular oxygen-producing organisms on early Earth.

25.3 The Origins of Eukaryotic Cells

Present-day eukaryotic cells have several interrelated characteristics that distinguish them from prokaryotes: (1) the separation of DNA and cytoplasm by a nuclear envelope; (2) the presence in the cytoplasm of membrane-bound compartments with specialized metabolic and synthetic functions—mitochondria, chloroplasts, the endoplasmic reticulum (ER), and the Golgi complex, among others; and (3) highly specialized motor (contractile) proteins that move cells and internal cell parts. In this section we discuss how eukaryotes most probably evolved from associations of prokaryotes. As you read, note that there are more good theories for the origin of eukaryotes than there are good data.

The Endosymbiotic Theory Proposes That Mitochondria and Chloroplasts Evolved from Ingested Prokaryotes

The **endosymbiotic theory,** put forward by Lynn Margulis at the University of Massachusetts, Amherst, proposes that the membranous organelles of eukaryotic cells, the mitochondria and chloroplasts, may each have originated from mutualistic (mutually advantageous) relationships between two prokaryotic cells **(Figure 25.5).** The theory proposes that the following events occurred:

1. **Generation of endosymbionts.** Mitochondria began to develop when photosynthetic and nonphotosynthetic prokaryotes coexisted in an oxygen-rich atmosphere. The nonphotosynthetic prokaryotes fed themselves by ingesting organic molecules from their environment. These prokaryotes included both anaerobes, which are unable to use oxygen as the final acceptor for electron transfer, and aerobes, which are fully capable of using oxygen. Only the aerobes could fully exploit the energy stored in organic molecules, but predatory anaerobes could capture that energy by eating aerobic cells. These anaerobic prokaryotes had become efficient predators, and lived by ingesting other cells. Among the ingested cells were some aerobic prokaryotes; instead of being digested, some of them persisted in the cytoplasm of the predators and continued to respire aerobically in their new location. They had become **endosymbionts,** organisms that live symbiotically within a host cell. The cytoplasm of the host anaerobe, formerly limited to the use of organic molecules as final electron acceptors, was now home to an aerobe capable of carrying out the much more efficient transfer of electrons to oxygen. From an evolutionary

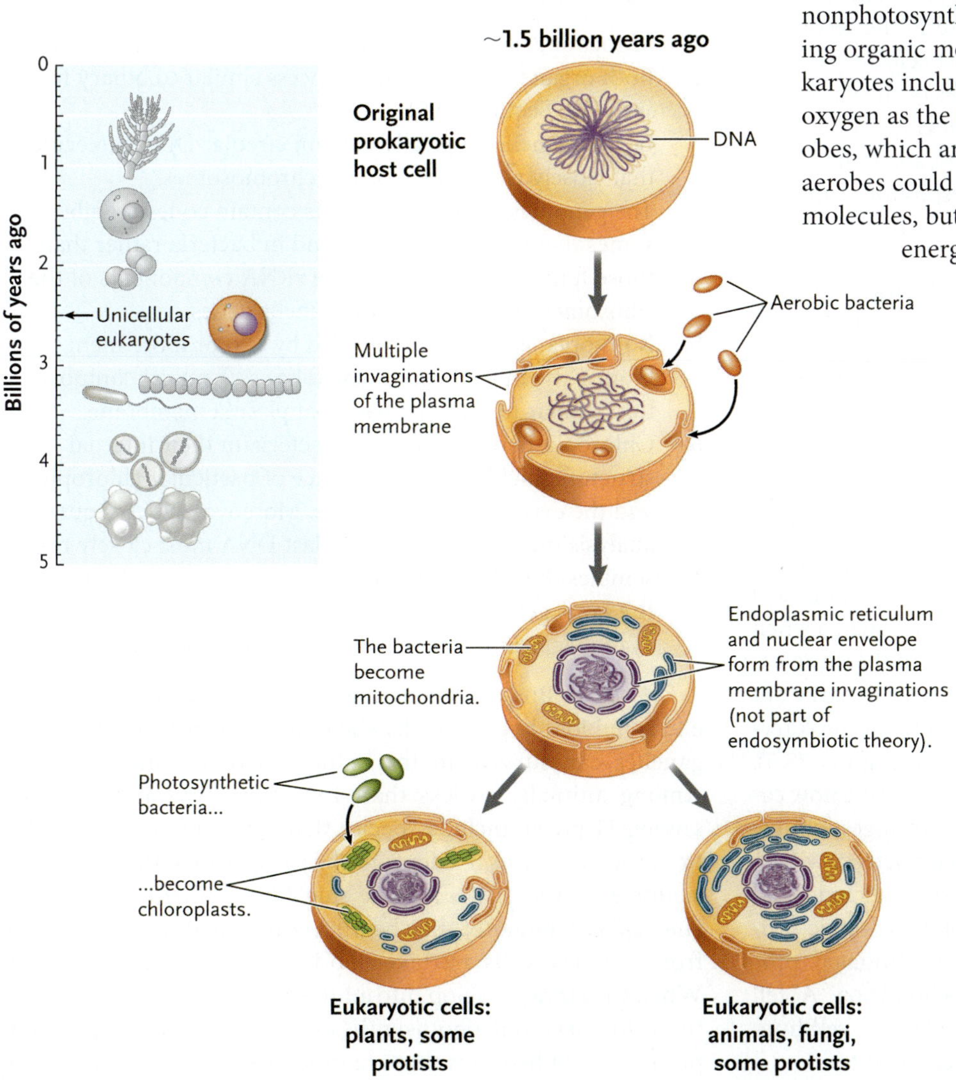

FIGURE 25.5 The endosymbiotic theory. The mitochondrion is thought to have originated from an aerobic prokaryote that lived as an endosymbiont within an anaerobic prokaryote. The chloroplast is thought to have originated from a photosynthetic prokaryote that became an endosymbiont within an aerobic cell that had mitochondria.

perspective, cells that had engulfed respiring bacterial cells capable of cellular respiration would have been at a considerable advantage compared to other cells, so natural selection would have favored their persistence. They could ingest simple carbohydrates from their environment and metabolize them as a source of energy.

2. **Formation of the nucleus and other membranous structures.** As a part of the transition to a true eukaryotic cell, the cell also evolved to acquire other membranous structures, the major ones being the nuclear envelope, the endoplasmic reticulum (ER), and the Golgi complex. Endocytosis, the process of infolding of the plasma membrane (see Figure 5.15), is thought to be responsible for the evolution of these structures. Researchers think that, in cell lines leading from prokaryotes to eukaryotes, pockets of the plasma membrane formed during endocytosis may have extended inward and surrounded the nuclear region. Some of these membranes fused around the DNA, forming the nuclear envelope and, hence, the nucleus. The remaining membranes formed vesicles in the cytoplasm that gave rise to the ER and the Golgi complex **(Figure 25.6).**
3. **Transfer of many functions from endosymbiont to the host cell.** Next, many functions duplicated in the aerobic endosymbiont were taken over by the host cell. As part of this transfer of function, most of the genes of the aerobe moved to the cell nucleus and became integrated into the host cell's DNA. At the same time, the host anaerobe became dependent for its survival on the respiratory capacity of the symbiotic aerobe. The ingested aerobe presumably benefited as well, because the host cell brought in large quantities of food molecules to be oxidized. This gradual process of mutual adaptation culminated in transformation of the cytoplasmic aerobe into a mitochondrion. The first eukaryotic cells had appeared, the ancestors of all modern-day eukaryotes.

The endosymbiotic theory also proposes that a similar mechanism led to the appearance of the membrane-bound **plastids** (the general term for chloroplasts and related organelles, both photosynthetic and nonphotosynthetic: see Section 5.4) some time after mitochondria evolved. A plastid originated when an aerobic cell that had a mitochondrion and were therefore capable of efficient cellular respiration, but was unable to carry out photosynthesis, ingested one or more photosynthetic prokaryotes resembling present-day cyanobacteria (see Figure 25.5). Cells that had engulfed photosynthetic prokaryotes were now capable of photosynthesis and no longer needed to ingest simple carbohydrates for energy. That is, the engulfed bacteria within them produced the carbon dioxide from carbon dioxide and water, with energy provided by sunlight. A photosynthetic prokaryote gradually changed into a plastid by evolutionary processes similar to those that produced a mitochondrion. A cell with both a plastid and a mitochondrion founded the cell lines that gave rise to the modern eukaryotic algae and plants.

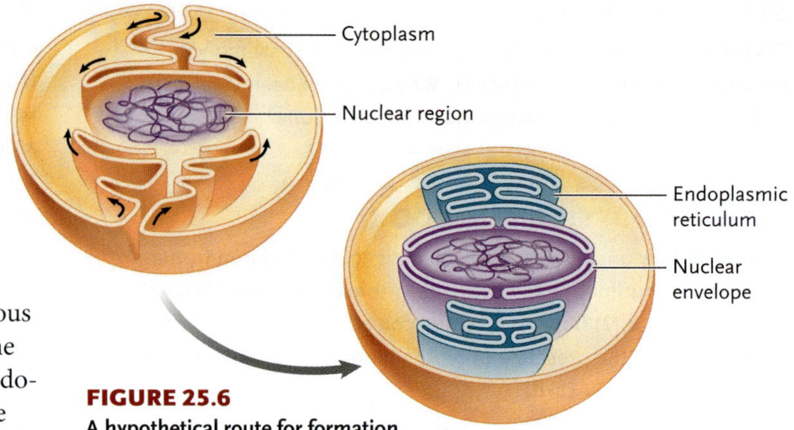

FIGURE 25.6
A hypothetical route for formation of the nuclear envelope and endoplasmic reticulum, through segments of the plasma membrane that were brought into the cytoplasm by endocytosis.

Several Lines of Evidence Support the Endosymbiotic Theory

A lot of evidence supports the endosymbiotic theory, including the following:

1. Mitochondria and plastids are similar in size to bacteria.
2. Both organelles divide by a process similar to binary fission (see Section 10.5).
3. Both organelles typically contain circular DNA molecules that closely resemble bacterial chromosomes.
4. The genomes of both organelles contain codes for ribosomes that resemble those found in bacteria rather than those found in eukaryotes. The rRNA components of the ribosomes are encoded by organelle genes.
5. Both organelles are surrounded by two or more membranes, the innermost of which has a chemical composition similar to that of a bacterial plasma membrane.
6. Chloroplasts resemble cyanobacteria in their internal structure, including the presence of particular chlorophylls and the existence of thylakoids. Moreover, DNA sequence analysis indicates that chloroplast DNA most closely resembles the DNA in modern cyanobacteria, with many of the same genes, supporting the conclusion that an ancient cyanobacterium was the ancestor of chloroplasts.

Another line of evidence supports a key assumption of the endosymbiont hypothesis by showing that engulfed cells or organelles can survive in the cytoplasm of the ingesting cell. Among animals, no less than 150 living genera, distributed among 11 phyla, include species that contain eukaryotic algae or cyanobacteria as residents in the cytoplasm of their cells. For example, larvae of the marine slug *Elysia* initially contain no chloroplasts, but after they begin feeding on algae, chloroplasts from the algal cells are taken up into the cells lining the gut. When the larvae develop into adult snails, the chloroplasts continue to carry out photosynthesis in their new location and produce carbohydrates that are used by the snails. Some of the

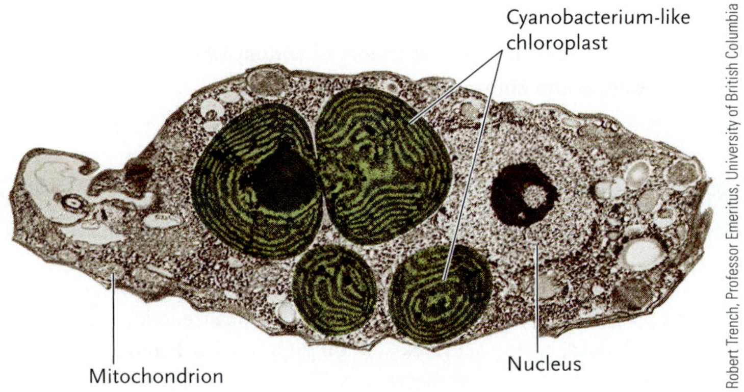

FIGURE 25.7 ***Cyanophora paradoxa*, a protist with chloroplasts that closely resemble cyanobacteria without cell walls.**

chloroplast proteins synthesized during this time are encoded by genes in the slug's nuclear genome. The uptake of functional chloroplasts has also been observed among the protists (see Chapter 27); **Figure 25.7** shows a protist with chloroplasts that closely resemble cyanobacteria.

How long did it take for evolutionary mechanisms to produce fully eukaryotic cells? The oldest known convincing microfossil eukaryotes are 1.5 billion years old. If prokaryotic cells first evolved some 3.5 billion years ago, it took up to 2 billion years for eukaryotic cells to evolve from prokaryotes (see Figure 25.1). If so, this long interval probably reflects the complexity of the adaptations leading from prokaryotic to eukaryotic cells. Of course, it is possible that eukaryotic cells evolved more quickly, and we have yet to find the supporting evidence.

Eukaryotic Cells May Have Evolved from a Common Ancestral Line Shared with Archaeans

The system of classification that has gained acceptance among biologists, and the one used in this book, groups all living organisms into three domains. One domain, Eukarya, contains the eukaryotes. The second domain, Bacteria, includes one of two groups of prokaryotes, the bacteria, which consists of both photosynthesizing and nonphotosynthesizing species. The third domain, Archaea, contains the other group of prokaryotes, the archaeans, many of which inhabit extreme environments, including highly saline environments and hot springs.

There is little question that the three domains originated from a common ancestral cell line. However, the events leading from this common ancestry to the three domains of life remain unclear. The most difficult questions surround the role of the archaeans in both bacterial and eukaryotic evolution.

Archaeans have some features that are typical of bacteria, including a circular DNA genome in a nucleoid region without a surrounding nuclear envelope, and no membrane-bound organelles in the cytoplasm. However, the archaeans also have some features that are typically eukaryotic, including the presence of introns (see Section 15.3) in some of their genes. The archaeans also have some characteristics that are unique to their domain, including features of gene and rRNA sequences, and features of cell wall and plasma membrane structure that are found nowhere else among living organisms. The characteristics shared by archaeans and eukaryotes suggest that their roots may lie in a common ancestral line that split off from the line leading to bacteria (see Figure 24.17). At some point, this ancestral line split into the lines leading to Domain Archaea and Domain Eukarya.

Multicellular Eukaryotes Probably Evolved in Colonies of Cells

The first eukaryotes were unicellular. They are the ancestors of the present-day diversity of unicellular eukaryotes. Multicellular eukaryotes evolved from unicellular eukaryotes and then diverged to produce multiple eukaryotic lineages. Molecular clock analysis indicates the first multicellular eukaryote likely arose between 800 and 1,000 million years ago, while the first fossil records (of small algae) are from 600 to 800 million years ago.

According to the prevalent theory, multicellular eukaryotes arose by the aggregation of cells of the same species into a colony. The ability to act in a coordinated way probably increased the capacity of colonies to adapt to changes in the environment. Subsequently, differentiation of cells into various specialized cell types with distinct functions produced organisms with a wider range of capabilities and adaptability. Cell differentiation in a colony would have required cell signals that affected gene expression. That is, because each cell in the colony has the same genome, the development of specific functions (phenotypes) would require intracellular signals that would change the program of gene regulation. (Gene regulation in eukaryotes is discussed in Section 16.4.) Over time, as genomes evolved, the division of function among cells led to the evolution of the tissues and organ systems of complex eukaryotes.

Multicellularity evolved several times in early eukaryotes, producing a number of lineages of algae as well as the ancestors of present day fungi, plants, and animals.

Life May Have Been the Inevitable Consequence of the Physical Conditions of the Primitive Earth

Some researchers maintain that the evolution of life on our planet was an inevitable outcome of the initial physical and chemical conditions established by Earth's origin, among them a reducing atmosphere (at least in some locations), a size that generates moderate gravitational forces, and a distance from the Sun that results in average surface temperatures between the freezing and boiling points of water. Given the same conditions and sufficient time, according to these scientists, it is inevitable that life has evolved or is evolving now on other planets in the universe.

The chapters that follow in this unit trace the course of evolution and its products after eukaryotic cells were added to the prokaryotes already living on Earth. Among prokaryotes, evolution established two major groups, the bacteria and archaea; among eukaryotes, further evolution established the protists, fungi, plants, and animals. The survey of life's diversity begins in the next chapter with a description of present-day bacteria and archaeans.

STUDY BREAK 25.3

What are the key points of the theory of endosymbiont origins for mitochondria and chloroplasts?

UNANSWERED QUESTIONS

What was the first polymer of life?

As discussed in this chapter, many researchers favor RNA for this role. This is because it both self-replicates and can catalyze chemical reactions, and is neatly connected with the contemporary life based on nucleic acids and proteins. There are, however, several problems with this hypothesis. One of them is that synthesis of RNA and its building blocks, nucleotides, is quite difficult under primordial conditions. To circumvent this difficulty, several researchers proposed that other genetic polymers that are simpler to synthesize might have preceded RNA. For example, Albert Eschenmoser from the Swiss Federal Institute of Technology (ETH) in Zurich replaced the difficult-to-synthesize ribose portion of nucleotides by the six-membered sugar pyranose. These polymers are stable and form a double helix. Most recently, Philipp Holliger and his group at the MRC Laboratory of Molecular Biology, Cambridge, U.K. showed that genetic information can be transferred between DNA and its analogs containing different sugars, xeno-nucleic acids. Another proposal is that the initial complement of nucleic acid bases was different than A, U, G, and C. This proposal is motivated by the poor stability of cytosine in water. Although we have no evidence that polymers alternative to contemporary nucleic acids were present on the early Earth, it is important to realize that such a possibility exists and might be used by life elsewhere.

The protein-first hypothesis is currently less popular, even though these polymers are excellent catalysts of chemical reactions, and their building blocks, amino acids, existed on prebiotic Earth and are found in relatively large quantities in meteorites. This is because there is no known mechanism for proteins to self-replicate. Some researchers speculate that a limited replication of proteins is possible. An alternative hypothesis is that replication of individual polymers was not necessary at the origin of life and, instead, the reproduction of protein functions in a population was initially sufficient. Currently, neither view has sufficient experimental support, but as we learn more about the structure and functions of small proteins major surprises might be in store.

THINK LIKE A SCIENTIST If RNA molecules were the first molecules capable of catalyzing biochemical reactions, why were they replaced by proteins?

Andrew Pohorille

Andrew Pohorille heads the NASA Center for Computational Astrobiology and Fundamental Biology at NASA's Ames Research Center. He is also professor of Chemistry and Pharmaceutical Chemistry at the University of California, San Francisco. For his work on the origin of life he was awarded the 2002 NASA Exceptional Scientific Achievement Medal.

REVIEW KEY CONCEPTS

To access the course materials and companion resources for this text, please visit www.cengagebrain.com.

25.1 The Formation of Molecules Necessary for Life

- Living cells are characterized by a boundary membrane, one or more nucleic acid coding molecules in a nuclear region, a system for using the coded information to make proteins, and a metabolic system providing energy for those activities.
- Oparin and Haldane independently hypothesized that life arose *de novo* under the conditions they thought prevailed on the primitive Earth, including a reducing atmosphere that lacked oxygen. Reduction reactions, fueled by natural energy acting on the primitive atmosphere, produced organic molecules. Random aggregations of these molecules were able to carry out primitive reactions which were characteristic of life and that gradually became more complex until life appeared. Chemistry simulation experiments support the hypothesis that organic molecules would form under these conditions (Figure 25.2).
- Present thinking is that early Earth's atmosphere was not reducing, but in fact was neutral (neither reducing nor oxidizing). This has caused skepticism about Oparin and Haldane's hypothesis. One new theory proposes that life developed near hydrothermal vents in the sea floor.

Animation: Milestones in the history of life

Animation: Miller's reaction chamber experiment

25.2 The Origin of Cells

- Organic molecules produced in early Earth's environment by chance formed aggregates that became membrane-bound in protocells, primitive cell-like structures with some of the properties of life. Protocells may have been the precursors of cells (Figure 25.3).
- Next, living cells may have developed from protocells by the development of several critical components, in particular a system based on nucleic acids that could store and pass on the information required to make proteins.
- Subsequently, fully cellular life evolved, with a nuclear region containing DNA and the mechanisms for copying its information into RNA messages; a cytoplasmic region containing systems for utilizing energy and systems for translating RNA messages into proteins; a membrane separating the cell from its surroundings; and a reproductive system duplicating the informational molecules and dividing them among daughter cells.
- The first living cells were prokaryotes. Eventually, some early cells developed the capacity to carry out photosynthesis using water as an electron donor; the oxygen produced as a by-product accumulated and the oxidizing character of Earth's atmosphere increased. From this time on organic molecules produced in the environment were quickly broken down by oxidation, and life could arise only from preexisting life, as in today's world.

Animation: Building blocks of life

25.3 The Origins of Eukaryotic Cells

- According to the endosymbiotic theory, mitochondria evolved from ingested prokaryotes that were capable of using oxygen as final electron acceptors; chloroplasts evolved from ingested cyanobacteria (Figure 25.5).
- Eukaryotic structures such as the ER, Golgi complex, and nuclear envelope appeared through infoldings of the plasma membrane as a part of endocytosis (Figure 25.6).
- Multicellular eukaryotes probably evolved by differentiation of cells of the same species that had aggregated into colonies. Multicellularity evolved several times, producing lineages of several algae and ancestors of fungi, plants, and animals.

Animation: Eukaryotic evolution (formerly "endosymbiont theory")

UNDERSTAND & APPLY

Test Your Knowledge

1. Earth was formed ___ years ago, whereas the oldest known living cell formed about ___ years ago.
   a. 400×10^3; 3.6×10^6.
   b. 4.6×10^9; 1.0×10^9.
   c. 3.8×10^9; 4.6×10^7.
   d. 4.6×10^9; 3.5×10^9.
   e. 2.0×10^9; 600×10^6.
2. Which of the following is *not* a characteristic of all living organisms?
   a. They replicate genetic information and convert the information into proteins.
   b. They pass genetic information between generations.
   c. They have an internal metabolic system that provides energy for biological activity.
   d. They use external energy to drive internal reactions that require energy.
   e. They use mitochondria for synthesis of ATP.
3. Evidence for the appearance of the following is separated by the greatest amount of time:
   a. nonlife: prokaryotes.
   b. photosynthetic prokaryotes: first eukaryotes.
   c. nonphotosynthetic bacteria: photosynthetic bacteria.
   d. unicellular eukaryotes: multicellular eukaryotes.
   e. first prokaryotic cells: first eukaryotic cells.
4. According to the Oparin–Haldane hypothesis, the atmosphere when life began was believed to be composed primarily of the following substances that contain the maximum number of electrons:
   a. H_2O, N_2, and CO_2.
   b. H_2, H_2O, NH_3, and CH_4.
   c. H_2O, N_2, O_2, and CO_2.
   d. O_2 and no H_2.
   e. H_2 only.
5. The Miller–Urey experiment:
   a. was based on the understanding that the atmosphere was oxidizing.
   b. was able to synthesize amino acids and macromolecules from reduced gases.
   c. did not require much energy or a continuous energy source to keep synthesizing.
   d. did not require water to produce organic molecules.
   e. used free oxygen as a reactant.
6. An unknown organism was found in a park. It was unicellular, had no nuclear membrane around its DNA, and contained no mitochondria and no chloroplasts. It is most likely a:
   a. eukaryote.
   b. vertebrate or plant.
   c. bacteria or archaean.
   d. plant or fungi.
   e. virus.
7. Hydrothermal vents are theorized as sources for the origin of life because:
   a. the temperature of the water around them supports most life.
   b. most organic molecules undergo dehydration synthesis at high temperatures.
   c. the amino acids degrade in the colder water soon after synthesis.
   d. water is needed by living things.
   e. reducing conditions and abundant chemicals exist.
8. For the transition of protocells to living cells, which of the following probably had to occur?
   a. DNA had to code for cell membranes.
   b. Proteins had to be aggregated by RNA.
   c. Ribozymes had to adsorb onto clays.
   d. H_2O, which is essential for the chemistry of all biological systems, had to become available to cells via metabolism and dehydration synthesis.
   e. Informational molecules, such as ribozymes, had to evolve.
9. As part of the evolution of eukaryotic cell, endocytosis, the process of infolding of the plasma membrane, led to the formation of:
   a. chromosomes.
   b. the cell wall.
   c. ribosomes.
   d. the nuclear envelope.
   e. microtubules.
10. Which of the following is *not* part of the evidence supporting the endosymbiotic theory? Both mitochondria and plastids:
   a. are each the size of many bacterial cells.
   b. have inner membranes with a chemical composition similar to that of a bacterial plasma membrane.
   c. have ribosomes that are similar to bacterial ribosomes.
   d. contain circular DNA.
   e. have DNA similar to nuclear DNA.

Discuss the Concepts

1. What evidence supports the idea that life originated through inanimate chemical processes?
2. Explain, in terms of hydrophilic and hydrophobic interactions, how protocells might have formed in water from aggregations of lipids, proteins, and nucleic acids.
3. What conditions would likely be necessary on a planet located elsewhere in the universe for life to originate in a way similar to that which occurred on Earth?
4. Most scientists agree that life on Earth can arise only from preexisting life, but also that life could have originated spontaneously on the primordial Earth. Can you reconcile these seemingly contradictory statements?

Design an Experiment

Suppose you discover a hot spring-fed pool on a remote mountain that has never before been explored by humans. In the pool you find a cellular life form that appears to be prokaryotic. What experiments would you do to distinguish between the alternative hypotheses that this organism evolved on Earth from ancestral prokaryotes, or is descended from a life form that arrived at that location in a meteorite?

Interpret the Data

Studies of ancient rocks and fossils can provide insight into many of the events in Earth's history, including changes in its atmosphere that have taken place over geologic time. The **Figure** shows how asteroid impacts and the composition of the atmosphere have changed over time. Use this figure and information in the chapter to answer the following questions.

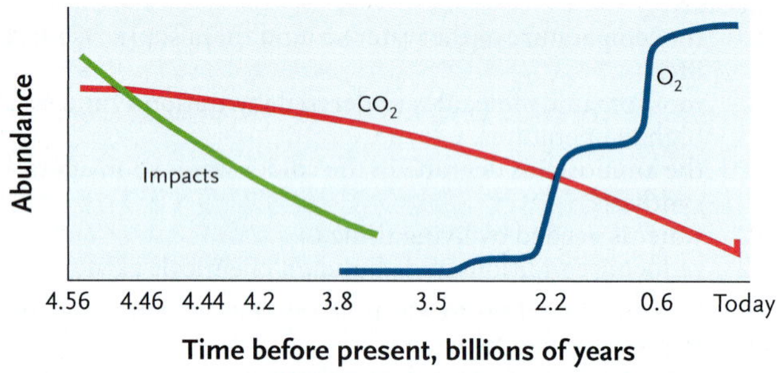

FIGURE **Changes over geologic time in asteroid impacts (green), atmospheric carbon dioxide concentration (pink), and oxygen concentration (blue).**

1. Which occurred first, a decline in asteroid impacts or a rise in the atmospheric level of oxygen?
2. How do modern levels of carbon dioxide and oxygen compare to those at the time when the first cells arose?
3. Which is now more abundant, oxygen or carbon dioxide?

Apply Evolutionary Thinking

In the evolution unit, you learned how changes in the environment can foster evolutionary changes in biological systems. How have changing biological systems influenced the evolution of changes in Earth's physical environment?

Viruses and Subviral Agents

24

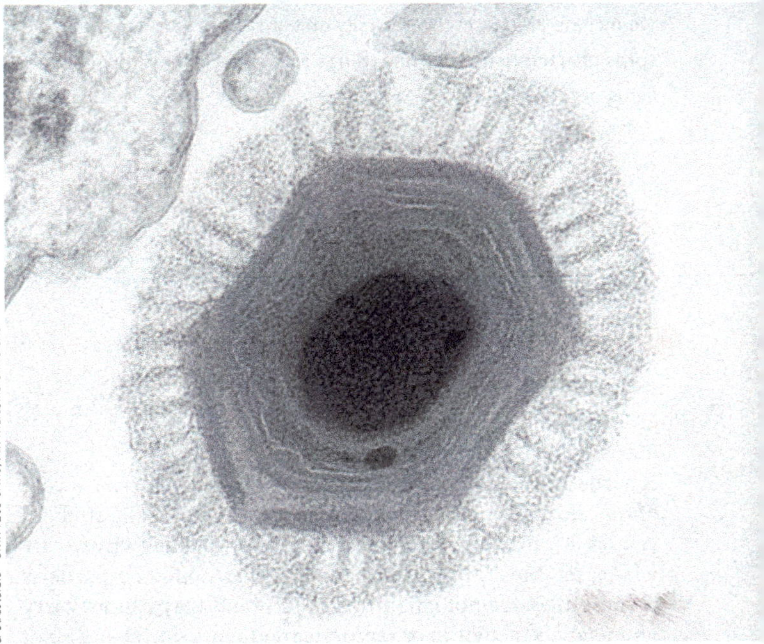

Chantal Abergel, IGS-IMM, CNRS-AMU nformation Génomique & Structurale - UMR 7256 CNRS, Aix-Marseille Université

EM of *Megavirus chilensis*. Megavirus is a giant virus, the size of a very small bacterium (mycoplasma). The giant viruses discovered to date infect amoebas.

During the late 1800s, botanists searched for the cause of tobacco mosaic disease, which stunts the growth of tobacco plants and gives the infected tobacco leaves a spotted, mosaic appearance. The investigators found that they could transmit this disease to healthy plants by daubing their leaves with the sap of diseased plants. In 1892, Dmitri Ivanowsky, a Russian botanist, showed that the sap was still infective after it had been passed through porcelain filters designed to filter out all known bacteria. A few years later, in 1898, Martinus Beijerinck, a Dutch microbiologist, provided evidence that the agent that caused tobacco mosaic disease had many characteristics of a living organism. However, it appeared that the infective agent could reproduce only within a living cell. Beijerinck named the infective agent "virus" (from the Latin word *virus,* which means "poison"). Tobacco mosaic virus was the first virus to be recognized.

Early in the 20th century, scientists discovered other infective agents, like those responsible for tobacco mosaic disease, that could cause disease in animals or kill bacteria. These pathogens were so small that they could not be seen with the light microscope. They also passed through filters that removed bacteria. Curiously, they could not be grown in laboratory cultures unless living cells were present. The development of the electron microscope in the 1930s made it possible to see viruses for the first time.

Most of the viruses that infect animals, plants, and bacteria were identified during the second half of the 20th century. Viruses cause many plant diseases and are responsible for billions of dollars in crop losses each year. You may recognize some of the many human viral diseases, such as rabies, influenza, herpes, and acquired human immunodeficiency syndrome (AIDS). You have probably received vaccinations for many common childhood viral diseases, including polio, rubella (German measles), mumps, and chickenpox.

Viruses make up a large component of Earth's biomass. Their huge numbers make up for their small size and weight. Viruses influence many ecological processes. For example, they kill large amounts of marine biomass each day, contributing significantly to the recycling of nutrients. Viruses adapt to their environments through evolution. They also influence the biodiversity of other organisms. Viruses can transfer their own genes into the genetic material of their hosts. They can also transfer eukaryotic genes from

KEY CONCEPTS

24.1 A virus is a small particle consisting of a nucleic acid core surrounded by a protein coat; to reproduce, a virus must infect a living cell.

24.2 Viruses can be classified based on their host range, what type of nucleic acid they have, and whether the nucleic acid is single-stranded or double-stranded.

24.3 In a lytic reproductive cycle, a virus uses the host cell's molecular machinery to replicate itself, destroying the host cell in the process. In a lysogenic cycle, the viral genome becomes integrated into the host DNA and is then called a "provirus."

24.4 Viruses infect the cells of all kinds of organisms; they cause serious diseases in plants and animals.

24.5 Three different hypotheses suggest how viruses may have originated and evolved.

24.6 Subviral agents, which include satellites, viroids, and prions, are smaller and simpler than viruses.

one organism to another, crossing species boundaries. The discovery of giant viruses (see photograph) has raised new questions about the structure, genomes, origin, and phylogenetic relationships of viruses.

In this chapter we examine the characteristics and diversity of viruses as well as the smaller subviral agents, viroids and prions. We include viruses in the diversity unit of *Biology* because they have some characteristics of living things and because they impact all living organisms.

24.1 THE STATUS AND STRUCTURE OF VIRUSES

LEARNING OBJECTIVES

1 Contrast a virus with a cellular organism.
2 Describe the structure of a virus.

A **virus** is a very small infective agent that consists of a core of nucleic acid and is dependent on a living host. The study of viruses is called *virology*, and biologists who study viruses are *virologists*. Most biologists view viruses as nonliving particles because they are not composed of cells and they cannot carry on metabolic activities or reproduce on their own. They do not have the components necessary to carry on cellular respiration or to synthesize proteins and other molecules.

Viruses do contain the nucleic acids necessary to make copies of themselves. They replicate by invading living cells and commandeering their metabolic machinery. To multiply, a virus must *infect* a cell in which it can replicate. Thus, viruses are *obligate intracellular parasites;* they survive only by using the resources of a *host cell*, the cell the virus invades. Viruses infect all types of organisms, including bacteria, archaea, protists, plants, fungi, and animals. Some viruses even infect other viruses. Interestingly, viruses evolve by natural selection. For these reasons, most biologists view viruses as being at the edge of life, although a few argue that they are very simple life-forms.

Viruses are very small

Most known viruses are very small, ranging in size from 20 to 300 nm. (Recall that a nanometer is one-thousandth of a micrometer.) The poliovirus is about 30 nm in diameter (about the size of a ribosome). If we could line these viruses up end to end, it would take almost a million of them to span 1 inch! A poxvirus that causes smallpox can measure up to 300 nm long and 200 nm wide.

The paradigm that all viruses are very small and simple, with very small genomes, has been challenged by the discovery of *giant* viruses. *Mimivirus* was discovered in 2004, followed by *Megavirus* in 2011, and Pandoraviruses in 2013. One of the Pandoraviruses approaches 1 μm in length and is about 0.5 μm wide, larger than many bacteria and some eukaryotic cells. These giant viruses were discovered by French researchers in seawater off the coast of Chile and in a freshwater pond in Australia.

A virus consists of nucleic acid surrounded by a protein coat

The nucleic acid core of the virus is surrounded by a protein coat called a **capsid.** Microbiologists use the term *virion* to refer to the complete virus particle that is outside a host cell in a dormant state. The virion is the form in which the virus moves from the cell in which it was produced to a new host cell in which it can replicate its genome. However, here we will use the term *virus* in a broad sense, even when referring to virions.

A typical virus contains *either* deoxyribonucleic acid (DNA) or ribonucleic acid (RNA), not both. In 2012, however, researchers discovered an RNA–DNA hybrid virus in a high temperature acidic California lake. Further studies have found three other similar RNA–DNA hybrids in marine environments.

The nucleic acid of a virus can be single-stranded or double-stranded. Thus, a virus can have single-stranded (ss) DNA, double-stranded (ds) DNA, ssRNA, or dsRNA. As we will discuss, the type of nucleic acid is important in classifying viruses. The virus genome typically consists of 5000 to more than 100,000 bases or base pairs (depending on whether it is single-stranded or double-stranded). Again, the giant viruses are exceptions. *Megavirus* has seven transfer RNAs and some metabolic genes never before identified in any other virus. Researchers have sequenced the double-stranded DNA genomes of two species of *Pandoravirus* and reported that they range from 1.9 to 2.5 million base pairs. The species with the largest genome has 2556 proposed protein-coding genes (only 6% of which match known proteins).

The capsid is a protective protein coat

The capsid consists of protein subunits called *capsomers*. The capsomers determine the shape of the virus. Capsids generally are helical, polyhedral, or a combination of both shapes. Helical viruses, such as the tobacco mosaic virus (TMV), appear as long rods or threads (FIG. 24-1a). Its capsid is a hollow cylinder made up of proteins that form a groove into which the RNA fits.

Polyhedral viruses, such as the adenoviruses (which cause a number of human illnesses, including some respiratory infections), appear somewhat spherical (FIG. 24-1b). Its capsomers are organized in equilateral triangles. The capsid of a large virus can consist of several hundred capsomers. The most common polyhedral structure is an *icosahedron*, a structure with 20 identical surface faces (each face is a triangle). The human immunodeficiency virus (HIV) that causes AIDS is an enveloped virus (see next section) (FIG. 24-1c).

Some viruses have both helical and polyhedral components. Viruses that infect bacteria are called *bacteriophages*, or simply *phages* (FIG. 24-1d). The T4 phage that infects the bacterium *Escherichia coli* consists of a polyhedral "head" attached to a helical "tail." Many phages have this shape. Many phages also have tail fibers that attach to the host cell.

Some viruses are surrounded by an envelope

Some viruses, called *enveloped viruses*, have an outer membranous envelope that surrounds the capsid. Typically, the virus

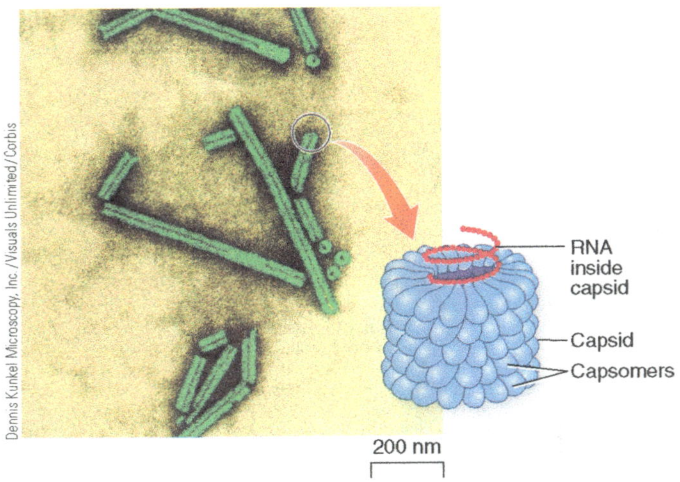

(a) TEM of tobacco mosaic virus. This rod-shaped virus has a helical arrangement of capsid proteins.

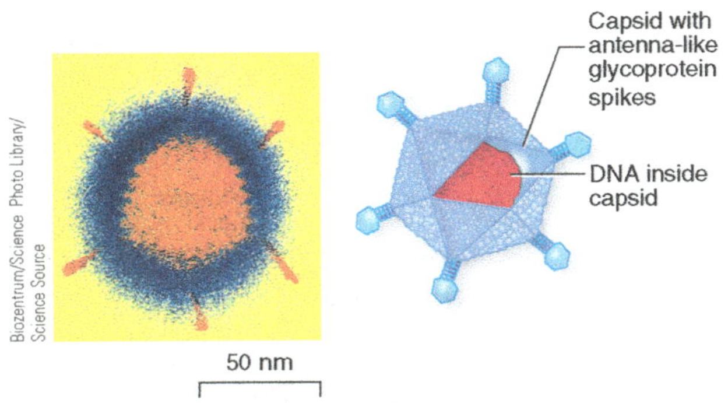

(b) Color-enhanced TEM of an adenovirus. The capsid is composed of 252 subunits (visible as tiny ovals) arranged into a 20-sided polyhedron. Twelve of the subunits have projecting glycoprotein spikes that permit the virus to recognize host cells.

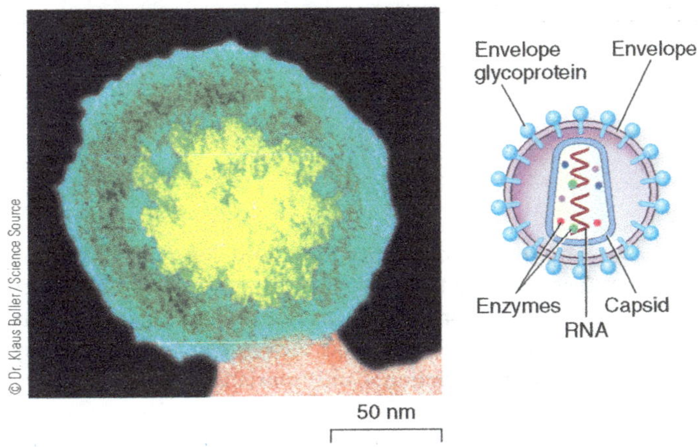

(c) Color-enhanced TEM of HIV, the virus that causes AIDS. The virus is leaving a host cell (*pink*). The virus is enclosed in an envelope (*green*) derived from the host's plasma membrane. Viral proteins project from the envelope. The capsid of the virus is shown in *yellow*.

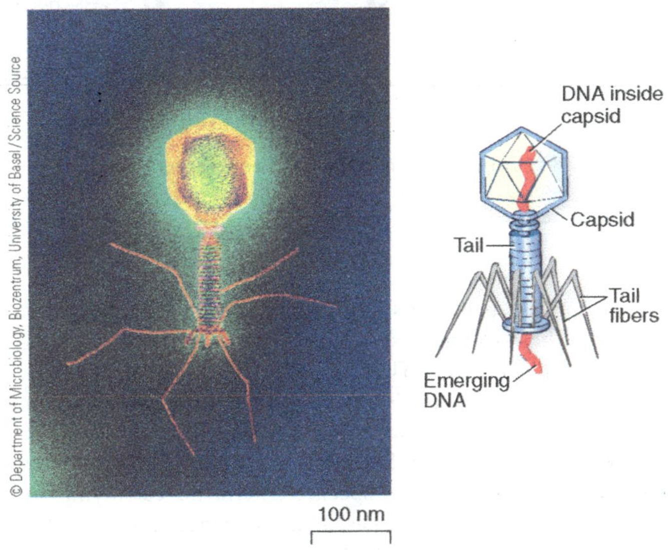

(d) Color-enhanced TEM of T4 bacteriophage. This virus has a polyhedral head and a helical tail. The virus attaches to the cell wall of the bacterial host by its tail fibers.

Figure 24-1 ***Animation*** **The structure of a virus**

A virus consists of a DNA or RNA core surrounded by a protein coat called a capsid. The capsid is made up of protein subunits called capsomers. Some viruses have an outer membranous envelope surrounding the capsid.

acquires the envelope from the host cell's plasma membrane as it leaves the host cell (FIG. 24-2). Interestingly, while inside the host cell, the virus synthesizes certain proteins and inserts them into the host's plasma membrane. Thus, the viral envelope consists of phospholipids and proteins of the host's plasma membrane as well as distinctive proteins produced by the virus itself. Some viruses produce envelope glycoproteins that extend out from the envelope as spikes. As we will discuss, these glycoprotein spikes can be very important in the virus's interaction with the host cell.

CHECKPOINT 24.1

- CONNECT *What characteristics of a living organism are absent in a virus?*
- *What are the structural components of a virus?*

24.2 CLASSIFICATION OF VIRUSES

LEARNING OBJECTIVE

3 Identify three characteristics used to classify viruses.

Viruses present a taxonomic challenge to biologists because they do not have the characteristics that define living organisms (see Chapter 1). They are not cellular, they do not carry on metabolic activities, and they reproduce only by taking over the reproductive machinery of other cells. Viruses do not manufacture proteins and so do not have distinctive rRNA. For these reasons, viruses are not classified in any of the three domains.

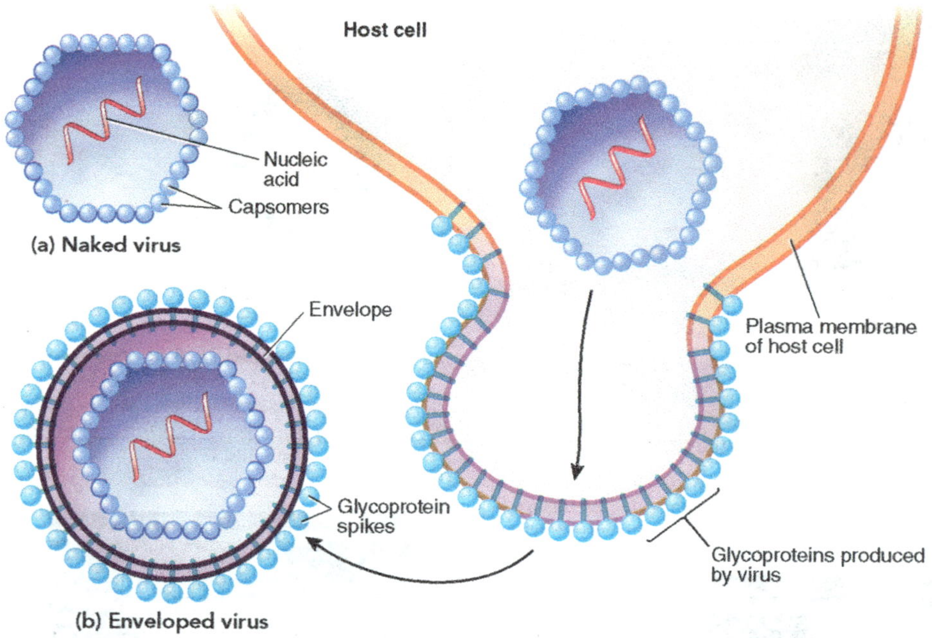

Figure 24-2 Comparison of naked and enveloped viruses

As the virus leaves the host cell, the host's plasma membrane wraps around the virus, forming the envelope. The envelope contains proteins produced by the virus.

Viruses can be classified based on their **host range,** the types of host species a specific virus (or other type of infectious agent) can infect. They may be referred to as plant viruses, animal viruses, bacterial viruses, and so on. Viruses are more formally classified into taxa from species to orders. The International Committee on Taxonomy of Viruses (ICTV), a group of virologists, decides on specific criteria for classifying and naming viruses. Recently, based on host range and other characteristics, the ICTV classified viruses into 7 orders, 96 families, 420 genera, and more than 2600 species. Virus family names include the suffix *-viridae.* Note that this classification system is not a traditional Linnaean system; it does not assign viruses to domains, kingdoms, or phyla.

The Baltimore classification system classifies viruses based on the type of nucleic acid the virus contains, whether the nucleic acid is single-stranded or double-stranded, and how mRNA is produced. Other traits considered in viral classification are the size and shape of the virus, the presence of an envelope, and the method by which the virus is transmitted from host to host.

CHECKPOINT 24.2

- *What are three characteristics used to classify viruses?*

24.3 VIRAL REPLICATION

LEARNING OBJECTIVES

4 Characterize bacteriophages.
5 Contrast a lytic cycle with a lysogenic cycle.

As we have discussed, viruses reproduce, but only within the complex environment of the living host cells they infect. Viruses use their genetic information to force their host cells to replicate their viral nucleic acid and make the proteins they need. They take over the transcriptional and translational mechanisms of the host cell.

Bacteriophages infect bacteria

Much of our knowledge of viruses has come from studying **bacteriophages** ("bacteria eaters"), more simply referred to as **phages.** These viruses can be cultured easily within living bacteria in the laboratory. Microbiologists have identified more than 5000 phages. Bacteriophages are among the most complex viruses (see Fig. 24-1d). Their most common structure consists of a long nucleic acid molecule (usually dsDNA) coiled within a polyhedral head. Most have a tail, which may be contractile and may function in penetration of the host cell.

Phages have been clinically used to treat infection for close to a century. Although abandoned in Western countries after the discovery of sulfa drugs and antibiotics in the 1940s, they have continued to be commonly used in member countries of the former Soviet Union to treat patients. Antibiotics, which have long been more dependable and easier to use, are increasingly less effective than in the past due to the widespread and escalating problem of bacterial resistance. Phage therapy, alone and paired with antibiotics, is again a focus of research in the United States as well as in Eastern Europe. Phages are being engineered to specifically destroy particular bacteria due to their ability to target specific host species. Scientists are also genetically engineering phages so that bacteria will be slower to evolve resistance to them.

Phage therapy also has applications in dentistry, veterinary medicine, agriculture, and food science. For example, certain phages can kill deadly strains of *E. coli* in cattle. These bacteria do not appear to make cattle ill but can cause illness and death in people who eat contaminated, undercooked ground beef.

Viruses replicate inside host cells

Viruses infect bacterial, plant, and animal cells in basically similar ways. We will focus on phage infection of bacteria because this process is best understood. The reproductive cycle of viruses begins with a virus coming into contact with a host cell. The virus typically attaches to the surface of the host cell. The viral nucleic acid must enter the host cell and synthesize the components it needs to reproduce itself. Then viral components are assembled, and viruses are released from the cell, ready to invade other cells. Two types of viral reproductive cycles are lytic and lysogenic cycles.

Lytic reproductive cycles destroy host cells. In a **lytic cycle,** the virus lyses (destroys) the host cell. When the virus infects a susceptible host cell, it forces the host to use its metabolic machinery to replicate viral particles. Viruses that have only a lytic cycle are described as *virulent,* which means that they cause disease and often death.

Five steps are typical in lytic viral reproduction (FIG. 24-3):

1. **Attachment (or adsorption).** The virus attaches to specific receptors on the host cell. This process ensures that the virus infects only its specific host.
2. **Penetration.** The virus penetrates the host plasma membrane and moves into the cytoplasm. Many viruses that infect animal cells enter the host cell intact. Some phages inject only their nucleic acid into the cytoplasm of the host cell; the capsid remains on the outside.
3. **Replication and synthesis.** The viral genome contains all the information necessary to produce new viruses. Once inside a host cell, the virus degrades the host-cell nucleic acid. The virus then uses the molecular machinery of the host cell to replicate its own nucleic acid and produce viral proteins. Many antiviral drugs interfere with replication of viral nucleic acid.
4. **Assembly.** The newly synthesized viral components are assembled into new viruses.
5. **Release.** Assembled viruses are released from the cell. Generally, lytic enzymes, produced by the phage late in the replication process, destroy the host plasma membrane. Phage release typically occurs all at once and results in rapid cell lysis. In contrast, animal viruses are often released slowly or bud off from the plasma membrane.

Once released, the viruses infect other cells, and the process begins anew. The time required for viral reproduction, from attachment to the release of new viruses, varies from less than 20 minutes to more than 1 hour.

How do bacteria protect themselves from phage infection? You may recall from Chapter 15 that bacteria produce *restriction enzymes,* enzymes that cut up the foreign DNA of the phage. This action prevents the phage DNA from duplicating. The

KEY POINT

Viruses reproduce by seizing control of the metabolic machinery of a host cell. In a lytic infection, the viruses destroy the host cell.

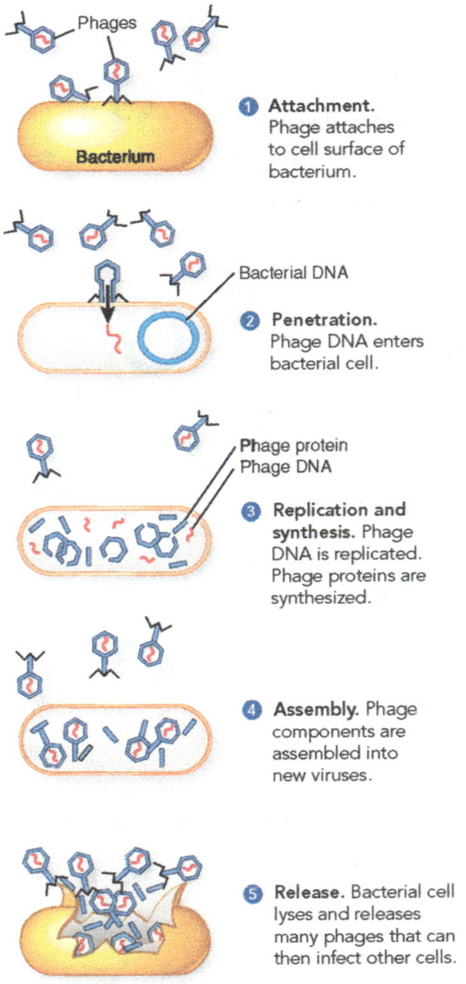

(a) The sequence of events in a lytic infection.

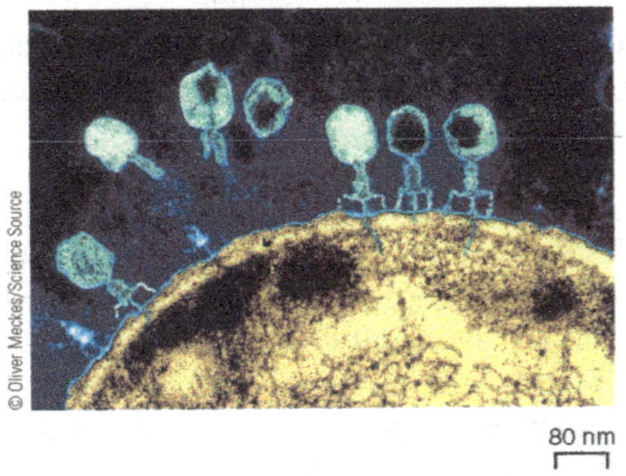

(b) Color-enhanced TEM of phages infecting a bacterium, *Escherichia coli.*

Figure 24-3 ***Animation*** **The sequence of events in a lytic infection**

Virulent bacteriophages kill their host cell during each cycle of infection.

PREDICT Would the phage components be able to infect other bacteria if they were released prior to assembly?

KEY POINT

In a lysogenic cycle, temperate phages integrate their nucleic acid into the host genome. The prophage replicates when the bacterial DNA replicates.

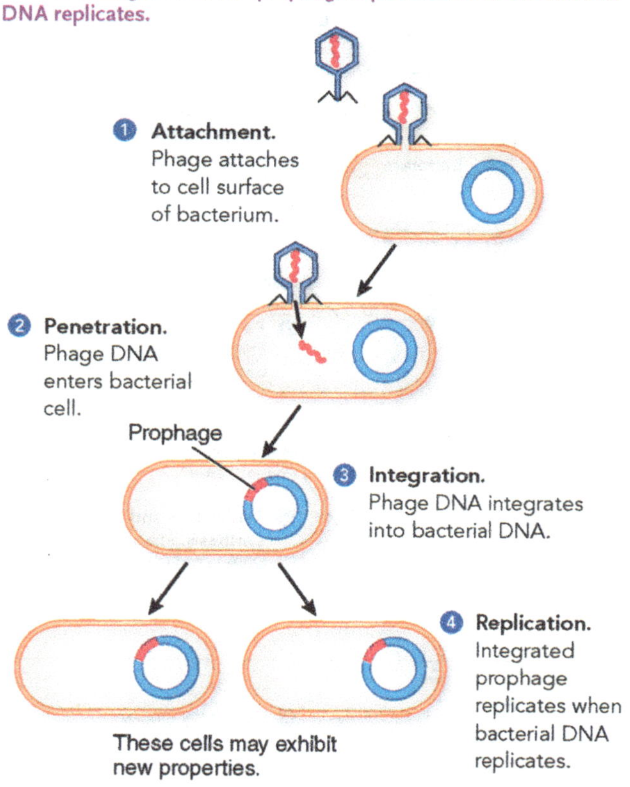

Figure 24-4 *Animation* **The sequence of events in a lysogenic cycle**

Temperate phages integrate their nucleic acid into the DNA of their bacterial host. The integrated viral DNA is called a prophage. The bacterium is called a lysogenic cell.

PREDICT How do temperate phages benefit from integrating into the bacterial DNA?

bacterial cell protects its own DNA by slightly modifying it after replication so that the restriction enzyme does not recognize the sites it would cut.

In lysogenic cycles temperate viruses integrate into the host DNA **Temperate phages** do not immediately destroy their hosts. These phages alternate between a lytic cycle and a lysogenic cycle. In a **lysogenic cycle,** the viral genome becomes integrated into the host bacterial DNA. The integrated virus is called a **prophage,** or **provirus.** When the bacterial DNA replicates, the prophage also replicates (FIG. 24-4). In this way, the prophage is passed on to new generations of the original infected cell. The viral genes that code for viral structural proteins may be repressed indefinitely. The temperate phage lambda (designated by the Greek letter λ) is a model organism used to study temperate viruses.

Certain external conditions (such as ultraviolet light and X-rays) cause temperate viruses to revert to a lytic cycle and then destroy their host. Sometimes temperate viruses become lytic spontaneously.

Bacterial cells carrying prophages are called *lysogenic cells.* Such bacterial cells may exhibit new properties. This type of change is called **lysogenic conversion.** An interesting example involves the bacterium *Corynebacterium diphtheriae,* which causes diphtheria. Two strains of this species exist, one that produces a toxin (and causes diphtheria) and one that does not. The only difference between these two strains is that the toxin-producing bacteria are infected by a specific temperate phage. The phage DNA codes for the powerful toxin that causes the symptoms of diphtheria. Similarly, the bacterium *Clostridium botulinum,* which causes botulism, a serious form of food poisoning, is harmless unless it contains certain prophage DNA that induces synthesis of the toxin.

CHECKPOINT 24.3

- VISUALIZE *What are the steps in a lytic cycle? Draw diagrams to illustrate your answer.*
- *How is a lysogenic cycle different from a lytic cycle?*

24.4 VIRAL DISEASES

LEARNING OBJECTIVES

6 Contrast viral infection of plants and animals, and identify specific diseases caused by viruses.

7 Describe the reproductive cycle of a retrovirus, such as human immunodeficiency virus (HIV).

Sir Peter Medawar, a winner of the 1960 Nobel Prize in Physiology or Medicine, defined a *virus* as "a piece of bad news wrapped in protein." Viruses that are **pathogens** are responsible for some very serious diseases, including Ebola hemorrhagic fever, rabies, influenza (flu), and AIDS. However, although viruses must infect cells to reproduce, most known viruses do *not* cause disease. Here we focus on viral infection of plants and animals.

Viruses cause serious plant diseases

Viruses cause many important plant diseases and are responsible for billions of dollars in agricultural losses and lower crop quality worldwide each year. Infected crops almost always produce lower yields than noninfected crops.

Viral infections do not usually kill plants, but they do stunt their growth; cause changes in the shape of the foliage; and may cause spots, streaks, or mottled patterns on leaves, flowers, or fruits (FIG. 24-5). Most plant viruses have capsids, but not envelopes. The genome of most plant viruses consists of ssRNA. Plant viruses are typically named according to the type of host plant they infect and their effects on the plant. Tobacco mosaic virus, described at the beginning of this chapter, was the first virus ever discovered.

Insects are important vectors of plant disease. As they feed on plant tissues, aphids, leaf-hoppers, and many other insects spread viral diseases among plants. Soil-borne nematodes and protozoa are vectors of viral infections in plant roots.

Because plants have thick cell walls, viruses cannot penetrate plant cells unless the cells are damaged. Plant viruses can be transmitted from one generation to the next through infected seeds or by asexual propagation (Fig. 24-5). Once a plant is infected, the virus spreads through the plant body by passing through *plasmodesmata*, cytoplasmic channels that connect adjacent plant cells.

There are no known cures for most viral diseases of plants, so infected plants are commonly burned. Many agricultural scientists are focusing their efforts on preventing viral disease by developing virus-resistant strains of important crop plants.

© iStockphoto.com/sjtoh

(a) Virus-streaked tulip. The virus that causes this relatively harmless disease affects pigment formation in the petals.

© Nigel Cattlin/Alamy

(b) Pepper leaves infected with tobacco mosaic virus. The leaf is characteristically mottled with light green areas.

Figure 24-5 **Plant diseases caused by viruses**

Viruses cause serious diseases in animals

Hundreds of different viruses infect animals, including humans. Animal viruses cause hog cholera, foot-and-mouth disease, canine distemper, and certain types of cancer (such as feline leukemia and cervical cancer). Viruses cause chickenpox, herpes simplex (one type causes genital herpes), mumps, rubella (German measles), rubeola (measles), rabies, warts, infectious mononucleosis, influenza, viral hepatitis, Ebola hemorrhagic fever, and AIDS (TABLE 24-1 and *Inquiring About: Influenza and Other Emerging Diseases*). Most humans suffer from two to six viral infections each year, including common colds and gastrointestinal disorders.

Norwalk virus is responsible for most worldwide viral gastroenteritis outbreaks and epidemics, affecting more than 200 million and killing more than 200,000 people each year. Norwalk virus is transmitted directly from person to person and indirectly by fecal contamination of water and food. The virus is also spread when infected people sneeze or cough, sending particles into the air; that is, virus particles become aerosolized. Infected individuals can shed viral particles for many weeks after symptoms abate. Norwalk virus infections are a serious concern, especially in closed communities, such as health care institutions, prisons, and cruise ships where people are in prolonged close contact. The genetically diverse strains of the Norwalk virus species make up the noroviruses.

Human activity, including social factors such as urbanization, global travel, and war, contributes to epidemics of infectious disease. Living conditions, including sanitation, nutrition, physical stress, level of health care, and sexual practices, are important factors in the spread of disease. In the United States and other highly developed countries, infectious disease accounts for about 4% to 8% of deaths compared with death rates of 30% to 50% in developing regions. Even at the level of our current knowledge about viruses and epidemiology, just how prepared are we to contain a particularly virulent virus? For example, how well are we containing HIV (the virus that causes AIDS), which continues to infect and kill millions of people worldwide every year?

Since the September 11, 2001, terrorist attacks in the United States, bioterrorism has become a critical concern worldwide. **Bioterrorism** is the intentional use of microorganisms or toxins derived from living organisms to cause death or disease in humans, animals, or plants on which humans depend. Terrorists could conceivably initiate epidemics of smallpox, anthrax, plague, yellow fever, Ebola hemorrhagic fever, and other potentially fatal diseases. (Note that anthrax and plague are caused by bacteria; see Table 25-5.) The quest for rapid identification and effective treatments and vaccines for these diseases has taken on new urgency.

Animal viruses have varied structure and life cycles

Most viruses cannot survive very long outside a living host cell, so their survival depends on their being transmitted from one animal to another. However, their host range may be quite limited because attachment to a host cell is very specific. The type of attachment proteins on the surface of a virus determines what type of cell it can infect. Receptors typically vary with each species and sometimes with each type of tissue. Thus, many human viruses can infect only humans because the viral attachment proteins combine only with receptor sites found on human cell membranes. The measles virus and poxviruses infect many types of human tissue because their attachment proteins combine with receptor sites on a variety of cells. In contrast, the poliovirus attaches to specific types of human cells, such as those that line the digestive tract and motor neurons of the brain and spinal cord.

Some viruses, such as the adenoviruses, have fibers that project from the capsid and adhere to complementary receptors

TABLE 24-1 Some Viruses That Infect Vertebrates

GROUP	DISEASES CAUSED	CHARACTERISTICS
DNA VIRUSES WITH ENVELOPE		
Poxviruses	Smallpox, cowpox,* monkeypox, and economically important diseases of domestic fowl	**dsDNA;** large, complex viruses; replicate in the cytoplasm of the host cell
Herpesviruses	Cold sores (herpes simplex virus type 1); genital herpes, a sexually transmitted disease (herpes simplex virus type 2); chickenpox and shingles (herpes varicella–zoster virus); infectious mononucleosis and Burkitt's lymphoma (Epstein–Barr virus)	**dsDNA;** medium to large, enveloped viruses; replicate in the host nucleus†
DNA VIRUSES WITH NO ENVELOPE		
Adenoviruses	Respiratory tract disorders (e.g., sore throat, tonsillitis), conjunctivitis, and gastrointestinal disorders are caused by more than 40 types of adenoviruses in humans; other varieties infect other animals	**dsDNA;** replicate in the host nucleus
Papovaviruses	Human warts and some degenerative brain diseases; some cancers, including cervical cancer‡‡	**dsDNA**
Parvoviruses	Infections in dogs, swine, arthropods, rodents; gastroenteritis in humans (transmitted by consumption of infected shellfish)	**ssDNA;** some require a helper virus to multiply
RNA VIRUSES WITH ENVELOPE		
Togaviruses	Rubella (German measles)	**ssRNA** that can serve as mRNA; large diverse group of medium-sized enveloped viruses; many transmitted by arthropods
Orthomyxoviruses	Influenza (flu) in humans and other animals	**ssRNA** that serves as template for mRNA synthesis; medium-sized viruses that often exhibit projecting glycoprotein spikes
Paramyxoviruses	Rubeola (measles) and mumps in humans; distemper in dogs	**ssRNA;** resemble orthomyxoviruses but somewhat larger
Rhabdoviruses	Rabies	**ssRNA**
Coronaviruses	Upper respiratory infections; SARS	**ssRNA;** largest known RNA virus
Flaviviruses	Yellow fever; West Nile virus; hepatitis C (the most common reason for liver transplants in the United States)	**ssRNA**
Filoviruses	Hemorrhagic fever, including that caused by the Ebola virus	**ssRNA**
Bunyaviruses	St. Louis encephalitis; hantavirus pulmonary syndrome (caused by Sin Nombre virus, a hantavirus)	**ssRNA**
Retroviruses	AIDS; some types of cancer	**ssRNA** viruses that contain reverse transcriptase for transcribing the RNA genome into DNA; two identical molecules of ssRNA
RNA VIRUSES WITH NO ENVELOPE		
Picornaviruses	Polio (poliovirus); hepatitis A (hepatitis A virus); intestinal disorders (enteroviruses); common cold (rhinoviruses); aseptic meningitis (coxsackievirus, echovirus)	**ssRNA** that can serve as mRNA; diverse group of small viruses
Reoviruses	Vomiting and diarrhea; encephalitis	**dsRNA**
Noroviruses (This group made up of strains of Norwalk virus)	Most common cause of viral gastroenteritis in humans	**ssRNA**

*The vaccinia (cowpox) virus is used to produce genetically engineered vaccines.
†These viruses frequently cause latent infections; some cause tumors.
‡‡The virus SV40 has been used as a vector to transport genes into cells.

on the host cell. Other viruses, such as those that cause herpes and rabies, are surrounded by a lipoprotein envelope with projecting glycoprotein spikes that attach to a host cell. The influenza virus has rodlike, glycoprotein spikes that project from its envelope. These spikes bind with specific receptors present only on cells lining the respiratory tract of certain vertebrates, including wild and domestic birds, horses, swine, and humans. Influenza vaccine prevents attachment by stimulating the host's antibodies to cover the glycoprotein spikes projecting from the virus.

Viruses have several ways to penetrate animal cells (FIG. 24-6). After attachment to a host-cell receptor, some enveloped viruses fuse with the animal cell's plasma membrane. The entire virus, including the capsid and nucleic acid, is released into the animal cell. Other viruses enter the host cell by *endocytosis.* In this process the plasma membrane of the animal cell invaginates, forming a membrane-enclosed vesicle that contains the virus. Endocytosis is advantageous to the virus because the endocytotic vesicle delivers the virus deep into the cytosol.

inquiring about **Influenza and Other Emerging Diseases**

How do new strains of the influenza virus re-emerge? **Emerging diseases,** those new to the human population—such as AIDS, severe acute respiratory syndrome (SARS), Ebola hemorrhagic fever, eastern equine encephalitis, and West Nile virus—often appear suddenly. Many new, continual, or re-emerging pathogens can strike globally.

Re-emerging diseases are those that have been almost eradicated and then suddenly recur, causing an epidemic, sometimes in a new geographic area. Many familiar diseases, such as influenza, malaria, tuberculosis, and bacterial pneumonias, continue to infect large numbers of people and increasingly reappear in forms that are resistant to drug therapy. Drug-resistant pathogens are a major challenge to public health and other health care professionals.

Historically, new viral strains have claimed many human lives. In 1918, for example, an influenza (flu) pandemic killed more than 20 million people throughout the world. Flu pandemics also occurred in 1957 and 1968. Each year new strains of influenza virus evolve and become infectious. If the new combination of viral genes is unfamiliar to the human immune system, the virus may spread easily, resulting in a pandemic. According to the Centers for Disease Control and Prevention, influenza kills about 36,000 individuals in the United States every year.

New strains of influenza are thought to infect one species, such as a species of birds. At first, they are not able to spread among mammals. However, because viruses can evolve, the avian strain can mutate and become virulent. The avian virus can also exchange RNA with a virus that has the genetic material necessary to spread among swine or other mammals. Swine can be susceptible to both avian and human strains of the virus. Inside a swine's body, the viruses can exchange bits of RNA and develop the ability to infect humans. When an animal disease crosses the species barrier and infects humans, it is called a **zoonotic disease.** (We discuss zoonosis further in Chapter 25.) Reassortment of genetic material can also occur inside a human host, producing a viral strain that can spread from person to person (see figure). H1N1 influenza evolved in this way.

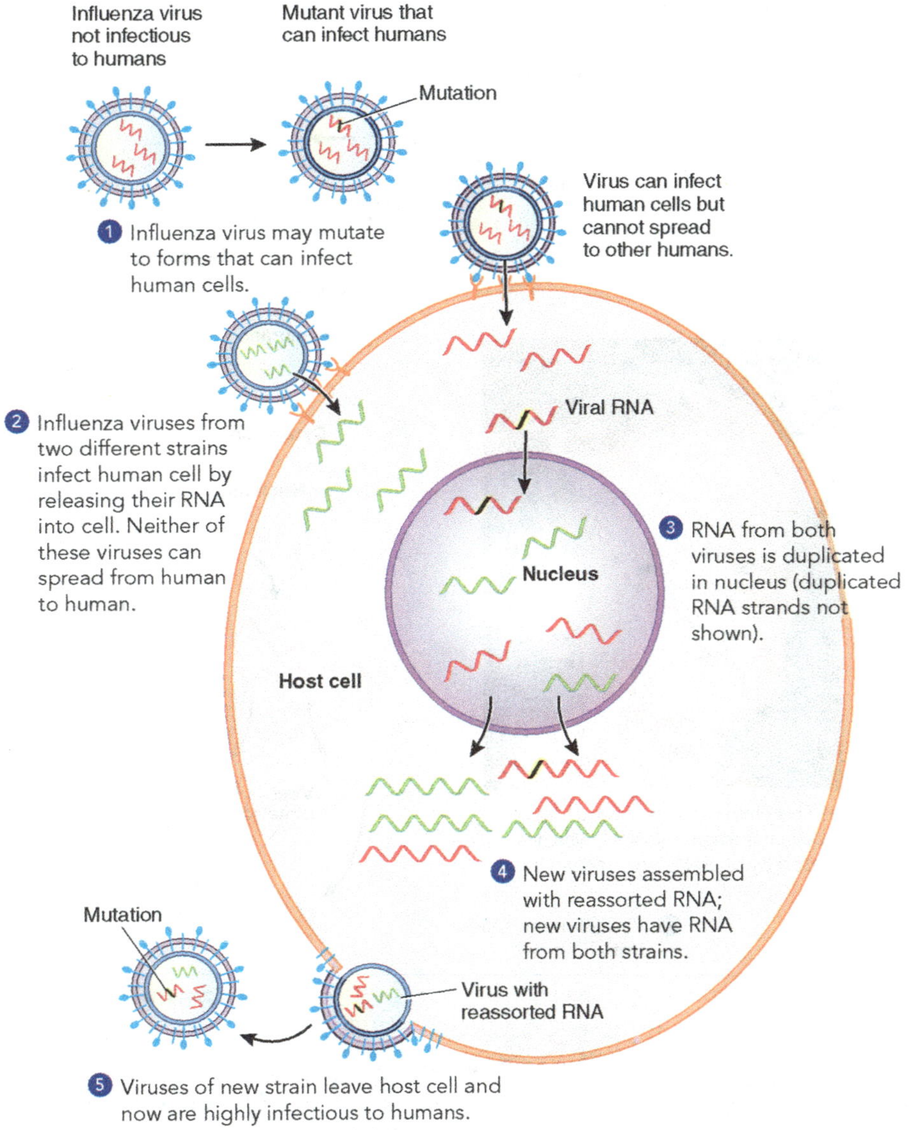

An influenza virus such as H1N1 evolves as it mutates and its RNA reassorts in the bodies of birds, swine, and humans. Eventually, it becomes virulent.

In DNA animal viruses, replication of viral DNA and protein synthesis are similar to the processes by which the host cell would normally carry out its own DNA replication and protein synthesis. In most RNA viruses, RNA synthesis takes place with the help of an RNA-dependent RNA polymerase.

Retroviruses are RNA viruses that have a DNA polymerase called **reverse transcriptase,** which transcribes the RNA genome into a DNA intermediate (FIG. 24-7). This DNA becomes integrated into the host DNA by an enzyme also carried by the virus. Copies of the viral RNA are synthesized as the incorporated DNA is transcribed by host RNA polymerases. The **human immunodeficiency virus (HIV)** that causes AIDS is a retrovirus. Some retroviruses can cause cancer directly or by interacting with other cancer-causing viruses.

After viral genes are transcribed, the viral structural proteins are synthesized. The capsid is produced, and then assembly of new virus particles takes place. Finally, release occurs.

KEY POINT

Viruses bind to specific receptor proteins on the animal cell's plasma membrane. Enveloped viruses then enter the cell by membrane fusion.

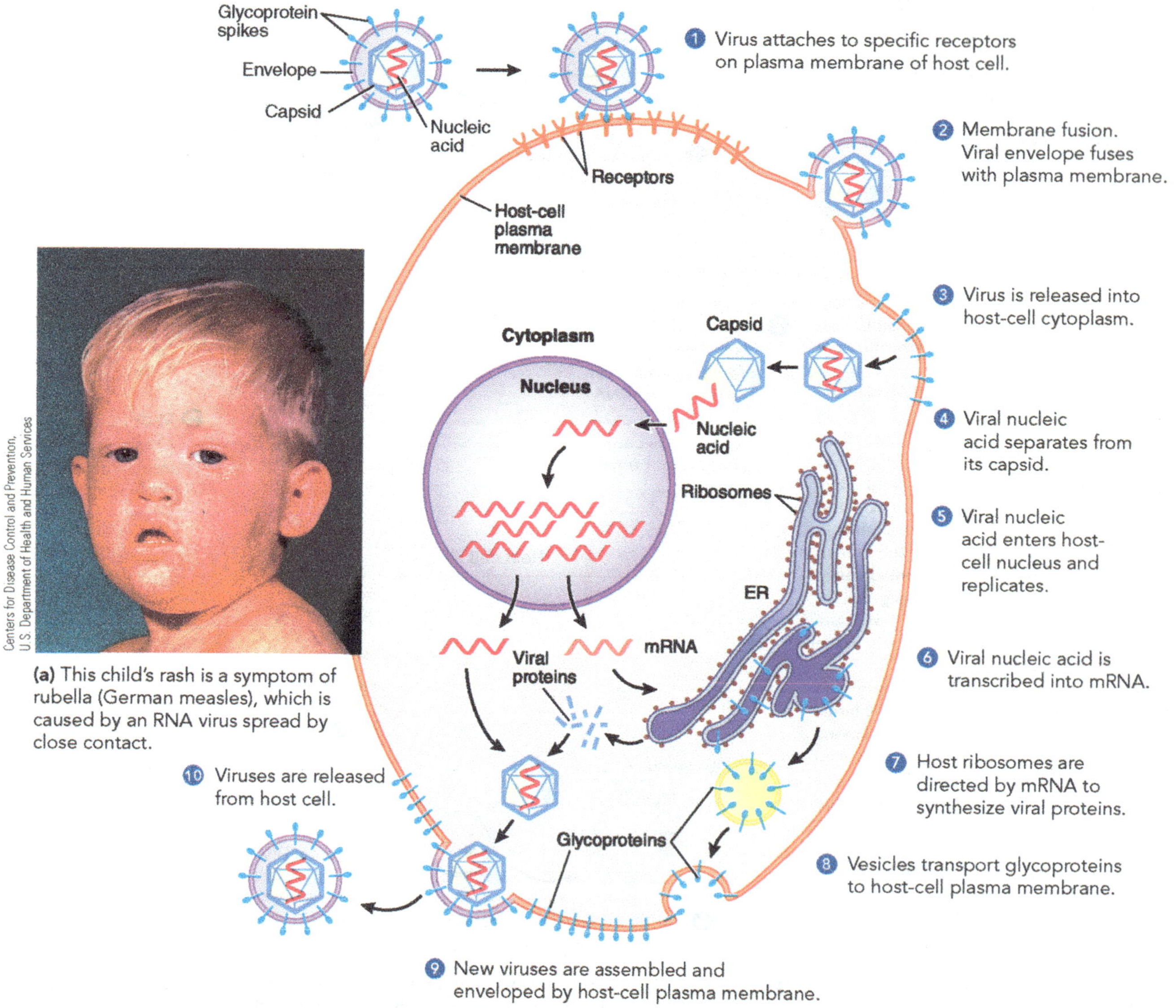

(a) This child's rash is a symptom of rubella (German measles), which is caused by an RNA virus spread by close contact.

(b) An enveloped virus enters an animal cell by fusing with its plasma membrane. After it makes copies of itself, new viruses are released. A new envelope forms around each virus as it leaves the host cell.

Figure 24-6 ***Animation*** **Viral infection in animals**

CONNECT Why can't new viruses be completely assembled in the nucleus of the host?

Viruses that do not have an outer envelope exit by cell lysis. The plasma membrane ruptures, releasing many new viral particles. Enveloped viruses obtain their lipoprotein envelopes by picking up a fragment of the host plasma membrane as they leave the infected cell (see Figs. 24-6b and 24-7).

Viral proteins damage the host cell in several ways. These proteins may alter the permeability of the plasma membrane or may inhibit synthesis of host nucleic acids or proteins. Viruses sometimes damage or kill their host cells by their sheer numbers. A poliovirus can produce 100,000 new viruses within a single host cell!

Viruses quickly become resistant to antiviral drugs

Antibiotics are specific for fighting bacteria; they do not kill viruses. Fortunately, researchers have developed antiviral drugs. These drugs inhibit the development, or replication, of many types of RNA and DNA viruses. Unfortunately, viruses rapidly develop resistance, making antiviral drug design and therapy a

KEY POINT

Retroviruses use reverse transcriptase to transcribe their RNA into DNA; the viral DNA becomes part of the host DNA. When activated, viral DNA uses host enzymes to transcribe viral RNA.

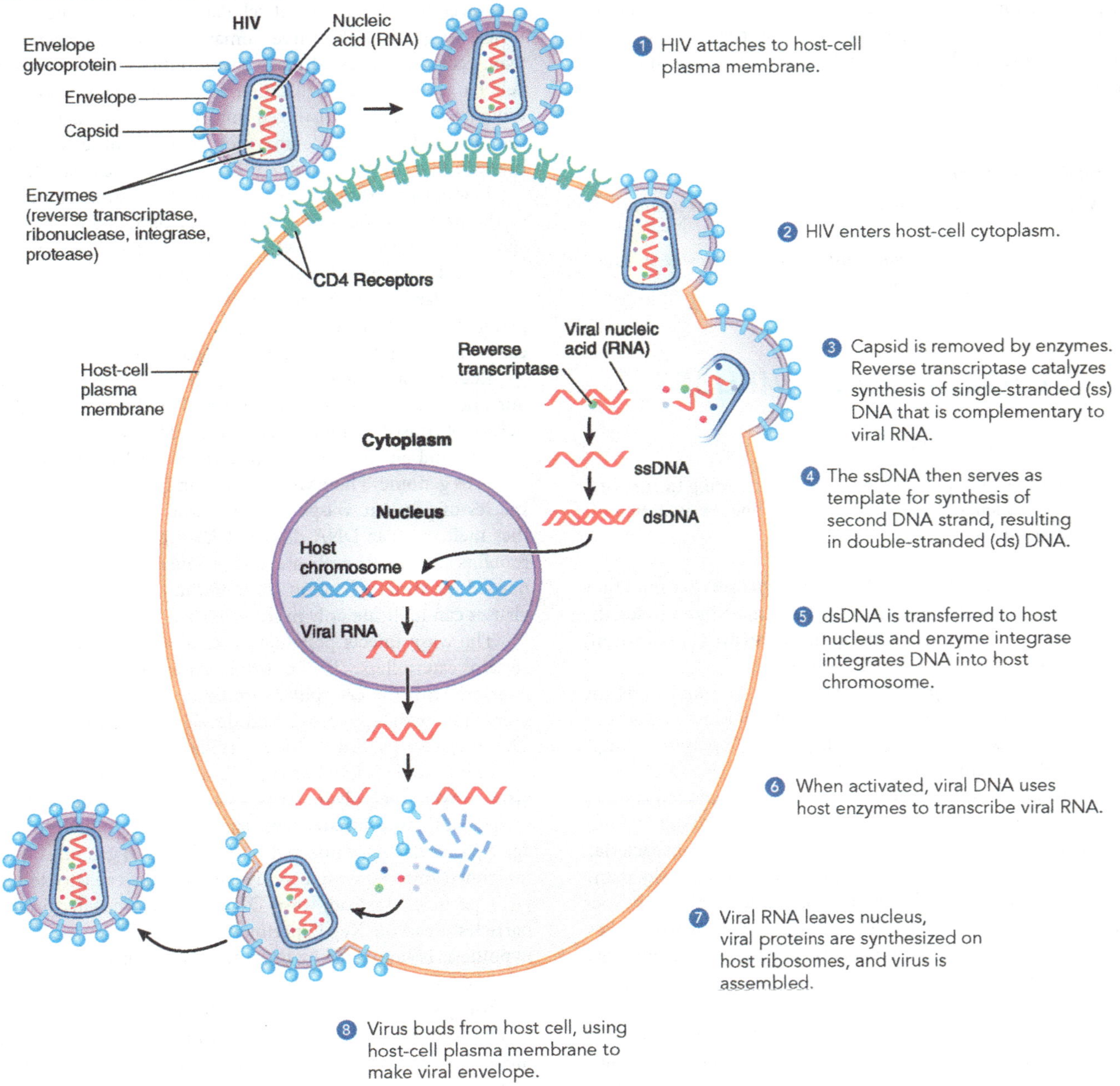

Figure 24-7 ***Animation*** **Life cycle of HIV, the retrovirus that causes AIDS**

HIV infects T helper cells, specialized cells of the host's immune system. The virus attaches to protein receptors, known as CD4, on the plasma membrane of T helper cells. HIV has two identical single-stranded RNA molecules.

CONNECT What advantage does the retrovirus that causes AIDS have compared to an RNA virus?

constantly changing, competitive endeavor. For example, amantadine, which inhibits penetration or uncoating of viral nucleic acids, has been effective in treating patients with influenza. However, most strains of influenza virus have become resistant to amantadine, perhaps as a result of common use of the drug in poultry feed in certain areas of Asia, such as China.

Another type of antiviral drug (e.g., Tamiflu) inhibits a viral enzyme (neuraminidase) necessary for the virus to leave the host cell. Certain viral strains are becoming resistant to Tamiflu. A single mutation that occurs spontaneously is apparently responsible for this resistance. Many more antiviral drugs are being designed or are currently in clinical trials. Some inhibit viral attachment to host cells, and others interfere with replication of viral nucleic acid.

CHECKPOINT 24.4

- *How do viruses infect plant cells?*
- *What are the mechanisms by which viruses enter animal cells?*
- VISUALIZE *Draw a diagram illustrating how retroviruses, such as HIV, make copies of themselves.*

24.5 EVOLUTION OF VIRUSES

LEARNING OBJECTIVE

8 Trace the evolutionary origin of viruses according to current hypotheses and describe research regarding the evolution of polydnaviruses.

Scientists have been debating the origin of viruses ever since they were discovered. Currently, there are three main hypotheses: the progressive (or escape) hypothesis, the regressive (or reduction) hypothesis, and the virus-first hypothesis.

According to the *progressive hypothesis*, also known as the *escape hypothesis*, viruses may have originated as mobile genetic elements such as *transposons* (see Chapter 13) or *plasmids* (small, circular DNA fragments discussed in Chapters 15 and 25). Such fragments could have escaped from one cell and entered another cell through damaged cell membranes. According to this hypothesis, viruses may trace their origin to animal, plant, bacterial, or archaeal cells. Their multiple origins may explain why many viruses are species specific; perhaps they infect only those species that are closely related to the organisms from which they originated. This hypothesis is supported by the genetic similarity between some viruses and their host cells—a closer similarity than exists between one type of virus and another.

The *regressive hypothesis*, also called the *reduction hypothesis*, asserts that viruses are remnants of cellular organisms and evolved from small cells that were parasites in larger cells. Genes that they did not need, like those for protein synthesis, were gradually lost through evolution. This hypothesis is supported by certain bacteria (chlamydia and rickettsia) that are able to reproduce only inside the cells of their hosts (see Table 25-4).

The regressive hypothesis is also supported by the discovery that some giant viruses have genes that encode components for protein translation. Proponents of the regressive hypothesis suggest that these giant viruses evolved from a free-living, complex cellular ancestor. Later, they became parasitic and gradually lost some of the genes necessary for protein synthesis. This hypothesis explains how viruses could have existed before their present-day hosts evolved.

The *virus-first hypothesis* states that viruses predate or coevolved with their current cellular hosts even before the life-forms assigned to the three domains diverged. One research team recently suggested that viruses initially existed in a pre-cellular world as self-replicating units. Gradually, these small units became more organized and more complex. Eventually they evolved to produce enzymes for the synthesis of membranes and cell walls, leading to the origin of the first cells.

Evidence for this hypothesis comes from similarities found in the protein structures of some viral capsids and in genetic similarities between some viruses that infect archaea and some that infect bacteria. Molecular biologists studying this hypothesis consider it improbable that these similarities evolved independently. This evidence suggests that viruses diverged very early—before archaea, bacteria, or eukaryotes.

Recent studies on the origin of polydnaviruses contribute to our understanding of viral evolution. **Polydnaviruses** are particles that consist of multiple circles of dsDNA encased in capsid proteins and an envelope. Each circle of DNA contains part of the virus genome. These viruses are found in ovary cells of many species of parasitic wasps. The polydnaviruses are unusual in that their circular DNA does not have genes for making the proteins needed to replicate and produce new viruses. Genes needed for viral replication are found in the wasp genome; the viruses can replicate only in the wasp ovary cells.

The wasp injects polydnaviruses along with her eggs into certain caterpillars. The polydnaviruses express toxins that interfere with the caterpillar's immune defenses and development. The wasp eggs hatch and develop inside the caterpillar. The young wasps feed on the caterpillar.

Biologists wondered whether the polydnaviruses were really viruses. Were the polydnaviruses simply particles derived from wasp genes? An alternative hypothesis was that millions of years ago a particular virus infected wasps. The virus genome became integrated into the wasp genome and lost the capacity to enter virus particles. Instead, wasp DNA was incorporated into the particles. Read the Key Experiment, FIGURE 24-8, to learn which hypothesis is supported by the research of Jean-Michel Drezen, Annie Bézier, and their colleagues.

Today's viruses may have originated multiple times, through many mechanisms, or by a mechanism not yet discovered. No matter how viruses came to be and how they continue to evolve, they have important roles in evolution. They mutate constantly, producing many new alleles. They replicate rapidly, and some of their genes are incorporated into the genomes of host cells through horizontal gene transfer. In fact, viruses may provide a major source of new genes for their host cells.

CHECKPOINT 24.5

- CONNECT *How could viruses have evolved before their hosts? Explain your answer.*
- CONNECT *In what way are polydnaviruses mutualistic partners with certain wasps?*

KEY EXPERIMENT

What is the origin of polydnaviruses?

HYPOTHESIS: Polydnaviruses evolved from ancient viruses that infected wasps.

EXPERIMENT: Annie Bézier, Jean-Michel Drezen, and their colleagues searched for virus-related genes in the wasp genome. The team analyzed DNA from three different wasp species. They compared nucleotide sequences with sequences of public databases containing sequences of all organisms and viruses studied.

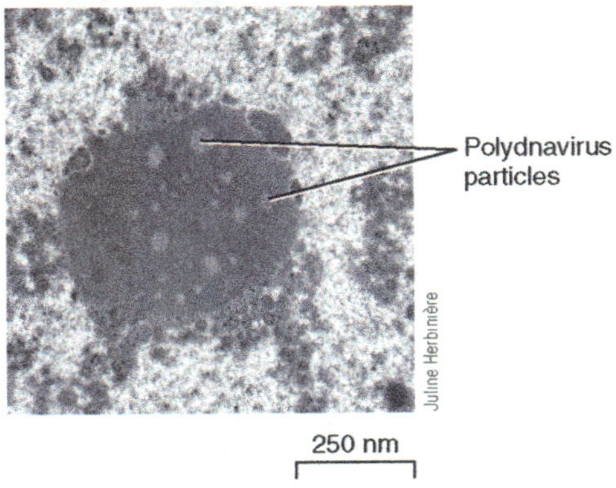

RESULTS AND CONCLUSION: The researchers identified 22 genes expressed in wasp ovaries where viral particles are produced. These genes are present in the wasp genome and resemble key genes of nudiviruses, a family of insect viruses. The investigators found that many of these genes code for proteins that are part of the polydnavirus particles. These genes have been conserved in several different wasp families that have polydnaviruses.

The research team concluded that the polydnavirus evolved from a nudivirus that infected wasps millions of years ago. In time, the virus genome became incorporated into the wasp genome. Proteins needed for viral replication are now part of the wasp's DNA, and the virus can replicate only in the wasp's ovaries.

At the same time, the wasps depend on the virus. The wasps inject their eggs into certain caterpillars along with polydnaviruses. The viruses depress the caterpillar's immune defenses, allowing the wasp larvae to develop in the caterpillar. Thus, a mutualistic relationship has evolved between the wasp and the polydnavirus.

The investigators suggested that the polydnavirus acts as a gene vector, transporting large chunks of DNA to the caterpillar. Might learning more about this mechanism have clinical applications?

SOURCE: Bézier, Annie, et al., "Polydnaviruses of Braconid Wasps Derive from an Ancestral Nudivirus." *Science*, Vol. 323, No. 5916, Feb. 13, 2009, pp. 926–930.

Figure 24-8 Exploring the evolution of polydnaviruses

This TEM shows polydnavirus particles in the nucleus of a cell in a wasp ovary. The polydnavirus particles are located at the periphery of an electron-dense region thought to be a "viral factory."

PREDICT What would happen if the polydnavirus mutated and no longer depressed the caterpillar's immune defenses?

24.6 SUBVIRAL AGENTS

LEARNING OBJECTIVE

9 Compare satellites, viroids, and prions.

Subviral agents are infective agents that are smaller and simpler than viruses. Subviral agents include satellites, viroids, and prions.

Satellites depend on helper viruses

Satellites are subviral agents that depend on co-infection of a host cell with a helper virus. The satellite, which consists of either DNA or RNA, cannot reproduce without the helper virus. The agent that causes hepatitis D is a satellite that can reproduce only when the hepatitis B virus is also present.

Sputnik is a satellite that infects Mimivirus, which in turn infects an amoeba. Sputnik is dependent on the replication and assembly machinery set up by Mimivirus. Some researchers now refer to Sputnik as a *virophage*, a word not yet completely defined by many virologists. The term implies that Sputnik, and perhaps other satellites, impairs the reproduction of Mimivirus, its helper virus.

Viroids are short, single strands of naked RNA

In 1971, Theodor Otto Diener, a plant pathologist, discovered an infective agent in potatoes that he named a **viroid.** Since then, about 30 species of viroids have been identified. Smaller than a virus, a viroid consists of a very short, circular, single strand of naked RNA (only 250 to 400 nucleotides). The viroid has no protective protein coat and no associated proteins to assist in duplication. Its RNA serves as a template that is copied by host RNA polymerases. Viroids are extremely hardy and are able to resist heat and ultraviolet radiation because of the condensed folding of their RNA.

Many viroids that cause plant diseases—including potato spindle tuber viroid, apple scar skin viroid, avocado sunblotch viroid, and tomato chlorotic dwarf viroid—have been identified. Viroids cause stunted or distorted growth and sometimes kill the plant. For example, cadang-cadang disease has killed more than 30 million coconut palm trees in the Philippines. Viroids are transmitted by infected pollen or seeds.

Viroids are generally found within the host-cell nucleus and appear to interfere with gene regulation. Molecular biologists have hypothesized that viroids mirror gene sequences in their hosts and silence critically important host genes. The viroid's ssRNA replicates, forming dsRNA. The plant's defense response cleaves this viroid RNA, producing small interfering RNAs (siRNAs), discussed in Chapters 13 and 14. However, the viroid siRNAs then cause host ribonucleases to selectively cleave host mRNAs with complementary base sequences. This action inactivates host mRNA and silences specific host genes. The viroids themselves are resistant to RNA silencing.

Prions are protein particles

Motivated by the death of a patient from Creutzfeldt–Jakob disease (CJD), a degenerative brain disease, Stanley Prusiner, professor of neurology and biochemistry at the University of California, San Francisco, began his studies of prions in the early 1970s. Prusiner discovered that the infective agent was not affected by radiation (which typically mutates nucleic acids), and he could not find DNA or RNA in the particles. In 1982, he named the infective agent **prion,** for "proteinaceous infectious particle."

Prusiner's hypothesis that an infectious agent could cause disease without nucleic acids was not readily accepted because there was no known mechanism by which this action could occur. However, Prusiner and other researchers continued to study prions and discovered how they can cause cells to malfunction. In 1997, Prusiner was awarded the Nobel Prize in Physiology or Medicine for his discovery of prions.

Prusiner and others have shown that animals have a gene that encodes a normally harmless protein known as PrP. Normal forms of the protein are found on the surfaces of brain cells and many other types of cells. The PrP protein, which consists of 208 amino acids, helps neurons transport copper and may play a role in the sense of smell in animals. Researchers have discovered that normal PrP interacts with the peptides associated with Alzheimer's disease.

Sometimes, the PrP protein folds into a different shape, an insoluble form that can cause disease. This misfolded protein is the prion. Mutations in the gene that encodes the PrP protein increase the risk that the protein will misfold and become a prion. The prion then somehow induces other PrP molecules to misfold into the pathogenic form (FIG. 24-9). Prions apparently can aggregate and accumulate in the brain and in certain other tissues and cause serious damage.

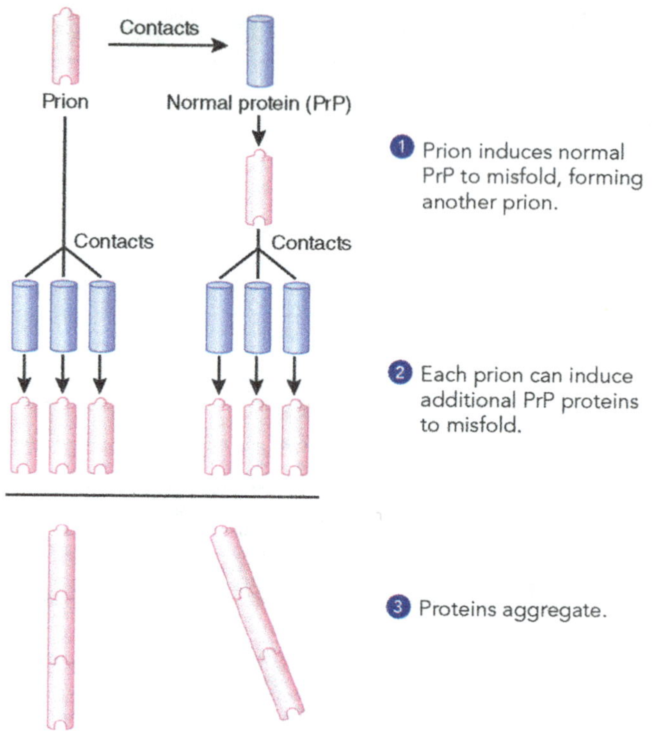

Figure 24-9 A model for how prion populations expand

Prions contact normal PrP proteins and induce them to misfold and become prions. Each new prion can then contact additional PrP molecules and induce them to misfold, thus expanding the prion population. Prions can form clumps by aggregating.

Prions are found in the brains of patients with *transmissible spongiform encephalopathies (TSEs)*. This group of fatal degenerative brain diseases has been identified in birds and mammals. These diseases are called TSEs because when infected, the brain appears to develop holes and becomes somewhat spongelike. The TSEs are in some ways like viral illnesses. However, there is a long latency period (about 5 years in cows and up to 10 years in humans) between contracting the infection and developing symptoms of the illness. Also, the TSEs have proven very difficult to treat. Genetically engineered mice that lack the prion protein gene are immune to TSE infection.

The oldest known prion disease is *scrapie* in sheep and goats. When infected, animals lose coordination, become irritable, and itch so severely that they scrape off their wool or hair. Bovine spongiform encephalopathy (BSE) is a related prion disease popularly referred to as "mad cow disease" because some diseased cattle become aggressive. In the 1990s, BSE became epidemic in cattle in the United Kingdom.

More than 200 people have died from a human variety of BSE, providing evidence that the disease is transmissible from cow to human. The human disease is called vCJD because it is a variant of CJD, which is caused by the transformation of PrP proteins into prions. The infectious agent of vCJD has been recovered from infected human neural tissue and appears similar to the prion that causes scrapie in sheep. New regulations have been instituted in many countries to safeguard the food supply. Human-to-human transmission of vCJD has been associated with tissue and organ transplants and with transfusion with contaminated blood.

Chronic wasting disease, an illness related to mad cow disease, has spread among deer and elk populations in North America. Studies are under way to determine whether chronic wasting disease can infect livestock or humans. Feline spongiform encephalopathy is a prion disease of domestic and captive felines. With this disease, large deposits of a misfolded protein accumulate in the liver and spleen.

Like viruses, prions exhibit variability. Different strains of prions may differ from one another in the conformation of their proteins, in their properties, and in the diseases they cause. Prions are different from viruses in that they can arise spontaneously, mostly as a result of mutation. Researchers are studying ways to block prions from forming. They are also looking for ways to stimulate cells to destroy prions. Early diagnosis is critical, and research for an assay, or test, to detect prion disease is ongoing.

CHECKPOINT 24.6

- CONNECT *What do viroids and prions have in common? How are they different?*

SUMMARY: FOCUS ON LEARNING OBJECTIVES

24.1 The Status and Structure of Viruses *(page 496)*

1 Contrast a virus with a cellular organism.

- A **virus** is a very small infective agent that consists of a core of nucleic acid and is dependent on a living host. Viruses contain the nucleic acids necessary to make copies of themselves, but to reproduce they must invade living cells and commandeer their metabolic machinery.

2 Describe the structure of a virus.

- A virus is a subcellular particle consisting of a DNA or RNA genome surrounded by a protein coat, called a **capsid.** In some viruses the capsid is surrounded by an outer envelope.

24.2 Classification of Viruses *(page 497)*

3 Identify three characteristics used to classify viruses.

- Viruses are classified based on **host range,** the types of host species they can infect; the type of nucleic acid they contain; and whether the nucleic acid is single-stranded or double-stranded. Other factors used to classify viruses include size and shape, presence of an envelope, and method of transmission from host to host.

24.3 Viral Replication *(page 498)*

4 Characterize bacteriophages.

- **Bacteriophages,** or **phages,** are complex viruses that infect bacteria. They typically consist of a long molecule of dsDNA coiled within a polyhedral head.

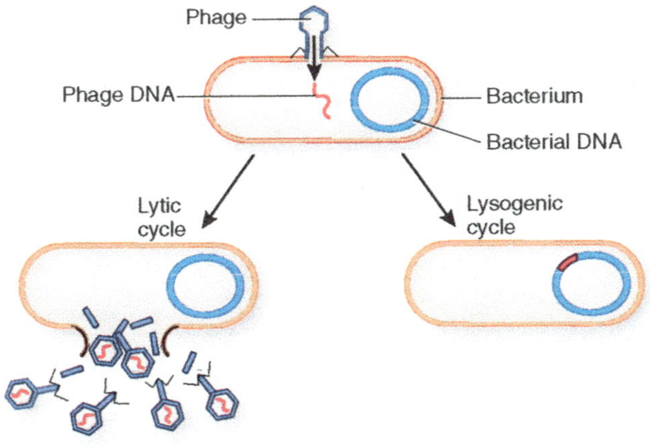

© Cengage Learning

5 Contrast a lytic cycle with a lysogenic cycle.

- In a **lytic cycle,** the virus destroys the host cell. The five steps in a lytic cycle are *attachment* to the host cell, *penetration* of viral nucleic acid into the host cell, *replication* of the viral nucleic acid, *assembly* of newly synthesized components into new viruses, and *release* from the host cell.
- The nucleic acid of some phages becomes integrated into the DNA of its bacterial host. The phage is then called a **prophage.** A **temperate virus** alternates between a lytic and a **lysogenic cycle.** In a lysogenic cycle, the genome of a temperate virus is replicated along with replication of the host's DNA without causing death of the host cell.
- Bacterial cells that carry prophages are lysogenic cells. In **lysogenic conversion** bacterial cells containing certain temperate viruses exhibit new properties.

24.4 Viral Diseases *(page 500)*

6 Contrast viral infection of plants and animals, and identify specific diseases caused by viruses.

- Most plant viruses are ssRNA viruses that do not have envelopes. Plant viruses penetrate plants through damaged cells and are also transmitted by infected seeds. Viruses can be spread among plants by insect vectors. Viruses spread through the plant via plasmodesmata.
- Viruses enter animal cells by membrane fusion or by endocytosis. Diseases caused by DNA viruses include herpes, respiratory infections, and gastrointestinal disorders. RNA viruses cause influenza, upper respiratory infections, AIDS, and some types of cancer.

7 Describe the reproductive cycle of a retrovirus, such as human immunodeficiency virus (HIV).

- **Retroviruses,** such as HIV, use **reverse transcriptase** to transcribe their RNA genome into a DNA intermediate that becomes integrated into the host DNA. Copies of the viral RNA are then synthesized.

24.5 Evolution of Viruses *(page 506)*

8 Trace the evolutionary origin of viruses according to current hypotheses and describe recent research regarding the evolution of polydnaviruses.

- According to the *progressive,* or *escape, hypothesis,* viruses may have originated as mobile genetic elements that could have escaped and moved from one cell and entered another through damaged cell membranes. Viruses may trace their origin to animal cells, plant cells, bacterial cells, or archaeal cells.
- The *regressive,* or *reduction, hypothesis* asserts that viruses are remnants of cellular organisms; they evolved from free-living, complex cellular ancestors. As they became parasitic, the viruses gradually lost some of the genes necessary for protein synthesis. Genes they no longer needed, like those for protein synthesis, were gradually lost through evolution.
- According to the *virus-first hypothesis,* viruses predate or coevolved with their current cellular hosts. Viruses may have initially existed as self-replicating units. As they evolved, viruses produced enzymes for the synthesis of membranes and cell walls, leading to the formation of the first cells.
- Evidence suggests that **polydnaviruses** evolved from viruses that infected wasps. They have become mutualistic partners with their wasp hosts.

24.6 Subviral Agents *(page 507)*

9 Compare satellites, viroids, and prions.

- Satellites, viroids, and prions are smaller than viruses. Satellites reproduce only with the help of a helper virus.
- A **viroid** consists of a short strand of RNA with no protein coat. Many viroids cause plant diseases. A **prion** consists only of protein. Prions cause fatal degenerative brain diseases, such as *transmissible spongiform encephalopathies (TSEs).*

TEST YOUR UNDERSTANDING

Know and Comprehend

1. The genome of a virus consists of (a) DNA (b) RNA (c) prions (d) DNA and RNA (e) DNA or RNA
2. The capsid of a virus consists of (a) protein subunits (b) nucleic acid (c) helical lipids (d) a carbohydrate envelope (e) RNA and lipid
3. Viruses that kill host cells are (a) lysogenic (b) lytic (c) viroids (d) prophages (e) temperate
4. In lysogenic conversion (a) bacterial cells may exhibit new properties (b) the host cell dies (c) prions sometimes convert to viroids (d) reverse transcriptase transcribes DNA into RNA (e) lytic viruses become temperate
5. The types of host species a particular virus can infect are referred to as its (a) assembly (b) commensal cohort (c) host range (d) prophage factor (e) virulence factor
6. Plant viruses (a) typically cause rapid death of infected plants (b) are mainly retroviruses (c) can be killed with antibiotics (d) are transmitted throughout the plant body by insects (e) can be transmitted from one generation to the next through infected seeds
7. According to the progressive, or escape, hypothesis, viruses (a) appeared before the three domains diverged (b) came from mobile genetic elements and had multiple origins (c) evolved from early plant cells (d) are bits of nucleic acid that escaped from animal cells (e) evolved from cells that were parasites in larger cells
8. Prions (a) consist of RNA with no protein coat (b) are misfolded proteins (c) cause several important plant diseases (d) are the infective agents of several emerging diseases (e) consist of proteins that stimulate host RNA to produce DNA

Apply and Analyze

9. Arrange the following list into the correct sequence for viral reproduction:
   1. penetration 2. assembly 3. replication 4. attachment 5. release
   (a) 1, 2, 3, 4, 5 (b) 5, 2, 3, 4, 1 (c) 4, 1, 3, 2, 5 (d) 4, 1, 2, 3, 5 (e) 3, 1, 2, 4, 5
10. Arrange the following list into the correct sequence for part of the cycle of a retrovirus:
   1. dsDNA integrated into host DNA 2. viral proteins synthesized on host ribosomes 3. viral DNA uses host enzymes to transcribe viral RNA 4. reverse transcriptase catalyzes synthesis of ssDNA 5. synthesis of second DNA strand
   (a) 5, 2, 1, 3, 4 (b) 5, 2, 3, 4, 1 (c) 4, 5, 1, 3, 2 (d) 4, 1, 2, 3, 5 (e) 2, 1, 3, 4, 5
11. **VISUALIZE** What does this diagram illustrate? Complete the labels.

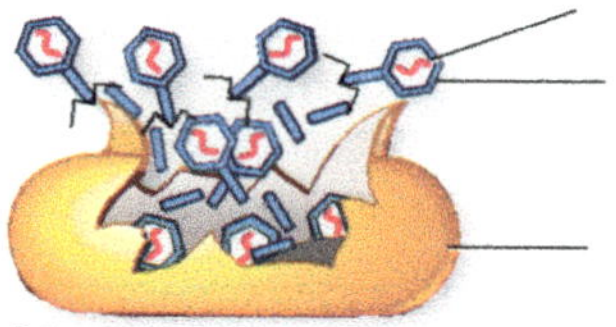

© Cengage Learning

12. **CONNECT** How are viral genomes different from those of plant or animal cells?
13. **CONNECT** How do viroids and prions differ from viruses?

Evaluate and Synthesize

14. **EVOLUTION LINK** Discuss the coevolution of polydnaviruses and their mutualistic partners. (See the discussion and Key Experiment on polydnaviruses.)
15. **EVOLUTION LINK** Based on what you have learned about viruses, present an argument for one of the three hypotheses of virus origin.
16. **SCIENCE, TECHNOLOGY, AND SOCIETY** What is the continuous challenge that the influenza virus presents to scientists and public health officials? How can they use technology and work together to meet this challenge?

aplia To access course materials, such as Aplia and other companion resources, please visit **www.cengagebrain.com.**

The bacterium *Clostridium butyricum*, one of the *Clostridium* species that produces the toxin botulin (colorized TEM).

26

STUDY OUTLINE

26.1 Prokaryotic Structure and Function

26.2 The Domain Bacteria

26.3 The Domain Archaea

Prokaryotes: Bacteria and Archaea

Why it matters . . . You wait in line at a fast-food restaurant, eagerly anticipating your hamburger. In the back of your mind you may worry because, not too many years ago, people were sickened, and a few even died, because their fast-food hamburgers were contaminated by a pathogenic strain of the bacterium *Escherichia coli,* the normally harmless bacteria that inhabit our intestinal tract. Since then, fast-food restaurants have cooked their hamburgers well beyond the point required to kill any lurking pathogenic *E. coli* or other pathogenic species of bacteria.

The bacterium *E. coli* is a prokaryote, an organism lacking a true nucleus. Prokaryotes, the topic of this chapter, are the smallest organisms in the world **(Figure 26.1).** Few species are more than 1 to 2 μm long; from 500 to 1,000 of them would fit side by side across the dot above this letter "i."

Prokaryotes are small, but their total collective mass (their *biomass*) on Earth may be greater than that of all plant life. They colonize every niche on Earth that supports life, meaning essentially everywhere. Huge numbers of bacteria inhabit surfaces and cavities of the human body, including the skin, the mouth and nasal passages, the large intestine, and the vagina. Collectively, bacteria in and on the human body outnumber the body's own cells by about tenfold.

On the basis of molecular phylogenetics analyses, biologists classify prokaryotes into two of the three domains of life, the **Archaea** and the **Bacteria** (see Section 24.7 and Figure 24.16). The third domain, the **Eukarya,** includes all eukaryotes. Bacteria are the prokaryotic organisms most familiar to us, including many types responsible for diseases of humans and other animals and many other types found in a wide variety of ecosystems. Many members of the domain Archaea (*arkhaios* = ancient) live under conditions so extreme, including high salinity, acidity, or temperature, that their environments cannot be tolerated by other organ-

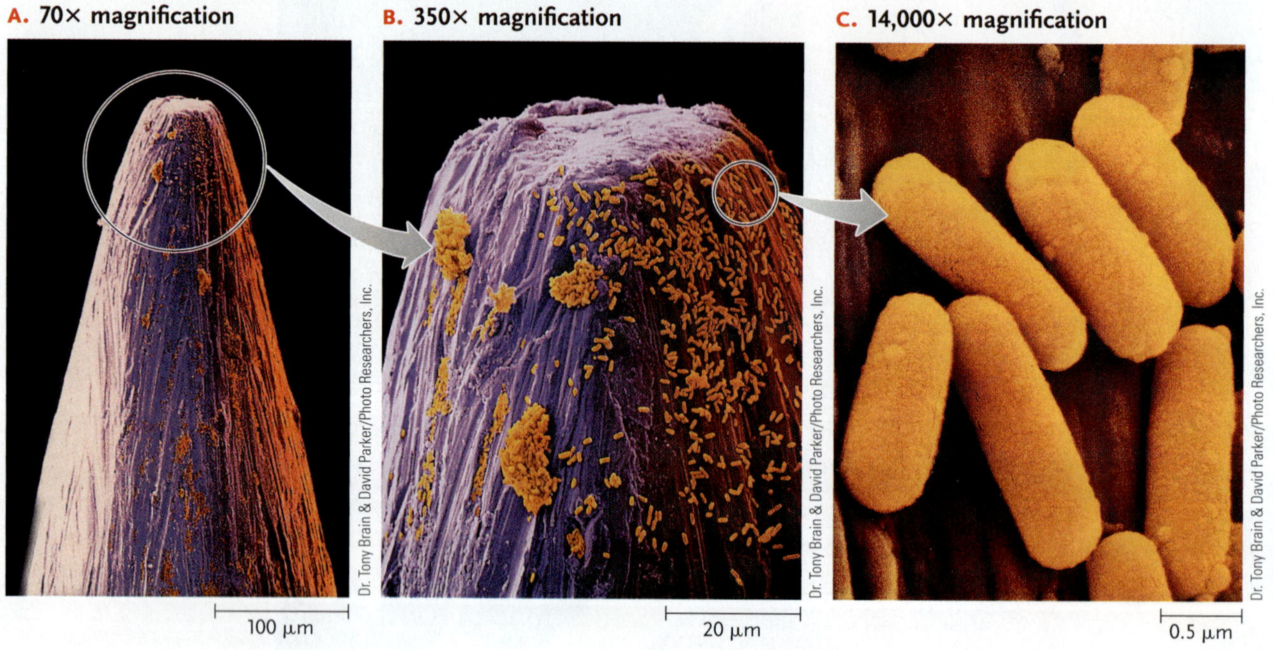

FIGURE 26.1 ***Bacillus*** **bacteria on the point of a pin.**

isms, including bacteria. In general, archaeans are less well studied, and therefore less well understood compared to bacteria.

As a group, prokaryotes have a wide range of metabolic capabilities. Their metabolic activities are crucial for maintenance of the biosphere. In particular, prokaryotes are the key players in the life-sustaining recycling of the elements carbon, nitrogen, and oxygen. For example, prokaryotes (along with other organisms) break down organic material in dead plants and animals, releasing carbon dioxide that is used for plant growth. In addition, prokaryotes are necessary to make nitrogen available to most forms of life. And a significant amount of the oxygen in the atmosphere originates from bacterial photosynthesis. An illustration of prokaryotes' importance is Biosphere 2, an attempt by scientists to build a completely closed ecosystem in Arizona. The attempt failed, in part because the researchers did not have a complete enough understanding of the activities of the microorganisms in the soil. Through respiration by soil microorganisms, the oxygen level in the Biosphere structure decreased to lower-than-expected levels and the ecosystem could no longer sustain itself. This small-scale example illustrates the essential role of prokaryotes in enabling life of all forms to exist.

Prokaryotes also have specific impacts on the lives of humans. Among other things, they are important for the production of certain foods; they carry out chemical reactions that are of importance in industry; they are used for the production of pharmaceutical products; they cause diseases; and they are used for bioremediation of polluted sites.

In Chapter 25 you learned how prokaryotic cells likely were produced from protocells. Evolution has generated the diversity of prokaryotes on Earth today. In this chapter you will learn about the structure and function of prokaryotes, and about the diverse organisms in the evolutionary branches of Bacteria and Archaea.

26.1 Prokaryotic Structure and Function

Prokaryotes show great diversity in their ability to colonize areas that can sustain life. Their cells are small and, while prokaryotes show a greater structural diversity than eukaryotes, prokaryotic cells are not as complex in organization as eukaryotic cells. For instance, although prokaryotes do not have a membrane-bound nucleus or organelles, their DNA and some proteins are localized in particular places. They vary in how their cell membrane is protected, and some species have specialized surface structures that protect them from their environment or that enable them to move actively. Prokaryotes also show great diversity in the ways they obtain energy and in their metabolic activities.

The diversity of prokaryotes has arisen through rapid adaptation to their environments as a result of evolution by natural selection. Genetic variability in prokaryotic populations, the basis for this rapid adaptation, derives largely from mutation and to a lesser degree from the transfer of genes between organisms by horizontal gene transfer, that is, by transformation, transduction, and conjugation (see Chapter 17). Since prokaryotes have much shorter generation times than eukaryotes and smaller genomes (roughly 1,000 times smaller than an average eukaryote), prokaryotes have roughly 1,000 times more mutations per gene, per unit time, per individual, than do eukaryotes. Furthermore, prokaryotes typically have much larger population sizes than eukaryotes, which contributes to their

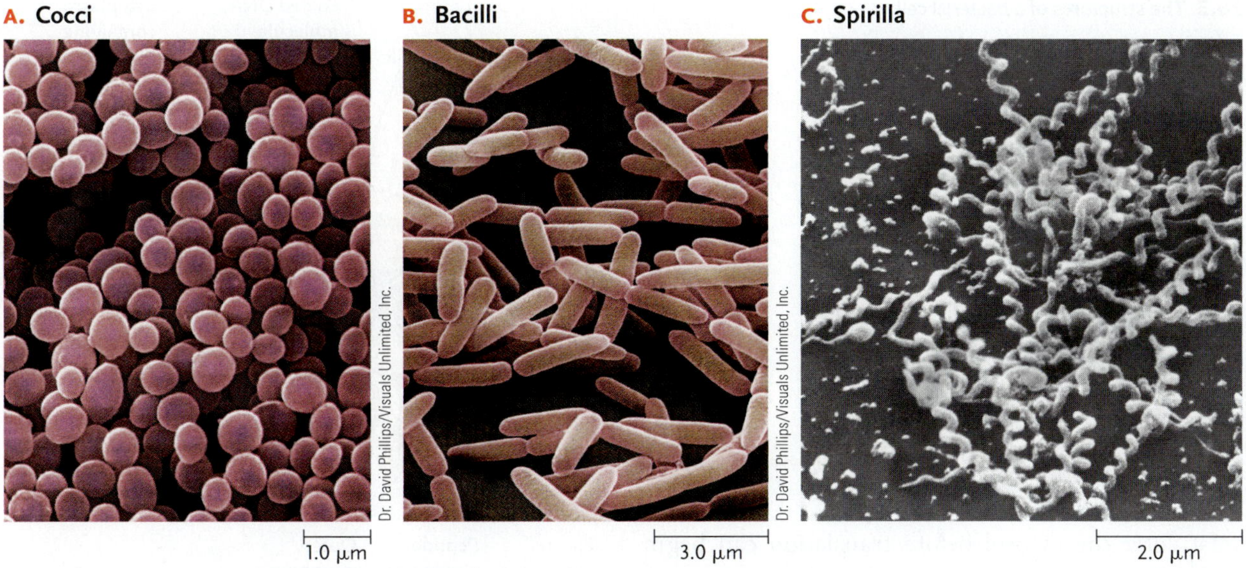

FIGURE 26.2 Common shapes among prokaryotes. (A) Scanning electron microscope (SEM) image of cocci of *Micrococcus* bacteria; **(B)** SEM image of bacilli of *Salmonella* bacteria; **(C)** SEM image of spirilla of *Spiroplasma* bacteria.

greater genetic variability. In short, prokaryotes have an enormous capacity to adapt, and this has been the key to their evolutionary success.

Prokaryotes Are Different in Structure and Organization from Eukaryotic Cells

Like eukaryotic cells, prokaryotic cells are bounded by a plasma membrane. Prokaryotic cells also have a cell wall, which provides reinforcement and protection. Within the cell, the genetic material of prokaryotes is localized to a non-membrane enclosed area of the cell, rather than being in a membrane-bound nucleus as in eukaryotes. Many prokaryotic species contain few if any internal membranes; they lack, therefore, the membrane-bound organelles characteristic of eukaryotic cells, including mitochondria, endoplasmic reticulum, Golgi complex, and chloroplasts. With few exceptions, the reactions carried out by these organelles in eukaryotes are distributed between the cytoplasmic solution and the plasma membrane in prokaryotes.

As you learned in Chapter 5, both prokaryotic and eukaryotic cells have cytoskeletons. Prokaryotic cytoskeletons play important roles in creating and maintaining the proper shape of cells, in cell division and, in certain prokaryotes, in determining polarity of the cells.

Although prokaryotic cells appear relatively simple according to this description, their simplicity is deceptive. Most can use a variety of substances as energy and carbon sources, and they can synthesize almost all of their required organic molecules from simple inorganic raw materials. In many respects, prokaryotes are more biochemically versatile than eukaryotes.

The following specifics of prokaryotic structure illustrate some of the diversity seen in this group of organisms.

SHAPE Three shapes are common among prokaryotes: spheres, rods, and spirals **(Figure 26.2).** The spherical prokaryotes are **cocci** (singular, *coccus*) (*kokkos* = berry). Cylindrical or rod-shaped prokaryotes are **bacilli** (singular, *bacillus*) (*bacillus* = small staff or rod). The spiral prokaryotes are the **spirilla** (singular, *spirillum*) (*speira* = coil, twist, wreath), which are twisted helically like a corkscrew, and **vibrios** (*vibrare* = to move quickly to and fro), which are curved and commalike. Among the prokaryotes of all structural types are some that live singly and others that link into chains or aggregates of cells.

INTERNAL STRUCTURES The genome of most prokaryotes consists of a single, circular DNA molecule called the *prokaryotic chromosome.* A few bacterial species, for example the causative agent of Lyme disease *(Borrelia burgdorferi),* have a linear chromosome. Genome sequencing projects have shown that the range of genome sizes among bacteria and archaeans is about 20-fold, with the smallest genome of any organism that can be cultured in the laboratory, that of *Mycoplasma genitalium,* being about 580,000 bp.

In all prokaryotes, the chromosome is packed into a central area of the cell called the **nucleoid.** The nucleoid has no nucleolus, and it has no boundary membranes equivalent to the nuclear envelope of eukaryotes (**Figure 26.3;** and see Sections 5.2 and 5.3). The arrangement of the genetic material in prokaryotes permits the translation of mRNA transcripts of protein-coding genes to begin while transcription is in progress. This is possible because the mRNA is immediately accessible to ribosomes, the cellular structures upon which translation takes place. Simultaneous transcription and translation is not possible in eukaryotes because the ribosomes are in the cytosol and the DNA is in the nucleus. Transcription and mRNA processing must be completed in the nucleus, and the

FIGURE 26.3 **The structures of a bacterial cell.**

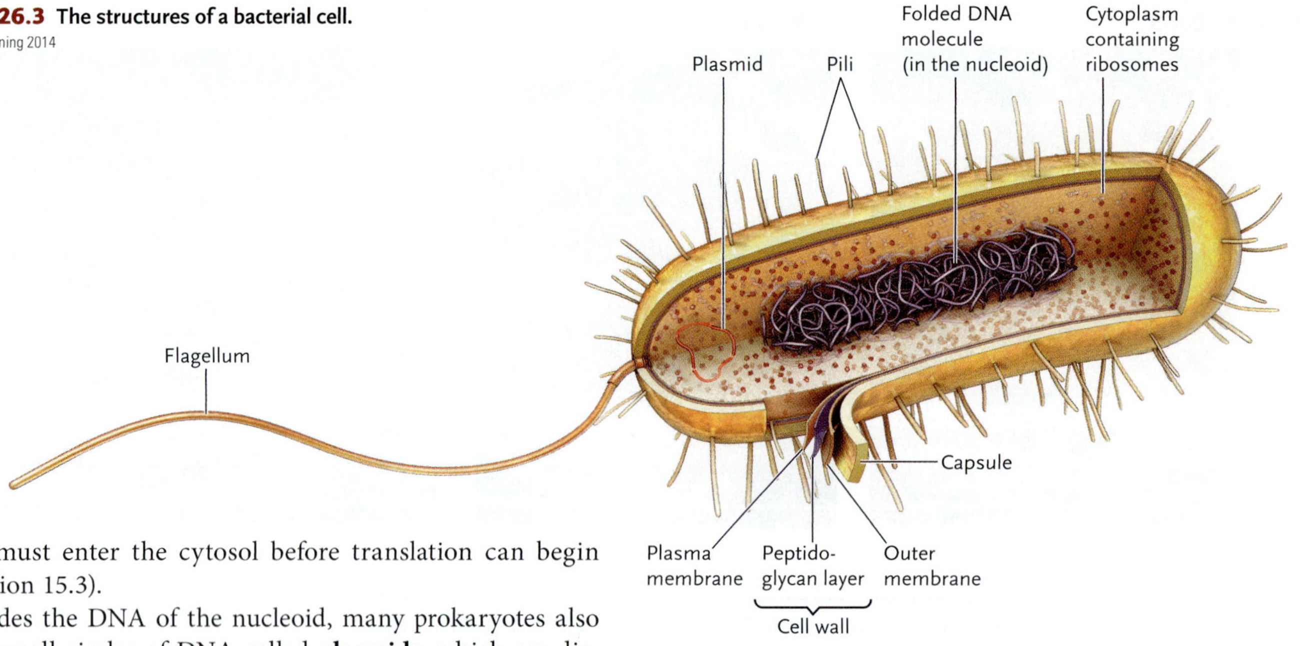

mRNA must enter the cytosol before translation can begin (see Section 15.3).

Besides the DNA of the nucleoid, many prokaryotes also contain small circles of DNA called **plasmids,** which are distributed in the cytoplasm. The plasmids, which often contain genes with functions that supplement those in the nucleoid, contain a replication origin that allows them to replicate along with the nucleoid DNA and be passed on during cell division (see Section 14.5).

Prokaryotic ribosomes are smaller than eukaryotic ribosomes and contain fewer proteins and RNA molecules. Archaeal ribosomes resemble those of bacteria in size, but differ in structure. Scientists have demonstrated that, with some differences in detail, bacterial ribosomes carry out protein synthesis by the same mechanisms as those of eukaryotes (see Section 15.4). Interestingly, protein synthesis in archaeans is a combination of bacterial and eukaryotic processes, with some unique archaeal features. As a result, antibiotics that stop bacterial infections by targeting ribosome activity do not stop protein synthesis of archaeans.

Some prokaryotes are capable of photosynthesis. Photosynthetic bacteria and archaeans have complex layers of intercellular membranes formed by invaginations of the plasma membrane on which photosynthesis takes place. Those membranous structures correspond to those that carry out photosynthesis in plants, but they are organized differently.

The cytoplasm of many prokaryotes also contains storage granules holding glycogen, lipids, phosphates, or other materials. The stored material is used as an energy reserve or a source of building blocks for synthetic reactions.

PROKARYOTIC CELL WALLS The plasma membrane of prokaryotes must withstand both high intracellular osmotic pressures and the action of natural chemicals in the environment that have detergent properties. Most prokaryotes have one or more layers of materials coating the plasma membrane to form a cell wall that provides the necessary protection, although the structure of the wall is quite different in bacteria and archaeans.

Bacteria typically are surrounded by a cell wall. The primary structural molecule of bacterial cell walls is **peptidoglycan,** a polymer formed from a polysaccharide backbone tied together by short polypeptides. (Polysaccharides are described in Section 3.2.) Peptidoglycan varies in chemical structure among different bacterial species.

Differences in bacterial cell wall composition are of clinical importance. In 1882, Hans Christian Gram, a Danish physician, developed a staining method to distinguish two types of bacteria in body fluids, each of which could cause pneumonia. In this **Gram stain technique,** an investigator treats bacteria with the dye crystal violet and then with iodine, which fixes the dye to the cell wall. Next the bacteria are washed with alcohol, and then treated with a second stain, either fuchsin or safranin. Bacteria that appear purple after these steps have retained the crystal violet stain; they are **Gram-positive.** Bacteria that appear pink after these steps have lost the crystal violet stain in the alcohol wash and are stained pink with the second dye; they are **Gram-negative.** (Gram-positive cells also react with the second dye, but the stain does not affect the color imparted by the crystal violet.)

The staining difference reflects differences in the cell walls of the bacteria **(Figure 26.4).** The cell wall of typical Gram-positive bacteria consists of a thick peptidoglycan layer (see Figure 26.4A). In contrast, the cell wall of typical Gram-negative bacteria consists of a thin layer of peptidoglycan (see Figure 26.4B). Outside of the thin cell wall is an additional boundary membrane, called the **outer membrane,** which covers the peptidoglycan layer. The outer membrane contains **lipopolysaccharides,** assembled from lipid and polysaccharide subunits found nowhere else in nature. The outer membrane protects Gram-negative bacteria from potentially damaging substances in their environment. For example, the outer membrane of *E. coli* protects it from the detergent effects of bile re-

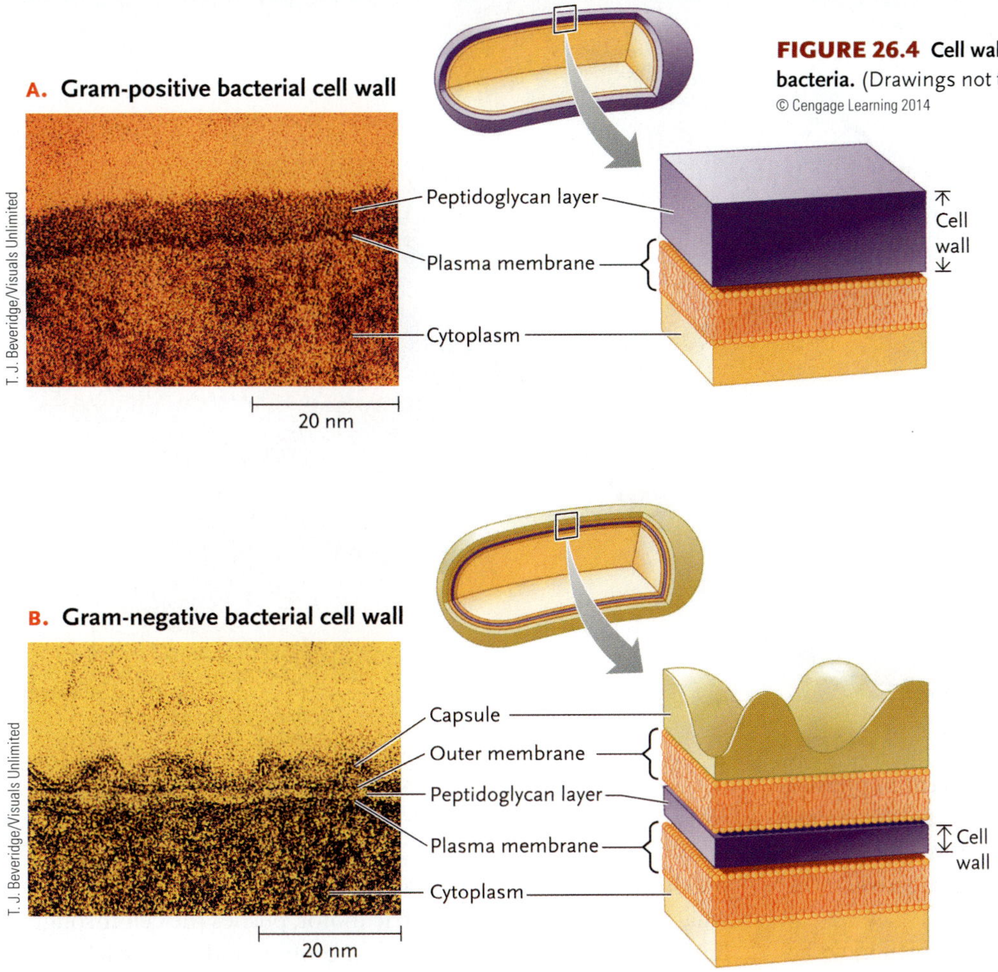

FIGURE 26.4 Cell wall structure in Gram-positive and Gram-negative bacteria. (Drawings not to scale.)

leased into the intestinal tract, which otherwise would lyse ("break open") the bacterium and kill it.

Rapidly distinguishing between Gram-positive and Gram-negative bacteria is important for determining the first line of treatment for bacterial-caused human diseases. Most pathogenic bacteria are Gram-negative species; their outer membrane protects them against the body's defense systems and blocks the entry of drugs such as antibiotics. For example, the antibiotic penicillin blocks new bacterial cell wall formation by inhibiting peptidoglycan crosslinking. The weakened cell wall soon leads to the death of the bacterium. Penicillin is effective against Gram-positive pathogens, but it is less effective against Gram-negative pathogens because their outer membrane inhibits entry of the antibiotic.

The walls of many Gram-positive and Gram-negative bacteria are coated with an external layer of polysaccharides called the **glycocalyx,** a slime coat typically composed of polysaccharides. When the glycocalyx is diffuse and loosely associated with the cells, it is a **slime layer;** when it is gelatinous and more firmly attached to the cells, it is a **capsule** **(Figure 26.5).** Depending on the species, the capsule ranges from a layer that is thinner than the cell wall to many times thicker than the entire cell. The glycocalyx helps protect cells from physical damage and desiccation, and to the effects of some antibiotics, and may enable a cell to attach to a surface.

In many bacteria, the capsule prevents bacterial viruses and molecules such as enzymes, antibiotics, and antibodies from reaching the cell surface. In many pathogenic bacteria, the presence or absence of the protective capsule differentiates virulent ("infective") from nonvirulent ("noninfective") forms. For example, normal *Streptococcus pneumoniae* bacteria are capsulated and are virulent, causing severe pneumonia in humans and other mammals. Mutant *S. pneumoniae* without capsules are nonvirulent and can easily be eliminated by the body's immune system if they are injected into mice or other animals (see Section 14.1).

Archaeans differ from bacteria by having some unique plasma membrane and cell wall characteristics. The lipid molecules in archaean plasma membranes have a chemical bond between the hydrocarbon chains and glycerol that is unlike that found in the plasma membranes of all other organisms. The difference is significant because the exceptional linkage is more resistant to disruption, making the plasma membranes of archaeans more tolerant of the extreme environmental conditions in which many of these organisms live.

The cell walls of some archaeans are assembled from molecules related to the peptidoglycans, but that have different molecular components and bonding structure. Others have walls assembled from proteins or polysaccharides instead of peptido-

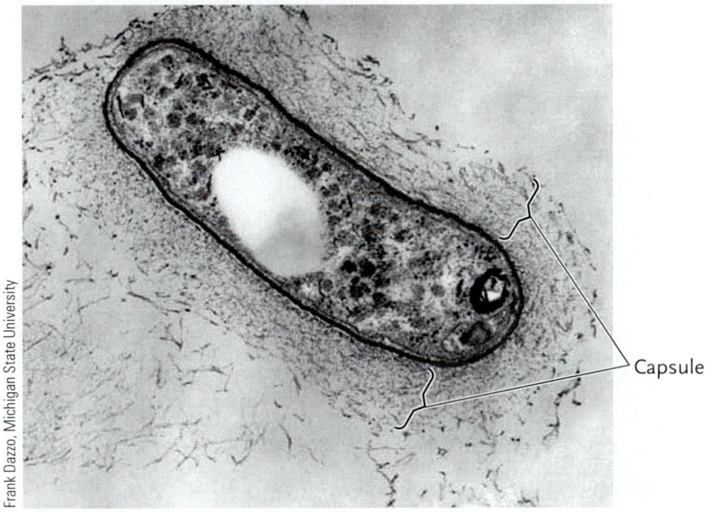

FIGURE 26.5 The capsule surrounding the cell wall of *Rhizobium*, a Gram-negative, soil-dwelling bacterium.

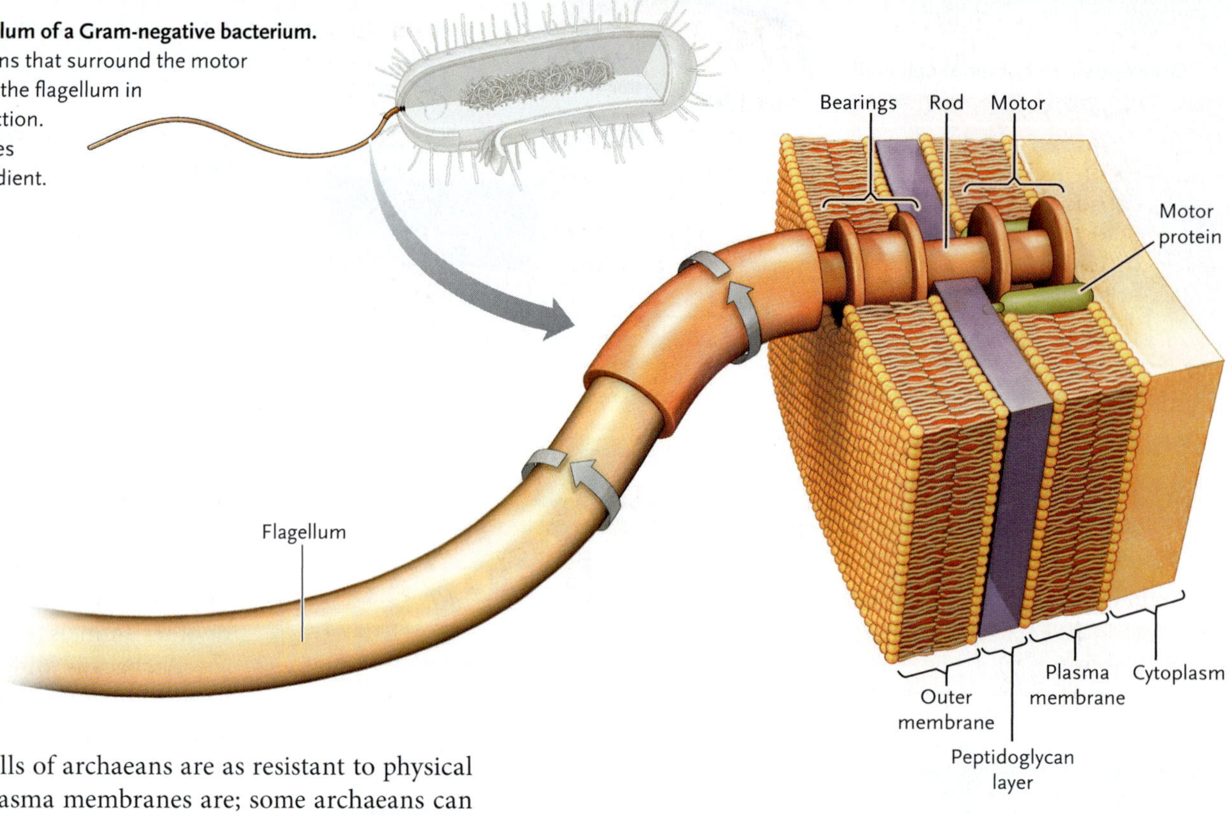

FIGURE 26.6 A flagellum of a Gram-negative bacterium. The ring of motor proteins that surround the motor provide torque to rotate the flagellum in a counterclockwise direction. Energy for rotation comes from a proton (H^+) gradient.

glycans. The cell walls of archaeans are as resistant to physical disruption as the plasma membranes are; some archaeans can be boiled in strong detergents without disruption. Different archaeans stain as either Gram-positive or Gram-negative.

FLAGELLA AND PILI Many bacteria and archaeans can move actively through liquids and across wet surfaces. The most common mechanism for movement involves the action of **flagella** (singular, *flagellum*) (*flagellum* = whip) extending from the cell wall **(Figure 26.6)**. These flagella are much smaller and simpler than the flagella of eukaryotic cells and contain no microtubules (eukaryotic flagella are discussed in Section 5.3).

A bacterial flagellum consists of a helical fiber of protein that rotates in a socket in the cell wall and plasma membrane, much like the propeller of a boat. The rotation, produced by what is essentially a tiny electric motor, pushes the cell through liquid. The motor is powered by a proton gradient, which creates an electrical repulsion that makes the flagellum rotate. In some bacteria, flagella group together in bundles that rotate to propel the cell.

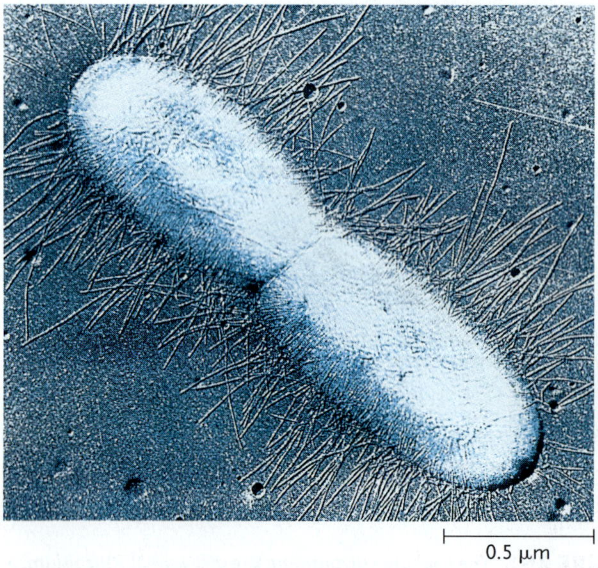

FIGURE 26.7 Pili extending from the surface of a dividing *E. coli* bacterium.

Archaeal flagella are analogous, not homologous, to bacterial flagella. That is, they carry out the same function, but they differ in their structure and mechanisms of action. The genes for the two types of flagellar systems are also different.

Some bacteria and archaeans have rigid, hairlike shafts of protein called **pili** (singular, *pilus*) extending from their cell walls **(Figure 26.7)**. Among bacteria, pili are characteristic primarily of Gram-negative bacteria; relatively few Gram-positive bacteria produce these structures. Recognition proteins at the tips of pili allow bacterial cells to adhere to other cells. One type, called *sex pili,* allows bacterial cells to adhere to each other as a prelude to conjugation, a joining of two cells that leads to transfer of genetic material from one to the other (see Section 17.1). Other types help bacteria to bind to animal cells. For example, *Neisseria gonorrhoeae,* the Gram-negative bacterium that causes gonorrhea, has pili that allow it to attach to cells of the throat, eye, urogenital tract, or rectum in humans.

Prokaryotes Have the Greatest Metabolic Diversity of All Living Organisms

All organisms take in carbon and energy in some form, but prokaryotes show the greatest diversity in their modes of securing these resources **(Figure 26.8)**. Prokaryotes are classified on the

Carbon source		Energy source: Oxidation of molecules*	Energy source: Light
	CO_2	**CHEMOAUTOTROPH** Found in some bacteria and archaeans; not found in eukaryotes	**PHOTOAUTOTROPH** Found in some photosynthetic bacteria, in some protists, and in plants
	Organic molecules	**CHEMOHETEROTROPH** Includes some bacteria and archaeans, and also in protists, fungi, animals, and plants	**PHOTOHETEROTROPH** Found in some photosynthetic bacteria

*Inorganic molecules for chemoautotrophs and organic molecules for chemoheterotrophs.

FIGURE 26.8 Modes of nutrition among Bacteria and Archaea. All four modes of nutrition occur in the Bacteria, with chemoheterotrophs as the most common type. Among the Archaea, chemoautotrophs are most common, while other archaeans are chemoheterotrophs.

basis of the source of carbon atoms they use. Some prokaryotes are **autotrophs** (*auto* = self; *trophe* = nourishment), meaning that they, like plants, obtain carbon from CO_2 (which is considered to be an inorganic molecule). Others are **heterotrophs,** meaning that they, like humans and other animals, obtain carbon from organic molecules. Prokaryotic heterotrophs obtain carbon from the organic molecules of living hosts, or from organic molecules in the products, wastes, or remains of dead organisms.

Prokaryotes are also classified according to the source of the energy they use to drive biological activities. **Chemotrophs** (*chemo* = chemical; *trophe* = nourishment) obtain energy by oxidizing inorganic or organic substances, while **phototrophs** obtain energy from light. Combining carbon source and energy gives us the following four types (see Figure 26.8):

1. **Chemoautotrophs:** Prokaryotic chemoautotrophs obtain energy by oxidizing inorganic substances such as hydrogen, iron, sulfur, ammonia, nitrites, and nitrates and use CO_2 as their carbon source. They use the electrons they remove in the oxidations to make organic molecules by reducing CO_2 or to provide the energy for ATP synthesis (using an electron transfer system embedded in the plasma membrane). Chemoautotrophy occurs widely among prokaryotes, including many bacteria and most archaeans, but is not found among eukaryotes.
2. **Chemoheterotrophs:** Prokaryotic chemoheterotrophs oxidize organic molecules as their energy source and obtain carbon in organic form. They include most of the bacteria that cause disease in humans, domestic animals, and plants and many bacteria that are responsible for decomposing matter. They are the largest prokaryotic group in terms of numbers of known species.
3. **Photoautotrophs:** Photoautotrophs are photosynthetic organisms that use light as their energy source and CO_2 as their carbon source. They include several groups of bacteria, for example, the *cyanobacteria,* as well as plants and many protists. The cyanobacteria use water (H_2O) as their source of electrons for reducing CO_2. Some other types of phototrophic bacteria use sulfur or sulfur compounds.
4. **Photoheterotrophs:** Photoheterotrophs use light as their ultimate energy source and preexisting organic carbon molecules as reducing agents, but they cannot use CO_2. Photoheterotrophs are limited to three subgroups of *proteobacteria.*

Prokaryotes Differ in Whether Oxygen Can Be Used in Their Metabolism

Prokaryotes also differ in how their metabolic systems function with respect to oxygen (see Chapter 8). **Aerobes** require oxygen for cellular respiration (in other words, oxygen is the final electron acceptor for that process). **Obligate aerobes** cannot grow without oxygen. **Anaerobes** do not require oxygen to live. **Obligate anaerobes** are poisoned by oxygen, and survive either by fermentation, in which organic molecules are the final electron acceptors (see Section 8.5), or by a form of respiration in which inorganic molecules such as nitrate ions (NO_3^-) or sulfate ions (SO_4^{2-}) are used as final electron acceptors. **Facultative anaerobes** use O_2 when it is present, but under anaerobic conditions, they live by fermentation.

Prokaryotes Fix and Metabolize Nitrogen

Nitrogen is a component of amino acids and nucleotides and, hence, is of vital importance for the cell. Prokaryotes are able to metabolize nitrogen in many forms. For example, a number of bacteria and archaeans are able to reduce atmospheric nitrogen (N_2, the major component of Earth's atmosphere) to ammonia (NH_3), a process called **nitrogen fixation.** The ammonia is ionized quickly to ammonium (NH_4^+), which the cell then uses in biosynthetic pathways to produce nitrogen-containing molecules such as amino acids and nucleic acids. Nitrogen fixation is an exclusively prokaryotic process and is the only means of replenishing the nitrogen sources used by most microorganisms and by all plants and animals. In other words, all organisms use nitrogen that is fixed by bacteria. Examples of nitrogen-fixing bacteria are some of the cyanobacteria and *Azotobacter* species, which are free-living, and *Rhizobium* species, which are in symbiotic relationships with plants (see Chapter 35).

Not all bacteria convert fixed nitrogen directly into organic molecules. Some bacteria carry out **nitrification,** the conversion of ammonium (NH_4^+) to nitrate (NO_3^-). This is carried out in two steps by two types of *nitrifying bacteria.* One type of nitrifying bacterium (for example, *Nitrosomonas* species) converts ammonium to nitrite (NO_2^-), while the other (for example, *Nitrobacter* species) converts nitrite to nitrate. Because of this specialization, both types of nitrifying bacteria are usually present in soils and water, with some converting ammonium to nitrite and others using that nitrite to produce nitrate. The nitrate can be used by plants and fungi to incorporate nitrogen into organic molecules. Animals obtain nitrogen in organic form by eating other organisms.

In sum, nitrification makes nitrogen available to many other organisms, including plants and animals and bacteria that cannot metabolize ammonia. You will learn more about nitrogen metabolism in connection with the nitrogen cycle (see Chapter 53). The metabolic versatility of the prokaryotes is one factor that accounts for their abundance and persistence on the planet; another factor is their impressive reproductive capacity.

Prokaryotes Pass Genes to Progeny by Binary Fission or, Rarely, from Cell to Cell by Horizontal Gene Transfer

In prokaryotes, reproduction occurs by binary fission in which a parent cell divides into two daughter cells that are exact genetic copies of the parent (see Figure 10.17). In this process, genetic material moves from generation to generation, that is, by descent.

Conjugation, in which two parent cells join or "mate," occurs in some bacterial and archaeal species. Conjugation depends on genes carried by a plasmid that replicates separately from the prokaryotic chromosome. Usually only the plasmid is passed on during conjugation, but in some bacteria, the plasmid integrates into the chromosome of the host so that host genes transfer from one parent (donor) to the other (recipient). Genetic recombination then occurs. The recombinant cell divides to produce daughter cells that differ in genetic information from either parent. The movement of genetic material between organisms by means other than descent, as is the case with conjugation, is **horizontal gene transfer.** Horizontal gene transfer by conjugation between bacterial cells is described in Section 17.1.

A small number of bacteria can produce an **endospore,** so-called because it develops *within* the cell **(Figure 26.9).** The endospore, which typically develops when environmental conditions become unfavorable, is a dormant cell containing the bacterial genome and some of the cytoplasm. It is inactive metabolically and highly resistant to heat, desiccation, and attack by enzymes or other chemical agents. In the first step of endospore formation, binary fission cuts the parent cell into two parts of unequal size. The larger cell then envelops the smaller one (which contains the genome) and surrounds it with a tough, chemically resistant protein coat; the smaller cell develops into the endospore. Rupture of the larger cell releases the endospore to the environment. If environmental conditions become favorable for growth, the spore germinates: it becomes permeable, water enters the cell, its surface coat breaks, and the cell is released in a metabolically active form.

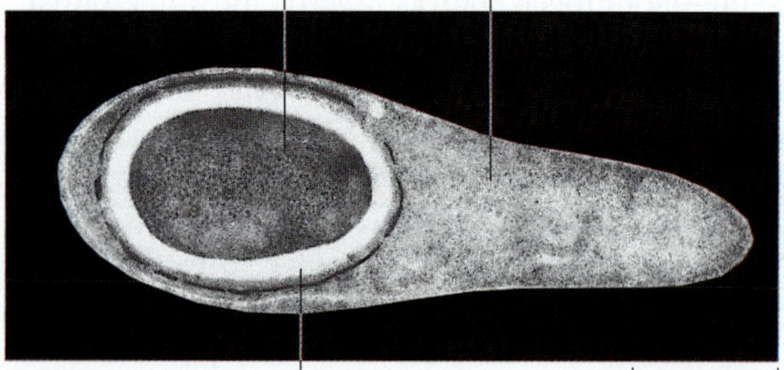

FIGURE 26.9 A developing endospore within a cell of the bacterium *Clostridium tetani*, a dangerous pathogen that causes tetanus.

No one is certain how long endospores can survive. There are claims that endospores survive for thousands to millions of years, but the data are controversial. They certainly can survive a very long time.

In Nature, Bacteria May Live in Multicellular Associations

Researchers grow prokaryotes as individuals in liquid cultures or as isolated colonies on solid media. The results from studies using pure cultures have been crucial in developing an understanding of, among many other things, the nature of the genetic material, DNA replication, gene expression, and gene regulation. But, in nature, bacteria typically are not in pure cultures. Instead, there are mixed populations of bacterial species, perhaps also with archaeal and eukaryotic microbial species. Thus, information from pure cultures in the laboratory may not apply to populations of the same species in nature.

In addition, while prokaryotes are unicellular organisms, in nature prokaryotes may form *multicellular associations,* either of the same or mixed species. Those multicellular associations may function essentially as single cells do, or they may undergo some differentiation so that cells within the association carry out different functions. For example, the photoautotrophic cyanobacteria may live singly or form multicellular associations of various kinds, including filaments, sheets, or balls of cells. In the case of the cyanobacterium *Anabaena spiroides,* the cells associate into filaments in which cells differentiate into three different types: vegetative cells, *heterocysts,* and spores (see Figure 26.16B). Photosynthesis occurs only in the vegetative cells. Nitrogen starvation induces heterocysts to form. These cells lose their ability to perform photosynthesis and specialize in nitrogen fixation. Organic acids leak from the heterocysts and attract heterotrophic bacteria, which consume the oxygen surrounding the cell, thereby creating the anaerobic conditions necessary for nitrogen fixation. The third cell type, the spore, is a resting stage that can withstand adverse environmental conditions and then germinate to produce a new filament.

Another type of prokaryotic multicellular association is a **biofilm,** which consists of complex aggregation of microorganisms (either one or multiple species) attached, in most cases, to a surface. For instance, biofilms may be found on lake surfaces, on rocks in freshwater or marine environments (making the rocks slippery), surrounding plant roots and root hairs, and on animal tissues such as the intestinal mucosa and teeth (human dental plaque is a biofilm).

Biofilms may be beneficial or detrimental to humans. On the beneficial side, for example, biofilms on solid supports are used in sewage treatment plants for processing organic matter before the water is discharged. For example, biofilms adhere to many kinds of surgical equipment and supplies, including catheters and synthetic implants such as pacemakers and artificial joints. When pathogenic bacteria are involved, biofilms can cause infections which are difficult to treat, because pathogenic bacteria in a biofilm are up to 1,000 times more resistant to antibiotics than are the same bacteria in pure liquid cultures. Other examples of medical conditions resulting from the activities of biofilms include middle-ear infections, bacterial endocarditis (an infection of the heart's inner lining or the heart valves), and Legionnaire's disease (an acute respiratory infection caused by breathing in pieces of biofilms containing the pathogenic bacterium *Legionella*).

As already stated, biofilms can involve a single species, or multiple species of microorganism. They can form on a variety of organic and inorganic surfaces, and they can assume many different forms and carry out many different functions for different species. Biofilm formation may be induced by environmental signals that vary depending on the species, including temperature, pH, oxygen level, and the presence of particular amino acids.

Figure 26.10 illustrates a common pattern in the formation of many types of biofilms. The numbers in the following list correspond to the steps in the figure.

1. The species-specific environmental signal induces marked changes in the pattern of gene expression compared to that of free bacteria. In effect, the prokaryote becomes a significantly different organism. This observation is especially important in the case of pathogenic bacteria, since most research on the control of those bacteria is done with liquid cultures. (*Insights from the Molecular Revolution* presents the results of a study showing the gene expression changes in a gum-disease-causing bacterial pathogen.) The challenge is to devise new treatment strategies for biofilm-caused diseases.
2. The genetically reprogrammed bacteria attach to a surface, for example by pili or lipopolysaccharides, and secrete molecules that cause other free bacteria to attach and form a monolayer.
3. The bacteria attaching to the surface begin communicating with one another by sending and receiving chemical signals in a process called **quorum sensing.** The chemical signaling molecules are synthesized and secreted continuously by individual cells, but it is only when the population of cells reaches a certain level (a "quorum") that the concentration of the signal is high enough to be sensed. Recognition of the signal triggers more genetic changes that cause the bacteria to bind more strongly to the surface and to each other, producing small colonies.
4. The bacteria produce an extracellular matrix of polysaccharide polymers and entrapped organic and inorganic materials. The matrix enables the biofilm to mature, leading to complex three-dimensional forms in which channels supply the bacteria with nutrients for growth. Maturation of the biofilm may also involve trapping other microorganisms, including other bacterial species, algae, fungi, and/or protists, thereby producing diverse microbial communities.

Much remains to be learned about how bacteria form a biofilm, how the change in gene expression during the transition is regulated, and how they interact within the biofilm. Particularly for pathogenic bacteria, having the knowledge of genetic changes will be important for informing scientists about treating a biofilm-associated infection.

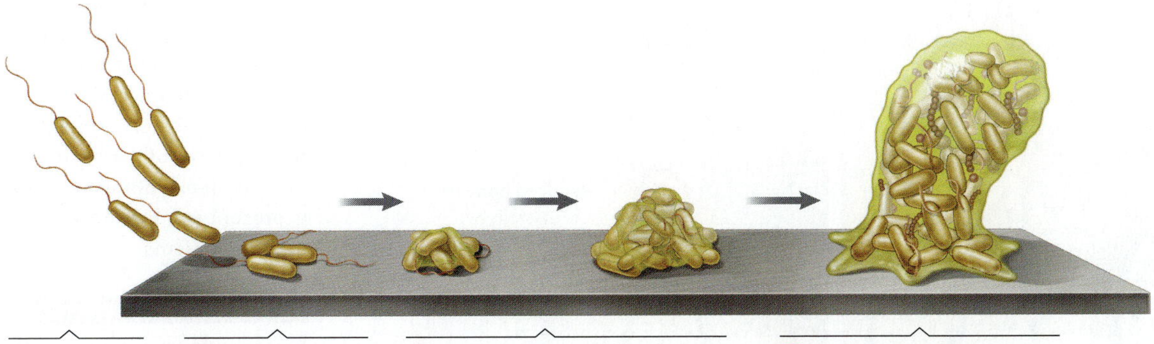

1 A specific environmental signal changes gene expression in free bacteria.

2 Bacteria attach to a surface coated with polysaccharides and glycoproteins, which results in more free bacteria attaching. A monolayer forms.

3 Bacteria attach more firmly to the surface and form small colonies.

4 The bacteria produce an extracellular matrix that enables the biofilm to mature, producing its three-dimensional shape.

FIGURE 26.10 Steps in the formation of many types of biofilms.

Gums the Word: Gene expression changes in an oral cavity pathogen associated with transition from free-living to biofilm

Gingivitis, a nondestructive periodontal disease, is inflammation of the gums. The gums are part of the periodontal tissue; that is, the tissue that supports the teeth. Commonly gingivitis develops due to plaque—bacterial biofilms—on the surfaces of teeth. Untreated, gingivitis can progress to periodontitis, which is a destructive periodontal disease affecting tooth-supporting tissues **(Figure)**. Without treatment, periodontitis can result in loosening and then loss of teeth.

Periodontitis is the leading cause of tooth loss in industrial nations, and one of the most frequently occurring infectious diseases in humans. *Porphyromonas gingivalis* is a Gram-negative pathogenic bacterium that is present in dental plaque. It is strongly implicated in the onset of periodontitis and the destructive progression of that disease. However, little is known about the physiology of *P. gingivalis* or about gene regulatory changes that occur in the biofilm that could relate to pathogenicity.

Research Question

How does gene expression differ in *P. gingivalis* in free-living cells versus cells in a biofilm?

Experiment

Eric Reynolds and his colleagues at Melbourne Dental School, The University of Melbourne, and Monash University in Australia answered this question using a genomics-based approach. That is, as was discussed in Section 18.3 and Figure 19.8, DNA microarray analysis can be used to compare, at the whole genome level, which genes are active under different conditions as well as determining the relative amount of expression. For their experiments, the researchers grew *P. gingivalis* in culture in fermenters to maintain anaerobic conditions. In one culture the bacterium was grown in a planktonic state (a culture of single cells), and in a second culture it was grown as a biofilm. Cells were harvested from each culture, RNA was extracted from the cells and from the mRNAs present, cDNA copies were made, incorporating a green fluorescent label for the cDNA derived from one culture and a red fluorescent label for the cDNA derived from the other culture. The labeled cDNAs were then mixed and added to a DNA microarray (DNA chip) containing DNA spots corresponding to all known genes of *P. gingivalis.* Hybridization of a cDNA to a spot was detected by laser analysis; the fluorescence indicated whether a particular cDNA hybridized and, therefore, whether the gene corresponding to that cDNA was active. More specifically, the results showed whether some genes were active only in free-living cells, only in biofilm cells, or active in both. For the latter, the relative amount of expression was quantified by the green and red fluorescence levels in the spot.

Results

The researchers found that 377 genes of the genome (18% of the total number of genes) were differentially expressed when the bacterium is grown as a biofilm. Of those genes, 191 were up-regulated, meaning they were expressed more in the biofilm compared with the planktonic cells, and the other 186 were down-regulated. Among those up-regulated were genes involved in transport (including signal transduction pathways), and among those down-regulated were genes involved in cell membrane synthesis, DNA replication, and energy production.

Conclusion

Taken together, the gene expression results suggest that the bacterium transitions to a decreased rate of cell replication and, hence of rate of growth, when it becomes a biofilm. The researchers argue that the changes in gene expression are involved in the adaptive response of the bacterium to growth in the biofilm state. How the changes in gene expression relate to the pathogenicity of *P. gingivalis* when it is growing in a biofilm is presently unknown. One caveat to consider here is that the experiments were done in pure cultures of the bacterium, whereas in the mouth exhibiting gingivitis, *P. gingivalis* is present in a biofilm with other microorganisms.

THINK LIKE A SCIENTIST Outline in a general way the molecular events that must occur for *P. gingivalis* to make the transition from planktonic to biofilm growth.

Source: A. W. Lo et al. 2009. Comparative transcriptomic analysis of *Porphyromonas gingivalis* biofilm and planktonic cells. *BMC Microbiology* 9:18.

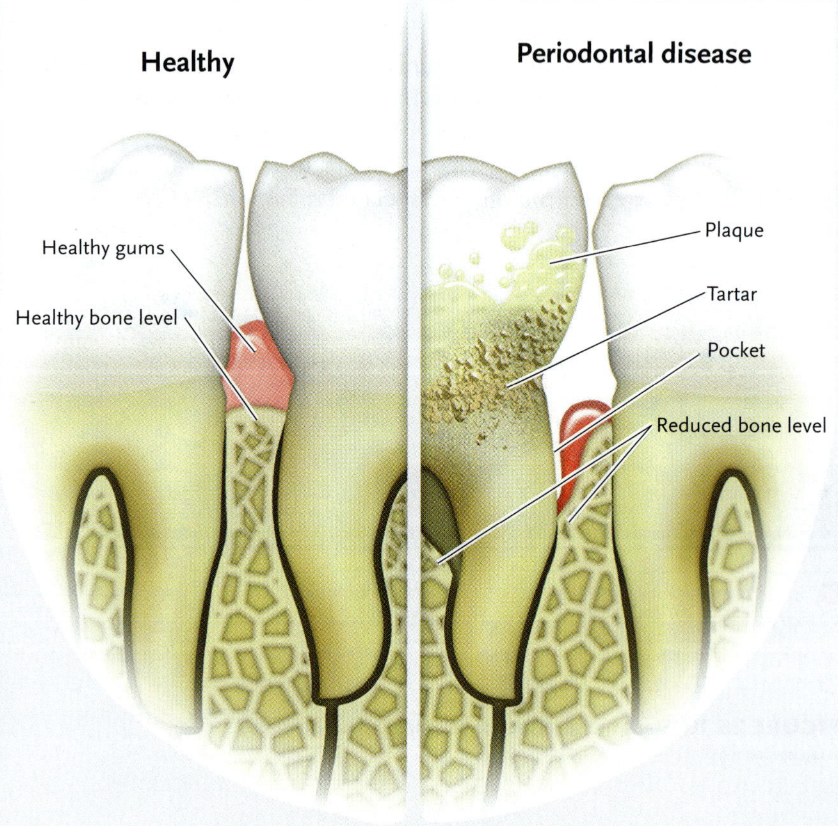

FIGURE Healthy gum tissue (left) and gum tissue with periodontal disease (right).

STUDY BREAK 26.1

1. What distinguishes a prokaryotic cell from a eukaryotic cell?
2. What is the difference between a chemoheterotroph and a photoautotroph?
3. What is the difference between an obligate anaerobe and a facultative anaerobe?
4. What is the difference between nitrogen fixation and nitrification? Why are nitrogen-fixing prokaryotes important?
5. What is a biofilm? Give an example of a biofilm that is beneficial to humans and one that is harmful.

THINK OUTSIDE THE BOOK

On your own or collaboratively, research a study in which genomic experiments have shown marked changes in gene expression pattern when a bacterium forms a biofilm. Outline how the experiment was done, and summarize the results of the experiments.

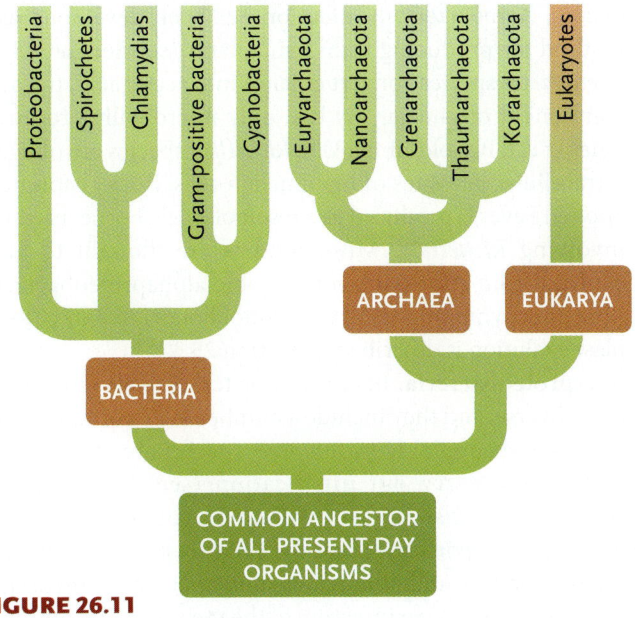

FIGURE 26.11
An abbreviated and simplified phylogenetic tree of prokaryotes.

26.2 The Domain Bacteria

Figure 26.11 presents a simplified phylogenetic tree of the domains Bacteria and Archaea (the Eukarya is the third domain of life). This tree is a model based on molecular phylogenetics analysis of DNA sequences of a number of prokaryotes, including hundreds of complete genome sequences. (Molecular phylogenetics is discussed in Chapter 24.) It shows five major evolutionary groups; complete trees show several more groups. As more sequence data become available, undoubtedly the details of the tree in terms of the species represented will be refined and altered.

Note that phylogenetic tree construction in prokaryotes is complicated by the phenomenon of horizontal gene transfer. That is, phylogenetic tree construction involves using molecular sequence comparisons to hypothesize how observed variations in genes may be explained in terms of evolutionary lineages. An important assumption in such analysis is vertical inheritance, namely that genes are passed from parents to offspring. However, much more so for prokaryotes than for eukaryotes, evolution of genes also involves horizontal gene transfer (see Section 26.1). For scientists trying to reconstruct the phylogenetic history of a group of organisms on the basis of gene sequences, horizontal gene transfer muddies the waters significantly. That is, suppose a gene moved from ancient bacterium A to unrelated ancient bacterium B on a plasmid, and then subsequently moved to bacterium B's chromosome. Based on studying the sequences of the gene in present-day descendants, an investigator might conclude that the two bacterial species are closely related, whereas genome level studies could show them to be distantly related. The bottom line is that creating one Tree of Life that depicts the evolution of all life forms accurately and completely likely is not possible and, especially relevant to the prokaryotes, we may never understand completely how the earliest prokaryotes and eukaryotes evolved.

We discuss the major groups of the domain Bacteria in this section and of the domain Archaea in the next section.

The Five Major Evolutionary Groups of Bacteria Have Diverse Properties

In this section we discuss the five major evolutionary groups of bacteria shown in Figure 26.11—the proteobacteria, the spirochetes, the chlamydias, the Gram-positive bacteria, and the cyanobacteria.

PROTEOBACTERIA The *proteobacteria* are a highly diverse group of Gram-negative bacteria that consist of at least 384 genera and 1,300 species. They exhibit a range of metabolic capabilities, including chemoautotrophy, chemoheterotrophy, and photoautotrophy. Some are aerobes, whereas others are anaerobes. Based on comparisons of rRNA sequences, the proteobacteria are divided into five subgroups, the Alphaproteobacteria, the Betaproteobacteria, the Gammaproteobacteria, the Deltaproteobacteria, and the Epsilonproteobacteria.

- **Alphaproteobacteria.** The alphaproteobacteria show considerable metabolic diversity. This subgroup includes many species that form symbiotic relationships with plants and animals, as well as some pathogens. For example, *Rhizobium* (see Figure 25.5) and *Agrobacterium* species infect the roots of leguminous plants to form nodules in which nitrogen fixation occurs. *Agrobacterium* infection has been developed as a method for producing transgenic plants (see Section 18.2). *Acetobacter* species convert ethanol into acetic acid, the acid component of vinegar. By law, only microbially produced vinegar can be sold in the U.S. *Rickettsia* species are obligate intracellular parasites of mammalian cells. Rocky Mountain Spotted Fever is produced as a result of a tick-borne infection involving *Rickettsia*. Mitochondria are thought to have evolved from endosymbiotic aerobic alphaproteobacteria. (The endosymbiotic theory for mitochondrion and chloroplast evolution is described in Section 25.3.)

- **Betaproteobacteria.** Betaproteobacteria are also metabolically diverse, and they include a number of important pathogens. For example, *Nitrosomonas* species are chemoautotrophs that carry out nitrification (see Section 26.1). *Bordetella pertussis* causes whooping cough in children, *Neisseria gonorrhoeae* causes the sexually transmissible disease gonorrhea, and *Neisseria meningitidis* causes meningococcal meningitis, an infection of the meninges (membranes that cover the brain and spinal cord).
- **Gammaproteobacteria.** The gammaproteobacteria are the largest subgroups of the proteobacteria. This subgroup is well known for facultative anaerobes that inhabit mammalian colons, such as *Escherichia coli* (see Figure 26.7). The subgroup also contains some important pathogens such as *Salmonella enterica typhi* (causes typhoid fever) (see Figure 26.2B), *Vibrio cholerae* (causes cholera), *Legionella* (causes legionellosis, one form of which is Legionnaire's disease), *Shigella dysenteriae* (causes dysentery), and *Haemophilus influenzae* (causes ear infections and meningitis, but not influenzae). Some gammaproteobacteria are photosynthetic; typically they are anaerobic bacteria that use photosynthesis to produce carbohydrates using chlorophyll, but do not produce oxygen.
- **Deltaproteobacteria.** The deltaproteobacteria include a group of anaerobes that are involved in the cyclic interchanges of sulfur-containing compounds in the biosphere. Other species have complex life cycles that include multicellular developmental forms. The best-studied examples are the slime-secreting myxobacteria ("slime bacteria"). Myxobacteria glide actively over the slime, typically traveling in swarms of cells that stay together using an intercellular signaling system. Enzymes secreted by the swarms digest "prey"—other bacteria, primarily—that become stuck in the slime. When environmental conditions become unfavorable, as when soil nutrients or water are depleted, myxobacteria form a *fruiting body* **(Figure 26.12)**, which contains clusters of spores. When the fruiting body bursts, the spores disperse and form new colonies.
- **Epsilonproteobacteria.** This subgroup consists of a small group of bacteria that inhabit the animal digestive tract. The epsilonproteobacteria includes *Campylobacter* and *Helicobacter* species, some of which are pathogenic. *C. jejuni* is a major cause of food-borne intestinal disease, and *H. pyloris* is the most common form of peptic ulcers **(Figure 26.13).**

Hans Reichenbach, Gesellschaft for Biotechnologische Forschung, Braunsweig, Germany

FIGURE 26.12 The fruiting body of *Chondromyces crocatus*, a myxobacterium, a member of the Deltaproteobacteria subgroup of the Proteobacteria. Cells of this species collect together to form the fruiting body.

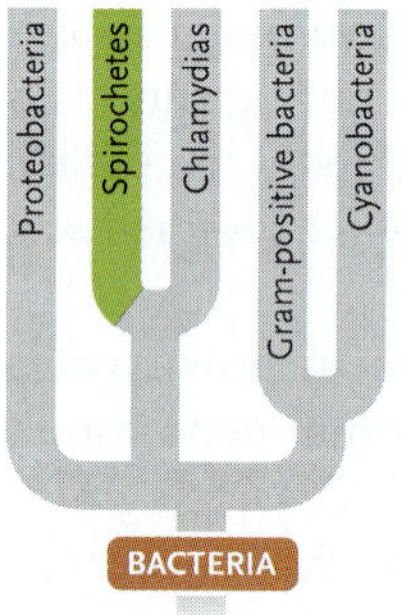

SPIROCHETES The spirochetes are Gram-negative bacteria with helically spiraled bodies and an unusual form of movement in which their flagella, which are embedded in the region between the outer membrane and the cell membrane, cause the entire cell to twist in a corkscrew pattern. Some spirochetes are pathogenic to humans. For example, *Treponema pallidum* causes syphilis, and *Borrelia burgdorferi* causes Lyme disease. Beneficial spirochetes in termite intestines aid in the digestion of plant fiber.

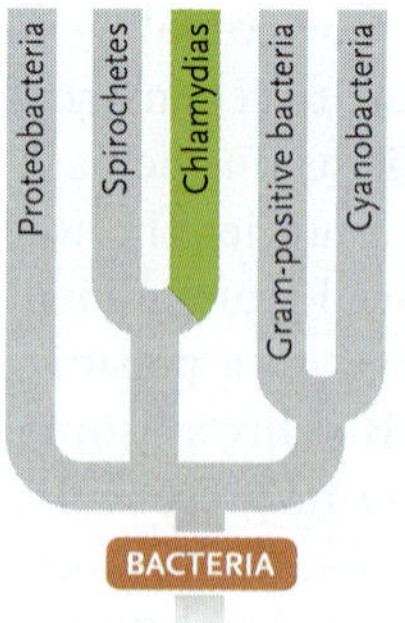

CHLAMYDIAS The chlamydias are structurally unusual among bacteria because, although they are Gram-negative and have cell walls with a membrane outside of them, they lack peptidoglycan. All the known chlamydias are obligate intracellular parasites, meaning that they can only reproduce within other cells. Chlamydias cause various diseases in animals. For example, *Chlamydia trachomatis* **(Figure 26.14)** is responsible for chlamydia, one of the most common sexually transmitted infections of the urinary and reproductive tracts of humans.

FIGURE 26.13 Experimental Research

Demonstration of the Bacterial Cause of Peptic Ulcers

Question: Do bacteria cause gastric and duodenal ulcers?

Experiment: A gastric ulcer (also called a peptic ulcer) is an erosion of the mucosal layer of the stomach that can potentially penetrate into the deeper layers of the stomach wall. Pepsin and HCl in the stomach irritate the exposed stomach wall, causing extreme pain. Similar ulcers, called duodenal ulcers, can form in the duodenum. In 1982, physicians "knew" that gastric ulcers were caused by excessive amounts of stomach acid that was caused by stress, smoking, personality issues, or susceptibility genes. They treated an ulcer with various kinds of antacids or drugs to inhibit acid production in the stomach. Typically the ulcer symptoms returned after treatment was stopped. Prior to that time, various researchers had observed the presence of spiral or curved bacteria in biopsies taken from the majority of patients with gastric and duodenal ulcers. The bacteria were considered either to be contamin ants or not related to ulcer formation.

In 1982, Barry Marshall and J. Robin Warren of Royal Perth Hospital, Perth, Western Australia, tested the hypothesis that bacteria caused peptic ulcers. They took biopsies of peptic ulcers and duodenal ulcers and assayed for the presence of bacteria by Gram staining and culturing on Petri plates.

Results: The results of the biopsies, published in 1984, are shown in the **Table**.

Association of Bacteria with Biopsy Samples

Biopsy Appearance	Total Samples	Number (%) Associated with Bacteria
Gastric ulcer	22	18 (77%)
Duodenal ulcer	13	13 (100%)
Total	35	31 (89%)

The researchers' results showed that almost all cases of gastric or duodenal ulcers were associated with bacteria. The bacteria had a curved appearance, but the species was unknown at the time. It is now known to be the epsilonproteobacterium *Helicobacter pylori.*

Helicobacter pylori

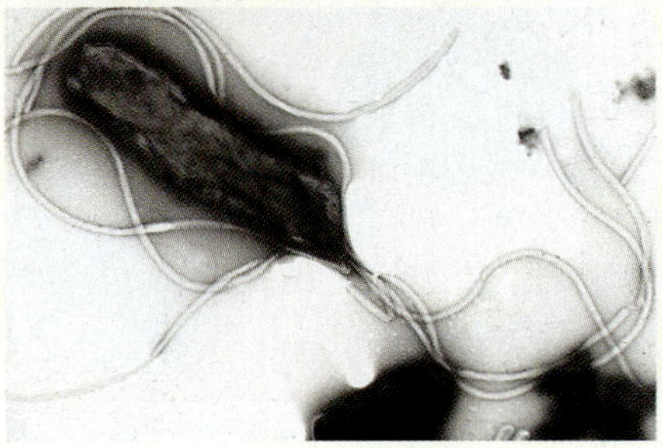

Yutaka Tsutsumi, M.D. Professor Department of Pathology Fujita Health University School of Medicine

Marshall and Warren's results were met with a great deal of skepticism, and most physicians were not persuaded that gastric and duodenal ulcers were bacterial diseases. How could it be shown that *H. pylori* can cause gastric and duodenal ulcers? The accepted way to prove that a bacterium is a pathogen is to fulfill Koch's postulates, a set of four conditions proposed in the nineteenth century by Robert Koch. The postulates are as follows:

1. The bacteria must be present in every case of the disease.
2. The bacteria must be isolated from the infected host and grown in pure culture.
3. The specific disease must be induced when a pure culture of the bacteria is introduced into a healthy host.
4. The bacteria must be isolated from the now experimentally infected host.

Postulates 3 and 4 had not been fulfilled in the study. Attempts to use piglets to fulfill them failed. Therefore, in a bold and highly unusual move, Marshall decided to try to fulfill postulates 3 and 4 himself. He drank 200 mL of an *H. pylori* suspension, then fasted the rest of the day. On the fifth to eighth day after drinking the bacteria, Marshall began feeling very nauseated to the point of vomiting. On the tenth day, a colleague performed a biopsy, the analysis of which showed the presence of *H. pylori* cells. *Helicobacter pylori* was proven to be a pathogen.

Conclusion: Marshall and Warren's experiments had shown that *Helicobacter pylori* can cause gastric and duodenal ulcers. This bacterium probably accounts for more than 80% of cases of such ulcers in humans in the United States. In *H. pylori*-caused ulcers, the effective treatment is an antimicrobial agent such as an antibiotic. Marshall and Warren received a Nobel Prize in 2005 "for their discovery of the bacterium *Helicobacter pylori* and its role in gastritis and peptic ulcer disease."

THINK LIKE A SCIENTIST Marshall developed a bacterial infection after drinking a solution of *Helicobacter.* Speculate on the origin of *Helicobacter* in patients who developed gastric or duodenal ulcers and were positive for the bacterium.

Sources: B. J. Marshall, J. A. Armstrong, D. B. McGechie, and R. J. Glancy. 1985. Attempt to fulfill Koch's postulates for pyloric Campylobacter. *Medical Journal of Australia* 142:436–439; B. J. Marshall and J. R. Warren. 1984. Unidentified curved bacilli in the stomach of patients with gastritis and peptic ulceration. *The Lancet* 1(8390):1311–1315.

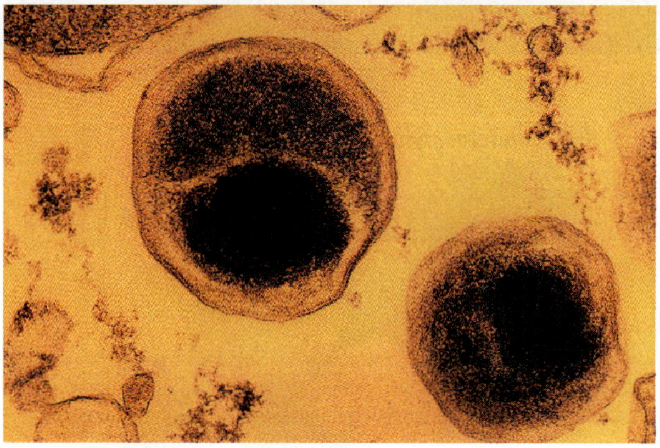

David M. Phillips/Visuals Unlimited

FIGURE 26.14 Cells of *Chlamydia trachomatis* inside a human cell. This bacterium is a major cause of infectious eye and genital diseases in humans.

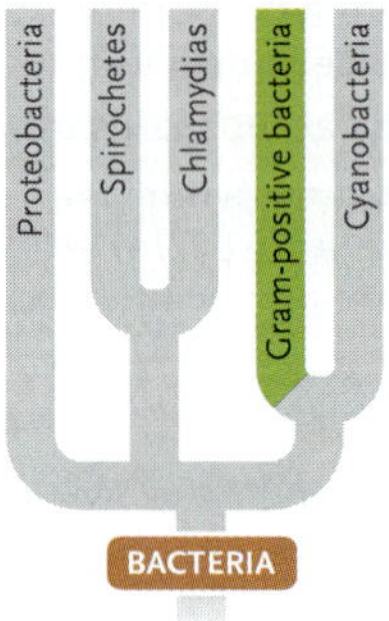

GRAM-POSITIVE BACTERIA The large group of Gram-positive bacteria contains many species that live primarily as chemoheterotrophs. One species, *Bacillus subtilis* (see Figure 26.1), is studied by biochemists and geneticists almost as extensively as is *E. coli.* A number of Gram-positive bacteria are pathogenic to humans, including *Bacillus anthracis,* which causes anthrax; *Staphylococcus aureus,* which causes a range of illnesses including pimples, abscesses, toxic shock syndrome, pneumonia, meningitis and sepsis (an infection state involving a whole-body inflammation; commonly referred to as blood poisoning); and *Streptococcus pyogenes* **(Figure 26.15),** which is a cause of streptococcal pharyngitis ("strep throat"), scarlet fever, and rheumatic fever. Some Gram-positive bacteria are beneficial; *Lactobacillus* species, for example, carry out the lactic acid fermentation used in the production of pickles, sauerkraut, yogurt, and some other fermented foods. One unusual group of bacteria, the mycoplasmas, is classified in the group Gram-positive bacteria by molecular studies even though they lack a cell wall and, therefore, they react negatively to the Gram stain. Some species are pathogenic to humans; *Mycoplasma pneumonia,* for instance, causes a form of pneumonia.

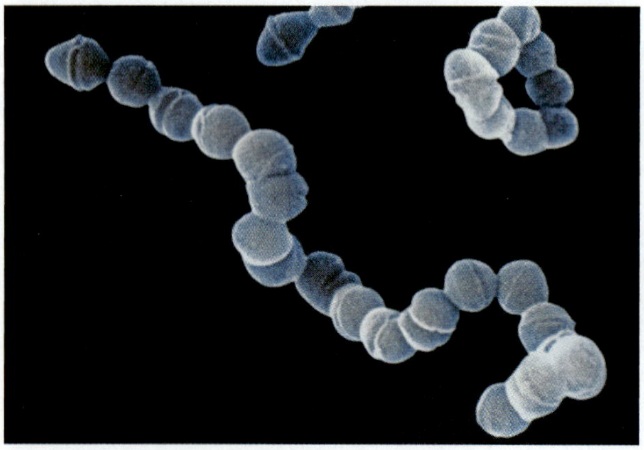

David M. Phillips/Visuals Unlimited

FIGURE 26.15 *Streptococcus* bacteria, a genus of Gram-positive bacteria, forming a long chain of cells, which is typical of the many species in this genus.

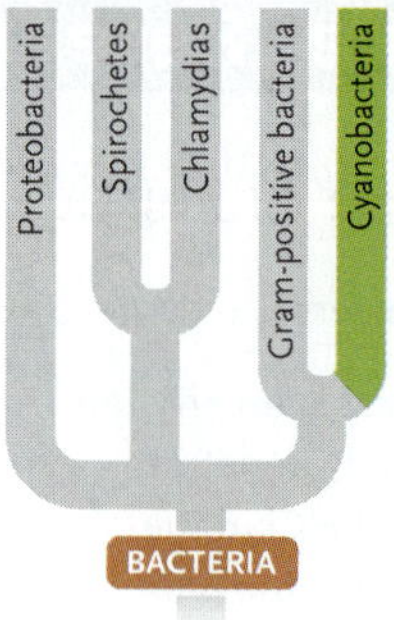

CYANOBACTERIA The cyanobacteria **(Figure 26.16)** are Gram-negative photoautotrophs that have a blue-green color and carry out photosynthesis using the same pathways as eukaryotic algae and plants, and use the same chlorophyll that plants have as their primary photosynthetic pigment. They release oxygen as a by-product of photosynthesis.

The direct ancestors of present-day cyanobacteria were the first organisms to use the water-splitting reactions of photosynthesis. As such, they were critical to the appearance of abundant oxygen in the atmosphere, which allowed the evolutionary development of aerobic organisms (see Section 25.2). Chloroplasts probably evolved from early cyanobacteria that were incorporated into the cytoplasm of primitive eukaryotes, which eventually gave rise to the algae and higher plants (see Section 25.3). Besides releasing oxygen, present-day cyanobacteria help fix nitrogen into organic compounds in aquatic habitats and also in lichens that involve a cyanobacterium and a filamentous fungus living together symbiotically (see Chapter 30). (Most lichens involve a green alga and a filamentous fungus.)

A. A population of cyanobacteria covering the surface of a pond

Dr. Jeremy Burgess/SPL/Photo Researchers, Inc.

B. Chains of cyanobacterial cells. Some cells in the chains form spores. The heterocyst is a specialized cell that fixes nitrogen.

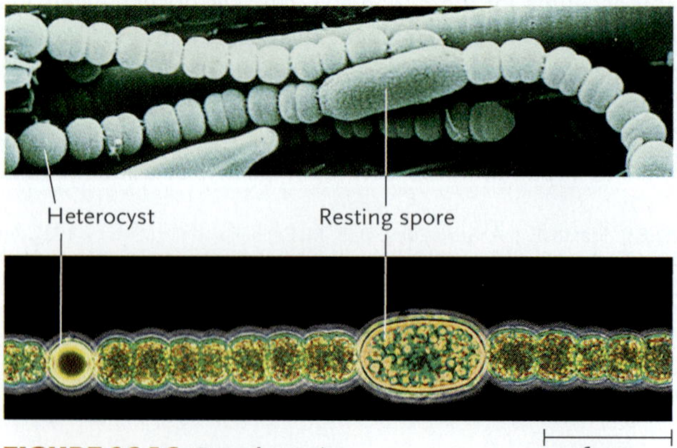

Tony Brian/SPL/Photo Researchers, Inc.

P. W. Johnson and J. McN. Sieburth, University of Rhode Island/Biological Photo Service

FIGURE 26.16 Cyanobacteria.

Bacteria Cause Diseases by Several Mechanisms

As you have just learned, some bacteria are pathogenic. Pathogenic bacteria vary in the ways they cause diseases. A number of bacterial lineages produce **exotoxins,** toxic proteins that are secreted from the bacterium or are released from the cell when it lyses, and interfere with the biochemical processes of body cells in various ways. For example, the exotoxin of the Gram-positive bacterium *Clostridium botulinum* is found as a contaminant in poorly preserved foods, and causes *botulism.* The botulism exotoxin is one of the most poisonous substances known: a few nanograms of the toxin can cause illness, and a few hundred grams could kill every human on Earth. It acts by interfering with the transmission of nerve impulses. The muscle paralysis produced by the exotoxin can be fatal if the muscles that control breathing are affected. The botulinum exotoxin, under the brand name Botox, is used in low doses for the cosmetic removal of wrinkles, in the treatment of migraine headaches, and of some other medical conditions.

Some other bacteria cause disease through **endotoxins.** Endotoxins are lipopolysaccharides released from the outer membrane surrounding the cell walls when the bacterium dies and lyses. Endotoxins are natural components of the outer membrane of all Gram-negative bacteria, which include the gammaproteobacteria *E. coli, Salmonella* species, and *Shigella* species. The lipopolysaccharides cause disease by overstimulating the host's immune system, often triggering inflammation (see Chapter 45). Endotoxin release has different effects, depending on the bacterial species and the site of infection, and include typhoid or other fevers, diarrhea, and, in severe cases, organ failure and death. For example, *Salmonella enterica typhi,* the cause of typhoid fever, enter the human intestines and penetrate the intestinal wall, eventually ending up in the lymph nodes. There they multiply, and some of the cells die and lyse, releasing endotoxins into the bloodstream. This both triggers the host's immune response and causes sepsis. Without treatment, the condition can progress to multiple organ system failure and, eventually, death.

Some bacteria release **exoenzymes,** enzymatic proteins that digest plasma membranes and cause cells of the infected host to rupture and die. Exoenzymes may also digest extracellular materials such as collagen, causing connective tissue diseases. Some exoenzymes attack red or white blood cells, leading to anemias, impairment of the immune response, or interference with blood clotting. Among the bacteria that release exoenzymes are particular species of the Gram-positive genera, *Streptococcus* (see Figure 26.15), *Staphylococcus,* and *Clostridium* (see Figure 26.9). Necrotizing fasciitis (flesh-eating disease), the spectacularly destructive and rapid degeneration of subcutaneous tissues in the skin, is caused by an exoenzyme released by *Streptococcus pyogenes, Staphylococcus aureus,* and some other bacteria.

Some of the ill effects of bacteria have little to do with exotoxins, endotoxins, or exoenzymes, but are caused purely by the body's responses to infection. The severe pneumonia caused by *Streptococcus pneumoniae,* for example, results from massive accumulation of fluid and white blood cells in the lungs in response to the infection. The white blood cells have little effect on the bacteria, however, because of the bacterial cell's protective capsule. As the fluid, white blood cells, and bacteria continue to accumulate, they block air passages in the lungs and severely impair breathing.

Pathogenic Bacteria Commonly Develop Resistance to Antibiotics

Antibiotics are routinely used to treat bacterial infections. These substances, produced as defensive molecules by some bacteria and fungi, or by chemical synthesis, kill or inhibit the growth of other microbial species. For example, streptomycins, produced by soil bacteria, block protein synthesis in their target cells. Penicillins, which are produced by fungi, prevent formation of covalent bonds that hold bacterial cell walls together, weakening the wall and causing the cells to rupture.

Many pathogenic bacteria develop resistance to antibiotics through mutations that allow them to break down the drugs or otherwise counteract their effects (see *Why It Matters,* Chapter 21). Resistance is also acquired by horizontal gene transfer when resistance genes on plasmids transfer between bacteria by conjugation, or when DNA is brought into pathogens by other pathways such as transformation and transduction (see Section 17.1). Taking antibiotics routinely in mild doses, or failing to complete a prescribed dosage, contribute to the development of resistance by selecting strains that can survive in the presence of the drug. Overprescription of antibiotics for colds and other virus-caused diseases does not affect the virus, but can result in resistance among bacteria present in the body. Antibacterial agents that may result in resistance are also commonly included in such commercial products as soaps, detergents, and deodorants. Resistance is a form of evolutionary adaptation; antibiotics alter the bacterium's environment, conferring a reproductive advantage on those strains best adapted to the altered conditions.

The development of resistant strains has made tuberculosis, cholera, typhoid fever, gonorrhea, "staph," and other diseases caused by bacteria difficult to treat with antibiotics. For example, as recently as 1988, drug-resistant strains of the Gram-positive bacterium *Streptococcus pneumoniae,* which causes pneumonia, meningitis, and middle-ear infections, were practically unheard of in the United States. Now, resistant strains of *S. pneumoniae* are common and increasingly difficult to treat with antibiotics.

In this section, you have seen that bacteria thrive in nearly every habitat on Earth, including the human body. However, some members of the second prokaryotic domain, the Archaea, the subject of the next section, live in habitats that are too forbidding even for the Bacteria.

STUDY BREAK 26.2

1. **What is an exotoxin, an endotoxin, and an exoenzyme, and how do they differ with respect to how they cause disease?**
2. **By what mechanisms can pathogenic bacteria develop resistance to antibiotics?**

> **THINK OUTSIDE THE BOOK**
> **Individually or collaboratively, research the Internet to identify an example of a phylogenetic tree for bacteria that does not have a traditional treelike form. What does the alternate tree form say about the evolution of the group?**

26.3 The Domain Archaea

The first-studied archaeans were found in extreme environments, such as hot springs, hydrothermal vents on the ocean floor, and salt lakes **(Figure 26.17).** For that reason, these prokaryotes were called *extremophiles* ("extreme lovers"). Subsequently archaeans have also been found living in normal environments; like bacteria, these are *mesophiles.*

Many archaeans are chemoautotrophs that obtain energy by oxidizing inorganic substances, while others are chemoheterotrophs that oxidize organic molecules. Recent research has identified some archaeans as possible human pathogens, for example a species associated with tooth infections.

Research has shown that archaeans have some eukaryotic features, some bacterial features, and some features that are unique to the group (also discussed in Section 25.3). **Table 26.1** compares the characteristics of Bacteria, Archaea, and Eukarya.

A. **Red-purple color produced by archaeans in highly saline water**

iStockphoto.com/David Crane

B. **Bright colors produced in hot, sulfur-rich water by the oxidative activity of archaeans that convert H_2S to elemental sulfur**

Sascha Burkard/Shutterstock.com

FIGURE 26.17 **Typically extreme archaean habitats.**

Molecular Studies Reveal Five Evolutionary Branches in the Archaea

Based on differences between the rRNA coding sequences in their genomes, the domain Archaea is divided into five groups (see Figure 26.11). Two major groups, the **Euryarchaeota** and

TABLE 26.1 Characteristics of the Bacteria, Archaea, and Eukarya

Characteristic	Bacteria	Archaea	Eukarya
DNA arrangement	Single, circular in most, but some linear and/or multiple	Single, circular	Multiple linear molecules
Chromosomal proteins	Prokaryotic histonelike proteins	Five eukaryotic histones	Five eukaryotic histones
Genes arranged in operons	Yes	Yes	No
Nuclear envelope	No	No	Yes
Mitochondria	No	No	Yes
Chloroplasts	No	No	Yes
Peptidoglycans in cell wall	Present	Present but modified, or absent	Absent
Membrane lipids	Unbranched; linked by ester linkages	Branched; linked by ether linkages	Unbranched; linked by ester linkages
RNA polymerase	One type	Multiple types	Multiple types
Ribosomal proteins	Prokaryotic	Some prokaryotic, some eukaryotic	Eukaryotic
First amino acid placed in proteins	Formylmethionine	Methionine	Methionine
Aminoacyl–tRNA synthetases	Prokaryotic	Eukaryotic	Eukaryotic
Cell division proteins	Prokaryotic	Prokaryotic	Eukaryotic
Proteins of energy metabolism	Prokaryotic	Prokaryotic	Eukaryotic
Sensitivity to chloramphenicol and streptomycin	Yes	No	No

the **Crenarchaeota,** contain many species of archaeans that have been cultured and examined in the laboratory. Each of the other three groups, the **Nanoarchaeota,** the **Thaumarchaeota,** and the **Korarchaeota,** is represented by one to a few living species, with others in the group recognized by related rRNA coding sequences in DNA from environmental samples.

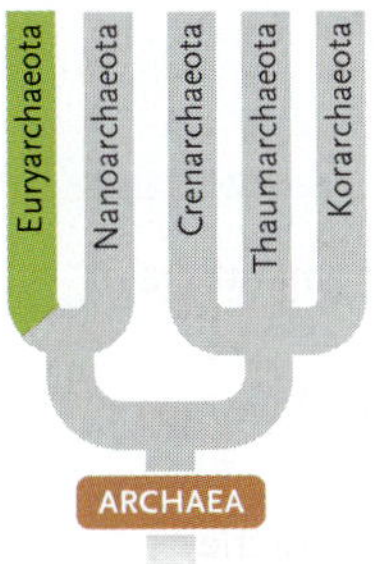

EURYARCHAEOTA The Euryarchaeota are found in different extreme environments. They include methanogens, which produce methane; extreme halophiles, which live in high concentrations of salt; and some extreme thermophiles, which live under high-temperature conditions.

Methanogens ("methane generators") live in reducing environments **(Figure 26.18),** and represent about one-half of all known species of archaeans. All known methanogens, for example *Methanococcus* and *Methanobacterium* species, belong to the Euryarchaeota. Methanogens are obligate anaerobes, meaning they are killed by oxygen. They are found in the anoxic ("oxygen-lacking") sediments of, for instance, swamps and marshes, as well as in more moderate environments, such as the rumen of cattle and sheep; the large intestine of dogs and humans; and the hindguts of insects such as termites and cockroaches. Methanogens generate energy by converting at least ten different substrates such as carbon dioxide and hydrogen gas, methanol, or acetate into methane gas (CH_4), which is released into the atmosphere.

Halophiles are salt-loving organisms. Extreme halophilic Archaea live in highly saline (salty) environments such as the Great Salt Lake or the Dead Sea, and on foods preserved by salting. Moreover, they *require* a minimum NaCl concentration of about 1.5 M (about 9% solution) to live, and they can live in a fully saturated solution (5.5 M, or 32%). All known extreme halophilic Archaea belong to the Euryarchaeota. Most are aerobic chemoheterotrophs; they obtain energy from sugars, alcohols, and amino acids using pathways similar to those of bacteria. Examples are *Halobacterium* and *Natrosobacterium* species.

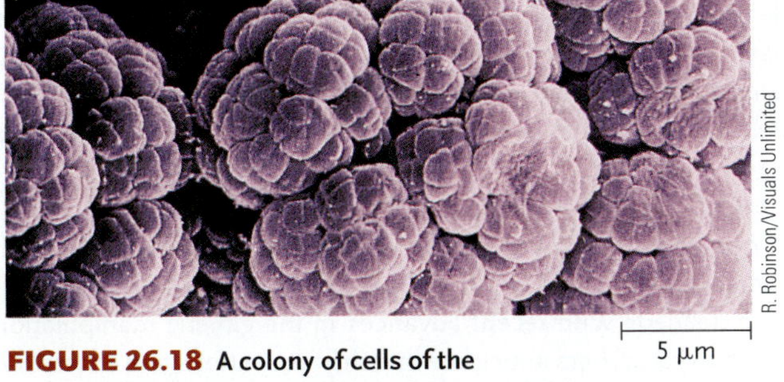

FIGURE 26.18 A colony of cells of the methanogenic archaean *Methanosarcina*, which live in the sulfurous, waterlogged soils of marshes and swamps.

Extreme thermophiles live in extremely hot environments. Archaeans of this type live in thermal areas such as ocean-floor hydrothermal vents and hot springs such as those in Yellowstone National Park. Their optimal temperature range for growth is 70° to 95°C, approaching the boiling point of water. By comparison, no eukaryotic organism is known to live at a temperature higher than 60°C. Some extreme thermophiles, such as *Pyrophilus* species and *Thermoplasma* species, are members of the Euryarchaeota. Some of them, such as *Pyrophilus* species, are obligate anaerobes, while others, such as *Thermoplasma* species, are facultative anaerobes that grow on a variety of organic compounds.

Two extreme thermophiles in this group have proved to be valuable for research. Recall that the polymerase chain reaction (PCR) uses a thermostable DNA polymerase (see Section 18.1). Two commercially available thermostable DNA polymerases for PCR are Vent DNA Polymerase, derived from *Thermococcus litoralis,* and Pfu DNA polymerase, derived from *Pyrococcus furiosus.*

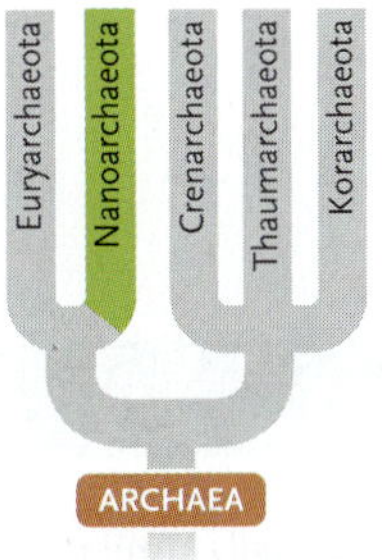

NANOARCHAEOTA The group Nanoarchaeota consists of only one known living species, *Nanoarchaeum equitas;* other members are represented only by sequences detected in environmental samples. This extreme thermophile was discovered in a hydrothermal vent off the coast of Iceland. Analysis of its genome shows that it lacks genes for several metabolic pathways, including those for most nucleotides, amino acids, and lipids. It survives by being a symbiont of a thermophilic member of the Crenarchaeota.

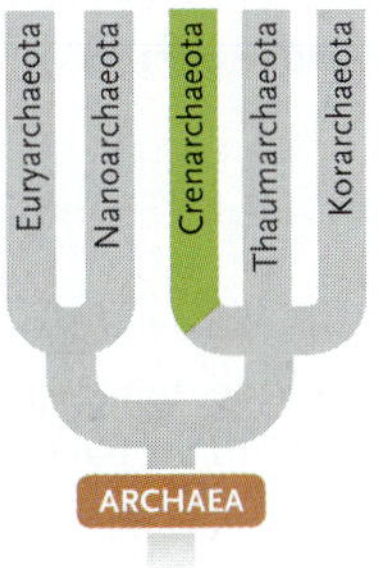

CRENARCHAEOTA The group Crenarchaeota contains most of the extreme thermophiles. Their optimal temperature range of 75° to 105°C is higher than that of the Euryarchaeota. Most are unable to grow at temperatures below 70°C. The most thermophilic member of the group, *Pyrolobus,* grows optimally at 106°C, but dies below 90°C. *Pyrolobus* species lives in ocean-floor hydrothermal vents where the pressure makes it possible to have temperatures above 100°C, the boiling point of water on Earth's surface.

Also within this group are **psychrophiles** ("cold loving"), organisms that grow optimally in cold temperatures in the range of −10 to −20°C. These organisms are found mostly in the Antarctic and Arctic oceans, which are frozen most of the year, and in the intense cold of ocean depths.

Mesophilic members of the Crenarchaeota comprise a large part of plankton in cool, marine waters where they are food sources for other marine organisms. As yet, no individual species of these archaeans has been isolated and characterized.

Crenarchaeota archaeans exhibit a wide range of metabolism with regard to oxygen, and include obligate anaerobes, facultative anaerobes, and aerobes.

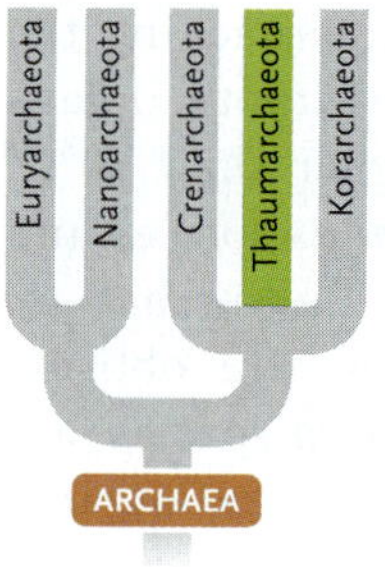

THAUMARCHAEOTA The Thaumarchaeota are a newly designated group of the Archaea consisting, at present, of four living species; other members are represented by sequences found in environmental samples. All living representatives are mesophilic, chemoautotrophic ammonia oxidizers. The establishing archaean for this group is *Cenarchaeum symbiosum.*

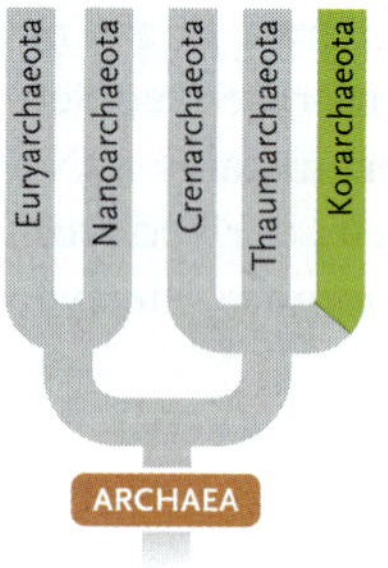

KORARCHAEOTA The group Korarchaeota consists of one living species, *Candidatus Korarcheum cryptofilum;* other members are represented in sequences in environmental samples. The living member is a thermophile; it was isolated from Obsidian Pool at Yellowstone National Park (a similar pool is shown in Figure 26.17B). The Korarchaeota diverged very early in the evolution of the Archaea.

From the highly varied prokaryotes, we now turn to eukaryotes, a group which does not equal the metabolic diversity or the conquest of extreme environments we have seen in the Bacteria and Archaea, but which does display new levels of physiological and behavioral complexity. The next six chapters discuss the eukaryotic kingdoms of protists, plants, fungi, and animals.

STUDY BREAK 26.3

1. **What distinguishes members of the Archaea from members of the Bacteria and Eukarya?**
2. **How does a methanogen obtain its energy? In which group or groups of Archaea are methanogens found?**
3. **Where do extreme halophilic archaeans live? How do they obtain energy? In which group or groups of Archaea are the extreme halophiles found?**
4. **What are extreme thermophiles and psychrophiles?**

UNANSWERED QUESTIONS

As you learned in this chapter, archaeans are morphologically similar to bacteria, yet they have a number of features that are more related to eukaryotes than to bacteria. More specifically, the DNA replication and transcription machineries of archaeans are related to those of eukaryotes and distinct from those found in bacteria. Also, recent work has revealed that, in some archaeans, aspects of cell division have close parallels in human cells but not in bacteria.

How is chromosome replication accomplished in species of the archaean genus *Sulfolobus*?

Analysis of the available genome sequences of archaeans show that the machinery that archaea use to replicate their DNA is, like the transcription machinery, related to that of eukaryotes and distinct from the bacterial apparatus. Recall from Chapter 14 that bacteria have a single replication origin per chromosome, whereas eukaryotes have many replication origins per chromosome. Some archaeans replicate their chromosomes from a single origin, but a number of species have multiple replication origins per chromosome. For example, species in the *Sulfolobus* genus (hyperthermophiles of the Crenarchaeota group of the domain Archaea) have three replication origins in their single chromosome. All three of these origins are used in every cell in every cell cycle, indicating that there is a sophisticated control system that ensures that each origin is used once, and only once, per cell division cycle. The nature of this control system is currently unknown.

The function of a replication origin is to recruit proteins to unwind double-stranded DNA, which exposes single-stranded DNA for use as templates. The single-stranded regions produced at an origin are extended by a DNA helicase (see Section 14.3), which, in archaeans, is called the MCM helicase. This archaean helicase has been the subject of intense structural and biochemical study, and much is now known about its form and function. However, the way in which it is recruited to replication origins remains poorly understood. Finally, with three replication origins being used in every cell for every cell division cycle, a total of six replication forks are simultaneously used for DNA replication (two from each origin). How these replication forks are resolved when they collide on the circular chromosome to produce two copies of the archaeal chromosome is currently not known.

How does cell division take place in the archaean *Sulfobolus*?

Once replication is complete, *Sulfolobus* keeps two replicated chromosomes paired for a time prior to their segregation at cell division. Until recently, the cell division machinery used by *Sulfolobus* and relatives in the Crenarchaeota was unknown. Nearly all bacteria, members of other archaeal groups, and eukaryotes have proteins belonging to the tubulin and actin families of cytoskeletal proteins, which play integral roles in cell division. However, *Sulfolobus* and its relatives lack these proteins yet they divide by binary fission. Recent work has established that cell division in *Sulfolobus* uses the so-called ESCRT machinery. Remarkably, the ESCRT proteins are conserved throughout the Eukarya, where they play roles in a wide variety of membrane processes. Cell biological and genetic studies have revealed that the ESCRT system plays a key role in *Sulfolobus* cell division, hinting at the ancestral role of this conserved machinery. Many questions remain unanswered, including: how are the *Sulfolobus* chromosomes segregated prior to cell division, and how is the site of cell division defined? With recent advances in the genetic manipulation of archaeans, researchers anticipate that these questions and others will be readily addressed.

THINK LIKE A SCIENTIST The ability to replicate DNA is fundamental to the propagation of all life on the planet. Yet it is apparent that bacteria use a different machinery to replicate their DNA than that employed by archaeans and eukaryotes. How can two distinct machineries for such a fundamentally important process have evolved?

Stephen D. Bell

Rachel Y. Samson

Stephen D. Bell and **Rachel Y. Samson** work at the Sir William Dunn School of Pathology, Oxford University. Steve is the Professor of Microbiology and his research group is focused on investigating transcription, DNA replication and cell division in the archaeal domain of life. Rachel joined Steve's group in 2007 and is researching the molecular and genetic basis of archaeal cell division. To learn more about their research, go to: http://users.path.ox.ac.uk/~sbell/Welcome.html.

REVIEW KEY CONCEPTS

To access the course materials and companion resources for this text, please visit www.cengagebrain.com.

26.1 Prokaryotic Structure and Function

- Three shapes are common in prokaryotes: spherical, rodlike, and spiral (Figure 26.2).
- Prokaryotic genomes typically consist of a single, circular DNA molecule packaged into the nucleoid. Many prokaryotic species also contain plasmids, which replicate independently of the main DNA (Figure 26.3).
- Gram-positive bacteria have a cell wall consisting of a thick peptidoglycan layer. Gram-negative bacteria have a thin peptidoglycan layer. The thin cell wall is surrounded by an outer membrane (Figure 26.4).
- A polysaccharide capsule, or slime layer, surrounds many bacteria. This sticky, slimy layer both protects the bacteria and helps them adhere to surfaces (Figure 26.5).
- Some prokaryotes have flagella, corkscrew-shaped protein fibers that rotate like propellers, and pili, protein shafts that help bacterial cells adhere to each other or to eukaryotic cells (Figure 26.7).
- Prokaryotes show great diversity in their modes of obtaining energy and carbon. Chemoautotrophs obtain energy by oxidizing inorganic substrates and use carbon dioxide as their carbon source. Chemoheterotrophs obtain both energy and carbon from organic molecules. Photoautotrophs are photosynthetic organisms that use light as a source of energy and carbon dioxide as their carbon source. Photoheterotrophs use light as a source of energy and obtain their carbon from organic molecules (Figure 26.8).
- Some prokaryotes are capable of nitrogen fixation, the conversion of atmospheric nitrogen to ammonia; others are responsible for nitrification, the two-step conversion of ammonium to nitrate.
- Prokaryotes normally reproduce by binary fission. Some prokaryotes are capable of conjugation, in which part of the DNA of one cell is transferred to another cell.
- In nature, prokaryotes may live in multicellular associations either of the same or mixed species. The associations may function as single cells do, or differentiate to produce cells with different functions. Examples are filaments of cyanobacteria, and biofilms, complex aggregations of microorganisms on a surface (Figure 26.10). Biofilms have both detrimental and beneficial consequences; they can harm human health, but they also can be effective in, for example, bioremediation.

Animation: Gram staining

Animation: Metabolic diversity of prokaryotes

Animation: Typical prokaryotic cell

26.2 The Domain Bacteria

- Bacteria are divided into several evolutionary branches (Figure 26.11).
- The proteobacteria are a highly diverse group of Gram-negative bacteria that exhibit a range of metabolic capabilities, including chemoautotrophy, chemoheterotropy, and photoautotrophy. Some are aerobic, others are anaerobes. Five subgroups of proteobacteria are recognized. Free-living proteobacteria include *Escherichia coli*, species that fix nitrogen, and the spore-forming myxobacteria (Figure 26.12).
- The chlamydias are Gram-negative, intracellular parasites that cause various diseases in animals. They have cell walls with an outer membrane, but they lack peptidoglycans (Figure 26.14).
- The spirochetes are spiral-shaped bacteria that are propelled by twisting movements produced by the rotation of flagella.
- The Gram-positive bacteria are primarily chemoheterotrophs and include many pathogenic species (Figure 26.15).
- The cyanobacteria are Gram-negative photoautotrophs that carry out photosynthesis and release oxygen as a by-product (Figure 26.16).
- Bacteria cause disease through exotoxins, endotoxins, and exoenzymes.
- Pathogenic bacteria may develop resistance to antibiotics through mutation of their own genes, or by acquiring resistance genes or plasmids from other bacteria.

26.3 The Domain Archaea

- Members of the domain Archaea have some features that are like those of bacteria, other features that are like those of eukaryotes, and some other features that are uniquely archaean (Table 26.1).
- The archaean plasma membrane contains unusual lipid molecules. The cell walls of archaeans consist of distinct molecules similar to peptidoglycans, or of protein or polysaccharide molecules.
- The Archaea are classified into five groups: the Euryarchaeota, which include the methanogens, the extreme halophiles, and some extreme thermophiles; the Nanoarchaeota, the only living member of which is an extreme thermophile; the Crenarchaeota, which contain most of the archaean extreme thermophiles, as well as psychrophiles and mesophiles; the Thaumarchaeota, all of which are mesophilic chemoautotrophs; and the Korarchaeota, the only living member of which is a thermophile.

UNDERSTAND & APPLY

Test Your Knowledge

1. A urologist identifies cells in a man's urethra as bacterial. Which of the following descriptions applies to the cells?
   a. They have sex pili, which give them motility.
   b. They have flagella, which allow them to remain in one position in the urethral tube.
   c. They are covered by a capsule, which enables them to multiply quickly.
   d. They are covered by pili, which keep them attached to the urethral walls.
   e. They contain a peptidoglycan cell wall, which gives them buoyancy to float in the fluids of the urethra.
2. A bacterium that uses nitrites as its only energy source was found in a deep salt mine. It is a:
   a. chemoautotroph.
   b. parasite.
   c. photoautotroph.
   d. heterotroph.
   e. photoheterotroph.
3. The ___ are all oxygen-producing photoautotrophs.
   a. spirochetes
   b. chlamydias
   c. cyanobacteria
   d. Gram-positive bacteria
   e. proteobacteria
4. At the health center, a fecal sample was taken from a feverish student. Organisms with corkscrew-like flagella and no endomembranes but with cell walls that lack peptidoglycan were isolated as the cause for the illness. These organisms probably belong to the group:
   a. chlamydias.
   b. spirochetes.
   c. Euryarchaeota.
   d. Cyanobacteria.
   e. Archaea.
5. Which of the following is *not* a property of an endospore?
   a. resistant to boiling—must be autoclaved to be killed
   b. metabolically inactive
   c. can potentially survive millions of years
   d. provides a method to preserve bacterial DNA under harsh conditions
   e. is the portion of the bacterial cell that undergoes conjugation
6. Bacterial cells are generally thought to act independently. An exception to this is bacterial cells involved in:
   a. biofilms.
   b. photosynthesis.
   c. peptidoglycan layering.
   d. toxin release.
   e. facultative anaerobic metabolism.
7. Penicillin, an antibiotic, inhibits the formation of cross-links between sugar groups in peptidoglycan. Bacteria treated with penicillin should be:
   a. aerobic.
   b. anaerobic.
   c. Gram-negative.
   d. Gram-positive.
   e. flagellated.
8. The best choice when using/prescribing antibiotics is to:
   a. increase the dosage when the original amount does not work.
   b. determine the kind of bacterium causing the problem.
   c. stop taking the antibiotic when you feel better but the prescription has not run out.
   d. ask the doctor to prescribe a drug as a precaution for an infection you do not have.
   e. choose soaps that are labeled "antibacterial."
9. Archaeans have:
   a. proteins of energy metabolism like bacteria, cell division proteins like bacteria, and mitochondria like eukaryotes.
   b. multiple types of RNA polymerases like eukaryotes, chromosomes like bacteria, and formylmethionine as the first amino acid placed in proteins.
   c. operons like bacteria, multiple types of RNA polymerases like eukaryotes, and cell division proteins like bacteria.
   d. no sensitivity to streptomycin, histones like eukaryotes, and no operons.
   e. histones like eukaryotes, a single type of RNA polymerase like bacteria, and operons like bacteria.
10. Methanogens, obligate anaerobes that generate methane, are:
   a. members of the Proteobacteria.
   b. members of the Cyanobacteria.
   c. members of the Spirochetes.
   d. members of the Euryarchaeota.
   e. members of the Korarchaeota.

Discuss the Concepts

1. The digestive tract of newborn chicks is free of bacteria until, for example, they eat food that has been exposed to the feces of adult chickens. The ingested bacteria establishes a population in the digestive tract that is beneficial for the digestion of food. However, if *Salmonella* are present in the adult feces, this bacterium, which can be pathogenic for humans who ingest it, may become established in the digestive tracts of the chicks. To eliminate the possibility that *Salmonella* might become established, should farmers feed newborn chicks a mixture of harmless known bacteria from a lab culture, or a mixture of unknown fecal bacteria from healthy adult chickens? Design an experiment to answer this question.
2. Investigators in Australia found that mats of pond scum formed by the bacterium *Botyrococcus braunii* decayed into a substance resembling crude oil when the ponds dried up. Formulate a hypothesis explaining how this process may have contributed to Earth's oil deposits.

Design an Experiment

You receive a sample of an unidentified prokaryote. Design an experiment to see if it is a bacterium or an archaean.

Interpret the Data

The proteobacteria in the genus *Pseudomonas* are important spoilers of food. The rate at which food spoils is dependent upon temperature. The data in the **Figure** show the relationship between the square root of the growth rate (how fast the population is growing per day; unitless) and the temperature of the culture.

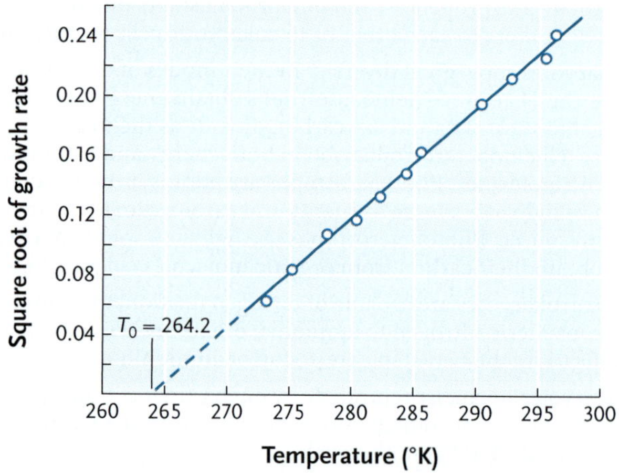

D. A. Ratkowsky et al. 1982. Relationship between temperature and growth rate of bacterial cultures. *Journal of Bacteriology* 149:1–5. Journal of bacteriology by American Society for Microbiology; Society of American Bacteriologists. Reproduced with permission of American Society for Microbiology (etc.) in the format Journal via Copyright Clearance Center.

1. Convert the temperature of the data point at the lowest temperature from °K to °C. Why are there no data below this temperature?
2. Does typical refrigeration (temperature lowered to 4°C) appear sufficient to limit growth rate compared to room temperature (20°C)?

Source: D. A. Ratkowsky et al. 1982. Relationship between temperature and growth rate of bacterial cultures. *Journal of Bacteriology* 149:1–5.

Apply Evolutionary Thinking

Does our understanding of horizontal gene transfer affect our application of the biological species concept to bacteria?

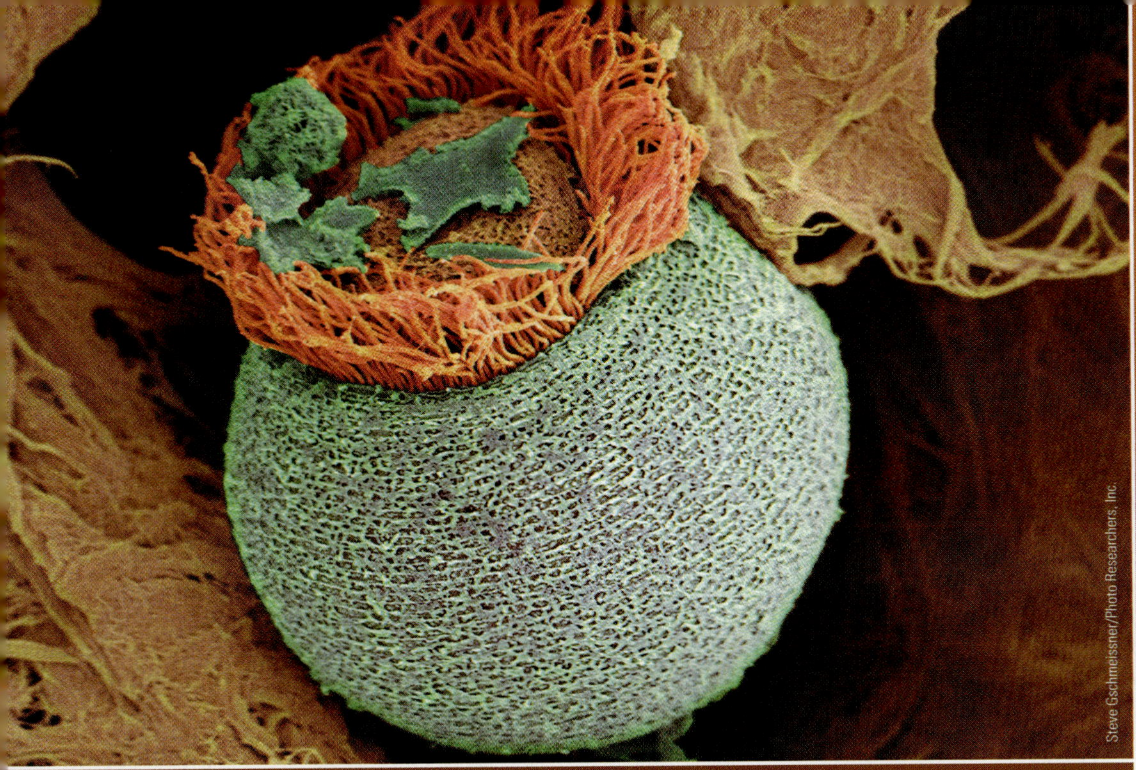

A ciliated protozoan, a type of protist (colorized SEM). This protozoan lives in water, feeding on bacteria and decaying organic matter.

27

STUDY OUTLINE

27.1 What Is a Protist?
27.2 The Protist Groups

Protists

Why it matters . . . Go for a swim just about anywhere in the natural world and you will share the water with multitudes of diverse organisms called protists. Like their most ancient ancestors, almost all of these eukaryotic species are aquatic. Structurally, single-celled protists are the simplest of all eukaryotes. Although most are microscopic in size, many have had or have a significant impact on the world. For example, the protist *Phytophthora infestans,* a water mold also referred to as a downy mildew, infects valuable crop plants such as potatoes. Potatoes were one of the main food staples in Europe in the nineteenth century, and *P. infestans* caused a lot of damage to potato crops in many countries. In Ireland, for instance, the growing seasons between 1845 and 1860 were cool and damp. Year after year, *P. infestans* spores spread along thin films of water on the plants. Late blight, a rotting of plant parts, became epidemic. In this so-called Irish potato famine, an estimated one million people starved to death or died of typhoid fever (a secondary effect), and another million people fled to other countries.

Today, related species threaten forests in the United States, Europe, and Australia. For example, when conditions favored its growth, *Phytophthora ramorum* started an epidemic of sudden oak death in California, during which tens of thousands of oak trees died. As the common name suggests, infected trees die rapidly. The first sign of infection is a dark red-to-black sap oozing from the bark surface. The pathogen has now jumped to madrones, redwoods, and certain other trees and shrubs. Cascading ecological changes resulting from tree death caused by this pathogen will reduce sources of food and shelter for forest species.

In this chapter, we describe several lineages of protists, a diverse array of eukaryotic organisms that are clearly not a monophyletic group. Also known as *protoctists,* protists are the products of the varied early branching of eukaryotic evolution. They are abundant on Earth and play key ecological, economic, and medical roles in the world's biological communities.

27.1 What Is a Protist?

Protists are easily the most varied of all Earth's creatures. **Figure 27.1** shows a number of protists, illustrating their great diversity. Protists include both microscopic single-celled and large multicellular organisms. They may inhabit aquatic environments, moist soils, or the bodies of animals and other organisms, and they may live as predators, photosynthesizers, parasites, or decomposers. The extreme diversity of the group has made the protists difficult to classify.

Protists Are Eukaryotes Other Than Fungi, Land Plants, and Animals

The one reasonable certainty about protist classification is that they are not prokaryotes, fungi, plants, or animals. Because protists are eukaryotes, the boundary between them and prokaryotes is clear and obvious. Unlike prokaryotes, protists have a true nucleus, with multiple, linear chromosomes and cytoplasmic organelles, including mitochondria (in most but not all species), endoplasmic reticulum, Golgi complex, and chloroplasts (in some species). They reproduce asexually by mitosis or sexually by meiosis and union of sperm and egg cells, rather than by binary fission as do prokaryotes.

Figure 27.2 shows the phylogenetic relationships between protists and other eukaryotes. This tree represents a consensus based on molecular data, including phylogenomics analysis, which compares sequences of whole genomes, or large segments of genomes. The tree shows the most ancient bifurcation of eukaryotes into the **Unikonta (unikonts),** eukaryotes with a single flagellum, and **Bikonta (bikonts),** eukaryotes with two flagella. Further evolution produced several categories of eukaryotes. As you can see in Figure 27.2, all eukaryotic organisms except animals, land plants, and fungi—three groups that arose from protist ancestors—are protists. Although some protists have features that resemble those of the fungi, land plants, or animals, several characteristics are distinctive. For instance, cell wall components in protists differ from those of the fungi (molds and yeasts). In contrast to land plants, protists lack highly differentiated structures equivalent to true roots, stems, and leaves; they also lack the protective structures that encase developing embryos in plants. Protists are distinguished from animals by their lack of highly differentiated structures such as limbs and a heart, and by the absence of features such as nerve cells, complex developmental stages, and an internal digestive tract. Protists also lack collagen, the characteristic extracellular support protein of animals.

The classification scheme for protists has been challenging to devise because of the diversity of these organisms. Changes and refinements are likely as more sequence data are accumulated. What is unequivocally clear, though, is that the protists are a polyphyletic assemblage of eukaryotic organisms. As Figure 27.2 shows, many protists are no more closely related to each other than they are to the fungi, plants, or animals.

For simplicity, we will refer to these organisms as protists in this book, with the understanding that it is a collection of largely unrelated organisms placed together for convenience. We will refer to the major evolutionary clusterings as groups (see Figure 27.2).

You may have heard two broad terms historically associated with large assemblages of protists, *protozoa* ("first animal") (also called *protozoans*) and *algae*. Neither of these assemblages corresponds to a taxonomic group of protists, however. Protozoa is a name for a diverse assemblage of hetero-

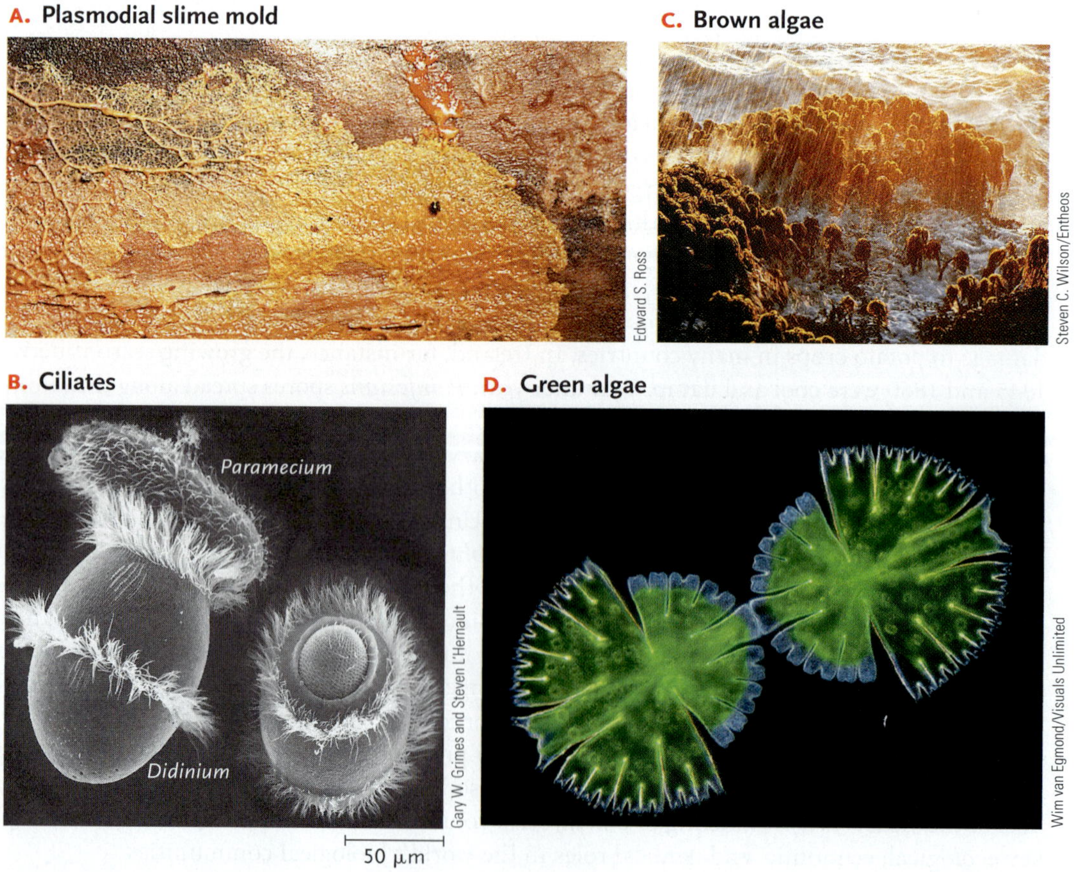

FIGURE 27.1 A sampling of protist diversity. (A) *Physarum*, a plasmodial slime mold (yellow shape, lower part of figure) migrating over a rotting log. **(B)** *Didinium*, a ciliate, consuming another ciliate, *Paramecium*. **(C)** *Postelsia palmaeformis* (the sea palm), a brown alga, thriving in the surf pounding a California coast. **(D)** *Micrasterias*, a single-celled green alga, here shown dividing in two.

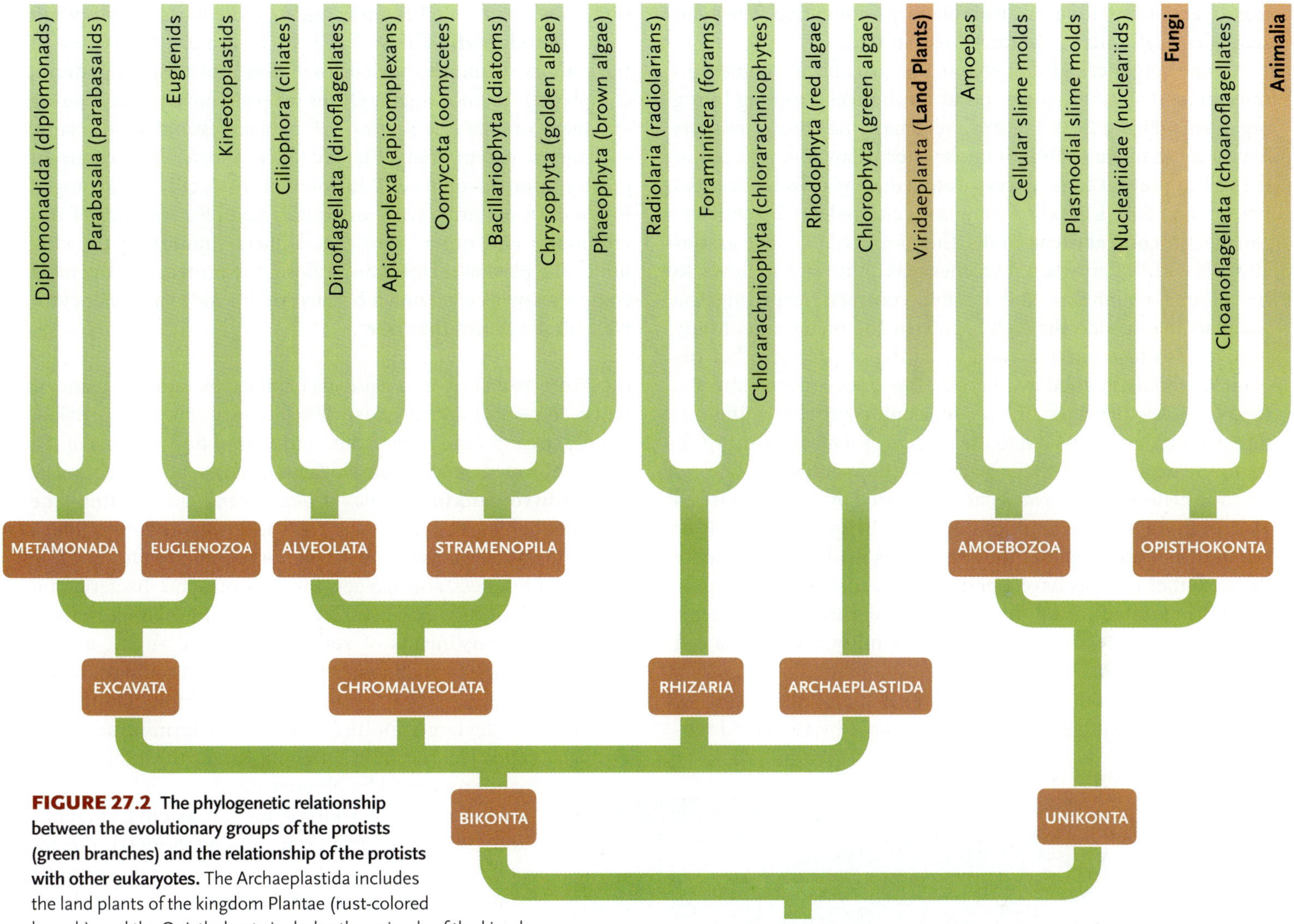

FIGURE 27.2 The phylogenetic relationship between the evolutionary groups of the protists (green branches) and the relationship of the protists with other eukaryotes. The Archaeplastida includes the land plants of the kingdom Plantae (rust-colored branch), and the Opisthokonta includes the animals of the kingdom Animalia and the fungi of the kingdom Fungi (rust-colored branches).

trophic protists, formerly considered to be animals, that are mobile and feed either by ingesting prey or absorbing organic molecules, whereas algae is a name for several sets of photosynthetic protists. The term "algae" will be used in this chapter in association with specific taxonomic groups of protists.

Protist Diversity Is Reflected in Their Metabolism, Reproduction, Structure, and Habitat

As you might expect from the broad range of organisms included, protists are highly diverse in metabolism, reproduction, structure, and habitat.

METABOLISM Almost all protists are aerobic organisms that live either as heterotrophs—by obtaining their organic molecules from other organisms—or as autotrophs—by producing organic molecules for themselves by photosynthesis. Among the heterotrophs, some protists obtain organic molecules by directly ingesting part or all of other organisms and digesting them internally. Others absorb organic molecules from their environment. A few protists can live as either heterotrophs or autotrophs.

REPRODUCTION Reproduction may be asexual by mitosis or sexual by meiotic cell division and formation of gametes. In protists that reproduce by both mitosis and meiosis, the two modes of cell division are combined into a life cycle that is highly distinctive among the different protist groups.

STRUCTURE Many protists live as single cells or as **colonies** in which individual cells show little or no differentiation and are potentially independent. Within colonies, individuals use cell signaling to cooperate on tasks such as feeding or movement. Some protists are large multicellular organisms, in which cells are differentiated and completely interdependent. For example, seaweeds are multicellular marine protists that include the largest and most differentiated organisms of the group; their structures include a holdfast to secure the organism to rocks, leaflike fronds, and, in some cases, an air bladder for flotation. The giant kelp of coastal waters rival forest trees in size.

Some single-celled and colonial protists have complex intracellular structures, some found nowhere else among living organisms **(Figure 27.3).** For example, many freshwater protists have a mechanism that maintains water balance in and out of the cell to prevent lysis. Excess water entering cells by osmosis (see Section 6.3) is handled using a specialized cytoplasmic organelle, the **contractile vacuole.** The contractile vacuole gradually fills with water; when it reaches maximum size it moves to the plasma membrane and forcibly contracts, expelling the water to the outside through a pore in the membrane. Many protists also have **food vacuoles** that digest prey or other organic material engulfed by the cells. Enzymes secreted into the food vacuoles digest the organic molecules; any remaining undigested matter is expelled to the outside by a mechanism similar to the expulsion of water by contractile vacuoles.

The cells of some protists are supported by an external cell wall, or by an internal or external shell built up from organic or mineral matter; in some, the shell takes on highly elaborate forms. Other protists have a **pellicle,** a layer of supportive protein fibers located inside the cell, just under the plasma membrane, which provides strength and flexibility instead of a cell wall (see Figure 27.5).

Almost all protists have structures that provide motility at some time during their life cycle. Some move by amoeboid motion, in which the cell extends one or more lobes of cytoplasm called **pseudopodia** ("false feet"; see Figure 27.19). The rest of the cytoplasm and the nucleus then flow into a pseudopodium, completing the movement. Other protists move by the beating of flagella or cilia (see Section 5.3; cilia are essentially the same as flagella, except that cilia are often shorter and occur in greater numbers on a cell). In some protists, cilia are arranged in complex patterns, with an equally complex network of microtubules and other cytoskeletal fibers supporting the cilia under the plasma membrane. Among the protists are the most complex single cells known because of the wide variety of cytoplasmic structures they have.

HABITAT Protists live in aqueous habitats, including aquatic or moist terrestrial locations such as oceans, freshwater lakes, ponds, streams, and moist soils, and within host organisms. In bodies of water, small photosynthetic protists collectively make up the **phytoplankton** (*phytos* = plant; *planktos* = drifting), the abundant organisms that capture the energy of sunlight in nearly all aquatic habitats. These photosynthetic protists provide organic substances and oxygen for heterotrophic bacteria and protists and for the small crustaceans and animal larvae that are the primary constituents of **zooplankton** (*zoe* = life, usually meaning animal life); although protists are not animals, biologists often include them among the zooplankton. The phytoplankton and the larger multicellular protists forming seaweeds collectively account for about half of the total organic matter produced by photosynthesis.

In the moist soils of terrestrial environments, protists play important roles among the detritus feeders that recycle matter from organic back to inorganic form. In their roles in phytoplankton, zooplankton, and as detritus feeders, protists are enormously important in the world ecosystem.

Protists that live in host organisms are parasites, obtaining nutrients from the host. Indeed, many of the parasites that have significant effects on human health are protists, causing diseases such as malaria, sleeping sickness, and giardiasis.

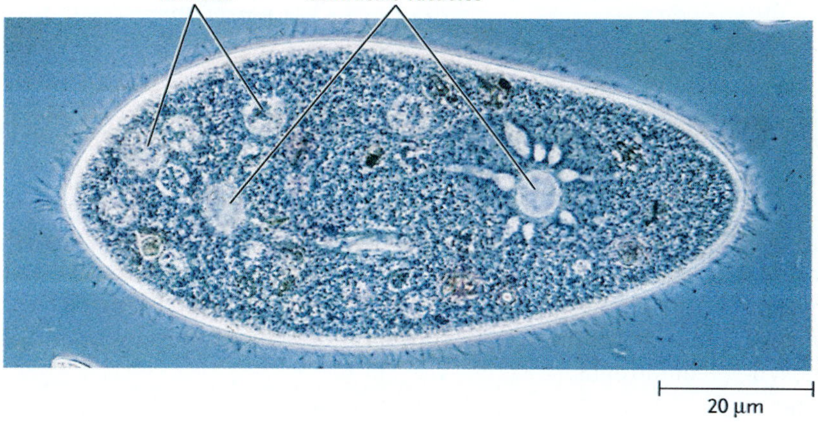

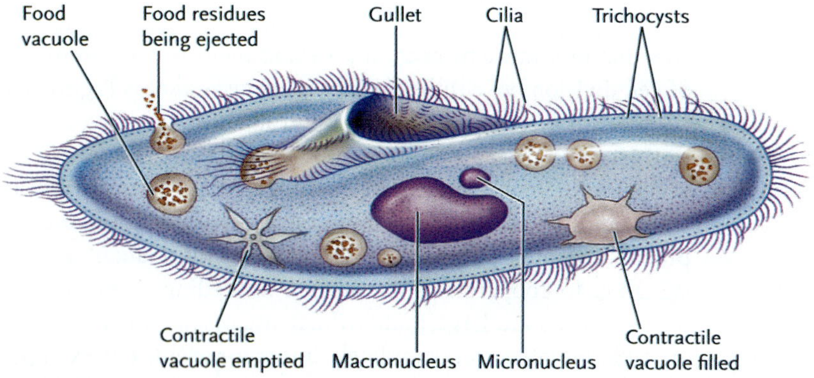

FIGURE 27.3 A ciliate, *Paramecium*, showing the cytoplasmic structures typical of many protists.

STUDY BREAK 27.1

What distinguishes protists from prokaryotes? What distinguishes protists from fungi, land plants, and animals?

THINK OUTSIDE THE BOOK

Phylogenomics employs comparisons of genome sequences or the sequences of many genes simultaneously to construct phylogenetic trees. Use the Internet or research literature to outline how recent phylogenomic analysis has provided new insights into the evolutionary relationships among the protists.

27.2 The Protist Groups

This section considers the biological features of each of the groups of protists in Figure 27.2.

The Excavates Include Protists with Unique Flagella and Protists with Modified Mitochondria

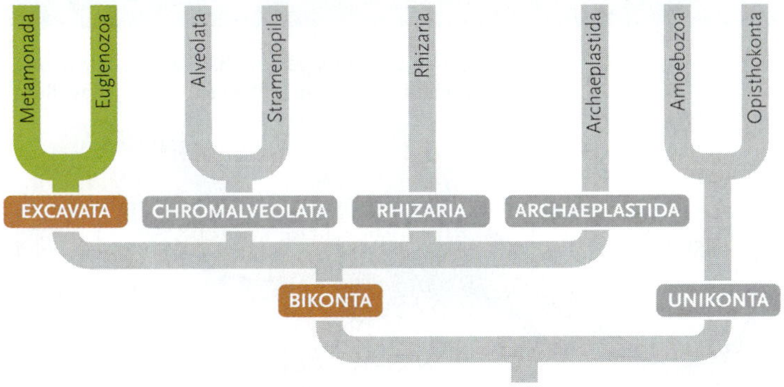

All members of the **Excavata** are single-celled animal parasites that have greatly reduced mitochondria, or organelles derived from mitochondria, and move by means of flagella; most have a scooped out (excavated) feeding apparatus on the ventral surface of the cell. Due to the lack of functional mitochondria, excavates are incapable of aerobic respiration and are limited to glycolysis as an ATP source. However, the nuclei of excavates contain genes derived from mitochondria, meaning that the ancestors of these protists probably had mitochondria. The changes in their mitochondria may have occurred as an adaptation to the parasitic way of life, in which oxygen is in short supply. Two major subgroups of the Excavata are the **Metamonada** and **Euglenozoa.**

METAMONADA **Metamonads** consist of the Diplomonadida and Parabasala. *Diplomonad* cells have two nuclei and move by means of multiple freely beating flagella. In addition to lacking mitochondria, they also lack a clearly defined endoplasmic reticulum and Golgi complex. The best-known representative of the group, *Giardia lamblia* **(Figure 27.4A),** infects the mammalian intestinal tract, inducing severe diarrhea and abdominal cramps. *Giardia* is spread by water that is contaminated with feces, in which resistant cysts of the protist can be present in large numbers.

Parabasalids also have freely beating flagella and, in addition, they have a sort of fin called an **undulating membrane,** formed by a flagellum buried in a fold of the cytoplasm. The buried flagellum allows parabasalids to move through thick and viscous fluids. Among the Parabasala are the trichomonads, including *Trichomonas vaginalis* **(Figure 27.4B),** which is responsible for infections of the urinary and reproductive tracts in both men and women. The infective trichomonad is passed from person to person primarily, but not exclusively, by sexual intercourse. It lives in the vagina in women and in the urethra of both sexes. The infection is usually symptomless in men, but in women *T. vaginalis* can cause severe inflammation and irritation of the vagina and vulva. It is easily cured by drugs.

A. *Giardia lamblia*

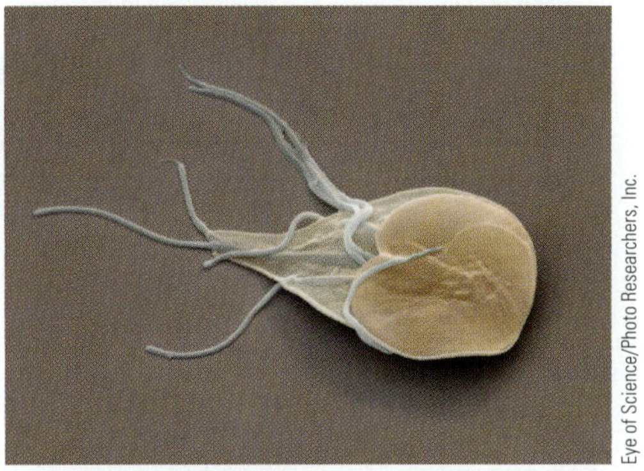

B. *Trichomonas vaginalis*

FIGURE 27.4 Metamonads of the excavates. (A) A diplomonad, *Giardia lamblia,* that causes intestinal disturbances. **(B)** A parabasalid, *Trichomonas vaginalis,* that causes a sexually transmitted disease, trichomoniasis.

EUGLENOZOA **Euglenozoans** include about 1,800 species, almost all single-celled, highly motile cells that swim by means of flagella. Their mitochondria are characterized by disc-shaped cristae (inner mitochondrial membranes). While most are photosynthetic, some are facultative heterotrophs; some can even alternate between photosynthesis and life as a heterotroph. The two major subgroups of the Euglenozoa are the *euglenids* and the *kinetoplastids.*

Euglenids. Euglenids are free-living protists with anterior flagella. The best-known species is *Euglena gracilis* **(Figure 27.5).** With the exception of a few marine species, the euglenids inhabit freshwater ponds, streams, and lakes. Most are autotrophs that carry out photosynthesis by the same mechanisms as plants, using the same photosynthetic pigments, including chlorophylls *a* and *b* and β-carotene. Many of the photosynthetic euglenids, including *E. gracilis,* can also live as heterotrophs by absorbing organic molecules through the plasma membrane. Some euglenids lack chloroplasts and live entirely as heterotrophs.

E. gracilis and other euglenids have a profusion of cytoplasmic organelles, including a contractile vacuole and, in photosynthetic species, chloroplasts (see Figure 27.5). Instead of an external cell wall, the euglenids have a spirally grooved pellicle formed from transparent, protein-rich material. Most of the photosynthetic euglenids, including *E. gracilis,* have an *eyespot*

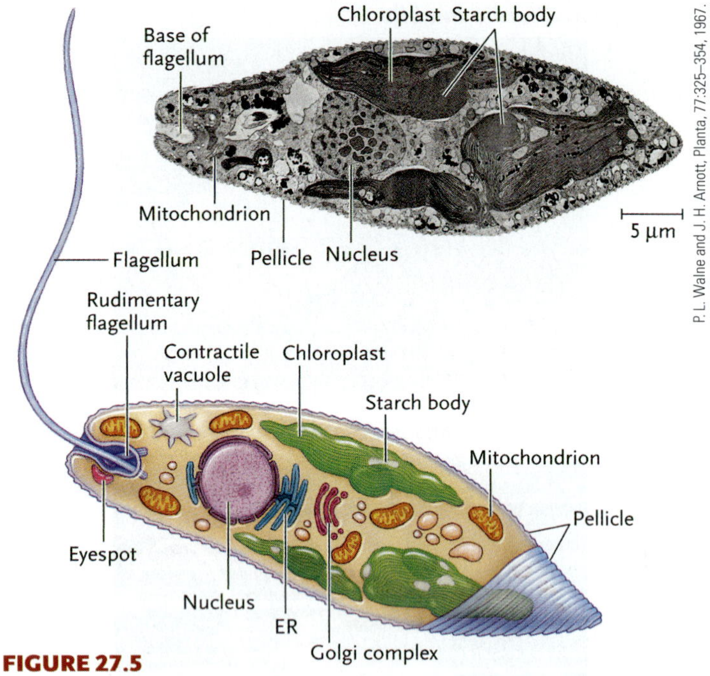

P. L. Walne and J. H. Arnott, Planta, 77:325–354, 1967.

FIGURE 27.5

Body plan and an electron micrograph of *Euglena gracilis*. The plane of section in the electron micrograph has cut off all but the base of the flagellum.

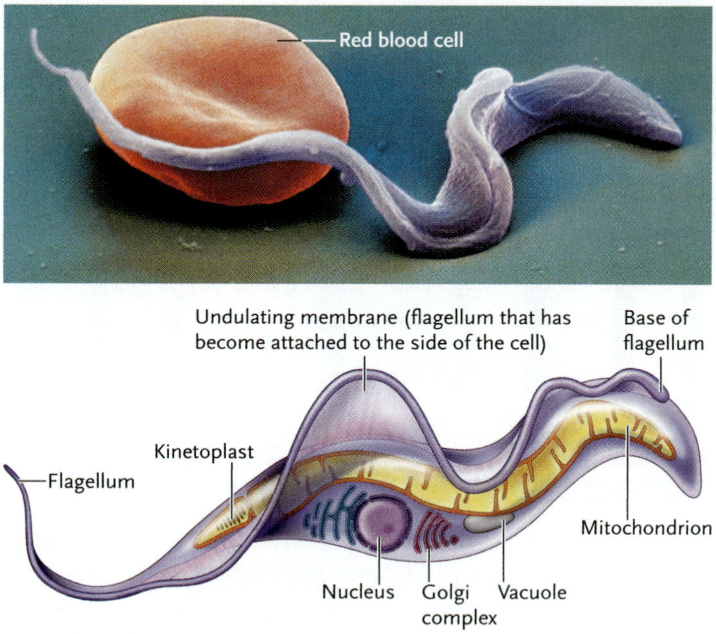

Oliver Meckes/Photo Researchers, Inc.

FIGURE 27.6 ***Trypanosoma brucei*, the parasitic kinetoplastid that causes African sleeping sickness.**

containing carotenoid pigment granules in association with a light-sensitive structure. The eyespot is part of a sensory mechanism that stimulates cells to swim toward moderately bright light or away from intensely bright light so that the organism is positioned in light conditions for optimal photosynthetic activity. The cells swim by whiplike movements of flagella that extend from a pocketlike depression at one end of the cell. Most have two flagella, one rudimentary and short, the other long.

Kinetoplastids. Kinetoplastids are a group of nonphotosynthetic, heterotrophic cells that live as animal parasites **(Figure 27.6).** Their name reflects the structure of the single mitochondrion their cells, which contains a large DNA–protein deposit called a *kinetoplast.* Most kinetoplastids have two flagella, which are used for movement. In some cases, as in Figure 27.6, one of the flagella is attached to the side of the cell, forming an undulating membrane that is often used to enable the organism to glide along or attach to surfaces.

The kinetoplastids include the trypanosomes, responsible for several diseases afflicting millions of humans in tropical regions. *Trypanosoma brucei* (see Figure 27.6) causes African sleeping sickness, transmitted from one host to another by bites of the tsetse fly. Early symptoms include fever, headaches, rashes, and anemia. Untreated, the disease damages the central nervous system, leading to a sleeplike coma and eventual death. The disease has proved difficult to control because the same trypanosome infects wild mammals, providing an inexhaustible reservoir for the parasite. Other trypanosomes, also transmitted by insects, cause Chagas disease, which can lead to severe brain and heart damage, and leishmaniasis, which can produce disfiguring skin sores and ulcers.

The Chromalveolates Are a Heterogeneous Group of Protists with a Range of Forms and Life Styles

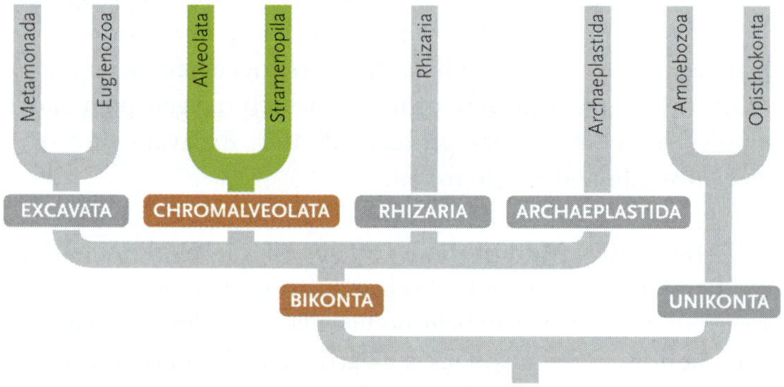

The **Chromalveolata** consists of the **Alveolata** and the **Stramenopila.**

ALVEOLATA The **alveolates** are so called because they have small, flattened, membrane-bound vesicles called *alveoli* (*alvus* = belly) in a layer just under the plasma membrane. The alveolates include two motile, primarily free-living groups, the **Ciliophora** and **Dinoflagellata,** and a nonmotile, parasitic group, the **Apicomplexa.**

Ciliophora. Ciliophora—the **ciliates**—include nearly 10,000 known species of primarily single-celled but highly complex heterotrophic organisms that swim by means of cilia (see Figures 27.1B and 27.3). Essentially any sample of pond water or bottom mud contains a wealth of these creatures.

The organisms in the Ciliophora have many highly developed organelles, including a mouthlike gullet lined with cilia;

structures that exude mucins, toxins, or other defensive and offensive materials from the cell surface; contractile vacuoles; and complex systems of food vacuoles. A pellicle reinforces cell shape. A complex cytoskeletal network of microtubules and other fibers anchors the cilia just below the pellicle and coordinates the ciliary beating. The cilia can stop and reverse their beating in synchrony, allowing ciliates to stop, back up, and turn if they encounter negative stimuli.

The ciliates are the only eukaryotes that have two types of nuclei in each cell: one or more small nuclei called *micronuclei,* and a single larger *macronucleus* (see Figure 27.3). A **micronucleus** is a diploid nucleus that contains a complete complement of genes. It functions primarily in cellular reproduction, which may be asexual or sexual. The number of micronuclei present depends on the species. The **macronucleus** develops from a micronucleus, but loses all genes except those required for basic "housekeeping" functions of the cell and for ribosomal RNAs. These DNA sequences are duplicated many times, greatly increasing its capacity to transcribe the mRNAs needed for these functions, and the rRNAs needed to make ribosomes.

In asexual reproduction by mitosis, both types of nuclei replicate their DNA, divide, and are passed on to daughter cells. **Figure 27.7** illustrates sexual reproduction of *Paramecium,* which is initiated when two cells **conjugate** and a cytoplasmic bridge forms between the cells.

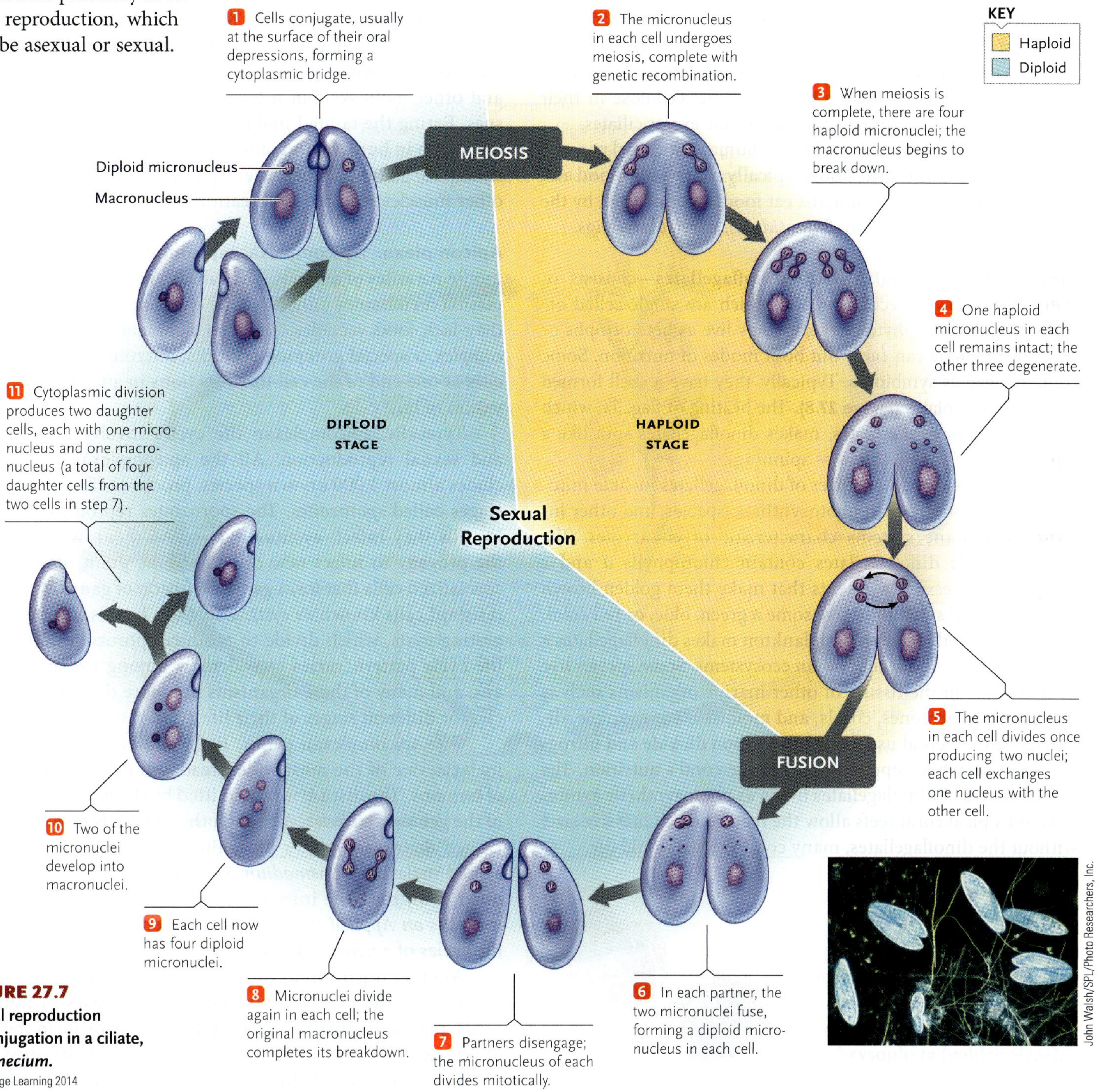

FIGURE 27.7
Sexual reproduction by conjugation in a ciliate, *Paramecium.*

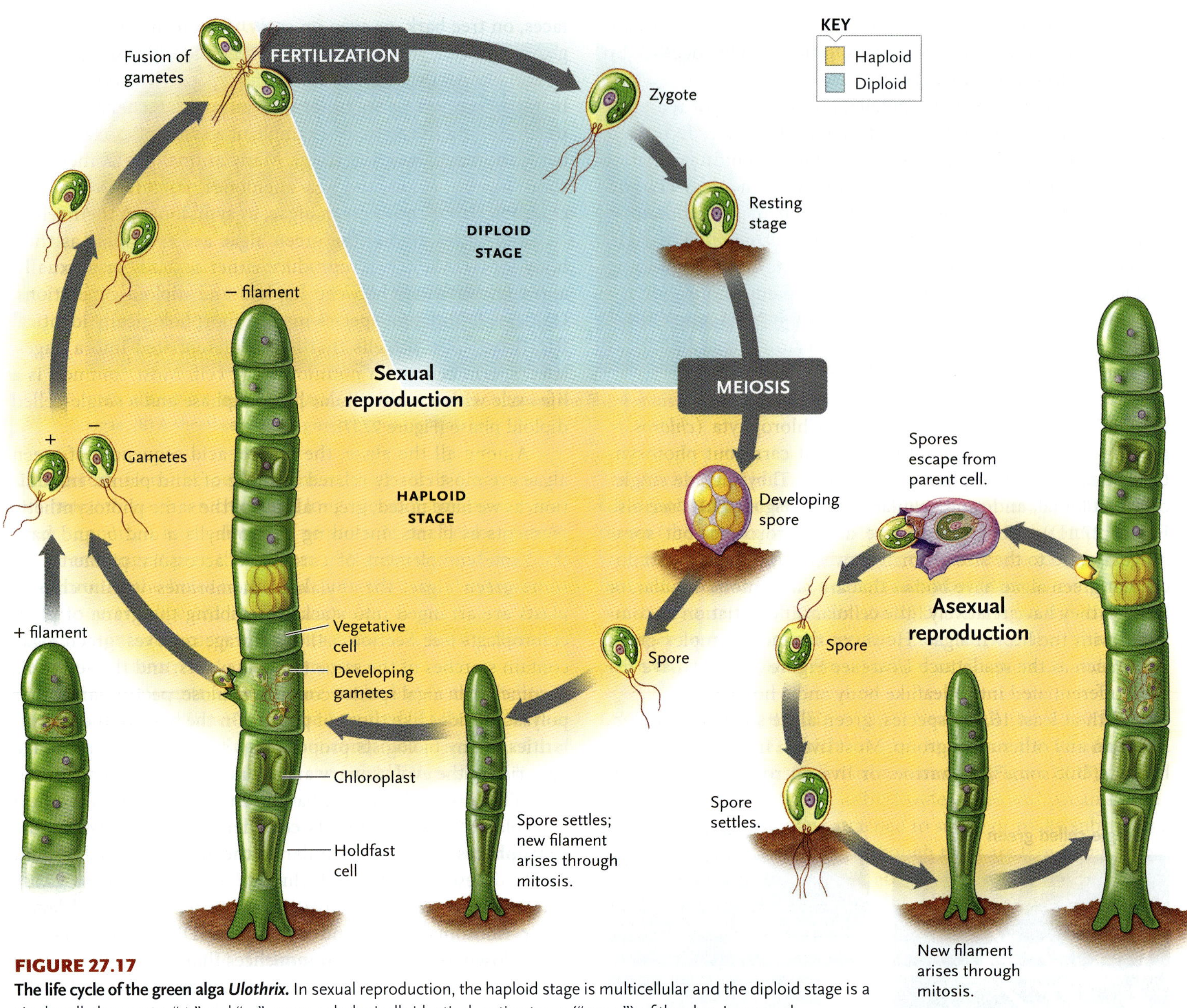

FIGURE 27.17

The life cycle of the green alga *Ulothrix*. In sexual reproduction, the haploid stage is multicellular and the diploid stage is a single cell, the zygote. "+" and "−" are morphologically identical mating types ("sexes") of the alga. In asexual reproduction, all stages are haploid.

FIGURE 27.18 **The charophyte *Spirogyra*, representative of a group of green algae that may have given rise to the plant kingdom.**

The **Unikonta—unikonts**—include amoebas with unsupported pseudopods, protists closely related to animals or fungi, and animals and fungi themselves. The unikonts consist of the **Amoebozoa** and the **Opisthokonta.**

AMOEBOZOA The **Amoebozoa** include most of the amoebas (others are in the Opisthokonta and the Rhizaria) as well as the cellular and plasmodial slime molds. All members of this group use pseudopods for locomotion and feeding for all or part of their life cycles.

Amoebas. Amoebas of the Amoebozoa are single-celled organisms that are abundant in marine and freshwater environments and in the soil. They use pseudopods for locomotion and

FIGURE 27.11 **Experimental Research**

The Evolutionary History of Diatoms as Revealed by Genome Analysis

Question: What do diatom genomes reveal about their evolutionary history?

Experiment: Diatoms are members of the chromalveolates. There are two major classes of diatoms, the radially symmetrical centrics and the bilaterally symmetric pennates. The earliest fossil centrics date to 180 million years ago, whereas the earliest fossil pennates date to 90 million years ago. Researchers have long been interested in how the two types of diatoms have evolved and diverged.

The genome sequence of a pennate diatom, *Phaeodactylum tricornutum,* was reported in 2008, and the genome sequence of a centric diatom, *Thalassiosira pseudonana,* was reported in 2004. Researchers involved in the 2008 report compared the genome sequences of the two diatoms, and compared the diatom genome sequences with the genome sequences of three metazoans, two plants, three green algae, one red alga, two fungi, and ten other chromalveolates.

Results: The table below presents the results of the genome sequence comparisons. In the table, three types of genes are distinguished:

- *Core genes:* Genes found in all of the eukaryotic groups with which the diatom sequences were compared.
- *Diatom-specific genes:* Genes found in both diatoms, but not in the other organisms.
- *Unique genes:* Genes found only in one or other of the two diatoms.

	Diatom species	
	P. tricornutum	*T. pseudonana*
Genome size (Mb[1])	27.4	32.4
Number of genes	10,402	11,776
Core genes	3,523	4,332
Diatom-specific genes	1,328	1,407
Unique genes	4,366	3,912

[1] Mb = 1 megabase = 10^6 base pairs

Notably, the results show that the two diatoms have only about 57% of their genes in common, illustrating a high degree of divergence since the split to the two diatom lineages.

The researchers made some other observations from the genome sequence comparisons:

- 171 of their chloroplast genes are of red algal origin; 108 of these genes are shared by the two species.
- 587 genes are of prokaryotic origin; 56% of these genes are found in both diatoms. These genes are presumed to have been obtained by horizontal gene transfer, the movement of genetic material between organisms other than by descent (see Sections 17.1 and 26.1). These genes represent more than 5% of the total genes of these diatoms, an unusually high percentage for a eukaryote. Most of the genes can be traced to proteobacteria, cyanobacteria, and various archaea. This observation is interpreted to mean that there has been a long-term association between diatoms and bacteria that has led to the transfer of useful genes.
- The diatom-specific genes are evolving faster than other genes in these diatoms or in eukaryotes in general. Although the exact reason for this fact is unknown, it is thought that it may correlate with the extensive diversification of diatoms.

Conclusion: Analysis of the two diatom genome sequences has provided insights into the evolutionary history of diatoms. The analysis provides interesting comparisons with other eukaryotic organisms, and supports the hypothesis that chromalveolates evolved from symbiosis between a photosynthetic red alga and a heterotrophic host (a concept discussed further later in the chapter). The evidence for the extent of horizontal gene transfer, and for the increased rate of evolution of diatom-specific genes, may help explain the high degree of diversity of diatoms and their adaptations for their life in the oceans.

THINK LIKE A SCIENTIST What would you do to obtain a better understanding of the evolutionary history of diatoms?

Sources: C. Bowler et al. 2008. The *Phaeodactylum* genome reveals the evolutionary history of diatom genomes. *Nature* 456:239–244; E. V. Armbrust et al. 2004. The genome of the diatom *Thalassiosira pseudonana:* ecology, evolution, and metabolism. *Science* 306:79–86.

produce flagellated gametes that lose their shells and fuse in pairs to form a zygote. The zygote grows to normal size before secreting a completely new shell with full-size top and bottom halves.

The shells of diatoms are common in fossil deposits. In fact, more diatoms are known as fossils than as living species—some 35,000 extinct species have been described as compared with 7,000 living species. For about 180 million years, the shells of diatoms have been accumulating into thick layers of sediment at the bottom of lakes and seas. Since diatoms store food as oil, fossil diatoms may be a source of oil in many oil deposits. **Figure 27.11** presents the results of genome analysis experiments that provide some insights into the evolution of diatoms.

Grinding the fossilized shells into a fine powder produces *diatomaceous earth,* which is used in abrasives and filters, as an insulating material, and as a pesticide. Diatomaceous earth kills crawling insects by abrading their exoskeleton, causing them to dehydrate and die. Insect larvae are killed in the same way. Insects also die when they eat the powder but larger animals, including humans, are unaffected by it.

Chrysophyta. **Chrysophyta—golden algae**—mostly are colonial forms in which each cell of the colony bears a pair of flagella **(Figure 27.12A).** The golden algae have glassy shells, but in the form of plates or scales rather than in the candy-box form of the diatoms.

FIGURE 27.12 Golden and brown algae. **(A)** A microscopic, swimming colony of *Synura,* a golden alga. Each cell bears two flagella, which are not visible in this light micrograph. **(B)** A brown alga, *Macrocystis.* Gas bladders keep the blades floating. **(C)** The holdfast, stemlike stipes, and leaflike blades, as seen in another brown alga, the sea palm *Polstelsia palmaeformis.*

Nearly all chrysophytes are autotrophs and carry out photosynthesis using pathways similar to those of plants. Their color is due to a brownish carotenoid pigment, fucoxanthin, which masks the green color of the chlorophylls. Golden algae are important in freshwater habitats and in "nanoplankton," a community of marine phytoplankton composed of huge numbers of extremely small cells. During the spring and fall, "blooms" of golden algae can give a fishy taste and brownish color to the water.

Phaeophyta. Phaeophyta—brown algae (*phaios* = brown)—are photosynthetic autotrophs that range from microscopic forms to giant kelps that reach lengths of 50 m or more (Figure 27.1C and **Figure 27.12B and C**). Their color is also due to fucoxanthin. Their cell walls contain cellulose and a mucilaginous polysaccharide, alginic acid.

Nearly all of the 1,500 known phaeophyte species inhabit temperate or cool coastal marine waters. The kelps form vast underwater forests; fragments of these algae litter the beaches in coastal regions where they grow. Great masses of another brown alga, *Sargassum,* float in an area of the mid-Atlantic Ocean called the Sargasso Sea, which covers millions of square kilometers between the Azores and the Bahamas.

Kelps are the largest and most complex of all protists. Their tissues are differentiated into leaflike *blades,* stalklike *stipes,* and rootlike *holdfasts* that anchor them to the bottom. Hollow, gas-filled bladders give buoyancy to the stipes and blades and help keep them upright. The stalks of some kelps contain tubelike vessels, similar to the vascular elements of plants, which distribute dissolved sugars and other products of photosynthesis rapidly throughout the body of the alga.

Typically, life cycles among the brown algae are complex and in many species consist of alternating haploid and diploid generations **(Figure 27.13).** The large structures recognized as kelps and other brown seaweeds are diploid **sporophytes,** so called because they give rise to haploid spores by meiosis. The spores germinate and divide by mitosis to form a small, independent, haploid **gametophyte** generation, which gives rise to haploid gametes. Variations occur in smaller brown algae, including some life cycles in which the sporophytes and gametophytes are the same size and some in which the gametophyte is larger than the sporophyte.

The alginic acid in brown algal cell walls, called **algin** when extracted, is an essentially tasteless and nontoxic substance used to thicken such diverse products as ice cream, pudding, salad dressing, jellybeans, cosmetics, paper, and floor polish. Brown algae are also harvested as food crops and as fertilizers.

The Rhizarians Consist of Amoebas with Filamentous Pseudopods

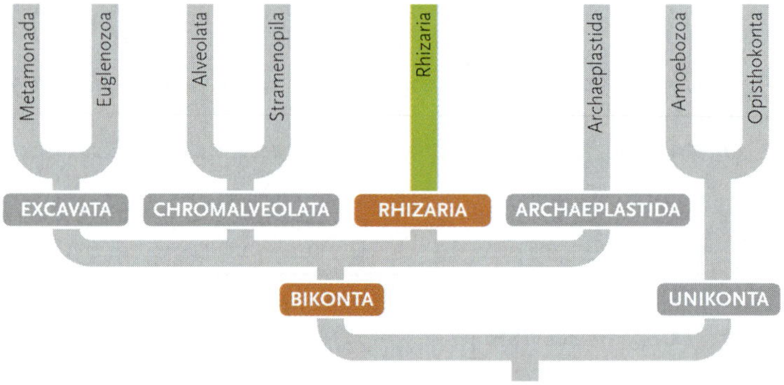

Amoeba is a descriptive term for a single-celled protist that moves by means of temporary cellular projections called pseudopods. Several major groups of protists contain amoebas, which are similar in form but are not all closely related. The amoebas of the **Rhizaria** produce stiff, filamentous pseudopodia, and many produce hard outer shells, also called *tests.* We consider here two heterotrophic groups of rhizarian amoebas, the **Radiolaria** and the **Foraminifera,** and a third, photosynthesizing group, the **Chlorarachniophyta.**

RADIOLARIA **Radiolaria—radiolarians**—are marine protists that are distinguished by axopods, slender, raylike strands of cytoplasm supported internally by long bundles of microtubules. They engulf prey organisms that stick to the axopods and digest them in food vacuoles.

Radiolarians secrete a glassy internal skeleton from which the axopods project **(Figure 27.14A and B).** Just outside the skeleton, the cytoplasm is crowded with frothy vacuoles and lipid droplets, which provide buoyancy. The skeletons of dead radiolarians sink to the bottom of the ocean and become part of the sediment. Over time, they harden into sedimentary rocks that form an important part of the geological record.

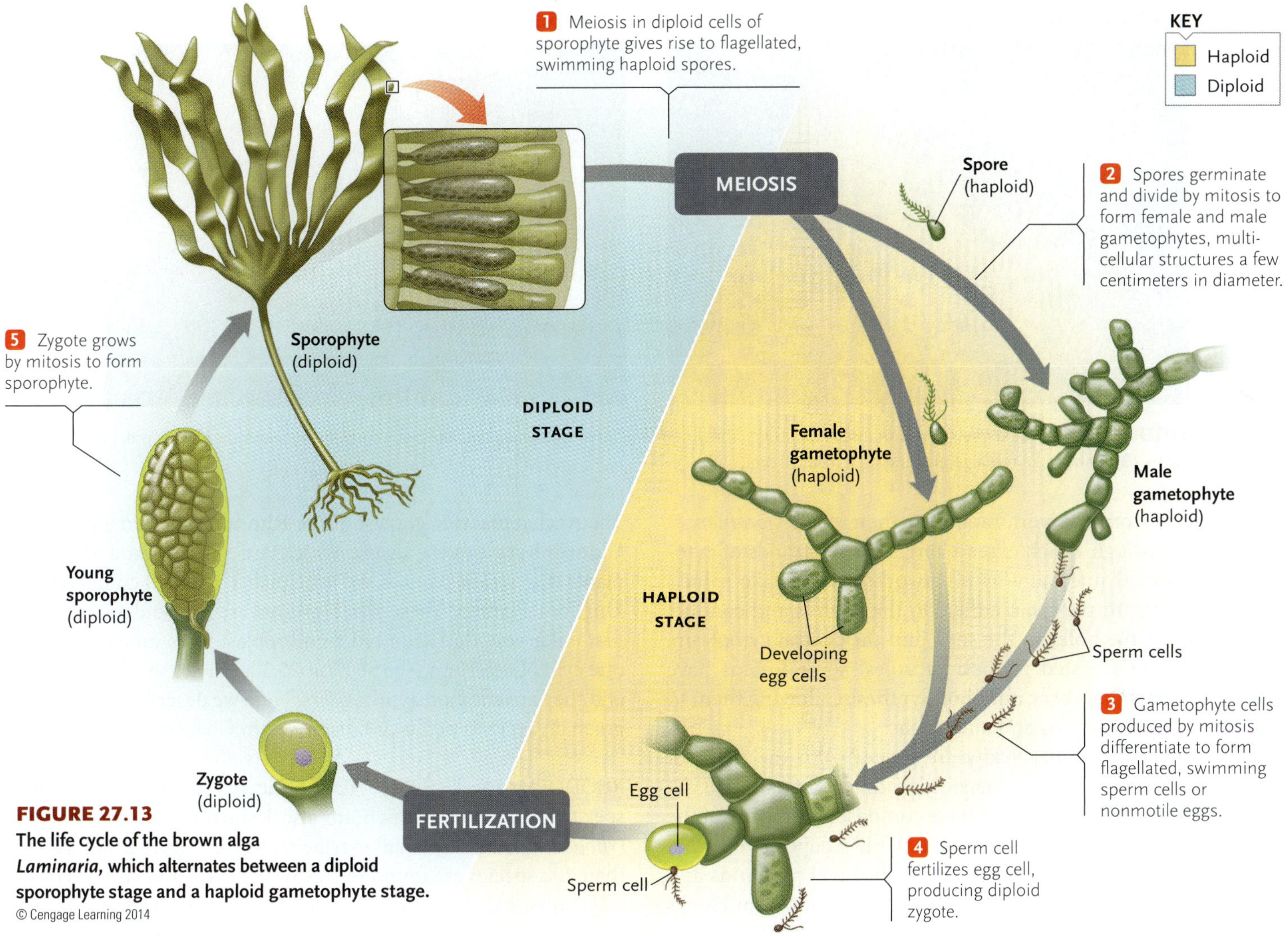

FIGURE 27.13
The life cycle of the brown alga *Laminaria*, which alternates between a diploid sporophyte stage and a haploid gametophyte stage.

FORAMINIFERA: FORAMS **Foraminifera—forams—**are also marine protists. Their shells consist of organic matter reinforced by calcium carbonate **(Figure 27.14C and D)**. Most foram shells are chambered, spiral structures that, although microscopic, resemble those of mollusks. Forams are identified and classified primarily by the form of the shell; about 250,000 species are known. Some species are planktonic, but they are most abundant on sandy bottoms and attached to rocks along coasts. Their

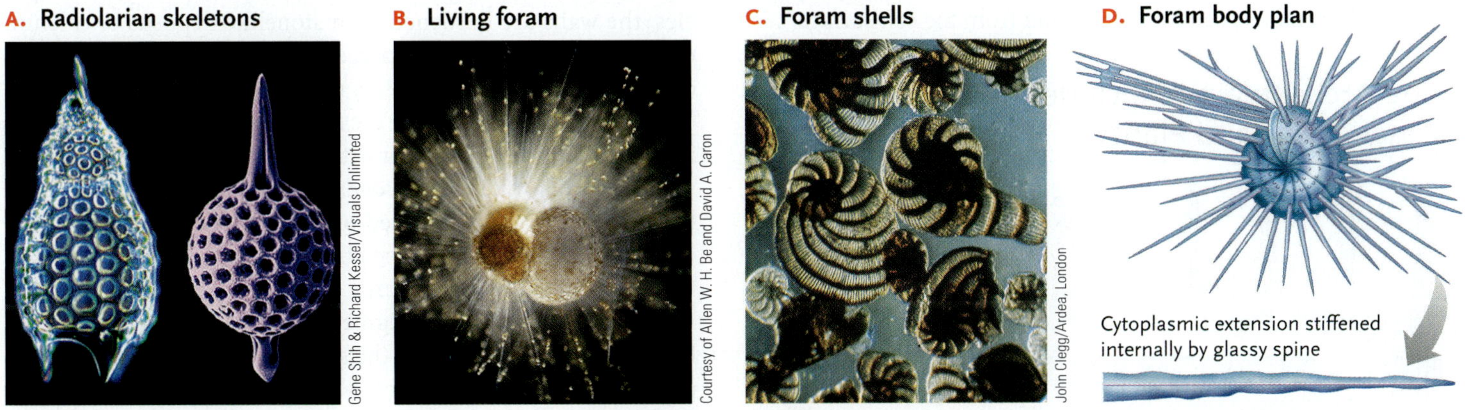

FIGURE 27.14 **Radiolarians and forams. (A)** The internal skeletons of two radiolarian species, possibly *Pterocorys* and *Stylosphaera*. Bundles of microtubules support the cytoplasmic extensions of the radiolarians. **(B)** A living foram, showing the cytoplasmic strands extending from its shell. **(C)** Empty foram shells. **(D)** The body plan of a foram. Needlelike, glassy spines support the cytoplasmic extensions of the forams.

FIGURE 27.15 **Red algae. (A)** *Antithamnion plumula*, showing the filamentous and branched body form most common among red algae. **(B)** A sheetlike red alga growing on a tropical reef.

name comes from the perforations in their shells (*foramen* = little hole), through which extend long, slender strands of cytoplasm supported internally by a network of needlelike spines. The forams engulf prey that adhere to the strands and conduct them through the holes in the shell into the central cytoplasm, where they are digested in food vacuoles. Some forams have algal symbionts that carry out photosynthesis, allowing them to live as both heterotrophs and autotrophs.

Marine sediments typically are packed with the shells of dead forams. The sediments may be hundreds of feet thick; the White Cliffs of Dover in England are composed primarily of the shells of ancient forams. Most of the world's deposits of limestone and marble contain foram shells; the great pyramids and many other monuments of ancient Egypt are built from blocks cut from fossil foram deposits. Because distinct species lived during different geological periods, they are widely used to establish the age of sedimentary rocks containing their shells.

CHLORARACHNIOPHYTA **Chlorarachniophyta—chlorarachniophytes**—are green, photosynthetic amoebas that also engulf food. They contain chlorophylls *a* and *b*, but they are phylogenetically distinct from other chlorophyll *b*-containing eukaryotes. Many filamentous pseudopodia extend from the cell surface.

The Archaeplastids Include the Red and Green Algae, and Land Plants

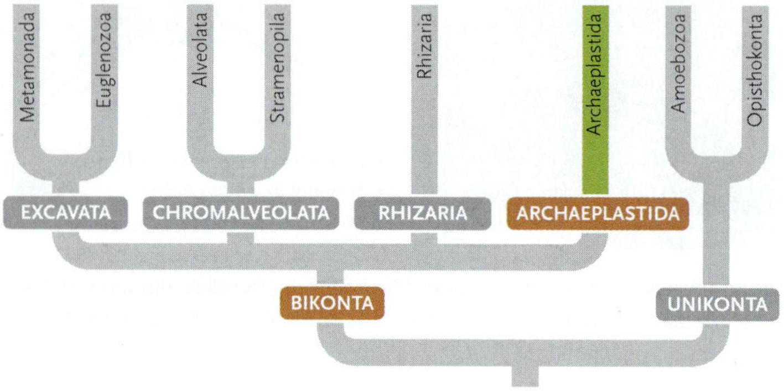

The **Archaeplastida** consist of the **Rhodophyta** (red algae) and **Chlorophyta** (green algae), which are protists, and the land plants (the *viridaeplantae,* or "true plants"), which comprise the kingdom Plantae. These three groups are descended from the first eukaryote that acquired a chloroplast from an endosymbiotic cyanobacterium (see Section 25.3 and later in this chapter), and they are all photosynthesizers. Here we describe the red and green algae; land plants are discussed in Chapters 28 and 29.

RHODOPHYTA: THE RED ALGAE Nearly all the 4,000 known species of red algae, which are also known as the Rhodophyta (*rhodon* = rose), are small marine seaweeds **(Figure 27.15).** Fewer than 200 species are found in freshwater lakes and streams or in soils. Most red algae grow attached to sandy or rocky substrates, but a few occur as plankton. Although most are free-living autotrophs, some are parasites that attach to other algae or plants.

Red algae are typically multicellular organisms, with plantlike bodies that are composed of interwoven filaments. The base of the body is differentiated into a holdfast, which anchors it to the bottom or other solid substrate, and into stalks with leaflike plates. Their cell walls contain cellulose and mucilaginous pectins that give them a slippery texture. In some species, the walls are hardened with stonelike deposits of calcium carbonate. Many of the red algae with stony cell walls resemble corals and occur with corals in reefs and banks.

Although most red algae are reddish in color, some are greenish purple or black. The color differences are produced by accessory pigments, mainly *phycobilins,* which absorb green wavelengths of light and reflect red ones, thereby masking the green color of their chlorophylls. The phycobilins are unusual photosynthetic pigments with structures similar to the ring structure of hemoglobin. The accessory pigments of some red algae make them highly efficient in absorbing the shorter wavelengths of light that penetrate to the ocean depths, allowing them to grow at deeper levels than any other algae. Some red algae live at depths to 260 m if the water is clear enough to transmit light to these levels.

Red algae have complex reproductive cycles involving alternation between diploid sporophytes and haploid gametophytes. No flagellated cells occur in the red algae; instead, gametes are released into the water to be brought together by random collisions in currents.

Extracts containing the mucilaginous pectins of red algal cell walls are widely used in industry and science. Extracted **agar** is used as a moisture-preserving, inert agent in cosmetics and baked goods, as a setting agent for jellies and desserts, and as a solidifying agent for culture medium in the laboratory. **Carrageenan,** extracted from the red alga *Eucheuma,* is used to thicken and stabilize paints, dairy products such as pudding and ice cream, and many other creams and emulsions.

Some red algae are harvested as food in Japan and China. *Porphyra,* one of these harvested algae, is used in sushi bars as the *nori* wrapped around fish and rice.

CHLOROPHYTA: THE GREEN ALGAE Chlorophyta (*chloros* = green)—green algae—are autotrophs that carry out photosynthesis using the same pigments as plants. They include single-celled, colonial, and multicellular species (**Figure 27.16;** see also Figure 27.1D). Most green algae are microscopic, but some range upward to the size of small seaweeds. Although the multicellular green algae have bodies that are filamentous, tubular, or leaflike, they have relatively little cellular differentiation as compared with the brown algae. However, the most complex green algae, such as the sea lettuce *Ulva* (see Figure 27.16C), have tissues differentiated into a leaflike body and a holdfast.

With at least 16,000 species, green algae show more diversity than any other algal group. Most live in freshwater aquatic habitats, but some are marine, or live on rocks and soil surfaces, on tree bark, or even on snow. The green, slimy mat that grows profusely in stagnant pools and ponds, for example, consists of filaments of a green alga. A few species live as symbionts in other protists or in fungi and animals. Lichens (see Figure 30.15) are the primary example of a symbiotic relationship between green algae and fungi. Many animal phyla, including some marine snails and sea anemones, contain green algal chloroplasts, or entire green algae, as symbionts in their cells.

Life cycles among the green algae are as diverse as their body forms. Many can reproduce either sexually or asexually, and some alternate between haploid and diploid generations. Gametes in different species may be morphologically identical flagellated cells, or cells that have differentiated into a flagellated sperm cell and a nonmotile egg cell. Most common is a life cycle with a multicellular haploid phase and a single-celled diploid phase **(Figure 27.17).**

Among all the algae, the nucleic acid sequences of green algae are most closely related to those of land plants. In addition, as we have noted, green algae use the same photosynthetic pigments as plants, including chlorophylls *a* and *b,* and have the same complement of carotenoid accessory pigments. In some green algae, the thylakoid membranes within chloroplasts are arranged into stacks resembling the grana of plant chloroplasts (see Section 5.4). As storage reserves, green algae contain starches of the same types as plants, and the cell walls of some green algal species contain cellulose, pectins, and other polysaccharides like those of plants. On the basis of these similarities, many biologists propose that some ancient green alga gave rise to the evolutionary ancestors of modern-day plants.

What green alga might have been the ancestor of modern land plants? Many biologists consider a group known as the **charophytes** to be most similar to the algal ancestors of land plants. These organisms, including *Spirogyra* **(Figure 27.18),** *Nitella,* and *Coleochaete,* live in freshwater ponds and lakes. Their ribosomal RNA and chloroplast DNA sequences are more closely related to plant sequences than those of any other green alga. Further, the new cell wall separating daughter cells in charophytes is formed through development of a cell plate, by a mechanism closely similar to that of plants (see Section 10.2). The body form is distinctly plantlike, with a stemlike axis on which whorls of leaflike blades occur at intervals.

A. Single-celled green alga

Wim Van Egmond/Visuals Unlimited

B. Colonial green alga

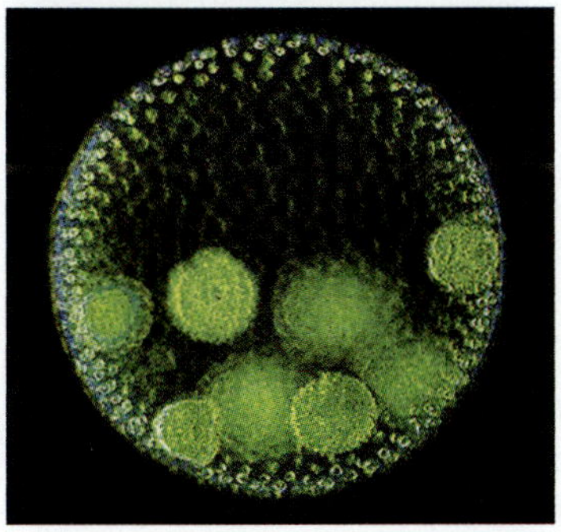

© Birke/AgeFotostock

C. Multicellular green alga

Marevision/Getty Images

FIGURE 27.16 Green algae. **(A)** A single-celled green alga, *Acetabularia,* which grows in marine environments. Each individual in the cluster is a large, multinucleate cell with a rootlike base, stalk, and cap. **(B)** A colonial green alga, *Volvox.* Each green dot in the spherical wall of the colony is a potentially independent, flagellated cell. Daughter colonies can be seen within the parent colony. **(C)** A multicellular green alga, *Ulva,* common to shallow seas around the world.

The Unikonts Include Protists That Are Closely Related to Animals and Fungi

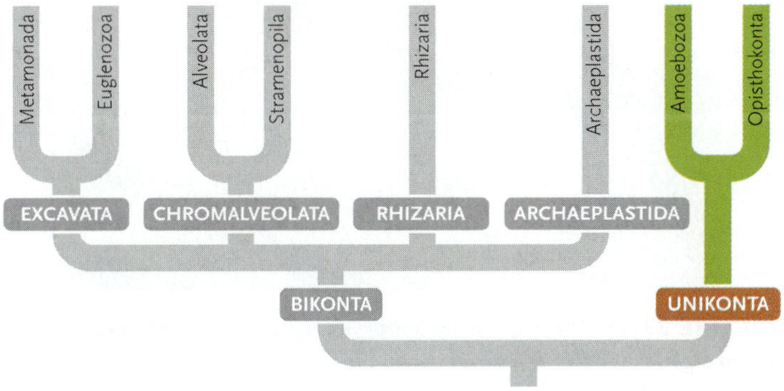

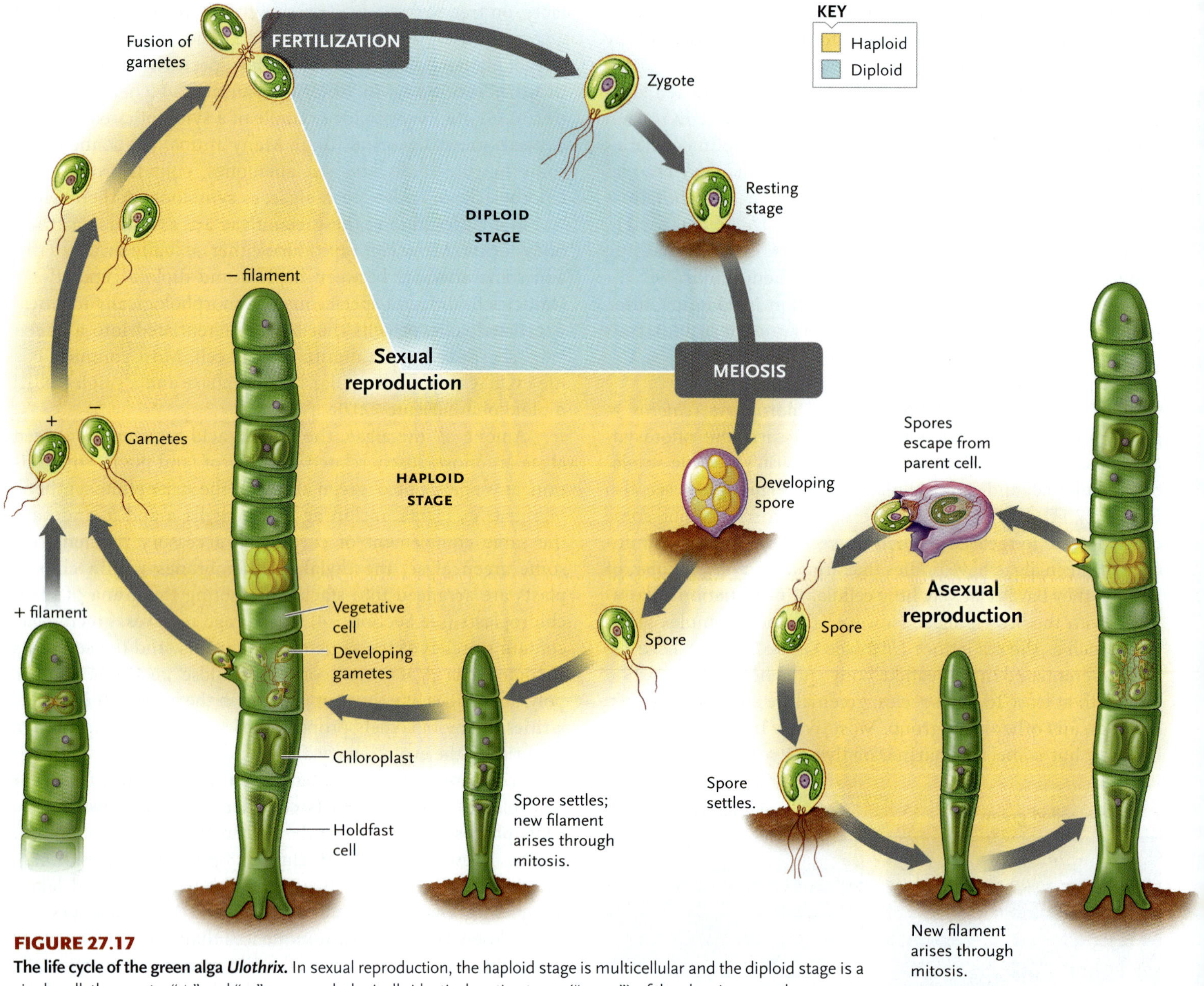

FIGURE 27.17

The life cycle of the green alga *Ulothrix*. In sexual reproduction, the haploid stage is multicellular and the diploid stage is a single cell, the zygote. "+" and "−" are morphologically identical mating types ("sexes") of the alga. In asexual reproduction, all stages are haploid.

FIGURE 27.18 **The charophyte *Spirogyra*, representative of a group of green algae that may have given rise to the plant kingdom.**

The **Unikonta—unikonts**—include amoebas with unsupported pseudopods, protists closely related to animals or fungi, and animals and fungi themselves. The unikonts consist of the **Amoebozoa** and the **Opisthokonta.**

AMOEBOZOA The **Amoebozoa** include most of the amoebas (others are in the Opisthokonta and the Rhizaria) as well as the cellular and plasmodial slime molds. All members of this group use pseudopods for locomotion and feeding for all or part of their life cycles.

Amoebas. Amoebas of the Amoebozoa are single-celled organisms that are abundant in marine and freshwater environments and in the soil. They use pseudopods for locomotion and feeding. The pseudopods extend and retract at any point on their body surface and are unsupported by any internal cellular organization. This type of pseudopod—called a *lobose* ("lobe-like") *pseudopod*—distinguishes these amoebas from those in the Rhizaria, which have stiff, supported pseudopods. As a result of their pseudopod activity, and the ability to flatten or round up, these amoebas have no fixed body shape. A number of species are parasites, but most species feed on bacteria, other protists, and bits of organic matter. The ingested matter is enclosed in food vacuoles and digested by enzymes secreted into the vacuoles. Any undigested matter is expelled to the outside by fusion of the vacuole with the plasma membrane. Their reproduction is entirely asexual, through mitotic divisions.

The most-studied amoeba of the amoebozoans is *Amoeba proteus* **(Figure 27.19)**. Its natural habitat is in freshwater ponds and streams. Another member, *Acanthamoeba,* which lives in the soil, is widely used as a source of actin and myosin for scientific studies of amoeboid motion and cytoplasmic streaming.

The parasitic amoebas include some 45 species that infect the human digestive tract, one in the mouth and the rest in the intestine. One of the intestinal parasites, *Entamoeba histolytica,* causes amoebic dysentery. Cysts of this amoeba contaminate water supplies and soil in regions with inadequate sewage treatment. When ingested, a cyst breaks open to release an amoeba that feeds and divides rapidly in the digestive tract. Enzymes released by the amoebas destroy cells lining the intestine, producing the ulcerations, painful cramps, and debilitating diarrhea characteristic of the disease. Amoebic dysentery afflicts millions of people worldwide; in less-developed countries, it is a leading cause of death among infants and small children. Other parasitic amoebas cause less severe digestive upsets.

Slime Molds. **Slime molds** are heterotrophic protists that, at some stage of their life cycle, exist as individuals that move by amoeboid motion but the remainder of the time exist in more complex forms. They live on moist, rotting plant material such as decaying leaves and bark. The cells engulf particles of dead organic matter, and also bacteria, yeasts, and other microorganisms, and digest them internally. At one stage of their life cycles, they differentiate into a funguslike, stalked structure called a **fruiting body,** which forms spores by either asexual or sexual reproduction. Some species are brightly colored in hues of yellow, green, red, orange, brown, violet, or blue. The two major evolutionary lineages of slime molds, the *cellular slime molds* and the *plasmodial slime molds,* differ in cellular organization.

Cellular slime molds exist primarily as individual cells, either separately or as a coordinated mass. Among the 70 or so species of cellular slime molds, *Dictyostelium discoideum* is best known. **Figure 27.20** shows asexual reproduction (steps 1–5) and sexual reproduction (steps 6–7) in *D. discoideum.*

Plasmodial slime molds exist primarily as a large composite mass, the **plasmodium,** in which individual nuclei are suspended in a common cytoplasm that is surrounded by a single plasma membrane. (Do not confuse this with *Plasmodium,* the genus of apicomplexans that causes malaria.) There are about 500 known species of plasmodial slime molds. The main phase of the life cycle, the plasmodium (see Figure 27.1A), flows and feeds as a single huge amoeba—a single cell that contains thousands to millions or even billions of diploid nuclei surrounded by a single plasma membrane. Typically, a plasmodium, which may range in size from a few centimeters to more than a meter in diameter, moves in thick, branching strands connected by thin sheets. The movements occur by cytoplasmic streaming, driven by actin microfilaments and myosin (see Section 5.3).

At some point, often in response to unfavorable environmental conditions, fruiting bodies form at sites on the plasmodium. At the tips of the fruiting bodies, nuclei become enclosed in separate cells, each surrounded by its own plasma membrane and cell wall. Depending on the species, either chitin or cellulose may reinforce the walls. These cells undergo meiosis, forming haploid, resistant spores that are released from the fruiting bodies and carried about by water or wind. If they reach a favorable environment, the spores germinate to form flagellated or unflagellated gametes, depending on the species, that fuse to form a diploid zygote. The zygote nucleus then divides repeatedly without an accompanying division of the cytoplasm, forming many diploid nuclei suspended in the common cytoplasm of a new plasmodium.

Both the cellular and plasmodial slime molds, particularly *Dictyostelium* (cellular) and *Physarum* (plasmodial; see Figure 27.1A), have been of great interest to scientists because of their ability to differentiate into fruiting bodies with stalks and spore-bearing structures. This differentiation is much simpler than the complex developmental pathways of other eukaryotes, providing a unique opportunity to study cell differentiation at its most fundamental level. One such study, examining the role of cAMP (cyclic AMP) in differentiation, is described in *Insights from the Molecular Revolution.*

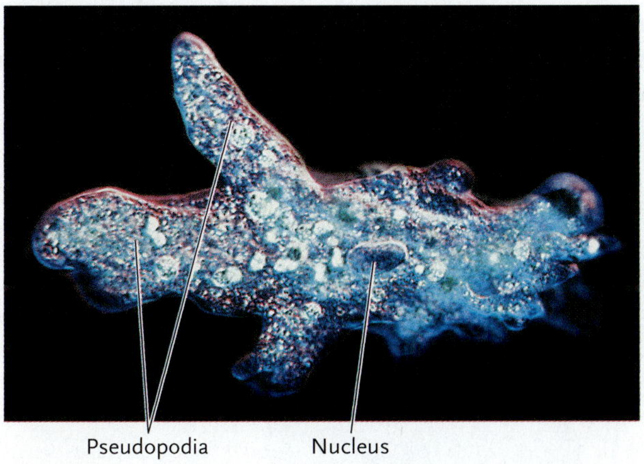

Michael Abbey/Visuals Unlimited, Inc.

FIGURE 27.19 ***Amoeba proteus*** **of the Amoebozoa, perhaps the most familiar protist.**

OPISTHOKONTA The opisthokonts (*opisthen* = behind or posterior) are a broad group of eukaryotes that includes the protist groups, the choanoflagellates and the nucleariids. If present at some stage of the life cycle, the single flagellum is located at the posterior.

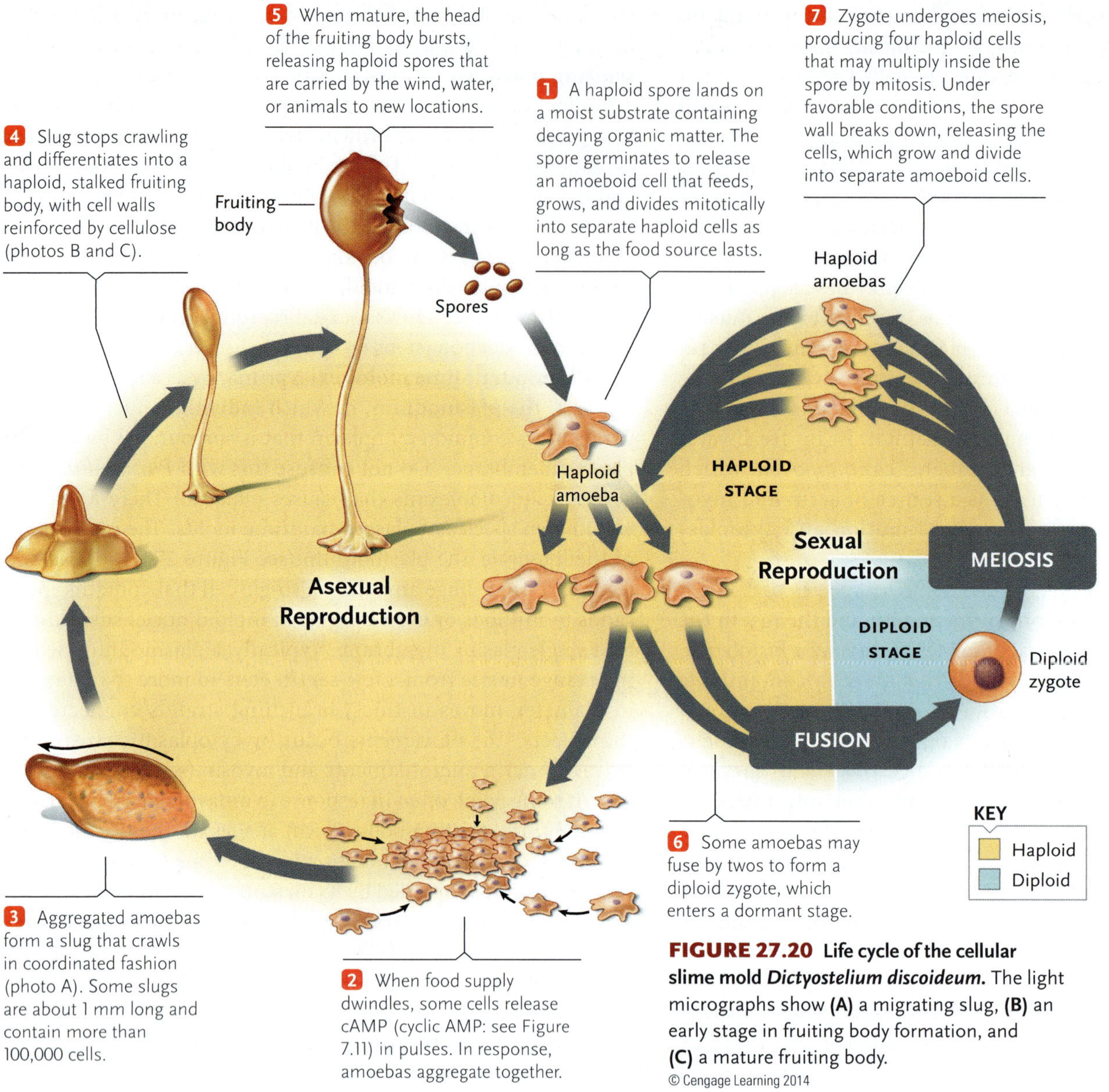

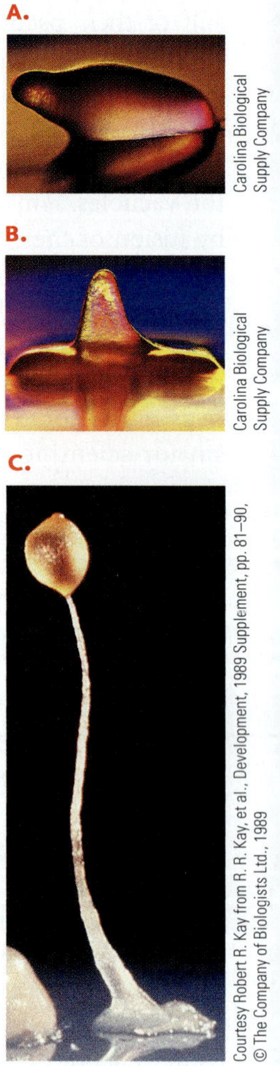

FIGURE 27.20 Life cycle of the cellular slime mold *Dictyostelium discoideum*. The light micrographs show **(A)** a migrating slug, **(B)** an early stage in fruiting body formation, and **(C)** a mature fruiting body.

Choanoflagellata (*choanos* = collar)—**choanoflagellates**—are named for a collar of closely packed microvilli that surrounds the single flagellum by which these protists move and take in food **(Figure 27.21).** The collar resembles an upside-down lampshade. There are about 150 species of choanoflagellates. They live in fresh and marine waters. Some species are mobile, with the flagellum pushing the cells along, as is the case with animal sperm, in contrast to most flagellates, which are pulled by their flagella. Most choanoflagellates, though, are *sessile;* that is, attached via a stalk to a surface. A number of species are colonial with a cluster of cells on a single stalk.

Choanoflagellates have the same basic structure as choanocytes (collar cells) of sponges, and they are similar to collared cells that act as excretory organs in organisms such as flatworms and rotifers. These morphological similarities, as well as molecular sequence comparison data, indicate that an ancestor of a present-day choanoflagellate is likely to have been the ancestor of animals.

Nucleariidae—**nucleariids**—are heterotrophic, predominantly spherical amoebae with radiating, fine pseudopods that are not supported by microtubules. Molecular sequence data show that nucleariids are more closely related to fungi than to animals (see Chapter 30).

FIGURE 27.21 A choanoflagellate.

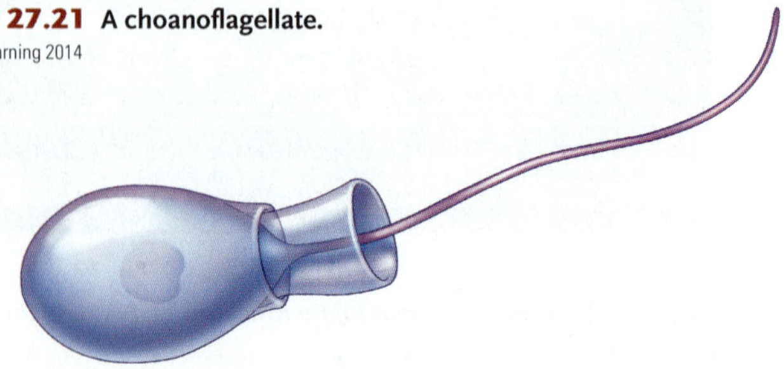

Getting the Slime Mold Act Together: What molecules regulate development of *Dictyostelium*?

Development of differentiated structures can be followed at its simplest level in slime molds. In the cellular slime mold *Dictyostelium discoideum,* the aggregation of individual cells leading to differentiation begins when unfavorable living conditions induce some cells to secrete cyclic AMP (cAMP). Other *Dictyostelium* cells move toward the regions of highest cAMP concentration and aggregate into the slug stage. Further pulses of cAMP trigger differentiation into a stalk and spores.

The extracellular cAMP binds a cell surface G-protein–coupled receptor, activating it. This leads to activation of adenylyl cyclase within the cell, which generates cAMP. This intracellular cAMP activates a cAMP-dependent protein kinase (PKA; see Section 7.4). The PKA, which is active only when cAMP is present, adds phosphate groups to target proteins in the cells. The target proteins, activated or deactivated by addition of the phosphate groups, trigger cellular developmental processes that lead to slug formation and differentiation of the stalk and spores.

Research Question

Which is more important to the process of development in *Dictyostelium,* cAMP or PKA?

Experiments

Adam Kuspa and his graduate student Bin Wang at Baylor College of Medicine, Houston, Texas, set out to answer this question. Their experimental approach is shown in the **Figure.**

Results

The researchers found that the cAMP-deficient cells with the modified PKA gene aggregated into slugs when the cultures were deprived of food (in this case, bacteria). The slugs then differentiated normally into fruiting bodies. No cAMP could be detected in the cells.

Conclusions

The results indicated that activated PKA by itself can trigger all the steps in the developmental pathway from individual cells to slugs. Thus the requirement for cAMP in normal slime mold development is primarily or exclusively to stimulate the PKA. And because development can proceed with active PKA alone, this protein kinase is probably more central to the growth and differentiation processes of the slime mold than is cAMP.

These results are of significance because both cAMP and cAMP-dependent protein kinases are also active in animal development and intercellular signaling, including that of humans and other mammals.

THINK LIKE A SCIENTIST In triggering the developmental response described, extracellular cAMP acts as a signal molecule that, by binding to and activating a G-protein–coupled receptor in the plasma membrane, triggers the production of the intracellular cAMP molecules which activate the protein kinase. Review Chapter 7 and outline the molecular steps involved in this sequence of events.

Source: B. Wang and A. Kuspa. 1997. *Dictyostelium* development in the absence of cAMP. *Science* 277:251–254.

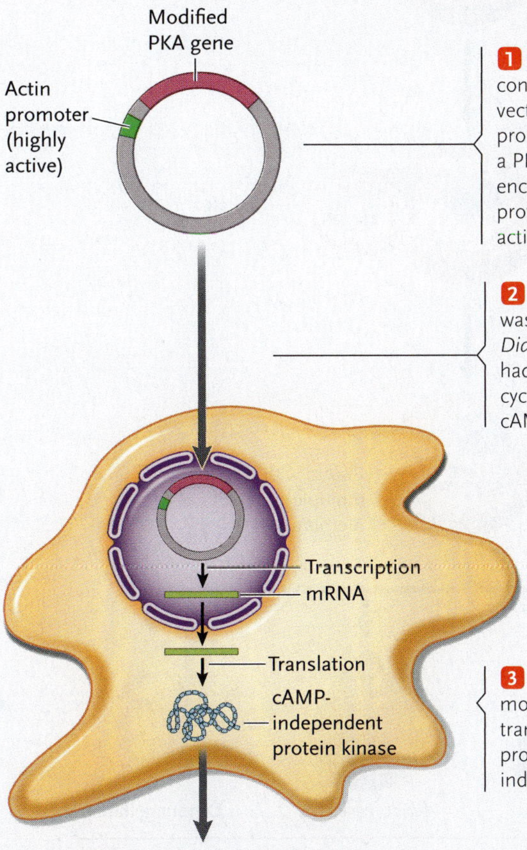

1 The researchers constructed an expression vector with a highly active promoter for transcribing a PKA gene modified to encode a cAMP-independent protein kinase—an enzyme active in the absence of cAMP.

2 Expression plasmid was introduced into *Dictyostelium* cells, which had a mutated adenylyl cyclase gene, so no cAMP could be made.

3 Transcription of the modified PKA gene and translation of the mRNA produced the cAMP-independent PKA.

4 The researchers tested whether cells with the modified gene aggregated into slugs when deprived of food.

In Several Protist Groups, Plastids Evolved from Endosymbionts

We have encountered chloroplasts in a number of eukaryotic organisms in this chapter: red algae, green algae, land plants, euglenids, dinoflagellates, stramenopiles, and chlorarachniophytes. How did these chloroplasts evolve?

In Section 25.3 we discussed the endosymbiotic theory for the origin of eukaryotes. In brief, an anaerobic prokaryote ingested an aerobic prokaryote, which survived as an endosymbi-

FIGURE 27.22 The origin and distribution of plastids among the eukaryotes by primary and secondary endosymbiosis.

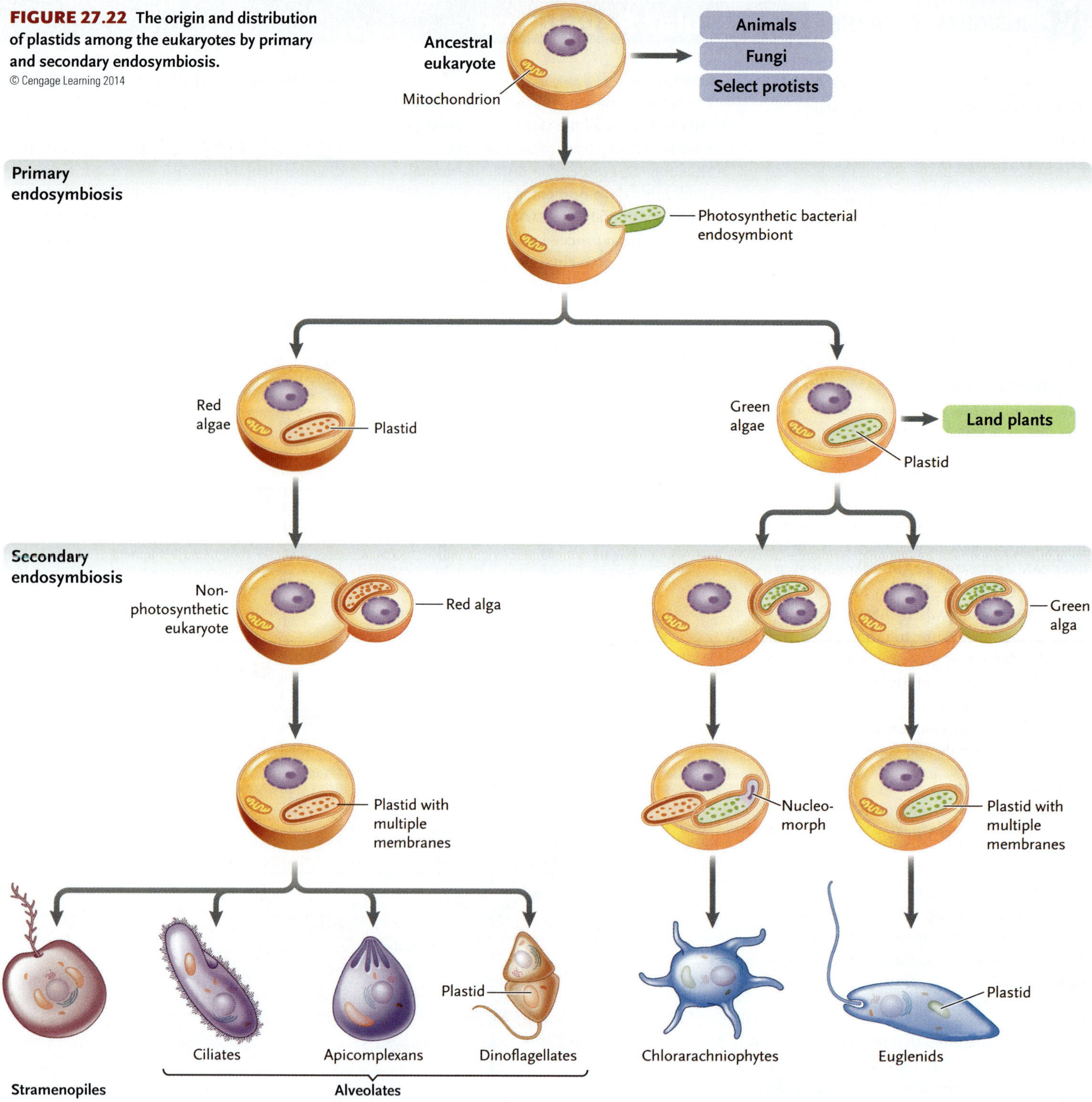

ont (see Figure 25.5). Over many generations, the endosymbiont became an organelle, the mitochondrion, which was incapable of free living, and the result was a true eukaryotic cell. Cells of animals, fungi, and some protists derive from this ancestral eukaryote. The addition of plastids (the general term for chloroplasts and related organelles) through further endosymbiotic events produced the cells of all photosynthetic eukaryotes, including land plants, algae, and some other protists.

Figure 27.22 presents a model for the origin of plastids in eukaryotes through two major endosymbiosis events. First, in a single **primary endosymbiosis** event perhaps 600 million years ago, a eukaryotic cell engulfed a photosynthetic cyanobacterium (a photosynthetic prokaryote, remember). In some such cells, the cyanobacterium was not digested, but instead formed a symbiotic relationship with the engulfing host cell—it became an endosymbiont. Over time the symbiont lost genes no longer required for independent existence, and most of the remaining genes migrated from the prokaryotic genome to the host's nuclear genome. The symbiont had become an organelle—a chloroplast. All plastids subsequently evolved from this original chloroplast. Evidence for the single origin of plastids comes from a variety of sequence comparisons, including recent sequencing of the genomes of key protists: a red alga and a diatom.

The first photosynthesizing eukaryote was essentially an ancestral single-celled alga. The chloroplasts of the Archaeplastida—the red algae, green algae, and land plants—result from evolutionary divergence of this organism. Their chloroplasts, which originate from primary endosymbiosis, have two membranes, one from the plasma membrane of the engulfing eukaryote and the other from the plasma membrane of the cyanobacterium.

At least three **secondary endosymbiosis** events led to the plastids in other protists (see Figure 27.22). In each case, a nonphotosynthetic eukaryote engulfed a photosynthetic eukaryote, and new evolutionary lineages were produced. In one of these events, a red alga ancestor was engulfed and became an endosymbiont. In models accepted by a number of scientists, the transfer of functions that occurred over evolutionary time led to the chloroplasts of the stramenopiles, ciliates, and dinoflagellates. And from the same photosynthetic ancestor, loss of chloroplast functions occurred in the lineage of the Apicomplexa, which have a remnant plastid. In an independent event, a nonphotosynthetic eukaryote engulfed a green alga ancestor. Subsequent evolution in this case produced the euglenids. In a different event, a similar endosymbiosis involving a green alga led to the chlorarachniophytes. In these protists, the chloroplast is contained still within the remnants of the original symbiont cell, with a vestige of the original nucleus (the nucleomorph) also present.

Note that secondary endosymbiosis has produced plastids with additional membranes acquired from the new host, or series of hosts. For example, euglenids have plastids with three membranes, while chlorarachniophytes have plastids with four membranes (see Figure 27.22). Sequencing the genomes of the chlorarachniophyte's nucleus, chloroplast, and vestigial nucleus is providing interesting information about the early endosymbiosis event that generated these organisms.

In sum, the protists are a highly diverse and ecologically important group of organisms. Their complex evolutionary relationships, which have long been the subject of contention, are now being revised as new information is discovered, including more complete genome sequences. A deeper understanding of protists is also contributing to a better understanding of their recent descendants, the fungi, plants, and animals. We turn to these descendants in the next four chapters, beginning with the fungi.

STUDY BREAK 27.2

1. **What is the evidence that the Excavates, which lack mitochondria, derive from ancestors that had mitochondria rather than from ancestors that were in lineages that never contained mitochondria?**
2. **In primary endosymbiosis, a nonphotosynthetic eukaryotic cell engulfed a photosynthetic cyanobacterium. How many membranes surround the chloroplast that evolved?**

UNANSWERED QUESTIONS

What was the first eukaryote?

Since prokaryotes precede eukaryotes in the fossil record, we assume that eukaryotes arose after prokaryotes. The first eukaryote would have been some sort of protist—a single-celled organism with a nucleus and some rudimentary organelles, perhaps even a half-tamed mitochondrion. One approach to identifying which of the surviving protists is the most ancient has been to infer evolutionary trees from gene sequence data. To determine the earliest branching eukaryote, these trees need to include the prokaryotes. But herein lies the problem—prokaryotes are very distant, evolutionarily speaking, from even the simplest eukaryotes, and the mathematical models used to construct evolutionary trees are not yet up to the job. Initially, these models suggested that some protist parasites, like the excavates *Giardia* and *Trichomonas,* might be the most ancient eukaryotes, and this idea fit nicely with the fact that these protists lacked mitochondria. Indeed, for a time it was thought that the excavates might actually have diverged from the eukaryotic branch of life before the establishment of mitochondria. However, we know now that *Giardia* and *Trichomonas* did have mitochondria initially. The latest research shows that they even have a tiny relic of the mitochondrion, though exactly what it does in these oxygen-shunning parasites remains to be figured out. Thus, trees depicting *Giardia* and *Trichomonas* at the base of the great expansion of eukaryotic life must be viewed with some caution—these protists might be the surviving representatives of the earliest cells with a nucleus, but they might not be. We simply need better methods for identifying just what the first eukaryotes were like.

How many times did plastids arise by endosymbiosis?

For many years researchers thought that the green algae, plants, and red algae were the only organisms to have primary endosymbiosis-derived plastids. However, a second, independent primary endosymbiosis has been recently discovered in which a shelled amoeba has captured and partially domesticated a cyanobacterium. This organism, known as *Paulinella,* is a vital window into the process by which autotrophic eukaryotes first arose some 600 million years ago. *Paulinella* has tamed the cyanobacterium sufficiently to have it divide and segregate in coordination with host cell division, but the endosymbiont is still very much a cyanobacterium and has undergone little of the modification and streamlining we see in the red or green algal plastids.

After a primary endosymbiosis was established, the second chapter in plastid acquisition could take place. Secondary endosymbiosis involves a eukaryotic host engulfing and retaining a eukaryotic alga. Essentially, secondary endosymbiosis can convert a heterotrophic organism into an autotroph by hijacking a photosynthetic cell and putting it to work as a solar-powered food factory. Secondary endosymbiosis results in plastids with three or four membranes, and we know that it occurred at least three times—once for the euglenids, once for the chlorarachniophytes, and once for the chromalveolates (a grouping of stramenopiles and alveolates). We can even tell what kind of endosymbiont was involved by the biochemistry and genetic makeup of the plastid: a green alga for euglenids and chlorarachniophytes, and a red alga for chromalveolates. The number of secondary endosymbioses is hotly debated, largely because not all protistologists support the existence of chromalveolates. Some contend that there were multiple, independent enslavements of different red algae to produce the dinoflagellates, stramenopiles, and apicomplexans. Understanding these events is crucial to confirming or refuting the proposed chromalveolate "supergroup."

A nice example of secondary endosymbiosis-in-action was recently discovered by Japanese scientists who found a flagellate, *Hatena*, with a green algal endosymbiont. *Hatena* has not yet assumed control of endosymbiont division and has to get new symbionts each time it divides, so it appears to be at a very early stage in establishing a relationship. We also want to know how secondary endosymbioses proceed because they have been a major driver in eukaryotic evolution. The stramenopiles, for instance, are the most important ocean phytoplankton and are key to ocean productivity and global carbon cycling. Knowing exactly how they got to be autotrophs in the first place is fundamental to understanding the world we live in.

THINK LIKE A SCIENTIST

1. What does the research examining the history of the earliest eukaryotes and endosymbioses reveal about evolution and the process of science?
2. Given that parasitic species are often considered to have free-living ancestors, and therefore are more evolutionarily recent than a closely related free-living form, is it possible that prokaryotes evolved from a single-celled eukaryote?

Courtesy of Geoff McFadden

Geoff McFadden is a professor of botany at the University of Melbourne. He studies the early evolution of eukaryotes, especially the origin and evolution of plastids and mitochondria. You can learn more about McFadden's research by visiting http://homepage.mac.com/fad1/McFaddenLab.html.

REVIEW KEY CONCEPTS

To access the course materials and companion resources for this text, please visit www.cengagebrain.com.

27.1 What Is a Protist?

- Protists are eukaryotes that differ from fungi in having motile stages in their life cycles and distinct cell wall molecules. Unlike plants, they lack true roots, stems, and leaves. Unlike animals, protists lack collagen, nerve cells, and an internal digestive tract, and they lack complex developmental stages (Figures 27.1 and 27.2).
- Protists are aerobic organisms that live as autotrophs or heterotrophs, or by a combination of both nutritional modes. Some are parasites or symbionts living in or among the cells of other organisms.
- Protists live in aquatic or moist terrestrial habitats, or as parasites within animals as single-celled, colonial, or multicellular organisms, and range in size from microscopic to some of Earth's largest organisms.
- Reproduction may be asexual by mitotic cell divisions, or sexual by meiosis and union of gametes in fertilization.
- Many protists have specialized cell structures including contractile vacuoles, food vacuoles, eyespots, and a pellicle, cell wall, or shell. Most are able to move by means of flagella, cilia, or pseudopodia (Figure 27.3).

27.2 The Protist Groups

- The most ancient bifurcation separates eukaryotes into the Unikonta and Bikonta. Protist groups are found in both the unikonts and bikonts (Figure 27.2).
- The Excavata includes the Metamonada, protists with unique flagella, and the Euglenozoa, protists with modified mitochondria.
- Metamonads consist of diplomonads and parabasalids, which have flagellated, single cells that lack mitochondria (Figure 27.4).
- Euglenozoans are almost all single-celled, autotrophic or heterotrophic (some are both), motile protists that swim using flagella. The free-living, photosynthetic forms—the euglenids—typically have complex cytoplasmic structures, including eyespots. The parasitic, nonphotosynthetic forms—the kinetoplastids—have characteristic mitochondrial structures (Figures 27.5 and 27.6).
- The Chromalveolata, protists with heterogeneous forms and life styles, includes the Alveolata and the Stramenopila.
- The alveolates include the ciliates, dinoflagellates, and apicomplexans. The ciliates swim using cilia and have complex cytoplasmic structures and two types of nuclei, the micronucleus and macronucleus. The dinoflagellates swim using flagella and are primarily marine organisms; some are photosynthetic. The apicomplexans are nonmotile parasites of animals (Figures 27.7 and 27.8).
- The stramenopiles include the funguslike groups, the Bacillariophyta (diatoms), Chrysophyta (golden algae), and Phaeophyta (brown algae). For most stramenopiles, flagella occur only on reproductive cells. Many Oomycota grow as masses of microscopic hyphal filaments and secrete enzymes that digest organic matter in their surroundings. Diatoms are single-celled organisms covered by a glassy silica shell; golden algae are colonial forms; brown algae are primarily multicellular marine forms that include large seaweeds with extensive cell differentiation (Figures 27.9–27.13).
- Rhizaria, amoebas with filamentous pseudopods supported by internal cellular structures, include the Radiolaria, the Foraminifera, and the Chlorarachniophyta. Many rhizarians produce hard outer shells.
- Radiolaria (radiolarians) are primarily marine organisms that secrete a glassy internal skeleton. They feed by engulfing prey that adhere to their axopods. Foraminifera (forams) are marine, single-celled organisms that form chambered, spiral shells containing calcium. They engulf prey that adhere to the strands of cytoplasm extending from their shells. Chlorarachniophytes engulf food using their pseudopodia (Figure 27.14).
- The Archaeplastida include the Rhodophyta (red algae) and Chlorophyta (green algae), as well as the land plants.
- Red algae are typically multicellular, primarily photosynthetic organisms of marine environments, with plantlike bodies composed of interwoven filaments. They have complex life cycles including alternation of generations, with no flagellated cells at any stage (Figure 27.15).
- Green algae are single-celled, colonial, and multicellular species that live primarily in freshwater habitats and carry out photosynthesis by mechanisms similar to those of plants; all produce flagellated gametes (Figures 27.16–27.18).

- The Unikonta includes the Amoebozoa, protists with unsupported pseudopods, and Opisthokonta, protists closely related to animals and fungi, and animals and fungi themselves.
- Amoebozoans include most amoebas and two heterotrophic slime molds, cellular (which move as individual cells) and plasmodial (which move as large masses of nuclei sharing a common cytoplasm). The amoebas in this group are heterotrophs abundant in marine and freshwater environments and in the soil. They move by extending pseudopodia (Figures 27.19 and 27.20).
- The opisthokonts are a broad group of eukaryotes that includes the choanoflagellates and the nucleariids. Choanoflagellates are characterized by a collar of microvilli surrounding a single flagellum. An ancestor of a present-day choanoflagellate is considered likely to have been the ancestor of animals (Figure 27.21). Nucleariids are spherical amoeba with fine pseudopods. They are more closely related to fungi than to animals.
- Several groups of protists, as well as land plants, contain chloroplasts. Present-day chloroplasts and other plastids result from endosymbiosis events that took place millions of years ago: in a primary endosymbiosis event, a eukaryotic cell engulfed a cyanobacterium, which became an endosymbiont. Over time, the symbiont became an organelle, the chloroplast. This first photosynthesizing organism was a green alga. Evolutionary divergence produced the red algae, green algae, and land plants. By secondary endosymbiosis, in which a nonphotosynthetic eukaryote engulfed a photosynthetic eukaryote, the various photosynthetic protists were produced (Figure 27.22).

Animation: Amoeboid motion

Animation: Green alga life cycle

Animation: Ciliate conjugation

Animation: Body plan of *Euglena*

Animation: Apicomplexan life cycle

Animation: Red alga life cycle

Animation: Cellular slime mold life cycle

UNDERSTAND & APPLY

Test Your Knowledge

1. Which of the following is a characteristic of protists that is also found in at least one other group?
   a. division by binary fission
   b. multicellular structures
   c. complex developmental stages
   d. peptidoglycan cell walls
   e. organelles and reproduction by meiosis/mitosis
2. Which of the following is *not* found among the protist groups?
   a. life cycles
   b. contractile vacuoles
   c. pellicles
   d. collagen
   e. pseudopodia
3. A member of this group can cause urinary infections and be spread by sexual intercourse. The group is characterized by a flagellum buried in a fold of the cytoplasm that allows the organisms to move through thick and viscous fluids of humans. The group is:
   a. Ciliophora.
   b. Apicomplexa.
   c. Amoebozoa.
   d. Parabasala.
   e. Diplomonadida.
4. When *Paramecium* conjugate:
   a. cytoplasmic division produces four daughter cells, each having two micronuclei and two macronuclei.
   b. one haploid micronucleus in each cell remains intact; the other three degenerate. The micronucleus of each cell divides once, producing two nuclei, and each cell exchanges one nucleus with the other cell. In each partner the two micronuclei fuse, forming a diploid zygote micronucleus in each cell.
   c. the micronucleus of each of the disengaged partners divides meiotically. Macronuclei divide again in each cell, and the original micronucleus breaks down. Each cell has two haploid micronuclei; one of the macronuclei develops into a micronucleus.
   d. the mating cells join together at opposite sites of their oral depression.
   e. the micronucleus in each cell undergoes mitosis. When mitosis is complete, there are four diploid macronuclei; the micronucleus then breaks down.
5. The group Diplomonadida is characterized by:
   a. a mouthlike gullet and hairlike surface. *Paramecium* is an example.
   b. flagella and a lack of mitochondria. *Giardia* is an example.
   c. nonmotility, parasitism, and sporelike infective stages. *Toxoplasma* is an example.
   d. switching between autotrophic and heterotrophic lifestyles. *Euglena* is an example.
   e. large protein deposits and movement by two flagella, which are part of an undulating membrane. *Trypanosoma* is an example.
6. The greatest contributors to protist fossil deposits, and probable source of oil in many oil deposits, are:
   a. Oomycota.
   b. Chrysophyta.
   c. Bacillariophyta.
   d. Sporophyta.
   e. Alveolata.
7. The group with the distinguishing characteristic of gas-filled bladders and a cell wall composed of alginic acid, which is used by humans to thicken such diverse products as ice cream and floor polish, is:
   a. Chrysophyta.
   b. Phaeophyta.
   c. Oomycota.
   d. Bacillariophyta.
   e. none of the preceding.
8. *Plasmodium* is transmitted to humans by the bite of a mosquito *(Anopheles)*. Its life cycle is characterized by spores, gametes, and cysts that can "hide" in human cells. This infective protist belongs to the group:
   a. Apicomplexa.
   b. Archaeplastida.
   c. Dinoflagellata.
   d. Oomycota.
   e. Ciliophora.
9. Tripping on a rotten log, a hunter notices a mass resembling mucus that appears to be moving slowly toward what appear to be brightly colored fruiting bodies. The organisms in the mass are:
   a. amoebas in the group Rhizaria.
   b. slime molds.
   c. red algae.
   d. green algae.
   e. charophytes.

10. Endosymbioses that lead to the evolution of euglenoids and, separately, the evolution of chlorarachniophytes were the result of the combining of:
   a. two ancestral nonphotosynthetic prokaryotes.
   b. two ancestral photosynthetic prokaryotes.
   c. a nonphotosynthetic eukaryote with a photosynthetic eukaryote.
   d. a photosynthetic prokaryote with a nonphotosynthetic eukaryote.
   e. mitochondria with an already established plastid.

Discuss the Concepts

1. You decide to vacation in a developing country where sanitation practices and standards of personal hygiene are inadequate. Considering the information about protists covered in this chapter, would you consider it safe to drink water in that country? What treatments could make the water safe to drink? What kinds of foods might be best avoided? What kinds of preparation might make foods safe to eat?
2. The overreproduction of dinoflagellates, which produces red tides, is sometimes caused by fertilizer runoff into coastal waters. The red tides kill countless aquatic species, birds, and other wildlife. Would you consider drastic cutbacks in the use of fertilizers as a means to lessen the red tides? Why?

Design an Experiment

Design an experiment to demonstrate whether the flagellated protist *Euglena* is phototropic, that is, is attracted to and moves toward light. Also propose a follow-up experiment (on the assumption of a positive result) to determine the wavelength range and light intensity range sufficient to cause phototropic movement.

Interpret the Data

Marine diatoms are a crucial part of marine productivity because they are food for many organisms. Their productivity is dependent on how well they photosynthesize. The **Figure** shows the response of photosynthesis in the marine diatom *Skeletonema costatum* to light.

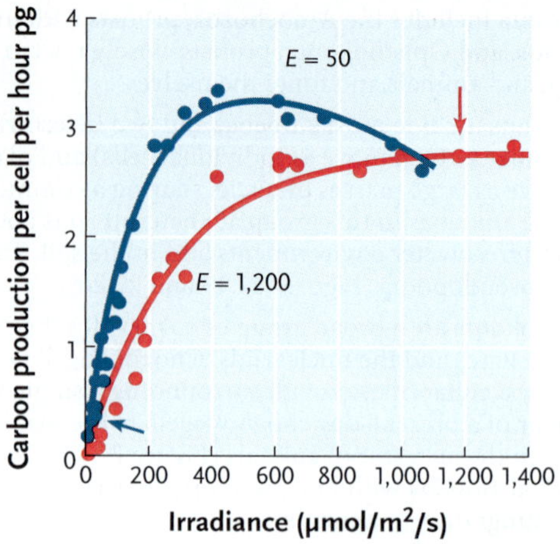

T. Anning et al. 2000. Photoacclimation in the marine diatom *Skeletonema costatum. Limnology and Oceanography* 45(8):1807–1817. Reprinted by permission of the Association for the Sciences of Limnology and Oceanography.

Photosynthesis is expressed on the *y* axis as carbon production in picograms (pg) per cell per hour, and light is expressed on the *x* axis as the number of photons of irradiance that can activate photosynthesis (0 mmol/m^2/s would be complete darkness, and 2,000 mmol/m^2/s would be full sunlight at midday). The blue symbols show a cell culture grown for five days at 50 mmol/m^2/s of irradiance (the treatment irradiance), with an arrow showing the level of photosynthetic production at the end of this period. The red symbols show a different population of cells from the same culture grown for five days at 1,200 mmol/m^2/s of irradiance (the treatment irradiance), with an arrow showing the level of photosynthetic production. Both cultures were then exposed to all of the possible irradiance conditions for 30-minute periods.

1. In each light treatment, at what irradiance was the photosynthetic rate maximized? Was this the same rate as the treatment irradiance? Why or why not?
2. The response of photosynthesis to irradiance at low irradiance is linear. The rate of this linear increase is the same as the organism's photosynthesis efficiency with respect to light. Which treatment has higher photosynthesis efficiency?

Source: T. Anning et al. 2000. Photoacclimation in the marine diatom *Skeletonema costatum. Limnology and Oceanography* 45(8):1807–1817.

Apply Evolutionary Thinking

Use the Internet to research why studies of a molecular sensor, receptor tyrosine kinase (see Section 7.3), supports the hypothesis that a choanoflagellate type of protist is the ancestor of animals. Summarize your findings.

Ferns, familiar representatives of the ancient seedless plant lineage Pteridophyta.

28

Seedless Plants

STUDY OUTLINE

28.1 The Transition to Life on Land

28.2 Bryophytes: Nonvascular Land Plants

28.3 Seedless Vascular Plants

Why it matters . . . Ages ago, along the edges of the ancient supercontinent Laurentia, the only sound was the rhythmic crash of waves breaking in the distance. There were no birds or other animals, no plants with leaves rustling in the breeze. In the preceding eons, oxygen-producing photosynthetic cells had come into being and had gradually changed Earth's atmosphere. Solar radiation had converted much of the oxygen into a dense ozone layer—a shield against lethal doses of ultraviolet radiation, which had kept early organisms below the water's surface. Now, they could populate the land.

Cyanobacteria may have been the first to adapt to intertidal zones and then to spread into the freshwater environment of shallow, coastal streams. Later, green algae and fungi made the same journey. Some 700 to 800 mya (million years ago), green algae living near the edge of a pond or stream became the ancestors of modern plants. From molecular and morphological evidence we know that a series of adaptive shifts in certain freshwater green algae gave rise to the clade we recognize as the **Kingdom Plantae.** Today the plant kingdom—sometimes called the Viridiplantae, or "green plants"—encompasses at least 260,000 living species, organized in this textbook into 10 phyla. The vast majority of these extant species are plants in which embryos develop protected inside a seed during the reproductive part of the life cycle. Chapter 29 focuses on such seed plants, including the flowering plants and conifers that dominate natural land ecosystems.

The first land plants lacked seeds, however, and their lineages persist today in the seedless groups we survey in this chapter, including mosses, horsetails, and ferns **(Figure 28.1).** As seedless plants arose, a variety of adaptive changes were crucial for the transition to a fully terrestrial life. Eons of natural selection provided solutions to problems such as avoiding desiccation, physically supporting the plant body in air, obtaining nutrients from soil, and reproducing sexually in environments where water would not be available for dispersal of eggs and sperm. With time, plants

FIGURE 28.1 Representatives of seedless plants. (A) *Sphagnum* moss growing at the edge of a stream. Mosses evolved relatively soon after plants made the transition to land. **(B)** Fern grotto. **(C)** Horsetails.

evolved features that not only addressed these problems but also provided access to nearly every land environment. Those ecological opportunities opened the way for the dramatic radiation of plant species—and of animal species that eventually came to include our own.

All plants are multicellular. Most also are autotrophs, the primary producers that carry out photosynthesis, using solar energy, carbon dioxide, water, and dissolved minerals to manufacture their own food. Together with photosynthetic bacteria and "protists," plant tissues provide the nutritional foundation for nearly all communities of life.

28.1 The Transition to Life on Land

Land plants and green algae share several fundamental traits: they have cellulose in their cell walls, they store energy captured during photosynthesis as starch, and their light-absorbing pigments include both chlorophyll *a* and chlorophyll *b*. Like some other green algae described in Chapter 27, the lineage from which land plants evolved were adapted to freshwater habitats such as ponds and stream edges, and probably had invaded land by about 450 mya. There is widespread agreement that these algae belonged to the phylum Charophyta. Comparison of 129 gene sequences from 40 plant taxa points to the charophyte order Zygnematales as the sister clade—the closest relatives—of land plants. Modern representatives of this clade include filamentous green algae in the genus *Spirogyra* **(Figure 28.2).** A key next step in exploring this proposed relationship will be gathering more information on charophyte genomes, including one or more full genome sequences.

The fossil record is sketchy in pinpointing when the first truly terrestrial plants appeared. One reason for the dearth of information is that the earliest land plants lacked hard parts (such as wood) that fossilize well. A British and Arab research team working in the Middle East found fossilized tissue and spores from what appears to be a land plant in rocks dated to about 475 mya. If the remains indeed are from a plant, they represent the earliest known plant fossils. Even in more recent deposits, the most common finds of possible plant parts are microscopic bits and pieces. Obvious leaves, stems, roots, and reproductive parts seldom occur together, or if they do, it can be difficult to determine if the fossilized bits all belong to the same individual. Whole plants are extremely rare. Adding to the challenge, more than a few chemical and structural adaptations to life on land arose independently in several plant lineages. Consequently, a fossil may have some but not all the features of modern land plants, making it difficult to determine whether a given specimen was actually terrestrial, aquatic, or a transitional form. Despite these problems, plant scientists have been able to gain insight into a variety of innovations and overall trends in plant evolution.

Early Biochemical and Structural Adaptations Enhanced Plant Survival on Land

Because land environments pose very different challenges than aquatic environments, evolution in land plants produced a range of adaptations essential to survival on dry land. For instance, unlike algae in their aquatic habitats, plants living—and reproducing—on land had to have protection against drying out. The earliest land plants may have benefited from an inherited ability to make **sporopollenin,** a resistant polymer that surrounds the zygotes of modern charophytes. In land plants, capsules called **sporangia** ("spore chamber"; singular, *sporangium*) produce reproductive spores, and sporopollenin is a major component of the thick sporangium wall that protects spores from drying and other damage. Other reproductive adaptations in plants included multicellular chambers that protect developing gametes,

FIGURE 28.2 A filamentous green alga in the genus *Spirogyra*. The genus was named for the spiral arrangement of chloroplasts in the algal cells.

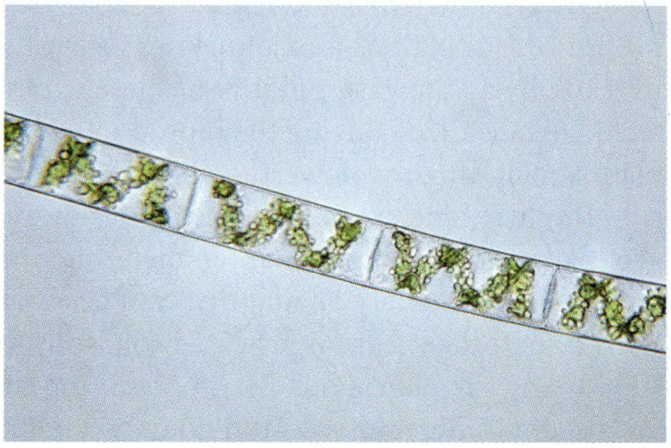

and a dependent, multicellular embryo that is sheltered inside tissues of a parent plant. Botanists use the term **embryophyte** (*phyton* = plant) as a synonym for land plants because all land plants produce a protected embryo during their reproductive cycle.

Some of the first land plants also evolved an outer waxy layer called a **cuticle,** which slows water loss, helping to prevent desiccation **(Figure 28.3A).** Another adaptation in most land plant lineages was the presence of **stomata** (singular, *stoma; stoma* = mouth), tiny openings in the cuticle-covered surfaces **(Figure 28.3B).** A stoma is formed by a pair of cells that can change shape and so open up or close a space between them, and stomata became the main route for plants to take up carbon dioxide and control water loss by evaporation. The next unit describes stomata more fully.

If you had been a botanist about 470 million years ago, you would have noted two general categories of land plants. One included **nonvascular plants**—for historical reasons also called **bryophytes**—which lacked well-developed internal tubes for transporting water and organic compounds such as sugars. Mosses are in this category. As far as we know, nonvascular plants are not monophyletic; that is, the various phyla described shortly do not share a recent ancestor. In fact, they are extremely diverse. The other general category encompassed **vascular plants,** more closely related lineages in which specialized transport tissues had arisen. One of these tissues, called **xylem,** contained stacked water-conducting cells called *tracheids,* which is why vascular plants collectively are also known as **tracheophytes.** Another vascular tissue, called **phloem,** is an adaptation for transporting sugars and other metabolic products.

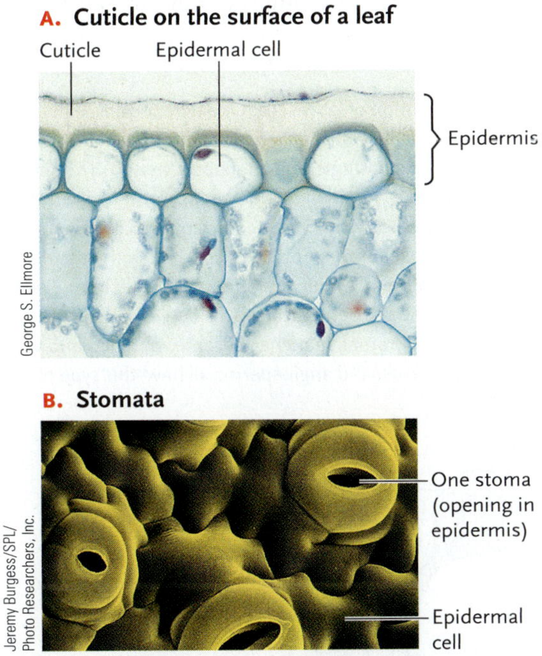

FIGURE 28.3 Land plant adaptations for limiting water loss. (A) A waxy cuticle, which covers the epidermis of land plants and helps reduce water loss. **(B)** Surface view of stomata in the epidermis (surface layer of cells) of a leaf. Stomata allow carbon dioxide to enter plant tissues and oxygen to leave while controlling loss of water.

In addition to vascular tissue, another adaptation that roughly correlated with the divergence of nonvascular and vascular plants was the capacity to synthesize **lignin,** a tough, rather inert polymer that strengthens the secondary walls of certain types of plant cells and thus helps vascular plants to grow taller and stay erect on land, giving photosynthetic tissues better access to sunlight. Vascular plants also have **apical meristems,** regions of unspecialized dividing cells near the tips of shoots and roots. Meristem tissue is the foundation for a vascular plant's extensively branching stem parts. Chapter 33 describes in detail how descendants of such unspecialized cells differentiate and form all mature plant tissues. As you'll read shortly, although mosses are classified as nonvascular, they have apical meristems and some other adaptive similarities to vascular plants.

Vascular Tissue Was an Adaptation for Transporting Substances within a Large Plant Body

The Latin *vas* means duct, and vascular plants have specialized tissues made up of cells arranged in lignified, tubelike structures that branch throughout the plant body, conducting water and solutes. As you've just read, the vascular tissue called xylem distributes water—including dissolved mineral ions—through plant parts. The other vascular tissue, phloem, distributes sugars manufactured during photosynthesis. Chapter 34 explains how tracheids and other specialized cells of xylem and phloem perform these key internal transport functions.

Most of the plants you are familiar with—flowering plants, conifers, ferns—are vascular plants. Supported by cells with lignin and serviced by a well-developed vascular system, the body of a plant can grow large. Extreme examples are the giant redwood trees of the northern California coast, some of which are more than 300 feet tall. By contrast, nonvascular plants lack lignified cells, and have very simple internal transport systems, or none at all. This is why modern nonvascular plants generally are small.

Shoot and Root Systems Were Adaptations for Support and Nutrition

A nonvascular plant has no true stems, structures that are fundamental adaptations for support of an erect plant body. In vascular plants, the evolution of sturdy stems became the basis of a large, aerial *shoot system.* This adaptive shift was correlated with the capacity to synthesize lignin.

The first unquestioned fossils of a vascular plant are specimens in the extinct genus *Cooksonia.* As **Figure 28.4A** shows, this plant had leafless "naked" stems with simple, forked branches of equal size. Sporangia formed at the branch tips and simple xylem wove through the stems. Other species formerly lumped into the genus *Cooksonia* were early vascular plants

A. Fossil of *Cooksonia*

B. Cross section of *Rhynia gwynne-vaughnii*

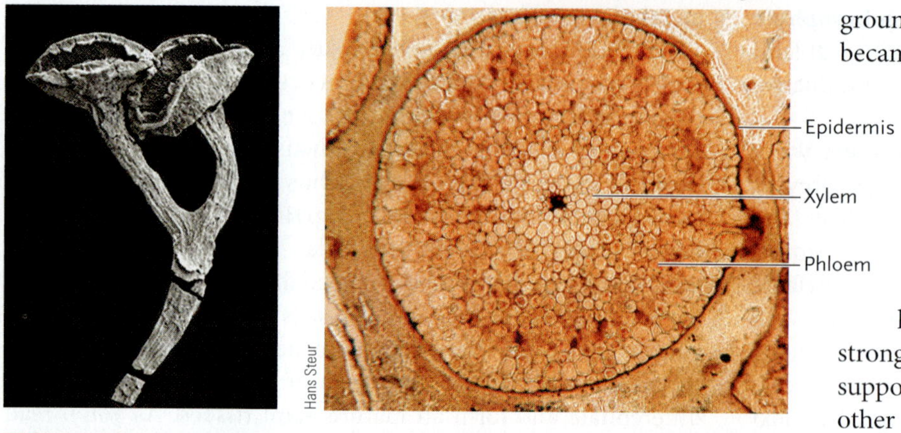

FIGURE 28.4 Fossils from early vascular plants. (A) *Cooksonia,* which dates to about 420 mya. *Cooksonia* stems lacked leaves and were probably less than 3 cm long. The cup-shaped structures at the top of the stems produced reproductive spores. **(B)** Cross section of the stem of a rhyniophyte, *Rhynia gwynne-vaughnii,* about 3 mm in diameter. This fossil was embedded in chert approximately 400 mya. Still visible in it are traces of the transport tissues xylem and phloem, along with other specialized tissues.

called *rhyniophytes* (because numerous specimens are from rock outcrops near Rhynie, Scotland). *Cooksonia* and rhyniophyte fossils all date to about 425 mya and show similar features, including vascular tissue **(Figure 28.4B).** Most likely none of these species had roots, which are organs specialized to anchor a vascular plant and take up water and minerals. Instead these ancestral forms probably were supported physically by a **rhizome**—a horizontal, modified stem that can penetrate a substrate and anchor the plant. Cells called *rhizoids,* which can help anchor a plant and take up water, may have extended into the soil.

UPRIGHT GROWTH: LIGNIFIED SHOOT SYSTEMS Above ground, the simple stems of early land plants gradually became more specialized, evolving into the **shoot systems** we see in all living vascular plants. Shoot systems have stems and leaves that arise from apical meristems and that absorb light energy from the sun and carbon dioxide from the air. Lignin became incorporated in the water-conducting tracheids of xylem. Its added mechanical strength in these and other types of cells provided vascular plants with adaptive advantages. For instance, a strong, internal scaffold composed of tracheids could support upright, branching stems bearing leaves and other photosynthetic structures—and so help increase the surface area for intercepting sunlight. Also, reproductive structures such as sporangia borne on aerial stems might serve as platforms for more efficient launching of spores from the parent plant.

ENHANCED PHOTOSYNTHESIS: LEAVES WITH STOMATA AND VEINS Structures we think of as "leaves" arose several times during plant evolution. Leaves are modifications of stems and fall into two general categories. *Microphylls* are narrow and have a single strand of vascular tissue, or **vein.** *Megaphylls* are broad leaves with multiple veins. **Figure 28.5** illustrates the basic steps of possible evolutionary pathways for these leaf forms. Among the lycophytes (see Section 28.3), microphyll-like parts may have evolved as flaplike outgrowths of the plant's main vertical stem (see Figure 28.5A). Club mosses, the most common living lycophytes, still have this type of leaf. In contrast, megaphylls are thought to have arisen from modi-

A. Development of microphylls as an offshoot of the main vertical axis

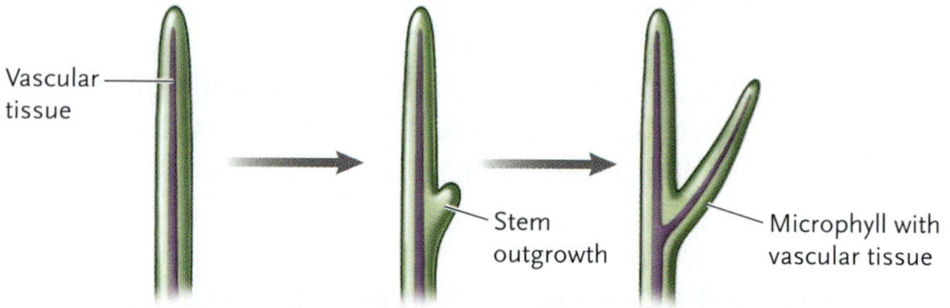

B. Development of megaphylls in a branching pattern

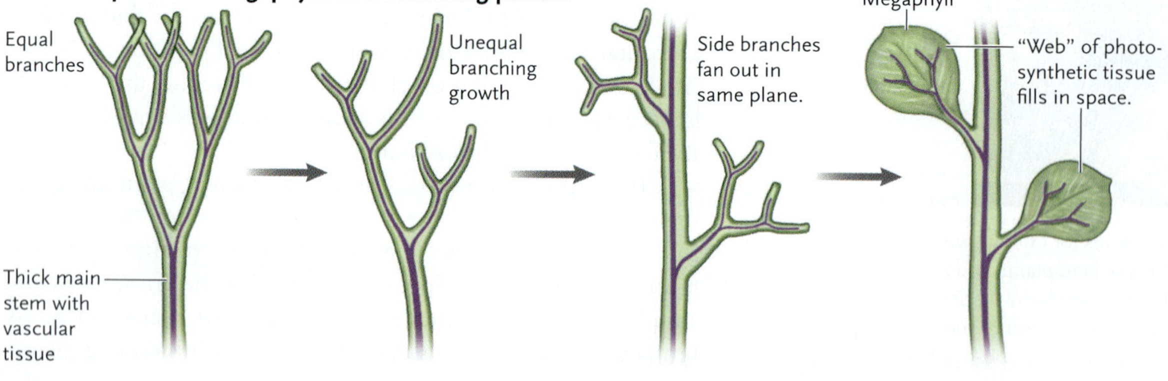

FIGURE 28.5 Evolution of leaves. (A) One type of early leaflike structure, the microphyll, may have evolved as an offshoot of the plant's main vertical axis. Each leaf contained a single vein (transport vessel). The seedless vascular plants known as lycophytes (club mosses) still have this type of leaf. **(B)** In other seedless vascular plants, the form of leaf called a megaphyll could have arisen in a series of steps that began when the main stem evolved a branching growth pattern. In this scenario, small side branches then fanned out and photosynthetic tissue filled the space between them, becoming the leaf blade. With time the small branches became modified into veins. Ferns, horsetails, gymnosperms, and angiosperms all have this type of leaf.
© Cengage Learning 2014

fied branches, and by the late Paleozoic era (about 350 mya), they were appearing in several plant groups. Figure 28.5B shows a simplified scenario for this adaptive shift: a slender stem branches into two roughly equal sections in a recurring pattern (called dichotomous branching). As slender branched sections are overtopped by more robust ones, the sections flatten out and a web of tissue having cells with chloroplasts then fills in the gap between branches. Over evolutionary time the tissue gains stomata, which provide access to carbon dioxide for photosynthesis, and a simple network of veins that transport water.

There is a strong correlation between the density of venation in a plant's leaves and the plant's capacity for photosynthesis—and, ultimately, for producing energy for growth. Timothy Brodribb and Taylor Feild, plant physiologists who compared the leaves of more than 500 living and extinct flowering plants, documented a dramatic increase in the number of leaf veins over the course of angiosperm evolution. They discovered that by about 100 mya, leaves of later evolving flowering plant species had up to 10 times more veins than leaves of more ancient species. Brodribb and Feild calculated that even a tripling in the number of veins would have increased the capacity of photosynthesis by 100%. Their findings suggest strongly that the evolution of large leaves with numerous stomata and a dense system of veins was a major factor in the remarkable proliferation of flowering plants that has occurred over the past 150 million years.

SUPPORT AND NUTRITION: ROOTS As upright growth emerged as a basic feature of vascular plants, it required a means of anchoring aerial parts in the soil, as well as effective strategies for obtaining water and soil nutrients. The evolution of **roots**—anchoring organs that also absorb water and nutrients—fulfilled these requirements. Frustratingly, fossils of early vascular plants are not very helpful in this regard and we are left with more questions than answers. Exactly when roots appeared is uncertain, and it is quite possible that the feature arose independently in various groups. At some point, ancestral vascular plants did come to have roots, and ultimately vascular plants developed whole specialized **root systems,** which generally consist of underground, cylindrical absorptive structures with a large surface area that favors the rapid uptake of soil water and dissolved minerals.

In the Plant Life Cycle, the Diploid Phase Became Dominant

As early plants moved into drier habitats, their life cycles also were modified considerably. You may recall that in sexually reproducing organisms, meiosis in diploid cells produces haploid (n) reproductive cells (see Chapter 11). These cells may be gametes—sperm or eggs—or they may be **spores,** which can give rise to a new haploid individual asexually, without mating.

All sexually reproducing organisms form both diploid and haploid cells, but what happens after each cell type forms varies greatly. In sexually reproducing animals, for example, a multicellular body develops only when diploid cells divide by mitosis. Haploid cells do not divide but become gametes. In plants, by contrast, mitosis occurs in both diploid and haploid cells, so plants essentially have two multicellular phases, one diploid and one haploid. This phenomenon, found in all plants, is called **alternation of generations (Figure 28.6).** The diploid generation produces haploid spores and is called a **sporophyte** ("spore producer"). The haploid generation produces gametes and is called a **gametophyte** ("gamete producer").

Some green algae show alternation of generations, but most are entirely haploid organisms. They produce many swimming gametes, and their zygotes do not develop a multicellular diploid phase. The first land plants had long-lived haploid gametophytes and short-lived diploid sporophytes. The bryophytes we consider later in this chapter still show this life cycle pattern. As plant adaptations to terrestrial life accumulated, the haploid gametophyte phase became physically smaller and less complex and had a shorter life span, while just the opposite occurred with the diploid sporophyte phase. Accordingly, in mosses and other bryophytes—all nonvascular plants—the sporophyte is a little larger and longer-lived than in green algae, and in vascular plants the sporophyte clearly is larger and more complex and lives much longer than the gametophyte **(Figure 28.7).** When you look at a pine tree, for example, you see a large, long-lived sporophyte.

The sporophyte generation begins after fertilization, when the resulting zygote grows mitotically into a multicellular, diploid organism. Its body will eventually develop sporangia, which produce spores.

The haploid phase of the plant life cycle begins in the sporangia. There, meiosis produces haploid spores. The spores divide by mitosis and give rise to multicellular haploid gametophytes. A gametophyte's function is to produce, nourish, and protect the forthcoming generation. These functions were linked to major changes in gametophyte structure as different groups of land plants evolved.

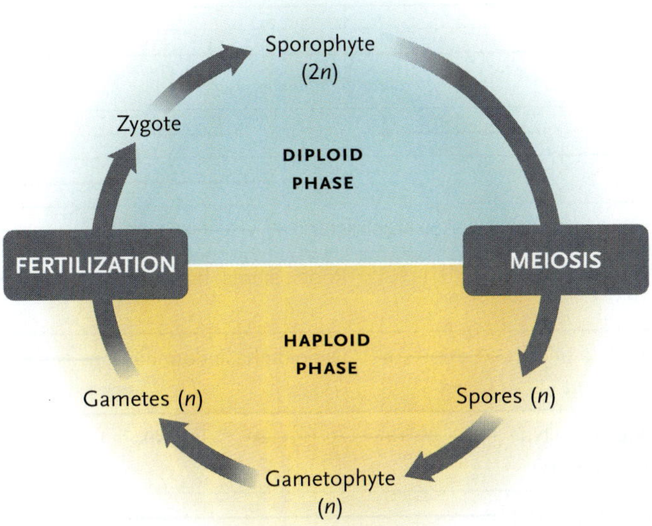

FIGURE 28.6 Overview of the alternation of generations, the basic pattern of the plant life cycle. The relative dominance of haploid and diploid phases is different for different plant groups (compare with Figure 28.7).

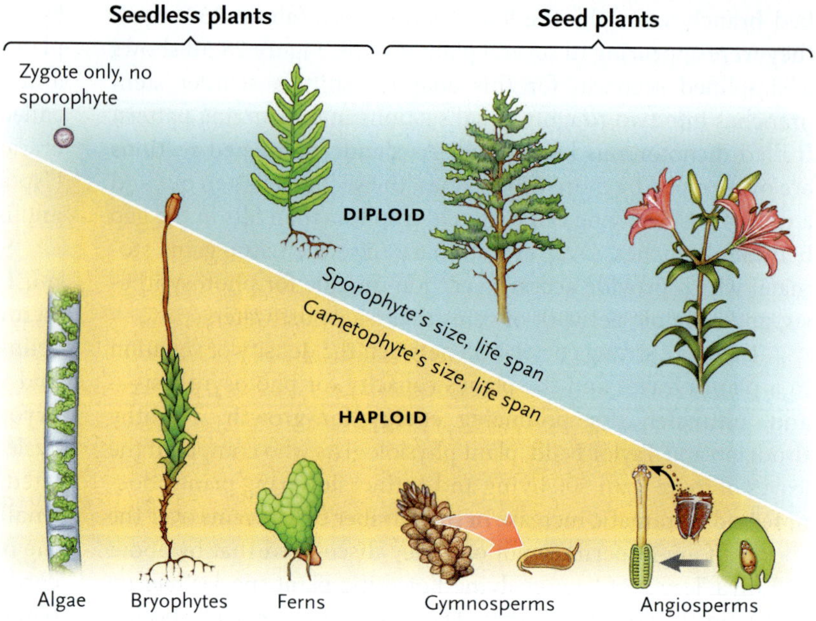

FIGURE 28.7 Evolutionary trend from dominance of the gametophyte (haploid) generation to dominance of the sporophyte (diploid) generation, represented here by existing species ranging from a charophyte alga *(Spirogyra)* to a flowering plant. This trend developed as early plants were colonizing habitats on land. In general, the sporophytes of vascular plants are larger and more complex than those of bryophytes, and their gametophytes are reduced in size and complexity. In this diagram ferns represent seedless vascular plants.

Some Vascular Plants Evolved Separate Male and Female Gametophytes

During sexual reproduction in plants, meiosis produces spores. When a plant makes only one type of spore, it is said to be **homosporous** ("same spore"). A gametophyte that develops from such a spore is bisexual—it can produce both sperm and eggs.

The sperm have flagella and are motile, for they have to swim through liquid water in order to encounter female gametes. Other vascular plants, including gymnosperms and angiosperms, are **heterosporous.** They produce two types of spores in two different types of sporangia, and those spores develop into small, sexually different gametophytes. The smaller spore type develops into a male gametophyte—a *pollen grain.* The larger one develops into a female gametophyte, in which eggs form and fertilization occurs.

So many new fossils appear in Devonian rocks that paleobotanists (who study fossil plants) have thus far been unable to determine which fossil lineages gave rise to modern plant phyla. Regardless, as each major lineage came into being, its characteristic adaptations included modifications of existing structures and functions (**Figure 28.8** and **Table 28.1**). The next sections fill out this general picture with respect to the seedless plants. We begin with bryophytes, nonvascular seedless plants that most clearly resemble the plant kingdom's algal ancestors.

TABLE 28.1 Trends in Plant Evolution

Traits derived from algal ancestor: cell walls with cellulose, energy stored in starch, two forms of chlorophyll (*a* and *b*); possibly, sporopollenin in spore wall.

Bryophytes	Ferns and Their Relatives	Gymnosperms	Angiosperms	Functions in Land Plants
Cuticle →	→	→	→	Protection against water loss, pathogens
Stomata →	→	→	→	Regulation of water loss and gas exchange (CO_2 in, O_2 out)
Nonvascular →	Vascular →	→	→	Internal tubes that transport water, nutrients
	Lignin →	→	→	Mechanical support for vertical growth
	Apical meristem →	→	→	Branching shoot system
	Roots, stems, leaves →	→	→	Enhanced uptake, transport of nutrients and enhanced photosynthesis
Haploid phase dominant →	Diploid phase dominant →	→	→	Genetic diversity
One spore type (homospory) →	→	Two spore types (heterospory) →	→	Promotion of genetic diversity
Motile gametes →	→	→	Nonmotile gametes →	Protection of gametes within parent body
Seedless →	→	Seeds →	→	Protection of embryo

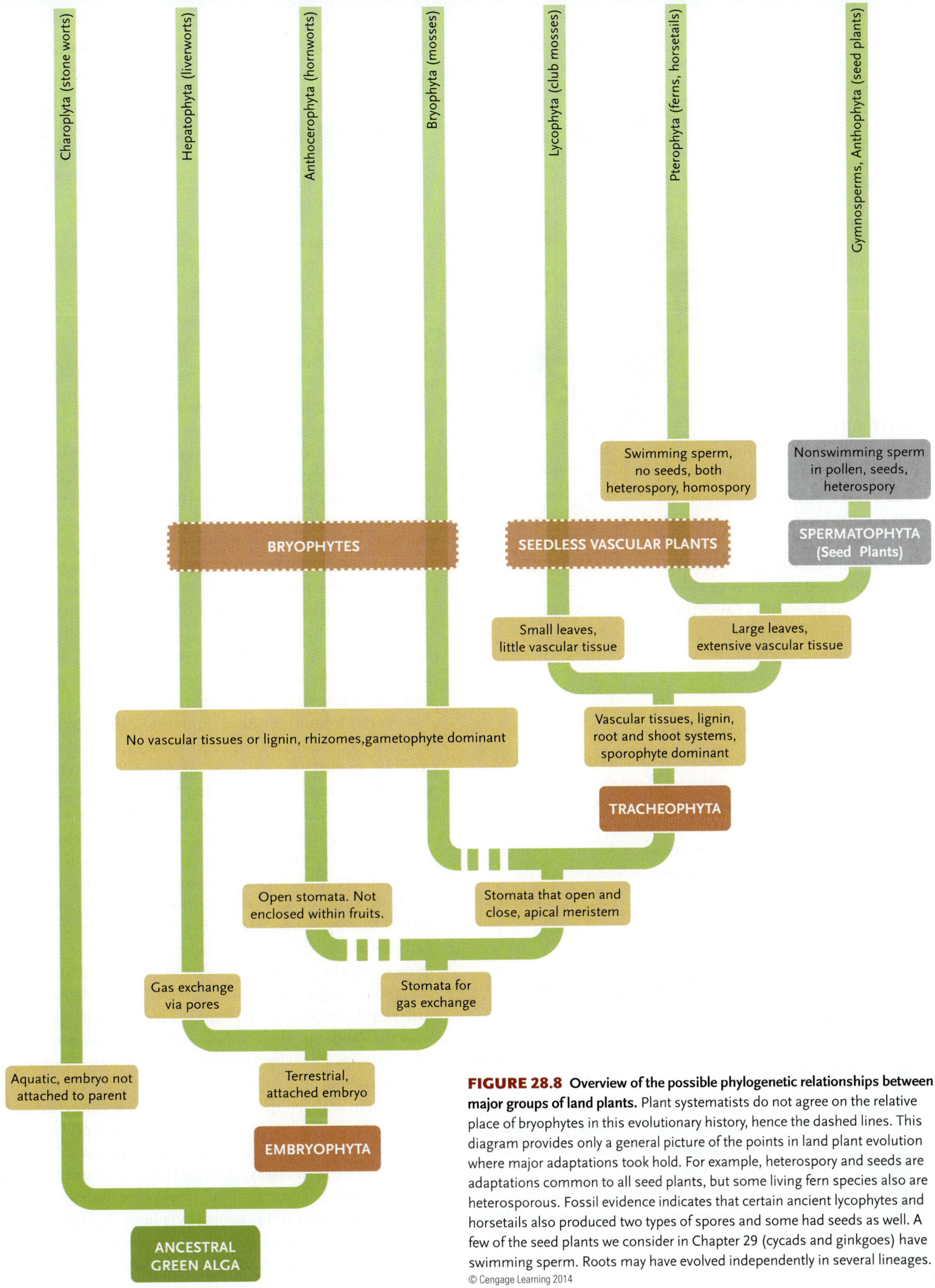

FIGURE 28.8 Overview of the possible phylogenetic relationships between major groups of land plants. Plant systematists do not agree on the relative place of bryophytes in this evolutionary history, hence the dashed lines. This diagram provides only a general picture of the points in land plant evolution where major adaptations took hold. For example, heterospory and seeds are adaptations common to all seed plants, but some living fern species also are heterosporous. Fossil evidence indicates that certain ancient lycophytes and horsetails also produced two types of spores and some had seeds as well. A few of the seed plants we consider in Chapter 29 (cycads and ginkgoes) have swimming sperm. Roots may have evolved independently in several lineages.

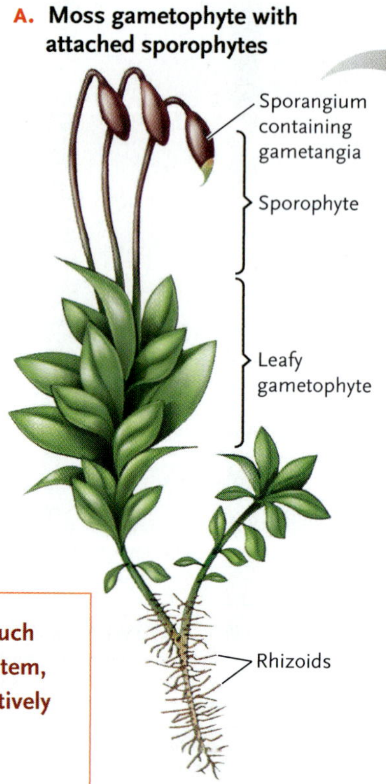

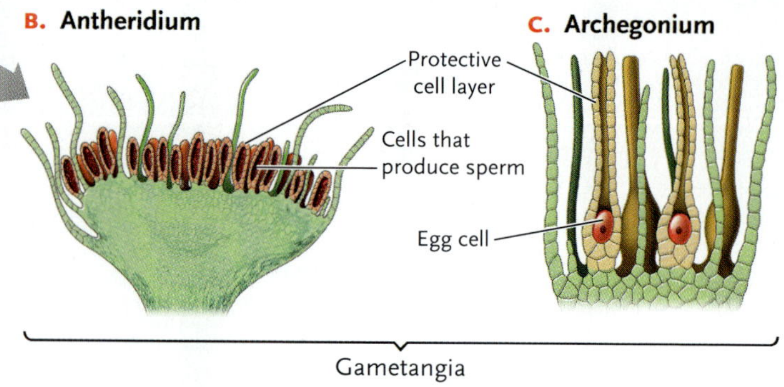

FIGURE 28.9 Multicellular structures enclosing plant gametes, a bryophyte innovation. **(A)** The gametophyte and sporophyte phases of the moss *Mnium*. In this species the gametangia are embedded in tissue of the gametophyte. In some other bryophytes the gametangia are attached on the gametophyte's surface. The two types of moss gametangia are the **(B)** antheridium, containing cells from which sperm arise, and the **(C)** archegonium, containing an egg cell. When fertilized, the egg cell gives rise to sporophytes.

STUDY BREAK 28.1

1. **How did plant adaptations such as a root system, a shoot system, and a vascular system collectively influence the transition to terrestrial life?**
2. **Describe the difference between homospory and heterospory, and explain how heterospory paved the way for other reproductive adaptations in land plants.**

THINK OUTSIDE THE BOOK

In 2009 researchers studying seaweeds off the coast of California reported discovering a species of red alga with lignin in its tissues. The project's lead scientist was Patrick Martone, now at the University of British Columbia. Use the Internet to learn more about Martone's work. Why was the discovery of lignin in a red alga so surprising, and what are its potential evolutionary implications?

28.2 Bryophytes: Nonvascular Land Plants

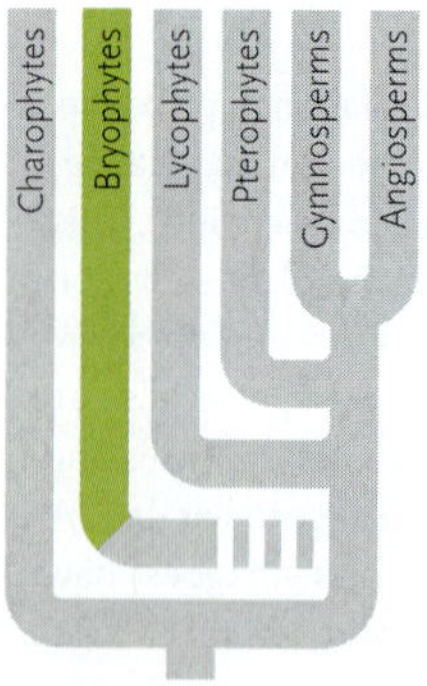

The bryophytes (*bryon* = moss)—liverworts, hornworts, and mosses—have a combination of traits that allow them to bridge aquatic and land environments. Bryophytes lack a well-developed system for conducting water, and while some can withstand desiccation, it is not surprising that they commonly grow on wet sites along creek banks or on rocks just above running water; in bogs, swamps, or the dense shade of damp forests; and on moist tree trunks or rooftops. Some species are **epiphytes** (*ep* = upon)—they grow independently (that is, not as a parasite) on another organism or some other moist place, ranging from the splash zone just above high tide on rocky shores to edges of snowbanks.

In general, bryophytes are strikingly algalike. They produce flagellated sperm that must swim through water to reach eggs, and they do not have a complex vascular system (although some have a primitive type of conducting tissue). Bryophytes do have parts that are rootlike, stemlike, and leaflike. However, the "roots" are rhizoids, and bryophyte "stems" and "leaves" did not evolve from the same structures as vascular plant stems and leaves did. (That is, stems and leaves are not homologous in the two groups.) Also, as already mentioned, bryophyte tissues do not contain lignified cells. The absence of this strengthening material and the lack of internal pipelines for efficient nutrient transport partly account for bryophytes' small size—typically less than 20 cm long—and for their tendency to grow sprawled along surfaces instead of upright.

Notwithstanding these limitations, bryophytes are clearly adapted to land. Along with their leaflike, stemlike, and fibrous, rootlike organs, sporophytes of some species have a water-conserving cuticle and stomata. Like most plants, bryophytes also have both sexual and asexual reproductive modes. And as is true of all plants, the life cycle has both gametophyte (n) and sporophyte ($2n$) phases, though the sporophyte is tiny and lives only a short time. **Figure 28.9** shows the green, leafy gametophyte of a moss plant, with miniscule diploid sporophytes attached to it by slender stalks. Bryophyte gametophytes produce gametes sheltered within a layer of protective cells called a **gametangium** (plural, *gametangia*). The gametangia in which bryophyte eggs form are flask-shaped structures called **archegonia** (*archi* = first, *gonos* = seed). Flagellated sperm form in rounded gametangia called **antheridia** (*antheros* = flowerlike). The sperm swim through a film of water on the plant to the archegonia and fertilize eggs. Each fertilized egg gives rise to a diploid embryo sporophyte, which stays attached to the gametophyte, produces spores—and the cycle repeats.

Despite their similarities to more complex plants, bryophytes are unique in several ways. Unlike in vascular species, the gametophyte is much larger than the sporophyte and obtains its nutrition independently of the sporophyte body. In fact, the comparatively tiny sporophyte remains attached to the gametophyte and depends on the gametophyte for much of its nutrition.

FIGURE 28.10 The bryophyte *Marchantia*, the only liverwort to produce (A) male and (B) female gametophytes on separate plants. *Marchantia* also reproduces asexually by way of **(C)** gemmae, multicellular vegetative bodies that develop in tiny cups on the plant body. Gemmae can grow into new plants when splashing raindrops transport them to suitable sites.

The position of bryophytes in plant evolution is still an open question. The basic bryophyte body plan may be similar to the ancestral condition from which higher plants evolved. In another view, bryophytes are a side shoot of evolution, completely separate from the path that led to vascular plants. The fossil record does not help much in resolving the issue. Although the 475-million-year-old spores noted earlier resemble modern liverwort spores, the first undisputed bryophyte fossils appear in late-Devonian rocks 350 mya, after vascular plants were already on the scene.

Despite questions raised by recent fossil finds, most current molecular, biochemical, cellular, and morphological evidence supports the view that bryophytes are not a monophyletic group. Instead, the various bryophytes evolved as separate lineages, in parallel with vascular plants. The relationships are far from resolved, however. For example, molecular evidence can be interpreted to mean that liverworts diverged early on from the lineage that led to all other land plants, with hornworts diverging later and mosses later still. Until new discoveries and interpretive work clarify this picture, the classification scheme in this chapter places liverworts, hornworts, and mosses in separate phyla. Our survey of nonvascular plants begins with the liverworts and hornworts, the simplest of the group, and concludes with mosses—plants that not only are more familiar to most of us, but whose structure and physiology more closely resemble that of vascular plants.

Liverworts May Have Been the First Land Plants

Liverworts make up the phylum Hepatophyta, and early herbalists thought that these small plants were shaped like the lobes of the human liver (*hepat* = liver; *wort* = herb). The resemblance might be vague to modern eyes: many of the 6,000 species of liverworts consist of a flat, branching, ribbonlike plate of tissue closely pressed against damp soil. This simple body, called a *thallus* (plural, *thalli*) is the gametophyte generation. Threadlike rhizoids anchor the gametophytes to their substrate. About two-thirds of liverwort species have leaflike structures, and some have stemlike parts. None have stomata, the openings that regulate gas exchange in most other land plants, although some species do have pores that open and close. Mitochondrial gene sequence data show that liverworts lack a few features (three introns) that are present in other bryophytes and in vascular plants. This finding has bolstered the hypothesis that liverworts were probably the first land plants.

In species of the liverwort genus *Marchantia* **(Figure 28.10)**, male and female gametophytes are separate plants. Male plants produce antheridia and female plants produce archegonia on specialized stalked organs (see Figure 28.10A, B). The sperm released from an antheridium must swim through surface water to encounter an egg inside an archegonium of a female gametophyte. After fertilization, a small, diploid sporophyte develops inside the archegonium, matures there, and produces haploid spores by meiosis. During meiosis, *Marchantia* sex chromosomes segregate, so some spores have the male genotype and others the female genotype. As in other liverworts, the spores develop inside jacketed sporangia that split open to release the spores. A spore that is carried by air currents to a suitable location germinates and gives rise to a haploid gametophyte, which is either male or female. *Marchantia* also can reproduce asexually by way of **gemmae** (*gem* = bud), small cell masses that form in cuplike growths on a thallus (see Figure 28.10C). Gemmae can grow into new thalli when rainwater splashes them out of the cups and onto an appropriately moist substrate.

Hornworts Have Both Plantlike and Algalike Features

About 100 species of hornworts make up the phylum **Anthocerophyta.** Simple stomata, which are permanently open, appear in this group. Like some liverworts, a hornwort gametophyte has a flat thallus, but the sporangium of the sporophyte phase is long and pointed, like a horn **(Figure 28.11).** The sporangia split into two or three ribbonlike sections when they release spores. Sexual reproduction occurs in basically the same way as in liverworts: free-swimming sperm fertilize eggs, which give rise to the sporophytes. Hornworts sometimes reproduce asexually by fragmentation as pieces of a thallus break off, form rhizoids, and develop into new individuals. Many hornwort species have cell features in common with green algae, including the presence in each cell of a single large chloroplast that contains protein bodies called pyrenoids. No other group of land plants has this feature, and some biologists have speculated that the

FIGURE 28.11 **The hornwort *Anthoceros*.** The base of each long, slender sporophyte is embedded in the flattened, leafy gametophyte.

distinction of "first land plant" should be assigned not to liverworts but to hornworts instead.

Mosses Most Closely Resemble Vascular Plants

Chances are that you have seen, touched, or sat on at least several of the approximately 10,000 species of mosses, and the use of the name **Bryophyta** for this phylum underscores the fact that mosses are the best-known bryophytes. They also are structurally and functionally most similar to the vascular plants we will consider in following sections. Their spores, produced by the tens of millions in sporangia, give rise to threadlike, haploid gametophytes that grow into the familiar moss plants, which often form tufts or carpets of vegetation on the surface of rocks, soil, or bark.

The moss life cycle, diagrammed in **Figure 28.12,** begins when a haploid (n) spore lands on a wet soil surface. After the

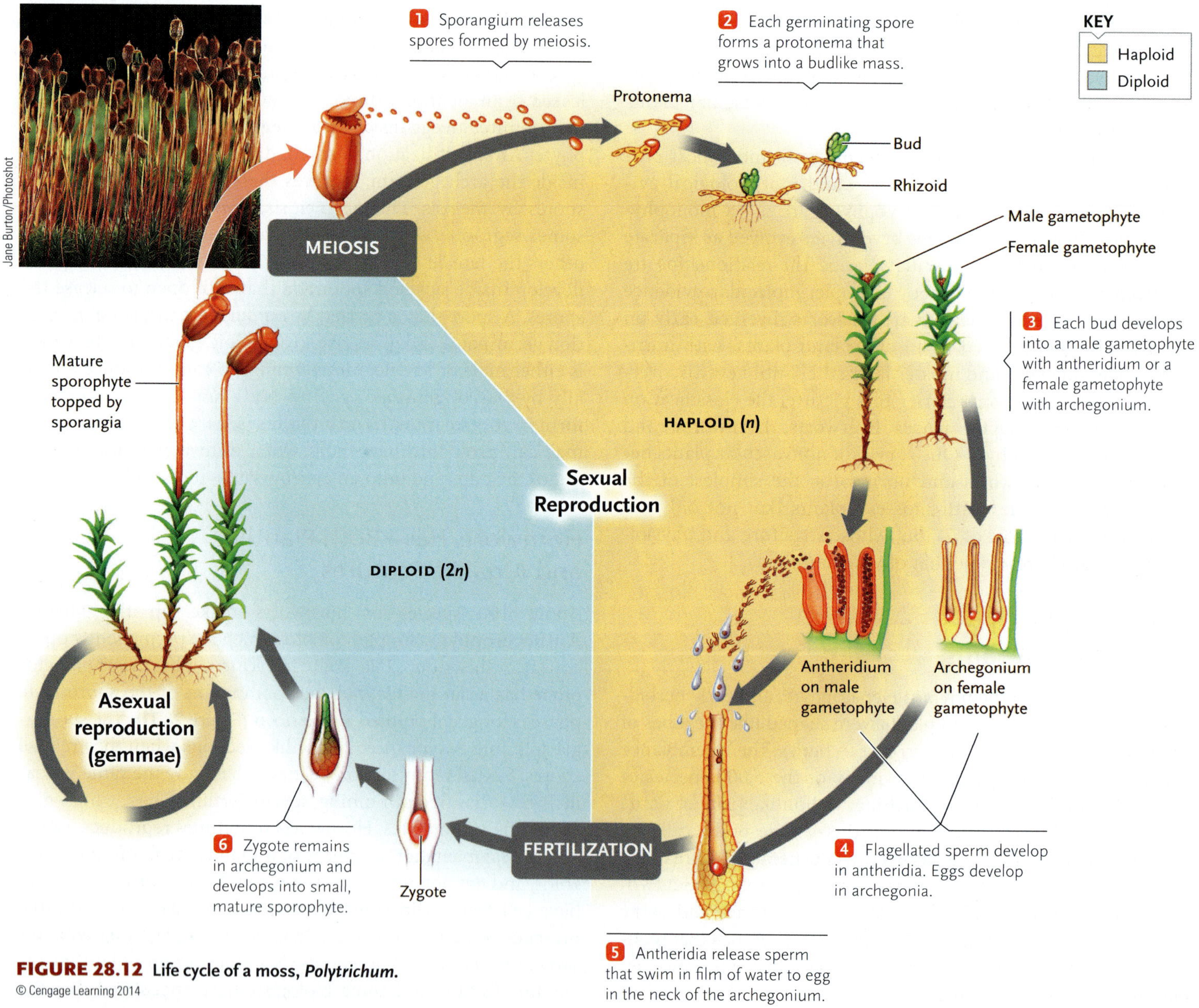

FIGURE 28.12 **Life cycle of a moss, *Polytrichum*.**

spore germinates, it elongates and branches into a filamentous web of tissue called a **protonema** ("first thread"), which can become dense enough to color the surface of soil, rocks, or bark visibly green. After several weeks of growth, the budlike cell masses on a protonema develop into leafy, green gametophytes anchored by rhizoids. A single protonema can be extremely prolific, producing bud after bud—and in this way giving rise to a dense clone of genetically identical gametophytes. Leafy mosses also may reproduce asexually by gemmae produced at the surface of rhizoids as well as on aboveground parts.

When a leafy moss is sexually mature, gametangia develop on its gametophytes and gametes form in them. In some moss genera, plants are unisexual and produce male *or* female gametangia—antheridia at the tips of male gametophytes and archegonia at the tips of female gametophytes. In other genera, plants are bisexual and produce both antheridia and archegonia. Propelled by a pair of flagella, sperm released from antheridia swim through a film of dew or rainwater and down a channel in the neck of the archegonium, attracted by a gradient of a chemical secreted by each egg. Fertilization produces the new sporophyte generation inside the archegonium, in the form of diploid zygotes that develop into small, mature sporophytes, each consisting of a sporangium on a stalk. Moss sporophytes may eventually develop chloroplasts and nourish themselves by photosynthesis, but initially they depend on the gametophytes for food. And even after a moss sporophyte begins photosynthesis, it still must obtain water, carbohydrates, and some other nutrients from the gametophyte.

Certain moss gametophytes are structurally complex, with features somewhat similar to those of higher plants. For example, mosses have stomata that open and close, and some species have a central strand of primitive conducting tissue. One kind of this tissue is made up of elongated, thin-walled, dead and empty cells that conduct water. These specialized cells, called hydroids, have oblique end walls that sometimes are partly dissolved or perforated with pores. Experiments with dyes show that water moves through them, as it does in similar xylemlike arrangements in vascular plants (see Chapters 33 and 34). In a few mosses, the water-conducting cells are surrounded by sugar-conducting tissue resembling the phloem of vascular plants.

Mosses and other bryophytes are important both ecologically and economically. As colonizers of bare land, their small bodies trap particles of organic and inorganic matter, helping to build soil on bare rock and stabilizing soil surfaces with a biological crust in harsh places like coastal dunes, inland deserts, and embankments created by road construction. Some hornworts harbor mutualistic nitrogen-fixing cyanobacteria, and so increase the amount of nitrogen available to other plants. In arctic tundras, bryophytes constitute as much as half the biomass, and they are crucial components of the food web that supports animals in that ecosystem. People have long used *Sphagnum* and other absorbent "peat" mosses (which typically grow in bogs) for everything from diapering babies and filtering whiskey to increasing the water-holding capacity of garden soil. Peat moss also is used as a fuel; each day the Rhode generating station in Ireland, one of several in that nation, burns 2,000 metric tons of peat to produce electricity.

In the next section we turn to the vascular plants, which have specialized tissues that can transport water, minerals, and sugars. Without the capacity to move these substances efficiently throughout the plant body, large sporophytes could not have evolved on land. Unlike bryophytes, modern vascular plants are monophyletic—all groups are probably descended from a common ancestor.

STUDY BREAK 28.2

1. **Give some examples of bryophyte features that bridge aquatic and terrestrial environments.**
2. **How do specific aspects of a moss plant's anatomy resemble those of vascular plants?**

28.3 Seedless Vascular Plants

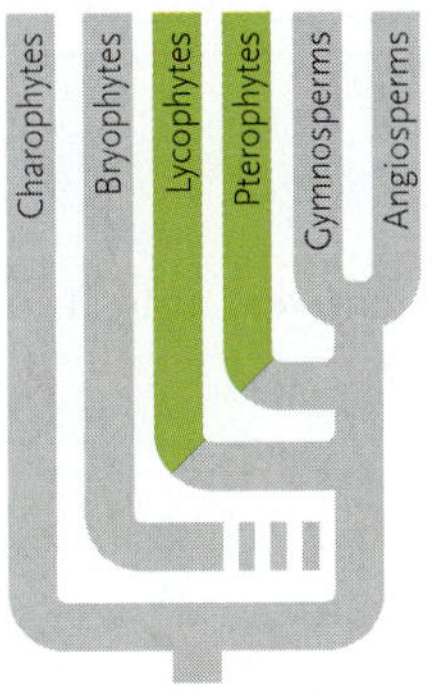

The first vascular plants, which did not "package" their embryos inside protective seeds, were the dominant plants on Earth for almost 200 million years, until seed plants became abundant. The fossil record shows that seedless vascular plants were well established by the late Silurian, some 427 mya, and they flourished until the end of the Carboniferous, about 250 mya. Some living seedless vascular plants have certain bryophyte-like traits, whereas others have some characteristics of seed plants. On one hand, like bryophytes, seedless vascular plants reproduce sexually by releasing spores, and they have swimming sperm that require free water to reach eggs. On the other hand, as in seed plants, the sporophyte of a seedless vascular plant separates from the gametophyte at a certain point in its development and has well-developed vascular tissues (xylem and phloem). Also, the sporophyte is the larger, longer-lived stage of the life cycle and the gametophytes are very small. Some gametophytes even lack chlorophyll. **Table 28.2** summarizes these characteristics and gives an overview of seedless vascular plant features.

Seedless vascular plants once encompassed a huge number of diverse species of trees, shrubs, and herbs. In the late Paleozoic era, they were Earth's dominant vegetation. Their remains formed most of the vast coal deposits that literally fueled the Industrial Revolution and continue to be major global energy sources. Some lineages have endured to the present, but collectively these survivors total fewer than 14,000 species. The taxonomic relationships between various lines are still under active investigation, and comparisons of gene sequences from the genomes in plastids, cell nuclei, and sometimes mitochondria are revealing previously unsuspected links between some of them. In this book we assign seedless vascular plants to two phyla, the

TABLE 28.2 Seedless Plant Phyla and Major Characteristics

Phylum	Common Name	Number of Species*	Common General Characteristics
Bryophytes: Nonvascular plants. Gametophyte dominant, free water required for fertilization, cuticle and stomata present in some.			
Hepatophyta	Liverworts	6,000	Leafy or simple flattened thallus with pores, rhizoids; spores in capsules; cuticle. Moist, humid habitats.
Anthocerophyta	Hornworts	100	Simple flattened thallus, rhizoids; stomata; hornlike sporangia. Moist, humid habitats.
Bryophyta	Mosses	10,000	Feathery or cushiony thallus, some have hydroids; spores in capsules. Moist, humid habitats; colonizes bare rock, soil, or bark.
Seedless vascular plants: Sporophyte dominant, swimming sperm, free water required for fertilization, cuticle and stomata present in all.			
Lycophyta	Club mosses	1,000	Simple leaves, true roots; most species have sporangia on sporophylls. Mostly wet or shady habitats.
Pterophyta	Ferns, whisk ferns, horsetails	13,000	*Ferns:* Finely divided leaves, woody stems in tree ferns; sporangia in sori. Habitats from wet to arid. *Whisk ferns:* Branching stem from rhizomes; sporangia on stem scales. Tropical to subtropical habitats. *Horsetails:* Hollow photosynthetic stem, scalelike leaves with silica in cell walls, sporangia in strobili. Swamps, disturbed habitats.

*Numbers of species are approximate.

Lycophyta (club mosses and their close relatives) and the Pterophyta (ferns, whisk ferns, and horsetails).

Early Seedless Vascular Plants Flourished in Moist Environments

The extinct plant genus *Cooksonia* (see Figure 28.4) probably was one of the earliest ancestors of modern seedless vascular plants. It and similar plants dominated the mudflats and swamps of the Devonian period (408–360 mya). Like other members of its extinct phylum (Rhyniophyta), *Cooksonia* species were small, rootless, and leafless, but their simple stems had a central core of xylem, an arrangement seen in many existing vascular plants. Although this and other now-extinct phyla came and went, ancestral forms of both modern phyla of seedless vascular plants appeared. In botanical terms, the earliest seedless vascular plants were "herbs"—that is, they did not have woody, lignified tissue. By the start of the ensuing Carboniferous period, however, the small, herbaceous Devonian plants had given rise to larger shrubby species and to trees—long-lived plants with a single main stem containing woody tissue. Other features of some of these early trees may have included bark, roots, leaves, and even seeds.

Carboniferous forests were lush, swampy places dominated by members of the phylum **Lycophyta,** a group in which roots evolved—an innovation that provided both firm anchoring in the soil and enhanced access to soil water and nutrients. One example is *Lepidodendron,* which had broad, straplike leaves and sporangia near the ends of the branches **(Figure 28.13A).** It also had xylem and several other types of tissues that are typical of all modern vascular plants, although probably not in the same proportions as seen today.

Carboniferous forests also were home to representatives of the phylum **Pterophyta,** including ferns such as *Medullosa* and giant horsetails such as *Calamites*—huge treelike plants that could have a trunk 30 cm in diameter. The strong, upright stems were attached to a system of rhizomes. Ferns populated the forest understory. Some early seed plants were present as well, including now-extinct fernlike plants, called seed ferns, which bore seeds at the tips of leaves **(Figure 28.13B).**

When leaves, branches, and old trees in swampy Carboniferous landscapes fell to the ground, they became buried in anaerobic sediments. Over geologic time, these buried remains were compressed and fossilized, and today they form much of the world's coal reserves. This is why coal is called a "fossil fuel," and the Carboniferous period is called the Coal Age. Characterized by a humid climate over much of the planet, and by the dominance of seedless vascular plants, the Carboniferous period continued for 150 million years, ending when climate patterns changed during the Paleozoic era (550–250 mya).

Most modern seedless vascular plants are ferns, and like their ancestors they are confined largely to wet or humid environments because they require external water for reproduction. Except for whisk ferns, their gametophytes do not rely on vascular tissues for water transport, and male gametes swim through water to reach eggs. In the wild, the few vascular seedless plants that are adapted to dry environments such as deserts can reproduce sexually only when seasonal rains make surface water available.

A. **Lycophyte tree *(Lepidodendron)***

B. **Artist's depiction of a Coal Age forest**

FIGURE 28.13 **Reconstruction of the lycophyte tree *Lepidodendron* and its environment.** **(A)** Fossil evidence suggests that *Lepidodendron* grew to be about 35 m tall with a trunk 1 m in diameter. **(B)** Artist's depiction of a Coal Age forest.

Modern Lycophytes Are Small and Have Simple Vascular Tissues

Unlike their ancient relatives, the most familiar of the 1,000 or so living species of lycophytes are small, ground-hugging species. They include spike mosses, quillworts, and the club mosses shown in **Figure 28.14A**—members of genera such as *Lycopodium* that grow on forest floors. A club moss sporophyte has stems and roots that contain a small amount of the vascular tissue xylem. Near the tips of stems are small, green microphyll-type leaves called **sporophylls,** with a single sporangium tucked at the base of each one. A cluster of sporophylls forms a conical **strobilus** (plural, *strobili*). In some species the sporangia release haploid spores produced by meiosis **(Figure 28.14B)**. If a spore germinates (sometimes several years after it is released), it forms a free-living gametophyte that differs markedly from the sporophyte. Ranging in size from nearly invisible to several centimeters, the gametophyte easily becomes buried under decomposing plant litter. There rhizoids attach it to a substrate. It cannot photosynthesize, and instead obtains nutrients by way of mycorrhizae, a symbiotic interaction in which a plant's rhizoids or roots associate with a particular fungus. Although many lycophyte species are homosporous—gametophytes are bisexual and produce both eggs and sperm—some are heterosporous. Regardless, as with ancestral lycophytes, the sperm require water in which they can swim to the eggs. After fertilization, the life cycle comes full circle as the zygote develops into a diploid embryo that grows into a sporophyte.

Ferns, Whisk Ferns, Horsetails, and Their Relatives Make Up the Diverse Phylum Pterophyta

Second only to the flowering plants, the phylum Pterophyta (*pteron* = wing) contains a large and diverse group of vascular plants—the 13,000 or so species of ferns, whisk ferns, and horsetails. Most ferns, including some that are popular houseplants, are native to tropical and temperate regions. Some floating species are less than 1 cm across, while some tropical "tree ferns" grow to 25 m tall. Other species are adapted to life in arctic and alpine tundras, salty mangrove swamps, and semi-arid deserts.

A. ***Lycopodium* sporophyte**

B. **Fossilized lycophyte spore**

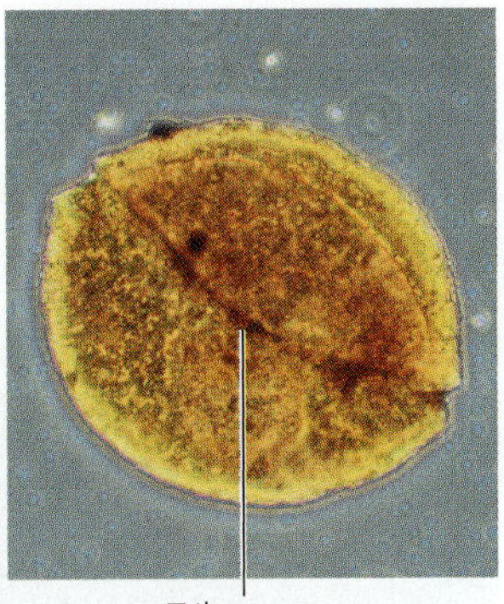

FIGURE 28.14 **Lycophytes.** **(A)** *Lycopodium* sporophyte, showing the conelike strobili in which spores are produced. **(B)** A fossilized lycophyte spore bearing a characteristic Y-shaped mark called a trilete scar.

COMPLEX ANATOMICAL FEATURES IN FERNS The familiar plant body of a fern is the sporophyte phase **(Figure 28.15).** It produces an aboveground clump of fern leaves, called fronds. Often finely divided and featherlike, and containing multiple strands of vascular tissue, fronds are an example of large, macrophyll-type leaves, a trait ferns share with seed plants. A typical frond has a well-developed epidermis with chloroplasts in the epidermal cells and stomata on the lower surface. Young fronds are tightly coiled, and as they emerge above the soil these "fiddleheads" (so named because they resemble the scroll of a violin) unroll and expand. Some people enjoy collecting and eating the unfurled fiddleheads of edible (noncarcinogenic) species.

Except for tropical tree ferns, the stems of most ferns are underground rhizomes. The stem's vascular system is organized into a complex, interconnecting network of bundles, each having a core of xylem surrounded by phloem. Roots descend along the length of the rhizomes. Remarkably, a rhizome can live for centuries, growing at its tip and extending outward horizontally through the soil, sometimes over a considerable area. In most ferns, the fronds arise from nodes positioned along the rhizome. A **node** is the point on a stem where one or more leaves are attached.

In most fern species, the sporophyte produces sporangia on the lower surface or margin of some leaves. Often, several sporangia are clustered into a rust-colored **sorus** (plural, *sori*). Sori may be exposed or they may be protected with a flap of tissue. Each sporangium is a delicate case, shaped rather like an old-fashioned pocket watch and covered by a layer of epidermal cells. In the layer, a row of thick-walled cells called the **annulus** ("ring") nearly encircles the sporangium.

Inside the sporangium, haploid spores arise by meiosis. Meanwhile, the sporangium slowly dries, and in the process

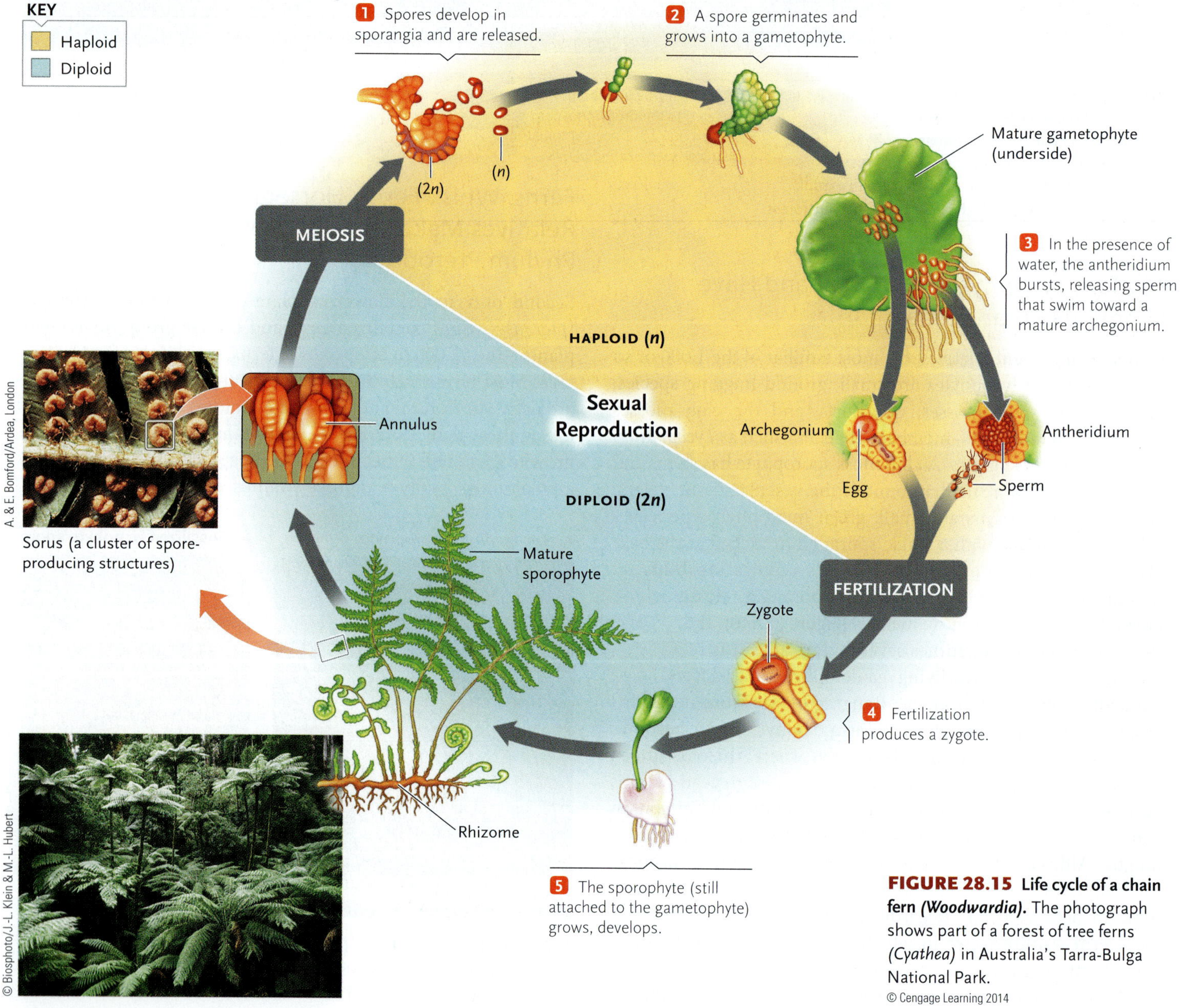

FIGURE 28.15 Life cycle of a chain fern *(Woodwardia).* The photograph shows part of a forest of tree ferns *(Cyathea)* in Australia's Tarra-Bulga National Park.

the annulus contracts. Eventually the force of the contracting annulus rips open the sporangium, which snaps back on itself, flinging out the mature spores. In this way fern spores can be dispersed up to 2 m away from the parent plant. Wind may carry them much farther: on board the *Beagle,* Charles Darwin collected fern spores hundreds of miles from shore.

A germinating spore develops into a gametophyte, which is typically a small, heart-shaped plant anchored to the soil by rhizoids. Both antheridia and archegonia are present on the lower surface of each gametophyte, where moisture is trapped. Inside an antheridium is a globular packet of haploid cells, each of which develops into a helical sperm cell with many flagella. When water is present, the antheridium bursts, releasing the sperm. If mature archegonia are nearby, the sperm swim toward them, drawn by a chemical attractant that diffuses from the neck of the archegonium, which is open when free water is present.

After a sperm fertilizes an egg, the diploid zygote begins dividing and developing into an embryo, which at this stage obtains nutrients from the gametophyte. In a short time, however, the embryo develops into a young sporophyte that is larger than the gametophyte and has its own green leaf and a root system. The sporophyte now is nutritionally independent, and the parent gametophyte degenerates and dies.

FEATURES OF EARLY VASCULAR PLANTS IN WHISK FERNS The whisk ferns and their relatives are represented by two genera, *Psilotum* (pronounced si-LOW-tum) and *Tmesipteris* (may-SIP-ter-is), with only about 10 species in all. These rather uncommon plants grow in tropical and subtropical regions, often as epiphytes. In the United States the range for *Psilotum* species **(Figure 28.16)** includes Hawaii, Gulf Coast states such as Florida and Louisiana, and parts of the West.

Whisk fern sporophytes are up to 60 cm tall and resemble the extinct *Cooksonia.* Like those early vascular plants, they lack roots and leaves. Instead, small scales adorn an upright branching stem that ascends from a horizontal rhizome system anchored by rhizoids. The rhizoids have mycorrhizal fungi associated with them, which provide enhanced access to some nutrients.

A whisk fern's stem is structurally and functionally multifaceted. Its epidermal cells carry out photosynthesis, whereas its core has the transport tissues xylem and phloem and other anatomical features of more complex vascular plants. Sporangia rest atop some of the stem scales. Inside them, meiotic divisions of specialized cells produce haploid spores.

HORSETAILS, REMNANTS OF AN ANCIENT PAST Only fifteen horsetail species in a single genus, *Equisetum,* have survived to the present **(Figure 28.17)**. Unlike towering ancestors such as *Calamites,* most modern horsetails grow less than a meter high, in moist soil along streams and in disturbed habitats, such as roadsides and beds of railroad tracks. Their sporophytes typically have underground rhizomes along with roots that anchor the rhizome to the soil. Whorls of scalelike leaves are arranged around a photosynthetic stem that is stiff and gritty from the silica that horsetails accumulate in their tissues. They came to be known popularly as "scouring rushes" because American pioneers used them to scrub out pots and pans.

Equisetum sporangia are borne in strobili on highly specialized stem structures quite different from the sporophylls of club mosses. Each stalked spore-bearing structure in a strobilus resembles an umbrella and is attached at right angles to a main axis. Haploid spores develop in sporangia attached near the edge of the "umbrella's" underside, and air currents disperse them.

FIGURE 28.16 Sporophytes of a whisk fern *(Psilotum)*, a seedless vascular plant. Three-lobed sporangia occur at the ends of stubby branchlets; inside the sporangia, meiosis gives rise to haploid spores.

FIGURE 28.17 A species of *Equisetum*, the horsetails. (A) Vegetative stem. **(B)** Strobili, which bear sporangia. **(C)** Close-up of sporangia and associated structures on a strobilus.

INSIGHTS FROM THE Molecular Revolution

The Powerful Genetic Toolkit for Studying Plant Evolution

Unlike most other eukaryotes, plants have three distinct sets of genes—in the cell nucleus, in mitochondria, and in chloroplasts. Chloroplast DNA, or cpDNA, has been especially useful for evolutionary studies, particularly the chloroplast *rbcL* gene. Mutations of the gene have occurred slowly, at about one-fourth to one-fifth the rate of genes in the nucleus. As a result, the DNA sequences of *rbcL* genes of different species diverge less than those of most other plant genes. Further, there are no introns—noncoding sequences—interrupting the coding sequence of the *rbcL* gene. Researchers can compare *rbcL* DNA from different species base by base, with no need to subtract introns. At the same time, the *rbcL* genes of different species are different enough to allow researchers to assemble evolutionary trees based on the degree of sequence variation.

Research Question

What were the major branch points in plant evolution?

Experiments

Studies using cpDNA have helped fuel several fundamental shifts in our understanding of branch points in plant evolution. For example, together with gene sequence data from nuclear DNA, analysis of *rbcL* genes provided evidence that charophyte green algae were the evolutionary forerunners of land plants. Similarly, in the late 1990s an international research team led by Yin-Long Qiu at the University of Massachusetts at Amherst correlated the loss of introns from two mitochondrial genes with the hypothesis that the first land plants were liverworts. Qiu and his colleagues carried out a genetic survey of more than 350 land plants representing all major lineages. They discovered that the noncoding sequences were present in all other bryophytes and all major lines of vascular plants, but were absent in liverworts, green algae, and all other eukaryotes. Analysis of *rbcL* sequences in various plant groups support the findings. Data from cpDNA and mtDNA analyses also underlie the hypothesis that, as land plants evolved, the ancient relatives of club mosses (lycophytes) were the forerunners of other vascular plants. Clearly, these varied molecular tools, and cpDNA in particular, are helping plant scientists explore evolutionary relationships across the whole spectrum of the Kingdom Plantae.

Conclusions

The evidence from cpDNA and mtDNA analysis points to liverworts as the first land plants, and ancient relatives of club mosses as the forerunners of other vascular plants.

THINK LIKE A SCIENTIST Suppose a paleobotanist came across an intact fossil of a small, previously unknown lycophyte having rather well developed vascular tissue. Might this discovery significantly alter our understanding of early vascular plant evolution?

Source: Y.-L. Qiu et al. 1998. The gain of three mitochondrial introns identifies liverworts as the earliest land plants. *Nature* 394:671–674.

They must germinate within a few days to produce gametophytes, which are free-living plants about the size of a small pea.

STUDY BREAK 28.3

1. **Compare and contrast the lycophyte and bryophyte life cycles with respect to the sizes and longevity of gametophyte and sporophyte phases.**
2. **In ferns, whisk ferns, and horsetails, what kinds of structures fulfill the roles of roots and leaves?**
3. **How does the life cycle of a horsetail differ from that of a fern?**

THINK OUTSIDE THE BOOK

Seed ferns, an extinct group mentioned briefly in this section, are assigned to the extinct phylum Pteridospermophyta. They were nearly wiped out by the Permian–Triassic mass extinction event that occurred about 251 mya and became extinct soon thereafter. Seed-bearing conifers living at the time fared better, and conifers were the dominant land plants for at least another 100 million years. Using Internet or library resources, develop a plausible explanation for why the seed fern and conifer lineages had such differing evolutionary fates.

UNANSWERED QUESTIONS

Why is the homosporous lifestyle correlated with unusually large numbers of chromosomes?

My laboratory at the University of Arizona is interested in how plant genomes evolve, and, in particular, how changes in chromosome number and organization occur. In the plant kingdom, and in fact across all eukaryotes, homosporous ferns and lycophytes have the largest numbers of chromosomes: on average $n = 57.05$. How this strikingly large average chromosome number correlates with reproductive strategies in homosporous lycophytes and ferns has long been a mystery to plant evolutionary biologists. In general, chromosome numbers may increase due to dysploidy—a change in the number of single chromosomes without a change in the overall number of genes—or by polyploidy (duplication of a whole genome). Decades of research have uncovered only rare examples of dysploidy in plants, but polyploidy is much more common. This observation has fueled hypotheses about why high chromosome numbers are correlated with homospory.

One possible explanation is that the capacity of a single spore to produce both sperm and egg may support higher rates of polyploidy. During polyploidy, the new zygote with duplicated chromosomes is nearly always reproductively isolated from its parental, diploid relatives. This means that for polyploid lineages to survive and establish clades with higher numbers of chromosomes, they must be able to reproduce. If a polyploid individual arises and is the only one of its kind, then the

reproductive assurance of having homosporous gametophytes will aid in the new line's initial survival. In contrast, a heterosporous plant would need to have a megaspore and a microspore near each other for reproduction, instead of just a single spore.

Alternatively, it may be that having more chromosomes relieves potential deleterious effects of the extreme inbreeding possible in homosporous ferns and lycophytes. Assuming that the overall number of genes remains similar, a greater number of chromosomes will increase independent assortment and thus effectively increase the recombination rate. Whether increased polyploidy, selection for increased independent assortment, or some combination of both is driving the evolution of high chromosome numbers in homosporous lycophytes and ferns remains an open question. New genomic tools are being applied to study the genomes of these groups to give us new insights into how the diversity of plant lifestyles has influenced the diversity of plant genomes.

THINK LIKE A SCIENTIST How would you expect changes in chromosome number to influence recombination and independent assortment? How are changes in chromosome number achieved without creating problems during meiosis? Would you expect all chromosomes to be equally likely to be duplicated or lost?

Courtesy of Michael Barker

Michael S. Barker is an Assistant Professor of Ecology and Evolutionary Biology at the University of Arizona in Tucson. His research focuses on the evolution genomics of plant diversity. To learn more about Barker's work, go to http://barkerlab.net/

REVIEW KEY CONCEPTS

To access the course materials and companion resources for this text, please visit www.cengagebrain.com.

28.1 The Transition to Life on Land

- Plants are thought to have evolved from green algae between 425 and 490 million years ago (Figure 28.2).
- Key adaptations in early land plants include an outer cuticle that helps prevent desiccation, tissues containing lignified cells, spores protected by a wall containing sporopollenin, multicellular chambers that protect developing gametes, and an embryo sheltered inside a parent plant (Figure 28.3).
- Land plant evolution also included the development of vascular tissues, root systems, and shoot systems, including stems containing lignified cells and leaves having stomata; a shift from dominance by a long-lived, larger haploid gametophyte to dominance of a long-lived, larger diploid sporophyte; and a shift from homospory to heterospory with separate male and female gametophytes (Figures 28.5–28.8).

Animation: The importance of alternation of generations

Animation: Evolutionary tree for plants

28.2 Bryophytes: Nonvascular Land Plants

- Existing nonvascular land plants, or bryophytes, include the liverworts (Hepatophyta), hornworts (Anthocerophyta), and mosses (Bryophyta). Liverworts may have been the first land plants.
- Bryophytes release spores and produce flagellated sperm that swim through free water to reach eggs. They lack a vascular system; true roots, stems, and leaves; and lignified tissue. A larger, dominant gametophyte (n) phase alternates with a small, fleeting sporophyte ($2n$) phase (Figures 28.9–28.12).

Animation: Fern life cycle

Animation: Marchantia, a liverwort

Animation: Moss life cycle

28.3 Seedless Vascular Plants

- Existing seedless vascular land plants include the lycophytes (club mosses), whisk ferns, horsetails, and ferns. They release spores and have swimming sperm, but have well-developed vascular tissues. The sporophyte is the larger, longer-lived stage of the life cycle and develops independently of the small gametophyte.
- Club mosses (Lycophyta) have sporangia clustered at the bases of specialized leaves called sporophylls. Clusters of sporophylls form a strobilus ("cone"). Haploid spores germinate to form small gametophytes. The life cycle is similar in ferns, whisk ferns, and horsetails (Pterophyta) (Figures 28.13–28.17).
- Ferns are the most diverse group of seedless vascular plants. Most species do not have aboveground stems, only leaves that arise from nodes along an underground rhizome. Fern leaves typically have well-developed stomata, and the vascular system consists of bundles, each with xylem surrounded by phloem. Sporangia on sporophylls ("fronds") release spores that develop into gametophytes. Sexual reproduction produces a much larger, long-lived sporophyte.

UNDERSTAND & APPLY

Test Your Knowledge

1. Which of the following is *not* an evolutionary trend among plants?
   a. developing vascular tissues
   b. becoming seedless
   c. having a dominant diploid generation
   d. producing nonmotile gametes
   e. producing two types of spores
2. As plants made the evolutionary transition to a terrestrial existence, they benefited from adaptations that:
   a. increased the motility of their gametes on dry land.
   b. flattened the plant body to expose it to the sun.
   c. reduced the number and distribution of roots to prevent drying.
   d. provided mechanisms for gaining access to nutrients in soil.
   e. allowed stems and leaves to absorb water from the atmosphere.

3. Which of the following is *not* generally thought to be accurate with respect to the evolution of leaves?
   a. Leaves arose more than once during land-plant evolution.
   b. Microphylls may have arisen as modifications of the main vertical stem.
   c. Macrophylls may have come about as modifications of stem branches.
   d. The evolution of higher-density leaf venation correlates with increased capacity for photosynthesis.
   e. Only macrophyll-type leaves, which have stomata and a network of veins, are observed in extant plant species.
4. Which is the correct matching of phylum and plant group?
   a. Hepatophyta: liverworts
   b. Anthocerophyta: mosses
   c. Bryophyta: horsetails
   d. Lycophyta: ferns
   e. Pterophyta: hornworts
5. A homeowner noticed moss growing between bricks on his patio. Closer examination revealed tiny brown stalks with cuplike tops emerging from green leaflets. These brown structures were:
   a. the sporophyte generation.
   b. the gametophyte generation.
   c. elongated haploid reproductive cells.
   d. archegonia.
   e. antheridia.
6. In addition to having simple vascular tissue, seedless vascular plants differ from all bryophytes in that:
   a. they do not require free water for fertilization.
   b. the sporophyte generation is dominant.
   c. most do not require a moist, humid, or shaded habitat.
   d. the group is much more diverse, with many more species overall.
   e. most have a feathery or flattened thallus.
7. Which feature(s) do ferns share with all other land plants?
   a. sporophyte and gametophyte life cycle stages
   b. gametophytes supported by a thallus
   c. dispersal of spores from a sorus
   d. asexual reproduction by way of gemmae
   e. water uptake by means of rhizoids
8. In comparing bryophytes and seedless vascular plants, which of the following statements is not accurate?
   a. In general, bryophyte traits more closely resemble those of algae.
   b. We only see true leaves, stems, and roots in seedless vascular plants.
   c. Both groups have lignified cells in some body parts.
   d. Bryophytes lack complex conducting tissues (xylem and phloem).
   e. Seedless vascular plants have sexual and asexual reproductive modes, but in bryophytes only asexual reproduction has been observed.
9. Based solely on numbers of species, the most successful seedless plants today are:
   a. horsetails.
   b. ferns.
   c. liverworts.
   d. mosses.
   e. club mosses.
10. Which of the following is/are true with respect to seedless vascular plants?
   a. Ferns are the most numerous species in this group.
   b. Lycophyte species grow the largest of all modern seedless vascular plants.
   c. Seedless vascular plants, including trees, shrubs, and herbaceous ("nonwoody") species dominated Earth's terrestrial environments for nearly 200 million years.
   d. The strobilus, where spore-producing parts sporangia are located, is a reproductive adaptation found in all seedless vascular plants.
   e. In the fern life cycle, the young sporophyte is dependent on the gametophyte, but quickly develops into an independent, larger plant with its own root system and leaves where photosynthesis occurs.

Discuss the Concepts

1. Working in the field, you discover a fossil of a previously undescribed plant species. The specimen is small and may not be complete; the bits you have do not include any floral parts. What sorts of observations would you need in order to classify the fossil as a seedless vascular plant with reasonable accuracy? What evidence would you need in order to distinguish between a fossil lycopod and a fern?
2. Compare the size, anatomical complexity, and degree of independence of a moss gametophyte and a fern gametophyte. Which one is the most protected from the external environment? Which trends in plant evolution does your work on this question bring to mind?

Design an Experiment

A moss species that you study occurs in lush abundance in the moist, shaded forest floor in a grove of coast redwoods *(Sequoia sempervirens)* in Northern California. When one of the massive old trees abruptly topples over, there is a dramatic increase in the amount of light and air movement in a sizable patch of the moss' habitat. This is a rare opportunity for you to study the effects on the moss of a sudden shift in the environmental conditions in which it grows. Devise an experimental plan to guide your research, including hypotheses about one or more ways the change may affect the species' growth and/or life cycle.

Interpret the Data

Green plants, green algae, red algae, and microscopic freshwater algae called glaucophytes are all primary photosynthetic eukaryotes—they have plastids, including chloroplasts that carry out photosynthesis. Biologists have long accepted the hypothesis that plastids arose by way of endosymbiosis of a cyanobacterium during the evolution of lineages leading to primary photosynthetic eukaryotes (see Sections 25.3 and 27.2). Until recently, however, there was no unequivocal molecular evidence to support this assumption. If the assumption is correct, then a phylogenetic tree based on molecular data should support the prediction that all primary photosynthetic eukaryotes are descended from a common ancestor. Naiara Rodríguez-Ezpeleta at the University of Montreal led a team that tested this prediction. They already knew that proteins associated with plastids in cyanobacteria and photosynthetic eukaryotes share some amino acid sequences. In the experiment of interest here, they sequenced the amino acids in 143 proteins encoded by nuclear genes of species representing all the primary photosynthetic groups. They used a computerized statistical method called bootstrapping to analyze the large number of different data sets from the selected species. Basically, bootstrapping produced a series of phylogenetic trees based on comparisons of the amino acid sequences in the target proteins from each species. It then compared the trees to see where overlaps occurred, and generated a summary tree like the one in **Figure A.**

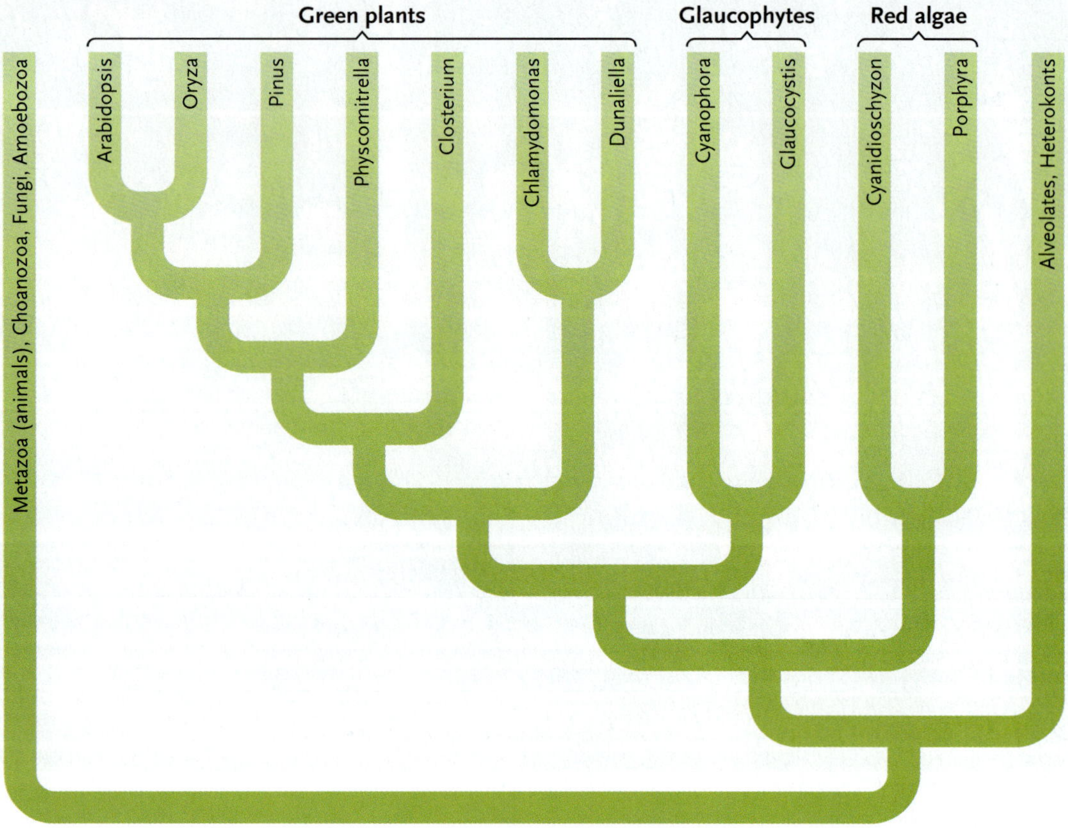

FIGURE A **Phylogeny based on a dataset of 143 nuclear encoded proteins.**

1. Does Figure A support or contradict the hypothesis that a single endosymbiosis event correlated with the evolution of organisms containing plastids?
2. Do the data support the prediction that glaucophytes and green plants are sister groups—that is, that they are more closely related to each other than to other groups in the phylogeny?

Source: N. Rodriguez-Ezpeleta et al. 2005. Monophyly of primary photosynthetic eukaryotes: green plants, red algae, and glaucophytes. *Current Biology* 15:1325–1330.

Apply Evolutionary Thinking

One proposed bryophyte phylogeny considers hornworts and vascular plants (such as lycophytes and ferns) to be sister clades (see Figure A). Among other factors, a shared morphological feature supports this hypothesis: hornwort and vascular plant sporophytes both show indeterminate growth—once a sporophyte develops, it grows for the remainder of the plant's life. Indeterminate growth has not been observed in other bryophyte lineages. Can you think of a different evolutionary explanation for the growth pattern shared by hornworts and vascular plants?

29

A temperate forest with representatives of the two clades of seed plants—conifers (gymnosperms), and flowering plants (angiosperms).

STUDY OUTLINE

29.1 Gymnosperms: The "Naked Seed" Plants

29.2 Angiosperms: Flowering Plants

29.3 Frontiers of Plant Genome Research

Seed Plants

A. **A ponderosa pine**

Robert Potts, California Academy of Sciences

B. **An orchid**

dashingstock/Shutterstock.com

C. **Cotton plants**

Steven Frame/Shutterstock.com

Why it matters . . . Chances are, *Gossypium hirsutum* is one of the few truly indispensable plants in your life. This angiosperm, commonly called upland or Mexican cotton, produces the most widely used plant fiber in the world. While questions remain about where and when *Gossypium* species arose, archeological finds in places as distant as Peru, Pakistan, and Egypt show that humans have been using cotton in diverse ways for thousands of years. Today *G. hirsutum* accounts for about 90% of commercially grown cotton worldwide; its fibers are raw materials in a vast gamut of items ranging from jeans, T-shirts, and virtually every other sort of clothing, to bandages, rugs, bedding, and upholstery. Botanically the fibers are long, cellulose-rich "hairs" that develop from the protective coat around a cotton plant's seeds **(Figure 29.1)**. The subset of cotton fibers called "lint" can be spun into strong thread.

FIGURE 29.1 **Some representative seed plants. (A)** A ponderosa pine, *Pinus ponderosa*. This species and other conifers belonging to the phylum Coniferophyta represent the gymnosperms. **(B)** An orchid *(Cattleya rojo)*, a showy example of a flowering plant. **(C)** Cotton plants (*Gossypium* spp.), producers of the most widely used plant fiber on Earth.

TABLE 29.1	Seed Plant Phyla and Major Characteristics		
Phylum	**Common Name**	**Number of Species***	**Common General Characteristics**
Gymnosperms: Vascular plants with "naked" seeds. Nonmotile sperm arise from male gametophytes in pollen grains.			
Cycadophyta	Cycads	185	Shrubby or treelike with palmlike leaves, pithy stems; male and female strobili on separate plants. Widely distributed in warm climates.
Ginkgophyta	Ginkgo	1	Woody-stemmed tree, deciduous fan-shaped leaves. Male, female structures on separate plants. Temperate areas of China.
Gnetophyta	Gnetophytes	70	Shrubs or woody vines; one has strappy leaves. Male and female strobili on separate plants. Limited to deserts, tropics.
Coniferophyta	Conifers	550	Mostly evergreen, woody trees and shrubs with needlelike or scalelike leaves; male and female cones usually on same plant.
Angiosperms (Anthophyta): Woody or herbaceous plants with flowers, and seeds protected inside fruits; nearly all land habitats, some aquatic.			
Major groups: Magnoliids, Monocots, Eudicots			
Magnoliids	Magnolias, laurels, avocado, black peppers, and others	8,000+	Pollen grains have a single groove; some species with three of more cotyledons.
Monocots	Grasses, palms, lilies, orchids, and others	60,000+	Pollen grains have a single groove; one cotyledon. Parallel-veined leaves common.
Eudicots	Most fruit trees, roses, cabbages, melons, beans, potatoes, and others	200,000+	Pollen grains have three grooves. Most species have two cotyledons; net-veined leaves common.

*Numbers of species are approximate.

Cotton is just one example of the extraordinary benefits human societies gain from seed plants. It's easy to tick off many more, starting with wood, paper, and crucial food crops like rice, corn, and wheat. Beyond meeting human needs, seed plants are the main sources of nutrients and other materials essential to the functioning of nearly all land ecosystems. Technically seed plants are spermatophytes ("seed plant"), and seed plants as a whole belong to a grouping called the Spermatophyta. They are subdivided into gymnosperms—conifers and their relatives—and angiosperms, the flowering plants (Table 29.1). In this chapter we begin our survey of seed plants with the gymnosperms, the first to appear on the evolutionary stage.

29.1 Gymnosperms: The "Naked Seed" Plants

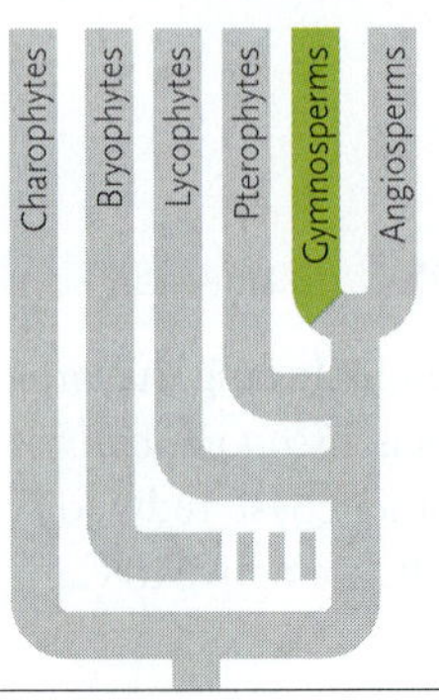

Gymnosperms are the conifers and their relatives. The earliest fossils identifiable as gymnosperms are found in Devonian rocks dating to about 390 mya (million years ago). By the Carboniferous (360–300 mya), when ferns and other seedless plants were dominant, many lines of gymnosperms had also arisen, and the first true conifers appeared. These radiated during the Permian period (300–250 mya), and the Mesozoic era that followed, 251 to 65 mya, was the age not only of the dinosaurs but of the gymnosperms as well.

The evolution of gymnosperms involved sweeping changes in plant structures related to reproduction. As a prelude to our survey of modern gymnosperms, we begin by considering some of these innovations, which opened new adaptive options for land plants.

Major Adaptations in Growth and Reproduction Occurred as Gymnosperms Evolved

Early gymnosperm evolution was marked by several innovations, all of which facilitated the radiation of gymnosperms into various land environments. These changes included increasing size and major reproductive adaptations—the ovule, pollen and pollination, and the seed. These adaptations helped spark the diversification of plants in the Devonian period, 416 to 359 mya. Although the fossil record is sketchy, it provides glimpses of the order and types of changes that accompanied this shift.

A NEW MERISTEM FOR SECONDARY GROWTH By about 390 mya, in the mid Devonian, a group called *progymnosperms* had evolved from seedless vascular plants. Progymnosperms did not have seeds, but they did have megaphyll-type leaves (with branching veins; refer to Figure 28.5). Some also had woody

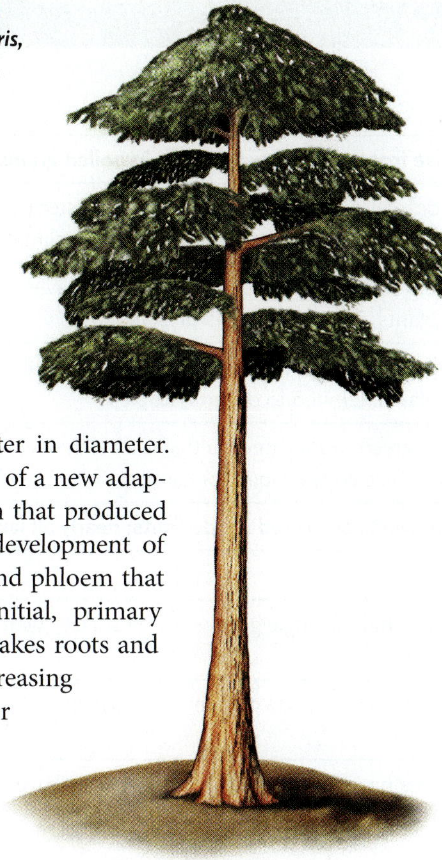

FIGURE 29.2 Fossil-based reconstruction of *Archaeopteris*, a Devonian progymnosperm.

trunks more than a meter in diameter. Such girth was the result of a new adaptation, a lateral meristem that produced *secondary growth*—the development of woody layers of xylem and phloem that takes place once the initial, primary plant body forms and makes roots and stems sturdier by increasing their diameter. (Chapter 33 discusses secondary growth in detail.) **Figure 29.2** shows an artist's rendering of the progymnosperm *Archaeopteris,* a small tree with a tapering trunk that was common over parts of Europe and North America and is thought to have reached a height of about 6 m.

OVULES, POLLEN, AND SEEDS The word *gymnosperm* is derived from the Greek *gymnos,* meaning naked, and *sperma,* meaning seed. Although progymnosperms were seedless, fossil evidence suggests that seed-bearing gymnosperm lineages probably arose from them. For example, fossils reveal that species in the progymnosperm group that included *Archaeopteris* were heterosporous: they produced two types of spores (megaspores and microspores) that were released from structurally different sporangia. Some progymnosperm fossils indicate that megaspores were retained instead of released, and were surrounded by a layer of tissue called an *integument*—an arrangement that has been interpreted as the foundation for an ovule that would develop into the gymnosperm seed. Fossils dating to the late Devonian confirm the existence of (at least) two seed-bearing gymnosperm groups: early conifers, and a now-extinct lineage of small, woody, treelike species called *seed ferns,* so named because they had leaves resembling fern fronds.

Other changes, perhaps happening simultaneously, resulted in the evolution of pollen grains from microspores. As described next, pollen grains have hard walls and fossilize relatively well. Some amazingly well-preserved specimens reveal that as the Devonian period came to a close, several different types of gymnosperm pollen grains existed.

POLLEN AND OVULES: SHELTER FOR SPORES Unlike bryophytes and seedless vascular plants, gymnosperm sporophytes do not disperse their spores. The sporophyte produces haploid spores by meiosis, but it retains these spores inside reproductive structures where they give rise to multicellular haploid gametophytes. A **pollen grain** is a male gametophyte, which produces sperm. Each pollen grain is produced from a microspore that typically has walls reinforced with the polymer sporopollenin. All but a few gymnosperms have nonmotile sperm. Usually, two of these nonswimming cells develop inside each pollen grain—very different from the flagellated, swimming sperm of algae and seedless plants.

An **ovule** is a structure in a sporophyte in which a female gametophyte develops, complete with an egg. Physically connected to the sporophyte and surrounded by the ovule's protective layers, a female gametophyte no longer faces the same risks of predation or environmental assault that can threaten a free-living gametophyte.

Pollination is the transfer of pollen to female reproductive parts via air currents or on the bodies of animal pollinators. Pollen and pollination were enormously important adaptations for gymnosperms, because the shift to nonswimming sperm along with a means for delivering them to female gametes meant that reproduction no longer required liquid water. The only gymnosperms that have retained swimming sperm are the cycads and ginkgoes described shortly, which have relatively few living species and are restricted to a smattering of native habitats.

SEEDS: PROTECTING AND NOURISHING PLANT EMBRYOS A **seed** is the structure that forms when an ovule matures after a pollen grain reaches it and a sperm fertilizes the egg. It consists of three basic parts: (1) the embryo sporophyte; (2) tissues around it containing carbohydrates, proteins, and lipids that nourish the embryo until it becomes established as a plantlet with leaves and roots; and (3) a tough, protective outer seed coat **(Figure 29.3).** This complex structure makes seeds ideal packages for sheltering an embryo from drought, cold, or other adverse conditions. As a result, seed-making plants enjoy a tremendous survival advantage over species that simply release spores to the environment. Encased in a seed, the embryo also can be transported far from its parent, as when ocean currents carry coconut seeds ("coconuts" protected in large, buoyant fruits) hundreds of kilometers across the sea. As discussed in Chapter 36, some plant embryos housed in seeds can remain dormant for months or years before environmental conditions finally stimulate them to germinate and grow.

Modern Gymnosperms Include Conifers and a Few Other Groups

Today there are about 800 gymnosperm species, including the largest and oldest living land plants. Giant sequoias *(Sequoiadendron giganteum)* may grow to more than 80 m, and coast redwoods *(Sequoia sempervirens)* to more than 112 m. Both are native to Cali-

A. Cutaway view of the seed of a pine, a gymnosperm

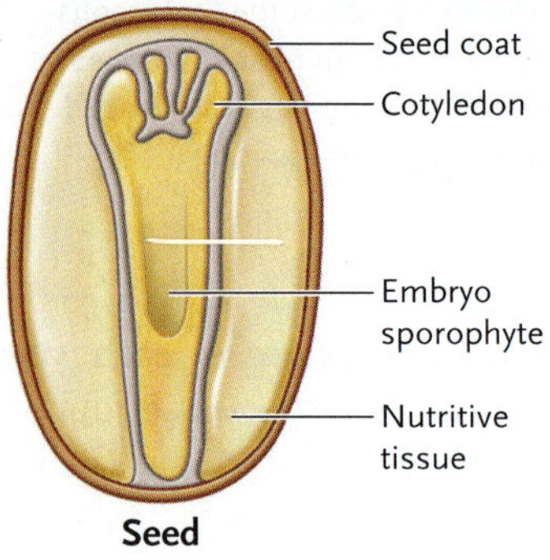

B. Examples of angiosperm seeds, which are contained inside fruits

Seeds of the coconut palm (*Cocos nucifera*) are contained within a fibrous, husklike fruit.

Vasiliy Koval/Shutterstock.com

Buttery, edible flesh (botanically, tissue of the ovary) encloses the seed of avocado (*Persea americana*).

Denis Tabler/Shutterstock.com

FIGURE 29.3 Seeds.

fornia. Tree-ring analysis revealed that one specimen of another California native, the bristlecone pine *(Pinus aristata),* has survived almost 5,000 years, making it the world's oldest known continuously standing tree. The sporophytes of nearly all gymnosperms are sizable trees or shrubs, although a few are woody vines. The most widespread and familiar gymnosperms are the conifers (Coniferophyta). Others are the cycads (Cycadophyta), ginkgoes (Ginkgophyta), and gnetophytes (Gnetophyta).

Economically, gymnosperms, particularly conifers, are vital to human societies. They are sources of lumber, paper pulp, turpentine, and resins, among other products. They also have huge ecological importance. Their habitats range from tropical forests to deserts, but gymnosperms are most dominant in the cool-temperate zones of the northern and southern hemispheres. They flourish in poor soils where flowering plants do not compete as well. In North America, for example, conifer forests cover more than one-third of the continent's landmass—although in some areas, logging has significantly reduced the once-lush forest cover. Our survey of living gymnosperms begins, however, with the cycads, ginkgoes, and gnetophytes—the latter two groups remnants of lineages that have all but vanished from the modern scene.

Carlton Ray/Science Photo Library/Photo Researchers, Inc.

FIGURE 29.4 The cycad *Encephalartos*. Note the large, terminal female cone and fernlike leaves.

Cycads Are Restricted to Warmer Climates

During the Mesozoic era (251–65 mya), the **Cycadophyta** (*kykas* = palm), or cycads, flourished along with the dinosaurs. This lineage probably arose from ancient seed ferns. About 185 species have survived to the present, confined to the tropics and subtropics.

At first glance, you might mistake a cycad for a small palm tree **(Figure 29.4).** Some cycads have massive, cone-shaped strobili (clusters of sporophylls) that bear either pollen or ovules. Air currents or beetles transfer pollen from male plants to the developing gametophyte on female plants. Poisonous alkaloids that may help deter insect predators occur in various cycad tissues. In tropical Asia, some people consume cycad seeds and flour made from cycad trunks, but only after the toxic compounds have been rinsed away. Much in demand from fanciers of unusual plants, cycads in some countries are uprooted and sold in what amounts to a black-market trade—greatly diminishing their numbers in the wild.

Ginkgoes Are Limited to a Single Living Species

The phylum **Ginkgophyta** may also have arisen from seed ferns, though evidence supporting this hypothesis is debated. The phylum has one living species, the ginkgo (maiden-hair) tree *(Ginkgo biloba),* which grows wild today only in warm-temperate forests of central China. Ginkgo trees are large, diffusely branching trees with characteristic fan-shaped leaves **(Figure 29.5)** that turn a brilliant yellow in autumn. Nursery-propagated male trees often are planted in cities because they are resistant to insects, disease, and air pollutants. The female trees are equally pollution-resistant, but their fleshy seeds synthesize butyric acid, which has a notoriously foul odor.

Gnetophytes Include Simple Seed Plants with Intriguing Features

The phylum Gnetophyta contains three genera—*Gnetum, Ephedra,* and *Welwitschia*—that together include about 70 species. Moist, tropical regions are home to about 30 species of *Gnetum,* which includes both trees and leathery leafed vines (lianas). About 35 species of *Ephedra* grow in desert regions of the world **(Figure 29.6A–B).**

A. Ginkgo tree

©Joyce Photographics/Photo Researchers, Inc.

B. Fossil and modern gingko leaves

Albert J. Copley/AgeFotostock

anaken2012/Shutterstock.com

C. Male cone

William Ferguson

D. Ginkgo seeds

Kingsley R. Stern

FIGURE 29.5 ***Ginkgo biloba.*** **(A)** A ginkgo tree. **(B)** Fossilized ginkgo leaf compared with a leaf from a living tree. The fossil formed at the Cretaceous–Tertiary boundary. Even though 65 million years have passed, the leaf structure has not changed much. **(C)** Pollen-bearing cones and **(D)** fleshy-coated seeds of the *Ginkgo*.

Of all the gymnosperms, *Welwitschia* strikes many people as the most bizarre. It grows in the hot deserts of southern and western Africa. The bulk of the plant is a deep-reaching taproot, with the only exposed part a woody disk-shaped stem that bears cone-shaped strobili and leaves. The plant never produces more than two strap-shaped leaves, which split lengthwise repeatedly as the plant grows older, producing a rather scraggly pile **(Figure 29.6C)**.

Although gnetophytes are structurally and functionally simpler than most other seed plants, recent studies of sexual reproduction mechanisms in *Gnetum* and *Ephedra* species uncovered a two-step process of fertilization—which is a hallmark of angiosperms, the most advanced seed plants. This discovery raised some provocative evolutionary questions, even leading some investigators to propose that ancient gnetophytes may have given rise to flowering plants. This hypothesis has since been contradicted by molecular findings, such as those arrived at by a research team at the Academia Sinica in Taiwan (China). When the team compared 65 nuclear rRNA sequences from ferns, gymnosperms, and angiosperms, their analysis supported the hypothesis that cycads and ginkgoes represent the earliest gymnosperm lineage, with a divergent lineage of gnetophytes and conifers arising later. The team found no molecular evidence for a link between the Gnetophyta and angiosperms. The cladogram in **Figure 29.7** shows these proposed evolutionary relationships.

Conifers Are the Most Common Gymnosperms

About 80% of all living gymnosperm species belong to the phylum **Coniferophyta,** or conifers ("cone-bearers"). Conifer sporophytes are longer-lived, and anatomically and morphologically more complex, than any plant sporophyte we have discussed so far. For example, each year most of the 550 conifer species produce a new layer of woody secondary tissue (which you can see illustrated in Figures 33.24 and 33.26); in some species, such as the sequoias and Douglas fir *(Pseudotsuga menziesii)*, the girth of mature trees can easily exceed 6 m, and some individuals measure more than 15 m in diameter. Conifers also form woody reproductive cones, and most species are woody trees or shrubs with needlelike or scalelike leaves, which are anatomically adapted to arid habitats. For instance, needles have a thick cuticle, sunken stomata, and a fibrous epidermis, all traits that reduce the loss of water vapor.

Most conifers are evergreens. That is, although they shed old leaves, often in autumn, they retain enough leaves so that they still look "green," unlike deciduous species like maples,

A. *Ephedra* plant

Edward S. Ross

B. *Ephedra* male cones

William Ferguson

FIGURE 29.6 **Gnetophytes. (A)** Sporophyte of *Ephedra*, with a close-up of **(B)** its pollen-bearing male cones. Female cones develop on separate plants. **(C)** Sporophyte of *Welwitschia mirabilis*, with seed-bearing cones.

C. *Welwitschia* plant with female cones

Fletcher & Baylis/Photo Researchers, Inc.

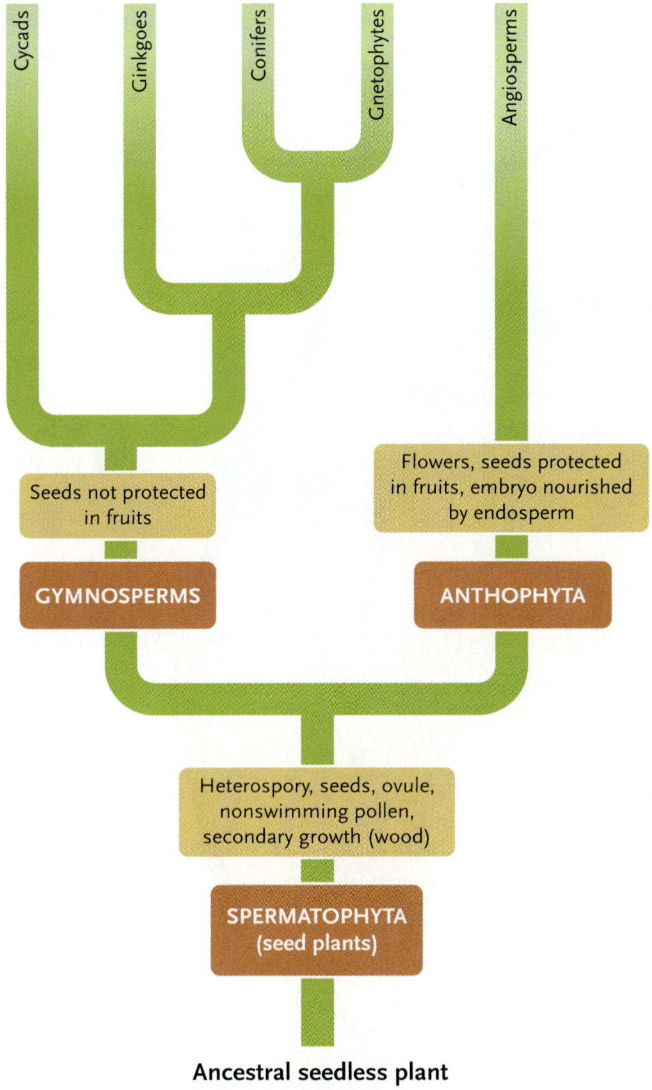

FIGURE 29.7 Overview of the possible phylogenetic relationships between the lineages of seed plants. This diagram provides only a general picture of the points in seed plant evolution where major adaptations took hold. For example, heterospory and seeds are shown as adaptations common to all seed plants, but some living fern species also are heterosporous. Fossil evidence indicates that certain ancient lycophytes and horsetails also produce two types of spores and some had seeds as well. Cycads and ginkgoes are unlike other gymnosperms in that they have swimming sperm.

which shed *all* their leaves as winter approaches. Familiar conifer examples are the pines, spruces, firs, hemlocks, junipers, cypresses, and redwoods. Conifers produce pollen in clusters of small strobili and ovules in larger, woody ones. Both types of strobili are often referred to as cones. Seeds develop on the shelf-like scales of the female cones.

Pines and many other gymnosperms produce resins, a mix of organic compounds that are by-products of metabolism. Resin accumulates and flows in long resin ducts through the wood, inhibiting the activity of wood-boring insects and certain microbes. Pine resin extracts are the raw material of turpentine and (minus the volatile terpenes) the sticky rosin used to treat violin bows.

We know a great deal about the pine life cycle **(Figure 29.8)**, so it is a convenient model for gymnosperm life cycles. All but 1 of the 93 pine species are trees (*Pinus mugo,* native to high elevations in Europe, is a shrub). The male cones (strobili) are relatively small and delicate, only about 1 cm long, and are borne on the lower branches. Each one consists of many small scales, which are specialized leaves (called sporophylls) attached to the cone's axis in a spiral. Two sporangia, called *microsporangia,* develop on the underside of each scale. Inside the microsporangia, spore "mother cells" called microsporocytes undergo meiosis and give rise to haploid **microspores.** Each microspore then undergoes mitosis to develop into a winged pollen grain—an immature male gametophyte. At this stage the pollen grain consists of four cells, two that will degenerate and two that will function later in reproduction.

Young female cones develop higher in the tree, at the tips of upper branches. The cone scales bear ovules. Inside each ovule is a spore mother cell called a megasporocyte. Unlike microsporocytes, the megasporocyte in an ovule undergoes meiosis only when conditions are right. It then produces four haploid spores called **megaspores.** Only one megaspore survives, however, and it develops slowly, becoming a mature female gametophyte only when pollination is underway. In a pine, the process takes well over a year. The mature female gametophyte is a small oval mass of cells with several archegonia at one end, each containing an egg.

Each spring, air currents lift vast numbers of pollen grains off male cones—by some estimates, a single pine may spew billions. The extravagant numbers ensure that at least some pollen grains will land on female cones. The process is not as random as it might seem: studies have shown that the contours of female cones create air currents that can favor the "delivery" of pollen grains near the cone scales. After pollination, the pollen grain develops into a *pollen tube* that grows toward the female spore mother cell. As it does, sperm form in the tube and stimulate maturation of the female gametophyte and the production of eggs. When a pollen tube reaches an egg, the stage is set for fertilization, the formation of a zygote, and early development of the plant embryo. Often, fertilization occurs months to a year after pollination. Once an embryo forms, a pine seed—which, recall, includes the embryo, female gametophyte tissue, and seed coat—eventually is shed from the cone. The seed coat protects the embryo from drying out, and the female gametophyte tissue serves as its food reserve. This tissue makes up the bulk of a "pine nut."

STUDY BREAK 29.1

1. **What are the four major reproductive adaptations that evolved in gymnosperms?**
2. **What are the basic parts of a seed, and how is each one adaptive?**
3. **Describe some features that make conifers structurally more complex than other gymnosperms.**

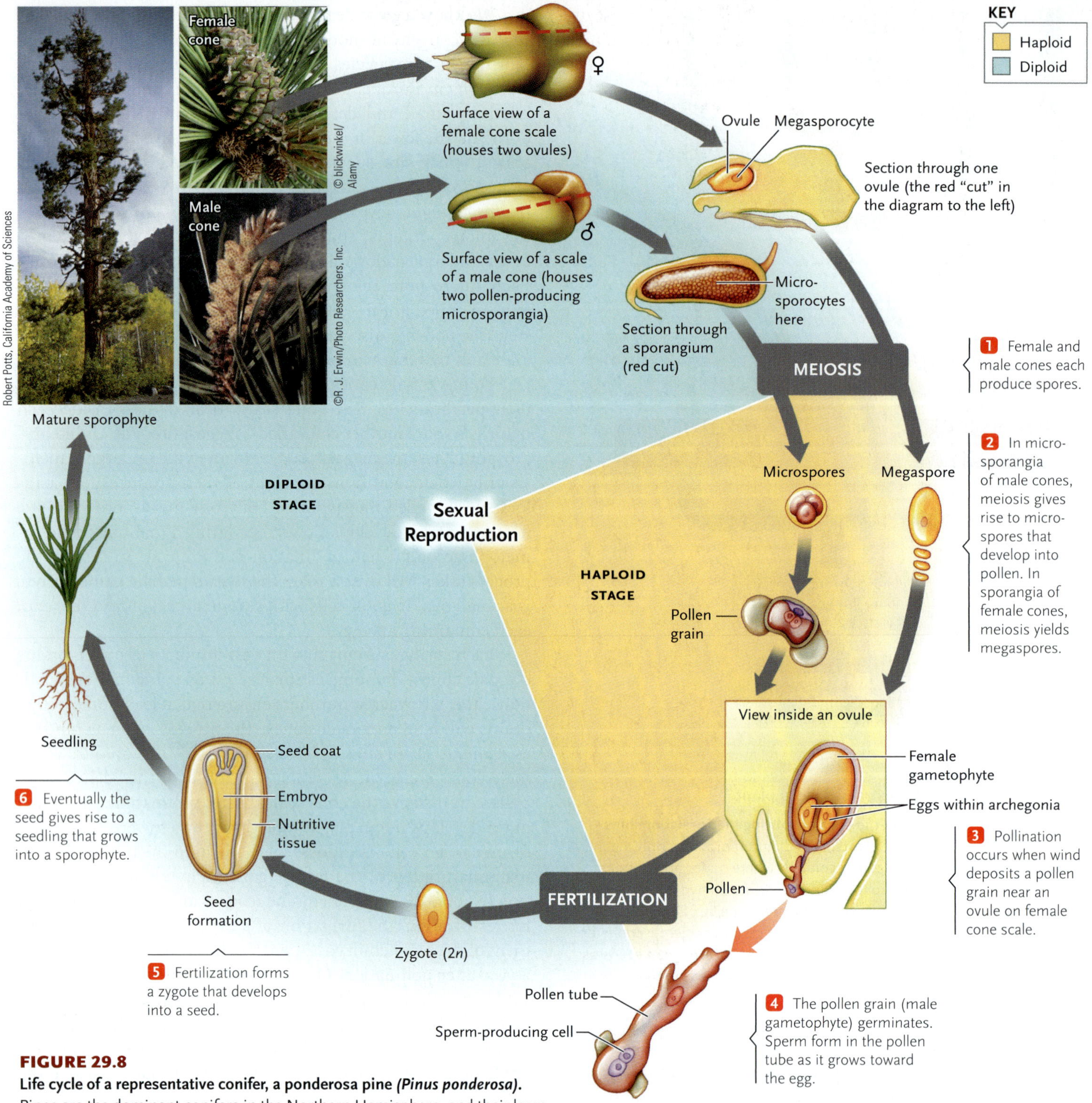

FIGURE 29.8

Life cycle of a representative conifer, a ponderosa pine *(Pinus ponderosa)*. Pines are the dominant conifers in the Northern Hemisphere, and their large sporophytes provide a heavily exploited source of wood.

THINK OUTSIDE THE BOOK

Seed ferns are assigned to the extinct phylum Pteridospermophyta. They were nearly wiped out by the Permian–Triassic mass extinction event that occurred about 251 mya and became extinct soon thereafter. Seed-bearing conifers living at the time fared better, and conifers were the dominant land plants for at least another 100 million years. Using Internet or library resources, develop a plausible explanation for why the seed fern and conifer lineages had such differing evolutionary fates.

A. Flowering plants in a desert

Marketa Mark/Shutterstock.com

B. Alpine angiosperms

BILL COSTER/Still Pictures

C. Triticale, a grass

Pietus/Shutterstock.com

D. A parasitic angiosperm

©Peter F. Zika/Visuals Unlimited

FIGURE 29.9 Flowering plants. Diverse photosynthetic species are adapted to nearly all environments, ranging from **(A)** deserts to **(B)** snowlines of high mountains. **(C)** Triticale, a hybrid grain derived from parental stocks of wheat *(Triticum)* and rye *(Secale)*, is one example of the various grasses utilized by humans. **(D)** The parasitic flowering plant Indian pipe *(Monotropa uniflora)* having no chlorophyll of its own, obtains food by associating with mycorrhizae, which are in turn associated with the roots of photosynthetic plants.

29.2 Angiosperms: Flowering Plants

Of all plant phyla, the flowering plants, or angiosperms, are the most widespread today. At least 260,000 species are known (**Figure 29.9** shows a few examples), and researchers regularly discover new ones in previously unexplored regions of the tropics. The word angiosperm means "enclosed seed." It refers to the fact that a modified leaf called a **carpel** surrounds and protects the ovules and later, the seeds of angiosperms. Carpels are parts of flowers, reproductive structures that are a key defining feature of angiosperms. Another defining angiosperm feature is the fruit, which surrounds the angiosperm embryo and aids seed dispersal.

In addition to having flowers and fruits, angiosperms are the most ecologically diverse plants on Earth, growing on dry land and in wetlands, in fresh water and the seas. They range in size from tiny duckweeds about 1 mm long to towering *Eucalyptus* trees more than 100 m tall. Most are free-living photosynthesizers. Others lack chloroplasts and obtain nutrients through associations with fungi.

The Fossil Record Reveals Little about the Origin of Flowering Plants

The evolutionary origin of angiosperms has confounded plant biologists for well over a hundred years. Charles Darwin famously called it the "abominable mystery," because flowering plants seem to appear suddenly in the fossil record, without a fossil sequence that clearly links them to any other plant groups. The oldest well-documented fossil specimens date back about 125 million years to the early Cretaceous. Discovered in China, these remarkable fossils show complex and strikingly modern-looking plants that have leaves, stems, fruits, and seeds **(Figure 29.10)**. Two species have been unearthed and have been assigned to the genus *Archaefructus*, representing a newly discovered, extinct angiosperm group.

The fossil record has yet to reveal obvious transitional organisms connecting either seedless vascular plants or gymnosperms to flowering plants.

As with gymnosperms, attempts to reconstruct the earliest flowering plant lineages from morphological, developmental, and biochemical characteristics have produced conflicting classifications and family trees. Some paleobotanists hypothesize

FIGURE 29.10 Fossil of *Archaefructus sinensis,* thought to have been an early flowering plant. The sketch shows what this small, possibly aquatic plant may have looked like.

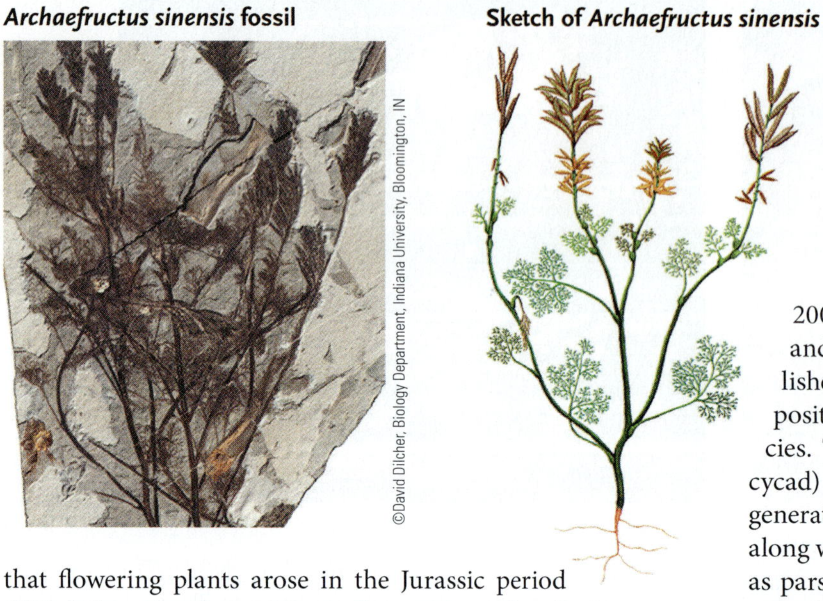

FIGURE 29.11 Eudicot pollen grain. Eudicot pollen grains have three slitlike grooves, only two of which are visible here. Pollen made by all other seed plants, including monocots, have just one groove.

that flowering plants arose in the Jurassic period (200–145 mya); others propose they arose earlier, in the Triassic (250–200 mya) from now-extinct gymnosperms or from seed ferns.

As the Mesozoic era ended and the modern Cenozoic era began, great extinctions occurred within both plant and animal kingdoms. Gymnosperms declined, and dinosaurs disappeared. Flowering plants, mammals, and social insects flourished, radiating into new environments. Today we live in what has been called "the age of flowering plants."

Angiosperms Are Subdivided into Several Clades, Including Monocots, Eudicots, and Various Basal Angiosperms

Angiosperms are assigned to the phylum **Anthophyta,** a name that derives from the Greek *anthos,* meaning flower. The great majority of angiosperms are classified either as monocots or eudicots. **Monocots** are distinguished by the morphology of their embryos, which have a single seed leaf called a cotyledon ("cuplike hollow"). **Eudicots** ("true dicots"), which generally have two cotyledons, are set apart from other angiosperms by the structure of their pollen grains, which have three grooves **(Figure 29.11).** By contrast, the pollen of monocots and all other seed plants (including more than 8,500 species once lumped with eudicots under the term "dicots") have only a single groove. Paleobotanists use this clear structural difference not only to help establish the general type of plant that produced fossil pollen, but also what types of plants were present in fossil deposits of a particular age or geographic location.

Although most angiosperms can fairly easily be categorized as either monocots or eudicots, figuring out the appropriate classification for other angiosperms is an ongoing challenge and an extremely active area of plant research. Genetic and phylogenetic studies are playing a pivotal role in this arena. For example, in 2007, Robert K. Jansen at the University of Texas in Austin and 15 colleagues in the United States and Germany published an ambitious phylogenetic study of 76,583 nucleotide positions in 81 plastid genes from the genomes of 64 plant species. Three species of gymnosperms (a pine, a ginkgo, and a cycad) served as the outgroup. The researchers used their data to generate alternative phylogenetic hypotheses that they evaluated along with six previously published ones, using approaches (such as parsimony) outlined in Chapter 24. Their statistical analysis provided the strongest support for a phylogenetic tree summarized in **Figure 29.12.** In addition to eudicots and monocots, this

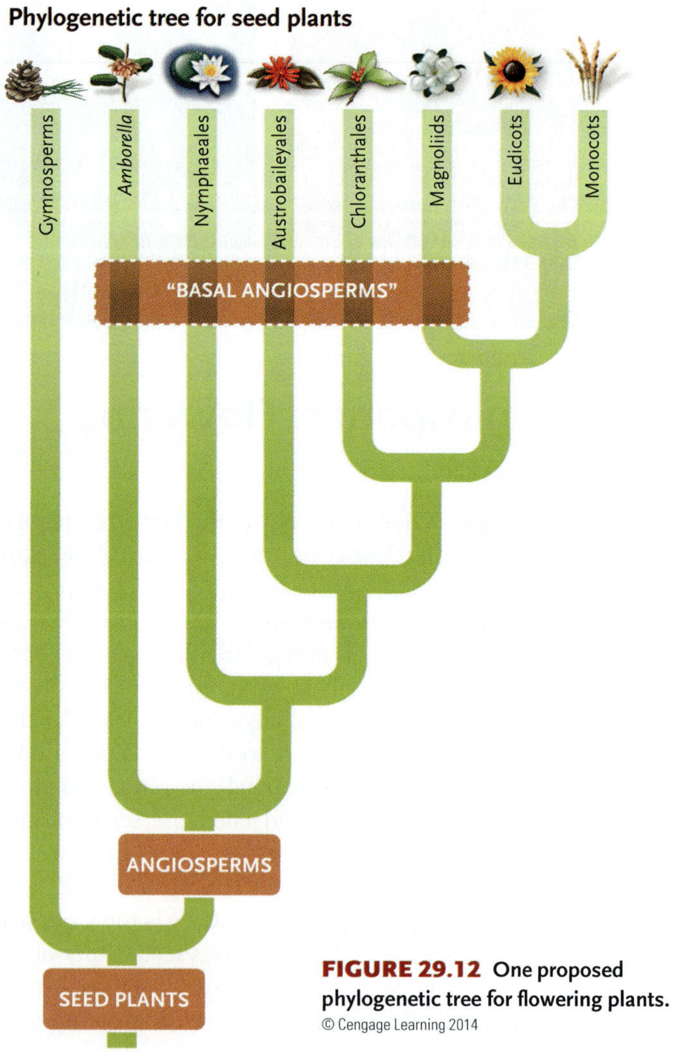

FIGURE 29.12 One proposed phylogenetic tree for flowering plants.

A. Southern magnolia *(Magnolia grandiflora)*, a magnoliid B. Star anise *(Illicium floridanum)* C. Yellow pond lily *(Nuphar polysepala)*, a water lily D. *Amborella*

FIGURE 29.13 Representatives of basal angiosperm clades. (A) Southern magnolia *(Magnolia grandiflora)*, a magnoliid. **(B)** Star anise *(Illicium floridanum)*. **(C)** Yellow pond lily *(Nuphar polysepala)*. **(D)** *Amborella trichopoda*.

proposed tree encompasses five clades of **basal angiosperms** thought to represent the earliest branches of the flowering plant lineage **(Figure 29.13)**. The **magnoliids** includes magnolias (see Figure 29.13A), laurels, and avocados. The **Chloranthales** includes plants that are the sources of spices such as peppercorns, nutmeg, and cinnamon. **Austrobaileyales** includes star anise and its close relatives (see Figure 29.13B), while water lilies are in the **Nymphaeales** (see Figure 29.13C). The Jansen group's work and several other analyses have identified the shrub *Amborella trichopoda* (see Figure 29.13D) as the closest living relative of the first flowering plants—hence *Amborella's* divergence near the base of the phylogenetic tree **(Figure 29.14)**. *Amborella* is found only in cloud forests of the South Pacific island of New Caledonia. Morphological differences also help make it the current leading candidate for "most ancient" living basal angiosperm. *Amborella's* small white flowers and vascular system are structurally simpler than those of other angiosperms, and its female gametophyte differs as well.

Figure 29.15A gives some idea of the variety of living monocots, which include grasses, palms, lilies, and orchids. The world's major crop plants (wheat, corn, rice, rye, sugarcane, and barley) are domesticated grasses, and all are monocots. There are at least 60,000 species of monocots, including 10,000 grasses and 20,000 orchids. Eudicots are even more diverse, with nearly 200,000 species **(Figure 29.15B)**. They include flowering shrubs and trees, most nonwoody (herbaceous) plants, and cacti. **Figure 29.16** shows the life cycle of a monocot, a lily. The life cycle of a typical eudicot is described in detail in the next unit, which focuses on the structure and function of flowering plants.

Many Factors, Including Developmental Plasticity, Contributed to the Adaptive Success of Angiosperms

Throughout the panorama of land plant evolution, no group can match angiosperms in terms of one fundamental attribute: their *developmental plasticity*. No other plant group even comes close to equaling the sheer numbers of angiosperm species and the astonishing diversity of the sizes, shapes, habitats, growth habits, and nutrition modes of flowering plants. Whatever changes resulted in the evolution of the angiosperm lineage from one rooted in a gymnosperm heritage, they must have included a capacity to undergo (and survive) genetic changes that created new options for growth, development, and functioning. Here we consider just a small subset of basic shifts that helped distinguish angiosperms from gymnosperms.

MORE EFFICIENT TRANSPORT OF WATER AND NUTRIENTS Where gymnosperms have only one type of water-conducting cell (tracheids), angiosperms have an additional, more specialized type (called vessel elements). As a result, an angiosperm's xylem vessels move water more rapidly from roots to shoot parts. Also, modifications in angiosperm phloem tissue allow it to more efficiently transport sugars produced in photosynthesis through the plant body.

ENHANCED NUTRITION AND PHYSICAL PROTECTION FOR EMBRYOS Other changes in angiosperms made it more likely that reproduction would succeed. For example, a two-step *double fertilization* process in the seeds of flowering plants gives rise to both an embryo and a unique nutritive tissue (called endosperm) that nourishes the embryonic sporophyte. The ovule containing a female gametophyte is enclosed within an **ovary,** which develops from a carpel and shelters the ovule against desiccation and against attack by herbivores or pathogens. In turn, ovaries develop into the fruits that house angiosperm seeds. As noted earlier, a fruit not only protects seeds, but helps disperse them—for instance, when an animal eats a fruit, seeds may pass through the animal's gut none the worse for the journey and be released in a new location in the animal's feces. Above all, angiosperms have flowers, the unique reproductive organs that you will read much more about in the next unit.

COEVOLUTION WITH ANIMAL POLLINATORS The evolutionary success of angiosperms correlates not only with the adaptations just described, but also with efficient mechanisms of transferring pollen to female reproductive parts. While a conifer depends on air currents to disperse its pollen, angiosperms coevolved with pollinators—insects, bats, birds, and other animals that withdraw pollen from male floral structures (often while

FIGURE 29.14 Experimental Research

Analyzing Phytochromes to Explore Angiosperm Evolution

Question: Did adaptive changes in light-sensitive pigments play a role in the early evolution of flowering plants?

Experiment: Many paleobotanists consider that the first angiosperms arose in the understory of moist land habitats dominated by large Mesozoic gymnosperms and ferns—a habitat not too unlike that where *Amborella* lives today. In this shaded environment, young plants that could gain a roothold despite the lack of bright light might have a distinct survival advantage. To investigate the possibility that strong selection pressure favoring adaptation to dim light played a part in angiosperm evolution, researchers looked at the genes coding for pigment proteins called *phytochromes* that allow plants to detect light of red and far-red wavelengths—wavelengths that predominate in dim light.

An activated phytochrome launches a signaling pathway that may stimulate or inhibit some aspect of plant growth. In seed plants, a small family of *PHY* genes encodes five main phytochromes, A through E, each with a distinct function. The nucleotide sequences of two of the genes, *PHYA* and *PHYC*, are about 50% identical. Plant scientists hypothesize that duplication of an ancestral *PHY* gene occurred before angiosperms arose. *PHYA* and *PHYC* evidently are the descendants of this duplicated ancestral gene. *PHYC* is sensitive to relatively bright light but apparently doesn't respond to dim light and *PHYA* is highly sensitive to dim light and is inactivated by bright light.

To investigate the hypothesis that strong selection pressure drove the divergence of these two genes in ancestral angiosperms, researchers obtained amino acid sequence data for key functional domains of the phytochrome proteins encoded by *PHYA* and *PHYC* in 45 plant species. Most species in the sample were angiosperms; several conifers represented the presumed ancestral gene. Analysis of the data focused both on the number of substitutions in the targeted phytochrome amino acid sequences, as well as on the biochemical effects of the substitutions. Among other information, the analysis produced a phylogenetic tree that displayed genetic divergences as branch points.

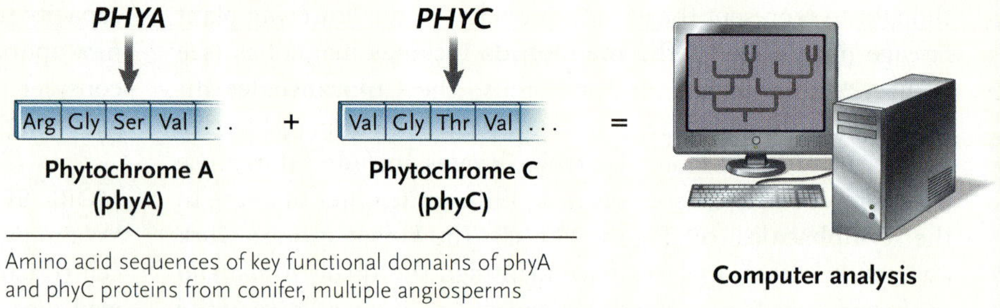

Results: In the tree branch leading from the presumed ancestral *PHY* gene to *PHYA*, 32 amino acid substitutions occurred. Of these, 11 could be best interpreted as shifts associated with selection pressure. In the branch leading to *PHYC*, only seven substitutions occurred; four were best interpreted as associated with selection pressure.

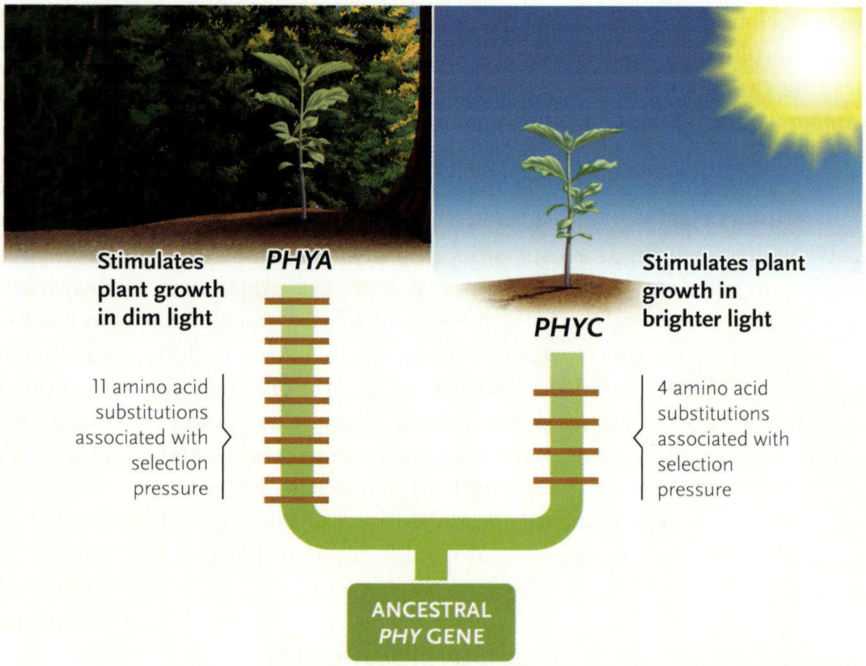

Conclusion: These findings support the hypothesis that diverging molecular characteristics of *PHYA* and *PHYC* correlated with diverging functional features of the phytochromes the genes encode. Evidently, *PHYA* evolved under strong selection pressure. The availability of phytochrome phyA may have allowed early angiosperm seedlings to grow in mostly dim light conditions—an adaptive change that may have fostered the establishment of angiosperms in the shade of Mesozoic forests.

Amino acid substitutions in key functional domains of phyA and phyC proteins from conifer, multiple angiosperms.

THINK LIKE A SCIENTIST What might we learn by examining the *PHY* genes of a moss and a fern, and comparing the results to those of this study?

Source: Sarah Mathews, J. Gordon Burleigh, and Michael J. Donoghue. 2003. Adaptive evolution in the photosensory domain of phytochrome A in early angiosperms. *Molecular Biology and Evolution* 20:1087–1097.

FIGURE 29.15 Examples of monocots and eudicots.

A. Representative monocots

Wheat *(Triticum)* | Tulips *(Tulipa)* | Eastern prairie fringed orchid *(Platanthera leucophaea)*

B. Representative eudicots

Rose *(Rosa)* | Yellow bush lupine *(Lupinus arboreus)* | Cherry *(Prunus)* | Claret cup cactus *(Echinocereus triglochidratus)*

obtaining nectar) and inadvertently transfer it to female reproductive parts. **Coevolution** occurs when two or more species interact closely in the same ecological setting. A heritable change in one species affects selection pressure operating between them, and the other species evolves as well. Over time, plants that came to have distinctive flowers, scents, and sugary nectar coevolved with animals that could take advantage of the rich food source.

In general, a flower's reproductive parts are positioned so that visiting pollinators will brush against them. In addition, many floral features correlate with specific pollinators. For example, reproductive parts may be located above nectar-filled floral tubes the same length as the feeding structure of a preferred pollinator. Nectar-sipping bats **(Figure 29.17A)** and moths forage by night. They pollinate intensely sweet-smelling flowers with white or pale petals that are more visible than colored petals in the dark. Long, thin mouthparts of moths and butterflies reach nectar in narrow floral tubes or flora spurs. The Madagascar hawkmoth uncoils a mouthpart the same length—an astonishing 22 cm—as a narrow floral spur of an orchid it pollinates, *Angraecum sesquipedale* **(Figure 29.17B).** Red and yellow flowers attract birds **(Figure 29.17C),** which have good daytime vision but a poor sense of smell. Hence bird-pollinated plants do not squander metabolic resources to make fragrances. By contrast, flowers of species that are pollinated by beetles or flies may smell like rotten meat, dung, or decaying matter. Daisies and other fragrant flowers with distinctive patterns, shapes, and red or orange components attract butterflies, which forage by day.

Bees see ultraviolet light and visit flowers with sweet odors and parts that appear to humans as yellow, blue, or purple **(Figure 29.17D).** Produced by pigments that absorb ultraviolet light, the colors form patterns called "nectar guides" that attract bees—which may pick up or "drop off" pollen during the visit. Here, as in our other examples, flowers contribute to the reproductive success of plants that bear them.

STUDY BREAK 29.2

1. **How has the relative lack of fossils of early angiosperms affected our understanding of this group?**
2. **Describe two basic features that distinguish monocots from eudicots, and give some examples of species in each clade.**
3. **List at least three adaptations that have contributed to the evolutionary success of angiosperms as a group.**

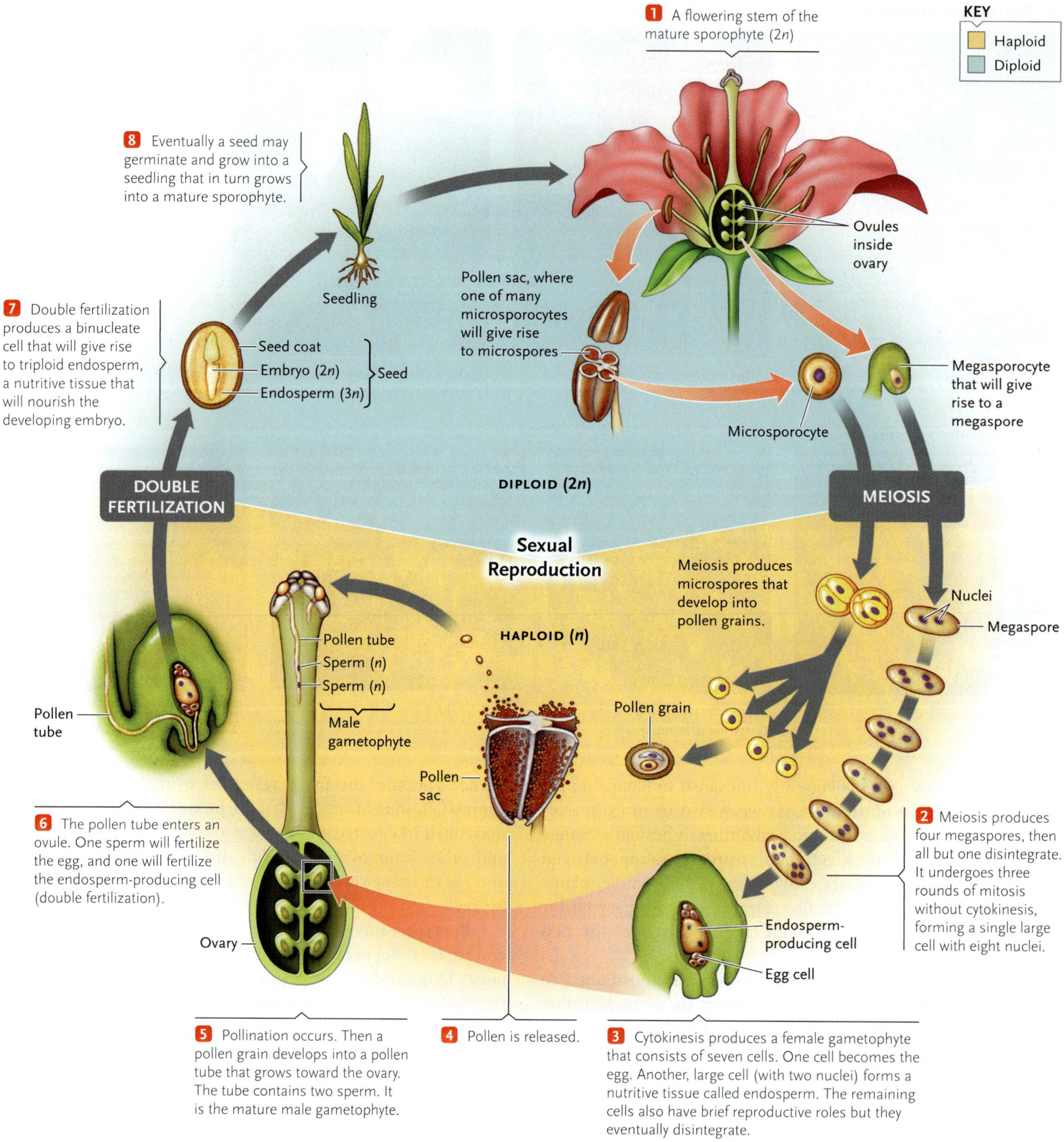

FIGURE 29.16 Life cycle of a flowering plant. This diagram shows a generalized life cycle. Notice that it includes double fertilization: the male gametophyte delivers two sperm to an ovule. One sperm fertilizes the egg, forming the embryo, and the other fertilizes the endosperm-producing cell, which nourishes the embryo. Figure 36.5 shows the life cycle of a eudicot.

FIGURE 29.17 Coevolution of flowering plants and animal pollinators. The colors and configurations of some flowers, and the production of nectar or odors, have coevolved with specific animal pollinators. **(A)** At night, nectar-feeding bats sip nectar from flowers of the giant saguaro *(Carnegia gigantea)*, transferring pollen from flower to flower in the process. **(B)** The hawkmoth *Xanthopan morgani praedicta* has a proboscis long enough to reach nectar at the base of the equally long floral spur of the orchid *Angraecum sesquipedale.* **(C)** A Bahama woodstar hummingbird *(Calliphlox evelynae)* sipping nectar from a hibiscus blossom *(Hibiscus)*. The long, narrow bill of hummingbirds coevolved with long, narrow floral tubes. **(D)** Under ultraviolet light, the bee-attracting pattern of a gold-petaled marsh marigold becomes visible to human eyes.

29.3 Frontiers of Plant Genome Research

Today some of the most exciting research in all of biology involves studies exploring links between genetic changes and key evolutionary transitions in plant form and functioning. A case in point is research on the gene *LFY,* which encodes the regulatory protein LEAFY (Chapter 16 discusses regulatory proteins in detail). The LEAFY protein typically controls expression of several genes by binding to the genes' control sequences. All land plants carry the *LFY* gene, but its effects on phenotype vary markedly in different plant groups. In mosses, which arose almost 400 million years ago, the LEAFY protein regulates growth throughout the plant. In ferns and gymnosperms, which arose later, LEAFY controls growth in a subset of tissues. In angiosperms, LEAFY regulates gene expression only in the particular type of meristem tissue that gives rise to flowers (a topic of Chapter 36). Curious about the evolutionary shift from a general to a specific effect, Alexis Maizel and his team at the Max Planck Institute for Developmental Biology in Germany compared *LFY* sequences and their corresponding proteins in 14 species, including a moss, ferns, gymnosperms, and the angiosperms *Arabidopsis* (thale cress) and snapdragon. Remarkably, they discovered that the evolutionary honing of the effects of the LEAFY protein correlated with only a handful of changes in the base sequence of the *LFY* gene. Each change affected how—or if—the LEAFY protein regulated the expression of a given gene. Over time, LEAFY took on its highly specific, crucial role in angiosperms, helping to direct the developmental events that produce flowers.

As more and more whole plant genomes are sequenced, plant scientists are able to analyze and compare the resulting data in revealing ways. One area of basic interest is the relationship between large-scale shifts in genomes and the evolution of fundamental characteristics of major lineages. *Insights from the Molecular Revolution* describes an example: the comparison of the genomes of a tree species, black cottonwood *(Populus trichocarpa)* and the small weed and model research organism thale cress *(Arabidopsis thaliana).*

Many plant genome-sequencing efforts—such as the one now underway to obtain the full genome sequence of *Amborella*—have a corollary goal of developing improved sequencing methods that will advance future opportunities for research. Sequencing of the genomes of globally important agricultural plants—including wheat *(Triticum),* tomato *(Solanum)* corn *(Zea mays),* rice *(Oryza sativa),* and soybeans *(Glycine max)*—is already providing plant geneticists with major new insights into traits such as growth regulation, disease and pest resistance, and tolerance to drought.

Jump Starting Studies on Plant Genome Evolution

Research Question

Can comparing genome sequences provide clues to the evolution of contrasting adaptations in woody and nonwoody plants?

Experiments

The black cottonwood (*Populus trichocarpa*, **Figure**) is a fast-growing tree native to western North America used widely in the manufacture of high-grade paper, plywood, and other products. It also has a relatively small nuclear genome of about 45,550 protein-coding genes located on 19 chromosomes. These and other characteristics made *P. trichocarpa* the species of choice for an international effort to obtain the first complete gene sequence of a tree. Led by tree geneticist Gerald A. Tuskan of Oak Ridge National Laboratory, 34 researchers worked to sequence the *P. trichocarpa* genome and then compare it to that of the nonwoody plant *Arabidopsis thaliana* (thale cress)—the only other complete eudicot genome sequence then available.

The sequencing used DNA from a single female genotype and employed a whole-genome shotgun approach. In this method, a genome is broken into random, overlapping fragments. After the nucleotide sequence of each fragment is determined, a computer assembles the sequences. In this project, subsequent gene mapping produced a draft sequence for the whole genome with genes assigned to specific chromosomes.

The study produced some provocative findings. For example, genetic events that can have major adaptive significance—such as large-scale chromosome rearrangements and nucleotide substitutions—have occurred much more slowly in *P. trichocarpa* than in *A. thaliana*. Said another way, the "molecular clock" described in Section 24.6 ticks more slowly in *P. trichocarpa*. About 110 mya, a duplication event in *P. trichocarpa* apparently coincided with the divergence of the *P. trichocarpa* and *A. thaliana* lineages from a common ancestor. Then, roughly 45 million years later, about 8,000 pairs of *P. tricocarpa* genes descended from the previous duplication were duplicated again. Tuskan and his colleagues hypothesized that this duplication event had a major impact, affecting more than 90% of the genome. Another key finding was that, while both *P. trichocarpa* and *A. thaliana* have genes that influence disease resistance and membrane transporters, in *P. trichocarpa* evolution has resulted in more multiple-gene families for such traits. The transporters include those that function in hormonal responses governing processes such as the development of vascular tissues that transport water and nutrients. Genes governing wood development (secondary growth) are present in *P. trichocarpa* but not in *A. thaliana*.

FIGURE Black cottonwood, *Populus trichocarpa*.

Conclusions

Comparing the draft *P. trichocarpa* genome sequence with that of *A. thaliana* pointed to a wide array of opportunities for studying how plant gene families arise and how genomes may become reorganized in the course of evolution. The research also provides a broader basis for investigating the genetics of disease resistance, development of vascular tissues, and the formation of wood.

THINK LIKE A SCIENTIST This study focused in part on comparing and contrasting genes for particular traits in two species with very different life histories—a tree and a tiny herbaceous plant. What other factors could be explored to learn more about the differences the study revealed?

Source: G. A. Tuskan et al. 2006. The genome of black cottonwood, *Populus trichocarpa* (Torr. & Gray). *Science* 313(5793):1596–1604.

To date, more than 30 plant genomes have been partially or fully sequenced. In addition to the species already noted, these genomes range from cucumber, potato, and cannabis (marijuana) to cultivated apple, orange, and cacao (the source of chocolate). As the genes of many more plant species are sequenced and correlated with evidence from comparative morphology and the fossil record, we can expect a steady stream of new insights about the adaptations and evolutionary journey of all major plant lineages. It will probably not be long before we have a much deeper understanding of the evolutionary paths by which leaves, flowers, efficient vascular systems and many other plant characteristics have come into being.

In Chapter 30 a very different group of organisms, the fungi, takes center stage. Although many fungal species seem superficially plantlike, biologists today are avidly exploring evolutionary links between fungi and animals.

STUDY BREAK 29.3

1. **How do studies of the plant regulatory protein LEAFY help elucidate changing gene functions during plant evolution?**
2. **Briefly explain some benefits to be gained from having the entire sequences of numerous plant genomes.**

UNANSWERED QUESTIONS

What is the role of gene duplication in the evolution of plant diversity?

What is the genetic basis of the origin of new and complex structures, such as the flower, during the course of evolution? One important factor is gene duplication, thought to be one of the driving forces behind the increase in organismal complexity that we see with evolution. When a plant gene is duplicated, initially there are two identical copies with identical functions (called "redundancy"); often over time one of the copies either is eliminated or becomes nonfunctional, so that the original condition of one gene is restored. But sometimes, through the process of mutation and sequence divergence, the two copies take on different functions. They may divide the functions of the original single gene between them ("subfunctionalization"), or one of the copies may take on entirely new functions, leaving the other copy to perform the original function ("neofunctionalization"). It is this last possibility—the origin of new gene functions after duplication—that is thought to provide the raw material for the origin of new plant structures such as the flower.

Like all gymnosperms, the ancestors of angiosperms produced separate cones with male and female reproductive structures, whereas flowers produce both male and female reproductive organs surrounded by a novel structure unique to flowers, the sterile perianth (sepals and petals). What genetic changes occurred in proto-angiosperms that allowed for the development of the complex bisexual flower and the perianth? Of course we cannot look at the genomes of the extinct angiosperm ancestors, however, by comparing the genomes of extant gymnosperms and angiosperms we can identify key flower development genes that are found only in angiosperms. In *Arabidopsis thaliana*, *APETALA1 (AP1)* and *SEPALLATA (SEP)* genes are required for proper flower formation. If *AP1* genes are eliminated, the plants will not form flowers. If *SEP* genes are eliminated, the plants form structures similar to flowers but with all the floral organs replaced by tiny leaves. *AP1* and *SEP* genes are found only in flowering plants; combined with the observation that these genes are required to form flowers, this suggests that they may have played a role in the evolution of the flower.

AP1 and *SEP* genes appear to have arisen by way of two duplications from a third gene group *(AGL6)* that is found in both gymnosperms and angiosperms. Along with the B- and C-function genes of the ABC model, these genes belong to the MADS-box family of transcription factors, members of which play key roles throughout plant development. Repeated duplication during the course of plant evolution has led to an increase in the number of these regulatory genes from one in algae to over 100 in *Arabidopsis;* the proliferation of these key developmental regulators may be one of the driving forces behind the increase in complexity from algae to angiosperms, and is quite likely to have played a role in the origin of the flower. Thus, duplications in these genes led to the origin in gymnosperms of the B- and C-function genes, which in gymnosperms appear to play a role in specifying reproductive organ identity that is carried into the angiosperms. In contrast, later duplications, coinciding with the origin of the angiosperms, led to the establishment of the *AP1* and *SEP* gene lineages. These angiosperm-specific genes are required for flower formation, and may have been critical, in particular, in the origin of the flower-specific perianth.

THINK LIKE A SCIENTIST If the A function of the ABC model is not found in any species outside the mustard family, explain how the "BCE" model can still account for the formation of four different types of floral organs.

Amy Litt

Amy Litt is Director of Plant Genomics and Cullman Curator at the New York Botanical Garden, where she also earned her Ph.D. Her main interests lie in the evolution of plant form and how changes in gene function during the course of plant evolution have produced novel plant forms and functions—particularly new flower and fruit morphologies. Learn more about her work at http://sciweb.nybg.org/science2/Profile_106.asp.

REVIEW KEY CONCEPTS

To access the course materials and companion resources for this text, please visit www.cengagebrain.com.

29.1 Gymnosperms: The "Naked Seed" Plants

- Gymnosperms (conifers and their relatives), together with angiosperms (flowering plants), are the seed-bearing vascular plants. During the Mesozoic era (248–65 mya), gymnosperms were the dominant land plants (Figures 29.1–29.3).
- Major reproductive adaptations in gymnosperms include pollination, the ovule, and the seed. Gymnosperm seeds mainly protect, nourish, and help disperse embryonic sporophytes (Figures 29.4–29.7).
- Gymnosperm phyla include cycads, ginkgos, and gnetophytes, but conifers are the dominant living group and the conifer life cycle is well understood. An ovule is a sporangium in which a large spore gives rise to a female gametophyte, which is attached to and protected by the sporophyte. Male gametophytes arise from a smaller spore type. Pollination and fertilization do not require water. Fertilization occurs with the ovule and ultimately results in a seed. Pollination takes place via air currents or animal pollinators (Figures 29.8 and 29.9).

Animation: Eudicot seed development

Animation: Pine life cycle

29.2 Angiosperms: Flowering Plants

- Angiosperms (Anthophyta) have dominated the land for more than 100 million years and currently are the most diverse plant group. The two main angiosperm clades are monocots and eudicots. Clades of basal angiosperms are represented by: magnolias and their relatives (magnoliids); peppercorns and their relatives (Chloranthales); the star anise group (Austrobaileyales), water lilies (Nymphaeales), and *Amborella*, a single species thought to be the most basal living angiosperm (Figures 29.10–29.14).

- Angiosperms have the most efficient plant vascular system (xylem and phloem). Reproductive adaptations include a protective ovary around the ovule, endosperm, fruits that aid seed dispersal, the complex organs called flowers, and the coevolution of flower characteristics with the structural and/or physiological characteristics of animal pollinators (Figures 29.15–29.18).

  Animation: Eudicot life cycle

  Animation: Eudicots and monocots

29.3 Frontiers of Plant Genome Research

- Sequencing and analysis of plant genomes has begun to yield important insights into plant evolution.
- Genome analysis also is increasingly being used in studies of the genetic basis of important agricultural traits, such as disease resistance and drought tolerance.

UNDERSTAND & APPLY

Test Your Knowledge

1. Which of the following was *not* a gymnosperm reproductive innovation?
   a. seeds
   b. pollen grains
   c. the ovule
   d. embryonic cotyledons
   e. rudimentary "proto flowers" at the tips of cones
2. Gymnosperm evolution is also notable for which of the following adaptive feature or features?
   a. pollination by means of water
   b. secondary growth by way of a lateral meristem
   c. no major increase in body size over ancestral seedless plants
   d. liberation of free microspores as part of the life cycle
   e. a seasonal loss of all leaves
3. In which modern gymnosperm groups do we see swimming sperm, an ancient reproductive trait?
   a. Ginkgoes and cycads
   b. Conifers and ginkgoes
   c. Gnetophytes and cycads
   d. Conifers and cycads
   e. Gnetophytes and gingkoes
4. The gymnosperm group distinguished by having the greater part of its bulk underground is:
   a. Coniferophyta.
   b. Ginkophyta.
   c. Cycadophyta.
   d. Gnetophyta.
5. Angiosperms and gymnosperms share which of the following reproductive characteristic(s)?
   a. pollination
   b. ovules protected within an ovary
   c. megaspores and microspores produced by meiosis
   d. seeds
   e. a dominant sporophyte generation
6. Which of the following are defining reproductive features of angiosperms?
   a. flowers
   b. seeds protected by a seed coat
   c. sperm transfer by way of a growing pollen tube
   d. development of fruit
   e. development of embryo-nourishing tissue within seeds
7. The angiosperm clade distinguished by having pollen with three grooves is:
   a. monocots.
   b. magnoliids.
   c. the star anise group.
   d. *Amborella*.
   e. eudicots.
8. Which is *not* true of eudicots?
   a. Many species produce grains humans use for food.
   b. The group includes a large number of flowering shrubs.
   c. Most nonwoody plants are eudicots.
   d. There are probably at least three times as many eudicots as monocots.
   e. Eudicots generally have more seed leaves than monocots do.
9. Developmental plasticity evidently has been a hallmark of angiosperm evolution. Which of the following is *not* an example of this?
   a. diversity of growth habits and habitats
   b. larger size than most gymnosperms
   c. unparalleled numbers of species
   d. more diverse shapes of the plant body
   e. wide range of nutrition modes
10. Adaptations that correlate with the adaptive success of angiosperms include:
   a. more rapid water and nutrient transport.
   b. protection of the ovule inside an ovary.
   c. seed dispersal by way of fruits.
   d. coevolution with animals that pollinate flowers.
   e. nourishment of developing embryos by tissues of the female gametophyte.

Discuss the Concepts

1. Suggest adjustments in the angiosperm life cycle that would better suit plants to some future world where environments were generally hotter and more arid. Do the same for a colder and wetter environment.
2. Working in the field, you discover a fossil of a previously undescribed plant species. The specimen is small and may not be complete; the parts you have do not include any floral organs. What sorts of observations would you need in order to classify the fossil as a vascular seed plant with reasonable accuracy? What evidence would you need in order to distinguish between a fossil monocot and eudicot?
3. Modern humans emerged about 100,000 years ago. How accurate is it to state that our species has lived in the Age of Wood? Explain.
4. Compare the size, anatomical complexity, and degree of independence of a Douglas fir female gametophyte and a dogwood female gametophyte. Which one is the most protected from the external environment? Which trend(s) in plant evolution does your work on this question bring to mind?

Design an Experiment

You are studying mechanisms that control the development of flowers, and your research to date has focused on eudicots. A colleague has suggested that you broaden your analysis to include representative basal angiosperms. Outline the rationale for this expanded approach and indicate which additional species or group(s) you plan to include. Discuss the type(s) of data you plan to gather and why you feel the information will make your study more complete.

Interpret the Data

In an inversion, a chromosome segment breaks and then reattaches to the chromosome with the order of genes reversed (see Section 13.3). Inversions disrupt recombination during meiosis, and can have evolutionary consequences. Some plant scientists have proposed a "local adaptation" mechanism in which reproductive isolation—the hallmark of speciation—arises after an inversion yields one or more altered phenotypes that confer a selective advantage in the local environment. There was not much real-world support for this hypothesis until a field study conducted by Duke University researchers David B. Lowery and John H. Willis documented local adaptation and reproductive isolation in populations of the yellow monkeyflower, *Mimulus guttatus*. In this species, which is native to a large swath of western North America, there are two distinct ecotypes. Annuals (AN) complete the life cycle in a single growing season and live in inland habitats where soil moisture is scarce in summer. Perennials (PE) survive for at least two growing seasons and live where soil is moist year-round. Lowery and Willis identified a chromosomal inversion that closely tracked this life history difference. They also found evidence that the inversion contributes to several reproductive isolating barriers, including the time frame for flowering (which is when pollination and fertilization occur). In the experiments, AN and PE plants were grown in test plots in both habitats. In addition, to help ensure that findings reflected the influence of the inversion, Lowery and Willis used laboratory breeding techniques to obtain seedlings having the AN or PE arrangement of genes in a variety of genetic backgrounds. The Key indicates colors associated with each set of test plants. **Figure A** graphs the cumulative proportion of *M. guttatus* plants surviving to flower at the inland growing site. **Figure B** graphs the cumulative proportion of plants that survived at the coastal field site. **Figure C** graphs the cumulative proportion of plants that survived at the inland field site.

Key:

Purple: Seedlings of wild inland annuals (AN)

Teal: Seedlings of wild coastal perennials (PE)

Blue: Lab-bred plants having a PE gene arrangement in an AN genetic background

Orange: Lab-bred plants having the AN arrangement in an AN background

Green: Lab-bred plants having an AN arrangement in a PE genetic background

Red: Lab-bred plants having a PE arrangement in a PE background

A.

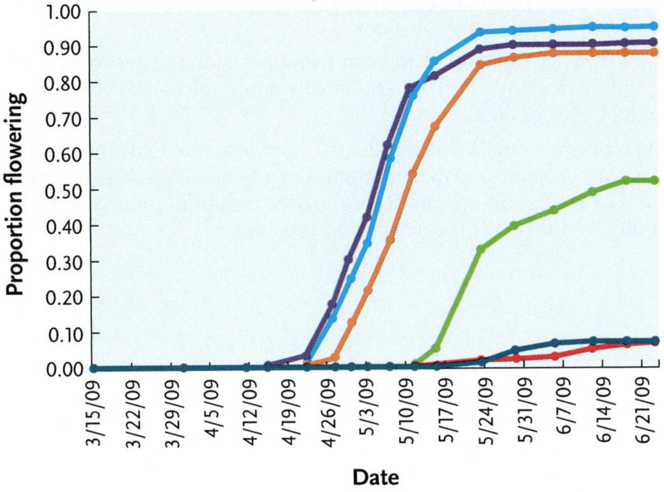

B.

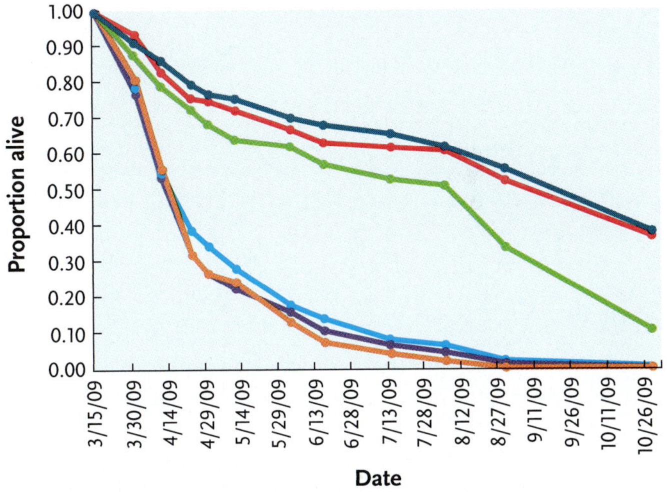

C.

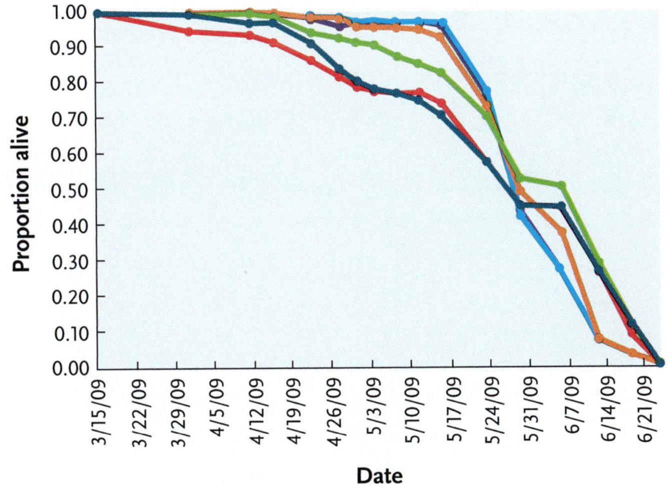

Lowry DB, Willis JH (2010). A Widespread Chromosomal Inversion Polymorphism Contributes to a Major Life-History Transition, Local Adaptation, and Reproductive Isolation. *PLoS Biol* 8(9): e1000500. doi:10.1371/journal.pbio.1000500

1. In Figure A, the graphed data suggest that the AN/PE genotypes are associated with reproductive isolation related to flowering time. Explain why this is so.
2. What general conclusion can you draw from the data graphed in Figures B and C? Propose an explanation for the survivorship patterns in Figure C.

Source: Lowry DB, Willis JH (2010). A Widespread Chromosomal Inversion Polymorphism Contributes to a Major Life-History Transition, Local Adaptation, and Reproductive Isolation. *PLoS Biol* 8(9): e1000500. doi:10.1371/journal.pbio.1000500

Apply Evolutionary Thinking

Evolutionary biologist Spencer C. H. Barrett has written that the reproductive organs of angiosperms are more varied than the equivalent structures of any other group of organisms. Explain why you agree or disagree with his view.

Mushroom-forming fungi of the genus *Mycena*, which are commonly found on decaying wood.

30

Fungi

STUDY OUTLINE

30.1 General Characteristics of Fungi

30.2 Major Groups of Fungi

30.3 Fungal Associations

Why it matters . . . In most natural environments, decay is everywhere—rotting leaves, moldering branches, perhaps the disintegrating carcass of an insect or a small mammal. Gradually, this organic matter is broken down and its elements are recycled in a long-term process that has a huge impact on terrestrial ecosystems. For example, each year the cycling of nutrients returns at least 85 billion tons of carbon, in the form of carbon dioxide, to the atmosphere. Chief among the recyclers are the curious and, to many eyes, beautiful organisms of the **Kingdom Fungi**—about 70,000 described species of molds, mushroom-forming fungi, yeasts, and their relatives **(Figure 30.1)**, and an estimated 1.5 million more that are yet to be described.

Fungi are eukaryotes, most are multicellular, and all are heterotrophs, obtaining their nutrients by breaking down organic molecules that other organisms have synthesized. Molecular evidence suggests that fungi were present on land at least 500 million years ago, and possibly much earlier. In the course of the intervening millennia, evolution equipped fungi with a remarkable ability to break down organic matter, ranging from living and dead organisms and organic wastes to your groceries, clothing, paper and wood, even photographic film. Along with heterotrophic bacteria, they have become Earth's premier decomposers.

Fungi collectively also are a major cause of plant diseases, and a host of species causes disease in humans and other animals. Some even produce carcinogenic toxins. On the other hand, 90% of plants obtain needed minerals by way of a symbiotic relationship with a fungus. Humans have harnessed the metabolic activities of certain fungi to obtain substances ranging from cheeses and wine to therapeutic drugs such as penicillin and the immunosuppressant cyclosporin. And, as you know from previous chapters, species such as the yeast *Saccharomyces cerevisiae* and the mold *Neurospora crassa* are pivotal model organisms in studies of DNA structure and function, and *S. cerevisiae* has also been important in the development of genetic engineering methods.

FIGURE 30.1 Examples of fungi that hint at the rich diversity within the Kingdom Fungi.

Despite their profound impact on ecosystems and other life forms, most people are acquainted with fungi mainly as edible treats or the invisible but annoying causes of skin infections like athlete's foot. This chapter introduces mycology, the study of fungi (*mykes* = mushroom, *logia* = study). We begin with general characteristics of this kingdom and then discuss its major groups.

30.1 General Characteristics of Fungi

We begin our survey of fungi by examining how fungi differ from other forms of life, how fungi obtain nutrients, and the adaptations for reproduction and growth that enable fungi to spread far and wide through the environment.

Fungi May Be Single-Celled or Multicellular

Two basic body forms, single-celled and multicellular, emerged as the lineages of fungi evolved. A single-celled form is called a **yeast.** A cottony mesh of threadlike filaments makes up multicellular fungi. Some fungal species alternate between yeast and multicellular forms at different stages of the life cycle. Regardless, a rigid wall usually surrounds the plasma membrane of fungal cells. Generally the polysaccharide **chitin** provides this rigidity, the same function it serves in the external skeletons of insects and other arthropods.

A multicellular fungus exploits food sources by way of its slender filaments, which branch repeatedly as they grow over or into organic matter. Each filament is a **hypha** (*hyphe* = web) (plural, *hyphae*), and the combined mass of hyphae is called a **mycelium** (plural, *mycelia*) **(Figure 30.2A).** Hyphae generally are tubular. In most multicellular fungi, cross walls called *septa* (*saeptum* = partition, or fence) (singular, *septum*) divide hyphae into cell-like compartments that contain organelles. In a relatively few multicellular species, nuclei in growing hyphae divide without cytokinesis, so no septa develop and organelles share a common cytoplasm **(Figure 30.2B).** Such fungi are said to be **coenocytic,** which means "contained in a shared vessel."

Depending on the species, hyphal cells may have more than one nucleus, and septa have pores that permit nuclei and other organelles to move between hyphal cells. These passages also allow cytoplasm to extend from one hyphal cell to the next, throughout the whole mycelium. By a mechanism called **cytoplasmic streaming,** cytoplasm can flow through the hyphae, carrying nutrients from food-absorbing parts of the fungus to reproductive structures or other nonabsorptive parts. While mycologists commonly consider fungi with cell-like compartments to be "multicellular," they do so with the understanding that the term has a special meaning for these organisms.

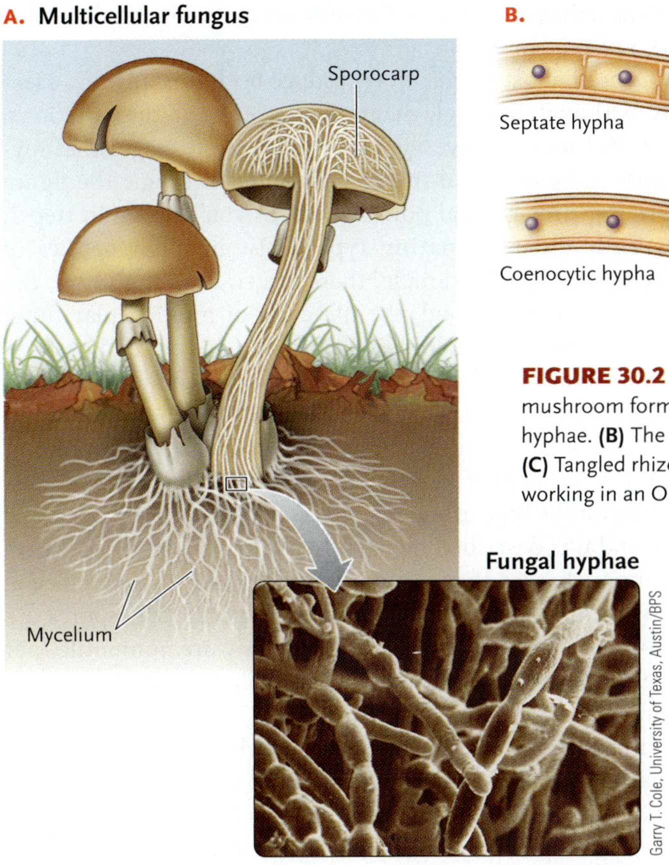

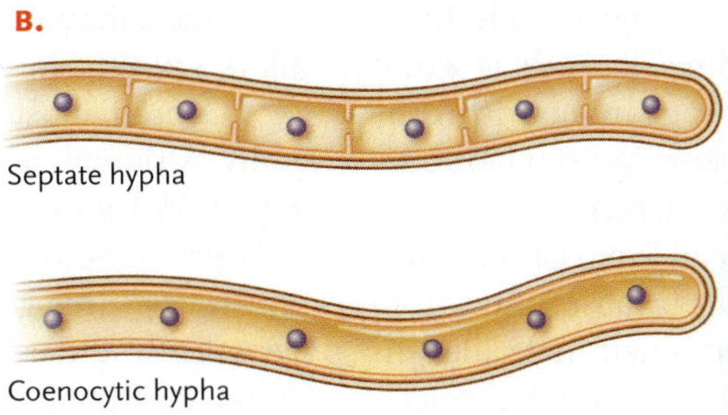

FIGURE 30.2 Fungal mycelia, hyphae, and rhizomorphs. (A) Sketch of the mycelium of a mushroom forming fungus, which consists of branching septate hyphae. *Inset:* Micrograph of fungal hyphae. **(B)** The two basic forms of fungal hyphae: septate (left) and coenocytic, or aseptate (right). **(C)** Tangled rhizomorphs of *Armillaria ostoyae,* the honey mushroom that was uncovered by biologists working in an Oregon forest. Rhizomorphs are modified hyphae that typically extend over roots or through crevices between a tree's bark and the underlying wood. They are features of some species in the phylum Basidiomycota described in Section 30.2.

A multicellular fungus grows larger as its hyphae elongate and branch. Each hypha elongates at its tip as new wall polymers (delivered by vesicles) are incorporated and additional organelles and cytoplasm are synthesized. As the new hyphae elongate and branch, an extensive mycelium can form quickly. Each forming branch fills with cytoplasm that includes new nuclei produced by mitosis. Although the rapid branching of hyphae is what allows multicellular fungi to grow aggressively—sometimes increasing in mass many times over within a few days—researchers have only recently gained the tools to explore the mechanisms that underlie this phenomenon. Studies spurred by the sequencing of the genome of *Neurospora crassa* suggest that multiple steps involving a variety of genes and their interacting protein products determine where and when a new branch arises. Given that the rapid growth of fungal mycelia has such a tremendous impact in ecosystems and in the progression of fungal disease in plants and animals, this topic is a significant area of mycological research.

Beyond their role in nutrient transport, aggregations of hyphae are the structural foundation for all other parts that arise as a multicellular fungus develops. For example, in many fungi a subset of hyphae interweave tightly, becoming prominent reproductive structures sometimes called *sporocarps* (see Figure 30.2A). Grocery store mushrooms are examples. But while a mushroom or some analogous structure may be the most conspicuous part of a given fungus, it usually represents only a small fraction of the organism's total mass. The rest, often invisible to the unaided eye, penetrates the food source the fungus is slowly digesting. In some fungi, modified hyphae form rootlike variations on mycelial threads called *rhizomorphs.* These cordlike structures both anchor the fungus to its existing substrate and provide a means for rapidly colonizing new food sources. Dense tangles of the strands can so damage susceptible tree roots that large patches of forest trees die. In a famous example, U.S. Forest Service scientists found that rhizomorphs of a single individual of the parasitic honey mushroom *Armillaria ostoyae* **(Figure 30.2C)** cover an area of about 880 hectares (roughly 1,665 football fields) in an eastern Oregon forest. By one estimate, the mass of tangled strands extends an average of 1 m deep and nearly 6,000 m across, making the fungus perhaps one of the largest organisms on Earth.

Fungi Obtain Nutrients by Extracellular Digestion and Absorption

Some major selection pressures have shaped the adaptations by which fungi obtain nutrients. As heterotrophs, fungi must secure nutrients by breaking down organic substances formed by other organisms. Nearly all fungi are terrestrial, but unlike other land-dwelling heterotrophs (such as animals), fungi do not move about in their environment. They also lack mouths or appendages for seizing and handling food items. Instead, fungi have a very different suite of adaptations for obtaining nutrients.

Experiments have shown that most species of fungi can synthesize nearly all their required nutrients from a few raw materials, including water, some minerals and vitamins (especially B vitamins), and a sugar or some other organic carbon source. For many species, carbohydrates in dead organic matter are the carbon sources, and fungi with this mode of nutrition are called *saprobes* (*sapros* = rotten). Other fungi are parasites, which extract carbohydrates from tissues of a living host, harming it in the process. Most fungi that parasitize living plants produce hyphal branches called **haustoria** (*haustor* = drinker), which penetrate the walls of a host plant's cells and channel nutrients back to the fungal body. Parasitic fungi include those responsible for many devastating plant diseases, such as wheat rust and Dutch elm disease. Still other fungi are nourished by plants with which they have a mutually beneficial symbiotic association, a topic we return to later in this chapter.

Regardless of their nutritional mode, all fungi gain the raw materials to build and maintain their cells by absorption from the environment. Fungi can absorb many small molecules directly, and they gain access to other nutrients through extracellular digestion. In this process, a fungus synthesizes digestive enzymes and packages them in secretory vesicles via the pathway described in Section 15.4. When released to the outside, the enzymes digest nearby organic matter, breaking down larger molecules into absorbable fragments. Fungal species differ in the particular digestive enzymes they synthesize, so a substrate that is a suitable food source for one species may be unavailable to another. Although there are exceptions, in nature fungi typically thrive only in moist environments where they can directly absorb water, dissolved ions, simple sugars, amino acids, and other small molecules. When some of a mycelium's hyphal filaments contact a food source, growth is channeled toward it.

Various parts of a hypha absorb nutrients. Small atoms and molecules pass readily through the tips, and then transport mechanisms move them through the underlying plasma membrane. Large organic molecules, such as the carbohydrate cellulose (see Section 3.2), *cannot* directly enter any part of a fungus. To feed on such substances, a fungus must secrete hydrolytic enzymes that break down the large molecules into smaller, absorbable subunits. Depending on the size of the subunit, further digestion may occur inside fungal cells.

As described earlier, fungal adaptations for extracellular digestion make them masters of the decay so vital to terrestrial ecosystems. For instance, in a single autumn, one elm tree can shed 400 pounds of withered leaves; and in a tropical forest, a year's worth of debris may total 60 tons per acre. Without the metabolic activities of saprobic fungi and other decomposers such as bacteria, natural communities would rapidly become buried in their own detritus. As fungi digest dead tissues of other organisms, they also make a major contribution to the recycling of chemical elements from those tissues, including key nutrients such as nitrogen and phosphorus. But the prime example of this recycling virtuosity involves carbon. The respiring cells of fungi and other decomposers give off carbon dioxide, liberating carbon that would otherwise remain locked in the tissues of dead organisms. Each year this activity recycles a vast amount of carbon to plants, the primary producers of nearly all ecosystems on Earth.

All Fungi Reproduce by Way of Spores, but Other Aspects of Reproduction Vary

Biologists have observed striking reproductive variations in fungi, differences that are part of what makes them fascinating to study. As you will learn in the next section, fungi have traditionally been classified on the basis of their reproductive characteristics, although today evidence from molecular analysis also plays a prominent role.

Overall, most fungi can reproduce both sexually and asexually. Although no single diagram can depict all the variations, **Figure 30.3** gives an overview of the life cycle stages that mycologists have observed in several groups of fungi. The figure illustrates three general points. First, as you can see in step 1, genetically different **mating types** exist in many species of fungi (in the figure example, there are two mating types, denoted + and −). Second, the life cycle of multicellular fungi typically involves a haploid (n) stage (step 1 is the end point of this stage), a diploid ($2n$) stage (step 5), and between them a dikaryotic ("two nuclei") stage in which the fungus forms hyphae (and a mycelium) that have two haploid nuclei, making them $n + n$—neither strictly haploid *nor* diploid (step 3). Depending on the type of fungus, this stage may be long lasting or brief, and it is described more fully later in this section. Third, all fungi, whether they are multicellular or in a single-celled (yeast) form, can reproduce via **fungal spores.** The spores are microscopic, and in all but one group they are nonmotile (not propelled by flagella). Each spore is a walled single cell or multicellular structure that is dispersed from the parent body, often via wind or water. The spores of single-celled fungi form inside the parent cell, then escape when the cell wall bursts. In multicellular fungi, spores arise in or on specialized hyphal structures and may develop thick walls that help them withstand cold temperatures or drying out after they are released.

Reproductive spores represent a crucial fungal adaptation. Most fungi are opportunists, obtaining energy by exploiting unpredictable food sources in the environment. Spores can go dormant for long periods, then germinate and grow when environmental conditions become favorable. Being lightweight, fungal spores are easily disseminated by air or water—increasing the potential for finding suitable substrates. And releasing vast numbers of spores, as some fungi do, improves odds that at least a few spores will germinate and produce a new individual.

In nature generally, opportunistic organisms are adapted to rapidly take advantage of new food sources. Fungi that can quickly degrade simple sugars and starches often are among the first decomposers to exploit a new source of food. Competition for such resources is stiff, but once fungal spores encounter potential food and favorable conditions, they can quickly develop into new individuals that simultaneously feed and rapidly make more spores.

Many opportunistic fungi develop rapidly and reproduce before the food source is depleted. A common trade-off for speed, however, is small, even microscopic, body size. Larger species of fungi are often adapted to exploit food sources such as cellulose and lignin (a complex polymer in the walls of many plant cells), which their predecessors may have lacked the enzymatic machinery to digest efficiently. Some of these fungi may produce huge mycelia (and reproductive sporocarps such as mushrooms) by extracting nutrients from dead trees that contain enough organic material to sustain an extended period of growth.

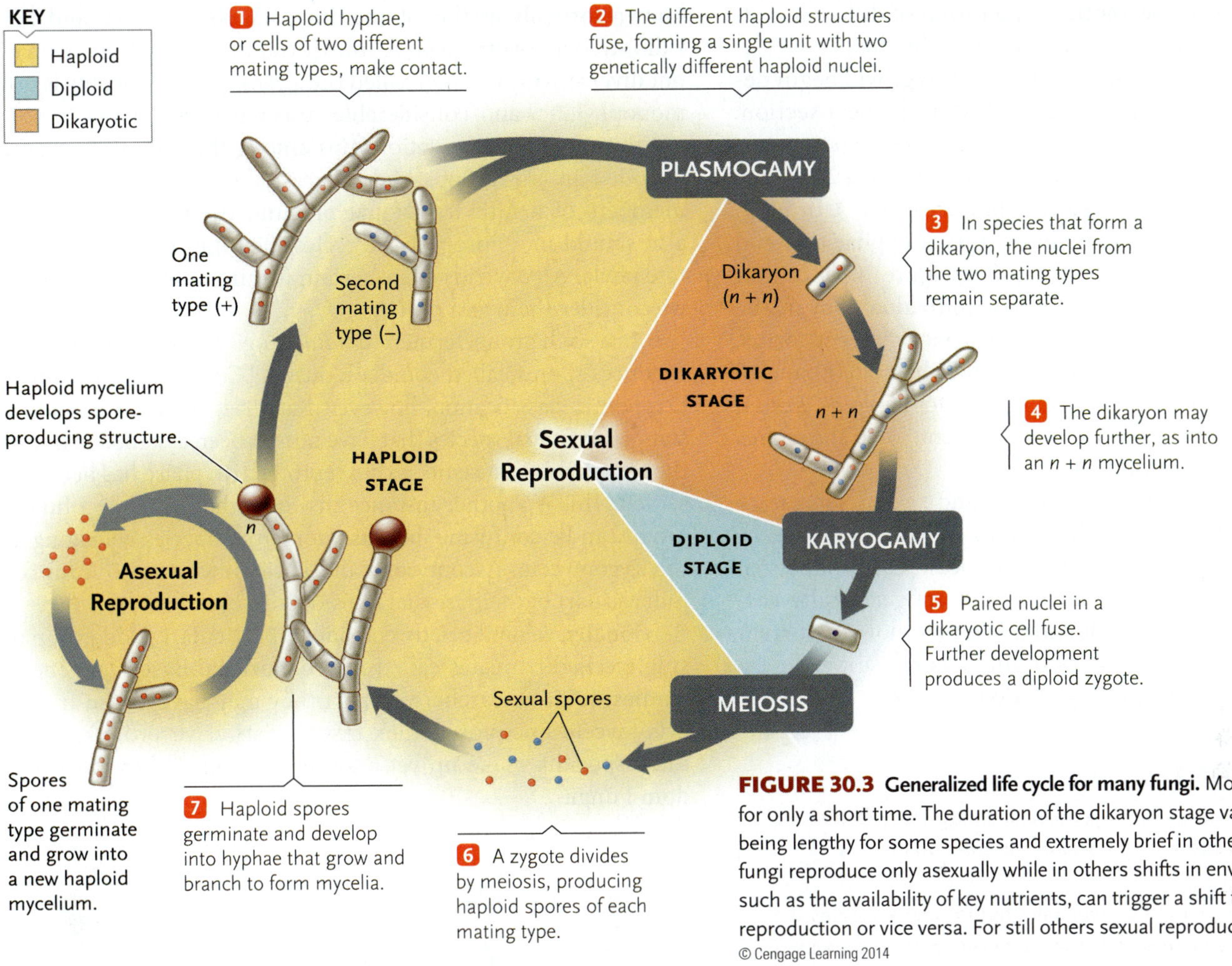

FIGURE 30.3 Generalized life cycle for many fungi. Most fungi are diploid for only a short time. The duration of the dikaryon stage varies considerably, being lengthy for some species and extremely brief in others. Some types of fungi reproduce only asexually while in others shifts in environmental factors, such as the availability of key nutrients, can trigger a shift from asexual to sexual reproduction or vice versa. For still others sexual reproduction is the norm.

FEATURES OF ASEXUAL REPRODUCTION IN FUNGI When a fungus produces spores asexually (see Figure 30.3), it may disperse billions of them into the environment. Some fungi (including many yeasts) also can reproduce asexually by budding or fission, or, in multicellular types, when fragments of hyphae break away from the mycelium and grow into separate individuals. In still others, the fungus produces hyphal fragments *or* asexual spores, depending on environmental factors. Asexual reproductive strategies all result in new individuals that are essentially clones of the parent fungus. They can be viewed as adaptations for reproductive efficiency, because the alternative—sexual reproduction—requires the presence of a suitable partner and generally involves several more steps.

The asexual stage of many multicellular fungi—including the pale gray fuzz you might see on berries or bread—is often called a **mold.** The term can be confusing if you are trying to keep track of taxonomic groupings. For example, the water molds and slime molds described in Chapter 27 are protists, although they were grouped with fungi until additional research revealed their true evolutionary standing. The mold visible on an overripe raspberry is actually a fungal mycelium with aerial structures bearing sacs of haploid spores at their tips.

FEATURES OF SEXUAL REPRODUCTION IN FUNGI Although asexual reproduction is the norm, quite a few fungi shift to sexual reproduction when environmental conditions (such as a lack of nitrogen) or other influences dictate. Recall from Chapter 11 that in sexual reproduction two haploid cells unite, and in most species fertilization—the fusion of two gamete nuclei to form a diploid zygote nucleus—soon follows. In fungi, however, the partners in sexual union can be two hyphae, two gametes, or other types of cells; the particular combination depends on the species involved. And in sharp contrast to most other life forms, many days, months, or even years may pass between the time fertilization gets underway and when it is completed.

During the initial sexual stage, called *plasmogamy* (*plasma* = a formed thing, *gamos* = marriage), the cytoplasms of two genetically different partners fuse. The resulting new cell, a **dikaryon** ("two nuclei"), contains two haploid nuclei, one from each parent. A dikaryon itself is not haploid (the condition of having one set of chromosomes) because it contains two nuclei. But neither is it diploid, because the nuclei are not fused. So, to be precise, we say that a dikaryon has an $n + n$ nuclear condition.

For example, one form of plasmogamy occurs when hyphal cells of two different mating types fuse. The uniting hyphae belong to mycelia of different individuals of the same species that happen to grow near one another. The fusion of different mating types ensures genetic diversity in new individuals.

Once a dikaryon forms, when the next stage will begin depends on the type of fungus, as described in the next section. Sooner or later, however, a second phase of fertilization unfolds: the nuclei in the dikaryotic cell fuse to make a $2n$ zygote nucleus. This process is **karyogamy** ("nuclear union"). In fungi that form mushrooms, it occurs in the tips of hyphae that end on the underside of a mushroom cap (see Figure 30.1). In animals, a zygote is the first cell of a new individual, but in the world of fungi the zygote has a different fate. After it forms, meiosis converts the zygote nucleus into four haploid (n) nuclei. Those nuclei commonly are packaged into haploid "sexual spores" that vary genetically from each parent. Then the spores are released to the environment.

To sum up, in most fungi asexual and sexual spores are haploid, and both can germinate into haploid individuals. However, asexual spores are genetically identical products of asexual reproduction, while sexual spores are genetically varied products of sexual reproduction. We turn now to current ideas on the evolutionary history of fungi, and a survey of the major taxonomic groups in this kingdom.

STUDY BREAK 30.1

1. **What features distinguish the two basic fungal body forms?**
2. **What is a fungal spore, and how does it function in reproduction?**
3. **Fungi reproduce sexually or asexually, but for many species the life cycle includes an unusual stage not seen in other organisms. What is this genetic condition, and what is its role in the life cycle?**

THINK OUTSIDE THE BOOK

Section 19.2 introduced computer-based methods of genome sequence determination and annotation. Complete genome sequences are now available for several species of fungi. Using the Internet or other resources, find out which sequences are available and describe potential research or benefits scientists are hoping to realize from this knowledge.

30.2 Major Groups of Fungi

The evolutionary origins and lineages of fungi have been obscure ever since the first mycologists began mulling over the characteristics of this group. With the advent of molecular techniques, these topics have become extremely active and exciting areas of biological research that may shed light on fundamental events in the evolution of all eukaryotes. Not surprisingly when so much new information is coming to light, mycologists hold a wide range of views on how different groups arose and may be related. Even so, there is wide agreement on three phyla of fungi, known formally as the Glomeromycota, Ascomycota, and Basidiomycota **(Figure 30.4)**. A fourth traditional phylum, a small but diverse array of fungi called the Zygomycota, probably is not monophyletic, and considerable research is underway to sort out the evolutionary relationships among the lineages typically included in the group. Molecular studies suggest that some members of a fifth traditional phylum, the Chytridiomycota, also should more properly be reclassified into two other, separate phyla. All chytrids share certain features, however, and here we consider them as a unit.

In a sixth group, termed *conidial* fungi, asexual reproduction produces spores called conidia ("dust"). "Conidial" is not a true taxonomic classification, however. Rather, it serves as a holding station for fungal species that have not yet been assigned to one of the five phyla because no sexual reproductive phase has been observed. This is another instance in which the name for a fungal group can be confusing, because numerous species belonging to the Zygomycota, Ascomycota, and Basidiomycota also form conidia as part of their asexual reproductive cycle.

Finally, we will briefly consider a particularly odd group of single-celled parasites called **microsporidia.** Based on genetic studies, some researchers suspect they may have arisen from a zygomycete ancestor. Other mycologists have proposed that microsporidia make up yet another phylum within the Kingdom Fungi.

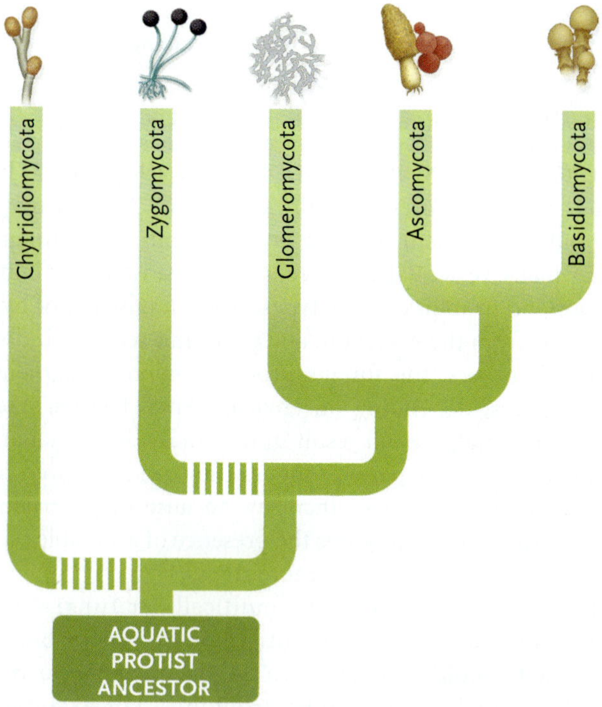

FIGURE 30.4 A phylogeny of fungi. This scheme represents a widely accepted view of the general relationships between major groups of fungi, but it may well be revised as new molecular findings provide more information. The dashed lines indicate that within two traditional groups, the chytrids and zygomycetes, lines of descent of some subgroups are not well understood and the phyla are probably not monophyletic.

Fungi Arose from an Amoeba-like Flagellated Protist

For many years fungi were classified as plants, partly because many species seem to have a plantlike appearance. Recent comparisons of RNA and DNA sequences, and the discovery of chitin in fungal cells, all indicate that fungi and animals are more closely related to each other than they are to most other eukaryotes (see *Insights from the Molecular Revolution*). Phylogenetic studies, including extensive gene sequence analysis, suggest that the lineages of both fungi and animals arose from single-celled, flagellated protists in the clade Opisthokonta (Section 27.2). One such protist, a single-celled choanoflagellate, evidently gave rise to the lineage of animals, while molecular data show that fungi are more closely related to **nucleariids,** single-celled amoebas belonging to a different line of opisthokont protists Among other implications, these findings suggest that multicellular forms of animals and fungi evolved independently.

In general, soft-bodied organisms do not fossilize well, so the fossil record of possible early fungi is sparse. Although traces of what may be fossil fungi exist in rock formations nearly 1 billion years old, the oldest fossils that we can confidently assign to the Kingdom Fungi appear in strata laid down about 500 million years ago.

TABLE 30.1 Summary of Fungal Phyla

Phylum	Body Type	Key Feature	
Chytridiomycota and other chytrids	One to several cells	Motile spores propelled by flagella; usually asexual	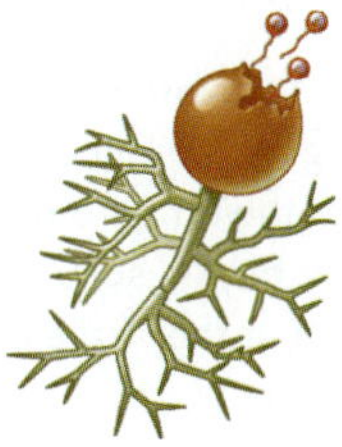
Zygomycota (zygomycetes)	Hyphal	Sexual stage in which a resistant zygospore forms for later germination	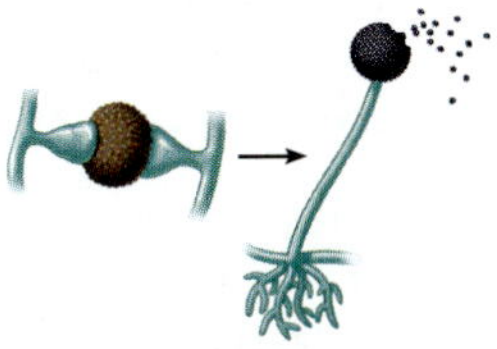
Glomeromycota (glomeromycetes)	Hyphal	Hyphae associated with plant roots, forming arbuscular mycorrhizae	
Ascomycota (ascomycetes)	Hyphal	Sexual spores produced in sacs called asci	
Basidiomycota (basidiomycetes)	Hyphal	Sexual spores (basidiospores) form in basidia of a prominent fruiting body (basidiocarp)	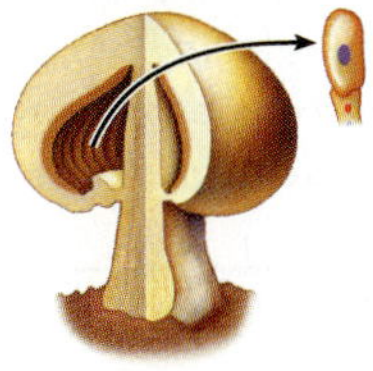

Once They Appeared, Fungi Radiated into Several Major Lineages

Most likely, the first fungi were aquatic, like their protist ancestors. When other kinds of organisms began to colonize land, they may well have brought fungi along with them. For example, researchers have discovered what appear to be mycorrhizae—symbiotic associations of a fungus and a plant—in fossils of some of the earliest known land plants (see Chapter 28). The final section of this chapter examines the nature of mycorrhizae more fully.

Over time, fungi diverged into the strikingly diverse groups that we consider in the rest of this section **(Table 30.1).** As the lineages diversified, different adaptations associated with reproduction arose. For this reason, mycologists traditionally assigned fungi to phyla according to the type of structure that houses the final stages of sexual reproduction and releases sexual spores. These features can still be useful indicators of the phylogenetic standing of a fungus, although now the powerful tools of molecular analysis are bringing many revisions to our understanding of the evolutionary journey of fungi. Our survey begins with chytrids, which probably most closely resemble the fungal kingdom's flagellated protist ancestors.

Chytrids Produce Motile Spores That Have Flagella

There are about a thousand species of chytrids, most assigned to the phylum Chytridiomycota. Chytrids are the only fungi that produce motile spores, propelled by flagella. Nearly all chytrids are microscopic **(Figure 30.5A),** and mycologists have recently begun paying significant attention to them, in part because their characteristics strongly suggest that the lineages in this group arose near the beginning of fungal evolution. Research interest is also keen because the chytrid *Batrachochytrium dendrobatidis* is responsible for a disease epidemic that has wiped out an estimated two-thirds of the species of harlequin frogs *(Atelopus)* of the American tropics **(Figure 30.5B and C).** The epidemic has correlated with the rising average temperature in the frogs' habitats, an increase credited to global warming. The warmer environment provides optimal growing temperatures for the chytrid pathogen.

Most chytrids are aquatic, although a few live as saprobes in soil, feeding on decaying organic matter; as parasites on plants and some animals, or even as symbiotic partners in the gut of cattle and some other herbivores. Wherever a chytrid

There Was Probably a Fungus among Us

What is the taxonomic relationship between the fungi and other eukaryotes? The relationships of the fungi to protists, plants, and animals are buried so far back in evolutionary history that they have proved difficult to reconstruct.

Research Question

At the Kingdom level, which other organisms are the closest relatives of fungi?

Experiment

Relying on morphological comparisons, taxonomists classified fungi as most closely related to protists or plants until an investigation of ribosomal RNA (rRNA) sequences forced significant revision of this hypothesis.

Patricia O. Wainwright, Gregory Hinkle, Mitchell L. Sogin, and Shawn K. Stickel of Rutgers University and the Marine Biology Laboratory, Woods Hole, carried out the analysis by comparing sequences of 18S rRNA, an rRNA molecule that forms part of the small ribosomal subunit in eukaryotes (see Section 15.4). The investigators compared the 18S rRNA sequences of fungi to the 18S rRNA molecules of certain sponges, ctenophores, and cnidarians, of plants, and of several protists, including protozoans and algae. They used a computer program that sorts the rRNA sequences into related groups under the assumption that species with the greatest similarities in 18S rRNA sequence are most closely related. Analyses of the sequence information in several different combinations yielded the same family tree each time **(Figure).**

Conclusion

The family tree placed animals as the branch most closely related to fungi and indicates that the two groups probably share a common protist ancestor not shared with algae or plants. Other investigators have cited similarities in biochemical pathways in fungi and animals, such as pathways that make the amino acid hydroxyproline, the protein ferritin (which combines with iron atoms), and the polysaccharide chitin, which is a primary constituent of both fungal cell walls and arthropod exoskeletons. As described in the main text, subsequent molecular studies have further refined our understanding of the early origins of both fungi and animals.

THINK LIKE A SCIENTIST This study compared the 18S rRNA sequence across widely different groups of organisms. In terms of molecular analysis, what is a logical next step in seeking new data that may support the research findings?

Source: Patricia O. Wainwright, Gregory Hinkle, Mitchell L. Sogin, and Shawn K. Stickel, 1993. Monophyletic origins of the Metazoa: An evolutionary link with Fungi. *Science* 260:340–342.

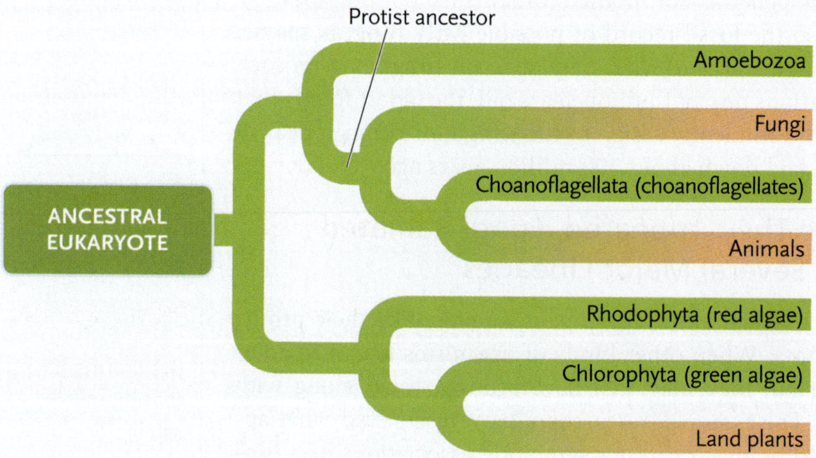

lives, reproduction requires at least a film of water through which the spores can swim.

A chytrid's entire life cycle may unfold within a few days. For most species, asexual reproduction occupies much of this brief span. Although individuals initially exist as a vegetative (nonreproductive) phase, the fungus soon shifts into a reproductive mode. Flagella-bearing spores are released and each swims until it comes to rest on a substrate and a tough cyst forms around it. Under proper conditions, it will soon germinate and launch the life cycle anew.

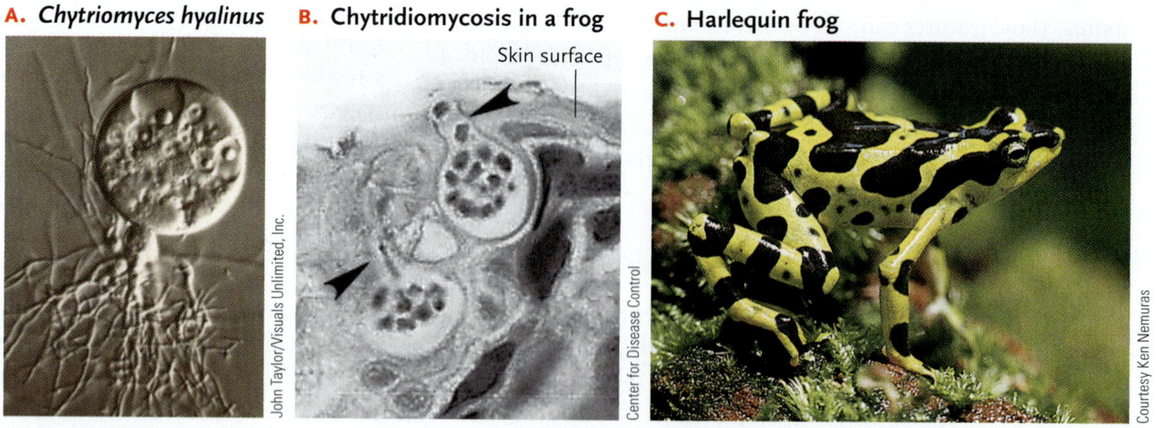

FIGURE 30.5 Chytrids. **(A)** *Chytriomyces hyalinus*, one of the few chytrids that reproduces sexually. **(B)** Chytridiomycosis, a fungal infection, shown here in the skin of a frog. The two arrows point to flask-shaped cells of the parasitic chytrid *Batrachochytrium dendrobatidis*, which has devastated populations of harlequin frogs **(C)**.

A few chytrids reproduce sexually. Mycologists have observed a remarkable variety of sexual modes, but in all of them spores of different mating types unite. Karyogamy directly follows plasmogamy to produce a 2*n* zygote. This cell may form a mycelium that produces sporangia, or it may directly give rise to either asexual or sexual spores.

Zygomycetes Form Zygospores for Sexual Reproduction

Fewer than a thousand species are assigned to the phylum Zygomycota—fungi that reproduce sexually by way of structures called *zygospores* **(Figure 30.6)**. What zygomycetes lack in numbers, however, they make up for in impact on other organisms. Many zygomycetes are saprobes that live in soil, feeding on plant detritus. There, their metabolic activities release mineral nutrients in forms that plant roots can take up. Some zygomycetes are parasites of insects (and even other fungi), and some wreak havoc on human food supplies, spoiling stored grains, bread, fruits, and vegetables such as sweet potatoes. Others, however, have become major partners in commercial enterprises, where they are used in manufacturing products that range from industrial pigments to pharmaceuticals.

Most zygomycetes have aseptate hyphae, unlike the other multicellular fungi. Mycelia of many zygomycetes may occur in either a + or − mating type (Figure 30.6, step 1), and the nuclei of the two mating types are equivalent to gametes. Each strain secretes steroidlike hormones that can stimulate the development of sexual structures in the complementary strain and cause sexual hyphae to grow toward each other. When +

FIGURE 30.6 Life cycle of the bread mold *Rhizopus stolonifer*, a zygomycete. Asexual reproduction is common, but different mating types (+ and −) also reproduce sexually. In both cases, haploid spores form and give rise to new mycelia.

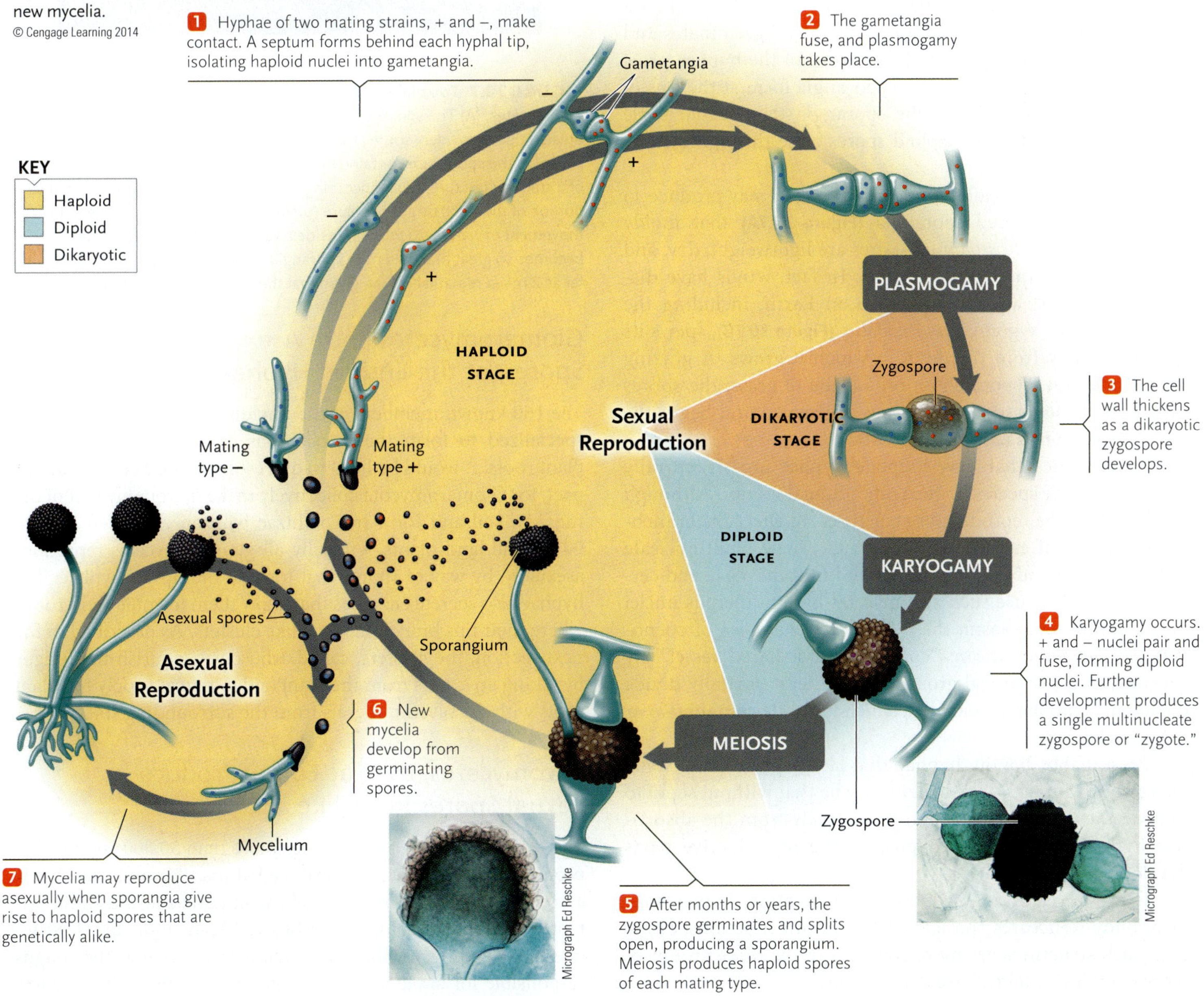

and − hyphae come into close proximity, a septum forms behind the tip of each hypha, producing a terminal **gametangium** that contains haploid nuclei. When the gametangia of the two strains touch, cellular enzymes digest the wall between them, yielding a single large, thin-walled cell that contains many nuclei from both parents (steps 2 and 3). In other words, plasmogamy has occurred, and this new cell is a dikaryon. Gradually an inner wall forms, thickens, and hardens. This structure, with the multinucleate cell inside it, is a **zygospore,** and gives this fungal group its scientific name. It becomes dormant and can stay dormant for months or years.

Karyogamy follows plasmogamy and eventually the diploid zygospore breaks dormancy. Then, meiosis produces a stalked sporangium (step 5). The sporangium contains haploid spores of each mating type, which are released to the outside. When a spore later germinates, it produces either a + or a − mycelium, and the sexual cycle can continue.

Like other fungi, however, zygomycetes usually reproduce asexually, as shown at the lower left in Figure 30.6. When a haploid spore lands on a favorable substrate, it germinates and gives rise to a branching mycelium. Some of the hyphae grow upward, and saclike, thin-walled sporangia form at the tips of these aerial hyphae. Inside the sporangia the asexual cycle comes full circle as new haploid spores arise through mitosis and are released.

The black bread mold, *Rhizopus stolonifer,* may produce so many charcoal-colored sporangia **(Figure 30.7A)** that moldy bread looks black. *R. stolonifer* spores are lightweight, dry, and readily wafted away by air currents. In fact, winds have dispersed them just about everywhere on Earth, including the Arctic. Another zygomycete, *Pilobolus* **(Figure 30.7B),** spews its sporangia away from the dung in which it grows. A grazing animal may eat a sporangium on a blade of grass; the spores then pass through the animal's gut unharmed and begin the life cycle again in a new dung pile.

Zygomycetes that have aseptate hyphae are structurally simpler than the species in most other fungal groups. Although septa wall off the reproductive structures, in effect the branching mycelium of each fungus is a single, huge, multinucleate cell—the same body structure as found in some algae and certain protists. Because such zygomycetes have numerous nuclei in a common cytoplasm, these fungi are examples of coenocytic species which means "contained in a shared vessel." By contrast, in other fungal groups septa at least partially divide the hyphae into individual cells, which typically contain two or more nuclei.

Presumably, having hyphae that lack septa confers some selective advantages. One benefit may be that without septa to impede the flow, nutrients can move freely from the absorptive hyphal tips to other hyphae where reproductive parts develop.

In zygomycetes, aggregations of "cooperating" hyphae may form body structures specialized for certain functions. However, such structures are more common in the three groups of more complex fungi that we consider next.

A. **Sporangia of *Rhizopus stolonifer***

B. **Sporangia (dark sacs) of *Pilobolus***

FIGURE 30.7 **Two of the numerous strategies for spore dispersal by zygomycetes. (A)** The sporangia of *Rhizopus stolonifer,* shown here on a slice of bread, release powdery spores that are easily dispersed by air currents. **(B)** In *Pilobolus,* the spores are contained in a sporangium (the dark sac) at the end of a stalked structure. When incoming rays of sunlight strike a light-sensitive portion of the stalk, turgor pressure (pressure against a cell wall due to the movement of water into the cell) inside a vacuole in the swollen portion becomes so great that the entire sporangium may be ejected outward as far as 2 m—a remarkable feat, given that the stalk is only 5 to 10 mm tall.

Glomeromycetes Form Asexual Spores, at the Ends of Hyphae

The 160 known members of the phylum Glomeromycota are all specialized to form the associations called mycorrhizae with plant roots. It would be hard to overestimate their ecological impact, for Glomeromycota collectively make up roughly half of the fungi in soil and form mycorrhizae with an estimated 80% to 90% of all land plants. Virtually all glomeromycetes reproduce asexually, by way of spores that form at the tips of hyphae. The hyphae also secrete enzymes that allow them to enter plant roots, where their tips branch into treelike clusters. As described in the next section, the clusters, called arbuscules, nourish the fungus by taking up sugars from the plant and in return supply the plant roots with dissolved minerals from the surrounding soil.

Ascomycetes, the Sac Fungi, Produce Sexual Spores in Saclike Asci

The phylum Ascomycota includes more than 30,000 species that produce reproductive structures called *asci* **(Figure 30.8).** A few ascomycetes prey on agricultural insect pests and thus have potential for use as "biological pesticides." Many more are destructive plant pathogens, including *Venturia inaequalis,* the fungus responsible for apple scab, and *Ophiostoma ulmi,* which causes

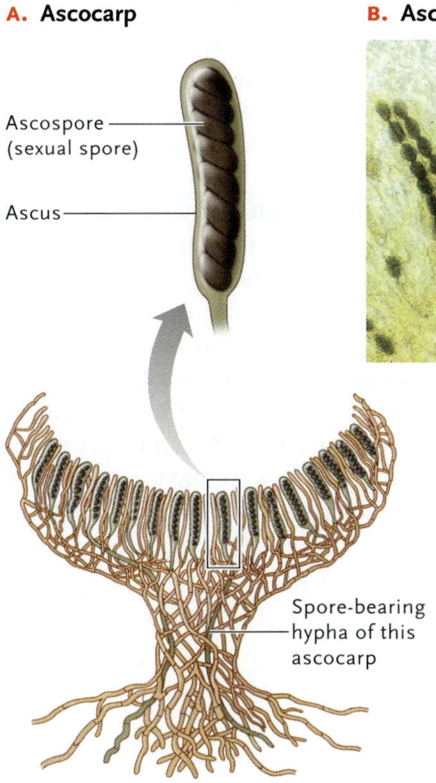

FIGURE 30.8 A few of the ascomycetes, or sac fungi. The examples shown are multicellular species that form ascocarps as reproductive structures. **(A)** A cup-shaped ascocarp, composed of tightly interwoven hyphae. The spore-producing asci occur inside the cup. **(B)** Asci on the inner surface of an ascocarp. **(C)** The velvety appearance the Scarlet cup fungus *(Sarcoscypha)* is due to the thousands of asci in its ascocarp. **(D)** True morels *(Morchella esculenta),* a prized edible fungus. **(E)** The ascomycete *Venturia inequalis* may grow on apple fruits and leaves, causing apple scab disease. A fungicide can help control the disease, but a severe infection can result in unmarketable fruit.

Dutch elm disease. Ascomycete pathogens on humans range from those that cause nuisance infections like athlete's foot and ringworm, to those that are potentially lethal. *Claviceps purpurea,* a parasite on rye and other grains, causes ergotism, a disease marked by vomiting, hallucinations, convulsions, and, in severe cases, gangrene and even death. Species of *Aspergillus* grow in damp grain or peanuts; over time their metabolic wastes, known as aflatoxins, can cause cancer in people who eat the poisoned food. By contrast, certain species of *Penicillium* **(Figure 30.9A)** are the source of the penicillin family of antibiotics, while others produce the aroma and distinctive flavors of Camembert and Roquefort cheeses. Gourmet delicacies such as truffles *(Tuber melanosporum)* and the succulent true morel *Morchella esculenta* come from ascomycetes as well. The orange bread mold *Neurospora crassa* has been important in genetic research, including the elucidation of the one gene–one enzyme hypothesis (see Section 15.1). A few ascomycetes even show trapping behavior, ensnaring small worms that they then digest **(Figure 30.9B).**

Although yeasts and filamentous fungi with a yeast stage in the life cycle occur in all fungal groups except chytrids and glomeromycetes, many of the best-known yeasts are ascomycetes. *Candida albicans* **(Figure 30.10)** commonly infects mucous membranes, especially of the mouth (where it causes a disorder called thrush) and the vagina. *Saccharomyces cerevisiae,* which

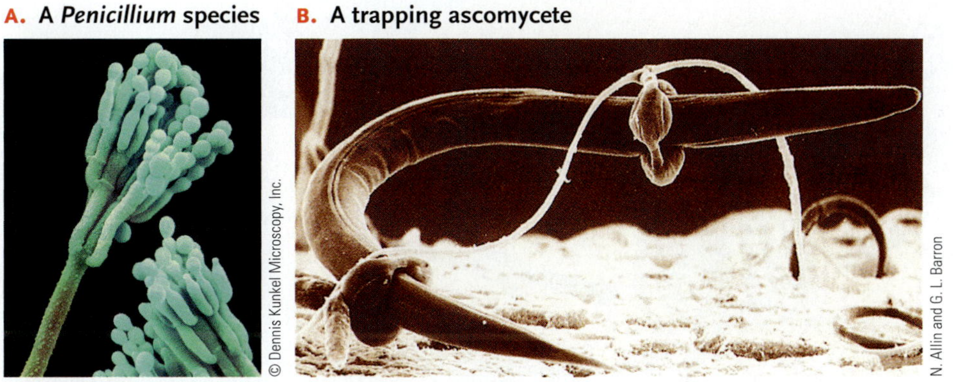

FIGURE 30.9 Other ascomycete representatives. (A) *Eupenicillium.* Notice the rows of conidia (asexual spores) atop the structures that produce them. **(B)** Hyphae of *Arthrobotrys dactyloides,* a trapping ascomycete, form nooselike rings. When the fungus is stimulated by the presence of a prey organism, rapid changes in ion concentrations draw water into the hypha by osmosis. The increased turgor pressure shrinks the "hole" in the noose and captures this nematode. The hypha then releases digestive enzymes that break down the worm's tissues.

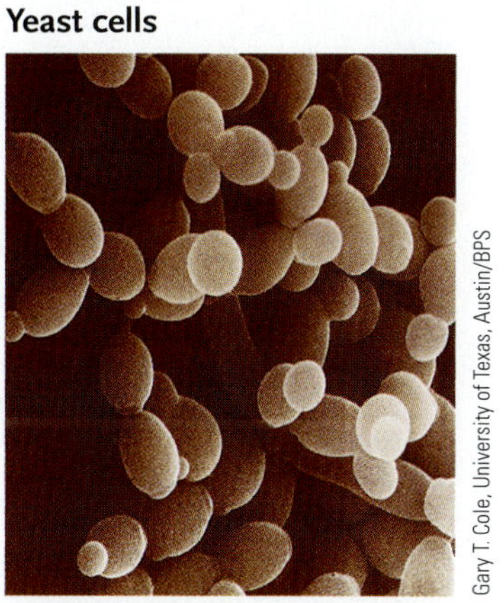

FIGURE 30.10 ***Candida albicans,* cause of yeast infections of the mouth and vagina.**

produces the ethanol in alcoholic beverages and the carbon dioxide that leavens bread, has also been a model organism for genetic research. By one estimate, it has been the subject of more genetic experiments than any other eukaryotic microorganism. Yeasts commonly reproduce asexually by fission or budding from the parent cell, but many also can reproduce sexually after the fusion of two cells of different mating types (analogous to the mating types described earlier). Many ascomycete yeasts are found naturally in the nectar of flowers and on fruits and leaves. At least 1,500 species have been described, and mycologists suspect that thousands more are yet to be identified.

Tens of thousands of ascomycetes, however, are not yeasts. They are multicellular, with tissues built up from septate hyphae. Although septa do slow the flow of nutrients (which, recall, can cross septa through pores), they also confer advantages. For example, septa present barriers to the loss of cytoplasm if a hypha is torn or punctured, whereas in an aseptate zygomycete, fluid pressure may force out a significant amount of cytoplasm before a breach can be sealed by congealing cytoplasm. In ways that are not well understood, septa can also limit the damage from toxins that are secreted by competing fungi.

As with zygomycetes, certain hyphae in ascomycetes are specialized for asexual reproduction. Many ascomycetes produce asexual spores called *conidia* (*konis* = dust) (singular, *conidium*). In some of these species, the conidia form in chains that elongate from modified hyphal branches called **conidiophores.** In other ascomycetes, the conidia pinch off from the hyphae in a series of "bubbles," a bit like a string of detachable beads. Either way, an ascomycete can form and release spores much more quickly than a zygomycete can. Each newly formed conidium contains a haploid nucleus and some

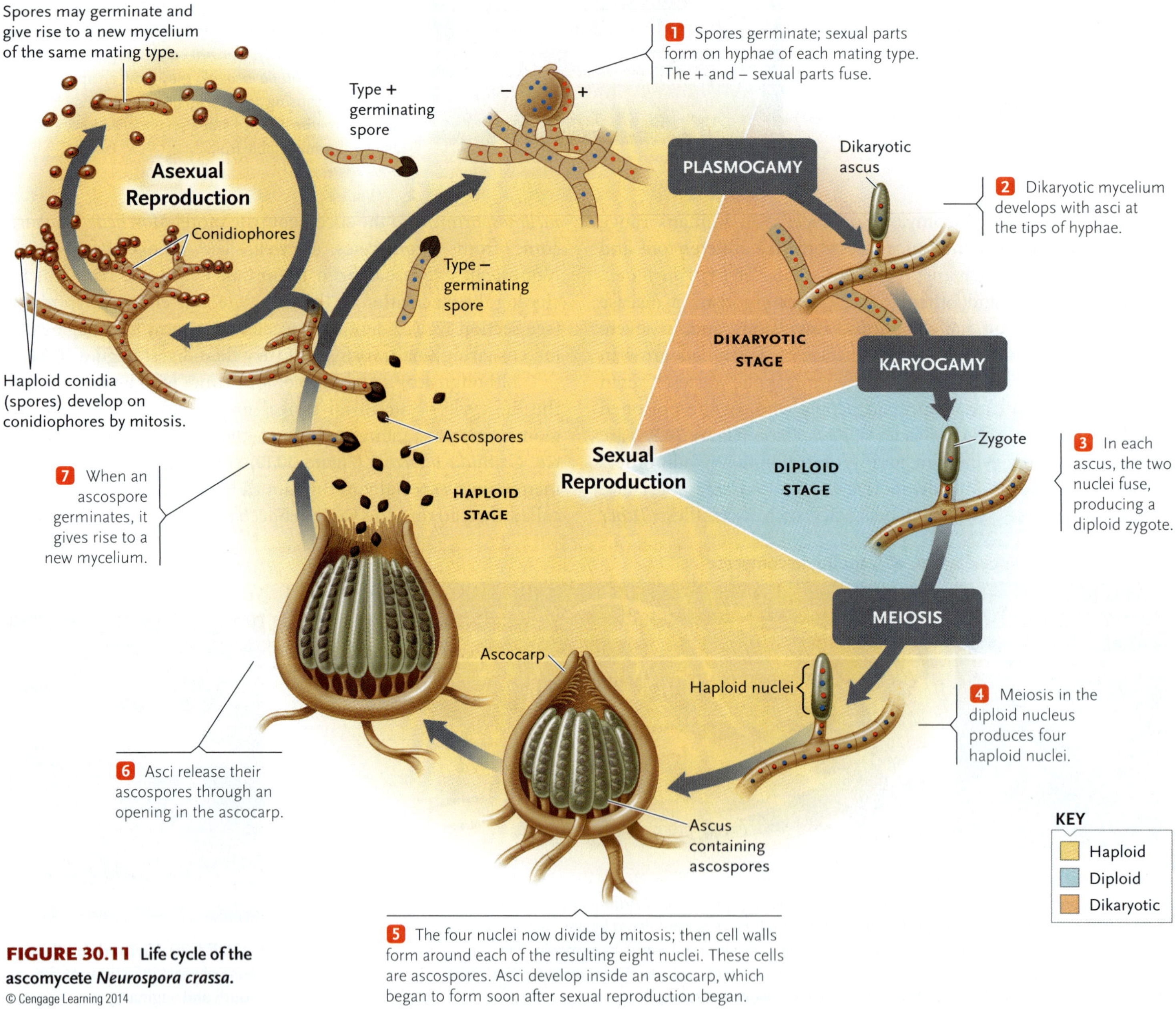

FIGURE 30.11 Life cycle of the ascomycete *Neurospora crassa.*

of the parent hypha's cytoplasm. Conidia and conidiophores of some ascomycete species are visible as the white powdery mildew that attacks grapes, roses, grasses, and the leaves of squash plants.

Ascomycetes can also reproduce sexually and are commonly termed *sac fungi* because most of the events that generate haploid sexual spores occur in saclike cells called *asci* (*askos* = leather bag) (singular, *ascus*). In *Neurospora crassa* and other complex ascomycetes, reproductive bodies called **ascocarps** bear or contain the asci. Some ascocarps resemble globes, others flasks or open dishes. In *N. crassa,* an ascocarp begins to develop when a conidium from a hypha of one mating type fuses with a hypha of the opposite mating type. **Figure 30.11,** step 1). Plasmogamy then takes place, with the details differing from species to species. During plasmogamy, the fused sexual structures give rise to dikaryotic hyphae. Asci form at the hyphal tips. Inside them, karyogamy takes place, producing a diploid zygote nucleus. It divides by meiosis, producing four haploid nuclei. In yeasts and some other ascomycetes, cell division stops at this point, but in *N. crassa* and in many other species, a round of mitosis ensues and results in eight nuclei. Regardless, the nuclei, other organelles, and a portion of cytoplasm then are incorporated into ascospores that may germinate on a suitable substrate and continue the life cycle.

Basidiomycetes, the Club Fungi, Form Sexual Spores in Club-Shaped Basidia

The 30,000 or so species of fungi in the phylum Basidiomycota include the mushroom-forming species, shelf fungi, coral fungi, bird's nest fungi, stinkhorns, smuts, rusts, and puffballs **(Figure 30.12).** Basidiomycetes are known commonly as club fungi because their spore-producing cells, called *basidia* (*basis* = base or foundation), usually are club shaped. Some species have enzymes for digesting cellulose and lignin and are important decomposers of woody plant debris. A surprising number of basidiomycetes, including the prized edible oyster mushrooms *(Pleurotus ostreatus),* also can trap and consume bacteria and small animals such as nematodes by secreting paralyzing toxins or gluey substances that immobilize the prey. This adaptation gives the fungus access to a rich source of molecular nitrogen, an essential nutrient that is often scarce in terrestrial habitats.

Many basidiomycetes take part in vital mutualistic associations with the roots of forest trees, as discussed later in this chapter. Others, the rusts and smuts, are parasites that cause serious diseases in wheat, rice, and other plants.

Amanita muscaria, the fly agaric mushroom (see Figure 30.12D), has been used as a fly poison. In humans one of its side effects is euphoria, which is why *A. muscaria* was used in

FIGURE 30.12 Representative basidiomycetes, or club fungi. (A) The light red coral fungus *Ramaria.* **(B)** The shelf fungus *Laetiporus.* **(C)** The white-egg bird's nest fungus *Crucibulum laeve.* Each tiny "egg" contains spores. Raindrops splashing into the "nest" can cause "eggs" to be ejected, thereby spreading spores into the surrounding environment. **(D)** The fly agaric mushroom *Amanita muscaria,* which is toxic to humans. **(E)** The scarlet hood *Hygrophorus.*

religious rituals of ancient societies in Central America, Russia, and India. (Less pleasant side effects can include powerful nausea.) Other species of this genus, including the death cap mushroom *Amanita phalloides,* produce deadly toxins. In eukaryotic cells, the *A. phalloides* toxin, a short peptide called α-amanitin, halts the transcription of protein-coding genes by RNA polymerase II (see Section 15.2), thereby inhibiting protein synthesis. Ingesting as little as 5 mg of the toxin can trigger violent vomiting and diarrhea within 8 to 24 hours. Later, kidney and liver cells start to degenerate; without intensive medical care, death can follow within a few days.

In a few basidiomycetes, reproduction is usually asexual, by budding or shedding a fragment of a hypha. One is *Cryptococcus neoformans,* which causes a form of meningitis in humans. In general, however, basidiomycetes do reproduce sexually, producing large numbers of haploid sexual spores. **Figure 30.13** shows the life cycle of a typical basidiomycete.

Basidia typically develop on a **basidiocarp,** which is the reproductive body of the fungus. A basidiocarp consists of tight clusters of hyphae; the feeding mycelium is buried in the soil or decaying wood. The shelflike bracket fungi on trees are basidiocarps, and about 10,000 species of club fungi produce the basidiocarps we call mushrooms. Each is a short-lived reproductive body consisting of a stalk and a cap. Basidia develop on sheetlike "gills" on the underside of the cap. The basidia undergo meiosis to produce microscopic, haploid **basidiospores** (Figure 30.13, inset) that disperse throughout the environment.

When a basidiospore lands on a suitable food source, it germinates and gives rise to a haploid mycelium. Compatible mating types growing near each other may undergo plasmog-

FIGURE 30.13 Generalized life cycle of the basidiomycete *Agaricus arvensis,* a close relative of the common button mushroom sold in grocery stores. During the dikaryotic stage, *A. arvensis* cells contain two genetically different nuclei, shown here in different colors. *Inset:* Micrograph showing basidia and basidiospores.

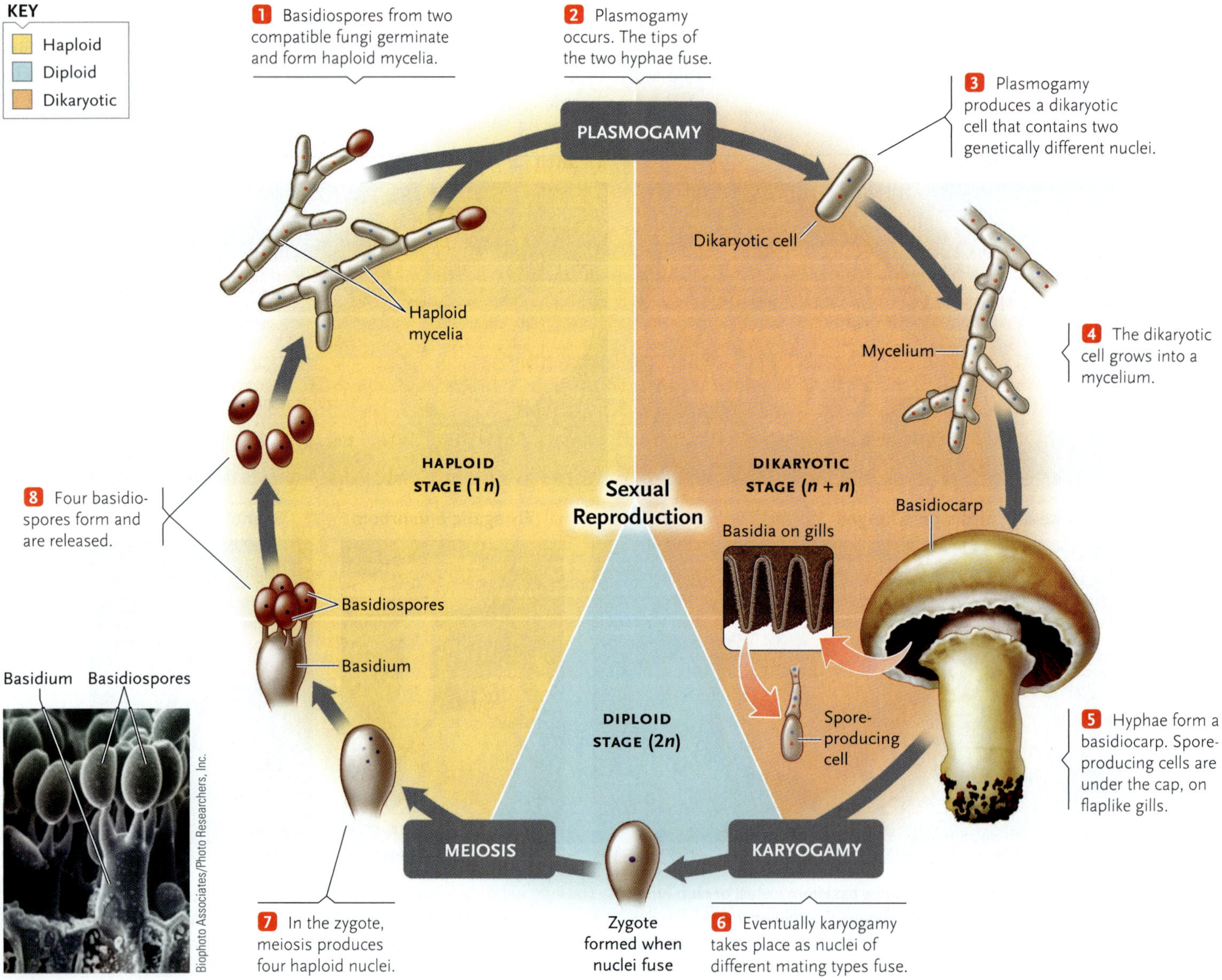

amy. The resulting mycelium is dikaryotic, its cells containing one nucleus from each mating type. The dikaryotic stage of a basidiomycete is the feeding mycelium, which can grow for years—a major departure from an ascomycete's short-lived dikaryotic stage. Accordingly, a basidiomycete has many more opportunities for producing sexual spores, and the mycelium can give rise to reproductive bodies many times.

After an extensive mycelium develops, and when environmental conditions such as moisture are favorable, basidiocarps grow from the mycelium and develop basidia. At first, each basidium in the mushroom or other reproductive body is dikaryotic, but then the two nuclei undergo karyogamy, fusing to form a diploid zygote nucleus. The zygote exists only briefly; meiosis soon produces haploid basidiospores, which are wafted away from the basidium by air currents. Basidia can produce huge numbers of spores—for many species, estimates run as high as 100 million spores *per hour* during reproductive periods, day after day.

Squirrels and many other small animals may eat mushrooms almost as soon as they appear, but in some species the underlying mycelium can live for many years. As such a mycelium grows, specialized mechanisms during cell division maintain the dikaryotic condition and the paired nuclei in each hyphal cell.

Conidial Fungi Are Species for Which No Sexual Phase Is Known

As noted earlier, fungi generally are classified on the basis of their structures for sexual reproduction. When a sexual phase is absent or has not yet been detected, the fungal species is lumped into a convenience grouping, the conidial fungi (recall that conidia are asexual spores). Other historical names for this grouping are "imperfect fungi" and deuteromycetes.

When researchers discover a sexual phase for a conidial fungus, or when molecular studies establish a clear relationship to a sexual species, the conidial fungus is reassigned to the appropriate phylum. Thus far, some have been classified as basidiomycetes, but most conidial fungi have turned out to be ascomycetes.

Microsporidia Are Single-Celled Sporelike Parasites

There are more than 1,200 species of the single-celled parasites called **microsporidia.** They are known to infect insects including honeybees and grasshoppers, and vertebrates including fish and humans—especially individuals with compromised immune systems. Microsporidia are rather mysterious organisms. Physically they resemble spores **(Figure 30.14)**, but they lack mitochondria and have other puzzling characteristics. Molecular studies suggest that they are related to zygomycetes, and some researchers have proposed that the group may have lost many typical fungal features as it evolved a highly specialized parasitic lifestyle.

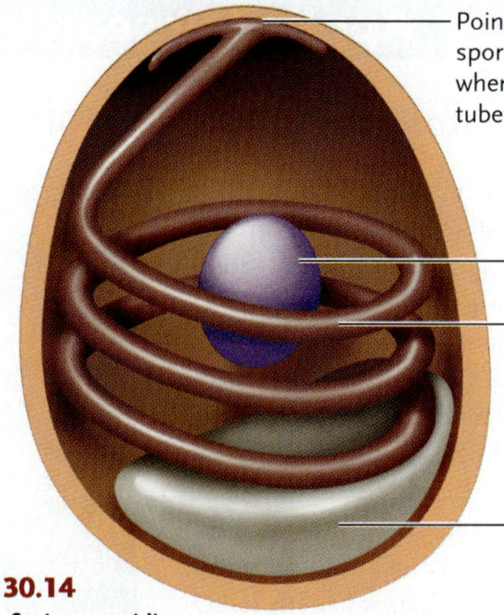

FIGURE 30.14
Structure of microsporidia.
When a spore germinates, its vacuole expands and forces the coiled "polar tube" outward and into a nearby, soon-to-be host cell. The nucleus and cytoplasm of the parasite enter the host through the tube, launching developmental steps that lead to the development of more microsporidia inside the host.

STUDY BREAK 30.2

1. **Name the major phyla and phylum-level categories of the Kingdom Fungi and describe the reproductive adaptations that distinguish each one.**
2. **In terms of structure, which are the simplest fungal groups? The most complex?**
3. **Describe some ways, positive or negative, that members of each fungal phylum interact with other life forms.**

THINK OUTSIDE THE BOOK

Humans are biologically attractive hosts for a wide range of fungi that colonize epidermal tissues—skin, hair, and nails. These fungal species, often described medically as dermatophytes, are categorized as anthrophilic, zoophilic, and geophilic species. Investigate this topic and make a list of the skin disorders associated with each type, as well as common sources of infection. What basic nutrition source do all dermatophytes share?

30.3 Fungal Associations

Many fungi are partners in mutually beneficial interactions with plants and other photosynthesizers, with insects, and even with mammals—associations that play major roles in the functioning of ecosystems. A **symbiosis** is a state such as parasitism or mutualism in which two or more species live together in close association. Chapter 54 discusses general features of symbiotic associations more fully; here we are interested in some examples of the symbioses fungi form with photosynthetic partners—cyanobacteria, green algae, and plants.

FOCUS ON Applied Research

Lichens as Monitors of Air Pollution's Biological Damage

Lichens have become reliable pollution-monitoring devices all over the world—in some cases, replacing costly electronic monitoring stations. Different species are vulnerable to specific pollutants. For example, *Ramalina* lichens are damaged by nitrate and fluoride salts. Elevated levels of sulfur dioxide (a major component of acid rain) cause old man's beard *(Usnea trichodea)* to shrivel and die, but strongly promote the growth of a crusty European lichen, *Lecanora conizaeoides.* The sensitivity of yellow *Evernia* lichens to SO_2 enabled the scientist who discovered its damage at remote Isle Royale in Michigan to point the finger northward to coal-burning furnaces at Thunder Bay, Canada. Conversely, healthy lichens on damaged trees of Germany's Black Forest lifted suspicion from French coal-burning power plants and allowed investigators to identify the true source of the tree damage: nitrogen oxides from automobile exhausts. The result was Germany's first auto emission standards, which went into effect in the 1990s.

Robert W. Domm/Visuals Unlimited/Getty Images

Usnea (old man's beard), a pendent (hanging) lichen.

A Lichen Is an Association between a Fungus and a Photosynthetic Partner

Thousands of ascomycetes and a few basidiomycetes form the kind of symbiosis called a lichen. Technically, a **lichen** is a single vegetative body that results from an association between a fungus and a photosynthetic partner. The fungal partner in a lichen, called the **mycobiont,** usually makes up about 90% of the whole. The other 10% is the photosynthetic partner, called the **photobiont.** Only about 100 photosynthetic species serve as photobionts. Most frequently, these are green algae of the genus *Trebouxia* or cyanobacteria of the genus *Nostoc.*

Lichens often live in harsh, dry microenvironments, including on bare rock and wind-whipped tree trunks. Yet lichens have vital ecological roles and important human uses. Lichens secrete acid that eats away at rock, breaking it down and facilitating the development of new soil. Some paleobiologists (who study ancient life forms) have suggested that lichens may have been some of the earliest land organisms, covering bare rocks during the Ordovician period (roughly 500 million–425 million years ago). In this scenario, millennia of decaying lichens would have created the first soils in which the earliest land plants could grow. Today, lichens continue to enhance the survival of other life forms. For instance, in arctic tundra, where plants are scarce, reindeer and musk oxen can survive by eating lichens. Insects and some other invertebrates also consume lichens, and they are nest-building materials for many birds and small mammals. People have derived dyes from lichens; they are even a component of garam masala, an ingredient in Indian cuisine. Some environmental chemists monitor air pollution by observing lichens, most of which cannot grow in heavily polluted air (see *Focus on Applied Research*).

Because lichens are composite organisms, it may seem odd to talk of lichen "species." Biologists do give lichens binomial names, however, based on the characteristics of the mycobiont. More than 13,500 lichens are recognized, each one a unique combination of a particular species of fungus and one or more species of photobiont. The relationship often begins when a fungal mycelium contacts a free-living cyanobacterium, algal cell, or both. The fungus parasitizes the photosynthetic host cell, sometimes killing it. If the host cell can survive, however, it multiplies in association with the fungal hyphae. The result is a tough, pliable body called a **thallus,** which can take a variety of forms **(Figure 30.15A).** Short, specialized hyphae penetrate photobiont cells of the thallus, which become the fungus's sole source of nutrients. Often, the mycobiont of a lichen absorbs up to 80% of the carbohydrates the photobiont produces.

Benefits for the photobiont are less clear-cut, in part because the drain on nutrients hampers its growth and reproduction. In one view, many and possibly most lichens are parasitic symbioses in which the photobiont does not receive equal benefit. On the other hand, it is relatively rare to find a lichen's photobiont species living independently in the same conditions under which the lichen survives. Studies have also revealed that at least some green algae do clearly benefit from the relationship. Such algae are sensitive to desiccation and intense ultraviolet radiation. Sheltered by a lichen's fungal tissues, a green alga can thrive in locales where alone it would perish.

As you might expect with such a communal life form, reproduction has its quirky aspects. In lichens that involve an ascomycete, the fungus produces ascospores that are dispersed by the wind. The spores germinate to form hyphae that may colonize new photosynthetic cells and so establish new symbioses. A lichen itself can also reproduce in at least two ways. In some types, a section of the thallus detaches and grows into a new lichen. In about one-third of lichens, specialized regions of the thallus give rise asexually to reproductive cell clusters called **soredia** (*soros* = heap) (singular, *soredium*). Each cluster includes both algal and hyphal cells **(Figure 30.15B).** As the lichen grows, the soredia detach and are dispersed by water, wind, or passing animals.

FIGURE 30.15 Lichens. (A) Sketch of a cross section through the thallus of the lichen *Lobaria verrucosa.* **(B)** The soredia, which contain both hyphae and algal cells, are a type of dispersal fragment by which lichens reproduce asexually. **(C)** Crustose lichen. **(D)** Erect, branching lichen, *Cladonia rangiferina.*

Mycorrhizae Are Symbiotic Associations of Fungi and Plant Roots

A **mycorrhiza** ("fungus root") is a mutualistic symbiosis in which fungal hyphae associate intimately with plant roots. Mycorrhizae greatly enhance the plant's ability to extract various nutrients, especially phosphorus and nitrogen, from soil (see Chapter 35).

In **endomycorrhizae,** the fungal hyphae penetrate the walls of root cells. This kind of association occurs on the roots of nearly all flowering plants, and in most cases a glomeromycete is the fungal partner. The branched hyphae of endomycorrhizae are called arbuscules **(Figure 30.16),** and glomeromycetes are sometimes referred to as arbuscular mycorrhizal fungi.

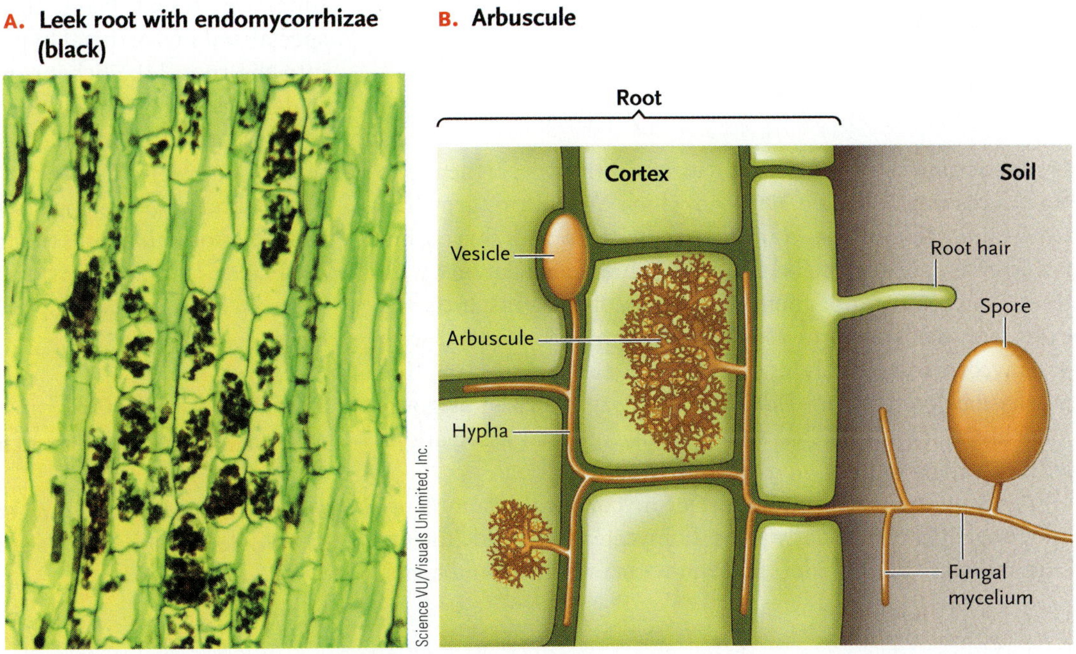

FIGURE 30.16 Endomycorrhizae. (A) In this instance, the roots of leeks are growing in association with the glomeromycete *Glomus versiforme* (longitudinal section). Notice the arbuscules that have formed as the fungal hyphae branched after entering the leek root **(B).**

A. Lodgepole pine

B. Mycorrhiza

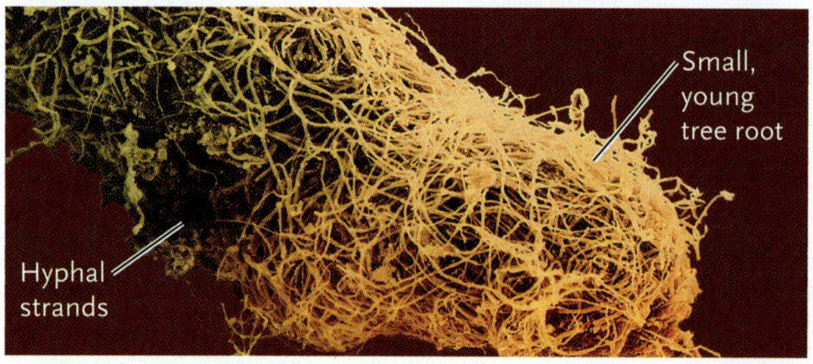

FIGURE 30.17 Ectomycorrhizae. (A) Seedling of pine, *Pinus contorta*, longitudinal section. Notice the extent of the mycorrhiza compared with the above-ground portion of the seedling, which is only about 4 cm tall. **(B)** Mycorrhiza of a hemlock tree.

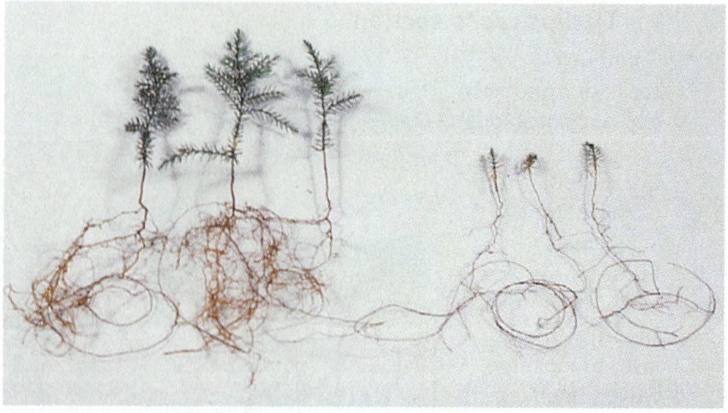

FIGURE 30.18 Effect of mycorrhizal fungi on plant growth. The six-month-old juniper seedlings on the left were grown in sterilized low-phosphorus soil inoculated with a mycorrhizal fungus. The seedlings on the right were grown under the same conditions but without the fungus.

Basidiomycetes are the usual fungal partners in **ectomycorrhizae (Figure 30.17)**, in which hyphal tips grow between and around the young roots of trees and shrubs but never enter the root cells. Ectomycorrhizal associations with trees—often several of them—are common. For instance, the extensive root system of a mature pine may be studded with ectomycorrhizae involving dozens of fungal species. The musky-flavored truffles *(Tuber melanosporum)* prized by gourmets are ascomycetes that form ectomycorrhizal associations with oak trees (genus *Quercus*).

Orchids are partners in a unique mycorrhizal relationship. The fungal partner, usually a basidiomycete, lives inside the orchid's tissues and provides the plant with a variety of nutrients. In fact, seeds of wild orchids germinate, and seedlings survive, only when such mycorrhizae are present.

In general, mycorrhizae represent a "win-win" situation for the partners. The fungal hyphae absorb carbohydrates made by the plant, along with some amino acids and perhaps growth factors as well. The growing plant in turn absorbs mineral ions made accessible to it by the fungus. Collectively, the fungal hyphae have a huge surface area for absorbing mineral ions from a large volume of the surrounding soil. Dissolved mineral ions accumulate in the hyphae when they are plentiful in the soil and are released to the plant when they are scarce. This service is a survival boon to a great many plants, especially species that cannot readily absorb mineral ions, particularly phosphorus **(Figure 30.18).** For plants that inhabit soils poor in mineral ions, such as in tropical rain forests, mycorrhizal associations are crucial for survival. Likewise, in temperate forests, species of spruce, oak, pine, and some other trees die unless mycorrhizal fungi are present. Plants that live in dry habitats often rely on specialized mycorrhizal hyphae that serve as conduits for water into the root. Like lichens, mycorrhizae are highly vulnerable to damage from pollutants, especially acid rain.

Mycorrhizae have a long evolutionary history. Fossils show that endomycorrhizae were common among ancient land plants, and some biologists have speculated they might have been key for enhancing the transport of water and minerals to the plants. In that scenario, endomycorrhizae may have played a crucial role in allowing plants to make the transition to life on land.

STUDY BREAK 30.3

1. **Explain what a lichen is and how each partner contributes to the whole.**
2. **Describe the biological and ecological roles of mycorrhizae.**
3. **How do endomycorrhizae and ectomycorrhizae differ?**

UNANSWERED QUESTIONS

Do different fungal species respond differently to environmental changes?

Ecologists Shigeo Yachi and Michel Loreau formulated the "biological insurance" hypothesis, which predicts that ecosystem functions will be maintained more reliably in natural communities having species that perform the same general ecological function (such as serving as decomposers) but that differ in their responses to environmental changes. When environmental conditions shift in such communities, a species that fills the same ecological role as another but is better suited to the new conditions can then "take over" for a species better adapted to the former conditions. Changes in the makeup of fungal communities resulting from climate change or other ecological perturbations could have important consequences for ecosystem functions such as the availability of soil nutrients, storage of carbon in organic material, and plant growth. For instance, experiments in which fungi were exposed to above-normal concentrations of nitrogen have demonstrated pronounced shifts in the composition of communities of mycorrhizal and saprobic fungi. Studies of CO_2 effects have shown changes in fungal community composition when the fungi were exposed to elevated CO_2 treatments, although several longer-term studies indicate that these differences wane over time. Gaining a better understanding of the mechanisms underlying these patterns and linking changes in community composition to changes in how ecosystems function are important directions for further research.

THINK LIKE A SCIENTIST Ectomycorrhizal fungal communities often exhibit high local diversity, where the root systems of single trees may contain dozens—and a single-host forest stand may contain hundreds—of different fungal symbionts. What factors might allow multiple species of fungi to coexist on a single tree? What factors might favor the dominance of one symbiont over another? How might a better understanding of these phenomena be beneficial to conservation planning?

Courtesy of Todd Osmundson

Todd Osmundson is a postdoctoral researcher at the University of California, Berkeley. His research areas include systematics, biodiversity, ecology, and conservation of fungi and microbes, with an emphasis on ectomycorrhizal basidiomycete fungi. Learn more about Osmundson's work at http://nature.berkeley.edu/~ tosmunds.

REVIEW KEY CONCEPTS

To access the course materials and companion resources for this text, please visit www.cengagebrain.com.

30.1 General Characteristics of Fungi

- Fungi are key decomposers contributing to the recycling of carbon and some other nutrients. They occur as single-celled yeasts or multicellular filamentous organisms.
- The fungal mycelium consists of filamentous hyphae that grow throughout the substrate the fungus feeds on (Figure 30.2). A wall containing chitin surrounds the plasma membrane, and in most species septa partition the hyphae into cell-like compartments.
- Pores in septa permit cytoplasm and organelles to move between hyphal cells. Aggregations of hyphae form all other tissues and organs of a multicellular fungus.
- Fungi gain nutrients by extracellular digestion and absorption. Saprobic species feed on nonliving organic matter. Parasitic types obtain nutrients from tissues of living organisms. Many fungi are partners in symbiotic relationships with plants.
- All fungi may reproduce via spores generated either asexually or sexually (Figure 30.3). Some types also may reproduce asexually by budding or fragmentation of the parent body.
- Sexual reproduction usually has two stages. First, in plasmogamy, the cytoplasms of two haploid cells fuse to become a dikaryon containing a haploid nucleus from each parent. Later, in karyogamy, the nuclei fuse and form a diploid zygote. Meiosis then generates haploid spores.

30.2 Major Groups of Fungi

- The main phyla of fungi are the Chytridiomycota (which have motile spores), Zygomycota (zygospore-forming fungi), Glomeromycota, Ascomycota (sac fungi), and Basidiomycota (club fungi) (Figure 30.4). The phyla traditionally have been distinguished mainly by the structures of sexual reproduction. When a sexual phase cannot be detected or is absent from the life cycle, the specimen is assigned to an informal grouping, the conidial fungi.
- Chytrids usually are microscopic. They are the only fungi that produce motile, flagellated spores. Many are parasites (Figure 30.5).
- Zygomycetes have aseptate hyphae and are coenocytic, with many nuclei in a common cytoplasm. They sometimes reproduce sexually by way of hyphae that occur in 1 and 2 mating types; haploid nuclei in the hyphae function as gametes. Further development produces the zygospore, which may become dormant.
- When the zygospore of a zygomycete breaks dormancy, it produces a stalked sporangium containing haploid spores of each mating type, which are released (Figures 30.6 and 30.7).
- Glomeromycetes form a distinct type of endomycorrhizae in association with plant roots. They reproduce asexually, by way of spores that form at the tips of hyphae.
- Most ascomycetes are multicellular (Figure 30.9). In asexual reproduction, chains of haploid asexual spores called conidia elongate or pinch off from the tips of conidiophores (modified aerial hyphae; Figure 30.10).

- In sexual reproduction of ascomycetes, haploid sexual spores called ascospores arise in saclike cells called asci. In the most complex species, reproductive bodies called ascocarps bear or contain the asci. Ascospores can give rise to a new haploid mycelium (Figures 30.8 and 30.11).
- Most basidiomycete species reproduce only sexually. Club-shaped basidia develop on a basidiocarp and bear sexual spores on their surface. When dispersed, these basidiospores may germinate and give rise to a haploid mycelium (Figure 30.13).
- Microsporidia are single-celled sporelike parasites of arthropods, fish, and humans (Figure 30.14).

Animation: Club fungus life cycle

Animation: Rhizopus

Animation: How a Fungus Makes a Living

30.3 Fungal Associations

- Many ascomycetes and a few basidiomycetes enter into symbioses with cyanobacteria or green algae to produce the communal life form called a lichen, which has a tough, pliable body called a thallus. The algal cells supply the lichen's carbohydrates, most of which are absorbed by the fungus.
- In some lichens a section of the thallus may detach and grow into a new individual. In others, specialized regions of the thallus give rise asexually to reproductive soredia that include both algal and hyphal cells (Figure 30.15).
- In the symbiosis called a mycorrhiza, fungal hyphae make mineral ions and sometimes water available to the roots of a plant partner. The fungus in turn absorbs carbohydrates, amino acids, and possibly other growth-enhancing substances from the plant (Figures 30.16–30.18).
- In endomycorrhizae, the fungal hyphae (usually of a glomeromycete) penetrate the cells of the root. With ectomycorrhizae, hyphae form a netlike structure that surrounds the cortical cells of young roots but hyphae do not enter root cells; the usual fungal partner is a basidiomycete.

Animation: Lichens

UNDERSTAND & APPLY

Test Your Knowledge

1. Which of the following statements does *not* reflect current understanding of phylogenetic relationships and features among fungi?
   a. Lineages leading to fungi diverged from those leading to plants about 1 billion years ago.
   b. Chytrids are probably a paraphyletic grouping and may most closely resemble the earliest fungi.
   c. There are four broadly accepted fungal phyla, the Zygomycota, Glomeromycota, Ascomycota, and Basidiomycota.
   d. As fungal lineages diversified, each evolved distinctive reproductive adaptations.
   e. Conidial fungi do not belong to a recognized fungal phylum.
2. Which of the following events is/are a necessary part of a typical asexual cycle in fungal reproduction?
   a. formation of a dikaryon
   b. hyphae developing into a mycelium
   c. formation of a diploid zygote
   d. plasmogamy, which occurs when hyphae fuse at their tips
   e. production and release of large numbers of spores
3. A trait common to all fungi is:
   a. reproduction via spores.
   b. parasitism.
   c. septate hyphae.
   d. a dikaryotic phase inside a zygospore.
   e. plasmogamy after an antheridium and ascogonium come into contact.
4. The chief characteristic used to classify fungi into the major fungal phyla is:
   a. nutritional dependence on nonliving organic matter.
   b. recycling of nutrients in terrestrial ecosystems.
   c. adaptations for obtaining water.
   d. features of reproduction.
   e. cell wall metabolism.
5. At lunch George ate a mushroom, some truffles, a little Camembert cheese, and a bit of moldy bread. Which of the following groups was *not* represented in the meal?
   a. Basidiomycota
   b. Ascomycota
   c. conidial fungi
   d. chytrids
   e. Zygomycota
6. Which of the following fungal reproductive structures are diploid?
   a. basidiocarps
   b. ascospores
   c. conidia
   d. gametangia
   e. zygospores
7. A mushroom is:
   a. the food-absorbing region of an ascomycete.
   b. the food-absorbing region of a basidiomycete.
   c. a reproductive structure formed only by basidiomycetes.
   d. a specialized form of mycelium not constructed of hyphae.
   e. a collection of saclike cells called asci.
8. A zygomycete is characterized by:
   a. usually, aseptate hyphae.
   b. mostly sexual reproduction.
   c. absence of + and − mating types.
   d. the tendency to form mycorrhizal associations with plant roots.
   e. a life cycle in which karyogamy does not occur.
9. Which of the following statements apply/applies to lichens?
   a. It is a fungus that breaks down rock to provide nutrients for an alga.
   b. It colonizes bare rocks and slowly degrades them to small particles.
   c. It spends part of the life cycle as a mycobiont and part as a fungus.
   d. It is an association between a basidiomycete and an ascomycete.
   e. It is an association between a photobiont and a fungus.
10. In a college greenhouse, a new employee observes fuzzy mycorrhizae in the roots of all the plants. Destroying no part of the plants, she carefully removes the mycorrhizae. The most immediate result of this "cleaning" is that the plants cannot:
    a. carry out photosynthesis.
    b. absorb water through their roots.
    c. transport water up their stems.
    d. extract as much nitrogen from soil.
    e. store carbohydrates in their roots.

Discuss the Concepts

1. A mycologist wants to classify a specimen that appears to be a new species of fungus. To begin the classification process, what kinds of information on body structures and/or functions must the researcher obtain in order to assign the fungus to one of the major fungal groups?
2. In a natural setting—a pile of horse manure in a field, for example—the sequence in which various fungi appear illustrates ecological succession, which is the replacement of one species by another in a community (see Chapter 52). The first fungi to appear are the most efficient opportunists, for they can form and disperse spores most rapidly. In what order would you expect representatives from each division of fungi to appear on the manure pile? Why?
3. As the text noted, conifers, orchids, and some other types of plants cannot grow properly if their roots do not form associations with fungi, which provide the plant with minerals such as phosphate and in return receive carbohydrates and other nutrients synthesized by the plant. In some instances, however, the plant receives proportionately more nutrients than the fungus does. Even so, biologists still consider this to be a mycorrhizal association. Explain why you agree or disagree.
4. Humans are fundamentally diploid organisms. Explain how this state of affairs compares with the fungal life cycle, then compare the two general life cycles in light of the two groups' overall reproductive strategies.

Design an Experiment

Experiments on the orange bread mold *Neurospora crassa,* an ascomycete, were pivotal in elucidating the concept that each gene encodes a single enzyme. As *N. crassa* ascospores arise through meiosis and then mitosis in an ascus, each ascospore occupies a particular position in the final string of eight spores the ascus contains:

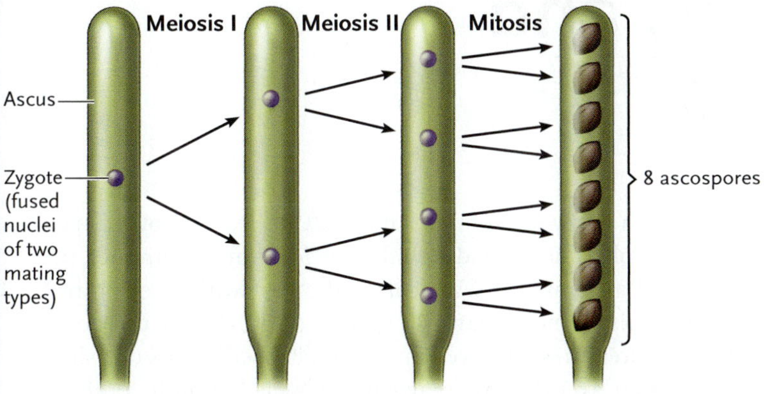

This quirk of ascospore development was extremely useful to early geneticists, because it vastly simplified the task of figuring out which alleles ended up in particular ascospores following meiosis. Recalling genetics topics discussed in Chapter 11, why was the analysis easier?

Interpret the Data

The fungus *Cryptococcus neoformans,* a basidiomycete yeast, is a dangerous pathogen in humans who have a weakened immune system, such as HIV/AIDS patients. The disease it causes, cryptococcosis, produces severe symptoms ranging from a pneumonia-like illness to encephalitis. *C. neoformans* thrives in dark places such as soil, dung, and caves, and light inhibits its sexual reproduction by preventing the development of a dikaryon stage. The recent sequencing of its genome has spurred efforts to identify genes that control this inhibition. Alexander Idnurm and Joseph Heitman of Duke University Medical Center performed a series of experiments using wild-type and mutant *C. neoformans* specimens. As described in Section 30.2, a dikaryon develops when haploid hyphae of two mating types fuse, forming the $n + n$ structure. In the wild *C. neoformans* has two mating types, denoted α and **a.** In both, the normal gene BWC encodes the light-based inhibition response; there is no inhibition in strains having the mutant *bwc1.* In one series of experiments, Idnurm and Heitman tested how efficiently fusion events occurred in crosses between matings of wild-type parental hyphae (denoted +) and mutant hyphae (denoted by Δ) that were exposed to different light wavelengths. Colored bars in the **Figure** indicate the color (wavelength) of the light stimulus; black denotes darkness, and red denotes dim light at the red/far-red end of the visible spectrum. Green and blue represent wavelengths toward the other end of the visible spectrum. To interpret this graph, you must read the mating combinations vertically.

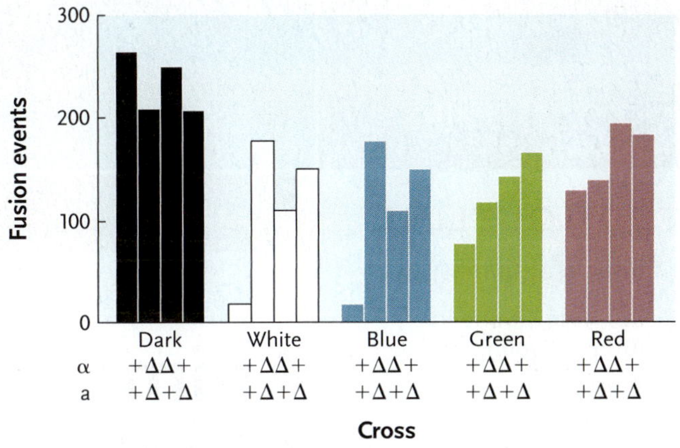

A. Idnurm and J. Heitman. 2005. Light controls growth and development via a conserved pathway in the fungal kingdom. *PLoS Biology* 3:e95.

1. In the dark, did matings in which both partners were either wild-type or mutant forms produce a different response to the varying light stimuli than crosses of a wild-type and a mutant?
2. Within the other categories of light stimuli, how did the makeup of the different mating pairs correlate with the response?

Source: A. Idnurm and J. Heitman. 2005. Light controls growth and development via a conserved pathway in the fungal kingdom. *PLoS Biology* 3:e95.

Apply Evolutionary Thinking

The hypothesis that fungi are more closely related to animals than to plants is supported by studies of fungus genomes. For instance, scientists have documented striking similarities in the structure of many fungal and human genes—similarities that may be especially important in medicine. One mycologist, John Taylor of the University of California, Berkeley, suggests that a close biochemical relationship between fungi and animals may explain why fungal infections are typically so resistant to treatment, and why it has proven rather difficult to develop drugs that kill fungi without damaging their human or other animal hosts. Dozens of fungal genomes have been sequenced, including those of several medically important species. If you are a researcher working to develop new antifungal drugs, how could you make use of this growing genetic understanding? Using Internet resources, can you find examples of antifungal drugs that exploit biochemical differences between animals and fungi?

31

STUDY OUTLINE

31.1 What Is an Animal?

31.2 Key Innovations in Animal Evolution

31.3 An Overview of Animal Phylogeny and Classification

31.4 Animals without Tissues: Parazoa

31.5 Eumetazoans with Radial Symmetry

31.6 Lophotrochozoan Protostomes

31.7 Ecdysozoan Protostomes

Weaver ants *(Oecophylla longinoda)* carry a leaf to repair their nest in Papua New Guinea.

Animal Phylogeny, Acoelomates, and Protostomes

Why it matters . . . In 1909, a lucky fossil hunter named Charles Walcott tripped over a rock on a mountain path in British Columbia, Canada. Under the force of his hammer, the rock split apart, revealing the discovery of a lifetime. As Walcott and other workers examined rocks at this locality, they soon found fossils of more than 120 species of previously unknown animals from the Cambrian period. These creatures had lived on the muddy sediments of a shallow ocean basin; about 530 million years ago, an underwater avalanche buried them in a rain of silt that was eventually compacted into finely stratified shale. Over millions of years, the shale was uplifted by tectonic activity and incorporated into the mountains of western Canada. It is now known as the Burgess Shale formation.

Some animals in the Burgess Shale were truly bizarre **(Figure 31.1).** For example, *Opabinia* was about as long as a tube of lipstick; it had five eyes on its head and a grasping organ that it may have used to capture prey. No living animals even remotely resemble *Opabinia.* The smaller *Hallucigenia* sported seven pairs of large spines on one side and seven pairs of soft organs on the other. Recent research suggests that *Hallucigenia* may belong in the phylum Onychophora, described in Section 31.7. Nevertheless, most species of the Burgess Shale left no descendants that are still alive today. Thus, this remarkable assemblage of fossils provides a glimpse of some evolutionary novelties that—whether through the action of natural selection or just plain bad luck—were ultimately unsuccessful.

Other animal lineages have shown much greater longevity. Zoologists have described nearly 2 million living species in the kingdom **Animalia.** The familiar **vertebrates,** animals with a back-

Opabinia

Dr. Chip Clark, National Museum of Natural History, Smithsonian Institution

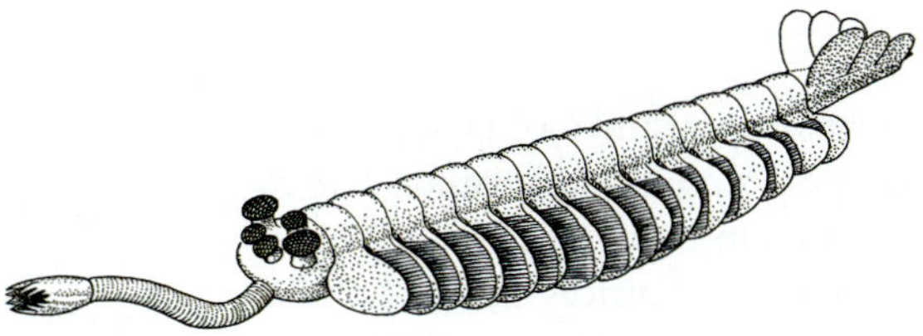

Hallucigenia

Dr. Chip Clark, National Museum of Natural History, Smithsonian Institution

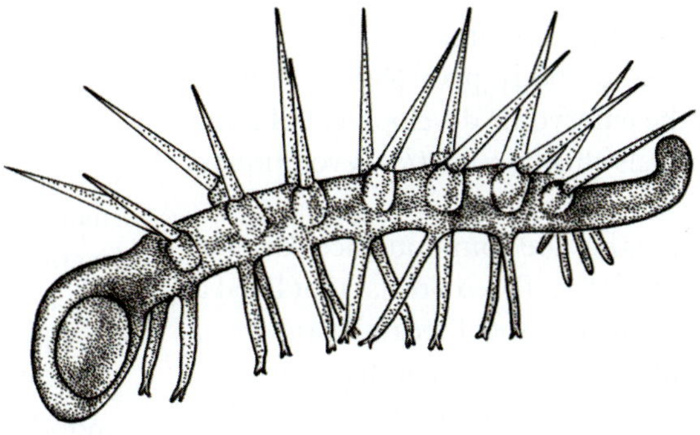

FIGURE 31.1 Animals of the Burgess Shale. *Opabinia* had five eyes and a grasping organ on its head. *Hallucigenia* had seven pairs of spines and soft protuberances.

bone, encompass only a small fraction (about 47,000) of that total. The overwhelming majority of animals fall within the descriptive grouping of **invertebrates,** animals without a backbone.

The remarkable evolutionary diversification of animals resulted from their ability to consume other organisms as food and, for most groups, their ability to move from one place to another. Today, animals are important consumers in nearly every environment on Earth. Their diversification has been accompanied by the evolution of specialized tissues and organ systems as well as complex behaviors.

In this chapter, we introduce the general characteristics of animals and a phylogenetic hypothesis about their evolutionary history and classification. We also survey some of the major invertebrate phyla; a *phylum* is an ancient monophyletic lineage with a distinctive body plan. In Chapter 32 we examine the deuterostome clade, which includes the vertebrates and their nearest invertebrate relatives.

31.1 What Is an Animal?

Biologists recognize the Kingdom Animalia as a monophyletic group that is easily distinguished from the other kingdoms.

All Animals Share Certain Structural and Behavioral Characteristics

Animals are eukaryotic, multicellular organisms. Their cells lack cell walls, a trait that differentiates them from plants and fungi. Because the individual cells of most animals are similar in size, very large animals like elephants have many more cells than small ones like fleas. In large animals, most cells are far from the body surface, but specialized tissues and organ systems deliver nutrients and oxygen to them and carry wastes away.

All animals are **heterotrophs:** they eat other organisms to acquire energy and nutrients. Food is ingested (eaten), then digested (broken down), and its breakdown products are absorbed by specialized tissues. Animals use oxygen to metabolize the food they eat through the biochemical pathways of aerobic respiration, and most store excess energy as glycogen, oil, or fat.

All animals are **motile**—able to move from place to place—at some time in their lives. They travel through the environment to find food or shelter and to interact with other animals. Most familiar animals are motile as adults. However, in some species, such as mussels and barnacles, only the larval stages are motile; they eventually settle down as **sessile** (unable to move from one place to another) adults. The advantages of motility have fostered the evolution of locomotor structures, including fins, legs, and wings that are powered by muscles, which are specialized contractile tissues that move individual body parts. Most animals also have sensory and nervous systems that allow them to receive, process, and respond to information about the environment through which they are traveling.

Animals reproduce either asexually or sexually; in some groups they switch from one mode to the other. Sexually reproducing species produce short-lived, haploid **gametes** (eggs and sperm), which fuse to form diploid **zygotes** (fertilized eggs). Animal life cycles generally include a period of development during which mitosis transforms the zygote into a multicelled **embryo;** an embryo develops into a reproductively immature juvenile or a free-living **larva,** which becomes a reproductively mature adult. Larvae often differ markedly from adults and may occupy different habitats and consume different foods.

Animals Probably Arose from a Colonial Choanoflagellate Ancestor

An overwhelming body of morphological and molecular evidence indicates that all animal phyla had a common ancestor. For example, all animals share similarities in their cell-to-cell junctions and the molecules in their extracellular matrices (see Section 5.5) as well as similarities in the structure of their ribosomal RNAs.

In 1874 the German embryologist Ernst Haeckel proposed that the common ancestor of all animals was a colonial, hollow, ball-shaped organism with unspecialized flagellated cells. According to his hypothesis, its cells became specialized for particular functions, and a developmental reorganization produced a double-layered, sac-within-a-sac body plan **(Figure 31.2)**. As you will see in Chapter 50, the embryonic development of many living animals roughly parallels this hypothetical evolutionary transformation.

Most biologists now agree that the common ancestor of all animals was probably a colonial, flagellated protist that lived at least 700 million years ago, during the Precambrian era. It may have resembled the minute, sessile choanoflagellates that live in both freshwater and marine habitats today (see Figure 27.21). Several lines of morphological and molecular evidence support this hypothesis. First, living choanoflagellates are almost identical in structure to the "collar cells" of sponges, the most ancient living clade of animals; similar cells have also been discovered in other distantly related animal phyla, such as flatworms and echinoderms. Second, DNA sequence data suggest that choanoflagellates and animals are sister clades. Moreover, researchers have identified several genes for cell adhesion and cell signaling that are present in both choanoflagellates and animals.

STUDY BREAK 31.1

1. What characteristics distinguish animals from plants and fungi?
2. How does the ability of animals to move through the environment relate to their acquisition of nutrients and energy?

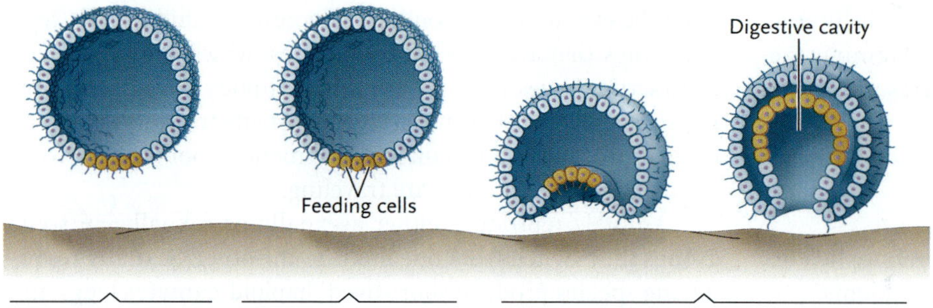

FIGURE 31.2 Animal origins. Many biologists agree that animals arose from a colonial, flagellated protist in which cells became specialized for specific functions and a developmental reorganization produced two cell layers. The cell movements illustrated here are similar to those that occur during the early development of many animals, as described in Chapter 50.

31.2 Key Innovations in Animal Evolution

Once established, the animal lineage diversified quickly into an amazing array of body plans. Before the development of molecular sequencing techniques, biologists identified several key morphological innovations that they used to develop hypotheses about the evolutionary relationships of the major animal groups.

Tissues and Tissue Layers Appeared Early in Animal Evolution

The presence or absence of **tissues**—groups of cells that share a common structure and function—divides the animal kingdom into two distinct branches. One branch, the sponges, or Parazoa (*para* = alongside, *zoon* = animal), lacks tissues. All other animals, collectively grouped in the Eumetazoa (*eu* = true, *meta* = in common with), have tissues.

During the development of eumetazoans, embryonic tissues form as either two or three concentric **primary cell layers** (described further in Chapter 50). The innermost layer, the **endoderm** (generally illustrated in yellow), eventually develops into the lining of the gut (digestive system) and, in some animals, respiratory organs. The outermost layer, the **ectoderm** (generally illustrated in blue), forms the external body surface and nervous system. Between the two, the **mesoderm** (generally illustrated in red) forms the muscles of the body wall and most other structures between the gut and the body surface. Some animals have a **diploblastic** body plan that includes only two layers, endoderm and ectoderm. However, most animals are **triploblastic,** having all three primary cell layers.

Most Animals Exhibit Either Radial or Bilateral Symmetry

The most obvious feature of an animal's body plan is its shape **(Figure 31.3)**. Most animals are **symmetrical;** in other words, their bodies can be divided by a plane into mirror-image halves. The rare exceptions include most sponges, which have irregular shapes and are therefore **asymmetrical.**

Eumetazoans exhibit one of two body symmetry patterns. The Radiata includes two phyla, Cnidaria (hydras, jellyfishes, and sea anemones) and Ctenophora (comb jellies), which have **radial symmetry.** Their body parts are arranged regularly around a central axis, like the spokes on a wheel. Thus, any cut down the long axis of a hydra divides it into matching halves. Radially symmetrical animals are usually sessile or slow moving and receive sensory input from all directions.

All other eumetazoan phyla fall within the Bilateria, animals that have **bilateral symmetry.** In other words, only a cut along the midline from head to tail divides them

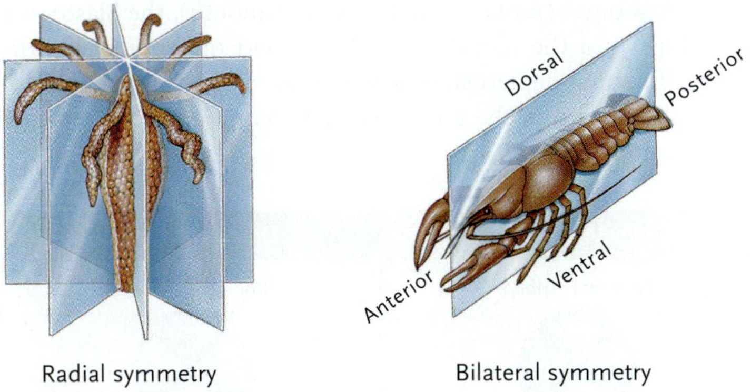

FIGURE 31.3 Patterns of body symmetry. Most animals have either radial or bilateral symmetry.

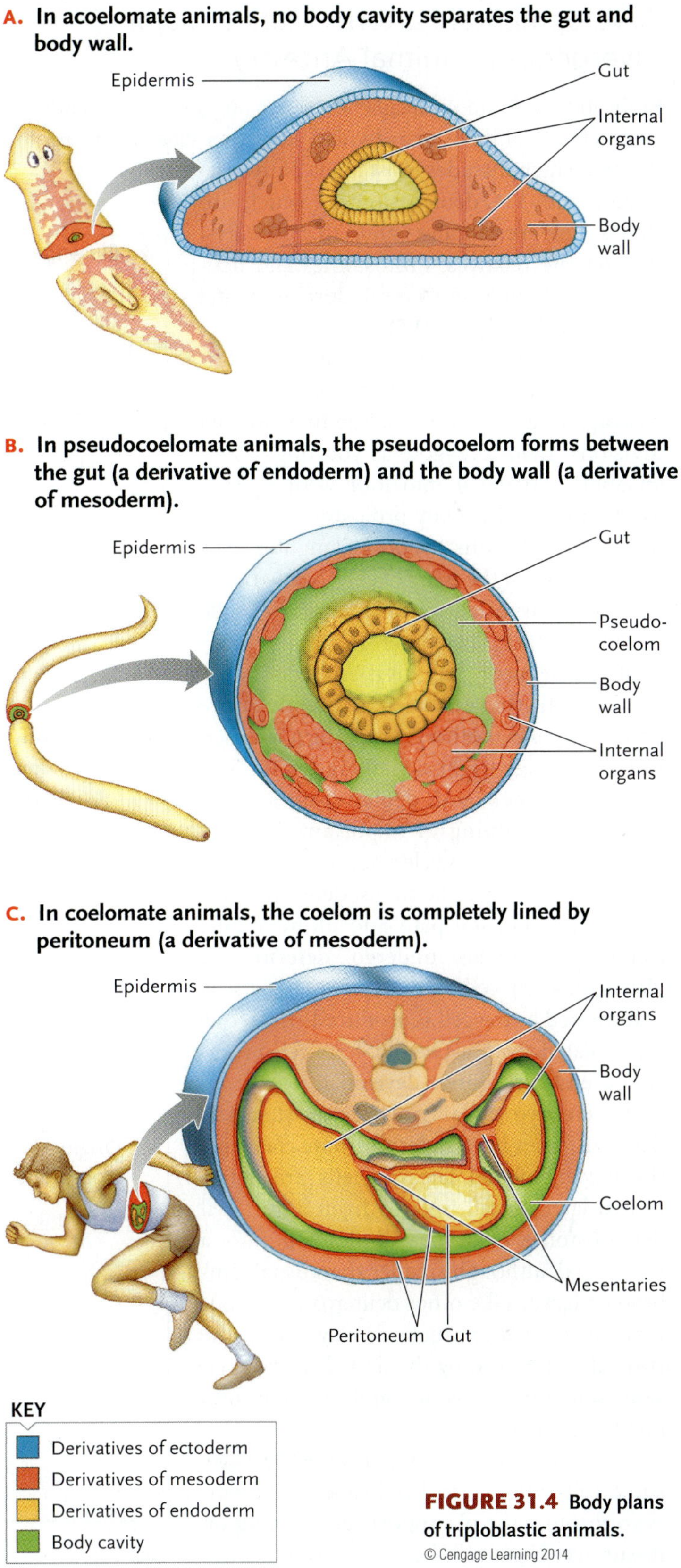

FIGURE 31.4 Body plans of triploblastic animals.

into left and right sides that are essentially mirror images of each other. Bilaterally symmetrical animals also have **anterior** ("front") and **posterior** ("back") ends as well as **dorsal** ("upper") and **ventral** ("lower") surfaces. As these animals move through the environment, the anterior end encounters food, shelter, or enemies first. Thus, in bilaterally symmetrical animals, natural selection also favored **cephalization,** the development of an anterior head where sensory organs and nervous-system tissue are concentrated.

Many Animals Have Body Cavities That Surround Their Internal Organs

The body plans of many bilaterally symmetrical animals include a body cavity that separates the gut from the muscles of the body wall **(Figure 31.4). Acoelomate** animals (*a* = not, *koilomat* = hollow), such as flatworms (phylum Platyhelminthes), lack such a cavity; a continuous mass of tissue, derived largely from mesoderm, packs the region between the gut and the body wall (see Figure 31.4A). **Pseudocoelomate** animals (*pseudo* = false), including the roundworms (phylum Nematoda) and wheel animals (phylum Rotifera), have a **pseudocoelom,** a fluid- or organ-filled space between the gut and the muscles of the body wall, that is partially lined by tissues derived from mesoderm (see Figure 31.4B). Internal organs lie within the pseudocoelom and are bathed by its fluid. **Coelomate** animals have a **coelom,** a fluid-filled body cavity, also between the gut and body wall, that is completely lined by the **peritoneum,** a thin tissue derived from mesoderm (see Figure 31.4C). Membranous extensions of the inner and outer layers of the peritoneum, the **mesenteries,** surround the internal organs and suspend them within the coelom.

Biologists describe the body plan of pseudocoelomate and coelomate animals as a "tube within a tube"; the digestive system forms the inner tube, the body wall forms the outer tube, and the body cavity lies between them. The body cavity separates internal organs from the body wall, allowing the organs to function independently of whole-body movements. The fluid within the cavity also protects delicate organs from mechanical damage. And, because the volume of the body cavity is fixed, the incompressible fluid within it serves as a **hydrostatic skeleton,** which provides support. In some animals muscle contractions can shift the fluid, changing the animals' shape and allowing them to move from place to place (see Section 43.2).

Developmental Patterns Mark a Major Divergence in Animal Ancestry

Embryological and molecular evidence suggests that bilaterally symmetrical animals are divided into two clades: the protostomes, which includes most phyla of invertebrates, and the deuterostomes, which includes the vertebrates and their nearest invertebrate relatives. Protostomes and deuterostomes differ in several developmental characteristics **(Figure 31.5)**.

Shortly after fertilization, an egg undergoes a series of mitotic divisions called **cleavage** (see Section 50.2). The first two cell divisions divide a zygote as you might slice an apple, cutting it into four wedges from top to bottom. In many protostomes, subsequent cell divisions produce daughter cells that lie *between* the pairs of cells below them; this pattern is called **spiral cleavage** (left side of Figure 31.5A). In deuterostomes, by contrast, subsequent cell divisions produce a mass of cells that are stacked directly above and below one another; this pattern is called **radial cleavage** (right side of Figure 31.5A).

Protostomes and deuterostomes often differ in the timing of important developmental events. During cleavage, certain genes are activated at specific times, determining a cell's developmental path and ultimate fate. Many protostomes undergo **determinate cleavage:** each cell's developmental path is determined as the cell is produced. Thus, one cell isolated from a two- or four-cell protostome embryo cannot develop into a functional embryo or larva. By contrast, many deuterostomes have **indeterminate cleavage:** the developmental fates of cells are determined later. A cell isolated from a four-cell deuterostome embryo will develop into a functional, although smaller than usual, embryo or larva. Like other deuterostomes, humans have indeterminate cleavage; thus, the two cells produced by the first cleavage division sometime separate and develop into identical twins.

As development proceeds after cleavage, some cells migrate to the inside of the embryo through the **blastopore,** an opening on its surface. These cells become the endoderm, which forms the **archenteron,** the developing gut (see Figure 31.5B). Later in development, a second opening at the opposite end of the embryo transforms the pouchlike archenteron into a digestive tube (see Figure 31.5C).

In protostomes (*protos* = first, *stoma* = mouth), the blastopore develops into the mouth, and the second opening forms the anus. In some deuterostomes (*deuteros* = second), the blastopore develops into the anus, and the second opening becomes the mouth.

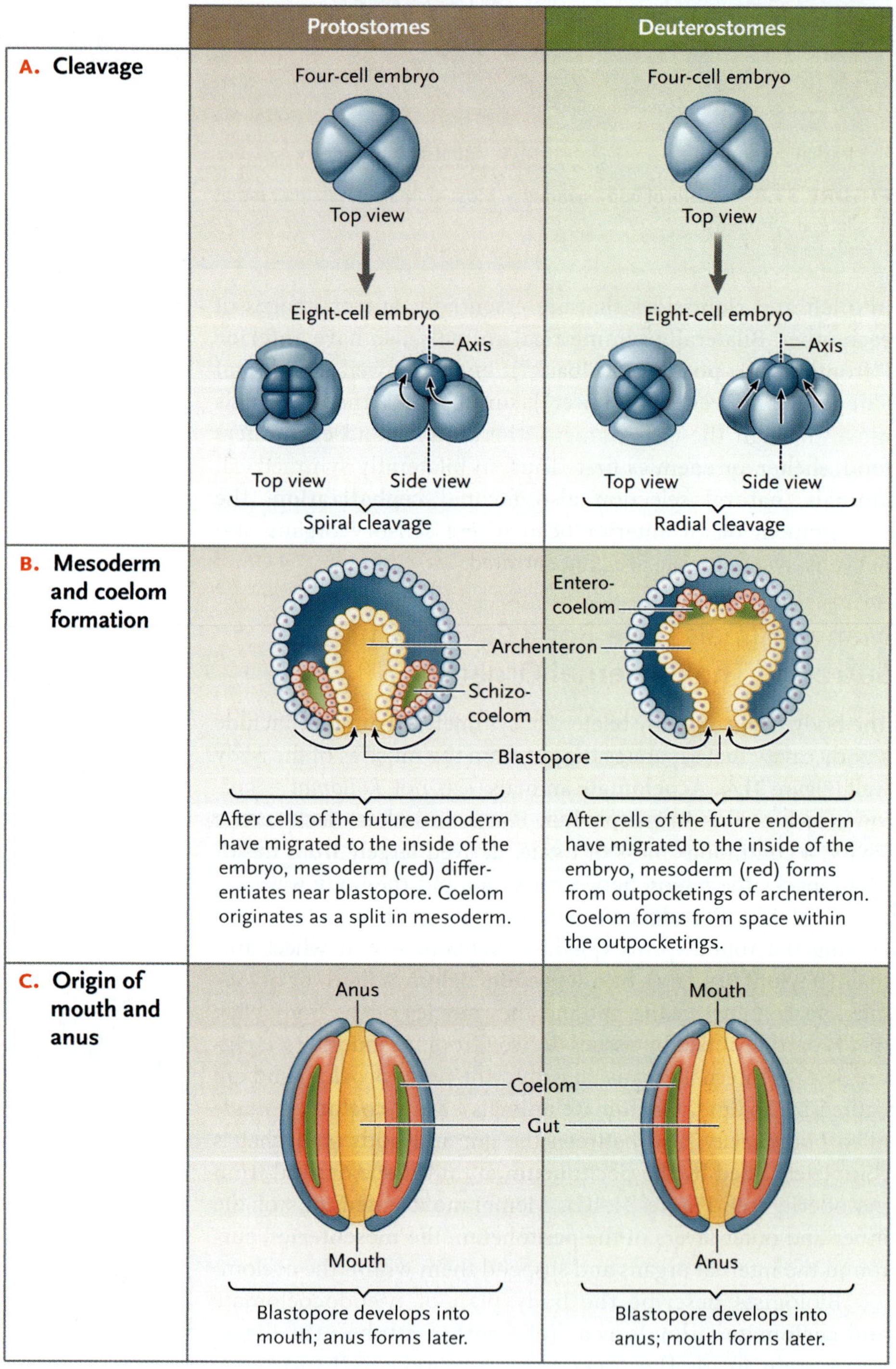

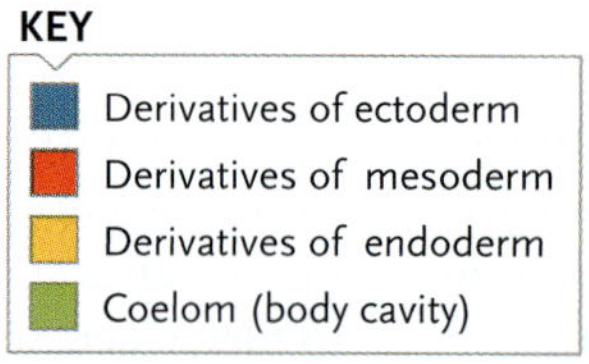

FIGURE 31.5 Protostomes and deuterostomes. The two clades of coelomate animals differ in **(A)** cleavage patterns, **(B)** the origin of mesoderm and the coelom, and **(C)** the origin of the mouth and the anus.

Protostomes and deuterostomes also differ in the origin of mesoderm and the coelom (see Figure 31.5B). In most protostomes, mesoderm originates from a few specific cells near the blastopore. As the mesoderm grows and develops, it splits into inner and outer layers. The space between the layers is called a **schizocoelom** (*skhizein* = split). In deuterostomes, mesoderm often forms from outpocketings of the archenteron. The space pinched off from the archenteron by the outpocketing mesoderm is called an **enterocoelom** (*enteron* = intestine).

Several other characteristics also differ in adult protostomes and deuterostomes. For example, the central nervous system of protostomes is generally positioned on the ventral side of the body, and their brain surrounds the anterior section of the digestive tract. By contrast, the nervous system and brain of deuterostomes lie on the dorsal side of the body.

Segmentation Divides the Bodies of Some Animals into Repeating Units

Some phyla in both the protostome and deuterostome clades exhibit varying degrees of **segmentation,** the production of body parts as repeating units. During development, segmentation first arises in the mesoderm, the middle tissue layer that produces most of the body's bulk. In humans and other vertebrates, we see evidence of segmentation in the vertebral column ("backbone"), ribs, and associated muscles, such as the "six-pack abs" that sit-ups accentuate. In some animals, segmentation is also reflected in structures derived from the endoderm and ectoderm. For example, the ringlike pattern on an earthworm or a caterpillar matches the underlying segments.

Segmentation provides several advantages. In markedly segmented animals, such as earthworms and their relatives, each segment may include a complete set of important organs, including respiratory surfaces and parts of the nervous, circulatory, and excretory systems. Thus, a segmented animal may survive damage to the organs in one segment, because those in other segments perform the same functions. Segmentation also improves control over movement, especially in wormlike animals. Each segment has its own set of muscles, which can act independently of those in other segments. Thus, an earthworm can move its anterior end to the left while it swings its posterior end to the right. Similarly, the segmented backbone and body wall musculature of vertebrates allow greater flexibility of movement than would unsegmented structures.

STUDY BREAK 31.2

1. **What is a tissue, and what three primary tissue layers are present in the embryos of most animals?**
2. **What type of body symmetry do humans have?**
3. **What is the functional significance of the coelom?**
4. **What does having a segmented body allow some animals to do?**

31.3 An Overview of Animal Phylogeny and Classification

For many years, biologists used the morphological innovations and embryological patterns described earlier to trace the phylogenetic history of animals. These efforts were sometimes hampered by the difficulty of identifying homologous structures in different phyla and by morphological data that led to contradictory interpretations. More recently, researchers have used molecular sequence data to reanalyze animal relationships. Although biologists now recognize nearly 40 animal phyla, in this chapter we focus on 14 phyla, many of which include substantial numbers of species, that represent a diversity of body forms and ways of life **(Table 31.1).** In Chapter 32, we consider three deuterostomes phyla, which represent a distinct clade within the animals.

TABLE 31.1 Classification of the Animal Phyla Described in This Book

Major Clade	Phylum	Known Diversity (Approximate Number of Living Species Described)
Parazoa	Porifera	9,000
Eumetazoa		
Radiata	Cnidaria	9,000
Bilateria	Ctenophora	150
Protostomia		
Lophotrochozoa	Ectoprocta	5,000
	Brachiopoda	300
	Phoronida	20
	Platyheminthes	20,000
	Rotifera	1,800
	Nemertea	1,100
	Mollusca	100,000
	Annelida	16,000
Ecdysozoa	Nematoda	25,000
	Onychophora	100
	Arthropoda	1,150,000
Deuterostomia	Echinodermata	6,000
	Hemichordata	100
	Chordata	65,000

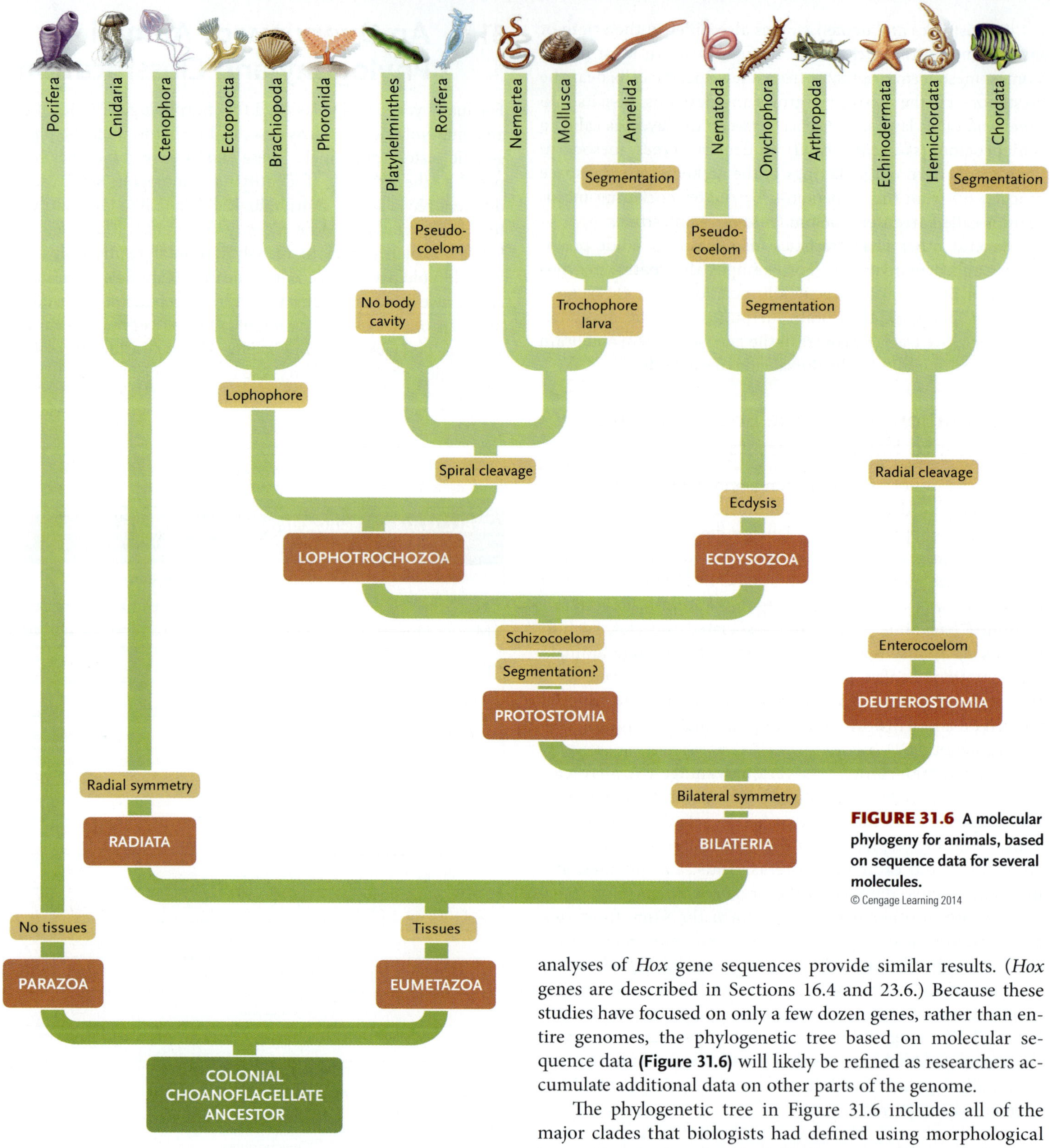

FIGURE 31.6 A molecular phylogeny for animals, based on sequence data for several molecules.

Molecular Phylogenetics Has Refined Our Understanding of Animal Phylogeny

Molecular phylogenies of animal relationships have usually been based on the nucleotide sequences in small subunit ribosomal RNA and mitochondrial DNA (see Chapter 15). Recent analyses of *Hox* gene sequences provide similar results. (*Hox* genes are described in Sections 16.4 and 23.6.) Because these studies have focused on only a few dozen genes, rather than entire genomes, the phylogenetic tree based on molecular sequence data **(Figure 31.6)** will likely be refined as researchers accumulate additional data on other parts of the genome.

The phylogenetic tree in Figure 31.6 includes all of the major clades that biologists had defined using morphological innovations and embryological characters. For example, molecular data confirm the distinctions between the Parazoa and the Eumetazoa and between the Radiata and the Bilateria. They also confirm the separation of the deuterostome phyla from all others within the Bilateria.

However, the molecular phylogeny groups many other phyla—including the acoelomate animals, pseudocoelomate animals, protostomes, and a few others—into one taxon, the

Protostomia. This group is, in turn, subdivided into two major clades, the Lophotrochozoa and the Ecdysozoa, which were not previously recognized. The name Lophotrochozoa (*lophos* = crest, *trochos* = wheel) refers to both the lophophore, a feeding structure found in three phyla (illustrated in Figure 31.15), and the trochophore, a type of larva found in annelids and mollusks (illustrated in Figure 31.23). The name Ecdysozoa (*ekdysis* = a stripping or casting off) refers to the cuticle or external skeleton that species within this group secrete and periodically molt (or "cast off") when they experience a growth spurt or begin a different stage of the life cycle (illustrated in Figure 31.34); the molting process is called **ecdysis.**

The Molecular Phylogeny Reveals Some Surprising Patterns in the Evolution of Key Morphological Innovations

Phylogenetic trees contain explicit hypotheses about evolutionary change, and the molecular phylogeny has forced biologists to reevaluate the evolution of several important morphological innovations. For example, traditional phylogenies based on morphology and embryology usually inferred that the absence of a body cavity, the acoelomate condition, was an ancestral state and that the presence of a body cavity, the pseudocoelomate or coelomate condition, was derived. But the molecular tree provides a very different view. Because most protostome phyla have a schizocoelom, the molecular phylogeny suggests that this trait evolved in the common ancestor of all protostomes. If that hypothesis is correct, then the acoelomate condition of flatworms (phylum Platyhelminthes) represents the evolutionary *loss* of the schizocoelom, *not* an ancestral condition. Similarly, the molecular phylogeny hypothesizes that the pseudocoelom evolved independently in rotifers (Lophotrochozoa, phylum Rotifera) and in roundworms (Ecdysozoa, phylum Nematoda). Thus, according to the molecular phylogeny, the pseudocoelomate condition of these organisms is the product of convergent evolution.

Phylogenies based on embryological and morphological characters also suggested that the segmented body plan of several protostome phyla was inherited from a segmented common ancestor and that segmentation arose independently in chordates by convergent evolution. The molecular phylogeny in Figure 31.6, by contrast, suggests that segmentation evolved independently in *three* clades—segmented worms (Lophotrochozoa, phylum Annelida), arthropods and velvet worms (Ecdysozoa, phyla Arthropoda and Onychophora), and chordates (Deuterostomia, phylum Chordata)—rather than in just two. However, in 2010, Nicolas Dray of the Centre Génétique Moléculaire du CNRS, and colleagues from other European research institutes, demonstrated that the development of segmentation in one species of annelid requires expression of the Hedgehog signaling pathway, which also plays a prominent role in the development of segmentation in arthropods. Thus, the genetic basis of segmentation may have originated in the common ancestor of annelids and arthropods. If this conclusion is correct, it is consistent with the traditional view that segmentation was present in the common ancestor of all protostomes. Further research should confirm or refute this hypothesis and reveal more about the evolutionary history of this key morphological innovation in animals.

Despite these surprising findings about morphological evolution, biologists embrace the hypothesis provided by molecular sequence studies. In the future, new data will undoubtedly foster fascinating discussions, heated disputes, and revisions of the phylogeny. As a student, you may be understandably frustrated by the lack of consensus among experts and an ever-changing phylogeny for animals. But these disagreements highlight the uniqueness of science as a way of knowing the natural world through the process of collecting evidence and rigorously challenging accepted hypotheses. The phylogenetic tree based on molecular sequence data is truly revolutionary, and the dust has yet to settle on the disagreements that new analyses provoke.

STUDY BREAK 31.3

1. **Which major groupings of animals defined on the basis of morphological characters have been confirmed by molecular sequence studies?**
2. **What type of body cavity is ancestral within the Deuterostomes?**

31.4 Animals without Tissues: Parazoa

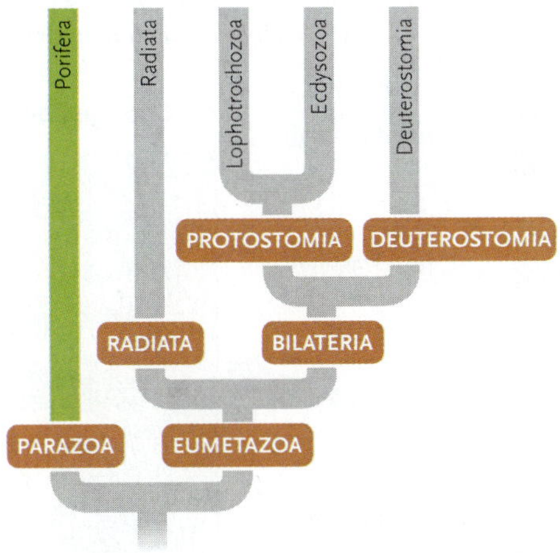

Parazoa is the basal animal clade that is the sister group to Eumetazoa. It includes just one group of animals, the sponges, phylum Porifera (meaning "pore bearers").

Sponges lack true tissues: during development, their cells do not form the complex layers typical of other phyla. Mature sponges are sessile, and their shapes are less fixed than those of other animals, because mobile cells allow them to change shape in response to local conditions **(Figure 31.7).** Sponges have been abundant since the Cambrian, especially in shallow coastal areas. Most of the 9,000 living species are marine and range in size from 1 cm to 2 m at maturity.

FIGURE 31.7 Asymmetry in sponges. Sponges, such as this *Haliclona* species from the Philippines, often have an asymmetrical shape.

Sponges have simple body plans **(Figure 31.8)**. Flattened cells form an outer layer, the **pinacoderm.** The inner surface of saclike sponges is lined by collar cells, **choanocytes,** each equipped with a beating flagellum and a surrounding "collar" of modified microvilli. Amoeboid cells wander through the gelatinous **mesohyl** between the two layers; they secrete a supporting skeleton of a fibrous protein and *spicules,* small needlelike structures of calcium carbonate or silica (see Figure 31.8D). The natural sponges that are used for bathing and washing are the fibrous remains of the bath sponge (*Spongia* species), which lacks mineralized parts.

A sponge's body is an elaborate system for filtering food particles from the surrounding water. Water flows through pores in the pinacoderm into a central chamber, the **spongocoel,** and then out of the sponge through one or more openings called **oscula** (singular, *osculum*). The beating flagellae of the choanocytes maintain a constant flow of water, and contractile pore cells *(porocytes)* adjust the flow rate. Even a small sponge may filter as much as 20 liters of water per day. The choanocytes capture suspended particles and microorganisms from the water and pass this food to mobile amoeboid cells, which carry nutrients to cells of the pinacoderm.

Most sponges are **hermaphroditic:** individuals produce both sperm and eggs. Sperm are released into the spongocoel and then out into the environment through oscula; eggs remain in the mesohyl, where sperm from other sponges, drawn in with water, fertilize them. Zygotes develop into flagellated larvae that are expelled to fend for themselves. Surviving larvae attach to substrates and undergo **metamorphosis** (a reorganization of form) into sessile adults. Some sponges also reproduce asexually; small fragments break off an adult and grow into new sponges. Many species also produce *gemmules,* clusters of cells with a resistant covering that allows them to survive unfavorable conditions; gemmules germinate into new sponges when conditions improve.

STUDY BREAK 31.4

1. **What type of body symmetry do sponges exhibit?**
2. **How does a sponge gather food from its environment?**

FIGURE 31.8 The body plan of sponges. Most sponges have **(A)** simple body plans and **(B)** relatively few cell types. **(C)** Beating flagella on the choanocytes create a flow of water through incurrent pores, into the spongocoel, and out through the osculum. **(D)** Venus' flower basket, a marine sponge, has spicules of silica fused into a rigid framework.

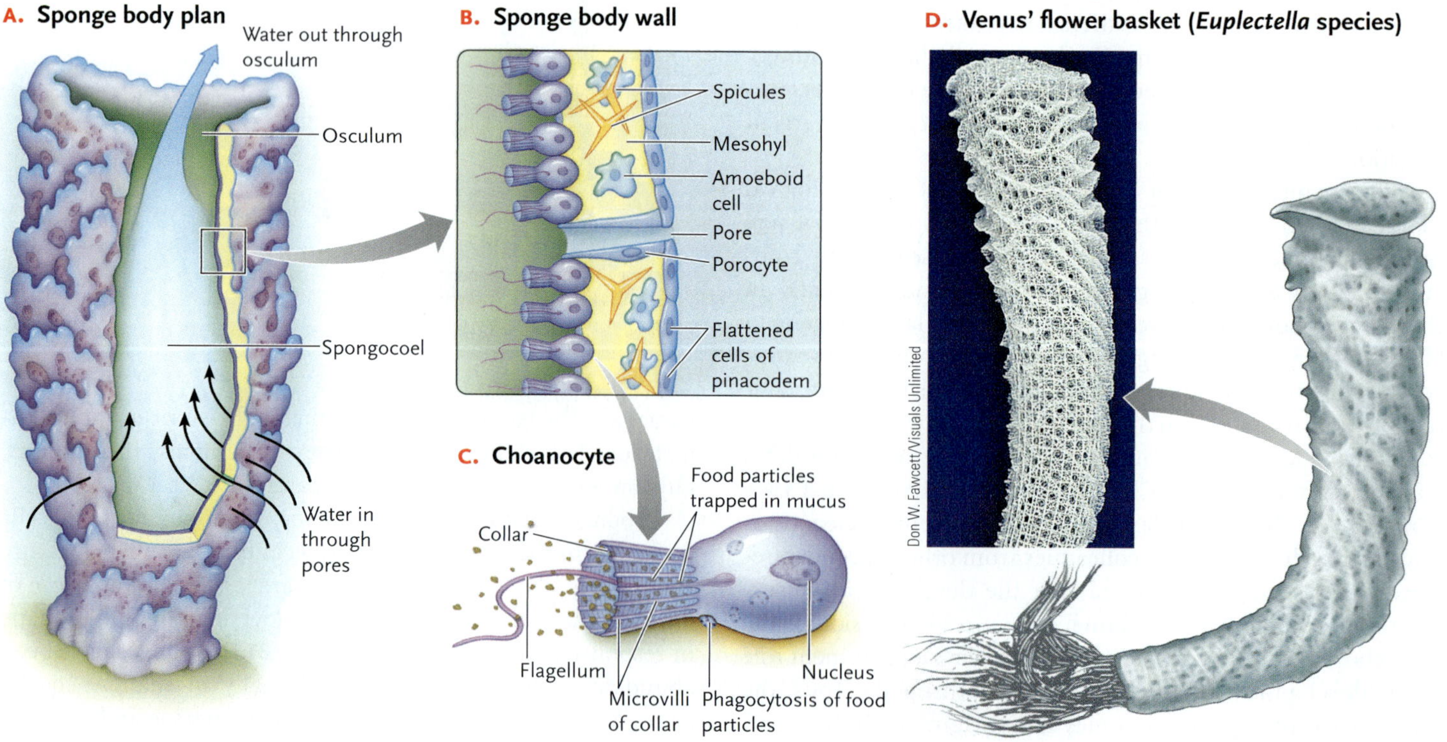

31.5 Eumetazoans with Radial Symmetry

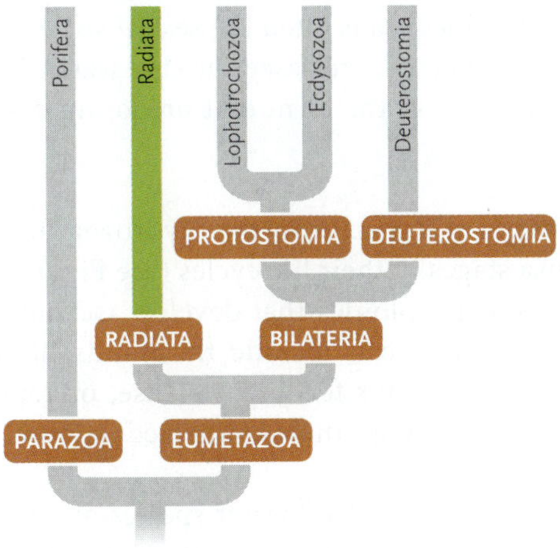

Unlike sponges, eumetazoans have true tissues, which develop from distinct layers in the embryo. Working together, the cells of a tissue perform complex functions beyond the capacity of individual cells. For example, nerve tissue transmits information rapidly through an animal's body, and epithelial tissue forms barriers that surround the body and line body cavities. In this section, we describe eumetazoans with radial symmetry. This symmetry enables them to sense stimuli from all directions, an effective adaptation for life in open water.

Two phyla of soft-bodied organisms, Cnidaria and Ctenophora, have radial symmetry and tissues, but they lack organ systems and a coelom. Species in both phyla possess a **gastrovascular cavity** that serves both digestive and circulatory functions. It has a single opening, the mouth. Gas exchange and excretion occur by diffusion because no cell is far from a body surface.

The radiate phyla have a diploblastic body plan with only two tissue layers, the inner *gastrodermis* (an endoderm derivative) and the outer *epidermis* (an ectoderm derivative). Most species also possess a gelatinous *mesoglea* (*mesos* = middle, *glia* = glue) between the two layers. The mesoglea contains widely dispersed amoeboid cells.

Cnidarians Use Nematocysts to Stun or Kill Prey

Nearly all of the 9,000 species in the phylum Cnidaria (*knide* = stinging nettle, a plant with irritating surface hairs) live in the sea **(Figure 31.9).** Their body plan is organized around a saclike gastrovascular cavity; the mouth is ringed with tentacles, which push food into it. Cnidarians may be vase-shaped, upward-pointing **polyps** or bell-shaped, downward-pointing **medusae** (see Figure 31.9A). Most polyps attach to a substrate at the *aboral* (opposite the mouth) end; medusae are unattached and float.

Cnidarians are the simplest animals that exhibit a division of labor among specialized tissues (see Figure 31.9B and C). (Sponges have specialized cells, but no tissues.) The gastrodermis includes gland cells and phagocytic nutritive cells, both of which contain circular muscle fibrils. Gland cells secrete enzymes into the gastrovascular cavity for the extracellular digestion of food, which is then engulfed by nutritive cells and exposed to intracellular digestion. The epidermis includes nerve

FIGURE 31.9 The cnidarian body plan. (A) Cnidarians exist as either polyps or medusae. **(B)** The body of both forms is organized around a gastrovascular cavity, which extends all the way to the aboral end of the animal. **(C)** The two tissue layers in the body wall, the gastrodermis and the epidermis, include a variety of cell types.

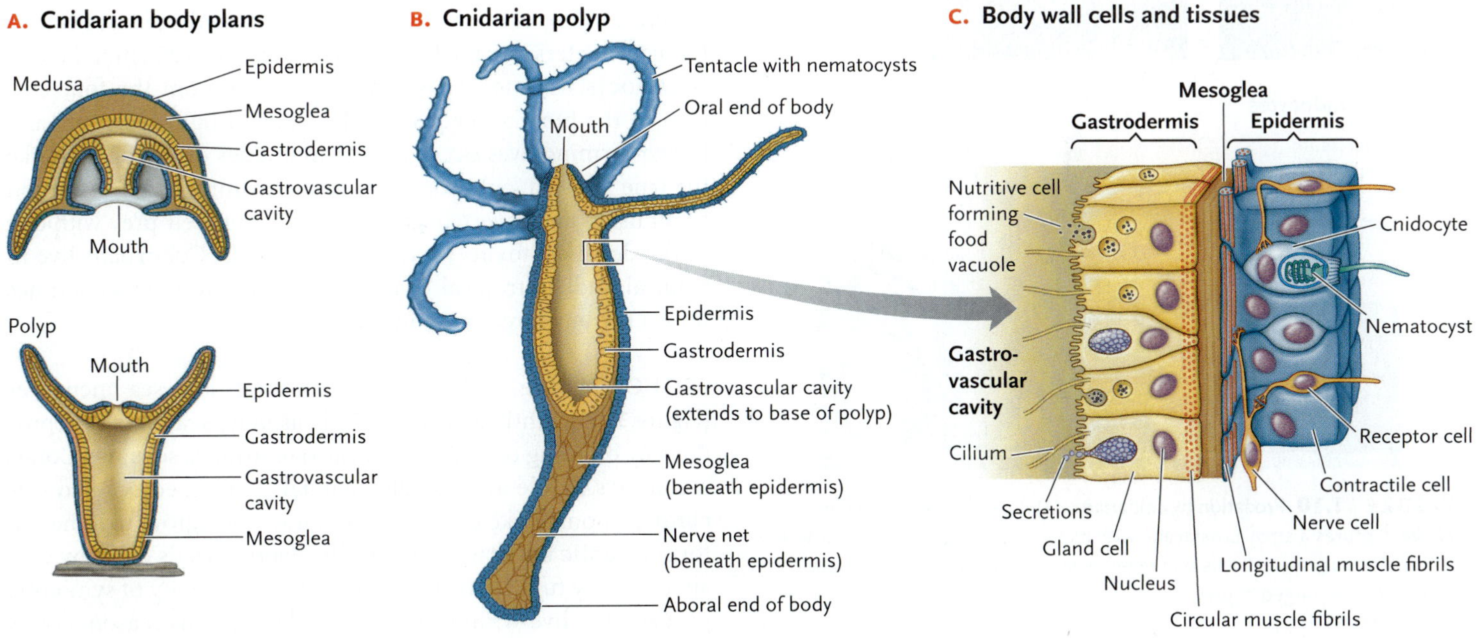

cells, sensory cells, contractile cells, and cells specialized for prey capture. A layer of acellular mesoglea separates the gastrodermis from the epidermis.

Cnidarians prey on crustaceans, fishes, and other animals. The epidermis includes unique cells, **cnidocytes,** each armed with a stinging **nematocyst** **(Figure 31.10).** The nematocyst is an encapsulated, coiled thread that is fired at prey or predators, sometimes releasing a toxin through its tip. Discharge of nematocysts may be triggered by touch, vibrations, or chemical stimuli. The toxin can paralyze small prey by disrupting nerve cell membranes. The painful stings of some jellyfishes and corals result from the discharge of nematocysts.

Cnidarians engage in directed movements by contracting muscle fibrils in cells of the epidermis and gastrodermis. In medusae, the mesogleal jelly serves as a deformable skeleton against which contractile cells act. Rapid contractions narrow the bell, forcing out jets of water that propel the animal. Polyps use their water-filled gastrovascular cavity as a hydrostatic skeleton. When some cells contract, fluid within the chamber is shunted about, changing the body's shape and moving it in a particular direction.

The **nerve net,** which threads through both tissue layers, coordinates responses to stimuli, but it has no central control organ or brain. Impulses initiated by sensory cells are transmitted in all directions from the site of stimulation.

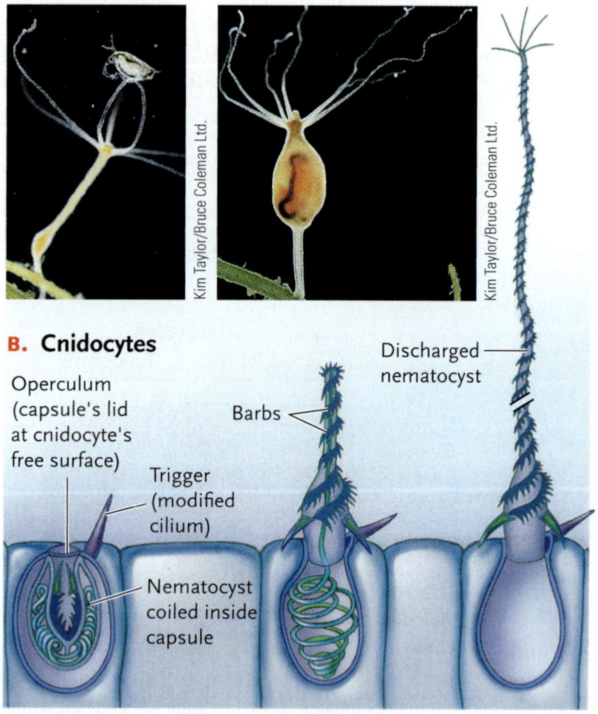

FIGURE 31.10 Predation by cnidarians. (A) A polyp of a freshwater *Hydra* captures a small crustacean with its tentacles and swallows it whole. **(B)** Cnidocytes, special cells on the tentacles, encapsulate nematocysts, which are discharged at prey.

Many cnidarians exist in only the polyp or the medusa form, but some have a life cycle that alternates between them **(Figure 31.11).** In the latter type, the polyp often produces new individuals asexually from buds that break free of the parent (see Figure 49.2). The medusa is often the sexual stage, producing sperm and eggs, which are released into the water. The four clades of Cnidaria differ in the form that predominates in the life cycle.

HYDROZOA Most of the species in the Hydrozoa have both polyp and medusa stages in their life cycles (see Figure 31.11). The polyps form sessile colonies that develop asexually from one individual. A colony can include thousands of polyps, which may be specialized for feeding, defense, or reproduction. They share food through their connected gastrovascular cavities.

Unlike most hydrozoans, freshwater species of *Hydra* (see Figure 31.10A) live as solitary polyps that attach temporarily to rocks, twigs, and leaves. Under favorable conditions, hydras reproduce by budding. Under adverse conditions, they produce eggs and sperm; the zygotes, which are encapsulated in a protective coating, develop and grow when conditions improve.

SCYPHOZOA The medusa stage predominates in the Scyphozoa **(Figure 31.12A),** which are commonly known as jellies or jellyfishes. Scyphozoans range from 2 cm to more than 2 m in diameter. Nerve cells near the margin of the bell control their tentacles and coordinate the rhythmic activity of contractile cells, which move the animal. Specialized sensory cells are clustered at the edge of the bell: **statocysts** sense gravity and **ocelli** are sensitive to light. Scyphozoan medusae are either male or female; they release gametes into the water where fertilization takes place.

CUBOZOA The 20 known species of box jellyfish, the Cubozoa **(Figure 31.12B),** exist primarily as cube-shaped medusas only a few centimeters tall; the largest species grows to 25 cm in height. Nematocyst-rich tentacles grow in clusters from the four corners of the boxlike medusa, and groups of light receptors and image-forming eyes occur on the four sides of the bell. Unlike the true jellyfish, cubozoans are active swimmers. They feed on small fishes and invertebrates, immobilizing their prey with one of the deadliest toxins produced by animals. Cubozoans live in tropical and subtropical coastal waters, where they sometimes pose a serious threat to swimmers.

ANTHOZOA The Anthozoa includes corals and sea anemones **(Figure 31.13).** Anthozoans exist only as polyps, and often reproduce by budding or fission; most also reproduce sexually. Corals are always sessile and usually colonial. Most species build calcium carbonate skeletons, which sometimes grow together to form gigantic underwater reefs. The energy needs of many corals are partly fulfilled by the photosynthetic activity of symbiotic protists that live within the corals' cells. For this reason, corals

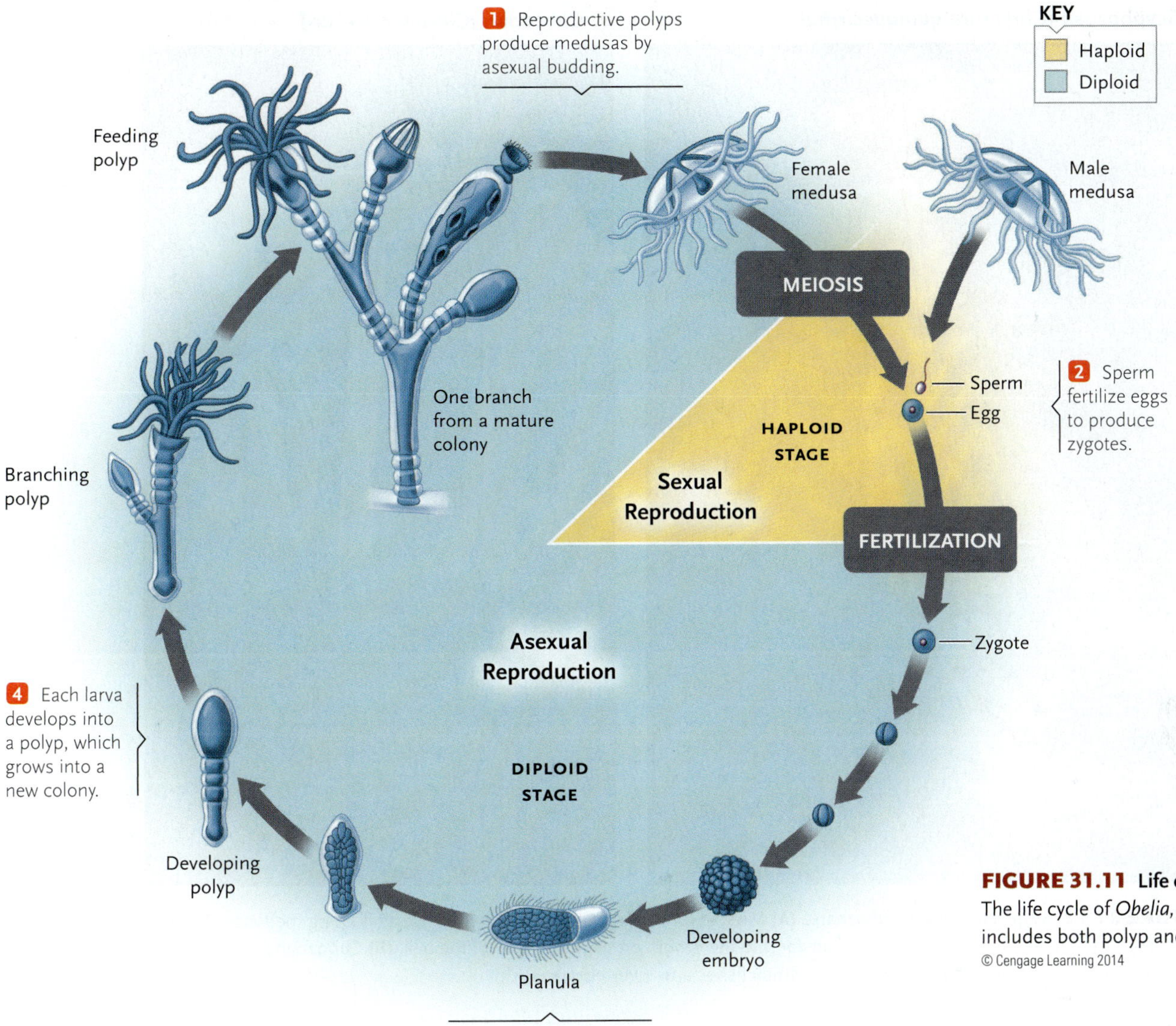

FIGURE 31.11 Life cycle of a *hydrozoan*. The life cycle of *Obelia*, a colonial hydrozoan, includes both polyp and medusa stages.

are restricted to shallow water where sunlight can penetrate. Sea anemones are soft-bodied, solitary polyps, ranging from 1 cm to 10 cm in diameter. Many species occupy shallow coastal waters. Most species are essentially sessile, but some move by crawling slowly or by using their gastrovascular cavity as a hydrostatic skeleton.

Ctenophores Use Tentacles to Feed on Microscopic Plankton

Like the cnidarians, the 150 species of comb jellies in the phylum Ctenophora (*ktenos* = comb, *-phoros* = bearing) have radial symmetry, mesoglea, and feeding tentacles. However, they differ from cnidarians in significant ways: they lack nematocysts; they expel some waste through anal pores opposite the mouth; and certain tissues appear to be of mesodermal origin. These transparent, and often luminescent, animals range in size from a few millimeters to a few meters **(Figure 31.14)**. They live primarily in coastal regions of the oceans.

Ctenophores move by beating cilia arranged on eight longitudinal plates that resemble combs. They are the largest animals to use cilia for locomotion, but they are feeble swimmers. Nerve cells connected to the cilia coordinate the animals' movements, and a gravity-sensing statocyst helps them maintain an upright position. They capture microscopic plankton with their two tentacles, which have specialized cells that discharge sticky filaments; food particles are ingested as the food-laden tentacles are drawn across the mouth. Ctenophores are hermaphroditic, producing gametes in cells that line the gastrovascular cavity. Eggs and sperm are expelled through the mouth or from special pores, and fertilization occurs in the open water.

STUDY BREAK 31.5

1. How do cnidarians capture, consume, and digest their prey?
2. Which group of cnidarians has only a polyp stage in its life cycle?
3. What do ctenophores eat, and how do they collect their food?

A. Scyphozoan *(Chrysaora quinquecirrha)*

B. Cubozoan *(Chironex fleckeri)*

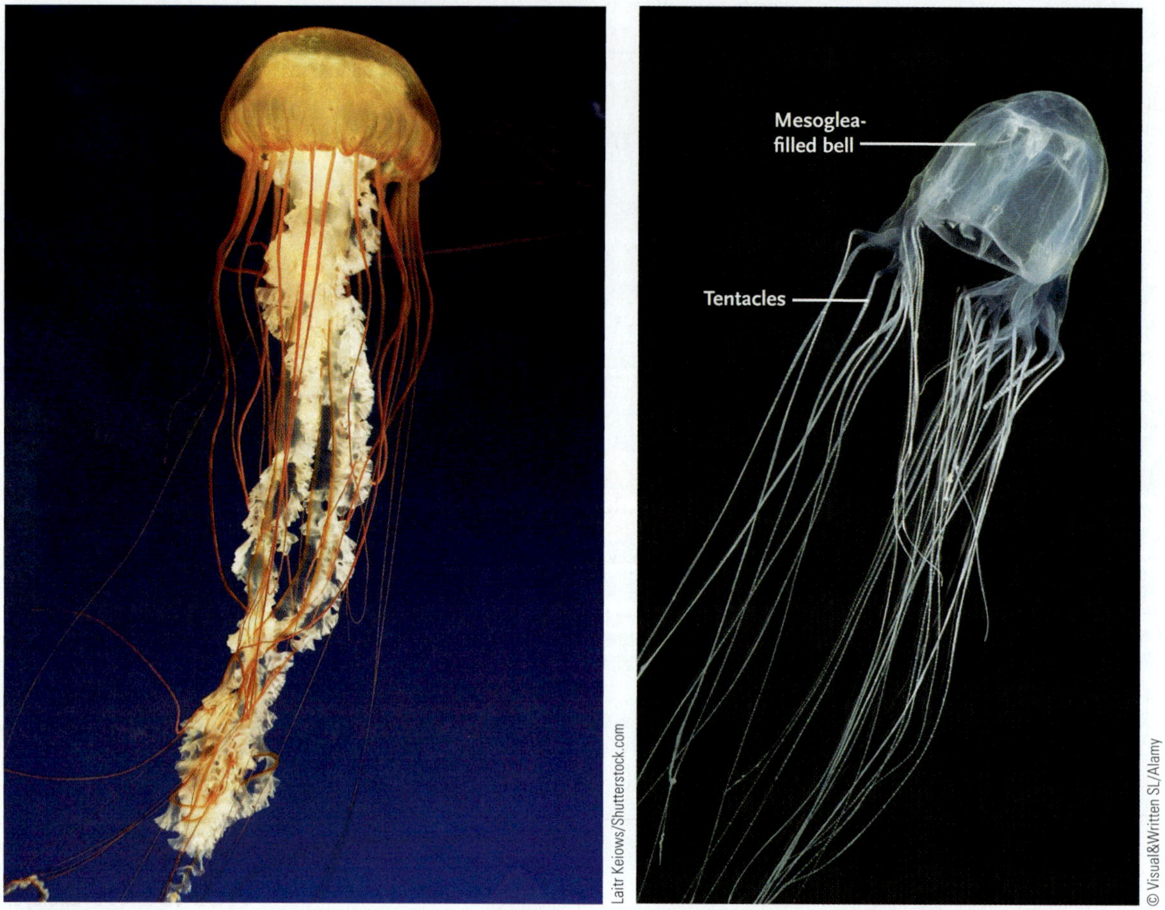

FIGURE 31.12 Scyphozoans and cubozoans. (A) Most scyphozoans, like the sea nettle, live as floating medusae. Their tentacles trap prey, and the long oral arms transfer it to the mouth on the underside of the bell. **(B)** Cubozoans, like the sea wasp, are strong swimmers that actively pursue small fishes and invertebrates.

A. Staghorn coral *(Acropora cervicornis)*

B. Sea anemone *(Urticina lofotensis)* escape behavior

FIGURE 31.13 Anthozoans. (A) Many corals are colonial, and their polyps build a hard skeleton of calcium carbonate. The skeletons accumulate to form coral reefs in shallow tropical waters (see Figure 51.29B). **(B)** A white-spotted sea anemone detaches from its substrate to escape from a predatory sea star.

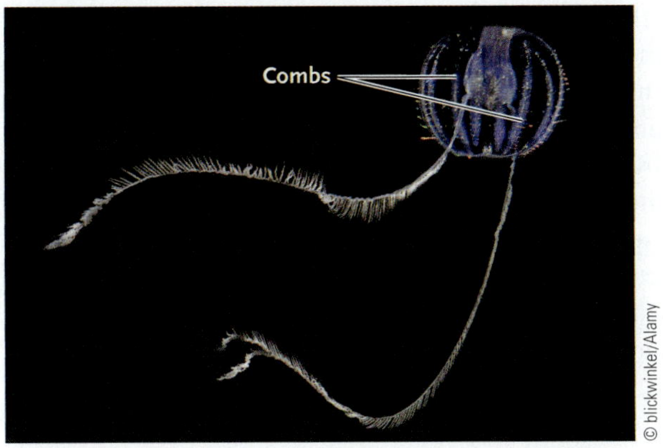

FIGURE 31.14 Ctenophores. Comb jellies, like this sea gooseberry *(Pleurobrachia pileus)* from the Gulf of Maine, collect microscopic prey on their long, sticky tentacles and then wipe the food-laden tentacles across their mouths.

FIGURE 31.15 Lophophorate animals. Although the lophophorate animals differ markedly in appearance, they all use a lophophore—the feathery structures in the photos—to acquire food.

31.6 Lophotrochozoan Protostomes

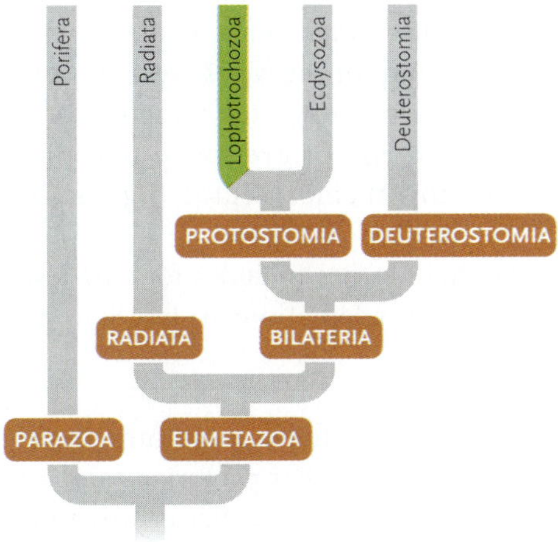

The remaining organisms described in this chapter fall within the group Bilateria: they have bilateral symmetry and triploblastic development (that is, their embryological development includes three tissue layers). Bilaterians also have organ systems, structures that include two or more tissue types that perform specific functions. Most of them also possess a coelom or pseudocoelom. With bilateral symmetry and sensory organs concentrated at the anterior end of the body, most bilaterians engage in highly directed, often rapid movements in pursuit of food, mates, or to escape danger. And their complex organ systems accomplish tasks more efficiently than simple tissues. For example, animals that have a tubular digestive system surrounded by a space (the coelom or pseudocoelom) use muscular contractions of the digestive system to move ingested food past specialized epithelia that break it down and absorb the breakdown products.

Molecular analyses group eight of the most diverse phyla into the monophyletic Lophotrochozoa, one of the two main protostome clades (see Figure 31.6).

The Lophophorate Phyla Share a Distinctive Feeding Structure

Three small groups of aquatic, coelomate animals—the phyla Ectoprocta, Brachiopoda, and Phoronida—possess a **lophophore,** a circular or U-shaped fold with one or two rows of hollow, ciliated tentacles surrounding the mouth **(Figure 31.15).** Molecular sequence data as well as the lophophore suggest that these phyla share a common ancestry.

The lophophore, which looks like a crown of tentacles at the anterior end of the animal, serves as a site for gas exchange and waste elimination as well as for food capture. Most lophophorates are sessile suspension-feeders as adults: movement of cilia on the tentacles brings food-laden water toward the lophophore, the tentacles capture small organisms and debris, and the cilia transport them to the mouth. The lophophorates have a complete digestive system, which is U-shaped in most species, with the anus lying outside the ring of tentacles.

PHYLUM ECTOPROCTA The Ectoprocta (sometimes called Bryozoa) are tiny colonial animals that mainly occupy marine habitats (see Figure 31.15A). They secrete a hard covering over their soft bodies and feed by extending the lophophore through a hole. Each colony, which may include more than a million individuals, is produced asexually by a single animal. Ectoprocta colonies are permanently attached to solid substrates, where they form encrusting mats, bushy upright growths, or jellylike blobs. Nearly 5,000 living species are known.

PHYLUM BRACHIOPODA The Brachiopoda, or lampshells, have two calcified shells that develop on the animal's dorsal and ventral sides (see Figure 31.15B). Most species attach to substrates with a stalk that protrudes through one of the shells. The lophophore is held within the two shells, and the animal feeds by opening its shell and drawing water over its tentacles. Although only 300 species of brachiopods live today, more than 30,000 extinct species are known from the fossil record, mostly from Paleozoic seas.

PHYLUM PHORONIDA The 20 or so species of phoronid worms vary in length from a few millimeters to 25 cm (see Figure 31.15C). They usually build tubes in soft ocean sediments or on hard substrates, and feed by protruding the lophophore from the top of the tube. The animal can withdraw into the tube when disturbed. Phoronida reproduce both sexually and by budding.

Flatworms Have Digestive, Excretory, Nervous, and Reproductive Systems, but Lack a Coelom

The 20,000 flatworm species in the phylum Platyhelminthes (*platys* = flat; *helminth-* = worm) live in aquatic and moist terrestrial habitats. Like cnidarians, flatworms can swim or float in water, but they are also able to crawl over surfaces. They range from less than 1 mm to more than 60 cm in length, but most are just a few millimeters thick. Free-living species eat live prey or decomposing carcasses, whereas parasitic species consume the tissues of living hosts.

Like the radiate phyla, flatworms are acoelomate, but they have a complex structural organization that reflects their triploblastic construction **(Figure 31.16)**. Endoderm derivatives line the gastrovascular (digestive) cavity with cells specialized for the chemical breakdown and absorption of ingested food. Mesoderm, the middle tissue layer, produces muscles and reproductive organs. Ectoderm produces a ciliated epidermis, the nervous system, and the *flame cell system*. Flame cells function in osmoregulation; they regulate the concentrations of salts and water within body fluids, allowing free-living flatworms to live in freshwater habitats. Flatworms do not have circulatory or respiratory systems, but, because all cells of their dorsoventrally (top-to-bottom) flattened bodies are near an interior or exterior surface, diffusion supplies them with nutrients and oxygen.

The flatworm nervous system includes two or more longitudinal ventral nerve cords interconnected by numerous smaller nerve fibers, like rungs on a ladder. An anterior **ganglion,** a concentration of nervous system tissue that serves as a primitive brain, integrates their behavior. Most free-living species have *ocelli,* or eye spots, that distinguish light from dark and tiny chemoreceptor organs that sense chemical cues.

The phylum Platyhelminthes includes four clades, defined largely by their anatomical adaptations to free-living or parasitic habits.

TURBELLARIA Most free-living flatworms (Turbellaria) live in the sea **(Figure 31.17)**, but the familiar planarians and a few others live in fresh water or on land. Turbellarians swim by undulating the body wall musculature, or they crawl across surfaces by using muscles and cilia to glide on mucous trails produced by the ventral epidermis.

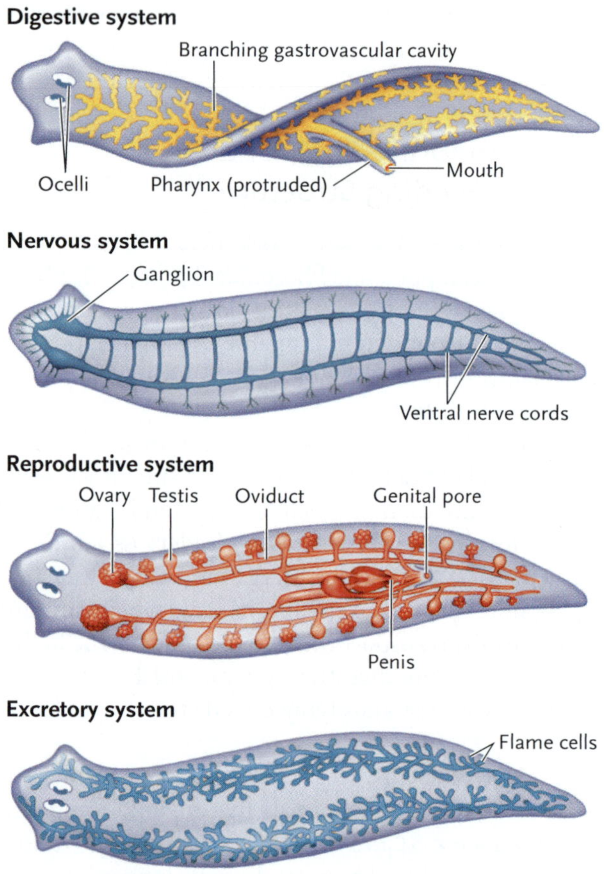

FIGURE 31.16 Flatworms. Species in the phylum Platyhelminthes, exemplified by a freshwater planarian, have well-developed digestive, excretory, nervous, and reproductive systems. Because flatworms are acoelomate, their organ systems are embedded in a solid mass of tissue between the gut and the epidermis.

FIGURE 31.17 Turbellaria. Some marine turbellarians, such as *Pseudoceros dimidiatus,* are brightly colored.

FIGURE 31.18 **Trematoda.** The hermaphroditic Chinese liver fluke *(Opisthorchis sinensis)* uses a well-developed reproductive system to produce thousands of eggs.

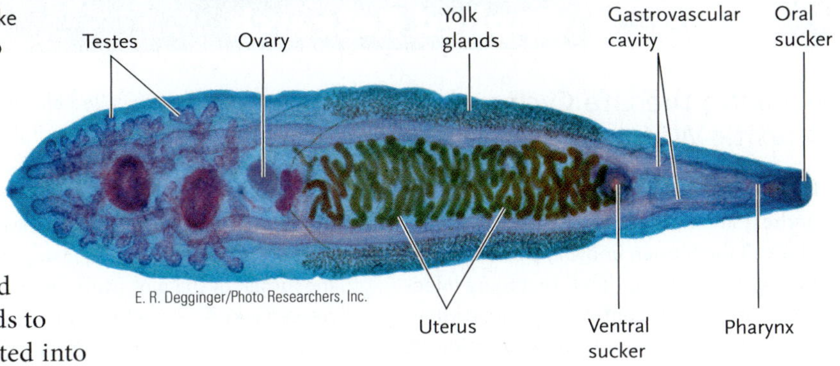

The gastrovascular cavity in free-living flatworms is similar to that of cnidarians. Food is ingested and wastes are eliminated through a single opening, the mouth, located on the ventral surface. Food passes from the mouth to a muscular **pharynx** that leads to the digestive cavity (see Figure 31.16). Chemicals secreted into the saclike cavity digest the food into particles; then cells throughout the gastrovascular surface engulf the particles and subject them to intracellular digestion. In some species, the digestive cavity is highly branched, increasing the surface area for digestion and absorption.

Nearly all turbellarians are hermaphroditic, with complex reproductive systems (see Figure 31.16). When they mate, each partner functions simultaneously as a male and a female. Many free-living species also reproduce asexually by separating their anterior half from their posterior half. Both halves subsequently regenerate the missing parts.

TREMATODA AND MONOGENOIDEA Flukes (Trematoda and Monogenoidea) are parasites that obtain nutrients from tissues of a living host **(Figure 31.18).** Most adult trematodes are **endoparasites,** living in the gut, liver, lungs, bladder, or blood vessels of vertebrates. Monogenes are **ectoparasites,** attaching to the gills or skin of aquatic vertebrates. Flukes are structurally specialized for a parasitic existence. They use suckers or hooks to attach to hosts, and a tough outer covering protects them from chemical attack. They produce numerous eggs that can readily infect new hosts. Monogene flukes usually have simple life cycles with a single host species. Trematodes, by contrast, have complex life cycles and multiple hosts. Humans suffer potentially fatal infections by many flukes, as discussed in *Focus on Applied Research.*

CESTODA Tapeworms (Cestoda) develop, grow, and reproduce within the intestines of vertebrates **(Figure 31.19).** Through evolution, they have lost their mouths and digestive systems and absorb nutrients directly through their body wall. Tapeworms have a specialized structure, the *scolex,* with hooks and suckers that attach to the host's intestine; like the flukes, they also have a protective covering resistant to digestive enzymes.

Most of a tapeworm's body, which may be up to 20 m long, consists of a series of segmental structures, *proglottids,* that contain little more than male and female reproductive systems. New proglottids are generated near the scolex; older proglottids, each carrying as many as 80,000 fertilized eggs, break off from the tapeworm's posterior end, and leave the host's body in feces. Tapeworms are important parasites on humans, pets, and livestock.

Rotifers Are Tiny Pseudocoelomates with a Jawlike Feeding Apparatus

Most of the 1,800 species in the phylum Rotifera (*rota* = wheel; *ferre* = to carry) live in fresh water **(Figure 31.20).** All are microscopic—about the size of a ciliate protist—but they have well-developed digestive, reproductive, excretory, and nervous systems as well as a pseudocoelom. In some habitats, rotifers make up a large part of the zooplankton, tiny animals that float in open water.

Rotifers use coordinated movements of cilia, arranged in a wheel-like *corona* around the head, to propel themselves in the environment. Cilia also bring food-laden water to their mouths. Ingested microorganisms are conveyed to the *mastax,* a toothed grinding organ, and then passed to the stomach and intestine. Rotifers have a **complete digestive tract:** food enters through the mouth, and undigested waste is voided through the anus.

The life history patterns of some rotifers are adapted to the ever-changing environments in small bodies of water. During

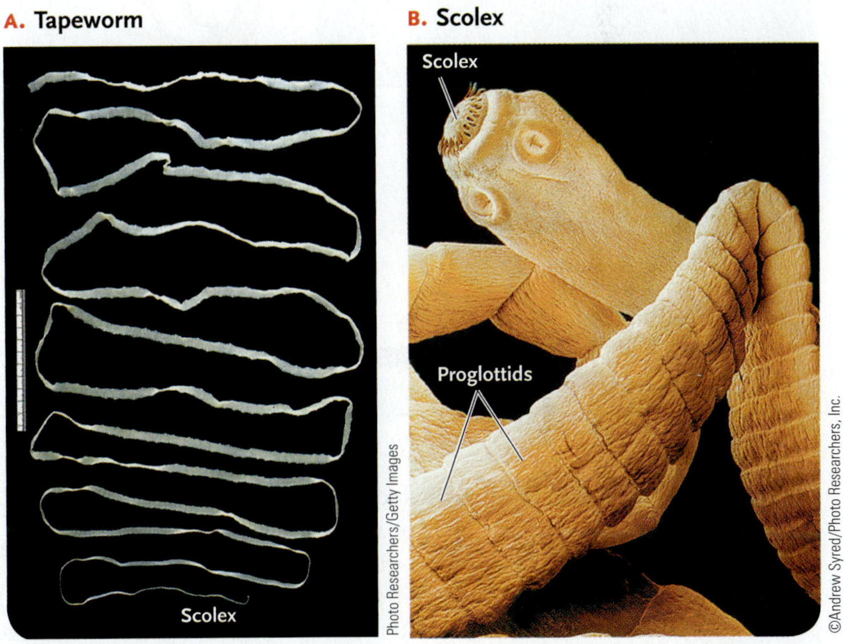

FIGURE 31.19 **Cestoda. (A)** Tapeworms have long bodies comprised of a series of proglottids that each produce thousands of fertilized eggs. **(B)** The anterior end is a scolex with hooks and suckers that attach to the host's intestinal wall.

Breaking the Life Cycles of Parasitic Worms

Many parasitic flatworms (phylum Platyhelminthes) and roundworms (phylum Nematoda) call the human body home, frequently causing disfiguring or life-threatening infections. The effort to control or eliminate these infections often begins when parasitologists or public health researchers study the worms' life cycles and ecology. This approach is successful because the worms often have more than one host: humans may be the *primary host,* harboring the sexually mature stage of the parasites' life cycles, but other animals serve as *intermediate hosts* to the larval stages. If researchers can learn the details of a parasite's life cycle, they can identify ways to cut it short before the parasite infects a human host.

Roughly 200 million people, most of them in sub-Saharan Africa or in Asia, suffer from *schistosomiasis,* a disease caused by several species of flatworms called blood flukes (Trematoda). Japanese blood flukes *(Schistosoma japonicum)* mature and mate in blood vessels of the human intestine **(Figure 1).** Sharp spines on their eggs rupture the blood vessels, releasing the eggs into the lumen of the gut, from which they are passed with feces. When the infected human waste enters standing fresh water, the eggs hatch into ciliated larvae, which burrow into some aquatic snail species. The larvae feed on the snail's tissues and reproduce asexually. Their offspring leave the snail and, when they contact human skin, bore inward to a blood vessel. They eventually reach the intestine, where they complete their complex life cycle and produce fertilized eggs.

Infected humans mount an immune response against flukes, but it is always a losing battle. Severe infections cause coughs, rashes, pain, and eventually diarrhea, anemia, and permanent damage to the intestines, liver, spleen, bladder, and kidneys. Death often results. Drug therapy can reduce the symptoms and limit the spread of these parasites, but schistosomiasis is most common in countries where people have limited access to medical care. Research has demonstrated that the disease can be controlled by proper sanitation and the elimination of the snails that serve as intermediate hosts.

The nematodes called filarial worms (*Wuchereria* and *Brugia*) cause another debilitating infection. These large roundworms (up to 10 cm long) live in the human lymphatic system, where they obstruct the normal flow of lymphatic fluid to the bloodstream. Female worms release first-stage larvae, which are acquired by mosquitoes, the intermediate host, when they feed on human blood. The larvae develop into second-stage larvae in mosquitoes, and when a mosquito bites another human, it may transmit those larvae to a new host. If victims experience severe filarial worm infection, their lymphatic vessels can be so obstructed that surrounding tissues swell grotesquely, a condition known as *elephantiasis* **(Figure 2).** Public health programs that reduce or eliminate mosquito populations lower the incidence of this disease.

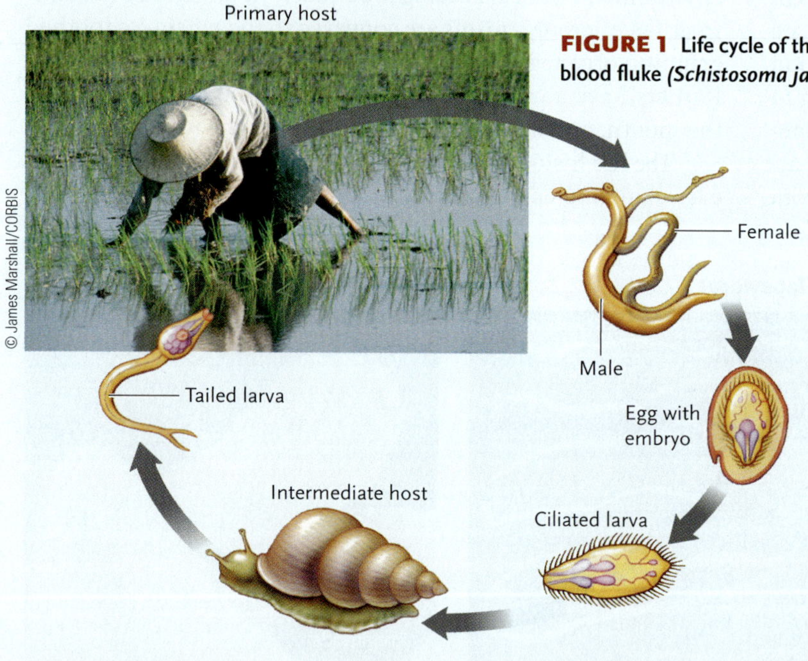

FIGURE 1 Life cycle of the Japanese blood fluke *(Schistosoma japonicum).*

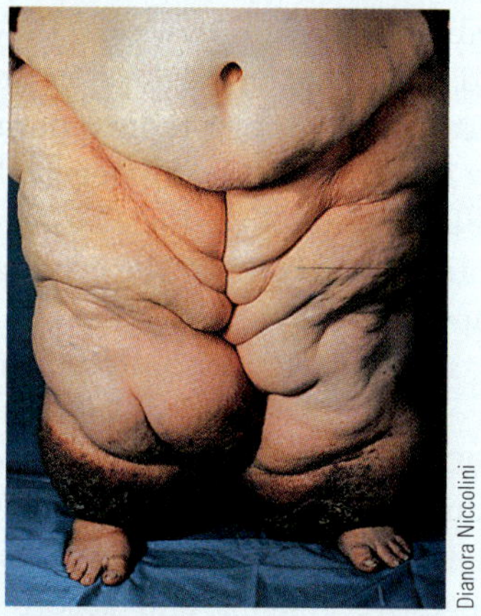

FIGURE 2 Elephantiasis caused by *Wuchereria bancrofti.*

most months, rotifer populations include only females that reproduce by **parthenogenesis,** a form of asexual reproduction in which unfertilized eggs develop into diploid females (see Section 49.1). When environmental conditions deteriorate, females produce eggs that develop into haploid males. The males fertilize haploid eggs to produce diploid female zygotes. The fertilized eggs have durable shells and food reserves to survive drying or freezing.

Ribbon Worms Use a Proboscis to Capture Food

The 1,100 species of ribbon worms or proboscis worms (phylum Nemertea) vary from less than 1 cm to 30 m in length **(Figure 31.21).** Most species are marine, but a few occupy moist, terrestrial habitats. Although the often brightly colored ribbon worms superficially resemble free-living flatworms, their body

FIGURE 31.20 **Phylum Rotifera. (A)** Despite their small size, rotifers have complex body plans and organ systems. **(B)** This rotifer is laying eggs.

A. Rotifer *(Philodina roseola)* body plan

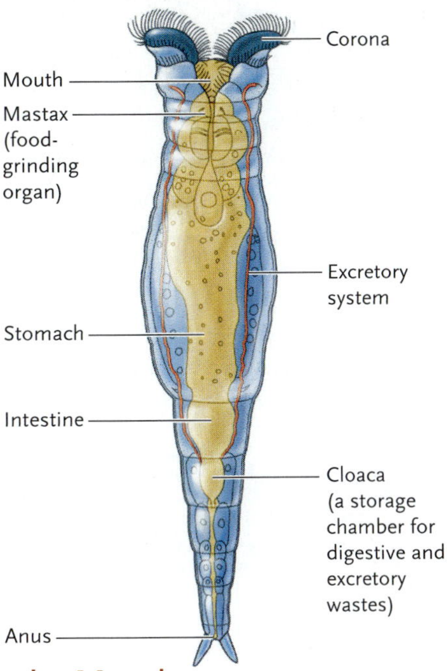

B. Rotifer (*Philodina* species)

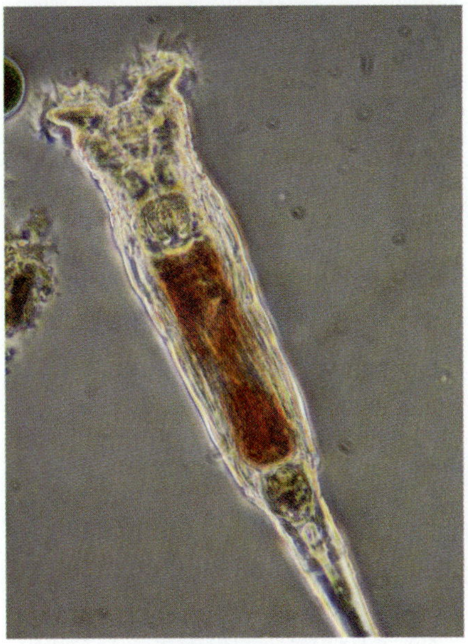

plans are more complex. First, they possess both a mouth and an anus; thus, they have a complete digestive tract. Second, nemerteans have a circulatory system in which fluid flows through **circulatory vessels** that carry nutrients and oxygen to tissues and remove wastes (see Section 44.1). Finally, they have a muscular, mucus-covered proboscis, a tube that can be everted (turned inside out) through the mouth or a separate pore to capture prey. The proboscis is housed within a chamber, the *rhynchocoel,* which is unique to this phylum.

Mollusks Have a Muscular Foot and a Mantle That Secretes a Shell or Aids in Locomotion

Most of the 100,000 species in the phylum Mollusca (*mollis* = soft)—including clams, snails, octopuses, and their relatives—are marine. However, many clams and snails occupy freshwater habitats, and some snails live on land. Mollusks vary in length from clams less than 1 mm to the giant squids, which can exceed 18 m.

In mollusks, the body is divided into three regions: the visceral mass, the foot, and the head **(Figure 31.22).** The **visceral mass** contains digestive, excretory, and reproductive systems and the heart. The muscular **foot** often provides the major means of locomotion. In more active groups, the **head** is well-defined and carries sensory organs and a brain. The mouth often includes a toothed **radula,** which scrapes food into small particles or drills through the shells of prey. Many mollusks are covered by a protective shell of calcium carbonate secreted by the **mantle,** which comprises one or two folds of the body wall that often enclose the visceral mass. The mantle also defines a space, the *mantle cavity.* In aquatic species, the mantle cavity houses the *gills,* delicate respiratory structures with enormous surface area. In most mollusks, cilia on the mantle and gills generate a steady flow of water into the mantle cavity.

The large size of mollusks requires a circulatory system to maintain cells that are far from the body surface. Most mollusks have an **open circulatory system** in which **hemolymph,** a bloodlike fluid, leaves the circulatory vessels and bathes tissues directly (see Figure 44.3A). Hemolymph pools in spaces called *sinuses,* and then drains into vessels that carry it back to the heart.

The sexes are usually separate, although many snails are hermaphroditic. Fertilization may be internal or external. The

FIGURE 31.21 **Phylum Nemertea. (A)** The flattened, elongated bodies of ribbon worms are often brightly colored. **(B)** Ribbon worms have a complete digestive system and a specialized cavity, the rhynchocoel, that houses a protrusible proboscis.

A. Ribbon worm (*Lineus* species)

B. Ribbon worm anatomy

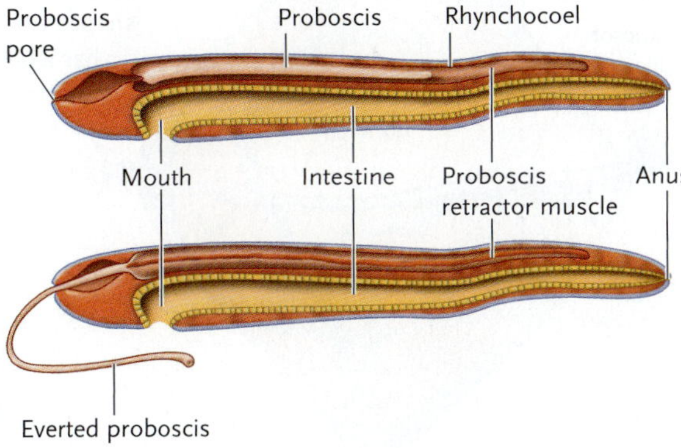

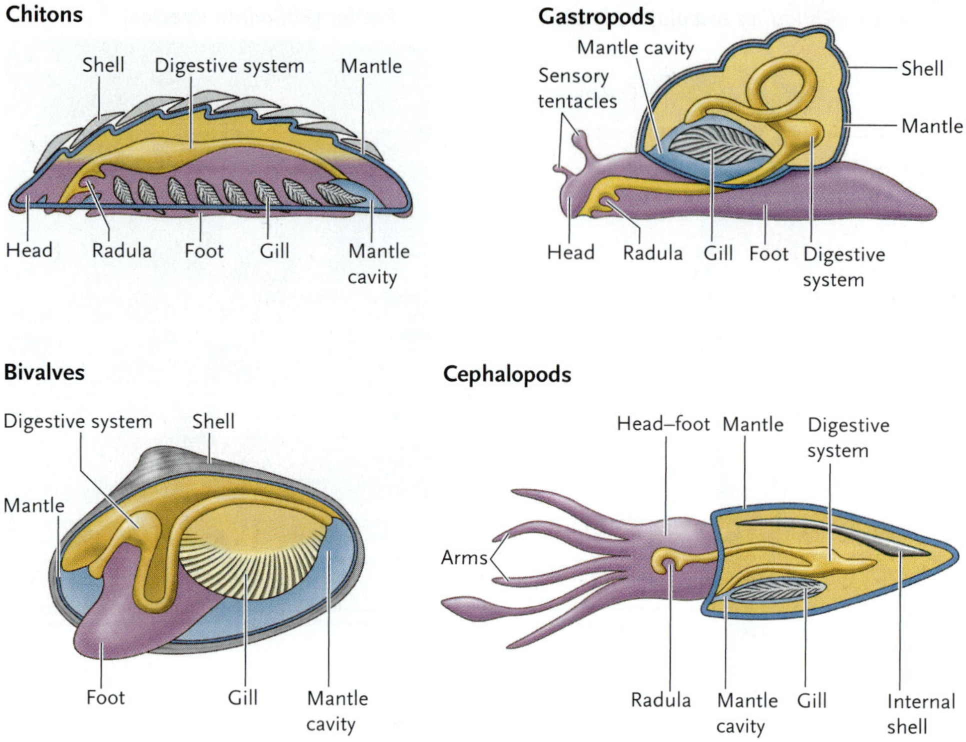

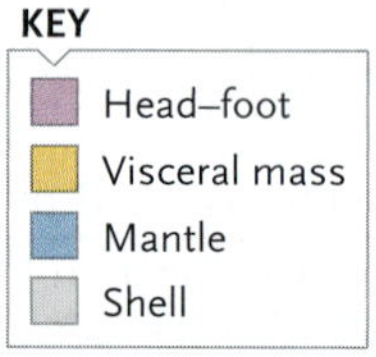

FIGURE 31.22 Molluscan body plans. The bilaterally symmetrical body plans of most mollusks include a muscular head–foot, a visceral mass, and a mantle. Bivalves have no head, and cephalopods have arms on the head–foot.

zygotes of marine species often develop into free-swimming, ciliated *trochophore* larvae **(Figure 31.23),** typical of both this phylum and the phylum Annelida, which we describe next. In some mollusks, the trochophore develops into a second larval stage, called a *veliger,* before metamorphosing into an adult. Some snails as well as octopuses and squids have direct development: embryos develop into miniature replicas of the adults.

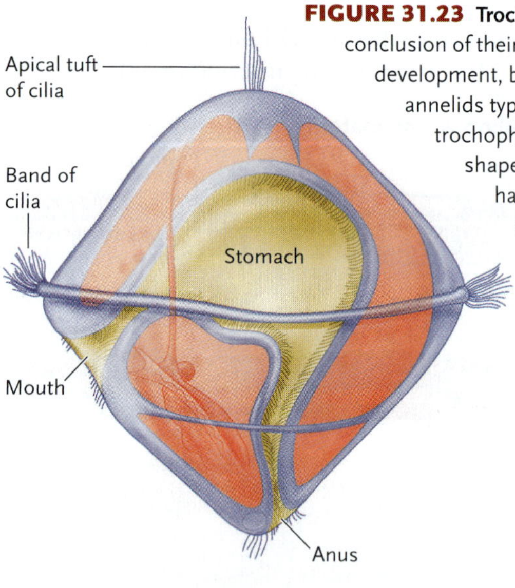

FIGURE 31.23 Trochophore larva. At the conclusion of their embryological development, both mollusks and annelids typically pass through a trochophore stage. The top-shaped trochophore larva has a band of cilia just anterior to its mouth.

Mollusca includes eight clades. Here we examine the four that are most commonly encountered.

POLYPLACOPHORA The chitons (Polyplacophora: *poly* = many; *plax* = flat surface) are sedentary mollusks that graze on algae along rocky marine coasts. The oval, bilaterally symmetrical body has a dorsal shell divided into eight plates that allow it to conform to irregularly shaped surfaces **(Figure 31.24).** When a chiton is disturbed or exposed to strong wave action, the muscles of its broad foot maintain a tenacious grip, and the mantle's edge functions like a suction cup to hold fast to the substrate.

GASTROPODA Snails and slugs (Gastropoda: *gaster* = belly; *podos* = foot) are the largest molluscan group, numbering 40,000 species **(Figure 31.25).** Aquatic species use gills to acquire oxygen, but in terrestrial species a modified mantle cavity functions as an air-breathing lung. Gastropods feed on algae, vascular plants, or animal prey. Some are scavengers, and a few are parasites.

The visceral mass of most snails is housed in a coiled or cone-shaped shell that is balanced above the rest of the body, much as you balance a backpack full of books (see Figure 31.25A and B). Most shelled snails undergo **torsion,** a curious realignment of body parts that is independent of shell coiling. Muscle contractions and differential growth twist the visceral mass and mantle 180° relative to the head–foot. This rearrangement moves the mantle cavity forward so that the head can be withdrawn into the shell in times of danger. It also brings the gills, anus, and excretory openings above the head—a potentially messy configuration, were it not for cilia that sweep away wastes.

FIGURE 31.24 Polyplacophora. Chitons live on rocky shores, where they use their foot and mantle to grip rocks and other hard substrates. This chiton *(Tonicella lineata)* lives in California.

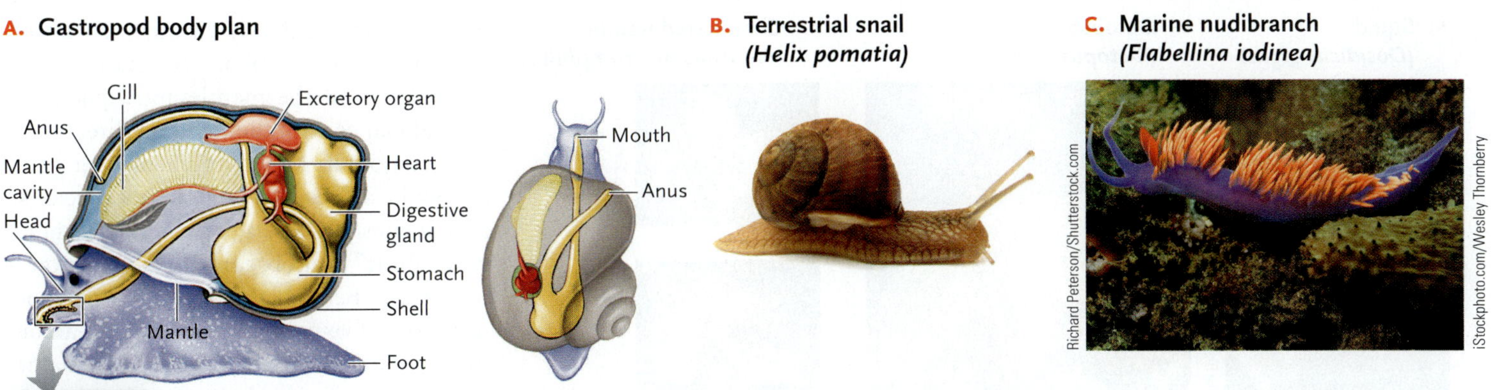

FIGURE 31.25 Gastropoda. (A) Most gastropods have a coiled shell that houses the visceral mass. A developmental process called torsion causes the digestive and excretory systems to eliminate wastes into the mantle cavity, near the animal's head. **(B)** The edible snail is a terrestrial gastropod. **(C)** Nudibranchs, like this Spanish shawl nudibranch, are shell-less marine snails.

Some gastropods, including terrestrial slugs and colorful nudibranchs (sea slugs), are shell-less, a condition that leaves them vulnerable to predators (see Figure 31.25C). Some nudibranchs consume cnidarians and then transfer undischarged nematocysts to projections on their dorsal surface, where these "borrowed" stinging capsules provide protection.

The nervous and sensory systems of gastropods are well developed. Tentacles on the head include chemical and touch receptors; the eyes detect changes in light intensity but don't form images.

BIVALVIA The clams, scallops, oysters, and mussels (Bivalvia: *bi* = two; *valva* = folding door) are restricted to aquatic habitats. They are enclosed within a pair of shells, hinged together dorsally by an elastic ligament **(Figure 31.26)**. Contraction of one or two **adductor muscles** closes the shell by pulling the two sides together; when the muscles relax, the elastic ligament opens the shell by pulling them apart (see Figure 31.26A). Although some bivalves are tiny, giant clams of the South Pacific can be more than 1 m across and weigh 225 kg (see Figure 31.26B).

Adult mussels and oysters are sessile and permanently attached to hard substrates, but many clams are mobile and use their muscular foot to burrow in sand or mud. Some bivalves, such as young scallops, swim by rhythmically clapping their valves ("shells") together, forcing a current of water out of the mantle cavity. The "scallops" that we eat are the scallops' well-developed adductor muscles.

Bivalves have a reduced head, and they lack a radula. Part of the mantle forms two tubes called *siphons* (see Figure 31.26C). Beating of cilia on the gills and mantle carries water into the mantle cavity through the incurrent siphon and out through the excurrent siphon. Incurrent water carries dissolved oxygen and food particles to the gills, where oxygen is absorbed. Mucus strands on the gills trap the food, which is then transported by cilia to *palps,* where final sorting takes place; edible tidbits are carried to the mouth. The excurrent water carries away metabolic wastes and feces.

Despite their sedentary existence, bivalves have moderately well-developed nervous systems; sensory organs that detect chemicals, touch, and light; and statocysts to sense their orientation. When they encounter pollutants, many bivalves stop pumping water and close their shells. When confronted by a predator, some burrow into sediments or swim away.

FIGURE 31.26 Bivalvia. (A) Bivalves are enclosed in a hinged two-part shell. Part of the mantle forms a pair of water-transporting siphons. **(B)** Giant clams in the South Pacific *(Tridachna gigas)* sometimes weigh hundreds of kilograms. **(C)** The geoduck is a clam with enormous muscular siphons.

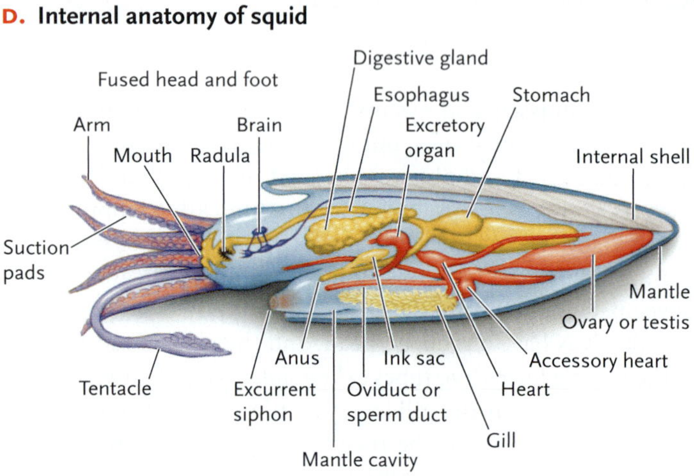

FIGURE 31.27 Cephalopoda. **(A)** Squids and **(B)** octopuses are the most familiar cephalopods. **(C)** The chambered nautilus and its relatives retain an external shell. **(D)** Like other cephalopods, the squid body includes a fused head and foot; most organ systems are enclosed by the mantle.

CEPHALOPODA The octopuses, squids, and nautiluses (Cephalopoda: *kephale* = head) are active marine predators, including the fastest and most intelligent invertebrates **(Figure 31.27).** They vary in length from a few centimeters to 18 m. Giant squids, the largest invertebrates known, may be the source of "sea monster" stories.

The cephalopod body has a fused head and foot (see Figure 31.27D). The head comprises the mouth and eyes. The ancestral "foot" forms a set of arms and tentacles, equipped with suction pads, adhesive structures, or hooks. Cephalopods use these structures to capture prey and a pair of beaklike jaws to bite or crush it. Venomous secretions often speed the captive's death. Some species use their radula to drill through the shells of other mollusks.

Cephalopods have a highly modified shell. Octopuses have no remnant of a shell at all. In squids and cuttlefishes, it is reduced to a stiff internal support. Only the chambered nautilus and its relatives retain an external shell.

Squids are the most mobile cephalopods, moving rapidly by a kind of jet propulsion. When muscles in the mantle relax, water is drawn into the mantle cavity. When they contract, a jet of water is squeezed out through a funnel-shaped excurrent siphon. By manipulating the position of the mantle and siphon, the animal can control the rate and direction of its locomotion. When threatened, a squid can make a speedy escape. Many species simultaneously release a dark fluid, commonly called "ink," that obscures their direction of movement. Being highly active, cephalopods need lots of oxygen, and they alone among the mollusks have a **closed circulatory system:** hemolymph is confined within the walls of hearts and blood vessels, providing increased pressure to the vascular fluid. Moreover, accessory hearts speed the flow of hemolymph through the gills, enhancing the uptake of oxygen and the release of carbon dioxide.

Compared with other mollusks, cephalopods have large and complex brains. Giant nerve fibers connect the brain with the muscles of the mantle, enabling quick responses to food or danger. Their image-forming eyes are similar to those of vertebrates. Cephalopods are also highly intelligent. Octopuses, for example, learn to recognize objects with distinctive shapes or colors, and they can be trained to approach or avoid them.

Cephalopods have separate sexes and elaborate courtship rituals. Males store sperm within the mantle cavity and use a specialized tentacle to transfer packets of sperm into the female's mantle cavity, where fertilization occurs. Fertilized eggs, wrapped in a protective jelly, are attached to objects in the environment. The young hatch with an adult body form.

Annelids Exhibit a Serial Division of the Body Wall and Some Organ Systems

The 16,000 species of segmented worms (phylum Annelida: *anellus* = little ring) occupy marine, freshwater, and moist terrestrial habitats. Terrestrial annelids eat organic debris; aquatic species consume algae, microscopic organisms, detritus, or other animals. They range from a few millimeters to several meters in length.

The annelid body is highly segmented: the body wall muscles and some organs—including respiratory surfaces, parts of the nervous, circulatory, and excretory systems, and the coelom itself—are divided into similar repeating units **(Figure 31.28).** Body segments are separated by transverse partitions called **septa.** The digestive system and major blood vessels are not segmented and run the length of the animal.

The body wall muscles of annelids have both circular and longitudinal layers (see Figure 31.28). Alternate contractions of these muscle groups allow annelids to make directed movements, using the pressure of the fluid in the coelom as a hydrostatic skeleton. All annelids except leeches also have chitin-reinforced bristles, called **setae** (sometimes written *chaetae*), which protrude outward from the body wall. Setae anchor the worm against the substrate, providing traction.

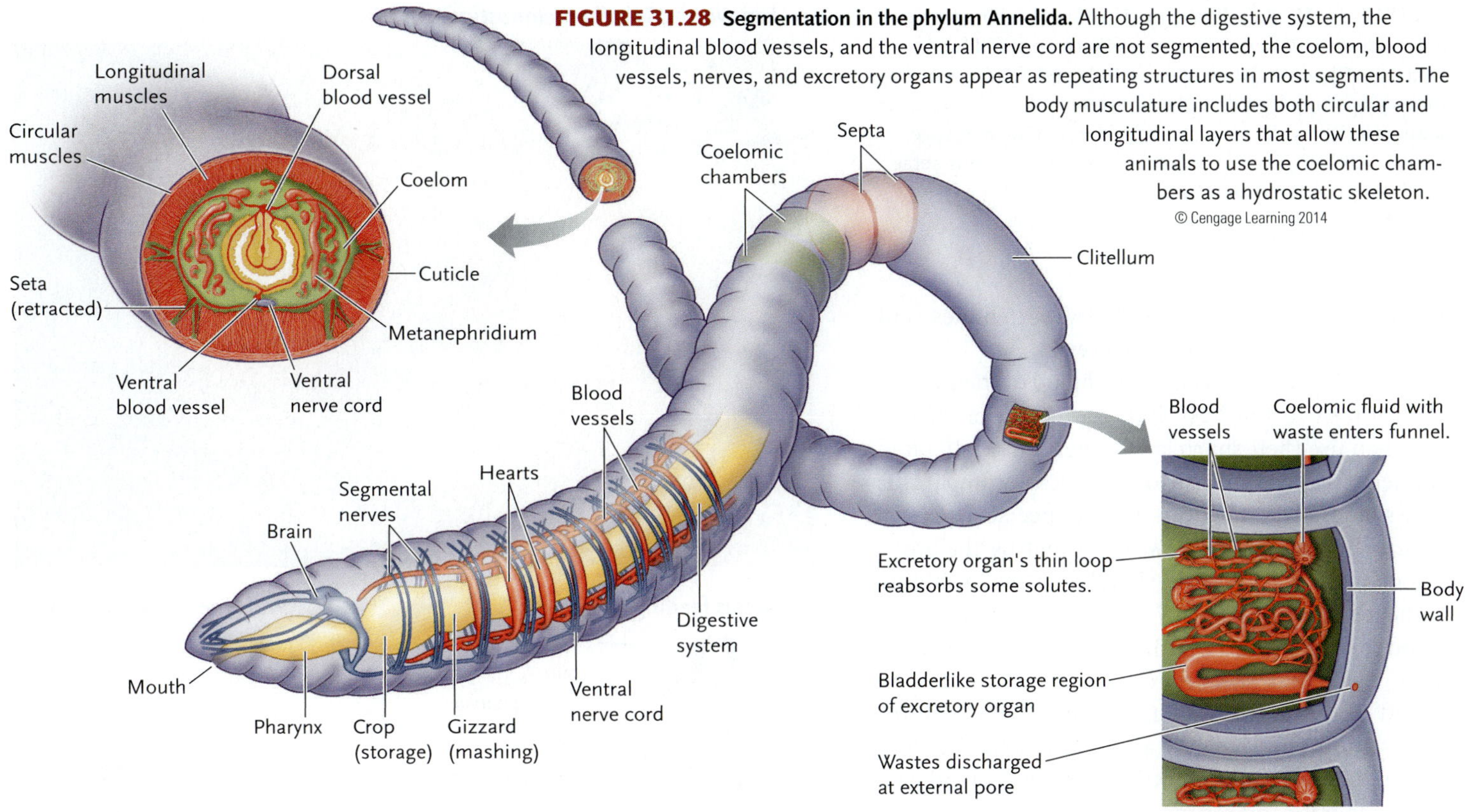

FIGURE 31.28 Segmentation in the phylum Annelida. Although the digestive system, the longitudinal blood vessels, and the ventral nerve cord are not segmented, the coelom, blood vessels, nerves, and excretory organs appear as repeating structures in most segments. The body musculature includes both circular and longitudinal layers that allow these animals to use the coelomic chambers as a hydrostatic skeleton.

Annelids have a complete digestive system and a closed circulatory system. However, they lack a discrete respiratory system; oxygen and carbon dioxide diffuse through the skin. The excretory system is composed of pairs of **metanephridia** that usually occur in all body segments posterior to the head. The nervous system is highly developed, with ganglia (local control centers) in every segment, a simple brain in the head, and sensory organs that detect chemicals, moisture, light, and touch.

Most freshwater and terrestrial annelids are hermaphroditic, and worms exchange sperm when they mate. Newly hatched worms have an adult morphology. Some terrestrial annelids also reproduce asexually by fragmenting and regenerating missing parts. Marine annelids usually have separate sexes, and release gametes into the sea for fertilization. The zygotes develop into trochophore larvae that add segments, gradually assuming an adult form.

Annelida includes three clades, each of which is largely restricted to one environment.

POLYCHAETA The bristle worms (Polychaeta: *chaite* = bristles or hair) are primarily marine **(Figure 31.29).** Many live under rocks or in tubes constructed from mucus, calcium carbonate secretions, grains of sand, or small shell fragments. Their setae project from well-developed **parapodia** (singular, *parapodium* = closely resembling a foot), fleshy lateral extensions of the body wall used for locomotion and gas exchange. In more mobile species, sense organs are concentrated on a well-developed head.

Crawling or swimming polychaetes are often predatory; they use sharp jaws in a protrusible muscular pharynx to grab small invertebrate prey. Other species graze on algae or scavenge organic matter. A few tube dwellers draw food-laden water into the tube by beating their parapodia; most others collect food by extending feathery, ciliated, mucus-coated tentacles.

OLIGOCHAETA Most oligochaete worms (Oligochaeta: *oligos* = few) are terrestrial **(Figure 31.30),** but they are restricted to moist habitats because they quickly dehydrate in dry air or soil. They range in length from a few millimeters to more than 3 m. Terrestrial oligochaetes, the earthworms, are nocturnal, spending their days in burrows that they excavate. Aquatic species live in mud or detritus at the bottom of lakes, rivers, and estuaries.

Earthworms are scavengers on decomposing organic matter. In his book *The Formation of Vegetable Mould through the Action of Worms,* Darwin noted that earthworms can ingest their own weight in soil every day. He calculated that a typical population of 16,000 worms per hectare (10,000 sq m) consumes more than 20 tons of soil in a year. This impressive activity aerates soil and makes nutrients available to plants by mixing the subsoil with the topsoil. Earthworms have complex organ systems (see Figure 31.28), and they sense light and touch at both ends of the body. In addition, they have moisture receptors, an important adaptation in organisms that must stay damp to allow gas exchange across the skin.

HIRUDINEA The leeches (Hirudinea: *hirudo* = leech) are mostly freshwater parasites. They have dorsoventrally flattened, tapered bodies with a sucker at each end. Although the body wall is segmented, the coelom is reduced and not partitioned by septa.

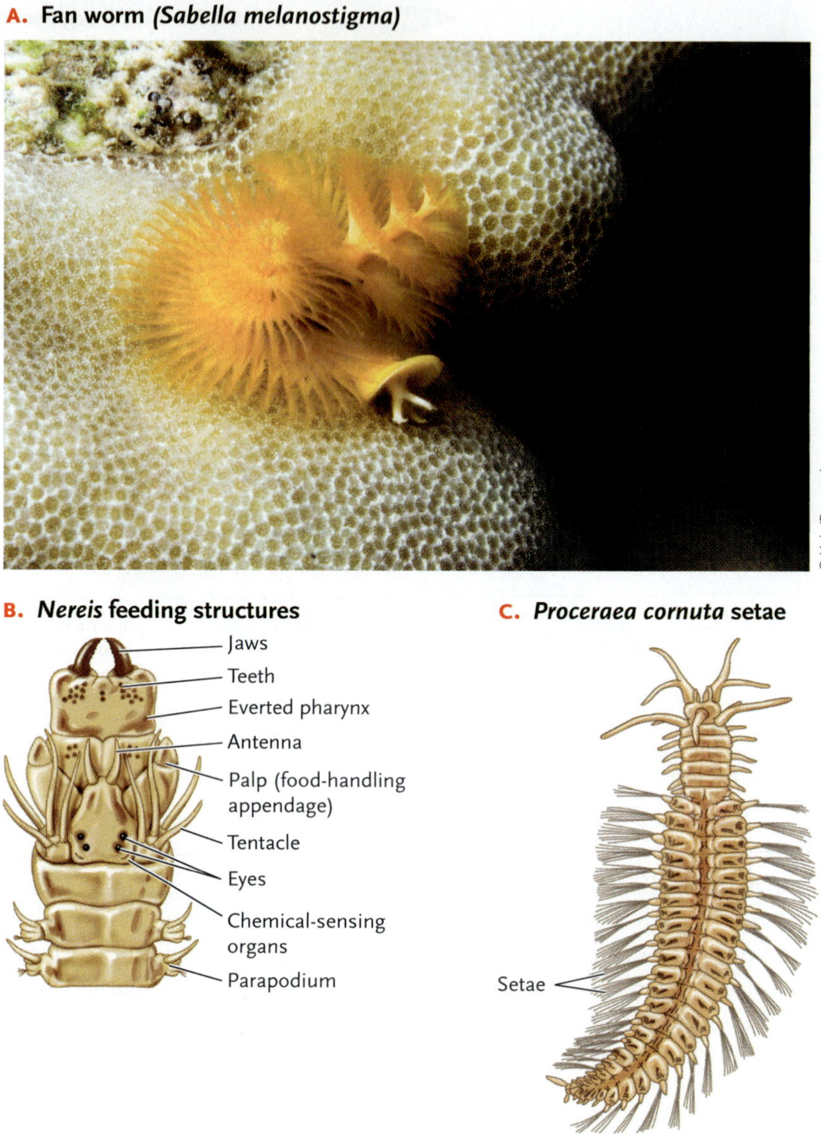

FIGURE 31.29 Polychaeta. (A) The tube-dwelling fan worm has mucus-covered tentacles that trap small food particles. **(B)** Some polychaetes actively seek food; when they encounter a suitable tidbit, they evert their pharynx, exposing sharp jaws that grab the prey and pull it into the digestive system. **(C)** Many marine polychaetes have numerous setae, which they use for locomotion.

Many leeches are ectoparasites of vertebrates, but some attack small invertebrate prey.

Parasitic leeches feed on the blood of their hosts. Most attach to the host with the posterior sucker, and then use their sharp jaws to make a small, often painless, triangular incision. A sucking apparatus draws blood from the prey, while a special secretion, *hirudin,* maintains the flow by preventing the host's blood from coagulating. Leeches have a highly branched gut that allows them to consume huge blood meals **(Figure 31.31).** For centuries, doctors used medicinal leeches *(Hirudo medicinalis)* to "bleed" patients; today, surgeons use them to drain excess fluid from tissues after reconstructive surgery, reducing swelling until the patient's blood vessels regenerate and resume this function.

STUDY BREAK 31.6

1. **What organs systems are present in free-living flatworms (Turbellaria)? Which of these organ systems is absent in tapeworms (Cestoda)?**
2. **What characteristic reveals the close evolutionary relationship of ectoprocts, brachiopods, and phoronid worms?**
3. **What anatomical structures and physiological systems allow squids and other cephalopods to be much more active than other types of mollusks?**
4. **Which organ systems exhibit segmentation in most annelid worms?**

FIGURE 31.30 Oligochaeta. Earthworms *(Lumbricus terrestris)* generally move across the ground surface at night.

THINK OUTSIDE THE BOOK

Search the records of your local health department, either online or in person, to discover whether people in your town or city routinely suffered from parasitic worm infections in the past. What public health measures stopped the outbreak?

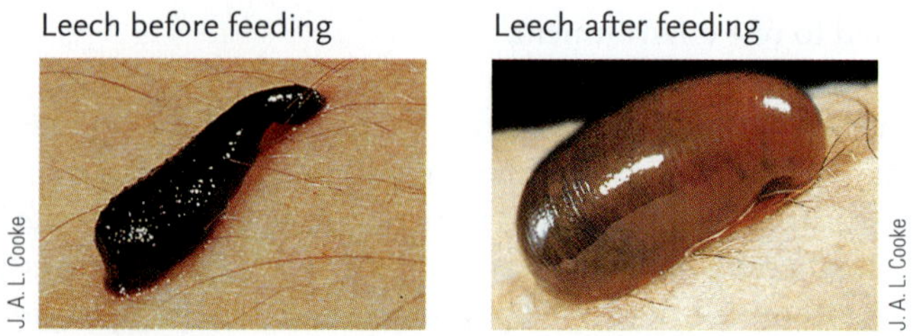

FIGURE 31.31 Hirudinea. Parasitic leeches consume huge blood meals, as shown by these before and after photos of a medicinal leech *(Hirudo medicinalis)*. Because suitable hosts are often hard to locate, gorging allows a leech to take advantage of any host it finds.

31.7 Ecdysozoan Protostomes

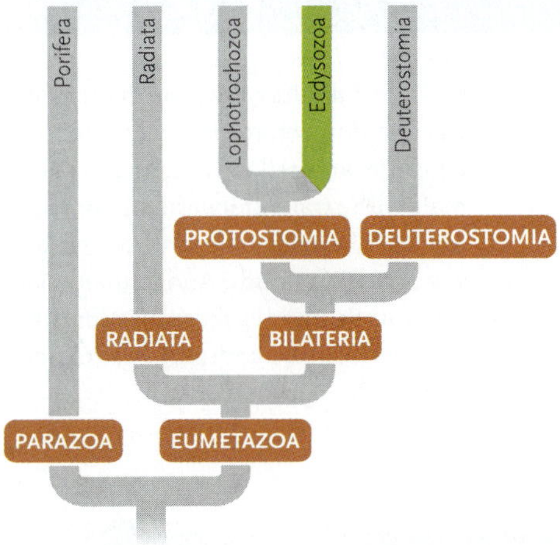

The three phyla in the protostome group Ecdysozoa all have an external covering that they shed periodically. The outer covering protects these animals from harsh environmental conditions, and it helps parasitic species resist host defenses. Although many of these animals live in either aquatic or moist terrestrial habitats, a tough exoskeleton allows one group, the insects, to thrive on dry land and in the air.

Nematodes Are Unsegmented Worms Covered by a Flexible Cuticle

Roundworms (phylum Nematoda: *nema* = thread) are perhaps the most abundant animals on Earth. A cupful of rich soil, a dead earthworm, or a rotting fruit may contain thousands of them. Although 25,000 species have been described, experts estimate that more than half a million exist. Many nematodes are almost microscopically small, but some species are a meter or more long. They occupy nearly every freshwater, marine, and terrestrial habitat on Earth, consuming detritus, microorganisms, plants, or animals.

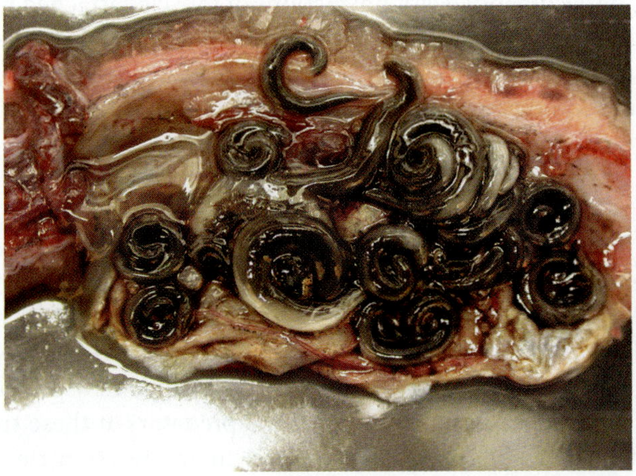

FIGURE 31.32 Phylum Nematoda. Some roundworms, like these *Anguillicola crassus* in the swim bladder of an eel, are parasites of plants or animals. Others are important consumers of dead organisms in most ecosystems.

The roundworm body is cylindrical and usually tapered at both ends **(Figure 31.32)**. None of the cells have cilia or flagella. Roundworms are covered in a tough but flexible, water-resistant **cuticle,** which must be molted periodically; the cuticle is replaced by the underlying epidermis as the worm grows. The cuticle prevents the animal from dehydrating in dry environments, and, in parasitic species, it resists attack by acids and enzymes in a host's digestive system. Beneath the cuticle and epidermis, a layer of longitudinal muscles extends the length of the body. Alternating contractions of these muscles on a nematode's dorsal and ventral sides push against the substrate and propel the worm forward, usually with a thrashing motion.

The adults of one soil-dwelling species, *Caenorhabditis elegans,* are transparent and contain fewer than 1,000 cells. As *Focus on Model Research Organisms* explains, biologists have studied this worm inside and out.

Nematodes reproduce sexually, and the sexes are separate in most species. In some, internal fertilization produces many thousands of fertile eggs per day. The eggs of many species can remain dormant if environmental conditions are unsuitable.

Because of their great numbers, nematodes have enormous ecological, agricultural, and medical significance. Free-living species are responsible for decomposition and nutrient recycling in many habitats. Parasitic nematodes attack the roots of plants, causing tremendous crop damage. Some species parasitize animals, including humans. Although some, like pinworms *(Enterobius),* are more of a nuisance than a danger, others, like trichinas or filarial worms, can cause serious disease, disfigurement, or even death (see *Focus on Applied Research* on page 686). More than 1 billion people worldwide suffer from debilitating and life-threatening nematode infections.

Velvet Worms Have Segmented Bodies and Numerous Unjointed Legs

The 100 or so living species of velvet worms (phylum Onychophora: *onyx* = claw) live under stones, logs, and forest litter in the tropics and in moist temperate habitats in the southern hemisphere. They range in size from 15 mm to 15 cm and feed on small invertebrates.

Onychophorans have a flexible cuticle, segmented bodies, and numerous pairs of unjointed legs **(Figure 31.33)**. Like the annelids, they have pairs of excretory organs in most segments. But unlike annelids, they have an open circulatory system, a specialized respiratory system, relatively large brains, jaws, and tiny claws on their feet. Many produce live young, which, in some species, are nourished within a uterus. Fossil evidence indicates that the onychophoran body plan has not changed much over the last 500 million years.

FOCUS ON Model Research Organisms

Caenorhabditis elegans

Researchers studying the tiny, free-living nematode *Caenorhabditis elegans* have made many recent advances in molecular genetics, animal development, and neurobiology. It is so popular as a model research organism that most workers simply refer to it as "the worm."

Several attributes make *C. elegans* a model research organism. It has an adult size of about 1 mm and thrives on cultures of *E. coli* or other bacteria; thus, thousands can be raised in a culture dish. It is hermaphroditic and often self-fertilizing, which allows researchers to maintain pure genetic strains. It completes its life cycle from egg to reproductive adult within 3 days at room temperature. Furthermore, stock cultures can be kept alive indefinitely by freezing them in liquid nitrogen or in an ultra-cold freezer set to minus 80°C. Researchers can therefore store new mutants for later research without having to clean, feed, and maintain active cultures.

Best of all, the worm is anatomically simple **(Figure)**; an adult contains just 959 cells (excluding the gonads). Having a fixed cell number is relatively uncommon among animals, and developmental biologists have made good use of this trait. The eggs, juveniles, and adults of the worm are completely transparent, and researchers can observe cell divisions and cell movements in living animals with straightforward microscopy techniques. There is no need to kill, fix, and stain specimens for study. And virtually every cell in the worm's body is accessible for manipulation by laser microsurgery, microinjection, and similar approaches.

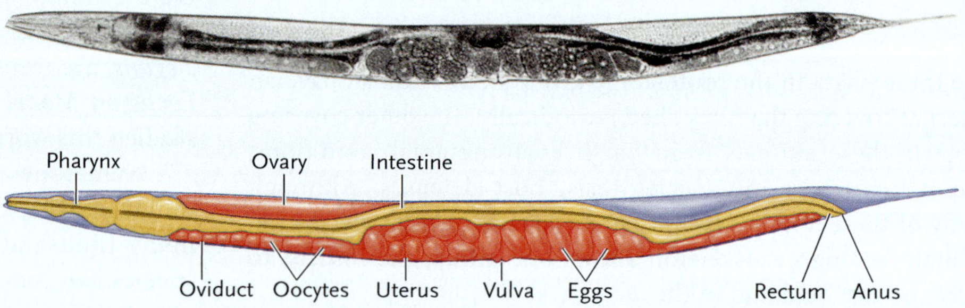

FIGURE 31.33 Phylum Onychophora. Species in the small phylum Onychophora, like this *Peripatus* species from the Andes Mountains of South America, have segmented bodies and unjointed appendages.

Arthropods Are Segmented Animals with a Hard Exoskeleton and Jointed Appendages

The 1.15 million known species of arthropods (phylum Arthropoda: *arthron* = joint) include more than half the animal species on Earth—and only a fraction of the living arthropods have been described. This huge clade includes insects, spiders, crustaceans, millipedes, centipedes, the extinct trilobites, and their relatives.

Arthropods have a segmented body encased in a rigid **exoskeleton.** This external covering is made of chitin, a mix of polysaccharide fibers glued together with glycoproteins, as well as waxes and lipids that block the passage of water. In some marine groups, such as crabs and lobsters, it is hardened with calcium carbonate. The exoskeleton probably first evolved in marine species, providing protection against predators. In terrestrial habitats, it provides support against gravity and protection from dehydration, which contributes to the success of insects in even the driest places on Earth. The exoskeleton is especially thin and flexible at the joints between body segments and in the appendages. Contractions of muscles attached to the inside of the exoskeleton move individual body parts like levers, allowing highly coordinated movements and patterns of locomotion that are more precise than those in soft-bodied animals with hydrostatic skeletons.

Although the exoskeleton has obvious advantages, it is nonexpandable and therefore could limit growth of the animal. But, like other Ecdysozoa, arthropods grow and periodically develop a soft, new exoskeleton beneath the old one, which they shed in the complex process of ecdysis **(Figure 31.34).** After shedding the old exoskeleton, aquatic species swell with water and terrestrial species swell with air before the new one hardens. They are especially vulnerable to predators at these times. "Soft-shelled" crabs, prized as food in many countries, are crabs that have recently molted.

As arthropods evolved, body segments became fused in various ways, reducing the overall number of segments. Each region of the body, along with its highly modified paired appendages, is specialized, but the structure and function vary

FIGURE 31.34 Ecdysis in insects. Like all other arthropods, this cicada *(Graptopsaltria nigrofuscata)* sheds its old exoskeleton as it grows and when it undergoes metamorphosis into a winged adult.

greatly among groups. In insects (see Figure 31.43), which have three body regions, the **head** includes a brain, sensory structures, and some sort of feeding apparatus. The segments of the **thorax** bear walking legs and, in some insects, wings. The **abdomen** includes much of the digestive system and sometimes part of the reproductive system.

The coelom of arthropods is greatly reduced, but another cavity, the *hemocoel,* is filled with bloodlike hemolymph. The heart pumps the hemolymph through an open circulatory system, bathing tissues directly.

Arthropods are active animals and require substantial quantities of oxygen. Different groups have distinctive mechanisms for gas exchange, because oxygen cannot cross the impermeable exoskeleton. Marine and freshwater species, like crabs and lobsters, rely on diffusion across gills. The terrestrial groups—insects and spiders—have developed unique and specialized respiratory systems.

High levels of activity also require intricate sensory structures. Many arthropods are equipped with a highly organized central nervous system, touch receptors, chemical sensors, **compound eyes** that include multiple image-forming units, and in some, hearing organs.

Arthropod systematics is an active area of research, and scientists are currently using molecular, morphological, and developmental data to reexamine relationships within this immense group. As *Insights from the Molecular Revolution* explains, hypotheses about the phylogeny and classification of arthropods are in a state of flux. Some researchers even argue for splitting them into four or more phyla. We follow the traditional definition of five *subphyla,* partly because this classification adequately reflects arthropod diversity and partly because no alternative hypothesis has been widely adopted by experts.

SUBPHYLUM TRILOBITA The trilobites (subphylum Trilobita: *tri* = three, *lobos* = lobe), now extinct, were among the most numerous animals in shallow Paleozoic seas. They disappeared in the Permian mass extinction, but the cause of their demise is unknown. Most trilobites were ovoid, dorsoventrally flattened, and heavily armored, with two deep longitudinal grooves that divided the body into the one median and two lateral lobes for which the group is named **(Figure 31.35).** Their segmented bodies were organized into three body regions: the head, which included a pair of sensory **antennae** (chemosensory organs) and two compound eyes, and a thorax and an abdomen, both of which bore pairs of walking legs.

The position of trilobites in the fossil record indicates that they were among the earliest arthropods. Thus, biologists are confident that their three body regions and numerous unspecialized appendages—one pair per segment—represent ancestral traits within the phylum. As you will learn as you read about the other four subphyla, the subsequent evolution of the different arthropod groups included dramatic remodeling of the major body regions as well as modifications of the ancestral, unspecialized paired appendages into highly specialized structures that perform different functions.

SUBPHYLUM CHELICERATA In spiders, ticks, mites, scorpions, and horseshoe crabs (subphylum Chelicerata: *chela* = claw, *keras* = horn), the first pair of appendages, the **chelicerae,** are fanglike structures used for biting prey. The second pair of appendages, the *pedipalps,* serve as grasping organs, sensory organs, or walking legs. All chelicerates have two major body regions, the **cephalothorax** (a fused head and thorax) and the abdomen. The group originated in shallow Paleozoic seas, but most living species are terrestrial. They vary in length from less than a millimeter to 20 cm; all are predators or parasites.

FIGURE 31.35 Subphylum Trilobita. Trilobites, like *Olenellus gilberti,* bore many pairs of relatively undifferentiated appendages.

Unscrambling the Arthropods: Morphological versus molecular characteristics?

For years, biologists disagreed about the evolutionary relationships among the major clades of arthropods. Some researchers grouped hexapods and myriapods together (**Figure,** *Traditional tree*) because they share certain morphological characteristics, which may indicate a common ancestry: unbranched appendages, a tracheal system for gas exchange, Malpighian tubules for excretion, and one pair of antennae. However, other researchers noted that hexapods and crustaceans shared a different set of important morphological characters: jawlike mandibles on the fourth head segment, similar compound eyes, comparable development of the nervous system, and similarities in the structure of thoracic appendages.

Research Question

Given the variety of morphological characteristics that hexapods share with myriapods and those that they share with crustaceans, what are the evolutionary relationships among these major clades of arthropods?

Experiment

A molecular study by Jeffry L. Boore and his colleagues at the University of Michigan lends support to the hypothesis that hexapods may indeed be more closely related to crustaceans than to myriapods. Boore and his coworkers compared the arrangement of genes in the mitochondrial DNA of representative arthropod species. Mitochondrial DNA (mtDNA) was chosen for study because it has been subject to many random rearrangements during the evolutionary history of the arthropods, producing wide variations in the order and placement of the 36 or 37 genes typically present. As long as the rearrangements do not disrupt gene function, they appear to have little or no effect on fitness; therefore, they may not be affected by natural selection. As Boore and his colleagues noted, the large number of possible rearrangements of the mtDNA genes makes it unlikely that the same gene order would appear by chance in any two groups; thus, groups with the same arrangement are likely to share a common ancestor in which the rearrangement first appeared.

The investigators sequenced the mtDNA of a chelicerate, two crustaceans, and a myriapod (listed in the figure caption). They also sequenced the mtDNA of a mollusk and an annelid, animals that are only distantly related to arthropods, to provide outgroup comparisons. They compared these sequences with those in 14 other invertebrates, including four hexapods and one crustacean, already obtained by other workers.

Results

The locations of two genes encoding leucine-carrying tRNAs provide the strongest clue to the evolutionary relationships of these organisms. These genes are positioned next to each other in the mtDNA of the chelicerate, the mollusk, and the annelid. This arrangement appears to reflect the ancestral gene order present in the common ancestor of annelids, mollusks, and arthropods. However, in the hexapods and the crustaceans studied, one of the leucine tRNA genes is located at a different position, between two genes coding for electron transfer proteins. Moreover, this rearrangement does not appear in the myriapod examined. It is extremely unlikely that the same translocation of the leucine tRNA gene occurred independently in hexapods and crustaceans. Instead, this gene rearrangement probably occurred in an ancestor shared by hexapods and crustaceans after this ancestor diverged from the lineage that includes the chelicerates and myriapods.

Conclusion

The shared gene arrangement in insects and crustaceans and the absence of a matching arrangement in myriapods suggests that hexapods and crustaceans are closely related, and that hexapods and myriapods are not, in spite of their morphological similarities (see **Figure,** *New tree*). Other investigators, who compared the sequences of ribosomal RNA genes in a variety of arthropod species, reached the same conclusions independently.

THINK LIKE A SCIENTIST If we were able to sequence the DNA of a trilobite, do you think that the locations of the two genes encoding leucine-carrying tRNAs would more closely match the locations of these genes in the mtDNA of chelicerates or of crustaceans?

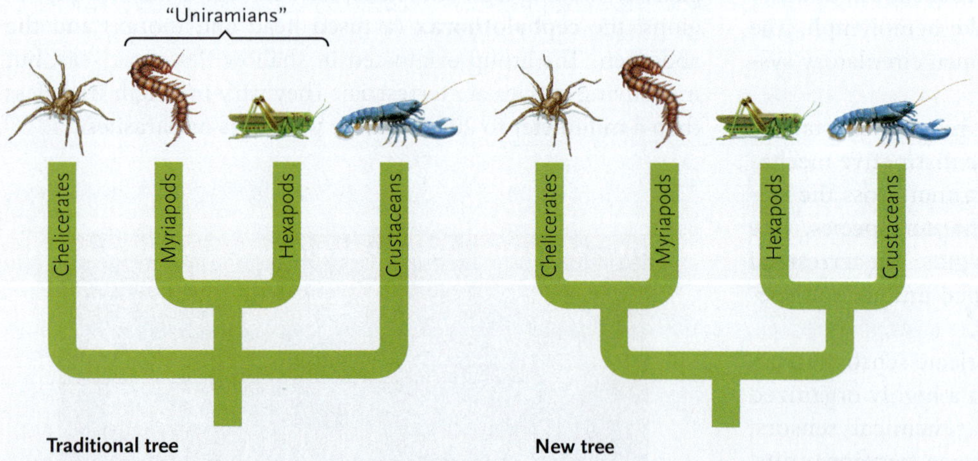

FIGURE The traditional phylogenetic tree for arthropods (left) differs from that suggested by the research described in this box. The DNA sequences used to construct the new phylogenetic tree (right) were obtained from the chelicerate *Limulus,* the myriapod *Thyrophygus,* the hexapod *Drosophila,* and the crustaceans *Daphnia* and *Homarus.*

Source: J. L. Boore et al. 1995. Deducing the pattern of arthropod phylogeny from mitochondrial DNA rearrangements. *Nature* 376:163–165.

A. Wolf spider (*Lycosa* species)

Cathy Keifer/Shutterstock.com

B. Spider anatomy

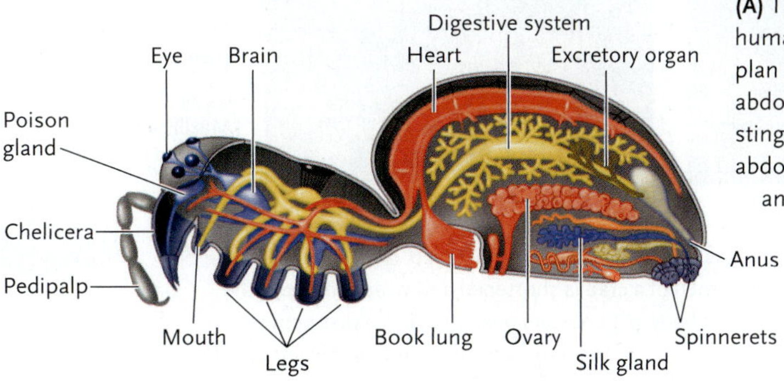

FIGURE 31.36 Subphylum Chelicerata, subgroup Arachnida. **(A)** The wolf spider is harmless to humans. **(B)** The arachnid body plan includes a cephalothorax and abdomen. **(C)** Scorpions have a stinger at the tip of the segmented abdomen. Many protect their eggs and young. **(D)** House dust mites, shown in a scanning electron micrograph, feed on microscopic debris.

C. Scorpion (*Tityus* species)

Ingo Arndt/Getty Images

D. House dust mite *(Dermatophagoides pteronyssinus)*

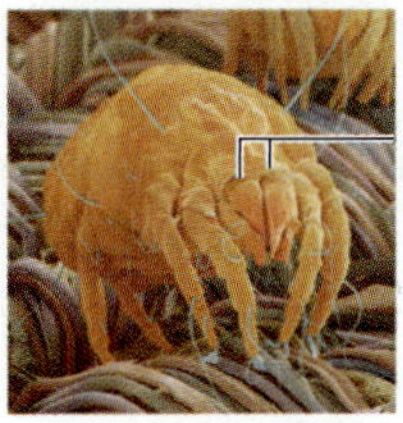

Andrew Syred/Photo Researchers, Inc.

The spiders, scorpions, mites, and ticks (Arachnida) represent the vast majority of chelicerates **(Figure 31.36).** Arachnids have four pairs of walking legs on the cephalothorax and highly modified chelicerae and pedipalps. In some spiders, males use their pedipalps to transfer packets of sperm to females. Scorpions use them to shred food and to grasp one another during courtship. Many predatory arachnids have excellent vision, provided by simple eyes on the cephalothorax. Scorpions and some spiders also have unique pocketlike respiratory organs, called **book lungs,** which are derived from abdominal appendages.

Spiders, like most other arachnids, subsist on a liquid diet. They use their chelicerae to inject paralyzing poisons and digestive enzymes into prey and then suck up the partly digested tissues. Many spiders are economically important predators, helping to control insect pests. Only a few are a threat to humans. The toxin of a black widow *(Latrodectus mactans)* causes paralysis, and the toxin of the brown recluse *(Loxosceles reclusa)* destroys tissues around the site of the bite.

Although many spiders hunt actively, others capture prey on silken threads secreted by **spinnerets,** which are modified abdominal appendages. Some species weave the threads into complex, netlike webs. The silk is secreted as a liquid, but quickly hardens on contact with air. Spiders also use silk to make nests, to protect their egg masses, as a safety line when moving through the environment, and to wrap prey for later consumption.

Most mites are tiny, but they have a big impact. Some are serious agricultural pests that feed on the sap of plants. Others cause mange (patchy hair loss) or painful and itchy welts on animals. House dust mites cause allergic reactions in many people. Ticks, which are generally larger than mites, are blood-feeding ectoparasites that often transmit pathogens, such as the bacteria that cause Rocky Mountain spotted fever and Lyme disease.

The subphylum Chelicerata also includes five species of horseshoe crabs (Merostomata), an ancient lineage with a morphology that has not changed much over its 350-million-year history **(Figure 31.37).** Horseshoe crabs are carnivorous bottom feeders in shallow coastal waters. Beneath their characteristic shell, they have one pair of chelicerae, a pair of pedipalps, four pairs of walking legs, and a set of paperlike gills that are derived from ancestral walking legs.

SUBPHYLUM CRUSTACEA The shrimps, lobsters, crabs, and their relatives (subphylum Crustacea: *crusta* = crust or hard shell) represent a lineage that emerged more than 500 million years ago. They are abundant in marine and freshwater habitats. A few species, such as sowbugs and pillbugs, live in moist, sheltered terrestrial environments. In many crustaceans two, or even all three, of the arthropod body regions—head, thorax, and abdomen—are fused; a fused cephalothorax and a separate abdomen is an especially common pattern. The edible "tail" of a lobster, shrimp, or crayfish is actually a highly muscularized abdomen. In some, the exoskeleton includes a

Darlyne A. Murawski/Getty Images

FIGURE 31.37 Marine chelicerates. Horseshoe crabs, like *Limulus polyphemus,* are included in the Merostomata.

A. Crab (*Ocypode* species)

EcoPrint/Shutterstock.com

B. Lobster (*Homarus americanus*)

Herve Chaumeton/Agence Nautr

C. Lobster external anatomy

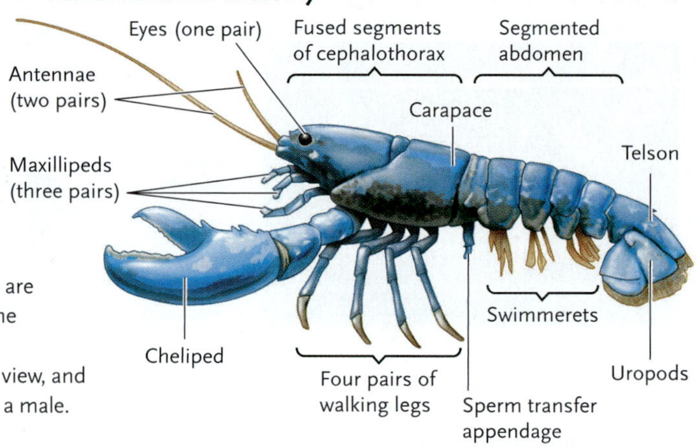

FIGURE 31.38 Decapod crustaceans. (A) Crabs, like this ghost crab, and **(B)** lobsters are typical decapod crustaceans. The abdomen of a crab is shortened and wrapped under the cephalothorax, producing a compressed body. **(C)** Lobsters bear 19 pairs of distinctive appendages; only appendages on the left side of the animal are illustrated in this lateral view, and one pair of mandibles and two pairs of maxillae are not shown. The animal illustrated is a male.

carapace, a protective covering that extends backward from the head. Crustaceans vary in size from water fleas less than 1 mm long to lobsters that can grow to 60 cm in length and weigh as much as 20 kg.

Crustaceans generally have five characteristic pairs of appendages on the head **(Figure 31.38).** Most have two pairs of sensory antennae and three pairs of mouthparts. The latter include one pair of *mandibles,* which move laterally to bite and chew, and two pairs of *maxillae,* which hold and manipulate food. Numerous paired appendages posterior to the mouthparts vary among groups.

Most crustaceans are active animals that exhibit complex patterns of movement. Their activities are coordinated by elaborate sensory and nervous systems, including chemical and touch receptors in the antennae, compound eyes, statocysts on the head, and sensory hairs embedded in the exoskeleton throughout the body. The nervous system is similar to that of annelids, but the ganglia are larger and more complex, allowing a finer level of motor control. High levels of activity require substantial amounts of oxygen, and larger species have complex, feathery gills tucked beneath the carapace. Activity also produces abundant metabolic wastes that are excreted by diffusion across the gills or, in larger species, by glands located in the head.

The sexes are typically separate, and courtship rituals are often complex. Eggs are usually brooded on the surface of the female's body or beneath the carapace. Many have free-swimming larvae that, after undergoing a series of molts, gradually assume an adult form.

The subphylum includes so many different body plans that it is usually divided into six major groups with numerous subgroups. The crabs, lobsters, and shrimps (Decapoda, meaning "10 feet," which is a subgroup of the Malacostraca) number more than 10,000 species. The vast majority of decapods are marine, but a few shrimps, crabs, and crayfishes occupy freshwater habitats. Some crabs also live in moist terrestrial habitats, where they scavenge dead vegetation, clearing the forest floor of debris.

All decapods exhibit extreme specialization of their appendages. In the American lobster, for example, each of the 19 pairs of appendages is different (see Figure 31.38C). Behind the antennae, mandibles, and maxillae, the thoracic segments have three pairs of *maxillipeds,* which shred food and pass it up to the mouth, a pair of large *chelipeds* (pinching claws), and four pairs of walking legs. The abdominal appendages include a pair specialized for sperm transfer (in males only), *swimmerets* for locomotion and for brooding eggs, and *uropods,* a pair of appendages that, combined with the *telson,* the tip of the abdomen, form a fan-shaped tail. If any appendage is damaged, the animal can autotomize ("drop") it and begin growing a new one before its next molt.

Representatives of several crustacean groups—fairy shrimps, amphipods, water fleas, krill, ostracods, and copepods **(Figure 31.39)**—live as plankton in the upper waters of oceans and lakes. Most are only a few millimeters long, but are present in huge numbers. They feed on microscopic algae or detritus and are themselves food for larger invertebrates, fishes, and some suspension-feeding marine mammals like the baleen whales. Planktonic crustaceans are among the most abundant animals on Earth.

Adult barnacles (Cirripedia, meaning "hairy footed," a subgroup of the Maxillopoda) are sessile, marine crustaceans that

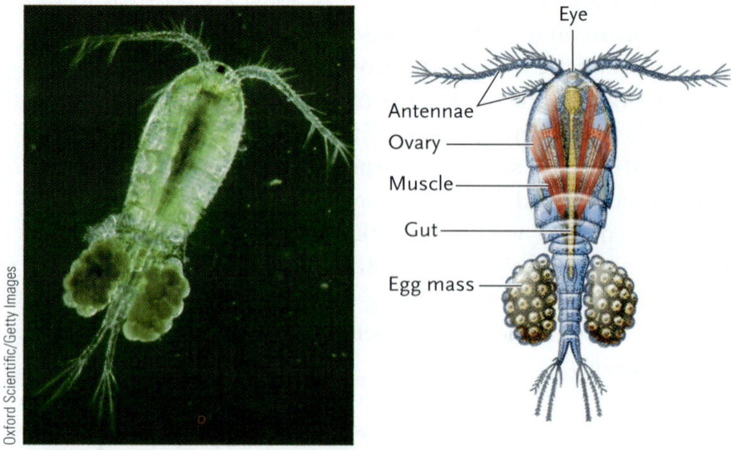

FIGURE 31.39 Copepods. Tiny crustaceans, like these copepods (*Cyclops* species), occur by the billions in freshwater and marine plankton.

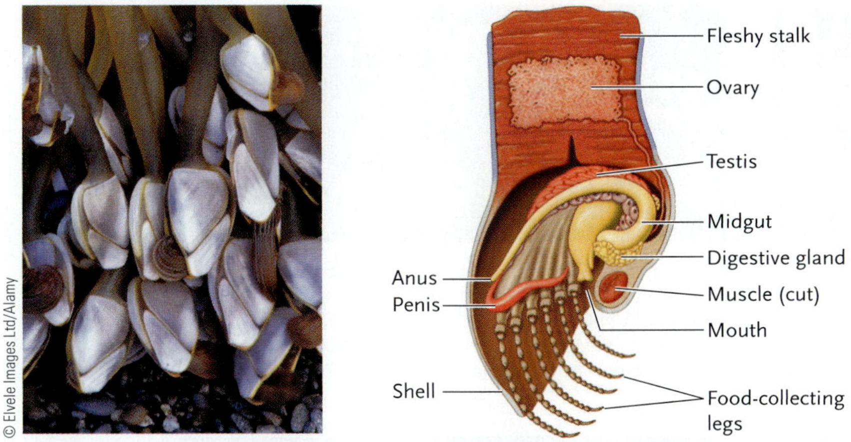

FIGURE 31.40 Barnacles. Gooseneck barnacles *(Lepas anatifera)* attach to the underside of floating debris. Like other barnacles, they open their shells and extend their feathery legs to collect particulate food from seawater.
© Cengage Learning 2014

live within a strong, calcified cup-shaped shell **(Figure 30.40).** Their free-swimming larvae attach permanently to substrates—rocks, wooden pilings, the hulls of ships, the shells of mollusks, even the skin of whales—and secrete the shell, which is actually a modified exoskeleton. To feed, barnacles open the shell and extend six pairs of feathery legs. The beating legs capture microscopic plankton and transfer it the mouth. Unlike most crustaceans, barnacles are hermaphroditic.

SUBPHYLUM MYRIAPODA The centipedes (Chilopoda) and millipedes (Diplopoda) are classified together in the subphylum Myriapoda (*myrias* = ten thousand, *pod-* = feet). Myriapods have two body regions, a head and a segmented trunk **(Figure 31.41).** The head bears one pair of antennae, and the trunk bears one (centipedes) or two (millipedes) pairs of walking legs on most of its many segments. Myriapods are terrestrial, and many species live under rocks or dead leaves. Centipedes are fast and voracious predators, using powerful toxins to kill their prey; they generally feed on invertebrates, but some eat small vertebrates. The bite of some species is harmful to humans. Although most species are less than 10 cm long, some grow to 25 cm. The millipedes are slow but powerful herbivores or scavengers. The largest species attain a length of nearly 30 cm. Although they lack a poisonous bite, they curl into a ball and exude noxious liquids when disturbed.

SUBPHYLUM HEXAPODA In terms of sheer numbers, diversity, and the range of habitats they occupy, the 1,000,000 or more species of insects and their closest relatives (subphylum Hexapoda, *hexapoda* = six feet) are the most successful animals on Earth. They were among the first animals to colonize terrestrial habitats, where most species still live. The oldest insect fossils date from the Devonian, 380 million years ago. Insects have one pair of antennae on the head, a pair of mandibles for feeding, and unbranched appendages. Insects are generally small, ranging from 0.1 mm to 30 cm in length. The lineage is divided into about 30 subgroups **(Figure 31.42).** The insect body plan always includes a head, thorax, and abdomen **(Figure 31.43).** The head is equipped with multiple mouthparts, a pair of compound eyes, and one pair of sensory antennae. The thorax has three pairs of walking legs and often one or two pairs of wings. Insects are the only invertebrates capable of flight. Their wings, which are made of lightweight but durable sheets of chitin and sclerotin, arise embryonically from the body wall; unlike the wings of birds and bats, insect wings are not derived from ancestral appendages.

Studies in evolutionary developmental biology have begun to unravel the genetic changes that fostered certain aspects of the insect body plan. For example, the *Distal-less* gene (*Dll* for short) is a highly conserved tool-kit gene (see Section 23.6) that triggers the development of appendages in all sorts of animals—the legs of chickens, the fins of fishes, the parapodia of polychaete worms, and the diverse appendages of arthropods. All arthropods also have a gene called *Ultrabithorax* (*Ubx* for short). It is one of the *Hox* genes that control the overall body plans of animals. In insects, the *Ubx* gene contains a unique mutation, not found in other arthropods, that causes the protein for which it codes to repress *Dll,* thereby preventing the forma-

A. Millipede (*Spirobolus* species)

B. Centipede (*Scolopendra* species)

FIGURE 31.41 Millipedes and centipedes. (A) Millipedes feed on living and decaying vegetation. They have two pairs of walking legs on most segments. **(B)** All centipedes are voracious predators, feeding on invertebrates and small vertebrates. Centipedes have one pair of walking legs per segment.

A. Silverfish (Thysanura, *Lepisma saccharina*) are primitive wingless insects.

B. Dragonflies, like the flame skimmer (Odonata, *Epitheca cynosura*), have aquatic larvae that are active predators; adults capture other insects in mid-air.

C. Male praying mantids (Mantodea, *Mantis religiosa*) are often eaten by the larger females during or immediately after mating.

D. This rhinoceros beetle (Coleoptera, *Lucanus cervus*) is one of more than 250,000 beetle species that have been described.

E. Fleas (Siphonoptera, *Hystrichopsylla dippiei*) have strong legs with an elastic ligament that allows these parasites to jump on and off their animal hosts.

F. Crane flies (Diptera, *Tipula* species) look like giant mosquitoes, but their mouthparts are not useful for biting other animals; the adults of most species live only a few days and do not feed at all.

G. The luna moth (Lepidoptera, *Actias luna*), like other butterflies and moths, has wings that are covered with colorful microscopic scales.

H. Like many other ant species, fire ants (Hymenoptera, *Solenopsis invicta*) live in large cooperative colonies. Fire ants—named for their painful sting—were introduced into southeastern North America, where they are now serious pests.

FIGURE 31.42 Insect diversity. Insects are grouped into about 30 clades, 8 of which are illustrated here.

tion of appendages wherever *Ubx* is expressed. And because insects express *Ubx* in their abdomen, they do not grow abdominal appendages. All other arthropods, which have the ancestral, nonrepressing form of the *Ubx* gene, have appendages in the posterior region of their body. Thus, one mutation in a *Hox*-family gene has fostered the evolution of a highly distinctive morphological trait in insects—having legs on the thorax, but not on the abdomen.

Insects exchange gases through a specialized **tracheal system,** a branching network of tubes that carry oxygen from small openings in the exoskeleton to tissues throughout the body (see Figure 46.5). Insects excrete nitrogenous wastes through specialized **Malpighian tubules** that transport wastes to the digestive system for disposal with feces (see Figure 48.6). Both of these organ systems are unique among animals. Insect sensory systems are diverse and complex. Besides image-forming compound eyes (see Figure 41.10), many insects have light-sensing ocelli on their heads. Many also have hairs, sensitive to touch, on their antennae, legs, and other regions of the body. Chemical receptors are particularly common on the legs, allowing the identification of food. And many groups of insects have hearing or-

FIGURE 31.43 The insect body plan. Insects have distinct head, thorax, and abdomen. Of all the internal organ systems, only the dorsal blood vessel, ventral nerve cord, and some muscles are strongly segmented.

External anatomy of a grasshopper

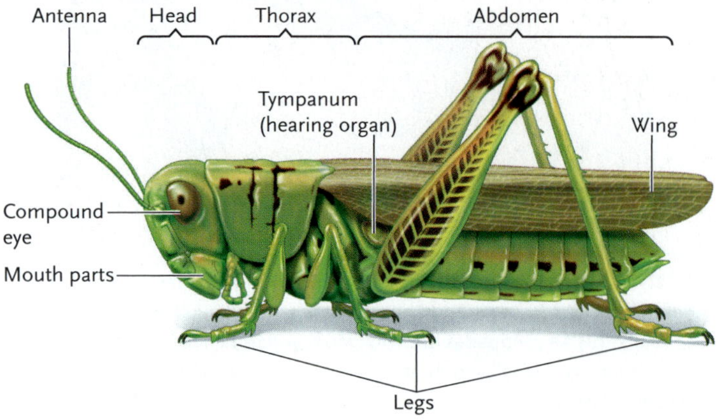

Internal anatomy of a female grasshopper

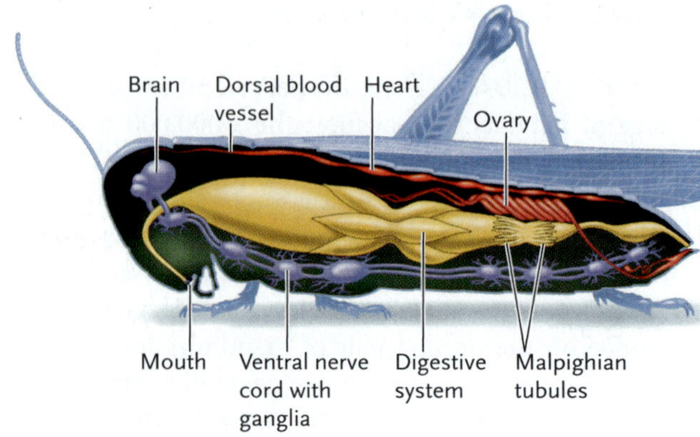

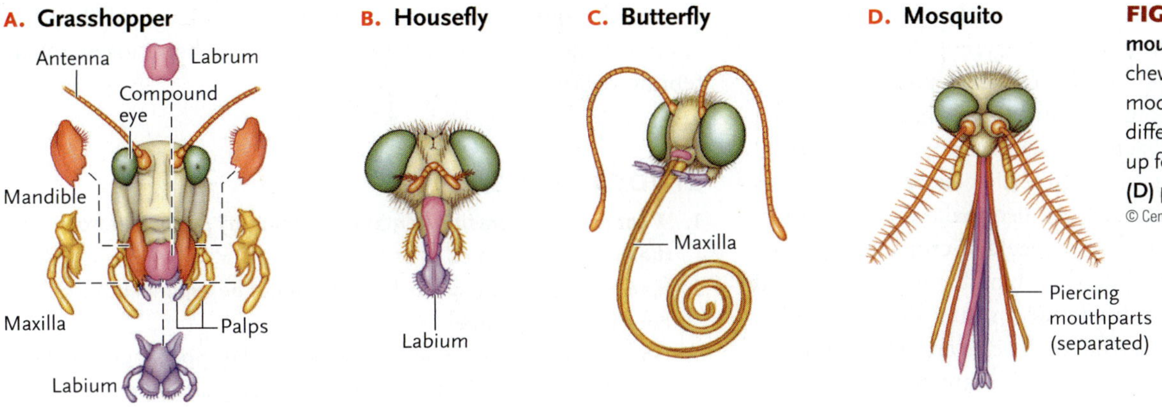

FIGURE 31.44 Specialized insect mouthparts. The **(A)** ancestral chewing mouthparts have been modified by evolution, allowing different insects to **(B)** sponge up food, **(C)** drink nectar, and **(D)** pierce skin to drink blood.

gans to detect predators and potential mates. The familiar chirping of crickets, for example, is a mating call emitted by males that may repel other males and attract females.

As a group, insects feed in every conceivable way and on most other organisms. Species that eat plants, such as grasshoppers, have a pair of rigid mandibles, which chew food before it is ingested. Behind the mandibles is a pair of maxillae, which may also aid in food acquisition. Insects also have inflexible upper and lower lips, the *labrum* and *labium,* respectively. A tongue-like structure, just dorsal to the labium in chewing insects, houses the openings of the salivary glands. But evolution has modified this ancestral mandibulate pattern in numerous ways **(Figure 31.44).** In houseflies, the mouthparts are adapted for sopping up liquid food. In butterflies and moths, the mouthparts include a long proboscis to drink nectar. And in some biting flies, like mosquitoes and blackflies, the mouthparts have evolved into piercing structures.

After it hatches from an egg, an insect passes through a series of developmental stages called *instars.* Several hormones control development and ecdysis, which marks the passage from one instar to the next. Insects exhibit one of three basic patterns of postembryonic development **(Figure 31.45).** Primitive, wingless species simply grow and shed their exoskeleton without undergoing major changes in morphology. Other species undergo **incomplete metamorphosis;** they hatch from the egg as a *nymph,* which lacks functional wings. In many species, such as grasshoppers (Orthoptera), the nymphs resemble the adults. In other insects, such as dragonflies (Odonata), the aquatic nymphs are morphologically very different from the adults.

Most insects undergo **complete metamorphosis:** the larva that hatches from the egg differs greatly from the adult. Larvae and adults often occupy different habitats and consume different food. The larvae (caterpillars, grubs, or maggots) are often worm-shaped, with chewing mouthparts. They grow and molt several times, retaining their larval morphology. Before they transform into sexually mature adults, they spend a period of time as a sessile **pupa.** During this stage, the larval tissues are drastically reorganized. The adult that emerges is so different from the larva that it is often hard to believe that they are of the same species. Moths, butterflies, beetles, and flies are examples of insects with complete metamorphosis. Their larval stages specialize in feeding and growth, whereas the adults are adapted for dispersal and reproduction. In some species, the adults never feed, relying on the energy stores accumulated during the larval stage.

FIGURE 31.45 Patterns of postembryonic development in insects.

The 240-million-year history of insects has been characterized by innovations in morphology, life cycle patterns, locomotion, feeding, and habitat use. Their well-developed nervous systems govern exceptionally complex patterns of behavior, including parental care, a habit that reaches its zenith in the colonial social insects, the ants, bees, and wasps (see Chapter 56). The factors that contribute to the insects' success also make them our most aggressive competitors. They destroy vegetable crops, stored food, wool, paper, and timber. They feed on blood from humans and domesticated animals, sometimes transmitting disease-causing pathogens as they do so. Nevertheless, insects are essential members of terrestrial ecological communities. Many species pollinate flowering plants, including important crops. Many others attack or parasitize species that are harmful to human activities. And most insects are a primary source of food for other animals. Some even make useful products like honey, beeswax, and silk.

In the next chapter we consider the lineage of deuterostomes, which includes the vertebrates and their closest invertebrate relatives.

STUDY BREAK 31.7

1. What part of a parasitic nematode's anatomy protects it from the digestive enzymes of its host?
2. If an arthropod's rigid exoskeleton cannot be expanded, how does the animal grow?
3. How do the number of body regions and the appendages on the head differ among the four subphyla of living arthropods?
4. How do the life stages differ between insects that have incomplete metamorphosis and those that have complete metamorphosis?

UNANSWERED QUESTIONS

What are the evolutionary relationships among the invertebrate clades?

If we step back from our vertebrate and terrestrial biases, invertebrates are the dominant form of life on Earth. Many students of natural history are overwhelmed by the sheer numbers of organisms that one encounters, particularly in the aquatic environment, and the challenge of categorizing them taxonomically. Sorting them out in terms of their ecological roles is equally daunting. The problem we face is that invertebrates do it all—they are predators, herbivores, parasites, detritivores, and the primary symbiotic organisms on Earth. This adaptability of form and function is a fascinating hallmark of the animal way of life, but it poses a legion of questions, many still unanswered. Recent advances in genetic technology have advanced our understanding, but there is much left to do.

In the past two decades, our ability to compare genetic information from various invertebrate groups has led to a remarkable reshuffling of the long-established categories used to classify these organisms. The first categories to be eliminated were groups based on superficial phenotypic resemblances, such as the "pseudocoelomates," which had plagued student understanding of diversity. Today, we have a much deeper knowledge of the evolutionary relationships of these organisms. Perhaps the most exciting discovery is that much of the diversity we see is not the product of slow changes in protein-coding gene sequences, but rather the result of variations in the timing and location of the expression of genes that affect development. Evo-devo, the melding of evolutionary and developmental biology—mostly made possible by intensive studies of two model invertebrates, *Caenorhabditis elegans* and *Drosophila melanogaster*—has revealed that changes in the expression of relatively simple sets of genes have brought about the myriad forms of life we see among the invertebrates. As systematists incorporate these new discoveries in their analyses, our understanding of the evolutionary relationships among the invertebrates will surely change.

What is the genetic basis of the diversity of form and function observed among invertebrates?

Because of advances in genetics research, we are on the cusp of being able to answer some fundamental questions. How does an animal's body develop either radial or bilateral symmetry? How does an organism develop a head with a concentration of nervous system tissue, and how do all the exquisite sensory systems associated with a big brain develop? From where do the respiratory pigments, which increase the capacity of the hemolymph or blood to carry oxygen, increasing an organism's capacity for activity, arise? How does the immune system develop, and what genetic change fosters the quantum leap from a nonspecific defense system to one that responds specifically to foreign invaders? More practically, are there genetic switches that we can manipulate? Can we make blood-feeding invertebrates, such as mosquitoes, or parasitic species, like tapeworms or filarial worms, innocuous? Is it possible to use genetic engineering to reduce the ability of the mosquito *Anopheles gambiae* (the "deadliest organism on Earth") to transmit malaria? Will it be possible to forestall or reverse the global decline of coral reefs, one of the richest habitats on Earth?

The answers to these and many more basic and applied questions lie within a deep knowledge and understanding of the invertebrates and the roles they play on our planet. If one looks at invertebrates as dynamic systems, as rich sources of clues to life on Earth, the questions they pose easily provide a lifetime of investigation and reward.

THINK LIKE A SCIENTIST

1. Based on your reading of this chapter, provide two examples of how recent genetic comparisons of invertebrates have confirmed groupings based on morphology and embryology.
2. Based on the content of this chapter, which genes would you study if you wanted to learn about the genetic basis of diversity among invertebrates?

William S. Irby

William S. Irby is an associate professor in the Department of Biology at Georgia Southern University in Statesboro. His research focuses on the ecology and evolution of blood-feeding behavior in mosquitoes. To learn more about Irby's work, go to http://www.bio.georgiasouthern.edu/amain/fac-list.html.

REVIEW KEY CONCEPTS

To access the course materials and companion resources for this text, please visit www.cengagebrain.com.

31.1 What Is an Animal?

- Animals are eukaryotic, multicellular, heterotrophs that are motile at some time in their lives.
- Animals probably arose from a colonial flagellated ancestor during the Precambrian era (Figure 31.2).

31.2 Key Innovations in Animal Evolution

- All animals except sponges have tissues, which are organized into either two or three tissue layers.
- Although some animals exhibit radial symmetry, most exhibit bilateral symmetry (Figure 31.3).
- Acoelomate animals have no body cavity. Pseudocoelomate animals have a body cavity between the derivatives of endoderm and mesoderm. Coelomate animals have a body cavity that is entirely lined by derivatives of mesoderm (Figure 31.4).
- Two lineages of bilaterally symmetrical animals differ in developmental patterns (Figure 31.5). Most protostomes exhibit spiral, determinate cleavage, and their coelom forms from a split in a solid mass of mesoderm. Most deuterostomes have radial, indeterminate cleavage, and the coelom usually forms within outpocketings of the primitive gut.
- Four animal phyla exhibit segmentation.

Animation: Types of body cavities

Animation: Directional terms and planes of symmetry

Animation: Developmental differences between protostomes and deuterostomes

31.3 An Overview of Animal Phylogeny and Classification

- Analyses of molecular sequence data have refined our view of animal evolutionary history (Figure 31.6). The molecular phylogeny recognizes some major clades that had been identified on the basis of morphological and embryological characters. Sponges are grouped in the Parazoa. All other clades are grouped in the Eumetazoa. Among the Eumetazoa, the Radiata includes animals with two tissue layers and radial symmetry, and the Bilateria includes animals with three tissue layers and bilateral symmetry.
- Bilateria is further subdivided into Protostomia and Deuterostomia. The new phylogeny divides the Protostomia into the Lophotrochozoa and the Ecdysozoa.
- The molecular phylogeny suggests that ancestral protostomes had a coelom and that segmentation arose more than once.

31.4 Animals without Tissues: Parazoa

- Sponges (phylum Porifera) are asymmetrical animals with limited integration of cells in their bodies (Figure 31.7).
- The body of many sponges is a water-filtering system (Figure 31.8). Flagellated choanocytes draw water into the body and capture particulate food.

Animation: Body plan of a sponge

31.5 Eumetazoans with Radial Symmetry

- The two major radiate phyla, Cnidaria and Ctenophora, have two well-developed tissue layers with a gelatinous mesoglea between them (Figure 31.9). Members of both phyla lack organ systems. All are aquatic.
- Cnidarians capture prey with tentacles and stinging nematocysts (Figures 31.10, 31.12, and 31.13). Their life cycles may include polyps, medusae, or both (Figure 31.11).
- Ctenophores use long tentacles to capture particulate food and use rows of cilia for locomotion (Figure 31.14).

Animation: Cnidarian body plans

Animation: Nematocyst action

Animation: Cnidarian life cycle

31.6 Lophotrochozoan Protostomes

- The taxon Lophotrochozoa includes eight phyla.
- Three small phyla (Ectoprocta, Brachiopoda, and Phoronida) use a lophophore to feed on particulate matter (Figure 31.15).
- Free-living flatworm species (phylum Platyhelminthes) have well-developed digestive, excretory, reproductive, and nervous systems (Figures 31.16 and 31.17). Parasitic species attach to their animal hosts with suckers or hooks (Figures 31.18 and 31.19).
- The rotifers (phylum Rotifera) are tiny and abundant inhabitants of freshwater and marine ecosystems (Figure 31.20). Movements of cilia in the corona control their locomotion and bring food to their mouths.
- The ribbon worms (phylum Nemertea) are elongate and often colorful animals with a proboscis housed in a rhynchocoel (Figure 31.21).
- Mollusks (phylum Mollusca) have fleshy bodies that are often enclosed in a hard shell. The molluscan body plan includes a head, a foot, the visceral mass, and a mantle (Figures 31.22 and 31.24–31.27).
- Segmented worms (phylum Annelida) generally exhibit segmentation of the coelom and of the muscular, circulatory, excretory, respiratory, and nervous systems. They use the coelom as a hydrostatic skeleton for locomotion (Figures 31.28–31.31).

Animation: Clam body plan

Animation: Crab life cycle

Animation: Cuttlefish body plan

Animation: Earthworm body plan

Animation: Insect development

Animation: Insect head parts

Animation: Molluscan classes

Animation: Planarian organ systems

Animation: Roundworm body plan

Animation: Tapeworm life cycle

31.7 Ecdysozoan Protostomes

- The taxon Ecdysozoa includes three phyla that periodically shed their cuticle or exoskeleton.
- Roundworms (phylum Nematoda) feed on decaying organic matter or parasitize plants or animals (Figure 31.32). They move by alternately contracting longitudinal muscles on the dorsal and ventral sides of the body wall.

- The velvet worms (phylum Onychophora) have segmented bodies and unjointed legs (Figure 31.33). Some species bear live young, which develop in a uterus.
- The segmented bodies of the arthropods (phylum Arthropoda) have specialized appendages for feeding, locomotion, or reproduction. Arthropods shed their exoskeleton as they grow or enter a new stage of the life cycle (Figure 31.34). They have an open circulatory system, a complex nervous system, and, in some groups, highly specialized respiratory and excretory systems.
- Arthropods are divided into five subphyla. The extinct trilobites (subphylum Trilobita), with three-lobed bodies and relatively undifferentiated appendages, were abundant in Paleozoic seas (Figure 31.35). Chelicerates have a cephalothorax and abdomen; appendages on the head serve as pincers or fangs and pedipalps (Figures 31.36 and 31.37). Crustaceans have a carapace that covers the cephalothorax as well as highly modified appendages, including antennae and mandibles (Figures 31.38–31.40). Myriapods have a head and an elongate, segmented trunk (Figure 31.41). Hexapods have three body regions, three pairs of walking legs on the thorax, and three pairs of feeding appendages on the head (Figures 31.42–31.44). Insects exhibit three patterns of postembryonic development (Figure 31.45).

UNDERSTAND & APPLY

Test Your Knowledge

1. Which of the following characteristics is *not* typical of most animals?
   a. heterotrophic
   b. sessile
   c. bilaterally symmetrical
   d. multicellular
   e. motile at some stage of life cycle
2. A body cavity that separates the digestive system from the body wall but is *not* completely lined with mesoderm is called a:
   a. schizocoelom.
   b. mesentery.
   c. peritoneum.
   d. pseudocoelom.
   e. hydrostatic skeleton.
3. Protostomes and deuterostomes typically differ in:
   a. their patterns of body symmetry.
   b. the number of germ layers during development.
   c. their cleavage patterns.
   d. the size of their sperm.
   e. the size of their digestive systems.
4. The nematocysts of cnidarians are used primarily for:
   a. capturing prey.
   b. detecting light and dark.
   c. courtship.
   d. sensing chemicals.
   e. gas exchange.
5. Which organ system is absent in flatworms (phylum Platyhelminthes)?
   a. nervous system
   b. reproductive system
   c. circulatory system
   d. digestive system
   e. excretory system
6. Which part of a mollusk secretes the shell?
   a. visceral mass
   b. radula
   c. trochophore
   d. head–foot
   e. mantle
7. What is the major morphological innovation seen in annelid worms?
   a. a complete digestive system
   b. image-forming eyes
   c. a respiratory system
   d. an open circulatory system
   e. body segmentation
8. Which phylum includes the most abundant animals in soil?
   a. Nematoda
   b. Rotifera
   c. Mollusca
   d. Annelida
   e. Brachiopoda
9. Which body region of an insect bears the walking legs?
   a. head
   b. carapace
   c. abdomen
   d. thorax
   e. trunk
10. Ecdysis refers to a process in which:
   a. bivalves use siphons to pass water across their gills.
   b. arthropods shed their old exoskeletons.
   c. cnidarians build skeletons of calcium carbonate.
   d. rotifers produce unfertilized eggs.
   e. squids escape from predators in a cloud of ink.

Discuss the Concepts

1. Many invertebrate species are hermaphroditic. What selective advantages might this characteristic offer? In what kinds of environments might it be most useful?
2. People who eat raw clams and oysters harvested from sewage-polluted waters often develop mild to severe gastrointestinal infections. These mollusks are suspension feeders. Develop a hypothesis about why people who eat them raw may be at risk.
3. On a voyage to the ocean bottom, a biologist discovers a worm that appears to be new to science. What characteristics of this animal should the biologist examine to determine whether or not she has discovered a previously undescribed phylum?
4. What are the relative advantages and disadvantages of radially symmetrical and bilaterally symmetrical body plans?

Design an Experiment

Design an experiment to test the hypothesis that the cuticle of parasitic nematodes protects them from the acids and enzymes present in the digestive systems of their hosts. Your design must include both experimental and control treatments.

Interpret the Data

A research team led by Andreas Michalsen of the Kliniken Essen, a hospital in Germany, compared the relative effectiveness of "leech therapy" and the topical application of diclofenac, an antiinflammatory drug, for the treatment of pain caused by osteoarthritis in the thumb. Leech saliva contains more than 30 biologically active compounds, including some with antiinflammatory effects. For the group receiving drug therapy, the patients applied the drug to their thumbs at least twice a day for up to 30 days. For the group receiving leech therapy, researchers applied two or three leeches to the affected joint and allowed them to feed until they were satiated; the average feeding time was 50 minutes. Patients in the leech therapy group received just one treatment. The researchers asked patients to score their pain (0 = no pain; 100 = severe pain) before treatment (day 0 in the **Figure**) and again one week, one month, and two months after treatment. The results are presented in the accompanying bar chart. Which treatment had the greater effect in reducing the pain of arthritis patients in this study? Did leech therapy have a prolonged effect?

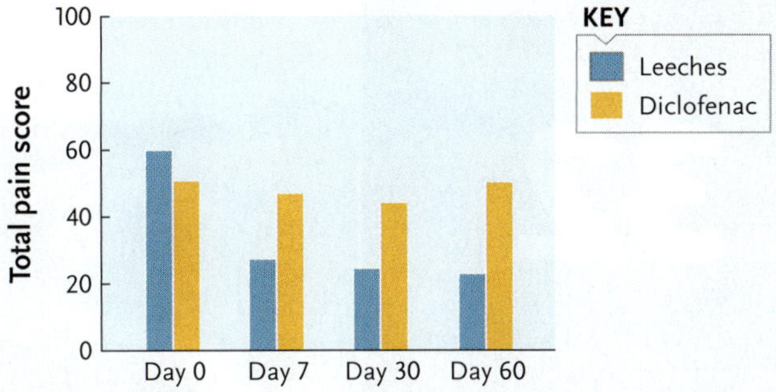

Source: A. Michalsen et al. 2008. Effectiveness of leech therapy in women with symptomatic arthrosis of the first carpometacarpal joint: A randomized controlled trial. *Pain* 137: 452–459.

Apply Evolutionary Thinking

Many insects have a larval stage that is morphologically different from the adult and that feeds on different foods. What selection pressures may have fostered the evolution of a life cycle with such distinctive life stages? Your answer should address the different biological activities that characterize each life cycle stage.

32

Snow monkeys *(Macaca fuscata)*. These snow monkeys, which have the northernmost distribution of any nonhuman primate, are soaking in a hot spring in Japan.

STUDY OUTLINE

32.1 Invertebrate Deuterostomes

32.2 Overview of the Phylum Chordata

32.3 The Origin and Diversification of Vertebrates

32.4 "Agnathans": Hagfishes and Lampreys, Conodonts and Ostracoderms

32.5 Gnathostomata: The Evolution of Jaws

32.6 Tetrapoda: The Evolution of Limbs

32.7 Amniota: The Evolution of Fully Terrestrial Vertebrates

32.8 Testudines: Turtles

32.9 Living Lepidosaurs: Sphenodontids and Squamates

32.10 Living Archosaurs: Crocodilians and Birds

32.11 Mammalia: Monotremes, Marsupials, and Placentals

32.12 Nonhuman Primates

32.13 The Evolution of Humans

Deuterostomes: Vertebrates and Their Closest Relatives

Why it matters . . . In 1798, naturalists at the British Museum skeptically probed a curious specimen that had been sent from Australia. The furry creature—about the size of a housecat—had webbed front feet, a ducklike bill, and a flat, paddlelike tail **(Figure 32.1).** The scientists eagerly searched for evidence that a prankster had stitched together parts from wildly different animals, but they found no signs of trickery and soon accepted the duck-billed platypus *(Ornithorhynchus anatinus)* as a genuine zoological novelty.

Further study has revealed that the platypus is even stranger than those scientists could have imagined. Like other mammals, the platypus is covered with hair, and females produce milk that the offspring lick off the fur on their mother's belly. But like turtles and birds, a platypus has no teeth, and it reproduces by laying eggs instead of giving birth to its offspring. And like turtles, birds, lizards, snakes, and crocodilians, it has a cloaca, a multipurpose chamber through which it releases feces, urine, and eggs. Scientists had never before seen such a weird combination of traits, and they didn't quite know what to make of them.

Studies of the platypus under natural conditions have helped biologists make sense of its characteristics. The platypus inhabits streams and lagoons in Australia and Tasmania. It rests in streamside burrows during the day, but at night it slips into the water to hunt for invertebrates. Its dense fur keeps its body warm and dry under water, and its tail serves both as a rudder and as a storehouse for energy-rich fat. It uses its bill to scoop up food and the horny pads that line its jaws to grind up prey. While underwater, the platypus clamps shut its eyes, ears, and nostrils, relying on roughly 800,000 sensory receptors in its bill to detect the movements and weak electrical discharges of nearby prey.

The platypus, with its strange combination of characteristics, illustrates the remarkable diversity of adaptations that enable vertebrates—animals with backbones—to occupy nearly every

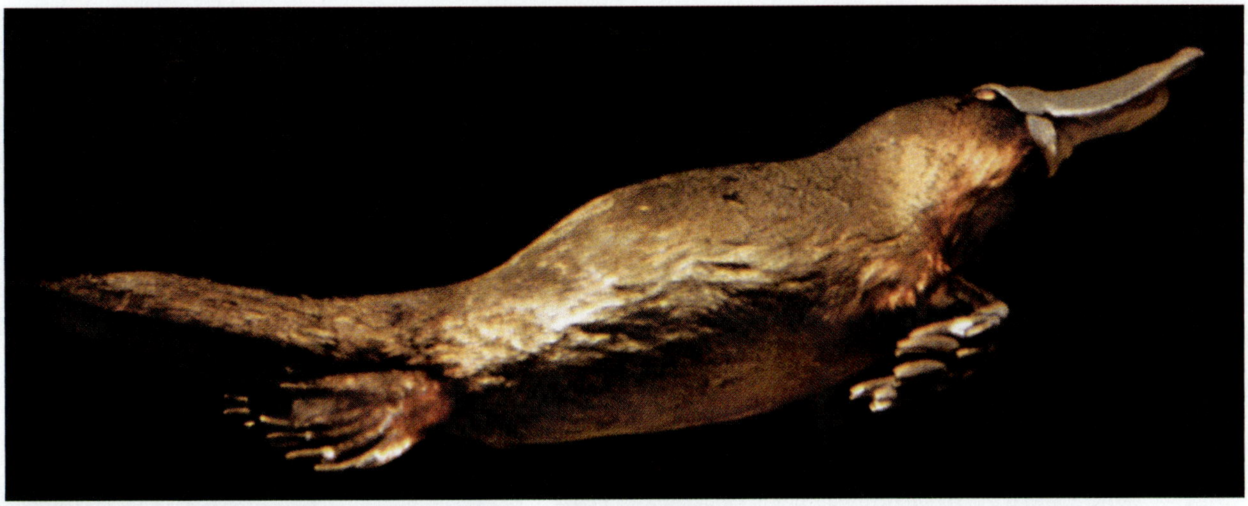

Tom McHugh/Photo Researchers, Inc.

FIGURE 32.1 A puzzling animal. Because of its strange mixture of traits, the platypus *(Ornithorhynchus anatinus)* amazed the first European zoologists who saw it.

habitat on Earth. Despite the platypus's mixed characteristics, biologists eventually classified it as a member of the mammal lineage because, like all other mammals, it has hair on its body and produces milk to nourish its offspring. Today biologists know that it is one of just a few remaining survivors of an early lineage of egg-laying mammals.

In this chapter, we survey the Deuterostomia, a monophyletic lineage of animals that dates to the Paleozoic. The deuterostomes are defined by features of early embryological development and molecular sequence data (see Chapter 31). There are three living phyla of deuterostomes; we briefly consider two phyla of invertebrate deuterostomes before focusing on the Phylum Chordata, which includes a few thousand species of invertebrates as well as many living species of vertebrates.

32.1 Invertebrate Deuterostomes

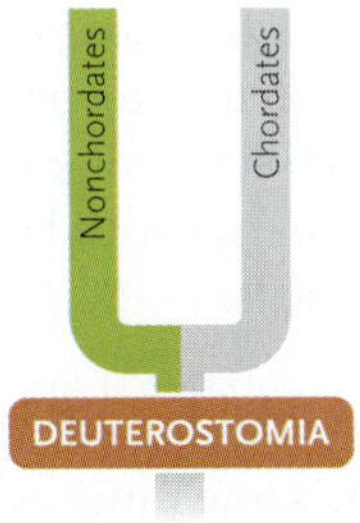

Deuterostome body plans have been so modified by evolution that a casual observer would not readily group the two phyla of invertebrate deuterostomes—Echinodermata and Hemichordata—together with the Phylum Chordata. However, embryological and molecular analyses agree that all three are indeed closely related.

Echinoderms Have Secondary Radial Symmetry and an Internal Skeleton

The phylum Echinodermata (*echinos* = spiny, *derma* = skin) includes roughly 6,000 species of sea stars, sea urchins, sea cucumbers, brittle stars, and sea lilies. These slow-moving or sessile, bottom-dwelling animals are important herbivores and predators in shallow coastal waters and the ocean depths. Although the phylum was diverse in the Paleozoic, only a remnant of that fauna remains. Living species vary in size from less than 1 cm to more than 50 cm in diameter.

Echinoderms develop from a bilaterally symmetrical, free-swimming larva. But as a larva develops, it assumes a secondary radial symmetry, often organized around five rays or "arms" **(Figure 32.2).** Many echinoderms have an *oral surface,* with the mouth facing the substrate, and an *aboral surface* facing in the opposite direction. Virtually all echinoderms have an internal skeleton made of calcium-stiffened *ossicles* that develop from mesoderm. In some groups, fused ossicles form a rigid container called a *test.* In most, spines or bumps project from the ossicles.

The internal anatomy of echinoderms is unique among animals **(Figure 32.3).** They have a well-defined coelom and a complete digestive system (see Figure 32.3A), but no excretory or respiratory systems, and most have only a minimal circulatory system. In many, gases are exchanged and metabolic wastes eliminated through projections of the epidermis and peritoneum near the base of the spines. Given their radial symmetry, there is no head or central brain; the nervous system is organized around nerve cords that encircle the mouth and branch into the rays. Sensory cells are abundant in the skin.

Echinoderms have a unique locomotor system, the *water vascular system,* which consists of fluid-filled canals (see Figure 32.3B). In a sea star, for example, water enters the system through the *madreporite,* a sievelike plate on the aboral surface. A short tube connects it to the *ring canal,* which surrounds the esophagus. The ring canal branches into five *radial canals* that extend into the arms. Each radial canal is connected to numerous *tube feet* that protrude through holes in the ossicles (see Figure 32.3C). Each tube foot has a mucus-covered, suckerlike tip and a small muscular bulb, the *ampulla,* which lies inside the body. When an ampulla contracts, fluid is forced into the tube foot, causing it to lengthen and attach to the substrate. The tube foot then contracts, pulling the animal along. As the tube foot

FIGURE 32.2
Echinoderm diversity. Echinoderms exhibit secondary radial symmetry, usually organized as five rays around an oral–aboral axis.

A. Asteroidea: This sea star *(Fromia milleporella)* lives in the intertidal zone.

B. Ophiuroidea: A brittle star *(Ophiothrix swensonii)* perches on a sponge.

C. Echinoidea: A sea urchin *(Strongylocentrotus purpuratus)* grazes on algae.

D. Holothuroidea: A sea cucumber *(Cucumaria miniata)* extends its tentacles, which are modified tube feet, to trap particulate food.

E. Crinoidea: A feather star *(Oxycomanthus bennetti)* feeds by catching small particles with its numerous arms.

shortens, water flows back into the ampulla, and the tube foot releases its grip on the substrate. The tube foot can then take another step forward, reattaching to the substrate. Although each tube foot has limited strength, the coordinated action of hundreds or thousands of them is so strong that they can hold an echinoderm to a substrate even against strong wave action.

Echinoderms have separate sexes, and most reproduce by releasing gametes into the water. Radial cleavage is so clearly apparent in the transparent eggs of some sea urchins that they are commonly used for demonstrations of cleavage in introductory biology laboratories. A few echinoderms also reproduce asexually by splitting in half and regenerating the missing parts; some can regenerate body parts lost to predators.

Echinoderms are divided into six groups, one of which, the sea daisies (Concentricycloidea), was discovered only in 1986. These small, medusa-shaped animals occupy sunken, waterlogged wood in the deep sea. Here we describe the five other groups, which are more diverse and better known.

ASTEROIDEA Sea stars (Asteroidea, from *asteroeides* = starlike) live from rocky shorelines to depths of 10,000 m. The asteroid body plan consists of a central disk surrounded by 5 to 20 radiating "arms" (see Figure 32.2A), with the mouth centered on the oral surface. The ossicles of the endoskeleton are not fused, permitting flexibility of the arms and disk. Small pincers, **pedicellariae,** at the base of short spines remove debris that falls onto the animal's surface (see Figure 32.3C). Many sea stars feed on soft-bodied invertebrates and small fishes. Other species consume bivalve mollusks: after using their tube feet to pry apart the two shells, they evert their stomachs and secrete digestive enzymes onto the mollusk's flesh. Some sea stars are destructive predators of corals, endangering many reefs.

OPHIUROIDEA Brittle stars and basket stars (Ophiuroidea, from *ophioneos* = snakelike) occupy roughly the same range of habitats as sea stars. Their bodies have a well-defined central disk and slender, elongate arms that are sometimes branched (see Figure 32.2B). Ophiuroids can crawl fairly swiftly across substrates by moving their arms in coordinated fashion. As their common name implies, the arms are delicate and easily broken, an adaptation that allows them to escape from predators with only minor losses. Brittle stars feed on small prey, suspended plankton, or detritus that they extract from muddy deposits.

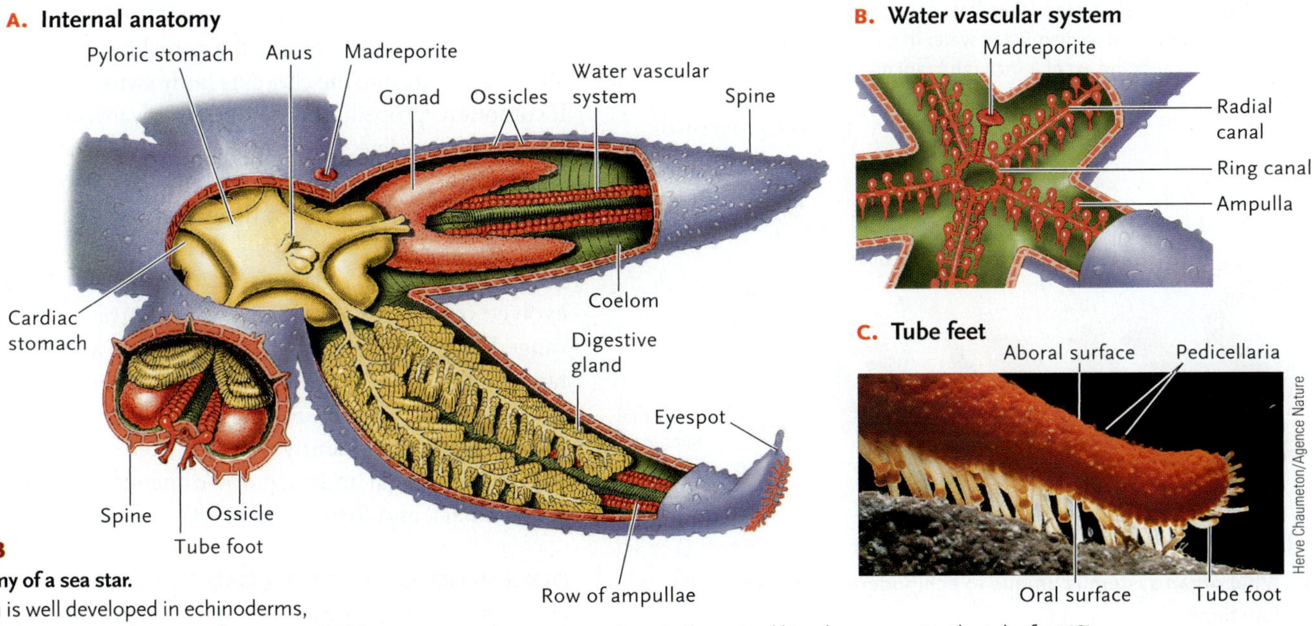

FIGURE 32.3

Internal anatomy of a sea star.

(A) The coelom is well developed in echinoderms, as illustrated by this cutaway diagram of a sea star. **(B)** The water vascular system, unique in the animal kingdom, operates the tube feet **(C)**, which are responsible for locomotion. Note the pedicellariae on the upper surface of the sea star's arm **(C)**.

ECHINOIDEA Sea urchins and sand dollars (Echinoidea, *ekhinos* = porcupine) lack arms altogether (see Figure 32.2C). Their ossicles are fused into solid tests, which provide excellent protection but restrict flexibility. The test is spherical in sea urchins and flattened in sand dollars. Five rows of tube feet, used primarily for locomotion, emerge through pores in the test. Most echinoids have movable spines, some with poison glands that protect them from predators; a jab from certain tropical species can cause severe pain and inflammation to a careless swimmer. Echinoids graze on algae and other organisms that cling to marine surfaces. In the center of an urchin's oral surface is a five-part nipping jaw that is controlled by powerful muscles. Some species damage kelp beds, disrupting the habitat of young lobsters and other crustaceans.

HOLOTHUROIDEA Sea cucumbers (Holothuroidea, from *holothourion* = water polyp) are elongate animals that lie on their sides on the ocean bottom (see Figure 32.2D). Although they have five rows of tube feet, their endoskeleton is reduced to widely separated microscopic plates. The body, which is elongated along the oral–aboral axis, is soft and fleshy, with a tough, leathery covering. Modified tube feet form a ring of tentacles around the mouth, which points to the side or upward. Some species secrete a mucous net that traps plankton or other food particles. Other species extract food from bottom sediments. Many sea cucumbers exchange gases through an extensively branched *respiratory tree* that arises from the rectum, the part of the digestive system just inside the anus at the aboral end of the animal. A well-developed circulatory system distributes oxygen and nutrients to tissues throughout the body.

CRINOIDEA Sea lilies and feather stars (Crinoidea, from *krinon* = lily) are the surviving remnants of a fauna that was diverse and abundant 500 million years ago (see Figure 32.2E). Most species occupy marine waters of medium depth. The central disk and mouth point upward rather than toward the substrate. Between five and several hundred branched arms surround the disk; new arms are added as a crinoid grows larger. The branches of the arms are covered with tiny mucus-coated tube feet, which trap suspended microscopic organisms. The sessile sea lilies have the central disk attached to a flexible stalk that can reach a meter in length. Adult feather stars can swim or crawl weakly, attaching temporarily to substrates.

Acorn Worms Use A Pharynx with Branchial Slits to Acquire Food and Oxygen

The approximately 100 species of acorn worms (phylum Hemichordata, *hemi* = half, *chorda*, referring to the phylum Chordata) are sedentary marine animals that live in U-shaped tubes or burrows in coastal sand or mud. Their soft bodies, which range from 2 cm to 2 m in length, are organized into an anterior proboscis, a tentacled collar, and an elongate trunk **(Figure 32.4)**. They use their muscular, mucus-coated proboscis to construct burrows and trap food particles. Acorn worms also have pairs of **branchial slits** (sometimes called *pharyngeal slits* or *gill slits*) in the pharynx, the part of the digestive system just posterior to the mouth. Beating cilia create a flow of water, which enters the pharynx through the mouth and exits through these openings. As water passes through, suspended food is trapped and shunted into the digestive system, and gases are exchanged across the partitions between the slits. This coupling of feeding and respiration reflects the evolutionary relationship between hemichordates and chordates, the phylum that we consider next.

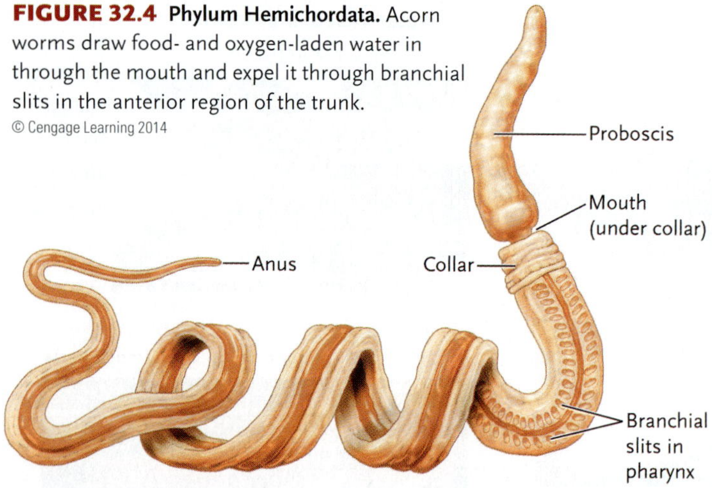

FIGURE 32.4 Phylum Hemichordata. Acorn worms draw food- and oxygen-laden water in through the mouth and expel it through branchial slits in the anterior region of the trunk.

STUDY BREAK 32.1

1. **What organ system is unique to echinoderms, and what is its function?**
2. **How does a perforated pharynx enable hemichordates to acquire food and oxygen from seawater?**

32.2 Overview of the Phylum Chordata

The phylum Chordata contains three subphyla: two lineages of invertebrates, Cephalochordata and Urochordata, and the diverse lineage of vertebrates, Vertebrata.

Key Morphological Innovations Distinguish Chordates from Other Deuterostome Phyla

Chordates are distinguished from other deuterostomes by a set of key morphological innovations: a *notochord, segmental muscles in the body wall and tail, a dorsal hollow nerve chord,* and a *perforated pharynx* **(Figure 32.5).** These structures foster higher levels of activity, unique modes of aquatic locomotion, and more efficient feeding and oxygen acquisition.

NOTOCHORD Early in chordate development, mesoderm that is dorsal to the developing digestive system forms a **notochord** (*noton* = the back; *chorda* = string). This flexible rod, constructed of fluid-filled cells surrounded by tough connective tissue, supports the embryo from head to tail. The notochord later forms the skeleton of invertebrate chordates. Body wall muscles are anchored to the notochord, and when these muscles contract, the notochord bends, but it does not shorten. As a result, the chordate body swings left and right during locomotion, propelling the animal forward; unlike annelids and other nonchordate invertebrates, the chordate body does not shorten when the animal is moving. Remnants of the notochord persist as gelatinous disks in the backbones of adult vertebrates.

SEGMENTAL BODY WALL AND TAIL MUSCLES Chordates evolved in water, and they swim by contracting segmentally arranged blocks of muscles in the body wall and tail. The chordate tail, which is posterior to the anus, provides much of the propulsion in aquatic species. Segmentation allows each muscle block to contract independently; waves of contractions pass down one side of the animal and then down the other, sweeping the body and tail back and forth in a smooth and continuous movement.

DORSAL HOLLOW NERVE CORD The central nervous system of chordates is a hollow nerve cord on the dorsal side of the animal (see Chapter 40). By contrast, most nonchordate invertebrates have solid nerve cords on the ventral side. In vertebrates, an anterior enlargement of the nerve cord forms the brain; in invertebrates, the anterior concentration of nervous system tissue is described as a *ganglion.*

PERFORATED PHARYNX Like the hemichordates described above, most chordates have outpocketings, perforations, or slits in the pharynx during some stage of the animal's life cycle. These paired openings originated as exit holes for water that carried particulate food into the mouth. Invertebrate chordates exchange gases with water as it passes by the walls of the pharynx, acquiring oxygen and eliminating carbon dioxide. Invertebrate chordates and fishes retain a perforated pharynx throughout their lives. In most air-breathing terrestrial vertebrates, the outpocketings or slits are present only during embryonic development and in larvae.

Invertebrate Chordates Are Small, Marine Suspension Feeders

Two subphyla of invertebrate chordates exhibit the basic chordate body plan in its simplest form.

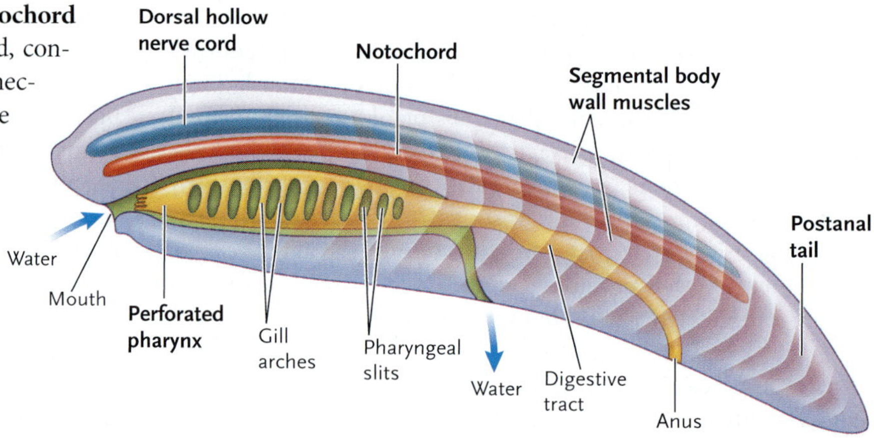

FIGURE 32.5 Diagnostic chordate characters. Chordates have a notochord; a muscular postanal tail; a segmental body wall and tail muscles; a dorsal, hollow nerve cord; and a perforated pharynx.

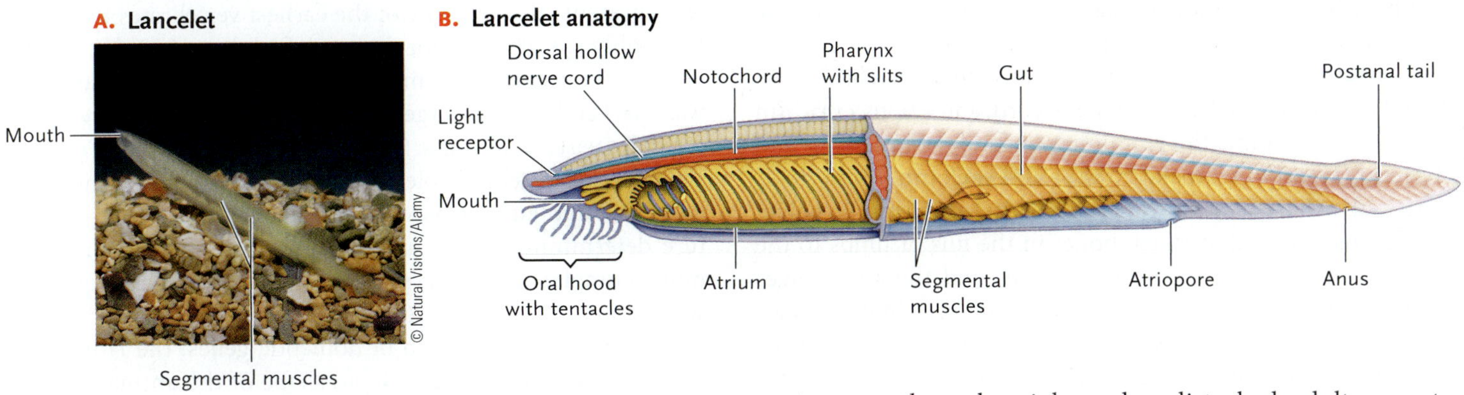

FIGURE 32.6 Cephalochordates. (A) The unpigmented skin of adult lancelets reveals their segmental body wall muscles. A cutaway view **(B)** illustrates their internal anatomy.

SUBPHYLUM CEPHALOCHORDATA The 28 lancelet species (subphylum Cephalochordata, *kephale* = head) occupy warm, shallow marine habitats where they lie mostly buried in sand **(Figure 32.6).** Although generally sedentary, they have well-developed body wall muscles and a prominent notochord. Most species are included in the genus *Branchiostoma* (formerly *Amphioxus*). Lancelet bodies, which are 5 to 10 cm long, are pointed at both ends like the double-edged surgical tools for which they are named. Adults have light receptors on the head as well as chemical sense organs on tentacles that grow from the **oral hood.** Lancelets use cilia to draw food-laden water through hundreds of pharyngeal slits; water flows into the atrium and is expelled through the **atriopore.** Most gas exchange occurs across the skin.

SUBPHYLUM UROCHORDATA The 2,500 species of tunicates, sometimes called sea squirts (subphylum Urochordata, from *oura* = tail), float in surface waters or attach to substrates in shallow marine habitats. The sessile adults of many species secrete a gelatinous or leathery "tunic" around their bodies and squirt water through a siphon when disturbed; adults grow to several centimeters **(Figure 32.7).** In the most common group of sea squirts (Ascidiacea), the swimming larvae possess the defining chordate features. Larvae eventually attach to substrates and transform into sessile adults. During metamorphosis, they lose most traces of the notochord, dorsal nerve cord, and tail, and their basketlike pharynx enlarges. In adults, beating cilia pull water into the pharynx through an **incurrent siphon.** A mucous net traps particulate food, which is carried, with the mucus, to the gut. Water passes through the pharyngeal slits, enters a chamber called the **atrium,** and is expelled—along with digestive wastes and carbon dioxide—through the **atrial siphon.** Oxygen is absorbed across the walls of the pharynx.

Vertebrates Possess Several Unique Tissues, Including Bone and Neural Crest

The most distinctive anatomical characteristic of the subphylum Vertebrata is an internal skeleton that provides structural support for muscles and protection for the nervous system and other organs. The skeleton—and the muscles attached to it—also enable most vertebrates to move rapidly through the environment. A vertebrate's skeleton is composed of many separate, bony elements. Indeed, vertebrates are the only animals

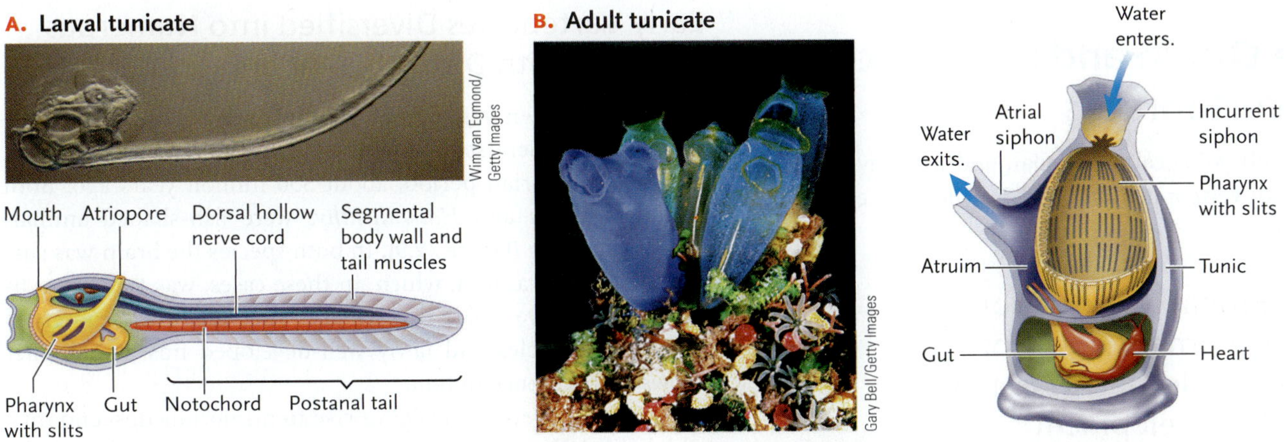

FIGURE 32.7 Urochordates. (A) A tadpolelike tunicate larva metamorphoses into **(B)** a sessile adult. After a larva attaches to a substrate at its anterior end, the tail, notochord, and most of the nervous system are recycled to form new tissues. Slits in the pharynx multiply, the mouth becomes the incurrent siphon, and the atriopore becomes the atrial siphon.

that have **bone,** a connective tissue in which living cells secrete the mineralized matrix that surrounds them (see Figure 38.4D). The **vertebral column,** made up of individual **vertebrae,** surrounds and protects the dorsal nerve cord, and a bony **cranium** surrounds the brain. The cranium, vertebral column, ribs, and sternum (breastbone) make up the **axial skeleton.** Most vertebrates also have a **pectoral girdle** anteriorly and a **pelvic girdle** posteriorly that attach bones in the fins or limbs to the axial skeleton. Bones of the two girdles and the appendages constitute the **appendicular skeleton.** One vertebrate lineage, Chondrichthyes, has lost its bone over evolutionary time; its skeleton is made of cartilage, a dense but flexible connective tissue that is often a developmental precursor of bone (see Section 38.2).

Vertebrates also possess a unique cell type, **neural crest,** which is distinct from endoderm, mesoderm, and ectoderm. Neural crest cells arise next to the developing nervous system, but later migrate throughout a vertebrate's body. They ultimately contribute to many uniquely vertebrate structures, including parts of the cranium, teeth, sensory organs, cranial nerves, and the medulla (that is, the interior part) of the adrenal glands.

Finally, the brains of vertebrates are much larger and more complex than those of invertebrate chordates. Moreover, the vertebrate brain is divided into three regions—the forebrain, midbrain, and hindbrain—each of which governs distinct nervous system functions (see Section 40.1).

STUDY BREAK 32.2

1. **On a field trip to a lake, a college student captures a worm-shaped animal with segmented body-wall muscles. While examining the specimen in the laboratory the following day, she determines that the main nerve cord runs along the ventral side of the animal. Is this animal a chordate?**
2. **What structures distinguish vertebrates from invertebrate chordates?**

32.3 The Origin and Diversification of Vertebrates

Biologists use embryological, molecular, and fossil evidence to trace the origin of vertebrates and to chronicle their evolutionary diversification.

Vertebrates Probably Arose from an Invertebrate Chordate Ancestor through the Duplication of Genes That Regulate Development

Recent genetic sequence studies suggest that vertebrates are more closely related to urochordates than to cephalochordates. Nevertheless, the fishlike form of adult lancelets probably better represents the anatomy of the earliest vertebrates than does the highly specialized morphology of adult tunicates. The evolution of vertebrates from an invertebrate chordate ancestor was marked by the emergence of neural crest, bone, and other typically vertebrate traits. What genetic changes were responsible for these remarkable developments? Biologists now hypothesize that an increase in the number of homeotic—structure determining—genes may have made the development of more complex anatomy possible. (Homeotic genes are described in Sections 16.4, 20.3, and 23.6.)

In animals, one group of homeotic genes, the *Hox* genes, influences the three-dimensional shape of the animal and the locations of important structures—such as eyes, wings, and legs—particularly along the head-to-tail axis of the body. *Hox* genes are arranged on the chromosomes in a particular order, forming what biologists call the *Hox* gene complex (see Section 23.6 and Chapter 36). Each gene in the complex governs the development of particular structures. Animal groups with the simplest structure, such as cnidarians, have two *Hox* genes. Those with more complex anatomy, such as insects, have 10. Chordates have as many as 13 or 14. Thus, lineages with many *Hox* genes generally have more complex anatomy than do those with fewer *Hox* genes.

Molecular analyses also reveal that the entire *Hox* gene complex was duplicated several times in the evolutionary history of vertebrates, producing multiple copies of all its genes **(Figure 32.8).** The cephalochordate *Branchiostoma* has just one *Hox* gene complex, but the most primitive living vertebrates, the jawless hagfishes described later, have two. All vertebrates that possess jaws, a derived characteristic, have at least four sets, and some fishes have as many as seven. Evolutionary developmental biologists hypothesize that after the wholesale duplication of *Hox* genes and other tool-kit genes, the original copies of these genes maintained their ancestral functions, but the duplicate copies assumed *new* functions, directing the development of novel structures, such as the vertebral column and jaws.

Early Vertebrates Diversified into Numerous Lineages with Distinctive Adaptations

The oldest known vertebrate fossils were discovered in the late 1990s, when scientists in China described several species from the early Cambrian period, about 550 million years ago. Both *Myllokunmingia* and *Haikouichthys* were fish-shaped animals about 3 cm long **(Figure 32.9).** In both species the brain was surrounded by a cranium, which, in these cases, was formed of fibrous connective tissue or cartilage. They also had segmented body-wall muscles and fairly well-developed fins, but neither shows any evidence of bone.

The early vertebrates gave rise to numerous descendants, which varied greatly in anatomy, physiology, and ecology. New feeding mechanisms and locomotor structures were often crucial to their success. Today, vertebrates occupy nearly every habitat and feed on virtually all other organ-

CLOSER LOOK — FIGURE 32.8 *Hox* genes and the evolution of vertebrates.

The *Hox* genes in different animal groups appear to be homologous, indicated here by their color and position in the complex.

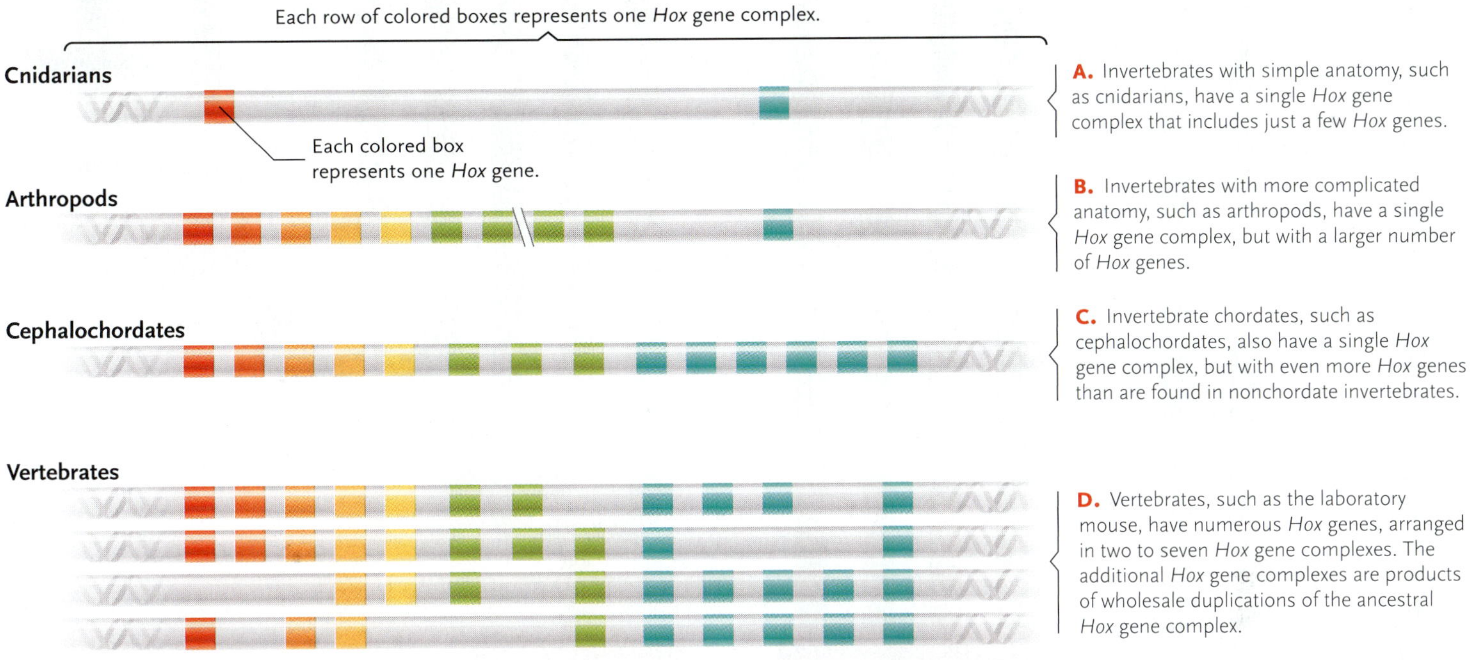

SUMMARY Vertebrates have many more individual *Hox* genes than most invertebrates do because the entire *Hox* gene complex was duplicated in the vertebrate lineage. The additional copies of *Hox* genes probably evolved to specify the development of uniquely vertebrate characteristics, such as the cranium, vertebral column, and neural crest cells.

THINK LIKE A SCIENTIST Given the information available to you in this figure, what change in overall body plan accompanied the evolution of an increased number of *Hox* genes *within* the *Hox* gene complex.

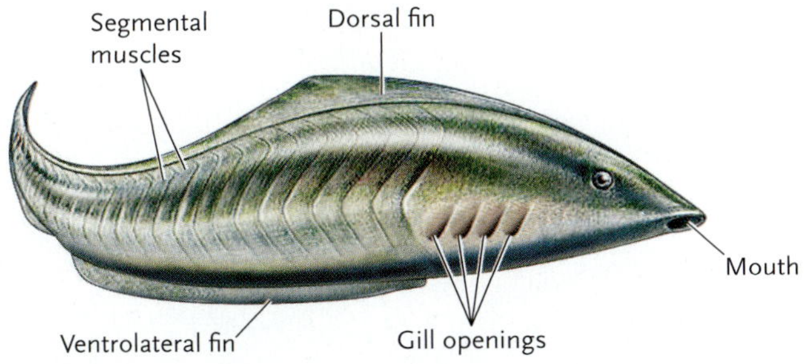

FIGURE 32.9 An early vertebrate. *Myllokunmingia,* one of the earliest vertebrates yet discovered, had no bones; it was about 3 cm long.

isms. Here we briefly introduce the major vertebrate lineages **(Figure 32.10)**.

Although biologists use four key morphological innovations—a cranium, vertebrae, bone, and neural crest cells—to identify vertebrates, these structures did not arise all at once. Instead, they appeared somewhat independently of one another as new groups arose. Some researchers and textbooks present a phylogeny and classification that places the "vertebrates" (animals that have vertebrae) within a larger lineage called the "craniates" (animals that have a cranium). But only one small group, the hagfishes (Myxinoidea, described later), has a cranium but no vertebrae, and some recent molecular analyses do not support its separation from the other vertebrates. Thus, for the sake of simplicity, we describe organisms that possessed any of the four key innovations as "vertebrates."

Several groups of early jawless vertebrates are *described* as "agnathans" (*a* = not, *gnathos* = jaw), but they do not form a monophyletic group (that is, one that includes an ancestor and all of its descendants). Although most became extinct in the Paleozoic era, two ancient lineages, Myxinoidea and Petromyzontoidea, still live today. All other vertebrates possess moveable jaws; they are members of the monophyletic lineage **Gnathostomata** ("jawed mouth"). Some early clades of jawed fishes, the Acanthodii and Placodermi, are now extinct. But three lineages of jawed fishes are still abundant in aquatic

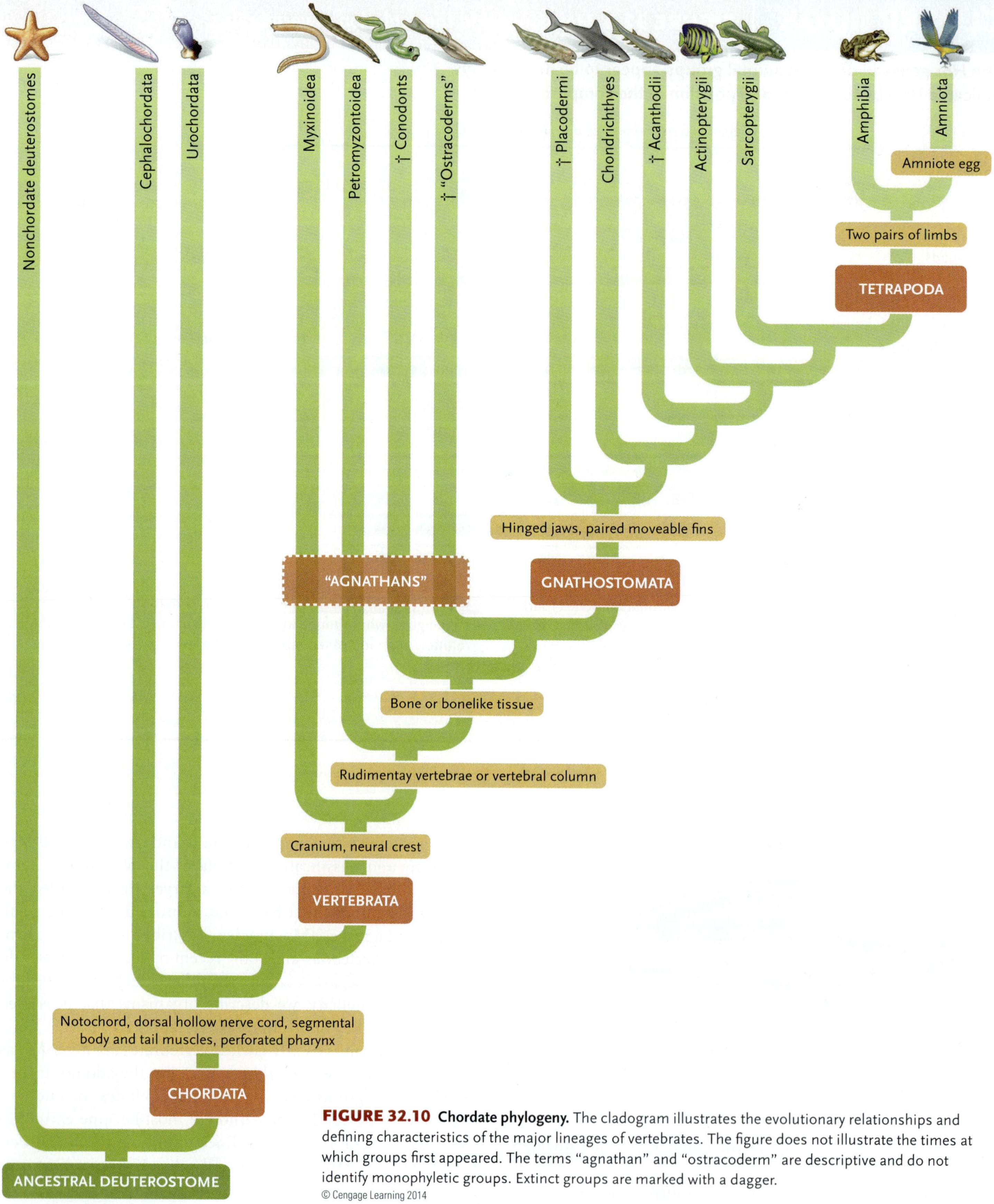

FIGURE 32.10 Chordate phylogeny. The cladogram illustrates the evolutionary relationships and defining characteristics of the major lineages of vertebrates. The figure does not illustrate the times at which groups first appeared. The terms "agnathan" and "ostracoderm" are descriptive and do not identify monophyletic groups. Extinct groups are marked with a dagger.

habitats: the Chondrichthyes, such as sharks and skates, have cartilaginous skeletons; the Actinopterygii and Sarcopterygii have bony skeletons. All jawless vertebrates and jawed fishes are restricted to aquatic habitats, and they use gills to extract oxygen from the water that surrounds them.

The Gnathostomata also includes the monophyletic lineage **Tetrapoda** (*tetra* = four, *pod* = foot); most tetrapods use four limbs for locomotion. Many tetrapods are semiterrestrial or terrestrial, although some, like sea turtles and porpoises, have secondarily returned to aquatic habitats. Adult tetrapods generally use air-breathing lungs for gas exchange. Within the Tetrapoda, one lineage, the amphibians, includes animals, such as frogs and salamanders, that typically need standing water to complete their life cycles. Another lineage, the **Amniota,** comprises animals with specialized eggs that can develop on land. Since their appearance, the amniotes diversified into lineages that include mammals, turtles, lizards, snakes, alligators, and birds. We consider the detailed evolutionary history of the amniotes in Section 32.7.

STUDY BREAK 32.3

1. **How do the *Hox* genes of vertebrates differ from those of cephalochordates?**
2. **Which of the taxonomic groups Amniota, Gnathostomata, and Tetrapoda includes the largest number of species? Which includes the fewest?**

32.4 "Agnathans": Hagfishes and Lampreys, Conodonts and Ostracoderms

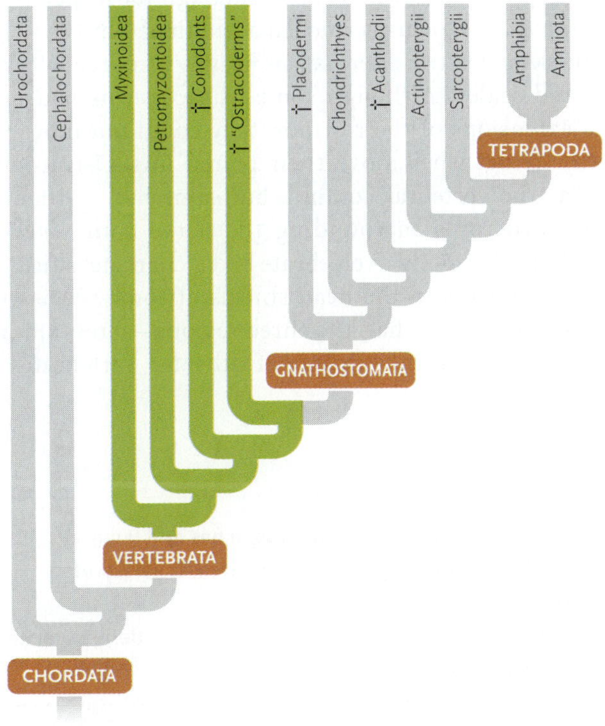

The earliest vertebrates lacked jaws, but they had a muscular pharynx, which they used to suck edible tidbits into their mouths. Although they do not form a monophyletic group, we describe four groups of jawless vertebrates together because their ancestral jawless condition distinguishes them from the monophyletic Gnathostomata. The two living groups of agnathans as well as species that flourished in the Paleozoic vary greatly in size and shape as well as in the number of vertebrate characters they possess.

Hagfishes and Lampreys Are the Living Descendants of Ancient Agnathan Lineages

Two apparently separate lineages of jawless vertebrates, hagfishes (Myxinoidea) and lampreys (Petromyzontoidea), still live today. Both have skeletons composed entirely of cartilage. Although scientists have found no fossilized hagfishes or lampreys from the early Paleozoic era, the absence of both jaws and bone in these groups suggests that their lineages arose early in vertebrate history, before the evolution of bone. Hagfishes and lampreys have a well-developed notochord, but no true vertebrae or paired fins, and their skin has no scales. Individuals grow to a maximum length of about 1 m **(Figure 32.11).**

The axial skeleton of the 60 living species of hagfishes includes only a cranium and a notochord; it has no specialized structures surrounding the dorsal nerve cord. Some biologists

FIGURE 32.11 Living agnathans. (A) Two groups of jawless fishes, hagfishes and lampreys, survive today. **(B)** Lampreys use a toothed oral disk to attach to a host and feed on its blood and soft tissues.

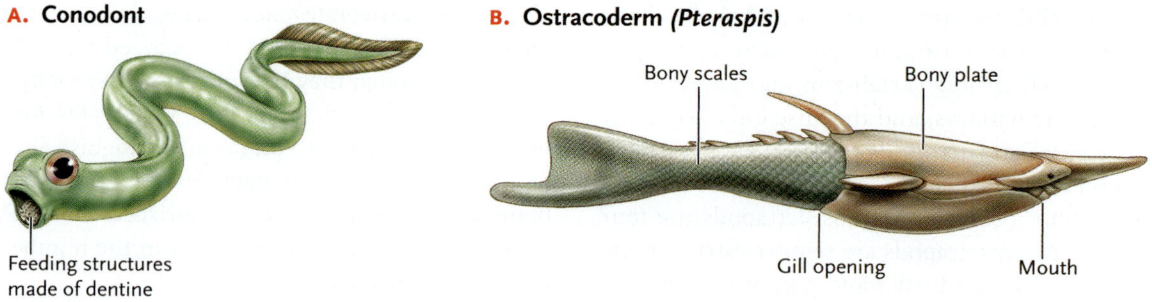

FIGURE 32.12 Extinct agnathans. (A) Conodonts were elongate, soft-bodied animals with bonelike feeding structures in the mouth and pharynx. **(B)** Some ostracoderms had large bony plates on the head and small bony scales on the rest of the body; this species was about 6 cm long.

do not even include hagfishes among the Vertebrata, because they lack any sign of vertebrae. Hagfishes are marine scavengers that burrow in sediments on continental shelves. They feed on invertebrate prey and on dead or dying fishes. In response to predators, they secrete an immense quantity of sticky, noxious slime; when no longer threatened, a hagfish ties itself into a knot and wipes the slime from its body. Hagfish life cycles are simple and lack a larval stage.

The 40 or so living species of lampreys have traces of an axial skeleton. Their notochord is surrounded by dorsally pointing cartilages that partially cover the nerve cord; many biologists hypothesize that this arrangement reflects an early stage in the evolution of the vertebral column. Most lamprey species are parasitic as adults. They have a circular mouth surrounded by a sucking disk with which they attach to a fish or other vertebrate host; they feed on a host's body fluids after rasping through its skin. In most species, sexually mature adults migrate from the ocean or a lake to the headwaters of a stream, where they lay eggs and then die. Their suspension-feeding larvae, which resemble adult cephalochordates, burrow into mud and develop for as long as seven years before undergoing metamorphosis and migrating to the sea or a lake to live as parasitic adults.

Conodonts and Ostracoderms Were Early Jawless Vertebrates with Bony Structures

Mysterious bonelike fossils, most less than 1 mm long, have long been known in oceanic rocks dating from the early Paleozoic era through the early Mesozoic era. Called **conodont** ("cone tooth") elements, these abundant fossils were once described as the support structures of marine algae or the feeding structures of ancient invertebrates. However, recent analyses of their mineral composition reveal that they were made of dentine, a bonelike component of vertebrate teeth. In the 1980s and 1990s, several research teams discovered fossils of intact conodont animals with these elements in place.

Conodonts were elongate, soft-bodied animals; most were 3 to 10 cm long. They had a notochord, cranium, segmental body wall muscles, and large, moveable eyes **(Figure 32.12A).** The conodont elements at the front of the mouth were forward pointing, hook-shaped structures that apparently functioned to collect food; those in the pharynx were stouter, suitable for crushing small items that had been consumed. Paleontologists now classify conodonts as vertebrates—the earliest vertebrates with bonelike structures.

An assortment of jawless fishes, representing several evolutionary lineages and collectively described as **ostracoderms** (*ostrakon* = shell), was abundant from the Ordovician through the Devonian periods **(Figure 32.12B).** Like their invertebrate chordate ancestors, ostracoderms used their pharynx to extract small food particles from mud and water. However, the ostracoderms' muscular pharynx enabled them to *suck* mud and water into their mouths, providing a much stronger flow than the cilia-driven currents of invertebrate chordates. The greater flow rate allowed ostracoderms to collect food more rapidly. It also supported a larger body size: although most species were small, some ostracoderms reached a length of 2 m.

The skin of ostracoderms was heavily armored with bony plates and scales. Although some ostracoderms had paired lateral extensions of their armor, they could not move them the way living fishes move their paired fins. Ostracoderms lacked a true vertebral column, but some had rudimentary support structures surrounding the nerve cord. They also had other distinctly vertebrate-like characteristics. For example, imprints in the head shields of some species indicate that their brains had the three regions—forebrain, midbrain, and hindbrain—typical of all later vertebrates (see Section 40.1).

STUDY BREAK 32.4

1. **What characteristics of the living hagfishes and lampreys suggest that their lineages arose very early in vertebrate evolution?**
2. **What traits in conodonts and ostracoderms are derived relative to those in hagfishes and lampreys?**

FIGURE 32.13
The evolution of jaws.
© Cengage Learning 2014

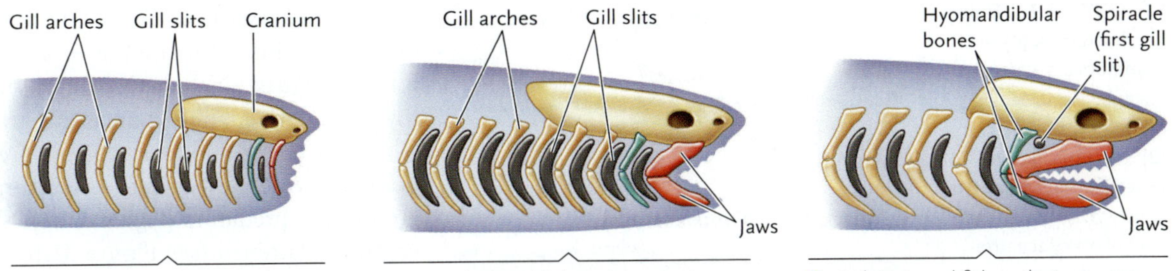

A. Jaws evolved from gill arches in the pharynx of jawless fishes.

B. In early jawed fishes, the upper jaw was firmly attached to the cranium.

C. In later jawed fishes, the jaws were supported by the hyomandibular bones, which were derived from a second pair of gill arches.

32.5 Gnathostomata: The Evolution of Jaws

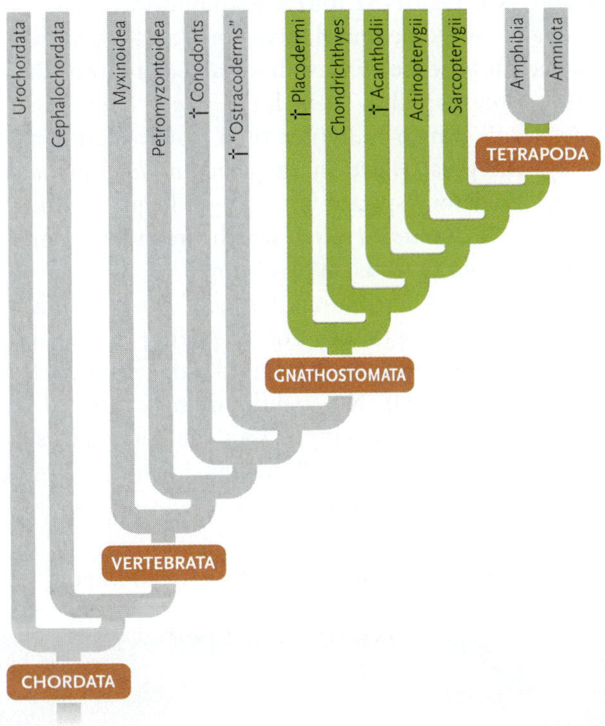

The first gnathostomes were jawed fishes. Key derived traits made their feeding and locomotion more efficient than those of their ancestors.

Jawed Fishes First Appeared in the Paleozoic Era

The renowned anatomist and paleontologist Alfred Sherwood Romer of Harvard University described the evolution of jaws as "perhaps the greatest of all advances in vertebrate history." Hinged jaws allow vertebrates to grasp, kill, shred, and crush large food items. Some living species also use their jaws for defense, for grooming, to construct nests, and to transport their young.

THE ORIGIN OF JAWS AND FINS Embryological evidence suggests that jaws evolved from paired **gill arches**—cartilaginous or bony supporting structures between the gill slits—in the pharynx of a jawless ancestor **(Figure 32.13).** One pair of ancestral gill arches formed bones in the upper and lower jaws, while a second pair was transformed into the hyomandibular bones that braced the jaws against the cranium. Nerves and muscles of the ancestral suspension-feeding pharynx control and move the jaws.

Innovative locomotor mechanisms have often appeared at roughly the same time as innovative feeding mechanisms in the vertebrate lineage; selection probably favored individuals that had the ability to pursue and capture a large range of animal prey as well as feed on them. Thus, many early jawed fishes also had fins. The earliest fins were folds of skin and moveable spines that stabilized locomotion and deterred predators. Moveable fins appeared independently in several lineages, and by the Devonian period, most fishes had unpaired (dorsal, anal, and caudal) and paired (pectoral and pelvic) fins **(Figure 32.14).**

EARLY JAWED FISHES In two early lineages of jawed fishes, spiny sharks and placoderms, the upper jaw was firmly attached to the cranium (see Figure 32.13B); their inflexible mouths simply snapped open and shut **(Figure 32.15).** Both groups also show evidence of an internal skeleton.

Spiny sharks (Acanthodii, *akantha* = thorn), which persisted from the late Ordovician through the Permian periods, were less than 20 cm long. Their small, light scales, streamlined bodies, well-developed eyes, large jaws, and numerous teeth suggest that they were fast swimmers and efficient predators. Most had a row of ventral spines and fins with internal skeletal support on each side of the body. Acanthodian anatomy suggests that they are closely related to the bony fishes living today.

Placoderms (Placodermi, *plax* = flat surface) appeared in the Silurian and diversified in the Devonian and Carboniferous

Andrey Burmakin/Shutterstock.com

FIGURE 32.14 Fish fins. Most fishes have both paired and unpaired fins.

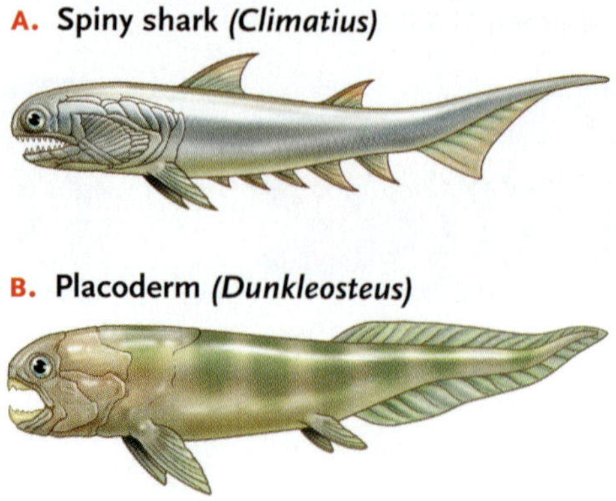

FIGURE 32.15 Early gnathostomes. (A) Most acanthodians were small, reaching a total length of no more than 20 cm. **(B)** Some placoderms were gigantic, up to 10 m in length. Although the jaws of many acanthodians had teeth, the jaws of placoderms had sharp, cutting edges.

periods, but they left no direct descendants. Some, like *Dunkleosteus*, reached a length of 10 m. The anterior part of the placoderm body was covered by heavy bony plates and the posterior part by smaller, bony scales. Their jaws had sharp cutting edges, but not separate teeth, and their paired fins had internal skeletons and powerful muscles.

Chondrichthyes Includes Fishes with Cartilaginous Endoskeletons

The 850 living species in the Chondrichthyes (*khondros* = cartilage, *ikhthys* = fish) have skeletons composed entirely of cartilage, which is much lighter than bone. In most vertebrates, the internal skeleton is first formed of cartilage, but the cartilage is gradually replaced by bone during embryonic development (see Section 38.2). Apparently, some mutation or mutations eliminated the second, bone-forming step in this process when this lineage arose. Because all fishes that lived before the appearance of Chondrichthyes had either bony armor or bony endoskeletons, biologists conclude that the absence of bone in Chondrichthyes is a derived, not an ancestral, trait.

Most living chondrichthyans are grouped in the Elasmobranchii, which includes the skates, rays, and sharks; nearly all are marine predators **(Figure 32.16)**. Skates and rays are dorsoventrally flattened (see Figure 32.16A). They swim by undulating their enlarged pectoral fins. Most are bottom dwellers that often lie partly buried in sand. They feed on hard-shelled invertebrates, which they crush with massive, flattened teeth. The largest species, the manta ray *(Manta birostris)*, which measures 6 m across, feeds on plankton in the open ocean. Some rays have electric organs that stun prey with as much as 200 volts.

Sharks (see Figure 32.16B) are among the ocean's dominant predators. Flexible fins, lightweight skeletons, streamlined bodies, and the absence of heavy body armor allow most sharks to pursue prey rapidly. Their large livers contain copious amounts of oil, which is lighter than water, increasing their buoyancy. The great white shark *(Carcharodon carcharias)* is the largest predatory species, attaining a length of 10 m. The whale shark *(Rhincodon typus)*, which grows to 18 m, is the largest fish known; it feeds on plankton.

Elasmobranchs—including sharks, skates, and rays—exhibit remarkable adaptations for acquiring and processing food. Their teeth develop in whorls under the fleshy parts of the mouth (Figure 32.16D). New teeth migrate forward as old, worn teeth break free. In many sharks, the upper jaw is loosely attached to the cranium, and it swings down during feeding. As the jaws open, the mouth spreads widely, sucking in large,

FIGURE 32.16 Chondrichthyes. (A) Skates and rays as well as **(B)** sharks are grouped in the Elasmobranchii. **(C)** Many shark egg cases include a large yolk that nourishes the developing embryo. **(D)** The teeth of elasmobranchs develop in whorls, with new teeth ready to migrate forward to replace older, worn teeth.

A. Sevruga sturgeon (*Accipenser stellatus*)

B. Long-nosed gar (*Lepisosteus osseus*)

FIGURE 32.17 Primitive actinopterygians. (A) Sturgeons and **(B)** gars are living representatives of early actinopterygian radiations.

hard-to-digest chunks of prey, which are swallowed intact. Although the elasmobranch digestive system is short, it includes a corkscrew-shaped **spiral valve,** which slows the passage of material and increases the surface area available for digestion and absorption.

Elasmobranchs also have well-developed sensory systems. In addition to vision and olfaction, they use **electroreceptors** to detect weak electric currents produced by other animals. And their **lateral-line system,** a row of tiny sensors in canals along both sides of the body, detects vibrations in water (see Figure 41.5).

Chondrichthyans exhibit numerous reproductive specializations. Males have a pair of organs, the **claspers,** on the pelvic fins, which help transfer sperm into the female's reproductive tract. Fertilization occurs internally. In many species, females produce large, yolky eggs with tough, leathery shells (see Figure 32.16C). Others retain the eggs within the oviduct until the young hatch. A few species nourish young within a uterus.

The Actinopterygii and Sarcopterygii Are Fishes with Bony Endoskeletons

In terms of diversity and sheer numbers, the fishes with bony endoskeletons—a cranium, vertebral column with ribs, and bones supporting their moveable fins—are the most successful of all vertebrates. The endoskeleton provides lightweight support, particularly compared with the heavy bony armor of ostracoderms and placoderms, and enhances their locomotor efficiency. Fishes with bony endoskeletons first appeared in the Silurian period and rapidly diversified into two lineages. The ray-finned fishes (Actinopterygii, *aktis* = ray, *pteron* = wing) have fins that are supported by thin and flexible bony rays. The fleshy-finned fishes (Sarcopterygii, from *sarco* = flesh) have fins that are supported by muscles and an internal bony skeleton. Ray-finned fishes have always been more diverse, and they vastly outnumber the fleshy-finned fishes today. The 21,000 living species of bony fishes occupy nearly every aquatic habitat and represent more than 95% of living fish species. Adults range from 1 cm to more than 6 m in length.

Bony fishes have numerous adaptations that increase their swimming efficiency. In many modern ray-finned fishes, a gas-filled **swim bladder** serves as a hydrostatic organ that increases buoyancy (see Figure 32.18A). The swim bladder is derived from an ancestral air-breathing lung that allowed early actinopterygians to gulp air, supplementing their gill respiration in aquatic habitats where dissolved oxygen concentration was low. The scales of most bony fishes are small, smooth, and lightweight. And their bodies are covered with a protective coat of mucus, which retards bacterial growth and smoothes the flow of water.

ACTINOPTERYGII The most primitive living actinopterygians, sturgeons and paddlefishes, have mostly cartilaginous skeletons **(Figure 32.17A).** These large fishes live in rivers and lakes of the northern hemisphere. Sturgeons feed on detritus and invertebrates; paddlefish consume plankton. Gars and bowfins are remnants of a more recent radiation **(Figure 32.17B).** They occur only in the eastern half of North America, where they feed on fishes and other prey. Gars are protected from predators by a heavy coat of bony scales.

Teleosts, the latest radiation of Actinopterygii, are the most diverse, successful, and familiar bony fishes. Evolution has produced a wide range of body forms **(Figure 32.18).** Teleosts have an internal skeleton made almost entirely of bone. On either side of the head, a flap of the body wall, the **operculum,** covers a chamber that houses the gills. Sensory systems generally include large eyes, a lateral-line system, sound receptors, chemoreceptive nostrils, and taste buds. Variations in jaw structure allow different teleosts to consume plankton, seaweed, invertebrates, or other vertebrates.

Teleosts exhibit remarkable feeding and locomotor adaptations. When some teleosts open their mouths, bones at the front of the jaws swing forward to create a circular opening. Folds of skin extend backward, forming a tube through which they suck food (see Figure 32.18F). Many also have symmetrical caudal fins, posterior to the vertebral column, which provide power for locomotion. And their pectoral fins lie high on the sides of the body, providing fine control over swimming. Some species use their pectoral fins for acquiring food, for courtship, and for care of eggs and young. Some teleosts even use them for crawling on land or gliding in air.

Most marine species produce small eggs that hatch into planktonic larvae. Eggs of freshwater teleosts are generally larger and hatch into tiny versions of the adults. In freshwater species, parents often care for their eggs and young, fanning oxygen-rich water over them, removing fungal growths, and

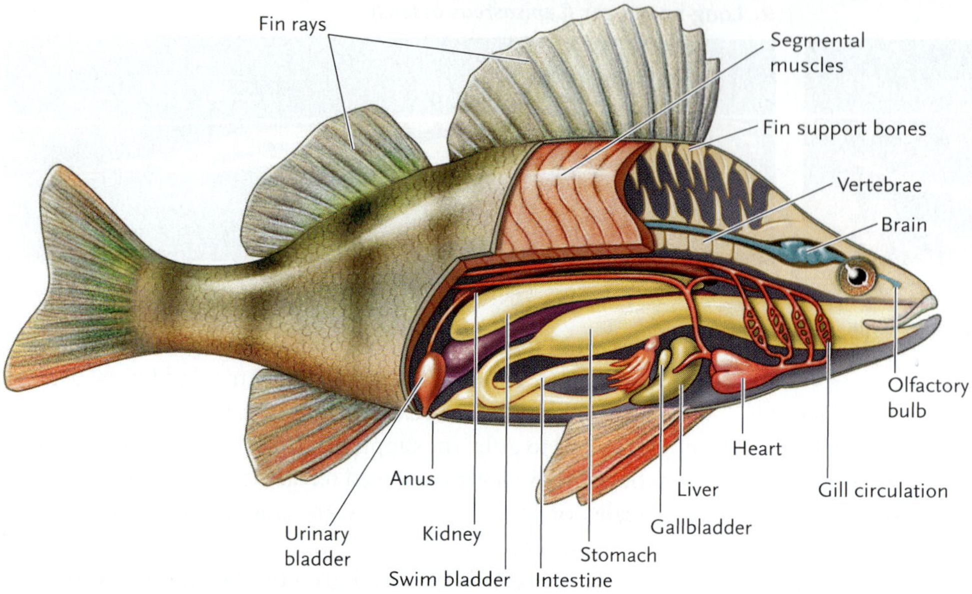

A. Teleost internal anatomy

B. Sea horses, like the thorny sea horse *(Hippocampus histrix)*, use a prehensile tail to hold on to substrates; they are weak swimmers.

C. The long, flexible body of a spotted moray eel *(Gymnothorax moringa)* can wiggle through the nooks and crannies of a reef.

D. Flatfishes, like this European flounder *(Platichthys flesus)*, lie on one side and leap at passing prey.

E. Open ocean predators, like the yellowfin tuna *(Thunnus albacares)*, have strong, torpedo-shaped bodies and powerful caudal fins.

F. Kissing Gouramis *(Helostoma temmincki)* extend their jaws into a tube that sucks food into the mouth.

FIGURE 32.18 Teleost diversity. Although all teleosts share similar internal features, their diverse shapes adapt them to different diets and types of swimming.

A. Coelacanth *(Latimeria chalumnae)*

Peter Scoones/Getty Images

B. African lungfish *(Protopterus annectens)*

Tom McHugh

FIGURE 32.19 Sarcopterygians. (A) Two species of lobe-finned fishes, coelacanths, still live in the oceans today. **(B)** The African lungfish is one of only six living lungfish species.

protecting them from predators. Some freshwater species, such as guppies, give birth to live young.

SARCOPTERYGII Two groups of fleshy-finned fishes (Sarcopterygii), the lobe-finned fishes and lungfishes, are now represented by only eight living species **(Figure 32.19).**

Although lobe-finned fishes were once thought to have been extinct for 65 million years, a living coelacanth *(Latimeria chalumnae)* was discovered in 1938 near the Comoros Islands, off the southeastern coast of Africa. We now know that a population of this meter-long fish lives at depths of 70 to 600 m, feeding on fishes and squid. Remarkably, a second population of coelacanths was discovered in 1998, when a specimen was found in an Indonesian fish market, 10,000 km east of the Comoros population. Based on analyses of its DNA, it is a distinct species *(Latimeria menadoensis).*

Lungfishes have changed relatively little over the last 200 million years. Six living species are distributed on southern continents. The Australian lungfishes, which live in rivers and pools, use their lungs to supplement gill respiration when dissolved oxygen concentration is low. The South American and African species, which live in swamps, use their lungs to collect oxygen during the annual dry season, which they spend encased in a mucus-lined burrow in the dry mud. When the rains begin, water fills the burrow and the fishes awaken from dormancy.

STUDY BREAK 32.5

1. What characteristics of sharks and rays make them more efficient predators than the acanthodians or placoderms?
2. How do the air bladder and fins of ray-finned bony fishes increase their locomotor abilities?
3. How do the lungs of lungfishes allow them to survive in stressful environments?

32.6 Tetrapoda: The Evolution of Limbs

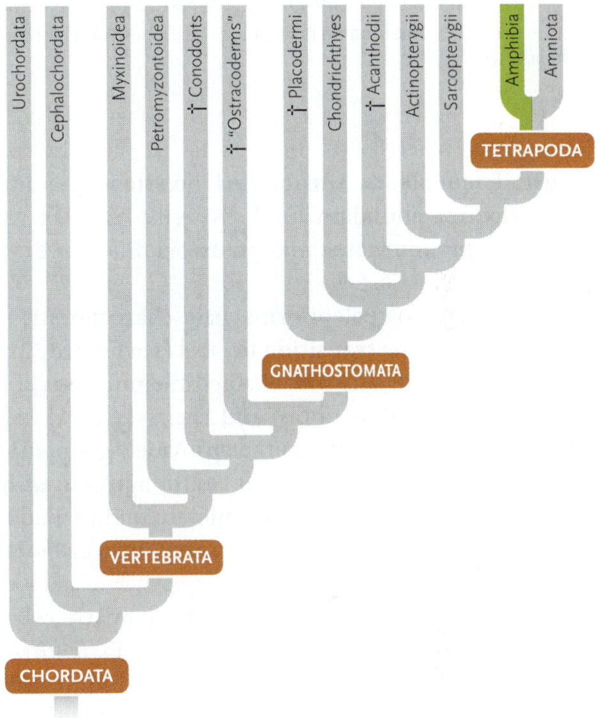

The fossil record suggests that tetrapods arose from a group of fleshy-finned fishes, the *osteolepiforms,* in the late Devonian period, about 380 million years ago. Osteolepiforms and early tetrapods shared several derived characteristics: both had curious infoldings of their tooth surfaces, a trait with unknown function, and the shapes and positions of bones of the skulls and appendages were similar **(Figure 32.20).**

Key Adaptations Facilitated the Transition to Land

Fishes are not adapted to live on land, and the first tetrapods faced serious environmental challenges. First, because air is less dense than water, it provides less support for an animal's body. Second, animals exposed to air inevitably lose body water by evaporation. Third, the sensory systems of fishes, which work well under water, do not function well in air. However, swampy, late-Devonian habitats also offered distinct

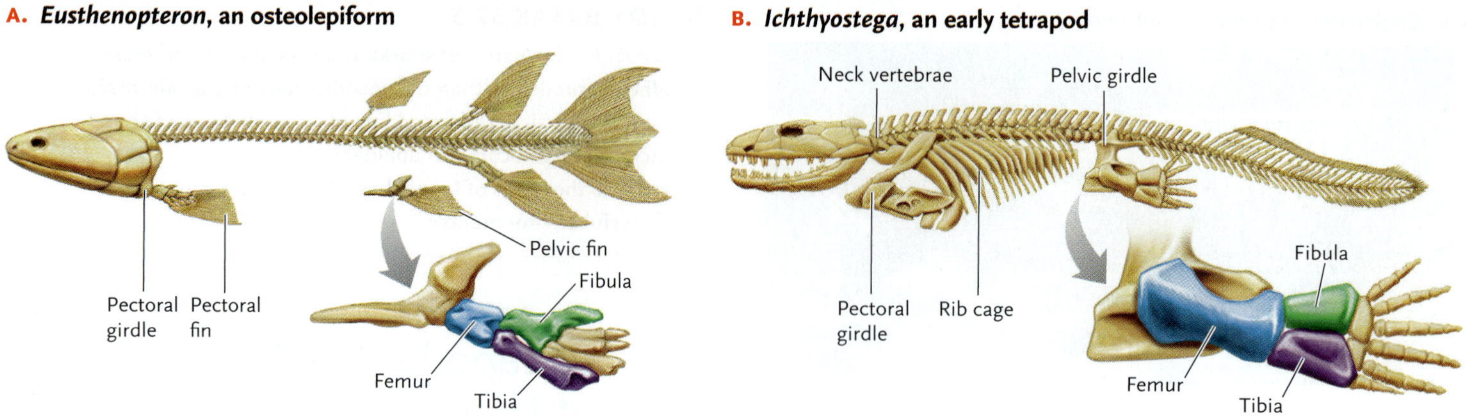

FIGURE 32.20 Evolution of tetrapod limbs. (A) The limb skeleton of osteolepiform fishes is homologous to that of **(B)** early tetrapods. Although *Ichthyostega* retained many fishlike characteristics, its pectoral girdle was completely freed from the cranium and it had a heavy rib cage. Fossils of its forefoot have not yet been discovered.

advantages. Land plants, soft-bodied invertebrates, and arthropods provided abundant food, oxygen was more readily available in air than in water, and predators did not yet live in these new habitats.

In some ways, osteolepiforms had characteristics that may have facilitated the transition to land (see Figure 32.20A). Most had strong, stout fins that enabled them to crawl on the muddy bottom of shallow pools, and their vertebral column included crescent-shaped bones that provided good support. They had nostrils leading to sensory pits that housed olfactory ("smell") receptors. And they almost certainly had lungs to augment gill respiration in the swampy, oxygen-poor waters where they lived.

The earliest tetrapod for which we have nearly complete skeletal data is the semiterrestrial, meter-long *Ichthyostega* (see Figure 32.20B), which lived about 365 million years ago. Compared with its fleshy-finned ancestors, *Ichthyostega* had a stronger vertebral column, sturdier girdles and appendages, a rib cage that protected its internal organs (including lungs), and a neck. In fishes the pectoral girdle is fused to the cranium; thus, fishes have no neck. But in *Ichthyostega,* several vertebrae separated the head from the body, allowing this early tetrapod to move its head to scan the environment and to capture food. However, *Ichthyostega* retained a fishlike lateral-line system, caudal fin, and scaly covering on its body.

Life on land also required changes in sensory systems. In fishes, for example, the body wall picks up sound vibrations and transfers them to sensory receptors directly. But sound waves are harder to detect in air. Early tetrapods developed a **tympanum,** a specialized membrane on either side of the head that is vibrated by airborne sounds. The tympanum connects to the **stapes,** a bone that is homologous to the hyomandibula, which had supported the jaws of fishes (see Figure 24.6). The stapes, in turn, transfers vibrations to the sensory cells of an inner ear.

Modern Amphibians Are Very Different from Their Paleozoic Ancestors

Most of the more than 6,000 species of living amphibians—including frogs, salamanders, and caecelians—are small, and their skeletons contain less bone than those of Paleozoic tetrapods like *Ichthyostega.* All living amphibians are carnivorous as adults, but the aquatic larvae of some species are herbivores.

Most living amphibians have a thin, scaleless skin, well supplied with blood vessels, that is a major site of gas exchange. Because gases must enter the body across a thin layer of water, the skin of most amphibians must remain moist, restricting them to aquatic or wet terrestrial habitats. Adults of some species also acquire oxygen through saclike lungs. The evolution of lungs was accompanied by modifications of the heart and circulatory system that increase the efficiency with which oxygen is delivered to body tissues (see Section 42.1).

The life cycles of many amphibians (*amphi* = of both kinds, *bios* = life) include both larval and adult stages. Eggs are laid and fertilized in water, where they hatch into larvae, such as the tadpoles of frogs, which eventually metamorphose into adults (see Figure 42.8). Although the larvae of many species are aquatic, adults may be aquatic, amphibious, or terrestrial. Some salamanders are paedomorphic; the larval stage attains sexual maturity without changing its form or moving to land. By contrast, some frogs and salamanders reproduce on land and skip the larval stage altogether. But even though they are terrestrial breeders, their eggs dry out quickly unless they are laid in moist places.

Modern amphibians are represented by three lineages **(Figure 32.21)**. Populations of practically all amphibian species have declined rapidly in recent years, probably because of exposure to environmental pollutants and fungal infections (see Chapter 55).

ANURA The 6,400 species of frogs and toads have short, compact bodies, and adults lack tails. Their elongate hind legs and webbed feet allow them to hop on land or swim. A few species

FIGURE 32.21 Living amphibians. (A) Frogs and toads have compact bodies and long hind legs; their larvae (inset), called tadpoles, are legless and rely on their tails for locomotion. **(B)** Salamanders and newts have an elongate body and four legs. **(C)** Caecelians are legless, burrowing amphibians.

are adapted to dry habitats, withstanding periods of drought by encasing themselves in mucous cocoons.

CAUDATA Most of the 580 species of salamanders and newts have an elongate, tailed body and four legs. They walk by alternately contracting muscles on either side of the body much the way fishes swim. Species in the most diverse group, the lungless salamanders, are fully terrestrial throughout their lives, using their skin and the lining of the throat for gas exchange.

GYMNOPHIONA The 170 species of caecelians (Gymnophiona, from *gymnos* = naked, *ophioneos* = snakelike) are legless burrowing amphibians with wormlike bodies. They occupy tropical habitats throughout the world. Unlike other modern amphibians, caecelians have small bony scales embedded in their skin. Fertilization is internal, and females give birth to live young.

STUDY BREAK 32.6

1. **For the first tetrapods, what were the advantages and disadvantages of moving onto the land?**
2. **What parts of the life cycle in most modern amphibians are dependent on water or very moist habitats?**

32.7 Amniota: The Evolution of Fully Terrestrial Vertebrates

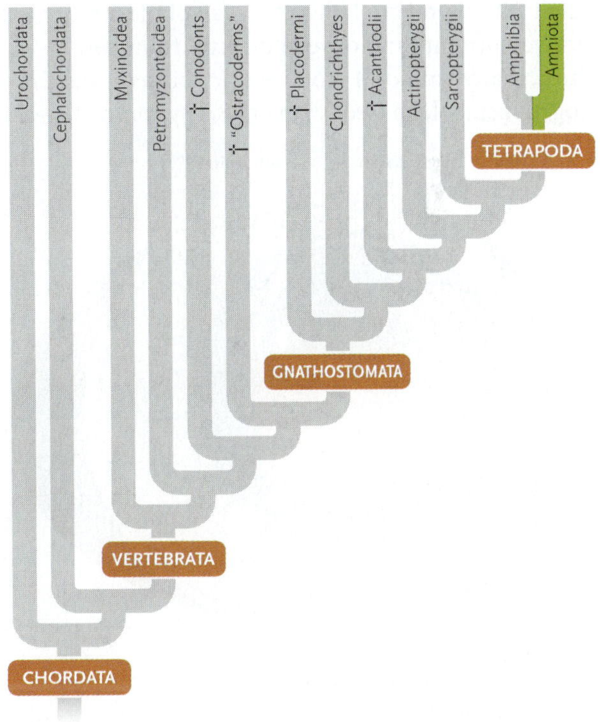

The amniote lineage arose about 320 million years ago, during the Carboniferous period, when seed plants and insects, which served as excellent food resources, began to occupy higher ground. The lineage is named for the amnion, a fluid-filled sac that surrounds the embryo during development.

Key Adaptations Allow Amniotes to Live a Fully Terrestrial Life

Based on the abundance and diversity of their fossils, amniotes were extremely successful, quickly replacing many nonamniote species in terrestrial habitats. Although the fossil record includes abundant skeletons of early amniotes, it provides little direct information about their soft body parts and physiology. For amniotes living today, three key adaptations allow them to live in dry habitats, freeing them from a dependency on moist surroundings and standing water. First, they have a tough, dry skin. The skin cells are filled with keratin and lipids, substances that are relatively impermeable to water. Thus, amniotes do not dehydrate in air as quickly as amphibians do.

Second, many amniotes produce an **amniote egg,** which can survive and develop on dry land. The eggs of modern reptiles and birds have four specialized membranes and a hard or leathery shell **(Figure 32.22).** The membranes protect the developing embryo and facilitate gas exchange and excretion; the shell, which is perforated by microscopic pores, mediates the exchange of air and water between the egg and its environment. The egg also includes generous supplies of **yolk,** the embryo's main energy source, and **albumin,** a source of nutrients and water. Compared with those of amphibians, amniote eggs are large; and, as amniotes lack a larval stage, the young hatch as miniature versions of the adult. In contrast to reptiles and birds, the eggs of virtually all mammals lack a shell. Instead, mammalian embryos, which have the same four membranes associated with them, implant in the wall of the mother's uterus and receive nutrients and oxygen directly from her.

Third, many amniotes produce uric acid as a waste product of nitrogen metabolism (see Chapter 48). By contrast, fishes and amphibians produce ammonium ions or urea, toxic materials that require lots of water to flush them from body tissues. Because uric acid is less toxic than these other compounds, it can be excreted as a semisolid paste, which conserves body water.

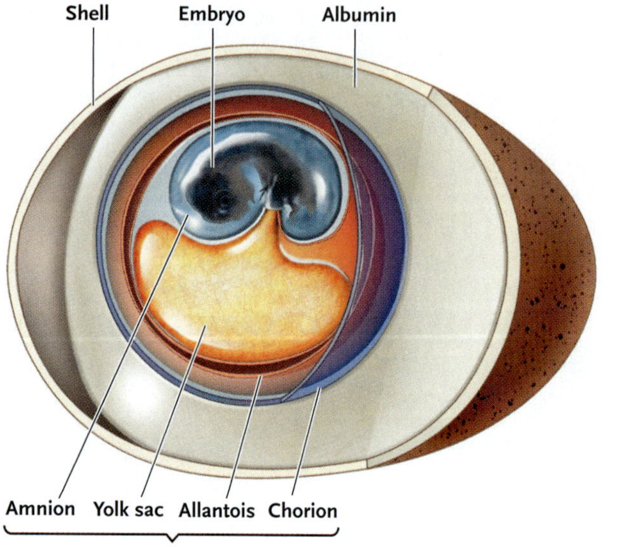

FIGURE 32.22 The amniote egg. A water-retaining egg with four specialized membranes (the amnion, allantois, chorion, and yolk sac) and a hard or leathery shell allowed amniotes and their descendants to reproduce in dry terrestrial environments.

The Fossil Records Suggests That Amniotes Diversified into Three Major Lineages

A traditional view, based largely on the morphology of extinct species, suggests that amniotes differentiated into three lineages during the Carboniferous and Permian periods: synapsids, anapsids, and diapsids **(Figure 32.23).** These three groups are distinguished by differences in the number of openings, called **temporal fenestrae** (*fenestra* = window), and the **bony temporal arches** that border them on the sides of the skull. In the animals that have them, temporal fenestrae provide space for large and powerful jaw muscles. We briefly describe the three traditionally recognized lineages and examine diversification within the diapsids before considering alternative views of amniote relationships.

SYNAPSIDA Nearly all researchers agree that **Synapsida** (*syn* = with; *apsis* = arch), a group of small terrestrial predators, was the earliest lineage to emerge from the ancestral amniotes. They had one temporal fenestra and one temporal arch on either side of the skull. Synapsids appeared late in the Permian period; mammals are their living descendants.

ANAPSIDA According to the traditional view, **Anapsida** ("no arch") was the second amniote lineage to arise; they lacked temporal fenestrae and temporal arches, a condition that scientists recognize as the ancestral amniote skull morphology. Traditionally, biologists hypothesized that turtles are the only living representatives of this lineage, because all extinct and living turtles lack temporal fenestrae and temporal arches.

DIAPSIDA Most Mesozoic amniotes belong to the third lineage, **Diapsida** ("two arches"); these animals had two temporal fenestrae and two temporal arches. Their living descendants include lizards and snakes, crocodilians, and birds. Morphological and genetic sequence data from living species confirm that the diapsids form a monophyletic lineage.

Diapsids Diversified Wildly during the Mesozoic Era

Fossil evidence and molecular sequence data from living species reveal that the early diapsids differentiated into two major lineages, the **Archosauromorpha** (*arkhon* = ruler, *sauros* = lizard, *morphe* = form) and the **Lepidosauromorpha** (*lepis* = scale), which differed in many skeletal characteristics.

The archosauromorphs (commonly called archosaurs), or "ruling reptiles," include crocodilians, pterosaurs, dinosaurs, and birds. Crocodilians, which first appeared during the Triassic period, have bony armor and a laterally flattened tail that propels

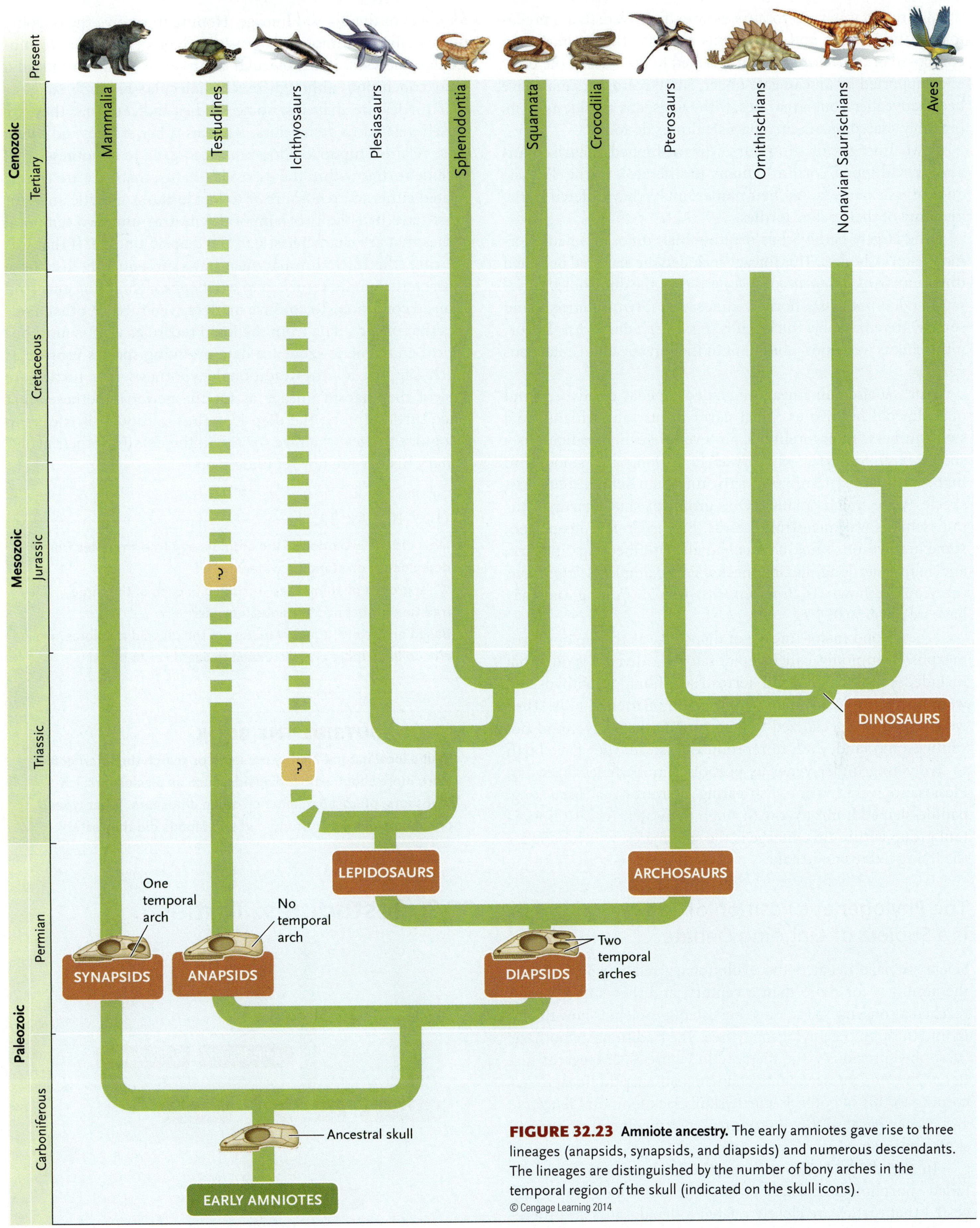

FIGURE 32.23 Amniote ancestry. The early amniotes gave rise to three lineages (anapsids, synapsids, and diapsids) and numerous descendants. The lineages are distinguished by the number of bony arches in the temporal region of the skull (indicated on the skull icons).

them through water. Pterosaurs, now extinct, were flying predators of the Jurassic and Cretaceous periods. Their wings were composed of thin sheets of skin attached to the sides of the body and supported by an elongate finger. Small pterosaurs may have been active fliers, but large ones, with wingspans as wide as 13 m, probably soared on air currents as vultures do today.

Two lineages of dinosaurs, "lizard-hipped" saurischians and "bird-hipped" ornithischians proliferated in the Triassic and Jurassic periods. As their names imply, they differed in the anatomy of their pelvic girdles.

The largely herbivorous ornithischian dinosaurs had enormous, stocky bodies. This lineage included the armored or plated dinosaurs (*Ankylosaurus* and *Stegosaurus*), the duck-billed dinosaurs *(Hadrosaurus),* horned dinosaurs *(Styracosaurus),* and some with remarkably thick skulls *(Pachycephalosaurus).* The ornithischians were most abundant in the Jurassic and Cretaceous periods.

The saurischian lineage included bipedal carnivores and quadrupedal herbivores. Most carnivorous saurischians were swift runners. Their forelimbs, however, were often ridiculously short. *Tyrannosaurus,* which was 15 m long and stood 6 m high, is the most familiar carnivorous saurischian, but most species were much smaller. One group of small carnivorous saurischians was ancestral to birds. By the Cretaceous period, some herbivorous saurischians had also attained gigantic size, and many had long, flexible necks. For example, *Apatosaurus* (previously known as *Brontosaurus*) was 25 m long and may have weighed 50,000 kg.

The second major lineage of diapsids was the lepidosauromorphs (commonly called lepidosaurs), a diverse group that included both marine and terrestrial animals. Ichthyosaurs were porpoise-like animals with laterally flattened tails. They were so highly specialized for marine life that they could not venture onto land, even to reproduce. Instead, they gave birth to live young under water as porpoises and whales do today. Plesiosaurs were marine, fish-eating creatures that used long, paddle-shaped limbs to row through the water. A third important group within this lineage is the squamates, which includes the living lizards and snakes.

The Phylogenetic Position of Turtles Is a Subject of Ongoing Debate

Scientists have debated the evolutionary relationships among the amniotes for more than a century, and they still have not reached a consensus. The most persistent problem is how turtles fit into the amniote phylogenetic tree. The traditional hypothesis (described previously and pictured in Figure 32.23) regards turtles as the living descendants of an ancient anapsid lineage. According to this hypothesis, turtle skulls lack temporal fenestrae and arches because this lineage retained the ancestral anapsid skull condition.

In the 1990s, however, researchers who conducted a detailed morphological analysis of some amniote groups suggested that turtles are closely related to lepidosaurs and therefore part of the diapsid lineage. Hoping to resolve the ongoing dispute about amniote relationships, biologists turned to molecular analyses. But molecular analyses have not produced a clear conclusion: although much molecular research suggests that turtles are diapsids, some studies indicate that they are closely related to archosaurs, whereas other studies indicate a close relationship to lepidosaurs. According to hypotheses that include turtles within the diapsids (that is, that they are closely related either to archosaurs *or* to lepidosaurs), a turtle ancestor must have had the two temporal fenestrae and two temporal arches that are characteristic of the diapsid lineage. If that was the case, the turtle lineage must have subsequently lost those characteristics and returned to what *appears to be* the ancestral anapsid condition. To confuse matters even further, other analyses that include data from the fossil record as well as morphological and genetic sequence data on living species provide as much support for the traditional hypothesis that turtles are part of the anapsid lineage as for the newer hypotheses that place turtles among the diapsids. Thus, although it is easy to recognize turtles when we see them, the details of their evolutionary history are not yet understood.

STUDY BREAK 32.7

1. **How did the evolution of the amniote egg free amniotes from a dependence on standing water?**
2. **What groups of animals are included in each of the three traditionally defined amniote lineages?**
3. **Based on the evolutionary history of the diapsid amniotes, are crocodilians more closely related to lizards or to birds?**

THINK OUTSIDE THE BOOK

Visit a local natural history museum or search the Internet to learn more about extinct diapsids, such as plesiosaurs, ichthyosaurs, pterosaurs, and non-avian dinosaurs. What types of habitats did they occupy, and what foods did they eat?

32.8 Testudines: Turtles

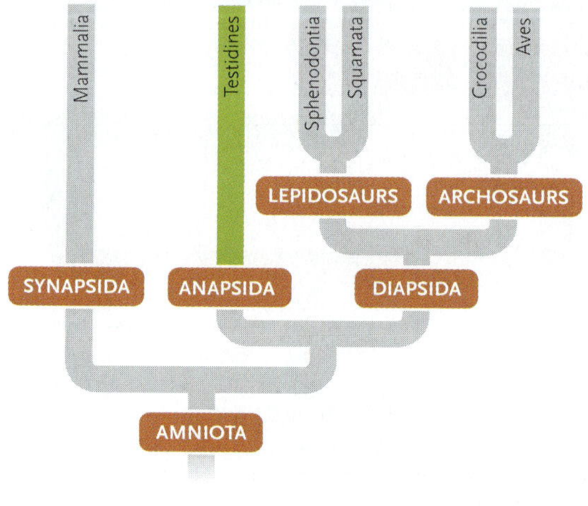

A. The turtle skeleton

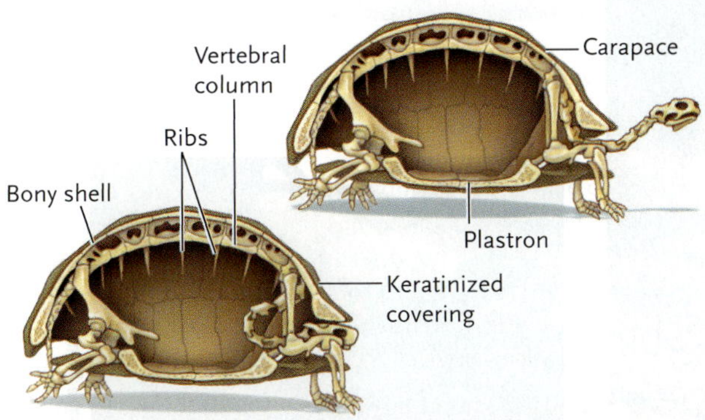

B. Eastern painted turtle (*Chrysemys picta*)

FIGURE 32.24 Testudines. (A) Most turtles can withdraw their heads and legs into a bony shell. **(B)** Aquatic turtles often bask in the sun to warm up. The sunlight may also help eliminate parasites that cling to the turtle's skin.

Turtles Have Bodies Encased in a Bony Shell

The turtle body plan, largely defined by a bony, boxlike shell, has changed little since the group first appeared during the Triassic period **(Figure 32.24).** The shell includes a dorsal **carapace** and a ventral **plastron.** A turtle's ribs are fused to the inside of the carapace, and, in contrast to other tetrapods, the pectoral and pelvic girdles lie *within* the ribcage. Large, keratinized scales cover the bony plates that form the shell.

The 250 living species of turtles occupy terrestrial, freshwater, and marine habitats. They range from 8 cm to 2 m in length. All species lack teeth, but they use a keratinized beak and powerful jaw muscles to feed on plants or animal prey. When threatened, most species retract into their shells. Many species are now highly endangered because adults are hunted for meat, their eggs are consumed by humans and other predators, and their young are collected for the pet trade.

STUDY BREAK 32.8

How does the overall structure of turtles distinguish them from other amniotes?

32.9 Living Lepidosaurs: Sphenodontids and Squamates

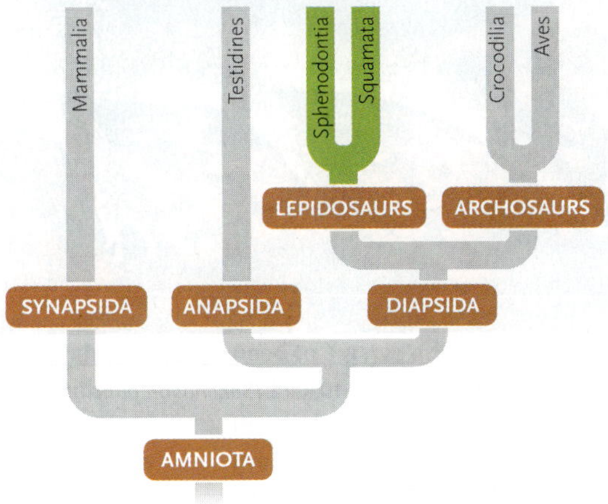

Two groups of lepidosaurs—Sphenodontida and Squamata—still live today. In addition to sharing certain skeletal characteristics, their bodies are encased in a dry, scaly skin.

Living Sphenodontids Are Remnants of a Diverse Mesozoic Lineage

The tuatara *(Sphenodon punctatus)* is one of two living representatives of the sphenodontids (*sphen* = wedge, *odont* = tooth), a diverse Mesozoic lineage **(Figure 32.25A).** These lizard-shaped animals survive on a few islands off the coast of New Zealand. Adults are about 60 cm long. They live in dense colonies, where males and females defend small territories using vocal and visual displays. They often share underground burrows with seabirds, feeding mainly on invertebrates and small vertebrates. They are primarily nocturnal and maintain low body temperatures during periods of activity.

Squamates—Lizards and Snakes—Are Covered by Overlapping, Keratinized Scales

The skin of lizards and snakes (Squamata, *squama* = scale) is composed of overlapping, keratinized scales that protect against dehydration. Squamates periodically shed their skin as they grow, much the way arthropods shed their exoskeletons (see Section 31.7). Most squamates regulate their body temperature behaviorally; they are active only when weather conditions are favorable, and they shuttle between sunny and shady places when they need to warm up or cool down (see Section 48.7).

Most of the 4,800 lizard species are less than 15 cm long **(Figure 32.25B).** However, the Komodo dragon *(Varanus komodoensis)* grows to nearly 3 m. Lizards occupy a wide range of habitats, but they are especially common in deserts and the tropics; one species *(Zootoca vivipara)* occurs within the Arctic Circle. Most lizards feed on insects, although some eat leaves or meat. The diverse tropical genus *Anolis* has be-

FIGURE 32.28
Bird diversity. Birds, the living feathered archosaurs, have diversified into lineages that are adapted to live in different habitats and feed on different foods.

A. The black-bowed albatross (Procellariiformes, *Thalassarche melanophrys*) has the long, thin wings typical of birds that fly great distances.

B. The saddle-billed stork (Ciconiiformes, *Ephippiorhynchus senegalensis*) searches for aquatic prey, which it captures with its long bill.

C. The bald eagle (Falconiformes, *Haliaeetus leucocephalus*) uses its sharp bill and talons to capture and tear apart prey.

D. A tawny frogmouth (Caprimulgiformes, *Podargus strigoides*) uses its wide mouth to capture flying insects.

E. A rufus hummingbird (Apodiformes, *Selasphorus rufus*) uses its long, thin bill to delve into flowers in search of nectar.

F. The black-capped chickadee (Passeriformes, *Poecile atricapillus*) uses its thin bill to probe for insects in dense vegetation.

species have a good sense of smell, which they use to find food. Migrating birds use polarized light, changes in air pressure, and Earth's magnetic field for orientation.

Most birds exhibit complex social behavior, including courtship, territoriality, and parental care. Many species communicate with vocalizations and visual displays to challenge other individuals or attract mates. Most raise their young in a nest, using body heat to incubate eggs. The nest may be a simple depression on a gravelly beach, a cup woven from twigs and grasses, or a feather-lined hole in a tree.

Many bird species embark on a semiannual, long-distance migration (see Section 56.3). The golden plover *(Pluvialis dominica),* for example, migrates 20,000 km twice each year. Migrations are a response to seasonal changes in climate. Birds travel toward the tropics as winter approaches; in spring, they return to high latitudes using seasonally available food sources.

STUDY BREAK 32.10

1. What anatomical and behavioral characteristics of crocodilians demonstrate their relatively close relationship to birds?
2. What specific adaptations allow birds to fly?
3. How do the structures of a bird's bill, wings, and feet reflect its dietary and habitat specializations?

32.11 Mammalia: Monotremes, Marsupials, and Placentals

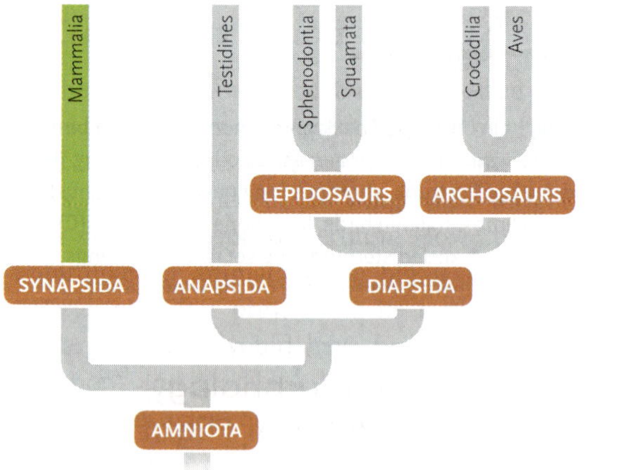

The synapsid lineage, which includes the living mammals, was the first group of amniotes to diversify broadly on land. Indeed, during the late Paleozoic era, medium- to large-sized synapsids were the most abundant predators in terrestrial habitats. One particularly successful and persistent branch of the synapsid lineage, the *therapsids,* exhibited many mammal-like characteristics in their legs, skulls, jaws, and teeth. And by the end of the

A. The turtle skeleton

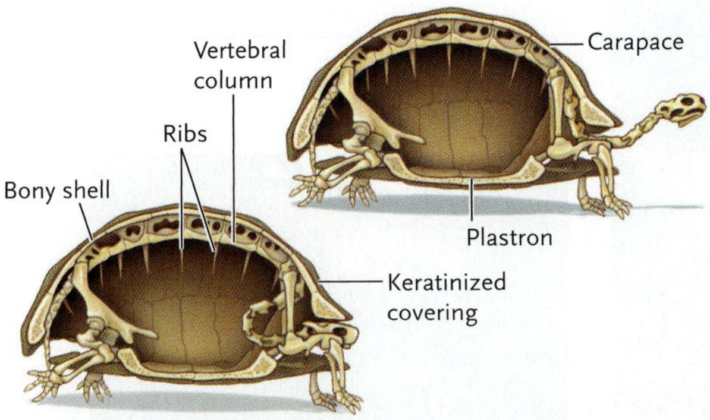

B. Eastern painted turtle (*Chrysemys picta*)

FIGURE 32.24 Testudines. (A) Most turtles can withdraw their heads and legs into a bony shell. **(B)** Aquatic turtles often bask in the sun to warm up. The sunlight may also help eliminate parasites that cling to the turtle's skin.

Turtles Have Bodies Encased in a Bony Shell

The turtle body plan, largely defined by a bony, boxlike shell, has changed little since the group first appeared during the Triassic period **(Figure 32.24).** The shell includes a dorsal **carapace** and a ventral **plastron.** A turtle's ribs are fused to the inside of the carapace, and, in contrast to other tetrapods, the pectoral and pelvic girdles lie *within* the ribcage. Large, keratinized scales cover the bony plates that form the shell.

The 250 living species of turtles occupy terrestrial, freshwater, and marine habitats. They range from 8 cm to 2 m in length. All species lack teeth, but they use a keratinized beak and powerful jaw muscles to feed on plants or animal prey. When threatened, most species retract into their shells. Many species are now highly endangered because adults are hunted for meat, their eggs are consumed by humans and other predators, and their young are collected for the pet trade.

STUDY BREAK 32.8

How does the overall structure of turtles distinguish them from other amniotes?

32.9 Living Lepidosaurs: Sphenodontids and Squamates

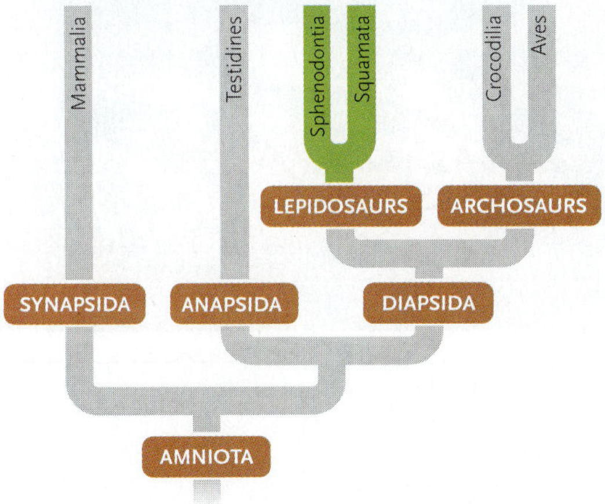

Two groups of lepidosaurs—Sphenodontida and Squamata—still live today. In addition to sharing certain skeletal characteristics, their bodies are encased in a dry, scaly skin.

Living Sphenodontids Are Remnants of a Diverse Mesozoic Lineage

The tuatara *(Sphenodon punctatus)* is one of two living representatives of the sphenodontids (*sphen* = wedge, *odont* = tooth), a diverse Mesozoic lineage **(Figure 32.25A).** These lizard-shaped animals survive on a few islands off the coast of New Zealand. Adults are about 60 cm long. They live in dense colonies, where males and females defend small territories using vocal and visual displays. They often share underground burrows with seabirds, feeding mainly on invertebrates and small vertebrates. They are primarily nocturnal and maintain low body temperatures during periods of activity.

Squamates—Lizards and Snakes—Are Covered by Overlapping, Keratinized Scales

The skin of lizards and snakes (Squamata, *squama* = scale) is composed of overlapping, keratinized scales that protect against dehydration. Squamates periodically shed their skin as they grow, much the way arthropods shed their exoskeletons (see Section 31.7). Most squamates regulate their body temperature behaviorally; they are active only when weather conditions are favorable, and they shuttle between sunny and shady places when they need to warm up or cool down (see Section 48.7).

Most of the 4,800 lizard species are less than 15 cm long **(Figure 32.25B).** However, the Komodo dragon *(Varanus komodoensis)* grows to nearly 3 m. Lizards occupy a wide range of habitats, but they are especially common in deserts and the tropics; one species *(Zootoca vivipara)* occurs within the Arctic Circle. Most lizards feed on insects, although some eat leaves or meat. The diverse tropical genus *Anolis* has be-

FIGURE 32.25 Living lepidosaurs.

Pete & Judy Morrin/Ardea London

A. Sphenodontia includes the tuatara *(Sphenodon punctatus)* and one other species.

Evgeny Murtola/Shutterstock.com

B. Unlike most geckos, the Madagascar giant day gecko *(Phelsuma madagascariensis)* is active during daylight hours.

Photo Researchers/Getty Images

C. A western diamondback rattlesnake *(Crotalus atrox)* of the American southwest bares its fangs with which it injects a powerful toxin into prey.

come a frequent subject of research, as described in *Focus on Model Research Organisms.*

The 2,900 species of snakes evolved from a lineage of lizards that lost their legs over evolutionary time (see Section 20.3). Streamlined bodies make snakes efficient burrowers or climbers **(Figure 32.25C).** Many subterranean species are only 10 or 15 cm long, but the giant constrictors may grow to 10 m. Unlike lizards, all snakes are predators that swallow prey whole. Snakes have thinner skull bones than their lizard ancestors did, and the bones are connected to each other by elastic ligaments that stretch remarkably, allowing some snakes to swallow food that is larger than their heads. Snakes also have well-developed sensory systems for detecting prey. The flicking tongue carries airborne molecules to sensory receptors in the roof of the mouth. Most snakes can detect vibrations on the ground, and some, like rattlesnakes, have heat-sensing organs (see Figure 41.22). Many snakes kill by constriction, which suffocates prey, and several groups produce toxins that immobilize, kill, and partially digest the prey before it is swallowed.

STUDY BREAK 32.9

1. **In addition to losing their legs over evolutionary time, how do snakes differ from their lizard ancestors?**
2. **How do the diets of lizards and snakes differ?**

32.10 Living Archosaurs: Crocodilians and Birds

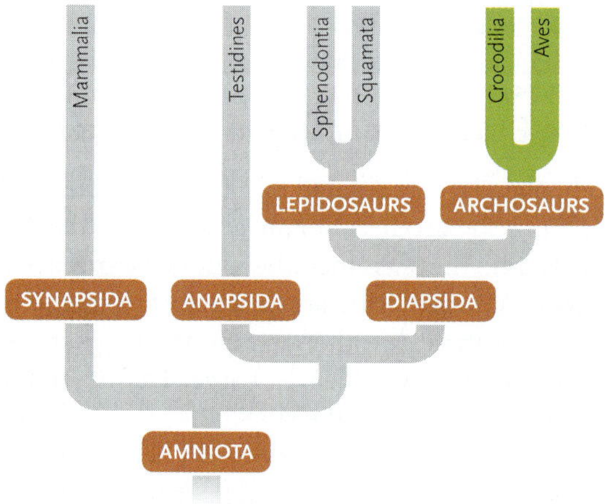

Two groups of archosaurs—Crocodilia and Aves—have survived since the Mesozoic era. Although they bear little resemblance to each other in their external morphology, similarities in their heart and respiratory structures as well as molecular sequence data confirm their close relationship.

Crocodilians Are Semiaquatic, Predatory Archosaurs

In addition to birds, the 21 living species of alligators and crocodiles (Crocodilia, from *crocodilus* = crocodile) are the remnants of a once-diverse archosaur lineage **(Figure 32.26).** The largest

Anolis Lizards of the Caribbean

The lizard genus *Anolis* has been a model system for studies in ecology and evolutionary biology since the 1960s, when Ernest E. Williams of Harvard University's Museum of Comparative Zoology first began studying it. With roughly 400 known species—and new ones being described all the time—*Anolis* is one of the most diverse vertebrate genera. Most anoles are less than 10 cm long, not including the tail, and many occur at high densities, making it easy to collect lots of data in a relatively short time. Male anoles defend territories, and their displays make them conspicuous even in dense forests.

Anolis species are widely distributed in South America and Central America, but nearly 40% occupy Caribbean islands. The number of species on an island is generally proportional to the island's size. Cuba, the largest island, has more than 50 species, whereas small islands have just one or two.

Studies by Williams and others suggest that the anoles on some large islands are the products of independent adaptive radiations. Eight of the 10 *Anolis* species now found on Puerto Rico probably evolved on that island from a common ancestor. Similarly, the seven *Anolis* species on Jamaica shared a common ancestor, which was different from the ancestor of the Puerto Rican species. The anole faunas on Cuba and Hispaniola are the products of several independent radiations on each island.

Williams discovered that these independent radiations had produced similar-looking species on different islands. He developed the concept of the *ecomorph,* a group of species that have similar morphological, behavioral, and ecological characteristics even though they are not closely related within the genus. Williams named the ecomorphs after the vegetation that they commonly used **(Figure).** For example, grass anoles are small, slender species that usually perch on low, thin vegetation. Trunk-ground anoles have chunky bodies and large heads, and they perch low on tree trunks, frequently jumping to the ground to feed. Although the grass anoles or the trunk-ground anoles on different islands are similar in many ways, they are not closely related to each other. Their resemblances are the products of convergent evolution.

Ecomorphs exist because evolutionary processes have accentuated the morphological differences among species that occupy different types of vegetation. Jon Losos of Harvard University has demonstrated that trunk-ground anoles, which have relatively long legs and tails, can run faster on wide surfaces and jump farther than species with relatively short legs. And in nature the trunk-ground anoles run and jump more frequently than the other ecomorphs do.

Different ecomorphs on an island use different parts of their habitats by choosing different perch sites (grass, tree trunks, rocks). When two or more species of the same ecomorph inhabit the same island, they occupy habitats with different temperature and shade conditions (see Figure). For example, in Puerto Rico, one species of trunk-ground anole *(Anolis gundlachi)* occupies cool, shady uplands; another *(Anolis cristatellus)* lives in warm, fairly open lowland habitats; and a third species *(Anolis cooki)* lives in desert habitats. Other species in Puerto Rico exhibit similar differences in their distributions. These differences in geographical distribution and habitat use presumably allow the different species to avoid competition with each other and gain access to the resources they need to survive and reproduce.

Evolutionary processes have also fostered physiological differences that reinforce the ecological separation established by the lizards' use of different habitats. For example, *A. cristatellus* maintains higher body temperatures than *A. gundlachi,* and neither is physiologically adapted to the environment of the other: *A. cristatellus* dies in the high altitude forests where *A. gundlachi* thrives, while *A. gundlachi* suffers heat stress at body temperatures that are typical for *A. cristatellus.*

Researchers throughout the Americas continue to explore the ecology and evolution of anoles. Some unravel their biogeography and systematic relationships; others focus on the ecology of populations and communities; still others study their social behavior or sensory physiology. With so many species distributed across hundreds of Caribbean islands, the lizard genus *Anolis* provides fertile ground for testing hypotheses about nearly every aspect of vertebrate biology.

A. cooki

Manuel Leal, Duke University

A. poncensis

Manuel Leal, Duke University

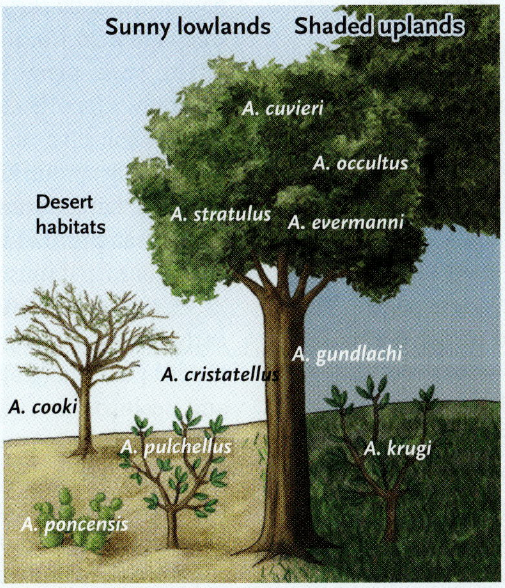

A. gundlachi

Manuel Leal, Duke University

A. krugi

Manuel Leal, Duke University

FIGURE 32.26 Living non-feathered archosaurs. Crocodilia includes semiaquatic predators, like this American alligator (*Alligator mississippiensis*) resting in the Okefenokee Swamp, Georgia.

species, the Australian saltwater crocodile (*Crocodylus porosus*), grows to 7 m. Crocodilians are aquatic predators that consume other vertebrates. Striking anatomical adaptations distinguish them from living lepidosaurs, including a four-chambered heart and a one-way flow of air through their lungs, characters that are homologous to those of birds.

American alligators (*Alligator mississippiensis*) exhibit strong maternal behavior, which also reflects their relationship to birds. Females guard their nests ferociously and help their young break out of their eggshells. The young stay close to the mother for about a year, feeding on scraps that fall from her mouth and living under her watchful protection.

Most alligator and crocodile species are highly endangered. Their habitats have been disrupted by human activities, and they have been hunted for meat and leather. Protection efforts have been extremely successful, however. American alligators, for example, have recovered from the brink of extinction.

Birds Have Key Adaptations That Reduce Body Weight and Power Flight

Birds (Aves) appeared in the Jurassic period as descendants of carnivorous, bipedal dinosaurs. Thus, they are full-fledged members of the archosaur lineage. Their evolutionary relationship to dinosaurs is evident in their skeletal anatomy, the scales on their legs and feet, and their posture when walking. However, powered flight gave birds access to new adaptive zones, setting the stage for their astounding evolutionary success **(Figure 32.27A)**.

The skeletons of birds are both lightweight and strong **(Figure 32.27B)**. For example, the endoskeleton of the frigate bird, which has a 1.5 m wingspan, weighs little more than 100 g, far less than its feathers. Most birds have hollow limb bones with small supporting struts that crisscross the internal cavities. Evolution has also reduced the number of separate bony elements in the wings, skull, and vertebral column (especially the tail), making the skeleton light and rigid. And all modern birds lack teeth, which are dense and heavy; they acquire food with a lightweight, keratinized bill. Many species have a long, flexible neck, which allows them to use their bills for feeding, grooming, nest-building, and social interactions.

The bones associated with flight are generally large. The forelimb and forefoot are elongate, forming the structural support for the wing. And most modern birds possess a **keeled sternum** (breastbone) to which massive flight muscles are attached **(Figure 32.27C)**. Not all birds are strong fliers, however; ostriches and other bipedal runners have strong, muscular legs but small wings and flight muscles (see Figure 20.2).

Like the skeleton, soft internal organs are modified in ways that reduce weight. Most birds lack a urinary bladder; uric acid paste is eliminated with digestive wastes. Females have only one ovary and never carry more than one mature egg; eggs are laid as soon as they are shelled.

All birds also possess **feathers (Figure 32.27D)**, sturdy, lightweight structures derived from scales in the skin of their ancestors. Each feather has numerous barbs and barbules with tiny hooks and grooves that maintain the feathers' structure, even during vigorous activity. Flight feathers on the wings provide lift, contour feathers streamline the surface of the body, and down feathers form an insulating cover close to the skin. Worn feathers are molted and replaced once or twice each year.

Other adaptations for flight allow birds to harness the energy needed to power their flight muscles. Their metabolic rates are eight to ten times higher than those of other comparably sized diapsids, and they process energy-rich food rapidly. A complex and efficient respiratory system (see Figure 46.7) and four-chambered heart (see Figure 44.5D) enable them to consume and distribute oxygen efficiently. As a consequence of high rates of metabolic heat production, birds maintain a high and constant body temperature (see Section 48.8).

Flying Birds Were Abundant by the Cretaceous Period

Although the earliest known bird, the pigeon-sized *Archaeopteryx*, had feathers, its skeleton was essentially that of a small dinosaur (see Figure 20.13). It had digits and claws on the forelimbs, teeth on its jaws, many bones in its wings and vertebral column, and only a poorly developed sternum. How could flight evolve in so unbirdlike an animal? Some biologists hypothesize that *Archaeopteryx* ran after prey, using its feathered wings like fly swatters. Larger wings would have provided extra lift when it jumped at prey, and gradual evolutionary modifications of the wing bones and muscles could have led to powered flight.

Crow-sized birds with full flight capability appeared by the early Cretaceous period. They had a keeled sternum and other modern skeletal features. The modern groups of wading birds and seabirds first appear in late Cretaceous rocks; fossils of other modern groups are found in slightly later deposits. Woodpeckers, perching birds, birds of prey, pigeons, swifts, the flightless ratites, penguins, and some other groups were all present by the end of the Oligocene; birds continued to diversify through the Miocene (see Table 23.1).

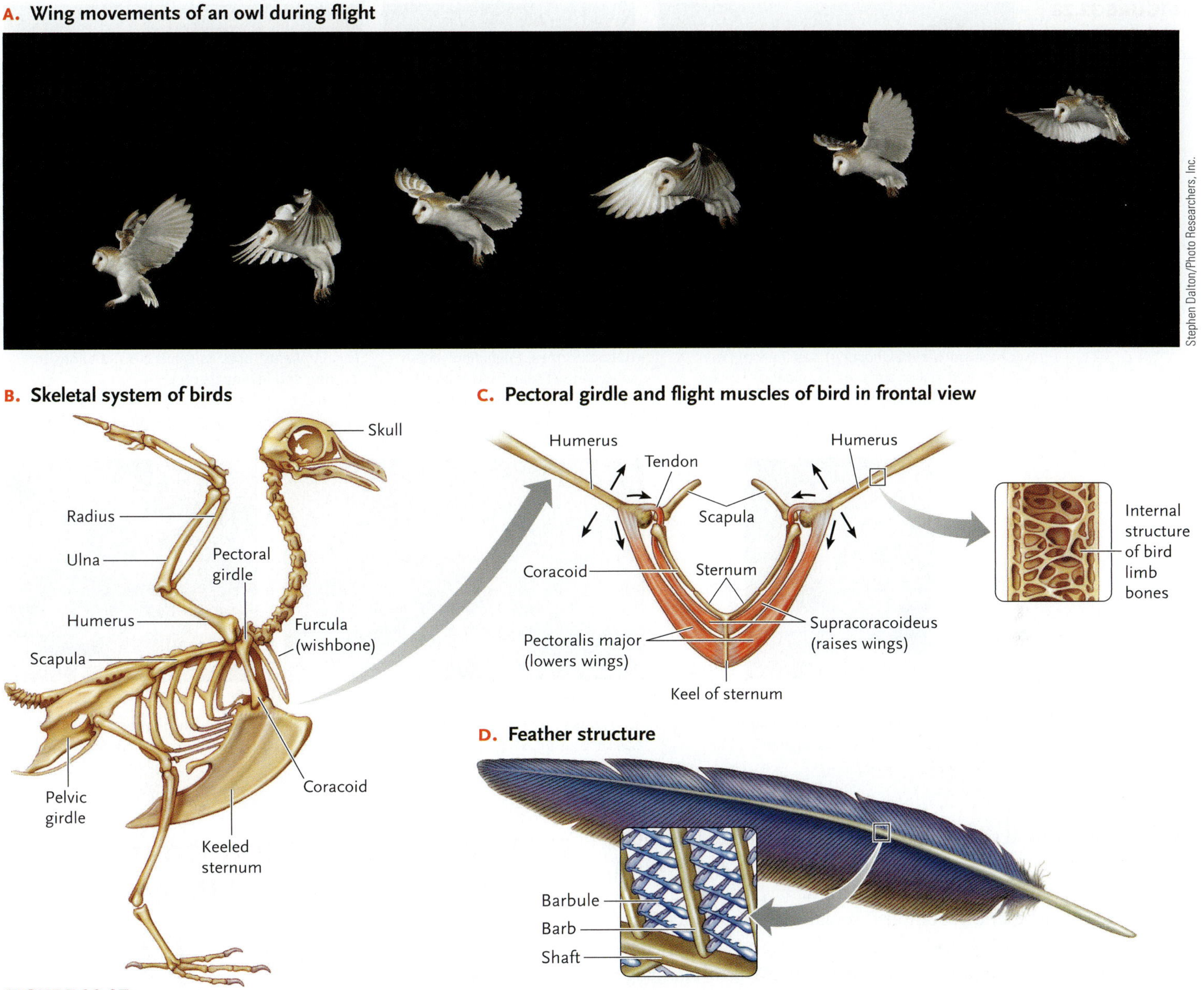

FIGURE 32.27

Adaptations for flight in birds. (A) The flapping movements of a bird's wing provide *thrust* for forward movement and *lift* to counteract gravity. **(B)** The bird skeleton includes a boxlike trunk, short tail, long neck, lightweight skull and beak, and well-developed limbs. In large birds, limb bones are hollow. **(C)** Two sets of flight muscles attach to a keeled sternum; one set raises the wings, and the other lowers it. **(D)** Flexible feathers form an airfoil on the wing surface.

Modern Birds Vary in Morphology, Diet, Habits, and Patterns of Flight

The 9,000 living bird species show extraordinary ecological specializations, but they share the same overall body plan. Living birds are traditionally classified into nearly 30 groups **(Figure 32.28).** They vary in size from the bee hummingbird *(Mellisuga helenae)* of Cuba, which weighs little more than 1 g, to the ostrich *(Struthio camelus),* which can weigh as much as 150 kg.

The structure of the bill usually reflects a bird's diet. Seed and nut eaters, such as finches and parrots, have deep, stout bills that crack hard shells. Carnivorous hawks and carrion-eating vultures have sharp bills to rip flesh, and nectar-feeding hummingbirds have slender bills to reach into flowers. The bills of ducks are modified to extract particulate matter from water, and many perching birds have slender bills to feed on insects.

Birds also differ in the structure of their feet and wings. Predators have large, strong talons ("claws"), whereas ducks and other swimming birds have webbed feet that serve as paddles. Long-distance fliers like albatrosses have narrow wings; those that hover at flowers, such as hummingbirds, have wide ones. The wings of some species, like penguins, are so specialized for swimming that they are incapable of aerial flight.

All birds have well-developed sensory and nervous systems, and their brains are proportionately larger than those of other diapsids of comparable size. Large eyes provide sharp vision, and most species also have good hearing, which nocturnal hunters like owls use to locate prey. Some vultures and other

FIGURE 32.28 **Bird diversity.** Birds, the living feathered archosaurs, have diversified into lineages that are adapted to live in different habitats and feed on different foods.

A. The black-bowed albatross (Procellariiformes, *Thalassarche melanophrys*) has the long, thin wings typical of birds that fly great distances.

B. The saddle-billed stork (Ciconiiformes, *Ephippiorhynchus senegalensis*) searches for aquatic prey, which it captures with its long bill.

C. The bald eagle (Falconiformes, *Haliaeetus leucocephalus*) uses its sharp bill and talons to capture and tear apart prey.

D. A tawny frogmouth (Caprimulgiformes, *Podargus strigoides*) uses its wide mouth to capture flying insects.

E. A rufus hummingbird (Apodiformes, *Selasphorus rufus*) uses its long, thin bill to delve into flowers in search of nectar.

F. The black-capped chickadee (Passeriformes, *Poecile atricapillus*) uses its thin bill to probe for insects in dense vegetation.

species have a good sense of smell, which they use to find food. Migrating birds use polarized light, changes in air pressure, and Earth's magnetic field for orientation.

Most birds exhibit complex social behavior, including courtship, territoriality, and parental care. Many species communicate with vocalizations and visual displays to challenge other individuals or attract mates. Most raise their young in a nest, using body heat to incubate eggs. The nest may be a simple depression on a gravelly beach, a cup woven from twigs and grasses, or a feather-lined hole in a tree.

Many bird species embark on a semiannual, long-distance migration (see Section 56.3). The golden plover *(Pluvialis dominica),* for example, migrates 20,000 km twice each year. Migrations are a response to seasonal changes in climate. Birds travel toward the tropics as winter approaches; in spring, they return to high latitudes using seasonally available food sources.

STUDY BREAK 32.10

1. **What anatomical and behavioral characteristics of crocodilians demonstrate their relatively close relationship to birds?**
2. **What specific adaptations allow birds to fly?**
3. **How do the structures of a bird's bill, wings, and feet reflect its dietary and habitat specializations?**

32.11 Mammalia: Monotremes, Marsupials, and Placentals

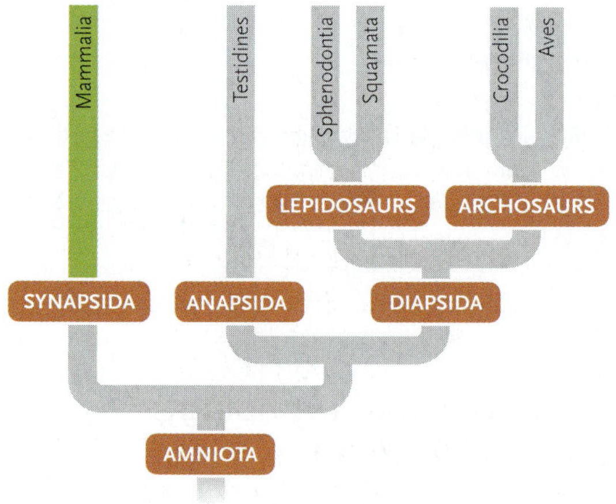

The synapsid lineage, which includes the living mammals, was the first group of amniotes to diversify broadly on land. Indeed, during the late Paleozoic era, medium- to large-sized synapsids were the most abundant predators in terrestrial habitats. One particularly successful and persistent branch of the synapsid lineage, the *therapsids,* exhibited many mammal-like characteristics in their legs, skulls, jaws, and teeth. And by the end of the

Triassic period, the earliest mammals—most of them no bigger than a rat—had emerged from therapsid ancestors. Several mammalian lineages coexisted with dinosaurs and other diapsids throughout the Mesozoic, but paleontologists hypothesize that most Mesozoic mammals were active only at night to avoid predatory dinosaurs, which were active during the day. Two mammalian lineages, the egg-laying Prototheria (or monotremes) and the live-bearing Theria (marsupials and placentals), survived the mass extinction that eliminated most dinosaurs at the end of the Mesozoic. The Theria diversified into the mammalian groups that are most familiar today.

Mammals Exhibit Key Adaptations in Anatomy, Physiology, and Behavior

Four sets of key adaptations fostered the success of mammals.

HIGH METABOLIC RATE AND BODY TEMPERATURE Like birds, mammals have high metabolic rates that release enough energy from food to maintain high activity levels and enough heat to maintain high body temperatures (see Section 48.8). An outer covering of fur and a layer of subcutaneous fat help retain body heat. Using metabolic heat to stay warm requires lots of oxygen, and mammals have a muscular organ, the diaphragm, that fills their lungs with air (see Figure 46.8). Four-chambered hearts and complex circulatory systems deliver oxygen to active tissues (see Figure 44.5D).

SPECIALIZATIONS OF THE TEETH AND JAWS Mammals also have anatomical features that allow them to feed efficiently. Ancestrally, mammals have four types of teeth (see Figure 47.18): flattened **incisors** nip and cut food; pointed **canines** pierce and kill prey; and two sets of cheek teeth, **premolars** and **molars,** grind and crush food. Moreover, teeth in the upper and lower jaws occlude (that is, fit together) tightly as the mouth is closed; thus, mammals can use their large jaw muscles to chew food thoroughly. The occlusion of premolars and molars allows some mammals to feed on exceptionally tough plant material, such as grasses and twigs.

PARENTAL CARE Mammals provide substantial parental care to their young. In most species, young complete development within a female's uterus, deriving nourishment through the **placenta,** a specialized organ that mediates the delivery of oxygen and nutrients (see Section 49.2). Females also have **mammary glands,** specialized structures that produce energy-rich milk, a watery mixture of fats, sugars, proteins, vitamins, and minerals. This perfectly balanced diet is the sole source of nutrients for newborn offspring.

COMPLEX BRAINS Finally, mammals have larger brains than other tetrapods of equivalent body size; the difference lies primarily in the **cerebral cortex,** the part of the forebrain responsible for information processing and learning (see Figure 40.6). Most mammals have a keen sense of olfaction; as *Insights from the Molecular Revolution* explains, a huge variety of olfactory receptor proteins send information to the parts of the cerebral cortex that allow mammals to sense and interpret a wide variety of odors. Extensive postnatal care provides opportunities for offspring to learn from older individuals. Thus, mammalian behavior is strongly influenced by past experience as well as by genetically programmed instincts.

The Major Groups of Modern Mammals Differ in Their Reproductive Adaptations

Biologists recognize a primary distinction between two lineages of modern mammals: the egg-laying Prototheria (*protos* = first, *therion* = wild beast), or monotremes, and the live-bearing Theria. The Theria, in turn, diversified into two sublineages: the

A. Short-nosed echidna *(Tachyglossus aculeatus)*

© FLPA/Alamy

B. Duck-billed platypus *(Ornithorhynchus anatinus)*

Jean Phillipe Varin/Jacana/Photo Researchers, Inc.

FIGURE 32.29 Monotremes. (A) The short-nosed echidna is terrestrial. **(B)** The duck-billed platypus raises its young in a streamside burrow.

FIGURE 32.30 Marsupials. An Eastern gray kangaroo *(Macropus giganteus)* carries her "joey" in her pouch.

Metatheria (*meta* = between), or marsupials, and the Eutheria (*eu* = true), or placentals, which also differ in their reproductive adaptations.

MONOTREMES The three living species of monotremes (Prototheria), which are limited to the Australian region, reproduce with a leathery-shelled egg **(Figure 32.29).** Newly hatched young lap up milk secreted by modified sweat glands on the mother's belly. The two species of echidnas, or spiny anteaters (*Tachyglossus aculeatus* and *Zaglossus bruijni*), feed on ants or termites. The duck-billed platypus *(Ornithorhynchus anatinus)* lives in burrows along riverbanks and feeds on aquatic invertebrates.

MARSUPIALS The 240 species of marsupials (Metatheria) have short gestation; the young are nourished through a placenta very briefly—sometimes only for 8 to 10 days—before birth. Newborns use their forelimbs to drag themselves across the mother's belly fur and enter her abdominal pouch, the **marsupium,** where they complete development attached to a teat. Marsupials are the dominant native mammals of Australia and a minor component of the South American fauna **(Figure 32.30);** only one marsupial species, the opossum *(Didelphis virginiana),* occurs in North America.

PLACENTALS The 4,000 species of placental mammals (Eutheria) are the dominant mammals today. They complete embryonic development in the mother's uterus, nourished through a placenta until they reach a fairly advanced stage of development. Some species, like humans, are helpless at birth, but others, such as horses, are quickly mobile.

Biologists divide eutherians into about 18 groups, traditionally identified as orders, only eight of which include more than 50 living species **(Figure 32.31).** Rodents (Rodentia) make up about 45% of eutherian species, and bats (Chiroptera) comprise another 22%. Our own group, Primates, is represented by fewer than 250 living species (less than 5% of all mammalian species), many of which are highly endangered.

Some eutherians have highly specialized locomotor structures. For example, whales and dolphins (Cetacea) are descended from terrestrial ancestors, but their appendages do not function on land; they are now restricted to aquatic habitats. By contrast, seals and walruses (Carnivora) feed underwater but rest and breed on land. Bats (Chiroptera) use wings for powered flight.

Although early mammals appear to have been insectivorous, the diets of modern eutherians are diverse. Odd-toed ungulates (*ungula* = hoof) like horses and rhinoceroses (Perissodactyla), even-toed ungulates like cows and camels (Artiodactyla), and rabbits and hares (Lagomorpha) feed on vegetation. Lions and wolves (Carnivora) consume other animals. Many hedgehogs, moles, and shrews (Lipotyphla) and bats eat insects, but some feed on flowers, fruit, and nectar. Many whales and dolphins prey on fishes and other animals, but some eat plankton. And some groups, including rodents and primates, feed opportunistically on both plant and animal matter.

STUDY BREAK 32.11

1. **During the Mesozoic era, why were most mammals active only at night?**
2. **Which key adaptations in mammals allow them to be active under many types of environmental conditions?**
3. **On what basis are the three major groups of living mammals distinguished?**

32.12 Nonhuman Primates

We now focus our attention on Primates, the mammalian lineage that includes humans, apes, monkeys, and their close relatives. The first Primates appeared early in the Eocene epoch, about 55 million years ago, in forested habitats in North America, Europe, Asia, and North Africa.

Key Derived Traits Enabled Primates to Become Arboreal, Diurnal, and Highly Social

Several derived traits allow primates to be arboreal (to live in trees rather than on the ground). For example, most primates have a more erect posture than other mammals, and they have flexible hip and shoulder joints, which allow a variety of locomotor activities. They can grasp objects with their hands and feet, because they have nails, not claws, on their fingers and toes; their fingertips are well endowed with sensory nerves that enhance the sense of touch. Unlike other mammals, most primates have an opposable big toe, which can touch the tips of other digits and the sole of the foot; many species also have an opposable thumb.

Most primates are diurnal (active during daylight hours), and, unlike most mammals, they rely more on vision than on their sense of smell (see *Insights from the Molecular Revolution*). Thus, they generally have short snouts and small olfactory lobes of the brain. Most species have forward-facing eyes with overlapping fields of vision, providing excellent depth perception, which comes in handy when moving through trees. Many species have color vision.

Primate brains—especially the regions that integrate information—are large and complex. As a result, they have an exceptional capacity to learn. Most species live in social groups; thus, young primates, which mature slowly, can interact with and learn from their elders and peers during an extended period of parental care. Females give birth to only one or two young at a time, allowing them to devote substantial attention to each offspring.

INSIGHTS FROM THE Molecular Revolution

Building a Better Nose: How did the sense of smell evolve in mammals?

The sense of olfaction (that is, smell) in vertebrates is made possible by olfactory receptor proteins (ORPs) in specialized cells in their nasal sinuses (see Section 41.5). Humans express hundreds of different ORPs, each of which binds specific airborne molecules, thereby allowing us to detect the presence of many different molecules. Hundreds of different genes—all of which are members of one large multi-gene family of g-protein-coupled receptors (see Chapters 7 and 20)—code for hundreds of different ORPs. The vertebrate sense of smell evolves when new genes arise through gene duplication and when old genes are lost as mutations turn them into pseudogenes (see Section 19.4).

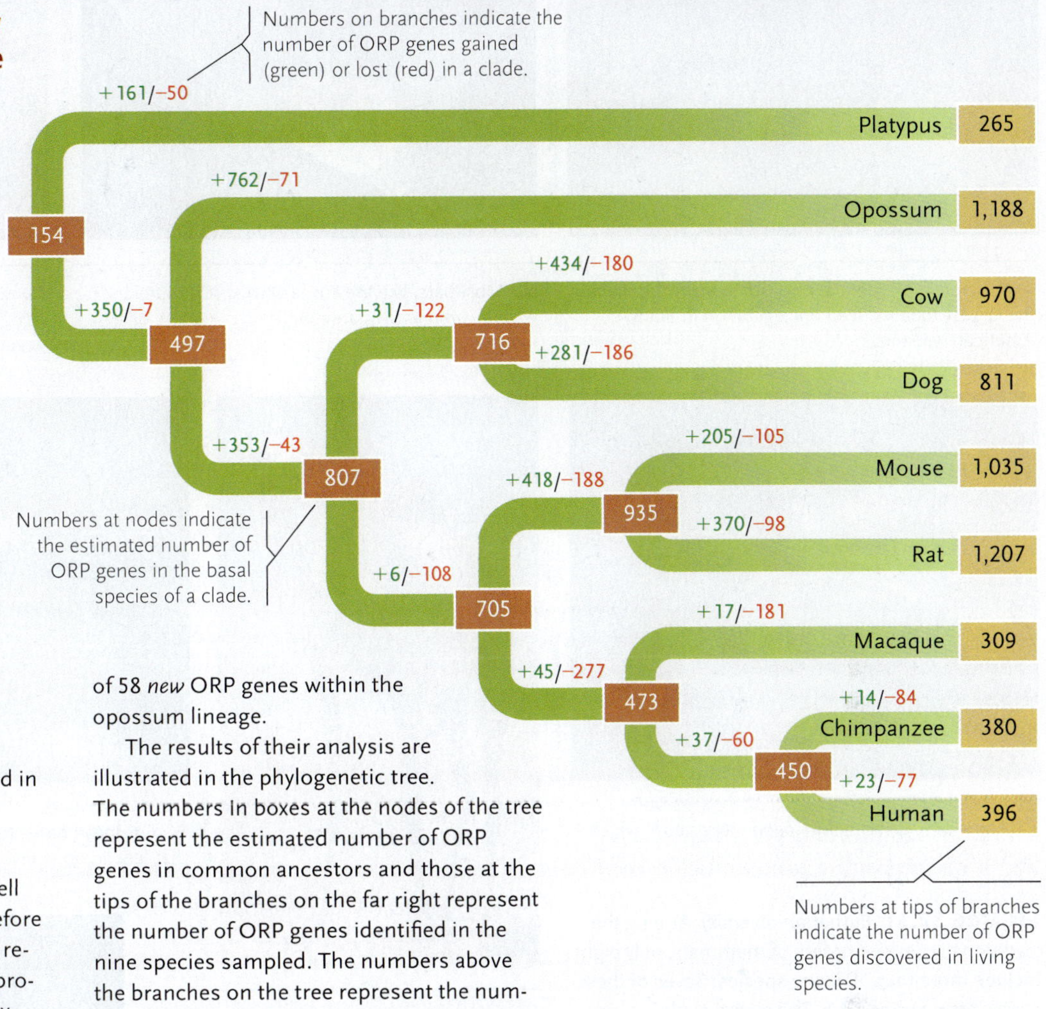

Research Question

What new olfactory receptor genes evolved in different groups of mammals?

Experiment

Mammals rely more on their sense of smell than do most other vertebrates. It is therefore reasonable to suppose that new olfactory receptor genes evolved in early mammals, producing their acute olfaction. To test this hypothesis, Yoshihito Niimura and coworkers searched a range of mammalian genomes for representatives of the multi-gene family coding for ORPs. Their work examined the genomes of one monotreme (platypus), one marsupial (opossum) and seven placentals (dog, cow, mouse, rat, macaque monkey, chimpanzee, and human).

The researchers made multiple pair-wise comparisons of the ORP gene sequences in the nine mammal species to construct phylogenetic trees of ORP genes; these phylogenetic trees illustrate which genes are more closely related to each other. By comparing the ORP gene phylogenetic trees with a phylogenetic tree for the nine mammal species, they were able to estimate how many new ORP genes had evolved in each mammalian lineage and how many ORP genes had been lost through mutation. For example, the opossum genome contains one family of 59 ORP genes that are more closely related to one another than they are to ORP genes in any of the other species. The most parsimonious explanation for this discovery is that a series of gene duplications led to the evolution of 58 *new* ORP genes within the opossum lineage.

The results of their analysis are illustrated in the phylogenetic tree. The numbers in boxes at the nodes of the tree represent the estimated number of ORP genes in common ancestors and those at the tips of the branches on the far right represent the number of ORP genes identified in the nine species sampled. The numbers above the branches on the tree represent the number of new ORP genes that evolved (in green) and the number of ORP genes lost to mutation (in red) within each lineage.

Many new ORP genes evolved in the lineage that was ancestral to both placental and marsupial mammals. After those clades diverged, new ORP genes continued to evolve in many lineages. The rodents (represented in the study by mouse and rat) experienced the evolution of a strikingly large number of new ORP genes. By contrast, the primates (represented in the study by macaque, chimpanzee, and human) experienced a net loss of ORP genes through mutation.

Conclusion

These findings reflect the relative importance of olfaction in different groups of mammals. The platypus is thought to rely less on olfaction than other mammals do, perhaps because they use electroreceptors in their bill to find food (see Section 41.7); this observation is consistent with the relatively few new ORP genes that evolved in that lineage. The evolution of nocturnal activity in early mammals is thought to have increased the need for olfactory information, an observation that is consistent with the evolution of many new ORP genes in the marsupial and placental lineages. In fact, of the 1,035 ORP genes in mice, at least 783 appeared after the emergence of mammals, suggesting that most of a mouse's ability to smell has been acquired during the past 200 million years. The numbers of ORP genes lost outpaced the numbers of ORP genes gained only among the primates, as members of that lineage began to rely increasingly on their sense of vision. A decreased reliance on olfaction is typical of all primates, including humans.

THINK LIKE A SCIENTIST Early primates were almost certainly arboreal, leaping or climbing from one branch to another in trees. How would this behavior favor the evolution of their increased reliance on vision and decreased reliance on olfaction?

Sources: Y. Niimura and M. Nei. 2007. Extensive gains and losses of olfactory receptor genes in mammalian evolution. *PLoS one* 2(8):e708. A. Matsui et al. 2010. Degeneration of olfactory receptor gene repertories in primates: no direct link to full trichromatic vision. *Mol. Biol. Evol.* 27:1192-1200. The figure represents a new analysis of data from both the 2007 and 2010 publications, kindly provided by Dr. Niimura.

Theo Allofs/Photonica/Getty Images Inc.

A. Capybaras (Rodentia, *Hydrochoerus hydrochaeris*), the largest rodents, feed on vegetation in South American wetlands.

J. Scott Altenbach, University of New Mexico

B. Most bats, like the Yuma Myotis (Chiroptera, *Myotis yumanensis*), are nocturnal predators of insects.

Vladimir Melnik/Shutterstock.com

C. Walruses (Carnivora, *Obodenus rosmarus*) feed primarily on marine invertebrates in frigid arctic waters.

Paul Nicklen/Getty Images

D. Narwhals (Cetacea, *Monodon monoceros*), which have one incisor that grows into a spiral tusk, may be a source of unicorn legends.

David Parker/SPL/Photo Researchers, Inc.

E. Arabian camels (Artiodactyla, *Camelus dromedarius*) use enlarged foot pads to cross hot desert sands.

iStockphoto.com/DAMIAN KUZDAK

F. European hares (Lagomorpha, *Lepus europaeus*) are herbivores that feed in open fields and grasslands.

© David Cole/Alamy

G. Common moles (Lipotyphla, *Talpa europaea*) feed largely on insects that they find underground.

FIGURE 32.31 Eutherian diversity. Among the traditionally defined orders of mammals, only eight include more than 50 living species. Seven of these clades are pictured here. The eighth clade, Primates, is described later in this chapter.

Living Primates Include Two Major Lineages

Primatologists recognize two lineages within the Primates **(Figure 32.32),** the Strepsirhini (*streptos* = twisted or turned, *rhinos* = nose) and the Haplorhini (*haplos* = single or simple).

STREPSIRHINI The 36 living species of Strepsirhini—lemurs, lorises, and galagos—possess many ancestral morphological traits, including moist, fleshy noses and eyes that are positioned somewhat laterally on their heads **(Figure 32.33).** Strepsirhines generally have short gestation periods and rapid maturation. Today, 22 lemur species survive on Madagascar, a large island off the east coast of Africa; they are ecologically diverse and range in size from 40 g to 7 kg. Some lemurs are arboreal, whereas others spend substantial time on the ground. The 12 species of lorises and galagos occupy tropical forest and subtropical woodlands in Africa, India, and Southeast Asia; they are all arboreal and nocturnal.

HAPLORHINI Most species in the Haplorhini—the familiar monkeys and apes—have many derived primate characteristics, including compact, dry noses, and forward-facing eyes.

However, five species of tarsiers, which are restricted to tropical forests on the islands of Southeast Asia, exhibit several ancestral traits: small body size (about 100 g), large eyes and ears, and two grooming claws on each foot **(Figure 32.34).** But they share the derived characteristics of dry noses and forward-facing eyes with the other haplorhines; DNA sequence data

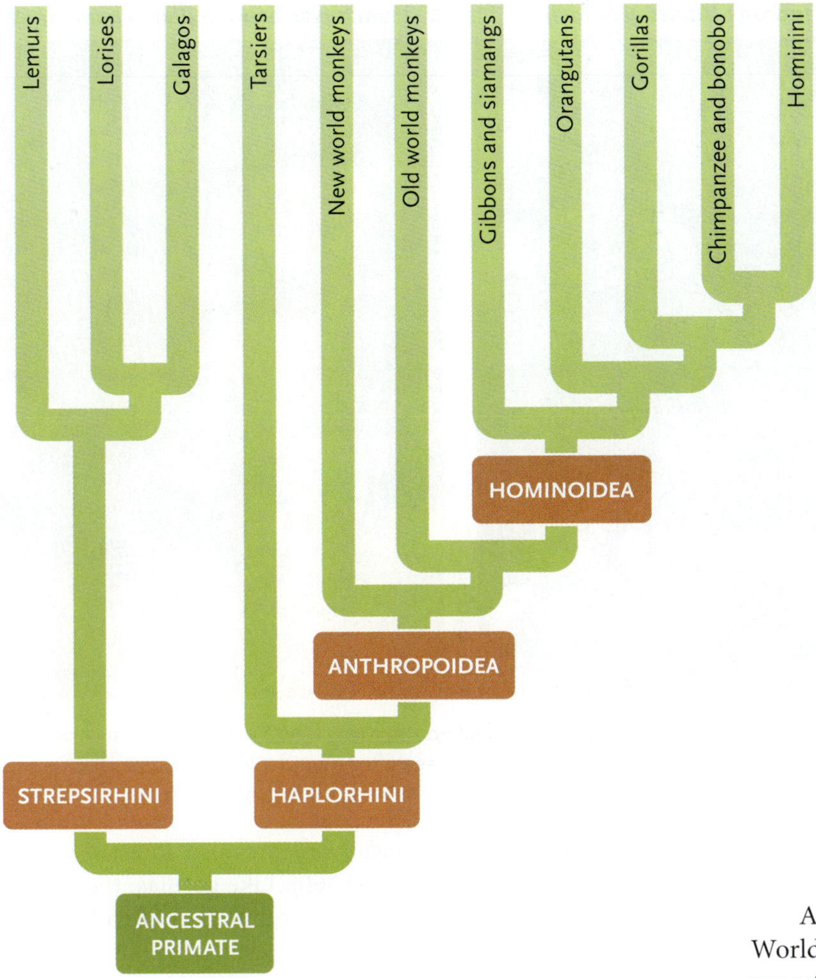

FIGURE 32.32 Primate phylogeny. A phylogenetic tree for the Primates illustrates the two main lineages: Strepsirhini and Haplorhini. Note that chimpanzees and bonobos are the closest living relatives of humans.

link them to the monkeys and apes and not to the strepsirhines described earlier.

The 130 or so species of monkeys, 13 species of apes, and humans constitute the monophyletic haplorhine lineage **Anthropoidea,** which probably arose in Africa. Fossils of a diverse and abundant radiation of forest-dwelling anthropoids, dating from the late Eocene epoch, have been discovered in northern Egypt. Continental drift then established long-term geographical and evolutionary separation of anthropoids in the New World and Old World **(Figure 32.35).**

By the middle of the Oligocene epoch, about 30 million years ago, the ancestors of the New World monkeys had arrived in South America and begun to diversify there. They probably rafted across the Atlantic, which was narrower at that time, on trees or other storm debris. New World monkeys now live in Central and South America (see Figure 32.35A). They range in size from tiny marmosets and tamarins (350 g), to hefty howler monkeys (10 kg). Most are exclusively arboreal and diurnal. The larger species may hang below branches by their arms, and some use a prehensile (grasping) tail as a fifth limb.

FIGURE 32.33 Strepsirhines. The ring-tailed lemur *(Lemur catta)* of Madagascar has ancestral primate characteristics, such as a long snout and wet nose.

Anthropoids diversified most spectacularly in the Old World, however, eventually giving rise to two lineages—one ancestral to Old World monkeys and the other to apes and humans. Although many people assume that the apes are descended from Old World monkeys, the fossil record contradicts that impression. The earliest hominoid ("ape") fossils date to

FIGURE 32.34 Tarsiers. Tarsiers *(Tarsius syrichta)* are classified as haplorhines, but they retain many ancestral characteristics.

the early Miocene, roughly 23 million years ago, but the oldest known Old World monkeys appeared several million years later.

Old World monkeys, which occupy habitats ranging from tropical rain forests to deserts in Africa and Asia, may grow as large as 35 kg (see Figure 32.35B). Many species are sexually dimorphic; in other words, males and females attain different adult sizes (see Section 21.3). Arboreal species use all four limbs for locomotion, but none has a prehensile tail. Some species, such as baboons, often walk or run on the ground.

Within the anthropoid lineage, the **Hominoidea** ("humanlike") is a monophyletic group that includes apes and humans. The climate of the early Miocene was wetter than it is today, and eastern Africa, where many early hominoid fossils are found, was covered with extensive forests. A shift to a cooler and drier climate in the middle Miocene, between 17 and 14 million years ago, converted dense forests into more open woodlands. Hominoids probably adopted a more terrestrial existence and shifted their diets to include leaves and hard foods. Miocene hominoids ranged in size from 4 kg to 80 kg. They occupied both forest and open woodland habitats; some may have been at least partially ground dwelling.

A. Spider monkey *(Ateles geoffroyi)* **B. Hamadryas baboon *(Papio hamadryas)***

Frans Lanting/Corbis

Joseph Van Os/Getty Images Inc.

FIGURE 32.35 New World and Old World monkeys. (A) Many New World monkeys, such as the spider monkey, have prehensile tails, which they use as a fifth limb. Old World monkeys lack prehensile tails, and many, such as **(B)** the Hamadryas baboon, are largely terrestrial.

Although hominoids are closely related to Old World monkeys, several characteristics distinguish them. Apes lack a tail, and the great apes (orangutans, gorillas, chimpanzees, and bonobos) are much larger than monkeys. Moreover, the posterior region of the vertebral column is shorter and more stable in apes. Apes also show more complex behavior.

The gibbons and siamangs, which live in tropical forests in Southeast Asia, are the smallest of the apes, ranging in weight from 6 to 11 kg. With extremely long arms and strong shoulders, they hang below branches by their arms and swing themselves forward, a pattern of locomotion called **brachiation (Figure 32.36A).** The much larger orangutan *(Pongo pygmaeus),* now restricted to forested areas on the islands of Borneo and Sumatra, can grow to 90 kg. Orangutans use both hands and feet to climb trees; they sometimes venture onto the ground on all fours.

Gorillas *(Gorilla gorilla),* which are currently restricted to two large central African forests, are the largest of the living primates. Males can weigh 180 kg; females are about half that size. Because of their size, gorillas spend most of their time on the ground. They often use "knuckle-walking" locomotion, leaning forward and supporting part of their weight on the backs of their hands. Gorillas are almost exclusively vegetarian.

Chimpanzees *(Pan troglodytes)* are also forest dwellers, weighing up to 45 kg **(Figure 32.36B).** Like gorillas, they spend most of their waking hours on the ground; they often knuckle-

A. Black-handed gibbon *(Hylobates agilis)* **B. Chimpanzee *(Pan troglodytes)***

Art Wolfe/Photo Researchers, Inc.

Kenneth Garrett/National Geographic Image Collection

FIGURE 32.36 Apes. (A) Small-bodied apes, such as the black-handed gibbon are agile brachiators that swing through the trees with ease. **(B)** Among the large-bodied apes, chimpanzees have opposable thumbs and big toes.

walk, but sometimes adopt a **bipedal** (two-legged) stance and swagger short distances. Groups of related males form loosely defined communities of up to 50 individuals, which may cooperate in hunts and foraging. Bonobos *(Pan paniscus),* sometimes called pygmy chimpanzees, are restricted to a small area in central Africa. Somewhat smaller than chimps, they have longer legs and smaller heads.

The Primates also includes humans *(Homo sapiens),* which occupy virtually all terrestrial habitats. Humans have adaptations that allow an upright posture and bipedal locomotion. They are ground-dwelling animals with extremely broad diets and complex social behavior.

STUDY BREAK 32.12

1. **What characteristics of primates allow them to spend a great deal of time in trees?**
2. **What is the lowest taxonomic group that includes monkeys, apes, and humans? What is the lowest taxonomic group that includes only apes and humans?**
3. **Which ape species spend the most time on the ground?**

32.13 The Evolution of Humans

Genetic analyses of living hominoid species indicate that African hominoids diverged into several lineages between 10 million and 5 million years ago; one lineage, the **hominins,** includes modern humans and our bipedal ancestors. (Until recently, biologists used the term "hominid" to describe humans and their bipedal ancestors, but that term now defines the group that includes humans, bipedal human ancestors, and the apes.)

Hominins First Walked Upright in East Africa about 6 Million Years Ago

Upright posture and bipedal locomotion are key adaptations that distinguish hominins from apes. Researchers infer these capabilities in early hominin fossils from the anatomy of the skull, spine, pelvis, knees, ankles, and feet **(Figure 32.37).** As a consequence of bipedal locomotion, the hands were no longer used for locomotor functions, allowing them to become specialized for other activities, such as tool use. Hominins also eventually developed larger brains.

Since the mid-1990s, paleontologists have uncovered an astounding array of fossils of hominin species that lived in East Africa and South Africa between 7 million and 1 million years ago **(Figure 32.38).** In 2000, researchers found 13 fossils of *Orrorin tugenensis* ("first man" in a local African language), a species that lived about 6 million years ago in East African forests. Shortly thereafter, researchers described the skull of an even older fossil, *Sahelanthropus tchadensis,* but because the pelvis and legs of this species are still unknown, scientists are unsure about its ability to walk bipedally. Indeed, the fossilized remains of most hominins are fragmentary.

In 1994, Tim White of the University of California, Berkeley and two Ethiopian colleagues made a spectacular discovery of early hominin fossils: they unearthed the remains of 36 individuals of *Ardipithecus ramidus* in the Awash Valley of central Ethiopia. The animals had lived in a closed-canopy woodland

FIGURE 32.37 Adaptations for bipedal locomotion. Differences in the posture, skeleton, and muscles of monkeys, great apes, and humans illustrate the anatomical changes that accompanied upright, bipedal locomotion. Evolutionary changes in the spine, pelvis, hip, knee, ankle, and foot were accompanied by changes in the sizes of leg muscles and their points of attachment to the bones they move.

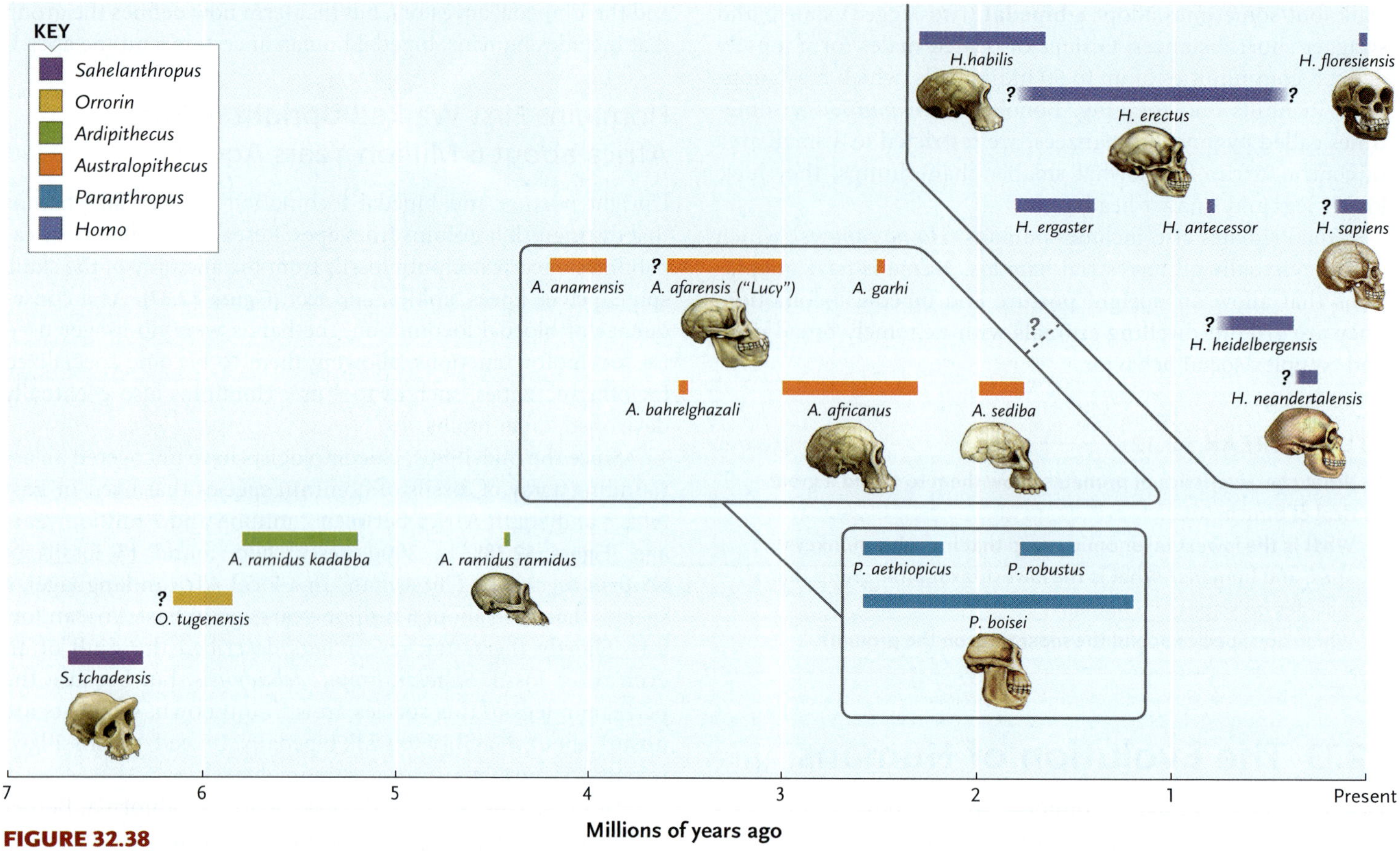

FIGURE 32.38

Hominin time line. Several species of hominins lived simultaneously at sites in eastern and southern Africa. The time line for each species and genus reflects the ages of known fossils. The question marks indicate uncertainty about the precise timing of the origin and extinction of various hominin species as well as evolutionary relationship between *Australopithecus* and *Homo*. Some of the skulls pictured are reconstructed from fragmentary fossils.

about 4.4 million years ago. The fossils were badly deformed and so fragile that they turned to dust in the researchers' hands. After the fossils were carefully excavated and transported to a laboratory at the National Museum of Ethiopia, an international and interdisciplinary team of 47 researchers labored for 15 years to analyze them. They published their findings in a series of 11 papers in *Science* magazine. The work was later hailed by the editors of *Science* as the "breakthrough of the year" in 2009.

Among the fossils that White's team uncovered, the researchers found 125 pieces of one female's skeleton, including bones from her arm and hand, pelvis, leg, ankle, and foot, lower jaw with teeth, and cranium **(Figure 32.39).** Nicknamed "Ardi," she had the body size and brain size of a modern chimpanzee; stood about 120 cm tall and weighed about 50 kg. Studies of Ardi's anatomy revealed that her face protruded less and her upper canine teeth were shorter and duller than those of modern chimpanzees. The upper parts of her pelvis are shorter and broader than those of living apes, and the anatomy of her hand suggests that she did not knuckle-walk as modern chimps and gorillas do. These observations suggest that she was able to walk bipedally on the ground, although perhaps not as well as later hominins. Nevertheless, Ardi's foot had an opposable big toe, and the bones of her fingers were long and curved; these characters suggest that she spent considerable time climbing trees, perhaps to escape predators, seek food, or build nests for sleeping. White and his team described Ardi as having an anatomy that was intermediate between that of chimps and later groups of hominins. Some researchers have taken issue with some of the interpretations that White's team published, partly because the fossils were in terrible condition when they were found and partly because, at the time of this writing, few other researchers have yet had the opportunity to study the fossils in detail.

Hominin fossils from 4.2 million to 1.2 million years ago are well known from many sites in East, Central, and South Africa. They are currently assigned to the genera *Australopithecus* (*australis* = southern, *pithekos* = ape) and *Paranthropus* (*para* = beside, and *anthropos* = human being). With their large faces, protruding jaws, and small skulls and brains, most of these hominins had an apelike appearance (see Figure 32.38). *Australopithecus anamensis,* which lived in East Africa around 4 million years ago, is the oldest known species. It had thick enamel covering on its teeth, a derived hominin characteristic; the structure of a fossilized leg bone suggests that it was bipedal.

Specimens of more than 60 individuals of *Australopithecus afarensis* have been found in northern Ethiopia, including about 40% of a female skeleton, named "Lucy" by its discover-

FIGURE 32.39 ***Ardipithecus.*** Scientists spent 15 years analyzing and reassembling this fairly complete skeleton of *Ardipithecus ramidus.*

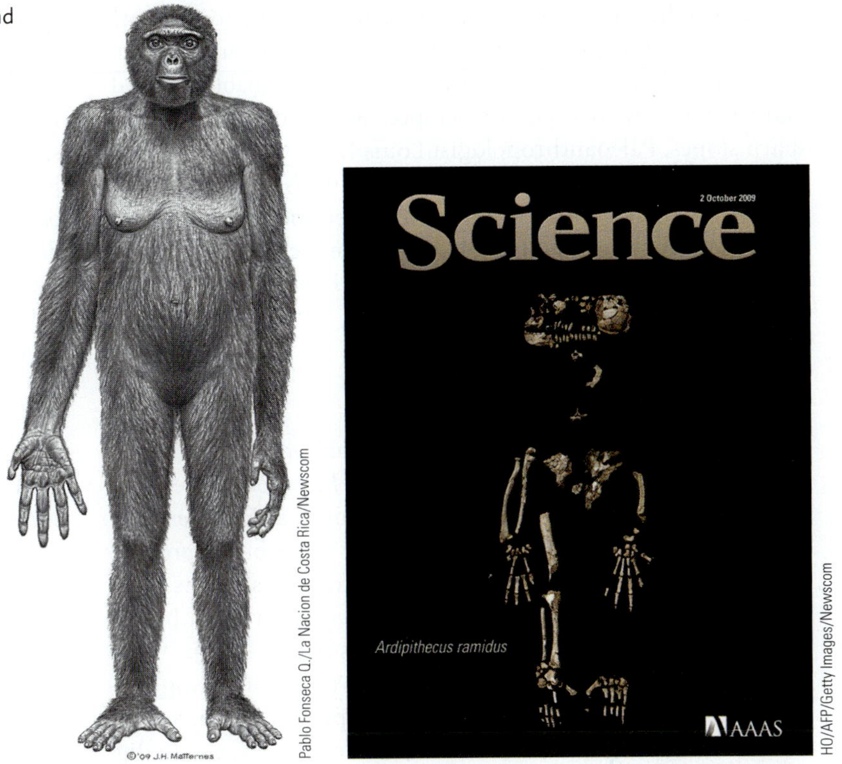

ers **(Figure 32.40)**. *A. afarensis* lived 3.5 million to 3 million years ago, but it retained several ancestral characteristics. For example, it still had moderately large and pointed canine teeth, and a relatively small brain. Males and females were 150 cm and 120 cm tall, respectively. Skeletal analyses suggest that *A. afarensis* was fully bipedal, a conclusion supported by fossilized footprints preserved in a layer of volcanic ash.

Other species of *Australopithecus* and *Paranthropus* lived in East Africa or South Africa between 3.7 million and 1 million years ago. Adult males ranged from 40 to 50 kg in weight and from 130 to 150 cm in height; females were smaller. Most species had deep jaws and large molars. Several species had a crest of bone along the midline of the skull, providing a large surface for the attachment of jaw muscles. These anatomical features suggest that they fed on hard food, such as nuts, seeds, and other vegetable products.

One species, *Australopithecus africanus,* known only from South Africa, had small jaws and teeth, indicating that it probably consumed a softer diet. Yet another species, *Australopithecus sediba,* first described in 2010 from fossils discovered in a South African cave, had even more derived characteristics than *A. africanus:* smaller teeth, longer legs, and a more vertically oriented pelvis. Researchers believe that *A. sediba* may be a descendant of *A. africanus* and an ancestor of the genus that includes modern humans. The phylogenetic relationships of the species classified as *Australopithecus* and *Paranthropus*—and their exact relationships to later hominins—are not yet fully understood. But most scientists agree that *Australopithecus* was ancestral to humans, which are classified in the genus *Homo.*

Homo Habilis Was Probably the First Hominin to Manufacture Stone Tools

Pliocene fossils of the earliest humans, which may have included several species, are fragmentary. They are also widely distributed in space and time, complicating analyses of their relationships. For the sake of simplicity, we describe them as belonging to one species, *Homo habilis* ("handy man").

From 2.3 million to 1.7 million years ago, *H. habilis* occupied the woodlands and savannas of eastern and southern Africa, sharing these habitats with various species of *Paranthropus.* The two genera are easy to tell apart because the brains of *H. habilis* were at least 20% larger, and they had larger incisors and smaller molars than their hominin cousins. Their diet included hard-shelled nuts and seeds as well as soft fruits, tubers, leaves, and insects. They may also have hunted small prey or scavenged carcasses left by large predators.

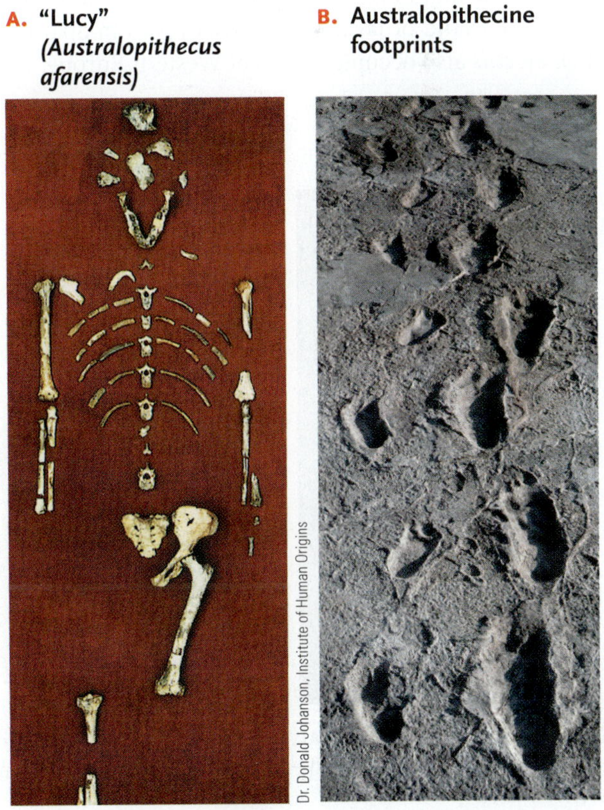

FIGURE 32.40 **Australopithecines.** **(A)** Researchers named the most complete fossil of *Australopithecus afarensis* "Lucy." **(B)** Mary Leakey discovered australopithecine footprints, made in soft, damp volcanic ash about 3.7 million years ago. The footprints indicate that australopithecines were fully bipedal.

Researchers have found numerous tools dating to the time of *H. habilis,* but they are not sure which species used them. Many of the hominin species alive at the time probably cracked marrowbones with rocks or scraped flesh from bones with sharp stones. Paleoanthropologist Louis Leakey was the first to discover evidence of tool *making* at East Africa's Olduvai Gorge, which cuts through a great sequence of sedimentary rock layers. The oldest tools at this site are crudely chipped pebbles, which were probably manufactured by *H. habilis.*

Homo Erectus Dispersed from Africa to Other Continents

Early in the Pleistocene epoch, about 1.8 million years ago, new species of humans appeared in East Africa. Most fossils are fragmentary. For convenience, we describe them all as *Homo erectus* ("upright man"), recognizing that they probably represent several species. One nearly complete skeleton suggests that *H. erectus* was taller than its ancestors, had a much larger brain, a thicker skull, and protruding brow ridges **(Figure 32.41).**

H. erectus made fairly sophisticated tools, including the hand axe (see Figure 32.41B), which they apparently used to cut food and other materials, to scrape meat from bones, and to dig for roots. *H. erectus* probably fed on both plants and animals; they may have hunted and scavenged animal prey. Archaeological data point to their use of fire to process food and to keep themselves warm.

The pressure of growing populations apparently forced groups of *H. erectus* to move out of Africa about 1.5 million years ago. They dispersed northward from East Africa into both northwestern Africa and Eurasia. Some moved eastward through Asia as far as the island of Java. Recent discoveries in Spain indicate that *H. erectus* also occupied parts of Western Europe.

FIGURE 32.41 ***Homo erectus.*** **(A)** A nearly complete skeleton of *Homo erectus* was discovered in Kenya. **(B)** Hand axes are frequently found at *H. erectus* sites.

Modern Humans Are the Only Surviving Descendants of *Homo Erectus*

Judging from its geographical distribution, *Homo erectus* was successful in many environments. Fossils from Africa, Asia, and Europe indicate that it gave rise to several species that first appeared more than 400,000 years ago. These early descendants, described as archaic humans, generally had a larger brain, rounder skull, and smaller molars than *H. erectus.* Like *H. erectus* before them, archaic humans left Africa and established populations in the Middle East, Asia, and Europe.

Some time later, 200,000 to 100,000 years ago, modern humans, *Homo sapiens* ("wise man"), arose in Africa. These modern humans also migrated into Europe and Asia and may have eventually driven the archaic humans to extinction, perhaps through competition for resources rather than hostile interactions.

ARCHAIC HUMANS The Neanderthals *(Homo neanderthalensis),* who occupied Europe and western Asia from 150,000 to 28,000 years ago, are the best-known archaic humans. Compared with modern humans, they had a heavier build, more-pronounced brow ridges, and slightly larger brains (see Figure 32.38). Neanderthals were culturally and technologically sophisticated. They made complex tools, including wooden spears, stone axes, and flint scrapers and knives. At some sites they built shelters of stones, branches, and animal hides, and they routinely used fire. They were successful hunters and probably consumed nuts, berries, fishes, and bird eggs. Some groups buried their dead, and they may have used rudimentary speech.

Researchers once classified Neanderthals as a subspecies of *H. sapiens,* but most scientists now recognize them as a separate species. In 1997 two teams of researchers, Matthias Kring and Svante Pääbo of the University of Munich and Anne Stone and Mark Stoneking of Pennsylvania State University, independently analyzed short segments of mitochondrial DNA (mtDNA) extracted from the fossilized arm bone of a Neanderthal. Unlike nuclear DNA, which individuals inherit from both parents, only mothers pass mtDNA to offspring. It does not undergo genetic recombination (see Section 13.5), and it has a high mutation rate, making it useful for phylogenetic analyses. Many scientists believe that mutation rates in mtDNA are fairly constant, allowing this molecule to serve as a molecular clock (see Section 24.6). Comparing the Neanderthal sequence with mtDNA from 986 living humans, the researchers discovered three times more differences between the Neanderthals and modern humans than between pairs of modern humans in their sample. These results suggest that Neanderthals and modern humans are different species that diverged from a common ancestor 690,000 to 550,000 years ago—hundreds of thousands of years before modern humans appeared.

In 2003, researchers discovered eight skeletons of a tiny hominin, tentatively described as *Homo floresiensis,* in a cave on the Indonesian island Flores. Remarkably, these individuals stood just 1 m tall and are estimated to have weighed about 25 kg. The researchers who discovered these remains have hypothesized that they represent a new archaic hominin species, a descendant of *H. erectus,* that lived in Southeast Asia until roughly 12,000 years ago. Some skeptics suggested that these individuals represent a population of modern humans who suffered from some genetic disease or condition that caused their small size. But subsequent analyses of their skeletal remains indicate that *H. floresiensis* branched off from *H. erectus* even earlier than *H. neanderthalensis.* Their diminutive stature may have resulted from selection for a reduction in size, a phenomenon observed in other island-dwelling mammals. Indeed, fossils of miniature elephants have been discovered nearby. The unresolved questions about this new archaic human include how and where it first evolved and how it may have interacted with modern humans who lived nearby.

MODERN HUMANS Modern humans *(Homo sapiens)* differ from Neanderthals and other archaic humans in having a slighter build, less-protruding brow ridges, and a more prominent chin. The earliest fossils of modern humans found in Africa and Asia are roughly 150,000 years old; those from the Middle East are 100,000 years old. Fossils from about 20,000 years ago are known from Western Europe, the most famous being those of the Cro-Magnon deposits in southern France. The widespread appearance of modern humans roughly coincided with the demise of Neanderthals in Western Europe and the Middle East 40,000 to 28,000 years ago. Although the two species apparently coexisted in some regions for thousands of years, until recently, we have had little concrete evidence about their interactions.

In May 2010, an international team of more than 50 scientists, led by Svante Pääbo, reported a draft sequence of the Neanderthal genome in the journal *Science.* The researchers collected genetic data from fossils of three Neanderthal females who had lived more than 38,000 years ago in what is now Croatia. They compared the Neanderthal DNA sequences to those of five living humans—one each from southern Africa, West Africa, China, Papua New Guinea, and France. Remarkably, the European and the Asians shared between 1% and 4% of their recently derived (that is, newly evolved) nuclear DNA sequences with Neanderthals. Subsequent research, published by another team of researchers in 2011, also found evidence of a small number of archaic human DNA sequences in some modern human populations in central Africa. These data suggest that modern humans interbred with archaic humans on several continents as recently as 35,000 years ago. Although the analyses are far from complete at the time of this writing, the genes that modern Europeans and Asians share with Neanderthals appear to be involved in cognitive development, skeletal development, and metabolism. Thus, there may well be a little Neanderthal in all of us, no matter where on Earth our recent ancestors lived.

Many researchers believe that all modern human populations are descended from an ancestral population that originated in Africa. DNA sequence data from modern humans confirm this hypothesis. In 1987, Rebecca Cann, Mark Stoneking, and Allan Wilson of the University of California, Berkeley, and their colleagues published an analysis of mtDNA sequences from more than 100 ethnically diverse humans on four continents. They found that contemporary African populations contain the greatest variation in mtDNA. One explanation for this observation is that neutral mutations have been accumulating in African populations longer than in others, marking the African populations as the oldest on Earth. They also found that all human populations contain at least one mtDNA sequence of African origin, suggesting an African ancestry for all modern humans. Cann and her colleagues named the ancestral population, which lived approximately 200,000 years ago, the "mitochondrial Eve."

Other researchers have examined genetic material that males inherit only from their fathers. In 1995, L. Simon Whitfield and his colleagues at Cambridge University published a study on an 18,000-base-pair sequence from the Y chromosome, which does not undergo recombination with the X chromosome. Because the sequence contains no genes, it should not be subject to natural selection. Thus, sequence variations should result only from random mutations, which can serve as a molecular clock. The researchers discovered only three sequence mutations among the five subjects examined—a surprising result, given that the sample included a European, a Melanesian, a South American, and two Africans. By contrast, a chimpanzee exhibited 207 differences from the human version of the sequence. Using a sophisticated statistical analysis, the Whitfield team calculated that the common ancestor of these five diverse humans, dubbed the "African Adam," lived between 37,000 and 49,000 years ago. The limited genetic diversity and relatively recent origin of a common ancestor clearly support the African origin of *H. sapiens.* Follow-up studies on the Y chromosomes of thousands of men from Africa, Europe, Asia, Australia, and the Americas have confirmed that all modern humans are the descendants of a single migration out of Africa.

Ever since Darwin first included humans in his analyses about the diversity of life, the study of human evolution has been a controversial subject. Some disputes arise because the fossil record for humans gives up its secrets slowly and in piecemeal fashion. Other disagreements result from the assumptions that researchers must make about the sizes and geographical ranges of ancient populations, the amount of gene flow they experienced, and how natural selection may have affected them. But intellectual disputes are routine in science, and they challenge researchers to refine their hypotheses and to test them in new ways. Questions about the details of human origins are at the center of one of the liveliest debates in evolutionary biology today; additional research will surely clarify the evolutionary history of our species.

STUDY BREAK 32.13

1. **What trait allows researchers to distinguish between apes and humans?**
2. **What evidence suggests that Neanderthals and modern humans represent two distinct species?**
3. **What evidence suggests that modern humans originated in Africa?**

UNANSWERED QUESTIONS

What causes the evolution of diversity?

In this chapter, you have read about the extensive diversity—in size, shape, color, structure, lifestyle, and habitats—of the vertebrate animals. Their mechanisms for maintaining themselves—including behaviors such as moving, feeding, reproducing, hiding, fighting, and sleeping—are equally varied. But nearly all vertebrates share some fundamental features. Recent research shows us that this generality is especially true of early development, but it is also apparent in many of the basic homeostatic mechanisms that vertebrates share—features of digestion and metabolism, respiration, and other characteristics regulated by the products of genetic networks and cascades. Biologists had long thought that once we understood the genetics of a variety of species, we would also understand the basis for their evolution and the maintenance of diversity. But now that biologists have sequenced a number of animal genomes, we have actually learned more about the genetic information that is shared by all animal species than we have learned about the genes that promote diversification. One of the great unanswered questions in biology, therefore, is how diversity arises and how it is maintained, given that so much genetic information is shared among even distantly related species. Equally important is why the same kinds of features evolved almost identically time after time in many unrelated lineages.

The ever-increasing body of genetic information is opening the "black box" of why there are so many species, and how they came to be. For example, we've long known that limbs evolved from fins, based on fossil evidence (which continues to accumulate, providing additional supportive evidence). But we haven't known what the *mechanisms* for forming a fin or a limb are, and how selection works to modify such structures. Now, with our knowledge of *Hox* and other tool-kit genes that control embryonic development, we understand how fins and limbs are formed. We can also experimentally manipulate development and perform selection experiments to test possible pathways of evolution—the "how they came to be."

Why do we see the recurrent, independent evolution of common themes?

A phenomenon that we often see, but don't yet fully understand, is why certain themes—such as body elongation, limblessness, and tooth modification—recur among distantly related vertebrates. The evolution of viviparity, live-bearing reproduction, is one such theme. In the vast majority of animals, reproduction occurs when a female lays her eggs in water and a male sprays sperm over them; typically both parents then abandon the fertilized eggs. But some species in many separate vertebrate lineages have evolved forms of viviparity. Cartilaginous and bony fishes exhibit diverse modes of embryonic nutrition, and in one group, the sea horses, it is the males that become pregnant. Amphibians also exhibit diverse patterns of viviparity: some have pregnant fathers and others have mothers that brood embryos in the skin of their backs, in their stomachs, or in their oviducts. And many squamate reptiles grow placentas similar to those of mammals. All mammals except monotremes are live-bearers with maternal nutrition. In fact, among the living vertebrate groups, birds are the only group that has not evolved the live-bearing habit. Can you think of some reasons why?

In some fishes, amphibians, and squamates, the mother supplies all the nutrition for the developing young through her investment in the egg; but in other species, females resorb their yolks and provide nutrients directly to offspring that are born as fully metamorphosed juveniles. Given that viviparity has evolved independently in approximately 200 vertebrate groups, it is not surprising that we see such a variety of reproductive patterns. Biologists are just beginning to understand the evolution of viviparity and to determine which patterns are shared among different evolutionary lineages and which are not. It appears that the hormonal basis for viviparity is similar in many groups, although the timing, the receptors involved, and the physiological responses vary. Researchers are now identifying candidate genes in the hope of unraveling the genetic networks that have fostered the different modes of viviparity. Ecological studies are revealing the interactions of potential selection regimes that influence reproductive modes. However, we still have much to learn about how genetics, development, physiology, and ecology interact in the evolution of diversity and common recurring themes.

THINK LIKE A SCIENTIST If viviparity has evolved multiple times in different vertebrate lineages, it must have been favored by natural selection. What might some of the benefits of viviparity be?

Marvalee H. Wake is a professor in the Department of Integrative Biology and the graduate school at the University of California, Berkeley. She studies vertebrate evolutionary morphology, development, and reproductive biology, with the goal of understanding evolutionary patterns and processes. To learn more about Dr. Wake's research, go to http://ibB.edu/research/interests/research_profile.php?person5236.

REVIEW KEY CONCEPTS

To access the course materials and companion resources for this text, please visit www.cengagebrain.com.

32.1 Invertebrate Deuterostomes

- Echinoderms have secondary radial symmetry, a five-part body plan, and a unique water vascular system that is used in locomotion and feeding (Figures 32.2 and 32.3).
- Hemichordates collect oxygen and particulate food from seawater that enters the mouth and exits the pharynx through the branchial slits (Figure 32.4).

32.2 Overview of the Phylum Chordata

- Chordates share several derived characteristics: a notochord; postanal tail; segmentation of body wall and tail muscles; a dorsal, hollow nerve cord; and a perforated pharynx at some stage of the life cycle (Figure 32.5).

- Two subphyla of invertebrate chordates use their perforated pharynx to collect particulate food (Figures 32.6 and 32.7).
- The subphylum Vertebrata includes animals with a bony endoskeleton, structures derived from neural crest cells, and a brain divided into three regions.

Animation: Chordate Innovations in the Animal Body Plan

Animation: Lancelet body plan

Animation: Tunicate body plan

Animation: Crocodile body plan

32.3 The Origin and Diversification of Vertebrates

- Vertebrates evolved from an invertebrate chordate ancestor, probably through duplication of the *Hox* gene complex (Figure 32.8).
- The earliest vertebrates were fish-shaped animals with only some vertebrate characteristics (Figure 32.9).
- Vertebrates diversified into numerous lineages (Figure 32.10). Gnathostomata includes all jawed vertebrates. Tetrapoda includes all lineages that ancestrally had four legs. Amniota includes groups descended from animals that produced an amniote egg.

Animation: Vertebrate evolution

32.4 "Agnathans": Hagfishes and Lampreys, Conodonts and Ostracoderms

- Living agnathans—hagfishes and lampreys—are jawless fishes that lack vertebrae and paired fins (Figure 32.11).
- Conodonts had bonelike elements in the pharynx. Ostracoderms were heavily armored, jawless fishes that sucked particulate food into their mouths (Figure 32.12).

32.5 Gnathostomata: The Evolution of Jaws

- Jaws arose through the evolutionary modification of gill arches, which supported the pharynx of ostracoderms (Figure 32.13). Fins arose at the same time (Figure 32.14).
- The first jawed fishes, Acanthodii and Placodermi, are now extinct (Figure 32.15). Chondrichthyans have a skeleton composed of cartilage (Figure 32.16). Actinopterygians and Sarcopterygians have a skeleton composed of bone. Actinopterygians have fins supported by bony rays (Figures 32.17 and 32.18). Sarcopterygians have fins supported by muscles and a bony endoskeleton (Figure 32.19).

Animation: Evolution of jaws

Animation: Cartilaginous fishes

Animation: Bony fish body plan

32.6 Tetrapoda: The Evolution of Limbs

- Tetrapods arose in the late Devonian (Figure 32.20). Key tetrapod adaptations include a strong vertebral column, girdles, limbs, and modified sensory systems.
- Modern amphibians are generally restricted to moist habitats. Their life cycles often include larval and adult stages. Caudata (salamanders) are elongate, tailed amphibians. Anura (frogs) have compact bodies, long legs, and no tails. Gymnophiona (caecelians) are legless burrowers (Figure 32.21).

Animation: Amniote egg

32.7 Amniota: The Evolution of Fully Terrestrial Vertebrates

- Key adaptations in amniotes, the first fully terrestrial vertebrates, included a water-resistant skin, amniote eggs (Figure 32.22), and the excretion of nitrogen wastes as uric acid.
- A traditional view suggests that early amniotes diversified into three lineages (anapsids, synapsids, and diapsids) distinguished by the number of temporal fenestrae and temporal arches on the skull (Figure 32.23).
- Diapsids split into two lineages: archosaurs and lepidosaurs.
- The phylogenetic position of turtles is the subject of disagreement.

32.8 Testudines: Turtles

- The turtle body plan includes a bony shell, with a dorsal carapace and ventral plastron (Figure 32.24).

Animation: Tortoise shell and skeleton

32.9 Living Lepidosaurs: Sphenodontids and Squamates

- Sphenodontids are remnants of a once-diverse lineage (Figure 32.25A).
- Squamates have skin composed of overlapping, keratinized scales (Figure 32.25B and C). Lizards consume a variety of foods, but all snakes are predators on other animals.

32.10 Living Archosaurs: Crocodilians and Birds

- Crocodilians are semiaquatic predators (Figure 32.26).
- Birds have adaptations that reduce their weight and generate power for flight (Figure 32.27).
- Modern birds exhibit adaptations of their bills, feet, wings, and behavior (Figure 32.28).

32.11 Mammalia: Monotremes, Marsupials, and Placentals

- Key adaptations of mammals include endothermy, which allows high levels of activity; modification of the teeth and jaws; extensive parental care of offspring; and large and complex brains.
- There are three major groups of mammals: the Prototheria, to which the monotremes belong, and two lineages of Theria, to which the marsupials and the placentals belong. The three groups differ in their reproductive patterns.
- Monotremes are restricted to Australia and New Guinea (Figure 32.29). Marsupials are abundant in Australia and occur in South America and North America (Figure 32.30). Most living mammals are placentals, occupying nearly all terrestrial and aquatic habitats (Figure 32.31).

32.12 Nonhuman Primates

- Key adaptations allow Primates to be arboreal and diurnal: upright posture and flexible limbs; good depth perception; and a large and complex brain.
- The Primates includes two major lineages (Figure 32.32). The Strepsirhini have many ancestral primate characteristics (Figure 32.33). The Haplorhini have many derived primate characteristics (Figures 32.34 and 32.35).
- Primates arose in forests about 55 mya (million years ago). The hominoid lineage, which includes apes and humans, arose in Africa about 23 mya (Figure 32.36).

32.13 The Evolution of Humans

- Hominins, the lineage that includes humans, arose in Africa between 10 mya and 5 mya. Hominin anatomy permits bipedal locomotion (Figure 32.37). Over time, hominins developed larger brains and tool-making behavior. Several genera of hominins occupied sub-Saharan Africa for several million years (Figures 32.38–32.40).
- *Homo habilis* was the first hominin species to make stone tools. *Homo erectus,* which arose in East Africa about 1.8 mya, made sophisticated stone tools (Figure 32.41).
- The early descendants of *H. erectus* left Africa in waves, populating Asia and Europe. Neanderthals, the best known of these groups, became extinct about 30,000 years ago.
- Modern humans, *Homo sapiens,* arose approximately 150,000 years ago and migrated out of Africa, eventually replacing archaic humans in Europe and Asia.

Animation: Primate skeletons

Animation: Genetic distance between human groups

UNDERSTAND & APPLY

Test Your Knowledge

1. Which phylum includes animals that have a water vascular system?
   a. Echinodermata
   b. Hemichordata
   c. Chordata
   d. Tetrapoda
   e. Amniota
2. Which of the following is *not* a characteristic of all chordates?
   a. notochord and postanal tail
   b. segmental body wall and tail muscles
   c. segmented nervous system
   d. dorsal hollow nerve cord
   e. perforated pharynx
3. Which group of vertebrates has adaptations that allow it to reproduce on land?
   a. agnathans
   b. tetrapods
   c. gnathostomes
   d. amniotes
   e. ichthyosaurs
4. Which group of fishes has the most living species today?
   a. sarcopterygians
   b. actinopterygians
   c. chondrichthyans
   d. acanthodians
   e. ostracoderms
5. Modern amphibians:
   a. closely resemble their Paleozoic ancestors.
   b. always occupy terrestrial habitats as adults.
   c. never occupy terrestrial habitats as adults.
   d. are generally larger than their Paleozoic ancestors.
   e. are generally smaller than their Paleozoic ancestors.
6. Which of the following key adaptations allows amniotes to occupy terrestrial habitats?
   a. the production of carbon dioxide as a metabolic waste product
   b. an unshelled egg that is protected by jellylike material
   c. a dry skin that is largely impermeable to water
   d. a lightweight skeleton with hollow bones
   e. feathers or fur that provide insulation against cold weather
7. Which of the following characteristics does *not* contribute to powered flight in birds?
   a. a lightweight skeleton
   b. efficient respiratory and circulatory systems
   c. enlarged forelimbs and a keeled sternum
   d. a high metabolic rate that releases energy from food rapidly
   e. scaly skin on the legs and feet
8. Which of the following characteristics did *not* contribute to the evolutionary success of mammals?
   a. extended parental care of young
   b. an erect posture and flexible hip and shoulder joints
   c. specializations of the teeth and jaws
   d. enlargement of the brain
   e. high metabolic rate and high body temperature
9. The Hominoidea is a monophyletic group that includes:
   a. apes and monkeys.
   b. apes only.
   c. humans and human ancestors.
   d. apes and humans.
   e. monkeys, apes, and humans.
10. Which of the following hominin species was the earliest?
   a. *Ardipithecus ramidus*
   b. *Australopithecus afarensis*
   c. *Homo habilis*
   d. *Homo erectus*
   e. *Homo neanderthalensis*

Discuss the Concepts

1. Most sharks and rays are predatory, but the largest species feed on plankton. Construct a hypothesis to explain this observation. How would you test your hypothesis?
2. When tetrapods first ventured onto the land, what new selection pressures did they face? What characteristics might have fostered the success of these animals as they made the transition from aquatic to terrestrial habitats?
3. Use a pair of binoculars to observe several species of birds that live in different types of environments, such as lakes and forests. How are their beaks and feet adapted to their habitats and food habits?
4. Imagine that you unearthed the complete fossilized remains of a mammal. How would you determine the food habits of this now extinct animal?
5. Many myths about human evolution are embraced by popular culture. Using the information you have learned about human evolution, argue against each of the following myths.
   a. Humans evolved from chimpanzees.
   b. Evolution occurred in a steady linear progression from primitive primate to anatomically modern humans.
   c. All human characteristics, such as bipedal locomotion and an enlarged brain, evolved simultaneously and at the same rate.

Design an Experiment

Walking along a rocky coast one day, you discover two small creatures—one lumpy and the other worm-shaped. What anatomical studies would you conduct to determine whether or not they are chordates? What genetic studies might provide supplementary evidence?

Interpret the Data

The phylogenetic tree for vertebrates depicted below was constructed from sequence data for two rRNA mitochondrial genes (12S and 16S). How do the results of this analysis compare with the traditional view of turtles as living representatives of an anapsid lineage?

Apply Evolutionary Thinking

Birds and crocodiles are both descended from an ancestral archosaur. What shared anatomical and behavioral characteristics reflect this common ancestry? Explain why dinosaurs, which were also members of the archosaur lineage, may have shared these traits as well. Review Figure 32.23 before formulating your answer.

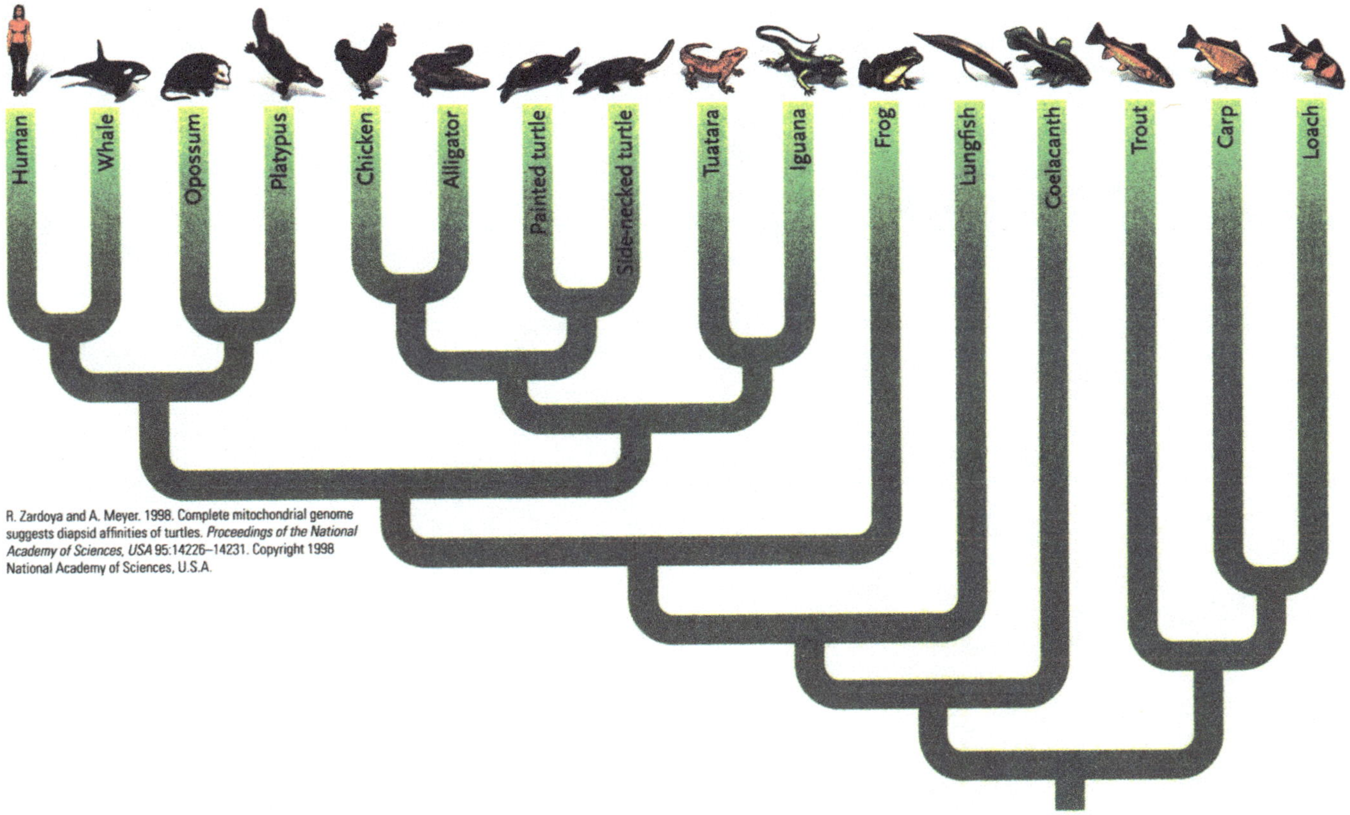

Source: R. Zardoya and A. Meyer. 1998. Complete mitochondrial genome suggests diapsid affinities of turtles. *Proceedings of the National Academy of Sciences, USA* 95:14226–14231.

Ted Kinsman/Photo Researchers, Inc.

The reproductive structures of an ornamental poppy *(Papaver rhoeas)*. Bright yellow male parts, where pollen grains develop, surround female parts that produce eggs. Within them fertilization occurs and seeds develop.

36

STUDY OUTLINE

36.1 Overview of Flowering Plant Reproduction

36.2 The Formation of Flowers and Gametes

36.3 Pollination, Fertilization, and Germination

36.4 Asexual Reproduction of Flowering Plants

36.5 Early Development of Plant Form and Function

Reproduction and Development in Flowering Plants

Why it matters . . . Chocolate, a confection beloved by humans with a sweet tooth, gets its start within the fruits of *Theobroma cacao,* a small flowering tree. *T. cacao* evolved in the undergrowth of tropical rainforests in Central America, where the Maya and Aztec peoples domesticated it and incorporated it into their religious and cultural life. Today cacao trees flourish on vast plantations in the tropical lowlands of Central and South America, the West Indies, West Africa, and New Guinea. Unlike most angiosperms, which produce flowers at the tips of floral shoots, *T. cacao* flowers grow directly from buds on the tree trunk. Small flying insects called midges pollinate the flowers. Pollination, in turn, is the first step toward fertilization of the eggs, and within about six months, large, podlike fruits develop from them **(Figure 36.1).** Each fruit contains from 20 to 60 seeds—the cacao "beans" that manufacturers process into cocoa, eating chocolate, and other products.

Angiosperms have elaborate reproductive systems housed in their flowers—systems that produce, protect, and nourish sperm, eggs, and developing embryos. As with *T. cacao,* the flowers of many species also attract animal pollinators that function in bringing sperm and egg together. Once a new individual forms and begins

© WILDLIFE GmbH/Alamy

Norman Chan/Shutterstock.com

FIGURE 36.1 Flowers and fruits growing from the trunk of a cacao tree *(Theobroma cacao)* in Central America. Each fruit is the mature ovary of a *T. cacao* flower.

to grow, finely regulated gene interactions guide the development of all a plant's parts.

Like the seeds of other flowering plants, cacao seeds result from sexual reproduction—the union of sperm and eggs that are produced by meiosis (see Section 11.1). As Section 11.2 describes in greater detail, this reproductive mode yields genetically diverse offspring. This diversity, in turn, provides an adaptive advantage in a changing environment, because it increases the chance that as change occurs—perhaps a shift in climate, or the appearance of a new predator or pathogen—at least some offspring will grow to maturity and reproduce successfully. *T. cacao* and many other plants are reproductively flexible, however. Under certain circumstances, they can also reproduce asexually, so that individuals of the new generation are genetically identical (clones) of their parents.

In the first three sections of this chapter, we will focus on sexual reproduction, which dominates the life cycle of flowering plants. We then consider asexual reproduction and conclude with a discussion of early plant development.

36.1 Overview of Flowering Plant Reproduction

In plants as in animals, sexual reproduction occurs when male and female haploid gametes unite to create a diploid fertilized egg. This fertilized egg—the zygote—then embarks on its gene-guided course of growth and development. As described in Chapter 10, animals grow mainly by mitosis, which increases the number of body cells, and the newly formed cells end up roughly the same size as the parent cell. Plants, however, grow larger by two mechanisms. Mitotic cell divisions in meristems generate new cells, but then the newly formed cells rapidly increase in size—especially length—expanding until they are much *larger* than the parent cell. Section 36.4 explains how this cell expansion occurs. Ultimately, as in animals, each daughter cell differentiates and becomes cell specialized for its particular function.

Once an angiosperm zygote has formed, early development generates an embryo enclosed within a seed. This embryo already contains early versions of the basic plant tissue systems, so the embryo technically is already a **sporophyte**—the diploid, spore-producing body of a plant described in Section 28.1. When most people look at a plant—say a cherry tree or a rose bush—what they think of as "the plant" is the sporophyte pictured at the top of **Figure 36.2.**

At some point during an angiosperm sporophyte's growth and development, one or more of its vegetative shoots undergo changes in structure and function and become reproductive shoots. These *floral shoots* will give rise to a flower or inflorescence (a group of flowers on the same floral shoot), and certain cells in the flowers will divide by meiosis. Unlike meiosis in animals, however, meiosis in plants does not yield haploid gam-

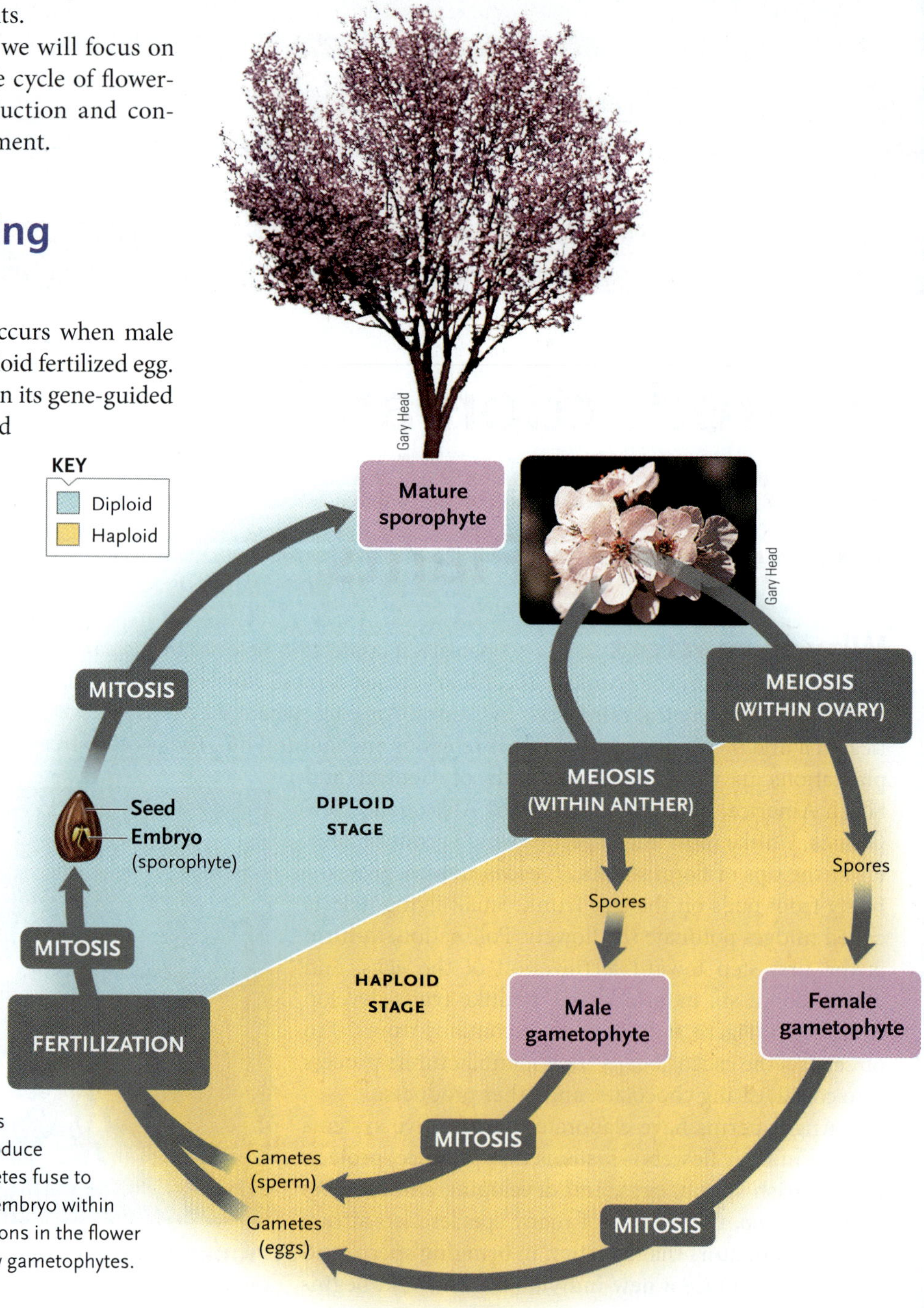

FIGURE 36.2 Overview of the flowering-plant life cycle, using the cherry *(Prunus)* as an example. This type of reproductive cycle, alternation of generations, has a haploid phase in which multicellular gametophytes produce gametes and a diploid phase that begins when two gametes fuse to form a zygote. This zygote develops into a multicellular embryo within a seed and then into a mature sporophyte. Meiotic divisions in the flower of the sporophyte produce spores, which give rise to new gametophytes.

etes. Instead, meiosis gives rise to haploid **spores,** walled cells that subsequently divide by mitosis. By way of these mitotic divisions, each spore grows into a multicellular—but still haploid—**gametophyte.** Although the term gametophyte means "gamete-producing plant," the gametophyte of a vascular plant is a tiny structure that produces haploid sex cells, the gametes, also by mitosis. Male gametophytes produce sperm cells, the male gametes of flowering plants; female gametophytes produce eggs. This division of a life cycle into a diploid, spore-producing generation and a haploid, gamete-producing one is called **alternation of generations.**

In virtually all plants, the gametophyte and sporophyte are strikingly different from one another in both function and structure. As Chapter 28 describes in detail, the course of plant evolution has been marked by a trend toward a larger and more nutritionally independent sporophyte. For instance, in bryophytes (mosses and liverworts), the sporophyte is usually smaller than the gametophyte, growing out of the gametophyte and obtaining nutrients from it (see Section 28.2). This relationship changed as vascular plants evolved. In ferns, which are seedless vascular plants, the gametophyte is much smaller than the sporophyte and is free-living for much of its life, and in most fern species the gametophyte nourishes itself by photosynthesis. In angiosperms and other seed plants, however, gametophytes are not independent "plants." Instead they are small structures that are retained *inside* the sporophyte for all or part of their lives. The female gametophyte of a flowering plant usually consists of only seven cells that are embedded in floral tissues, as you will read shortly. Male gametophytes consist of only three cells. They are released into the environment as pollen grains so small that they are measured in micrometers. The pollen grain matures when it reaches a compatible ovule. When fertilization ensues, an embryo develops within a seed.

As you will read later on in this chapter, sporophytes may also reproduce asexually by several means, all of which produce genetically identical offspring. We turn now to our consideration of sexual reproduction in angiosperms, beginning with the crucial step in which flowers develop.

STUDY BREAK 36.1

1. **What are the two "alternating generations" of plants?**
2. **How do these two life phases differ in structure and function?**

36.2 The Formation of Flowers and Gametes

Flowering marks a developmental shift for an angiosperm. Chemical signals—triggered in part by environmental cues such as day length and temperature—travel to the apical meristem of a shoot and trigger changes in cell activity there. The shoot stops its vegetative growth and undergoes further changes that transform it into a floral shoot capable of giving rise to floral organs.

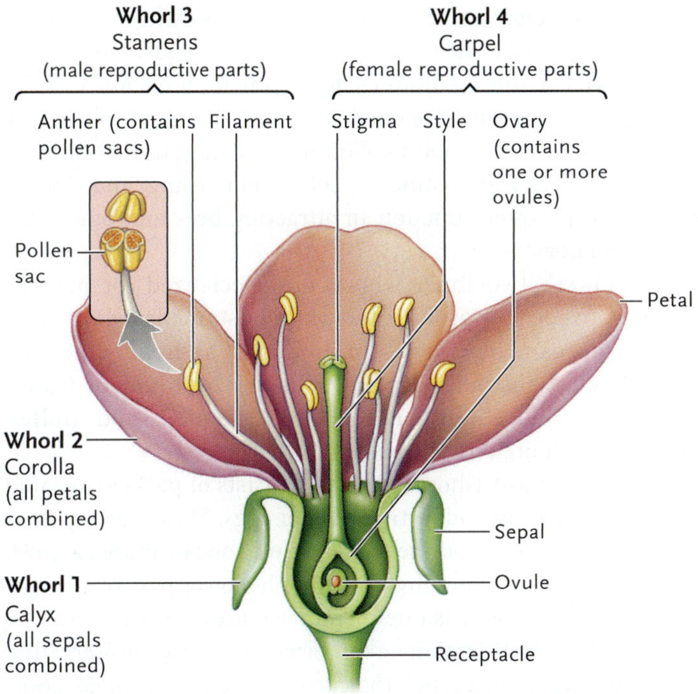

FIGURE 36.3 Structure of a cherry *(Prunus)* flower, with the four whorls indicated. Like the flowers of many angiosperms, it has a single carpel and several stamens. The anthers of each stamen produce haploid pollen. The stigma of the carpel receives the pollen, and the ovule inside the ovary contains the haploid egg. In this diagram, two of the cherry flower's five petals have been removed. For clarity, this diagram shows a single ovule, but as the text notes, flowers of many angiosperm species contain multiple ovules.

In Angiosperms, Flowers Contain the Organs for Sexual Reproduction

A flower develops from the end of the floral shoot, called the **receptacle.** In a "typical" flower, cells in the receptacle differentiate to produce four types of concentric tissue regions called *whorls* **(Figure 36.3).** Flowers with all four whorls are called **complete flowers (Figure 36.4A).** The two outer whorls consist of

FIGURE 36.4 Examples of complete and incomplete flowers. (A) Complete flower of a dog rose *(Rosa canina)*, which has all four floral whorls. **(B)** Incomplete flower of a squash plant *(Cucurbita)*—in this case, a "female" plant in which the stamen whorl is missing and the flowers contain only carpels.

nonfertile, vegetative structures. The outermost whorl (whorl 1), the **calyx,** is made up of leaflike **sepals.** The calyx is usually green, and, early in the flower's development, it encloses all the other parts, as in an unopened rose bud. The next whorl, the **corolla,** includes the **petals.** Corollas usually are the "showy" parts of flowers with distinctive colors, patterning, and shapes. These features often function in attracting bees and other animal pollinators.

A flower's two inner whorls are specialized for making gametes. Inside the corolla is the whorl of **stamens** (whorl 3), in which male gametophytes form. In almost all living flowering plant species, a stamen consists of a slender **filament** (stalk) capped by a bilobed **anther.** Each anther contains four **pollen sacs,** in which pollen develops.

The innermost whorl (whorl 4) consists of parts associated with the formation and fertilization of eggs. These parts, sometimes collectively called the *pistil,* include one or more **carpels,** in which female gametophytes form. The lower part of a carpel is the **ovary.** Inside it is one or more **ovules,** in which an egg develops and fertilization takes place. In many flowers that have more than one carpel, the carpels fuse into a single, common ovary containing multiple ovules. Typically, the carpel's slender **style** widens at its upper end, terminating in the **stigma,** which serves as a landing platform for pollen. Fused carpels may share a single stigma and style, or each may retain separate ones. The name angiosperm ("seed vessel") refers to the carpel.

The arrangement and number of whorl types varies in different species. In some species flowers lack one or more of the whorls, and thus botanists describe them as **incomplete flowers** **(Figure 36.4B).** Botanists also distinguish flowers on the basis of the sexual parts they contain. Most angiosperms produce **perfect flowers,** which have both kinds of sexual parts—that is, both stamens and carpels. **Imperfect flowers** have stamens or carpels, but not both. (Notice that all imperfect flowers are also incomplete because they lack one of the whorls.) Species with imperfect flowers are further divided according to whether individual plants produce both sexual types of flowers, or only one. In **monoecious** (*mono-* = one; *oikia* = house) species, such as oaks, each plant has some "male" flowers with only stamens and some "female" flowers with only carpels. In **dioecious** ("two houses") species, such as willows, a given plant produces flowers having only stamens or only carpels. With this basic angiosperm reproductive anatomy in mind, we can consider the processes by which male and female gametes come into being.

Pollen Grains Arise from Microspores in Anthers

Most of a flowering plant's reproductive life cycle, from production of sperm and eggs to production of a mature seed, takes place within its flowers. **Figure 36.5** summarizes this cycle as it unfolds in a perfect flower. The microspores that give rise to male gametophytes are produced in a flower bud's anthers (see Figure 36.5, left). The pollen sacs inside each anther hold diploid microsporocytes (or *microspore mother cells*); each microsporocyte undergoes meiosis and eventually produces four small haploid **microspores.** Like most plant cells, the microspores are walled, and inside its wall each microspore divides again, this time by mitosis. The result is an immature, haploid male gametophyte—a **pollen grain.**

Of the two nuclei produced by the mitotic division of a microspore, one again divides. After this second round of mitosis the male gametophyte consists of three cells—two sperm cells plus a third cell that controls the development of a **pollen tube** **(Figure 36.6A).** As you can see in Figure 36.5, step 5, when pollen lands on a stigma, this tube grows through the tissues of a carpel and carries the sperm cells to the ovary. A mature male gametophyte consists of the pollen tube and sperm cells—the male gametes. Once the pollen tube and sperm cells have formed, the stage is set for fertilization, which we discuss more thoroughly in Section 36.3.

The walls of pollen grains are hardened by the decay-resistant polymer *sporopollenin,* and are tough enough to protect the male gametophyte during the somewhat precarious journey from anther to stigma. These walls are so distinctive that the family to which a plant belongs often can be identified from pollen alone—based on the size and wall sculpturing of the grains, as well as the number of pores in the wall **(Figure 36.6B–D).** Because they withstand decay, pollen grains fossilize well and can provide revealing clues about the evolution of seed plants and the ecological communities that lived in the past.

Eggs and Other Cells of Female Gametophytes Arise from Megaspores

Meanwhile, in the ovary of a flower, one or more dome-shaped masses form on the inner wall. Each mass becomes an ovule (see Figure 36.5, right), which, if all goes well, develops into a seed. Only one ovule forms in the carpel of some flowers, such as the cherry. Dozens, hundreds, or thousands may form in the carpels of other flowers, such as those of a bell pepper plant (*Capsicum annuum*). At one end, the ovule has a small opening, called the **micropyle.**

Inside the cell mass, a diploid megasporocyte (or *megaspore mother cell*) divides by meiosis, forming four haploid **megaspores.** In most species, three of these megaspores disintegrate. The remaining megaspore enlarges and develops into the female gametophyte in a sequence of steps tracked in Figure 36.5.

First, three rounds of mitosis occur *without* cytoplasmic division; the result is a single cell with eight nuclei arranged in two groups of four. Next, one nucleus in each group migrates to the center of the cell; these two **polar nuclei** ("polar" because they migrate from opposite ends of the cell) may fuse or remain separate. The cytoplasm then divides, and a cell wall forms around the two polar nuclei, forming a single large endosperm mother cell (also called the *central cell*). A wall also forms around each of the other nuclei. Three of these walled nuclei

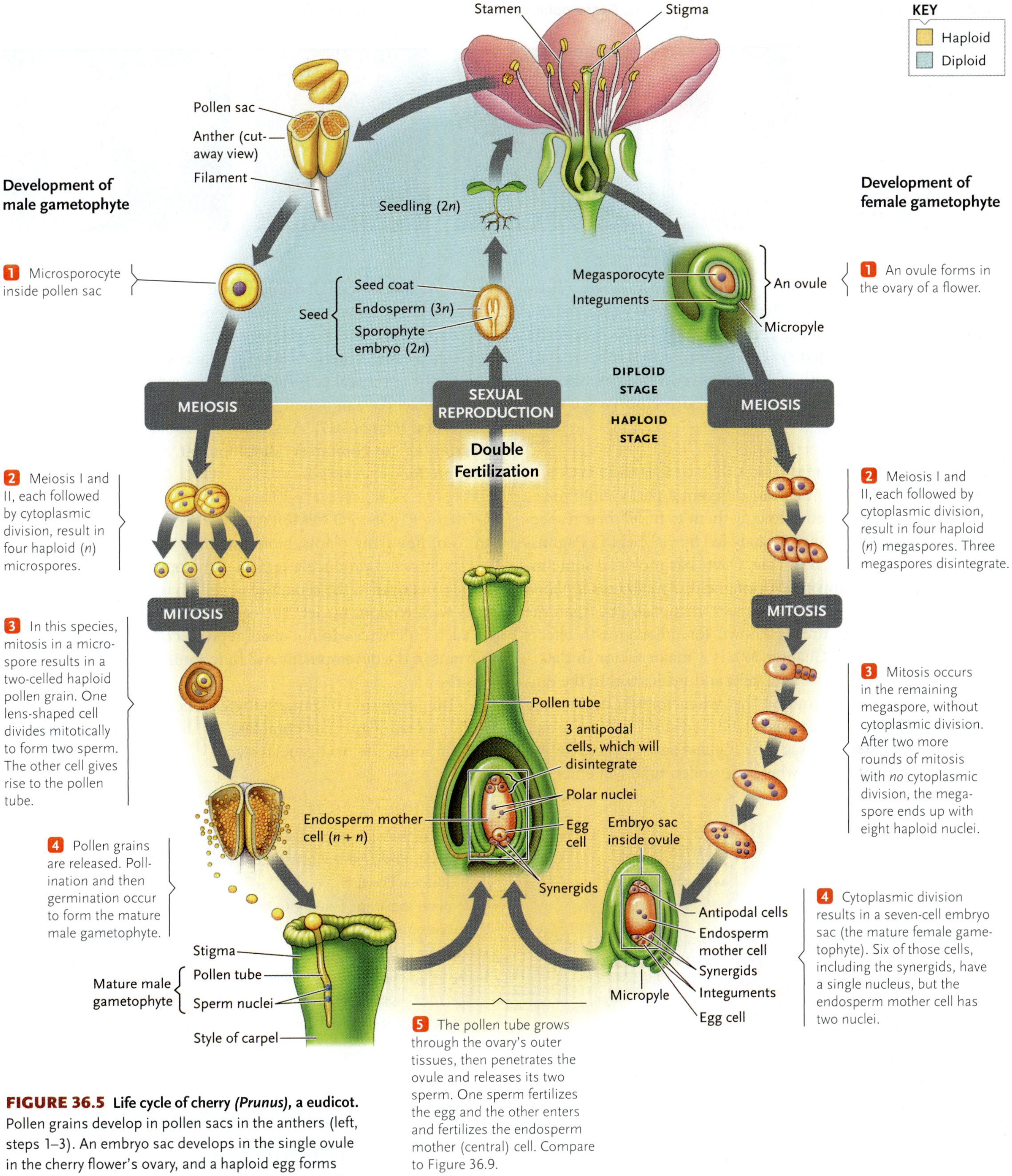

FIGURE 36.5 Life cycle of cherry *(Prunus)*, a eudicot. Pollen grains develop in pollen sacs in the anthers (left, steps 1–3). An embryo sac develops in the single ovule in the cherry flower's ovary, and a haploid egg forms within the embryo sac (right, steps 1–4). When a pollen grain is released and contacts the stigma (left, step 4), it germinates and produces a pollen tube containing two haploid sperm. The tube penetrates the stigma, then grows to the ovary and penetrates the ovule (step 5). Eventually double fertilization occurs, giving rise to an embryonic sporophyte and endosperm that become encased together in a seed coat.

FIGURE 36.6 Examples of a pollen tube and pollen grain diversity. (A) Scanning electron micrograph of a pollen tube (orange) extending from a germinating pollen grain of prairie gentian *(Gentiana)*. Three other SEMs show pollen grains from **(B)** a grass, **(C)** chickweed *(Stellaria)*, and **(D)** ragweed *(Ambrosia)* plants.

A. Pollen tube **B.** Grass pollen grains **C.** Chickweed pollen grain **D.** Ragweed pollen grains

become *antipodal cells,* which eventually disintegrate. Three others form a cluster (called the "egg apparatus") near the micropyle; one of them is an **egg cell** that may eventually be fertilized. The other two, called *synergids,* will have a role in fertilization. The eventual result of all these events is an **embryo sac** containing seven cells and eight nuclei. This embryo sac is the female gametophyte.

DISCOVERING HOW CHEMICAL CUES GUIDE EMBRYO SAC DEVELOPMENT How do the four different types of embryo sac cells become differentiated, allowing them to fulfill their respective reproductive functions? A study led by Gabriela C. Pagnussat at the University of California, Davis has provided some insights into this mechanism. Working with *Arabidopsis thaliana* plants, Pagnussat and her colleagues demonstrated that the plant hormone auxin, already known for other growth effects in plants (described in Chapter 37), is a major factor in determining the identity and roles of cells and nuclei within the embryo sac. The team determined that when mitosis begins in a megaspore, an auxin gradient is established along the future axis of the developing embryo sac. The highest auxin concentration occurs near the micropyle, where the pollen tube will enter the ovule. As mitotic cell divisions continue, each of the eight products develops as an egg, synergid, polar nucleus, or antipodal cell depending on its position within the auxin gradient. Thus the egg cell and synergids develop where auxin is most concentrated, the antipodal cells develop where auxin is lowest, and the polar nuclei develop where there is an intermediate auxin concentration **(Figure 36.7).** Auxin can be said to provide *positional information* for embryo sac development, a topic we return to in Section 36.5.

OTHER OPTIONS FOR EMBRYO SAC DEVELOPMENT In about a third of flowering plants, biologists have observed variations in the events that produce a female gametophyte. In lilies, for example, changes in the sequence of cell divisions produce several cells with triploid nuclei. The egg cell is not involved, however, so such differences do not affect reproduction. They may be important in the development and functioning of other embryonic tissues.

The formation of gametophytes sets the stage for events by which a seed plant may complete its life cycle. As we see next, pollination is the next crucial step.

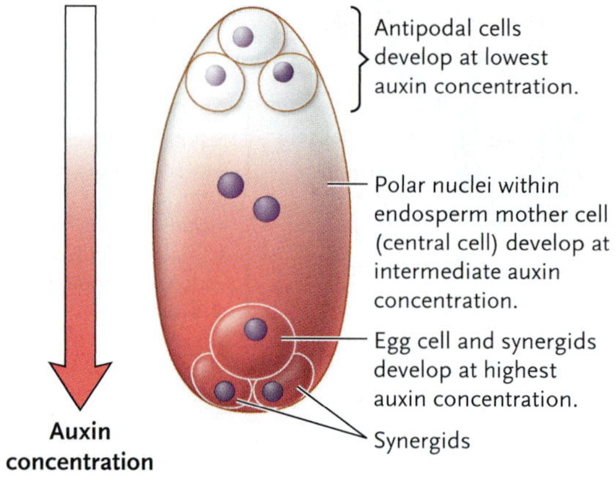

FIGURE 36.7 Model for patterning of the female gametophyte by an auxin gradient. A study by Gabriela Pagnussat and coworkers showed that the concentration of auxin (indicated in red) determined the fate of cells in the developing gametophyte. Where auxin was most concentrated, cells became specialized as synergids. At the lowest concentration, they became antipodal cells. Egg cells developed in between these two extremes.

STUDY BREAK 36.2

1. **What is the biological role of flowers, and what fundamental developmental shift must occur before an angiosperm can produce a flower?**
2. **Explain the steps leading to the formation of a mature male gametophyte, beginning with microsporocytes in a flower's anthers. Which structures are diploid and which haploid?**
3. **Trace the development of a female gametophyte, beginning with the megasporocyte in an ovule of a flower's ovary. Which structures are diploid and which haploid?**

36.3 Pollination, Fertilization, and Germination

The process by which plants produce seeds—which have the potential to give rise to new individuals—begins with **pollination,** when pollen grains make contact with the stigma of a flower. Air or water currents, birds, bats, insects, or other agents make the transfer. (Section 29.2 discussed the complex relationship between some flowering plants and their animal pollinators.)

Pollination is the first in a series of events leading to *fertilization,* the fusion of the haploid nuclei of an egg and sperm inside the flower's ovary. The resulting embryo and its ovule mature into a seed housing a young diploid sporophyte, and when the seed sprouts, or *germinates,* the sporophyte begins to grow. Before these events can unfold, however, certain "compatibility" requirements must be met.

Pollination Requires Compatible Pollen and Female Tissues

Even after pollen reaches a stigma, in most cases pollination and fertilization can take place only if the pollen and stigma are compatible. For example, if pollen from one species lands on a stigma from another, chemical incompatibilities usually prevent pollen tubes from developing.

In different angiosperm taxa, various mechanisms have evolved that limit or prevent inbreeding. For example, in monoecious species—in which, recall, individual plants produce flowers having either male *or* female reproductive parts—the stigmas of "female" flowers may be receptive to pollen a day or two before or after stamens of "male" flowers release pollen. In another mechanism, even when the sperm-bearing pollen and a stigma are "ready" at the same time, pollination may not lead to fertilization unless the pollen and stigma belong to genetically distinct individuals. In one such scenario, when pollen from a given plant lands on that plant's own stigma, a pollen tube may begin to develop, but stop before reaching the embryo sac. This **self-incompatibility** is a biochemical recognition and rejection process that prevents self-fertilization, and it apparently results from interactions between proteins encoded by *S* (self) genes.

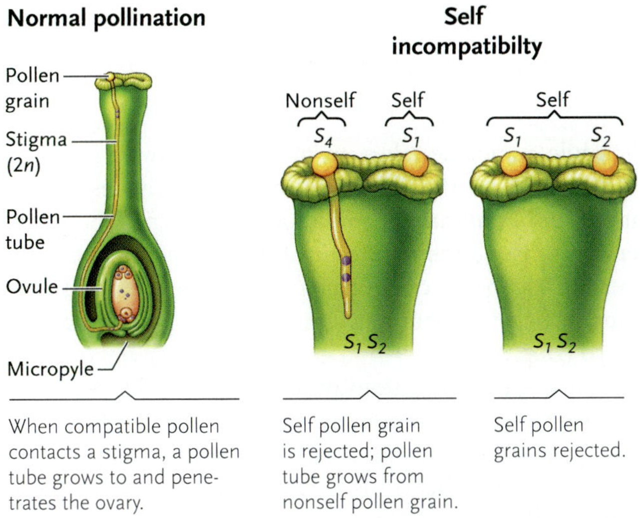

FIGURE 36.8 Self-incompatibility. When a pollen grain has an *S* allele that matches one in the stigma (which is diploid), the result is a biochemical response that prevents fertilization—in this illustration, by preventing the growth of a pollen tube.

Research has shown that *S* genes usually have multiple alleles—in some species there may be hundreds—and a common type of incompatibility occurs when pollen and stigma carry an identical *S* allele. The result is a biochemical signal that prevents proper formation of the pollen tube **(Figure 36.8).** For example, studies on plants of the mustard family have revealed that pollen contacting an incompatible stigma produces a protein that prevents the stigma from hydrating the relatively dry pollen grain, an essential step if the pollen tube is to grow. A wide range of self-incompatibility responses has been discovered, however. In cacao, for instance, when incompatible pollen contacts a stigma, a pollen tube grows normally but a hormonal response soon causes the flower to drop off the plant, preventing fertilization.

Self-incompatibility prevents inbreeding and promotes genetic variation, which is the raw material for natural selection and adaptation. Even so, many flowering plants do self-pollinate, either partly or exclusively, because that mode, too, has benefits in some circumstances. (Mendel's peas are a classic example.) For instance, "selfing" may help preserve adaptive traits in a population. It also reduces or eliminates a plant's reliance on wind, water, or animals for pollination, and thus ensures that seeds will form when conditions for cross-pollination are unfavorable, such as when pollinators or potential mates are scarce.

Double Fertilization Occurs in Flowering Plants

If a pollen grain lands on a compatible stigma, it absorbs moisture and germinates a pollen tube, which burrows through the stigma and style toward an ovule (see Figure 36.5 and **Figure 36.9A**). Chemical cues from the two synergid cells lying close to the egg cell help guide the pollen tube toward its destination. Before or during these events, the pollen grain's haploid sperm-producing cell divides by mitosis, forming two haploid sperm. When the pollen tube reaches the ovule, it enters through the micropyle and an opening forms in its tip. By this time one synergid has begun to die (an example of programmed cell death), and the two sperm are released into the disintegrating cell's cytoplasm. Experiments suggest that elements of the synergid's cytoskeleton guide the sperm cells onward, one to the egg cell and the other to the endosperm mother cell.

Next there occurs a remarkable sequence of events called **double fertilization (Figure 36.9B),** which has been observed only in flowering plants and (in a somewhat different version) in the gnetophyte *Ephedra* (see Section 29.2). Typically, one sperm nucleus fuses with the egg to form a diploid ($2n$) zygote. The other sperm nucleus fuses with the endosperm mother cell, forming a cell with a triploid ($3n$) nucleus. Tissues derived from that $3n$ cell are called **endosperm** (*endon* = within; *sperma* = seed). The endosperm nourishes the embryo and, in monocots, the seedling, until its leaves form and photosynthesis has begun.

The evolution of embryo-nourishing endosperm in angiosperms coincided with reduced size of the female gametophyte (see Figure 28.7). In other land plants, such as gymnosperms

CLOSER LOOK **FIGURE 36.9 Pollination and double fertilization.**

Step 1. When compatible pollen contacts a stigma, chemical signals guide a pollen tube toward the ovary. The tube contains two haploid sperm nuclei.

A. Pollination

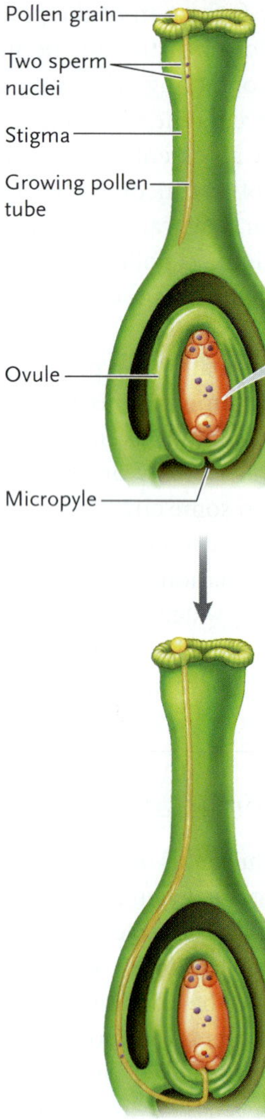

Step 2. The growing pollen tube penetrates the ovary, reaches the micropyle, and passes through it into the ovule. At this point the sperm nuclei are released.

B. Double fertilization

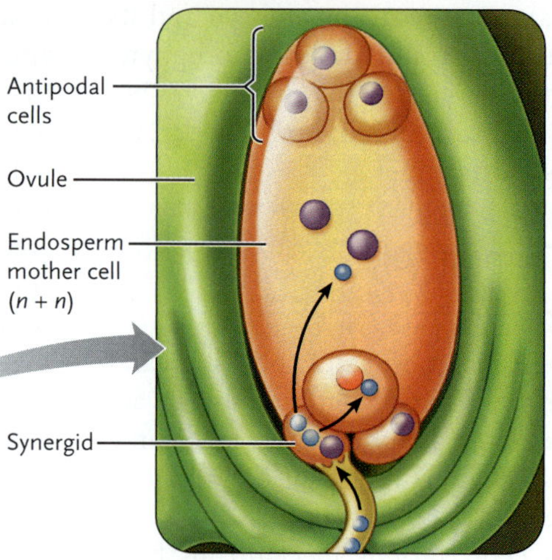

Step 1. One haploid sperm nucleus enters the endosperm mother (central) cell ($n + n$), forming a triploid ($3n$) cell from which nutritive endosperm develops. Later a seed will form as a seed coat encases the endosperm and embryonic sporophyte (see Figure 36.5).

Step 2. The other haploid sperm nucleus migrates through a synergid to the haploid egg. The egg is fertilized as its nucleus fuses with the sperm nucleus. The result is a diploid ($2n$) embryonic sporophyte.

SUMMARY Pollination of a compatible stigma triggers the growth of a pollen tube that delivers two haploid sperm nuclei to the ovule within an ovary. There one sperm nucleus fertilizes the egg and the other nucleus "fertilizes" the endosperm mother cell, triggering the development of endosperm. This process of double fertilization yields an embryonic sporophyte along with endosperm to nourish it.

THINK LIKE A SCIENTIST How can we learn more about chemical signals that guide the migration of a pollen tube toward the ovule?

and ferns, the gametophyte itself contains enough stored food to nourish the embryonic sporophytes. This observation suggests an adaptive advantage of double fertilization. While other seed plants expend energy on the development of a relatively large, nutrient-rich gametophyte that may or may not successfully contribute to reproduction, an angiosperm does not expend energy in filling an ovule with food reserves until after an egg has been fertilized.

The Embryonic Sporophyte Develops inside a Seed

When the zygote first forms, it starts to develop and elongate even before mitosis begins. For example, in shepherd's purse *(Capsella),* shown in **Figure 36.10,** most of the organelles in the zygote, including the nucleus, become housed in the top half of the cell, while a vacuole takes up most of the lower half (see Figure 36.10, step 1). The first round of mitosis divides the zygote into an upper *apical cell* and a lower *basal cell.* The apical cell then gives rise to the multicellular embryo, while most descendants of the basal cell form a simple row of cells, the **suspensor,** which transfers nutrients from the parent plant to the embryo (step 2).

The first apical cell divisions produce a globe-shaped structure attached to the suspensor. As they continue to grow, embryos of *Capsella* and other eudicots become heart-shaped (step 3). Each lobe of the "heart" is a developing cotyledon (seed leaf), which will provide nutrients for growing tissues in a germinating seedling. Typically, the two cotyledons absorb much of the nutrient-storing endosperm and become plump and fleshy. For instance, mature seeds of a sunflower *(Helianthus annuus)* have no endosperm at all. In some eudicots, however,

FIGURE 36.10 Stages in the embryonic development of shepherd's purse *(Capsella bursa-pastoris)*, a eudicot. The micrographs are not to the same scale. Figure 36.19 shows more detail of the development of early plant embryos.

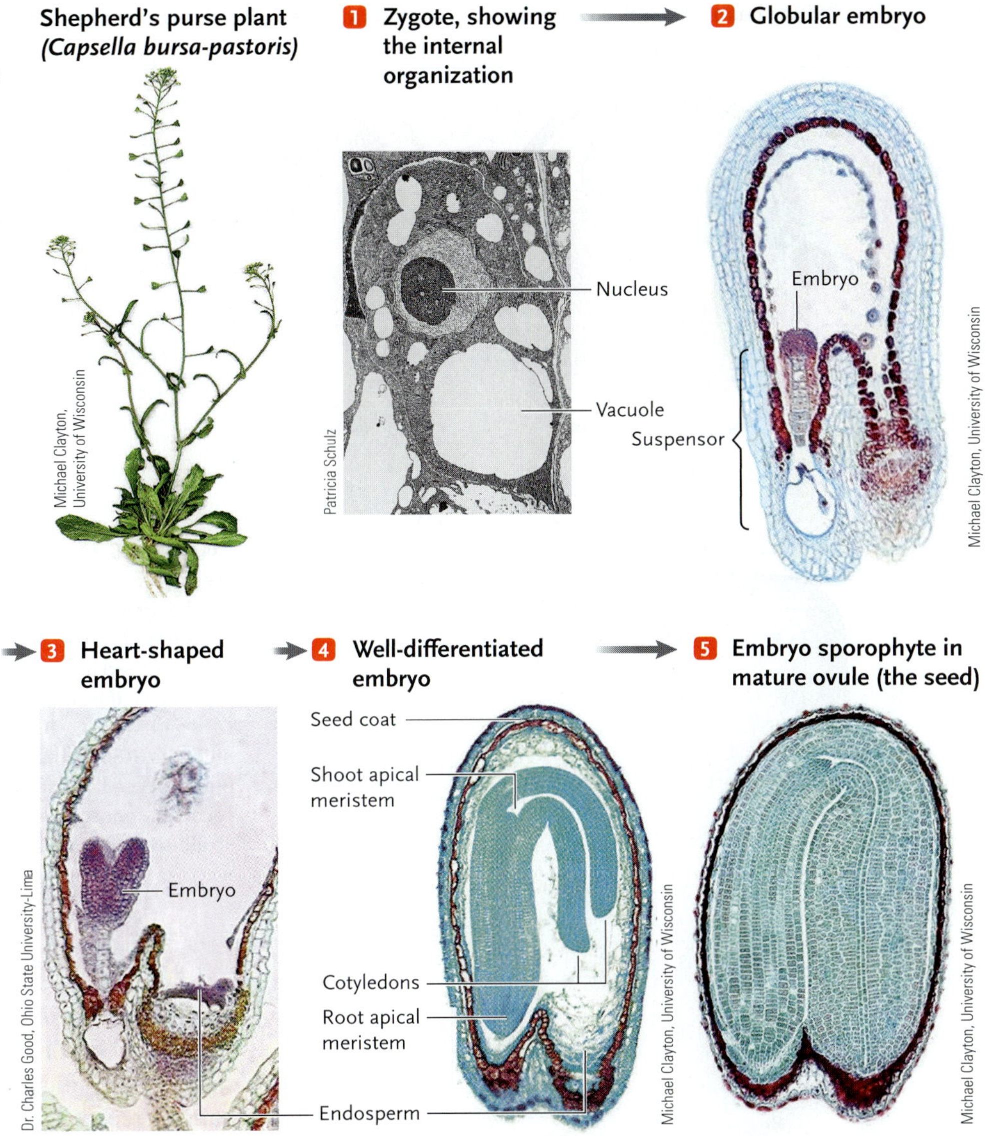

the cotyledons remain as slender structures; they produce enzymes that digest the seed's ample endosperm and transfer the liberated nutrients to the seedling. Monocots have one large cotyledon; in many monocot species, especially grasses such as corn and rice, the cotyledon absorbs the endosperm after germination, when the embryo inside the seed begins to grow.

A **seed** is a mature ovule. By the time the ovule is mature, its integument has developed into a protective **seed coat.** The embryo inside the seed has a lengthwise axis with a root apical meristem at one end and a shoot apical meristem at the other (steps 4 and 5).

Figure 36.11A and **Figure 36.11B** show the structural organization of seeds of two eudicots, the kidney bean *(Phaseolus vulgaris)* and the castor bean plant *(Ricinus communis).* The embryo makes up nearly all of both types of seeds. A mature kidney bean embryo has broad, fleshy cotyledons, and none of the seed's endosperm remains to nourish it. A castor bean embryo has much thinner cotyledons that are surrounded by endosperm, but in other ways the embryos are quite similar. The **radicle,** or embryonic root, is located near the micropyle, where the pollen tube entered the ovule prior to fertilization. The radicle attaches to the cotyledon at a region of cells called the **hypocotyl.** Beyond the hypocotyl is the **epicotyl,** which has the shoot apical meristem at its tip and which often bears a cluster of tiny foliage leaves, the **plumule.** At germination, when the root and shoot first elongate and emerge from the seed, the cotyledons are positioned at the first stem node with the epicotyl above them and the hypocotyl below them.

Deconstructing the seed of a monocot can be a bit more complex. To begin with, we see a great deal of variation in the cotyledons of monocots—some are large and plump with stored endosperm, while others (such as seeds of orchids and onions) are extremely small and contain little or no food stores for the embryo. **Figure 36.11C** shows the specialized seed of a grass—corn—which includes several structures not seen in the seeds of other angiosperm taxa. Housed inside a kernel (a type of fruit called a *grain*) a corn seed contains a large store of endosperm. The embryo has only one very large cotyledon, a shield-shaped mass called a **scutellum,** which absorbs nutrients from the endosperm. In monocot embryos generally, protective tissues blanket the root and shoot apical meristems, and in grasses the shoot apical meristem and plumule are covered by a **coleoptile,** a sheath of cells that protects them during upward growth through the soil. A similar covering, the **coleorhiza,** sheathes the radicle until it breaks out of the seed coat and enters the soil as the young plant's primary root.

Fruits Protect Seeds and Aid Seed Dispersal

Like the corn seed in its kernel, most angiosperm seeds are housed inside fruits, which provide protection and often aid seed dispersal. A **fruit** is a matured or ripened ovary. Usually, a fruit begins to develop after pollination and the subsequent fertilization of an egg in a flower's ovule. The start of ovule growth after fertilization is called "fruit set." The fruit wall, called the **pericarp,** develops from the ovary wall and can have several layers. Hormones in pollen grains provide the initial stimulus

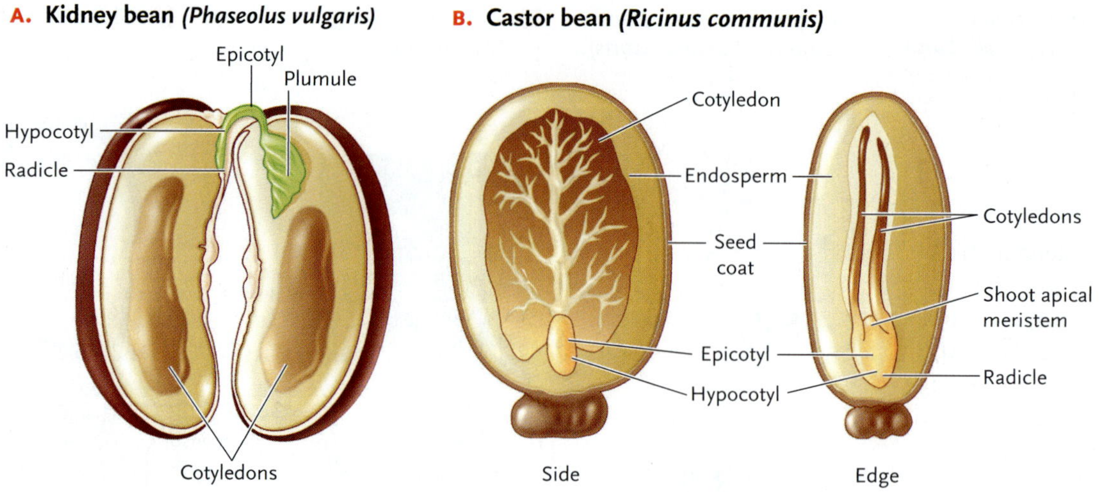

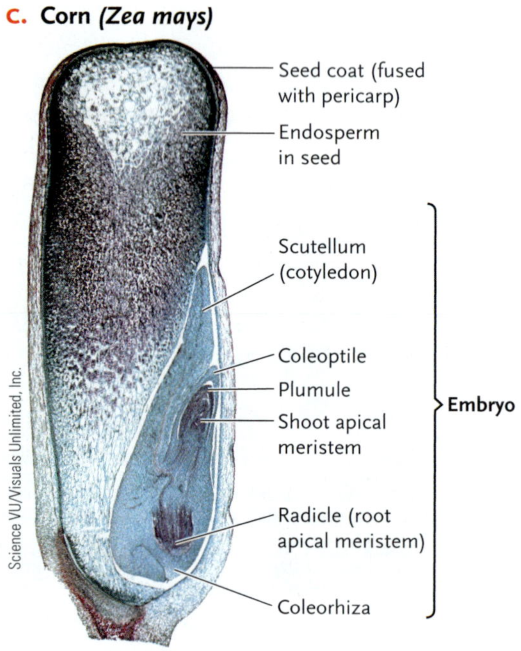

FIGURE 36.11 The structure of eudicot and monocot seeds. Eudicot seeds have two cotyledons, which store food absorbed from the endosperm, but the timing of this function varies in different species. **(A)** The cotyledons of a kidney bean *(Phaseolus vulgaris)* take up nutrients from endosperm while the seed develops, becoming fleshy. **(B)** In the castor bean *(Ricinus communis),* the endosperm is thick and the cotyledons are thin until the seed germinates, when the cotyledons begin to take up endosperm nutrients. **(C)** A kernel of corn *(Zea mays)* contains a representative monocot seed, which has been isolated here and is shown in longitudinal section. Monocot seeds have a single cotyledon, which develops into a shield-shaped scutellum that absorbs nutrients from endosperm.

that turns on the genetic machinery leading to fruit development; additional signals come from hormones produced by the developing seeds.

Fruits are extremely diverse, and biologists classify them into types based on combinations of structural features. A major defining feature is the nature of the pericarp, which may be fleshy (as in peaches) or dry (as in a hazelnut). A fruit also is classified according to the number of ovaries or flowers from which it develops. **Simple fruits** develop from a single ovary. In many of them, such as peaches, tomatoes, and the cacao fruits pictured in Figure 36.1, at least one layer of the pericarp is fleshy and juicy. Other simple fruits, including grains and nuts, have a thin, dry pericarp, which may be fused to the seed coat (like it is in corn). The garden pea *(Pisum sativa)* is a simple fruit, the peas being the seeds and the surrounding shell the pericarp. **Aggregate fruits** are formed from several ovaries in a single flower. Examples are raspberries and strawberries, which develop from clusters of individual ovaries. Strawberries also qualify as *accessory* fruits, in which floral parts in addition to each ovary become incorporated as the fruit develops. For instance, anatomically, the fleshy part of a strawberry is an expanded receptacle (the end of the floral shoot) and the strawberry fruits are the tiny, dry nubbins (called *achenes*) you see embedded in the fleshy tissue of each berry. *Multiple fruits* develop from several ovaries in multiple flowers. For example, a pineapple is a multiple fruit that develops from the enlarged ovaries of several flowers clustered together in an inflorescence. **Figure 36.12** shows examples of some different types of fruits.

Fruits not only protect seeds—they also aid seed dispersal in specific environments. For example, maple fruits have winglike extensions for dispersal (see Figure 36.12E). When the fruit drops, the wings cause it to spin sideways and downward and also can carry it away on a breeze. This aerodynamic property propels maple seeds to new locations, where they will not have to compete with the parent tree for water and minerals. Fruits also may have hooks, spines, hairs, or sticky surfaces, and they are ferried to new locations when they adhere to feathers, fur, or blue jeans of animals that brush against them. Fleshy fruits such as blueberries are nutritious food for many animals, and their seeds are adapted for surviving digestive enzymes in the animal gut. The enzymes remove just enough of the hard seed coats to increase the chance of successful germination when the seeds are expelled from the animal's body in feces.

A. **Peach *(Prunus)*, a fleshy simple fruit**

B. **Raspberry *(Rubus)*, an aggregate fruit**

C. **Strawberry *(Fragaria)*, an accessory fruit**

D. **Pineapple *(Ananas comosus)*, a multiple fruit**

E. **Maple *(Acer)*, a dry simple fruit**

FIGURE 36.12 **Fruits.** **(A)** Peach, a fleshy simple fruit. **(B)** Raspberry *(Rubus)*, an aggregate fruit. **(C)** Strawberry *(Fragaria ananassa)*, an accessory fruit that is also an aggregate fruit. **(D)** Pineapple *(Ananas comosus)*, a multiple fruit. **(E)** Winged fruits of maple *(Acer)* are dry simple fruits.

Seed Germination Continues the Life Cycle

A mature seed is essentially dehydrated. On average, water accounts for only about 10% of its weight—too little for cell expansion or metabolism. Germination gets underway when the seed begins to soak up water. The hydrated embryo then resumes growth. Ideally, a seed germinates when external conditions favor the survival of the embryo and growth of the new sporophyte. This timing is important, because once germination is underway the embryo loses the protection of the seed coat and other structures that surround it. Overall, the amount of soil moisture and oxygen, the temperature, day length, and other environmental factors influence when germination takes place.

In some species, the life cycle may include a period of seed **dormancy** (*dormire* = to sleep), in which the seed's metabolic activity is suspended. The conditions required for dormant seeds to germinate vary widely. Depending on the species, seeds may require minimum periods of daylight or darkness, repeated soaking, mechanical abrasion, or exposure to certain enzymes, the high heat of a fire, or a freeze–thaw cycle before they finally break dormancy. A control system based on the hormone abscisic acid, or ABA, underlies the shift from active metabolism to dormancy, and then reactivation and germination. When environmental conditions—perhaps too-cold temperature or drought—do not favor the survival of a potential seedling, seeds produce high levels of ABA. The hormone in turn stimulates synthesis of a transcription factor that turns on germination-inhibiting genes. When conditions eventually favor a seedling's survival, the genes switch off and dormancy ends. Kenji Miura of the University of Tsukuba in Japan and Mike Hasegawa at Purdue University recently discovered that this reversal occurs when a regulatory protein binds the transcription factor. Then, the germination-inhibiting genes are no longer expressed and germination can begin.

In some desert plants, hormones in the seed coat inhibit growth of a seedling until heavy rains flush them away. This adaptation ensures that seeds germinate only when there is enough water in the soil to support growth of the plant through the flowering and seed production stages before the soil dries again. Many plants in harsh environments, such as deserts, cycle from germination to growth, flowering, and seed development in the space of a few weeks and their offspring remain dormant as seeds until conditions once again favor germination and growth.

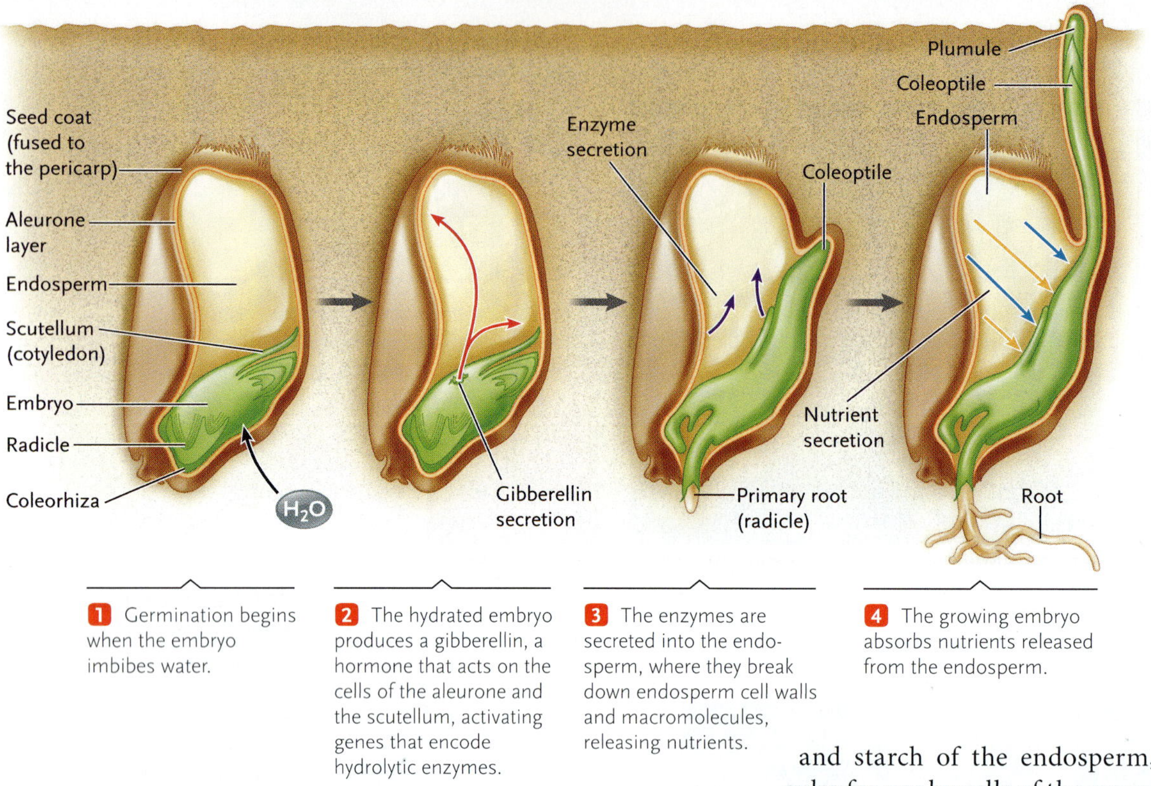

FIGURE 36.13 How food reserves are mobilized in a germinated seed within a grain of barley *(Hordeum vulgare)*, a monocot. Each grain is a fruit containing a single large seed.

Seeds of some species appear to remain viable for amazing lengths of time. Thousand-year-old lotus seeds *(Nelumbo lutea)* discovered in a dry lakebed have germinated trouble-free. And in one startling case, seeds of arctic lupine *(Lupinus arcticus)* were discovered along with fecal material in 10,000-year-old lemming burrows. When the seeds were thawed, they germinated readily as well.

Germination begins with **imbibition,** in which water molecules move into the seed, attracted to hydrophilic groups of stored proteins. As water enters, the seed swells, the coat ruptures, and the radicle begins its downward growth into the soil. Within this general framework, however, there are many variations among plants.

Once the seed coat splits, water and oxygen move more easily into the seed. Metabolism switches into high gear as cells divide and elongate to produce the seedling. Stable enzymes that were synthesized before dormancy become active; other enzymes are produced as the genes encoding them begin to be expressed. Among other roles, the increased gene activity and enzyme production mobilize the seed's food reserves in cotyledons or endosperm. Nutrients released by the enzymes sustain the rapidly developing seedling until its root and shoot systems are established.

The events of seed germination have been studied extensively in grains, and **Figure 36.13** illustrates them in barley. The seed's endosperm is separated from the pericarp by a thin layer of cells called the **aleurone.** As a hydrating seed imbibes water, the embryo produces a *gibberellin,* a hormone that stimulates aleurone cells to manufacture and secrete hydrolytic enzymes. Some of these enzymes digest components of endosperm cell walls; others digest proteins, nucleic acids, and starch of the endosperm, releasing nutrient molecules for use by cells of the young root and shoot. Although it is clear that nutrient reserves are also mobilized by metabolic activity in eudicots and in gymnosperms, the details of the process are not well understood.

Inside a germinating seed, embryonic root cells are generally the first to divide and elongate, giving rise to the radicle. When the radicle emerges from the seed coat as the primary root, germination is complete. **Figure 36.14** and **Figure 36.15** depict the stages of early development in a kidney bean, a eudicot, and in corn, a monocot. As the young plant grows, its development continues to be influenced by interactions of hormones and environmental factors, as you will read in Chapter 37.

Most plants give rise to large numbers of seeds because, in the wild, only a tiny fraction of seeds survive, germinate, and eventually grow into another mature plant. Also, flowers, seeds, and fruits represent major investments of plant resources. Asexual reproduction, discussed next, is a more "economical" means by which many plants can propagate themselves.

STUDY BREAK 36.3

1. **Explain the sequence of events in a flowering plant that begins with formation of a pollen tube and culminates with the formation of a diploid zygote and the 3*n* cell that will give rise to endosperm in a seed.**
2. **Early angiosperm embryos undergo a series of general changes as a seed matures. Summarize this sequence, then describe the structural differences that develop in the seeds of monocots and eudicots.**
3. **Germination begins when a seed imbibes water. What are the next key biochemical and developmental events that bring an angiosperm's life cycle full circle?**

2 Food-storing cotyledons are lifted above the soil surface when cells of the hypocotyl elongate. The hypocotyl becomes hook-shaped and forces a channel through the soil as it grows. At the soil surface, the hook straightens in response to light. For several days, cells of the cotyledons carry out photosynthesis; then the cotyledons wither and drop off. Photosynthesis is taken over by the first leaves that develop along the stem and later by foliage leaves.

3 Leaves break through the seed coat.

FIGURE 36.14 Stages in the development of a representative eudicot, the kidney bean *(Phaseolus vulgaris)*.

FIGURE 36.15 Stages in the development of a monocot, the corn plant *(Zea mays)*. The presence of a coleoptile over developing leaves is typical of grasses but not other monocots.

1 A corn grain germinates.

2 The young leaves are enclosed in a coleoptile, which protects them during upward growth through the soil. Adventitious roots develop from the first node at the base of the coleoptile.

3 The coleoptile emerges from the soil, and the primary root extends downward into it. The coleoptile and first foliage leaf of seedlings break through the soil surface.

THINK OUTSIDE THE BOOK

For many perennial plant species in temperate and high-latitude regions, dormancy is a major adaptation for surviving low winter temperatures. However, cold winters do not occur at tropical latitudes. Using library or Internet research sources, find out if there are environmental conditions that trigger automatic dormancy in wild-growing tropical plants. What implications do your findings have for gardeners in your area who want to grow tropical species year round?

FIGURE 36.16 ***Kalanchoe daigremontiana*, the mother-of-thousands plant.** Each tiny plant growing from the leaf margin can become a new, independent adult plant.

36.4 Asexual Reproduction of Flowering Plants

As we mentioned in Section 36.1, many flowering plants have the capacity for both sexual and asexual reproduction. Some flowering plants, including some citrus species and the grass variety known as Kentucky blue grass *(Poa pratensis),* can reproduce asexually through a mechanism called **apomixis,** in which there is no union of haploid maternal and paternal gametes. Typically, an embryo develops from an unfertilized egg (which is diploid due to a related shift in embryo sac development) or from diploid cells in the ovule tissue around the embryo sac. The resulting seed is said to contain a **somatic embryo,** which is genetically identical to the parent.

Another asexual alternative is **vegetative reproduction,** in which a new, genetically identical individual develops from a parent plant's nonreproductive tissue, usually a bit of meristematic tissue in a bud on the root or stem. Some of the stem modifications described in Section 33.3 function in this way. For instance, although strawberries can reproduce sexually by way of seeds, new roots and shoots also develop at each node along the stolons ("runners") a strawberry plant sends out. Likewise, new plantlets develop from "suckers" that sprout from the roots of blackberry bushes and "eyes" in the tubers of potatoes, or from buds on the short underground stems of onions and lilies.

Vegetative reproduction relies on an intriguing property of plants—namely, that many fully differentiated plant cells are **totipotent** (*totus* = all; *potens* = powerful). That is, they have the genetic potential to develop into a whole, fully functional plant. Under appropriate conditions, a totipotent cell can *dedifferentiate;* it returns to an unspecialized embryonic state, and the genetic program that guides the development of a new individual is turned on.

Vegetative Reproduction Is Common in Nature

Various plant species have developed different mechanisms for vegetative reproduction. In **fragmentation,** cells in a piece of the parent plant dedifferentiate and then can regenerate missing plant parts. Many gardeners have discovered to their frustration that a chunk of dandelion root left in the soil can rapidly grow into a new dandelion plant in this way. When a leaf falls or is torn away from a jade plant (*Crassula* species), a new plant can develop from meristematic tissue that is adjacent to the wound surface in the detached leaf. The "mother of thousands" plant, *Kalanchoe daigremontiana,* produces a wealth of seeds, while more or less simultaneously meristematic tissue in notches along the leaf margin gives rise to tiny plantlets **(Figure 36.16)** that eventually fall to the ground, where they can sprout roots and grow to maturity.

In wild plant species, most types of asexual reproduction result in offspring located near the parent. As already noted, these clonal populations lack the variability provided by sexual reproduction, variation that enhances the odds for survival when environmental conditions change. Yet asexual reproduction offers an advantage in some situations. It usually requires less energy than producing complex reproductive structures such as seeds and showy flowers to attract pollinators. Moreover, clones are likely to be well suited to the environment in which the parent grows. In addition, it may allow the rapid colonization of disturbed habitats, where seeds might be blown or washed away.

FIGURE 36.17 An orchard of grafted English walnut trees *(Juglans regia)* in California's Sacramento Valley. Even when the trees are several decades old, it is easy to distinguish their grafted scion and stock portions.

FIGURE 36.18 **Research Method**

Plant Cell Culture

Purpose: To determine if differentiated plant cells can be returned to a totipotent state.

Protocol: Fragments of tissue are excised from a plant and grown in a medium that contains all required nutrients. The procedure disrupts normal interactions among the excised cells. They dedifferentiate, forming an unorganized callus. When cells removed from the callus are grown in a medium containing nutrients and hormones, some develop into plantlets.

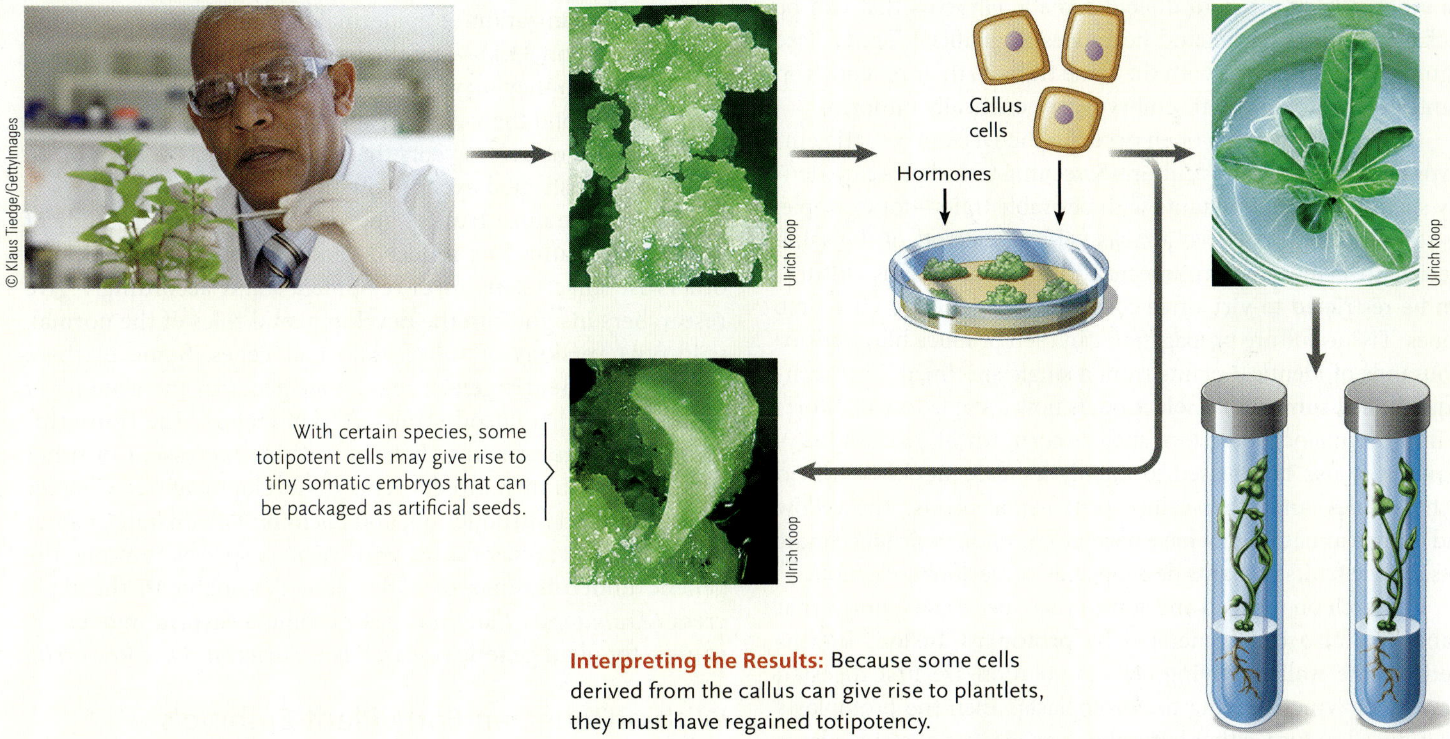

Interpreting the Results: Because some cells derived from the callus can give rise to plantlets, they must have regained totipotency.

Many Commercial Growers and Gardeners Use Artificial Vegetative Reproduction

For centuries, gardeners and farmers have used asexual plant propagation to grow particular crops and trees and some ornamental plants. They routinely use *cuttings,* pieces of stems or leaves, to generate new plants. Placed in water or moist soil, a cutting may sprout roots within days or a few weeks. Virtually the entire world supply of commercially sold bananas is harvested from plants grown from cuttings of a single variety called the "Cavendish banana." Because the tens of millions of plants that produce this banana are genetically uniform, the plants lack genetic diversity that would provide some measure of adaptive insurance against harmful fungi and other pests. As a result, by some estimates bananas are the most heavily pesticide-treated fruit in the global marketplace.

Fruit trees and wine grapes often are propagated by grafting a bud or branch from a plant with desirable fruit traits—the *scion*—and joining it to a root or stem from a plant with useful root traits—the *stock.* A grafted plant usually produces flowers and fruit identical to those of the scion's parent plant. The scion of a grafted wine grape or walnut variety may be chosen for the quality of its fruit, and the stock for its hardy, disease-resistant root system **(Figure 36.17).** Vegetative propagation can also be used to grow plants from single cells. Rose bushes and fruit trees from nurseries and commercially important fruits and vegetables such as Bartlett pears, McIntosh apples, Thompson seedless grapes, and asparagus come from plants produced vegetatively in tissue culture conditions that cause their cells to dedifferentiate to an embryonic stage.

VEGETATIVE PROPAGATION IN TISSUE CULTURE In groundbreaking experiments in the 1950s, Frederick C. Steward explored the totipotency of plant cells. Together with his coworkers at Cornell University, Steward propagated whole carrot plants in the laboratory by culturing carrot root phloem. Later researchers confirmed that almost any plant cell that has a nucleus and lacks a secondary cell wall may be totipotent.

The method of plant tissue culture Steward pioneered is simple in its general outlines **(Figure 36.18).** Bits of tissue are excised from a plant and grown in a nutrient medium. The procedure disrupts normal interactions between cells in the tissue sample, and the cells dedifferentiate and form an unorganized, white cell mass called a **callus.** When cultured with nutrients and growth hormones, some cells of the callus regain totipotency and develop into plantlets with roots and shoots.

Steward's work laid the foundation for large-scale commercial applications of plant tissue culture, as well as for a whole new field of research on *somatic embryogenesis* in plants. Single cells derived from a callus generated from shoot meristem are placed in a medium containing nutrients and hormones that promote cell differentiation. With some species, totipotent cells in the sample eventually give rise to diploid somatic embryos that can be packaged with nutrients and hormones in artificial "seeds" (see Figure 36.18). Endowed with the same traits as their parent, crop plants grown from somatic embryos are genetically uniform.

However, mutations often occur in the DNA of somatic embryos derived from callus culture. Screening techniques can identify such *somaclonal* mutants with desirable traits—for example, resistance to a disease that attacks wild-type plants of the same species. In plants that are infected with viruses, callus cultures can be restricted to virus-free cells and thus generate virus-free clones. Tissue culture propagation can then produce hundreds or thousands of identical plants from a single specimen. This technique, called **somaclonal selection,** is now a staple tool in efforts to improve major food crops, such as corn, wheat, rice, and soybeans. It is also being used to rapidly increase stocks of hybrid orchids, lilies, and other valued ornamental plants. The yellow and orange tomatoes that have become common in produce markets are the fruits of plants developed by somaclonal selection.

Research on tobacco and some other species has shown that plants can also be regenerated by **protoplast fusion.** In this method, the walls of living cells in solution are first digested away by enzymes, leaving the protoplasts. Then the protoplasts are induced to fuse, either by applying an electric current, a laser beam, or chemical additives to the solution, which briefly "loosen" the plasma membranes. The resulting cell (now $4n$, or tetraploid) is transferred to a solid nutrient medium and allowed to develop into a callus, then individual callus cells are stimulated to develop into embryos. If the fused protoplasts come from somatic cells of a single species, the embryos often grow into fertile plants. It has proven more difficult to grow fertile hybrids from fused protoplasts of different species and genera, probably because there are species-specific signals that govern key physiological functions. Even so, this method has produced the pomato, a cross between a potato and a tomato.

Regardless of how it comes into being, an embryonic sporophyte changes significantly as it begins the developmental journey toward maturity, when it will be capable of reproducing. Next we explore what researchers are learning about these developmental changes.

STUDY BREAK 36.4

1. **Describe three modes of asexual reproduction that occur in flowering plants.**
2. **What is totipotency, and how do methods of tissue culture exploit this property of plant cells?**

36.5 Early Development of Plant Form and Function

Unlike animals, plants have specialized body parts such as leaves and flowers that may arise from meristems throughout an individual's life—sometimes for thousands of years. Accordingly, in plants the biological role of embryonic development is not to generate the tissues and organs of the adult, but to establish a basic body plan—the root–shoot axis and the radial, "outside-to-inside" organization of epidermal, ground, and vascular tissues (see Section 33.1)—and the precursors of the primary meristems. Though they may sound simple, these fundamentals and the stages beyond them all depend on an intricately orchestrated sequence of molecular events that plant scientists are exploring through sophisticated experimentation.

One of the most fruitful approaches has been the study of plants with natural or induced gene mutations that block or otherwise affect steps in development—and accordingly give researchers insight into the developmental roles of the normal, wild-type versions of those abnormal genes. Some of these genes are **homeotic genes,** regulatory genes in the genome of an organism that encode transcription factors. The transcription factors are proteins that control the expression of other genes, which in turn direct events in development (see Chapter 16 for a general introduction, and *Focus on Research* in Chapter 33). While researchers work with various species to probe the genetic underpinnings of early plant development, the thale cress *(Arabidopsis thaliana)* has become a favorite model organism for plant genetic research (see *Focus on Basic Research*).

Within Hours, an Early Plant Embryo's Basic Body Plan Is Established

The entire *A. thaliana* genome has been sequenced, providing a powerful molecular "database" for determining how various genes contribute to shaping the plant body. Experimenters' ability to trace the expression of specific genes has shed considerable light on how the root–shoot axis is set and how the three basic plant tissue systems arise.

THE ROOT–SHOOT AXIS Shortly after fertilization gives rise to an *A. thaliana* zygote, the single cell divides. Electron microscopy shows that, as with the *Capsella* zygote described earlier, this first round of mitosis produces a small apical cell and a larger basal cell **(Figure 36.19A).** The apical cell receives the lion's share of the cytoplasm, while the basal cell receives the zygote's large vacuole and less cytoplasm. Researchers have confirmed that in this asymmetrical division of the zygote, the daughter cells receive different mixes of mRNAs—the gene transcripts that will be translated into proteins.

Translation of differing mRNAs produces proteins that include several transcription factors. As these factors trigger the expression of differing genes, distinct biochemical pathways unfold in the two cells in ways that establish the organization of the plant body. For example, a basal cell initially exports a

FOCUS ON Basic Research

Model Research Organisms: *Arabidopsis thaliana*

For plant geneticists, the little white-flowered thale cress, *A. thaliana* has attributes that make it a prime subject for genetic research. A tiny member of the mustard family Brassicaceae, *A. thaliana,* is revealing answers to some of the biggest questions in plant development and physiology.

Each plant grows only a few centimeters tall, so little laboratory space is required to house a large population. As long as *A. thaliana* is provided with damp soil containing basic nutrients, it grows easily and rapidly in artificial light. Like Mendel's peas, *A. thaliana* is self-compatible and self-fertilizing, and the flowers of a single plant can yield thousands of seeds per mating. Seeds grow to mature plants in just over a month and then flower and reproduce themselves in another 3 to 4 weeks. This permits investigators to perform desired genetic crosses and obtain large numbers of offspring having known, desired genotypes with relative ease. Individual *A. thaliana* cells also grow well in culture.

Arabidopsis thaliana

Elliot Meyerowitz

The *A. thaliana* genome was the first complete plant genome to be sequenced; at this writing researchers have identified approximately 28,000 genes arranged on five pairs of chromosomes. The genome contains relatively little repetitive DNA, so it is fairly easy to isolate *A. thaliana* genes, which can then be cloned using genetic engineering techniques. Cloned genes are inserted into bacterial plasmids and the recombinant plasmids transferred to the bacterial species *Agrobacterium tumefaciens,* which readily infects *A. thaliana* cells. Amplified by the bacteria, the genes and their protein products can be sequenced or studied in other ways.

Typically, researchers use chemical mutagens or recombinant bacteria to introduce changes in the *A. thaliana* genome. These mutants have become powerful tools for exploring molecular and cellular mechanisms that operate in plant development—for example, elucidation of the homeotic genes responsible for flower development described in this chapter. *A. thaliana* mutants are also being used to probe fundamental questions such as how plant cells respond to gravity, and what roles the pigments called phytochromes play in plant responses to light.

Several years ago an ambitious, multinational research effort was launched to determine the functions of all *A. thaliana* genes. This work is providing a comprehensive genetic portrait of a flowering plant—how each gene affects the functioning of not only individual cells but of the plant as a whole. The Arabidopsis Information Resource (TAIR) tallies the percentages of *A. thaliana* genes in different functional categories **(Figure).**

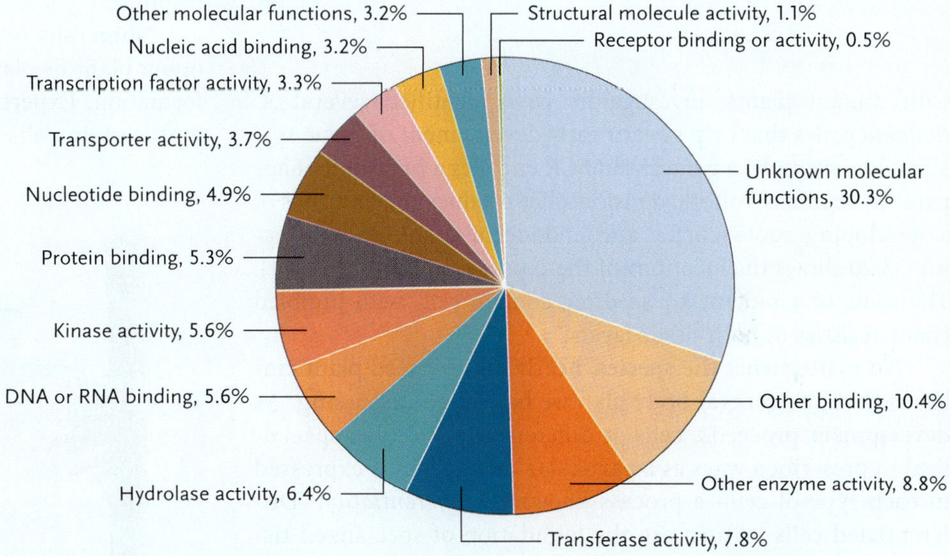

(Courtesy of the Arabidopsis Information Resource, 2009.)

FIGURE The percentages of *A. thaliana* genes that influence different functional categories.

signaling molecule (a hormone of the auxin family, discussed in Chapter 37) to the apical cell, and this sets in motion steps leading to the development of the various embryonic shoot features (Figure 36.19B). Later, gene expression and the flow of chemical signals shift in ways that promote the development of specific structures in the basal cell, including portions of the root apical meristem.

Several of the genes that influence root–shoot polarity have been identified, and when any of them is disrupted, the result can be a serious defect. For example, when an embryo receives two copies of a mutant gene called *gnom,* the embryo doesn't develop distinct root and shoot regions. Instead it remains a lumpy blob.

RADIAL ORGANIZATION OF TISSUE LAYERS A day or so after an *A. thaliana* egg cell is fertilized, the embryo consists of eight cells. Even at this early stage, both the root–shoot axis and the beginnings of tissue systems are present. When an embryo reaches the so-called torpedo stage, cells representing all three basic tissue systems are in place **(Figure 36.19C).** Again working

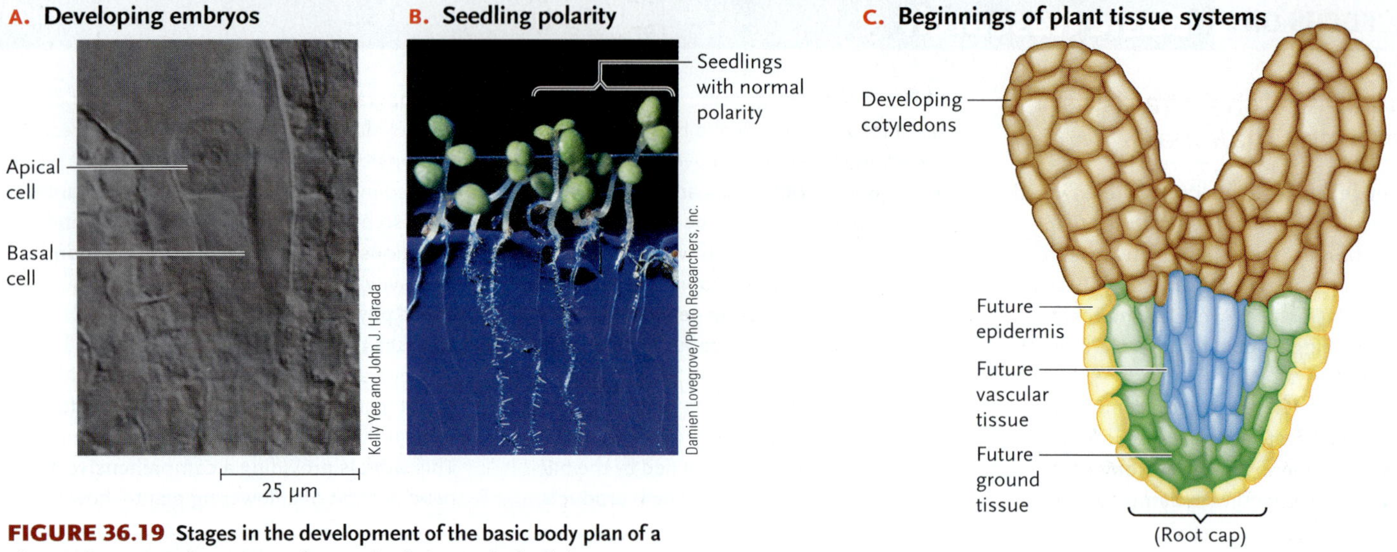

FIGURE 36.19 Stages in the development of the basic body plan of a plant embryo. (A) After a zygote forms, the first round of cell division produces an embryo with an apical cell that contains much of the zygote's cytoplasm, and a larger basal cell that receives the vacuole and less cytoplasm. The division allots different transcription factors to each cell and establishes the plant's root–shoot axis. **(B)** Normal *A. thaliana* seedlings, in which the root–shoot polarity has become established. **(C)** The approximate locations of early embryonic cells that are the forerunners of epidermal, ground, and vascular tissue systems, respectively.

with mutant plants, investigators have identified several *A. thaliana* genes that help govern early development of tissue systems. For example, a gene called *SCR* encodes a protein that apparently regulates mitotic divisions that produce the first cells of a developing root's cortex and endoderm tissue layers (Figure 33.20 shows the locations of these tissues in a mature root). The roots of a mutant *scr* seedling contain cells with jumbled characteristics of both tissue layers.

No matter what the species, nearly all new seed plant embryos have the general body plan we have been discussing. As development proceeds, cells at different sites become specialized in prescribed ways as a particular set of genes is expressed in each type of cell—a process known as *differentiation.* Differentiated cells in turn are the foundation of specialized tissues and organs, which come about through processes we consider next.

Key Developmental Cues Are Based on a Cell's Position

Although many of the specifics of development differ in animals and plants, one fundamental holds true for both: normal development produces ordered spatial arrangements of differentiated tissues. Examples in plants include root and shoot apical meristems at opposite ends of the root–shoot axis, the cotyledons that divide the shoot into an upper epicotyl and a lower hypocotyl, and the nested layers of vascular, ground, and epidermal tissue systems. Developmental biologists call this progressive ordering of parts **pattern formation,** and a wealth of research has shown that it is guided by the position of cells relative to one another. Like the information provided by a hormone gradient mentioned in Section 36.2, this positional information provides cues that "tell" cells where they are in the developing embryo and thus lay the groundwork for an appropriate genetic response.

Numerous researchers have explored how cells in a developing plant or plant part receive and respond to positional information. Experiments have demonstrated, for example, that only certain cells in the epidermis of an embryonic root will

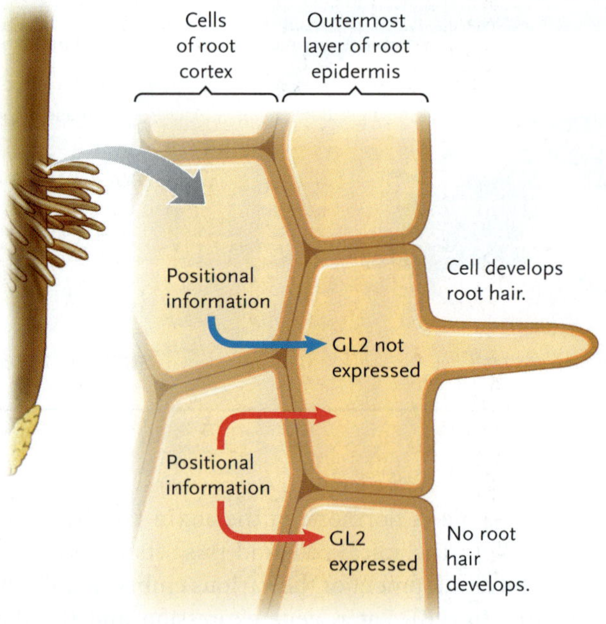

FIGURE 36.20 One model of how positional information influences the development of root hairs. In this model, the only epidermal cells that develop root hairs are those whose inner wall is in contact with two root cortex cells. Such positioning gives rise to signals that block the expression of the GL2 (GLABRA2) gene. When GL2 is expressed, a root epidermal cell will not develop a root hair.

INSIGHTS FROM THE Molecular Revolution

Trichomes: Window on development in a single plant cell

The delicate plant cell extensions called trichomes are helping to illuminate developmental processes that go on in a single plant cell as it differentiates—that is, as it acquires its ultimate specialized structure and function. In *A. thaliana,* each of these minute protuberances consists of a single cell with a branching tripartite pattern **(Figure 1).**

As a trichome differentiates, increases in size, and extends branches in different directions, its chromosomes—and the cell's DNA—are duplicated several times over without mitosis. As a result of the process, called *endoduplication,* the cell has multiple copies of its chromosomes. Experiments that isolate the effects of different mutants have helped confirm that the overall amount of DNA in the cell strongly influences the cell's structure, and also that several genes interact to determine the structure. One of these genes is called *TRY* (for *TRIPTYCHON*). When it is mutated, the affected plant's trichomes have a double complement of DNA and develop extra branches **(Figure 2).** But genes that influence endoduplication are only part of the story. Experiments with other mutants show that numerous other genes also help produce the characteristic three-pronged trichome branching. For example, when a gene called *STICHEL* is mutated, *sti* mutants don't develop any branches **(Figure 3).** The underlying reason for this phenotype is not yet understood.

These examples underscore how complex molecular interactions that affect multiple aspects of a cell's functioning ultimately determine a cell's form and function. Because the genes that operate in trichomes are also involved in the development of other types of plant cells, studying them promises to shed light on processes that generate differentiated cells throughout the plant body.

THINK LIKE A SCIENTIST In addition to being situated at the surface of leaf and stem epidermis, trichomes are transparent. Given these attributes, what other sorts of cell processes might mutant trichomes be used to study?

Source: Sven Schellman and Martin Hülskamp, 2005. Epidermal differentiation: Trichomes in *Arabidopsis* as a Model System. *Int. J. Dev. Biol.* 49:579–584.

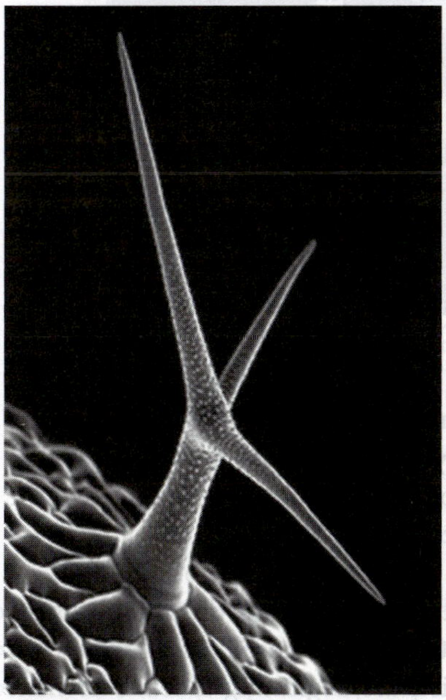

Jonathan Plett and Sharon Regan

FIGURE 1 Normal trichome from the epidermis of a leaf of *Arabidopsis thaliana.*

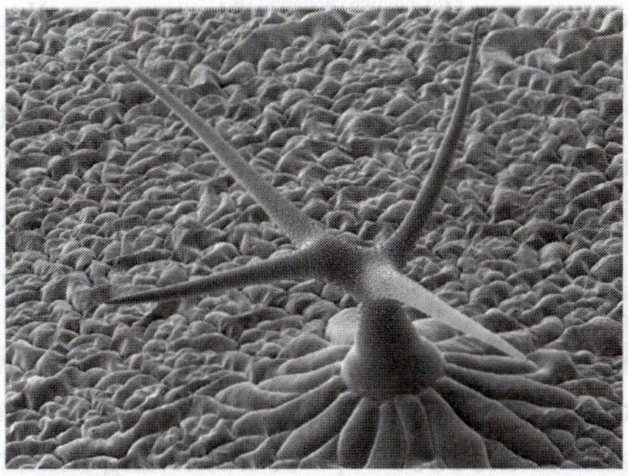

Dr. Stanley Flegler/Visuals Unlimited, Inc.

FIGURE 2 Abnormal trichome from a *try* mutant.

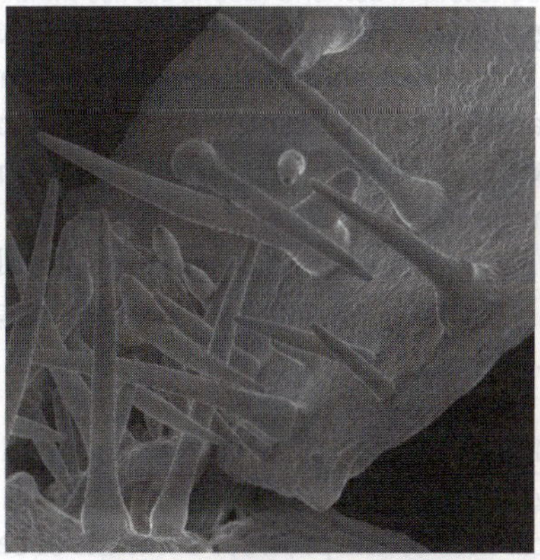

© Dr. Daniel Szymanski, Agronomy Department, Purdue University

FIGURE 3 The unbranched trichomes of an *sti* mutant.

give rise to root hairs, the type of trichomes that take up water and minerals from soil (see Section 33.2). These specialized root epidermal cells all share the same position with respect to the underlying root cortex—each abuts two cortical cells. By contrast, no root hair extension will develop from an epidermal cell that lines up against only one cortical cell. **Figure 36.20** diagrams one model of what happens next. In this scenario, one or more chemical signals may cross from cortical to epidermal cells by way of plasmodesmata. When an epidermal cell receives signals from a single cortex cell, a series of genes are expressed in a cascade of effects that culminate in the expression of a gene called GL2 (or GLABRA2). The product of GL2 blocks the formation of root hairs. If, on the other hand, an epidermal cell aligns with two cortex cells, it receives signals from both and the cascade of gene effects blocks expression of GL2—and a root hair develops. *Insights from the Molecular Revolution* gives more examples of ways that trichomes such as root hairs have become popular experimental models for studying the differentiation of plant cells.

Morphogenesis Shapes the Plant Body

As tissues of differentiated cells form in a plant embryo, the stage is set for different body regions to develop characteristic shapes and structures that correlate with their function. This

FIGURE 36.25 Early stages in leaf development. (A) Leaf primordia at the shoot tip of a member of *Coleus*, a genus that includes several popular house plants. **(B)** Diagram showing leaf primordia in different stages of development. See also Figure 36.13, which shows the progression of leaves that form during the early growth and development of a eudicot.

A. Leaf primordia of *Coleus*

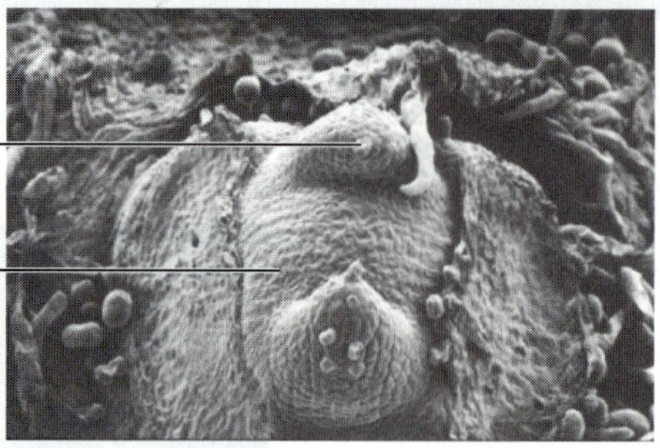

B. Early stages of leaf development

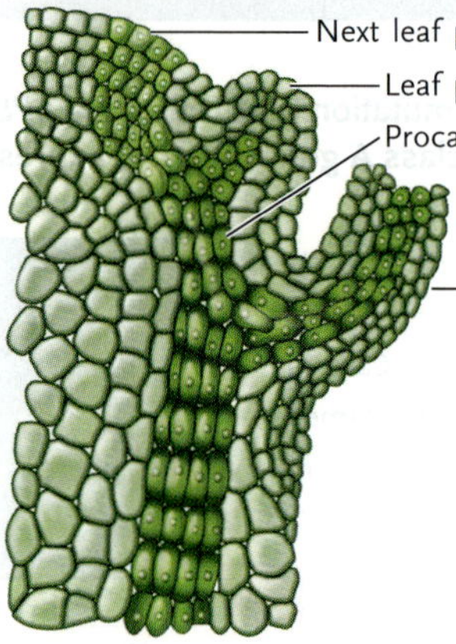

have used clonal analysis to study leaf development in garden peas, tomatoes, grasses, and tobacco, among others.

Like flowers, leaves arise through a developmental program that begins with gene-regulated activity in meristematic tissue. Hormones or other signals may arrive at target cells via the stem's vascular tissue, activating genes that regulate development. Studies show that small phloem sieve tubes penetrate a young leaf primordium almost immediately after it begins to bulge out from the underlying meristematic tissue, and xylem soon follows. The early phloem connections are especially vital to the leaf's survival, because the leaf does not begin photosynthesis until it attains one-third of its mature size.

A growing primordium becomes cone-shaped, wider at its base than at its tip. At a certain point, mitosis speeds up in cells along the flanks of the lengthening cone. In eudicots, the rapid cell divisions occur perpendicular to the surface and produce the leaf blade. In monocot grasses, which have long, narrow leaves, vertical "files" of cells develop as cells in the meristem at the base of the cone divide in a plane parallel to the surface. The ultimate shape of a leaf depends in part on variations in the plane and rate of these cell divisions.

Cells at the leaf tip, and those of xylem and phloem that service it, are the oldest in the leaf, and it is here that photosynthesis begins. Commonly, leaf tip cells also are the first to stop dividing. By the time a leaf has expanded to its mature size, all mitosis has ended and the leaf is a fully functional photosynthetic organ.

In nature, genes that govern leaf and flower development switch on or off in response to changing environmental conditions. In many perennials, new leaves begin to develop inside buds in autumn, then become dormant until the following spring, when external conditions favor further growth. Environmental cues stimulate the gene-guided production of hormones that travel through the plant in xylem and phloem, triggering renewed leaf growth and expansion. Leaves and other shoot parts also age, wither, and fall away from the plant as hormonal signals change. The far-reaching effects of plant hormones on growth and development are the subject of Chapter 37.

STUDY BREAK 36.5

1. **What is a homeotic gene? Give at least two examples of plant tissues such genes might govern in a species such as *A. thaliana*.**
2. **What are the two basic mechanisms of morphogenesis in plants? Describe the patterns of cell division by which a plant part (a) grows longer and (b) adds girth.**
3. **Summarize the gene-guided developmental program that gives rise to a leaf.**

UNANSWERED QUESTIONS

What are the chemical cues from female tissues that facilitate compatible pollination?

As you learned in this chapter, if a pollen tube lands on a compatible stigma, chemical cues produced by the female tissue then guide the pollen tube from the stigma to the embryo sac of an ovule. Several lines of evidence point to a role for both sporophytic and gametophytic tissues in proper guidance of pollen tubes to ovules. Recently, it was shown that cysteine-rich polypeptides (LUREs) secreted from synergid cells facilitate pollen tube entry into the ovules of *Torenia fournieri* (wishbone flower). Despite these advances, the identities of the cues released by other pistil tissues remain unknown. What hurdles have hampered the efforts to uncover pollen-tube navigation cues of even the known signaling events described above? First, the pistil tissue within which the pollen tube elongates to the ovule is comprised of several types of tissue, including stigma, style, and transmitting tract, and these tissues are not readily accessible. Second, analyzing the dynamic responses of pollen tubes is difficult given that pollen-tube navigation occurs well within opaque pistils. Third, it appears that multiple, stage-specific, short-range, and readily labile signals produced in minute quantities mediate pollen-tube guidance. However, recent development of highly sensitive global approaches and assays that directly monitor pollen-tube elongation offer hope that guidance cues will be uncovered sooner rather than later.

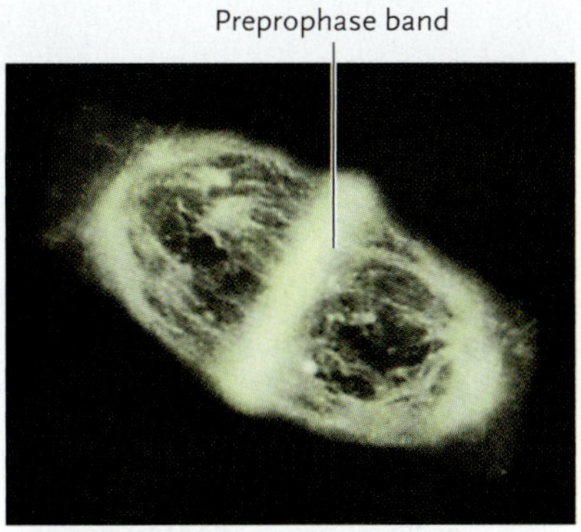

A. In preprophase of mitosis, the microtubules break their links to the plasma membrane and assemble into a thick preprophase band.

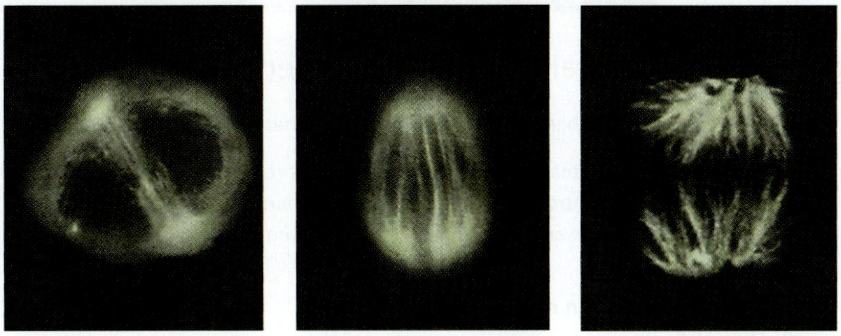

B. As cell division begins, the preprophase band disassembles and microtubules organize into a spindle.

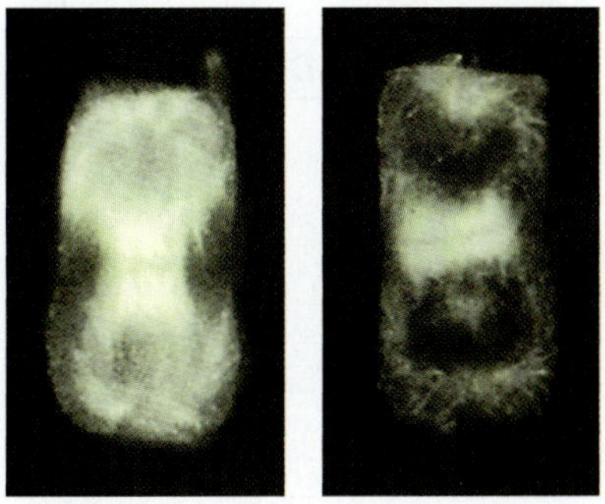

C. After division is complete, the spindle disassembles, leaving a layer of microtubules in the same plane as the preprophase band. The new cell wall and cell plate form there.

FIGURE 36.22 How the plane of cell division is determined in a plant cell. This series of micrographs shows events in onion *(Allium cepa)* root tip cells, starting with preprophase of mitosis.

FIGURE 36.23 Cell expansion and the orientation of cellulose microfibrils. In each cell, microtubules inherited from the parent cell are already oriented in prescribed patterns that govern how cellulose microfibrils will be oriented in the cell wall. Their orientation in turn governs the direction in which a cell can expand.

A. When microfibrils are oriented at random, the primary wall is elastic all over, so the cell can grow in all directions.

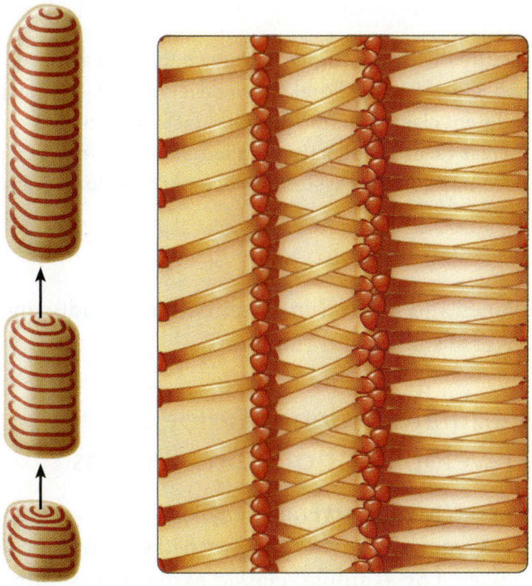

B. When microfibrils are oriented transversely, the cell can grow only longitudinally.

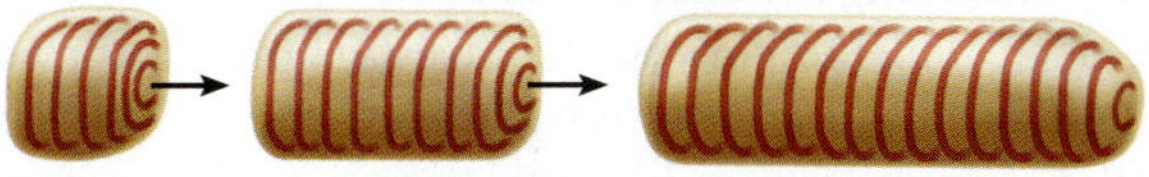

C. When microfibrils are oriented longitudinally, the cell can grow only laterally.

The direction of cell expansion depends on the orientation of the newly formed cellulose microfibrils **(Figure 36.23).** If the microfibrils are randomly oriented, the cell expands equally in all directions. If they are oriented at right angles to the cell's long axis, the cell expands lengthwise. And if new fibrils are deposited parallel to the long axis of the cell, the cell expands laterally.

PATTERNS OF CELL DIVISION DURING EARLY GROWTH Like the first mitotic division in an *A. thaliana* zygote, it's not uncommon for cell divisions in a growing plant to be asymmetrical, so that one daughter cell ends up with more cytoplasm than the other. The unequal distribution of cytoplasm means that the daughter cells differ in their composition and structure, and the differences affect how they interact with their neighbors during growth, even though all cells carry the same genes. Their cytoplasmic differences and interactions with one another trigger selective gene expression. Such events seal the developmental fate of particular cell lineages. Their descendant cells divide in prescribed

FIGURE 36.24 Experimental Research

Probing the Roles of Floral Organ Identity Genes

Question: What are the genetic mechanisms that govern the formation of the parts of a flower?

Experiments: Starting in the 1990s, teams of scientists at various research centers grew *A. thaliana* plants having mutated, inactivated versions of the genes that were suspected of controlling the proper development of floral organs. They compared the types and arrangements of floral organs in the test plants with the organs present in normal, wild-type *A. thaliana* flowers.

A. Floral organ whorls in normal *A. thaliana*

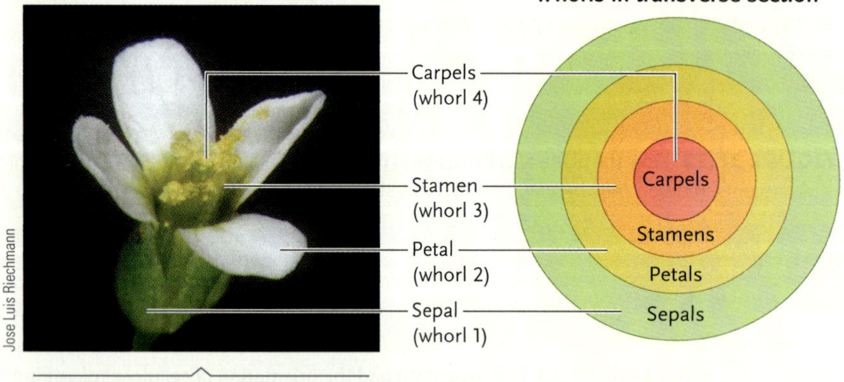

Normal arrangement of organs: carpels in whorl 4, stamens in whorl 3, petals in whorl 2, and sepals in whorl 1

B. Overlapping activity fields of floral organ identity genes

		C
	B	C
A	B	
A		
E	E	E

Results: Early experiments showed that at least three classes of homeotic genes—A, B, and C—encoded transcription factors that regulate different aspects of normal floral organ development. Later experiments by others revealed an E class of floral homeotic genes. Various combinations of the different classes of transcription factors produce a normal *A. thaliana* flower's whorls (parts A and B). Particular combinations and dosages of E-class factors, collectively called *SEPALLATA* or *SEP* genes, are required for the normal development of all four whorls. Mutant plants are a window onto this complex system for establishing floral organ identity (Figures C–F).

planes and expand in set directions, producing plant parts with diverse shapes and functions.

Regulatory Genes Guide the Development of Floral Organs

Research with several plant species has shed light on the genetic mechanisms that govern the formation of the parts of a flower. For example, classic experiments with *A. thaliana* carried out by Elliot Meyerowitz and his colleagues at the California Institute of Technology showed that *floral organ homeotic genes* regulate the development of the sepals, petals, stamens, and carpels in flowers.

The Meyerowitz team studied plants with various mutations in floral organs. By observing the effects of specific mutations on the structure of *A. thaliana* flowers, the investigators identified three classes of homeotic gene activity—which they named A, B, and C—that appeared to regulate different aspects of normal flower development in eudicots. The gene products are transcription factors. Subsequent studies by other scientists identified an E class of gene activity that apparently is an essential partner in the functioning of the other floral organ homeotic genes. Today there is widespread acceptance of this **ABCE model of floral organ patterning (Figure 36.24).** Expression of A, B, and C class genes overlaps and the products of E class genes function in all four whorls.

Abnormal floral patterns such as those in Figure 36.24C–F show how deactivation of floral organ homeotic genes can affect the development of flower parts. For example, a mutation that deactivates the A-class gene *APETALA2* produces a flower having both carpels and stamens in whorl 1. An intriguing finding is that the A and C activity classes normally oppose each other. When no A gene is expressed, C activity spreads into whorls where the A usually occurs, and vice versa. Also, the E-class genes are "redundant": the transcription factors they encode have such similar effects that an abnormal phenotype develops only when three or more of the genes are deactivated. Subsequent studies have examined many other floral homeotic genes, as well as the genes that control the various gene classes.

C. When mutation inactivates the *APETALA2* gene, class A genes are not expressed.

Stamens replace petals and carpels replace sepals. Organ identity in the other whorls does not change.

D. When mutation inactivates the *APETALA3* or *PISTILLATA* gene, class B genes are not expressed.

Carpels replace stamens and sepals replace petals. Organ identity in the other whorls does not change.

E. When mutation inactivates the *AGAMOUS* gene, class C genes are not expressed.

No carpels; instead petals develop in whorl 3 and a version of a floral meristem develops where whorl 4 would normally be. It gives rise to extra petals and sepals. Organ identity in other whorls does not change.

F. When mutation inactivates all four known *SEP* genes, class E genes are not expressed.

Leaflike structures develop but flower parts are poorly defined.

G. SEP4 mutant *A. thaliana* plant

Conclusions: In *A. thaliana,* ABCE activity genes underlie the development of a normal pattern of floral organs. The fields of activity overlap. Further studies have shown that A and C activity apparently counteract each other, and if one is absent the other can spread beyond the whorls where it normally appears. Dosage-dependent activity of E-class *SEP* genes is required for proper expression of all other organ identity genes. The SEP genes are redundant; that is, to some degree their products perform the same or similar functions. As a result, their role in establishing floral organ identity only became apparent when researchers developed mutants lacking functional copies of at least three of the genes.

THINK LIKE A SCIENTIST Do other flowering plants have a comparable system of floral organ identity genes?

Source: This box summarizes experimental studies by several research teams: D. Weigel and E. M. Meyerowitz, 1993. Activation of floral homeotic genes in *Arabidopsis*. *Science* 261, 1723–1726. J. L. Bowman, J. Alvarez, D. Weigel, E. M. Meyerowitz, and D. R. Smyth, 1993. Control of flower development in *Arabidopsis thaliana* by *APETALA1* and interacting genes. *Development* 119, 721–743. S. Pelaz, G. Ditta, S. Kohlami, E. Baumann and M. Yanofsky, 2000. B and C floral organ identity functions require *SEPALLATA* MADS-box genes. *Nature* 405, 200–203.

As the genes governing flower development are isolated, they can be cloned and their nucleotide sequences defined and manipulated. Such cloned genes already are of keen interest in plant genetic engineering, because food grains such as wheat and many other vital agricultural commodities come directly or indirectly from flowers.

Leaves Arise from Leaf Primordia in a Closely Regulated Sequence

A mature leaf may have many millions of differentiated cells organized into tissues such as epidermis and mesophyll. As described in Chapter 33, leaves develop from leaf primordia that arise just behind the tips of shoot apical meristems **(Figure 36.25)**.

Clonal analysis has opened a window on many aspects of plant development, including how leaf primordia originate and give rise to leaves. In this method, the investigator cultures meristematic tissue that contains a mutated embryonic cell having a readily observable trait, such as the absence of normal pigment. (In the laboratory, this kind of mutation can be induced by chemicals or radiation.) The unusual trait then serves as a marker that identifies the mutant cell's clonal descendants, making it possible to map the growing structure. Researchers

FIGURE 36.25 Early stages in leaf development. (A) Leaf primordia at the shoot tip of a member of *Coleus*, a genus that includes several popular house plants. **(B)** Diagram showing leaf primordia in different stages of development. See also Figure 36.13, which shows the progression of leaves that form during the early growth and development of a eudicot.

A. Leaf primordia of *Coleus*

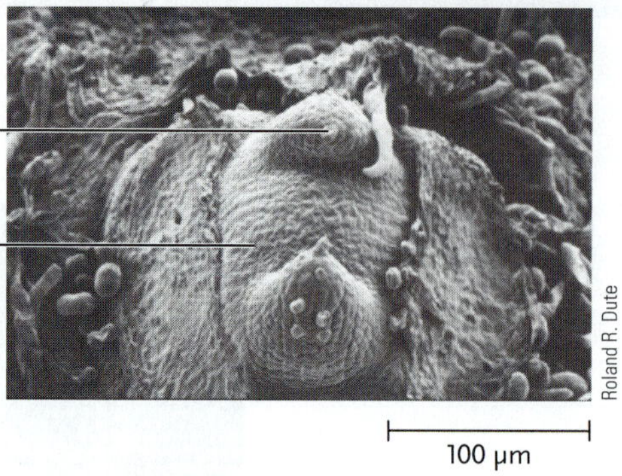

B. Early stages of leaf development

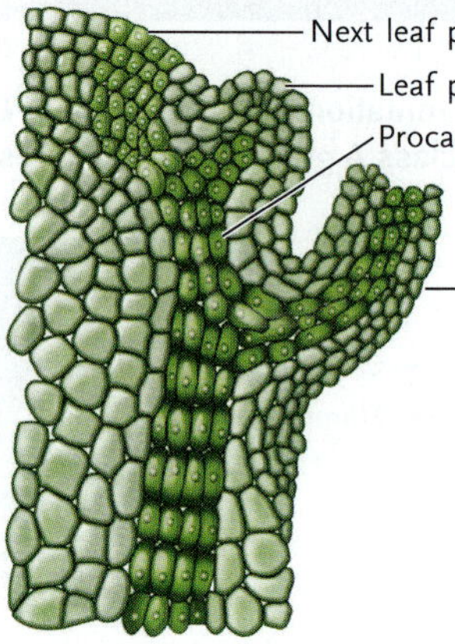

have used clonal analysis to study leaf development in garden peas, tomatoes, grasses, and tobacco, among others.

Like flowers, leaves arise through a developmental program that begins with gene-regulated activity in meristematic tissue. Hormones or other signals may arrive at target cells via the stem's vascular tissue, activating genes that regulate development. Studies show that small phloem sieve tubes penetrate a young leaf primordium almost immediately after it begins to bulge out from the underlying meristematic tissue, and xylem soon follows. The early phloem connections are especially vital to the leaf's survival, because the leaf does not begin photosynthesis until it attains one-third of its mature size.

A growing primordium becomes cone-shaped, wider at its base than at its tip. At a certain point, mitosis speeds up in cells along the flanks of the lengthening cone. In eudicots, the rapid cell divisions occur perpendicular to the surface and produce the leaf blade. In monocot grasses, which have long, narrow leaves, vertical "files" of cells develop as cells in the meristem at the base of the cone divide in a plane parallel to the surface. The ultimate shape of a leaf depends in part on variations in the plane and rate of these cell divisions.

Cells at the leaf tip, and those of xylem and phloem that service it, are the oldest in the leaf, and it is here that photosynthesis begins. Commonly, leaf tip cells also are the first to stop dividing. By the time a leaf has expanded to its mature size, all mitosis has ended and the leaf is a fully functional photosynthetic organ.

In nature, genes that govern leaf and flower development switch on or off in response to changing environmental conditions. In many perennials, new leaves begin to develop inside buds in autumn, then become dormant until the following spring, when external conditions favor further growth. Environmental cues stimulate the gene-guided production of hormones that travel through the plant in xylem and phloem, triggering renewed leaf growth and expansion. Leaves and other shoot parts also age, wither, and fall away from the plant as hormonal signals change. The far-reaching effects of plant hormones on growth and development are the subject of Chapter 37.

STUDY BREAK 36.5

1. **What is a homeotic gene? Give at least two examples of plant tissues such genes might govern in a species such as *A. thaliana*.**
2. **What are the two basic mechanisms of morphogenesis in plants? Describe the patterns of cell division by which a plant part (a) grows longer and (b) adds girth.**
3. **Summarize the gene-guided developmental program that gives rise to a leaf.**

UNANSWERED QUESTIONS

What are the chemical cues from female tissues that facilitate compatible pollination?

As you learned in this chapter, if a pollen tube lands on a compatible stigma, chemical cues produced by the female tissue then guide the pollen tube from the stigma to the embryo sac of an ovule. Several lines of evidence point to a role for both sporophytic and gametophytic tissues in proper guidance of pollen tubes to ovules. Recently, it was shown that cysteine-rich polypeptides (LUREs) secreted from synergid cells facilitate pollen tube entry into the ovules of *Torenia fournieri* (wishbone flower). Despite these advances, the identities of the cues released by other pistil tissues remain unknown. What hurdles have hampered the efforts to uncover pollen-tube navigation cues of even the known signaling events described above? First, the pistil tissue within which the pollen tube elongates to the ovule is comprised of several types of tissue, including stigma, style, and transmitting tract, and these tissues are not readily accessible. Second, analyzing the dynamic responses of pollen tubes is difficult given that pollen-tube navigation occurs well within opaque pistils. Third, it appears that multiple, stage-specific, short-range, and readily labile signals produced in minute quantities mediate pollen-tube guidance. However, recent development of highly sensitive global approaches and assays that directly monitor pollen-tube elongation offer hope that guidance cues will be uncovered sooner rather than later.

THINK LIKE A SCIENTIST

1. Do you think pollen-tube guidance signals from different plant species will be similar?
2. What are some of the practical applications of deciphering signals that mediate pollen-tube guidance?

Courtesy of Ravi Palanivelu

Ravi Palanivelu is an assistant professor in the Department of Plant Sciences at the University of Arizona. His current research focuses on the isolation and characterization of pollen-tube guidance signals during *Arabidopsis* reproduction, with the long-term goal of understanding the molecular basis of how cells communicate with each other. To learn more about Palanivelu's research, go to http://ag.arizona.edu/research/ravilab.

REVIEW KEY CONCEPTS

To access the course materials and companion resources for this text, please visit www.cengagebrain.com.

36.1 Overview of Flowering Plant Reproduction

- In most flowering plant species, a multicellular diploid sporophyte (spore-producing plant) stage alternates with a multicellular haploid gametophyte (gamete-producing plant) stage. The sporophyte develops roots, stems, leaves, and, at some point, flowers. This life cycle pattern is called alternation of generations (Figure 36.2).

Animation: Flowering plant life cycle

36.2 The Formation of Flowers and Gametes

- A flower develops at the tip of a floral shoot and has up to four whorls, which are supported by the receptacle. The calyx and corolla consist of the sepals and petals, respectively. The third whorl consists of stamens, and carpels make up the innermost whorl (Figure 36.3).
- The anther of a stamen contains sacs where pollen grains develop. Fertilization occurs if pollen lands on a compatible stigma, which contains an ovary in which eggs develop.
- A complete flower has both male and female reproductive parts. Monoecious species bear both types of flowers on each plant; in dioecious species the "male" and "female" flowers are on different plants (Figure 36.4).
- In pollen sacs, pollen grains (immature male gametophytes) develop inside microspores. In a pollen grain, one cell develops into two sperm cells (male gametes). Another cell produces the pollen tube (Figures 36.5 and 36.6).
- An ovule forms inside a carpel. In the ovule, meiosis produces megaspores, of which one survives and develops further, producing the female gametophyte: the seven-celled embryo sac, one cell of which is the haploid egg. A cell containing polar nuclei helps give rise to endosperm (Figure 36.7).

Animation: Parts of a flower

Animation: Floral structure and function

Animation: Microspores to pollen

Animation: Eudicot life cycle

36.3 Pollination, Fertilization, and Germination

- Upon pollination of a compatible stigma, the pollen grain resumes growth, a pollen tube develops, and mitosis of the male gametophyte's sperm-producing cell produces two sperm nuclei. In a double fertilization process, one sperm nucleus fuses with one egg nucleus to form a diploid ($2n$) zygote. The other sperm nucleus and the two polar nuclei formed in the embryo sac also fuse, forming a $3n$ cell that will give rise to triploid ($3n$) endosperm in the seed (Figures 36.8 and 36.9).
- After the endosperm forms, the ovule expands, and the embryonic sporophyte develops. This mature ovule is a seed. Inside it, the embryo has a lengthwise axis with a root apical meristem at one end and a shoot apical meristem at the other (Figure 36.10).
- Eudicot embryos have two cotyledons. The embryonic shoot consists of an upper epicotyl and a lower hypocotyl; also present is an embryonic root, the radicle. The single cotyledon of a monocot forms a scutellum that absorbs nutrients from endosperm (Figure 36.11).
- A fruit is a matured or ripened ovary. Fruits protect seeds and aid in their dispersal by animals, wind, or water. They may be simple, aggregate, or multiple, depending on the number of flowers or ovaries from which they develop. Fruits also vary in the characteristics of the pericarp that surrounds the seed (Figure 36.12).
- The seeds of many plants remain dormant until external conditions such as moisture, temperature, and day length favor the survival of the embryo and the development of a new sporophyte (Figures 36.13–36.15).

Animation: Plant development

Animation: Eudicot seed development

36.4 Asexual Reproduction of Flowering Plants

- Many flowering plants also reproduce asexually, with new plants arising by three general modes: by mitosis at nodes or buds along modified stems, by vegetative propagation (Figures 36.16 and 36.17), or by tissue culture from a parent plant's somatic (nonreproductive) cells (Figure 36.18).

36.5 Early Development of Plant Form and Function

- In plant sexual reproduction, development starts at fertilization. Early on, a new embryo acquires its root–shoot axis, and cells in different regions begin to become specialized for particular functions (Figure 36.19). In morphogenesis, body regions develop characteristic shapes and structures that correlate with their function (Figure 36.20).
- Dividing plant cells can synthesize a new cell plate in a different plane from the old one. Such changes establish the direction in which structures such as lateral roots, branches, and leaf and flower buds grow (Figures 36.21 and 36.22).
- Chemical signals that help guide morphogenesis appear to act on certain cells in meristematic tissue, activating homeotic genes that ultimately regulate cell division and differentiation (Figures 36.23 and 36.24).

Animation: Cell shapes

Animation: Flower formation

UNDERSTAND & APPLY

Test Your Knowledge

1. An angiosperm life cycle includes:
   a. meiosis within the male gametophyte to produce sperm.
   b. meiosis within the female gametophyte to produce eggs.
   c. meiosis within the ovary to produce megaspores.
   d. fertilization leading to development of microspores.
   e. fertilization leading to development of megaspores.
2. In a flower:
   a. the ovary contains the ovule.
   b. the stamens support the petals.
   c. the anther contains the megaspores.
   d. the carpel includes the sepals.
   e. the corolla includes the receptacle.
3. Double fertilization in a flower means:
   a. six sperm fertilize two groups of three eggs each.
   b. one sperm fertilizes the egg; a second sperm fertilizes the mother cell.
   c. one microspore becomes a pollen grain; the other microspore becomes a sperm-producing cell.
   d. one pollen grain can make sperm nuclei and a pollen tube.
   e. one sperm can fertilize two endosperm mother cells.
4. A seed is best described as a (an):
   a. epicotyl.
   b. endosperm.
   c. ovary.
   d. mature spore.
   e. mature ovule.
5. The primary root develops from the embryonic:
   a. epicotyl.
   b. hypocotyl.
   c. coleoptile.
   d. radicle.
   e. plumule.
6. Which of the following is *not* a step in the germination of a monocot seed?
   a. Enzymes secreted into the endosperm digest the endosperm cell wall and macromolecules.
   b. The embryo imbibes water and then produces gibberellin.
   c. The embryo absorbs nutrients released from the endosperm.
   d. Endosperm develops as a food reserve.
   e. Gibberellin acts on the cells of the aleurone and scutellum to encode hydrolytic enzymes.
7. A student cuts off a leaflet from a plant and places it in a glass of water. Within a week roots appear on the base of the cutting. A month later she places the growing cutting into soil and it grows to the full size of the "parent" plant. This is an example of:
   a. parthenocarpy.
   b. fragmentation.
   c. grafting.
   d. vegetative reproduction.
   e. tissue culture propagation.
8. Which of the following is *not* an example of pattern formation in developing plants?
   a. an epidermal cell receiving developmental signals from a cortical cell
   b. the loosening of the cell wall to allow the elongation of selected cells to reach mature size
   c. regulation by homeotic genes of the position of different flower parts
   d. oriented cell division that establishes the shape of an organ
   e. cell expansion that directs specific cells to undergo mitosis at a given time and place
9. During the development of a leaf:
   a. mitotic cell divisions occur on planes specific to different plant groups.
   b. xylem vessels are the first to penetrate the leaf primordium.
   c. the growing leaf primordium becomes wider at its base than at its tip.
   d. the leaf primordium bulges from the region behind the shoot apical meristem.
   e. all of the above occur during leaf development.
10. In spring a lone walnut tree in your backyard develops attractive white flowers, and by the end of summer roughly half the flowers have given rise to the shelled fruits we know as walnuts. Walnut trees are self-pollinating. Assuming that pollination was 100% efficient in the case of your tree, which of the following statements best describes your tree's reproductive parts?
    a. Its flowers are in the botanical category of "perfect" flowers.
    b. The tree is monoecious.
    c. The tree is dioecious.
    d. The tree has imperfect, monoecious flowers.
    e. Answers a and b together provide the best description of the tree's flowers.

Discuss the Concepts

1. A plant physiologist has succeeded in cloning a gene for pest resistance into petunia cells. How can she use tissue culture to propagate a large number of petunia plants that have the gene?
2. A large tree may have tens of thousands of shoot tips, and the cells in each tip can differ genetically from cells in other tips, sometimes substantially. Propose a hypothesis to explain this finding, and speculate about how it might be beneficial to the plant. How might this sort of natural variation be useful to human society?
3. Grocery stores separate displays of fruits and vegetables according to typical uses for these plant foods. For instance, bell peppers, cucumbers, tomatoes, and eggplants are in the vegetable section, while apples, pears, and peaches are displayed with other fruits. How does this practice relate to the biological definition of a fruit?

Design an Experiment

The developmental genetics of flowers are of keen interest in plant biotechnology, especially with regard to food plants such as wheat and rice. Outline a research program for a crop species that would exploit the genetics of flower development, including the effects of homeotic genes, to engineer a more productive variety.

Interpret the Data

As Chapter 37 discusses, the plant hormone ABA (abscisic acid) inhibits seed germination, while any of several GAs (gibberellins) stimulate it. For hundreds of plant species, exposure to a substance in wood smoke—one of a family of plant growth regulators called karrikins, or KARs—markedly improves the germination rate. When flames destroy the existing vegetation at a fire scene, more sunlight reaches the ground. In the presence of ample sunlight, KARs apparently promote germination by enhancing the expression of genes that encode GA.

The **Figure** below summarizes experiments by Australian researchers studying KAR activity in mutant *A. thaliana* plants (called ga1-3), which do not synthesize GA. The graph tracks germination activity after two groups of identical test seeds began imbibing water in a medium that contained the gibberellin GA_3. All seeds were exposed to identical levels of a karrikin called KAR_1. The red symbols designate seeds of mutant ABA-deficient plants that were supplied with supplementary ABA. The white symbols are ABA-deficient seeds that did not receive the supplement.

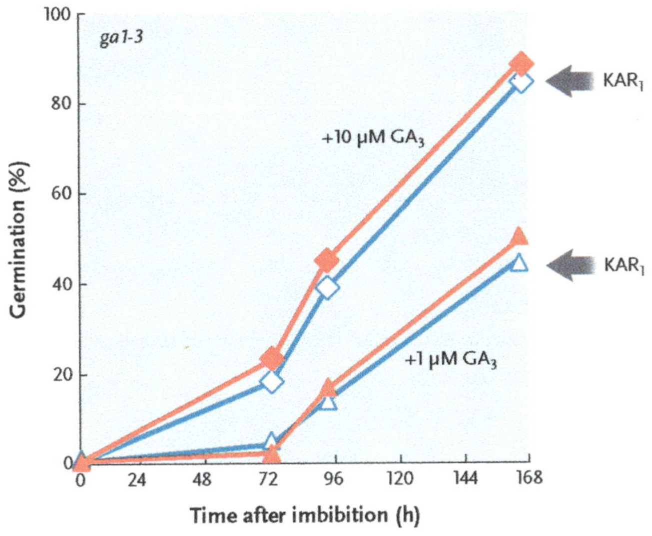

FIGURE Response of phytohormone mutants to KAR_1: Germination of *A. thaliana* ga1-3 mutants in the presence of 1 or 10 µM GA_3 6 1 µM KAR_1.

David C. Nelson, Julie-Anne Riseborough, Gavin R. Flematti, Jason Stevens, Emilio L. Ghisalberti, Kingsley W. Dixon and Steven M. Smith, 2009. Karrikins discovered in smoke trigger *Arabidopsis* seed germination by a mechanism requiring gibberellic acid synthesis and light. *Plant Physiology* 149:863–873. Plant physiology by AMERICAN SOCIETY OF PLANT PHYSIOLOGISTS Copyright 2009. Reproduced with permission of AMERICAN SOCIETY OF PLANT BIOLOGISTS in the format Textbook via Copyright Clearance Center.

Based on this graph:

1. What were the experimental variables in the germination experiment?
2. Did the presence or absence of ABA have a significant effect on the experiment's results?
3. Comparing the upper and lower data sets, what are the *two* basic observations you can make about the effects of the seed exposure to differing GA concentrations?

Source: David C. Nelson, Julie-Anne Riseborough, Gavin R. Flematti, Jason Stevens, Emilio L. Ghisalberti, Kingsley W. Dixon and Steven M. Smith, 2009. Karrikins discovered in smoke trigger *Arabidopsis* seed germination by a mechanism requiring gibberellic acid synthesis and light. *Plant Physiology* 149:863–873.

Apply Evolutionary Thinking

Botanists estimate that half or more of angiosperm species may be polyploids that arose initially through hybridization. *Polyploidy*—having more than a diploid set of the parental chromosomes—can result from nondisjunction of homologous chromosomes during meiosis, or when cytokinesis fails to occur in a dividing cell. *Hybridization* is the successful mating of individuals from two different species. Such an interspecific hybrid is likely to be sterile because it has uneven numbers of parental chromosomes, or because the chromosomes are too different to pair during meiosis. A sterile hybrid may reproduce asexually, however, and if by chance its offspring should become polyploid, that plant will be fertile because the original set of chromosomes will have homologs that can pair normally during meiosis. Explain why both the hybrid parent and fertile polyploid offspring may be considered a new species, and describe at least two ways in which this route to speciation differs from speciation in the animal kingdom.

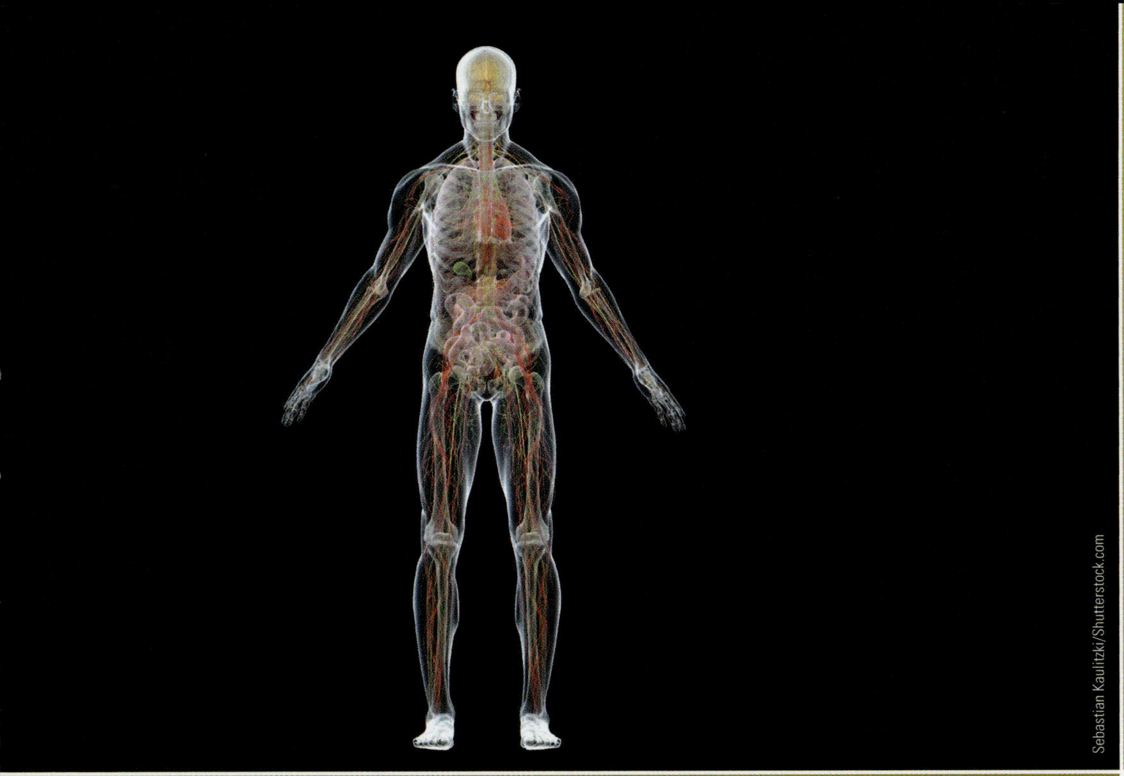

See-through image of a human male showing the major organs and the skeleton (computer rendering).

38

STUDY OUTLINE

38.1 Organization of the Animal Body
38.2 Animal Tissues
38.3 Coordination of Tissues in Organs and Organ Systems
38.4 Homeostasis

Introduction to Animal Organization and Physiology

Why it matters . . . On April 10, 1912, the RMS *Titanic* left Southampton, England, on her maiden voyage, bound for New York. A total of 2,223 people were on board, including 899 crew. Four days out, near midnight, the ship hit an iceberg. Some parts of the hull buckled, and rivets popped out below the waterline. As a result, the *Titanic* began to take on water. The captain ordered passengers and crew to the lifeboats. The ship's lifeboats could hold a maximum of 1,178 people, far fewer than the number of people on board. Two hours and forty minutes after striking the iceberg, the *Titanic* sank. Of the people on board, 1,517 died, most of them succumbing to hypothermia in the −2°C water. At that temperature, death likely occurred within 15 minutes. *Hypothermia* is a condition in which the core temperature of the body falls significantly below normal for a prolonged period. In humans, a drop in core temperature of a few degrees affects brain function and leads to confusion; continued hypothermia, as in the case of the *Titanic,* can lead to death.

Human body temperature is normally regulated within a narrow range so as to provide optimal conditions for cellular life. Body temperature is just one aspect of an animal's internal environment that must be regulated. The maintenance of the internal environment in a relatively stable state is called **homeostasis** (*homeo* = similar; *stasis* = standing or stopping). The processes and activities responsible for homeostasis are called **homeostatic control systems.** These control systems compensate both for the external environmental changes that a human or other animal encounters, such as differences in temperature and humidity, and for changes in their own body systems, such as the availability of nutrients for cells and tissues. As we saw for the unfortunate *Titanic* passengers and crew, the homeostatic control systems for temperature regulation are limited in their ability to maintain the stable state, after which they break down. Not only can animals die from hypothermia after prolonged exposure to the cold, they can also die from hyperthermia after prolonged exposure to excessive heat. Similarly, an imbalance between energy input (in the form of nutrients) and energy output can be lethal.

All animals have body systems for acquiring and digesting nutrients to provide energy for life, growth, reproduction, and movement. Other body systems seen in animals include the nervous system, endocrine system, muscular system, integumentary system, circulatory system, lymphatic system, respiratory system, excretory system, and reproductive system. Biologists are interested in the *structures* and *functions* of all of these systems. **Anatomy** is the study of the *structures* of organisms, and **physiology** is the study of their *functions*—the physicochemical processes of organisms.

In this unit, the structures and functions of the body systems of animals are discussed. In this introductory chapter to animal organization and physiology, we describe the organization of individual cells into tissues, organs, and organ systems, the major body structures that carry out animal activities. Our discussion continues with a look at how the functions of organ systems coordinate to accomplish homeostasis. The other chapters in this unit discuss the structures and functions of individual organ systems. Although the unit emphasizes vertebrates, with particular reference to human physiology, comparisons are made with invertebrates, to keep the structural and functional diversity of the animal kingdom in perspective and to understand the evolution of the structures and functions involved.

38.1 Organization of the Animal Body

The individual cells of animals must be surrounded by an aqueous solution that contains ions and molecules in particular concentrations to maintain osmotic balance and to supply energy and maintenance needs. Most animal cells also require oxygen to serve as the final acceptor for electrons removed in oxidative reactions. Also, waste molecules and other by-products of their metabolic activities, such as carbon dioxide, need to be released from the cells.

The evolution of multicellularity (see Section 25.3) made it possible for organisms to create an **internal environment** of fluid that supplies all the needs of individual cells so that multicellular organisms are able to occupy a wider range of **external environments** than is possible for single-celled organisms. For instance, dry terrestrial habitats likely would be lethal to single-celled organisms. Multicellular organisms can also become relatively large because their individual cells remain small enough to exchange ions and molecules with the internal fluid.

The evolution of multicellularity also allowed major life functions to be subdivided among specialized groups of cells. In animals, some groups of cells became specialized for movement, others for functions such as food capture, digestion, internal circulation of nutrients, excretion of wastes, and reproduction. Specialization greatly increases the efficiency by which animals carry out these functions.

In most animals, specialized groups of cells are organized into tissues, the tissues into organs, and the organs into organ systems **(Figure 38.1)**. A **tissue** is a group of cells with the same structure and function, working together as a unit to carry out one or more specialized activities; four types of tissues are shown in the figure. An **organ** integrates two or more different tissues into a structure that carries out a specific function. The eye, liver, and stomach are examples of organs. An **organ system** (also called a **body system**) coordinates the activities of two or more organs to carry out a major body function such as movement, digestion, or reproduction. The nervous system, for example, is the main regulatory system of the body and consists of the brain, spinal cord, peripheral nerves, and sensory organs.

STUDY BREAK 38.1

1. **What are some advantages for an organism being multicellular?**
2. **What is the difference between a tissue, an organ, and an organ system?**

38.2 Animal Tissues

Although the most complex animals may contain hundreds of distinct cell types, all can be classified into one of four basic tissue groups: *epithelial, connective, muscle,* and *nervous* (see Figure 38.1). Each tissue type is assembled from individual cells. The properties of those cells determine the structure and, therefore, the function of the tissue. More specifically, the structure and integrity of a tissue depends on the structure and organization of the cytoskeleton within the cell, the type and organization of the extracellular matrix (ECM) surrounding the cell, and the junctions holding cells together (see Section 5.5).

Junctions of various kinds link cells into tissues (see Figure 5.27). *Anchoring junctions* form buttonlike spots or belts that weld cells together. They are most abundant in tissues subject to stretching, such as skin and heart muscle. *Tight junctions* seal the spaces between cells, keeping molecules and even ions from leaking between cells. For example, tight junctions in the tissue lining the urinary bladder prevent waste molecules and ions from leaking out of the bladder into other body tissues. *Gap junctions* open channels between cells in the same tissue, allowing ions and small molecules to flow freely from one to another. For example, gap junctions between muscle cells help muscle tissue function as a unit.

Epithelial Tissue Forms Protective, Secretory, and Absorptive Coverings and Linings of Body Structures

Epithelial tissue (*epi* = over; *thele* = covering) consists of sheetlike layers of cells that are usually connected by tight junctions, with little ECM material between them **(Figure 38.2)**. Also called *epithelia* (singular, *epithelium*), these tissues cover body surfaces and the surfaces of internal organs, as well as line cavi-

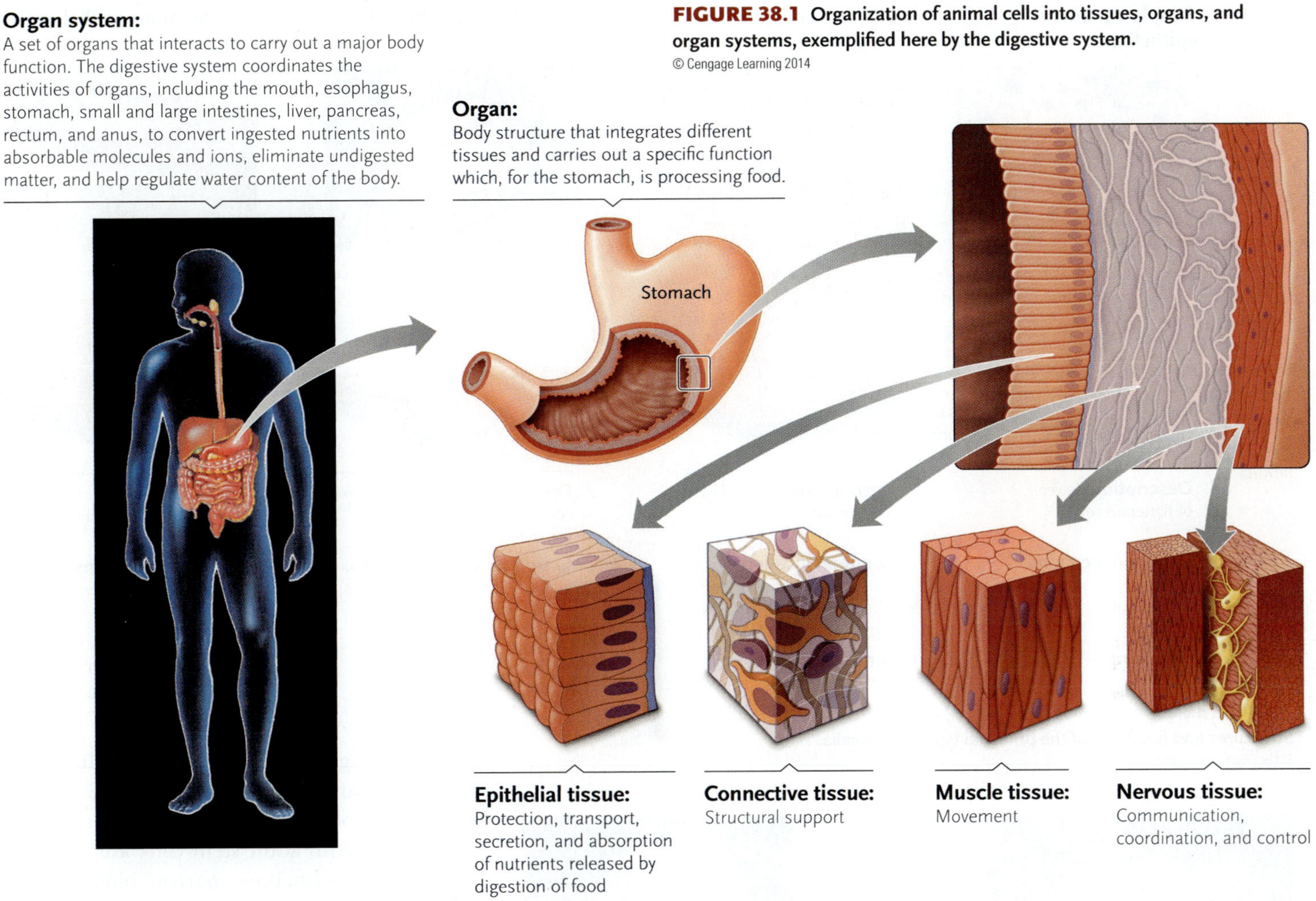

FIGURE 38.1 Organization of animal cells into tissues, organs, and organ systems, exemplified here by the digestive system.

ties and ducts within the body. They protect body surfaces from abrasion or from invasion by bacteria and viruses and secrete or absorb substances. For example, the epithelium covering a fish's gill structures serves as a barrier to bacteria and viruses and exchanges oxygen, carbon dioxide, and ions with the aqueous environment. In the epidermis of vertebrates, some epithelial cells contain a network of keratin, a family of fibrous proteins. Keratin forms protective structures: the fingernails, hair, claws, hooves, and horns of mammals, the feathers of birds, and the scales of reptiles and fish. In arthropods, the surface epithelium secretes a tough cuticle that forms a barrier to the environment and serves as the animal's skeleton.

Some epithelia, such as those lining the capillaries of the circulatory system, act as filters, allowing ions and small molecules to leak from the blood into surrounding tissues while barring the passage of blood cells and large molecules such as proteins.

Because epithelia form coverings and linings, they have an outer surface and an inner surface. The outer, **apical surface** may be exposed to water, air, or fluids within the body. The inner, **basal surface** adheres to a layer of ECM secreted by the epithelial cells called the **basal lamina,** which fixes the epithelium to underlying tissues, often connective tissues.

In internal cavities and ducts, the apical surface is often covered with *cilia* (see Section 5.3), which beat like oars to move fluids through the cavity or duct. The epithelium lining the oviducts in mammals, for example, is covered with cilia that generate fluid currents to move eggs from the ovaries to the uterus. In some epithelia, including the lining of the small intestine, the free surface is crowded with *microvilli,* fingerlike extensions of the plasma membrane that increase the area available for secretion or absorption.

TYPES OF EPITHELIA Epithelia are classified as *simple*—formed by a single layer of cells—or *stratified*—formed by multiple cell layers (see Figure 38.2). The shapes of cells within an epithelium may be *squamous* (mosaic, flattened, and spread out), *cuboidal* (shaped roughly like dice or cubes), or *columnar* (elongated, with the long axis perpendicular to the epithelial layer). Four principal types of epithelia are found in the body (see Figure 38.2).

The cells of some epithelia, such as those forming the skin and the lining of the intestine, divide constantly to replace worn and dying cells. New cells are produced through division of stem cells in the basal layer of the skin. *Stem cells* are undifferentiated cells that divide to produce either more stem cells or

A. **Simple squamous epithelium**

B. **Stratified squamous epithelium**

C. **Cuboidal epithelium**

D. **Columnar epithelium**

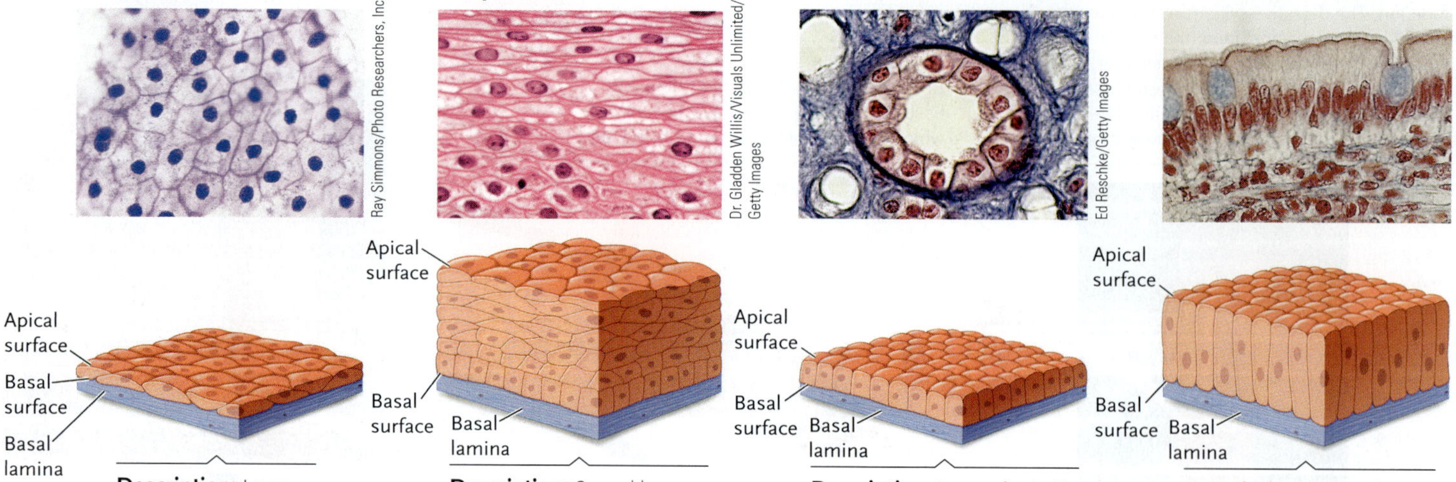

A. Description: Layer of flattened cells

Common locations: Blood vessel walls; air sacs of lungs

Function: Diffusion

B. Description: Several layers of flattened cells

Common locations: Skin and other surfaces subject to abrasion, such as the mouth, esophagus, and vagina

Function: Protection against abrasion; typically not involved in secretion or absorption

C. Description: Layer of cubelike cells; free surface may have microvilli

Common locations: Glands and tubular parts of nephrons in kidneys

Function: Secretion, absorption

D. Description: Layer of tall, slender cells; free surface may have microvilli

Common locations: Lining of gut and respiratory tract

Function: Secretion, absorption

FIGURE 38.2

Structures and functions of the principal types of epithelia.

differentiating cells that become specialized into one of the many cell types of the body. Stem cells are found in both adult organisms and embryos. Besides the skin, adult stem cells are found in tissues of the brain, bone marrow, blood vessels, skeletal muscle, and liver. (*Insights from the Molecular Revolution* describes an effort to culture embryonic stem cells as a source of replacements for damaged tissues and organs.)

GLANDS FORMED BY EPITHELIA Epithelia typically contain or give rise to cells that are specialized for secretion. Some of these secretory cells are scattered among nonsecretory cells within the epithelium. Others form structures called **glands,** which are derived from pockets of epithelium during embryonic development.

Some glands, called **exocrine glands** (*exo* = external; *crine* = secretion), remain connected to the epithelium by a duct, which empties their secretions at the epithelial surface. Exocrine secretions include mucus, saliva, digestive enzymes, sweat, earwax, oils,

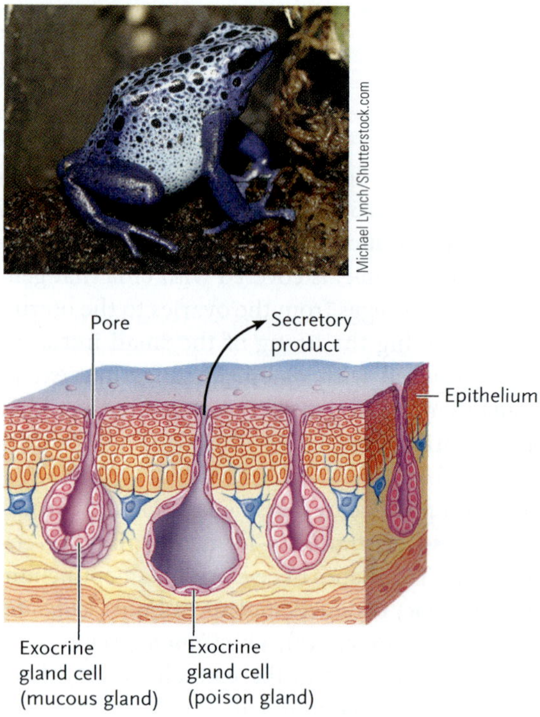

A. Examples of exocrine glands: The mucus- and poison-secreting glands in the skin of a blue poison dart frog

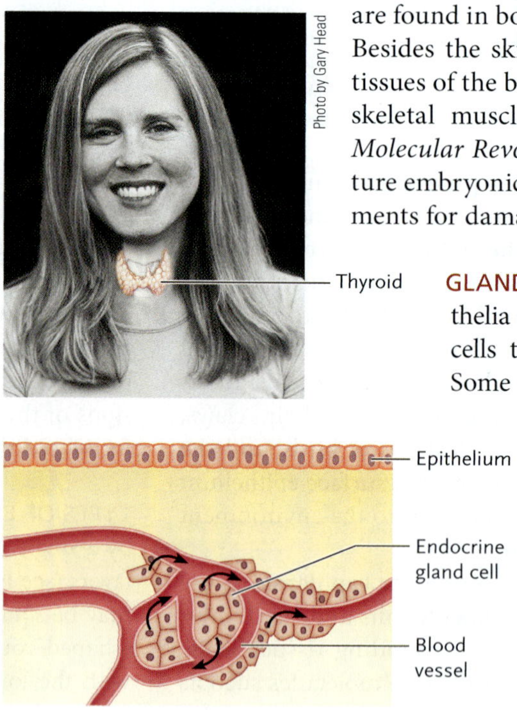

B. Example of an endocrine gland: The thyroid gland, which secretes hormones that regulate the rate of metabolism and other body functions

FIGURE 38.3 Exocrine and endocrine glands. The poison secreted by the blue poison dart frog *(Dendrobates azureus)* is one of the most lethal glandular secretions known.

Culturing Human Embryonic Stem Cells

Embryonic stem (ES) cells—cells derived from embryos—are totipotent, meaning that they have the potential to develop into any tissue. Human ES cells are highly useful for research and hold great promise for biomedical applications. If stem cells can be directed to become any kind of differentiated cells in the test tube, those cells can potentially be used for replacing dead or damaged tissue in the body with healthy cells. The key to the value of human ES cells in biomedical science is the ability to maintain them indefinitely in an undifferentiated state, that is, without becoming specialized.

Research Question

How can human ES cells be cultured?

Experiment

A few years ago, James A. Thomson and his coworkers at the University of Wisconsin and a collaborator at the Rambam Medical Center, Haifa, Israel, developed a successful method for culturing stem cells **(Figure).** Their use of mouse fibroblast cells derived from similar experiments in which nonhuman primate ES cells were successfully cultured. It was unknown at the outset if this experimental step would work with human cells.

Results

Inner cell mass cells multiplied on the fibroblasts, producing clumps of cells. As long as the clumps did not contain more than 50 to 100 cells, cells remained undifferentiated. To continue culturing the ES cells, some of the cells were taken and placed over a bed of mouse fibroblast cells in a fresh culture dish. Using this technique, the cell cultures were maintained for as long as 8 months in the laboratory with no signs of differentiation.

Critically, the researchers demonstrated by several tests that the cultured cells retained full stem cell function:

1. The cultured stem cells had characteristic surface molecules that change when the cells begin to differentiate into adult tissues.
2. Cultured stem cells contained fully active telomerase, an enzyme that helps maintain chromosomes at their normal length during rapid cycles of DNA replication and cell division (see Section 14.3). The enzyme becomes inactive in most cells as they differentiate into adult form.
3. Stem cells can differentiate into a wide range of tissues. When mice were injected with samples of cultured cells, the cells grew into balls of tissue that included skin cells, gut epithelium, cartilage, bone, smooth and striated muscle, and nerve cells. The cultured cells seemed to differentiate into adult tissues when stimulated to do so, and thus to have all the characteristics of stem cells.

Conclusions

The researchers demonstrated that human ES cells could be cultured indefinitely while retaining all their stem cell properties. This was pioneering research, the first human ES cell line to be isolated.

THINK LIKE A SCIENTIST The researchers observed that, when the ES cell cultures grew until they covered the culture dish and the cells began to pile up on one another, the cells differentiated spontaneously. Conceptually, what does that observation indicate?

Source: J. A. Thomson et al. 1998. Embryonic stem cell lines derived from human blastocysts. *Science* 282:1145–1147.

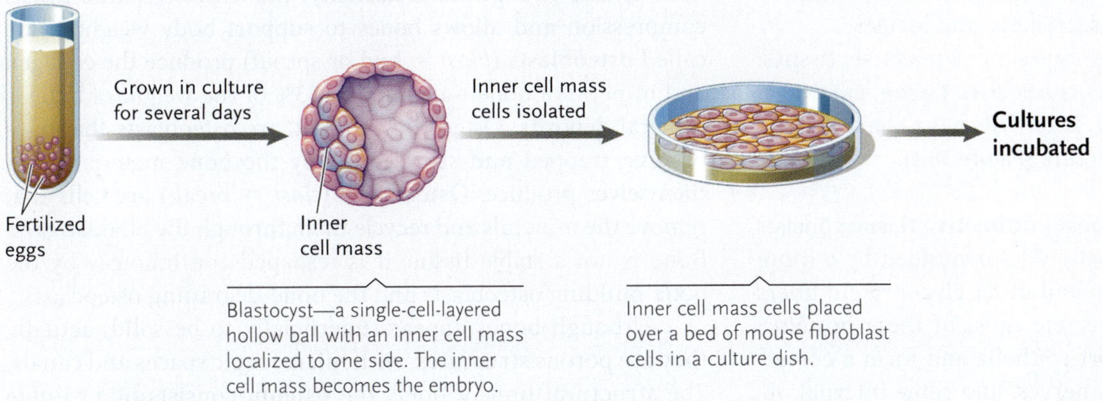

milk, and venom (**Figure 38.3A** shows an exocrine gland in the skin of a poisonous tree frog). Other glands, called **endocrine glands** (*endo* = inside), become suspended in connective tissue underlying the epithelium, with no ducts leading to the epithelial surface. These ductless glands, such as the pituitary gland, adrenal gland, and thyroid gland **(Figure 38.3B),** release their products—hormones (see Chapters 7 and 42)—directly into the interstitial fluid, to be picked up and distributed by the circulatory system.

Some glands act as both exocrine glands and endocrine glands. The pancreas, for instance, has an exocrine function of secreting pancreatic juice through a duct into the small intestine where it plays an important role in food digestion, and an endocrine function of secreting the hormones insulin and glucagon into the bloodstream to help regulate glucose levels in the blood.

Connective Tissue Supports Other Body Tissues

Most animal body structures contain one or more types of **connective tissues.** Connective tissues support other body tissues, transmit mechanical and other forces, and in some cases act as filters. They consist of cells that form networks or layers in and around body structures and that are separated by nonliving material, specifically the ECM secreted by the cells of the tissue (see Section 5.5). Many forms of connective tissue have more ECM material (both by weight and by volume) than cellular material.

The mechanical properties of a connective tissue depend on the type and quantity of its ECM. The consistency of the ECM ranges from fluid (as in blood and lymph), through soft and firm gels (as in tendons), to hard and crystalline (as in bone).

In most connective tissues, the ECM consists primarily of the fibrous glycoprotein **collagen** embedded in a network of proteoglycans—glycoproteins that are very rich in carbohydrates. Collagen is the most abundant family of proteins in animals; collagens occur in all Eumetazoa.

In bone, the glycoprotein network surrounding the collagen is impregnated with mineral deposits that produce a hard, yet still somewhat elastic, structure. Another family of glycoproteins, **fibronectin,** aids in the attachment of cells to the ECM and helps hold the cells in position.

In some connective tissues another rubbery protein, **elastin,** adds elasticity to the ECM—it is able to return to its original shape after being stretched, bent, or compressed. Elastin fibers, for example, help the skin return to its original shape when pulled or stretched, and give the lungs the elasticity required for their alternating inflation and deflation. Related to elastin, the protein **resilin** is found only in insects and some crustaceans and is the most elastic material known. Resilin is the basis for the jumping of fleas, crickets, and locusts.

Vertebrates have six major types of connective tissues: **loose connective tissue, fibrous connective tissue, cartilage, bone, adipose tissue,** and **blood.** Each type has a characteristic function correlated with its structure **(Figure 38.4).**

LOOSE CONNECTIVE TISSUE **Loose connective tissue** consists of sparsely distributed **fibroblast** cells surrounded by a more or less open network of collagen and other glycoprotein fibers **(Figure 38.4A);** the fibroblasts secrete most of these proteins. Loose connective tissues support epithelia and form a corset-like band around blood vessels, nerves, and some internal organs; they also reinforce deeper layers of the skin. Sheets of loose connective tissue, covered on both surfaces with epithelial cells, form the **mesenteries,** which hold the abdominal organs in place and provide lubricated, smooth surfaces that prevent chafing or abrasion between adjacent structures as the body moves.

FIBROUS CONNECTIVE TISSUE In **fibrous connective tissue,** fibroblasts are sparsely distributed among dense masses of collagen and elastin fibers that are lined up in highly ordered, parallel bundles **(Figure 38.4B).** The parallel arrangement produces maximum tensile strength and elasticity. Examples include **tendons,** which attach muscles to bones, and **ligaments,** which connect bones to each other at a joint. The cornea of the eye is a transparent fibrous connective tissue formed from highly ordered collagen molecules.

CARTILAGE **Cartilage** consists of sparsely distributed cartilage-producing cells called **chondrocytes,** surrounded by networks of collagen fibers embedded in a tough but elastic matrix of the glycoprotein *chondroitin sulfate* **(Figure 38.4C).** Elastin is also present in some forms of cartilage.

The elasticity of cartilage allows it to resist compression and stay resilient, like a piece of rubber. Bending your ear or pushing the tip of your nose, which are supported by a core of cartilage, shows the flexible nature of this tissue. In humans, cartilage also supports the larynx, trachea, and smaller air passages in the lungs. It forms the disks cushioning the vertebrae in the spinal column and the smooth, slippery capsules around the ends of bones in joints such as the hip and knee. Cartilage also serves as a precursor to bone during embryonic development; in sharks and rays and their relatives, almost the entire skeleton remains as cartilage in adults.

BONE The densest form of connective tissue, **bone,** forms the skeleton, which supports the body, protects softer body structures such as the brain, and contributes to body movements (see Chapter 43).

Mature bone consists primarily of cells called **osteocytes** (*osteon* = bone) embedded in an ECM containing collagen fibers and glycoproteins impregnated with *hydroxyapatite,* a calcium-phosphate mineral **(Figure 38.4D).** The collagen gives bone tensile strength and elasticity; the hydroxyapatite resists compression and allows bones to support body weight. Cells called **osteoblasts** (*blast* = bud or sprout) produce the collagen and mineral of bone—as much as 85% of the weight of bone is mineral deposits. Osteocytes, in fact, are osteoblasts that have become trapped and surrounded by the bone materials they themselves produce. **Osteoclasts** (*clast* = break) are cells that remove the minerals and recycle them through the bloodstream. Bone is not a stable tissue; it is reshaped continuously by the bone-building osteoblasts and the bone-degrading osteoclasts.

Although bones appear superficially to be solid, actually they are porous structures, with microscopic spaces and canals. The structural unit of bone, the **osteon,** consists of a minute central canal surrounded by osteocytes embedded in concentric layers of mineral matter (see Figure 38.4D). A blood vessel and extensions of nerve cells run through the central canal, which is connected to the spaces containing cells by very fine, radiating canals filled with interstitial fluid. The blood vessels supply nutrients to the cells with which the bone is built, and the nerve cells hook up the bone cells to the body's nervous system.

ADIPOSE TISSUE The connective tissue called *adipose tissue* mostly contains large, densely clustered cells called **adipocytes** that are specialized for fat storage **(Figure 38.4E).** It has little ECM. Adipose tissue also cushions the body and, in mammals, forms an especially important insulating layer under the skin.

A. Loose connective tissue

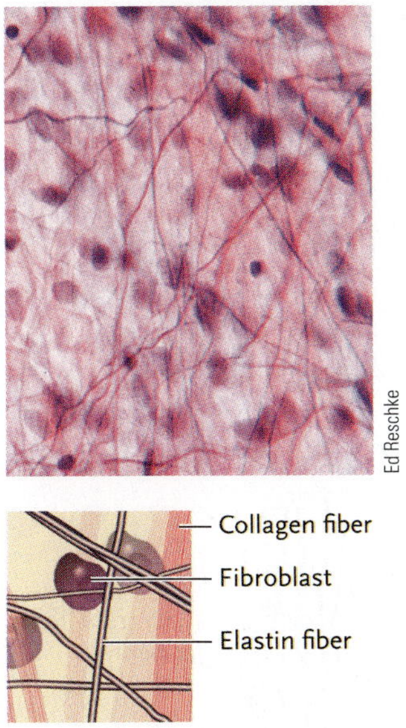

Description: Fibroblasts and other cells surrounded by collagen and elastin fibers forming a glycoprotein matrix

Common locations: Under the skin and most epithelia; around blood vessels, nerves, and some internal organs

Function: Support, elasticity, diffusion

B. Fibrous connective tissue

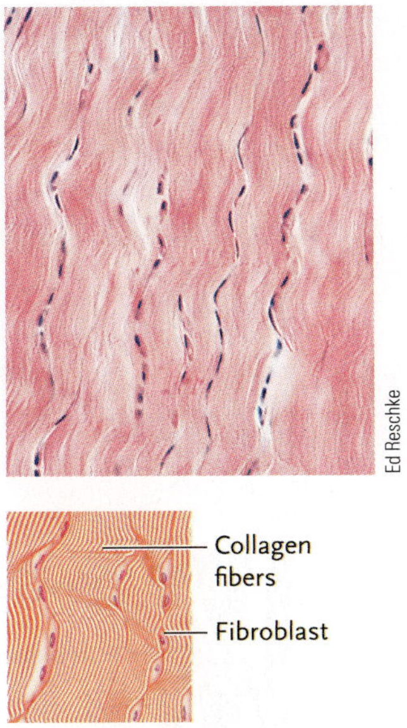

Description: Long rows of fibroblasts surrounded by collagen and elastin fibers in parallel bundles with a dense ECM

Common locations: Tendons, ligaments

Function: Strength, elasticity

C. Cartilage

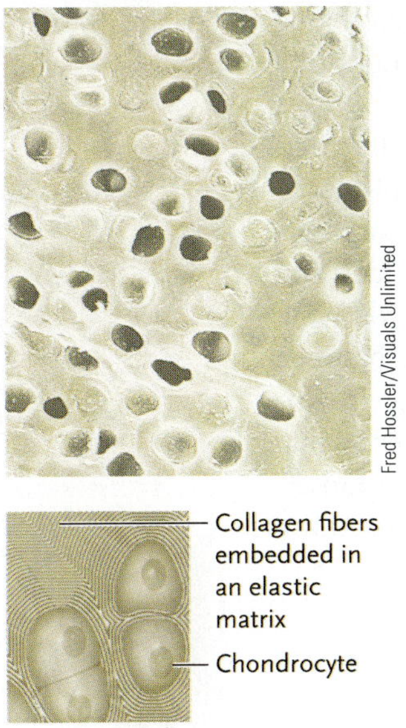

Description: Chondrocytes embedded in a pliable, solid matrix of collagen and chondroitin sulfate

Common locations: Ends of long bones, ears, nose, parts of airways, skeleton of vertebrate embryos

Function: Support, flexibility, low-friction surface for joint movement

D. Bone tissue

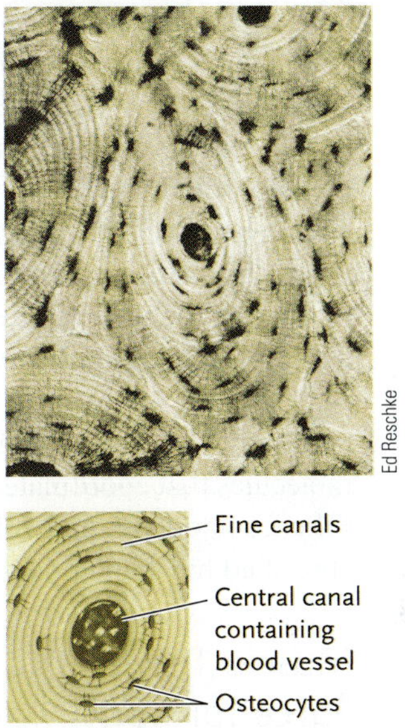

Description: Osteocytes in a matrix of collagen and glycoproteins hardened with hydroxyapatite

Common locations: Bones of vertebrate skeleton

Function: Movement, support, protection

E. Adipose tissue

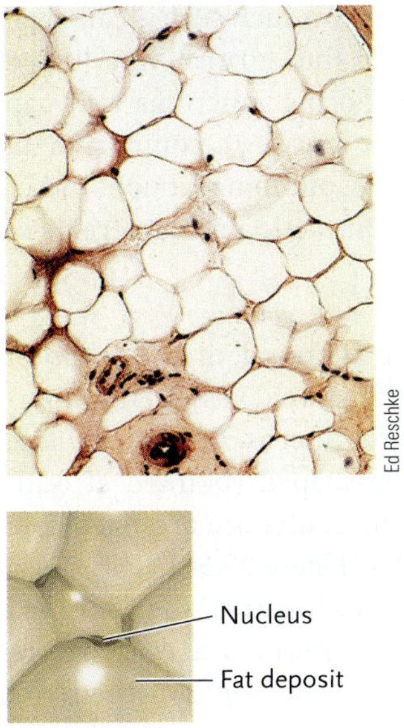

Description: Large, tightly packed adipocytes with little ECM

Common locations: Under skin; around heart, kidneys

Function: Energy reserves, insulation, padding

F. Blood

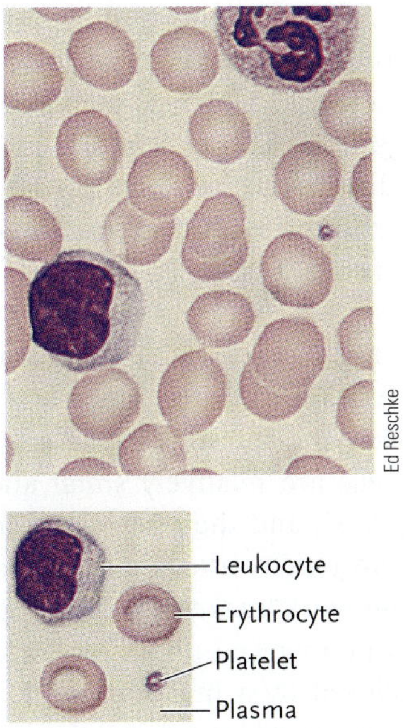

Description: Leukocytes, erythrocytes, and platelets suspended in plasma, a fluid ECM

Common locations: Circulatory system

Function: Transport of substances

FIGURE 38.4 Structures and functions of the six major types of connective tissues in vertebrates.

The animal body stores limited amounts of carbohydrates, primarily in muscle and liver cells. Excess carbohydrates are converted into the fats stored in adipocytes. The storage of chemical energy as fats offers animals a weight advantage. For example, the average human would weigh about 45 kg (100 pounds) more if the same amount of chemical energy was stored as carbohydrates instead of fats. Adipose tissue is richly supplied with blood vessels, which move fats or their components to and from adipocytes.

BLOOD **Blood (Figure 38.4F)** is considered a connective tissue because its cells are suspended in a fluid ECM, plasma. The straw-colored **plasma** is a solution of proteins, nutrient molecules, ions, and gases.

Blood contains two primary cell types, **erythrocytes** (red blood cells; *erythros* = red) and **leukocytes** (white blood cells; *leukos* = white). Erythrocytes are packed with hemoglobin, a protein that can bind and transport oxygen. There are several types of leukocytes—all help to protect the body against invading viruses, bacteria, and other disease-causing agents. The blood plasma also contains **platelets,** membrane-bound fragments of specialized blood cells, which take part in the reactions that seal wounds with blood clots.

Blood is the major transport vehicle of the body. It carries oxygen and nutrients to body cells, removes wastes and by-products such as carbon dioxide, and maintains the internal fluid environment, including the osmotic balance between cells and the interstitial fluid. Blood also transports hormones and other signal molecules that coordinate body responses. (The components and roles of blood are described in Chapter 44.)

Muscle Tissue Produces the Force for Body Movements

Muscle tissue consists of cells that have the ability to contract ("shorten"). The contractions, which depend on the interaction of two proteins—**actin** and **myosin**—move body limbs and other structures, pump the blood, and produce a squeezing pressure in organs such as the intestine and uterus. Three types of muscle tissues, *skeletal, cardiac,* and *smooth,* produce body movements in vertebrates **(Figure 38.5).**

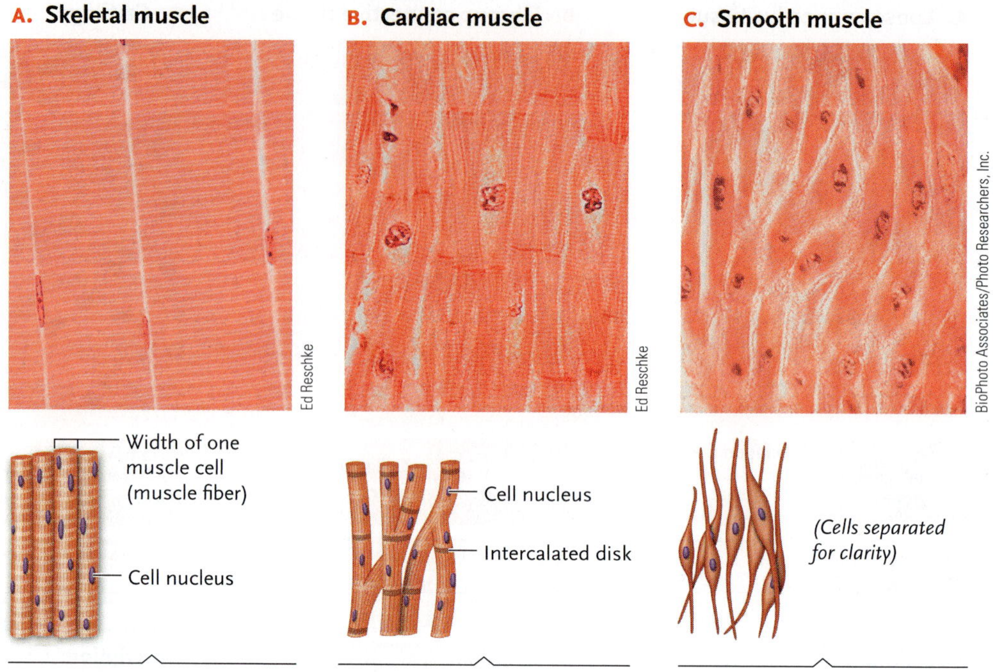

A. Description: Bundles of long, cylindrical, striated, contractile, multinucleate cells called muscle fibers

Typical location: Attached to bones of skeleton

Function: Locomotion, movement of body parts

B. Description: Interlinked network of short and branched cylindrical, striated cells stabilized by anchoring junctions and gap junctions

Location: Wall of heart

Function: Pumping of blood within circulatory system

C. Description: Loose network of contractile cells with tapered ends

Typical location: Wall of internal organs, such as stomach

Function: Movement of internal organs

FIGURE 38.5 Structures and functions of skeletal, cardiac, and smooth muscle.

SKELETAL MUSCLE **Skeletal muscle** is so called because typically it is attached by tendons to the skeleton. Skeletal muscle cells are also called **muscle fibers (Figure 38.5A).** These cells are multinucleate (contain many nuclei in the same cytoplasm) and are packed with actin and myosin molecules arranged in highly ordered, parallel units that give the tissue a banded or striated appearance when viewed under a microscope. Muscle fibers packed side-by-side into parallel bundles surrounded by sheaths of connective tissue constitute muscles, such as the biceps.

Skeletal muscle contracts in response to signals carried by the nervous system. The contractions of skeletal muscles, characteristically rapid and powerful, move body parts and maintain posture. In some cases disorganized muscle contractions (shivering) are used to release heat as a by-product of cellular metabolism. This heat helps mammals, birds, and some other vertebrates maintain their body temperatures when environmental temperatures fall. (Skeletal muscle is discussed further in Chapter 43.)

CARDIAC MUSCLE **Cardiac muscle** is the contractile tissue of the heart **(Figure 38.5B).** Cardiac muscle has a striated appearance because it contains actin and myosin molecules arranged like those in skeletal muscle. However, cardiac muscle cells are short and branched, with each cell connecting to several neighboring cells; the joining point between two such cells is called an **intercalated disk.** Cardiac muscle cells thus form an interlinked network, which is stabilized by anchoring junctions and gap junctions. This network enables heart muscle to contract in all directions, producing a squeezing or pumping action rather than the lengthwise, unidirectional contraction characteristic of skeletal muscle.

SMOOTH MUSCLE **Smooth muscle** is found in the walls of tubes and cavities in the body, including blood vessels, the stomach and intestine, the bladder, and the uterus. Smooth muscle cells are relatively small and spindle-shaped (pointed at both ends), and their actin and myosin molecules are arranged in a loose network rather than in bundles **(Figure 38.5C).** This loose network makes the cells appear smooth rather than striated when viewed under a microscope. Smooth muscle cells are enclosed by a mesh of connective tissue, and some of them are connected by gap junctions. The gap junctions allow the passage of current that makes smooth muscles contract as a unit, typically producing a squeezing motion. Although smooth muscle contracts more slowly than skeletal and cardiac muscles do, its contractions can be maintained at steady levels for a much longer time. These contractions move and mix the stomach and intestinal contents, constrict blood vessels, and push the infant out of the uterus during childbirth.

Nervous Tissue Receives, Integrates, and Transmits Information

Nervous tissue contains cells called **neurons** (also called *nerve cells*) that serve as lines of communication and control between body parts. Billions of neurons are packed into the human brain; others extend throughout the body. Nervous tissue also contains **glial cells** (*glia* = glue), which physically support and provide nutrients to neurons, provide electrical insulation between them, and scavenge cellular debris and foreign matter.

A neuron consists of a *cell body*, containing the nucleus and organelles, and two types of cell extensions, dendrites and axons **(Figure 38.6)**. *Dendrites* are usually highly branched, whereas *axons* are usually unbranched except at their terminals. Depending on the type of neuron and its location in the body, its axon may extend from a few micrometers or millimeters to more than a meter.

The cell body or the dendrites of a neuron receive chemical or electrical signals from other neurons. The signals are converted into an electrical wave that is transmitted along the axons to the axon terminals. Depending on the type of neuron, axons either convert the electrical signal at their terminals to a chemical signal that stimulates a response in nearby muscle cells, gland cells, or other neurons or transfers the electrical signal directly to another neuron through gap junctions. (Neurons and their organization in body structures are discussed further in Chapters 39, 40, and 41.)

All four major tissue types—epithelial, connective, muscle, and nervous—combine to form the organs and organ systems of animals. The next section depicts the major organs and organ systems of vertebrates, and outlines their main functions.

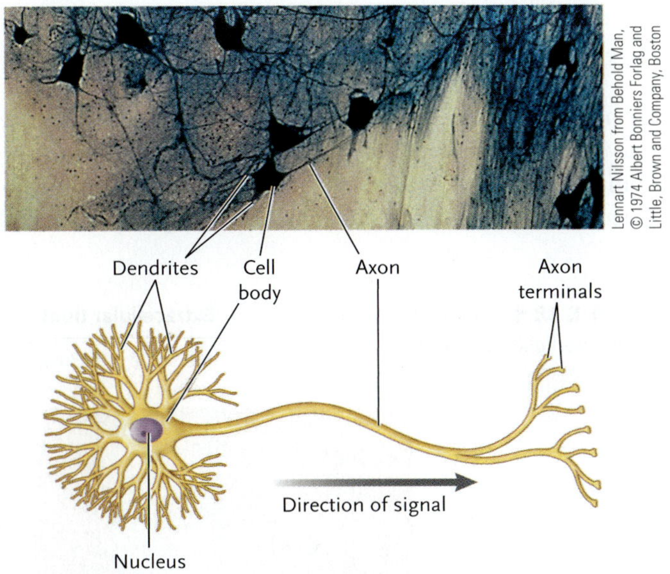

FIGURE 38.6 Neurons and their structure. The micrograph shows a network of motor neurons, which relay signals from the brain or spinal cord to muscles and glands.

STUDY BREAK 38.2

1. **Distinguish between exocrine and endocrine glands. What is the tissue type of each of these glands?**
2. **What are the six major types of connective tissue in vertebrates?**
3. **What three types of muscle tissue produce movements in the body?**

THINK OUTSIDE THE BOOK

Individually or collaboratively, determine the type or types of muscle tissues found in invertebrates and give examples of each type.

38.3 Coordination of Tissues in Organs and Organ Systems

In the tissues, organs, and organ systems of an animal, each cell engages in the basic metabolic activities that ensure its own survival, and performs one or more functions of the system to which it belongs. All vertebrates (and most invertebrates) have eleven major organ systems. **Figure 38.7** summarizes structures and functions of these organ systems, which are discussed in the rest of this unit.

The functions of all these organ systems are coordinated and integrated to accomplish collectively tasks that are vital to all animals, whether a flatworm, a fly, a salmon, a platypus, or a human. These functions include:

1. Acquiring nutrients and other required substances such as oxygen, coordinating their processing, distributing them throughout the body, and disposing of wastes.
2. Synthesizing the protein, carbohydrate, lipid, and nucleic acid molecules required for body structure and function.
3. Sensing and responding to changes in the environment, such as temperature, pH, and ion concentrations.
4. Protecting the body against injury or attack from other animals, and from viruses, bacteria, and other disease-causing agents.
5. Reproducing and, in many instances, nourishing and protecting offspring through their early growth and development.

Together these tasks maintain homeostasis, preserving the internal environment required for survival of the body. Homeostasis is the topic of the next section.

STUDY BREAK 38.3

What are the major functions of each of the eleven organ systems of the vertebrate body?

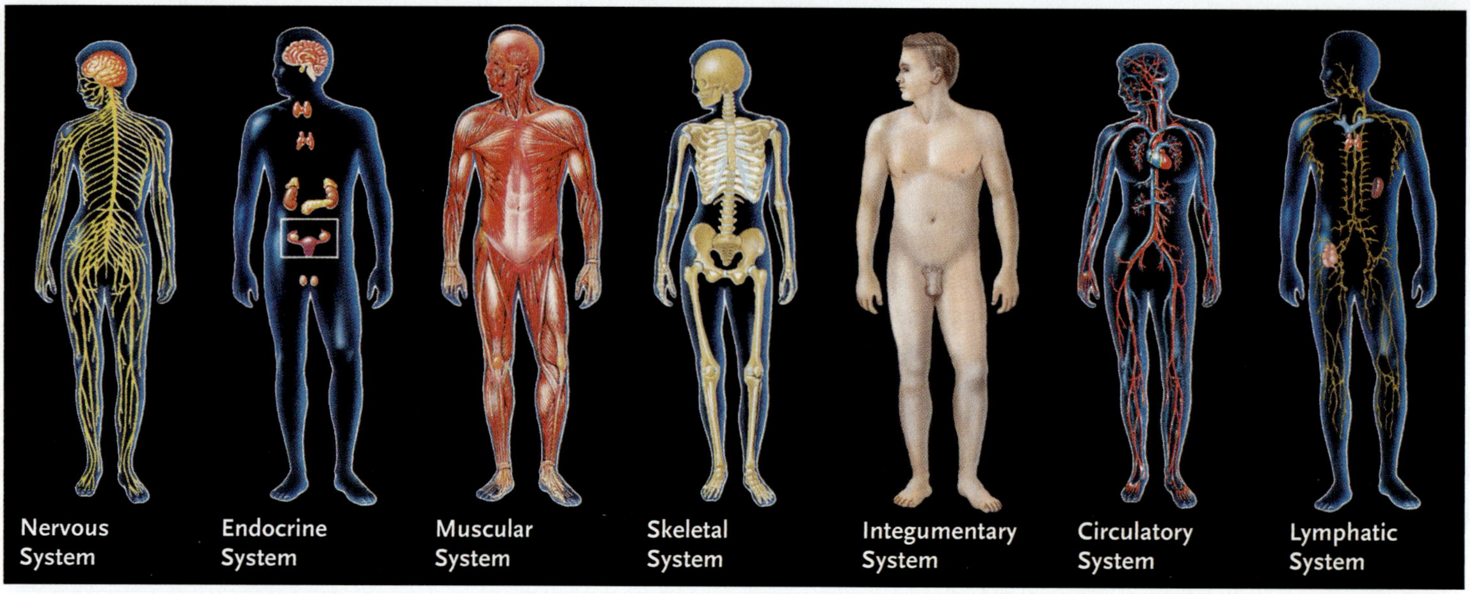

System	Main structures	Main functions
Nervous System	Brain, spinal cord, peripheral nerves, sensory organs	Principal regulatory system; monitors changes in internal and external environments and formulates compensatory responses; coordinates body activities
Endocrine System	Pituitary, hypothalamus, thyroid, adrenal, pancreas, and other hormone-secreting glands	Regulates and coordinates body activities through secretion of hormones
Muscular System	Skeletal, cardiac, and smooth muscle	Moves body parts; helps run bodily functions; generates heat; moves intestinal lumen contents
Skeletal System	Bones, tendons, ligaments, cartilage	Supports and protects body parts; provides leverage for body movements; stores minerals
Integumentary System	Skin, sweat glands, hair, nails	Covers external body surfaces and protects against injury and infection; helps regulate water content and body temperature
Circulatory System	Heart, blood vessels, blood	Distributes water, nutrients, oxygen, hormones, and other substances throughout body and carries away carbon dioxide and other metabolic wastes; helps stabilize internal temperature and pH
Lymphatic System	Lymph nodes, lymph ducts, spleen, thymus	Returns excess fluid to the blood; defends body against invading viruses, bacteria, fungi, and other pathogens as part of immune system (Other parts of the immune system, which are primarily cellular in nature, are not shown in this figure.)

FIGURE 38.7 Structures and functions of the organ systems of the human body.

38.4 Homeostasis

To live, cells of all organisms must take in nutrients and O_2 from the external environment and eliminate wastes such as CO_2 and particular chemicals to that external environment. A single-celled organism such as an amoeba is in direct contact with the external environment. Although most cells of a multicellular animal are isolated from direct contact with the external environment, those cells have the same needs of nutrient and O_2 input and waste elimination. Those needs are met by the internal environment, namely the aqueous **extracellular fluid (ECF) (Figure 38.8).** The ECF has two components:

- **Plasma,** the fluid portion of blood.
- **Interstitial fluid** (*inter* = between; *stitial* = that which stands), the fluid that surrounds the cells.

The ECF connects all cells to the external environment. Thus, no matter where a cell is within the body, it can make the exchanges essential to its life with the interstitial fluid. Particular organ systems enable these exchanges between the external and internal environments. The digestive system processes incoming food and transfers the resulting nutrients into the plasma of the blood. The nutrients reach all parts of the body by

FIGURE 38.8 Nature of the extracellular fluid (ECF).

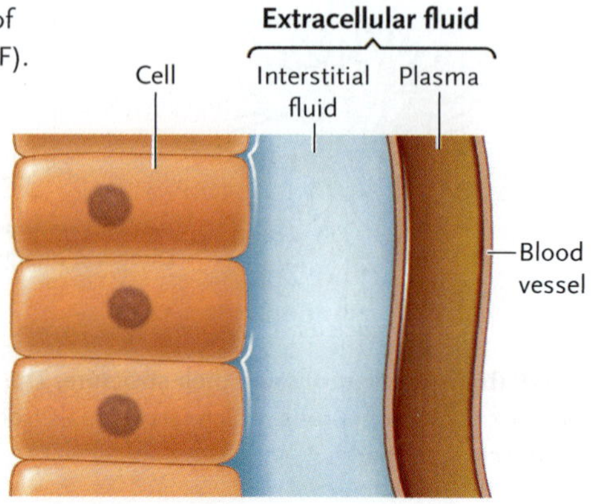

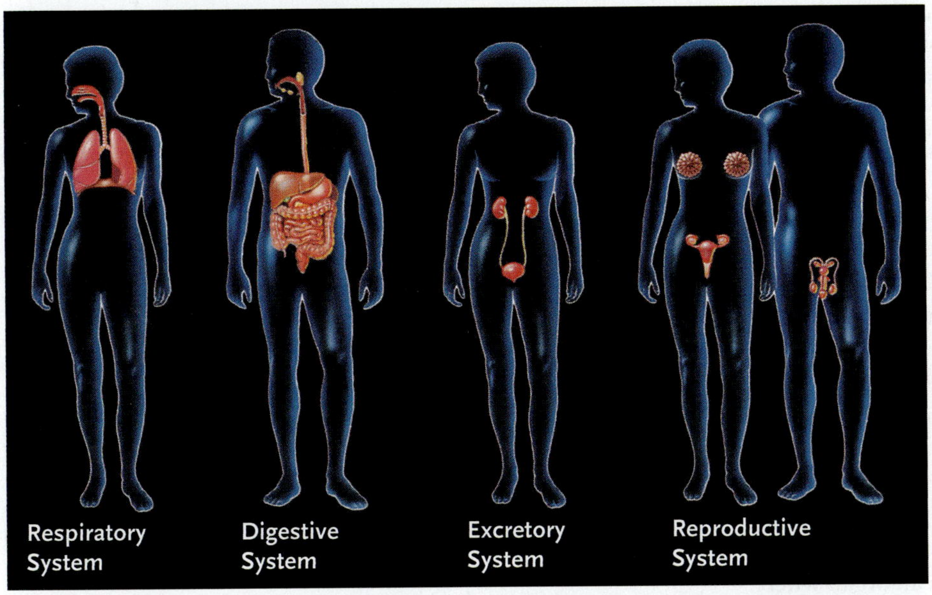

	Respiratory System	Digestive System	Excretory System	Reproductive System
Main structures:	Lungs, diaphragm, trachea, and other airways	Oral cavity, pharynx, esophagus, stomach, intestines, liver, pancreas, rectum, anus	Kidneys, bladder, ureter, urethra	*Female*: ovaries, oviducts, uterus, vagina, mammary glands *Male*: testes, sperm ducts, accessory glands, penis
Main functions:	Exchanges gases with the environment, including uptake of oxygen and release of carbon dioxide	Converts ingested matter into molecules and ions that can be absorbed into body; eliminates undigested matter; helps regulate water content	Removes and eliminates excess water, ions, and metabolic wastes from body; helps regulate internal osmotic balance and pH; helps regulate blood pressure	Maintains the sexual characteristics and passes on genes to the next generation

the action of the circulatory system, along with O_2 which enters the blood by the action of the respiratory system. The nutrients and O_2 in the plasma reach the interstitial fluid through the capillaries and, from there, they enter the cells as needed. Waste moves in the opposite direction: from the cells into the interstitial fluid, and then to the plasma. The respiratory system handles removal of CO_2, and the excretory system handles the metabolic wastes.

For optimal function of these systems, the composition and state of the ECF must be maintained within a narrow range so that cells have available the necessary nutrients and O_2 and wastes can be eliminated. Further, other aspects of the internal environment that are important for cellular (and therefore organismal) life, such as temperature, must also be regulated within a narrow range (especially in warm-blooded animals).

As introduced in *Why It Matters,* the maintenance of the internal environment in a relatively stable state is **homeostasis.** Note that, although the *stasis* part of homeostasis might suggest a static, unchanging process, homeostasis is actually a *dynamic* process, in which internal adjustments are made continuously to compensate for external changes. For example, internal adjustments are needed for homeostasis during exercise or hibernation. The factors controlled by homeostatic mechanisms all require energy that must be constantly acquired from the external environment.

Homeostatic control systems are found in all organisms. Humans and other mammals maintain a consistent internal environment, but many kinds of multicellular animals show much more variability in their internal environment than mammals do. Animals fall into two major categories in this regard: **regulators** maintain factors of the internal environment in a relatively constant state, and **conformers** have internal environments that match the external environment. For instance, mammals and birds are thermoregulators, meaning that they maintain their body temperature at a relatively constant value regardless of the temperature of the external environment. However, most animals—for example, fishes, reptiles, and insects—are thermoconformers, meaning that their body temperature matches that of the external environment. We will use mammals, and especially humans, as primary examples in the discussions throughout this unit.

Many Factors of the Internal Environment Are Regulated Homeostatically

Factors of the internal environment that are regulated homeostatically include:

1. *Nutrient concentration.* Energy production by cells requires a constant supply of nutrient molecules. The energy generated by catabolizing the nutrients is used for basic cellular processes and any specialized activities of the cell.
2. *Concentration of O_2.* Cellular respiration (see Chapter 8), the process that generates energy from catabolic reactions, requires a constant supply of O_2 for optimal productivity.
3. *Concentration of CO_2.* The CO_2 produced by the catabolic reactions of cellular respiration must be removed as waste or the ECF would become increasingly acidic.
4. *Concentration of waste chemicals.* Particular biochemical reactions in the cell generate products that would be toxic to the cell if not removed as waste.
5. *Concentration of water and NaCl.* The relative concentrations of NaCl and water in the ECF affect how much water enters or leaves a cell and hence the cell's volume (see Chapter 6). These concentrations must be regulated to maintain a cell volume that is optimal for function; swollen or shrunken cells are functionally impaired.
6. *pH.* Changes in pH of the ECF can adversely affect enzymatic activities within cells, as well as nerve cell functions.

7. *Volume and pressure of plasma.* Both the volume and pressure in the vessels of the plasma component of the ECF must be maintained at levels adequate to distribute the fluid throughout the body. This circulation is vitally important for supplying cells with their needs and removing their wastes.
8. *Temperature.* Body cells (of warm-blooded animals) function optimally within a fairly narrow temperature range. Outside of that range, chemical reactions change their rates and may be completely inhibited. If cells become too cold, the rates of enzymatic reactions decrease too much. If cells become too hot, structural and enzymatic proteins can be denatured and, therefore, become inactive.

Most Homeostatic Control Systems Operate Throughout the Body

Local homeostatic controls operate only within an organ where a change in the internal environment needs to be addressed. For example, when you exercise, a skeletal muscle increases its use of O_2, which reduces the concentration of O_2 in the muscle cells and sets up a requirement for increased O_2 uptake from the interstitial fluid. The local homeostatic control in this instance responds to the decreased O_2 concentration by triggering the relaxation of the smooth muscle in the walls of the blood vessels supplying the exercising muscle. As a result, the blood vessels dilate (expand in diameter), increasing the blood flow to the exercising muscle, which brings more O_2. The end result is maintenance of an optimal O_2 concentration for the exercising skeletal muscle.

Systemic homeostatic controls are initiated outside of an organ or organ system to control that organ's or organ system's activity. The body's two major regulatory systems, the nervous system and the endocrine system, are responsible for most systemic homeostatic control of organs and organ systems. This type of control enables the regulation of several organs or organ systems to be coordinated toward a common goal. For example, if blood pressure drops too low, the nervous system acts on the heart to increase contraction strength and on blood vessels to constrict them so as to increase pressure in the system. Blood pH is controlled by both the nervous and endocrine systems, blood glucose by the endocrine system, internal temperature by the nervous and endocrine systems, and oxygen and carbon dioxide concentrations by the nervous system.

Homeostasis Is Accomplished by Negative Feedback Control Systems

The primary mechanism of homeostasis is **negative feedback** **(Figure 38.9).** The following presents the components of a negative feedback control system maintaining homeostasis. In parentheses are descriptions of the specific homeostatic control system used by mammals and birds—warm-blooded vertebrates—to maintain body temperature within a relatively narrow range around a set point:

- **Stimulus**—an environmental change that triggers a response. (A change in body temperature beyond the normally controlled range around the normal temperature level.)
- **Sensor**—the body component that detects the environmental change. (Temperature-monitoring nerve cells throughout the body.)
- **Integrator**—a control center that receives information from the sensor and compares it with the **set point,** the normal level at which the condition being controlled is to be maintained. The integrator, typically part of the brain or endocrine system, sends out commands to correct any significant change from the set point detected. (The temperature control center in a region of the brain called the *hypothalamus* (see Chapter 40) detects changes in the temperature of the brain and the rest of the body, and compares it with a set point. For humans, the set point has a relatively narrow range centered at about 37°C.)

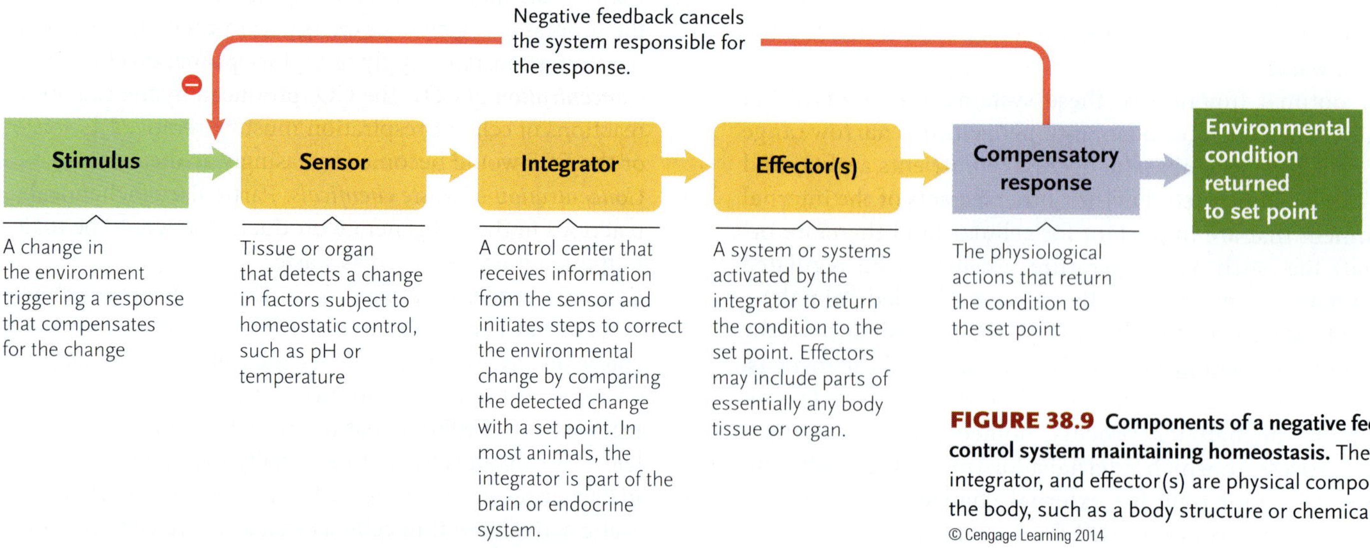

FIGURE 38.9 Components of a negative feedback control system maintaining homeostasis. The sensor, integrator, and effector(s) are physical components in the body, such as a body structure or chemical.

- **Effector**—a system that is activated by the integrator to bring the condition under control back to the set point. Effectors may include parts of essentially any body tissue or organ. (Physiological or behavioral responses that function to return body temperature to the set point.)

Let us consider the effectors for body temperature homeostasis in warm-blooded vertebrates in a little more detail. If the temperature falls below the lower limit, as was the case for the passengers and crew of the *Titanic* who ended up in the water (see *Why It Matters*), the hypothalamus activates effectors that constrict the blood vessels in the skin. The reduction in blood flow means that less heat is conducted from the blood through the skin to the environment, reducing heat loss from the skin. Other effectors may induce shivering, a physical mechanism to generate body heat. Also, integrating neurons in the brain, stimulated by signals from the hypothalamus, make us consciously sense a chill, which we may counteract behaviorally by, if possible, putting on more clothes or moving to a warmer area.

Conversely, if blood temperature rises above the set point by exposure to high external temperature or by vigorous exercise, the hypothalamus triggers effectors that dilate the blood vessels in the skin, increasing blood flow and heat loss from the skin. Other effectors induce sweating, which cools the skin and the blood flowing through it as the sweat evaporates. And again, through integrating neurons in the brain, we may consciously sense being overheated, which we may counteract behaviorally by shedding clothes, moving to a cooler location, stopping exercise, or taking a dip in a pool or a cool shower.

All mammals have similar homeostatic control systems that maintain or adjust body temperature. **Figure 38.10** illustrates how a dog responds to high environmental temperatures by the mechanisms we have discussed and, in addition, by panting. Birds also pant to reduce their body temperatures. Many terrestrial animals enter or splash water over their bodies to cool off.

The homeostatic control system for temperature can also be overwhelmed by prolonged exposure to a particularly high or low temperature. In either case, the end result will be death unless the external temperature returns to a more moderate value. As the *Titanic* story describes, only 15 minutes in about-freezing water can kill a person. Likewise, homeostatic control systems for other regulated factors can also be overwhelmed or be disrupted. For example, the inability to regulate blood glucose concentration homeostatically leads to the disease state of diabetes.

Whereas mammals and birds regulate their internal body temperature within a narrow range around a set point, certain other vertebrates regulate over a broader range. These vertebrates use other, less precise negative feedback mechanisms for their temperature regulation. Snakes and lizards, for example, respond behaviorally to compensate for variations in environmental temperatures. They may absorb heat by basking on sunny rocks in the cool early morning and move to cooler, shaded spots in the heat of the afternoon.

Some invertebrates, such as dragonflies, moths, and butterflies, use muscular contractions equivalent to shivering when their body temperature falls below the level required for flight. The shivering contractions warm the muscles to flying temperature. All of these physiological and behavioral responses depend on negative feedback mechanisms involving sensors, integrators, and effectors.

Animals Also Have Positive Feedback Mechanisms That Do Not Result in Homeostasis

Under certain circumstances, animals respond to a change in internal or external environmental condition by a **positive feedback** mechanism that intensifies or adds to the change. Such mechanisms, with some exceptions, do not result in homeostasis. Positive feedback mechanisms are less common than negative feedback mechanisms.

The birth process in mammals provides an example of positive feedback. During human childbirth, initial contractions

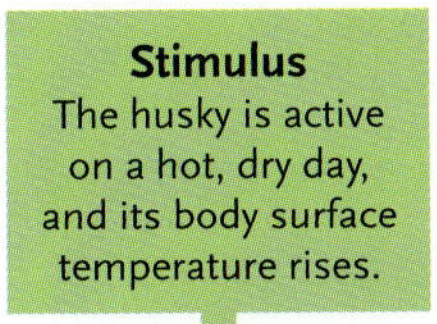

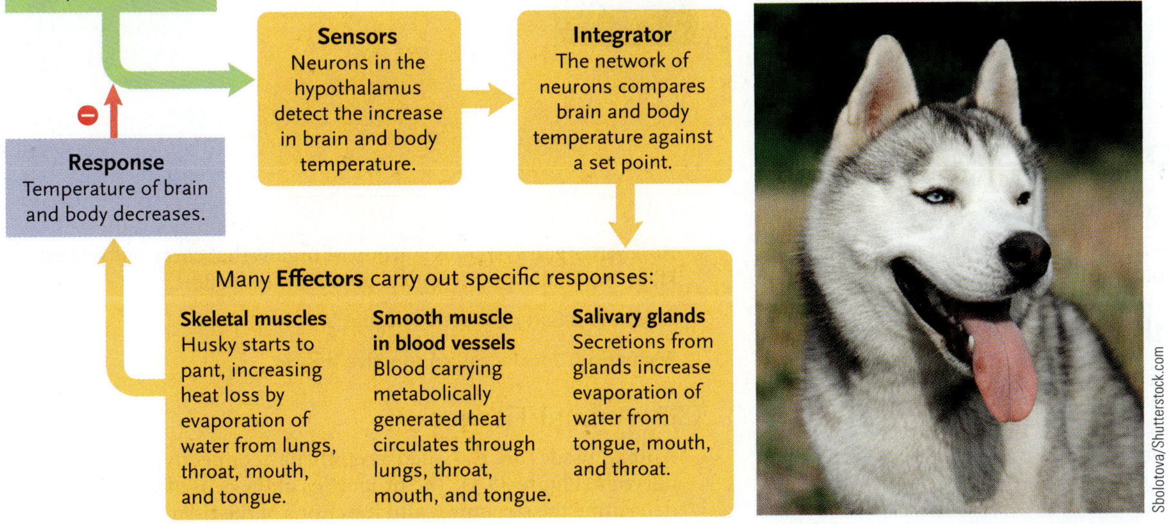

FIGURE 38.10
Homeostatic negative feedback control system maintaining the body temperature of a dog when environmental temperatures are high.

of the uterus push the head of the fetus against the cervix, the opening of the uterus into the vagina, causing the cervix to stretch. Sensors that detect the stretching signal the hypothalamus to stimulate the release of the hormone oxytocin from the pituitary gland. Oxytocin increases the uterine contractions, intensifying the squeezing pressure on the fetus and further stretching the cervix. The stretching results in more oxytocin release and stronger uterine contraction, repeating the positive feedback circuit and increasing the squeezing pressure until the fetus is pushed entirely out of the uterus.

In conclusion, we learned in this chapter about the various tissues and organ systems of the body, and of the involvement of organ systems in homeostasis. Next, we begin a series of chapters describing the organ systems in detail, starting with the nervous system.

STUDY BREAK 38.4

What are the components of a negative feedback control system that results in homeostasis?

UNANSWERED QUESTIONS

As described in the chapter, all animals have homeostatic control systems through which they maintain a consistent internal environment. The physiological processes underlying many of these control systems are shaped by evolution to adapt to daily changes in the external environment. In mammals, many physiological processes display 24-hour rhythms that are generated by an internal gene-based clock called a circadian clock (*circa* = approximately; *dies* = day). Some malfunctions of systems that are normally under homeostatic control also seem to exhibit a circadian cycle.

Why do so many strokes and heart attacks occur in the morning?

Strokes and heart attacks occur most frequently at a particular time of the day—in the morning—exhibiting a 24-hour, or circadian, rhythm. Circadian rhythms are generated through a molecular signaling mechanism called the circadian clock. A "central" circadian clock in the brain acts to control circadian rhythm in sleep-wake cycles. These same proteins are also found throughout the organs and cells of the body, including blood vessels (vascular clock). Recent work has shown that a broken clock also breaks down important physiologic signals, like nitric oxide to worsen heart attack, stroke, and vascular disease in animal models. (The role of nitric oxide on neuron function is discussed in Chapter 39.) Current research is addressing how circadian clock malfunctions contribute to both acute and chronic disease processes. In our laboratory, we are using mice with targeted disruption of circadian clock genes (knockout or mutant mice) and studying their response to cardiovascular stresses, including hypertension and vascular disease.

Location, location, location: Where in the body does the circadian clock malfunction to cause vascular disease?

Though heart attacks and strokes are the culmination of ongoing vascular disease, their onset is acute, often occurring in the morning around 8 A.M. In contrast, the progression to vascular disease is chronic, occurring in a time-span of years. Our laboratory and others have now demonstrated that the genes comprising the circadian clock can contribute to vascular disease. This finding suggests that, in addition to daily cycles, the circadian clock controls aspects of biology that are also important in the long-term. We think of circadian disorders as overt behavioral disorders that include poor sleep, night/shift work, and jet-lag, but there may be more to the picture. Recently, we have demonstrated that dysfunction of the circadian clock within local tissue (blood vessels) is important to confer disease, independent of the central clock. Thus, circadian malfunction in tissues outside the brain may contribute to diseases such as atherosclerosis, hypertension, diabetes, and even cancer. These basic science observations of circadian function in animals may lay the foundation for clinical studies in humans, which ultimately may change the way we understand and treat arteriosclerosis, hypertension, and heart attack.

THINK LIKE A SCIENTIST How might circadian rhythms have been important for survival in early organisms?

R. Daniel Rudic

R. Daniel Rudic is an Associate Professor in the Department of Pharmacology and Toxicology at the Medical College of Georgia. To learn more about his research on circadian rhythms and vascular biology, go to http://www.mcg.edu/som/phmtox/RudicLab/index.asp.

REVIEW KEY CONCEPTS

To access the course materials and companion resources for this text, please visit www.cengagebrain.com.

38.1 Organization of the Animal Body

- In most animals, cells are specialized and organized into tissues, tissues into organs, and organs into organ systems. A tissue is a group of cells with the same structure and function, working as a unit to carry out one or more activities. An organ is an assembly of tissues integrated into a structure that carries out a specific function. An organ system is a group of organs that carry out related steps in a major physiological process (Figure 38.1).

Animation: Soft connective tissues

Animation: Specialized connective tissues

38.2 Animal Tissues

- Animal tissues are classified as epithelial, connective, muscle, or nervous (Figure 38.1). The properties of the cells of these tissues determine the structures and functions of the tissues.

- Various kinds of junctions link cells in a tissue. Anchoring junctions "weld" cells together. Tight junctions seal the cells into a leak-proof layer. Gap junctions form direct avenues of communication between the cytoplasm of adjacent cells in the same tissue.
- Epithelial tissue consists of sheetlike layers of cells that cover body surfaces and the surfaces of internal organs and line cavities and ducts within the body (Figure 38.2).
- Glands are secretory structures derived from epithelia. They may be exocrine (connected to an epithelium by a duct that empties on the epithelial surface) or endocrine (ductless, with no direct connection to an epithelium) (Figure 38.3).
- Connective tissue consists of cell networks or layers and an extracellular matrix (ECM). It supports other body tissues, transmits mechanical and other forces, and in some cases acts as a filter (Figure 38.4).
- Loose connective tissue consists of sparsely distributed fibroblasts surrounded by an open network of collagen and other glycoproteins. It supports epithelia and organs of the body and forms a covering around blood vessels, nerves, and some internal organs.
- Fibrous connective tissue contains sparsely distributed fibroblasts in a matrix of densely packed, parallel bundles of collagen and elastin fibers. It forms high tensile-strength structures such as tendons and ligaments.
- Cartilage consists of sparsely distributed chondrocytes surrounded by a network of collagen fibers embedded in a tough but highly elastic matrix of branched glycoproteins. Cartilage provides support, flexibility, and a low-friction surface for joint movement (Figure 38.4).
- In bone, osteocytes are embedded in a collagen matrix hardened by mineral deposits. Osteoblasts secrete collagen and minerals for the ECM; osteoclasts remove the minerals and recycle them into the bloodstream (Figure 38.4).
- Adipose tissue consists of cells specialized for fat storage. It also cushions and rounds out the body and provides an insulating layer under the skin (Figure 38.4).
- Blood consists of a fluid matrix, the plasma, in which erythrocytes and leukocytes are suspended. The erythrocytes carry oxygen to body cells; the leukocytes produce antibodies and initiate the immune response against disease-causing agents (Figure 38.4).
- Muscle tissue contains cells that have the ability to contract forcibly (Figure 38.5). Skeletal muscle, containing long cells called muscle fibers, moves body parts and maintains posture.
- Cardiac muscle, which contains short contractile cells with a branched structure, forms the heart.
- Smooth muscle consists of spindle-shaped contractile cells that form layers surrounding body cavities and ducts.
- Nervous tissue contains neurons and glial cells. Neurons communicate information between body parts in the form of electrical and chemical signals (Figure 38.6). Glial cells support the neurons or provide electrical insulation between them.

Animation: Differences between cell and tissue types

Animation: Structure of an epithelium

Animation: Neuron structure and function

38.3 Coordination of Tissues in Organs and Organ Systems

- Organs and organ systems are coordinated to carry out vital tasks, including maintenance of internal body conditions; nutrient acquisition, processing, and distribution; waste disposal; molecular synthesis; environmental sensing and response; protection against injury and disease; and reproduction.
- In vertebrates and most invertebrates, the major organ systems that accomplish these tasks are the nervous, endocrine, muscular, skeletal, integumentary, circulatory, lymphatic, immune, respiratory, digestive, excretory, and reproductive systems (Figure 38.7).

Animation: Human organ systems

38.4 Homeostasis

- Homeostasis is the process by which animals maintain their internal environment—the extracellular fluid (Figure 38.8)—under conditions their cells can tolerate. It is a dynamic state, in which internal adjustments are made continuously to compensate for environmental changes.
- Multicellular organisms fall into two categories with respect to homeostatic regulation of the internal environment: regulators maintain factors of the internal environment in a relatively constant state, whereas conformers have internal environments that match the external environment.
- Homeostatically regulated factors of the internal environment include nutrient concentration; O_2 and CO_2 concentrations; concentrations of waste chemicals, water and NaCl; pH; and volume and pressure of plasma.
- Most homeostatic control systems operate systemically, whereas only some operate locally.
- Homeostasis is accomplished by negative feedback mechanisms that include a sensor, which detects a change in an external or internal condition; an integrator, which compares the detected change with a set point; and an effector or effectors, which return the condition to the set point if it has varied (Figure 38.9).
- Animals also have positive feedback mechanisms, in which a change in an internal or external condition triggers a response that intensifies the change, and typically does not result in homeostasis.

Animation: Negative feedback system

Animation: Feedback control of temperature

UNDERSTAND & APPLY

Test Your Knowledge

1. Which tissue type consisting of sheetlike layers of cells can both exchange oxygen and act as a barrier to bacteria?
   a. nervous
   b. epithelial
   c. connective
   d. muscle
   e. heart
2. Which of the following is a constant source of adult stem cells in a mammal?
   a. bone marrow
   b. pancreas
   c. basal lamina
   d. heart muscle
   e. kidney

3. A flexible, rubbery protein in connective tissue is called ____, whereas a more fibrous, less flexible glycoprotein is called ____.
   a. adipose; cartilage
   b. endocrine; exocrine
   c. sweat; hormones
   d. chondroitin sulfate; hydroxyapatite
   e. elastin; collagen
4. Adipose tissue:
   a. gives elasticity under epithelium.
   b. gives strength to tendons.
   c. insulates and is an energy reserve.
   d. provides movement, support, and protection.
   e. supports the nose and airways.
5. The bones of an elderly woman break more easily than those of a younger person. You would surmise that with aging, the cell type that diminishes in activity is the:
   a. osteocyte.
   b. osteoblast.
   c. osteoclast.
   d. chondrocyte.
   e. fibroblast.
6. Lifting weights will most increase the size of:
   a. skeletal muscle.
   b. smooth muscle.
   c. cardiac muscle.
   d. involuntary muscle.
   e. interlinked, branched muscle.
7. Which muscle types appear striated under a microscope?
   a. skeletal muscles only
   b. cardiac muscles only
   c. skeletal muscles and cardiac muscles
   d. smooth muscles only
   e. skeletal muscles and smooth muscles
8. Which of the following is *not* a homeostatic response?
   a. In a contest, a student eats an entire chocolate cake in 10 minutes, and his blood glucose level does not change dramatically.
   b. A jogger is sweating after a two-mile jog.
   c. The pupils in humans' eyes constrict when looking at a light.
   d. Physical activity increases carbon dioxide blood levels, which lowers blood pH, and increases breathing.
   e. Oxytocin is released by the hypothalamus during human childbirth.
9. When you exercise, a skeletal muscle increases its use of O_2, which reduces the concentration of O_2 in the muscle cells and sets up a requirement for increased O_2 uptake from the interstitial fluid. This is an example of:
   a. osmolarity.
   b. environmental sensing.
   c. integration.
   d. systemic homeostatic control.
   e. local homeostatic control.
10. The system that coordinates other organ systems is the:
   a. skeletal system.
   b. reproductive system.
   c. muscular system.
   d. nervous system.
   e. integumentary system.

Discuss the Concepts

1. Blood is often described as an atypical connective tissue. If you had to argue that blood is a connective tissue, what reasons would you include? What reasons would you include if you had to argue that blood is not a connective tissue?
2. What effect do you think a program of lifting weights would have on the bones of the skeleton? How would you design an experiment to test your prediction?
3. Positive feedback mechanisms are rare in animals compared with negative feedback mechanisms. Why do you think this is so?
4. Near the time of childbirth, collagen fibers in the connective tissue of the cervix break down, and gap junctions between the smooth muscle cells of the uterus increase in number. What do you think is the significance of these tissue changes?
5. Explain how, when driving, you control the car's speed by a typical negative feedback mechanism.

Design an Experiment

The regulation of temperature in mammals and birds is an example of homeostasis. Design an experiment to observe and measure processes involved in temperature homeostasis in sedentary versus athletic humans during exercise.

Interpret the Data

Vasopressin is an important regulator of blood pressure. Taurine has been shown to inhibit vasopressin release. An experiment was done to test the hypothesis that taurine acts as part of a negative feedback system that regulates vasopressin. The experiment measured the release of taurine from rat pituitary cells in the presence of vasopressin (10 nM) **(Figure)**. Are these data consistent with the hypothesis? Why?

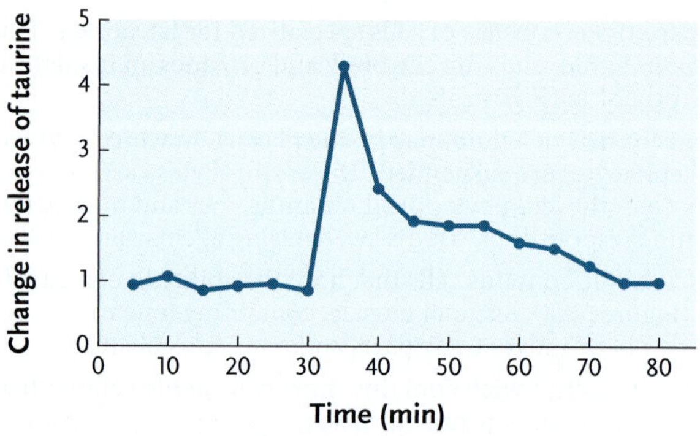

L. Rosso et al. 2004. Vasopressin-induced taurine efflux from rat pituicytes: a potential negative feedback for hormone secretion. *The Journal of Physiology* 554:731–742. Reprinted by permission of John Wiley and Sons.

Source: L. Rosso et al. 2004. Vasopressin-induced taurine efflux from rat pituicytes: a potential negative feedback for hormone secretion. *The Journal of Physiology* 554:731–742.

Apply Evolutionary Thinking

Large animals (like many of the vertebrates described in this chapter) often have more types of specialized tissues and organ systems than do small animals (such as flatworms). Why would natural selection favor the evolution of additional tissue types and organ systems in larger organisms?

Two North American bull elks contesting for cows. The shorter days of autumn trigger hormone production, battling, and reproductive behavior.

42

The Endocrine System

STUDY OUTLINE

42.1 Hormones and Their Secretion

42.2 Mechanisms of Hormone Action

42.3 The Hypothalamus and Pituitary

42.4 Other Major Endocrine Glands of Vertebrates

42.5 Endocrine Systems in Invertebrates

Why it matters . . . Every September, as the days grow shorter, bull elks *(Cervus canadensis)* begin to strut their stuff. Although they have grazed peacefully together at high mountain elevations from the Yukon to Arizona, they now become testy with each other. They also rasp at tree branches and plow the ground with their antlers. Soon, they descend to lower elevations, where the cow elks have been feeding in large nursery groups with their calves and yearlings.

The bulls move in among the cows and chase away the male yearlings. As part of the mating ritual, the bulls bugle, square off, strut, and circle; then they clash their antlers together, attempting to drive each other from the cows. The winning males claim harems of about 10 females each.

After the mating season ends, tranquility returns. The cows again graze in herds; the males form now-friendly bachelor groups that also feed quietly in the meadows. Eating is their major occupation, storing nutrients in preparation for the winter. The young will be born eight to nine months later, when summer returns.

The next year, the shortening days of late summer and fall again trigger the transition to mating behavior. Detected by the eyes and registered in the brain, reduced day length initiates changes in the secretion of long-distance signaling molecules called **hormones** (*horme* = to excite). Hormones released from one group of cells are transported through the circulatory system to other cells, their target cells, whose activities they change. Among the changes will be a rise in the concentration of hormones responsible for mating behavior.

We too are driven by our hormones. They control our day-to-day sexual behavior—sometimes as outlandish as that of a bugling bull elk—as well as a host of other functions, from the concentration of salt in our blood, to body growth, to the secretion of digestive juices. Along with the central nervous system, hormones coordinate the activities of multicellular life.

The best-known hormones are secreted by cells of the **endocrine system** (*endo* = within; *krinein* = to separate), although hormones actually are produced by many organ systems in the body. The endocrine system, like the nervous system, regulates and coordinates distant organs. The two systems are structurally, chemically, and functionally related, but they control different types of activities. The nervous system, through its high-speed electrical signals, enables an organism to interact rapidly with the external environment, while the endocrine system mainly controls activities that involve slower, longer-acting responses. Typical responses to hormones may persist for hours, weeks, months, or even years.

The mechanisms and functions of the endocrine system are the subjects of this chapter. As in other chapters of this unit, we pay particular attention to the endocrine system of humans and other mammals.

42.1 Hormones and Their Secretion

Cells signal other cells using neurotransmitters, hormones, and local regulators. Recall from Chapters 39, 40, and 41 that a neurotransmitter is a chemical released by an axon terminal at a synapse that affects the activity of a postsynaptic cell. Our focus in this chapter is hormones and local regulators.

The Endocrine System Includes Four Major Types of Cell Signaling

Four types of cell signaling occur in the endocrine system: classical endocrine signaling, neuroendocrine signaling, paracrine regulation, and autocrine regulation. In *classical endocrine signaling,* hormones are secreted into the extracellular fluid (ECF) by the cells of ductless secretory organs called **endocrine glands (Figure 42.1A).** (In contrast, *exocrine glands,* such as the sweat and salivary glands, release their secretions into ducts that lead outside the body or into the cavities of the digestive tract, as described in Section 38.2.) The hormones are circulated throughout the body in the blood and, as a result, most body cells are exposed to a wide variety of hormones. (The cells of the central nervous system are sequestered from the general circulatory system by the blood–brain barrier, described in Section 40.3.) Only *target cells* of a hormone, those with *receptor proteins* recognizing and binding that hormone, respond to it. Through these responses, hormones control such vital functions as digestion, osmotic balance, metabolism, cell division, reproduction, and development. The action of hormones may either speed up or inhibit these cellular processes. For example, growth hormone stimulates cell division, whereas glucocorticoids inhibit glucose uptake by most cells in the body.

Hormones are cleared from the body by enzymatic breakdown, mainly in the liver and kidneys, but also in target cells

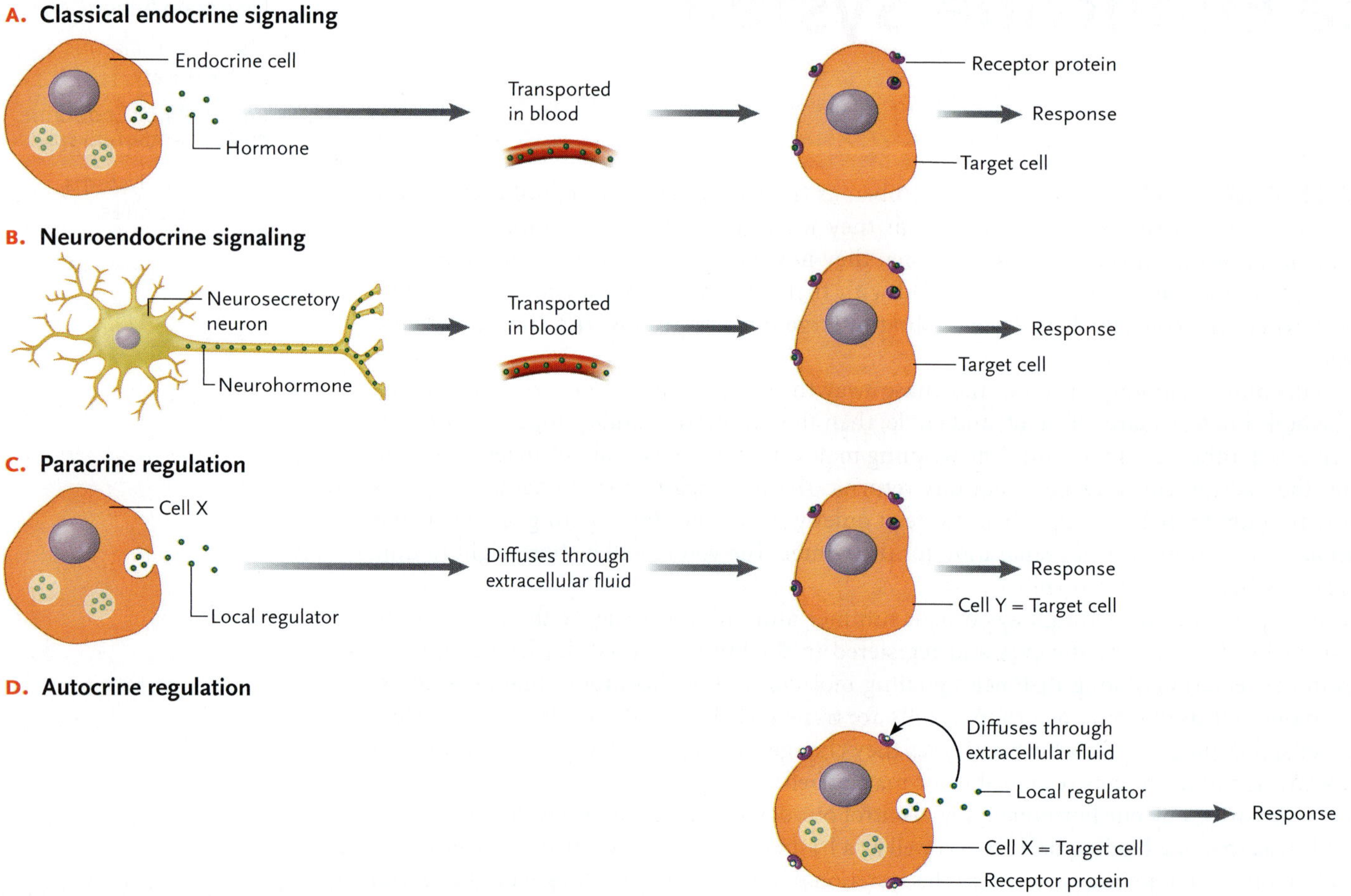

FIGURE 42.1 **The four major types of cell signaling in the endocrine system.**

themselves. Breakdown products are excreted in urine and feces. Depending on the hormone, the breakdown takes minutes to days.

In *neuroendocrine signaling,* specialized neurons called **neurosecretory neurons** release a hormone called a *neurohormone* into the circulatory system when appropriately stimulated **(Figure 42.1B).** The neurohormone is distributed by the circulatory system and elicits a response in target cells that have receptors for the hormone. Note that both neurohormones and neurotransmitters are secreted by neurons. An important distinction between neurohormones and neurotransmitters is that neurohormones are carried to target cells by the blood, whereas neurotransmitters act across a synaptic cleft (see Figure 39.6B). However, both neurohormones and neurotransmitters function in the same way—they cause cellular responses by interacting with specific receptors on target cells.

In *paracrine regulation,* a cell releases a signaling molecule that diffuses through the ECF and acts on nearby cells—regulation is *local* rather than at a distance, as is the case with hormones and neurohormones **(Figure 42.1C).** In some cases the local regulator acts on the same cells that produced it; this is called *autocrine regulation* **(Figure 42.1D).** For example, many of the growth factors that regulate cell division and differentiation act in both a paracrine and autocrine fashion.

Hormones and Local Regulators Can Be Grouped into Four Classes Based on Their Chemical Structure

More than 60 hormones and local regulators have been identified in humans. Many human hormones are either identical or very similar in structure and function to those in other animals, but other vertebrates as well as invertebrates have hormones not found in humans. Most hormones and local regulators fall into four molecular classes: amine, peptide, steroid, and fatty acid–derived molecules.

Amine hormones are involved in classical endocrine and neuroendocrine signaling. Most amine hormones are based on tyrosine. With one major exception, they are hydrophilic molecules, which diffuse readily into the blood and ECF. On reaching a target cell, they bind to receptors at the cell surface. The amine hormones include dopamine, epinephrine, and norepinephrine, already familiar as neurotransmitters released by some neurons (see Section 39.3). The exception is thyroxine, a hydrophobic amine hormone secreted by the thyroid gland. This hormone, which is based on a pair of tyrosines, passes freely through the plasma membrane and binds to a receptor inside the target cell, as do steroid hormones (see following discussion and Section 7.5).

The *peptide hormones* consist of amino acid chains, ranging in length from as few as 3 amino acids to more than 200. Some have carbohydrate groups attached to the amino acid chain. They are involved in classical endocrine signaling (for example, insulin) and neuroendocrine signaling (for example, gonadotropin-releasing hormone). Peptide hormones are released into the ECF, and from there they enter the blood. One large group of peptide hormones, the **growth factors,** regulates the division and differentiation of many cell types in the body. Many growth factors act in both paracrine and autocrine manners, as well as in classical endocrine signaling. Because they can switch cell division on or off, growth factors are an important focus of cancer research.

Steroid hormones are involved in classical endocrine signaling. All are hydrophobic molecules derived from cholesterol. They combine with hydrophilic carrier proteins to form water-soluble complexes that can diffuse through the ECF and enter the bloodstream. On contacting a cell, the hormone is released from its carrier protein, passes through the plasma membrane of the target cell, and binds to internal receptors in the nucleus or cytoplasm. Steroid hormones include aldosterone, cortisol, and the sex hormones. Steroid hormones may vary little in structure, but produce very different effects. For example, testosterone and estradiol, two major sex hormones responsible for the development of mammalian male and female characteristics, respectively, differ in little more than the presence or absence of a methyl group (see Figure 3.15).

Fatty acid-derived molecules are involved in paracrine and autocrine regulation. **Prostaglandins,** for example, are important as local regulators. Virtually every cell can secrete prostaglandins, and they are present at essentially all times. In semen, they enhance the transport of sperm through the female reproductive tract by increasing the contractions of smooth muscle cells, particularly in the uterus. During childbirth, prostaglandins secreted by the placenta work with a peptide hormone called oxytocin to stimulate labor contractions (see Sections 38.4 and 50.4). Other prostaglandins induce contraction or relaxation of smooth muscle cells in many parts of the body, including blood vessels and air passages in the lungs. When released as a product of membrane breakdown in injured cells, prostaglandins may also intensify pain and inflammation.

Many Hormones Are Regulated by Feedback Pathways

The secretion of many hormones is regulated by feedback pathways. Most of these pathways are controlled by negative feedback—that is, a product of the pathway inhibits an earlier step in the pathway (see Section 38.4). For example, in some mammals, secretion by the thyroid gland is regulated by a negative feedback loop **(Figure 42.2).** Secretory neurons in the hypothalamus secrete thyroid-releasing hormone (TRH) into the ECF and capillaries connecting the hypothalamus to the pituitary gland. In response, the pituitary releases thyroid-stimulating hormone (TSH) into the blood to stimulate the thyroid gland to release thyroid hormones. As the thyroid hormone concentration in the blood increases, it begins to inhibit TSH secretion by the pituitary; this action is the negative feedback step. The result is that secretion of the thyroid hormones is reduced.

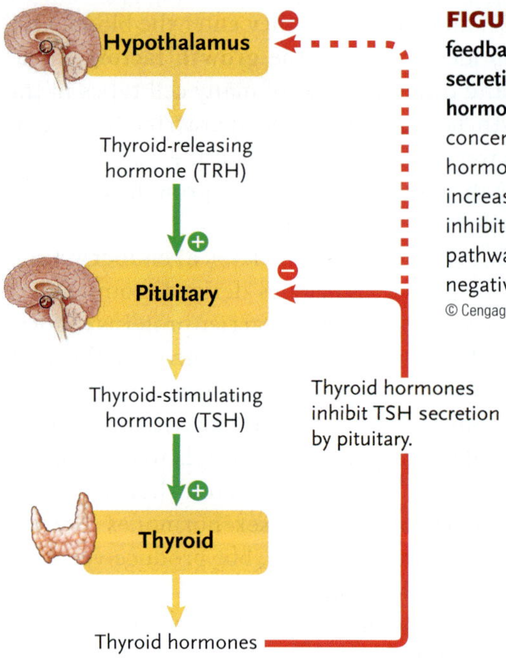

FIGURE 42.2 A negative feedback loop regulating secretion of the thyroid hormones. As the concentration of thyroid hormones in the blood increases, the hormones inhibit an earlier step in the pathway (indicated by the negative sign).

Body Processes Are Regulated by Coordinated Hormone Secretion

Although we will talk mostly about individual hormones in the remainder of the chapter, body processes are affected by more than one hormone. For example, the blood concentrations of glucose, fatty acids, and ions such as Ca^{2+}, K^{+}, and Na^{+} are regulated by the coordinated activities of several hormones secreted by different glands. Similarly, body processes such as oxidative metabolism, digestion, growth, sexual development, and reactions to stress are all controlled by multiple hormones.

In many of these systems, negative feedback loops adjust the levels of secretion of hormones that act in antagonistic ("opposing") ways, creating a balance that maintains homeostasis. For example, consider the regulation of fuel molecules such as glucose, fatty acids, and amino acids in the blood. We usually eat three meals a day and fast to some extent between meals. During these periods of eating and fasting, four hormone systems act in coordinated fashion to keep the fuel levels in balance: (1) insulin and glucagon, secreted by the pancreas; (2) growth hormone, secreted by the anterior pituitary; (3) epinephrine and norepinephrine, released by the sympathetic nervous system and the adrenal medulla; and (4) glucocorticoid hormones, released by the adrenal cortex.

The entire system of hormones regulating fuel metabolism resembles the fail-safe mechanisms designed by human engineers, in which redundancy, overlapping controls, feedback loops, and multiple safety valves ensure that vital functions are maintained at constant levels in the face of changing and even extreme circumstances.

STUDY BREAK 42.1

1. **Distinguish between a hormone and an exocrine secretion.**
2. **How is endocrine signaling different from autocrine signaling?**

42.2 Mechanisms of Hormone Action

Hormones control cell functions by binding to receptor molecules on or in their target cells. Small quantities of hormones can typically produce profound effects in cells and body functions as a result of **amplification** (see Figure 7.5). In amplification, binding of a hormone to a receptor triggers activation of many proteins, each of which then activates an even larger number of proteins for the next step in the cellular reaction pathway, and so on, increasing in magnitude for each step in the pathway. It has been estimated that, by amplification, a single molecule of epinephrine acting on a liver cell will liberate 10^6 molecules of glucose from stored glycogen.

Hydrophilic Hormones Bind to Surface Receptors, Activating Protein Kinases Inside Cells

Hormones that bind to receptor molecules in the plasma membrane—primarily hydrophilic amine and peptide hormones—produce their responses through signal transduction pathways (see Section 7.2). In brief, when a surface receptor binds a hormone, the receptor is activated and transmits the signal through the plasma membrane. Within the cell, the signal is transduced, changed into a form that causes the cellular response **(Figure 42.3A).** Typically, the reactions of signal transduction pathways involve protein kinases, enzymes that add phosphate groups to proteins. Adding a phosphate group to a protein may activate it or inhibit it, depending on the protein and the reaction. The particular response produced by a hormone depends on the kinds of protein kinases activated, the type of cell that can respond, and the types of target proteins they phosphorylate.

Two major types of surface receptors bind hydrophilic hormones: receptor tyrosine kinases and G-protein–coupled receptors. Receptor tyrosine kinases have a built-in protein kinase on the cytoplasmic side of the receptor (Figure 42.3A illustrates this type of receptor). Binding of a signal molecule to this type of receptor turns on the receptor's protein kinase, which activates the receptor. The activated receptor then initiates a signaling cascade within the cell, typically involving protein kinases, which phosphorylate target proteins, resulting in a cellular response. The cytoplasmic reactions that receptor tyrosine kinases control when they are activated are described in detail in Section 7.3. Insulin, a peptide hormone that lowers glucose concentration in the blood, elicits a cellular response using a signal transduction pathway initiated by activating a receptor tyrosine kinase. Insulin acts mainly by binding to receptors on liver cells, adipose tissue (fat), and skeletal muscle cells to stimulate glucose transport into cells, conversion of glucose to glycogen, and other metabolic activities.

G-protein–coupled receptors lack a built-in protein kinase. This type of receptor responds to the binding of a hormone by activating a G protein associated with the cytoplasmic end of the receptor. The activated G protein then activates an effector molecule, which then generates a second messenger molecule. Sec-

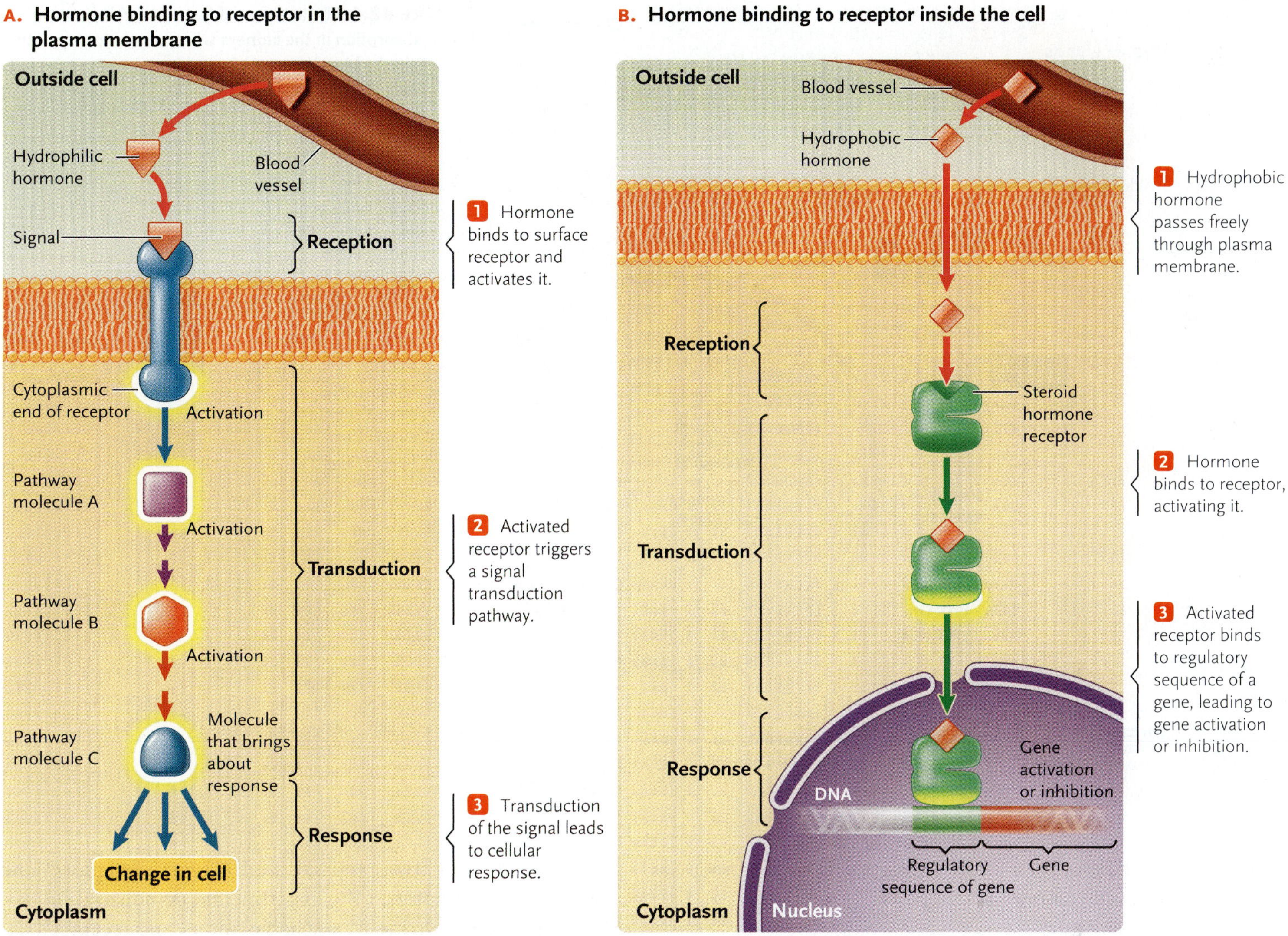

FIGURE 42.3 Outline of the reaction pathways activated by hormones that bind to receptor proteins in the plasma membrane (A) or inside cells (B). In both mechanisms, the signal—the binding of the hormone to its receptor—is transduced to produce the cellular response.

ond messengers activate protein kinases in the cell, which elicit the cellular response by phosphorylating target proteins. The cytoplasmic reactions that G-protein–coupled receptors control when activated are described in detail in Section 7.4. Glucagon, a peptide hormone that raises glucose concentration in the blood, elicits a cellular response using a signal transduction pathway initiated by activating a G-protein–coupled receptor. When glucagon binds to surface receptors on liver cells, it triggers the breakdown of glycogen stored in those cells into glucose. The glucose then is released into the circulatory system.

Hydrophobic Hormones Bind to Receptors Inside Cells, Activating or Inhibiting Genetic Regulatory Proteins

After passing through the plasma membrane, the hydrophobic steroid and thyroid hormones bind to internal receptors in the nucleus or cytoplasm (**Figure 42.3B,** and described in detail in Section 7.5). Binding of the hormone activates the receptor, which then binds to a regulatory sequence of specific genes. Depending on the gene, binding the regulatory sequence either activates or inhibits its transcription, leading to changes in protein synthesis that accomplish the cellular response. The characteristics of the response depend on the specific genes controlled by the activated receptors, and on the presence of other proteins that modify the activity of the receptor.

One of the actions of the steroid hormone aldosterone illustrates the mechanisms triggered by internal receptors (**Figure 42.4**). If blood pressure falls below optimal levels, aldosterone is secreted by the adrenal glands. The hormone affects only kidney cells that contain the aldosterone receptor in their cytoplasm. When activated by aldosterone, the receptor binds to the regulatory sequence of a gene, leading to the synthesis of proteins that increase reabsorption of Na^+ by the kidney cells. The resulting increase in Na^+ concentration in body fluids increases water retention and, with it, blood volume and pres-

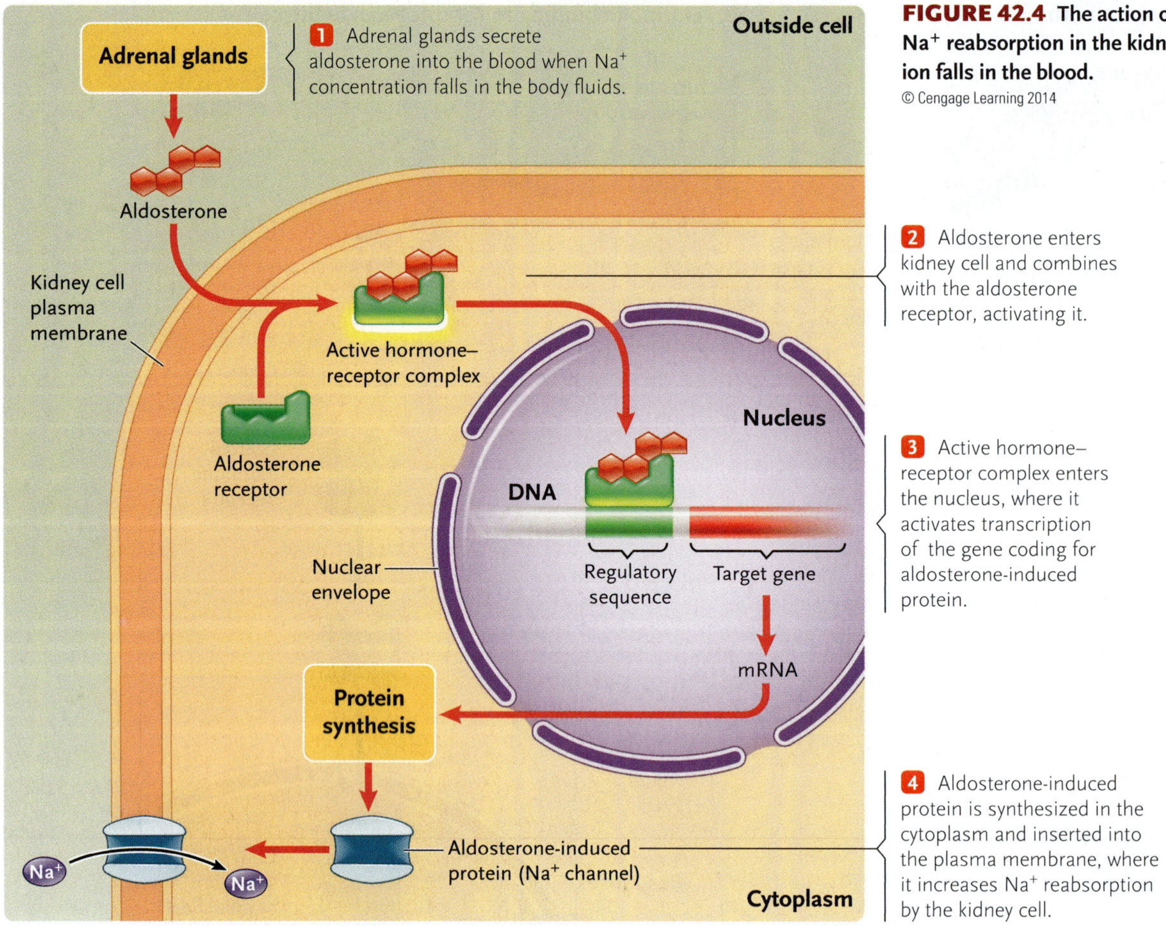

FIGURE 42.4 The action of aldosterone in increasing Na^+ reabsorption in the kidneys when concentration of the ion falls in the blood.

sure. These changes then feedback negatively on the processes that initiated aldosterone secretion.

Target Cells May Respond to More Than One Hormone, and Different Target Cells May Respond Differently to the Same Hormone

A single target cell may have receptors for several hormones and respond differently to each hormone. For example, vertebrate liver cells have receptors for the pancreatic hormones insulin and glucagon. Insulin increases glucose uptake and conversion to glycogen, which decreases blood glucose levels, while glucagon stimulates the breakdown of glycogen into glucose, which increases blood glucose levels.

Conversely, particular hormones interact with different types of receptors in or on a range of target cells. Different responses are then triggered in each target cell type because the receptors trigger different transduction pathways. For example, the amine hormone epinephrine secreted by the adrenal medulla prepares the body for handling stress (including dangerous situations) and physical activity. (Epinephrine is discussed in more detail in Section 42.4.) In mammals, epinephrine can bind to several different G-protein–coupled receptors (see Section 7.4) known as **adrenergic receptors** because of the adrenal origin of the hormones that bind to them. These receptors are categorized into two types: α-adrenergic receptors and β-adrenergic receptors. (The experimental demonstration that binding of epinephrine to a specific adrenergic receptor triggers a cellular response is described in **Figure 42.5.**) When epinephrine binds to an adrenergic receptor on a smooth muscle cell, such as that of a blood vessel, it triggers a response pathway that causes the cells to constrict, reducing circulation to peripheral organs. When epinephrine binds to β_1-adrenergic receptors on heart muscle cells, the contraction rate of the cells increases, which in turn enhances blood supply. When epinephrine binds to β_2-adrenergic receptors on liver cells, it stimulates the breakdown of glycogen to glucose, which is released from the cell. The overall effect of these, and a number of other, responses to epinephrine secretion is to supply energy to the major muscles responsible for locomotion—the body is now prepared for handling stress or for physical activity.

Different receptors binding hydrophobic hormones also may generate diverse responses. *Insights from the Molecular Revolution* (pp. 954–955) describes an investigation that tested the cellular responses produced by different receptors binding the same steroid hormone.

In summary, the mechanisms by which hormones work have four major features:

1. Only the cells that contain surface or internal receptors for a particular hormone respond to that hormone.

FIGURE 42.5 Experimental Research

Demonstration That Binding of Epinephrine to β-Adrenergic Receptors Triggers a Signal Transduction Pathway within Cells

Question: Is binding of epinephrine to β-adrenergic receptors necessary for triggering a signal transduction pathway within cells?

Experiment: Epinephrine triggers a signal transduction pathway. First, activation of adenylyl cyclase causes the level of the second messenger cAMP to increase, and then cAMP activates protein kinases in a signaling cascade that generates a cellular response (see Section 7.4). Richard Cerione and his colleagues at Duke University Medical Center studied whether the signal transduction pathway is stimulated by binding of epinephrine to β-adrenergic receptors.

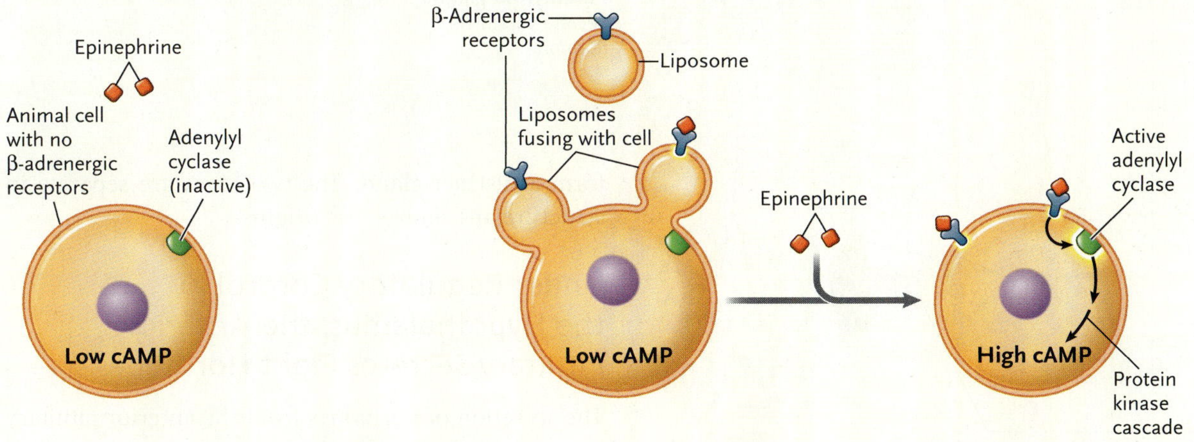

1. Epinephrine was added to animal cells lacking β-adrenergic receptors.

Result: No change occurred to the low level of cAMP in those cells. This result demonstrated that epinephrine alone was not able to trigger an increase in cAMP.

2. Liposomes—artificial spherical phospholipid membranes—containing purified β-adrenergic receptors were fused with the animal cells, and then epinephrine was added.

Result: When the liposomes fused with the animal cells, β-adrenergic receptors became part of the fused cell's plasma membrane. Then, adding epinephrine triggered synthesis of cAMP, resulting in high levels of cAMP in the cells. This result demonstrated that β-adrenergic receptors must be present in the membrane for epinephrine to trigger an increase in cAMP in the cell. The simplest interpretation was that epinephrine bound to the β-adrenergic receptors, activating adenylyl cyclase within the cell.

Conclusion: The cellular response depended on binding of the hydrophilic hormone to a specific plasma membrane–embedded receptor.

THINK LIKE A SCIENTIST How could this system be used to study what structural features of the receptor molecules are important for controlling the associated signal transduction pathway?

Source: R. A. Cerione et al. 1983. Reconstitution of the β-adrenergic receptors in lipid vesicles: Affinity chromatography-purified receptors confer catecholamine responsiveness on a heterologous adenylate cyclase system. *Proceedings of the National Academy of Sciences USA* 80:4899–4903.

2. Once bound by their receptors, hormones may produce a response that involves stimulation or inhibition of cellular processes through the specific types of internal molecules triggered by the hormone action.
3. Because of the amplification that occurs in both the surface and internal receptor mechanisms, hormones are effective in very small concentrations.
4. The response to a hormone differs among target organs.

In the next two sections, we discuss the major endocrine cells and glands of vertebrates. The locations of these cells and glands in the human body and their functions are summarized in **Figure 42.6** and **Table 42.1.** Peptide hormones secreted by other body regions, including the stomach and small intestine, the thymus gland, the kidneys, and the heart, will be described in the chapters in which these tissues and organs are discussed.

STUDY BREAK 42.2

1. **Compare and contrast the mechanisms by which glucagon and aldosterone cause their specific responses.**
2. **Explain how one type of target cell could respond to different hormones, and how the same hormone could produce different effects in different cells.**

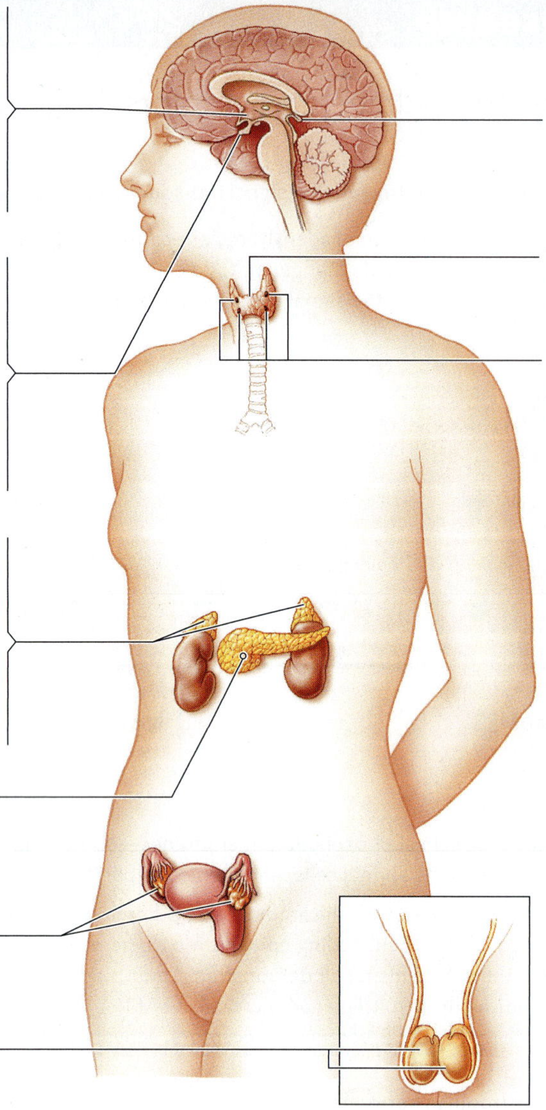

FIGURE 42.6 Major endocrine glands of the human body. Among the other organs that contain hormone-producing cells are the kidneys (see Section 44.2), the stomach and small intestine (see Section 47.4), and the heart (see Section 48.4).

42.3 The Hypothalamus and Pituitary

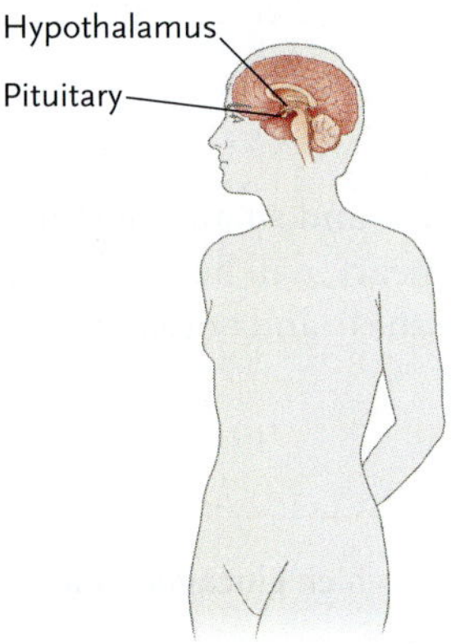

The hormones of vertebrates work in coordination with the nervous system. The action of several hormones is closely coordinated by the hypothalamus and the connected pituitary.

The hypothalamus is a region of the brain located in the floor of the cerebrum (see Section 40.3). The **pituitary gland,** consisting mostly of two fused lobes, is suspended just below it by a slender stalk of tissue that contains both neurons and blood vessels **(Figure 42.7).** The **posterior pituitary** contains axons and nerve endings of neurosecretory neurons that originate in the hypothalamus. The **anterior pituitary** contains nonneuronal endocrine cells that form a distinct gland. The two lobes are separate in structure and embryonic origins.

Under Regulatory Control by the Hypothalamus, the Anterior Pituitary Secretes Eight Hormones

The secretion of hormones from the anterior pituitary is controlled by peptide neurohormones called **releasing hormones (RHs)** and **inhibiting hormones (IHs),** released by the hypothalamus. These neurohormones are carried in the blood from the hypothalamus to the anterior pituitary in a *portal vein,* a special vein that connects the capillaries of the two glands. The portal vein provides a critical link between the brain and the endocrine system, ensuring that most of the blood reaching the anterior pituitary first passes through the hypothalamus.

RHs and IHs regulate the anterior pituitary's secretion of another group of hormones; those hormones in turn control many other endocrine glands of the body, and some body processes directly.

Secretion of hypothalamic RHs is controlled by neurons containing receptors that monitor the blood to detect changes in body chemistry and temperature. For example, TRH is secreted in response to a drop in body temperature. Input to the hypothalamus also comes through numerous connections from control centers elsewhere in the brain, including the brain stem and limbic system. Negative feedback pathways regulate secretion of the releasing hormones, such as the pathway regulating TRH secretion.

Under the control of the hypothalamic RHs, the anterior pituitary secretes six major hormones into the bloodstream: prolactin, growth hormone, thyroid-stimulating hormone, adrenocorticotropic hormone, follicle-stimulating hormone, and luteinizing hormone. Two additional hormones with more complex regulation, melanocyte-stimulating hormone (MSH) and endorphins, are also produced in pituitary tissue. **Prolactin (PRL)** influences reproductive activities and parental care in

TABLE 42.1	The Major Human Endocrine Glands and Hormones			
Secretory Tissue or Gland	**Hormones**	**Molecular Class**	**Target Tissue**	**Principal Actions**
Hypothalamus	Releasing and inhibiting hormones	Peptide	Anterior pituitary	Regulate secretion of anterior pituitary hormones
Anterior pituitary	Thyroid-stimulating hormone (TSH)	Peptide	Thyroid gland	Stimulates secretion of thyroid hormones and growth of thyroid gland
	Adrenocorticotropic hormone (ACTH)	Peptide	Adrenal cortex	Stimulates secretion of glucocorticoids by adrenal cortex
	Follicle-stimulating hormone (FSH)	Peptide	Ovaries in females, testes in males	Stimulates egg growth and development and secretion of sex hormones in females; stimulates sperm production in males
	Luteinizing hormone (LH)	Peptide	Ovaries in females, testes in males	Regulates ovulation in females and secretion of sex hormones in males
	Prolactin (PRL)	Peptide	Mammary glands	Stimulates breast development and milk secretion
	Growth hormone (GH)	Peptide	Bone, soft tissue	Stimulates growth of bones and soft tissues; helps control metabolism of glucose and other fuel molecules
	Melanocyte-stimulating hormone (MSH)	Peptide	Melanocytes in skin of some vertebrates	Promotes darkening of the skin
	Endorphins	Peptide	Pain pathways of PNS	Inhibit perception of pain
Posterior pituitary	Antidiuretic hormone (ADH)	Peptide	Kidneys	Raises blood volume and pressure by increasing water reabsorption in kidneys
	Oxytocin	Peptide	Uterus, mammary glands	Promotes uterine contractions; stimulates milk ejection from breasts
Thyroid gland	Calcitonin	Peptide	Bone	Lowers calcium concentration in blood
	Thyroxine and triiodothyronine	Amine	Most cells	Increase metabolic rate; essential for normal body growth
Parathyroid glands	Parathyroid hormone (PTH)	Peptide	Bone, kidneys, intestine	Raises calcium concentration in blood; stimulates vitamin D activation
Adrenal medulla	Epinephrine and norepinephrine	Amine	Sympathetic receptor sites throughout body	Reinforce sympathetic nervous system; contribute to responses to stress
Adrenal cortex	Aldosterone (mineralocorticoid)	Steroid	Kidney tubules	Helps control body's salt–water balance by increasing Na^+ reabsorption and K^+ excretion in kidneys
	Cortisol (glucocorticoid)	Steroid	Most body cells, particularly muscle, liver, and adipose cells	Increases blood glucose by promoting breakdown of proteins and fats
Testes	Androgens, such as testosterone*	Steroid	Various tissues	Control male reproductive system development and maintenance; most androgens are made by the testes
	Oxytocin	Peptide	Uterus	Promotes uterine contractions when seminal fluid is ejaculated into vagina during sexual intercourse

*Small amounts secreted by ovaries and adrenal cortex.

**Small amounts secreted by testes.

Continued

TABLE 42.1 The Major Human Endocrine Glands and Hormones—*Continued*

Secretory Tissue or Gland	Hormones	Molecular Class	Target Tissue	Principal Actions
Ovaries	Estrogens, such as estradiol**	Steroid	Breast, uterus, other tissues	Stimulate maturation of sex organs at puberty, and development of secondary sexual characteristics
	Prolactins, such as progesterone**	Steroid	Uterus	Prepare and maintain uterus for implantation of fertilized egg and the growth and development of embryo
Pancreas (islets of Langerhans)	Glucagon (alpha cells)	Peptide	Liver cells	Raises glucose concentration in blood; promotes release of glucose from glycogen stores and production from noncarbohydrates
	Insulin (beta cells)	Peptide	Most cells	Lowers glucose concentration in blood; promotes storage of glucose, fatty acids, and amino acids
Pineal gland	Melatonin	Amine	Brain, anterior pituitary, reproductive organs, immune system, possibly others	Helps synchronize body's biological clock with day length; may inhibit gonadotropins and initiation of puberty
Many cell types	Growth factors	Peptide	Most cells	Regulate cell division and differentiation
	Prostaglandins	Fatty acid	Various tissues	Have many diverse roles

**Small amounts secreted by testes.

vertebrates. In mammals, PRL stimulates development of the secretory cells of mammary glands during late pregnancy, and stimulates milk synthesis after a female mammal gives birth. Stimulation of the mammary glands and the nipples, as occurs during suckling, leads to PRL release.

Growth hormone (GH) stimulates cell division, protein synthesis, and bone growth in children and adolescents, thereby causing body growth. GH also stimulates protein synthesis and cell division in adults. For these actions, GH binds to target tissues, mostly liver cells, causing them to release **insulin-like growth factor (IGF),** a peptide that directly stimulates growth processes. GH also controls a number of major metabolic processes in mammals of all ages, including the conversion of glycogen to glucose and fats to fatty acids as a means of regulating their levels in the blood. In addition, GH stimulates body cells to take up fatty acids and amino acids and limits the rate at which muscle cells take up glucose. These actions help maintain the availability of glucose and fatty acids to tissues and organs between feedings; this is particularly important for the brain. In humans, deficiencies in GH secretion during childhood produce *pituitary dwarfs,* who remain small in stature through childhood and into adulthood. Overproduction of GH during childhood or adolescence, often due to a tumor of the anterior pituitary, produces *pituitary giants,* who may grow above 2.4 m (8 ft) in height.

The other four major hormones secreted by the anterior pituitary control endocrine glands elsewhere in the body. **Thyroid-stimulating hormone (TSH)** stimulates the thyroid gland to grow in size and secrete thyroid hormones. **Adrenocorticotropic hormone (ACTH)** triggers hormone secretion by cells in the adrenal cortex. **Follicle-stimulating hormone (FSH)** controls egg development and the secretion of sex hormones in female mammals, and sperm production in males. **Luteinizing hormone (LH)** regulates part of the menstrual cycle in human females and the secretion of sex hormones in males. FSH and LH are grouped together as **gonadotropins** because they regulate the activity of the gonads (ovaries and testes). The roles of the gonadotropins and sex hormones in the reproductive cycle are described in Chapter 49.

Melanocyte-stimulating hormone (MSH) and **endorphins** are two other hormones produced by the anterior pituitary. MSH is named because of its effect in some vertebrates on melanocytes, skin cells that contain the black pigment melanin. For example, an increase in secretion of MSH produces a marked darkening of the skin of fishes, amphibians, and reptiles. The darkening is produced by a redistribution of melanin from the centers of the melanocytes throughout the cells. In adult humans, the pituitary secretes little or no MSH.

Endorphins, peptide hormones produced by the hypothalamus and pituitary, are also released by the anterior pituitary (see Table 42.1). In the peripheral nervous system (PNS), endorphins act as neurotransmitters in pathways that control pain, thereby inhibiting the perception of pain.

The Posterior Pituitary Secretes Two Hormones into the Circulatory System

The neurosecretory neurons in the posterior pituitary secrete two peptide hormones, antidiuretic hormone and oxytocin, directly into the circulatory system (see Figure 42.7).

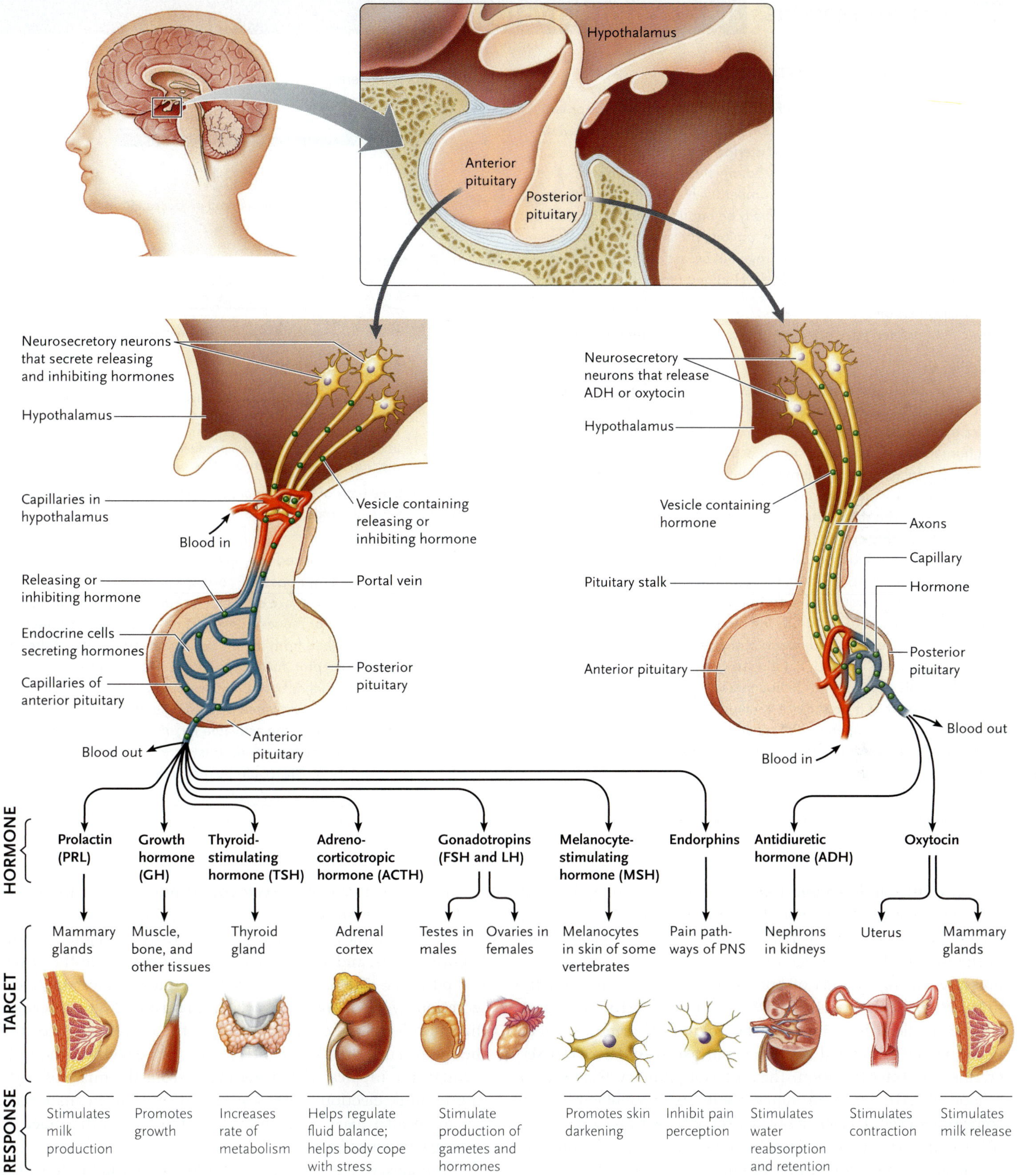

FIGURE 42.7 The hypothalamus and pituitary. Hormones secreted by the anterior and posterior pituitary are controlled by neurohormones released in the hypothalamus.

Two Receptors for Estrogens

Estrogens have many different effects on female sexual development, behavior, and the menstrual cycle. One negative effect is to stimulate the growth of tumors in breast and uterine cancer. This cancer-enhancing effect can be reduced by administering *antiestrogens,* estrogen-like chemicals that bind competitively to estrogen receptors and block the sites that would normally be bound by the hormone. The antiestrogen *tamoxifen,* for example, inhibits the growth of breast tumors by blocking the activity of estrogen in breast tissue, but patients receiving it are at increased risk of developing uterine cancer.

Research Question

How can tamoxifen have opposite effects in two different tissues?

Experiment

A group of investigators led by Thomas Scanlan and Peter Kushner at the University of California, San Francisco, joined by others at the Karolinska Institute and the Karo Bio Company in Sweden, had discovered that humans have two highly similar estrogen receptors, ERα and ERβ. Could differences between them account for the opposing effects of tamoxifen in breast and uterine tissues?

To find out, the researchers constructed two pairs of recombinant DNA plasmids **(Figure 1).** One pair consisted of either the ERα receptor gene or the ERβ receptor gene, adjacent to a promoter for continuous transcription of the receptor gene in human tissue culture cells that were not derived from breast cancer **(Figure 1A).** The other pair consisted of the firefly luciferase gene, which catalyzes a light-producing reaction, adjacent to one of two gene regulatory sequences, ERE or AP1, which act in estrogen-regulated systems **(Figure 1B).** In this experiment, the luciferase gene acts as a reporter gene, meaning that it provides a sensitive measure of the events occurring with the estrogen receptors and the gene regulatory sequences—luciferase has nothing to do with estrogen or estrogen activity.

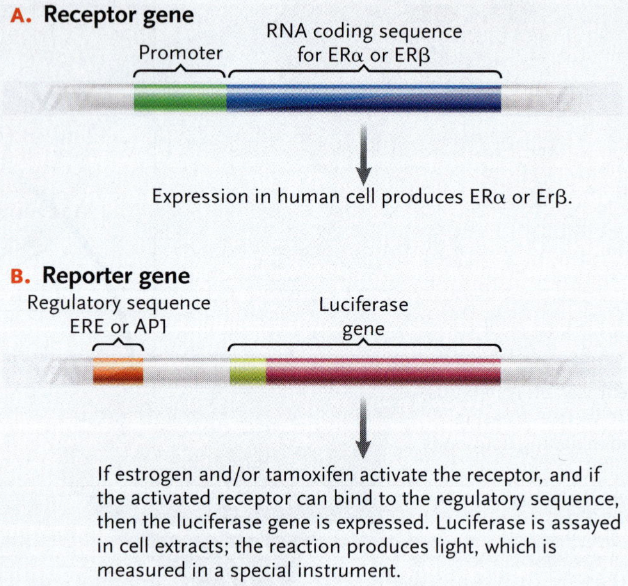

FIGURE 1 Schematic of the receptor genes and reporter genes on the recombinant DNA plasmids.

The researchers introduced one receptor gene plasmid and one reporter gene plasmid in four possible combinations into human cell lines that do not normally make estrogen receptors. Estrogen receptors are produced in all four of the resulting cell lines because the gene for the receptor is transcribed from the introduced plasmid. Two of the cell lines make the ERα receptor and two make the ERβ receptor.

With this experimental design, the researchers could test whether the ERα or ERβ receptor is activated by binding estrogen and/or tamoxifen, and also whether the activated receptor would bind to the luciferase plasmid. If these two conditions were met, luciferase would be synthesized, and its activity could be measured.

Antidiuretic hormone (ADH) stimulates kidney cells to absorb more water from urine, thereby increasing the volume of the blood. The hormone is released when sensory receptor cells of the hypothalamus detect an increase in the blood's osmotic pressure during periods of body dehydration or after a salty meal. Ethyl alcohol reduces ADH secretion, explaining in part why alcoholic drinks increase the volume of urine excreted. Nicotine and emotional stress, in contrast, stimulate ADH secretion and water retention. The posterior pituitary also releases a flood of ADH when an injury results in heavy blood loss or some other event triggers a severe drop in blood pressure. ADH helps maintain blood pressure by reducing water loss and also by causing small blood vessels in some tissues to constrict.

Hormones with structure and action similar to ADH are also secreted in fishes, amphibians, reptiles, and birds. In amphibians, these ADH-like hormones increase the amount of water entering the body through the skin and from the urinary bladder.

We have noted that **oxytocin** stimulates the ejection of milk from the mammary glands of a nursing mother. Stimulation of the nipples in suckling sends neuronal signals to the hypothalamus, and leads to release of oxytocin from the posterior pituitary. The released oxytocin stimulates more oxytocin secretion by a positive feedback mechanism. Oxytocin causes the smooth muscle cells surrounding the individual alveoli of the mammary glands to contract, forcibly expelling the milk through the nipples. The entire cycle, from the onset of suckling to milk ejection, takes less than a minute in mammals. Oxytocin also plays a key role in childbirth, as discussed in Sections 38.4 and 50.4.

In males, oxytocin is secreted into the seminal fluid by the testes. Like prostaglandins, when the seminal fluid is ejaculated into the vagina during sexual intercourse, oxytocin stimulates contractions of the uterus that aid movement of sperm through the female reproductive tract.

Results

The results are shown in **Figure 2.**

Conclusion

The experiments indicate that the previously baffling and opposing effects of tamoxifen on breast and uterine tissues occur because different estrogen receptors—ERα or ERβ—are present, and act on genes controlled by either the ERE or AP1 regulatory sequences. The results emphasize the fact that hormones can have distinct effects in different cell types depending on the types of receptors present.

THINK LIKE A SCIENTIST The researchers also performed the experiment of Figure 2D with cultured human breast cancer cells. The same general results were obtained, except that tamoxifen resulted in a significantly higher level of luciferase. Propose an explanation for this result.

Source: K. Paech et al. 1997. Differential ligand activation of estrogen receptors ERα and ERβ at AP1 sites. *Science* 277:1508–1510.

FIGURE 2 **Effects of estrogen and/or tamoxifen on luciferase expression in human cell lines containing a receptor gene plasmid and a reporter gene plasmid.**

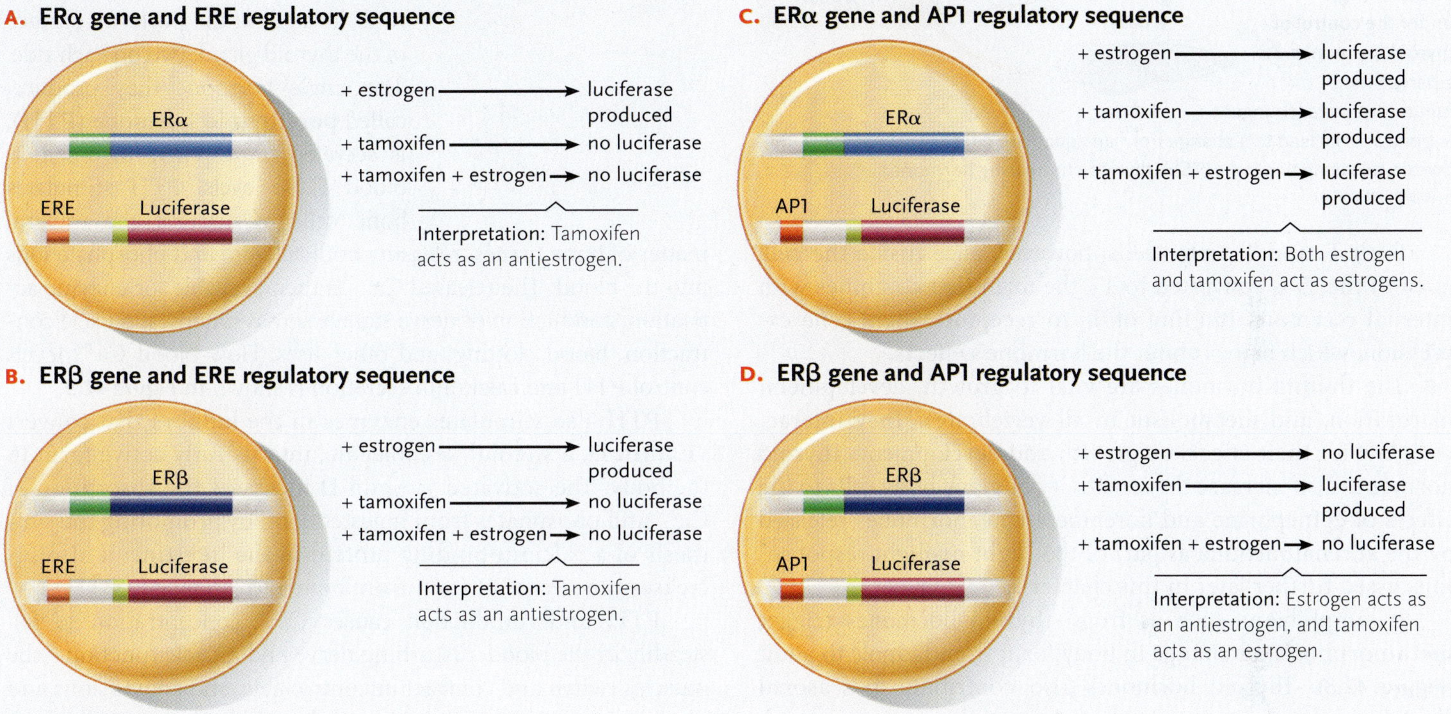

STUDY BREAK 42.3

1. **Summarize the functional relationship between the hypothalamus and the anterior pituitary gland.**
2. **What, in addition to the hormones they secrete, distinguishes the anterior and posterior pituitary glands?**

42.4 Other Major Endocrine Glands of Vertebrates

Besides the hypothalamus and pituitary, the body has seven major endocrine glands or tissues, many of them regulated by the hypothalamus–pituitary connection. These glands are the thyroid gland, parathyroid glands, adrenal medulla, adrenal cortex, gonads, pancreas, and pineal gland (shown in Figure 42.6 and summarized in Table 42.1).

The Thyroid Hormones Stimulate Metabolism, Development, and Maturation

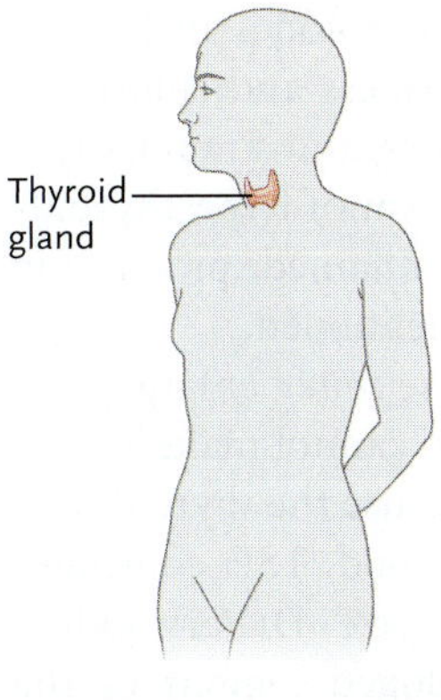

The **thyroid gland,** which is located in the front of the throat in humans, has a shape similar to that of a bowtie. It secretes the same hormones in all vertebrates. The primary circulating thyroid hormone, **thyroxine,** is known as T_4 because it contains four iodine atoms. The thyroid also secretes smaller amounts of a closely related hormone, **triiodothyronine** or T_3, which contains three iodine atoms. A supply of iodine in the diet is necessary for production of these hormones. Normally, their concentrations are kept at finely balanced levels in the blood by negative feedback loops such as that described in Figure 42.2.

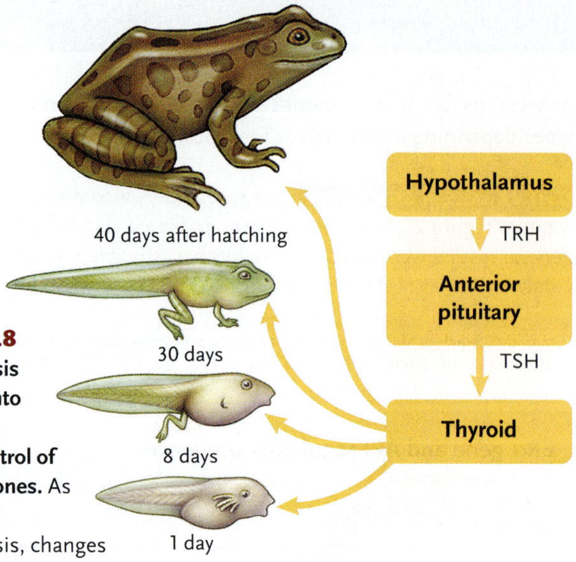

FIGURE 42.8 Metamorphosis of a tadpole into an adult frog, under the control of thyroid hormones. As a part of the metamorphosis, changes in gene activity lead to a change from an aquatic to a terrestrial habitat. TRH, thyroid-releasing hormone; TSH, thyroid-stimulating hormone.

Both T_4 and T_3 enter cells; however, once inside the cell, most of the T_4 is converted to T_3, the form that combines with internal receptors. Binding of T_3 to receptors alters gene expression, which brings about the hormone's effects.

The thyroid hormones are vital to growth, development, maturation, and metabolism in all vertebrates. They interact with GH for their effects on growth and development. Thyroid hormones also increase the sensitivity of many body cells to the effects of epinephrine and norepinephrine, hormones released by the adrenal medulla as part of the "fight-or-flight response" (discussed further later in this chapter).

In amphibians such as frogs, thyroid hormones trigger **metamorphosis,** or change in body form from tadpole to adult **(Figure 42.8).** Thyroid hormones also contribute to seasonal changes in the plumage of birds and coat color in mammals.

In human adults, low thyroid output, *hypothyroidism,* causes affected individuals to be sluggish mentally and physically; they have a slow heart rate and weak pulse, and often feel confused and depressed. Hypothyroidism in infants and children leads to cretinism, that is, stunted growth and diminished intelligence. Overproduction of thyroid hormones in human adults, *hyperthyroidism,* produces nervousness and emotional instability, irritability, insomnia, weight loss, and a rapid, often irregular heartbeat. The most common form of hyperthyroidism is *Graves' disease,* characterized by inflamed, protruding eyes in addition to the other symptoms mentioned.

Insufficient iodine in the diet can cause *goiter,* enlargement of the thyroid. Without iodine, the thyroid cannot make T_3 and T_4 in response to stimulation by TSH. Because the thyroid hormone concentration remains low in the blood, TSH continues to be secreted, and the thyroid grows in size. Dietary iodine deficiency has been eliminated in developed regions of the world by the addition of iodine to table salt.

In mammals, the thyroid also has specialized cells that secrete the peptide hormone **calcitonin.** The hormone lowers the level of Ca^{2+} in the blood by inhibiting the ongoing dissolution of calcium from bone. Calcitonin secretion is stimulated when Ca^{2+} levels in blood rise above the normal range and inhibited when Ca^{2+} levels fall below the normal range.

The Parathyroid Glands Regulate Ca^{2+} Level in the Blood

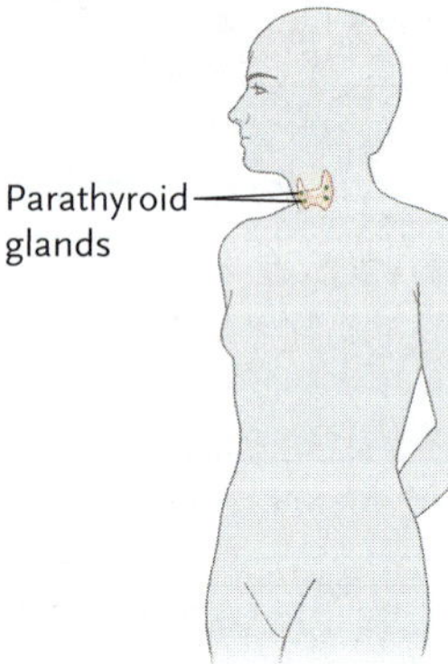

The **parathyroid glands** occur only in tetrapod vertebrates—amphibians, reptiles, birds, and mammals. Each is a spherical structure about the size of a pea. Mammals have four parathyroids located on the posterior surface of the thyroid gland, two on each side. The single hormone they produce, called **parathyroid hormone (PTH),** is secreted in response to a fall in blood Ca^{2+} levels. PTH stimulates bone cells to dissolve the mineral matter of bone tissues, releasing both calcium and phosphate ions into the blood. The released Ca^{2+} is then available for enzyme activation, conduction of nerve signals across synapses, muscle contraction, blood clotting, and other uses. How blood Ca^{2+} levels control PTH and calcitonin secretion is shown in **Figure 42.9.**

PTH also stimulates enzymes in the kidneys that convert **vitamin D,** a steroidlike molecule, into its fully active form in the body. The activated vitamin D increases the absorption of Ca^{2+} and phosphates from ingested food by promoting the synthesis of a calcium-binding protein in the intestine; it also increases the release of Ca^{2+} from bone in response to PTH.

PTH underproduction causes Ca^{2+} concentration to fall steadily in the blood, disturbing nerve and muscle function—the muscles twitch and contract uncontrollably, and convulsions and cramps occur. Without treatment, the condition is usually fatal, because the severe muscular contractions interfere with breathing. Overproduction of PTH results in the loss of so much calcium from the bones that they become thin and fragile. At the same time, the elevated Ca^{2+} concentration in the blood causes calcium deposits to form in soft tissues, especially in the lungs, arteries, and kidneys (where the deposits form kidney stones).

The Adrenal Medulla Releases Two "Fight-or-Flight" Hormones

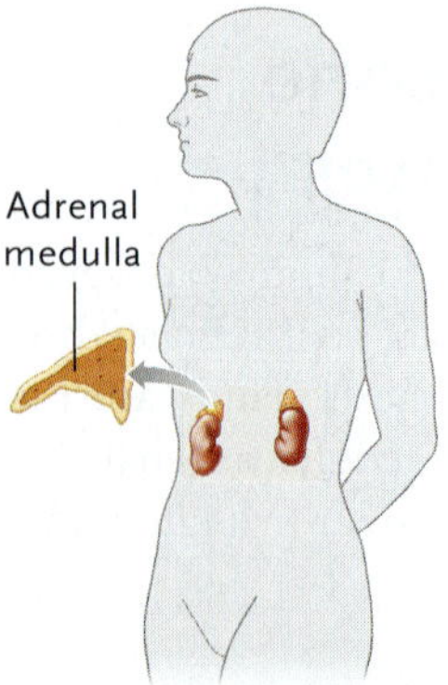

The adrenal glands (*ad* = to; *renes* = kidneys) of mammals consist of two distinct regions. The central region, the **adrenal medulla,** contains secretory cells of neural crest origin (see Chapter 50); the tissue surrounding it, the **adrenal cortex,** contains endocrine cells with a different developmental origin. The two regions secrete hormones with entirely different functions. Non-

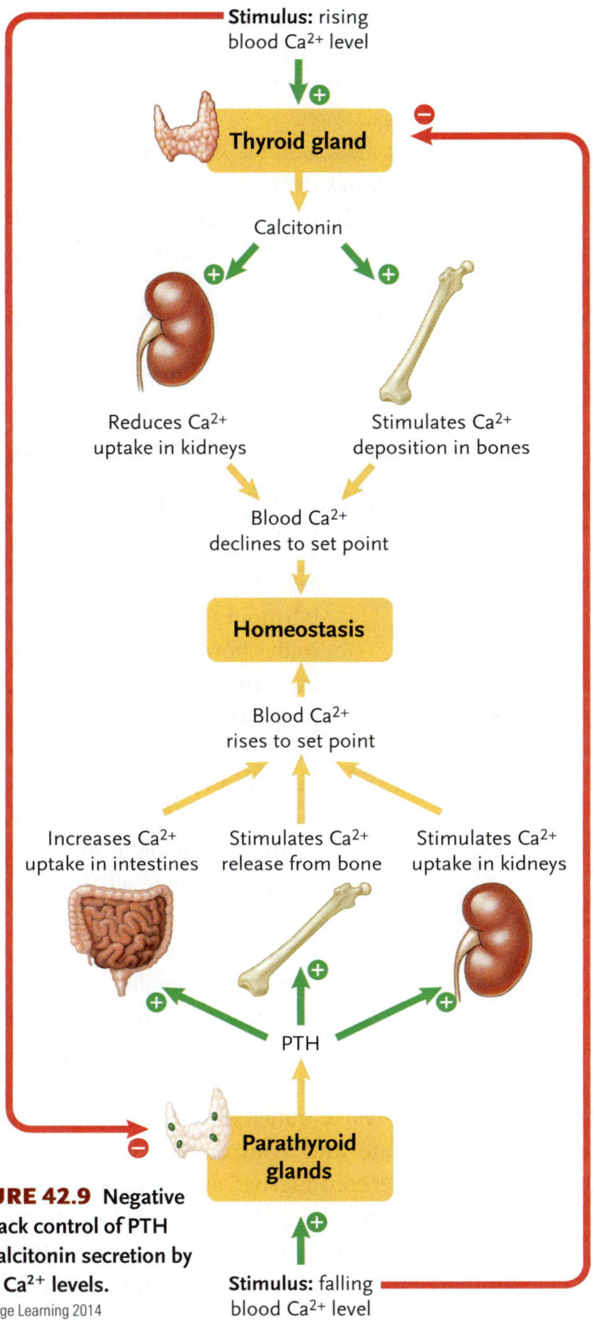

FIGURE 42.9 Negative feedback control of PTH and calcitonin secretion by blood Ca^{2+} levels.

mammalian vertebrates have glands equivalent to the adrenal medulla and adrenal cortex of mammals, but they are separate. Most of the hormones produced by these glands have essentially the same functions in all vertebrates. The only major exception is aldosterone, which is secreted by the adrenal cortex or its equivalent only in tetrapod vertebrates.

In most species, the adrenal medulla secretes two amine hormones, **epinephrine** and **norepinephrine,** which are **catecholamines,** chemical compounds derived from the amino acid tyrosine that circulate in the bloodstream. They bind to receptors in the plasma membranes of their target cells. (Epinephrine and norepinephrine are also secreted by some cells of the CNS and neurons of the sympathetic nervous system. In these cases, epinephrine and norepinephrine function as neurotransmitters in a broad array of brain and body functions; see Section 39.3.)

Epinephrine and norepinephrine are secreted from sympathetic neurons and the adrenal medulla when the body encounters stresses such as emotional excitement, danger (fight-or-flight situations), anger, fear, infections, injury, even midterm and final exams. Epinephrine in particular prepares the body for handling stress or physical activity. The heart rate increases. Glycogen and fats break down, releasing glucose and fatty acids into the blood as fuel molecules. In the heart, skeletal muscles, and lungs, the blood vessels dilate to increase blood flow. Elsewhere in the body, the blood vessels constrict, raising blood pressure, reducing blood flow to the intestines and kidneys, and inhibiting smooth muscle contractions, thereby reducing water loss and slowing digestive processes. Airways in the lungs also dilate, helping to increase the flow of air.

The effects of norepinephrine on heart rate, blood pressure, and blood flow to the heart muscle are similar to those of epinephrine. However, in contrast to epinephrine, norepinephrine causes blood vessels in skeletal muscles to constrict. This antagonistic effect is largely canceled out because epinephrine is secreted in much greater quantities.

No known human diseases are caused by underproduction of the hormones of the adrenal medulla, as long as the sympathetic nervous system is intact. Overproduction of epinephrine and norepinephrine, which can occur if there is a tumor in the adrenal medulla, leads to symptoms duplicating a stress response.

The Adrenal Cortex Secretes Two Groups of Steroid Hormones That Are Essential for Survival

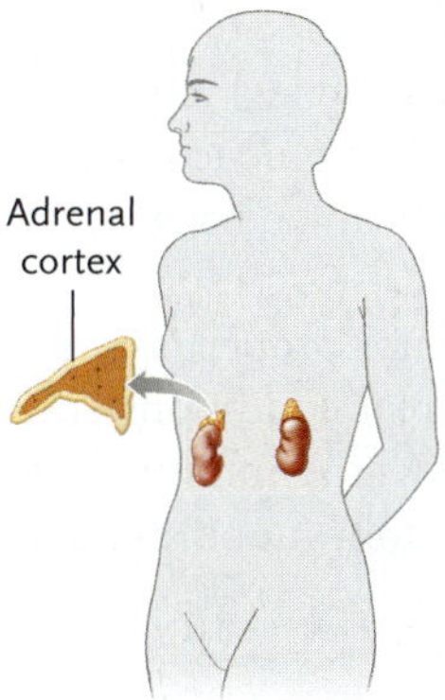

The adrenal cortex of mammals secretes two major types of steroid hormones: glucocorticoids help maintain the blood concentration of glucose and other fuel molecules, and mineralocorticoids regulate the levels of Na^+ and K^+ ions in the blood and ECF.

THE GLUCOCORTICOIDS The glucocorticoids help maintain glucose levels in the blood by three major mechanisms: (1) stimulating the synthesis of glucose from noncarbohydrate sources such as fats and proteins; (2) reducing glucose uptake by body cells except those in the central nervous system; and (3) promoting the breakdown of fats and proteins, which releases fatty acids and amino acids into the blood as alternative fuels when glucose supplies are low. The absence of down-regulation of glucose uptake to the CNS keeps the brain well supplied with glucose between meals and during periods of extended fasting. **Cortisol** is the major glucocorticoid secreted by the adrenal cortex.

Secretion of glucocorticoids is ultimately under control of the hypothalamus **(Figure 42.10)**. Low glucose concentrations

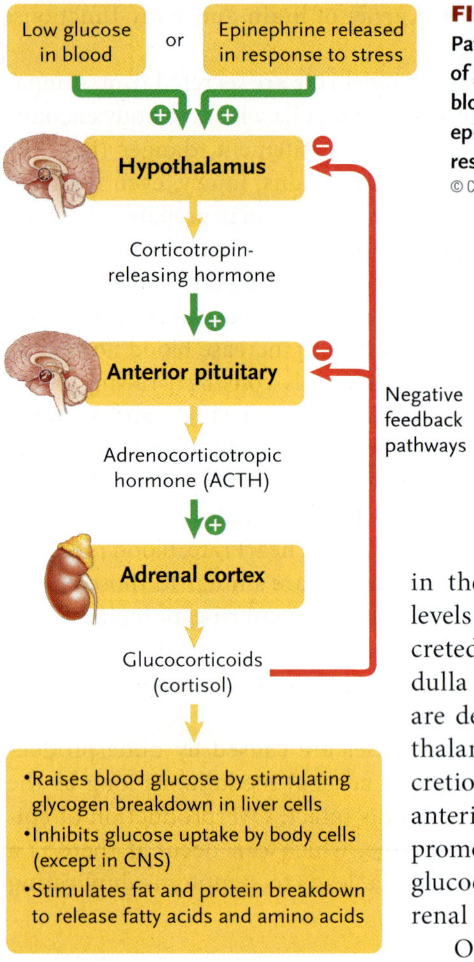

FIGURE 42.10 **Pathways linking secretion of glucocorticoids to low blood sugar and epinephrine secretion in response to stress.**

in the blood or elevated levels of epinephrine secreted by the adrenal medulla in response to stress are detected in the hypothalamus, leading to secretion of ACTH by the anterior pituitary. ACTH promotes the secretion of glucocorticoids by the adrenal cortex.

Overproduction of glucocorticoids makes blood glucose rise and increases fat deposition in adipose tissue and protein breakdown in muscles and bones. The loss of proteins from muscles causes weakness and fatigue; loss of proteins from bone, particularly collagens, makes the bones fragile and susceptible to breakage. Underproduction of glucocorticoids causes blood glucose concentration to fall below normal levels and diminishes tolerance to stress.

Glucocorticoids have antiinflammatory properties and, consequently, they are used clinically to treat conditions such as arthritis or dermatitis. They also suppress the immune system and are used in the treatment of autoimmune diseases such as rheumatoid arthritis.

THE MINERALOCORTICOIDS In tetrapods, the mineralocorticoids, primarily **aldosterone,** increase the amount of Na^+ reabsorbed from the fluids processed by the kidneys and absorbed from foods in the intestine. They also reduce the amount of Na^+ secreted by salivary and sweat glands and increase the rate of K^+ excretion by the kidneys. The net effect is to keep Na^+ and K^+ balanced at the levels required for normal cellular functions, including those of the nervous system. Secretion of aldosterone is tightly linked to blood volume and indirectly linked to blood pressure (see Section 42.2 and Chapter 48).

Moderate overproduction of aldosterone causes excessive water retention in the body, so that tissues swell and blood pressure rises. Conversely, moderate underproduction can lead to excessive water loss and dehydration. Severe underproduction is rapidly fatal unless mineralocorticoids are supplied by injection or other means.

The adrenal cortex also secretes small amounts of androgens, steroid sex hormones responsible for maintenance of male characteristics, which are synthesized primarily by the gonads. These hormones have significant effects only if they are overproduced, as can occur with some tumors in the adrenal cortex. The result is altered development of the primary or secondary sex characteristics.

The Gonadal Sex Hormones Regulate the Development of Reproductive Systems, Sexual Characteristics, and Mating Behavior

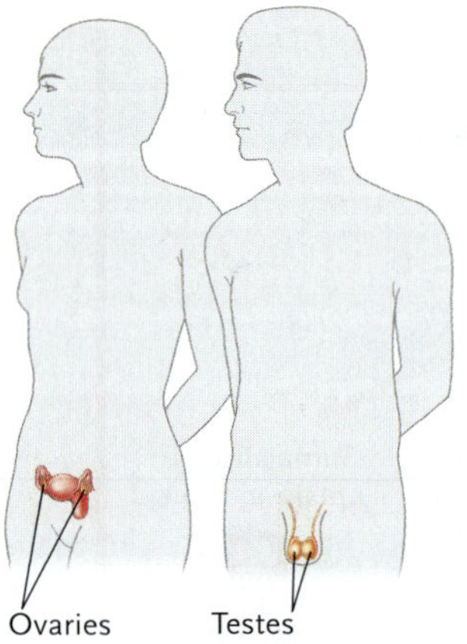

The **gonads,** the testes and ovaries, are the primary source of sex hormones in vertebrates. The steroid hormones they produce, the **androgens, estrogens,** and **progestins,** have similar functions in regulating the development of male and female reproductive systems, sexual characteristics, and mating behavior. Both males and females produce all three types of hormones, but in different proportions. Androgen production is predominant in males, whereas estrogen and progestin production is predominant in females. An outline of the actions of these hormones is presented here, and a more complete picture is given in Chapter 49.

The **testes** of male vertebrates secrete androgens, steroid hormones that stimulate and control the development and maintenance of male reproductive systems and masculine characteristics. The principal androgen is **testosterone,** the male sex hormone. In young adult males, a jump in testosterone levels stimulates puberty and the development of secondary sexual characteristics, including the growth of facial and body hair, muscle development, changes in vocal cord morphology, and development of normal sex drive. The synthesis and secretion of testosterone by cells in the testes is controlled by the release of luteinizing hormone (LH) from the anterior pituitary, which in turn is controlled by **gonadotropin-releasing hormone (GnRH)** secreted by the hypothalamus.

A large number of synthetic derivatives of androgens, known as **anabolic-androgenic steroids (AAS)** (also referred to as **anabolic steroids**), mimic the effects of androgens. They have been in the news over the years because of their use by bodybuilders and other athletes in sports in which muscular strength is important. *Focus on Basic Research* discusses the potential adverse effects of anabolic-androgenic steroids.

The **ovaries,** under the stimulatory influence of follicle-stimulating hormone (FSH), produce estrogens, steroid hor-

FOCUS ON **Basic Research**

Neuroendocrine and Behavioral Effects of Anabolic-Androgenic Steroids in Humans

Anabolic-androgenic steroids (AAS) are synthetic derivatives of the natural steroid hormone testosterone. They were designed to have potent anabolic (tissue building) activity and low androgenic (masculinizing) activity in therapeutic doses.

AAS are used for treating conditions such as delayed puberty and subnormal growth in children, as well as for therapy in chronic conditions such as cancer, AIDS, severe burns, liver and kidney failure, and anemias. Because of their anabolic effects, which include an increase in muscle mass, strength, and endurance, as well as acceleration of recovery from injuries, AAS are used by some athletes such as bodybuilders, weight lifters, baseball players, and football players. This use is actually abuse, because the doses typically administered for these purposes are far higher than therapeutic doses. AAS abuse is significant: in the early 1990s, about 1 million Americans had used or were using AAS to increase strength, muscle mass, or athletic ability.

Are AAS harmful at high doses? When researchers gave rodents doses of AAS comparable to those associated with human AAS abuse, they observed significant increases in aggression, anxiety, and sexual behaviors. These changes occur as a result of alterations in the neurotransmitters and other signaling molecules associated with those behaviors. All of these changes have been hypothesized to occur in human AAS abusers.

To study the effect of high doses of AAS on the human endocrine system, R. C. Daly and colleagues at the National Institute of Mental Health in Bethesda, Maryland, administered the AAS methyltestosterone (MT) to normal (medication-free) human volunteers over a period of time in an in-patient clinic. The subjects were examined for the effects of MT on pituitary–gonadal, pituitary–thyroid, and pituitary–adrenal hormones, and the researchers attempted to correlate endocrine changes with psychological symptoms caused by the MT.

The researchers found, for instance, that high doses of MT caused a significant decrease in the levels of gonadotropins and gonadal steroid hormones in the blood. At the same time, thyroxine and TSH levels increased. No significant increases were seen in pituitary–adrenal hormones.

The decrease in testosterone levels correlated significantly with cognitive problems, such as increased distractibility and forgetfulness. The increase in thyroxine correlated significantly with a rise in aggressive behavior, notably anger, irritability, and violent feelings. There were no changes in activities associated with pituitary–adrenal hormones—energy, disturbed sleep, and sexual arousal—as was expected by the lack of change in those hormones.

In sum, behavioral changes associated with high doses of an AAS suggest that AAS-induced hormonal changes may well contribute to the adverse behavioral and mood changes that occur during AAS abuse.

mones that stimulate and control the development and maintenance of female reproductive systems. The principal estrogen is **estradiol,** which stimulates maturation of sex organs at puberty and the development of secondary sexual characteristics. Ovaries also produce progestins, principally **progesterone,** the steroid hormone that prepares and maintains the uterus for implantation of a fertilized egg and the subsequent growth and development of an embryo. The synthesis and secretion of progesterone by cells in the ovaries is controlled by the release of LH from the anterior pituitary, which in turn is controlled by the same GnRH as in males.

The Pancreatic Islet of Langerhans Hormones Regulate Glucose Metabolism

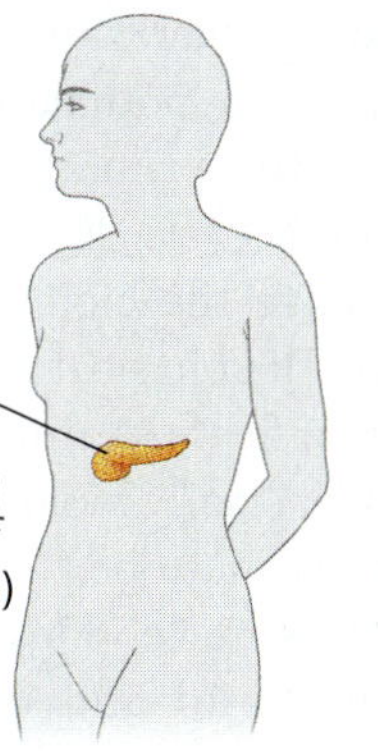

Most of the pancreas, a relatively large gland located just behind the stomach, forms an exocrine gland that secretes digestive enzymes into the small intestine (see Chapter 47). However, about 2% of the cells in the pancreas are endocrine cells that form the islets of Langerhans. Found in all vertebrates, the islets secrete the peptide hormones insulin and glucagon into the bloodstream.

Insulin and glucagon regulate the metabolism of fuel substances in the body. **Insulin,** secreted by *beta cells* in the islets, acts mainly on cells of skeletal muscles, liver cells, and adipose tissue (fat). (Brain cells do not require insulin for glucose uptake.) Insulin lowers circulating blood glucose, fatty acid, and amino acid levels and promotes their storage. That is, the actions of insulin include stimulation of glucose transport into cells, glycogen synthesis from glucose, uptake of fatty acids by adipose tissue cells, fat synthesis from fatty acids, and protein synthesis from amino acids. Insulin also inhibits glycogen degradation to glucose, fat degradation to fatty acids, and protein degradation to amino acids.

Glucagon, secreted by *alpha cells* in the islets, has effects opposite to those of insulin: it stimulates glycogen, fat, and protein degradation. Glucagon also promotes the use of amino acids and other noncarbohydrates as the input for glucose synthesis; this aspect of glucagon function operates during fasting. Negative feedback mechanisms that are keyed to the concentration of glucose in the blood control secretion of both insulin and glucagon to maintain glucose homeostasis **(Figure 42.11).**

Diabetes mellitus, a disease that afflicts more than 14 million people in the United States and at least 200 million people worldwide, results from problems with insulin production or action. The three classic diabetes symptoms are frequent urination, increased thirst (and consequently increased fluid intake), and increased glucose in the blood (hyperglycemia). Frequent urina-

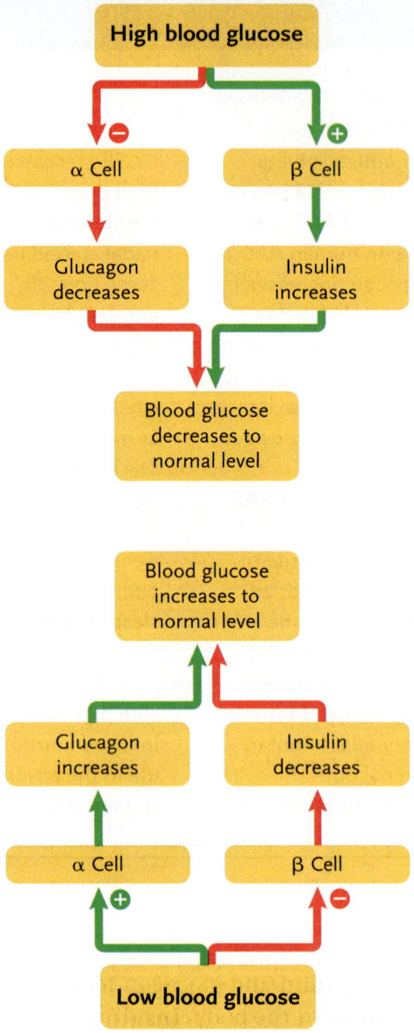

FIGURE 42.11 The action of insulin and glucagon in maintaining the concentration of blood glucose at an optimal level.

tion occurs because body cells are not stimulated to take up glucose, leading to abnormally high glucose concentration in the blood. Excretion of the excess glucose in the urine requires water to carry it, which causes increased fluid loss from the blood. Food intake is necessary to offset the negative energy balance resulting from the decreased glucose uptake by cells or else weight loss will occur. Two of these classic symptoms gave the disease its name: *diabetes* is derived from a Greek word meaning "siphon," referring to the frequent urination; and *mellitus,* a Latin word meaning "sweetened with honey," refers to the sweet taste of a diabetic's urine. (Before modern blood or urine tests were developed, physicians tasted a patient's urine to detect the disease.)

The disease occurs in two major forms called *type 1* and *type 2.* Type 1 diabetes, which occurs in about 10% of diabetics, results from insufficient insulin secretion by the pancreas. This type of diabetes is usually caused by an autoimmune reaction that destroys pancreatic beta cells. To survive, type 1 diabetics must receive regular insulin injections (typically, a genetically engineered human insulin called Humulin); careful dieting and exercise also have beneficial effects, because active skeletal muscles do not require insulin to take up and utilize glucose.

In type 2 diabetes, insulin is usually secreted at or above normal levels, but the target cells of affected people have significantly reduced responsiveness to the hormone compared with the cells of normal people. About 90% of patients in the developed world with type 2 diabetes are obese. A genetic predisposition can also be a factor. Most affected people can lead a normal life by controlling their diet and weight, exercising, and taking drugs that enhance insulin action or secretion.

Diabetes has long-term effects on the body. Its cells, unable to utilize glucose as an energy source, start breaking down proteins and fats to generate energy. The protein breakdown weakens blood vessels throughout the body, particularly in the arms and legs and in critical regions such as the kidneys and retina of the eye. The circulation becomes so poor that tissues degenerate in the arms, legs, and feet. At advanced stages of the disease, bleeding in the retina causes blindness. The breakdown of circulation in the kidneys can lead to kidney failure. In addition, in type 1 diabetes, acidic products of fat breakdown (ketones) are produced in abnormally high quantities and accumulate in the blood. The resulting lowering of blood pH can disrupt heart and brain function, leading to coma and death if the disease is untreated.

The Pineal Gland Regulates Some Biological Rhythms

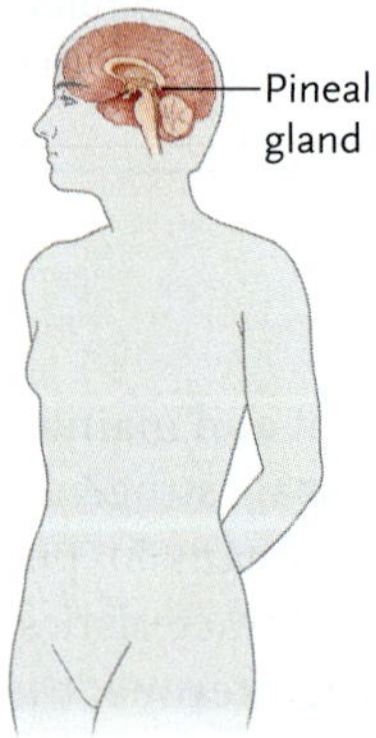

The **pineal gland** is found at different locations in the brains of vertebrates—for example, in mammals, it is at roughly the center of the brain, while in birds and reptiles, it is on the surface of the brain just under the skull. The pineal gland regulates some biological rhythms.

The earliest vertebrates had a third, light-sensitive eye at the top of the head, and some species, such as lizards and tuataras (New Zealand reptiles), still have an eyelike structure in this location. In most vertebrates, the third eye became modified into a pineal gland, which in many groups retains some degree of photosensitivity. In mammals, it is too deeply buried in the brain to be affected directly by light; nonetheless, specialized photoreceptors in the eyes make connections to the pineal gland.

In mammals, the pineal gland secretes a peptide hormone, **melatonin,** that helps to maintain daily biorhythms. Secretion of melatonin is regulated by an inhibitory pathway. Light hitting the eyes generates signals that inhibit melatonin secretion; consequently, the hormone is secreted most actively during periods of darkness. Melatonin targets a part of the hypothalamus called the *suprachiasmatic nucleus,* which is the primary biological clock coordinating body activity to a daily cycle. The nightly release of melatonin may help synchronize the biological clock with daily cycles of light and darkness. The physical and mental discomfort associated with jet lag may reflect the time required for melatonin secretion to reset a traveler's daily biological clock to match the period of daylight in a new time zone.

Melatonin also plays a role in other vertebrates. In some fishes, amphibians, and reptiles, melatonin and other hormones produce changes in skin color through their effects on *melanophores,* the pigment-containing cells of the skin. Skin color may vary with the season, the animal's breeding status, or the color of the background.

STUDY BREAK 42.4

1. What effect does parathyroid hormone have on the body?
2. What hormones are secreted by the adrenal medulla, and what are their functions?
3. What are the two types of hormones secreted by the adrenal cortex, and what are their functions?
4. To what molecular class of hormones do estradiol and progesterone belong, and where do they act in the body?

THINK OUTSIDE THE BOOK

Individuals who are allergic to bee stings are at risk for anaphylactic shock. Physicians recommend that such individuals carry an EpiPen, a container of epinephrine for immediate injection in case of a sting. Individually or collaboratively, use the Internet or research literature to determine what anaphylactic shock is and how it develops, and how an epinephrine injection is an effective treatment.

42.5 Endocrine Systems in Invertebrates

In even the simplest animals, such as the cnidarian *Hydra,* hormones produced by neurosecretory neurons control reproduction, growth, and development of some body features. In annelids, arthropods, and mollusks, endocrine cells and glands produce hormones that regulate reproduction, water balance, heart rate, and sugar levels.

Some hormones occur in related forms in invertebrates and vertebrates. For example, fruit flies, mollusks, and humans have insulin-like hormones and receptors, even though molecular studies suggest that their last common ancestor existed more than 800 million years ago. Both invertebrates and vertebrates secrete peptide and steroid hormones, but most of the hormones' structures differ between the two groups, and therefore most have no effect when injected into members of the other group. However, the reaction pathways stimulated by the hormones are the same in both groups, suggesting that these regulatory mechanisms appeared very early in animal evolution.

Hormones have been studied in detail in only a few invertebrate groups, with the most extensive studies focusing on regulation of metamorphosis in insects. Butterflies, moths, and flies undergo the most dramatic changes as they mature into adults. They hatch from the egg as a caterpillar-like *larva.* During the larval stage, growth is accompanied by one or more *molts,* in which an old exoskeleton is shed and a new one forms. The insect then enters a typically nonmotile stage, the *pupa,* in which the body forms a thick, resistant coating, and finally it transforms into an adult.

Three major hormones regulate molting and metamorphosis in insects: **prothoracicotropic hormone (PTTH),** a peptide hormone secreted by neurosecretory neurons in the brain; **ecdysone** (*ekdysis* = getting out), a steroid hormone secreted by the *prothoracic glands;* and **juvenile hormone (JH),** a terpenoid (a type of unsaturated hydrocarbon) secreted by the *corpora allata,* a pair of glands just behind the brain **(Figure 42.12).** The outcome of the molt depends on the level of JH. If it is high, the molt produces a larger larva; if it is low, the molt leads to pupation and the emergence of the adult.

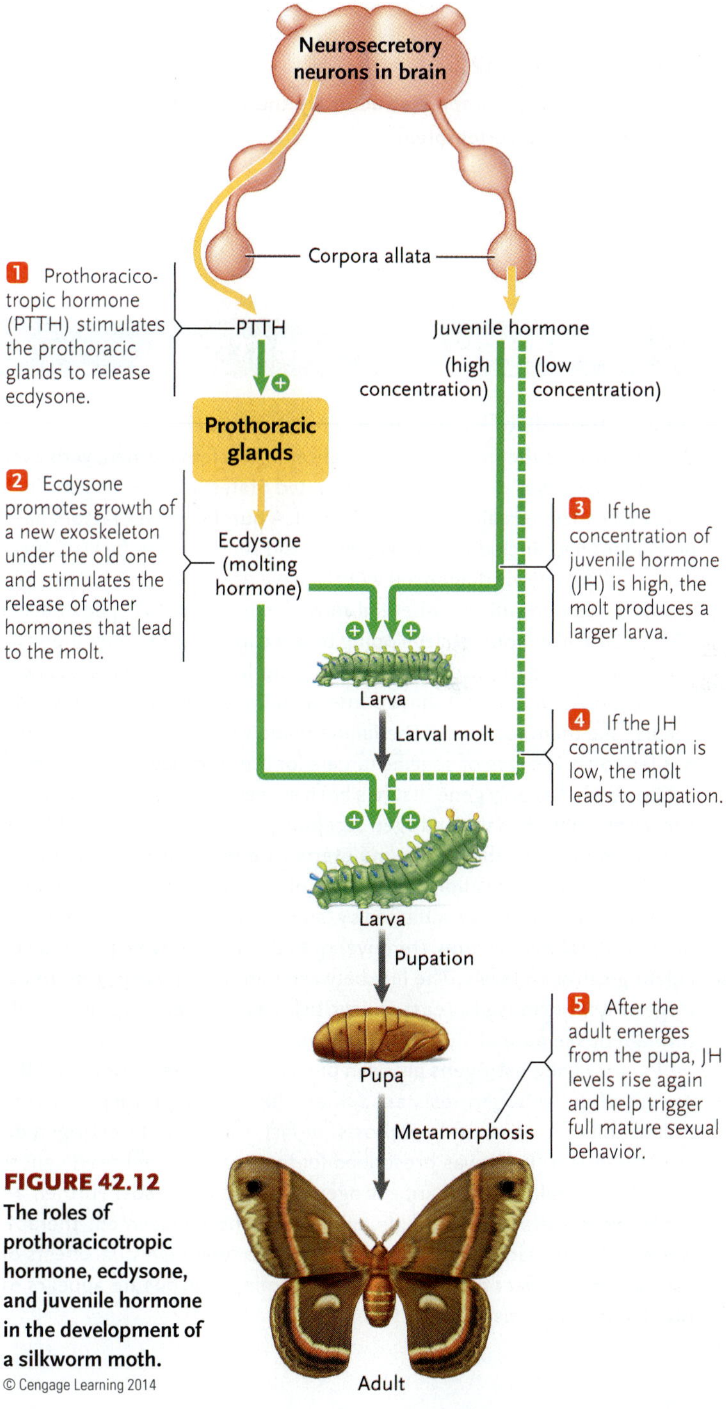

FIGURE 42.12 **The roles of prothoracicotropic hormone, ecdysone, and juvenile hormone in the development of a silkworm moth.**

Hormones that control molting have also been detected in crustaceans, including lobsters, crabs, and crayfish. Before growth reaches the stage at which the exoskeleton is shed, **molt-inhibiting hormone (MIH),** a peptide neurohormone secreted by a gland in the eye stalks, inhibits ecdysone secretion **(Figure 42.13).** As body size increases to the point requiring a molt, MIH secretion is inhibited, ecdysone secretion increases, and the molt is initiated.

In the next chapter, we discuss the structure and functions of muscles, and their interactions with the skeletal system to cause movement. Muscle function depends primarily on the action of the nervous system, but the endocrine system plays a role in the control of smooth muscle contraction.

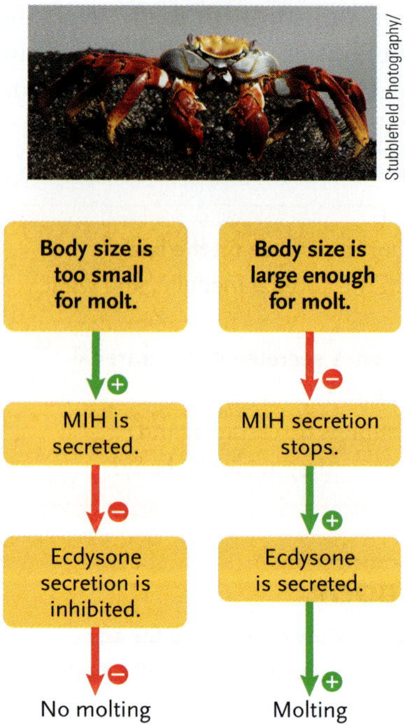

Stubblefield Photography/ Shutterstock.com

FIGURE 42.13 Control of molting by molt-inhibiting hormone (MIH), which is secreted by a gland in the eye stalks of crustaceans such as this crab.

STUDY BREAK 42.5

How do hormones compare structurally and functionally in invertebrates and vertebrates?

UNANSWERED QUESTIONS

What is the role of estrogens in breast cancer?

Breast cancer is the most common cancer in Western women, with over 230,000 new cases diagnosed in the United States in 2011. The etiology of breast cancer remains largely unknown. A number of risk factors have been identified, including family history and reproductive factors; however, ~80% of newly diagnosed women have no known risk factors. For example, women with an inherited mutation in the *BRCA1* gene have a 70% to 80% likelihood of developing breast cancer during their lifetime, yet only one-third of women with a strong family history of breast cancer carry a *BRCA1* mutation or have some other known mutation that would predispose them to cancer. The failure in identifying strong risk factors has led to an absence of specific targets for breast cancer prevention.

Currently, the only gene that has been successfully targeted for breast cancer prevention is the estrogen receptor (ER) gene. Estrogen induces cell proliferation in the breast and exerts its effects through binding to the ER, leading to activation or inhibition of several downstream signaling targets. ER activity can be inhibited by tamoxifen and other antiestrogens, or by bilateral ovariectomy (removal of both ovaries), which lowers circulating estrogen levels. The link between high lifetime exposure to estrogens and increased breast cancer risk has been established clearly through observational studies in women.

Paradoxically, estrogens also can prevent breast cancer or inhibit the growth of antiestrogen resistant tumors, likely through their ability to induce differentiation and apoptosis. In fact, the synthetic estrogen diethylstilbestrol (DES) was prescribed for the treatment of breast cancer until it was replaced with antiestrogens in the early 1980s. Further, an exposure to estradiol, as single agent hormone replacement therapy, prevents breast cancer and its metastasis. The seemingly dual effects of estrogen may reflect the critical role that timing of exposure appears to play in affecting breast cancer risk.

How does the timing of estrogenic exposures affect breast cancer risk?

Millions of pregnant women in the 1940s to 1970s took DES in the U.S. and Europe, initially to prevent miscarriage but also to aid various pregnancy-related conditions, such as morning sickness. It was subsequently proven ineffective and even dangerous. Most notable, young daughters exposed to DES *in utero* exhibited an unusually high incidence of clear cell vaginal adenocarcinoma (a malignant tumor), and the practice of treating pregnant women with DES stopped. The DES daughters have now reached the age when breast cancer is commonly detected (older than 40 years old), and the findings indicate that these daughters exhibit two times higher risk of developing breast cancer as well. Risk also is increased in mothers themselves.

Research from my laboratory at Georgetown University, using preclinical models of breast cancer, shows that maternal exposure to estrogenic dietary compounds during pregnancy, such as high fat diet, increases mammary cancer risk among female offspring. This increase is not limited to daughters but also is seen in granddaughters and great granddaughters, and mothers, without any further intervening exposures. Our studies suggest that these multigenerational effects are inherited via permanent changes in the epigenome (the set of epigenetic modifications in the genome), rather than through genetic changes, such as mutations. Genes which are epigenetically silenced in the mammary tissue from generation to generation following maternal dietary exposures during pregnancy include those that regulate proliferation of progenitor cells (they are daughters of stem cells) and their differentiation. The increased methylation of these genes, in turn, is proposed to be a hallmark of increased breast cancer susceptibility. Importantly, epigenetic changes are in principle reversible by compounds which demethylate silenced genes, and many such compounds are currently available and used to treat some cancers. Our preclinical studies suggest that these same compounds may also prevent *in utero* estrogen-exposed animals from developing mammary cancer.

The period when estrogens appear to be protective is around puberty, when the breast begins to grow. Pregnancy is characterized by 50- to 100-fold increase in circulating estrogen levels, and women who had their first child before age 20 have half the risk of developing breast cancer than women who were over 30 when their first child was born. Other estrogenic exposures at young adulthood also reduce lifetime breast cancer risk. Women who consumed high levels of soy foods during childhood, which contain the estrogenic isoflavone genistein, have a reduced risk of developing breast cancer. Similar findings have been obtained in preclinical animal models. Work done in my laboratory is investigating whether the protective effect of childhood genistein/soy intake is mediated by epigenetic changes that involve genes which may protect mammary epithelial cells from malignant transformation.

THINK LIKE A SCIENTIST What are the benefits and weaknesses of using *in vitro* (tissue-culture based) and *in vivo* (animal model) approaches versus performing human studies to understand the role of estrogen in breast cancer etiology.

Leena Hilakivi-Clarke

Leena Hilakivi-Clarke is a professor of Oncology at Georgetown University in Washington D.C. Her research involves understanding why timing of estrogenic exposures, including those originating from diet, determines whether they increase, reduce, or have no effect on breast cancer risk and risk of recurrence of this disease. Learn more about her research from http://clarkelabs.georgetown.edu/index.

REVIEW KEY CONCEPTS

To access the course materials and companion resources for this text, please visit www.cengagebrain.com.

42.1 Hormones and Their Secretion

- Hormones are substances secreted by cells that control the activities of cells elsewhere in the body. The cells that respond to a hormone are its target cells. The best-known hormones are secreted by the endocrine system.
- The endocrine system includes four major types of cell signaling: classical endocrine signaling, in which endocrine glands secrete hormones; neuroendocrine signaling, in which neurosecretory neurons release neurohormones into the circulatory system; paracrine regulation, in which cells release local regulators that diffuse through the ECF to regulate nearby cells; and autocrine regulation, in which cells release local regulators that regulate the same cells that produced them (Figure 42.1).
- Most hormones and local regulators fall into one of four molecular classes: amines, peptides, steroids, and fatty acids.
- Many hormones are controlled by negative feedback mechanisms (Figure 42.2).

42.2 Mechanisms of Hormone Action

- Hormones typically are effective in very low concentrations in the body fluids because of amplification.
- Hydrophilic hormones bind to receptor proteins embedded in the plasma membrane, activating them. The activated receptors transmit a signal through the plasma membrane, triggering signal transduction pathways that cause a cellular response. Hydrophobic hormones bind to receptors in the cytoplasm or nucleus, activating them. The activated receptors control the expression of specific genes, the products of which cause the cellular response (Figure 42.3).
- As a result of the types of receptors they have, target cells may respond to more than one hormone, or they may respond differently to the same hormone.
- The major endocrine cells and glands of vertebrates are the hypothalamus, pituitary gland, thyroid gland, parathyroid glands, adrenal medulla, adrenal cortex, testes, ovaries, islets of Langerhans of the endocrine pancreas, and pineal gland. Hormones are also secreted by endocrine cells in the stomach and intestine, thymus gland, kidneys, and heart. Most body cells are capable of releasing prostaglandins (Figure 42.6).

Animation: Major human endocrine glands

Animation: The Organs of the Endocrine System

Animation: Endocrine System

42.3 The Hypothalamus and Pituitary

- The hypothalamus and pituitary together regulate many other endocrine cells and glands in the body (Figure 42.7).
- The hypothalamus produces hormones (releasing hormones and inhibiting hormones) that control the secretion of eight hormones by the anterior pituitary: prolactin (PRL), growth hormone (GH), thyroid-stimulating hormone (TSH), adrenocorticotropic hormone (ACTH), follicle-stimulating hormone (FSH), luteinizing hormone (LH), melanocyte-stimulating hormone (MSH), and endorphins.
- The posterior pituitary secretes antidiuretic hormone (ADH), which regulates body water balance, and oxytocin, which stimulates the contraction of smooth muscle in the uterus as a part of childbirth and triggers milk release from the mammary glands during suckling of the young.

Animation: Anterior pituitary function

42.4 Other Major Endocrine Glands of Vertebrates

- The thyroid gland secretes the thyroid hormones and, in mammals, calcitonin. The thyroid hormones stimulate the oxidation of carbohydrates and lipids, and coordinate with growth hormone to stimulate body growth and development. Calcitonin lowers the Ca^{2+} level in the blood by inhibiting the release of Ca^{2+} from bone. In amphibians, such as the frog, thyroid hormones trigger metamorphosis (Figure 42.8).
- The parathyroid glands secrete parathyroid hormone, which stimulates bone cells to release Ca^{2+} into the blood. PTH also stimulates the activation of vitamin D, which promotes Ca^{2+} absorption into the blood from the small intestine (Figure 42.9).

- The adrenal medulla secretes epinephrine and norepinephrine, which reinforce the sympathetic nervous system in responding to stress. The adrenal cortex secretes glucocorticoids, which help maintain glucose at normal levels in the blood, and mineralocorticoids, which regulate Na^+ balance and ECF volume. The adrenal cortex also secretes small amounts of androgens (Figure 42.10).
- The gonadal sex hormones—androgens, estrogen, and progestins—play a major role in regulating the development of reproductive systems, sexual characteristics, and mating behavior.
- The islet of Langerhans cells of the endocrine pancreas secrete insulin and glucagon, which together regulate the concentration of fuel substances in the blood. Insulin lowers the concentration of glucose in the blood and inhibits the conversion of noncarbohydrate molecules into glucose. Glucagon raises blood glucose by stimulating glycogen, fat, and protein degradation (Figure 42.11).
- The pineal gland secretes melatonin, which interacts with the hypothalamus to set the body's daily rhythms.

Animation: Parathyroid hormone action

Animation: Hormones and glucose metabolism

42.5 Endocrine Systems in Invertebrates

- Hormones control development and function of the gonads, manage salt and water balance in the body fluids, and control molting in insects and crustaceans.
- Three major hormones—prothoracicotropic hormone (PTTH), ecdysone, and juvenile hormone (JH)—control molting and metamorphosis in insects. Hormones that control molting are also present in crustaceans (Figures 42.12 and 42.13).

UNDERSTAND & APPLY

Test Your Knowledge

1. Amine hormones are usually:
   a. hydrophilic when secreted by the thyroid gland.
   b. based on tyrosine.
   c. paracrine but not autocrine.
   d. not transported by the blood.
   e. repelled by the plasma membrane.
2. Prostaglandins would be best described as inducers of:
   a. male and female characteristics.
   b. cell division.
   c. nerve transmission.
   d. smooth muscle contractions.
   e. cell differentiation.
3. When the concentration of thyroid hormone in the blood increases, it:
   a. inhibits TSH secretion by the pituitary.
   b. stimulates TRH secretion by the hypothalamus.
   c. stimulates the pituitary to secrete TRH.
   d. stimulates the pituitary to secrete TSH.
   e. activates a positive feedback loop.
4. Which of the following statements about endocrine targeting and reception is correct?
   a. The idea that one hormone affects one type of tissue is illustrated when epinephrine binds to smooth muscle cells in blood vessels as well as to beta cells in heart muscle.
   b. The idea that one hormone affects one type of tissue is illustrated when epinephrine cannot activate both the receptors on liver cells and the beta receptors of heart muscle.
   c. The idea that a target cell can respond to more than one hormone is illustrated by a vertebrate liver cell responding to insulin and glucagon.
   d. The idea that a minute concentration of hormone can cause widespread effects demonstrates the specificity of cells for certain hormones.
   e. The idea that the response to a hormone is the same among different target cells is shown when different liver cells are activated by insulin.
5. The posterior pituitary secretes:
   a. hormones that control the hypothalamus.
   b. IGF, which simulates cell division and protein synthesis.
   c. ADH, which increases water absorption in the kidneys.
   d. oxytocin, which controls egg and sperm development.
   e. prolactin, which stimulates milk synthesis.
6. Blood levels of calcium are regulated directly by:
   a. insulin synthesized by the alpha cells of the pancreas.
   b. PTH made by the pituitary.
   c. vitamin D activated in the liver.
   d. prolactin synthesized by the anterior pituitary.
   e. calcitonin secreted by specialized thyroid cells.
7. If the human body is stressed, glucocorticoids:
   a. promote the breakdown of proteins in the muscles and bones.
   b. increase the amount of sodium reabsorbed from urine in the kidneys.
   c. decrease potassium secretion from the kidneys.
   d. decrease glucose uptake by cells in the nervous system.
   e. inhibit the synthesis of glucose from noncarbohydrate sources.
8. When blood glucose rises:
   a. alpha cells increase glucagon secretion.
   b. beta cells increase insulin secretion.
   c. urination decreases in a person with type 1 diabetes who has not recently received an insulin injection.
   d. glucagon stimulates the breakdown of amino acids into glycogen.
   e. target cells decrease their insulin receptors.
9. In mammals:
   a. the suprachiasmatic nucleus of the pineal gland controls both male and female reproductive systems.
   b. estradiol is produced by the hypothalamus to control ovulation.
   c. melatonin controls anabolic steroid production.
   d. GnRH stimulates LH to control testosterone production.
   e. progesterone increases the secretion of LH from the posterior pituitary.
10. Insect development is regulated by:
    a. ecdysone, a peptide secreted by the brain.
    b. juvenile hormone, a terpenoid secreted by the corpora allata near the brain.
    c. molt-inhibiting hormone, a steroid secreted by the prothoracic glands.
    d. prothoracicotropic hormone, a steroid secreted by the hypothalamus.
    e. melatonin, a peptide secreted by the brain in the larval stage.

Discuss the Concepts

1. A physician sees a patient whose symptoms include sluggishness, depression, and intolerance to cold. What disorder do these symptoms suggest?
2. A 20-year-old woman with a malignant brain tumor has her pineal gland removed. What kinds of side effects might this loss have?
3. In integrated pest management, a farmer uses a variety of tools to combat unwanted insects. These include applications of either hormones or hormone-inhibiting compounds to prevent insects from reproducing successfully. How might each of these hormone-based approaches disrupt reproduction?

Design an Experiment

The Environmental Protection Agency (EPA) defines *endocrine disruptors* as chemical substances that can "interfere with the synthesis, secretion, transport, binding, action, or elimination of natural hormones in the body that are responsible for the maintenance of homeostasis (normal cell metabolism), reproduction, development, and/or behavior." The chemicals, sometimes called environmental estrogens, come from both natural and human-made sources. A simple hypothesis is that endocrine disruptors act by mimicking hormones in the body. Many endocrine disruptors affect sex hormone function and, therefore, reproduction.

Examples of endocrine disruptors are the synthetic chemicals DDT (a pesticide), dioxins, and natural chemicals such as phytoestrogens (estrogen-like molecules in plants), which are found in high levels in soybeans, carrots, oats, onions, beer, and coffee.

Design an experiment to investigate whether a new synthetic chemical (pick your own interesting scenario) is an endocrine disruptor. (*Hint:* You probably want to work with a model organism.)

Interpret the Data

Researchers measured prolactin and growth hormone in the blood plasma of six subjects at regular intervals over a 24-hour period. The investigators were interested in determining if the levels of these two hormones cycle over a 24-hour period. The **Figure** presents the results of their experiments, with the *y* axis values showing the amount of each hormone expressed as a percentage of the 24-hour mean value, and each point showing the average of the six subjects' levels.

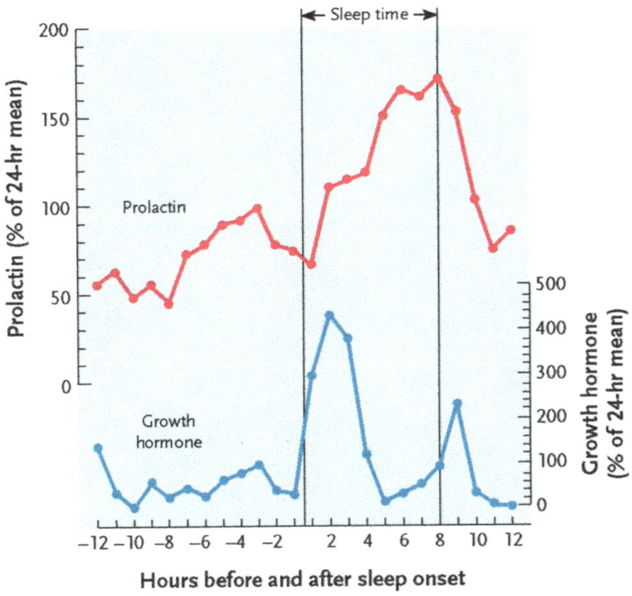

FIGURE Mean concentrations of prolactin and growth hormone in plasma, averaged for six subjects and expressed as a percentage of the average concentrations measured for the 24-hour period.

From J. F. Sassin et al. 1972. Human prolactin: 24-hour pattern with increased release during sleep. *Science* 177:1205–1207. Reprinted with permission from AAAS.

1. How did the level of growth hormone change over the 24-hour period and, in particular, how did it relate to the sleep period?
2. How did the level of prolactin change over the 24-hour period and, in particular, how did it relate to the sleep period?
3. Is there any consistent relationship between the patterns of prolactin and growth hormone in the plasma in the 24-hour period?
4. Do the results support the hypothesis that the changes in prolactin level over a 24-hour period are associated with the sleep–wake cycle?

Source: J. F. Sassin et al. 1972. Human prolactin: 24-hour pattern with increased release during sleep. *Science* 177:1205–1207.

Apply Evolutionary Thinking

Which endocrine system evolved earlier, endocrine glands or neurosecretory neurons? Support your conclusion with information obtained from online research.

Glomeruli, ball-like tufts of capillaries in the nephrons of a kidney. Glomeruli filter water and solutes from the blood as the first step in urine formation (colorized SEM).

48

STUDY OUTLINE

48.1 Introduction to Osmoregulation and Excretion

48.2 Osmoregulation and Excretion in Invertebrates

48.3 Osmoregulation and Excretion in Mammals

48.4 Regulation of Mammalian Kidney Function

48.5 Kidney Function in Nonmammalian Vertebrates

48.6 Introduction to Thermoregulation

48.7 Ectothermy

48.8 Endothermy

Regulating the Internal Environment

Why it matters . . . The nine crew of the World War II bomber *Lady Be Good* flew a night mission to Naples, Italy, from a base on the North African coast on August 4, 1943. Trouble during the flight forced the crew to turn back before reaching their destination. Navigational errors and a cloud layer led them to miss their home base and continue south over the Sahara Desert. Some 440 miles from the coastline, with the fuel running out, the crew members parachuted from the aircraft. In a few more minutes, the bomber crash-landed, leaving its crew miles behind.

The eight men who survived began a northward trek with only half a canteen of water among them, in desert heat that reached 130°F during the day. In a testimony to the physiological mechanisms that conserve water and cool the body, they continued onward for eight days. But then, one by one, they died as prolonged dehydration and the heat overwhelmed their homeostatic control systems for *water balance* (the equilibrium in inward and outward flow of water) and temperature regulation, and hyperthermia exceeded their capacity to survive. Rescue teams searched the desert for weeks after their disappearance, but no trace was found of the crew or their airplane.

The fate of the *Lady Be Good* remained unknown until 1958, when the aircraft was spotted from the air, sitting largely intact in the desert sands. A two-year search led to the remains of the crew, some of them more than a hundred miles north of the downed bomber. Diaries found among the scattered effects told the poignant story of the flight and the futile struggle against the fiery desert environment.

This story illustrates the trials of animal life under changing environmental conditions. Water and required nutrients may become more or less abundant. Temperatures may rise or fall. Through their homeostatic control systems, animals have an astounding capacity to compensate for fluctuating external conditions and to maintain the internal environment of their bodies within the relatively narrow limits that cells can tolerate (homeostasis is discussed in Chapter 38).

These limits, and the compensating mechanisms that maintain them, are the subjects of this chapter. First we examine **osmoregulation,** the control of water and ion balance, and the closely related topic of **excretion,** which helps maintain the body's water and ion balance while ridding the body of metabolic wastes. Then we consider **thermoregulation,** the control of body temperature.

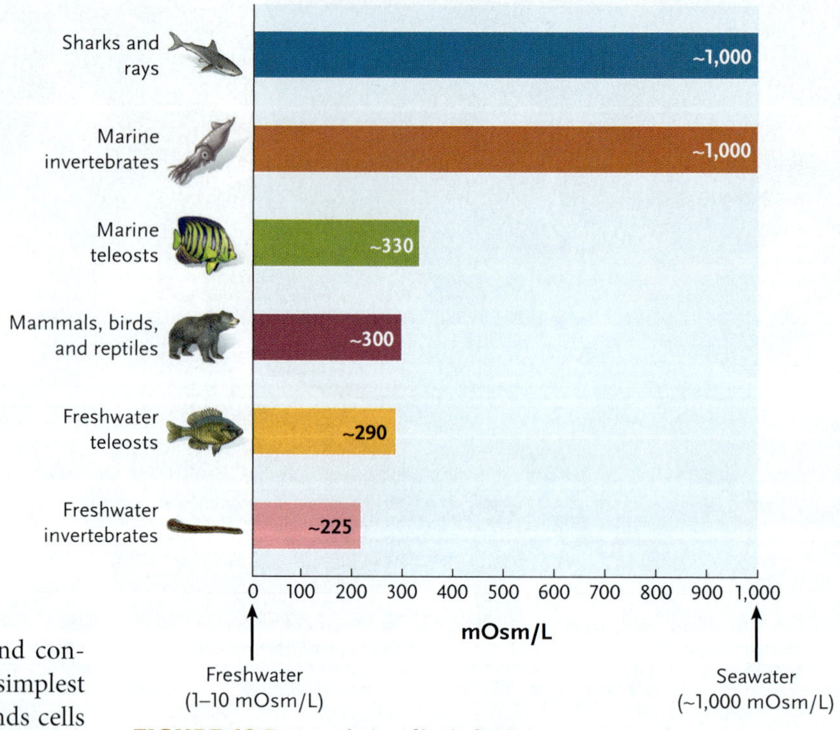

FIGURE 48.1 Osmolarity of body fluids in some animal groups. Reptiles include turtles, lizards, and snakes.

48.1 Introduction to Osmoregulation and Excretion

Living cells contain water, are surrounded by water, and constantly exchange water with their environment. For the simplest animals, the water of the external environment surrounds cells directly. For more complex animals, an aqueous extracellular fluid (ECF) surrounds the cells, and is separated from the external environment by a body covering. In animals with a circulatory system, the ECF includes both the interstitial fluid immediately surrounding cells and the plasma, the fluid portion of the blood, or other circulated fluid.

In this section, we review the mechanisms cells use to exchange water and solutes with the surrounding fluid through osmosis and diffusion.

Osmosis Is a Form of Passive Diffusion

In osmosis (see Section 6.3 and Figures 6.9 and 6.10), water molecules move across a selectively permeable membrane from a region where they are more highly concentrated to a region where they are less highly concentrated. The difference in water concentration is produced by differing numbers of solute molecules or ions on the two sides of the membrane. The side of the membrane with a *lower* solute concentration has a *higher* concentration of water molecules, so water will move osmotically to the other side, where water concentration is *lower.*

A key factor in osmosis for helping maintain differences in solute concentration on either side of biological membranes is the selective permeability of those membranes. Proteins are among the most important solutes in establishing the conditions that produce osmosis.

The total solute concentration of a solution, called its **osmolarity,** is measured in *osmoles*—the number of solute molecules and ions (in moles)—per liter of solution. Because the total solute concentration in the body fluids of most animals is less than 1 osmole, osmolarity is usually expressed in thousandths of an osmole, or *milliosmoles* (mOsm). As shown in **Figure 48.1**, the osmolarity of body fluids in humans and other mammals is about 300 mOsm/L; marine fishes and marine invertebrates have higher values and freshwater fishes and freshwater invertebrates have lower values.

Considering solutions on either side of a selectively permeable membrane, a solution of higher osmolarity is said to be *hyperosmotic* to a solution of lower osmolarity, and a solution of lower osmolarity is said to be *hypoosmotic* to a solution of higher osmolarity. If the solutions on either side of a membrane have the same osmolarity, they are said to be *isoosmotic.* Water moves across the membrane between solutions that differ in osmolarity (see Figure 6.9), whereas when two solutions are isoosmotic, no net water movement occurs.

Animals Use Different Approaches to Keep Osmosis from Swelling or Shrinking Their Cells

For metabolic stability, animals must keep their cellular fluids and ECFs isoosmotic. In some animals, called **osmoconformers,** the osmolarity of the cellular and extracellular solutions matches the osmolarity of the environment. Most marine invertebrates are osmoconformers. Other animals, called **osmoregulators,** use control mechanisms to keep the osmolarity of cellular and extracellular fluids the same, but at levels that may differ from the osmolarity of the surroundings. Most freshwater and terrestrial invertebrates, and almost all vertebrates, are osmoregulators.

For terrestrial animals, one of the greatest challenges to osmoregulation is the limited supply of water in the environment, as the crew of the *Lady Be Good* experienced.

Excretion Is Closely Tied to Osmoregulation

Control over osmolarity is maintained partly by removing certain molecules and ions from cells and body fluids and releasing them into the environment; thus, excretion is closely related to osmoregulation. Animals excrete H^+ ions to keep the pH of body fluids near the neutral levels required by cells for survival. They also excrete toxic products of metabolism, such as nitrogenous (nitrogen-containing) compounds resulting from the breakdown of proteins and nucleic acids, and breakdown products of poisons and toxins. Excretion of ions and metabolic products is accompanied by water excretion because water serves as a solvent for those molecules. Animals that take in large amounts of water may also excrete water to maintain osmolarity.

Microscopic Tubules Form the Basis of Excretion in Most Animals

Except in the simplest animals, minute tubular structures carry out osmoregulation and excretion **(Figure 48.2)**. The tubules are immersed in body fluids at one end (the *proximal end* of the tubules), and open directly or indirectly to the body exterior at the other end (the *distal end* of the tubules). The tubules are formed from a **transport epithelium**—a layer of cells with specialized transport proteins in their plasma membranes. The transport proteins move specific molecules and ions into and out of the tubule by either active or passive transport, depending on the particular substance and its concentration gradient.

Typically, the tubules function in a four-step process:

1. **Filtration.** Filtration is the nonselective movement of water and a number of solutes—ions and small molecules, but not large molecules such as proteins—into the proximal end of the tubules through spaces between cells. In animals with an open circulatory system, the water and solutes come from body fluids, with movement into the tubules driven by the higher pressure of the body fluids compared with the fluid inside the tubule. (The pressure here is the hydrostatic pressure of the plasma in the capillaries; that is, blood pressure.) In animals with a closed circulatory system, such as humans, the water and solutes from the capillary blood plasma exchange with the intratubular fluid, with the movement into the tubules similarly driven by hydrostatic pressure. (Open and closed circulatory systems are described in Section 44.1.)
2. **Tubular reabsorption.** Tubular reabsorption is a selective process in which some molecules (for example, glucose and amino acids) and ions are transported by the transport epithelium from the lumen of the tubule back into the ECF and eventually into the blood (in animals with closed circulatory systems) as the filtered solution moves through the excretory tubule.
3. **Tubular secretion.** Tubular secretion is a selective process in which specific small molecules and ions are transported from the ECF and blood into the tubules. Both tubular secretion and filtration play an important role in eliminating substances from the body fluid or blood. The difference between the two processes is that filtration is nonselective whereas tubular secretion is selective for substances transported.
4. **Release.** The fluid containing waste materials—urine—is released into the environment from the distal end of the tubule. In some animals, the fluid is stored in a bladder; in others, it is concentrated into a solid or semisolid form.

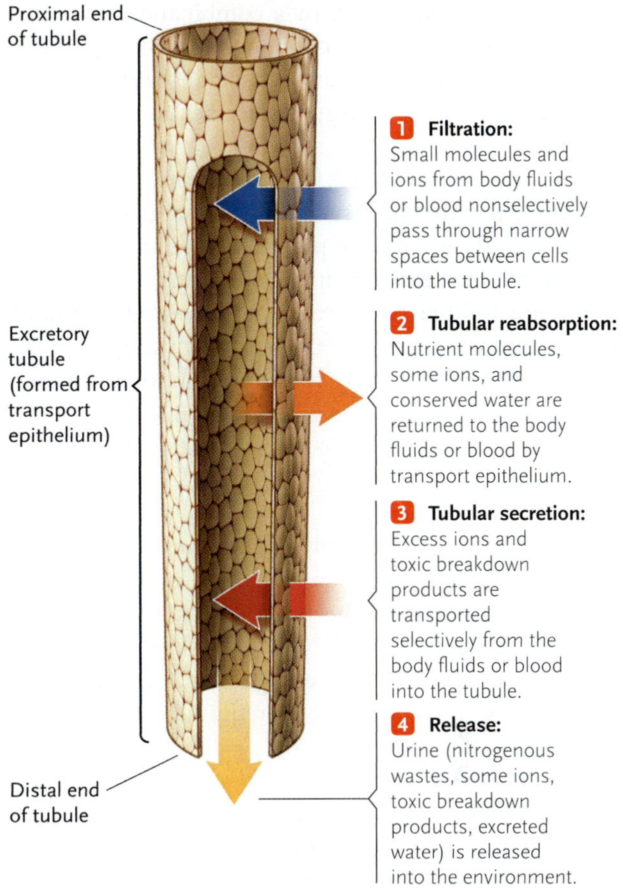

FIGURE 48.2 Common structures and operations of the tubules carrying out osmoregulation and excretion in animals. The tubules are typically formed from a single layer of cells with transport functions.

In all vertebrates and many invertebrates, the excretory tubules are concentrated in specialized organs, the *kidneys,* which are discussed in later sections.

Animals Excrete Nitrogen Compounds as Metabolic Wastes

The metabolism of ingested food is a source of both energy and molecules for the biosynthetic activities of an animal. Importantly, metabolism of ingested food produces water—called *metabolic water*—that is used in chemical reactions and is involved in physiological processes such as the excretion of wastes.

The nitrogenous products of the breakdown of proteins, amino acids, and nucleic acids are excreted by most animals as *ammonia, urea,* or *uric acid,* or a combination of these substances **(Figure 48.3).** The particular molecule or combination of molecules produced depends on a balance among toxicity, water conservation, and energy requirements.

AMMONIA Ammonia (NH_3) forms by a series of biochemical steps beginning with the removal of amino groups ($—NH_3^+$) from amino acids as a part of protein breakdown. Ammonia is readily soluble in water, but it is also highly toxic. Therefore, ammonia must either be excreted or converted to a nontoxic derivative. However, because of its toxicity, ammonia can be excreted from the body only in dilute solutions, making this path possible only in animals with a plentiful supply of water. Those animals include aquatic invertebrates, teleosts, and larval amphibians.

UREA All mammals, most amphibians, turtles, some marine fishes, and some terrestrial invertebrates combine ammonia with HCO_3^- and convert the product in a series of steps to *urea,* a soluble and relatively nontoxic substance. Although producing urea requires more energy than forming ammonia, excreting urea instead of ammonia requires only about 10% as much water.

URIC ACID Water is conserved further in some animals, including terrestrial invertebrates, lizards and snakes, and birds, by the formation of uric acid instead of ammonia or urea. Uric acid is nontoxic, and so insoluble that it precipitates in water as a crystal. (The white substance in bird droppings is uric acid.) The embryos of reptiles and birds also conserve water by forming uric acid, which is stored as a waste product.

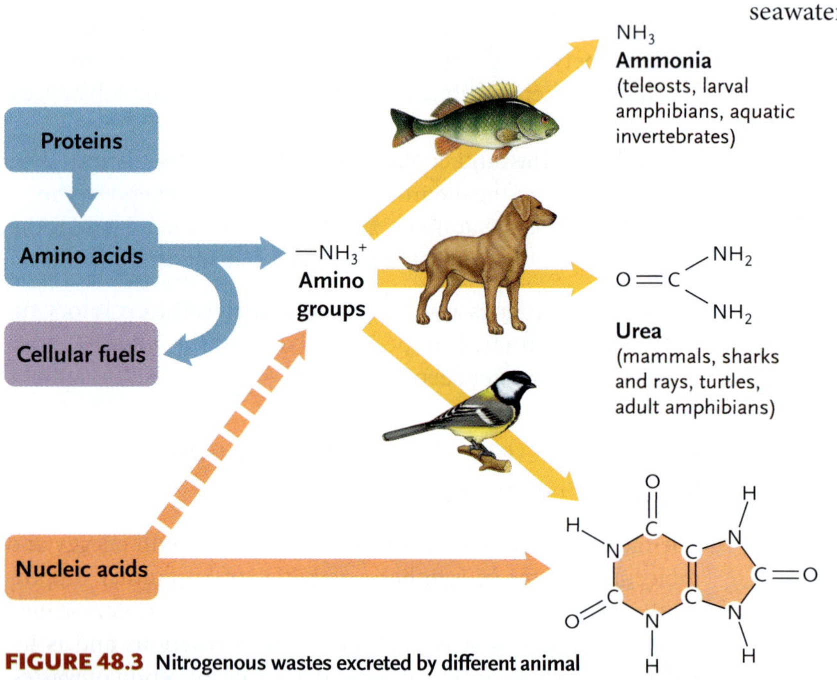

FIGURE 48.3 Nitrogenous wastes excreted by different animal groups. Although humans and other mammals primarily excrete urea, they also excrete small amounts of ammonia and uric acid.

Although making uric acid requires even more energy than making urea, molecule for molecule it contains four times as much nitrogen as ammonia. And, because uric acid precipitates from water, it can be excreted as a concentrated paste. These factors conserve about 99% of the water that would be required to excrete an equivalent amount of nitrogen as ammonia.

In the sections that follow, we look at the specifics of osmoregulation and excretion in different animal groups, beginning with the invertebrates.

STUDY BREAK 48.1

Define the terms *osmosis, osmolarity, hypoosmotic, osmoregulator,* and *transport epithelium.*

48.2 Osmoregulation and Excretion in Invertebrates

Except for the simplest groups, most invertebrates, whether osmoconformers or osmoregulators, carry out excretion by specialized excretory tubules.

Most Marine Invertebrates Are Osmoconformers; All Freshwater and Terrestrial Invertebrates Are Osmoregulators

Most marine invertebrates are osmoconformers. They release water, certain ions, and nitrogenous wastes—usually in the form of ammonia—directly from body cells to the surrounding seawater. The cells of these animals do not swell or shrink because the osmolarity of their intracellular and extracellular fluids and the surrounding seawater is the same, about 1,000 mOsm/L. Therefore, they do not have to expend energy to maintain their osmolarity. However, osmoconformers do expend energy to keep some ions, such as Na^+, at lower concentrations inside cells than in the surroundings.

By contrast, all freshwater invertebrates are osmoregulators because their cells could not survive if their internal ion concentrations were reduced to freshwater levels. These animals must expend energy to keep their internal fluids hyperosmotic to their surroundings. Although osmoregulation is energetically expensive, these invertebrates can live in more varied habitats than can osmoconformers.

Freshwater osmoregulators such as flatworms and mussels have cellular and extracellular fluids with an osmolarity higher than that of the external environment. This condition causes water to move constantly from the surroundings into their bodies. This excess water must be excreted, at a considerable cost in energy, to maintain the hyperosmotic state of their body fluids.

These animals must also obtain the salts required to keep their body fluids hyperosmotic to fresh water. The salts are obtained from foods, and by actively transporting salt ions from the water into their bodies (even fresh water contains some dissolved salts). This active ion transport occurs through the skin or gills.

Terrestrial annelids (earthworms), arthropods (insects, spiders and mites, millipedes, and centipedes), and mollusks (land snails and slugs) must obtain salts from their surroundings, usually in their foods. Although they do not have to excrete water entering by osmosis, they must constantly replace water lost from their bodies.

In Invertebrate Osmoregulators, Specialized Excretory Tubules Participate in Osmoregulation and Carry Out Excretion

Invertebrate osmoregulators typically use specialized tubules for carrying out excretion. Three common types of these specialized tubules are *protonephridia,* found in flatworms and larval mollusks; *metanephridia,* found in annelids and most adult mollusks; and *Malpighian tubules,* found in insects and other arthropods. Each processes body fluids in a different way.

PROTONEPHRIDIA The flatworm *Dugesia* provides an example of the simplest form of invertebrate excretory tubule, the **protonephridium** (*protos* = first; *nephros* = kidney). In *Dugesia,* two branching networks of protonephridia run the length of the body **(Figure 48.4).** The proximal branches of the tubule network end with a *flame cell* containing a bundle of cilia that extend into the tubule and beat to move fluid through the tubule. When the hemolymph, the invertebrate equivalent to blood, passes through the protonephridia, some molecules and ions are reabsorbed and others, including nitrogenous wastes, are secreted into the tubules. The urine resulting from this filtration system is released through pores at the distal ends of the tubules where they reach the body surface.

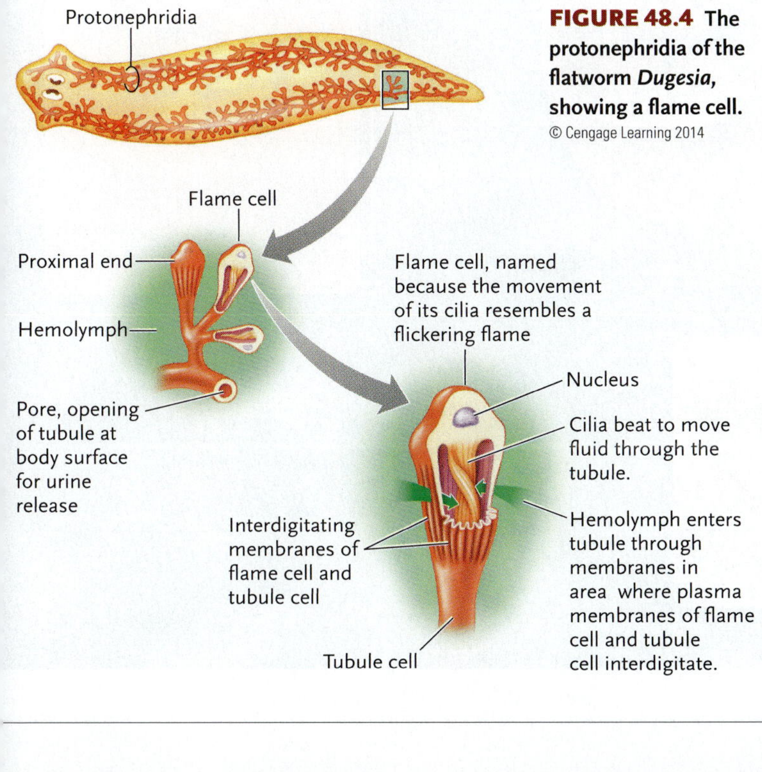

FIGURE 48.4 The protonephridia of the flatworm *Dugesia*, showing a flame cell.

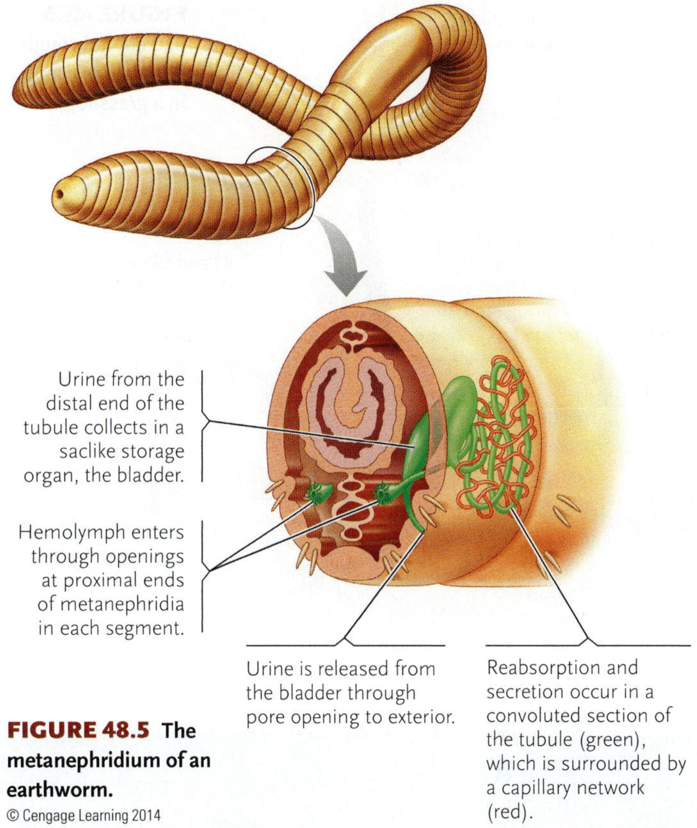

FIGURE 48.5 The metanephridium of an earthworm.

METANEPHRIDIA The excretory tubule of most annelids and adult mollusks, the **metanephridium** (*meta* = between), has a funnel-like proximal end surrounded with cilia that admits hemolymph. Like protonephridia, metanephridia are filtration systems. As hemolymph moves through the tubule, some molecules and ions are reabsorbed and other ions and nitrogenous wastes are secreted into the tubule and excreted from the distal end at the body surface.

Figure 48.5 shows the arrangement and operation of metanephridia in an earthworm. The proximal ends of a pair of metanephridia are located in each body segment, one on either side of the animal. Each tubule of the pair extends into the following segment, where it bends and folds into a convoluted arrangement surrounded by a network of blood vessels.

MALPIGHIAN TUBULES The excretory tubule of insects, the Malpighian tubule, has a closed proximal end that is immersed in the hemolymph **(Figure 48.6).** The distal ends of the tubules empty into the gut. Unlike other excretory systems, Malpighian tubules do not use pressure for the filtration step. Instead, they secrete K^+ into the lumen of the proximal segment, which makes the tubule fluid more positively charged and draws in Cl^- ions from the hemolymph surrounding the tubule. The accumulation of KCl makes the tubule fluid more concentrated, causing water from the hemolymph to move in by osmosis. Now organic wastes, in particular

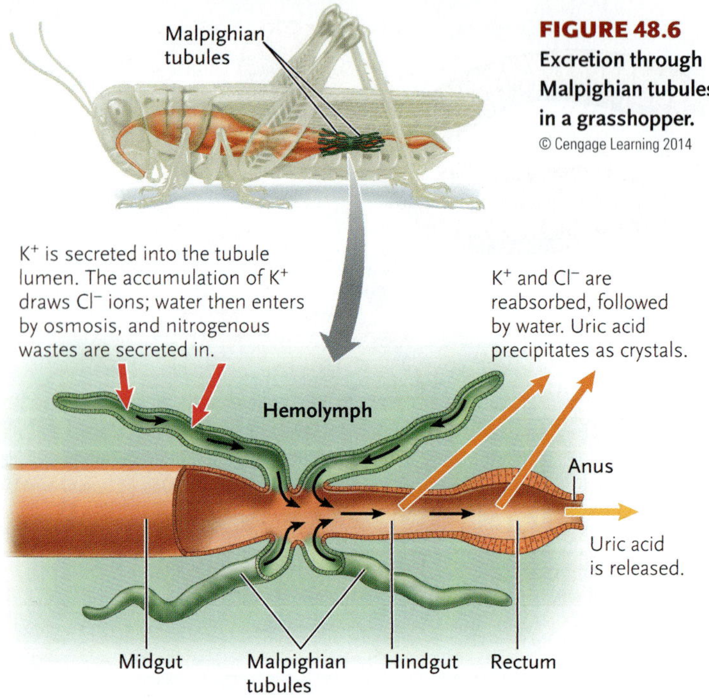

FIGURE 48.6 Excretion through Malpighian tubules in a grasshopper.

uric acid, are secreted into the tubule. The tubule empties into the gut. When the fluid reaches the hindgut and rectum, K^+ and Cl^- are reabsorbed, followed by water. As the water is removed, uric acid precipitates as crystals, which mix with the undigested matter in the rectum and are released with the feces. In this way, nitrogenous waste excretion is accomplished with minimal water loss.

STUDY BREAK 48.2

How are protonephridia, metanephridia, and Malpighian tubules different? In which animal groups are each of these excretory tubules found?

48.3 Osmoregulation and Excretion in Mammals

In all vertebrates, specialized excretory tubules contribute to osmoregulation and carry out excretion. The excretory tubules, called **nephrons,** are located in a specialized organ, the kidney.

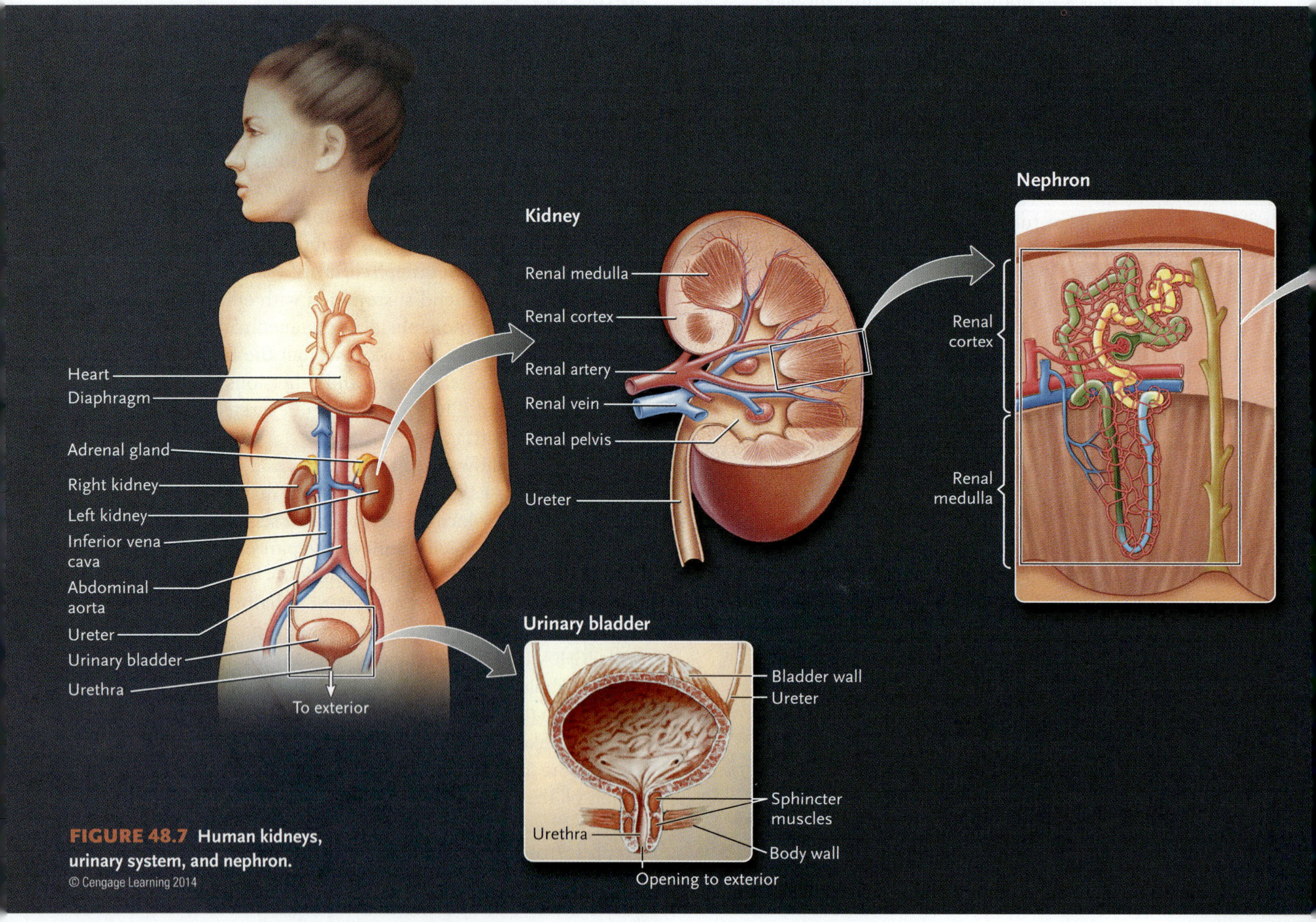

FIGURE 48.7 Human kidneys, urinary system, and nephron.

We begin our survey of vertebrate osmoregulation and excretion with a description of the structure and function of the mammalian kidney.

The Kidneys and Ureters, the Bladder, and the Urethra Form the Urinary System

Mammals have a pair of kidneys, located on either side of the vertebral column at the back of the abdominal cavity **(Figure 48.7).** Internally, the mammalian kidney is divided into an outer **renal cortex** surrounding a central region, the **renal medulla.**

The **renal artery** carries blood into the kidney, where metabolic wastes and excess ions are moved into urine. The filtered blood is routed away from the kidney by the **renal vein.** The urine leaving individual nephrons is processed further in **collecting ducts** and then drains into a central cavity in the kidney called the **renal pelvis.**

From the renal pelvis, the urine flows through a duct called the **ureter** to the **urinary bladder,** a storage sac located outside the kidneys. Urine leaves the bladder through the **urethra,** a tube that (in most mammals) opens to the outside. In human females, the opening of the urethra is just in front of the vagina; in males, the urethra opens at the tip of the penis. The two kidneys and ureters, the urinary bladder, and the urethra constitute the mammalian urinary system.

Two sphincter muscles control the flow of urine from the bladder to the urethra. In human infants, urination is an autonomic reflex triggered by stretch receptors in the bladder wall. When the bladder becomes full, the sphincters relax, smooth muscles in the bladder wall contract, and the urine is forced to the exterior. At about the age of two years, children learn to override the autonomic reflex by consciously contracting their sphincters until urination is convenient.

Mammalian Nephrons Are Differentiated into Regions with Specialized Functions

Each human kidney has more than a million nephrons. Nephrons are differentiated into regions that perform successive steps in excretion. At its proximal end, a human nephron forms the **Bowman's capsule,** an infolded region that cups around a ball of arterial capillaries called the **glomerulus** (see Figure 48.7). The capsule and glomerulus are located in the renal cortex. As blood flows through the glomerulus, plasma filters through the glomerular capillaries into Bowman's capsule; this process, called **glomerular filtration,** is the first step of urine formation.

Following Bowman's capsule, the nephron forms a **proximal convoluted tubule** that descends into the renal medulla in a U-shaped bend called the **loop of Henle** and then ascends again to form a **distal convoluted tubule.** The distal tubule drains the urine into a collecting duct that leads to the renal pelvis. As many as eight nephrons may drain into a single collecting duct. The combined activities of the proximal convoluted tubule, the loop of Henle, the distal convoluted tubule, and the collecting duct convert the filtrate entering the nephron into urine.

Unlike most capillaries in the body, the capillaries in the glomerulus do not lead directly to venules. Instead, they form another arteriole that branches into a second capillary network called the **peritubular capillaries.** These capillaries thread around the proximal and distal convoluted tubules and the loop of Henle. Molecules and ions that are reabsorbed from the filtrate are transferred indirectly between the nephron and the peritubular capillaries. First, the molecules or ions pass through the one-cell-thick wall of the tubule, then they diffuse through the interstitial fluid, and finally they pass into the capillary through its one-cell-thick wall.

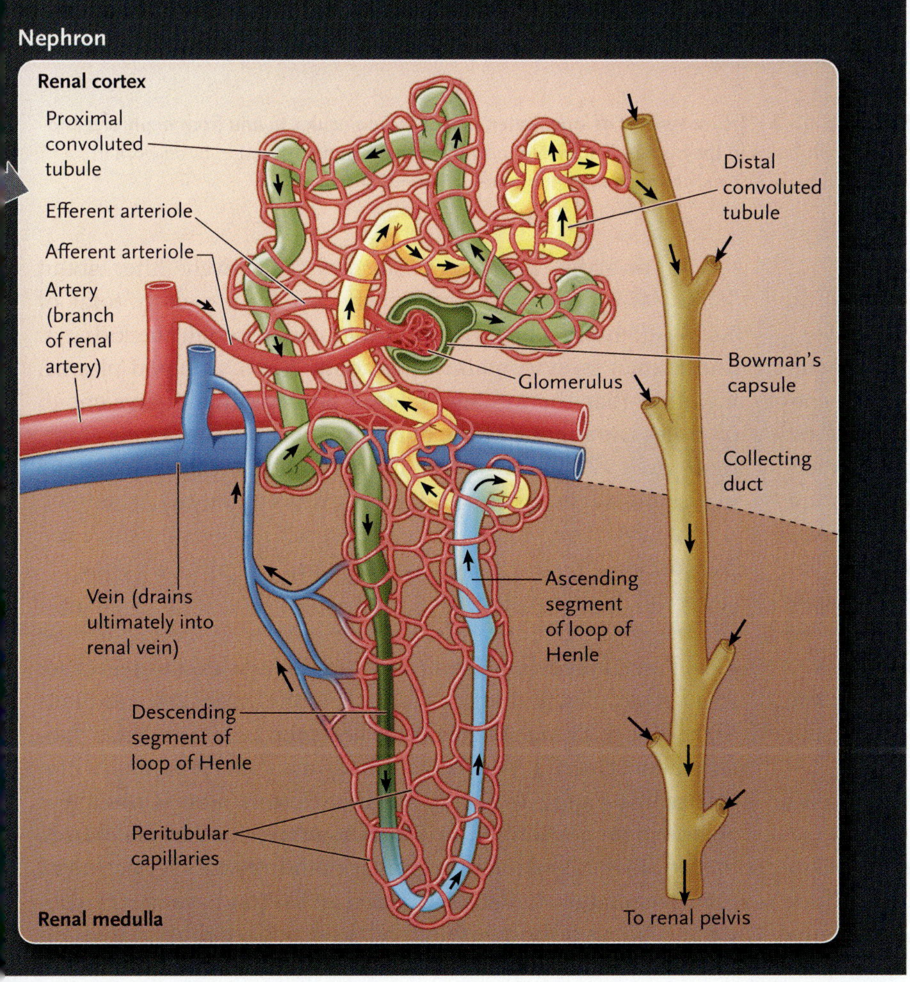

Mammalian Nephrons Interact with Surrounding Kidney Structures to Produce Hyperosmotic Urine

In mammals, urine is hyperosmotic to body fluids. All other vertebrates except for a few aquatic bird species produce urine that is hypoosmotic (or isoosmotic) to body fluids. Production of hyperosmotic urine is a water-conserving adaptation involving an interaction between the activities of the mammalian nephrons and the highly ordered structure of the mammalian kidney. Three features underlie this interaction:

1. The arrangement of the loop of Henle, which descends through the medulla and returns to the cortex again.
2. Differences in the permeability of successive regions of the nephron, established by a specific group of membrane transport proteins in each region.
3. A gradient in the concentration of molecules and ions in the interstitial fluid of the kidney, which increases gradually from the renal cortex to the deepest levels of the renal medulla.

These features interact to conserve nutrients and water, balance salts, and concentrate wastes for excretion from the body.

Three processes are involved in forming urine: *glomerular filtration, tubular reabsorption,* and *tubular secretion.* These processes are described in the next two subsections.

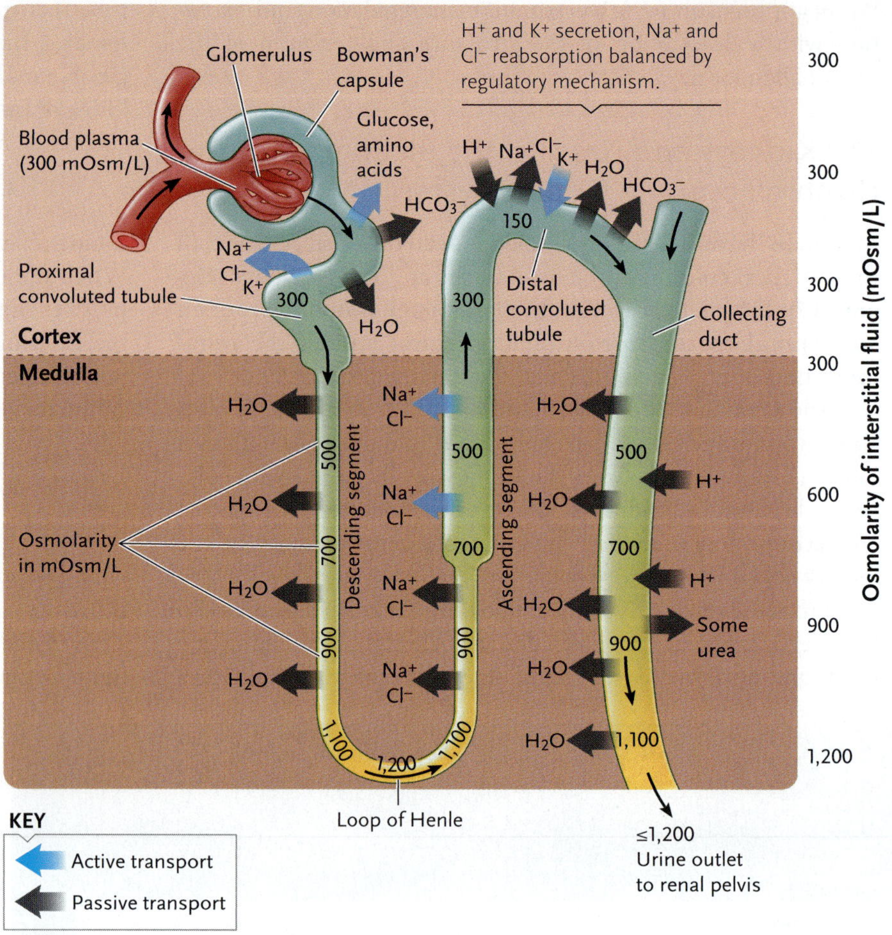

FIGURE 48.8 The movement of ions, water, and other molecules to and from nephrons and collecting tubules in the human kidney. Nephrons in other mammals and in birds work in similar fashion. The numbers are osmolarity values in mOsm/L.

Glomerular Filtration Begins the Process of Excretion

Glomerular filtration begins the process of excretion (shown in **Figure 48.8** and summarized in **Table 48.1**). The single layer of epithelial cells forming the glomerular capillaries is perforated by many pores that make the vessels much more permeable to water and solutes than other capillaries in the body. The higher pressure of the blood drives fluid containing ions, small nutrient molecules such as glucose and amino acids, and nitrogenous waste molecules (primarily urea) through the pores of the glomerular capillaries into Bowman's capsule. Blood cells and plasma proteins are too large to pass and are retained inside the capillaries.

The blood pressure driving fluid through the glomerular capillaries into Bowman's capsule is maintained because the diameter of the **afferent arteriole** delivering blood to the glomerulus is larger than that of the **efferent arteriole** that takes blood away from the glomerulus. That is, because more blood can enter the glomerulus through the afferent arteriole than can leave through the efferent arteriole, glomerular capillary pressure is maintained at a high level as a result of blood damming up in the capillaries.

In humans, Bowman's capsules collectively filter about 180 L (47.5 gallons) of fluid each day, from a daily total of 1,400 L (369.5 gallons) of blood that pass through the kidneys. The human body contains only about 2.75 L of blood plasma, meaning that the kidneys filter a fluid volume equivalent to 65 times the volume of the blood plasma each day. On average, more than 99% of the filtrate, mostly water, is reabsorbed in the nephrons, leaving about 1.5 L to be excreted daily as urine.

Reabsorption and Secretion Take Place in the Remainder of the Nephron

The fluid filtered into Bowman's capsule contains water, other small molecules, and ions at essentially the same concentrations as the blood plasma. As the filtrate flows through the tubules, substances of value are returned to the plasma in the peritubular capillaries (see Figure 48.7) through the selective process of tubular reabsorption. In addition, selective movement of substances from the blood of the peritubular capillaries into the tubules occurs by tubular secretion. Tubular secretion is a second avenue for substances to enter the tubules from the blood, the first being glomerular filtration. By the time the fluid reaches the distal end of

TABLE 48.1	Filtration, Reabsorption, and Secretion in Nephrons and Collecting Ducts			
Segment	Location	Permeability and Movement	Osmolarity of Filtrate and Urine	Result of Passage
Bowman's capsule	Cortex	Water, ions, small nutrients, and nitrogenous wastes move through spaces between epithelia	300 mOsm/L, same as surrounding interstitial fluid	Water and small substances, but not proteins, pass into nephron
Proximal convoluted tubule	Cortex	Na^+ and K^+ actively reabsorbed, Cl^- follows; water leaves through aquaporins; HCO_3^- reabsorbed into plasma of peritubular capillaries; glucose, amino acids, and other nutrients actively reabsorbed	300 mOsm/L	67% of ions, 65% of water, 50% of urea, and all nutrients return to interstitial fluid; pH maintained
Descending segment of loop of Henle	Cortex into medulla	Water leaves through aquaporins; no movement of ions or urea	From 300 mOsm/L at top to 1,200 mOsm/L at bottom of loop	Additional water returned to interstitial fluid
Ascending segment of loop of Henle	Medulla into cortex	Na^+ and Cl^- actively transported out; no entry of water; no movement of urea	From 1,200 mOsm/L at bottom to 150 mOsm/L at top of loop	Additional ions returned to interstitial fluid
Distal convoluted tubule	Cortex	K^+ and Na^+ secreted via active transport into urine; Na^+ and Cl^- reabsorbed; water moves into urine through aquaporins; HCO_3^- reabsorbed into plasma of peritubular capillaries	From 150 mOsm/L at beginning to 300 mOsm/L at junction with collecting duct	Ion balance, pH balance
Collecting ducts	Cortex through medulla, empties into renal pelvis	Water moves out via aquaporins; no movement of ions; some urea leaves at bottom of duct	From 300 mOsm/L to 1,200 mOsm/L at junction with renal pelvis	More water and some urea returned to interstitial fluid; some H^+ added to urine

the tubules and passes through the collecting ducts, reabsorption out of the tubules and secretion into them have markedly altered the concentrations of all components of the filtrate.

THE PROXIMAL CONVOLUTED TUBULE Reabsorption of water, ions, and nutrients into the interstitial fluid is the main function of the proximal convoluted tubule. Na^+/K^+ pumps in the epithelium of the proximal convoluted tubule move Na^+ and K^+ from the filtrate into the interstitial fluid surrounding the tubule (see Figure 48.8). The movement of positive charges sets up a voltage gradient that causes Cl^- ions to be reabsorbed from within the tubule with the positive ions. Specific active transport proteins reabsorb essentially all the glucose, amino acids, and other nutrient molecules from the filtrate into the interstitial fluid, making the filtrate hypoosmotic to the interstitial fluid surrounding the tubule. As a result, water moves from the tubule into the interstitial fluid by osmosis. The osmotic movement is aided by *aquaporins,* transport proteins that form passages for water molecules in the transport epithelium of the tubule cells. (To review aquaporins, see Section 6.2 and *Unanswered Questions* for Chapter 6.) The nutrients and water that entered the interstitial fluid move into the peritubular capillaries.

Tubular secretion moves some substances into the tubule, notably detoxified poisons are transferred by passive diffusion (detoxification takes place in the liver). Small amounts of ammonia are also secreted into the tubule.

In all, the proximal convoluted tubule reabsorbs about 67% of the Na^+, K^+, and Cl^- ions, 65% of the water, 50% of the urea, and essentially all the glucose, amino acids, and other nutrient molecules in the filtrate. The ions, nutrients, and water reabsorbed by the tubule are transported into the interstitial fluid, and then into capillaries of the peritubular network. Although half of the urea is reabsorbed, the constant flow of filtrate through the tubules keeps the concentration of nitrogenous wastes low in body fluids.

The proximal convoluted tubule has structural specializations that fit its function. The epithelial cells that make up its walls are carpeted on their inner surface by a brush border of microvilli. Like the brush border of epithelial cells in the small intestine (see Section 47.3), these microvilli greatly increase the surface area available for reabsorption and secretion.

THE DESCENDING SEGMENT OF THE LOOP OF HENLE The loop of Henle has dual functions: (1) to recapture water in the descending segment; and (2) to reabsorb salts in the ascending segment.

The filtrate leaving the proximal convoluted tubule enters the descending segment of the loop of Henle, where water is

INSIGHTS FROM THE **Molecular Revolution**

An Ore Spells Relief for Osmotic Stress

The osmolyte sorbitol is made from glucose in a reaction catalyzed by the enzyme *aldose reductase (AR)*. Placing renal cells in a hyperosmotic medium induces transcription of the AR gene.

Research Question

How does hyperosmotic stress activate the AR gene?

Experiments

To answer the question, Joan D. Ferraris and her colleagues at the National Institutes of Health in Bethesda, Maryland, cloned a DNA segment containing the AR gene and its upstream promoter and regulatory sequences from a rabbit genomic library (see Chapter 18) **(Figure A).** (Recall from Section 16.2 that the promoter, the sequence immediately upstream of the gene's RNA coding sequence, plays a basic role in initiating transcription of the gene, but does not contain regulatory sequences, which are located upstream of the promoter.)

The researchers then constructed a composite gene by fusing the upstream segment of the AR gene with the RNA coding sequence of the firefly luciferase gene **(Figure B).** The luciferase enzyme catalyzes a reaction that emits light as a product. By using light production as a measure of luciferase activity, the construct enabled the researchers to study the AR upstream region to identify the regulatory sequence required for activating transcription in cells exposed to a hyperosmotic medium.

The researchers transformed the composite gene into cultured rabbit renal cells, and then divided the cells into two groups. One group was placed in an isoosmotic medium (300 mOsm/L); the other group was placed in a medium made hyperosmotic (500 mOsm/L) by added NaCl. After a period of incubation, luciferase activity was measured in cell extracts. The cells in the hyperosmotic solution had several-fold higher luciferase activity than the cells in the isoosmotic solution, showing that the luciferase gene had been turned on under the high salt condition because of the presence of the regulatory sequence derived from the AR gene.

The next step was to identify the particular regulatory sequence. First, they identified the promoter of the AR gene by adding various lengths of fragments immediately upstream of the AR coding sequence to the luciferase gene. They found that a 208-bp sequence gave promoter activity **(Figure C);** that is, they saw a low level of transcription of luciferase in isoosmotic conditions, but that level of transcription did not change in hyperosmotic conditions. Next, the researchers broke the DNA upstream of the promoter into fragments, attached them one at a

reabsorbed. As this segment descends, it passes through regions of increasingly higher solute concentrations in the interstitial fluid of the medulla (see Figure 48.8). (The generation of this concentration gradient is described later.) As a result, more water moves out of the tubule by osmosis as the fluid travels through the descending segment.

Aquaporins in the descending segment allow the rapid transport of water. The outward movement of water concentrates the molecules and ions inside the tubule, gradually increasing the osmolarity of the fluid to a peak of about 1,200 mOsm/L at the bottom of the loop. This is the same as the osmolarity of the interstitial fluid at the bottom of the medulla.

THE ASCENDING SEGMENT OF THE LOOP OF HENLE The fluid then moves into the ascending segment of the loop of Henle, where Na^+ and Cl^- are reabsorbed into the interstitial fluid. As this segment ascends, it passes through regions of gradually lessening osmolarity in the medulla (see Figure 48.8). The ascending segment has membrane proteins that transport salt ions, but no aquaporins. Because water is trapped in the ascending segment, the osmolarity of the urine is reduced as the salt ions move out of the tubule.

By the time the fluid reaches the cortex at the top of the ascending loop, its osmolarity has dropped to about 150 mOsm/L. During the travel of fluid around the entire loop of Henle, water, nutrients, and ions have been conserved and returned to body fluids, and the total volume of the filtrate in the nephron has been reduced greatly. Urea and other nitrogenous wastes have been concentrated in the filtrate. Little secretion occurs in either the descending or ascending segments of the loop of Henle.

THE DISTAL CONVOLUTED TUBULE Additional water is recovered by osmosis from the fluid in the distal convoluted tubule. In response to hormones triggered by changes in the body's salt concentrations (described in Section 48.4), a varying amount of K^+ is secreted into the fluid, and varying amounts of Na^+ and Cl^- ions are reabsorbed. Bicarbonate ions are reabsorbed from the filtrate as in the proximal tubule.

The amounts of urea and other nitrogenous wastes remain the same. By the time the fluid—now urine—enters the collecting ducts at the end of the nephron, it is isoosmotic with blood plasma (about 300 mOsm/L) but very different in composition.

THE COLLECTING DUCTS The collecting ducts concentrate the urine. These ducts, which are permeable to water but not salt ions, descend from the cortex through the medulla of the kidney. As the ducts descend, they encounter the gradient of increasing solute concentration in the interstitial fluid of the medulla. This increase makes water move osmotically out of the ducts and greatly increases the concentration of the urine, which can become as high as 1,200 mOsm/L at the bottom of the medulla. Near the bottom of the medulla, the walls of the collecting ducts contain passive urea transporters that allow a portion of this nitrogenous waste to pass from the duct into the interstitial fluid. This urea adds significantly to the concentration gradient of solutes in the medulla.

time to the promoter–luciferase gene construct, and tested them for luciferase expression in isoosmotic and hyperosmotic conditions. Eventually, they identified the minimal sequence that activates the luciferase gene in cells placed in hyperosmotic medium; it proved to be the sequence 5′-CGGAAAATCAC-3′, beginning 1,105 bp upstream of the site where transcription begins. The investigators termed the activating sequence an *osmotic response element (ORE)*.

Conclusion

Molecular experiments identified for the first time the regulatory sequence responsible for activating the AR gene under hyperosmotic conditions. The result was a first step in elucidating the molecular mechanism for how hyperosmotic stress regulates the expression of genes involved in responding to that stress.

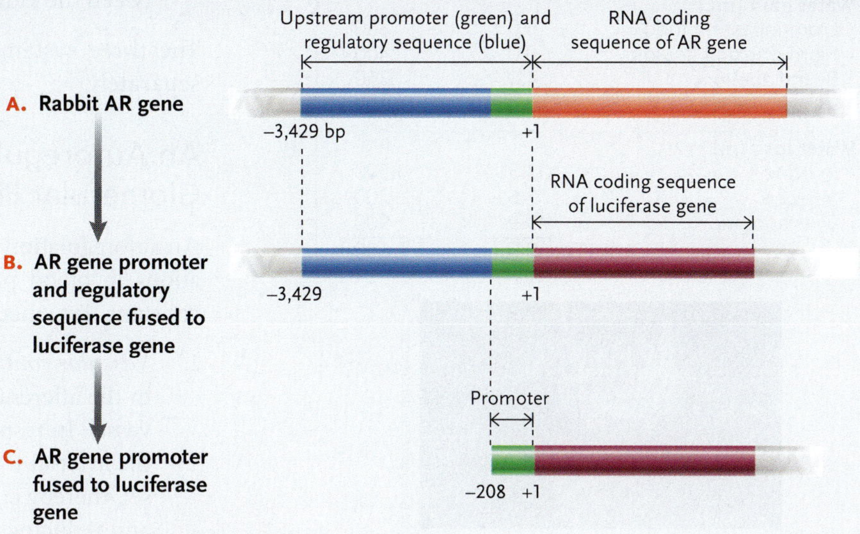

THINK LIKE A SCIENTIST Given its location with respect to the site where transcription begins, in what gene regulatory element is the ORE (osmotic response element) likely to be?

Sources: J. D. Ferraris et al. 1994. Cloning, genomic organization, and osmotic response of the aldose reductase gene. *Proceedings of the National Academy of Sciences, USA* 91:10742–10746; J. D. Ferraris et al. 1996. ORE, a eukaryotic minimal essential osmotic response element. *Journal of Biological Chemistry* 271:18318–18321.

At its maximum value of 1,200 mOsm/L, reached when water conservation is at its maximum, the urine at the bottom of the collecting ducts is about four times more concentrated than body fluids. It can also be as low as 50 to 70 mOsm/L when very dilute urine is produced in response to conditions such as excessive water intake.

The hyperosomotic state of the interstitial fluid toward the bottom of the medulla would damage the medulla cells if they were not protected against osmotic water loss. The protection comes from high concentrations of otherwise inert organic molecules called *osmolytes* in the cytoplasm of these cells. The osmolytes, primarily a sugar alcohol called *sorbitol*, raise the osmolarity of the cells to match that of the surrounding interstitial fluid. *Insights from the Molecular Revolution* describes research that identified the molecular controls leading to sorbitol production in kidney medulla cells and other cells subjected to osmotic stress.

Urine flows from the end of the collecting ducts into the renal pelvis, and then through the ureters into the urinary bladder where it is stored. From the bladder, urine exits through the urethra to the outside.

Terrestrial Mammals Have Additional Water-Conserving Adaptations

Terrestrial mammals have other adaptations that complement the water-conserving activities of the kidneys. One is the location of the lungs deep inside the body, which reduces water loss by evaporation during breathing (see Section 46.1). Another is a body covering of keratinized skin. Skin is nearly impermeable so that it almost eliminates water loss by evaporation, except for the controlled loss through evaporation of sweat in mammals with sweat glands.

Water-conserving adaptations reach their greatest efficiency in desert rodents such as the kangaroo rat **(Figure 48.9)**. The proportion of nephrons with long loops extending deep into the kidney medulla of kangaroo rats is very high, allowing them to excrete urine that is 20 times more concentrated than body fluids. Further, most of the water in the feces is absorbed in the large intestine and rectum. Lacking sweat glands, kangaroo rats lose little water by evaporation from the body surface. Much of the moisture in their breath is condensed and recycled by specialized passages in the nasal cavities. They stay in burrows during daytime, and come out to feed only at night.

About 90% of the kangaroo rat's daily water supply is generated from oxidative reactions in its cells. (Humans, by contrast, can make up only about 12% of their daily water needs from this source.) The remaining 10% of the kangaroo rat's water comes from its food. These structural and behavioral adaptations are so effective that a kangaroo rat can survive in the desert without ever drinking water.

Marine mammals, including whales, seals, and manatees, eat foods that are high in salt content and never drink fresh water. They are able to survive the high salt intake because they produce urine that is more concentrated than seawater. As a result, they are easily able to excrete all the excess salt they ingest in their diet.

	Kangaroo rat	Human
Water gain (mL)		
From ingesting food	6.0	850
From drinking liquids	0.0	1,400
By metabolism	54.0	350
	60.0	2,600
Water loss (mL)		
In urine	13.5	1,500
In feces	2.6	200
By evaporation	43.9	900
	60.0	2,600

FIGURE 48.9 A comparison of the sources of water for a human and a kangaroo rat (genus *Dipodomys*). Water conservation in the kangaroo rat is so efficient that the animal never has to drink water.

STUDY BREAK 48.3

1. **Describe the structure of a human nephron from the proximal end to the distal end.**
2. **The urine entering the collecting ducts at the end of the nephron has an osmolarity essentially the same as that of fluids in other parts of the body. How is the urine subsequently made more concentrated?**

THINK OUTSIDE THE BOOK

On your own or collaboratively, research the symptoms and possible causes of chronic renal disease (also called chronic kidney disease). Then outline the hypothesis and research approach for a study of chronic renal disease that uses the rat as the experimental organism.

48.4 Regulation of Mammalian Kidney Function

Mammalian excretory functions are integrated into overall body functions by three primary control systems, which link kidney functions to blood pressure, to the osmolarity and pH of body fluids, and to the body's water balance:

- An autoregulation system located entirely within the kidney keeps glomerular filtration constant during variations in arterial blood pressure, as when we move from sitting to standing.
- Two hormonal control systems compensate for excessive loss of salt and body fluids and adjust the rate of water uptake in the kidneys to compensate for excessive water intake or loss. These two hormonal systems regulate interactions between the kidneys and the rest of the body.

The three systems work together, but we will discuss them separately.

An Autoregulation System Controls Glomerular Filtration Rate

An autoregulation system responds almost instantly to small variations in blood pressure to keep the glomerular filtration rate constant. Two mechanisms are involved in the autoregulation:

1. *Vascular control of blood pressure.* Increased blood pressure in the afferent arteriole is detected by receptors in the vessel. In response to the stretching, the smooth muscles of the arterial wall contract, which constricts the blood vessel, thereby constraining the blood flow to the glomerulus and reducing the glomerular filtration rate. In the opposite condition of reduced blood pressure in the arteriole, the receptors signal the blood vessel to dilate (expand), thereby increasing blood flow into the glomerulus and, therefore, glomerular filtration rate, even though arterial blood pressure has dropped.
2. *Feedback control from the distal convoluted tubule to the juxtaglomerular apparatus.* The **juxtaglomerular apparatus** (*juxta* = near) is located at a point where the distal convoluted tubule contacts the afferent arteriole carrying blood to the glomerulus **(Figure 48.10, inset).** The feedback control mechanism is triggered by salt concentration in the distal convoluted tubule. If the glomerular filtration rate increases above normal, more fluid than normal is filtered, and a higher salt concentration results in the distal convoluted tubule. Specialized tubule cells in the juxtaglomerular apparatus monitor the salt level of the fluid flowing past them in the tubule. If the salt level rises, those cells secrete chemical messengers that act locally (by paracrine regulation; see Section 42.1 and Figure 42.1) on the adjacent afferent arteriole, causing it to constrict. As a result, glomerular blood flow is reduced and glomerular filtration rate decreases to normal. If the glomerular filtration decreases below normal, the specialized tubule cells detect the lower salt level that results, secretion of the chemical messengers from those cells stops and the afferent arteriole dilates, thereby allowing an increase in the glomerular filtration rate.

The Renin-Angiotensin-Aldosterone System (RAAS) Responds to Na^+ by Triggering Na^+ Reabsorption

Major changes in blood volume and pressure occur when the body loses or gains Na^+ in excessive amounts. Excessive Na^+ loss may result from prolonged and heavy sweating, repeated vomiting, severe diarrhea, or insufficient Na^+ uptake in the diet. The Na^+ loss reduces the osmolarity of body fluids, which causes less

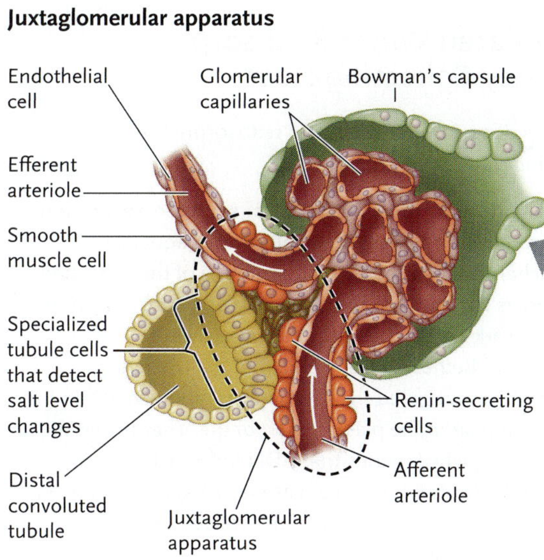

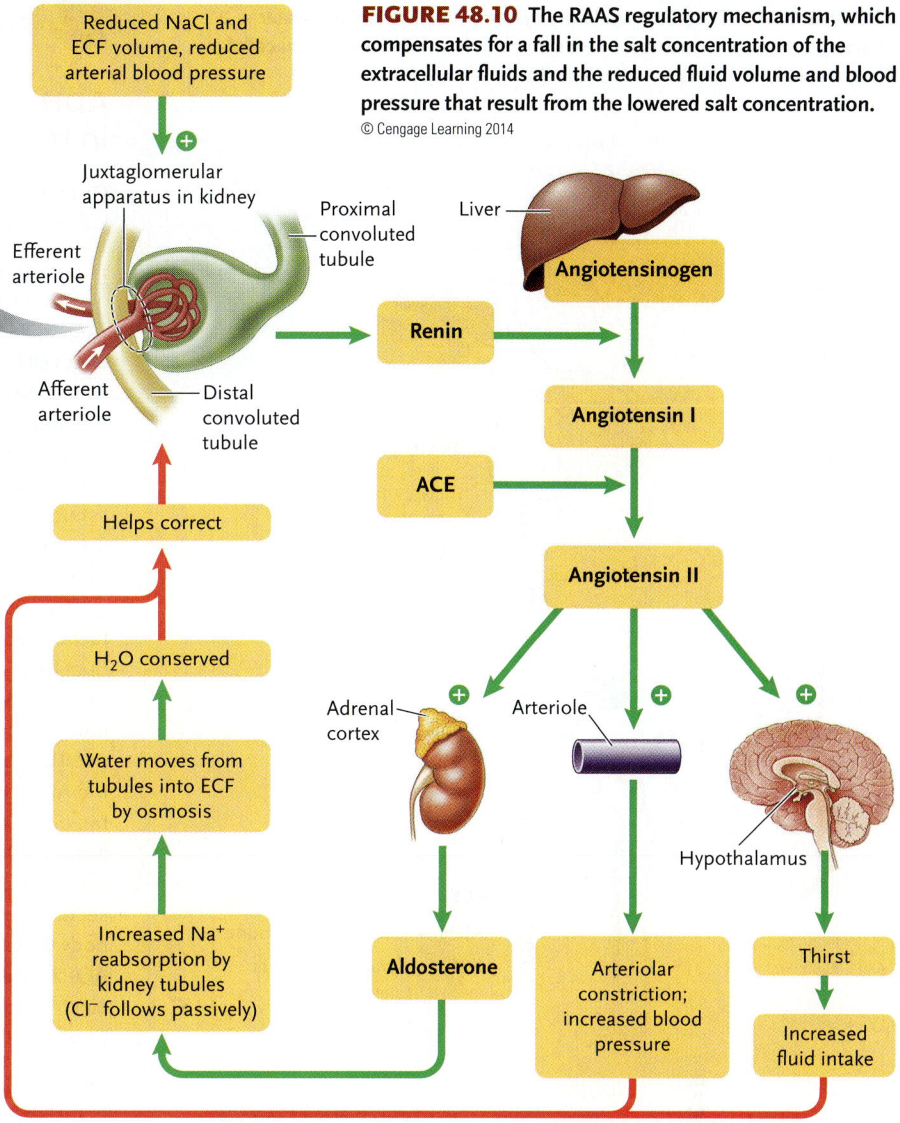

FIGURE 48.10 The RAAS regulatory mechanism, which compensates for a fall in the salt concentration of the extracellular fluids and the reduced fluid volume and blood pressure that result from the lowered salt concentration.

water to be reabsorbed in the kidneys. The water loss reduces the volume of blood and interstitial fluid and causes the blood pressure to drop. Excessive Na^+ intake in salty foods may have the opposite effects. The body must compensate for significant changes in Na^+.

The **renin-angiotensin-aldosterone system (RAAS)** is the most important hormonal system involved in regulating Na^+ **(Figure 48.10).** At normal body salt concentrations, the RAAS allows about 10 g of salt to be excreted in the urine each day. If excessive Na^+ is excreted, blood pressure and body fluid volume drop, and the glomerular filtration rate falls below levels that can be restored by the juxtaglomerular apparatus autoregulation. In response to this condition, the following events take place (see Figure 48.10):

- Cells in the juxtaglomerular apparatus secrete the enzyme **renin** into the bloodstream. (The RAAS also is activated to secrete renin when blood pressure or blood volume decreases independently of Na^+ levels, as in the case of a hemorrhage.)
- Renin cleaves the plasma protein *angiotensinogen* to produce **angiotensin I.**
- **Angiotensin-converting enzyme (ACE)** converts angiotensin I to **angiotensin II.**

Angiotensin II has three effects: (1) it raises blood pressure quickly by constricting arterioles in most parts of the body; (2) it stimulates synthesis of the steroid hormone **aldosterone** and its secretion from the adrenal cortex; and (3) it stimulates thirst so that more water is brought into the body. The aldosterone increases Na^+ reabsorption in the kidneys, which raises the osmolarity of body fluids. As a result, water moves from the tubules into the interstitial fluid, which conserves water. Angiotensin II may also stimulate secretion of **antidiuretic hormone (ADH)** (also called *vasopressin*) by the posterior pituitary (antidiuretic means "against urine output") (not shown in the figure). ADH increases water absorption in the kidneys. Overall, the combined effects of angiotensin act to return the blood pressure to normal.

In the opposite situation, when salt intake is too high, both body fluid volume and blood pressure rise above normal. Under these conditions, renin secretion is inhibited and, as a result, angiotensin production and aldosterone synthesis are not stimulated. The reduction in angiotensin lowers blood pressure by allowing arterioles to dilate; the reduction in aldosterone increases Na^+ loss in the urine by retarding the reabsorption of Na^+ and Cl^- from the kidney tubules.

Elevated blood pressure also stimulates specialized cells in the heart to release **atrial natriuretic factor (ANF),** a peptide hormone that inhibits renin release. ANF increases the filtration rate as well by dilating the arterioles that deliver blood to glomeruli and by inhibiting aldosterone release. As less Na^+ is reabsorbed and urine volume increases, both plasma volume and blood pressure return to normal.

FIGURE 48.11 The ADH regulatory system, which stimulates water reabsorption to compensate for a loss in the fluid volume of the extracellular fluids because of excessive water loss from the body.

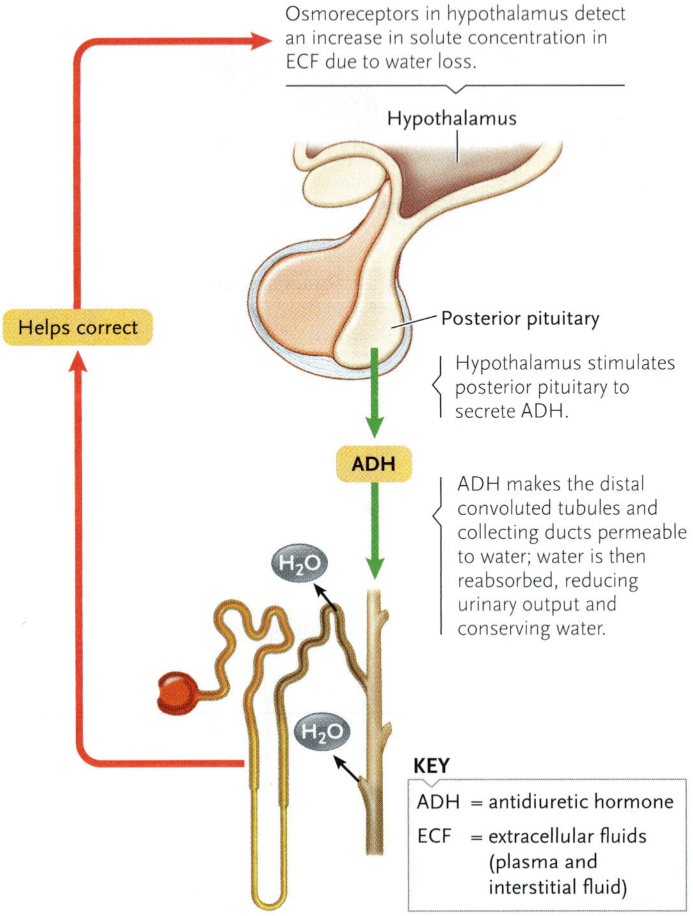

The Antidiuretic Hormone (ADH) System Also Regulates Osmolarity and Water Balance

You have just learned that the ADH system may be stimulated by angiotensin II of the RAAS in response to an increase in Na^+. The control centers of the hypothalamus that regulate ADH secretion (and, therefore, urinary output) and thirst (and, therefore, drinking) function together. That is, both ADH secretion and thirst are stimulated by a bodily water deficit and suppressed by bodily water excess. Logically, the same conditions that require reduced urinary output to conserve water also cause the sensation of thirst to replace water.

The ADH system regulates osmolarity and water balance—and, therefore, urinary output—by increasing water reabsorption in the kidneys without changing the usual excretion of salt **(Figure 48.11).** *Osmoreceptors* in the hypothalamus that are located near the ADH-secreting cells and the thirst center monitor control ADH secretion. (An **osmoreceptor** is a sensory receptor that monitors the osmolarity of the extracellular fluid bathing it, which, in turn, reflects the concentration of the entire internal environment.) When the osmoreceptors detect an increase in osmolarity (too little water) and the need to conserve water increases, both ADH secretion and thirst are stimulated. As a result, reabsorption of water in the distal convoluted tubules and collecting ducts increases so that urinary output is reduced and water is conserved while, at the same time, water intake (drinking) is stimulated. These actions restore depleted water stocks, and dilute

FIGURE 48.12 Experimental Resear

ADH-Stimulated Water Reabsorption in the Kidney Collecting Duct

Question: How does ADH cause water reabsorption in the kidney collecting ducts?

Experiment: The peptide hormone ADH causes the reabsorption of water in the kidney by increasing the number of aquaporin2 (AQP2) water channels in the plasma membrane of the collecting duct cells. AQP2 is the only aquaporin expressed exclusively in the collecting duct. Mark Knepper and his colleagues at the National Institutes of Health, Bethesda, Maryland, and the University of Aarhus, Denmark, investigated the mechanism by which ADH causes AQP2 to appear in the plasma membrane. They tested the "shuttle" hypothesis, which states that ADH induces the movement of AQP2 from intracellular vesicles (IVs) to the plasma membrane of cells on the lumen side of the collecting duct.

The researchers first attached sections of collecting duct from rat kidneys to a pipet and passed an ADH solution through the duct **(Figure 1).** They then counted the AQP2 in the IVs and in the plasma membrane of the lumen cells. To do this, they cut treated ducts into thin sections and added an antibody that specifically recognizes AQP2. The antibody had gold particles attached to it. Under an electron microscope, the electron-dense gold is easily visualized. Counting the gold particles therefore quantified the AQP2.

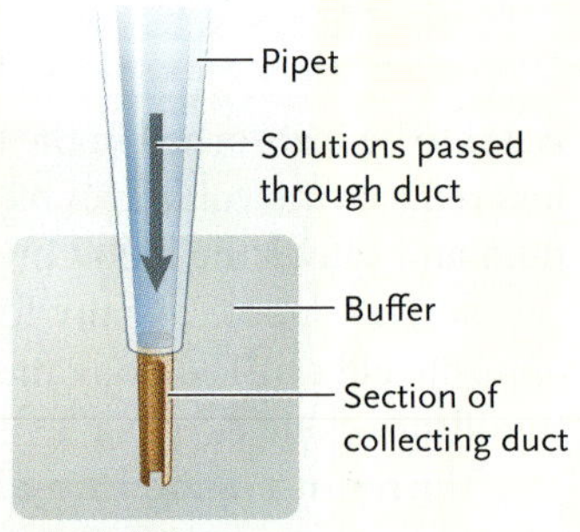

FIGURE 1

Results: The researchers measured the distribution of AQP2 in the absence of ADH, in the presence of ADH, and then after ADH had been washed away. Their results are expressed as the ratio of AQP2 in the plasma membrane of lumen cells to AQP2 in IVs:

Before ADH treatment: 0.32 ± 0.03

In presence of ADH: 0.38 ± 0.14

After ADH removal: 0.35 ± 0.04

Conclusion: ADH induces a change in distribution of AQP2, consistent with the shuttle hypothesis of AQP2 channels moving from IVs to the plasma membrane of lumen cells. The redistribution of AQP2 channels is reversible, as shown by the results after ADH removal.

Figure 2 shows the current molecular model for the ADH-induced redistribution of AQP2 channels in collecting duct epithelial cells.

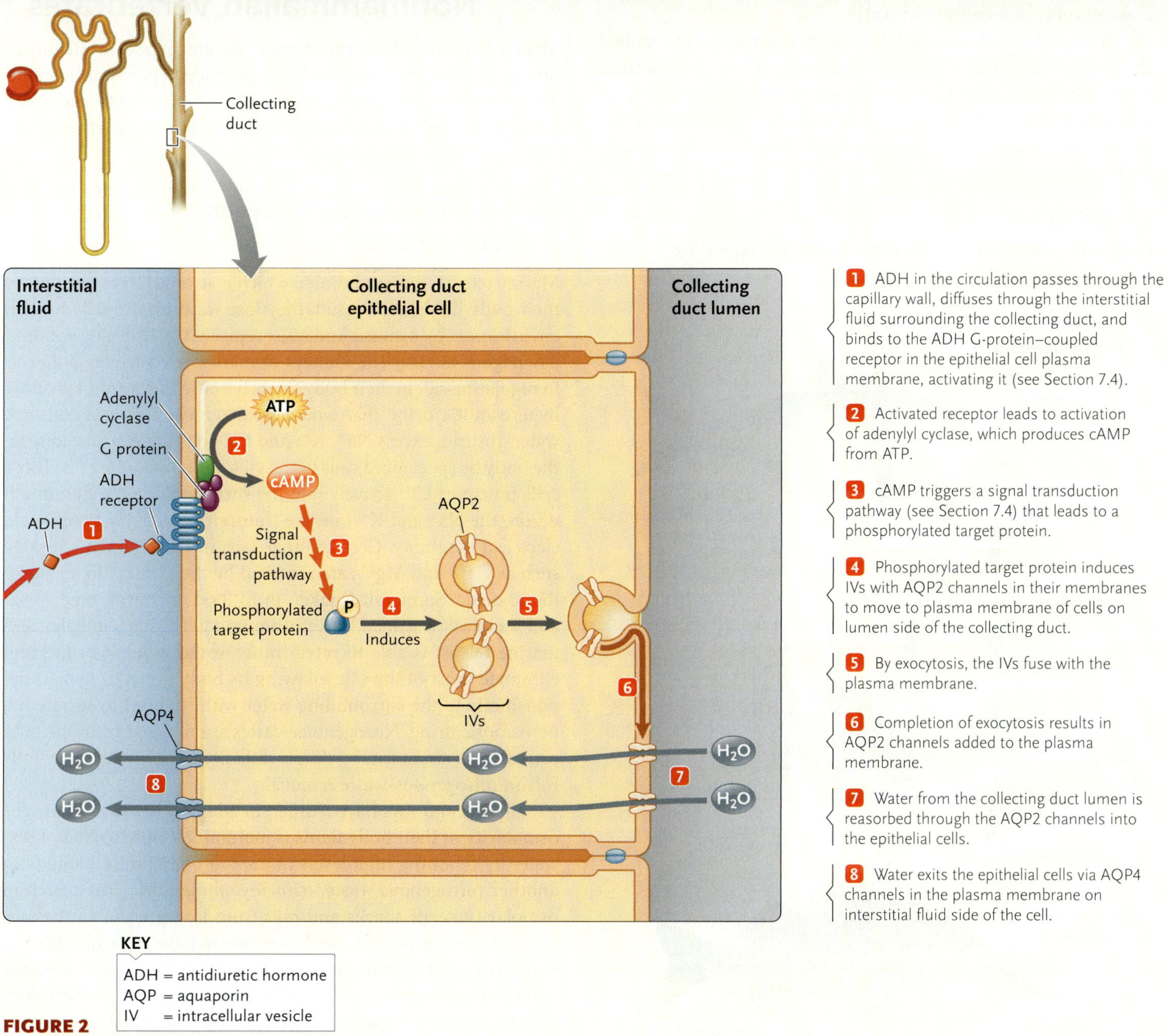

FIGURE 2

THINK LIKE A SCIENTIST In diabetes insipidus, the kidneys are unable to conserve water. One heritable form of the disease, central diabetes insipidus, can be treated by desmopressin, a synthetic replacement for ADH. Another heritable form of the disease, nephrogenic diabetes insipidus, does not respond to treatment with desmopressin. Hypothesize in which components shown in **Figure 2** might have been affected by the mutations in the two forms of the disease.

Source: S. Nielsen et al. 1995. Vasopressin increases water permeability of kidney collecting duct by inducing translocation of aquaporin-CD water channels to plasma membrane. *Proceedings of the National Academy of Sciences, USA* 92:1013–1017.

the solutes back to normal concentration. If, instead, there is excess water in the extracellular fluids, the osmolarity of those fluids drops below normal levels. The osmoreceptors are not stimulated so ADH secretion is reduced and this leads to increased urinary output. In addition, thirst is suppressed, reducing water intake.

Mechanistically, the action of ADH increases water reabsorption in the distal convoluted tubule and collecting ducts by promoting the insertion of more aquaporins into the epithelial membranes **(Figure 48.12).** Without ADH, water channels are retrieved from the distal convoluted tubule and collecting duct epithelial membranes (endocytosis of plasma membrane to produce intracellular vesicles; see Figure 48.12), and water reabsorption is reduced significantly.

Although the RAAS and ADH systems interact to regulate the body's water balance over a wide range of conditions, their regulatory mechanisms cannot compensate for water losses for more than a few days if water is unavailable. Dehydration becomes fatal when water loss amounts to about 12% of the normal fluid volume of the body.

STUDY BREAK 48.4

Outline the roles of the RAAS and ADH system in regulating mammalian kidney function.

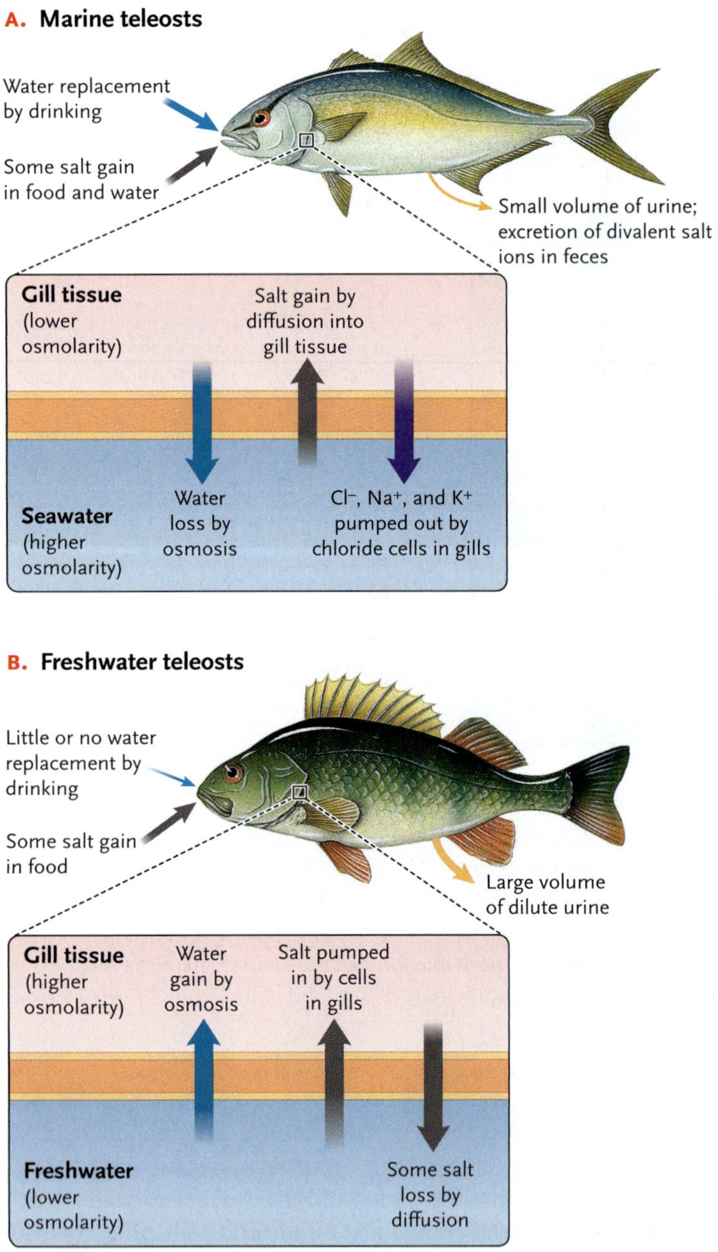

FIGURE 48.13 The mechanisms balancing the water and salt content of (A) marine teleosts and (B) freshwater teleosts.

48.5 Kidney Function in Nonmammalian Vertebrates

Most nonmammalian vertebrates secrete hypoosmotic urine; only a few species of aquatic birds produce hyperosmotic urine. The particular adaptations that maintain osmolarity and water balance among these animals vary depending on whether retention of water or salts is the major issue.

Marine Fishes Conserve Water and Excrete Salts

Marine teleosts live in seawater, which is strongly hyperosmotic to their body fluids. As a result, they lose water continually to their environment by osmosis and must replace it by drinking continually **(Figure 48.13A)**. The kidneys of marine teleosts play little role in regulating salt in their body fluids because they cannot produce hyperosmotic urine that would both remove salt and conserve water. Instead, excess Na^+, K^+, and Cl^- ions are eliminated from the body by specialized cells in the gills, called *chloride cells.* Those cells transport Cl^- actively into the surrounding seawater; in addition, the Na^+ and K^+ ions are transported actively to maintain electrical neutrality. Certain other ions in the ingested seawater, such as Ca^{2+} and Mg^{2+}, are removed by the kidneys in an isoosmotic urine. Some salt is taken in in food and swallowed water, and some salt is gained by diffusion into gill tissue. On balance, a marine teleost is able to retain most of the water it drinks and eliminate most of the salt, allowing its body fluids to remain hypoosmotic to the surrounding water with no need to secrete hyperosmotic urine. Nitrogenous wastes are released from the gills, primarily as ammonia, by simple diffusion. The kidneys play little role in nitrogenous-waste removal.

Sharks and rays have a different adaptation to seawater. The osmolarity of their body fluids is maintained close to that of seawater by retaining high levels of urea in body fluids, along with another nitrogenous waste, *trimethylamine oxide.* The match in osmolarity keeps sharks and rays from losing water to the surrounding sea by osmosis, and they do not have to drink seawater continually to maintain their water balance. Excess salts ingested with food are excreted in the kidney and by specialized secretory cells in a *rectal salt gland* located near the anal opening.

Freshwater Fishes and Amphibians Excrete Water and Conserve Salts

The body fluids of freshwater fishes and aquatic amphibians (no amphibians live in seawater) are hyperosmotic to the surrounding water, which usually ranges from about 1 to 10 mOsm/L. Water therefore moves osmotically into their tissues **(Figure 48.13B)**. Such animals rarely drink, and they excrete large volumes of dilute urine to get rid of excess water. In freshwater fishes, salt ions lost with the urine and by diffusion out of the gill tissue are replaced by salt in foods and by active transport of Na^+ and K^+ into the body by the gills; Cl^- follows to maintain electrical neutrality. Aquatic amphibians obtain salt in the diet and by active transport across the skin from the surrounding water. Nitrogenous wastes are excreted from the gills as ammonia in both freshwater fishes and aquatic amphibians.

Terrestrial amphibians must conserve both water and salt, which is obtained primarily in foods. In these animals, the kidneys secrete salt into the urine, causing water to enter the urine by osmosis. In the bladder, the salt is reclaimed by active transport and returned to body fluids. The water remains in the bladder, making the urine very dilute; during times of drought, it is reabsorbed as a water source. Terrestrial amphibians also have behavioral adaptations that help minimize water loss, such as seeking shaded, moist environments and remaining inactive during the day.

Larval amphibians, which are completely aquatic, excrete nitrogenous wastes from their gills as ammonia. Adult amphibians excrete nitrogenous wastes through their kidneys as urea.

Reptiles and Birds Excrete Uric Acid to Conserve Water

Terrestrial reptiles (lizards and snakes) conserve water by secreting nitrogenous wastes in the form of an almost water-free paste of uric acid crystals. Further water conservation occurs as the epithelial cells of the cloaca, the common exit for the digestive and excretory systems, absorb water from feces and urine before those wastes are excreted. Most birds conserve water by the same processes—they excrete nitrogenous wastes as uric acid and absorb water from the urine and feces in the cloaca. In reptiles, the scales covering the skin allow almost no water to escape through the body surface.

Reptiles and birds that live in or around seawater, including reptiles such as crocodilians, sea snakes, and sea turtles and birds such as seagulls, penguins, and pelicans, take in large quantities of salt with their food and rarely or never drink fresh water. These animals typically excrete excess salt through specialized *salt glands* located in the head **(Figure 48.14)**, which remove salts from the blood by active transport. The salts are secreted to the environment as a water solution in which salts are two to three times more concentrated than in body fluids. The secretion exits through the nostrils of birds and lizards, through the mouth of marine snakes, and as salty tears from the eye sockets of sea turtles and crocodilians. Neural and hormonal controls, essentially the same as those regulating osmolarity in mammals, control the rate of fluid secretion and its salt concentration.

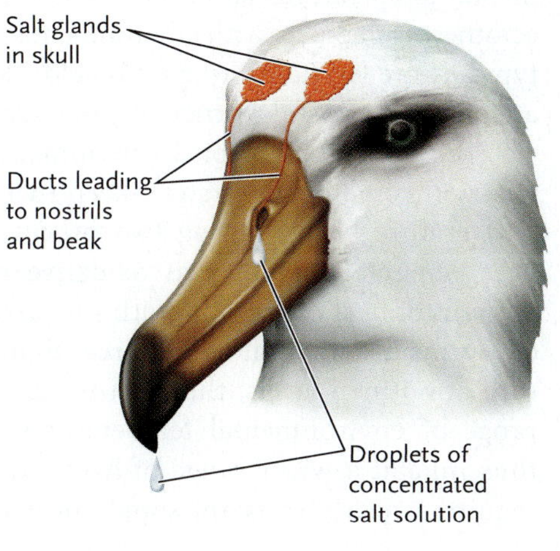

FIGURE 48.14 Salt glands in a bird living on a seacoast.

In sum, the adaptations described in this section permit excretion of toxic wastes and allow animals to maintain the salt concentration of body fluids at levels that keep cells from swelling or shrinking.

STUDY BREAK 48.5

1. **How do marine and freshwater teleosts differ in water, salt, and nitrogenous-waste regulation?**
2. **Reptiles and birds excrete nitrogenous wastes in the form of uric acid. Is there an advantage to doing this as opposed to the mammalian process of excreting nitrogenous wastes as urea?**

48.6 Introduction to Thermoregulation

Another vital challenge for animals is maintaining their internal environment at temperatures that can be tolerated by body cells. Animal cells can survive only within a temperature range from about 0°C to 45°C (32°F–113°F). Not far below 0°C, the lipid bilayer of a biological membrane changes from a fluid to a frozen gel, which disrupts vital cell functions, and ice crystals destroy the cell's organelles. As temperatures approach 45°C, most proteins and nucleic acids unfold from their functional form. Either condition leads quickly to cell death. Animals therefore usually maintain internal body temperatures somewhere within the 0°C to 45°C limits.

Temperature regulation—*thermoregulation*—is based on negative feedback pathways in which temperature receptors *(thermoreceptors)* detect changes from a temperature *set point*. Signals from the receptors trigger physiological and behavioral responses that return the temperature to the set point (thermoreceptors are discussed in Section 41.6; homeostatic control systems involving negative feedback mechanisms and set points are discussed in Section 38.4). All of the responses triggered by negative feedback mechanisms involve adjustments in the rate of heat-generating oxidative reactions within the body, coupled with adjustments in the rate of heat gain or loss at the body surface. The particular adaptations accomplishing these responses vary widely among species. And, although body temperature is regulated closely around a set point in all endotherms, the set point itself may vary over the course of a day and between seasons.

Thermoregulation Allows Animals to Reach Optimal Physiological Performance

Within the 0°C to 45°C range of tolerable temperatures, an animal's *organismal performance*—the rate and efficiency of its biochemical, physiological, and whole-body processes—varies

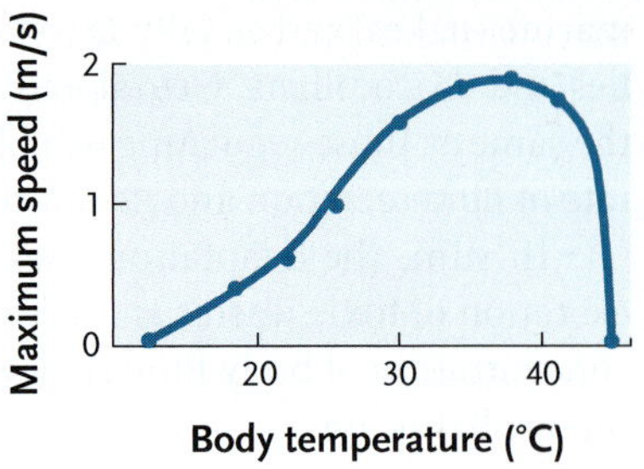

FIGURE 48.15 Body temperature and organismal performance. The maximum sprint speed of a lizard *(Agama stellio)* changes dramatically with body temperature.

greatly. For example, the speed at which the Middle Eastern lizard *Agama stellio* can sprint is low when the animal's body temperature is cold, rises smoothly with body temperature until it levels to a fairly broad plateau, and then drops off dramatically with further increases in body temperature **(Figure 48.15).** The temperature range that provides good organismal performance varies from one species to another.

Animals Exchange Heat with Their Environments by Conduction, Convection, Radiation, and Evaporation

As part of thermoregulation, animals exchange heat with their environments. As with all physical objects, heat flows into animals if they are cooler than their surroundings and flows outward if they are warmer. This heat exchange occurs by four mechanisms: *conduction, convection, radiation,* and *evaporation* **(Figure 48.16).**

Conduction is the flow of heat between atoms or molecules in direct contact. An animal loses heat by conduction when it contacts a cooler object and gains heat when it contacts an object that is warmer. **Convection** is the transfer of heat from a body to a fluid, such as air or water, that passes over its surface. The movement maximizes heat transfer by replacing fluid that has absorbed or released heat with fluid at the original temperature. **Radiation** is the transfer of heat energy as electromagnetic radiation. Any object warmer than absolute zero (−273°C) radiates heat; as the object's temperature rises, the amount of heat it loses as radiation increases as well. Animals also gain heat through radiation, particularly by absorbing radiation from the sun. **Evaporation** is heat transfer through the energy required to change a liquid to a gas. Evaporation of water from a surface is an efficient way to transfer heat; when the water in sweat evaporates from the body surface, the body cools down because heat is being transferred to the evaporated water in the surrounding air.

FIGURE 48.16 Heat flow to (in red) and from (in blue) a runner on a hot, sunny day. Unlike conduction, convection, and evaporation, which take place through the kinetic movement of molecules, electromagnetic radiation is transmitted through space as waves of energy.

All animals gain or lose heat by a combination of these four mechanisms. If you run on a hot summer day, for example, you lose heat by the evaporation of sweat from the skin, by convection as air flows over the skin, and by outward infrared radiation. You gain heat from internal biochemical reactions (especially oxidations), by absorbing infrared and solar radiation, and by conduction as the feet contact the hot ground. To maintain a constant body temperature, the heat gained and lost through these pathways must balance.

Ectothermic and Endothermic Animals Rely on Different Heat Sources to Maintain Body Temperature

Different animals use one of two major strategies to balance heat gain and loss. Animals that obtain heat primarily from the external environment are **ectotherms** (*ecto* = outside); those obtaining most of their heat from internal physiological sources are **endotherms** (*endo* = inside). All ectotherms generate at least some heat from internal reactions, however, and endotherms can obtain heat from the environment under some circumstances.

Virtually all invertebrates, fishes, amphibians, and lizards and snakes are ectotherms. These animals are described popularly as cold-blooded, although the body temperature of some, such as an active lizard, may be as high as or higher than ours on a sunny day. Ectotherms regulate body temperature by controlling the rate of heat exchange with the environment. Through behavioral and physiological mechanisms, they adjust body temperature toward a level that allows optimal physiological performance. However, most ectotherms are unable to maintain optimal body temperature when the temperature of their surroundings departs too far from that optimum, particularly when environmental temperatures fall. As a result, the body temperatures of ectotherms fluctuate with environmental temperatures, and they typically are less active when it is cold. Nevertheless, ectotherms are highly successful, particularly in warm environments.

The endotherms—birds, mammals, some fishes, sea turtles, and some invertebrates—keep their bodies at an optimal temperature by regulating two processes: (1) the amount of heat generated by internal oxidative reactions; and (2) the amount of heat exchanged with the environment. Because endotherms use internal heat sources to maintain body temperature at optimal levels, they can remain active over a broader range of environmental temperatures than ectotherms, and thus inhabit a wider range of habitats. However, endotherms require a nearly constant supply of energy to maintain their body temperatures. Because that energy is provided by food, endotherms typically consume much more food than ectotherms of equivalent size.

The difference between ectotherms and endotherms is reflected in their metabolic responses to environmental temperature **(Figure 48.17).** For example, the metabolic rate of a resting mouse *increases* steadily as the environmental temperature falls from 25°C to 10°C (77°F–50°F). This increase reflects the fact that to maintain a constant body temperature in a colder environment, endotherms must process progressively more food and generate more heat to compensate for their increased rate of heat loss. In this respect, an endotherm can be likened to a house in winter. To maintain a constant internal temperature, the homeowner must burn more oil or gas on a cold day than on a warm day.

By contrast, the metabolic rate of a resting lizard typically *decreases* steadily over the same temperature range. Because ectotherms do not maintain a constant body temperature, their biochemical and physiological functions, including oxidative reactions, slow down as environmental and body temperatures decrease. Thus, an ectotherm consumes and uses less energy when it is cold than when it is warm. This difference between ectotherms and endotherms is so fundamental that even samples of living tissue extracted from an ectotherm consume energy more slowly than equivalent samples from an endotherm.

Ectothermy and endothermy represent different strategies for coping with the variations in environmental temperature that all animals encounter; neither strategy is inherently superior to the other. Endotherms can remain fully active over a wide temperature range. Cold weather does not prevent them from foraging, mating, or escaping from predators, but it does increase their energy and food needs—and, to satisfy their need for food, they may not have the option of staying curled up safely in a warm burrow. Ectotherms do not have the capacity to be active when environmental temperatures drop too low; they move sluggishly and are unable to capture food or escape from predators. However, because their metabolic rates are lower under such circumstances, so are their food needs, and they do not have to look for food and expose themselves to danger to the extent that endotherms do.

In the next two sections, we examine in more detail how ectothermic and endothermic animals regulate their body temperatures.

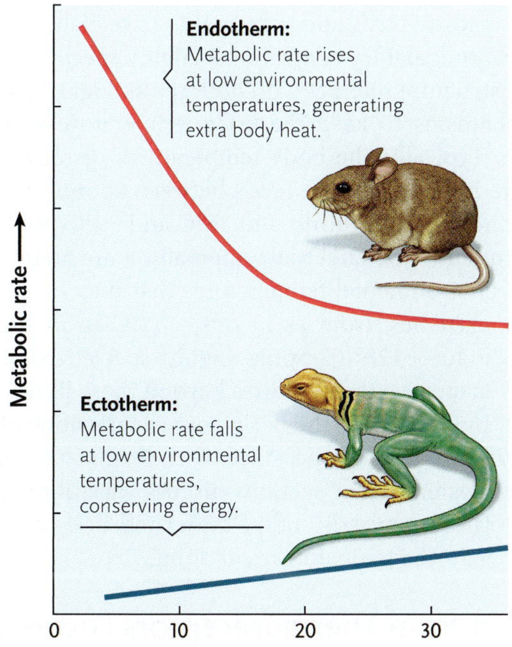

FIGURE 48.17 Metabolic responses of ectotherms and endotherms to cooling environmental temperatures. At any temperature, the metabolic rates of endotherms are always higher than those of ectotherms of comparable size.

STUDY BREAK 48.6

Distinguish between ectothermy and endothermy. Give one advantage and one disadvantage for each form of thermoregulation.

48.7 Ectothermy

Ectotherms vary widely in their ability to regulate internal body temperatures. For example, most aquatic invertebrates have such limited ability for thermoregulation that their body temperatures closely match those of the surrounding environment. These species live in or seek warm or temperate environments, where temperatures fall within a range that produces optimal physiological performance. Ectotherms with a greater ability for thermoregulation may occupy more varied habitats.

Ectotherms Are Found in All Invertebrate Groups

Most aquatic invertebrates are limited thermoregulators whose body temperature closely follows the temperature of their surroundings. However, even among these animals, some use behavioral responses to regulate body temperature. For example, a South American intertidal mollusk, *Echinolittorina peruviana,* is longer than it is wide. This animal orients itself as a means of thermoregulation. On sunny, summer days, it faces the sun, offering a smaller surface area for the sun's rays. On overcast summer days, or during the winter, it orients itself with a lateral side—which has the larger surface area—toward the sun's rays.

Invertebrates living in terrestrial habitats regulate body temperatures more closely. Many also use behavioral responses, such as moving between shaded and sunny regions, to regulate body temperature. Some winged arthropods, including bees, moths, butterflies, and dragonflies, use a combination of behavioral and heat-generating physiological mechanisms for thermoregulation. For example, in cool weather, these animals warm up before taking flight by vibrating the large flight muscles in the thorax, in a mechanism similar to shivering in humans.

Most Fishes, Amphibians, and Reptiles Are Ectotherms

Vertebrate ectotherms—fishes, amphibians, and reptiles (turtles, lizards, and snakes)—also vary widely in their ability for thermoregulation. Most aquatic species have a more limited thermoregulatory capacity than that found among terrestrial species, particularly the reptiles. Some fishes, however, are highly capable thermoregulators.

FISHES The body temperatures of most fishes remain within one or two degrees of their aquatic environment. However, many fishes use behavioral mechanisms to keep body temperatures at levels allowing good physiological performance. Freshwater species, for example, may use opportunities provided by the thermal stratification of lakes and ponds (see Figure 51.24). During hot summer days, they remain in deep, cool water, moving to the shallows to feed only during early morning and late evening when air and water temperatures fall.

AMPHIBIANS AND REPTILES The body temperatures of most amphibians also closely match environmental temperatures. Some terrestrial amphibians bask in the sun to raise their body temperature and seek shade to lower body temperature. However, basking can be dangerous to amphibians because they lose water rapidly through their permeable skin.

Thermoregulation is more pronounced among terrestrial reptiles. Some lizard species can maintain temperatures that are nearly as constant as those of endotherms. For small lizards, the most common behavioral thermoregulatory mechanism is shuttling between sunny (warmer) and shady (cooler) regions; in the deserts, lizards and other reptiles retreat into burrows during the hottest part of summer days. Some, such as the desert iguana *(Dipsosaurus dorsalis)*, lose excess heat by *panting*—rapidly moving air in and out of the airways. The air movement increases heat loss by convection and by evaporation of water from the respiratory tract.

Lizards also frequently adjust their posture to foster heat exchange with the environment and control the angle of their body relative to the rays of the sun. Snakes and lizards can often be found on large rocks and on roads on chilly nights, taking advantage of the heat retained by the stone or concrete.

Ectotherms Can Compensate for Seasonal Variations in Environmental Temperature

Many ectotherms undergo physiological changes, called **thermal acclimatization,** in response to seasonal shifts in environmental temperature. These changes allow the animals to attain good physiological performance at both winter and summer temperatures.

For example, in the summer a bullhead catfish (*Ameiurus* species) can survive water temperatures as high as 36°C (97°F), but it cannot tolerate temperatures below 8°C (46°F). In the winter, however, the bullhead cannot survive water temperatures above 28°C (82°F), but can tolerate temperatures near 0°C (32°F).

Another acclimatizing change involves the phospholipids of biological membranes (see *Focus on Basic Research* in Chapter 6). For example, membrane phospholipids have higher proportions of double bonds in carp living in colder environments than in carp living in warmer environments. The higher proportion of double bonds makes it harder for the membrane to freeze. A higher proportion of cholesterol also protects membranes from freezing.

When seasonal temperatures fall below 0°C, some ectotherms add molecules to their body fluids that act as antifreeze molecules to depress their freezing point and retard ice crystal formation. For example, antifreeze proteins allow fishes such as the winter flounder to remain active in seawater as cold as −1.8°C (−9°F) (see *Insights from the Molecular Revolution* in Chapter 51).

Ectotherms thus control body temperature primarily by regulating heat exchange with the environment; internal-heat generating mechanisms contribute to the control mechanisms in some species, but are never the primary source of body heat. The opposite conditions occur among endotherms: although these animals also regulate heat exchange with the environment, their primary sources of body heat are internal.

STUDY BREAK 48.7

1. **Describe two mechanisms an ectothermic animal can use to regulate its temperature.**
2. **What is thermal acclimatization?**

48.8 Endothermy

Endotherms—mostly birds and mammals—have elaborate and extensive thermoregulatory adaptations. Highly specialized features of body structure interact with both physiological and behavioral mechanisms to keep the body temperature within a narrow range. Typically, the body temperatures of fully active individuals are held constant at levels between about 39°C and 42°C (102°F and 108°F) in birds, and 36°C and 39°C (97°F and 102°F) in mammals. These internal temperatures are maintained in the face of environmental temperatures that may range over much greater extremes, from as low as −42°C to as high as +48°C (−45°F to +120°F). Some highly specialized endotherms can even survive temperatures beyond these limits.

We begin by describing the basic homeostatic control systems that maintain body temperature, with primary emphasis on the human system. Later sections discuss variations in the responses of other mammals and of birds, and daily and seasonal variations in the temperature set point.

Information from Thermoreceptors Located in the Skin and Internal Structures Is Integrated in the Hypothalamus

Thermoreceptors (discussed in Section 41.6) are found in various locations in the human body, including the **integument**

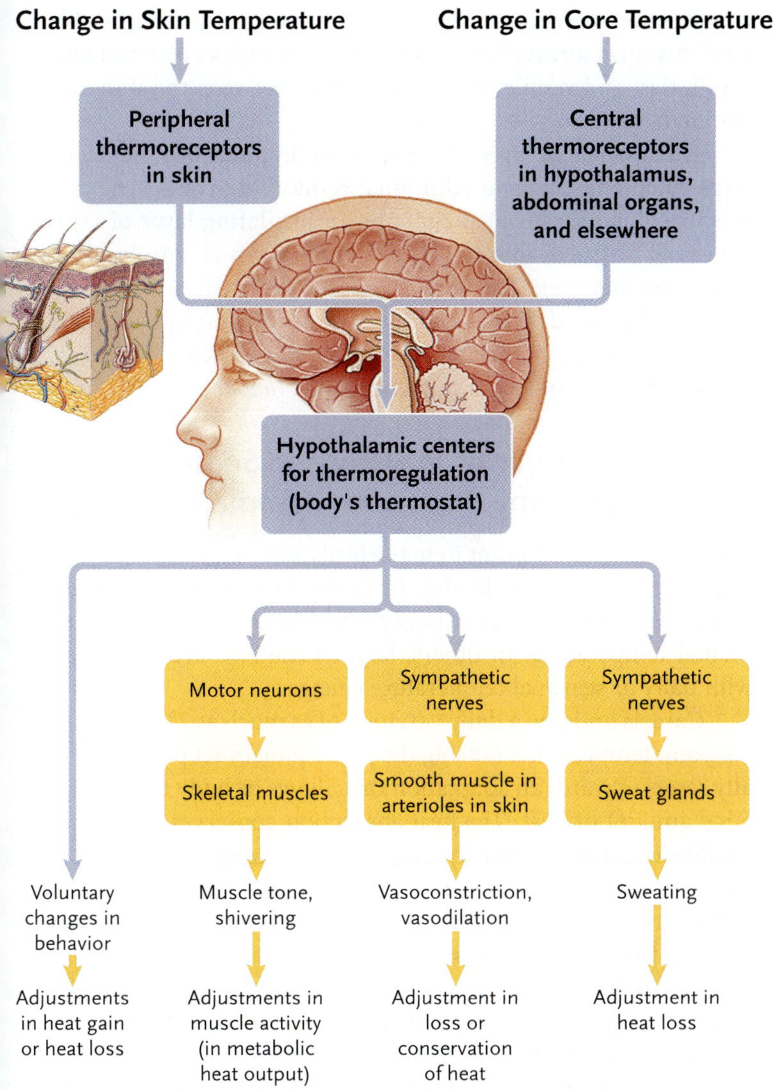

FIGURE 48.18 **The physiological and behavioral responses of humans and other mammals to changes in skin and core temperature.**

(skin; introduced in Chapter 38), spinal cord, and hypothalamus. Current evidence suggests the existence of five thermoreceptor types, each active in specific temperature ranges, beginning at less than 8°C and extending to above 52°C. Signals from the thermoreceptors are integrated in the hypothalamus and other regions of the brain to bring about compensating physiological and behavioral responses **(Figure 48.18).** The responses keep body temperature close to the set point, which varies normally in humans between 35.5°C and 37.7°C (96.0°F and 99.9°F) for the head and trunk. The appendages typically vary more widely in temperature; in freezing weather, for example, our arms, hands, legs, and feet are typically cooler in temperature than the body core—and the ears and nose especially so.

The hypothalamus was identified as a major thermoreceptor and response integrator in mammals by experiments in which various regions of the brain were heated or cooled with a temperature probe. Within the brain, only the hypothalamus produced thermoregulatory responses such as shivering or panting.

RESPONSES WHEN CORE TEMPERATURE FALLS BELOW THE SET POINT When thermoreceptors signal a fall in core temperature below the set point, the hypothalamus triggers compensating responses by sending signals through the autonomic nervous system. Among the immediate responses is constriction of the arterioles in the skin (vasoconstriction), which reduces the flow of blood to capillary networks in the skin. The reduced flow cuts down the amount of heat delivered to the skin and lost from the body surface. The reduction in flow is most pronounced in the skin covering the extremities, where blood flow may be reduced by as much as 99% when core temperature falls.

Another immediate response is contraction of the smooth muscles erecting the hair shafts in mammals and feather shafts in birds, which traps air in pockets over the skin, reducing convective heat loss. The response is minimally effective in humans because hair is sparse on most parts of the body—it produces the goose bumps you experience when the weather gets chilly. However, in mammals with fur coats or in birds, erection of the hair or feather shafts significantly increases the thickness of the insulating layer that covers the skin.

Immediate behavioral responses triggered by a reduction in skin temperature also help reduce heat loss from the body. Mammals may reduce heat loss by moving to a warmer locale, curling into a ball, or huddling together.

If these immediate responses do not return body temperature to the set point, the hypothalamus triggers further responses, most notably the rhythmic tremors of skeletal muscle we know as shivering. The heat released by the muscle contractions and the oxidative reactions powering them can raise the total heat production of the body substantially. At the same time, the hypothalamus triggers secretion of epinephrine (from the adrenal medulla) and thyroid hormones (see Section 42.4). These hormones increase heat production by stimulating the oxidation of fats and other fuels. The generation of heat by oxidative mechanisms in nonmuscle tissue throughout the body is termed **nonshivering thermogenesis.**

In human newborn babies and many other mammals, the most intense heat generation by nonshivering thermogenesis takes place in a specialized **brown adipose tissue** (also called **brown fat**) that can produce heat rapidly. Heat is generated by a mechanism that uncouples electron transport from ATP production in mitochondria (see Section 8.4); the heat is transferred throughout the body by the blood. Animals that hibernate or are active in cold regions, as well as the young of many others, contain brown adipose tissue. In most mammals, brown adipose tissue is concentrated between the shoulders in the back and around the neck. In human newborn babies, this tissue accounts for about 5% of body weight. Typically the tissue shrinks as humans age, until it is absent or essentially so in most adults. However, if exposure to cold is ongoing, as is the case with male Finlanders who work outside during the year, significant amounts of the tissue remain.

If none of these responses succeeds in raising body temperature to the set point, the result is **hypothermia,** a condition in which the core temperature falls below normal for a prolonged period. In humans, a drop in core temperature of only a few degrees affects brain function and leads to confusion; continued hypothermia can lead to coma and death (see *Why It Matters* in Chapter 38).

RESPONSES WHEN CORE TEMPERATURE RISES ABOVE THE SET POINT When core temperature rises above the set point, the hypothalamus sends signals through the autonomic system that trigger responses lowering body temperature (see Figure 38.10). As an immediate response, the signals relax smooth muscles of arterioles in the skin (vasodilation), increasing blood flow and with it, the heat lost from the body surface. In addition, in humans and other mammals with sweat glands, signals from the hypothalamus trigger the secretion of sweat, which absorbs heat as it evaporates from the surface of the skin.

Some endotherms, including dogs (which have sweat glands only on their feet) and many birds (which have no sweat glands), use panting as a major way to release heat (see Figure 38.10). These physiological changes are reinforced by behavioral responses such as seeking shade or a cool burrow, plunging into cold water, or taking a cold drink.

When the heat gain of the body is too great to be counteracted by these responses, **hyperthermia** results. An increase of only a few degrees above normal for a prolonged period is enough to disrupt vital biochemical reactions and damage brain cells. Most adult humans become unconscious if their body temperature reaches 41°C (106°F) and die if it goes above 43°C (110°F) for more than a few minutes.

The Skin Is an Organ of Heat Transfer in Birds and Mammals

The skin of birds and mammals is an organ of heat transfer. The arterioles delivering blood to the capillary networks of the skin constrict or dilate to control blood flow and, with it, the amount of heat transferred from the body core to the surface.

The outermost living tissue of human skin, the **epidermis,** consists of cells that grow and divide rapidly **(Figure 48.19),** becoming packed with fibers of a highly insoluble protein, *keratin* (see Section 5.3). When fully formed, the epidermal cells die and become compacted into a tough, nearly impermeable layer that limits water loss primarily to evaporation of the fluids secreted by the sweat glands.

The sweat glands and hair follicles are embedded in the layer below the epidermis. Called the **dermis,** it is packed with connective tissue fibers such as collagen, which resist compression, tearing, or puncture of the skin. The dermis also contains thermoreceptors and the dense networks of arterioles, capillaries, and venules that transfer heat between the skin and the environment.

The innermost layer of the skin, the **hypodermis,** contains larger blood vessels and additional reinforcing connective tissue. The hypodermis also contains an insulating layer of fatty tissue below the dermal capillary network, which ensures that heat flows between the body core and the surface primarily through the blood. The insulating layer is thickest in mammals that live in cold environments, such as whales, seals, walruses, and polar bears, in which it is known as *blubber.*

The Set Point Varies in Daily and Seasonal Rhythms in Many Birds and Mammals

The temperature set point in many birds and mammals varies in a regular cycle during the day. In some, the daily variations are relatively small and not obviously keyed to changes in environmental temperature. In others, larger variations are correlated with daily or seasonal temperature changes.

Camels undergo a daily variation of as much as 7°C (13°F) in set-point temperature. During the day, a camel's set point gradually resets upward, an adaptation that allows its body to absorb a large amount of heat. The heat absorption conserves water that would otherwise be lost by evaporation to keep the body at a lower set point. At night, when the desert is cooler, the thermostat resets again, allowing the body temperature to cool several degrees, releasing the excess heat absorbed during the day.

When the environmental temperature is cool, having a lowered temperature set point greatly reduces the energy required to maintain body temperature. In many animals, the lowered set point is accompanied by reductions in metabolic, nervous, and physical activity (including slower respiration and heartbeat), producing a sleeplike state known as **torpor.**

Many small animals and birds enter into **daily torpor,** a period of inactivity keyed to variations in daily temperature. Typically these animals expend more energy per unit of body

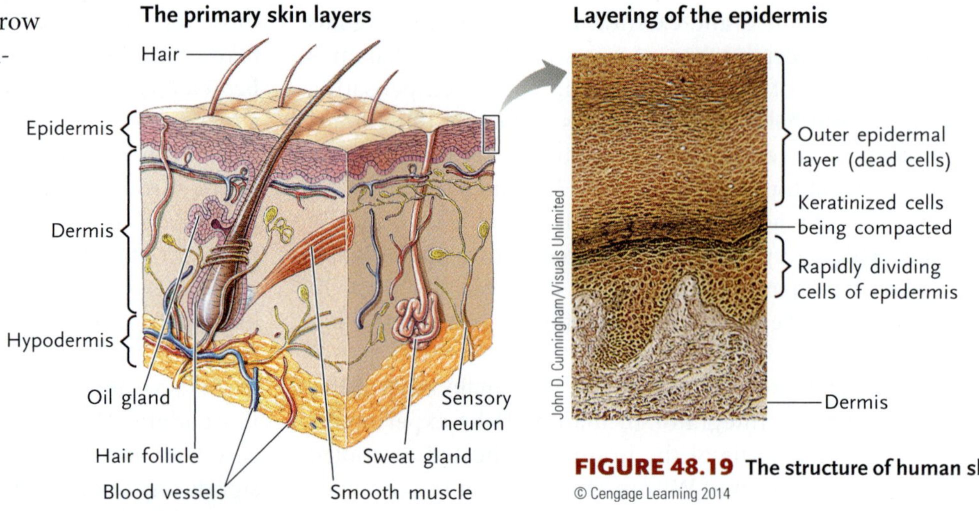

FIGURE 48.19 The structure of human skin.

weight to keep warm than larger animals, because the ratio of body surface to volume increases as body size decreases. Hummingbirds, for example, feed actively during the daytime, when their set point is close to 40°C (104°F). During the cool of night, however, the set point drops to as low as 13°C (55°F); this allows the birds to conserve enough energy to survive overnight without feeding. Some nocturnal animals, including bats and small rodents such as deer mice, become torpid in cool locations during daylight hours when they do not actively feed. At night, their temperature set point rises and they become fully active **(Figure 48.20)**.

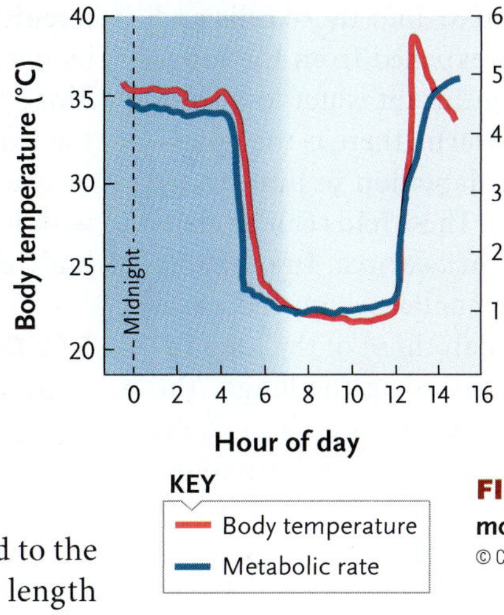

FIGURE 48.20 Cycle of daily torpor in a deer mouse *(Peromyscus maniculatus)*.

Many animals enter a prolonged state of torpor tied to the seasons, triggered in most cases by a change in day length that signals the transition between summer and winter. The importance of day length has been shown by laboratory experiments in which animals have been induced to enter seasonal torpor by changing the period of artificial light to match the winter or summer day length.

Extended torpor during winter, called **hibernation** (*hibernus* = relating to winter), greatly reduces metabolic expenditures when food is unobtainable. Typically, hibernators must store large quantities of fats to serve as energy reserves. The drop in body temperature during hibernation varies with the mammal. In some, such as hedgehogs, woodchucks, and squirrels, body temperature may fall by 20°C (36°F) or more. Body temperature even drops to near 0°C in some small hibernating mammals and, in the Arctic ground squirrel, the body supercools (goes to a below-freezing, unfrozen state) during hibernation, with body temperature dropping to about −3°C. Some ectotherms, including amphibians and reptiles living in northern latitudes and even some insects, also become torpid during winter.

Some mammals enter seasonal torpor during summer, called **estivation** (*aestivus* = relating to summer), when environmental temperatures are high and water is scarce. Some ground squirrels, for example, remain inactive in the cooler temperatures of their burrows during extreme summer heat. Many ectotherms, among them land snails, lungfishes, many toads and frogs, and some desert-living lizards, weather such climates by digging into the soil and entering a state of estivation that lasts throughout the hot dry season.

Some Animals Use Countercurrent Heat Exchangers to Retain Core Heat

Recall from Section 46.2 that many animals with internal gills use countercurrent exchange to maximize the amounts of O_2 and CO_2 exchanged with water. A countercurrent exchange system can also be used for heat exchange. **Figure 48.21** illustrates one type of *countercurrent heat exchanger.* Here, a set of arteries and veins is packed closely together in a region between the core and the periphery of the body, typically an extremity (one artery and one vein are shown in Figure 48.21). Warm blood from the core flows through arteries toward cold peripheral tissues and organs. In the countercurrent heat exchanger, that warm arterial blood flows next to cold blood from the body periphery that is in the veins. By conduction, heat moves from the artery to the vein and the now-warmer venous blood continues to the core.

Countercurrent heat exchanges of this type are found in many endotherms. In birds they are often in the legs, thereby limiting heat loss from the feet. In mammals, they are also often in the limbs; for example, in a dophin's fluke and flippers, and in the legs of an arctic wolf.

Another type of countercurrent exchange that helps to limit heat loss takes place in the nasal passages of birds and mammals. The air in the lungs of these animals is at 100% relative humidity because the cells of the lung capillaries must be

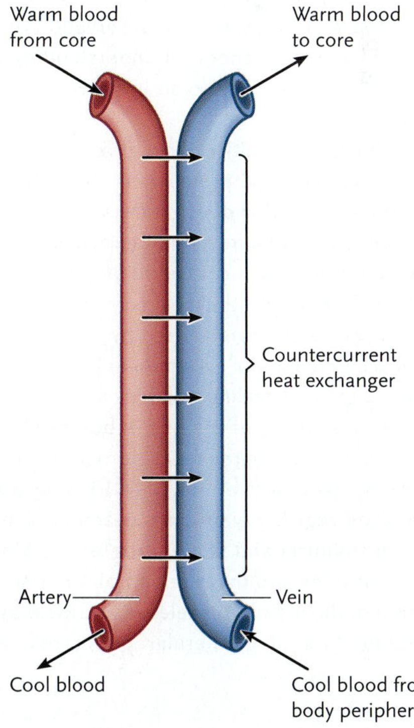

FIGURE 48.21 A countercurrent heat exchanger for conserving core heat.

moist to allow oxygen breathed in to pass into those cells and on into the pulmonary veins. The air expelled from the lungs therefore has the potential to cause significant water loss from the body. And, because the lung air is warm, there is the potential also for significant heat loss. An adaptation to limit water and heat loss is folds in the nasal cavity. These folds can be elaborate and represent a relatively large surface area. In outgoing breaths, some of the water in the air condenses on the nasal folds because the surfaces are cooler than those of the lung interior. Inhaled air is drier than the air leaving the lungs. The inhaled air evaporates some of the water on the nasal folds, which cools the folds for the next outgoing breath. This countercurrent exchange system involves opposing fluids separated by time rather than space as in the system discussed previously. This system is important for thermoregulation because it limits the loss of heat in expelled air. By this mechanism, mammals on average reclaim up to 45% of the water in expelled air, along with its heat content. Kangaroo rats reclaim up to 88% of their body water in this fashion.

STUDY BREAK 48.8

Describe how thermoreceptors and negative feedback pathways achieve temperature regulation in endotherms.

UNANSWERED QUESTIONS

How can we use genetics to understand how the kidney works in humans?

A variety of disorders of kidney function are inherited as Mendelian traits. For example, the disorder of renal growth known as polycystic kidney disease was recognized as a familial phenotype well before the structure of DNA was defined. Many other complex and common renal phenotypes (for example, blood pressure and susceptibility to the development of decreased kidney filtration function) also have genetic components.

The use of molecular cloning methods in the 1980s and 1990s led to enormous growth in our understanding of the specific transport proteins used by the kidney to fine-tune the extracellular environment. These discoveries were quickly followed by the elucidation of a large number of rare inherited disorders caused by mutations in the genes encoding many of these proteins. Careful observation of these human phenotypes, coupled with a biochemical understanding of the effects of these mutations, has in turn led to an improved understanding of basic renal physiology.

In the urine produced by a healthy kidney, protein is normally present only in very small amounts; little protein gets through the glomerular filtration barrier, and most of what does get through is reabsorbed in the renal tubule. By contrast, a large fraction of human kidney diseases are characterized by protein in the urine, a condition called proteinuria. Diabetic kidney disease, for example, the most common form of kidney disease in the Western world, is characterized by increased urine protein and the abnormal accumulation of extracellular protein in the renal glomerulus, leading to a gradual decline in kidney function.

For the past decade, most of my own laboratory's research has focused on inherited diseases of the renal glomerulus, using methods of molecular cloning (see Section 18.1) to identify disease genes. Mutations in the actin-regulatory proteins α-actinin-4 and INF2 both cause late onset proteinuria and slowly progressive kidney dysfunction. Based on our studies and the research of other labs, we believe that fine regulation of the actin cytoskeleton is critical to maintaining the complex architecture of glomerular podocytes, a specialized epithelial cell that forms the final part of the glomerular filter. When this architecture is disturbed, the podocytes become more easily injured.

Large amounts of protein (on the order of 3 g/day) are the central feature of a serious medical condition known as the nephrotic syndrome. By studying very early onset forms of the nephrotic syndrome, genes for proteins that are unique to the podocyte cell–cell junction have been shown to be mutated in these disorders. For example, the cell surface protein nephrin has been shown to play important structural and signaling functions. Despite the growing list of genes known to be important to normal filtering of the blood, much work needs to be done before we can answer the question of how the products of these genes interact to produce a functioning kidney.

Humans have many disadvantages as experimental organisms. We cannot do invasive experiments on them, we cannot do planned matings, and they take a long time to grow. Nonetheless, the careful study of renal phenotypes in humans followed by genetic analysis has led to a much better understanding of kidney disease and kidney function. It is important both for medicine and for understanding the kidney to use the information gained through genetic studies to inform biology.

THINK LIKE A SCIENTIST Kidney transplantation is a standard form of treatment for kidney failure. In some instances, the kidney disease that affected the original kidneys returns to involve the transplanted kidney. What does this say about the mechanism of the disease? What would this say about the mechanism of an inherited form of kidney disease?

Martin Pollak

Martin Pollak is Chief of Nephrology at the Beth Israel Deaconess Medical Center in Boston and Associate Professor of Medicine at Harvard Medical School. The focus of his laboratory is to learn more about the causes of kidney disease in patients and families by studying genetics. To learn more, visit http://www.fsgs.bwh.harvard.edu/index.html.

REVIEW KEY CONCEPTS

To access the course materials and companion resources for this text, please visit www.cengagebrain.com.

48.1 Introduction to Osmoregulation and Excretion

- Solute concentration is measured as osmolarity in milliosmoles per liter of solution (mOsm/L). A solution can be comparatively hyperosmotic, hypoosmotic, or isoosmotic to another solution. Water moving from a region of higher osmolarity to a region of lower osmolarity across a selectively permeable membrane occurs by osmosis.
- Osmoconformers allow the osmolarity of their body fluids to match that of the environment. Osmoregulators keep the osmolarity of body fluids different from that of the environment.
- Molecules and ions must be removed from the body to keep cellular and extracellular fluids isoosmotic. In most animals, ECF is filtered through tubules formed from a transport epithelium and released to the exterior of the animal as urine (Figure 48.2).
- Nitrogenous wastes are excreted as ammonia, urea, or uric acid, or as a combination of these substances (Figure 48.3).

48.2 Osmoregulation and Excretion in Invertebrates

- Most marine invertebrates are osmoconformers; they expend little or no energy on maintaining water balance. Freshwater and terrestrial invertebrates are osmoregulators. They must expend energy to excrete water that moves into their cells by osmosis.
- The cells of the simplest marine invertebrates exchange water and solutes directly with the surrounding seawater. More complex invertebrates have specialized excretory tubules (Figures 48.4–48.6).

48.3 Osmoregulation and Excretion in Mammals

- In mammals and other vertebrates, excretory tubules are concentrated in the kidney.
- The mammalian excretory tubule, the nephron, has a proximal end at which filtration takes place, a middle region in which reabsorption and secretion occur, and a distal end that releases urine. A network of capillaries surrounding the nephron takes up ions and water and other molecules absorbed by the nephron. The urine leaving individual nephrons is processed further in collecting ducts and then pools in the renal pelvis. From there it flows through the ureter to the urinary bladder, and through the urethra to the exterior of the animal (Figure 48.7).
- Glomerular filtration, a process in plasma of the blood filters through the glomerular capillaries into Bowman's capsule, is the first step of excretion. By the process of tubular reabsorption in the proximal convoluted tubule and loop of Henle, substances of value (Na^+, Cl^-, K^+, water, HCO_3^-, and nutrients) are returned to the plasma as the filtrate flows through the nephron tubules. By the process of tubular secretion, selective movement of substances (such as detoxified poisons) takes place from the blood of the peritubular capillaries. Subsequently, the concentration of salts is balanced between the urine and the interstitial fluid surrounding the nephron, and more water is reabsorbed (Figure 48.8 and Table 48.1).

Animation: Human urinary system

Animation: Urine Formation

Animation: Tubular reabsorption

48.4 Regulation of Mammalian Kidney Function

- The kidney's autoregulation system responds rapidly to small variations in blood pressure so as to keep the glomerular filtration rate constant. The autoregulation system involves: (1) vascular control of blood pressure and (2) a feedback control mechanism between the tubule and specialized cells in the juxtaglomerular apparatus (Figure 48.10).
- When blood volume and blood pressure drop, the hormones of the renin-angiotensin-aldosterone system (RAAS) raise blood pressure by stimulating arteriole constriction and increasing NaCl reabsorption in the kidneys (Figure 48.10).
- ADH, which increases water reabsorption, is released from the pituitary when osmoreceptors detect an increase in the osmolarity of body fluids (Figures 48.11 and 48.12).

48.5 Kidney Function in Nonmammalian Vertebrates

- Marine teleosts continually drink seawater to replace body water lost by osmosis to their hyperosmotic environment. Excess salts and nitrogenous wastes are excreted by the gills (Figure 48.13A).
- The body fluids of sharks and rays are isoosmotic with seawater. They do not lose water by osmosis, and do not drink seawater. Excess salts are excreted in the kidney and by a rectal salt gland.
- Body fluids of freshwater fishes and amphibians are hyperosmotic to their environment, and these animals must excrete the excess water that enters by osmosis. Body salts are obtained from food and, in fishes, through the gills (Figure 46.13B). Nitrogenous wastes are excreted from the gills of fishes and larval amphibians as ammonia and through the kidneys of adult amphibians as urea.
- Reptiles and birds conserve water by secreting nitrogenous wastes as uric acid and by absorbing water from urine and feces in the cloaca.

Animation: Diffusion, osmosis, and countercurrent systems

48.6 Introduction to Thermoregulation

- Animals must maintain body temperature at a level that provides optimal physiological performance. Heat flows between animals and their environment by conduction, convection, radiation, and evaporation (Figures 48.15 and 48.16).
- Ectothermic animals obtain heat energy primarily from the environment; endothermic animals obtain heat energy primarily from internal reactions (Figure 48.17).

Animation: Control of human body temperature

48.7 Ectothermy

- Ectotherms obtain heat energy externally and control body temperature primarily by physiological or behavioral methods of regulating heat exchange with the environment.
- Many animals undergo thermal acclimatization, a structural or metabolic change in the limits of tolerable temperatures as the environment alternates between warm and cool seasons.

48.8 Endothermy

- Endotherms obtain heat energy primarily from internal reactions and maintain body temperature over a narrow range by balancing internal heat production against heat loss from the body surface.
- Internal heat production is controlled by negative feedback pathways triggered by thermoreceptors. When deviations from the temperature set point occur, signals from the receptors bring about compensating responses such as changes in blood flow to

the body surface, sweating or panting, and behavioral modifications (Figure 48.18).

- The skin of endotherms is water-impermeable, reducing heat lost by direct evaporation of body fluids. The blood vessels of the skin regulate heat loss by constricting or dilating. A layer of insulating fatty tissue under the vessels limits losses to the heat carried by the blood. The hair of mammals and feathers of birds also insulate the skin. Erection of the hair or feathers reduces heat loss by thickening the insulating layer (Figure 48.19).
- The temperature set point in many birds and mammals varies in daily and seasonal patterns. During cooler conditions, a lowered set point is accompanied by torpor (Figure 48.20).
- Some animals use countercurrent heat exchangers to retain core heat. One form of exchanger is a set of arteries and veins arrayed close together in a region between the core and the body periphery. The opposing directions of fluid flow in the vessels allows heat to move from the warm blood in the arteries to warm up the cold blood in the veins which is heading toward the core. Another form of exchanger involves the nasal folds in birds and mammals. The folds provide an exchange mechanism to conserve body water and its heat content (Figure 48.21).

Animation: Human thermoregulation

Animation: Structure of human skin

UNDERSTAND & APPLY

Test Your Knowledge

1. Which of the following statements about osmoregulation is true?
   a. In freshwater invertebrates, salts move out of the body into the water because the animal is hypoosmotic to the water.
   b. A marine teleost has to fight gaining water because it is isoosmotic to the sea.
   c. Most land animals are osmoconformers.
   d. Vertebrates are usually osmoregulators.
   e. Terrestrial animals can regulate their osmolarity without expending energy.
2. One role of tubules in excretion is to:
   a. absorb H^+ ions to buffer body fluids.
   b. transport proteins across transport epithelium.
   c. reabsorb glucose and amino acids.
   d. move toxic substances from the filtrate into the cells composing the transport tubules.
   e. filter by maintaining a lower pressure in the fluid outside the tubule than inside it.
3. Products of metabolism in humans, as in:
   a. terrestrial amphibians, can include urea, which requires more energy to produce than ammonia.
   b. birds and reptiles, can include uric acid, which is nontoxic and excreted as a paste.
   c. sharks, are primarily excreted as ammonia.
   d. hydra, must be isoosmotic with the water ingested.
   e. other mammals, cannot be water as water comes only from what they drink.
4. Filtration and/or excretion can be performed by:
   a. ciliated metanephridia in insects.
   b. protonephridia containing flame cells in flatworms.
   c. a nephron and bladder in insects.
   d. Malpighian tubules on the segments of earthworms.
   e. the hindgut, which reabsorbs Na^+ and K^+ into the hemolymph of earthworms.
5. A mammalian nephron contains the:
   a. Bowman's capsule, which delivers the filtrate to the glomerulus.
   b. Bowman's capsule, which filters fluids, 99% of which will be excreted.
   c. proximal convoluted tubule, which moves Na^+ and K^+ into the filtrate of the interstitial fluids.
   d. proximal convoluted tubule, which reabsorbs K^+, Na^+, Cl^-, and H_2O.
   e. proximal convoluted tubule, which lacks microvilli to ease fluid movement through it.
6. Which of the following correctly describes a part of kidney function?
   a. Collecting ducts dilute urine because they are permeable to salt but not water.
   b. In the ascending loop of Henle, Na^+ and Cl^- move into the tubules because the osmolarity of the filtrate is increased.
   c. The descending loop of Henle receives filtrate from the ascending loop.
   d. The distal convoluted tubule pumps water into the tubule by active transport.
   e. The renal pelvis receives urine from the collecting ducts and carries it to the ureters.
7. Which of the following is an example of autoregulation of kidney function?
   a. The RAAS regulates Na^+ by secreting renin when blood pressure or blood volume decreases.
   b. The ADH system regulates water balance by decreasing water reabsorption and increasing excretion of salt.
   c. Receptors in the juxtaglomerular apparatus detect a higher salt concentration in the distal convoluted tubule and trigger constriction of the afferent arteriole to reduce glomerular filtration rate.
   d. ANF is released by the kidney to increase renin release.
   e. Angiotensin II lowers blood pressure by constricting arterioles.
8. Deficient water levels in humans are prevented by:
   a. osmoreceptors in the hypothalamus that detect decreases in salt concentrations.
   b. the hypothalamus stimulating the posterior pituitary to secrete a hormone that allows the collecting ducts and distal convoluted tubules to be permeable to water.
   c. inhibiting ADH, which causes a rise in osmolarity of the ECF.
   d. producing dilute urine.
   e. drinking alcohol, which stimulates aldosterone to raise the osmolarity of body fluids.
9. Which best exemplifies ectothermy?
   a. The metabolic rate increases as the temperature decreases.
   b. Body temperature remains constant when environmental temperatures change.
   c. Food demand increases when temperatures drop.
   d. Virtually all invertebrate groups are ectotherms.
   e. No vertebrate groups are ectotherms.
10. Unique to endotherms is:
    a. torpor.
    b. thermal acclimatization.
    c. a nonchanging body temperature.
    d. response to seasonal temperature changes.
    e. thermoregulation by a hypothalamus.

Discuss the Concepts

1. A urinalysis reveals glucose, urea, hemoglobin, and sodium. Which of these substances are abnormal in urine, and why?
2. As a person ages, nephron tubules lose some of their ability to concentrate urine. What is the effect of this change?
3. Shivering increases air movement over the body surface. What effect does this air movement have on heat conservation in the shivering animal?
4. What heat transfer processes might account for the change in body temperature when a mammal's body temperature undergoes daily variations?

Design an Experiment

Design experiments to show the role of fluid consumption in thermoregulation during endurance exercise.

Interpret the Data

Products labeled as "organic" fill an increasing amount of space on supermarket shelves. What does that label mean? A food that carries the USDA's organic label must be produced without pesticides such as malathion and chlorpyrifos that conventional farmers typically use on fruits, vegetables, and grains.

Does eating organic food significantly affect the level of pesticide residues in a child's body? Chensheng Lu of Emory University used urine testing to find out. For 15 days, the urine of 23 children (ages 3–11) was monitored for breakdown products of pesticides. During the first 5 days (phase 1), children ate their standard, nonorganic diet. For the next 5 days (phase 2), they ate organic versions of the same types of foods and drinks. Then, for the final 5 days (phase 3), the children returned to their nonorganic diet. The results are shown in the **Table.**

Levels of Metabolites (Breakdown Products) of Malathion and Chlorpyrifos in the Urine of Children Taking Part in a Study of Effects of an Organic Diet*

		Malathion metabolite		Chlorpyrifos metabolite	
Study phase	No. of samples	Mean (μg/L)	Maximum (μg/L)	Mean (μg/L)	Maximum (μg/L)
1. Nonorganic	87	2.9	96.5	7.2	31.1
2. Organic	116	0.3	7.4	1.7	17.1
3. Nonorganic	156	4.4	263.1	5.8	25.3

*The difference in the mean level of metabolites in the organic and inorganic phases of the study was statistically significant.

1. During which phase of the experiment did the children's urine contain the lowest level of the malathion metabolite?
2. During which phase of the experiment was the maximum level of the chlorpyrifos metabolite detected?
3. Did switching to an organic diet lower the amount of pesticide excreted by children?

Source: C. Lu et al. 2006. Organic diets significantly lower children's dietary exposure to organophosphorus pesticides. *Environmental Health Perspectives* 114:260–263.

Apply Evolutionary Thinking

Humans produce urea as an excretion product, whereas reptiles and birds produce uric acid. Indeed, human kidneys are not as efficient as those of reptiles and birds. What does this mean in an evolutionary sense?

49

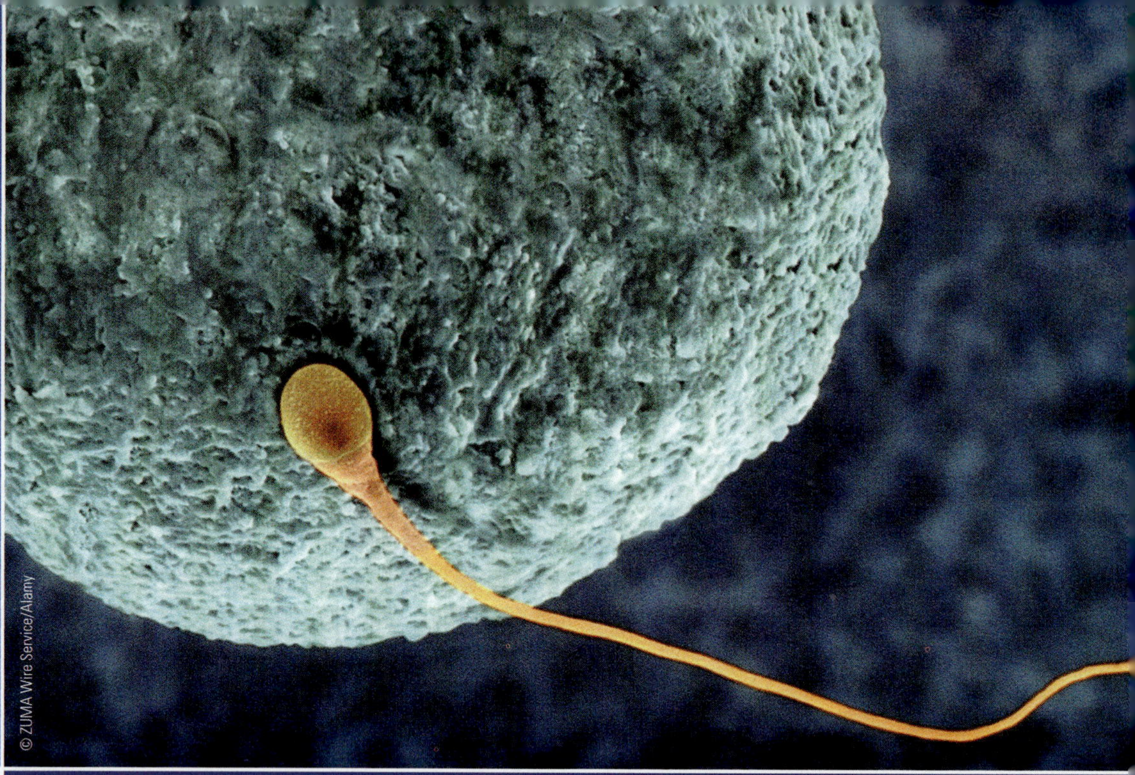

Fertilization, the union of gametes such as the sperm and egg shown here, initiates development of a new individual.

STUDY OUTLINE

49.1 Animal Reproductive Modes: Asexual and Sexual Reproduction

49.2 Cellular Mechanisms of Sexual Reproduction

49.3 Sexual Reproduction in Humans

49.4 Methods for Preventing Pregnancy: Contraception

Animal Reproduction

FIGURE 49.1 The palolo worm. Gametes are packed into segments of the tail section (in blue).

Why it matters . . . It is seven days after the October full moon and night is falling. All the inhabitants of the Samoan island of Tutuila who have access to a boat are gathered on the island's large lagoon. Some hold lanterns and look into the water; others have nets at the ready. They are awaiting the palolo worm *(Eunice viridis)* **(Figure 49.1),** which has appeared in the water as the moon rises on this same night of the lunar year for as long as the islanders can remember.

The moon rises and the excitement of the crowd rises. Then, there they are, thousands of blue and green worms, squirming in the water like animated spaghetti. The boaters scoop up the worms by the netful and dump them into buckets. When the buckets are full, the islanders glide toward the shore where steaming pots are waiting, for palolo worms are a delicacy that the islanders savor only once a year. For the islanders, a night of feasting, singing, and dancing follows as they cook and eat the worms.

The worms that come to the surface to delight the islanders are not complete individuals. They are tail sections about 10 to 20 cm long that break from adults after they become filled with eggs or sperm (see Figure 49.1). The adults are polychaete annelids that live in burrows in coral reefs of the Samoan and Fiji islands. These annelids develop tail segments once a year, just after the October full moon. On the seventh night following the full moon, the tails break off and swim to the surface, where—if Samoans do not net them first—they disintegrate and release eggs and sperm by the millions. The anterior ends of the worms, safe in their burrows, survive to produce tails the next year.

The story of the palolo worm concerns the *reproductive cycle* of an animal. There is a biological clock in the worms, timed by periods of moonlight, that precisely sets both the appearance of the mating swarm and, indirectly, the appearance of the islanders with their boats. Most animals show cycles in reproductive activity, and those cycles often are linked to the seasons.

The swarm of the palolo worms is only one of many adaptations that accomplish mating in animals. For animals that reproduce by eggs and sperm, the adaptations are as diverse as the number of species on Earth. This diversity allows individuals of the same species to find each other and unite eggs and sperm. Within the diversity, however, are underlying patterns that are shared by all animals.

Both the underlying patterns and the diversity of animal reproduction are the subjects of this chapter. We also discuss the development of eggs and sperm, and the union of egg and sperm that begins the development of a new individual. The next chapter continues with the events of development after eggs and sperm have united.

49.1 Animal Reproductive Modes: Asexual and Sexual Reproduction

Reproduction is part of a life cycle in which individuals grow, develop, and reproduce according to instructions encoded in DNA. Rather than survival of the individual, reproduction is the means of passing on the genes of an individual to new generations of the species. As such, it is among the most vital functions of living organisms.

Two basic modes of reproduction operate in the animal kingdom. In **asexual reproduction,** a single individual gives rise to offspring without fusion of **gametes** (egg and sperm); that is, there is no genetic input from another individual. In **sexual reproduction,** male and female parents produce offspring through the union of egg and sperm generated by meiosis (meiosis is discussed in Chapter 11).

Asexual Reproduction Produces Offspring with Genes from Only One Individual

Many aquatic invertebrates and some terrestrial annelids and insects reproduce asexually. Asexual reproduction is rare among vertebrates. In asexual reproduction, one or many cells of a parent develop directly into a new individual. In a few animals that undergo asexual reproduction, the cells taking part are genetically varied products of meiosis, but in most they are the products of mitosis. The offspring therefore are genetically identical to one another and to the parent; in other words, they are genetic clones of the parent. For this reason, asexual reproduction of this kind is also called **clonal reproduction.**

Genetic uniformity of offspring can be advantageous in environments that remain stable and uniform. Asexual reproduction tends to preserve gene combinations, producing individuals that are successful in such environments. Further, individuals do not have to expend energy to produce gametes or find a mate. Asexual reproduction can also bring reproductive advantages to individuals living in sparsely populated areas, or to sessile animals, which cannot move from place to place.

Asexual reproduction involving mitosis occurs in animals by three basic mechanisms: *fission, budding,* and *fragmentation.*

1. **Fission.** In **fission,** the parent separates into two or more offspring of approximately equal size. Planarians (flatworms), for instance, reproduce asexually by fission; depending on the species, they may divide by transverse or longitudinal fission.
2. **Budding.** In **budding,** a new individual grows and develops while attached to the parent. Sponges, tunicates, and some cnidarians reproduce asexually by this mechanism. The offspring may break free from the parent, or remain attached to form a *colony.* In the cnidarian *Hydra,* for example, an offspring buds and grows from one side of the parent's body and then detaches to become a separate individual **(Figure 49.2).**
3. **Fragmentation.** In **fragmentation,** pieces separate from the body of a parent and develop *(regenerate)* into new individuals. Many species of cnidarians, flatworms, annelids, and some echinoderms can reproduce by fragmentation.

Some animals produce offspring by the growth and development of an egg without fertilization. The offspring may be haploid or diploid depending on the species. This form of asexual reproduction is called **parthenogenesis** (*parthenos* = virgin, *genesis* = origin). Because the egg from which a parthenogenetic offspring is produced derives from meiosis in the female parent, the offspring are not genetically identical to the parent or to each other. (How chromosome segregation and genetic recombination during meiosis produce gametes with gene combinations different from the parent is described shortly.)

Parthenogenesis occurs in some invertebrates, including certain aphids, bees, and water fleas. In water fleas, for instance, the female can produce two types of eggs. One type of egg develops only if it is fertilized, whereas the other develops by parthenogenesis. The conditions set the switch for which form of reproduction occurs: sexual reproduction occurs when there is

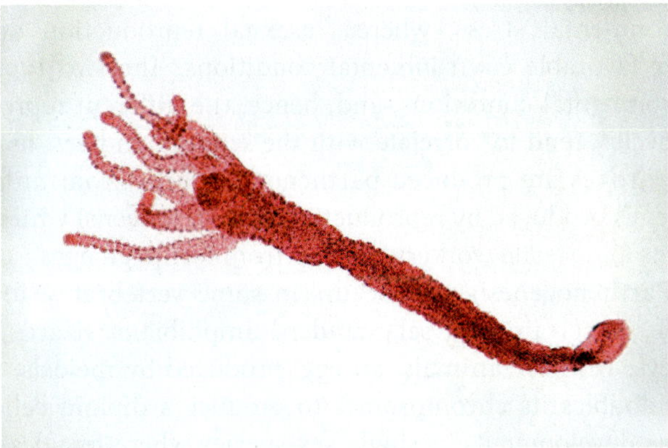

D. Kucharski & K. Kucharska/Shutterstock.com

FIGURE 49.2 Asexual reproduction by budding in *Hydra* (stained, visualized by light microscopy).

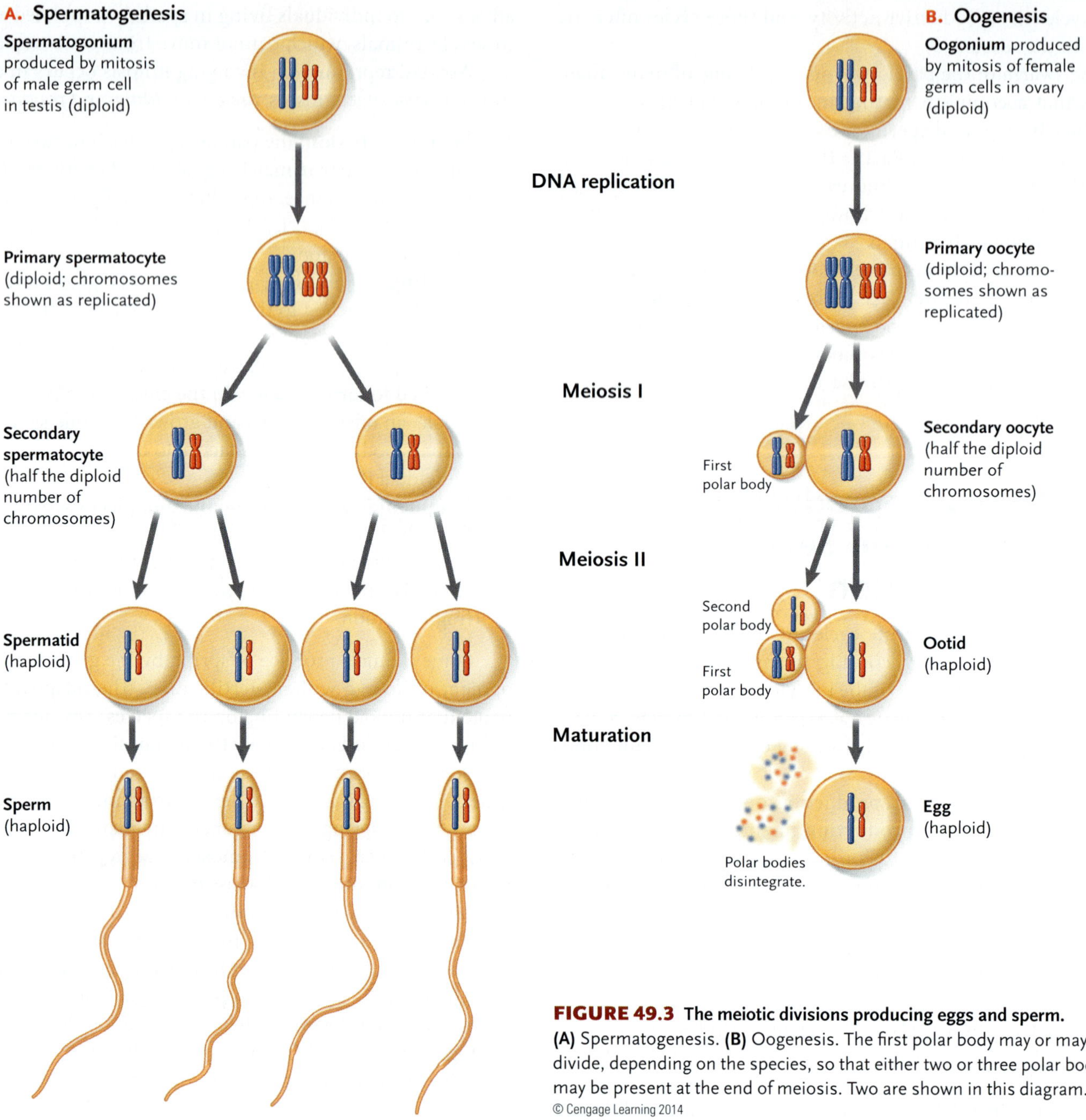

FIGURE 49.3 The meiotic divisions producing eggs and sperm. **(A)** Spermatogenesis. **(B)** Oogenesis. The first polar body may or may not divide, depending on the species, so that either two or three polar bodies may be present at the end of meiosis. Two are shown in this diagram.

environmental stress, whereas asexual reproduction occurs under favorable environmental conditions. The two types of environmental conditions, and, hence, the different reproductive cycles, tend to correlate with the seasons. In bees, haploid male drones are produced parthenogenetically from unfertilized eggs produced by reproductive females (queens) while new queens and sterile workers develop from fertilized eggs.

Parthenogenesis also occurs in some vertebrates, for example, in certain fish, salamanders, amphibians, lizards, and turkeys. In these animals, an egg, produced by meiosis, typically doubles its chromosomes to produce a diploid cell that begins development. In single-sex species where females have two identical sex chromosomes, the offspring are female, whereas in single-sex species where males have two identical sex chromosomes, the offspring are male. For instance, all whiptail lizards (*Cnemidophorus* species) are females, produced solely by parthenogenesis. These females go through the motions of mating and copulation with each other.

Sexual Reproduction Generates Diversity among Offspring

Animals reproduce sexually by the union of sperm and eggs produced by meiosis. The overriding advantage of sexual reproduction is the generation of genetic diversity among offspring. This diversity increases the chance that, in a changing environment, at least some offspring will grow and reproduce successfully. Diversity also increases the chance that offspring may be able to live and reproduce in environments previously unoccupied by the species.

Two mechanisms that are part of meiosis give rise to the genetic diversity in eggs and sperm: *genetic recombination* (see Section 11.2) and the *independent assortment* of chromosomes of maternal and paternal origin (see Section 12.1). Genetic recombination mixes the alleles of parents into new combinations within chromosomes; independent assortment results in random combinations of maternal and paternal chromosomes in gamete nuclei. Additional variability is generated at fertilization when eggs and sperm from genetically different individuals fuse together at random to initiate the development of new individuals. To these sources of variability are added random DNA mutations, which are the ultimate source of variability for both sexual and asexual reproduction.

The disadvantages of sexual reproduction include the expenditure of energy and raw materials in producing gametes and finding mates, and producing fewer offspring per parent. The need to find mates, for instance, can expose animals to predation and takes time from finding food and shelter and caring for existing offspring. Concerning offspring number, consider a female that reproduces asexually. All of her progeny will be females, each of whom will produce more offspring by asexual reproduction. However, in sexual reproduction, one half of the offspring of a mating pair are females, and one-half are males (on average). Therefore, only one-half of the progeny produced by sexual reproduction of a female produce offspring for the following generation.

We turn now to the mechanisms of sexual reproduction, which include both cellular and whole-organism activities. The next section discusses cellular mechanisms.

STUDY BREAK 49.1

What are the advantages and disadvantages of asexual reproduction? Of sexual reproduction?

49.2 Cellular Mechanisms of Sexual Reproduction

The cellular mechanisms of sexual reproduction are **gametogenesis,** the formation of male and female gametes, and **fertilization,** the union of gametes that initiates development of a new individual. The pairing of a male and a female for the purpose of sexual reproduction is **mating.**

Gametogenesis Involves the Coordinated Events of Meiosis and Sperm and Egg Development

Gametes in most animals form from **germ cells,** a cell line that is set aside early in embryonic development and remains distinct from the other, **somatic cells** of the body. During development, the germ cells collect in specialized gamete-producing organs, the **gonads**—the **testes** (singular, *testis*) in males and **ovaries** in females. Mitotic divisions of the germ cells produce **spermatogonia** in males and **oogonia** in females; these are the cells that enter meiosis to give rise to gametes by **spermatogenesis** in males and **oogenesis** in females **(Figure 49.3).** In humans, each gamete has only one of 2^{23} possible combinations of parental chromosomes. In some animals, the germ cells also give rise to families of cells that assist gamete development.

Meiosis reduces the number of chromosomes from the diploid level characteristic of somatic cells of the species, in which there are two copies of each chromosome, to the haploid level of gametes, in which there is one copy of each chromosome. The fusion of a haploid sperm and egg during fertilization restores the diploid number of chromosomes and produces a **zygote,** the first cell of a new individual.

SPERMATOGENESIS Spermatogenesis produces mature, haploid sperm cells, also called **spermatozoa** (singular, *spermatozoon*) or simply *sperm* **(Figure 49.4).** The sperm of most animal species are motile cells, driven through a watery medium by the whiplike beating of a flagellum that extends from the posterior end of the cell. During maturation from spermatid to sperm, most of the cytoplasm is lost, except for mitochondria, which surround the base of a flagellum. These mitochondria produce the ATP used as the energy source for flagellar beating. At the head of the sperm, a specialized secretory vesicle, the **acrosome,** forms a cap over the nucleus. The acrosome contains enzymes and other proteins that help the sperm attach to and penetrate the surface coatings of an egg of the same species.

OOGENESIS Oogenesis produces mature, haploid egg cells, also called **ova** (singular, *ovum*) or simply *eggs*. The eggs of all ani-

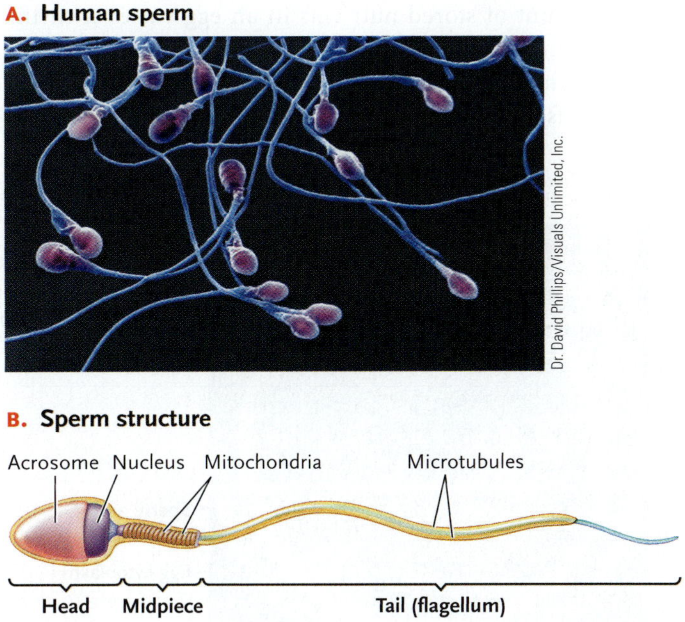

FIGURE 49.4 Spermatozoa. (A) Photomicrograph of human sperm. **(B)** Structure of a sperm.

mals are nonmotile cells, typically much larger than sperm of the same species. Whereas in spermatogenesis, all four products of meiosis develop into functional sperm, in oogenesis, only one of the cell products of meiosis develops into a functional egg **(Figure 49.5).** That egg retains almost all of the cytoplasm of the parent cell. The other products form nonfunctional cells called **polar bodies** (see Figure 49.3B). The unequal cytoplasmic divisions concentrate nutrients and other molecules required for development in the egg. In most species, the polar bodies eventually disintegrate and do not contribute to fertilization or embryonic development.

The oocytes of most animals do not complete meiosis until fertilization. In mammals, for example, oocytes stop developing at the end of the first meiotic prophase within a few weeks after a female is born. The oocytes remain in the ovary at this stage of development until the female is sexually mature. Then, one to several oocytes advance to the metaphase of the second meiotic division and are released from the ovary at intervals ranging from days to months, or at certain seasons, depending on the species. (In humans, some oocytes may therefore remain in prophase of the first meiotic division for perhaps 50 years until menopause.) As in other animals, meiosis is completed at fertilization to produce the fully mature egg. Mature eggs are the largest cell type of an animal species.

An egg typically has specialized features, which include: (1) stored nutrients and cytoplasmic determinants required for at least the early stages of embryonic development (see Section 16.4 for a discussion of cytoplasmic determinants); (2) egg coats of one or more kinds that protect the egg from mechanical injury and infection and, in some species, protect the embryo after fertilization; and (3) mechanisms that prevent the egg from being fertilized by more than one sperm cell (discussed shortly).

The amount of stored nutrients in an egg varies with the animal. Mammalian eggs are microscopic, containing few stored nutrients. In mammals, the embryo develops inside the mother and is supplied with nutrients by the mother's body. In contrast, the relatively large eggs of birds and reptiles contain all the nutrients required for complete embryonic development: the "yolk" contains the egg cell, and the "white" contains the nutrients. No matter what the size of an animal egg, however, most of the volume is cytoplasm, and the egg nucleus is microscopic or nearly so in all species.

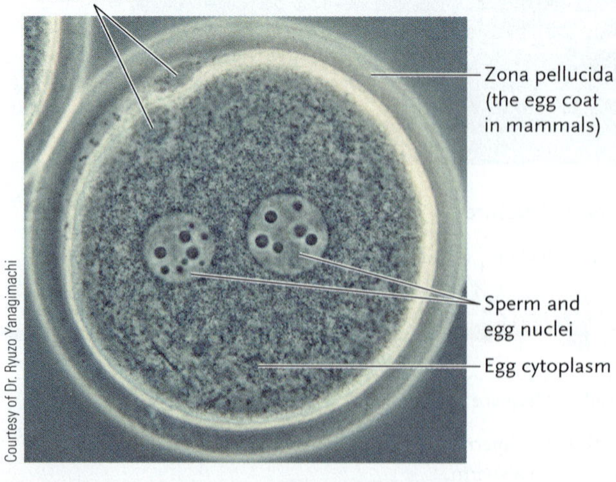

FIGURE 49.5 A mature hamster egg that has been fertilized.

Egg coats are surface layers added during oocyte development or fertilization in many species. The **vitelline coat,** called the **zona pellucida** in mammals (see Figure 49.5) is a gel-like matrix of proteins, glycoproteins, or polysaccharides located immediately outside the plasma membrane of the egg cell. Insect eggs have additional outer protein coats that form a hard, water-impermeable layer for preventing desiccation. Amphibians and some echinoderms have an additional outer egg jelly layer instead of a tough protein coat that protects the egg from drying out (see Figure 49.8).

In birds, reptiles, and one group of egg-laying mammals, the **monotremes,** the egg white, a thick solution of proteins, surrounds the vitelline coat. Outside the white is the *shell* of the egg, flexible and leathery in reptiles and mineralized and brittle in birds. Both the egg white and the shell are added while the egg—fertilized or not—is in transit through the **oviduct,** the tube through which the egg moves from the ovary to the outside of the body. In mammals, the egg is surrounded by **follicle cells** during its development. These cells, which grow from ovarian tissue, nourish the developing egg. They also make up part of the zona pellucida while the egg is in the ovary, and remain as a protective layer after it is released.

Fertilization Requires an Internal or External Aquatic Medium

Eggs and sperm are delivered from the ovaries and testes to the site of fertilization by oviducts in females and by sperm ducts in males; in many species, external accessory sex organs participate in the delivery. **Figure 49.6** shows examples of invertebrate and vertebrate reproductive systems. The nonmotile eggs move through the oviducts on currents generated by the beating of cilia lining the oviducts, or by contractions of the oviducts or the body wall.

Depending on the species, fertilization may be *external,* taking place in a watery medium outside the body of both parents, or *internal,* taking place in a watery fluid inside the body of the female. In **external fertilization,** which occurs in most aquatic invertebrates, bony fishes, and amphibians, sperm and eggs are shed into the surrounding water. The sperm swim until they collide with an egg of the same species. The process is helped by synchronization of female and male gamete release, and by the enormous quantities of gametes released. In some animals, such as sea urchins and amphibians, the sperm are attracted to the egg by diffusible attractant molecules released by the egg.

Most amphibians, even terrestrial species such as toads, mate in an aquatic environment. Frogs typically mate by a reflex response called *amplexus,* in which the male clasps the fe-

FIGURE 49.6 Some reproductive systems. (A) An insect, *Drosophila* (fruit fly). **(B)** An amphibian, a frog. **(C)** A mammal, a cat. Female systems are shown in blue, and male systems in yellow.

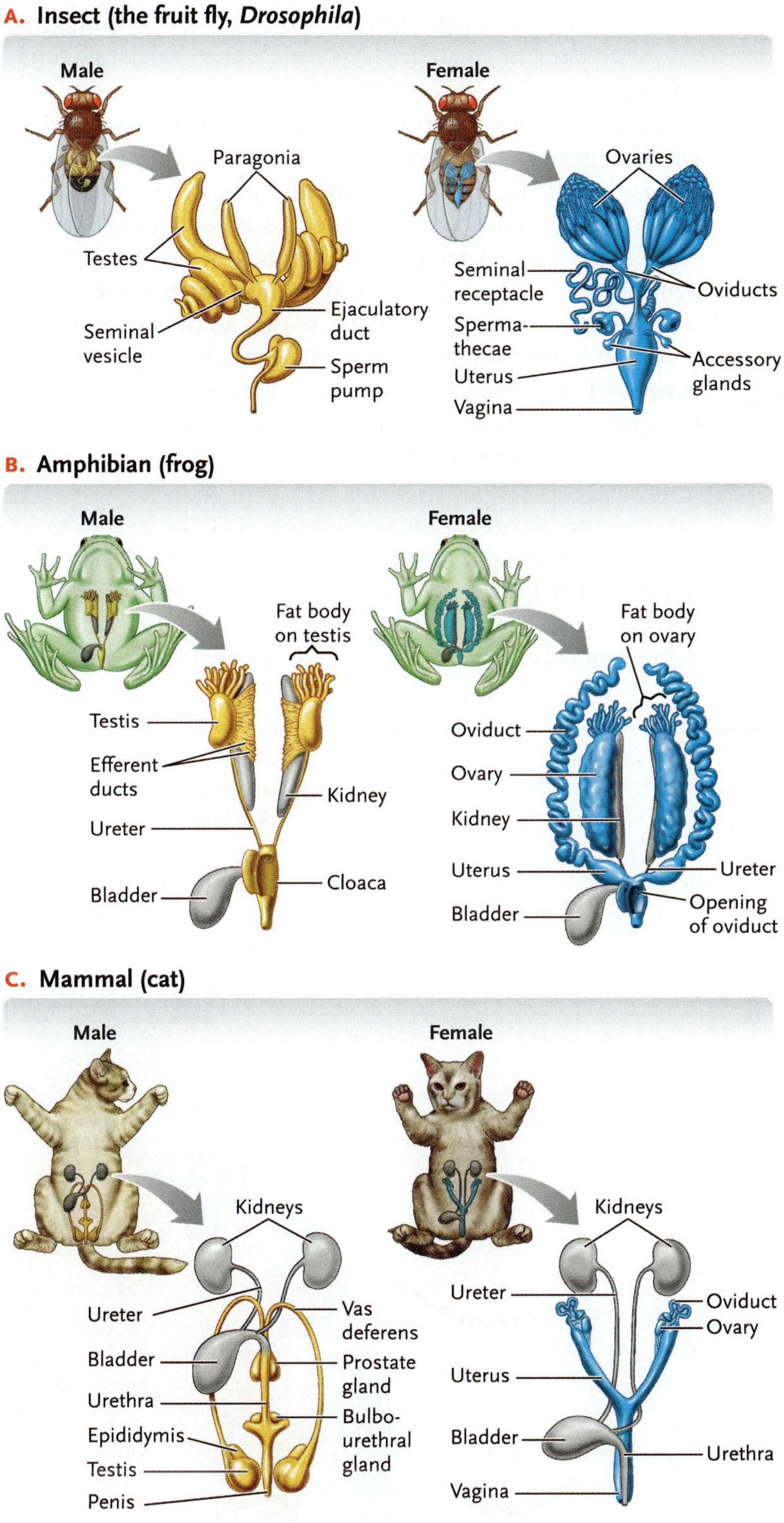

FIGURE 49.7 A male leopard frog *(Rana pipiens)* clasping a female in a mating embrace known as amplexus. The tight squeeze by the male frog stimulates the female to release her eggs, which can be seen streaming from her body, embedded in a mass of egg jelly. Sperm released by the male fertilize the eggs as they pass from the female.

Hans Pfletschinger

male tightly around the body with his forelimbs **(Figure 49.7).** The embrace stimulates the female to shed a mass of eggs into the water through the **cloaca**—the cavity in reptiles, birds, amphibians, and many fishes into which both the intestinal and genital tracts empty. As the eggs are released, they are fertilized by sperm released by the male.

Internal fertilization takes place in invertebrates such as annelids, some arthropods, and some mollusks, and in vertebrates such as reptiles, birds, mammals, some fishes, and some salamanders. In these animals, the sperm are released by the male close to or inside the entrance of the reproductive tract of the female. The sperm swim through fluids in the reproductive tract until they reach and fertilize each egg. In some species, molecules released by the egg attract the sperm to its outer coats. The physical act involving the introduction of the male's accessory sex organ (for example, penis) into a female's accessory sex organ (for example, vagina) to accomplish internal fertilization is known as **copulation.** Internal fertilization makes terrestrial life possible by providing the aquatic medium required for fertilization inside the female's body without the danger of gametes drying by exposure to the air.

Male reptiles, birds, and mammals have accessory sex organs that place sperm directly inside the reproductive tract of females, where fertilization takes place. In reptiles and birds, sperm fertilize eggs as they are released from the ovary and travel through the oviducts, before the shell is added. In mammals, the male's penis delivers sperm into the female's vagina. Unlike the cloaca of the reptiles and birds, which has both sexual and excretory functions, the vagina is specialized for reproduction. The introduced sperm swim into the tubular oviducts containing the eggs, and fertilization takes place.

Fertilization Involves Fusion of a Sperm and an Egg, Which Activates the Egg for Development

Once a sperm touches the outer surface of an egg of the same species, receptor proteins in the sperm plasma membrane bind the sperm to the vitelline coat or zona pellucida. In most animals, only a sperm from the same species as the egg can recognize and bind to the egg surface.

Species recognition is highly important in animals that carry out external fertilization, because the water surrounding the egg may contain sperm of many different species. It is less important in internal fertilization, because structural adaptations and behavioral patterns of mating usually limit sperm transfer from males to females of the same species.

FERTILIZATION **Figure 49.8** shows the steps of fertilization in the sea urchin, a well-studied organism for invertebrate development and reproduction, and an often-used model for human fertility research. The steps are similar in other organisms. The attachment of sperm to egg triggers a rapid series of events starting with the **acrosomal reaction,** in which enzymes in the acrosome are released from the sperm and digest a path through the egg coats. The sperm, with its tail still beating, follows the path until its plasma membrane touches and fuses with the plasma membrane of the egg. Fusion introduces the sperm nucleus into the egg cytoplasm and activates the egg to complete meiosis and begin development.

EGG ACTIVATION AND BLOCKS TO POLYSPERMY *Polyspermy,* the phenomenon in which more than one sperm fertilizes an egg, can be prevented by two mechanisms: a *fast block* within seconds of fertilization, and a *slow block* within minutes.

In many invertebrate species, such as the sea urchin, the fusion of egg and sperm opens ion channels in the plasma membrane of the egg, spreading a wave of electrical depolarization over the egg surface, much like the nerve impulse traveling along a neuron. The depolarization alters the egg plasma membrane's potential from negative to positive so that it cannot fuse with any additional sperm. Because it occurs within a few seconds after fertilization, the barrier set up by the wave of depolarization is called the **fast block to polyspermy** (see Figure 49.8, step 5).

In vertebrates, the wave of membrane depolarization following sperm–egg fusion is not as pronounced, and does not prevent additional sperm from fusing with the egg. However, any additional sperm nuclei entering the egg cytoplasm usually break down and disappear, so that only the first sperm nucleus to enter fuses with the egg nucleus.

In both invertebrates and vertebrates, fusion of egg and sperm triggers the release of stored calcium (Ca^{2+}) ions from the endoplasmic reticulum of the egg into the cytosol. The Ca^{2+} ions activate control proteins and enzymes that initiate intense metabolic activity in the fertilized egg, including a rapid increase in cellular oxidations and synthesis of proteins and other molecules.

The Ca^{2+} ions also trigger the **cortical reaction,** in which **cortical granules,** secretory vesicles just under the plasma membrane of the egg, fuse with the membrane and release their contents to the outside by exocytosis (see Figure 49.8, step 7; the process of exocytosis is described in Section 6.5). Enzymes released from the cortical granules alter the egg coats within minutes after fertilization, so that no further sperm can attach and penetrate to the egg. Once this barrier, termed the **slow block to polyspermy,** is set up, no further sperm can reach the egg plasma membrane in any animal species.

After the sperm nucleus enters the egg cytoplasm, microtubules move the sperm and egg nuclei together in the egg cytoplasm and they fuse. The chromosomes of the egg and sperm nuclei then assemble together and enter mitosis. The subsequent, highly programmed events of embryonic development that convert the fertilized egg into an individual capable of independent existence are described in the next chapter.

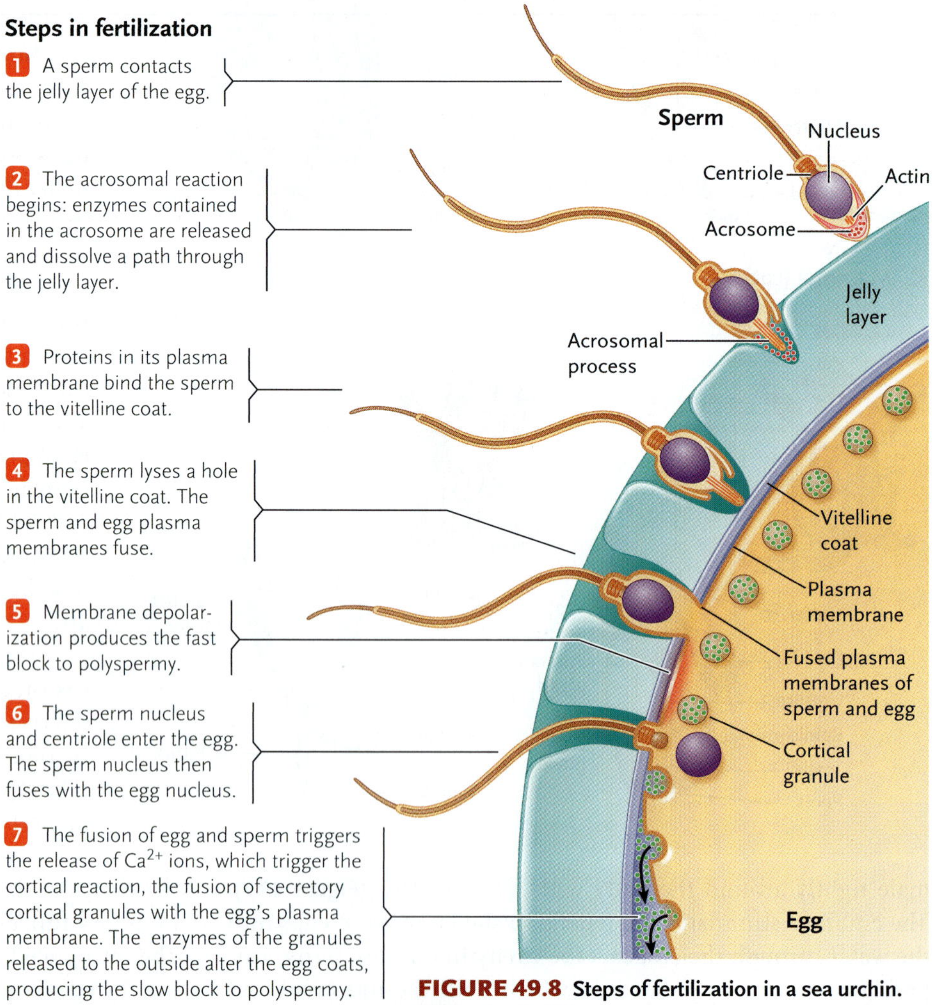

FIGURE 49.8 Steps of fertilization in a sea urchin.

Reproductive Systems May Be Oviparous or Viviparous in Animals with Internal Fertilization

In animals with internal fertilization, three major types of support for embryonic development have evolved: *oviparity,* meaning egg laying; *viviparity,* meaning giving birth to live offspring; and *ovoviparity,* meaning giving birth to live offspring that first hatch internally from eggs.

- **Oviparous** animals (*ovum* = egg, *parere* = to bring forth, to bear) lay eggs that contain the nutrients needed for development of the embryo outside the mother's body. Examples are insects, spiders, most reptiles, and birds. The only oviparous mammals are the *monotremes:* the echidnas (spiny anteaters—found in Australia and New Guinea) and the duck-billed platypus (*Ornithorhynchus anatinus*—found in Australia).
- **Viviparous** animals (*vivus* = alive) retain the embryo within the mother's body and nourish it during at least early embryo development. All mammals except the monotremes are viviparous. Viviparity is seen also in all other vertebrate groups except for the crocodiles, turtles, and birds. In viviparous animals, development of the embryo takes place in a specialized saclike organ, the **uterus** *(womb).* Among mammals, one group, called the *placental mammals* or *eutherians,* has a specialized temporary structure, the **placenta,** which connects the embryo to the uterus. The placenta facilitates the transfer of nutrients from the blood of the mother to the embryo and the movement of wastes in the opposite direction. Humans are an example of placental mammals. The other group of mammals, the *marsupials* or *metatherians,* originally were called nonplacental mammals because of a belief that they lacked a placenta. In fact, they do have a placenta, but it derives from a different tissue than that of eutherians and does not connect the embryo and the uterus. Instead it provides nutrients to the embryo from an attached membranous sac containing yolk for only the early stages of its development. In many metatherians, the embryo is then born at an early stage and crawls over the mother's fur to reach the **marsupium,** an abdominal pouch within which it attaches to nipples and continue its development. Kangaroos, koalas, wombats, and opossums are marsupials.
- **Ovoviviparous** animals retain fertilized eggs within the body and the embryo develops using the nutrients provided by the egg. There is no uterus or placenta involved. When development is complete, the eggs hatch inside the mother and the offspring are released to the exterior. Ovoviviparity is seen in some fishes, lizards, and amphibians, many snakes, and many invertebrates.

Hermaphroditism Is a Variation on Sexual Reproduction

Some animals have evolved modified mechanisms that they use as their normal sexual reproduction process. One of these mechanisms is **hermaphroditism** (from *Hermaphroditos,* the son of *Hermes* and *Aphrodite,* a Greek god and goddess), in which both mature egg-producing and mature sperm-producing tissue is present in the same individual. Most flatworms, earthworms, land snails, and numerous other invertebrates are hermaphroditic. In humans and other mammals, there are rare cases of individuals who have both testicular and ovarian tissues. They are not true hermaphrodites because they are not both male and female; rather, they are genetically either male or female, but with ambiguous genitalia. Hence, they are called *pseudohermaphrodites* rather than true hermaphrodites.

Most hermaphroditic animals do not fertilize themselves, because anatomical barriers prevent individuals from introducing sperm into their own body, or the egg and sperm mature at different times. The prevention of self-fertilization maintains the genetic variability of sexual reproduction.

Hermaphroditism takes two forms: **simultaneous hermaphroditism,** in which individuals develop functional ovaries and testes at the same time, and **sequential hermaphroditism,** in which individuals change from one sex to the other. The two earthworms shown in **Figure 49.9** provide a common example of simultaneous hermaphroditism. The only known vertebrate simultaneous hermaphrodites are hamlets (genus *Hypoplectrus*), a group of predatory sea basses. Sequential hermaphroditism is seen among a number of invertebrates (for example, the slipper shell *Crepidula fornicata,* a gastropod), and some ectothermic vertebrates, notably fishes

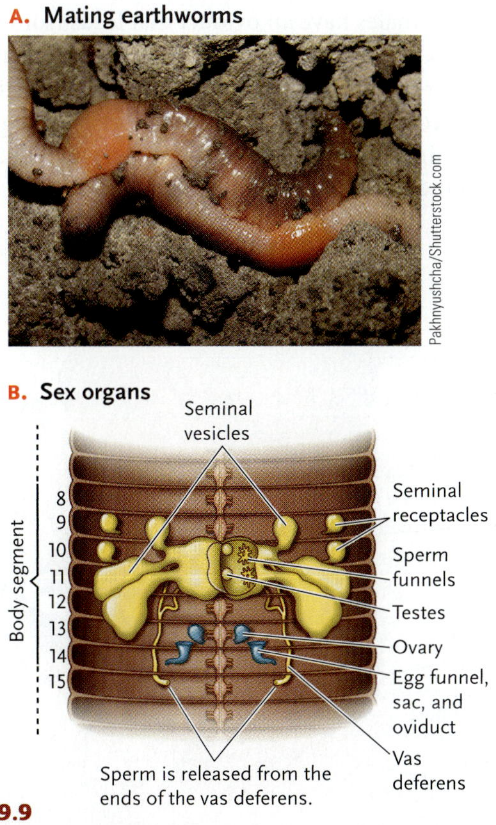

FIGURE 49.9

Simultaneous hermaphroditism in the earthworm. (A) Copulation by a mating pair of earthworms, in which each earthworm releases sperm that fertilize eggs in its partner. **(B)** Sex organs in the earthworm.

(for example, the clownfish, genus *Amphiprion*). In some species, the initial sex is male (as with the slipper shell and the clownfish), and in others it is female.

STUDY BREAK 49.2

1. What are egg coats, and what is their function? What egg coats do mammalian and bird eggs have?
2. How is the slow block to polyspermy brought about?

THINK OUTSIDE THE BOOK

Use the Internet or research literature to outline a model for how hermaphroditism may have evolved.

49.3 Sexual Reproduction in Humans

Except for structural details, human reproduction is typical of that of eutherian (placental) mammals. Internally, these mammals have a pair of gonads, either ovaries or testes. The gonads have a dual function in mammals, as they do in all vertebrates: they both produce gametes and secrete hormones that are responsible for sexual development and mating behavior (see Section 42.4). Males have ducts that carry sperm from the testes to the exterior. Females have an oviduct that leads from each ovary to the uterus, in which fertilized eggs implant and proceed through embryonic development. Nutrients from the mother and wastes from the embryo are exchanged through the placenta. After birth, the newborn offspring is nourished with milk secreted by the mother's mammary glands.

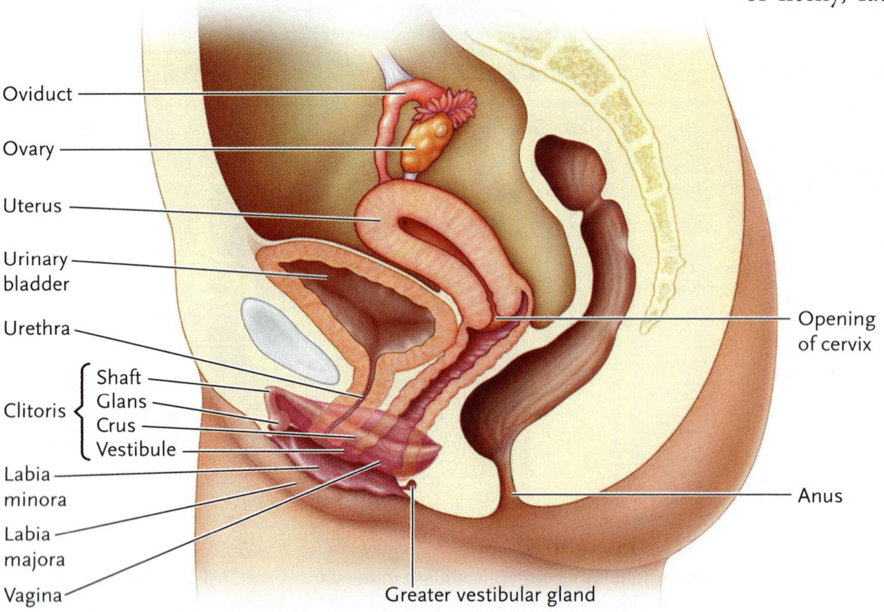

FIGURE 49.10 The reproductive organs of a human female.

In this section we survey reproductive structures and functions in humans as representative of eutherian mammals. Our story of human development continues in the next chapter, which traces the process from fertilization to birth.

Human Female Sexual Organs Function in Oocyte Production, Fertilization, and Embryonic Development

Human females have a pair of ovaries suspended in the abdominal cavity **(Figure 49.10).** An oviduct leads from each ovary to the uterus, which is a hollow, saclike organ with walls containing smooth muscle. The uterus is lined by the **endometrium,** which is formed by layers of connective tissue with embedded glands and is richly supplied with blood vessels. If an egg is fertilized and begins development, it must implant in the endometrium to continue developing. The lower end of the uterus, the **cervix,** opens into a muscular canal, the **vagina,** which leads to the exterior. Sperm enter the female reproductive tract via the vagina and, at birth, the baby passes from the uterus to the outside through the vagina.

At the birth of a female, each of her ovaries contains about 1 million oocytes, arrested at the end of the first meiotic prophase. Of these oocytes, about 200,000 to 400,000 survive until a female becomes sexually mature; about 400 are **ovulated**—released into the oviducts as immature eggs—during the lifetime of a woman. The egg is released into the abdominal cavity and is pulled into the nearby oviduct by the current produced by the beating of the cilia lining the oviduct. The cilia also propel the egg through the oviduct and into the uterus. Fertilization of the egg occurs in the oviduct.

The external female sex organs, collectively called the **vulva,** surround the opening of the vagina. Two folds of tissue, the **labia minora,** run from front to rear on either side of the opening to the vagina. These folds are partially covered by a pair of fleshy, fat-padded folds, the **labia majora,** which also run from front to rear on either side of the vagina. At the anterior end of the vulva, the labia minora join to partly cover the head of the **clitoris.** The rest of the clitoris is within the body. The clitoris contains erectile tissue and has the same embryonic origins as the penis. A pair of **greater vestibular glands,** with openings near the entrance to the vagina, secretes a mucus-rich fluid that lubricates the vulva. The opening of the urethra, which conducts urine from the bladder, is located between the clitoris and the vaginal opening. Most nerve endings associated with erotic sensations are concentrated in the clitoris, in the labia minora, and around the opening of the vagina. When a human female is born, a thin flap of tissue, the **hymen,** partially covers the opening of the vagina. This membrane, if it has not already been ruptured by physical exercise or other disturbances, is broken by the first sexual intercourse.

Ovulation in Human Females Occurs in a Monthly Cycle

Reproduction in human females is under neuroendocrine control, involving complex interactions between the hypothalamus, pituitary, ovaries, and uterus. Under this control, approximately every 28 days from puberty to menopause, a female releases an egg from one of her ovaries. The cyclic events in the ovary leading to ovulation are known as the **ovarian cycle.** This cycle is coordinated with the **uterine cycle,** or **menstrual cycle** (*menstruus* = monthly), events in the uterus that prepare it to receive the egg if fertilization occurs. While the cycle is commonly considered a monthly one, there is a lot of variation among human females with a range of about 15 to 45 days, and an average is about 28 to 29 days.

THE OVARIAN CYCLE The ovarian cycle produces a mature egg **(Figure 49.11).** The starting point for the cycle is a primary oocyte in prophase of meiosis division I. The beginning of the cycle is triggered by an increase in the release of **gonadotropin-releasing hormone (GnRH)** by the hypothalamus. This hormone stimulates the pituitary to release **follicle-stimulating hormone (FSH)** and **luteinizing hormone (LH)** into the bloodstream **(Figure 49.12A).** FSH stimulates 6 to 20 primary oocytes in the ovaries to be released from prophase of meiosis I and continue through the meiotic divisions. As the primary oocytes develop into secondary oocytes—which arrest in metaphase of meiosis II—they become surrounded by cells that form a **follicle** (day 2 of the cycle; **Figure 49.12B**). During this follicular phase, the follicle grows and develops and, at its largest size, becomes filled with fluid and may reach 12 to 15 mm in diameter. Usually only one follicle develops to maturity with release of the egg (secondary oocyte) by ovulation. If two or more follicles develop and their eggs are ovulated, multiple births of nonidentical siblings can result.

As the follicle enlarges, FSH and LH interact to stimulate the follicular cells to secrete **estrogens** (female sex hormones), primarily **estradiol** (see Section 42.4) **(Figure 49.12C).** Initially, the estrogens are secreted in low amounts; at this level, the estrogens have a *negative* feedback effect on the pituitary, inhibiting its secretion of FSH. As a result, FSH secretion declines briefly. However, estrogen secretion increases steadily, and its level peaks at about 12 days after follicle development begins (day 14 of cycle). The high estrogen level now has a *positive* feedback effect on the hypothalamus and pituitary, increasing the release of GnRH and stimulating the pituitary to release a burst of FSH and LH. The increased estrogen levels also convert the mucus secreted by the uterus to a thin and watery consistency, making it easier for sperm to swim through the uterus.

The burst in LH secretion stimulates the follicle cells to release enzymes that digest away the wall of the follicle, causing it to burst and release the egg (see Figure 49.12); this is ovulation. LH also initiates the last phase of the menstrual cycle, the *luteal phase.* That is, LH causes the follicle cells remaining at the surface of the ovary to grow into an enlarged, yellowish structure, the **corpus luteum** (*corpus* = body, *luteum* = yellow) (see Figure 49.11). Acting as an endocrine gland (see Section 42.1), the corpus luteum secretes several hormones: estrogens, large quan-

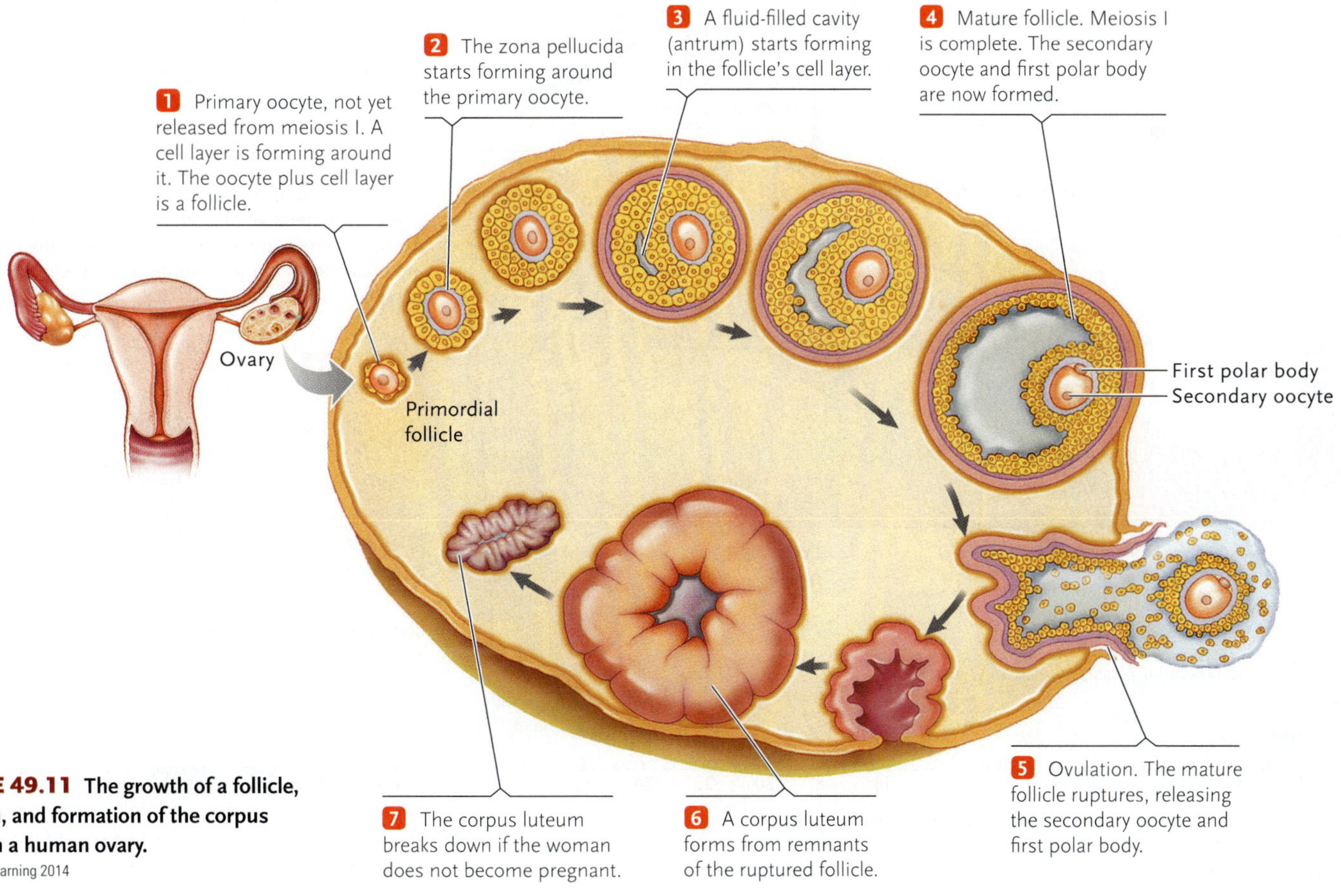

FIGURE 49.11 The growth of a follicle, ovulation, and formation of the corpus luteum in a human ovary.

CLOSER LOOK

FIGURE 49.12 The ovarian and uterine (menstrual) cycles of a human female.

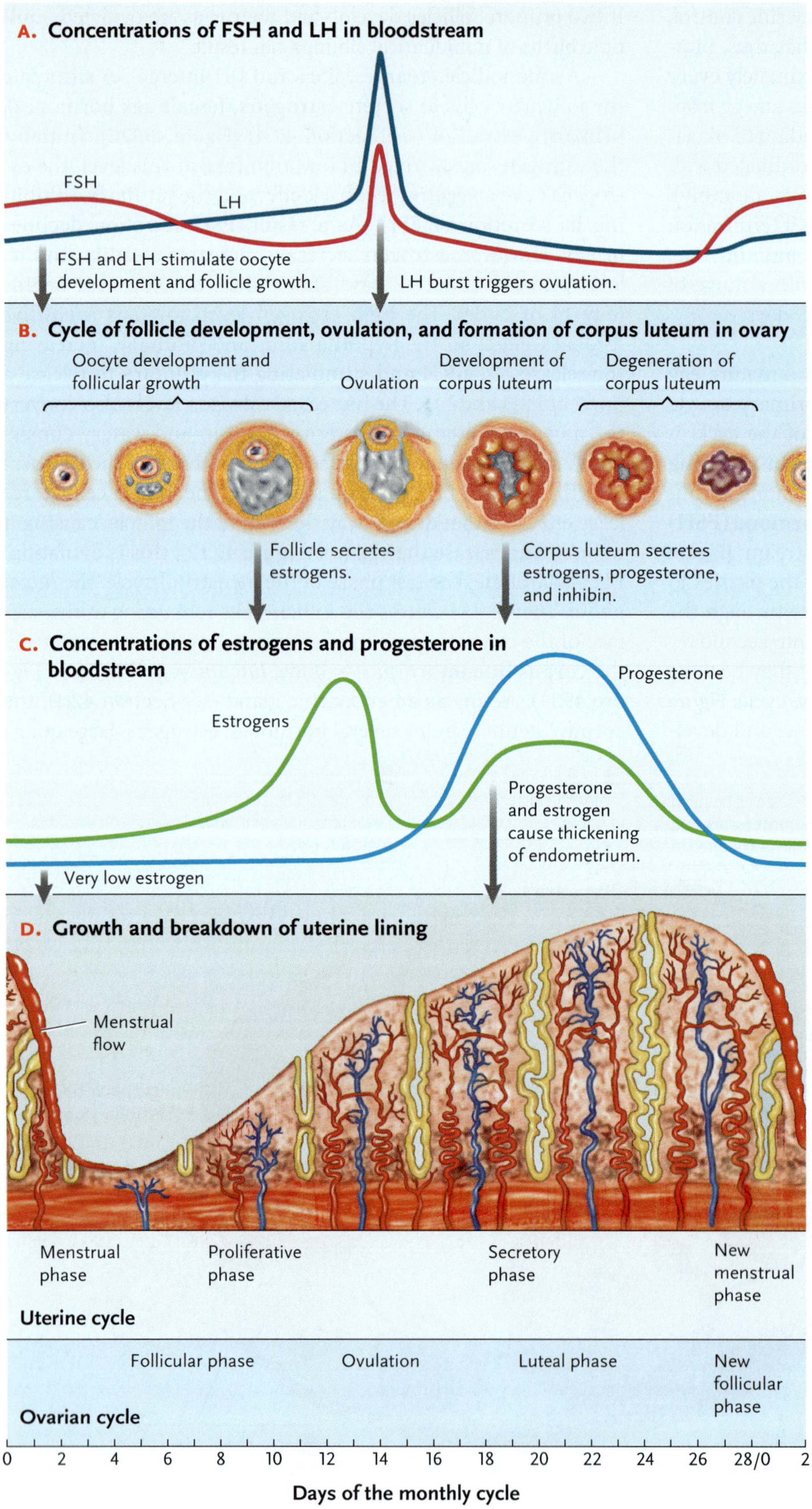

The ovarian cycle begins with an increase in the release of GnRH by the hypothalamus, which stimulates the release of FSH and LH from the anterior pituitary. The changes in concentrations of FSH and LH in the bloodstream play important roles in follicle development and ovulation.

In the follicular phase, FSH stimulates follicle growth. As the follicle enlarges, FSH and LH interact and stimulate cells of the follicle to secrete estrogens. Initially the estrogens inhibit secretion of FSH by the pituitary, causing a decrease in FSH concentration. When the estrogens reach a high concentration, they then stimulate the hypothalamus and pituitary to increase GnRH release, thereby stimulating the pituitary to produce a burst of LH and FSH. The burst in LH secretion causes ovulation and initiates the luteal phase, the development of the corpus luteum.

The concentrations of estrogens and progesterone are low during the follicular phase. The increase in estrogens late in that phase brings about ovulation and then their level drops. In the luteal phase, the corpus luteum secretes estrogens and large amounts of progesterone which stimulates growth of the uterine lining (the endometrium). In the absence of fertilization, the corpus luteum degenerates gradually and hormone levels fall bringing about menstruation.

The same hormones that control the ovarian cycle control the uterine cycle, which involves endometrium growth (proliferative phase), further uterine lining growth if there is no fertilization (the secretory phase), and break down of the endometrium (the menstrual phase).

SUMMARY The ovarian cycle produces an egg. First, GnRH released by the hypothalamus stimulates the anterior pituitary to release FSH and LH. FSH stimulates oocytes in the ovaries to initiate meiosis; one of them typically develops to maturity and becomes surrounded by cells that form a follicle. The follicle secretes estrogens, which stimulate an increase of FSH and LH concentrations in the blood. At about day 14, the LH stimulates the release of the egg from the follicle. The remainder of the follicle forms the corpus luteum, which secretes estrogens, progesterone, and inhibin.

THINK LIKE A SCIENTIST If the level of estrogens remains very low and does not increase as shown in the figure, what effect would that have on the menstrual cycle?

FIGURE 49.13 Experimental Research

Vocal Cues of Ovulation in Human Females

Question: Does the pitch of the female voice change in association with ovulation in human females?

Experiment: Primate females other than human females display well-characterized visual or olfactory cues that signal their reproductive state. Some research has documented particular cues associated with ovulation in humans; those cues relate to femininity and female attractiveness. Based on those observations, Gregory Bryant and Martie Haselton of the University of California, Los Angeles, hypothesized that vocal cues associated with female attractiveness would increase in frequency (pitch) over the menstrual cycle. A higher voice pitch is a signal associated with higher levels of female sex hormones, and correlates with being younger, both of which correlate with high fertility. To test their hypothesis, the researchers recruited 69 women with normal menstrual cycles and collected two sets of vocal samples from them saying "Hi, I'm a student at UCLA." One set was taken during a high-fertility phase of the cycle (at a time near ovulation) and the other set was taken during a low-fertility phase (at the luteal phase). The phase of the menstrual cycle at the time of vocal sampling was confirmed directly by hormonal tests. The samples were analyzed for the pitches of the voices and significant differences were determined using statistical methods.

Results: The average pitch of women's voices when the subjects said a simple sentence was significantly higher during the high-fertility phase of the menstrual cycle compared with during the low-fertility phase **(Figure).** The researchers confirmed in blind studies that the difference in frequency is readily detectable.

Conclusion: The change in vocal pitch seen for high-fertility (approaching ovulation) versus low-fertility phases of the menstrual cycle is seen as evidence for a cyclic fertility cue in the female human voice. Potentially the cue signals to males that females are in a high-fertility state.

THINK LIKE A SCIENTIST If you wanted to determine whether vocal cues for ovulation is evolutionarily conserved in mammals, what experiment(s) would you do?

Source: G. A. Bryant and M. G. Haselton. 2009. Vocal cues of ovulation in human females. *Biology Letters* 5:12–15.

tities of **progesterone,** and **inhibin.** Progesterone, a female sex hormone, stimulates growth of the uterine lining and inhibits contractions of the uterus. Both progesterone and inhibin have a negative feedback effect on the hypothalamus and pituitary. Progesterone inhibits the secretion of GnRH. Without GnRH, the pituitary does not release FSH and LH. FSH secretion from the pituitary is also inhibited directly by inhibin. The fall in FSH and LH levels diminishes the signal for follicular growth, and no new follicles begin to grow in the ovary.

If fertilization does not occur, the corpus luteum gradually degenerates as cells are destroyed by phagocytosis and blood supply is cut off. By about 10 days after ovulation, little tissue remains, meaning that estrogen, progesterone, and inhibin are no longer secreted. In the absence of progesterone, *menstruation* begins (described in the next section). As progesterone and inhibin levels decrease, FSH and LH secretion is no longer inhibited, and a new monthly cycle begins.

THE UTERINE (MENSTRUAL) CYCLE The hormones that control the ovarian cycle also control the uterine (menstrual) cycle **(Figure 49.12D),** keeping the processes connected physiologically. Day 0 of the monthly cycle in the figure is the beginning of follicular development in the ovary (see Figure 49.12B); in the uterus, this correlates with the time at which menstrual flow begins.

Menstrual flow results from the breakdown of the thickened endometrium. Blood and tissue breakdown products are released from the uterus to the outside through the vagina. When the flow ceases, at day 4 to 5 of the cycle, the endometrium begins to grow again—this is the *proliferative phase.* As the endometrium gradually thickens, the oocytes in both ovaries begin to develop further, eventually leading to ovulation at about 14 days after the beginning of the cycle, as already described. If fertilization does not take place, the uterine lining continues to grow for another 14 days after ovulation—this is the *secretory phase.* At the end of that time, the absence of progesterone results in the contraction of arteries supplying blood to the uterine lining, shutting down the blood supply and causing the lining to disintegrate. The menstrual flow begins. Contractions of the uterus, no longer inhibited by progesterone, help expel the debris. Prostaglandins released by the degenerating endometrium add to the uterine contractions, making them severe enough to be felt as the pain of "cramps," and also sometimes causing other effects such as nausea, vomiting, and headaches.

Menstruation—the menstrual flow—occurs only in human females and our closest primate relatives, gorillas and chimpanzees. In other mammals, the uterine lining is completely reabsorbed if a fertilized egg does not implant during the period of reproductive activity. The uterine cycle in those mammals is called the *estrous* cycle.

Hormonal tests can show where a woman is in her menstrual cycle. In nonhuman primates, various cues inform the male of the reproductive state of a female. **Figure 49.13** describes a research study demonstrating that vocal cues may signal the fertility state of human females.

Human Male Sexual Organs Function in Sperm Production and Delivery

Organs that produce and deliver sperm make up the male reproductive system **(Figure 49.14).** The testes are located outside the abdominal cavity. Sperm, which are produced by the testes, pass through tubules that enter the abdominal cavity and join with the urethra, the duct that carries urine from the bladder to an opening at the tip of the penis.

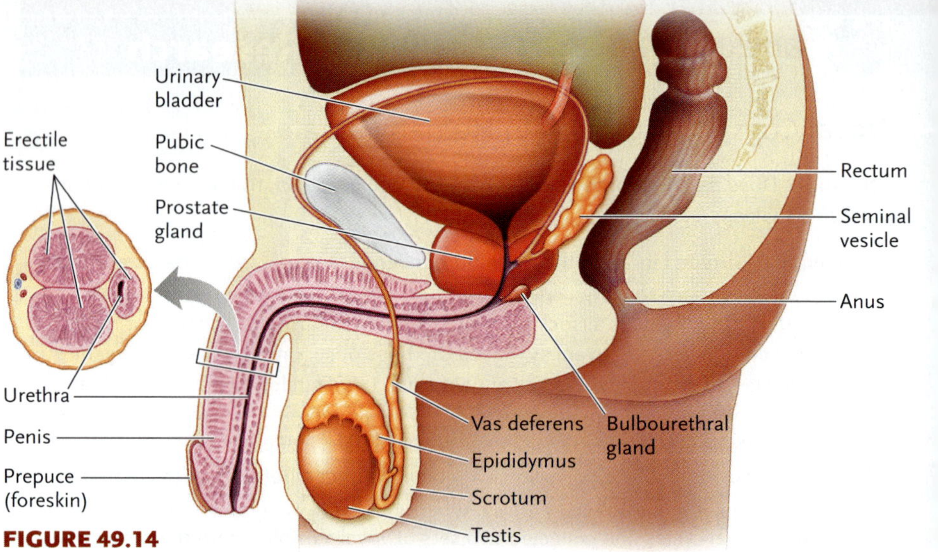

FIGURE 49.14
The reproductive organs of a human male.

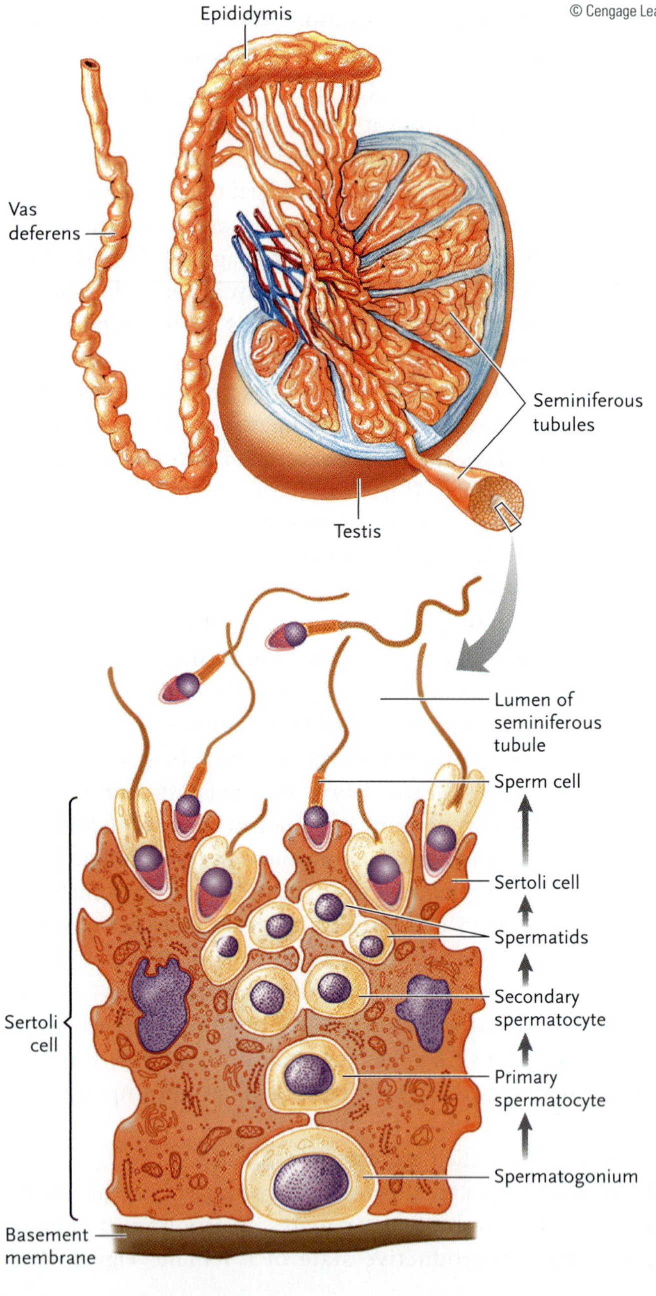

MALE REPRODUCTIVE STRUCTURES Human males have a pair of testes, suspended in the baglike **scrotum.** Suspension in the scrotum keeps the testes cooler than the body core, at a temperature that provides an optimal environment for sperm development. Some land mammals such as elephants and monotremes have relatively low body temperatures and have internal testes (testes carried within the body). A testis is packed with about 125 m of **seminiferous tubules,** in which sperm proceed through all the stages of spermatogenesis **(Figure 49.15).** The entire process, from spermatogonium to sperm, takes about 9 to 10 weeks. The testes produce about 130 million sperm each day.

Supportive cells called **Sertoli cells** completely surround the developing spermatocytes in the seminiferous tubules. They supply nutrients to the spermatocytes and seal them off from the body's blood supply. Other cells located in the tissue surrounding the developing spermatocytes, the **Leydig cells,** produce the male sex hormones, known as **androgens,** particularly **testosterone** (see Figure 49.15 and Section 42.4).

Mature sperm flow from the seminiferous tubules into the **epididymis,** a coiled storage tubule attached to the surface of each testis. Rhythmic muscular contractions of the epididymis move the sperm into a thick-walled, muscular tube, the **vas deferens** (plural, *vasa deferentia*), which leads into the abdominal cavity. Just below the bladder, the vasa deferentia empty into the urethra. During ejaculation, muscular contractions force the sperm into the urethra and out of the penis. The sperm are activated and become motile as they come in contact with alkaline secretions added to the ejaculated fluid by accessory glands.

Most of the interior of the penis is filled with three cylinders of spongelike tissue that become filled with blood and

FIGURE 49.15 **Structure of seminiferous tubules and the stages of spermatogenesis.** Spermatogonia are located nearest the outer wall and mature sperm cells nearest the tubule lumen. Sertoli cells surround the developing spermatocytes completely and protect them from attack by the immune system.

cause erection during sexual arousal. Although the human penis depends solely on engorgement of spongy tissue for erection, the males of many mammalian species, including bats, rodents, walruses, and most other primates, have a bone in the penis, the *baculum,* which helps maintain the penis in an erect state.

The penis ends in a soft, caplike structure, the **glans.** Most of the nerve endings producing erotic sensations are crowded into the glans and the region of the penile shaft just behind the glans. A loose fold of skin, the **prepuce** or **foreskin,** covers the glans (see Figure 49.14). In many cultures, the prepuce is removed for hygienic, religious, or other ritualistic reasons by the procedure called **circumcision** (*circum* = around, *caedere* = cut).

ACCESSORY GLANDS AND THE SEMEN About 150 million to 350 million sperm are released in a single ejaculation. Before they leave the body, these cells are mixed with the secretions of several accessory glands, forming the fluid known as **semen.** In humans, about two-thirds of the volume is produced by a pair of **seminal vesicles,** which secrete a thick, viscous liquid, the **seminal fluid,** into the vasa deferentia near the point where they join with the urethra. The seminal fluid contains prostaglandins that, when ejaculated into the female, trigger contractions of the female reproductive tract which help move the sperm into and through the uterus.

The large **prostate gland,** which surrounds the region where the vasa deferentia empty into the urethra, adds a thin, milky fluid to the semen. The alkaline prostate secretion, which makes up about one-third of the volume of the semen, raises the pH of the semen, and of the vagina, to about pH 6, the level of acidity best tolerated by sperm. The raised pH also activates motility of the sperm. As part of the prostate secretion, a fast-acting enzyme converts the semen to a thick gel when it is first ejaculated. The thickened consistency helps keep the semen from draining from the vagina when the penis is withdrawn. A second, slower-acting enzyme in the prostate secretion then gradually breaks down the semen clot and releases the sperm to swim freely in the female reproductive tract.

Finally, a pair of **bulbourethral glands** secretes a clear, mucus-rich fluid into the urethra before and during ejaculation. This fluid lubricates the tip of the penis and neutralizes the acidity of any residual urine in the urethra. In total, the secretions of the accessory glands make up more than 95% of the volume of semen; less than 5% is sperm.

Hormones Also Regulate Male Reproductive Functions

Many of the hormones regulating the menstrual cycle, including GnRH, FSH, LH, and inhibin, also regulate male reproductive functions. Testosterone, secreted by the Leydig cells in the testes, also plays a key role **(Figure 49.16).**

In sexually mature males, the hypothalamus secretes GnRH in brief pulses every 1 to 2 hours. The GnRH, in turn, stimulates the pituitary to secrete LH and FSH. LH stimulates the Leydig cells to secrete testosterone, which stimulates sperm production and controls the growth and function of male reproductive structures. FSH stimulates Sertoli cells to secrete a protein and other molecules that are required for spermatogenesis.

The concentrations of these hormones are maintained by feedback mechanisms. If the concentration of testosterone falls in the bloodstream, the hypothalamus responds by increasing GnRH secretion. If the concentration of testosterone rises too high, negative feedback mechanisms operate to return its concentration to an optimal level in the bloodstream (see Figure 49.16).

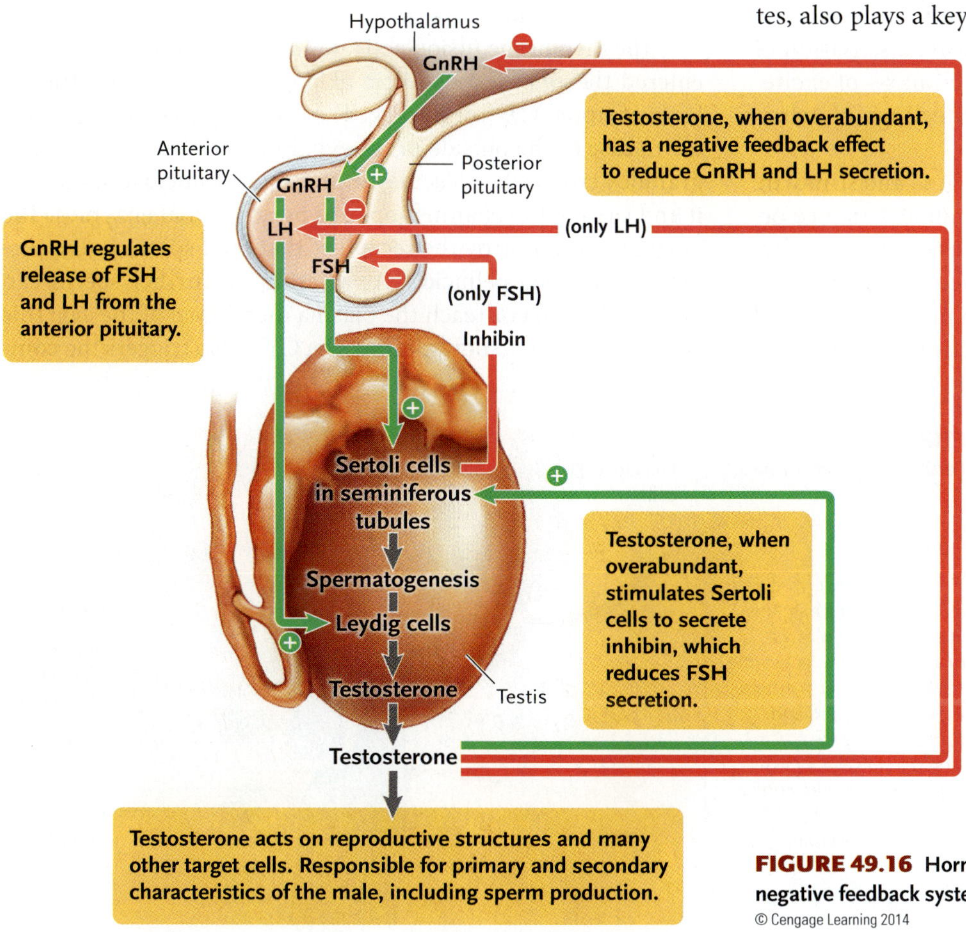

FIGURE 49.16 Hormonal regulation of reproduction in the male, and the negative feedback systems controlling hormone levels.

Human Copulation Follows a Typical Mammalian Pattern

When the male is sexually aroused, the sphincter muscles controlling the flow of blood to the spongy erectile tissue of the penis relax, allowing the tissue to become engorged with blood. (The penis is a hydrostatic skeleton structure; see Section 43.2.) As the spongy tissue swells, it maintains the pressure by compressing and almost shutting off the veins draining blood from the penis. The engorgement produces an erection in which the penis lengthens, stiffens, and enlarges. During continued sexual arousal, lubricating fluid secreted by the bulbourethral glands may be released from the tip of the penis.

Female sexual arousal results in enlargement and erection of the clitoris, in a process analogous to erection of the penis in males. The labia minora also become engorged with blood and swell in size, and lubricating fluid is secreted onto the surfaces of the vulva by the greater vestibular glands. In addition to these changes, the nipples become erect by contraction of smooth muscle cells, and the breasts swell due to engorgement with blood.

Insertion of the penis into the vagina and the thrusting movements of copulation lead to the reflex actions of ejaculation, including spasmodic contractions of muscles surrounding the vasa deferentia, accessory glands, and urethra. During ejaculation, the sphincter muscles controlling the exit from the bladder close tightly, preventing urine from being released from the bladder and mixing with the ejaculate. Ejaculation is usually accompanied by *orgasm,* a sensation of intense physical pleasure that is the peak—climax—of excitement for sexual intercourse, followed by feelings of relaxation and gratification.

The motions of copulation stretch the vagina and stimulate the clitoris. The stretching and stimulation can also induce orgasm in females. The vaginal stretching also stimulates the hypothalamus to secrete oxytocin, which induces contractions of the uterus. The contractions keep the sperm in suspension and aid their movement through the reproductive tract. Uterine contractions are also induced by the prostaglandins in the semen.

Sperm reach the site of fertilization in the oviducts within 30 minutes after their ejaculation into the vagina. Of the millions of sperm released in a single ejaculation, only a few hundred actually reach the oviducts. After orgasm, the penis, clitoris, and labia minora gradually return to their unstimulated size. Females can experience additional orgasms within minutes or even seconds of a first orgasm, but most males enter a *refractory period* that lasts for 15 minutes or longer before they can regain an erection and have another orgasm.

A Human Egg Can Be Fertilized Only in the Oviduct

A human egg can be fertilized only during its passage through the third of the oviduct nearest the ovary. If the egg is not fertilized during the 12- to 24-hour period that it is in this location, it disintegrates and dies. However, sperm do not swim randomly for a chance encounter with the egg. Rather, they first swim up the cervical canal to reach the oviduct, and then are propelled up the oviduct by contractions of its smooth muscles. Further, researchers have found evidence that eggs release chemical attractant molecules that the sperm recognize, causing them to swim directly toward the egg. (*Insights from the Molecular Revolution* describes some of this research.)

To reach the egg, the fertilizing sperm uses enzymes in its plasma membranes to penetrate the layer of follicle cells surrounding the egg, and then by the acrosomal reaction dissolves a path through the zona pellucida coating the egg surface **(Figure 49.17).** As soon as the first sperm cell reaches the egg, the sperm and egg plasma membranes fuse, and the sperm cell enters the cytoplasm of the egg. Although only one sperm fertilizes the egg, the combined release of acrosomal enzymes from many sperm greatly increases the chance that a complete channel will be opened through the zona pellucida. Partly for this reason, a low sperm count is often a source of male infertility. Low sperm count has a number of causes, including infection, heat, frequent intercourse, smoking, and excess alcohol consumption.

The membrane fusion activates the egg. The sperm that has entered the egg releases nitric oxide, which stimulates the release of stored Ca^{2+} in the egg. The Ca^{2+} triggers cortical granule release to the outside of the egg. Enzymes from the cortical granules crosslink molecules in the zona pellucida, hardening it and sealing the channels opened by acrosomal enzymes. The enzymes also destroy the receptors that bind sperm to the surface of the zona pellucida. As a result, no further sperm can bind to the zona or reach the plasma membrane of the egg; this is the slow block to polyspermy. The Ca^{2+} also triggers the com-

FIGURE 49.17
Fertilization in mammals.

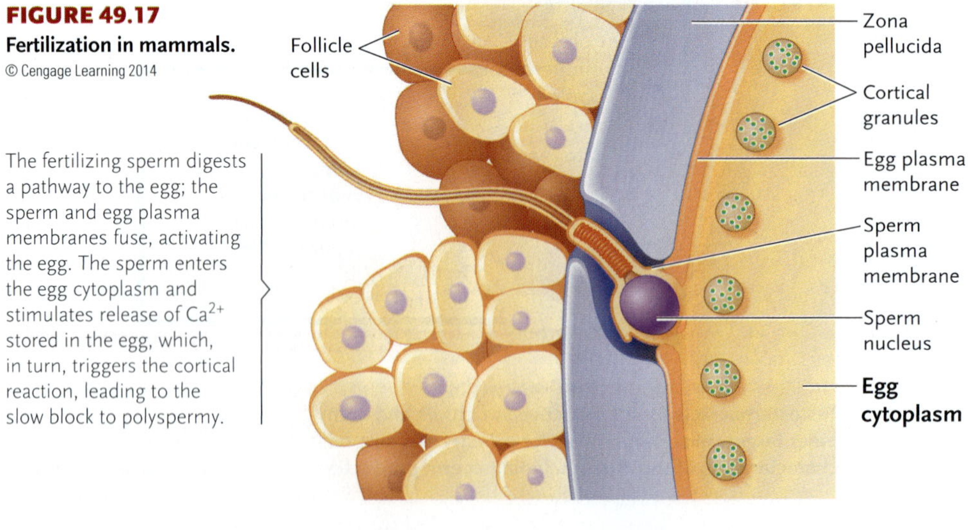

INSIGHTS FROM THE **Molecular Revolution**

Egging on the Sperm

Odorants are aroma molecules that can be specifically recognized ("smelled") by receptor-based mechanisms. In vertebrates, more than a thousand genes encode odorant receptors, most of them olfactory receptors associated with the senses of smell and taste. However, odorant receptors are also located on cell types that do not function in taste and smell, including sperm.

Research Question

Do sperm odorant receptors function in sperm–egg attraction?

Experiments

Marc Spehr and his colleagues at Ruhr University Bochum performed a number of molecular experiments to answer the question. They first performed experiments that showed that testicular tissue contains two active odorant receptor genes. One gene was for hOR17-2 (hOR = *h*uman olfactory receptor), which had been detected previously by others; and the other was for hOR17-4 (now called hOR1D2), which was not previously known to be active. They then molecularly cloned the gene for hOR17-4 and expressed it in cultured human embryonic kidney (HEK) cells. They found that an odorant chemical *bourgeonal* triggered a strong response in the genetically altered cells, but not in cells without the introduced gene. This result provided evidence that the product of the hOR17-4 gene does confer responsiveness to the odorant. They then tested untreated HEK cells for responsiveness to the odorant, and found that they did not respond, providing further evidence that the product of the odorant receptor gene was essential for the response.

In their final experiment, the researchers tested the response of human sperm cells in micropipettes to gradients of bourgeonal solutions.

Result

The sperm swam consistently toward the regions of highest concentration, and swam faster and more directly as the concentration increased. This experiment showed that human sperm can detect and respond to chemical attractants by swimming toward the source of the attractant.

Conclusion

An odorant receptor is on a sperm cell's surface and sperm will swim towards a chemical known to activate that receptor. Whether human eggs actually release such chemicals that act as sperm attractants remains to be determined. If so, the egg attractant detected by the hOR17-4 receptor is likely to resemble bourgeonal in chemical structure, because odorant receptors are highly specific in their responses.

As part of their research, the Spehr team found that another chemical, *undecanal*, strongly inhibits the binding of the hOR17-4 receptor to bourgeonal. With chemicals at hand that can both stimulate and eliminate sperm attraction, the system might provide a method for either contraception or procreation.

THINK LIKE A SCIENTIST A later study by other researchers demonstrated that male human subjects detected bourgeonal at significantly lower concentrations (mean value = 13 parts per billion) compared with female subjects (mean value = 26 parts per billion). What might this result mean?

Source: M. Spehr et al. 2003. Identification of a testicular odorant receptor mediating human sperm chemotaxis. *Science* 299:2054–2058.

pletion of meiosis of the egg (recall that, up to that point, it is a secondary oocyte arrested in meiosis II). The sperm and egg nuclei then fuse and the cell is now a zygote. Mitotic divisions of the zygote soon initiate embryonic development.

The first cell divisions of embryonic development take place while the fertilized egg is still in the oviduct. By about 7 days after ovulation, the embryo passes from the oviduct and implants in the uterine lining. During and after implantation, cells associated with the embryo secrete **human chorionic gonadotropin (hCG),** a hormone that keeps the corpus luteum in the ovary from breaking down. Excess hCG is excreted in the urine; its presence in urine or blood provides the basis of pregnancy tests.

The continued activity of the corpus luteum keeps estrogen and progesterone secretion at high levels, maintaining the uterine lining and preventing menstruation. The high progesterone level also thickens the mucus secreted by the uterus, forming a plug that seals the opening of the cervix from the vagina. The plug keeps bacteria, viruses, and sperm cells from further copulation from entering the uterus.

Later in development, about 10 weeks after implantation, the placenta takes over the secretion of progesterone, hCG secretion drops off, and the corpus luteum regresses. However, the corpus luteum continues to secrete the hormone *relaxin,* which inhibits contraction of the uterus until the time of birth is near.

Infertility Has Many Possible Causes

About 10% to 15% of couples in the United States are infertile. **Infertility** is defined as the inability for the female of the couple to get pregnant after 12 months of frequent, contraceptive-free intercourse. Infertility may result from a cause in either member of the couple, or a combination of factors involving both members of the couple.

MALE INFERTILITY Male infertility may result from a problem with the sperm or with the delivery of the sperm into the vagina. Problems with the sperm include low sperm concentration or altered sperm shape that affects mobility. For example, testosterone deficiency can lead to a reduced sperm count, and sexually transmitted diseases may temporarily affect sperm motility. In many cases, however, the root cause of low sperm count is not known. Problems with sperm delivery include sexual issues such as erectile dysfunction and blockage of the epididymis or ejaculatory ducts. Environmental factors such as a person's health or lifestyle can also affect male fertility. Such factors include obesity, being underweight, malnutrition, emotional stress, alcohol or drug dependency, exposure to pesticides or other chemicals, and tobacco smoking.

FEMALE INFERTILITY Female infertility may result from any of a variety of causes including physical or hormonal changes. For example, inflammation of the Fallopian tubes (most frequently caused by *Chlamydia* infection) can cause damage, thereby affecting movement of the egg through the tube. Hormonal defects such as low levels of the hypothalamus–pituitary controlled LH and FSH, or certain medications, can block ovulation. Hormone deficiencies such as this can result from various perturbations of normal activity of the hypothalamus and pituitary, including tumors, injury, and excessive exercise. Many of the environmental factors that cause female infertility are the same as those that cause male infertility. In addition, female athletes with vigorous training regimens, such as cycling and running, may experience dysfunctions of their menstrual cycles, a condition called *athletic menstrual cycle irregularities.* The dysfunctions include amenorrhea (no menstrual periods) and cycles of normal length but without ovulation.

STUDY BREAK 49.3

Outline the roles of follicle-stimulating hormone (FSH) and luteinizing hormone (LH) in the ovarian cycle of a human female.

THINK OUTSIDE THE BOOK

The concept of athletic menstrual cycle irregularities was just introduced. Use the Internet or research literature to learn about possible hormonal differences in female athletes and to determine whether they might be involved in athletic menstrual cycle irregularities.

TABLE 49.1 Pregnancy Rates for Birth Control Methods

Method	Lowest Expected Rate of Pregnancy[a]	Typical-Use Rate of Pregnancy[b]
Rhythm method	1%–9%	25%
Withdrawal	4%	19%
Condom (male)	3%	14%
Condom (female)	5%	21%
Diaphragm and spermicidal jelly	6%	20%
Vasectomy (male sterilization)	0.1%	0.15%
Tubal ligation (female sterilization)	0.5%	0.5%
Contraceptive pill (combination estrogen/progestin)	0.1%	5%
Contraceptive pill (progestin only)	0.5%	5%
Implant (progestin)	0.09%	0.09%
Intrauterine device (IUD) (copper T)	0.6%	0.8%

[a]Rate of pregnancy when the birth control method was used correctly every time.

[b]Rate of pregnancy when the method was used typically, meaning that it may not have been always used correctly every time.

Source: U.S. Food and Drug Administration, http://www.fda.gov/fdac/features/1997/conceptbl.html. Data reported in 1997 for effectiveness of methods in a 1-year period.

49.4 Methods for Preventing Pregnancy: Contraception

In human society, an unwanted pregnancy can be inconvenient at the least, or at the worst can have serious physical and social repercussions, particularly for the mother. Many methods exist for achieving **contraception**—the prevention of pregnancy—some old and others relatively new.

The oldest method of contraception is total abstinence from sex. Unfortunately, millions of years of animal evolution have stacked the cards against total abstinence by making the sex drive among the most powerful of compulsions. Millions of unwanted children attest to the failures of this method. Other methods of preventing pregnancy include techniques for: (1) preventing the sperm from reaching the site of fertilization; (2) preventing ovulation; or (3) interfering with implantation if fertilization does occur. **Table 49.1** lists the most common contraceptive techniques and their reliability, based on 1 year of use. You should refer to the table as we discuss each of the methods of contraception in the following sections. Note that, while we discuss each method individually, combinations of particular methods may be used to reduce further the chance of pregnancy.

Vasectomy and Tubal Ligation Are the Most Effective Methods for Preventing Fertilization

A natural technique for preventing fertilization is the *rhythm method,* which consists of avoiding intercourse during the time of the month when the egg can be fertilized. Because sperm can survive for as long as 5 days in the female reproductive tract, intercourse should be avoided from 5 days before ovulation and, for safety's sake, for another 4 or 5 days after ovulation. The method is difficult to apply because of the typical unpredictability of the time of ovulation (and the power of the sex drive).

Another natural method to prevent fertilization is *withdrawal*—starting sexual intercourse, but withdrawing the penis before ejaculation. Unfortunately, once ejaculation begins, it proceeds as a series of reflexes that is extremely difficult to interrupt; in addition, some sperm may be present in lubrication produced before ejaculation.

The *condom,* a thin, close-fitting sheath of latex, lambskin, or polyurethane worn over the penis, is one of the traditional methods of preventing ejaculated sperm from entering the vagina. Condoms made from latex may also provide a barrier to the transmission of disease between sexual partners (condoms made from natural skin do not block viruses such as HIV). Pouchlike "female condoms," inserted into the vagina, prevent ejaculated sperm from entering the uterus.

The *diaphragm* is a cuplike rubber device that blocks the cervix in females. (The similar *cervical cap* is smaller and fits more closely over the cervix.) Typically a spermicidal jelly or cream is also used. To be most effective, a diaphragm and the spermicidal jelly must be inserted no more than an hour before intercourse, and left in place for the recommended time afterward.

The *intrauterine device (IUD)* is a small plastic or copper device inserted into the uterus just inside the cervix. The IUD remains in place as a long-term preventive measure; depending on the type, a single IUD is approved for 5 to 10 years of use. It is not completely clear how an IUD works; it is thought to cause a mild inflammation of the uterine lining that interferes with sperm function so that fertilization is less likely to occur. IUDs may also interfere with implantation of a fertilized egg but this may be a smaller contribution to their contraceptive effect. The IUD is effective as long as it is not deflected from its correct position in the uterus; unfortunately, this may happen without warning or the user's awareness. A few women also experience unpleasant side effects from the IUD such as cramps, uterine infections, or excessive menstrual bleeding.

Fertilization can also be prevented surgically, by cutting and closing off either the vasa deferentia in males or the oviducts in females. In *vasectomy,* the procedure for males, an incision is made in the scrotum and each vas deferens is severed and tied off. After vasectomy, the seminal fluid is still produced and ejaculated, but it does not contain sperm. In *tubal ligation,* the procedure for females, the oviducts are cut and tied off, or seared with heat (cauterized) to close them. The ligation prevents eggs from being fertilized or reaching the uterus. Neither vasectomy nor tubal ligation interferes with the production of sex hormones by the ovaries or testes, or results in any change in sexual behavior. Both operations are highly effective in preventing pregnancy. Although they can be reversed, the procedures are difficult and not always successful.

The Oral Contraceptive Pill Is the Most Effective Method for Preventing Ovulation

The primary method used to prevent ovulation is the *oral contraceptive pill,* or simply "the pill," containing a combination of estrogen and *progestin* (a synthetic form of progesterone) or progestin alone. In this highly effective method, the pill is taken daily for 20 to 21 days after the end of the menstrual flow and then stopped (placebo pills are taken for the remaining days of the cycle to maintain the routine of pill taking) to allow menstruation, and then the next month's course is begun. If pregnancy is desired, the pill is simply not taken after the menstrual flow.

The pill works by inhibiting the secretion of FSH and LH by the pituitary; without these hormones, ovulation does not occur. When the pill is stopped after 20 to 21 days, the resulting drop in progestin concentration causes the uterine lining to break down and initiates the menstrual flow. Since ovulation does not occur, fertilization and pregnancy are not possible.

Most pregnancies among women taking the pill result from failure to take it on schedule—often simply by forgetting to take the pill for a day or two at the wrong time of the month. Some women, about one in four, experience unpleasant side effects, such as nausea, tenderness of the breasts, irritability, nervousness, or changes in skin color or texture. Modern versions of the pill have almost eliminated the more serious side effects, such as increased incidence of breast cancer and formation of blood clots. However, cigarette smoking significantly increases the risk of heart attacks and strokes for women taking the pill. This risk increases with age and with the number of cigarettes smoked per day.

The Morning-After Pill Blocks Ovulation or Fertilization

Whatever the method of birth control, its effectiveness is improved if sex partners are highly motivated to avoid pregnancy, and careful in its use. The effectiveness of condoms, for example, is greatly improved if the penis is withdrawn immediately after ejaculation (before the semen has time to spread under the condom and leak into the vagina).

Another method used to prevent pregnancy is the so-called *emergency contraception pill,* commonly referred to as the "morning-after pill." These pills are administered after intercourse has occurred as a means to prevent pregnancy. A high-dosage synthetic progestin emergency contraception pill called Plan B is available in the United States without prescription to women who are 18 or older. This pill is highly effective if taken within 72 hours after unprotected sexual intercourse. Pregnancy tests do not work until significantly after this time. Research data show that Plan B works by blocking ovulation or fertilization. It may also inhibit implantation of a fertilized egg by altering the endometrium, but it has no effect if the process of implantation has begun.

Another emergency contraception pill is *mifepristone (RU-486),* which contains a molecule that binds to and blocks progesterone receptors in the uterine lining. The blockage prevents the lining from responding to progesterone and causes it to break down (that is, a menstrual period is initiated), taking with it any embryo that may have implanted. Mifepristone is approved in the United States for terminating pregnancies up to 49 days postconception; the time period is longer in some foreign countries. It is available only by prescription.

In this chapter we have focused on animal reproduction up to the point of the fertilized egg. In the next chapter, we address the final stage of reproduction in sexually reproducing organisms, the development of a new individual from the fertilized egg.

STUDY BREAK 49.4

How does the oral contraceptive pill prevent pregnancy?

UNANSWERED QUESTIONS

What molecules are responsible for sperm recognition of egg coats?

Although fertilization is necessary for sexual reproduction, how sperm recognize eggs, particularly in mammals, has been an enigma. Fertilization success or failure may be attributed to how well complementary molecules on each gamete interact with each other. Furthermore, as discussed in Section 22.2, one of the prezygotic species-isolating mechanisms relies on the fact that gamete receptors from one species often do not recognize receptors from other species. However, in artificial environments (such as experimental situations), gametes from different species sometimes do recognize each other and form hybrids. These hybrids are usually explained by structural similarity of the molecules responsible for gamete recognition.

During fertilization in mammals, sperm must penetrate through several layers that surround the egg. Using motility and their own surface hyaluronidase, the sperm first move through the cumulus mass, a remnant of follicle cells (referred to as *cumulus cells*) and the sticky hyaluronate-containing matrix that cumulus cells produce. Next, the sperm bind to the zona pellucida, a tough acellular glycoprotein coat. In mammals, this coat is composed of three to four proteins, often referred to as ZP1, ZP2, ZP3, and ZP4. Mice, the mammal in which fertilization has been studied the most, produce ZP1, ZP2, and ZP3.

Starting in the 1980s, researchers have studied the function of zona pellucida glycoproteins. Early experiments studied the function of individual proteins using a competition assay. For this assay, zona pellucida proteins were dissolved and purified. Then, each was added individually to sperm to determine if it could occupy putative receptors on the plasma membrane of sperm cells and prevent the sperm from binding to cumulus-free eggs. Only ZP3 did this.

ZP3 also induces the acrosome reaction in sperm, which is necessary for sperm to penetrate the zona pellucida (discussed in Section 47.2). Subsequent research found that ZP3 bound to a form of the enzyme β-1,4-galactosyltransferase found on the sperm plasma membrane. Curiously, another ZP3 binding protein, sp56, was also found within the acrosome. These proteins may act sequentially to allow sperm to adhere to the zona pellucida (before and during or after the acrosome reaction).

Recent mouse genetic studies suggest that a more complex assembly of zona pellucida proteins is responsible for binding sperm. Other genetic studies indicate that additional receptors on sperm also contribute to sperm–zona pellucida adhesion. Further work is necessary to resolve the early biochemical data and the more recent results from genetic experiments. It is also important to clarify the role of other receptors that have been identified. Finally, recent evidence suggests that sperm begin the acrosome reaction in the cumulus cell mass prior to interaction with the zona pellucida. Regardless, it seems likely that fertilization includes a number or perhaps a sequence of molecular adhesive steps that must function properly for normal fertility and that there may be molecular redundancy in such a fundamental process.

How is membrane fusion regulated during the sperm acrosome reaction?

To release the acrosome, the outer acrosomal membrane fuses at hundreds of points with the plasma membrane overlying it. These fusion points expand, forming vesicles that are released, freeing the acrosomal contents. Acrosomal proteins are released in two phases. First, the more soluble acrosomal proteins are released. Then, acrosomal proteins that initially remained bound to a protein matrix within the acrosome are released. Regulated membrane fusion during secretion has been studied in most detail in neurons. In neurons, each secretory vesicle forms a single fusion point with the plasma membrane. Although membrane fusion during the acrosome reaction involves hundreds of fusion points, and secretory vesicles in somatic cells form just one, we found that proteins promoting membrane fusion in neuronal secretion at the synapse were also involved in the sperm acrosome reaction.

Proteins called SNAREs (from *S*oluble *N*-ethylmaleimide-sensitive factor *A*ttachment protein *RE*ceptor) promote membrane fusion in neuronal secretion. So-called "neuronal" SNAREs are also found in sperm near the acrosome. We have studied mice deficient in a protein called complexin that regulates SNARE function, and have found that sperm from these mice have defective acrosome reactions as well as markedly reduced fertility. Complexin appears to promote SNARE complex formation and then suspend the SNARE complex in an intermediate state in which membrane fusion can be completed readily by a Ca^{2+} influx into sperm.

Despite some molecular similarity to the neuronal process, membrane fusion in sperm is clearly unique. It is important to identify the important steps that prepare sperm for membrane fusion and to determine how SNARE function is regulated during the acrosome reaction. This should help understand sperm malfunctions that result in defects in the acrosome reaction, a frequent cause of male infertility.

THINK LIKE A SCIENTIST One of the procedures used in fertility clinics is called intracytoplasmic sperm injection (ICSI). A sperm is aspirated into a pipette, and the pipette is driven through the zona pellucida and egg membrane. The sperm is then injected directly into an egg, and the egg develops into an embryo. Predict the types of fertility problems for which this procedure could be used to produce offspring.

David Miller is Associate Professor in the Department of Animal Sciences at University of Illinois, Urbana–Champaign. His laboratory studies the molecular underpinnings of sperm maturation and fertilization. You can learn more about Miller's research by visiting: http://labs.ansci.illinois.edu/millerlab/.

REVIEW KEY CONCEPTS

To access the course materials and companion resources for this text, please visit www.cengagebrain.com.

49.1 Animal Reproductive Modes: Asexual and Sexual Reproduction

- In asexual reproduction, a single parent gives rise to offspring without genetic input from another individual. In sexual reproduction, offspring are produced by the union of gametes—eggs and sperm—from two parents.
- Asexual reproduction involving mitosis occurs in animals by fission, budding, or fragmentation (Figure 49.2). In parthenogenesis, a form of asexual reproduction, females produce eggs that develop without being fertilized.
- In sexual reproduction, genetic variability is produced by the meiotic processes of genetic recombination and independent assortment.

Animation: Spermatogenesis

49.2 Cellular Mechanisms of Sexual Reproduction

- Sexual reproduction includes two cellular processes, gametogenesis and fertilization, and a whole-organism process, mating. Gametogenesis is the formation of male and female gametes by meiotic cell division, followed by differentiation of the gametes; fertilization is the union of gametes that initiates development of new individuals (Figure 49.3).
- Gametogenesis takes place in the testes of males and in the ovaries of females. Sperm and eggs are delivered to the site of fertilization by sperm ducts in males and oviducts in females. External reproductive structures aid the delivery in many species.
- In male gametogenesis—spermatogenesis—each cell entering meiosis produces four haploid motile sperm cells. In female gametogenesis—oogenesis—each cell entering meiosis produces one haploid egg cell. The meiotic divisions of oogenesis concentrate almost all the cytoplasm in the single egg cell; the other division products are nonfunctional polar bodies (Figure 49.3).
- The egg contains stored nutrients and information required for at least the early stages of embryonic development. It is covered by one or more protective coats, and it has a mechanism that blocks additional sperm from entering after fertilization (Figure 49.5).
- Fertilization, which follows mating in most animals, may be external or internal. In external fertilization, sperm and eggs are shed into the surrounding water. In internal fertilization, sperm are released close to or inside the female reproductive ducts via copulation (Figure 49.6).
- When a sperm and egg touch during fertilization, their plasma membranes fuse, introducing the sperm nucleus into the egg cytoplasm. The sperm and egg nuclei then fuse to form a diploid zygote nucleus and initiate embryonic development (Figure 49.8).
- Oviparous animals lay eggs in which development of new individuals takes place outside the female's body. In viviparous animals, development takes place inside the female's body. In ovoviviparous animals, fertilized eggs are retained within the body while the embryo develops, the eggs hatch within the mother, and the offspring are then released from the body.
- In hermaphroditism, single individuals produce both mature egg-producing tissue and mature sperm-producing tissue (Figure 49.9).

Animation: Fertilization

49.3 Sexual Reproduction in Humans

- In females, eggs released from the ovaries travel through the oviducts to the uterus. The uterus opens into the vagina, the entrance for sperm and the exit for offspring during birth (Figure 49.10).
- The ovarian cycle produces an egg. The cycle begins with the release of GnRH by the hypothalamus, which stimulates the release of FSH and LH from the anterior pituitary. FSH stimulates oocytes in the ovaries to begin meiosis. One oocyte typically develops to maturity and is surrounded by cells that form a follicle (Figures 49.11 and 49.12).
- The enlarging follicle secretes estrogens, causing a burst in FSH and LH blood concentrations; at about day 14 of the cycle, the LH stimulates ovulation, the bursting of the follicle and the release of the egg. The remainder of the follicle forms the corpus luteum, which secretes estrogens, progesterone, and inhibin (Figures 49.11 and 49.12).
- Day 0 of the monthly uterine (menstrual) cycle correlates with the beginning of follicular development in the ovary and the beginning of the menstrual flow. Secretion of estrogen from the developing follicle stimulates the growth of a new endometrium. If fertilization does not occur, progesterone and inhibin maintain the endometrium until day 28 of the cycle, when the corpus luteum regresses. Without progesterone, the endometrium breaks down and is released as the menstrual flow (Figure 49.12).
- In males, sperm develop in seminiferous tubules in the testes and are released into the epididymis. When a male ejaculates, sperm travel from the epididymis to the vas deferens, and then through the urethra and the penis. The seminal vesicles, prostate gland, and bulbourethral glands add fluids to the sperm traveling to the outside (Figures 49.14 and 49.15).
- Sperm production in males is also controlled by LH and FSH. LH stimulates Leydig cells in the testes to secrete testosterone, which stimulates sperm production. FSH stimulates Sertoli cells in the testes to secrete molecules needed for spermatogenesis (Figure 49.16).
- During copulation, sperm are ejaculated into the vagina of the female. The sperm then swim through the female reproductive tract, aided by contractions of the oviduct and guided by molecules released by the egg. On contact with the egg in the oviduct, the acrosomes of sperm release enzymes that digest a path through the coats of the egg. When the fertilizing sperm contacts the egg, the sperm and egg plasma membranes fuse, releasing the sperm nucleus into the egg cytoplasm and activating the egg. The egg completes meiosis, and the sperm and egg nuclei fuse, producing the zygote (Figure 49.17).
- As the embryo implants, the hormone hCG sustains the corpus luteum, which continues to secrete estrogen and progesterone at high levels. These hormones maintain the uterine lining and prevent menstruation.
- Infertility has many causes, including physical problems, hormonal changes, or environmental factors.

Animation: Female Reproductive System: Anterior view of uterus

Animation: Female Reproductive System: Section of female pelvic area

Animation: Female reproductive system

Animation: Ovarian function

Animation: Hormones and the menstrual cycle

Animation: Male Reproductive System: Section of male pelvic area

Animation: Male reproductive system

49.4 Methods for Preventing Pregnancy: Contraception

- Methods of contraception work by preventing sperm from reaching the site of fertilization, by preventing ovulation, or by interfering with implantation (Table 49.1).
- Methods for preventing fertilization include the rhythm method, the condom, the diaphragm or cervical cap, the IUD, and a vasectomy or tubal ligation.
- The oral contraceptive pill prevents ovulation. It contains a combination of estrogen and the progesterone-like progestin, which inhibiting the secretion of FSH and LH and follicle formation.
- The morning-after pill blocks ovulation or fertilization, and may also inhibit implantation.

UNDERSTAND & APPLY

Test Your Knowledge

1. Asexual reproduction is most successful in:
   a. changing environments.
   b. sessile animals.
   c. densely settled populations.
   d. land animals.
   e. genetically varied individuals.
2. Which of the following processes does not increase genetic diversity?
   a. parthenogenesis
   b. random DNA mutations
   c. genetic recombination
   d. independent assortment
   e. random combinations of paternal and maternal chromosomes
3. Gametogenesis has parallel stages in egg and sperm formation. The stage in eggs that is equivalent to spermatids is the:
   a. primary oocyte.
   b. oogonium.
   c. ovum.
   d. ootid and polar bodies.
   e. secondary oocyte and polar body.
4. External fertilization provides a male with relative certainty of paternity and is found in:
   a. amphibians.
   b. birds.
   c. sharks.
   d. reptiles.
   e. mammals.
5. The slow block to polyspermy:
   a. is caused by a change in membrane potential from negative to positive.
   b. triggers the movement of Ca^{2+} from the cytosol to the endoplasmic reticulum.
   c. triggers a decrease in egg oxidation and protein synthesis.
   d. describes the fusion of egg and sperm nuclei.
   e. includes the fusion of cortical granules with the egg's plasma membrane.
6. Some placental animals provide nutrients to their embryos from an attached membranous yolk-containing sac. They are called:
   a. oviparous animals.
   b. ovoviviparous animals.
   c. metatherians.
   d. eutherians.
   e. mammals.
7. Which activity is a step in the ovarian cycle?
   a. FSH stimulates the pituitary to release GnRH.
   b. When FSH and LH levels fall, the corpus luteum shrinks and the uterine lining breaks down.
   c. Luteinizing hormone stimulates the uterus to make progesterone.
   d. Estrogen levels initially have a positive feedback effect on the pituitary, which is followed by higher estrogen levels causing negative feedback.
   e. A fully developed corpus luteum inhibits uterine lining growth.
8. During spermatogenesis in mammals, sperm travels from the:
   a. Sertoli cells past the epididymis and urethra, through the vas deferens to the prepuce.
   b. seminal vesicles past the prostate gland, through the glans and prepuce to the bulbourethral glands.
   c. vestibular glands past the Leydig cells, through the accessory glands and epididymis to the vas deferens.
   d. labia past the bulbourethral glands, through the vas deferens and urethra to the epididymis.
   e. seminiferous tubules past the Sertoli cells, through the epididymis and vas deferens to the urethra.
9. The secondary oocyte in humans is fertilized in the:
   a. uterus.
   b. vagina.
   c. oviduct.
   d. cervical canal.
   e. ovary.
10. The most effective method to prevent fertilization is:
    a. the oral contraceptive.
    b. the IUD.
    c. the morning-after pill.
    d. vasectomy in men and tubal ligation in women.
    e. the rhythm method.

Discuss the Concepts

1. Currently under development is an "antipregnancy vaccine" that stimulates a woman's immune system to develop antibodies against human chorionic gonadotropin (hCG). How would this method prevent pregnancy?
2. Men sometimes have reduced fertility because of *testicular varicoceles,* varicose veins in the testes in which blood pools. Based on what you now know of the conditions under which sperm develop properly, how do you think this condition might impair sperm development?
3. Spermatogenesis produces four sperm for each primary spermatocyte, but oogenesis produces only one egg for each primary oocyte. Why might these different outcomes be adaptive?

4. Sertoli cells protect spermatocytes from attack by antibodies during their development in the human male. What structures might protect the oocyte and egg from attack by antibodies in the human female?
5. Compare the advantages and disadvantages of sexual and asexual reproduction for an aphid and a parasitic worm.
6. It may be possible to develop a birth control drug that would prevent conception by interfering with fertilization. Outline the design for such a drug and explain exactly how it would work. (There may be more than one design that, in theory, would be effective.)

Design an Experiment

Design experiments to determine if, and at what dose, vitamin E can decrease menstrual cramping significantly.

Interpret the Data

Contamination of water by agricultural chemicals affects the reproductive function of some animals. Are there effects on humans? Epidemiologist Shanna Swan and her colleagues studied sperm collected from men in four cities in the United States **(Table)**. The men were partners of women who had become pregnant and were visiting a prenatal clinic, so all were fertile. Of the four cities, Columbia, Missouri, is located in the county with the most farmlands. New York City represents an area with no agriculture.

Data from a Study of Sperm Collected from Men Who Were Partners of Pregnant Women Who Visited Prenatal Health Clinics in One of Four Cities

	Location of clinic			
	Columbia, Missouri	Los Angeles, California	Minneapolis, Minnesota	New York, New York
Average age	30.7	29.8	32.2	36.1
Percent nonsmokers	79.5	70.5	85.8	81.6
Percent with history of STD*	11.4	12.9	13.6	15.8
Sperm count (million/mL)	58.7	80.8	98.6	102.9
Percent motile sperm	48.2	54.5	52.1	56.4

*STD stands for sexually transmitted disease.

1. Where did researchers record the highest and lowest sperm counts?
2. In which cities did samples show the highest and lowest sperm motility (ability to move)?
3. Aging, smoking, and STDs adversely affect sperm. Could differences in any of these variables explain the regional differences in sperm count?
4. Do these data support the hypothesis that living near farmlands can adversely affect male reproductive function?

Source: S. Swan et al. 2003. Geographic differences in semen quality of fertile US males. *Environmental Health Perspectives* 111:414–420.

Apply Evolutionary Thinking

The nematode species *Caenorhabditis elegans* and *C. briggsae* are both hermaphroditic. Phylogenetic evidence indicates that the last common ancestor of these two species had a normal male–female mechanism of reproduction. What does this evidence suggest about their hermaphroditism?

50

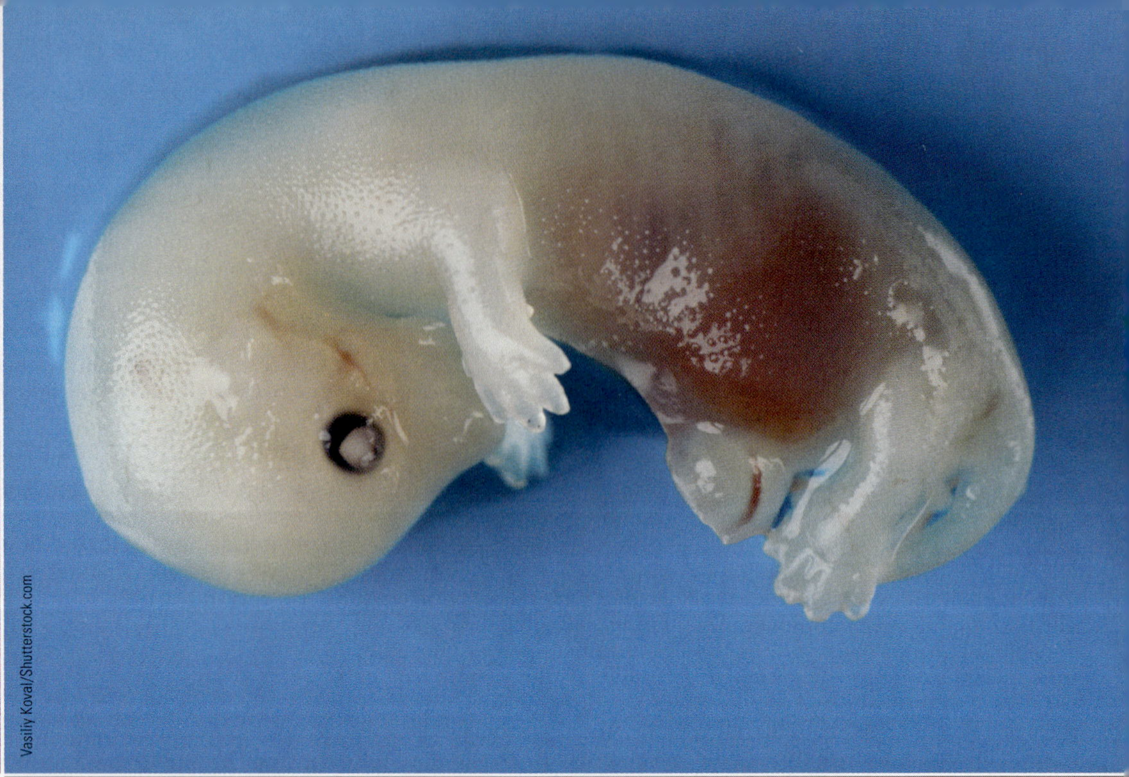

Mink embryo.

STUDY OUTLINE

50.1 Mechanisms of Embryonic Development

50.2 Major Patterns of Cleavage and Gastrulation

50.3 From Gastrulation to Adult Body Structures: Organogenesis

50.4 Embryonic Development of Humans and Other Mammals

50.5 The Cellular Basis of Development

Animal Development

Why it matters . . . The uterine contractions announcing birth are taking place at shorter intervals and with greater intensity. The mother-to-be endures the discomfort and apprehension with the knowledge that the child that has been growing in her body will soon come into the world. It formed from a fertilized egg, about the size of a period on this page, and grew through a program of cell divisions, complex cell migrations, and molecular interactions.

Her baby's development required no conscious attention on her part: human development, like that of all animals, is programmed to proceed inexorably from fertilized egg to free-living offspring. Even childbirth is the result of programmed events that, once started, normally move to conclusion without requiring deliberate input from the mother.

Over the course of its development, the baby's body formed all the organ systems required for independent existence and, at its birth, they are already working to sustain its life. Most astonishing, perhaps, is the baby's brain. It formed from a tube of nerve tissue that bulged outward and enlarged, continually adding nerve cells and connecting them into complex circuits—all as part of the automated events of development. Still, the human brain is unique only in the degree of its complexity and integrative capacity; the brains of other mammals are basically similar and develop through the same embryonic pathways.

The baby enters the outside world passing head first through the cervix, and then the vagina. It is a girl, who with further luck and good care will continue developing through childhood, puberty, adult life, and old age, all through programs built into her DNA. As part of these passages, she may bring her own child into the world.

People have tried since ancient times to understand how development and birth take place. The scientific quest began with Aristotle, who observed chick development and correctly interpreted the functions of the placenta and umbilical cord in humans. The investigators who followed Aristotle concentrated on describing developmental changes in **morphology,** which is the form or shape of an organism, or of a part of an organism. More recently, investigators began to trace the molecular underpinnings of the morphological events.

In this chapter we survey the results of these investigations. We take up the story of animal development where the previous chapter left off, with the fertilized egg. We continue with the early events leading from the fertilized egg to the primary tissues of the embryo, and then trace the development of organs from these tissues. Next, we describe human development as representative of the process in mammals. Then, we survey the cellular bases of these mechanisms. At the cellular level, the development of an adult animal from a fertilized egg involves cell division, in which more cells are produced by mitosis; **cell differentiation,** in which changes in gene expression establish cells with specialized structure and function; and **morphogenesis** ("form creation"), the generation of the body form of the animal as differentiated cells end up in their appropriate sites.

FIGURE 50.1 Body axes: anterior–posterior, dorsal–ventral, and left–right.

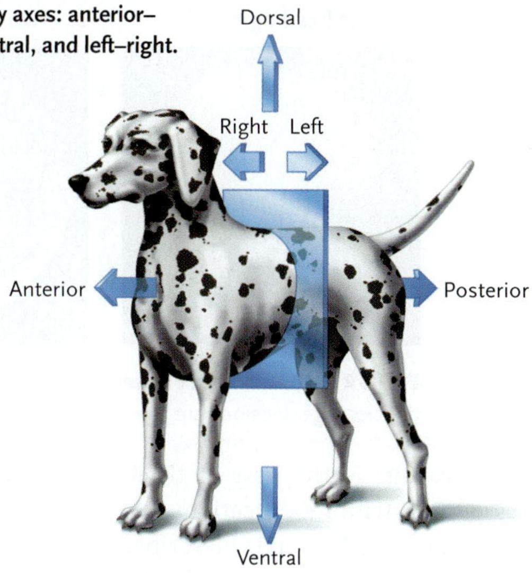

50.1 Mechanisms of Embryonic Development

Fertilization of an egg by a sperm cell produces a zygote. Embryonic development begins at this point and ultimately produces a free-living organism. All the instructions required for development are packed into the fertilized egg.

Developmental Information and Components for Growth Are Stored in the Fertilized Egg

Once the zygote is formed when egg and sperm nuclei fuse; mitotic divisions begin the developmental activity (see Section 49.2).

INFORMATION STORAGE IN THE EGG As introduced in Section 16.4, the initiation of development depends primarily on the DNA in the zygote nucleus, and on *cytoplasmic determinants,* that is, mRNA and protein molecules stored in the cytoplasm. Most of the mRNA and protein molecules are maternal in origin because the fertilizing sperm contributes essentially no cytoplasm to the zygote. The mRNAs and proteins direct the first stages of development up until the genes of the zygote become active.

OTHER COMPONENTS OF THE EGG In addition to cytoplasmic determinants, the egg cytoplasm also contains ribosomes and other cytoplasmic components required for protein synthesis and the early cell divisions of embryonic development. For example, the egg cytoplasm contains all the tubulin molecules required to form the spindles for early cell divisions. It also contains mitochondria, nutrients stored in granules in the yolk and in lipid droplets, and, in many animals, pigments that color the egg or regions of it.

Yolk contains nutrients. The eggs of insects, reptiles, and birds contain large amounts of yolk, which supplies all of the nutrients for development of the embryo. The eggs of placental mammals, in contrast, contain very little yolk, which supplies nutrients only for the earliest stages of development.

Depending on the species, the yolk may be concentrated at one end or in the center of the egg, or distributed evenly throughout the cytoplasm. Its distribution influences the rate and location of cell division during early embryonic development. Typically, cell division proceeds more slowly in the region of the egg containing the yolk. In the large, yolky eggs of birds and reptiles, cell division takes place only in a small, yolk-free patch at the surface of the egg.

Unequal distribution of yolk and other components in a mature egg is termed **polarity.** In most species, the egg nucleus is located toward one end of the egg. This end of the egg, called the **animal pole,** typically gives rise to surface structures and the anterior end of the embryo. The opposite end of the egg, the **vegetal pole,** typically gives rise to internal structures such as the gut and the posterior end of the embryo. Yolk, when unequally distributed in the egg cytoplasm, is most commonly concentrated in the vegetal half of the egg.

Fertilization of the egg launches the early events of development. As you follow these events, keep in mind that the developmental processes involved work together to produce the body plan of the adult. For a bilaterally symmetrical animal such as a human or a dog, for example, the adult must develop three body axes: the anterior–posterior axis, the dorsal–ventral (back–front) axis, and the left–right axis **(Figure 50.1).**

Cleavage, Gastrulation, and Organogenesis Are Early Events in Development

Soon after fertilization, the zygote begins **cleavage,** a series of mitotic divisions in which cycles of DNA replication and division occur *without* the production of new cytoplasm. As a result, the cytoplasm of the egg is partitioned into successively smaller cells without increasing the overall size or mass of the embryo **(Figure 50.2).** These cells are called **blastomeres** (*blastos* = bud or offshoot; *meros* = part or division). In the frog *Xenopus laevis,* for example, twelve cleavage divisions produce an embryo of about 4,000 cells.

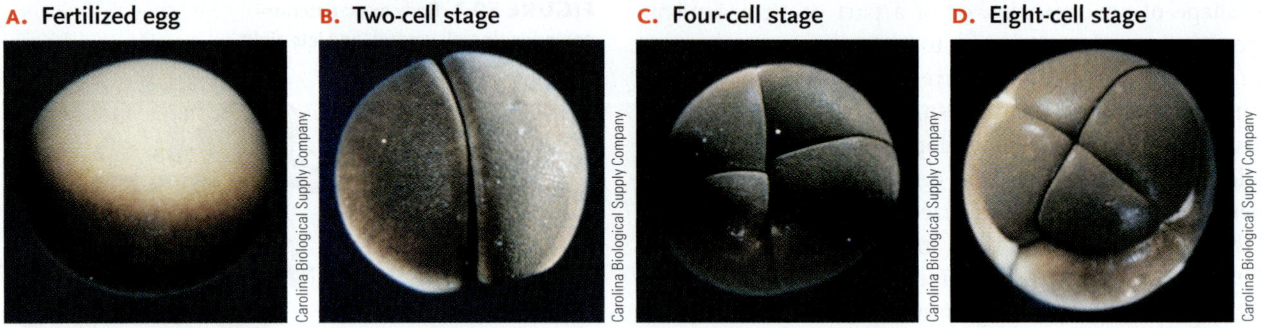

FIGURE 50.2 The first three cleavage divisions of a frog embryo, which convert the fertilized egg into the eight-cell stage. Note that the cleavage divisions cut the volume of the fertilized egg into successively smaller cells, called blastomeres.

Cleavage is the first of three major developmental processes that, with modifications, are common to the early development of most animals (described in detail for particular animals in the next section). Following cleavage, the second major process, **gastrulation,** produces an embryo with three distinct primary tissue layers. The third process, **organogenesis,** accomplishes the development of the major organ systems. At the end of organogenesis, the embryo has the body organization characteristic of its species. Gastrulation and organogenesis both involve the processes of cell division, cell migrations, and cell rearrangements. **Figure 50.3** outlines these stages in the life cycle of a frog.

FIGURE 50.3 Stages of animal development shown in a frog.

The cleavage divisions lead to three successive developmental stages that are common to the early development of most animals. The first stage, called a **morula** (*morula* = mulberry), is a solid ball or layer of blastomeres. As cleavage divisions continue, the ball or layer hollows out to form the second stage, the **blastula** (*ula* = small), in which the blastomeres enclose a fluid-filled cavity, the **blastocoel** (*koilos* = hollow) (see Figure 50.3).

Once cleavage is complete, the blastomeres undergo extensive cellular rearrangements. This morphogenetic process is called **gastrulation,** and during the process the embryo is termed a **gastrula** (*gaster* = gut or belly) (see Figure 50.3). The details of gastrulation differ among animal groups, but the result is the same for all: some surface cells of the blastula move to an interior position, and three primary cell layers of the embryo are produced. The three layers are the **germ layers:** the outer **ectoderm** (*ecto* = outside; *derma* = skin), the inner **endoderm** (*endo* = inside), and the **mesoderm** (*meso* = middle) between the ectoderm and the endoderm. Gastrulation establishes body pattern; that is, each tissue and organ of the adult animal originates in one of the three primary cell layers of the gastrula **(Table 50.1).**

As gastrulation proceeds, embryonic cells begin to differentiate: they become recognizably different in biochemistry, structure, and function. The developmental potential of the cells also becomes more limited than that of the fertilized egg from which they originated. That is, although a fertilized egg is **totipotent,** meaning that it is capable of producing all the various types of cells of the adult, progressively the cells produced become more specialized. That is, totipotent cells give rise to **pluripotent** cells, which can give rise to most adult cell types, and then pluripotent cells give rise to **multipotent** cells, which give rise to cells with particular functions. Thus, for example, a multipotent mesoderm cell may develop into muscle or bone but not normally into outside skin or brain.

The restriction of developmental potential does not occur, as was once thought, because the cells have lost all their genes except those for the structure and function of the cell type they will become. Rather, except in rare instances, the differentiating cells all contain complete genomes of the organism, but each type of cell has a different program of gene expression. Several definitive experiments supporting this conclusion were carried out several decades ago by Robert Briggs and Thomas King of Lankenau Hospital Research Institute in Philadelphia (now Fox Chase Cancer Center) and extended by John B. Gurdon of the University of Cambridge, United Kingdom. In a typical experiment, the nucleus of a fertilized frog egg was destroyed by ultraviolet light. A micropipette was then used to transfer a nucleus from a fully differentiated tissue, intestinal epithelium, to the enucleated egg. Some of the eggs receiving the transplanted nuclei subsequently developed into normal tadpoles and adult frogs. This outcome was possible only if the differentiated intestinal cells still retained their full complement of genes. This conclusion was extended to mammals in 1997 when Ian Wilmut and his colleagues successfully cloned a sheep—Dolly—starting with an adult cell nucleus (see Section 18.2). The constancy of the genome during development was also discussed in Section 16.4.

TABLE 50.1 Origins of Adult Tissues and Organs in the Three Primary Tissue Layers

Primary Tissue Layer	Adult Tissues and Organs
Ectoderm	Skin and its elaborations, including hair, feathers, scales, and nails; nervous system, including brain, spinal cord, and peripheral nerves; lens, retina, and cornea of eye; lining of mouth and anus; sweat glands, mammary glands, adrenal medulla, and tooth enamel
Mesoderm	Muscles; most of skeletal system, including bones and cartilage; circulatory system, including heart, blood vessels, and blood cells; internal reproductive organs; kidneys and outer walls of digestive tract
Endoderm	Lining of digestive tract, liver, pancreas, lining of respiratory tract, thyroid gland, lining of urethra, and urinary bladder

A Number of Processes Are Responsible for Development

Development in all animals is accomplished by a number of processes that are under genetic control but are influenced to some extent by the environment (for example, temperature affects the rate of cell division). The processes are:

1. Mitotic cell divisions.
2. Cell migrations.
3. **Selective cell adhesions,** in which cells make and break specific connections to other cells or to the extracellular matrix.
4. **Induction,** in which one group of cells (the inducer cells) causes or influences another nearby group of cells (the responder cells) to follow a particular developmental pathway (see Section 16.4).
5. **Determination,** in which the developmental fate of a cell is set, and it is committed to becoming a particular cell type. Typically, determination is the result of induction.
6. **Differentiation,** which follows determination, establishes a cell-specific developmental program in cells. Differentiation results in cell types with clearly defined structures and functions.
7. **Apoptosis,** programmed cell death, in which tissues no longer required for continued development of the organism are removed.

Examples of these mechanisms in the examples of development are discussed in the following three sections.

STUDY BREAK 50.1

1. How do cleavage divisions differ from cell division in an adult organism?
2. What are the primary cell layers of the embryo, and what process is responsible for producing them?

50.2 Major Patterns of Cleavage and Gastrulation

With the principles of early embryonic development established, we describe cleavage and gastrulation in three animal groups that have been models in *embryology* (the study of embryos and their development): sea urchins, amphibians, and birds. Later in the chapter, we describe cleavage and gastrulation in humans and other mammals, which resemble the pattern in birds.

Sea Urchin Gastrulation Follows a Symmetrical Pattern That Reflects an Even Distribution of Yolk

The sea urchin is an echinoderm, a type of deuterostome (see Section 32.1). Cleavage divisions first produce a solid mass of blastomeres, the morula (**Figure 50.4,** step 1). Continued divisions produce a blastula containing about a thousand cells (step 2).

Gastrulation begins at the vegetal pole of the blastula. As a result of induction, some cells in the middle of that region become elongated and cylindrical, causing the region to flatten and thicken. Then, some cells break loose and migrate into the blastocoel (step 3). These cells, the *primary mesenchyme cells* (mesenchyme means "middle juice"), move around inside the blastocoel, making and breaking adhesions, until eventually they attach along the ventral sides of the blastocoel. These cells eventually become mesoderm cells (see step 7), from which the larval skeleton is produced. Next, the flattened vegetal pole of the blastula invaginates, pushing gradually into the interior (steps 4 and 5). The invaginated region is the **archenteron** (*arche* = beginning; *enteron* = intestine or gut), which is lined with future endoderm cells, and its opening at the vegetal pole end is the **blastopore.** The archenteron forms the primitive gut of the sea urchin embryo. *Secondary mesenchyme cells* form at the top of the archenteron. These cells also eventually become mesoderm cells.

The secondary mesenchyme cells send out extensions that stretch across the blastocoel and contact the inside of the ectoderm (step 6). These extensions make tight adhesions and then contract, pulling the invaginated cell layer inward and thereby eliminating most of the blastocoel.

At this point the embryo has two complete cell layers. The outer layer remaining from the original blastula surface makes up the ectoderm of the embryo. The second, inner

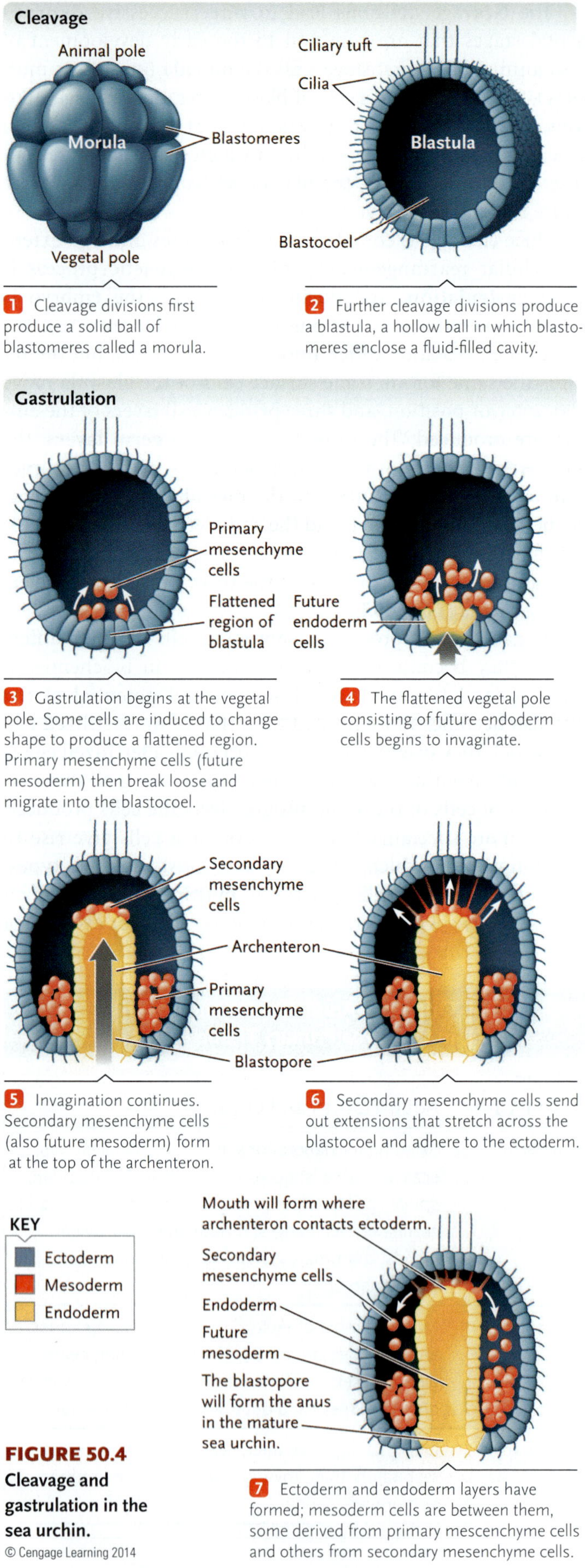

FIGURE 50.4 Cleavage and gastrulation in the sea urchin.

layer, derived from the cells forming the archenteron, makes up the endoderm. Mesodermal cells are also beginning to form a third layer, the mesoderm. Some are derived from the primary mesenchyme cells and others from the secondary mesenchyme cells that migrate into the space between ectoderm and endoderm (step 7). The mesodermal cells originating from secondary mesenchyme cells give rise to mesodermal organs of the sea urchin. When the mesoderm layer is complete, the embryo has three complete layers: ectoderm, mesoderm, and endoderm. At this point, cells within each layer begin to differentiate, as evidenced by the synthesis of different proteins in each layer.

As the ectoderm, mesoderm, and endoderm develop, the embryo lengthens into an ellipsoidal shape with the blastopore marking its posterior end. From this point on, organ systems differentiate through further cell division, cell migrations, selective cell adhesions, induction, and differentiation. The blastopore forms the anus; a mouth forms at the opposite, anterior end of the embryonic gut where the archenteron contacts and fuses with the ectoderm.

An archenteron and a blastopore are not present in the embryos of all organisms. For those that do have them, the blastopore gives rise to the anus or mouth of the embryo, depending on the animal group (see Section 31.2). In deuterostomes, which include echinoderms and chordates, the blastopore develops into the anus and the mouth forms at the opposite end of the embryonic gut formed by the archenteron. In the protostomes, which include annelids, arthropods, and mollusks, the blastopore develops into the mouth, and the anus forms at the opposite end of the embryonic gut.

Amphibian Cleavage and Gastrulation Are Influenced by an Unequal Distribution of Yolk

In amphibian eggs, such as those of frogs, yolk is concentrated in the vegetal half, which gives it a pale color. The animal half is darkly colored by a layer of pigment granules just below the surface. **Figure 50.5** shows the steps of fertilization and cleavage of a frog egg. The sperm typically fertilizes the egg in the animal half (step 1). After fertilization, the pigmented layer of cytoplasm rotates toward the site of sperm entry, exposing a crescent-shaped region of the underlying cytoplasm at the side opposite the point of sperm entry (step 2). This region, called the **gray crescent,** establishes the dorsal–ventral axis of the embryo, with the gray crescent marking the future dorsal side. The first cleavage division runs perpendicular to the long axis of the gray crescent and divides the crescent equally between the resulting two cells (step 3).

If one of the first two blastomeres does not receive gray crescent material, and the cells are separated experimentally, the cell without gray crescent divides to produce a disordered mass that stops developing. The cell receiving the gray crescent produces a normal embryo. Thus cytoplasmic material localized in the gray crescent is essential to normal development in frog embryos.

The second cleavage division is also in the vertical plane and at 90 degrees to the first cleavage division, producing the four-cell stage (step 4). At this point, the yolk in the vegetal half of the egg has slowed the progression of the first two cleavage furrows from going through that part of the egg. The third cleavage division is in the horizontal plane in the animal half of the egg and produces the eight-cell stage (step 5). By now the cleavage furrows of the first two cleavage divisions have continued all the way through the vegetal half of the egg. Embryo development continues with more divisions, first producing a morula (defined as a 16- to 64-cell stage in amphibians), and then a blastula (defined as at least 128 cells in amphibians) (step 6).

Gastrulation begins when cells from the animal pole move across the embryo surface and reach the region derived from the gray crescent (**Figure 50.6,** steps 1–2). This site is marked by a crescent-shaped depression rotated clockwise 90° called the **dorsal lip of the blastopore.** Cells changing shape and pushing inward from the surface in a process called **invagination** produce the depression. With continued inward migration of additional cells, the depression eventually forms a complete circle, which is the blastopore.

Cells now migrate into the blastopore by **involution,** a process in which the cells entering from the outer layer of the embryo spread over the internal surfaces of the remaining exterior cells (step 3). A consequence of involution is that the pigmented cell layer of the animal half expands to cover the entire surface of the embryo. The cells of the vegetal half are enclosed by the movement and show on the outside as a yolk plug in the blastopore (step 4). In amphibians, the blastopore gives rise to the anus.

Continued involution moves cells into the interior and upward (see steps 3 and 4), forming two layers that line the inside top half of the embryo. The uppermost of these layers is induced to become the dorsal mesoderm (see step 4). The layer beneath it, which contains cells originating from both the outer surface of the embryo and the yolky interior, becomes the endoderm (shown in yellow). The pigmented cells remaining at the surface of the embryo form the ectoderm (shown in blue). The ventral mesoderm begins to be induced near the vegetal pole.

As the mesoderm and endoderm form, the depression created by the inward cell migration gradually deepens and extends inward as the archenteron (see steps 3 and 4), which displaces the blastocoel. The cells of the three primary cell layers continue to increase in number by further cell migration and divisions as development proceeds.

During frog gastrulation, cells of the dorsal lip of the blastopore are inducer cells that control blastopore formation; if the cells in the dorsal lip are removed and transplanted elsewhere in the egg, they cause a second blastopore—and a second embryo—to form in this region (see Section 50.5).

The events of gastrulation in frogs thus include the same developmental processes as in sea urchins—cell divisions, cell migrations, selective adhesions, induction, and differentiation.

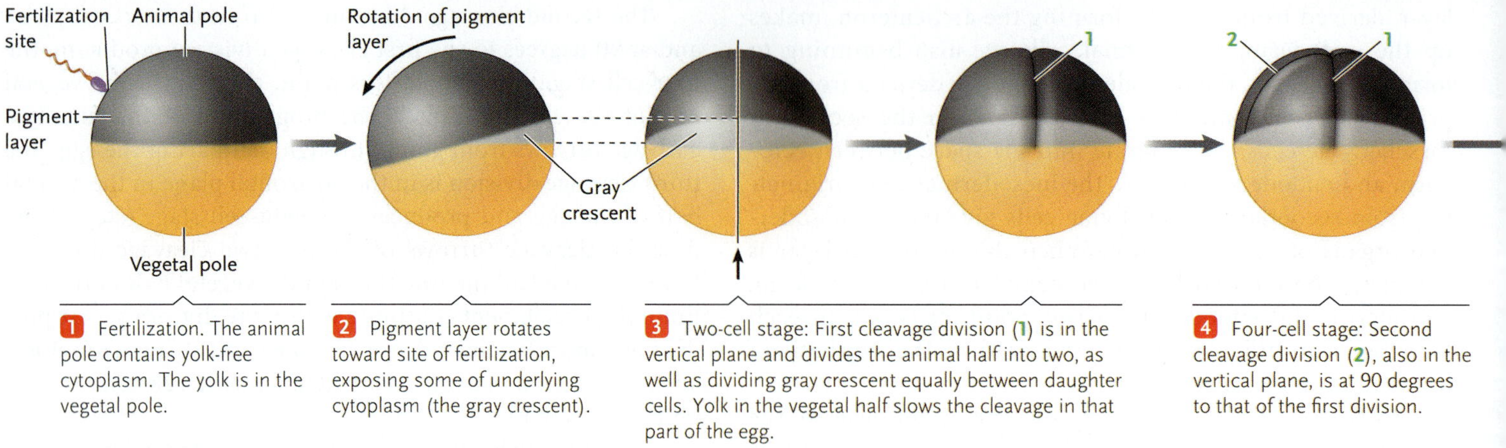

FIGURE 50.5 Fertilization and cleavage of a frog egg. The red numbers designate the cleavage divisions.

Gastrulation in Birds Proceeds at One Side of the Yolk

Gastrulation in amniotes (see Section 32.7) such as birds and reptiles is modified by the distribution of yolk, which occupies almost the entire volume of the egg. The portion of the cytoplasm that divides to give rise to the primary tissues of the embryo is confined to a thin layer at the egg surface. Although eggs of mammals (which are also amniotes) have little yolk, gastrulation follows a similar pattern, as discussed in Section 50.4.

CLEAVAGE AND GASTRULATION IN BIRDS The early cleavage divisions in birds produce a disclike layer of cells at the surface of the yolk called the **blastodisc** (**Figure 50.7,** step 1). When blastodisc formation is complete, the layer contains about 20,000 cells. The cells of the blastodisc then separate into two layers, called the **epiblast** (top layer) and **hypoblast** (bottom layer). The flattened cavity between them is the blastocoel (step 2).

Gastrulation begins as cells in the epiblast stream toward the midline of the blastodisc, thickening the epiblast in this region. The thickened layer—the **primitive streak**—begins forming in the posterior end of the embryo and extends toward the anterior end as more cells of the epiblast move into it (step 3). The primitive streak initially designates the future posterior end of the embryo, and by the time it has elongated fully, it has established the left and right sides of the embryo. The primitive streak forms on what will become the dorsal side of the embryo, with the ventral side below.

As the primitive streak forms, its midline sinks, forming the **primitive groove.** The primitive groove is a conduit for migrating cells to move into the blastocoel. The first cells to migrate through the primitive groove are epiblast cells (step 4), which will form the endoderm. Cells migrating laterally between the epiblast and the endoderm form the mesoderm. The epiblast cells left at the surface of the blastodisc form the ectoderm (see step 4). Thus all three of the primary tissue layers of

FIGURE 50.6 Gastrulation in a frog embryo.

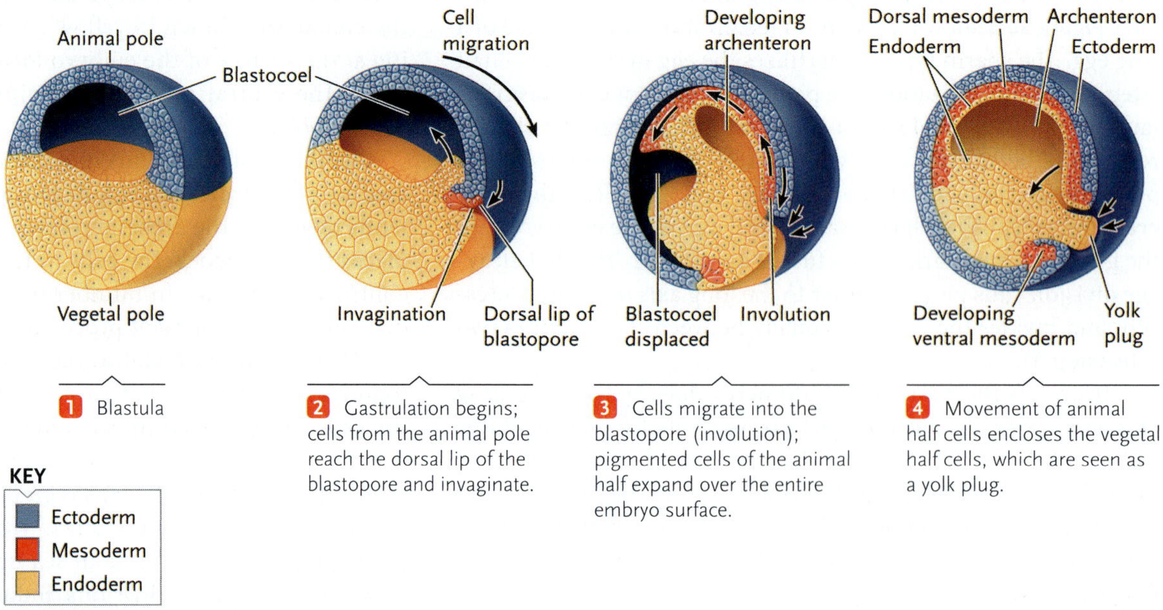

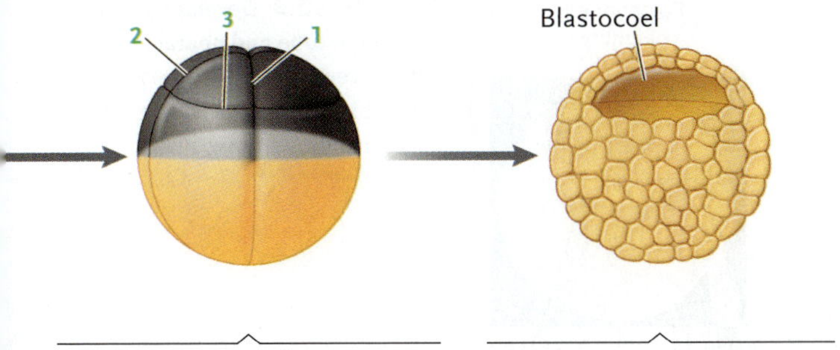

5 Eight-cell stage: Third cleavage division (3) is in the horizontal plane within the animal half of the cell. Divisions 1 and 2 have now continued through the vegetal half of the egg.

6 Blastula stage (cross section): More cleavage divisions produce the blastula, which contains at least 128 cells and has a fluid-filled blastocoel.

the chick embryo arise from the epiblast. Once the three layers are formed, gastrulation is complete.

Of the cells in the hypoblast, only a few near the posterior end of the embryo contribute directly to the embryo. These hypoblast cells form the *germ cells* that, later in development, migrate to the developing gonads and establish the cell line leading to eggs and sperm (see Section 49.2).

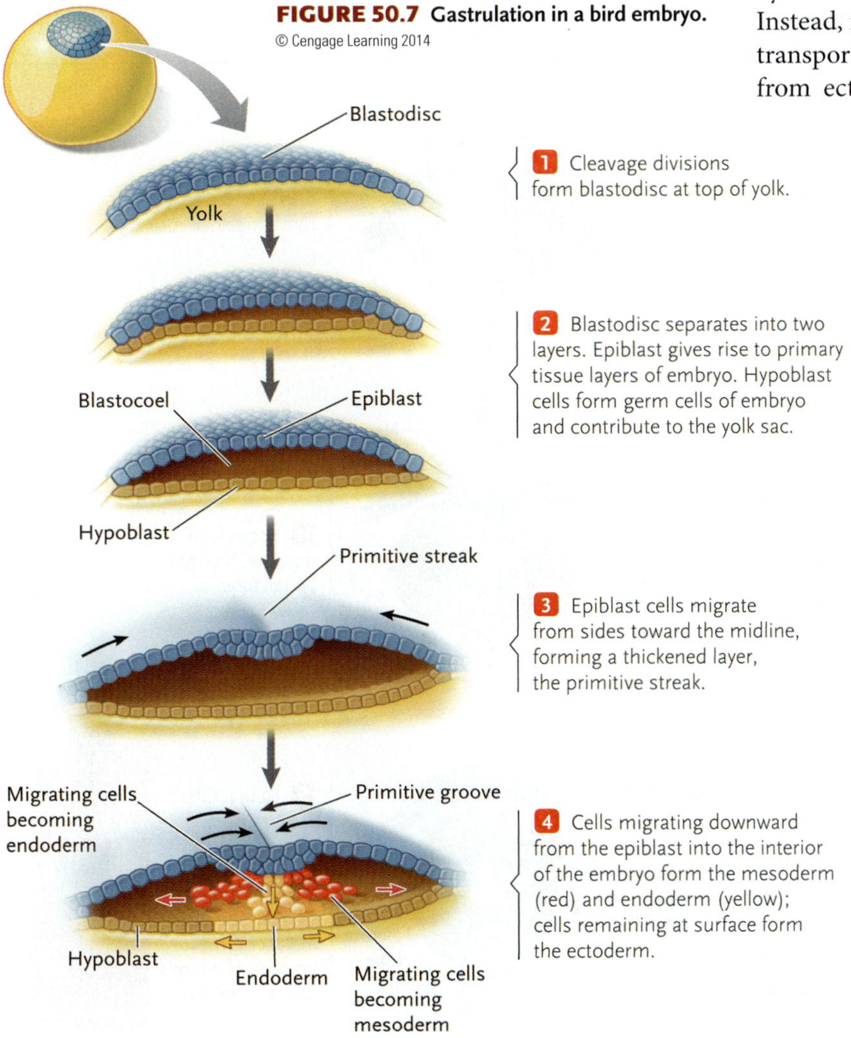

FIGURE 50.7 **Gastrulation in a bird embryo.**

FORMATION OF EXTRAEMBRYONIC MEMBRANES Each primary tissue layer of a bird embryo extends outside the embryo to form four **extraembryonic membranes (Figure 50.8),** which conduct nutrients from the yolk to the embryo, exchange gases with the environment outside the egg, and store metabolic wastes removed from the embryo. The **yolk sac** consists of extensions of mesoderm and endoderm that enclose the yolk. Although the yolk sac remains connected to the gut of the embryo by a stalk, yolk does not directly enter the embryo by this route. Instead, it is absorbed by blood vessels in the membrane, which transport the nutrients to the embryo. The **chorion,** produced from ectoderm and mesoderm, is the outermost membrane, which surrounds the embryo and yolk sac completely, and lines the inside of the egg shell. This membrane exchanges oxygen and carbon dioxide with the environment through the shell of the egg. The **amnion** is the innermost membrane, which closes over the embryo to form the *amniotic cavity.* The cells of the amnion secrete *amniotic fluid* into the cavity, which bathes the embryo and provides an aquatic environment in which it can develop. Reptilian and mammalian embryos are also surrounded by an amnion and amniotic fluid. The adaptation of providing the embryo with an aquatic environment made the development of fully terrestrial vertebrates possible. The evolutionary importance of the amnion to the fully terrestrial vertebrates is recognized by classifying them together as amniotes (see Section 32.7). A membrane derived from mesoderm and endoderm that has bulged outward from the gut forms a sac called the **allantois.** This sac closely lines the chorion and fills much of the space between the chorion and the yolk sac. The allantois stores nitrogenous wastes (primarily uric acid) removed from the embryo. In addition, the part of the allantoic membrane that lines the chorion forms a rich bed of blood capillaries that is connected to the embryo by arteries and veins. This circulatory system delivers carbon dioxide to the chorion and picks up the oxygen that is absorbed through the shell and chorion.

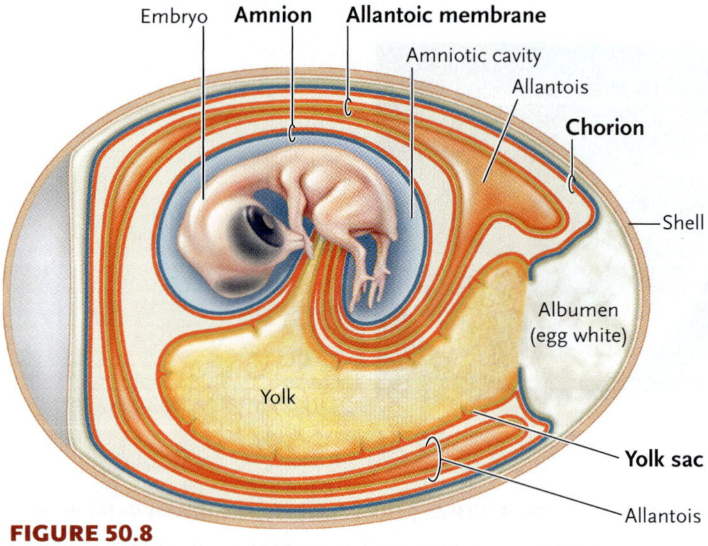

FIGURE 50.8
The four extraembryonic membranes in a bird embryo (in bold).

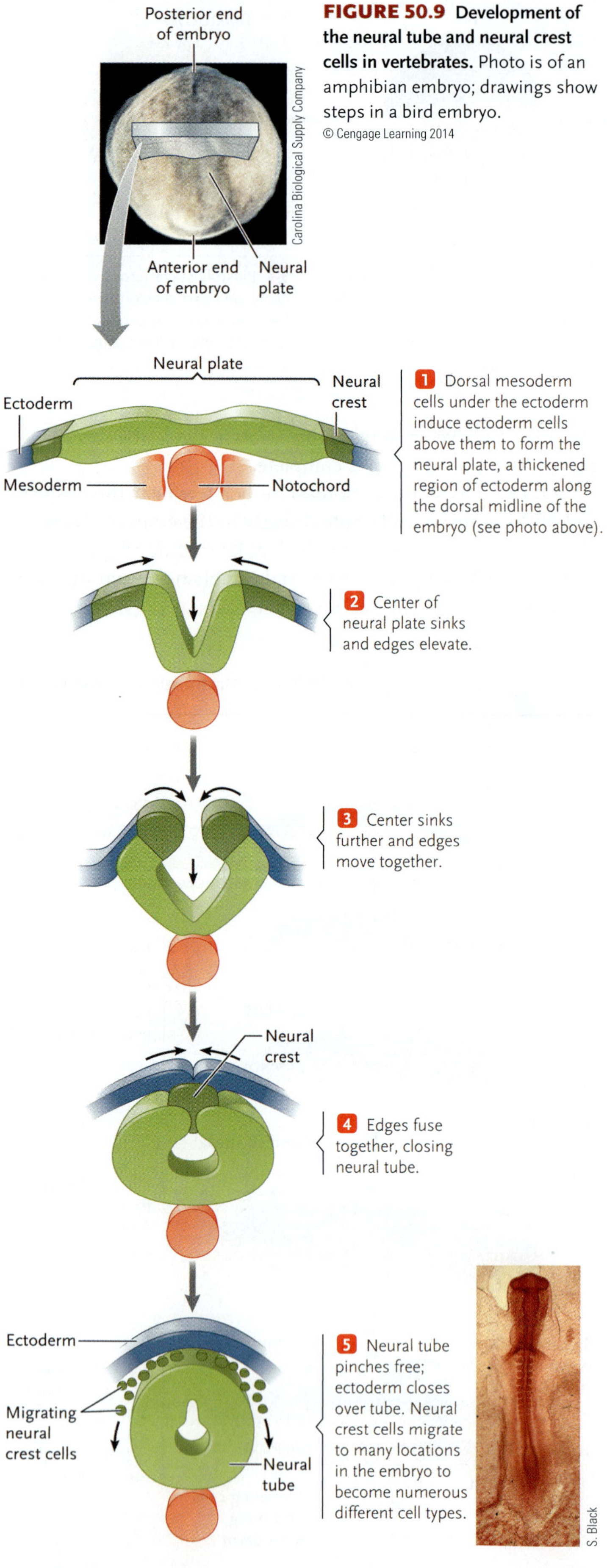

FIGURE 50.9 Development of the neural tube and neural crest cells in vertebrates. Photo is of an amphibian embryo; drawings show steps in a bird embryo.

STUDY BREAK 50.2

1. **What is the role of the gray crescent in amphibian development?**
2. **What evidence indicates that cells of the dorsal lip of the blastopore act as inducer cells?**
3. **What are the extraembryonic membranes in birds, and what are their functions?**

50.3 From Gastrulation to Adult Body Structures: Organogenesis

Following gastrulation, organogenesis—the process by which the ectoderm, mesoderm, and endoderm develop into organs—gives rise to an individual with the body organization characteristic of its species. Organogenesis involves the same mechanisms used in gastrulation—cell division, cell migrations, selective cell adhesion, induction, and differentiation—plus an additional mechanism, *apoptosis,* in which certain cells are programmed to die (apoptosis is also discussed in Section 45.2 and later in this section). As with other aspects of development, the details of organogenesis differ among animal groups. The frog and the chick are used here as examples.

The Nervous System Develops from Ectoderm

In vertebrates, organogenesis begins with **neurulation,** the development of the nervous system from ectoderm. Preliminary to neurulation, cells of the dorsal mesoderm form the **notochord,** a solid rod of tissue that extends the length of the embryo under the dorsal ectoderm. Dorsal mesoderm cells under the ectoderm then induce the ectoderm cells above them to thicken and flatten into a longitudinal band called the **neural plate** (**Figure 50.9,** step 1).

Next, the neural plate sinks downward along its midline (step 2), creating a deep, longitudinal groove. At the same time, the edges elevate along the sides of the neural plate. The groove becomes deeper and the edges move together (step 3). Next, the edges fuse together and close over the center of the groove, converting the neural plate into a **neural tube** that runs the length of the embryo (step 4). The neural tube then pinches off from the overlying ectoderm, which closes over the tube (step 5). The ectoderm that comprises the neural tube is called *neural ectoderm* (colored light green in figures) to distinguish it from surface ectoderm (colored blue in figures). The central nervous system, including the brain and spinal cord, develops directly from the neural tube.

During formation of the neural tube, cells of the **neural crest**—the region where the neural tube pinches off from the ectoderm—migrate to many locations in the developing embryo and become numerous different types of cells that contribute to a variety of organ systems. Neural crest cells are also ectodermal cells; they are colored dark green in Figure 50.9. (The neural crest is one of the defining features of vertebrates.) Some cells develop into cranial nerves in the head; others contribute to the bones of the inner ear and skull, the cartilage of facial structures, and the teeth. Still others form ganglia of the autonomic nervous system, peripheral nerves leading from the spinal cord to body structures, and nerves of the developing gut. Neural crest cells also move to the skin, where they form pigment cells, and to the adrenal glands, where they form the medulla of the kidney. The migration of neural crest cells contributes to the development of all vertebrates.

Other structures differentiate in the embryo while the neural tube is forming. On either side of the notochord, the mesoderm separates into blocks of cells called **somites,** spaced one after the other along both sides of the notochord **(Figure 50.10).** The somites give rise to the vertebral column, the ribs, the repeating sets of muscles associated with the ribs and vertebral column, and muscles of the limbs. The mesoderm outside the somites, which extends around the primitive gut (lateral mesoderm in Figure 50.10), splits into two layers, one covering the surface of the gut, and the other lining the body wall. The space between the layers is the coelom of the adult (see Section 31.2).

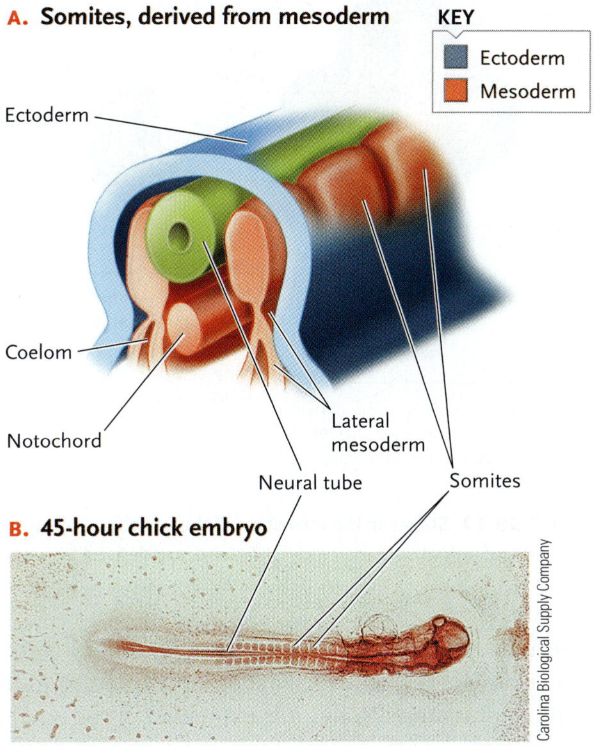

FIGURE 50.10 Later development of the mesoderm. (A) The somites develop into segmented structures such as the vertebrae, the ribs, and the musculature between the ribs. The lateral mesoderm gives rise to other structures, such as the heart and blood vessels and the linings of internal body cavities. **(B)** The somites in a 45-hour chick embryo.

Sequential Inductions and Differentiation Are Central to Eye Development

We now discuss the development of the eye to show how cellular mechanisms interact in organogenesis. Eyes develop by the same pathway in all vertebrates.

The brain forms at the anterior end of the neural tube from a cluster of hollow vesicles that swell outward from the neural tube (**Figure 50.11,** step 1). One paired set of vesicles, the *optic vesicles,* develop into the eyes. The figure depicts the optic vesicles in the brain of a frog embryo; the morphology of the forebrain, midbrain, and hindbrain in embryos differs among vertebrates.

The optic vesicles are derived from a specific region of neural ectoderm in the neural plate. The simultaneous expression of a specific set of transcription factors—Pax6, Six3, and Rx1—in the most anterior tip of the neural plate is responsible for optic vesicle development. The optic vesicles grow outward and contact the overlying surface ectoderm, where they induce a series of developmental responses in both tissues. The outer surface of the optic vesicle thickens and flattens at the region of contact and then pushes inward, transforming the optic vesicle into a double-walled *optic cup,* which eventually becomes the retina. The optic cup induces the surface ectoderm to thicken into a disclike swelling, the *lens placode* (step 2). The center of the lens placode invaginates—sinks inward—toward the optic cup, and its edges eventually fuse together, forming a ball of ectodermal cells, the *lens vesicle* (step 3).

Ectoderm now closes over the lens vesicle, which then detaches from the surface ectoderm (step 4). The lens vesicle becomes the developing lens, the cells of which begin to synthesize *crystallins,* which are transparent, lens-specific proteins that collect into clear, glassy deposits. The lens cells finally lose their nuclei and form the elastic, crystal-clear lens.

As the developing lens invaginates, it induces the overlying surface ectoderm to begin differentiation into the cornea. First, the overlying ectoderm secretes layers of collagen. Neural crest cells then migrate into the collagen layers and form new layers of cells that eventually mature into the clear cornea. Relatively little is known about the development of the cornea, but the

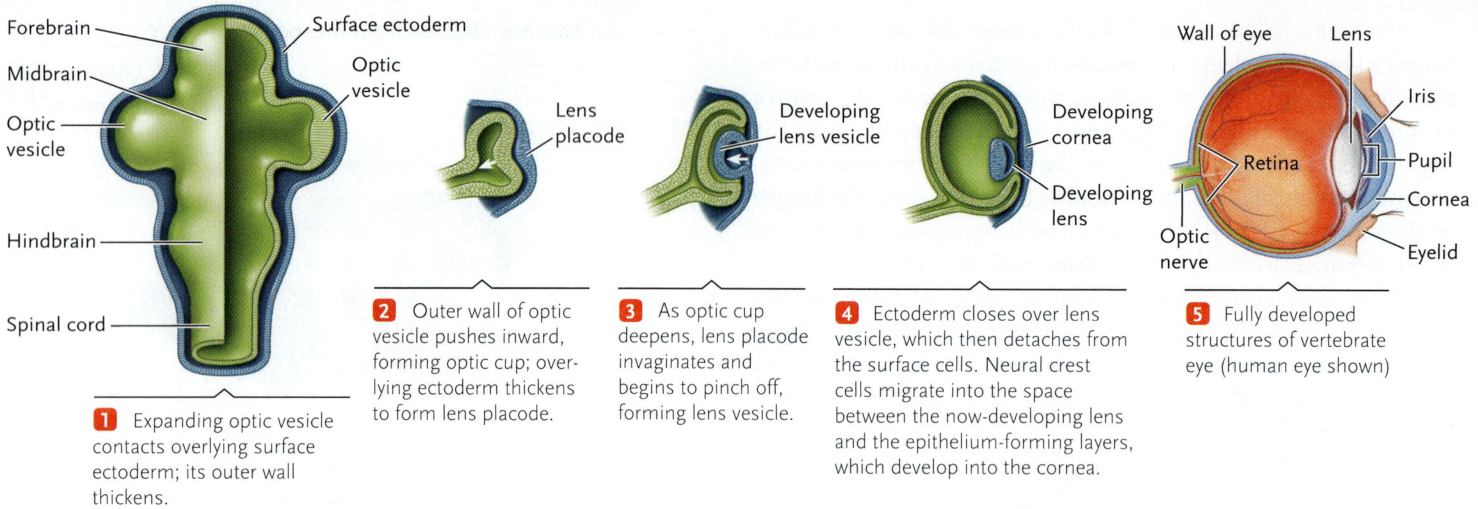

FIGURE 50.11 **Stages in the formation of the vertebrate eye from the optic vesicle of the brain and the surface ectoderm.** (The brain and spinal cord, colored light green, are derived from the neural tube, which consists of neural ectoderm.)

transcription factor Pax6 (see earlier discussion) is important for the process. The gene for Pax6 is not expressed in most tissues of adult mammals, but it is expressed in the adult cornea. The gene products that result from the action of Pax6 are important for healing wounds of the cornea.

Other cells contribute to accessory structures of the eye. For example, mesoderm and neural crest cells contribute to the reinforcing tissues in the wall of the eye and the muscles that move the eye. Figure 50.11, step 5, shows a fully developed vertebrate eye.

Many experiments have shown that the initial induction by the optic vesicle is necessary for development of the eye. For example, if an optic vesicle is removed before lens formation, the ectoderm fails to develop a lens placode and vesicle. Moreover, placing a removed optic vesicle under the ectoderm in other regions of the head causes a lens to form in the new location. Or, if the ectoderm over an optic vesicle is removed and ectoderm from elsewhere in the embryo is grafted in its place, a normal lens will develop in the grafted ectoderm, even though in its former location it would not differentiate into lens tissue.

Eye development also demonstrates differentiation. Ectoderm cells that are induced to form the lens of the eye synthesize crystallins; in other locations, ectoderm cells typically synthesize a different protein, *keratin,* as their predominant cell product. Keratin is a component of surface structures such as skin, hair, feathers, scales, and horns. In other words, as a response to induction by the optic vesicle, the genes of the ectoderm cells coding for crystallin are activated, whereas genes coding for keratin are not expressed.

STUDY BREAK 50.3

1. **What is the outcome of organogenesis?**
2. **What tissues or organs develop from the neural tube and neural crest cells?**

50.4 Embryonic Development of Humans and Other Mammals

The embryonic development of humans is representative of the placental mammals. In the uterus, the embryo is nourished by the placenta, which supplies oxygen and nutrients and carries away carbon dioxide and nitrogenous wastes.

Pregnancy or **gestation,** the period of mammalian development in the uterus, varies in different species. Larger mammals bearing larger young generally have longer gestation periods; for example, gestation takes 600 days in elephants and about 1 year in blue whales, versus 21 days in hamsters.

In humans, gestation takes about 38 weeks, which amounts to about 40 weeks from the beginning of the menstrual cycle in which fertilization takes place. Human gestation is divided into three **trimesters,** each approximately 3 months long.

The major developmental events in human gestation—cleavage, gastrulation, and organogenesis—take place during the first trimester. By the fourth week, the embryo's heart is beating, and by the end of the eighth week, the major organs and organ systems have formed. From this point until birth, the developing human is called a **fetus.** Only 5 cm long by the end of the first trimester, the fetus grows during the second and third trimesters to an average length of 50 cm and an average mass of 3.5 kg (or about 19.7 inches and 7.7 pounds). The period of gestation ends with birth.

Cleavage and Implantation Occupy the First 2 Weeks of Development

Human fertilization occurs when the egg is in the first third of the oviduct leading from the ovary to the uterus (see Section 49.3). After fertilization, cleavage divisions take place during passage of the developing embryo down the oviduct (also called the fallopian tube in mammals) and while it is still enclosed in the zona pellucida—the original coat of the egg **(Figure 50.12).**

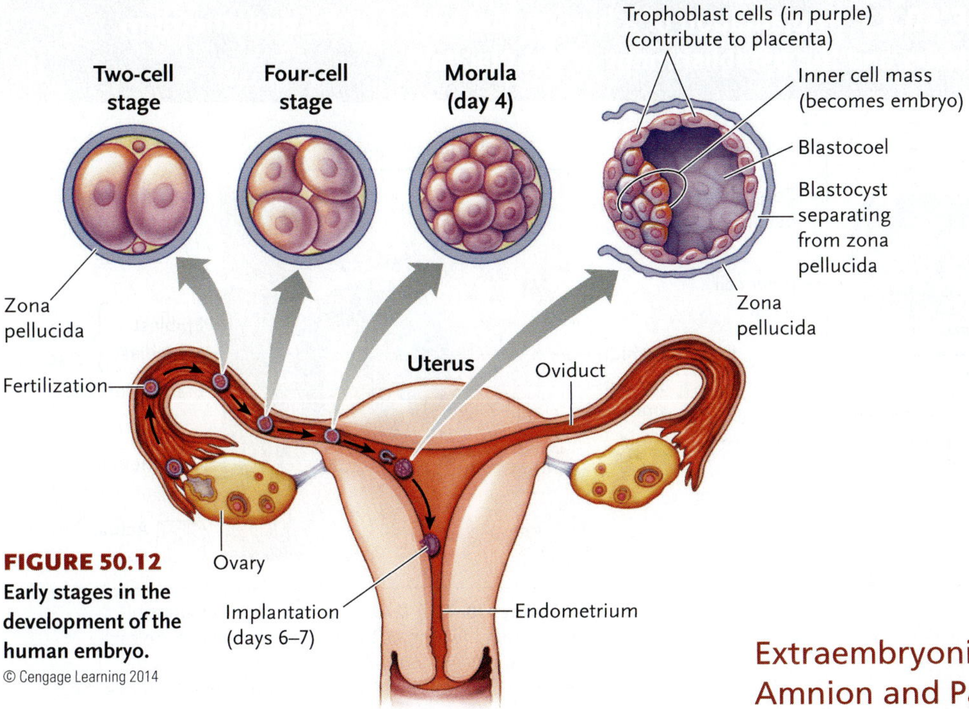

FIGURE 50.12 Early stages in the development of the human embryo.

By day 4, the morula, a 16- or 32-cell ball, has been produced. By the time the endometrium (uterine lining) is ready for implantation (about 7 days after ovulation; see Section 49.3), the morula has reached the uterus and has undergone further cell divisions and differentiation into a type of blastula called a **blastocyst,** which is a characteristic of mammalian development. At this time, the blastocyst is a single-cell-layered hollow ball that is made up of about 70 to 100 cells, and contains a fluid-filled cavity. The blastocoel, the fluid-filled cavity, has a dense mass of cells localized to one side called the **inner cell mass.** This inner cell mass gives rise to the embryo as well as the yolk sac, the allantois, and the amnion. The rest of the blastocyst becomes tissues that support the development of the embryo in the uterus. The outer single layer of cells of the blastocyst is the **trophoblast.**

Next, when the blastocyst has increased to an appropriate size, it breaks out of the zona pellucida and sticks to the endometrium on the side with the inner cell mass (**Figure 50.13,** step 1). Implantation begins when the trophoblast secretes proteases that digest pathways between endometrial cells. The trophoblast extends into the pathways, appearing like fingerlike projections into the endometrium. Continued digestion of endometrial cells generates a hole in the endometrium for the blastocyst and releases nutrients that the developing embryo can use.

The inner cell mass of the burrowing blastocyst separates into the *embryonic disc,* which consists of two distinct cell layers (see Figure 50.13, step 1). The layer farther from the blastocoel is the *epiblast,* which gives rise to the embryo proper, and the layer nearer the blastocoel is the *hypoblast,* which gives rise to part of the extraembryonic membranes. When implantation is complete, the blastocyst has completely burrowed into the endometrium and is covered by a layer of endometrial cells (step 2).

Mammalian Gastrulation and Neurulation Resemble the Reptilian–Bird Pattern

Gastrulation proceeds as in birds (see Figure 50.7), with the formation of a primitive streak in the epiblast. Some epiblast cells remain in place, becoming the ectoderm, whereas others enter the streak to form the endoderm and mesoderm. The ectoderm, mesoderm, and endoderm are located initially in three layers; from this initial arrangement, the endoderm folds to form the primitive gut, and becomes surrounded with ectoderm and mesoderm. Neurulation in human and other mammalian embryos takes place essentially as in birds (see Figure 50.9).

Extraembryonic Membranes Give Rise to the Amnion and Part of the Placenta

Next, a layer of cells separates from the epiblast along its top margin (see Figure 50.13, step 2). The fluid-filled space created by the separation becomes the amniotic cavity, and the layer of ectodermal cells forming its roof becomes the amnion, the extraembryonic membrane surrounding the cavity. The amnion expands until eventually it surrounds the embryo completely and suspends it in amniotic fluid. As in birds, the hypoblast develops into the yolk sac. However, in mammals, the mesoderm of the yolk sac gives rise to the blood vessels in the embryonic portion of the placenta.

Although the amnion is expanding around the embryonic disc, blood-filled spaces form in maternal tissue, and trophoblast cells grow rapidly around both the embryo and amnion to form the chorion (step 3). Then, a connecting stalk forms between the embryonic disc and the chorion, while the chorion begins to grow into the endometrium as fingerlike extensions called **chorionic villi** (step 4). When complete, the chorionic villi greatly increase the surface area of the chorion. Where these villi grow into the endometrium is the area of the future placenta. As the chorion develops, mesodermal cells of the yolk sac grow into it and form a rich network of blood vessels, the embryonic circulation of the placenta. The expanding chorion stimulates the blood vessels of the endometrium to grow into the maternal circulation of the placenta (step 5).

Within the placenta of humans, apes, monkeys, and rodents, the maternal circulation opens into spaces in which the maternal blood bathes the trophoblast layers of the placenta (step 6). The fetal capillaries are positioned next to the trophoblast so that there can be as few as two cell layers separating the maternal and fetal blood. (Different types of placentas are found in other mammals.) The embryonic circulation remains closed, however, so that the embryonic and maternal blood do not mix directly. This prevents the mother from developing an immune reaction against cells of the embryo, which, because it is geneti-

CLOSER LOOK

FIGURE 50.13 Implantation of a human blastocyst in the endometrium of the uterus and the establishment of the placenta.

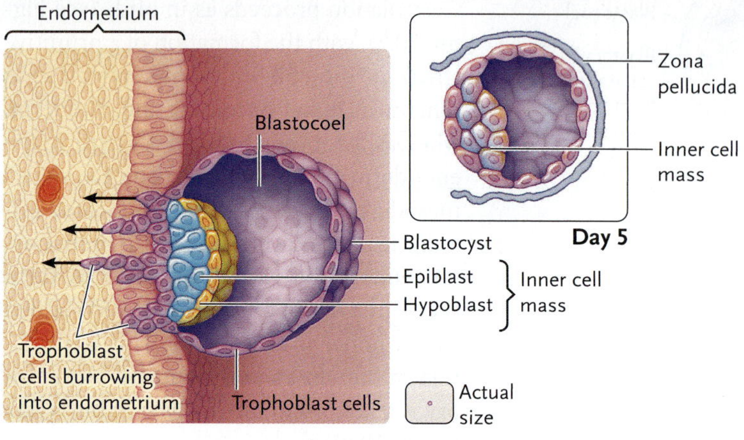

1 Days 6–7: Surface cells of the blastocyst attach to the endometrium and start to burrow into it. Implantation is under way.

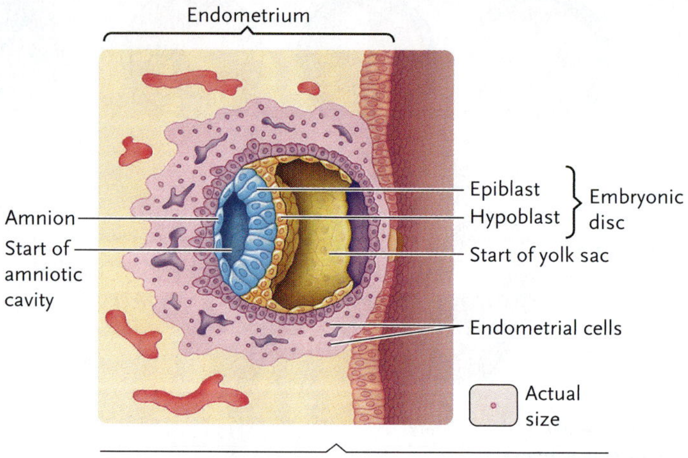

2 Days 10–11: A layer of epiblast cells separates, producing the amniotic cavity. The cells above the cavity become the amnion, which eventually surrounds the embryo. The hypoblast begins to form around the yolk sac.

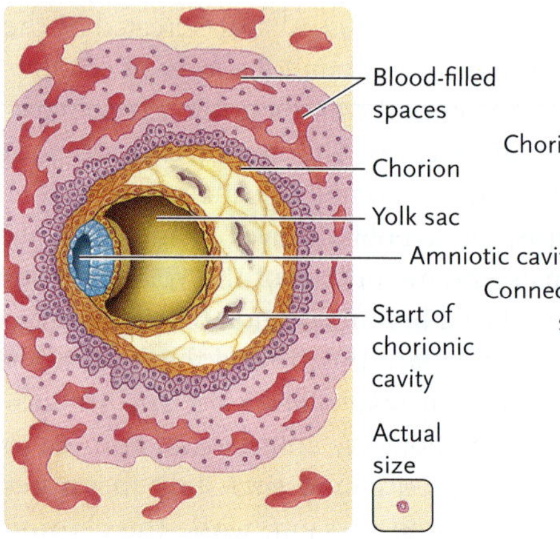

3 Day 12: Blood-filled spaces form in maternal tissue. The chorion forms, derived from trophoblast cells, and encloses the chorionic cavity.

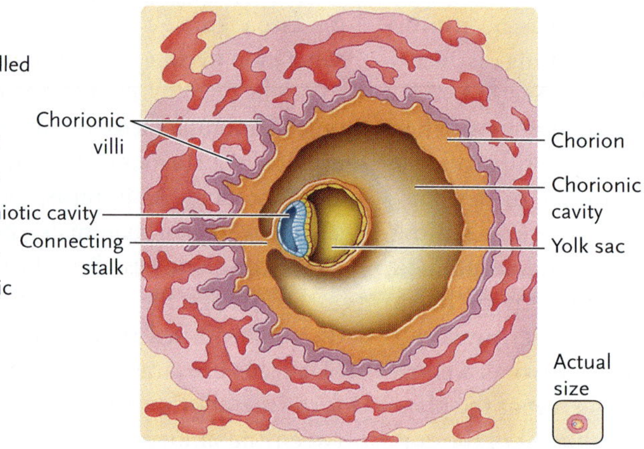

4 Day 14: A connecting stalk has formed between the embryonic disk and chorion. Chorionic villi, which will be features of a placenta, start to form.

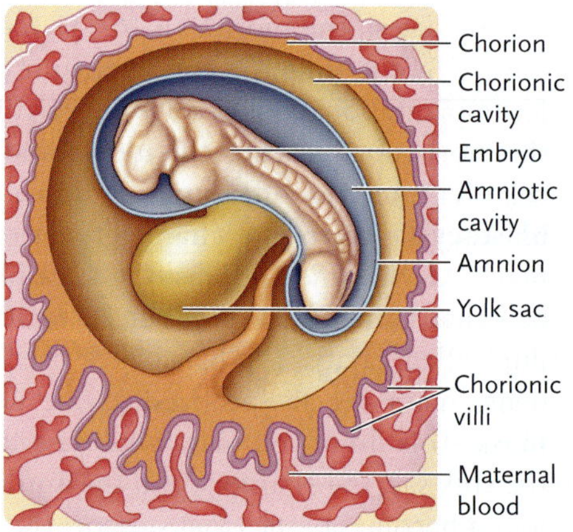

5 Day 25: The chorion continues to grow into the endometrium, producing the chorionic villi. The chorion growth stimulates blood vessels of the endometrium to grow into the maternal circulation of the placenta.

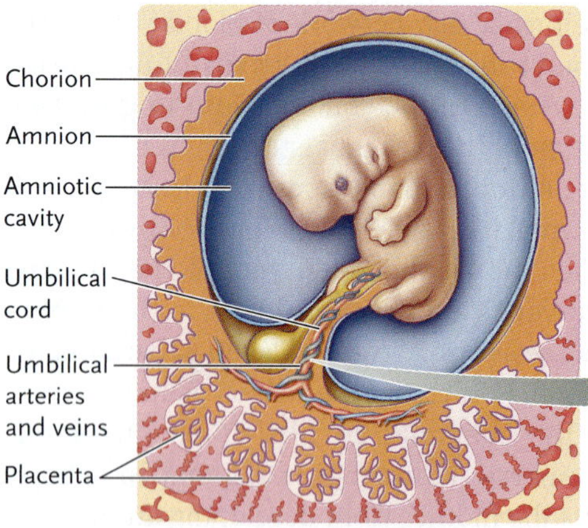

6 Day 45: Blood circulation has been established through the umbilical cord to the placenta.

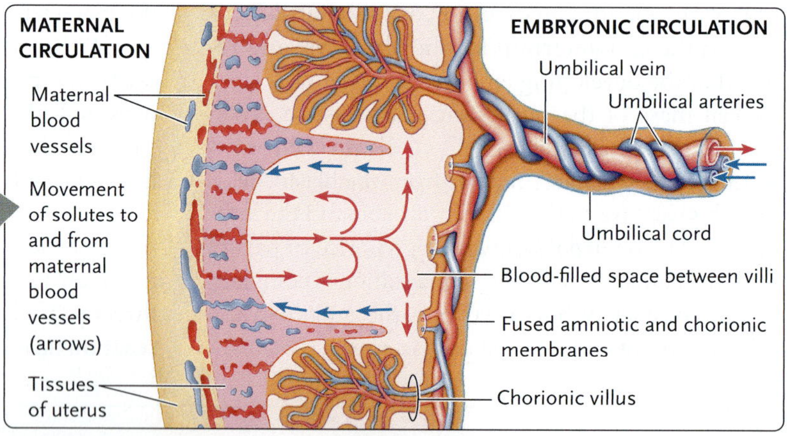

SUMMARY Implantation of a human blastocyst in the endometrium occurs with the first two weeks of development. When implantation is complete, the embryonic disc (epiplast plus hypoblast) is connected to the chorion by a stalk, chorionic villi have started to form, and an amniotic cavity is present. As development continues, the placenta forms and blood circulation is established through the umbilical cord of the embryo and the placenta. The blood circulation of the embryo is kept separate from the maternal circulation.

THINK LIKE A SCIENTIST Why is it important to keep the circulation of the embryo and fetus separate from that of the mother?

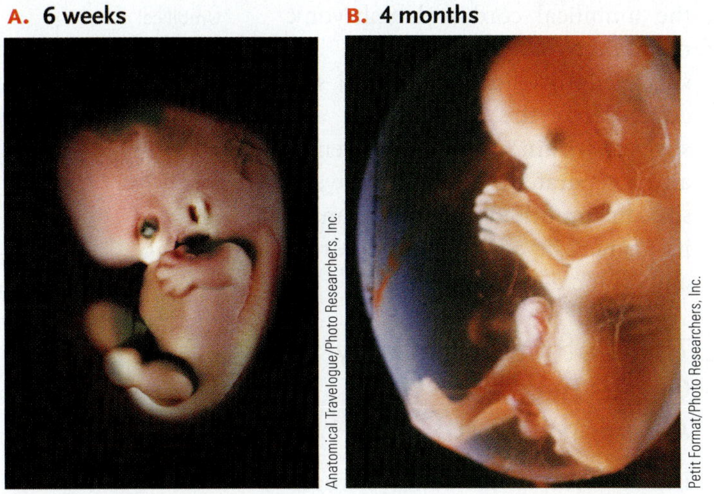

FIGURE 50.14 **The human embryo after 6 weeks (A) and 4 months (B) of development.**

cally different from the mother, may be recognized as foreign by the mother's immune system. Eventually, the placenta and its blood circulation grow to cover about a quarter of the inner surface of the enlarged uterus and reach the size of a dinner plate.

As the embryonic blood circulation develops, this connecting stalk between the embryo and placenta develops into the **umbilical cord,** a long tissue with blood vessels linking the embryo and the placenta. The vessels in the umbilical cord are derived from the extraembryonic membrane, the *allantois.* They conduct blood between the embryo and the placenta (shown in the inset for Figure 50.13, step 6).

Within the placenta, nutrients and oxygen pass from the mother's circulation into the circulation of the embryo. Besides nutrients and oxygen, many other substances taken in by the mother—including alcohol, caffeine, drugs, and toxins in cigarette smoke—can pass from mother to embryo. Carbon dioxide and nitrogenous wastes pass from the embryo to the mother, and are disposed of by the mother's lungs and kidneys.

If the presence of a genetic disease such as cystic fibrosis or Down syndrome is suspected, tests can be carried out on cells removed from the embryonic portion of the placenta or from the amniotic fluid, which contains cells derived from the embryo. The test using cells of the placenta is called *chorionic villus sampling;* the test using cells derived from the amniotic fluid is called *amniocentesis* (see Section 13.4). Chorionic villus sampling can be carried out as early as the eighth week, compared with 14 weeks for amniocentesis. Both tests carry some degree of risk to the embryo.

Further Growth of the Fetus Culminates in Birth

By the end of its fourth week, a human embryo is 3 to 5 mm long, which is 250 to 500 times the size of the zygote. It has a tail and **pharyngeal arches,** which are embryonic features of all vertebrates. The pharyngeal arches contribute to the formation of the face, neck, mouth, larynx, and pharynx. After 6 weeks, the embryo is about 13 mm long and has a more humanlike appearance with, for example, eyes and developing limbs apparent **(Figure 50.14A).** At 16 weeks, the fetus's bones and muscle tissue continue to form, producing a more complete skeleton; its skin begins to form; its limbs with fingers and toes have developed; its organ systems have formed; and its sex is identifiable **(Figure 50.14B).** At birth the baby will have grown to about 3 times the length and 17 times the weight of the fetus at this stage.

Figure 50.15 shows the hormonal events and associated physical events of birth. As the period of fetal growth comes to a close, the fetus typically turns so that its head is downward, pressed against the cervix. A steep rise in the levels of estrogen secreted by the placenta at this time causes cells of the uterus to express the gene for the receptor of the hormone *oxytocin.* The receptors become inserted into the plasma membranes of those cells. Oxytocin—which is secreted by the pituitary gland—binds to its receptor, triggering the smooth muscle cells of the uterine wall to contract and begin the rhythmic contractions of labor. These contractions mark the beginning of **parturition** (*parturire* = to be in labor), the process of giving birth.

The contractions push the fetus further against the cervix and stretch its walls (step 1). In response, stretch receptors in the walls send nerve signals to the hypothalamus, which responds by stimulating the pituitary to secrete more oxytocin. In turn, the oxytocin stimulates more forceful contractions of the uterus, pressing the fetus more strongly against the cervix, and further stretching its walls. The positive feedback cycle continues, steadily increasing the strength of the uterine contractions.

As the contractions force the head of the fetus through the cervix (step 2), the amniotic membrane bursts, releasing the amniotic fluid. Usually, within 12 to 15 hours after the onset of uterine contractions, the head passes entirely through the cervix. Once the head is through, the rest of the body follows quickly and the entire fetus is forced through the vagina to the exterior, still connected to the placenta by the umbilical cord (step 3).

After the baby takes its first breath, the umbilical cord is cut and tied off by the birth attendant. Contractions of the uterus continue, expelling the placenta and any remnants of the umbilical cord and embryonic membranes as the afterbirth. The short length of umbilical cord still attached to the infant dries and shrivels within a few days. Eventually, it separates entirely and leaves a scar, the **umbilicus** or navel, to mark its former site of attachment during embryonic development.

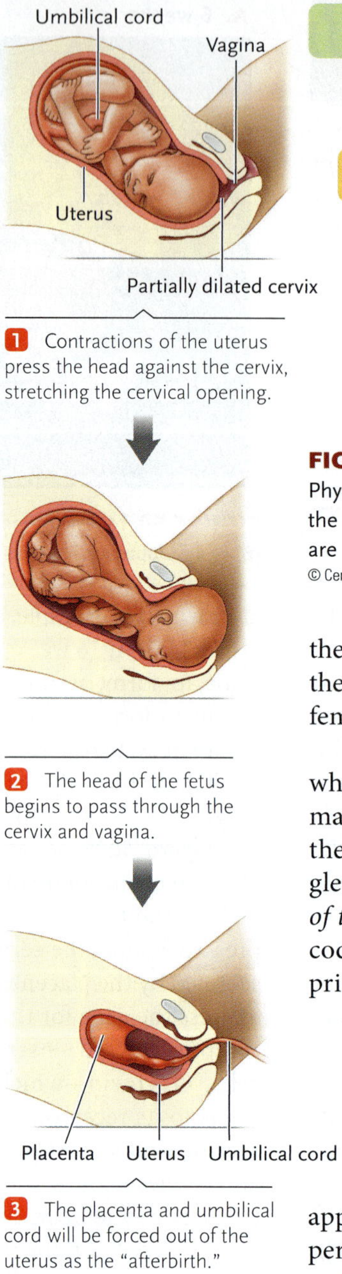

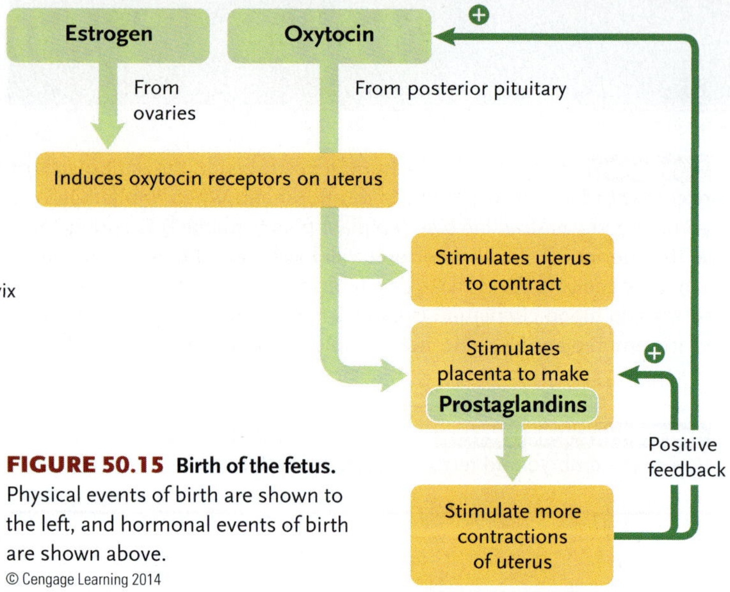

FIGURE 50.15 Birth of the fetus. Physical events of birth are shown to the left, and hormonal events of birth are shown above.

The Mother's Mammary Glands Become Active after Birth

Before birth of the fetus, estrogen and progesterone secreted by the placenta stimulate the growth of the mammary glands in the mother's breasts. However, the high levels of these hormones prevent the mammary glands from responding to *prolactin,* the hormone secreted by the pituitary that stimulates the glands to produce milk. After birth of the fetus and release of the placenta, the levels of estrogen and progesterone fall steeply in the mother's bloodstream, and the breasts begin to produce milk (stimulated by prolactin) and secrete it (stimulated by oxytocin).

Continued milk secretion depends on whether the infant is suckled by the mother. If the infant is suckled, stimulation of the nipples sends nerve impulses to the hypothalamus, which responds by signaling the pituitary to release a burst of prolactin and oxytocin. Hormonal stimulation of milk production and secretion continues as long as the infant is breastfed.

So far, we have followed the development of a generic human, but certain aspects of development differ depending on the offspring's sex. Next we look at the specifics of male and female development.

A Gene on the Y Chromosome Determines the Development of Male or Female Sex Organs

The gonads and their ducts begin to develop during the fourth week of gestation. Until the seventh week, male and female embryos have the same set of internal structures derived from mesoderm, including a pair of gonads **(Figure 50.16A).** Each gonad is associated with two primitive ducts, the **Wolffian duct** (also called the *mesonephric duct*) and the **Müllerian duct** (also called the *paramesonephric duct*), which lead to a cloaca. At this time, the gonads are *bipotential:* they can develop into either male or female gonads.

The presence or absence of a Y chromosome determines whether the bipotential gonads and internal ducts develop into male or female gonads and their associated internal ducts. If the fetus has the XY combination of sex chromosomes, a single gene on the Y chromosome, *SRY (Sex-determining **Region** of the **Y**),* becomes active in the seventh week. The protein encoded by the gene sets a molecular switch that causes the primitive gonads to develop into testes. The fetal testes then secrete two hormones, testosterone and the *anti-Müllerian hormone (AMH).* The testosterone stimulates development of the Wolffian ducts into the male reproductive tract, including the epididymis and vas deferens **(Figure 50.16B),** and seminal vesicles (see Figure 49.14). AMH causes the Müllerian ducts to degenerate and disappear. (*Insights from the Molecular Revolution* describes experiments that traced the activity of the *SRY* gene and its encoded protein in male development.) Testosterone additionally stimulates the development of the male genitalia.

If the fetus has the XX combination of chromosomes, no SRY protein is produced and the primitive gonads, under the influence of the estrogens and progesterone secreted by the placenta, develop into ovaries. The Müllerian ducts develop into the oviducts, uterus, and part of the vagina, and the Wolffian ducts degenerate and disappear **(Figure 50.16C).** The female sex hormones additionally stimulate the development of the female external genitalia.

Development Continues after Birth

Once fetal development is over, humans and other mammals, as well as most other animals, follow a prescribed course of further growth and development that leads to the adult, the sexually mature form of the species. In humans, the internal and external sexual organs mature and secondary sexual characteristics appear at puberty. Similar changes occur in most mammals. There are, in fact, many examples among different animal groups of developmental changes that take place after hatching or birth. In some cases, offspring hatch that are distinctly different in structure from the adult. Examples among invertebrates include insects such as *Drosophila* and butterflies, in which eggs hatch to produce larva that undergo metamorphosis into the adult. Frogs similarly hatch as tadpoles, which undergo metamorphosis to produce the adult.

We have now described embryonic development in animals from a morphological perspective. In the rest of the chapter, we focus on the cellular mechanisms that underlie development.

STUDY BREAK 50.4

1. **Distinguish between the roles of the trophoblast and inner cell mass of the blastocyst in mammalian development.**
2. **What hormone would you use to induce labor in a pregnant woman?**

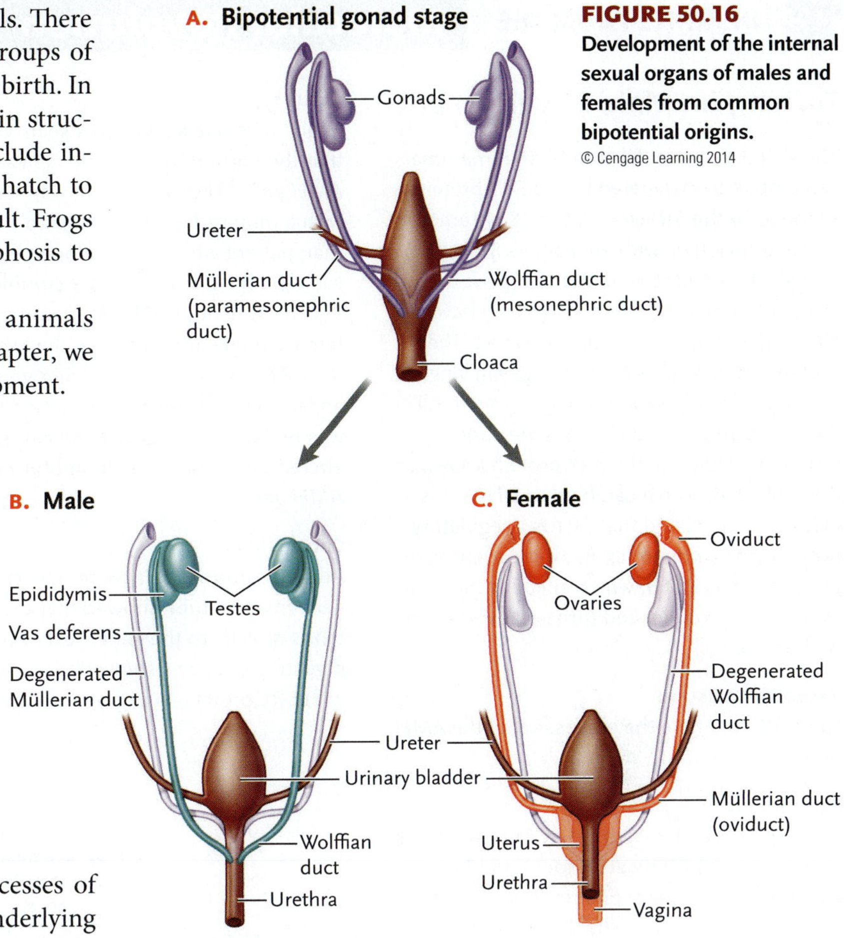

FIGURE 50.16 **Development of the internal sexual organs of males and females from common bipotential origins.**

50.5 The Cellular Basis of Development

In the preceding sections, you learned about the processes of development from a mainly structural point of view. Underlying those developmental processes are specific cellular and molecular events. In this section, you will learn about some of the cellular events that underlie the stages of development. The genetic and molecular regulation of development was discussed in Section 16.4.

Cell Division Varies in Orientation and Rate during Embryonic Development

Recall from earlier in the chapter that *morphogenesis* is the generation of the body form as differentiated cells end up in their appropriate positions. In animals, morphogenesis occurs by changes in cell shape, cell position, and cell adhesion. For embryo morphogenesis, the *orientation* and *rate* of mitotic cell division have special significance. Regulation of these two features of mitotic cell division occurs at all stages of development.

The orientation of cell division refers to the angles at which daughter cells are added to older cells as development proceeds. Orientation is determined by the location of a furrow that separates the cytoplasm after mitotic division of the nucleus (furrowing is discussed in Section 10.2). The furrow forms in alignment with the spindle midpoint. Therefore, when the spindle is centrally positioned in the cell, the furrow leads to symmetrical division of the cell. However, when the spindle is displaced to one end of the cell, the furrow leads to asymmetrical division of the cell into a smaller and a larger cell. Little is known about how spindle positioning is regulated.

The rate of cell division primarily reflects the time spent in the G_1 period of interphase (see Section 10.2); once DNA replication begins, the rest of the cell cycle is usually of uniform length in all cells of the same species. As an embryo develops and cells differentiate, the time spent in interphase increases and varies in length in different cell types. Therefore, different cell types proliferate at various rates, giving rise to tissues and organs with different cell numbers. Ultimately, the rate of cell division is under genetic control.

Frog egg cleavage provides examples of how both changes in orientation and rate of mitotic division affect development. The first two cleavages start at the animal pole and extend to the vegetal pole, producing four equal blastomeres (see Figure 50.2). The third cleavage occurs equatorially. However, because there is yolk in the vegetal region of the embryo, this cleavage furrow forms not at the equator but up higher toward the animal pole. The result is an eight-cell embryo with four small blastomeres in the animal region of the embryo, and four large blastomeres in the vegetal region. All the blastomeres then divide at the same rate up until the twelfth division. Thereafter, blastomeres in the animal region divide rapidly whereas blastomeres in the vegetal region divide more slowly because division is inhibited by yolk. As a result, the animal region becomes filled with many small cells, whereas the vegetal region contains a relatively small number of yolk-filled blastomeres.

Turning on Male Development

The switch to male development in mammalian embryos is triggered by the SRY protein encoded by the *SRY* gene on the Y chromosome. Individuals with an inactive SRY protein due to a mutation in the *SRY* gene develop into females, even though they have the XY combination of sex chromosomes. The mutant SRY proteins have changes in single amino acids or have a missing segment. All the single amino acid changes are concentrated in a region of the SRY protein known as the *HMG box,* which can bind to DNA. This discovery suggested that SRY is a regulatory protein that binds to regulatory sequences of genes such as *AMH,* which encodes the anti-Müllerian hormone, and turns on expression of those genes.

Research Question

Does SRY regulate the expression of the *AMH* directly?

Experiment

A group of investigators led by Michael Weiss at the Harvard Medical School and Massachusetts General Hospital in Boston tested whether SRY turns on the *AMH* gene directly.

For their experiments, the researchers attached the regulatory sequences that control transcription of the *AMH* gene to the amino acid-coding sequence of a gene encoding firefly luciferase **(Figure).** Firefly luciferase is an enzyme that catalyzes a reaction with the substrate luciferin to produce light (see *Insights from the Molecular Revolution,* Chapter 42). If SRY binds to the *AMH* gene regulatory sequence of the constructed gene and activates transcription, translation of the resulting mRNA produces luciferase. Luciferase activity is measured in cell extracts; the quantification of the amount of light produced in the reaction indicates the degree of activation of transcription caused by SRY. The data are then extrapolated to the effects of SRY on the complete *AMH* gene.

The researchers introduced the composite *AMH*–luciferase gene into embryonic cells removed from the developing gonads of XY rat embryos. A normal human *SRY* gene was then introduced into the gonad cells containing the artificial gene, and luciferase activity was measured.

Results

High luciferase activity was seen, confirming that the normal SRY protein activates the *AMH* gene. The experiment was repeated with a mutant *SRY* gene isolated from a human patient who had developed into a female even though she had the XY combination of sex chromosomes. The mutation involved altered a single amino acid in the HMG box of the SRY protein. When her *SRY* gene was added to the embryonic rat gonad cells, there was no luciferase activity, indicating that her altered SRY protein could not turn on the *AMH* gene.

Conclusions

Adding a normal SRY protein to the *AMH* gene in a test tube showed that the protein binds directly to the gene. Tests with DNA-digesting enzymes showed that combination with SRY protects a segment of the regulatory sequence of the *AMH* gene from attack by the enzymes. This protection indicated that SRY binds this sequence, as expected for a regulatory protein.

THINK LIKE A SCIENTIST The experiments described involved missense mutations in the HMG box-encoding region of the *SRY* gene. (Types of mutations are described in Section 15.5). Predict what result would have been obtained in this experiment with: (a) a nonsense mutation in the *SRY* gene upstream of the HMG-box encoding region; (b) a nonsense mutation downstream of that region; and (c) a frameshift mutation upstream of that region.

Source: C. M. Haqq et al. 1994. Molecular basis of mammalian sexual determination: activation of Müllerian inhibiting substance gene expression by SRY. *Science* 266:1494–1500.

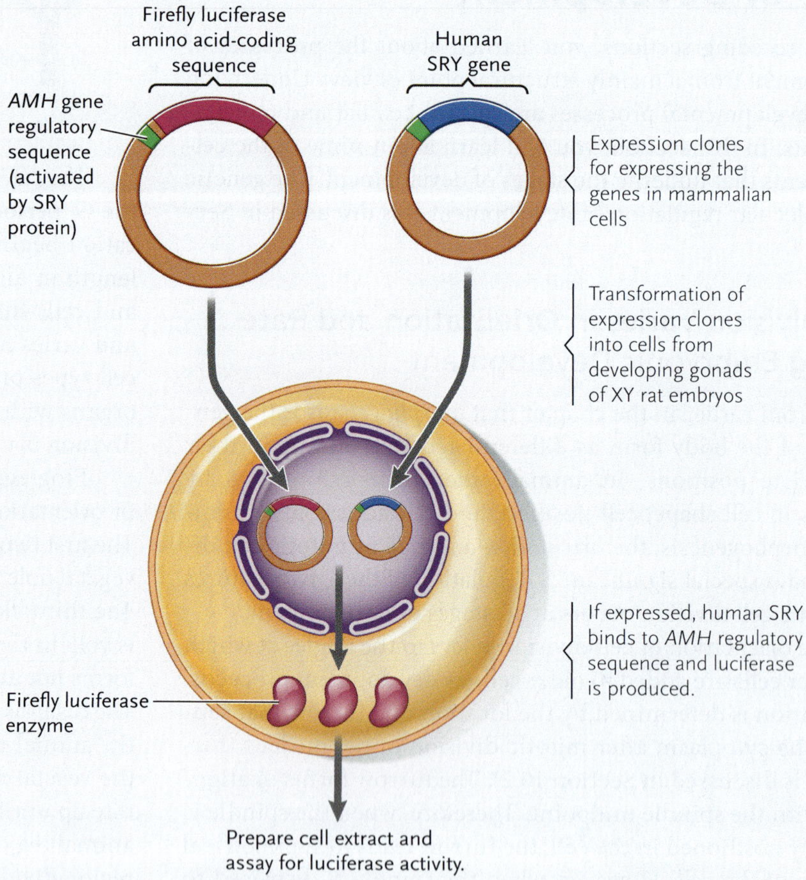

Cell-Shape Changes and Cell Migrations Play Important Roles in Development

We have seen that embryonic cells undergo changes in shape that generate cell migrations, such as the infolding of surface layers to produce endoderm or mesoderm. Entire cells also migrate during the embryonic growth of animals, both singly and in groups. Both the shape changes and the whole-cell migrations are produced by microtubules (powered by dyneins and kinesins) and microfilaments (powered by myosins) (see Section 5.3). Cell migrations are also produced by changes in the rate of growth or by the breakdown of microtubules and microfilaments. Generally speaking, changes in both cell shape and cell migration play important roles in cleavage, gastrulation, and organogenesis.

CHANGES IN CELL SHAPE Changes in cell shape typically result from reorganization of the cytoskeleton. For example, during the development of the neural plate in frogs (see Figure 50.9), the ectoderm flattens and thickens; that is, microtubules within cells in the ectoderm layer lengthen and slide farther apart, causing the cells to transition from a cubelike to a columnar shape **(Figure 50.17A)**.

Once formed, the neural plate sinks downward along its midline as a result of a change in cell shape from columnar to wedgelike **(Figure 50.17B)**. As the tops of the cells narrow, the entire cell layer is forced inward—it invaginates. How does this occur? Each wedge-shaped cell contains a group of microfilaments arranged in a circle at its top. The microfilaments slide over each other, tightening the ring like a drawstring and narrowing the top of the cell.

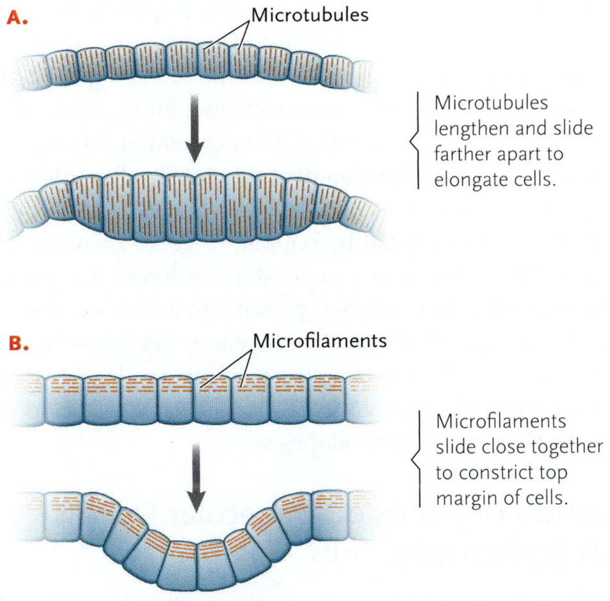

FIGURE 50.17 **The roles of microtubules (A) and microfilaments (B) in the changes in cell shape that produce developmental movements, such as neural plate formation.**

WHOLE-CELL MIGRATIONS Among the most striking examples of whole-cell migrations in embryonic development are the migrations of cells during gastrulation to form the three germ layers, that is, the ectoderm, mesoderm, and endoderm, and the often long-distance migrations of neural crest cells. These whole-cell migrations involve the coordinated activity of microtubules and microfilaments and large-scale reorganization of the cytoskeleton.

Cell migration proceeds in four stages:

1. *Polarization.* The front and back ends of a cell becomes defined by a polarization process. Polarization is triggered either by diffusible signals or by signals from the extracellular matrix (ECM). In response to a polarizing signal, the cell reorganizes its cytoskeleton so that the cell has a front and a back end.
2. *Protrusion of the front end of the cell.* Actin microfilaments assemble into long, parallel bundles, causing the front end of the cell to protrude.
3. *Adhesion.* In order to migrate, a cell must be anchored to a substrate that it can push on. Therefore, the cell at this stage adheres to the surrounding extracellular matrix. (Some molecular aspects of the adhesion process are described later.)
4. *Release of rear adhesion, and forward migration to a new adhesion point.* Adhesion of the cell to the substrate at its rear end are released, and the cell migrates forward and adheres to a new attachment point. Repetition of this stage results in continued forward movement in a fashion somewhat resembling how an amoeba moves (see Section 5.3 and Figure 6.15).

How do the cells know where to go? Typically, cells migrate over the surface of stationary cells in one of the embryo's layers. In many developmental systems, migrating cells follow tracks formed by molecules of the extracellular matrix that is secreted by the cells along the route over which they travel. An important track molecule is *fibronectin,* a fibrous, elongated glycoprotein of the extracellular matrix. Migrating cells can recognize and adhere to the fibronectin; in response, internal changes in the cells trigger movement in a direction based on the alignment of the fibronectin molecules. For instance, paths of fibronectin guide primordial germ cells to the gonads, and cells of the heart to the midline of the embryo. Experiments in which the formation of fibronectin pathways was prevented have shown how crucial those pathways are for embryonic development. For example, gastrulation does not occur in amphibians lacking fibronectin, and in mammals, mesoderm and neural tube development is defective in the absence of fibronectin.

Some migrating cells follow concentration gradients instead of molecular tracks. The gradients are created by the diffusion of molecules (often proteins) released by cells in one part of an embryo. Cells with receptors for the diffusing molecule follow the gradient toward its source, or move away from the source.

Selective Cell Adhesions Underlie Cell Migrations

Selective cell adhesion, the ability of an embryonic cell to make and break specific connections to other cells, is closely related to cell migration. As development proceeds, many cells break their initial adhesions and move, forming new adhesions in different locations. Final cell adhesions hold the embryo in its correct shape and form. Junctions of various kinds, including tight, anchoring, and gap junctions, reinforce the final adhesions (see Section 5.5).

The selective nature of cell adhesions was first demonstrated in a classic experiment by Johannes Holtfreter of the University of Rochester and his student P. L. Townes. The researchers removed pieces of ectoderm, mesoderm, and endoderm from living amphibian embryos in the neurulation stage, separated them into individual cells, and added the cells in various combinations to a culture medium. Initially the cells clumped together at random into a ball. After a few hours, they sorted themselves out and moved into arrangements resembling their normal locations in the gastrula **(Figure 50.18)**.

Further research has identified many cell surface proteins responsible for selective cell adhesions, including **cell adhesion molecules** (CAMs; see Section 5.5) and **cadherins** *(calcium-dependent adhesion molecules)*. As cells develop, different types of CAMs or cadherins may appear or disappear from their surfaces as they make and break cell adhesions. The changes reflect alterations in gene activity, often in response to molecular signals arriving from other cells. For example, in the neural plate stage of neurulation in the frog (see Figure 50.9), N-cadherin is on neural plate cells, keeping those cells together, whereas E-cadherin is on the adjacent ectodermal cells, keeping those cells together. The neural tube is produced when the neural plate cells separate from the ectodermal cells, whereas both cell types retain their respective cadherin type. The neural crest cells have neither cadherin bound to them, so they do not bind to each other and they disperse (as described earlier). However, if N-cadherin is expressed in the ectodermal cells through experimental manipulation, the forming neural tube does not separate from the flanking ectodermal cells because all of the cells are held together by N-cadherin.

As they begin to disperse, neural crest cells produce *integrins,* which insert into the plasma membrane. Recall from Section 5.5 that integrins are receptor proteins that span the plasma membrane. On the cytoplasmic side, integrins bind to actin microfilaments of the cytoskeleton. Integrins function to integrate changes outside and inside the cell by communicating changes in the extracellular matrix to the cytoskeleton. In the example here, the integrin receptors of the neural crest cells interact with proteins in the extracellular matrix in a way that causes cytoskeletal changes. The cytoskeletal changes move the cells along their migration pathways. More generally, integrins are involved in many key developmental processes.

The development of a new organism involves the cell shape changes, cell migrations, and cell adhesions we have just discussed and, in addition, the programmed emergence of particular types of cells, or cell lineages, as we will see next.

Fate Mapping Maps Adult Structures onto Regions of the Embryos from Which They Developed

From the early days of studying development, embryologists have focused on describing not only how embryos form and develop, but exactly how adult tissues and organs are produced from the cells of the embryo. Thus, an important goal in embryology is to trace cell lineages from embryo to adult. For most organisms, it is not possible to trace lineages at the individual cell level, primarily because of the complexity of the developmental process and the typical opacity of embryos. However, it has been possible to map adult or larval structures onto the region of the embryo from which each structure developed. This type of study is called *fate mapping,* and the result is called a **fate map.** Experimentally, fate mapping is done by following development of living embryos under the microscope, either using species in which the embryo is transparent, or by marking cells so they can be followed. Cells may be marked with vital dyes (dyes that do not kill cells), fluorescent dyes, or radioactive labels. Fate maps have been produced for a number of organisms, including the chick, *Xenopus,* and *Drosophila.*

In most cases a fate map is not detailed enough to relate how particular cells in the embryo gave rise to cells of the adult. The exception is the fate map of the nematode *Caenorhabditis elegans,* an organism that has a fixed, reproducible developmental pattern. This animal has a transparent body, and scientists have been able to map the fate—trace the **cell lineage**—of every somatic and germ-line cell as the zygote divides and the resulting embryo differentiates into the 959-cell adult hermaphrodite or the 1,031-cell adult male **(Figure 50.19)**. All somatic cells of the adult can be traced from five somatic *founder cells* produced during early development. Knowing the cell lineages of *C. elegans* has been a valuable tool for research into the genetic and molecular control of development in this organism, for mutants affecting development have an easily visible effect. Among the results of such studies, we have learned that not only are proteins encoded by regulatory genes involved in the control of development, but so are some miRNAs. For example, genetic manipulation makes it possible to knock out the functions of specific miRNA-coding genes. For most miRNA-coding genes, the knockouts do not affect development. However, for a few of the miRNA-coding genes, the knockouts cause defects in the timing of development.

Induction Depends on Molecular Signals Made by Inducing Cells

Recall from Section 50.1 that induction is the process by which a group of cells (the inducer cells) causes or influences a nearby group of cells (the responder cells) to follow a particular developmental pathway. Recall also that induction is the major pro-

FIGURE 50.18 **Experimental Research**

Demonstrating the Selective Adhesion Properties of Cells

Question: Do cells make specific connections to other cells?

Experiment: Johannes Holtfreter and P. L. Townes demonstrated that cells make specific connections to other cells, meaning that cells have selective adhesion properties.

Amphibian embryos of different species

1. Holtfreter and Townes separated ectoderm, mesoderm, and endoderm tissue from amphibian embryos soon after the neural tube had formed. They used embryos from amphibian species that had cells of different colors and sizes, enabling them to follow under the microscope where each cell type ended up. (The colors shown here are for illustrative purposes only.)

2. The researchers placed the tissues individually in alkaline solutions, which caused the tissues to break down into single cells.

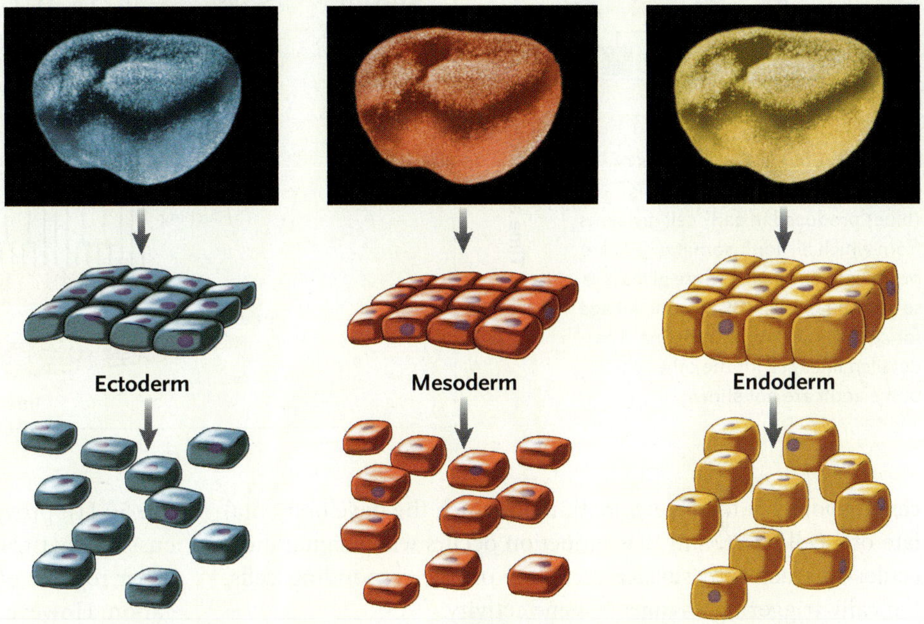

3. Holtfreter and Townes combined suspensions of single cells in various ways. Shown here are ectoderm + mesoderm, and ectoderm + mesoderm + endoderm. When the pH was returned to neutrality, the cells formed aggregates. Through a microscope, the researchers followed what happened to the aggregates.

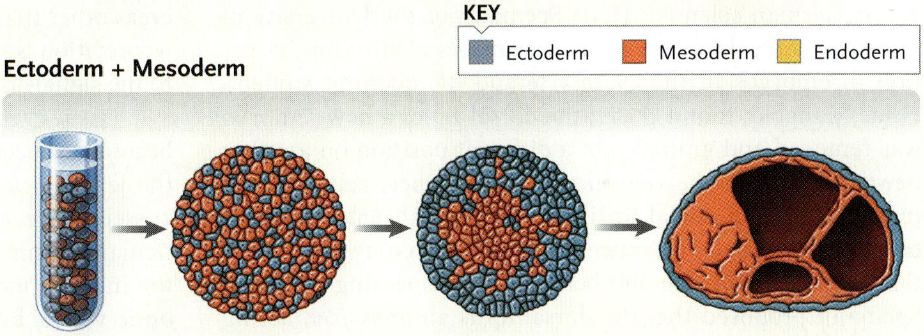

Ectoderm + Mesoderm + Endoderm

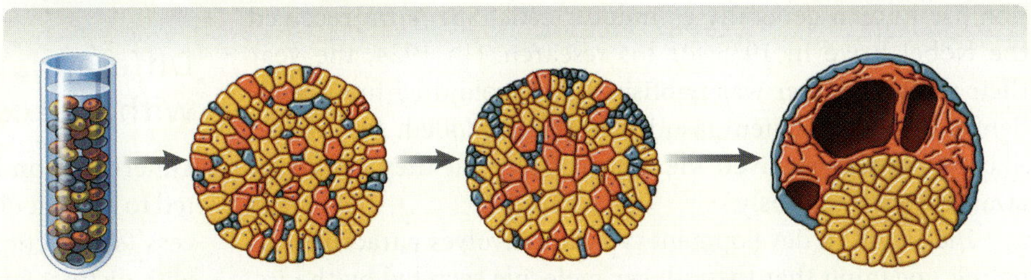

Results: In time the reaggregated cells sorted themselves with respect to cell type; that is, instead of the cell types remaining mixed, each cell type became separated spatially. In the ectoderm + mesoderm mixture, the ectoderm moved to the periphery of the aggregate, surrounding mesoderm cells in the center. In no case did the two cell types remain randomly mixed. In the ectoderm + mesoderm + endoderm mixture, cell sorting in the aggregates generated cell positions reflecting the positions of the cell types in the embryo. That is, the endoderm cells separated from the ectoderm and mesoderm cells and became surrounded by them. In the end, the ectoderm cells were located on the periphery, the endoderm cells were internal, and the mesoderm cells were between the other two cell types.

Conclusion: Holtfreter interpreted the results to mean that cells have selective affinity for each other—cells have selective adhesion properties. Specifically, he proposed that ectoderm cells have positive affinity for mesoderm cells but negative affinity for endoderm cells, whereas mesoderm cells have positive affinity for both ectoderm cells and endoderm cells. In modern terms, these properties result from cell surface molecules that give cells specific adhesion properties.

THINK LIKE A SCIENTIST As the main text discusses, subsequent research identified cadherins as one type of cell surface proteins responsible for selective cell adhesions. Suppose you have a cloned cDNA for a cadherin, which is proposed to be one type of cell surface protein responsible for selective cell adhesions. How could you use that cloned cDNA to test directly whether or not it has that role?

Source: P. L. Townes and J. Holtfreter. 1955. Directed movements and selective adhesion of embryonic amphibian cells. *Journal of Experimental Zoology* 128:53–120.

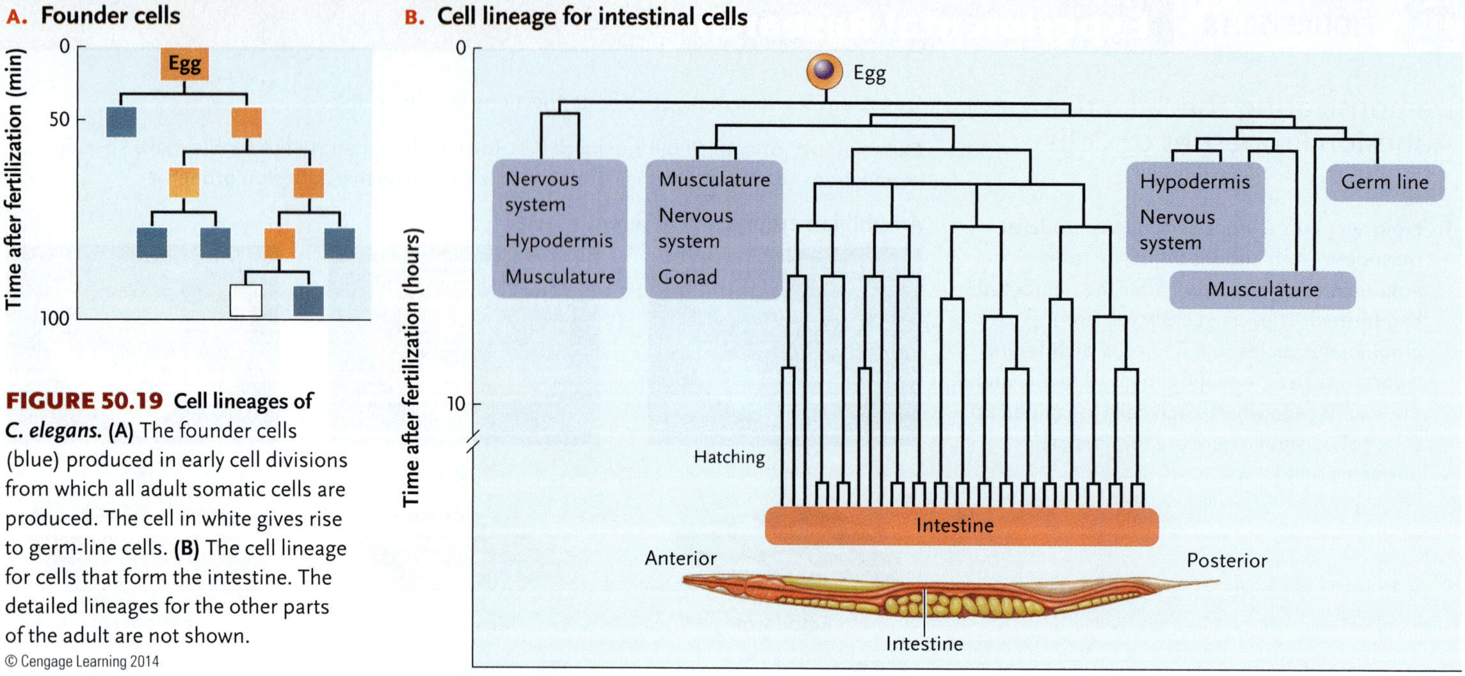

FIGURE 50.19 Cell lineages of *C. elegans*. (A) The founder cells (blue) produced in early cell divisions from which all adult somatic cells are produced. The cell in white gives rise to germ-line cells. **(B)** The cell lineage for cells that form the intestine. The detailed lineages for the other parts of the adult are not shown.

cess responsible for determination, in which the developmental fate of a cell is set, and that induction occurs when signal molecules interact with surface receptors on the responding cells, typically triggering changes in gene activity.

A German scientist, Hans Spemann of the University of Freiburg, carried out the first experiments identifying induction in embryos in the 1920s. He and his doctoral student, Hilde Mangold, found that if the dorsal lip of a newt embryo was removed and grafted into a different position on another newt embryo, on the ventral side for instance, cells moving inward from the dorsal lip induced a neural plate, a neural tube, and eventually an entire embryo to form in the new location **(Figure 50.20)**. On the basis of his pioneering research, Spemann proposed that the dorsal lip is an *organizer,* acting on other cells to alter their course of development. This action is now known as *induction,* and the cells responsible for induction are known generally as inducer cells. Spemann received the Nobel Prize in 1935 for his research. (In 1924, the year their research paper was published, Mangold died in an accident when her kitchen gasoline heater exploded. She would likely have also received the Nobel Prize, but they are never awarded posthumously.)

Induction in development typically involves paracrine signaling, meaning that the inducer molecule secreted by the inducer cell changes the cellular behavior of responder cells that are located nearby the inducer cells (see Section 42.1). For a cell to respond to the inducer molecule, it must have a receptor for that molecule. In addition, the responder cell must be able to respond to the specific inducer, a condition called *competence.* That is, simply having a receptor is not sufficient to generate a response; in addition, the responder cell must be expressing a genetic program that makes it competent to respond. Induction by the optic vesicle, discussed earlier in this chapter (see Figure 50.11), provides an example. You learned in the earlier discussion that transplanting the optic vesicle under ectoderm in other regions of the head induces a lens to form in the new location. However, if the optic vesicle is placed under ectoderm in areas other than the head, lens tissue does not develop. The interpretation is that only head ectoderm is competent to respond to the signals from the optic vesicle and produce a lens.

Tissues responding to inducing signals may themselves become inducers as development proceeds. For instance, once the lens has formed as a result of the induction events discussed above, the lens itself induces other tissues. In this particular example, the optic vesicle becomes a responder tissue for inducer molecules secreted by the lens. As a result, the optic vesicle becomes the optic cup (see Figure 50.11, step 2) and then the wall of the optic cup differentiates into the layers of the retina.

Differentiation Produces Specialized Cells without Loss of Genes

Differentiation is the process by which cells that have committed to a particular developmental fate by the determination process (see Section 50.1) now develop into specialized cell types with distinct structures and functions. As part of differentiation, cells concentrate on the production of molecules characteristic of the specific types. For example, 80% to 90% of the total protein that lens cells synthesize is crystallin.

Research into differentiation confirmed that as cells specialize, they retain all the genes of the original egg cell; except in rare instances, differentiation does not occur through selective gene loss. Several definitive experiments supporting this conclusion were carried out several decades ago by Robert Briggs and Thomas King of Lankenau Hospital Research Institute in Phila-

FIGURE 50.20 **Experimental Research**

Spemann and Mangold's Experiment Demonstrating Induction in Embryos

Question: Does induction occur in embryonic development?

Experiment: Hans Spemann and Hilde Mangold performed transplantation experiments with newt embryos.

1. The researchers grafted the dorsal lip of the blastopore from one newt embryo and onto a different position—the ventral side—of another embryo. The two embryos were from different newt species that differed in pigmentation, allowing the fate of the transplanted tissue to be followed easily during development.

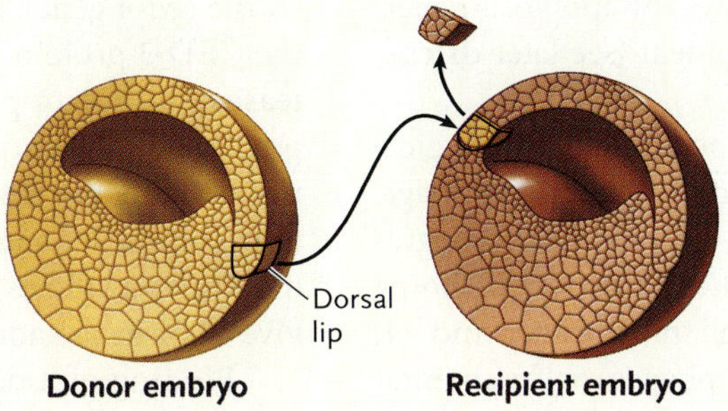

2. The embryo with the transplant was allowed to develop.

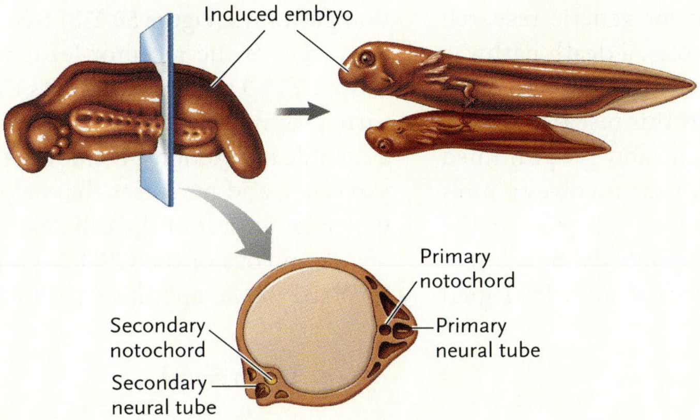

Result: At the ventral location on the recipient embryo where the dorsal lip of the blastopore was grafted, another embryo developed simultaneously with the recipient embryo. Eventually, two mature embryos were produced attached on their ventral surfaces.

Conclusion: The grafted dorsal lip of the blastopore induced a second gastrulation and subsequent development in the ventral region of the recipient embryo. The result demonstrated the ability of particular cells to induce the specific development of other cells.

THINK LIKE A SCIENTIST As a preliminary step, you transplanted cells from the upper tip of the primitive streak from one gastrulating chick embryo to an off-center site in another gastrulating chick embryo. The result was that, at the site of transplantation, a second set of dorsal structures formed in addition to the dorsal structures formed as a result of the recipient's primitive streak. The conclusion from this experiment is that the tip of the primitive streak induces dorsal structures. Now, if you transplant a primitive streak tip from a donor duck embryo to a recipient chick embryo, a second set of dorsal structures is also induced. Are these dorsal structures formed from duck cells or chick cells?

Source: H. Spemann and H. Mangold. 1924. Über induktion von Embryonalanlagen durch implantation artfremder Organisatoren. *Roux's Archives of Developmental Biology* 100:599–638. [Reprinted as English translation by V. Hamburger: Induction of embryonic primordia by implantation of organizers from a different species. *International Journal of Developmental Biology* 45:13–38 (2001).]

delphia (now Fox Chase Cancer Center), and extended by John B. Gurdon of the University of Cambridge, United Kingdom. In a typical experiment, the nucleus of a fertilized frog egg was destroyed by ultraviolet light. A micropipette was then used to transfer a nucleus from a fully differentiated tissue, intestinal epithelium, to the enucleated egg. Some of the eggs receiving the transplanted nuclei subsequently developed into normal tadpoles and adult frogs. This outcome was possible only if the differentiated intestinal cells still retained their full complement of genes. This conclusion was extended to mammals in 1997 when Ian Wilmut and his colleagues successfully cloned a sheep—Dolly—starting with an adult cell nucleus. (This experiment is described in Section 18.2.)

Apoptosis Is a Normal Part of Development

Induction and differentiation build complex, specialized organs from the three fundamental tissue types. Complementing these processes is *apoptosis*—a type of programmed cell death—a process that removes cells in a regulated way at a specific point in development. Apoptosis is an active process that is a normal part of development in animals. Programmed cell death processes have also been identified in plants (see Chapter 37). We focus here on animal development. (Apoptosis is also involved in removing cells that are damaged beyond repair, or cells that are infected with viruses. Those roles for apoptosis are not part of animal development and are not discussed here.)

An example of apoptosis is afforded by the nematode worm, *C. elegans*. In the worm, division of the fertilized egg eventually generates 1,090 cells, of which exactly 131 die by apoptosis at prescribed times to produce the 959 cells characteristic of an adult hermaphrodite. (*C. elegans* cell fates are summarized earlier in this section; see Figure 50.19.) In humans, for example, a developing fetus in the uterus produces about three times more neurons than is found at birth; the others are destroyed by apoptosis. Numerous other examples of apoptosis occur in animal development, including controlling the number of cells in particular tissues (such as neurons), generating complex organs (such as the heart and the retina), and removing structures not needed (such as mammary tissue in male mammals, the tails of frogs, and tissue to create the vaginal opening) and controlling the spacing and orientation of tissues (such as neuron spacing and orientation and the spaces between digits [fingers and toes]). Considering the last example, in early development, tissue fills in the spaces between what will become the digits. The removal of that tissue—webbing—by apoptosis generates the free digits found later in development (see later discussion for more details about this example).

Apoptosis is characterized by specific morphological changes in cells. These changes include: (1) cell shrinkage, which can also result in a loss of cell-to-cell contact; (2) degradation of chromatin (which includes the DNA); (3) blebbing (production of small bulges) of the cell membrane; and (4) phagocytosis of the whole cell by macrophages, or fragmentation of the cell, followed by phagocytosis of the fragments.

C. elegans was the model organism for genetic research that first delineated a pathway for apoptosis, a death pathway if you will. (Sydney Brenner, H. Robert Hurvitz, and John Sulson received a Nobel Prize in 2002 for this research on the "genetic regulation of organ development and programmed cell death.") The molecular basis of apoptosis involves a molecule called a death signal that binds to a receptor in the plasma membrane of a cell targeted for apoptosis. Binding of the death signal to the receptor activates it and, by signal transduction (see Chapter 7), leads to the activation of cellular proteins that kill the cell. In the absence of the death signal, the killing proteins remain inactive. Let us step through the two situations:

In the absence of a death signal, the membrane receptor is inactive **(Figure 50.21A)**. This allows a protein associated with the outer mitochondrial membrane, CED-9 (encoded by the *ced-9 cell d*eath gene) to bind to and inactivate CED-4 (encoded by the *ced-4* gene). CED-4 in active form is required to activate the CED-3 protein (encoded by the *ced-3* gene), which is a protease of a class of proteases called *caspases*. Active CED-3 caspase is needed to turn on the cell death program. But here, in the absence of a death signal, CED-9 inactivates CED-4 which then cannot activate CED-4, so no apoptosis occurs. Cells with molecular switches set in this way are those that normally survive to form the adult nematode.

Binding of a death signal to the receptor activates the receptor and the events that follow are typical of signal transduction pathways **(Figure 50.21B)** (see Sections 7.1–7.4). In this case, activation of the receptor leads to inactivation of CED-9. Because CED-9 no longer inhibits CED-4, CED-4 is active and, in turn, it activates the CED-3 protease. Activated CED-3 initiates a cascade of reactions, including the activation of other caspase proteases, and nucleases that degrade cell structures and chromosomes as part of the cell death program.

Homologs of the CED-3, CED-4, and CED-9 proteins are involved in the apoptosis pathway that is common to all ani-

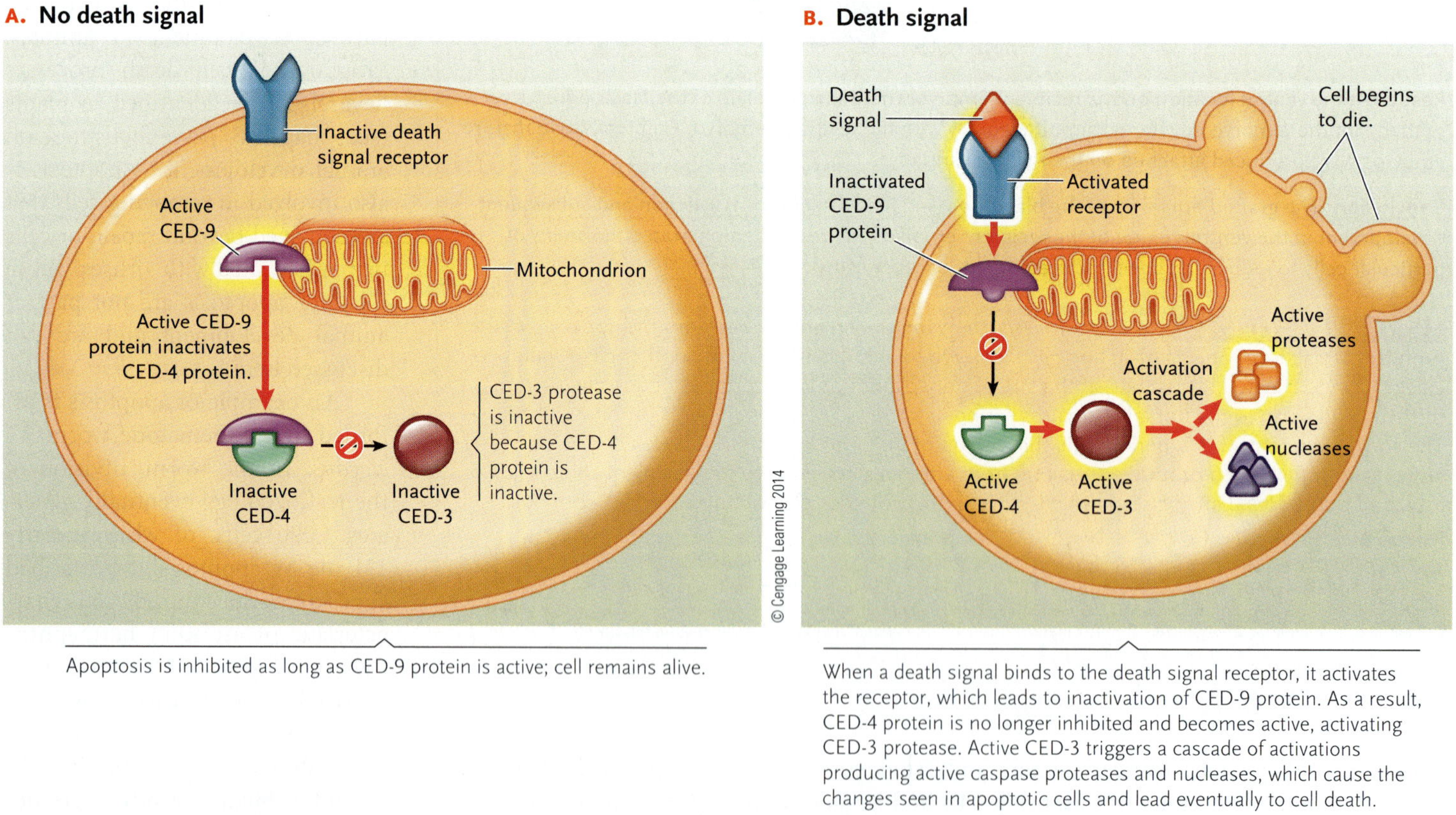

FIGURE 50.21 The molecular basis of apoptosis in the nematode, *C. elegans*. (A) In the absence of a death signal, no apoptosis occurs. **(B)** In the presence of a death signal, activation of CED-4 and CED-3 proteins triggers a pathway that leads to death of the cell.

mals that have been studied. Genetic studies have demonstrated their importance in development. For instance, worms deficient in CED-4 have 15% more cells than normal worms, yet they are viable. However, mice with deletions of their *ced-4* homologs late in embryonic development exhibit severe head and face anomalies and brain overgrowth. Mice with deletions of their *ced-3* homologs die around birth due to massive cell overgrowth in their nervous systems.

Let us consider the example of the role of apoptosis in the formation of digits in tetrapods in a little more detail. Some tetrapods, such as humans, mice, and chickens have separated digits, whereas others, such as ducks and a number of other bird species, frogs, bats, and otters, have webbing between digits. In all cases, early development produces a primordial limb with connected digits. The removal or not of the webbing by apoptosis then is under genetic control. Consider the chicken and the duck, which have highly similar feet except for the webbing in the duck. During development of the feet, bone morphogenic proteins (BMPs) are expressed in the tissue between the digits in both birds. BMPs signal the apoptosis events that result in the removal of the tissue between the digits, and those events produce the separated digits we see in the chicken. In the development of the duck foot, however, a protein called Gremlin is expressed in the tissue between the digits that inhibits the action of the BMPs. As a result, the duck has webbed feet. We can interpret these molecular data to mean that the evolution of web-footed birds likely involved the inhibition of BMP-mediated apoptosis in the tissue between the developing digits. The inhibition of BMP-mediated apoptosis by Gremlin is also involved in the development of the wing of the bat; the wing is webbing between the digits of the forefeet.

Recent research is showing that miRNAs are involved in apoptosis and, in fact, may be critical for its regulation in a number of developing tissues. For instance, in the frog, *Xenopus,* a particular miRNA is expressed during eye development in the neural layer of the retina and is required for correct formation of the eye structure. Inhibition of that miRNA results in a reduction in eye size, and that reduction occurs because of a significant increase in apoptosis in the retina. Research has shown that the miRNA involved negatively regulates the *ced-4* homolog, confirming its direct role in the regulation of apoptosis during normal development.

STUDY BREAK 50.5

1. **What are the key cellular events that contribute to morphogenesis in animals?**
2. **What is induction? What molecules are involved?**

THINK OUTSIDE THE BOOK

Use the Internet or research literature to outline evidence associating an increase in the level of N-cadherin with breast cancer.

UNANSWERED QUESTIONS

How do cells that comprise animals know what sex an animal should be as it grows up?

This should be an easy question to answer, because we know that in mammals it is the presence of the Y chromosome (containing the gene *SRY* discussed earlier in this chapter) that determines the development of testes in males. Testes then secrete testosterone and anti-Müllerian hormone (AMH), which stimulate Wolffian duct development and Müllerian duct regression. In females, ovaries develop because of the absence of SRY protein and the Müllerian ducts form. Steroid hormones secreted from the gonads then act to differentiate other tissues, including the brain. However, in nature there are many reports of animals that are gynandromorphic; that is, they have both female and male characteristics. These animals can be either bilaterally symmetrical or mosaics (containing a mix of male and female cells). Gynandromorphs have been seen in chickens, song birds, blue crabs, spiders, and insects, including butterflies and carpenter bees, as well as others. Some gynandromorphs that have been well studied include a song bird, the gynandromorphic Australian zebra finch, which had a testis and male plumage on the right side of the body and an ovary and female plumage on the left side of the body. Avian sex determination is chromosomal, and birds have a *ZW* sex-determination system that is reversed compared to the mammalian *XY* system. Female birds have heterogametic chromosomes *(ZW)* and males have homogametic chromosomes *(ZZ)*, and the sex-determining gene in birds is the *Z*-linked DMRT1 gene. The gynandromorphic zebra finch was studied by Dr. Arthur P. Arnold at the University of California, Los Angeles, where he and a team of researchers determined that the right side of the brain of the bird contained no *W* sex chromosomes while the left side contained both *Z* and *W* chromosomes. The fact that these animals develop naturally indicates that there are still many unanswered questions about sexual differentiation. What is not yet known is if the sex imposed on a developing cell comes entirely from hormonal signals originating from the outside of the cell, or if there are also signals coming from within the cells from the cell's own genome. Also, in vertebrates, how do sex chromosomes, acting on individual cells, play a role in the differences that develop between male and female tissues, including the brain?

What initiates childbirth?

You learned in this chapter that a series of hormonal and physical events are associated with childbirth. There is an increase in the secretion of oxytocin from the posterior pituitary gland and oxytocin receptors are found in the smooth muscle cells of the uterine wall. Oxytocin stimulates the muscle to begin the rhythmic contractions of labor marking the beginning of birth or parturition. This process has been well studied, but the initiation of labor is a complex process that has yet to be fully explained. What fetal-derived signals lead to the initiation of labor? How do these signals play a role in preterm labor? What roles do signals, such as prostaglandins, from the placenta play in regulating the timing of birth? What roles do the fetal hypothalamus and adrenal glands play in initiating childbirth? What roles do hormones, such as estradiol from the mother, or cortisol and corticotropin-releasing factor from the fetus, play in sending the signals to begin the process of parturition? One researcher who is making strides in this area is Louis J. Muglia M.D., Ph.D., from Washington University School of Medicine in St. Louis, Missouri. Dr. Muglia has been researching the timing of birth and the challenges to prevent premature births. His research has demonstrated that prostaglandins are essential for the initiation of parturition, and he has determined that the regulation of COX-1 is important for the initiation of parturition in mice. COX-1 (cyclooxygenase-1) is a protein that acts as an enzyme to speed up the production of prostaglandins in the stomach and the expression of the *COX-1* gene increases in the uterus during pregnancy. Identifying regulatory factors, such as COX-1, may provide critical information into timing the onset of parturition.

THINK LIKE A SCIENTIST Based on the first unanswered question, do you think there are mammals that are gynandromophs? Does your answer to this question alter the significance of the research?

Courtesy of Laura Carruth

Laura Carruth, an assistant professor of biology at Georgia State University, studies the genetic and hormonal factors that lead to sex differences in brain development. Her work focuses on model systems in songbirds (the Australian zebra finch) and mice. To learn more about Dr. Carruth's research, go to http://www2.gsu.edu/~wwwllc/.

REVIEW KEY CONCEPTS

To access the course materials and companion resources for this text, please visit www.cengagebrain.com.

50.1 Mechanisms of Embryonic Development

- Developmental information is stored in both the nucleus and cytoplasm of the fertilized egg. The mRNA and protein molecules that direct the first stages of development are the cytoplasmic determinants.
- The unequal distribution of yolk and other components makes eggs polar. The animal pole typically gives rise to surface structures and the anterior end of the embryo, whereas the vegetal pole typically gives rise to internal structures of the embryo such as the gut (Figure 50.2).
- Following fertilization, cleavage divisions produce the morula. The morula hollows out to form the blastula, which develops into the gastrula, the stage in which rearrangements of cells produce the ectoderm, mesoderm, and endoderm. Gastrulation establishes the body pattern, in that the organs and other structures of embryo arise from these three tissue layers (Figure 50.3; Table 50.1).
- Development proceeds as a result of cell division, cell movements, selective adhesions, induction, determination, and differentiation.

Animation: Body Planes

Animation: Leopard frog life cycle

50.2 Major Patterns of Cleavage and Gastrulation

- In sea urchin eggs, yolk is distributed evenly. As a result, cleavage divisions take place at the same rate in all regions of the embryo, and gastrulation follows a symmetrical pattern (Figure 50.4).
- In amphibian eggs, yolk is distributed unequally, with most in the vegetal pole. As a result, the rate of cell division is more rapid in the animal pole, and gastrulation shows an asymmetrical pattern (Figures 50.5 and 50.6).
- In bird and reptile embryos, the cleavage divisions give rise to a flat disc of cells at the top of the yolk, which divides into the epiblast and the hypoblast. In gastrulation, cells of the epiblast migrate to the interior to form the endoderm and the mesoderm. The epiblast cells left at the surface form the ectoderm (Figure 50.7).
- In birds and reptiles, the yolk sac, chorion, amnion, and allantoic membrane form from extensions of the primary tissue layers. These extraembryonic membranes conduct nutrients from the yolk to the embryo, exchange gases with the environment, and store metabolic wastes (Figure 50.8).

Animation: Process of gastrulation

Animation: Structure of the placenta

50.3 From Gastrulation to Adult Body Structures: Organogenesis

- In organogenesis, the three primary tissues give rise to the tissues and organs of the embryo. Organogenesis begins with neurulation, the development of the nervous system from ectoderm (Figures 50.9 and 50.10).
- The mesoderm splits into somites, which give rise to the vertebral column and to the muscles of the ribs, vertebral column, and limbs (Figure 50.10).
- Development of the eye from optic vesicles is illustrative of the inductions and differentiations common to organogenesis in vertebrates (Figure 50.11).

50.4 Embryonic Development of Humans and Other Mammals

- In humans, as in other placental mammals, cleavage divisions produce a morula that differentiates into a blastocyst. The blastocyst implants into the endometrium of the uterus, and its inner cell mass separates into the epiblast and hypoblast. The epiblast produces the ectoderm, mesoderm, and endoderm of the embryo (Figures 50.12 and 50.13, steps 1–2).
- Gastrulation, neurulation, differentiation of cell layers, and formation of extraembryonic membranes occur by mechanisms similar to those of bird and reptile embryos. Differentiation of ectoderm, mesoderm, and endoderm into their final tissues and organs is also similar in birds and reptiles.
- Extraembryonic membranes form in mammals by processes that are also similar to the reptilian–bird pattern. However, some of the membranes have altered functions, reflecting the minimal amount of yolk in mammalian embryos, and maintenance of the embryo by the placenta (Figure 50.13, steps 3–5).
- The placenta is connected to the embryo by the umbilical cord, which conducts blood between the embryo and the placenta (Figure 50.13, step 6).
- Fetal growth proceeds until birth, when the fetus is forced from the uterine cavity and through the vagina by contractions of the uterus, stimulated by oxytocin (Figures 50.14 and 50.15).
- The mother's mammary glands secrete milk once the offspring is born. Suckling by the offspring stimulates prolactin and oxytocin release from the pituitary, which stimulates milk production and secretion from the glands, respectively.

- Embryos develop internal male or female sex organs from the same primitive structures. The presence or absence of a Y chromosome, which carries the key *SRY* gene, determines whether the internal structures develop into male or female sexual organs (Figure 50.16).
- Most animals continue development after hatching or birth, leading to the adult, the sexually mature form of the species.

Animation: Hormonal control of sperm production

Animation: Cleavage and implantation

Animation: Female reproductive system

50.5 The Cellular Basis of Development

- Development in animals involves the regulation of specific cellular events, including cell division, cell migration, and cell adhesion.
- Cell division in development varies in orientation and rate.
- Cell movements in development occur through changes in cell shape or the migrations of entire cells. Shape changes are produced by microtubules or microfilaments. Cell migrations involve the activity of microtubules and microfilaments and large-scale reorganization of the cytoskeleton. The four stages of cell migration are polarization, protrusion of the front end of the cell, adhesion, and then release of adhesion at the rear of the cell and forward migration to a new adhesion point. In many developmental systems, migrating cells follow tracks formed by molecules of the extracellular matrix (Figure 50.17).
- Selective cell adhesions, which depend on surface glycoproteins including CAMs and cadherins, underlie many cell movements. The final cell adhesions hold the embryo in its correct shape and form (Figure 50.18).
- For some organisms, the origins of adult or larval structures have been mapped to regions of the embryo from which each structure derived (Figure 50.19).
- Induction results from the effects of signaling molecules of the inducing cells on the responder cells that are nearby the inducer cells. To change its cellular behavior, the responder cell must have a receptor for the inducer molecule and it must be competent to respond (Figure 50.20).
- In differentiation, cells change from embryonic form to specialized types with distinct structures and functions. Differentiation occurs by differential gene activation.
- Apoptosis—a type of programmed cell death—plays an important role in normal development by removing tissues present during development but not in the adult tissue or organ (Figure 50.21).

Animation: Embryonic induction

Animation: Active and Facilitated Diffusion

UNDERSTAND & APPLY

Test Your Knowledge

1. Major contributors to the axes of the animal body are the:
   a. sperm and egg cytoplasm.
   b. sperm and egg chromosomes.
   c. ribosomes and mitochondria.
   d. egg nucleus and yolk.
   e. pigments.
2. The process by which cells undergo mitosis without a corresponding increase in cytoplasm is called:
   a. polarity.
   b. cleavage.
   c. gastrulation.
   d. organogenesis.
   e. induction.
3. Which of the following mechanisms does *not* contribute to zygote development?
   a. meiosis
   b. mitosis
   c. selective cell adhesions
   d. determination
   e. induction
4. A major event during gastrulation is:
   a. the outward movement of cells at the dorsal lip of the blastopore.
   b. the displacement of the archenteron by the blastocoel.
   c. the formation of the coelom from the endoderm.
   d. the extension of ectoderm and endoderm to form the yolk sac.
   e. the development of ectoderm to form epidermal and neural tissues.
5. The presence of the gray crescent in a frog embryo:
   a. corresponds to the location of the future mouth.
   b. corresponds to the location of sperm entry into the egg.
   c. will cause the formation of a disordered mass of cells.
   d. contains concentrated amounts of darkly colored pigment granules from the animal pole.
   e. indicates that the egg has been fertilized.
6. To contribute to the formation of a nervous system:
   a. the neural crest develops into motor neurons.
   b. the neural tube is converted into a neural plate.
   c. the notochord induces the overlying ectoderm to become a neural plate.
   d. the roof of the archenteron induces the formation of the neural tube.
   e. somites give rise to the autonomic nervous system.
7. In mammalian development:
   a. the morula develops into a trophoblast.
   b. the chorionic villi allow the blastocyst to move down the oviduct.
   c. the allantois takes over the work of the amnion.
   d. the pharyngeal arches transform into the pharynx, larynx, and nasal cavities.
   e. prolactin stimulates parturition.
8. In the development of the female sex organs:
   a. all ducts in the 7-week embryo become Wolffian ducts.
   b. the *SRY* gene is activated.
   c. anti-Müllerian hormone is secreted.
   d. the Müllerian ducts develop into oviducts.
   e. the mother secretes oxytocin.
9. In the embryonic development of the eye:
   a. the optic vesicle cells permanently adhere to each other to prevent movement, whereas the optic cup cells are very motile.
   b. signals from the optic cup trigger surface receptors on the lens placode.
   c. gradients determine that the ectoderm overlying the lens vesicle develops into the optic vesicle.
   d. microtubules powered by myosins and microfilaments powered by dyneins move the eye components around in the head region.
   e. cadherins function in the presence of calcium to allow the lens placode and optic cup to break apart.

10. Which of the following statements about development is incorrect?
    a. As microtubules lengthen and slide apart, the shape of cells within the ectoderm changes.
    b. The extracellular matrix molecule fibronectin guides the movement of cells during development.
    c. The furrow formed during mitosis of cells in the developing embryo is centrally located to ensure all cells divide symmetrically.
    d. Cell surface cadherins regulate the ability of developing cells to adhere to one another.
    e. Integrins are responsible for transmitting information from the extracellular matrix to the cytoskeleton.

Discuss the Concepts

1. Experimentally, it is possible to divide an amphibian egg so that the gray crescent is wholly within one of the two cells formed. If the two cells are separated, only the cell with the gray crescent will form an embryo with a long axis, notochord, nerve cord, and back musculature. The other forms a shapeless mass of immature gut and blood cells. Propose an explanation of these outcomes.
2. Developmental biologist Lewis Wolpert once observed that "it is not birth, marriage or death, but gastrulation which is truly the most important time in your life." In what sense is he correct?
3. Arguably, in sexually reproducing animals, development begins when eggs and sperm form in the parents. In a paragraph, explain the rationale for this idea.
4. Investigators discovered a *Drosophila* protein that triggers development of the nerve cord on the ventral side of the embryos. When an mRNA encoding the protein was injected into cells on the ventral side of *Xenopus* embryos, dorsal structures were formed on the ventral side, including incomplete heads. What do these findings suggest about the evolution of embryonic development?

Design an Experiment

You learned in this chapter about the molecular basis of apoptosis in the nematode worm, *C. elegans*. In humans, the *BCL2* gene is a homolog of the worm *ced-9* gene. What experiment would convince you that the human BCL2 protein is functional homologous to the worm CED-9 protein?

Interpret the Data

People considering fertility treatments should be aware that such treatments raise the risk of multiple births and that multiple pregnancies are associated with an increased risk of some birth defects. The **Table** shows the results of Yiwei Tang's study of birth defects reported in Florida from 1996 to 2000. Tang compared the incidence of various defects among single and multiple births. She calculated the relative risk for each type of defect based on type of birth, and corrected for other differences that might increase risk: maternal age, income, race, previous adverse pregnancy experience, education, Medicaid participation during pregnancy, and the infant's sex and number of siblings. A relative risk of less than 1 means a defect occurs less often with multiple births than single births. A relative risk of greater than 1 means that multiples are more likely to have a defect.

Prevalence, per 10,000 Live Births, of Various Types of Birth Defects among Multiple and Single Births

	Prevalence of defect		Relative
	Multiples	Singles	Risk
Total birth defects	358.50	250.54	1.46
Central nervous system defects	40.75	18.89	2.23
Chromosomal defects	15.15	14.20	0.93
Gastrointestinal defects	28.13	23.44	1.37
Genital/urinary defects	72.85	58.16	1.31
Heart defects	189.71	113.89	1.65
Musculoskeletal defects	20.92	25.87	0.92
Fetal alcohol syndrome	4.33	3.63	1.03
Oral defects	19.84	15.48	1.29

1. What was the most common type of birth defect in the single-birth group?
2. Was that defect more or less common in the multiple-birth group than among single births?
3. Tang found that multiples have more than twice the risk of single newborns for one type of defect. Which type?
4. Does a multiple pregnancy increase the relative risk of chromosomal defects in offspring?

Source: Y. Tang et al. 2006. The risk of birth defects in multiple births: A population-based study. *Maternal and Child Health Journal* 10:75–81.

Apply Evolutionary Thinking

In this chapter, you have learned that cleavage and gastrulation differ among vertebrates that have different amounts of yolk in their eggs. What are some features of early development that have been conserved across vertebrate groups during their evolution?

Appendix A: Answers*

Chapter 1

Study Break

STUDY BREAK 1.1

1. The major levels in the hierarchy of life and some of their emergent properties are: cells—life; organisms—learning; populations—birth and death rates; communities, ecosystems, biosphere—diversity and stability.
2. Organisms use energy collected from the external environment for growth (including the production of new molecules and cells), maintenance and repair of body parts, and reproduction.
3. A life cycle is the series of structurally and functionally distinct developmental stages through which organisms pass.

STUDY BREAK 1.2

1. In artificial selection, humans selectively breed individuals with desirable heritable characteristics to enhance those traits in the next generation. In natural selection, genetically-based characteristics that increase survival and reproduction become more common in the next generation.
2. Random changes in DNA—mutations—may change the structure of proteins that contribute to the physical appearance and internal functions of an organism.
3. Being camouflaged may make an animal less likely to be noticed by a predator.

STUDY BREAK 1.3

1. In the cells of prokaryotic organisms, DNA is not separated from other parts of the cell. In the cells of eukaryotic organisms, DNA is enclosed within a nucleus.
2. Humans are classified in Domain Eukarya and Kingdom Animalia.

STUDY BREAK 1.4

1. A scientific hypothesis must be falsifiable. In other words, we must be able to imagine what sort of data would demonstrate that the hypothesis is incorrect.
2. The copper lizard models told the researchers how frequently lizards would perch in the sun just by chance and what the temperatures of nonthermoregulating lizards would be.
3. Model organisms are usually easy to maintain and study, and they have been so well studied that researchers already know a lot about their biology.
4. When scientists describe a set of ideas as a "theory," they recognize that the ideas have already withstood many scientific tests.

Think Like a Scientist

FIGURE 1.11

Animals are more closely related to fungi than they are to plants. According to the phylogenetic tree, animals and fungi shared a common ancestor more recently than did animals and plants.

FIGURE 1.15

To test the effects of two experimental variables simultaneously, you would need to design an experiment that had experimental and control groups for both variables (that is, fertilizer and water). One group might receive both fertilizer and water; a second group would receive fertilizer but no water; a third group would receive water but no fertilizer; and a fourth group would receive neither fertilizer nor water. You could then compare the results of the four treatments to see the effects of both experimental variables, alone and in combination, on your plants.

FIGURE 1.16

The temperature is not very variable among the shaded sites in this environment. Most of the models were in shaded sites, and the temperatures of the shaded sites are clustered around 19° to 20°C.

Test Your Knowledge

1. c 2. b 3. c 4. d 5. b 6. d 7. c 8. a 9. e 10. d

Interpret the Data

Anolis gundlachi does not regulate its body temperature. The graphs show that lizards perch in patches of sun at about the same rate as randomly positioned models and that lizards and models exhibit similar distributions of body temperatures. These data suggest that this lizard species perches at random with respect to environmental factors that might influence its body temperature.

Chapter 2

Study Break

STUDY BREAK 2.1

An element is a pure substance that consists of one type of atom. An atom is the smallest unit of an element that retains its chemical and physical properties. A molecule is a collection of atoms chemically combined in fixed numbers and ratios. Molecules can consist of the same atoms, as is seen for the two oxygen atoms in the molecule oxygen, or of different atoms, as in the combination of two hydrogen atoms and one oxygen atom in a molecule of water. Molecules with component atoms that are different, such as water, are compounds.

STUDY BREAK 2.2

1. Protons and neutrons are found in the nucleus of an atom. Electrons are found in orbitals located in energy levels (shells) that surround the nucleus.
2. Carbon-11 has six protons and five neutrons. Oxygen-15 has eight protons and seven neutrons.
3. The number of valence electrons—the electrons in the outermost shell of an atom—determines its chemical reactivity. If the outermost shell is not completely filled with electrons, the atom tends to be chemically reactive, whereas if that shell is completely filled, the atom is nonreactive.

STUDY BREAK 2.3

1. An ionic bond forms between atoms when those atoms gain or lose electrons completely. An example is the ionic bond in NaCl.
2. A covalent bond forms when atoms share a pair of valence electrons rather than gaining or losing them completely.
3. Electronegativity is a measure of an atom's attraction for the electrons it shares in a chemical bond with another atom. When electrons are shared equally, the atoms remain uncharged and the result is a nonpolar covalent bond. When electrons are shared unequally, one atom carries a partial negative charge and the other atom carries a partial positive charge. The molecule then has polarity, and the bond is a polar covalent bond.
4. In a chemical reaction, atoms or molecules interact to form new chemical bonds or break old ones. Atoms are added to or removed from molecules, or linkages of atoms in molecules are rearranged as a result of bond formation.

STUDY BREAK 2.4

1. Hydrogen bonds between neighboring water molecules produce a water lattice. The constant breakage and reformation of hydrogen bonds in the lattice allows water to flow easily. The polarity of water molecules also contributes to the properties of water; that is, in liquid water, the lattice resists invasion by other molecules unless the invading molecules also contain polar regions that can form competing attractions with water molecules. In that case, the water lattice opens, forming a cavity in which the polar or charged molecule can move. However, nonpolar molecules are unable to affect the water lattice. Hydrogen bonds also give water its unusual ability to resist changes in temperature by absorbing or releasing heat energy, its unusually high boiling point, and its unusually high internal cohesion and surface tension.
2. A solute is a dissolved substance. A solvent is a substance capable of dissolving another substance. A solution is a solute dissolved in a solvent. For example, salt (NaCl) is a solute that can dissolve in the solvent water.

STUDY BREAK 2.5

1. Acids are hydrogen ion (proton, H^+) donors; bases are proton acceptors. An acid dissociates in water to produce a hydrogen ion and an anion. Most bases dissociate in water to give hydroxide ions, which then accept protons to produce water.
2. Buffers act to control the pH of a solution. In living organisms, buffers keep the pH of body and cell fluids within a narrow range, enabling normal cell and body functions to occur. Outside of the normal pH range, the functions of proteins can be affected, thereby adversely affecting the functions of the organism.

Think Like a Scientist

UQ

Properties of natural soils can determine the flow pattern and how much carbon source can be delivered to bacteria, therefore controlling their capability of remediating uranium. In addition, the abundance and spatial distribution of certain minerals can affect the magnitude of energy sources for the bacteria and therefore the efficiency of uranium bioremediation.

Test Your Knowledge

1. d 2. e 3. a 4. b 5. a 6. d 7. d 8. c 9. e 10. b

Interpret the Data

1. About 50 minutes
2. (a) Using a pH of 5.9, H^+ concentration is 0.0000012589 (1.2589×10^{-6}) *M* and OH^- concentration is 0.00000000789433 (7.9433×10^{-9}) *M*. (b) Using a pH of 2.5, H^+ concentration is 0.00316228 (3.16228×10^{-3}) *M* and OH^- concentration is 0.0000000000031623 (3.1623×10^{-12}) *M*.
3. Using a pH of 2.5 for peak reflux and pH of 4.0 for clinical reflux, the H^+ concentration is 0.003 *M* greater, and OH^- concentration is 0.0000000000968377 (9.68377×10^{-11}) *M* less than a pH of 4.0.

Chapter 3

Study Break

STUDY BREAK 3.1

1. Organic molecules are molecules based on carbon. Hydrocarbons are a type of organic molecule that consists of carbon linked only to hydrogen atoms.
2. The maximum number of bonds that a carbon atom can form is four.
3. Carboxyl groups donate a hydrogen ion in water and therefore act as acids. Amino groups accept a hydrogen ion in water and therefore act as bases. Phosphate groups donate hydrogen ions in water and therefore act as acids.
4. In a dehydration synthesis reaction, components of water (—H and —OH) are removed. In hydrolysis, the components of a water molecule are added to functional groups as molecules are broken down into smaller subunits.

STUDY BREAK 3.2

A monosaccharide is the structural unit of carbohydrate molecules. Monosaccharides are simple sugars such as trioses, pentoses, and hexoses. Glucose, galactose, and fructose are hexoses. A disaccharide is a molecule assembled from two monosaccharides linked by a dehydration synthesis reaction. Lactose, sucrose, and maltose are disaccharides. A polysac-

*Abbreviations: IMR = Insights from the Molecular Revolution; UQ = Unanswered Questions charide is a polymer of monosaccharide subunits. The subunits are identical or different, depending on the particular polysaccharide. Glycogen, starch, cellulose, and chitin are polysaccharides.

STUDY BREAK 3.3

The three most common lipids found in living organisms are neutral lipids, phospholipids, and steroids. Most neutral lipids consist of a three-carbon backbone chain formed from glycerol, with each carbon linked to a fatty acid side chain. In the most common phospholipids, glycerol is the backbone, with two of its binding sites linked to fatty acids. The third binding site is linked to a polar phosphate group. Steroids have structures based on a framework of four carbon rings. Differences in side groups attached to the rings distinguish the different types of steroids.

STUDY BREAK 3.4

1. Differences in the side groups (R in the figures) give the amino acids their individual properties.
2. A peptide bond is the bond between the C of the carboxyl group of one amino acid and the N of the amino group of the adjacent amino acid (see Figure 3.19). The bond is formed in a dehydration synthesis reaction between an amino group of one amino acid and a carboxyl group of another amino acid.
3. Domains are distinct structural subdivisions in the final folded forms of proteins. They are the result of the amino acid sequence of the protein (the primary structure of the protein) and the secondary, tertiary, and quaternary (if more than one polypeptide is involved) structures of the protein.

STUDY BREAK 3.5

1. Nucleic acids are formed from nucleotide monomers. A nucleotide consists of a nitrogenous base, a five-carbon sugar, and a phosphate.
2. In DNA, the five-carbon sugar is deoxyribose; in RNA it is ribose. DNA has the pyrimidine nitrogenous base T (thymine), and RNA has U (uracil).

Think Like a Scientist

FIGURE 3.24

In the presence of a high concentration of urea, the formation of the four disulfide bridges of the native enzyme apparently only rarely occurs. As a result, only a low level of enzyme activity is measured in the solution. The other enzymes appear to have formed three-dimensional structures, but likely they contain nonnative disulfide linkages and therefore result in inactive enzymes. Overall, the result strengthens Anfinsen's conclusions because it means that particular disulfide linkages are responsible for the stable, and therefore active, form of the enzyme. If the enzyme instead forms with incorrect pairs of disulfide linkages, the structure is sufficiently different that enzyme activity is lost.

IMR

1. The enzyme that assembles DNA molecules in the bacterium *Escherichia coli* (see Figure 3.26B). The enzyme has two domains, one to assemble the DNA molecules (the polymerase domain) and the other to correct mistakes during DNA assembly (the exonuclease domain).
2. The method of analysis described in *Insights from the Molecular Revolution* was the comparison of domain structure and organization. Fundamentally, the domains consist of particular sequences of amino acids. The function of a domain depends on the sequence of amino acids plus the higher-order structures it forms within the protein. As a domain diverges through evolutionary time, core functional elements must stay the same or highly similar in order that the function of the domain be retained. At the same time, other parts of the domain can experience changes in amino acid sequence. Therefore, through bioinformatics-based computer analysis of proteins sharing a particular function, researchers can determine how the domain has changed over time and, thereby, construct a phylogenetic tree based on the changes.

UQ

In type 2 diabetes, SREBPs in the liver are activated, and this leads to overproduction of fatty acids and triglycerides. Inhibitors of SREBP processing might reduce plasma triglycerides and prevent their toxic effects on blood vessels.

Test Your Knowledge

1. a 2. d 3. c 4. e 5. c 6. d 7. b 8. d 9. b 10. a

Interpret the Data

1. LDL was highest in the saturated fats diet group, and HDL was lowest in the *trans* fatty acids diet group.
2. The *trans* fatty acid group had the highest LDL-to-HDL ratio.
3. From best to worst diet: *cis* fatty acids, saturated fats, *trans* fatty acids.

Chapter 4

Study Break

STUDY BREAK 4.1

1. Kinetic energy is the energy of motion, whereas potential energy is stored energy.
2. An isolated system exchanges neither matter nor energy with its environment. A closed system can exchange energy, but not matter, with its environment. An open system can exchange both energy and matter with its environment.

STUDY BREAK 4.2

1. The change in energy content of a system, and its change in entropy. Reactions tend to be spontaneous when the products have less potential energy than the reactants, and when the products are less ordered than the reactants.
2. The greater the negative value of ΔG, the further a reaction will proceed toward completion and therefore the greater the concentration of product molecules versus reactant molecules.
3. An exergonic reaction releases free energy—ΔG is negative because the products contain less free energy than the reactants. An endergonic reaction requires free energy from the surroundings to run—ΔG is positive because the products contain more free energy than the reactants.

   In a catabolic reaction, energy is released during the breakdown of a complex reactant molecule to a simpler product molecule, whereas in an anabolic reaction, energy is used to create a product molecule that is more complex than the reactant molecule.

   Individual reactions may be exergonic or endergonic. When a series of reactions (each of which may have a positive or a negative ΔG value) forms a metabolic pathway, the pathway is catabolic if the overall sum of individual reaction ΔG values is negative, and it is anabolic if the overall sum of individual reaction ΔG values is positive.

STUDY BREAK 4.3

1. ATP contains the five-carbon sugar ribose, with the nitrogenous base adenine linked to one of the carbons and a chain of three phosphate bonds to another carbon. The phosphate groups are closely associated with each other and their negative charges strongly repel each other, making the bonding arrangements unstable and storing potential energy. Hydrolysis of ATP to remove one or two of the three phosphates is a spontaneous reaction that relieves the repulsion and releases large amounts of free energy.
2. Many individual reactions found in living cells are not spontaneous because they have a positive ΔG. By joining such a reaction to another reaction with a large negative ΔG, the reaction can be completed. The combined reaction is called a coupled reaction. ATP participates in coupled reactions in the enzyme-driven process of energy coupling; that is, ATP comes into close contact with a reactant molecule in an endergonic reaction. When ATP is hydrolyzed, the terminal phosphate group is transferred to the reactant molecule, which makes that molecule less stable so that the reaction continues spontaneously.

STUDY BREAK 4.4

1. Enzymes accelerate reactions by reducing the activation energy of a reaction, the initial input of energy required to start a reaction. Enzymes lower activation energy by rearranging the atoms and bonds of the reacting molecules into the transition state, an activated state that is highly unstable. With relatively little change in energy, the transition state can move forward toward products or backward toward reactants.
2. No.

STUDY BREAK 4.5

1. There is a maximum rate at which an enzyme can combine with substrates and release products. Beyond that point, increasing substrate concentration does not increase the reaction rate any further.
2. In competitive inhibition, the inhibitor competes with the normal substrate molecule for binding to the active site of the enzyme, whereas in noncompetitive inhibition, the inhibitor does not compete directly with the substrate for binding to the active site.
3. As the temperature increases, the increasing kinetic motions of the enzyme's amino acid chains eventually disrupt the enzyme's three-dimensional structure, causing it to unfold and denature. At that point, there is no enzyme activity.

STUDY BREAK 4.6

A ribozyme is an RNA molecule that accelerates the rate of a biological reaction. A ribozyme qualifies as an enzyme because it remains unchanged after the reaction is complete; that is, it is a true catalyst.

Think Like a Scientist

IMR

Both the 5′ and 3′ constant regions are necessary for the active ribozyme structure to form.

UQ

The problem is that each nucleotide in the ribozyme would need to serve as template for polymerization. This means that the ribozyme would have to open up its own structure and inactivate its own catalytic site. Therefore, the existence of such a molecule seems very unlikely. However, it may be possible to design such a molecule if it has two catalytic centers, or if the crucial parts of the ribozyme do not need to template because they exist twice in the molecule.

Test Your Knowledge

1. c 2. d 3. b 4. c 5. e 6. a 7. a 8. b 9. d 10. e

Interpret the Data

1. ΔG is minus 5.5 minus 3.5 = −9 kcal/mol for the mutated polypeptide and minus 2.0 minus 4.0 = −6.0 kcal/mol for the parent polypeptide.
2. Binding of both of the polypeptides is exergonic and spontaneous, and the parent polypeptide would yield more free energy on binding.
3. No, because the ΔG is less negative, it would not be as energetically favorable.

Chapter 5

Study Break

STUDY BREAK 5.1

The plasma membrane is a bilayer of lipid and suspended protein molecules that bounds the cytoplasm of a cell. The lipid bilayer is hydrophobic; therefore, it is a barrier to the passage of water-soluble substances. The membrane has protein channels through which selected water-soluble substances are able to pass.

STUDY BREAK 5.2

The DNA of a prokaryotic cell is located in the nucleoid region, a central area of the cell that has no membrane around it to separate it from the cytoplasm. In most prokaryotes, the DNA is a folded mass. In its unfolded state, it is a circular DNA molecule.

STUDY BREAK 5.3

1. Most of the DNA of a eukaryotic cell is found within a nucleus located roughly in the center of the cell. The nucleus is bounded by the nuclear envelope, a membrane that separates its contents from the cytoplasm. The DNA is complexed with proteins and organized into several linear chromosomes.
2. The nucleolus is an area within the nucleus. It forms around the genes for rRNA in the chromosomes and is the location in which the information in those genes is copied into rRNA. The rRNA combines with proteins in the nucleolus to form the large and small ribosomal subunits. A large and a small ribosomal subunit function together in the cytoplasm to synthesize proteins.
3. The endomembrane system is a collection of organelles, membranous channels, and vesicles that form a major traffic network for the synthesis, distribution, and storage of proteins and other molecules. Its structure includes the endoplasmic reticulum (ER) and the Golgi complex. The rough ER has ribosomes on its outer surface. Proteins synthesized by those ribosomes enter the ER lumen, where they fold into their final shape and may be modified chemically. Then they are delivered to other regions of the cell within vesicles that pinch off from the rough ER. Most of the proteins made on rough ER go to the Golgi complex.

   The smooth ER consists of membranes that lack ribosomes. Functions of the smooth ER include synthesis of lipids that become parts of cell membranes.

   The Golgi complex is a stack of flattened, membranous sacs. The *cis* face of this organelle receives vesicles containing proteins released from the rough ER and continues their chemical modifications. The proteins are then sorted into vesicles that pinch off from the *trans* face of the Golgi complex, the side that faces the plasma membrane. Some of the released vesicles remain in the cytoplasm as storage vesicles of various types, whereas others, called secretory vesicles, release their contents to the outside of the cell.
4. A mitochondrion is an organelle enclosed by two membranes. The outer mitochondrial membrane is smooth and covers the outside of the organelle. The inner mitochondrial membrane is highly folded into cristae. Within the inner membrane is the mitochondrial matrix, which contains DNA and ribosomes. Most of the energy required for eukaryotic cellular activities is generated by reactions in the cristae and the matrix. Those reactions break down sugars, fats, and other fuel molecules into water and carbon dioxide, releasing energy mostly in the form of ATP.
5. The cytoskeleton is an internal cytoplasmic network of filaments and tubules composed mainly of actin and tubulin proteins. The function of the cytoskeleton is to maintain the shape of the cell, reinforce the plasma membrane, and organize internal structures. Changes in the cytoskeleton are responsible for movements of cell organelles, movements of parts of the cell, or movements of the whole cell.

STUDY BREAK 5.4

1. A chloroplast has two membranes: an outer boundary membrane and an inner boundary membrane. The latter, similar to the cristae of the mitochondrion, is highly folded. The two membranes enclose an inner compartment known as the stroma. Within the stroma is a membrane system composed of flattened, closed sacs called thylakoids. Chloroplasts are the sites of photosynthesis in plant cells. The thylakoid membranes contain chlorophyll and other molecules that absorb light energy and convert it into chemical energy. Enzymes in the stroma use the chemical energy to make carbohydrates and other complex organic molecules from water, carbon dioxide, and other simple inorganic precursors.
2. The tonoplast, the membrane surrounding the central vacuole, contains transport proteins that move substances into and out of the vacuole. Central vacuoles also store organic and inorganic salts, organic acids, sugars, storage proteins, pigments, and, in some cells, waste products. Chemical defense molecules are found in the central vacuoles of some plants.

STUDY BREAK 5.5

1. Anchoring junctions are spots or belts that run entirely around cells, effectively sticking adjacent cells together. Microfilaments (in adherens junctions) or intermediate filaments (in desmosomes) anchor the junction in the underlying cytoplasm. Tight junctions involve fusion of a network of junction proteins in the outer halves of the plasma membranes of adjacent cells, forming a tight seal that can keep even ions from moving between the cells. Gap junctions open direct channels between adjacent cells through which ions and small molecules can pass directly. The gap junctions are formed by aligned hollow protein cylinders in the plasma membranes of two cells.
2. The extracellular matrix (ECM) is a complex of proteins and polysaccharides secreted by the cells that it surrounds. Depending on the nature of the network of proteoglycans (carbohydrate-rich glycoproteins) in the ECM, the consistency ranges from soft and jellylike to hard and elastic.

Think Like a Scientist

FIGURE 5.11

Presumably a nuclear envelope-associated component recognizes the nuclear localization signal of proteins destined to enter the nucleus. A logical location for such a component is in association with the nuclear pore complex. Clearly, a very restricted region of a protein can be sufficient to direct nuclear localization. In the same vein, the transport system must be able to be highly selective in recognizing proteins with such a small nuclear localization signal.

FIGURE 5.17

The chemical is inhibiting a step in the pathway for protein secretion from the cell. Possible steps affected are the transport of the protein in vesicles from the ER to the Golgi complex, the production of secretory vesicles by budding from the Golgi membranes, or the release of the protein at the cell surface by exocytosis.

IMR

All the *Methanococcus* genes related to genes in other organisms would have had sequences similar just to those in bacteria rather than some similar to bacterial genes and others similar to eukaryotic genes.

UQ

Perturbing the activity of individual genes and proteins allows cell biologists to determine whether each plays a necessary role in a given cellular process, but does not indicate what mechanistic role the protein performs or how it cooperates with other proteins to perform its functions. On the other hand, reconstituting a process in a test tube allows cell biologists to build a functioning system from individual components and understand how each contributes to the system, but does not address the functional importance of each component in the context of the cell. Both approaches can be technically challenging, but reconstitution is often more challenging because it requires both identification of all of the important players and the isolation of the players in a functional form.

Test Your Knowledge

1. b 2. e 3. c 4. a 5. b 6. c 7. b 8. b 9. b 10. d

Interpret the Data

Most of the cathepsin activity occurs in the fraction containing the lysosomal marker enzymes (fraction 5), thus localizing the worm cathepsin activity to the lysosome.

Chapter 6

Study Break

STUDY BREAK 6.1

1. The fluid mosaic model proposes that the membrane consists of a fluid phospholipid bilayer in which proteins are embedded and float freely.
2. Integral proteins include transport proteins, receptor proteins, recognition proteins, and cell adhesion proteins. Peripheral proteins include microtubules, microfilaments, intermediate filaments, and proteins that link the cytoskeleton together.

STUDY BREAK 6.2

1. In passive transport, molecules and ions move across the membrane from the side with the higher concentration to the side with the lower concentration; that is, the difference in concentration provides the energy for passive transport. In active transport, molecules and ions move across the membrane from the side with the lower concentration to the side with the higher concentration—that is, against the concentration gradient. The energy for active transport comes from the hydrolysis of ATP.
2. The transport of substances through membranes based solely on molecular size and lipid solubility is simple diffusion. The diffusion of polar and charged molecules across membranes with the help of transport proteins is facilitated diffusion.

STUDY BREAK 6.3

1. Osmosis is the passive transport of water across a membrane. This movement follows concentration gradients. For osmosis to occur, there must be a selectively permeable membrane—that is, a membrane that will allow water molecules, but not molecules of the solute, to pass. As long as the solute is at different concentrations on the two sides of the membrane, water movement will occur; that is, it is not necessary for pure water to be present on one side of the membrane.
2. If animal cells are in a hypertonic solution, water molecules will move by osmosis from within the cells to the surrounding solution. If the outward movement of water exceeds the capacity of the cells to replace the lost water, the cells will shrink.

STUDY BREAK 6.4

1. Active transport is the movement of substances across membranes against their concentration gradients by pumps; the energy for active transport comes from ATP hydrolysis. In primary active transport, ATP hydrolysis directly drives the process; that is, the same protein that transports a substance also hydrolyzes ATP. In secondary active transport, ATP hydrolysis indirectly drives the process; that is, the transport proteins themselves do not hydrolyze ATP. Rather, the transporters use a favorable concentration gradient of ions, generated by primary active transport (when ATP hydrolysis is used), as their energy source for active transport of a different ion or molecule.
2. A membrane potential is a voltage difference across a membrane. Ion transport by membrane pumps contributes to this voltage difference. The sodium–potassium pump in the plasma membrane pushes three sodium ions out of the cell and two potassium ions into the cell with each turn of the pump. This leads to an accumulation of positive charges outside the membrane, causing the inside of the cell to become negatively charged with respect to the outside of the cell. In addition, an unequal distribution of ions across the membrane is created by passive transport. The electrical potential difference (voltage) across the plasma membrane is the membrane potential.

STUDY BREAK 6.5

1. In exocytosis, secretory vesicles in the cytoplasm contact and fuse with the plasma membrane, releasing their contents to the outside of the cell.
2. Endocytosis is a mechanism by which substances are brought into the cell from the exterior. The substances become trapped in pitlike depressions that bulge inward from the plasma membrane. The depression pinches off as an endocytic vesicle. In bulk-phase endocytosis, no binding by surface receptors is involved. Extracellular water is taken in together with any other molecules that are in solution in the water. This is the simplest form of endocytosis. In receptor-mediated endocytosis, molecules to be taken in become bound to the outer cell surface by receptor proteins. The receptor proteins are specific in that they recognize and bind only certain molecules from the solution that surrounds the cell. The molecules recognized are mostly proteins or other molecules carried by proteins. Once the receptors have bound their target molecules, the receptors collect into a coated pit, a depression in the plasma membrane. The pits, with the contained

Abbreviations: IMR = Insights from the Molecular Revolution; UQ = Unanswered Questions

target molecules, pinch off from the plasma membrane to form endocytic vesicles.

Think Like a Scientist

FIGURE 6.6

The protein synthesis inhibitors did not affect the pattern of the membrane proteins. Therefore, new protein synthesis is not necessary for generating the intermixed pattern of membrane proteins. The temperature experiment shows that intermixing depends on the temperature. This makes sense if the intermixing is a property of the membrane because, as you learned earlier in the chapter, the phospholipid bilayer can freeze into a semisolid, gel-like state at low temperatures. Logically, membrane proteins would not be able to move or they would only be able to move a little under those conditions. Together, these experiments support the fluid bilayer model and reject the new protein synthesis model.

FIGURE 6.11

The hydrolysis of ATP to ADP plus phosphate is key to the ability of the channel protein to transport ions. If the rate of hydrolysis of ATP was reduced drastically, the primary active transport pump would only be able to move the ion it transports at a much lower rate.

IMR

The result showed that unlabeled LDL competed with labeled LDL for binding to the cells. The interpretation was that there are specific binding sites for LDL on the cell surface and that their number is limited. The result supported the researchers' overall conclusion because it indicated that specific binding was occurring in the experiments rather than nonspecific binding; that is, there is a specific receptor for LDL.

UQ

Restricting the passage of protons aids in conservation of the membrane's electrochemical potential.

Test Your Knowledge

1. d 2. b 3. a 4. c 5. e 6. d 7. e 8. b 9. a 10. c

Interpret the Data

1. Cell set B is displaying resistance because these cells fail to accumulate paclitaxel. They are using active transport to move the anticancer drug back out across the cell membrane.
2. The drug treatment imatinib has a slight but significant effect on increasing the accumulation of the labeled paclitaxel in set A; however, a much more dramatic effect is seen using either of these drugs on cell set B. This suggests that in cell set B, the active transport mechanism is inhibited by the presence of these additional drugs.

Chapter 7

Study Break

STUDY BREAK 7.1

Specificity of a cellular response depends on the signal–receptor interaction. Specificity starts with the signal molecule; that is, the specific signal molecule is the messenger that elicits a specific cellular response. For example, the hormone epinephrine causes glucose to be released into the bloodstream. Specificity also depends on the target cells; that is, only target cells respond to the signal molecule because they exclusively have receptors for the signal molecule.

STUDY BREAK 7.2

1. Protein kinases are enzymes that add phosphate groups to other proteins. The result of phosphorylation is that the protein will be either stimulated or inhibited in its activity. The cellular responses of signal transduction pathways are produced through the actions of protein kinases.
2. Amplification is the phenomenon of an increase in the magnitude of each step of a signal transduction pathway. Amplification typically occurs because the proteins conducting each step of the pathway are enzymes; that is, each enzyme, when it is becomes activated, activates large numbers of molecules entering the next step of the pathway.

STUDY BREAK 7.3

1. For a receptor tyrosine kinase to become activated, the signal molecule first binds to the receptor, which then assembles into a dimer. The receptor adds phosphate groups to tyrosines on the cytoplasmic side of itself, which activates the receptor.
2. A fully activated receptor tyrosine kinase has phosphorylated tyrosines on each of its two monomers. A signaling protein that recognizes a phosphorylated tyrosine as well as surrounding parts of the polypeptide binds to the receptor. Depending on the signaling protein, the binding itself may activate the signaling protein, or it is activated by tyrosine phosphorylation catalyzed by the receptor. In its activated form, the signaling protein initiates a transduction pathway leading to a cellular response. A given receptor can initiate different responses because different combinations of signaling proteins can bind to the receptor.

STUDY BREAK 7.4

1. The first messenger in a G-protein–coupled receptor–controlled pathway is the extracellular signal molecule. When it binds to the G-protein–coupled receptor, it activates a site on the cytoplasmic side of the receptor; the activated receptor, in turn, activates the G protein next to it.
2. The effector is activated by the G protein. The effector is a plasma-associated enzyme that generates a nonprotein signal molecule called the second messenger. The second messenger leads to the activation of protein kinases leading to the cellular responses triggered by the signal molecule.
3. A main way the pathway is turned off is by the conversion of cAMP to 5′-AMP by phosphodiesterase. As long as the receptor is bound by the signal molecule, cAMP is being generated by the activated effector. The continued synthesis of cAMP balances the degradation of cAMP by phosphodiesterase, ensuring that the pathway continues to run. However, if the signal molecule no longer is bound to the receptor, the effector again becomes inactive, cAMP therefore is not generated, and existing cAMP is rapidly degraded by phosphodiesterase. As a result, the protein kinase cascade is shut down and no cellular responses occur.

STUDY BREAK 7.5

1. The steroid receptor is within the cell, whereas the receptor tyrosine kinase and G-protein–coupled receptors are in the membrane or associated with the membrane. Also, the activated steroid receptor directly activates genes, whereas the other two receptors, when active, are just the first steps in pathways that may or may not activate genes.
2. A steroid hormone brings about a specific cellular response because whether a cell responds to a steroid hormone depends on whether it has the internal receptor for the hormone. Then within the cells with the receptor, the specific genes that are controlled are those with regulatory sequences that are recognized by the activated receptor.

STUDY BREAK 7.6

Signal transduction pathways, cellular response systems triggered by cell adhesion molecules, and communication pathways that involve gap junctions between adjacent cells might be integrated in a cross-talk network.

Think Like a Scientist

FIGURE 7.2

The experiment showed that the second messenger was not a protein, or any other large molecule that would be denatured by boiling.

FIGURE 7.7

The transduction pathways controlled by the receptor tyrosine kinase would be active all the time, producing cellular responses in an uncontrolled manner—that is, in the absence of the signal molecule. Mutations such as this are known to be associated with some cancers.

FIGURE 7.9

There are two main possibilities. One is that the receptor would no longer recognize the first messenger signal molecule. In this case, the cellular response(s) controlled by the signal molecule would not occur. The other is that the receptor would be active even in the absence of the first messenger signal molecule. In this case, cellular response(s) would occur in an uncontrolled way. As you learned in the text, there are a very large number of G-protein–coupled receptors and they control a great many cellular processes. Thus, malfunctions of these receptors can have serious consequences. In fact, mutations in genes for G-protein–coupled receptors have been shown to be responsible for more than 30 human diseases.

IMR

You would test U0126 first in an animal model and, if the results were promising, then in humans in clinical trials. In fact, U0126 has been shown to reduce influenza virus propagation in lungs of infected mice.

UQ

1. Perhaps the male's behavior induces the release of dopamine or another relevant neurotransmitter onto specific neurons in the brain of the female, which contain progestin receptors and are involved in regulation of sexual behavior. The dopamine, in turn, may activate those receptors leading to neuronal changes resulting in the expression of sexual behavior.
2. To answer this question, you have to think of other situations in which the environment might cause the release of a neurotransmitter that then, via cross talk, activates a steroid receptor resulting in changes in behavior. Perhaps stimulation from pups activates neuronal steroid hormone receptors, resulting in changes in maternal behavior in a lactating mother. There are many other examples in which stimulation from the environment or another animal might, via neurotransmitter release, influence the function of a particular steroid hormone receptor (by activating it).

Test Your Knowledge

1. b 2. a 3. d 4. d 5. c 6. c 7. e 8. b 9. c 10. e

Interpret the Data

1. In normal cells the least amount of CFTR was found in vesicles, whereas the greatest amount was found in endoplasmic reticulum. In cells with the CF mutation the least amount of CFTR was found in Golgi bodies, whereas the greatest amount was found in endoplasmic reticulum.
2. The CFTR protein is found in almost the same amount in the endoplasmic reticulum of normal and CF cells.
3. The CFTR protein is held up in the endoplasmic reticulum.

Chapter 8

Study Break

STUDY BREAK 8.1

1. Oxidation is the removal of electrons from a substance; reduction is the addition of electrons to a substance.
2. Cellular respiration refers to the reactions in which oxygen is used as final electron acceptor; it includes the reactions that transfer electrons from organic molecules to oxygen and the reactions that make ATP. Oxidative phosphorylation is the process by which ATP is synthesized using the energy released by electrons as they are transferred to oxygen.

STUDY BREAK 8.2

1. The initial steps of glycolysis, which require 2 ATP, convert glucose to a phosphorylated derivative. The later steps, which release 4 ATP, remove electrons from the glucose derivatives and generate two molecules of pyruvate.
2. The redox reaction in glycolysis is the glyceraldehyde-3-phosphate (G3P) to 1,3-bisphosphoglycerate reaction (see Figure 8.7, step 6).
3. ATP is synthesized in glycolysis by substrate-level phosphorylation occurring in the 1,3-bisphosphoglycerate to 3-phosphoglycerate reaction (see Figure 8.7, step 7), and in the phosphoenolpyruvate (PEP) to pyruvate reaction (see Figure 8.7, step 10).
4. If excess ATP is present in the cytosol, it binds to and inhibits the activity of phosphofructokinase. As a result, the

concentration of fructose-1,6-bisphosphate, the product of the phosphofructokinase reaction, decreases, and the subsequent reactions of glycolysis are slowed or stopped. This is reversed when the ATP level in the cytosol decreases. In the end, this control mechanism helps prevent the needless oxidation of fuel molecules when the cell has an adequate supply of ATP.

STUDY BREAK 8.3

The three-carbon pyruvate molecules are transported from the cytosol into the mitochondria, where they are converted into two-carbon acetyl units through pyruvate oxidation. The citric acid cycle oxidizes the acetyl units completely to carbon dioxide with the transfer of electrons to NAD^+ or FAD.

STUDY BREAK 8.4

1. Each complex contains a unique combination of nonprotein carriers that pick up and release electrons.
2. The proton pumps push protons (H^+) from the mitochondrial matrix to the intermembrane compartment, increasing the proton concentration there. The resulting proton gradient produces an electrical gradient across the inner mitochondrial membrane with the matrix negatively charged with respect to the intermembrane compartment. The charge and proton concentration differences together provide energy for ATP synthesis in what is called proton-motive force.

STUDY BREAK 8.5

Fermentation occurs when oxygen is absent or limited. The electrons carried by the NADH produced in glycolysis are transferred to an organic molecule instead of the electron transfer system. In lactate fermentation, the end product of glycolysis, pyruvate, is converted to lactate. The lactate stores electrons temporarily, transferring them to the mitochondrial electron transfer system when the oxygen content of cells returns to normal. In alcoholic fermentation, pyruvate is converted into ethyl alcohol.

Think Like a Scientist

FIGURE 8.13

The H^+ concentration will increase in the intermembrane compartment. The increase will lead to an increase in ATP production.

FIGURE 8.15

In the light, ATP synthesis would occur and then, in the dark, ATP synthesis would stop. This was in fact observed by Racker and Stoeckenius.

IMR

The observation indicates that the two UCPs are members of the same family of proteins and that presumably they have evolved by diversification from a common ancestral gene.

UQ

One possible hypothesis: there is a direct cause and effect relationship between altered regulation of mitochondrial expression and AD. If a significant number of mice whose genes were not genetically altered were to develop AD, this hypothesis would have to be rejected.

Test Your Knowledge

1. a 2. b 3. a 4. e 5. c 6. b 7. e 8. a 9. d 10. a

Interpret the Data

1.

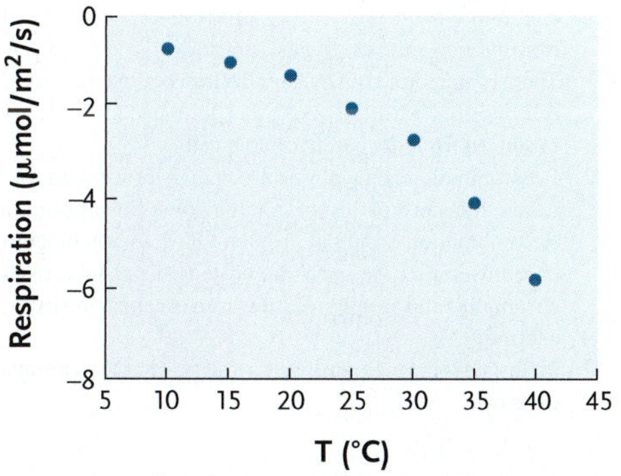

2. Two.
3. A curve represents the data best. A Q10 of 2 indicates a curve because the respiration rate is doubling for every 10°C increase in temperature. This type of relationship is an exponential increase.
4. Temperatures are predicted to increase, which would increase respiration according to these data. The data are insufficient to predict the actual increase in respiration since the plants would likely change their enzyme amounts or types (acclimate) in response to the rising temperature. Further, temperature is not the only factor that will likely change with rising CO_2. Other factors include precipitation, cloud cover, nutrient availability, etc.

Chapter 9

Study Break

STUDY BREAK 9.1

1. The two stages of photosynthesis are: (1) the light-dependent reactions, in which the energy of sunlight is absorbed and converted into chemical energy in the form of ATP and NADPH; and (2) the light-independent (dark) reactions, in which electrons carried by NADPH are used as a source of energy to convert carbon dioxide from inorganic to organic form.
2. In plants, photosynthesis takes place in the chloroplast. The light-dependent reactions are carried out on the thylakoid membranes and stromal lamellae. The light-independent reactions are carried out in the stroma.

STUDY BREAK 9.2

1. The chlorophyll *a* molecules in the antenna complexes are normal molecules of the pigment, consisting of a carbon ring structure with a magnesium atom bound at the center and an attached hydrophobic side chain. These chlorophyll *a* pigments absorb light. The chlorophyll *a* molecules in the reaction centers have modified light absorption properties that result from interactions with particular proteins of the photosystems. The two special chlorophyll *a* molecules of photosystem II are P680; those of photosystem I are P700. These pigment molecules capture light energy from the antenna complex pigments in the form of an excited electron that is passed to a primary acceptor molecule. That electron is passed to the electron transfer system.
2. The making of NADPH begins when electrons derived from water splitting are pushed to higher energies by light absorption in photosystem II. The high-energy electrons pass to a primary acceptor in photosystem II and then down an electron transfer system to P700 in photosystem I, losing energy along the way. Light energy absorbed by photosystem I again excites the electrons, which pass to different electron carriers, ending with ferredoxin. The ferredoxin transfers high-energy electrons to $NADP^+$, which is reduced to NADPH by $NADP^+$ reductase.
3. In the linear electron flow pathway, electrons run through the entire set of photosystems and electron carriers, producing both NADPH and ATP. In the cyclic electron flow pathway, electrons flow cyclically around photosystem I; photosystem II is not involved. The cycle of electrons is through the cytochrome complex and plastocyanin to photosystem I, to ferredoxin, but then back to the cytochrome complex rather than on to $NADP^+$ reductase. Only ATP is produced by this pathway.

STUDY BREAK 9.3

1. Rubisco catalyzes a reaction combining carbon dioxide with RuBP to form two molecules of 3-phosphoglycerate (3PGA). Rubisco, an enzyme unique to photosynthetic organisms, is the key enzyme for producing the world's food because it is responsible for carbon dioxide fixation, a process that ultimately provides organic molecules for most of the world's organisms. Rubisco is the key regulatory site of the Calvin cycle for the following reason: During the daytime, sunlight powers the light-dependent reactions, and the NADPH and ATP produced by those reactions stimulate rubisco, which, in turn, keeps the Calvin cycle running. In darkness, however, NADPH and ATP levels are low, and as a result, rubisco's activity is inhibited and the Calvin cycle slows down or stops.
2. For each carbon atom that is released from the Calvin cycle in a carbohydrate molecule, one carbon dioxide molecule must enter the cycle. Therefore, to produce a molecule containing 12 carbon atoms, 12 molecules of carbon dioxide must enter the cycle.

STUDY BREAK 9.4

1. Photorespiration uses oxygen and releases CO_2. It occurs when oxygen concentrations are high relative to CO_2 concentrations. In that condition, rubisco acts as a Rubisco rather than a carboxylase, catalyzing the combination of RuBP with O_2 rather than CO_2. The toxic products formed by this reaction cannot be used in photosynthesis and are eliminated from the plant as CO_2. Photorespiration uses energy to salvage the carbons from phosphoglycolate, which greatly reduces the efficiency of energy use in photosynthesis. This can be seen in the reduced growth of plants grown under photorespiration conditions.
2. In the C_4 pathway, carbon fixation involves the reaction of CO_2 with phosphoenolpyruvate (PEP) to produce a four-carbon molecule, oxaloacetate. The oxaloacetate is reduced to malate by electrons transferred from NADPH, and malate then is oxidized to pyruvate in a reaction releasing CO_2, which is used in the rubisco-catalyzed first step of the Calvin cycle. In C_4 plants, carbon fixation and the Calvin cycle occur in different cell types: carbon fixation in mesophyll cells, and the Calvin cycle in bundle sheath cells. This alternative method of carbon fixation minimizes photorespiration.
3. In C_4 plants, carbon fixation and the Calvin cycle occur in different cell types, mesophyll cells and bundle sheath cells, respectively. In CAM plants, carbon fixation and the Calvin cycle occur at different times, at night and during the day, respectively.

STUDY BREAK 9.5

The reactions of photosynthesis and cellular respiration are essentially the reverse of one another, with CO_2 and H_2O the reactants of photosynthesis and the products of cellular respiration. Phosphorylation reactions involving electron transfer systems are part of each process, namely photophosphorylation in photosynthesis and oxidative phosphorylation in cellular respiration. G3P is an intermediate in both pathways: in photosynthesis it is a product of the Calvin cycle, and in cellular respiration it is generated in glycolysis in the conversion of glucose to pyruvate. In photosynthesis, G3P is used for the synthesis of sugars and other fuel molecules, and in cellular respiration, it is part of the catabolism of sugars to simpler organic molecules.

Think Like a Scientist

FIGURE 9.10

Two.

IMR

There are two types of subunit, so there are two genes that encode the complete enzyme. The enzyme consists of eight copies each of the small subunit, encoded by one of the genes, and of the large subunit, encoded by the other gene.

UQ

Imagine two types of desert plants. Some, like the wild watermelon plants of the Kalahari, grow very rapidly, with very robust photosynthesis, as soon as rain falls. They produce as many seeds as they can, quickly before severe drought sets in. This strategy requires rapid photosynthesis, even at the risk of photodamage. Others, such as desert scrub plants like sage, persist throughout the drought. They may grow slowly, with highly protected photosynthesis, even during times of rain. If they grew too fast, their large leaf surface areas would result in high water loss during drought. In some invasive plant species, aggressive photosynthesis and high growth rates have a selective advantage, allowing them to outcompete their rivals for resources like sunlight or growth space. In these cases, high rates of photosynthesis may be an advantage even as they risk photodamage. Other plants may invest for the longer term, building long-lived resilient structures (leaves, wood, etc.). The large investment may render risky photosynthetic strategies less viable, instead preferring more "conservative" down-regulatory strategies. Still other plants may have to contend with lack of key nutrients, in which high photosynthetic rates or the need for rapid repair of the photosynthetic apparatus would be too "expensive" in terms of resources. Finally, humans have selected for traits in crop plants that are not related to photosynthetic yield—for example, tasty fruits, disease resistance, or short growing season. These traits often take precedence over photosynthetic efficiency.

Test Your Knowledge

1. d 2. b 3. a 4. c 5. e 6. a 7. b 8. d 9. d 10. b

Interpret the Data

1. When PPFD is zero, no photosynthesis is occurring, but respiration continues to occur. Based on the definition of net photosynthesis here and respiration from Chapter 8 *Interpret the Data,* respiration alone would have a negative value of CO_2 flux.

   The temperature for a respiration rate of 0.2 μmol/m^2/s is 25°C.

2.

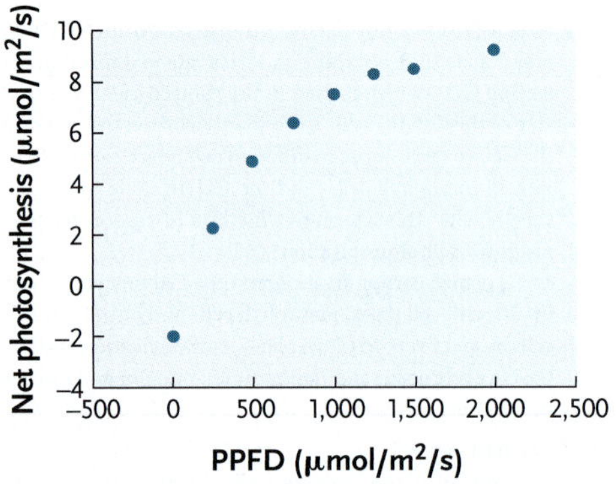

3. A curve represents the data best. Net photosynthesis is saturating with increasing PPFD, indicating that some other factor (such as the amount of photosynthetic enzymes) is limiting photosynthesis.
4. Temperatures are predicted to increase, which would increase respiration according to the data and thus lower net photosynthesis. These data are insufficient to predict the actual impact of CO_2 concentrations on net photosynthesis because the plants would likely change their enzyme amounts or types (acclimate) in response to the rising temperature. Further, temperature is not the only factor that will likely change with rising CO_2. Other factors include precipitation, cloud cover, nutrient availability, etc.

Chapter 10

Study Break

STUDY BREAK 10.1

1. (1) An elaborate master program of molecular checks and balances ensures an orderly and timely progression through the cell cycle. (2) The process of DNA synthesis replicates each DNA chromosome into two copies with almost perfect fidelity. (3) A structural and mechanical web of interwoven "cables" and "motors" of the mitotic cytoskeleton separates the DNA copies precisely into the daughter cells.
2. A linear DNA molecule complexed with proteins
3. In mitosis, DNA replication is followed by the equal separation of the replicated DNA molecules and their delivery to daughter cells. The process ensures that the two cell products of a division have the same DNA content and the same genetic information as the parent cell entering division has.

STUDY BREAK 10.2

1. In order, the stages of mitosis are prophase, prometaphase, metaphase, anaphase, and telophase.
2. Each eukaryotic chromosome has a specialized region known as a centromere. The centromere is where a complex of several proteins, called a kinetochore, forms. During mitosis, some spindle microtubules attach to each kinetochore. These connections determine the outcome of mitosis because they attach the sister chromatids of each chromosome to microtubules leading to the opposite spindle poles; during anaphase, the spindle separates sister chromatids and pulls them to opposite spindle poles. In brief, the centromeres are key to chromosome segregation during mitosis. Although not mentioned in the chapter, this is also apparent when problems occur in which a chromosome fragment without a centromere breaks off from a chromosome. The fragment without a centromere cannot connect to the spindle and, hence, is not segregated properly.
3. Joined sister chromatid pairs attach to kinetochore microtubules during prometaphase and begin their migration to the metaphase plate. The spindle microtubules are also necessary for segregating the sister chromatids to opposite poles of the cell during anaphase. If colchicine is present, no spindle will form and no kinetochore microtubules are present to attach to the sister chromatids. Therefore, the cell will be stuck in mitosis with the condensed pairs of sister chromatids in an unorganized array.

STUDY BREAK 10.3

1. In animal cells, all of which have centrosomes, the spindle forms through division of the cell center. As the dividing centrosome separates into two parts, the microtubules of the spindle form between them. Plant cells lack centrosomes. In plant cells, the spindle microtubules simply assemble around the nucleus. In either case, the microtubules assemble in a parallel array that creates two poles in the dividing cell.
2. Chromosomes (sister chromatids) move apart during anaphase. During the anaphase movements, the kinetochores move along the kinetochore microtubules, which become shorter as anaphase progresses. The nonkinetochore microtubules slide over each other, decreasing the degree of overlap and pushing the poles farther apart. The total distance traveled by the chromosomes is the sum of the two movements.

STUDY BREAK 10.4

1. A Cdk can become active only once it has complexed with a cyclin protein. Each cyclin is present only during a particular segment of the cell cycle, controlled by when it is synthesized and degraded. Thus, the period in the cell cycle when a particular Cdk is active depends on when its activating cyclin is present.
2. When the kinase of the Cdk is activated upon binding to a cyclin, it phosphorylates target proteins in the cell, regulating their activities. Those proteins play roles in initiating or regulating key events of the cell cycle, namely DNA replication, mitosis, and cytokinesis. The progression through the cell cycle is regulated, then, by a succession of cyclin–Cdk complexes, each of which has specific regulatory effects.
3. An oncogene is an altered gene in an organism that contributes to the development of cancer—that is, uncontrolled cell division. Some of the genes that become oncogenes encode components of the cyclin–Cdk system that regulates cell division, whereas others encode proteins that regulate gene activity, form cell-surface receptors, or make up elements of the systems controlled by the receptors.
4. Metastasis is when cells break loose from a tumor, spread through the body, and grow into new tumors in other body regions.

STUDY BREAK 10.5

1. Prokaryotic cell division begins with replication of the bacterial chromosome, starting with duplication of the origin of replication. Once the origin of replication is duplicated, the two origins actively migrate to the two ends of the cells, a process that separates the two replicating chromosomes in the cell. Division of the cytoplasm then occurs by means of a partition of cell wall material that grows inward until the cell is separated into two parts. The cytoplasmic division divides the replicated DNA molecules and cytoplasmic structures between the daughter cells.
2. Present in eukaryotic cell division, but absent from prokaryotic cell division, are the following: the process of mitosis; any form of microtubules for chromosome segregation; a spindle apparatus; cyclin/CDK control proteins.

Think Like a Scientist

FIGURE 10.4

The colchicine would block spindle formation. As a result the cells would not be able to complete mitosis. In fact, colchicine is used to get cells to arrest in metaphase when the chromosomes are at their most condensed. At that stage, the chromosomes are the easiest to visualize by microscopy (see Figure 10.7).

FIGURE 10.12

The result supporting the kinetochore microtubule movement hypothesis would have been the movement of the bleached region toward the pole.

FIGURE 10.14

The result indicates that there are factors in the G_2 nucleus that accelerate the progress of the S nucleus towards mitosis.

IMR

The result indicates that expression of K cyclin (which occurred when the inducer was added) is sufficient to induce S phase entry in quiescent (G_0) cells.

UQ

Although *E. coli* is easy to culture, there is no known prokaryotic equivalent of mitosis; mammalian cells are hard to culture and cannot be kept in long-term culture (see *Focus on Basic Research*). *Focus on Model Organisms* provides abundant rationale for *Saccharomyces* as a model organism for this type of research. Ubiquitin is given this name because it is indeed ubiquitous—present in almost the same form in essentially all eukaryotes (see Chapter 16).

Test Your Knowledge

1. c 2. e 3. e 4. b 5. d 6. d 7. a 8. b 9. b 10. c

Interpret the Data

1. Radium exposure does not have an immediate effect on mitosis; that is, for about two hours after initiating radium exposure, there is not a significant drop in the number of cells in the population undergoing mitosis. (The researchers interpreted this to mean that not only were cells completing mitosis, but other cells commenced mitosis.) However, after about two hours of radium exposure, mitosis in the cell population exhibits a sudden drop to zero. In other words, radium blocks mitosis after a long enough exposure. The corollary is that there was no evidence for radium stimulating mitosis.
2. The effect of radium exposure was not permanent. Mitosis resumed about two hours after radium was removed. (Importantly, this result also shows that the decrease in mitosis as a result of radium exposure was not because of death of the cells.)

Chapter 11

Study Break

STUDY BREAK 11.1

1. Mitosis produces daughter cells that are genetically identical to the parent cell. Either a haploid or a diploid cell can undergo mitosis. Meiosis starts with a diploid cell. There is one round of DNA replication but two rounds of cell division, with the result that four haploid cells are produced from the parent diploid cell.
2. Recombination is the physical exchange of segments between the chromatids of homologous chromosomes. Recombination occurs in prophase I, when homologous chromosomes have each duplicated to produce sister chromatids and are aligned fully in an organization called a tetrad.
3. Meiosis II is the meiotic division that is similar to a mitotic division.

STUDY BREAK 11.2

1. There are three ways in which sexual reproduction generates genetic variability. First, recombination, which involves the physical exchange of segments between homologous chromatids in prophase I of meiosis, generates new combinations of alleles. Second, the random separation of homologous chromosomes during meiosis generates genetic variability; that is, in metaphase I, for each homologous pair of chromosomes, one chromosome makes spindle connections leading to one pole and the other chromosome connects to the opposite pole. This process operates independently for each homologous pair of chromosomes; thus, for each meiosis random combinations of maternal and paternal chromosomes move to the poles during anaphase I. Third, the random joining of male and female gametes produces additional genetic variability.
2. The proportion of gametes that will have chromosomes that originate from the animal's female parent is: $(1/2)^6 = 1/64$.

STUDY BREAK 11.3

In animals, the diploid phase dominates the life cycle; mitotic divisions occur only in this phase. Meiosis in the diploid phase gives rise to products that develop directly into egg and sperm cells without undergoing mitosis.

In most plants, the life cycle alternates between haploid and diploid generations, both of which grow by mitotic divisions. Fertilization produces the diploid sporophyte generation; after growth by mitotic divisions, cells of the sporophyte undergo meiosis and produce haploid spores. The spores germinate and grow by mitotic divisions into the gametophyte generation. After growth of the gametophyte, cells develop directly into egg or sperm nuclei, which fuse in fertilization to produce the diploid sporophyte generation again.

Think Like a Scientist

IMR

The result suggests that CYP26B1 plays an essential role in germ cell development perhaps among all vertebrates, not just mammals.

UQ

The outcome depends on a variety of factors such as the extent of the rearrangement, its location in the genome and the particular organism or species being analyzed. However, very frequently, a phenomenon referred to as "synaptic adjustment" is observed. This allows the duplications and/or inversions to be accommodated (in many cases these just "loop out" from the fully paired homologs) and the homologs still succeed in synapsing. Occasionally, these rearrangements can also lead to nonhomologous synapsis and asynapsis.

Test Your Knowledge

1. b 2. a 3. d 4. c 5. b 6. b 7. d 8. a 9. b 10. c

Interpret the Data

1. At 20°C, one mother produced 21 male offspring; assuming 100 offspring, this would be 21%. This was the highest. The lowest was zero males produced by at least one mother at 20°C and 16-hour photoperiod. 15°C to 20°C seems to result in more male production than either higher or lower temperatures. No mothers produced male offspring at 25°C; some mothers produced no male offspring at 11°C. Less than 10% males were produced at 25°C or 11°C or at 20°C and 16-hour photoperiod.
2. At the moderate temperatures, slightly more males are produced with 12-hour days rather than 16-hour days. However, there is a wide range. If time has an effect, the effect is much less than the effect temperature has on the production of males.

NOTE: The authors concluded that the difference seen for photoperiod was not statistically significant, meaning that the difference seen could be merely because of chance.

Chapter 12

Study Break

STUDY BREAK 12.1

1. The numerical results approximate a 9:3:3:1 ratio. Therefore, both parents must be heterozygous for both genes. If we designate the alleles for one of the pairs of traits as *A* and *a*, and the alleles for the other pair of traits as *B* and *b*, each parent has the genotype *Aa Bb*.
2. A ratio of 1:1:1:1 is the typical outcome of a testcross involving a parent who is heterozygous for the alleles of two genes. Using the same allele symbols as in (1), the genotypes of the parents are *Aa Bb* and *aa bb;* this is a testcross.

STUDY BREAK 12.2

1. The color pattern involved is an incompletely dominant trait.
2. The fur colors here involve multiple alleles of a single gene. The allele symbols are *C* for wild type, c^{ch} for chinchilla, c^h for Himalayan, and *c* for albino, with dominance in the order $C \rightarrow c^{ch} \rightarrow c^h \rightarrow c$: that is, the *C* allele is completely dominant to the c^{ch} allele, the c^{ch} allele is completely dominant to the c^h allele, and so on. Therefore, we have these genotypes and phenotypes:
   C with c^{ch}, c^h, or c = agouti (CC, Cc^{ch}, Cc^h, Cc)
   c^{ch} with c^{ch}, c^h, or c = chinchilla ($c^{ch}c^{ch}$, $c^{ch}c^h$, $c^{ch}c$)
   c^h with c^h or c = Himalayan (c^hc^h, c^hc)
   c with *c* = albino *(cc)*

Think Like a Scientist

FIGURE 12.5

Purple-flowered F_2 plants are either *PP* or *Pp* in genotype. Only if *Pp* is selfed will you get purple-flowered and white-flowered progeny (it would be same as the $F_1 \times F_1$ self shown in the figure). There is a ratio of 1 *PP*:2 *Pp* plants as the Punnett square in the figure shows. Therefore, there is a 2 in 3 (two-thirds) chance that you will see both purple-flowered and white-flowered plants in the progeny.

FIGURE 12.8

This problem relates to the testcross shown in the figure. The ratio of the two types of progeny is approximately 1:1, a result that matches the Experiment 1 results in the figure. Therefore, the white-fat parent must have been heterozygous. The yellow-fat parent has to be homozygous recessive because of the information given about the fat-color phenotype. In short, the parents were *Ww* (white fat) × *ww* (yellow fat), which gives an expected 1:1 ratio of *Ww* (white fat):*ww* (yellow fat).

FIGURE 12.9

The cross would be *Rr Yy* × *Rr yy.* The progeny are determined by constructing a Punnett square and filling in the squares. For the *Rr Yy* parent, there are four types of gametes just as in the figure; that is, 1/4 *R Y*, 1/4 *R y*, 1/4 *r Y*, and 1/4 *r y*. For the *Rr yy* parent, there are two types of gametes; that is, 1/2 *R y* and 1/2 *r y*. This gives you 8 squares in this particular Punnett square with these progeny: 1/8 *RR Yy*, 1/8 *RR yy*, 1/8 *Rr Yy*, 1/8 *Rr yy*, 1/8 *Rr Yy*, 1/8 *Rr yy*, 1/8 *rr Yy*, and 1/8 *rr yy*. Collecting the genotypes with the same phenotypes together we get 3/8 round, yellow (1/8 *RR Yy* + 1/8 *Rr Yy* + 1/8 *Rr Yy*), 3/8 round, green (1/8 *RR yy* + 1/8 *Rr yy* + 1/8 *Rr yy*), 1/8 wrinkled, yellow (1/8 *rr Yy*), and 1/8 wrinkled, green (1/8 *rr yy*).

IMR

In Chapter 4, you learned that the substrate(s) for an enzyme interact with a small region of an enzyme called the *active site.* It is at the active site where enzyme catalysis occurs. The active site of an enzyme is formed by a particular array of amino acids in the polypeptide chain. Those amino acids have evolved to be optimal for enzyme activity. A mutation that alters an amino acid in or near the active site can affect the structure of the active site and, therefore, adversely affect the enzyme's catalytic activity. The extent of the deleterious effect depends on the particular amino acid substitution and the chemical changes brought about in the protein by the substitution.

UQ

If you sequence lots of species you will identify lots of regions of strong evolutionary constraint. If you look at many individuals within a species, you can look at the rate of fast-evolving alleles that might have responded to population-specific pressures or drift and start asking questions pertaining to the genome of the human. Ultimately, your choice will depend on the question you are trying to answer. If you are trying to understand developmental processes, cross-species comparisons are extremely useful, but if you are looking at immunity, maybe the within-species comparison is better.

Test Your Knowledge

1. (a) The *CC* parent produces all *C* gametes, and the *Cc* parent produces 1/2 *C* and 1/2 *c* gametes. All offspring would have colored seeds—half homozygous *CC* and half heterozygous *Cc*. (b) The *Cc* parent produces 1/2 *C* gametes and 1/2 *c* gametes, and the *cc* parent produces all *c* gametes. Half of the offspring are colored (1/2 *Cc*) and half are colorless (1/2 *cc*). (c) Both parents produce 1/2 *C* and 1/2 *c* gametes. Of the offspring, three-fourths would have colored seeds (1/4 *CC* + 1/2 *Cc*) and one-fourth would have colorless seeds (1/4 *cc*).
2. The genotypes of the parents are *Lele* and *lele*.
3. The taster parents could have a nontaster child, but nontaster parents are not expected to have a child who can taste PTC. The chance that they might have a taster child is 3/4. The chance of a nontaster child being born to the taster couple would be 1/4. Because each combination of gametes is an independent event, the chance of the couple having a second child, or any child, who cannot taste PTC is expected to be 1/4.
4. (a) All *A B*. (b) 1/2 *A B* + 1/2 *a B*. (c) 1/2 *A b* + 1/2 *a b*. (d) 1/4 *A B* + 1/4 *A b* + 1/4 *a B* + 1/4 *a b*.
5. (a) All *Aa BB*. (b) 1/4 *Aa Bb* + 1/4 *Aa bb* + 1/4 *aa Bb* + 1/4 *aa bb*. (c) 1/4 *AA BB* + 1/4 *AA Bb* + 1/4 *Aa BB* + 1/4 *Aa Bb*. (d) 1/4 *Aa Bb* + 1/8 *AA Bb* + 1/8 *Aa BB* + 1/8 *Aa bb* + 1/8 *aa Bb* + 1/16 *AA BB* + 1/16 *AA bb* + 1/16 *aa BB* + 1/16 *aa bb*.
6. (a) All *A B C*. (b) 1/4 *A B C* + 1/4 *A B c* + 1/4 *a B C* + 1/4 *a B c*. (c) 1/2 *A B c* + 1/2 *a B c*. (d) 1/8 *A B C* + 1/8 *A B c* + 1/8 *A b C* + 1/8 *A b c* + 1/8 *a B C* + 1/8 *a B c* + 1/8 *a b C* + 1/8 *a b c*.
7. Because the man can produce only 1 type of allele for each of the 10 genes, he can produce only 1 type of sperm cell with respect to these genes. The woman can produce 2 types of alleles for each of her 2 heterozygous genes, so she can produce 2 × 2 = 4 different types of eggs with respect to the 10 genes. In general, as the number of heterozygous genes increases, the number of possible types of gametes increases as 2^n, where n = the number of heterozygous genes.
8. Use a standard testcross; that is, cross the guinea pig with rough, black fur with a double recessive individual, *rr bb* (smooth, white fur). If your animal is homozygous *RR BB*, you would expect all the offspring to have rough, black fur.
9. One gene probably controls pod color. One allele, for green pods, is dominant; the other allele, for yellow pods, is recessive.
10. The cross *RR* × *Rr* will produce 1/2 *RR* and 1/2 *Rr* offspring. The cross *Rr* × *Rr* will produce 1/4 *RR*, 1/2 *Rr*, and 1/4 *rr* as combinations of alleles. However, the 1/4 *rr* combination is lethal, so it does not appear among the offspring. Therefore, the offspring will be born with only two types, *RR* and *Rr*, with twice as many *Rr* as *RR* in a 1:2 ratio (or 1/3 *RR* + 2/3 *Rr*).
11. The parental cross is *GG LeLe RR* × *gg lele rr*. All offspring of this cross are expected to be tall plants with green pods and round seeds, or *Gg Lele Rr*. When crossed, this heterozygous F_1 generation is expected to produce eight different phenotypes among the offspring: green–tall–round, green–dwarf–round, yellow–tall–round, green–tall–wrinkled, yellow–dwarf–round, green–dwarf–wrinkled, yellow–tall–wrinkled, yellow–dwarf–wrinkled, in a 27:9:9:9:3:3:3:1 ratio.

12. The genotypes are: bird 1, *Ff Pp;* bird 2, *FF PP;* bird 3, *Ff PP;* bird 4, *Ff Pp.*
13. Yes, it can be determined that the child is not hers, because the father must be AB to have both an A and B child with a type O wife; none of the woman's children could have type O blood with an AB father.
14. The cross is expected to produce white, tabby, and black kittens in a 12 : 3 : 1 ratio.
15. The mother is homozygous recessive for both genes, and the father must be heterozygous for both genes. The child is homozygous recessive for both genes. The chance of having a child with normal hands is 1/2, and that of having a child with woolly hair is 1/2. Using the product rule of probability, the probability of having a child with normal hands and woolly hair is $1/2 \times 1/2 = 1/4$.

Interpret the Data

Yes, the data support the hypothesis that the tolerance trait is dominant. Let us assign *T* as the symbol for the dominant allele for being tolerant, and *t* as the symbol for the recessive allele for being intolerant.

Cross 1: The crosses of true-breeding tolerant plants with true-breeding intolerant plants are $TT \times tt$. The F_1 offspring are *Tt*. If the tolerance trait is dominant, the $F_1 \times F_1$ crosses are genotypically $Tt \times Tt$, which produces an F_2 generation with a genotypic ratio of 1 *TT*:2 *Tt*:1 *tt*, which gives a phenotypic ratio of 3 alive (tolerant):1 dead (intolerant), which approximately matches the results.

Cross 2: As described for Cross 1, the F_1 plants have the *Tt* genotype. An F_1 × intolerant parent cross is $Tt \times tt$. The offspring from this cross are expected to have a genotypic ratio of 1 *Tt* :1 *tt*, which gives a phenotypic ratio of 1 alive:1 dead, which approximately matches the results.

Cross 3: An F_1 × tolerant parent cross is $Tt \times TT$. The offspring from this cross are expected to have a genotypic ratio of 1 *TT*:1 *Tt;* so all of the progeny are expected to be alive, which matches the results.

Chapter 13

Study Break

STUDY BREAK 13.1

The cross to use is the testcross. Here, the testcross would be *Aa Bb* × *aa bb*. A testcross is used so that you can follow the meiotic events in the dihybrid parent (including the consequences of crossing over between linked genes) because all of the gametes from the testcross parent carry recessive alleles for the genes in the cross. A testcross shows linkage when the ratio of 1 : 1 : 1 : 1 for the four possible phenotypes is not seen; that is, the 1 : 1 : 1 : 1 ratio result occurs when two genes assort independently. However, if two genes are linked, there will be excess of the two parental classes of progeny compared with the two recombinant classes.

STUDY BREAK 13.2

The differences between sex-linked inheritance and autosomal inheritance are seen clearly when reciprocal crosses are made and followed through to the F_2 generation. If sex-linked inheritance is involved, a cross of miniature-winged female × normal-winged male flies will give an F_1 generation of all normal-winged female and all miniature-winged male flies. (This result would not be found with autosomal inheritance. Instead, all F_1 flies—both males and females—would have normal wings.) Selfing the F_1 flies will give an F_2 generation with 1 : 1 normal-winged : miniature-winged flies in both sexes. (For autosomal inheritance, you would see a 3 : 1 ratio of normal-winged : miniature-winged flies in both sexes.)

In the reciprocal cross of true-breeding normal-winged female × miniature-winged male, the F_1 flies will all have normal wings if sex-linked inheritance is involved. (This result is the same as for autosomal inheritance.) Selfing the F_1 flies will give an F_2 generation in which all females will have normal wings and the males will be 1/2 normal-winged and 1/2 miniature-winged. (For autosomal inheritance, you would see a 3:1 ratio of normal-winged:miniature-winged flies in both sexes.)

In summary, reciprocal crosses show different segregation patterns of phenotypes for sex-linked inheritance and autosomal inheritance. The two modes of inheritance are easiest to distinguish in a cross of a mutant female × wild-type male because then, in the F_1 generation, all males show the mutant phenotype when sex-linked inheritance is involved.

STUDY BREAK 13.3

(a) Duplication of a chromosome segment occurs when a segment breaks from one chromosome and is inserted into its homolog.

(b) A Down syndrome individual results when nondisjunction of chromosome 21 during meiosis occurs (usually in females), producing gametes with two copies of chromosome 21 and one copy of every other chromosome. When such a gamete fuses with a normal gamete, the result is a zygote with three copies of chromosome 21 and two copies of the other chromosomes. This individual will have Down syndrome. Aneuploidy is the term for the condition of extra or missing chromosomes.

(c) A translocation occurs when a broken segment of a chromosome becomes attached to a different, nonhomologous chromosome.

(d) Polyploidy means that there are more sets of chromosomes than the typical diploid set. Polyploidy may result if the spindle fails to function properly in mitosis of cell lines leading to gametes. Cells affected in this way will have twice the normal number of sets of chromosomes. When meiosis subsequently occurs, the gametes produced will have two sets of chromosomes. Fusion of these gametes with, for instance, a gamete with one set of chromosomes will produce a zygote with three sets of chromosomes—a triploid cell.

STUDY BREAK 13.4

1. Autosomal recessive inheritance: For a child to exhibit an autosomal recessive trait, he or she must inherit one recessive allele from each parent. For autosomal recessive inheritance to explain Simpson syndrome in the family, the father must be homozygous for the Simpson syndrome allele, *ss,* and the mother must be heterozygous, *Ss.* The expectation would be that 1/2 of the children would be *Ss* and 1/2 would be *ss,* regardless of sex, and that is what is found. Therefore, on the assumption that the mother is heterozygous, the syndrome could be an autosomal recessive trait. Sex-linked recessive inheritance: One characteristic of sex-linked recessive inheritance is that affected females pass on the trait to all their sons. Here, we start with an affected male, and he would have to be X^sY if it is a sex-linked recessive trait. To explain the children, we would have to assume that the mother is heterozygous, X^SX^s. The cross of $X^SX^s \times X^sY$ is expected to give 1/2 females with the syndrome and 1/2 males with the syndrome, which is what is described. Therefore, the syndrome could be a sex-linked recessive trait.
2. Autosomal recessive inheritance: In pedigrees of autosomal recessive traits, the appearance of progeny with a trait when both parents do not have the trait is one common feature. If we assume that both parents are heterozygous, *Ww,* then the children can be explained; that is, you can get both wiggly-eared children (*ww,* expected frequency 1/4) and nonwiggly-eared children (*WW* or *Ww,* combined expected frequency 3/4), regardless of sex. Therefore, wiggly ears could be an autosomal recessive trait based on this family. Sex-linked recessive inheritance: Because a male individual has only one X chromosome, it is not possible for two nonwiggler parents to produce a wiggler daughter. To get a wiggler daughter, the male would have to be X^wY (wiggler) and the female would have to be X^WX^w (nonwiggler), which is not the case here. Therefore, we cannot conclude that ear wiggling is a sex-linked recessive trait based on this family.

STUDY BREAK 13.5

A mutant trait that shows cytoplasmic inheritance is caused by an alteration in the DNA of an organelle, either the mitochondrion or the chloroplast. A key property of cytoplasmic inheritance is that a trait is transmitted by a parent to all offspring, regardless of sex. The most common form of this is maternal inheritance, in which the progeny inherit the trait from their mother, paralleling the inheritance of mitochondria and mitochondrial DNA from the female parent and not from the male parent. This pattern of inheritance would not be seen for genes on chromosomes in the nucleus (as explained in Chapter 11).

Think Like a Scientist

FIGURE 13.2

The two largest classes in the progeny of the testcross would be red, vestigial and purple, normal—that is, the parental phenotypes. In this case the parental phenotypes are not a wild-type flies and flies with both mutations, but flies each with just one of the mutant phenotypes and the other phenotype wild-type.

The two smallest classes in the progeny of the testcross would be wild-type flies and flies that are both purple and vestigial.

FIGURE 13.8

The woman with normal vision must have at least one c^+ allele. One of her X chromosomes is inherited from her father, who is color blind and therefore *c*Y in genotype. Therefore, the woman is heterozygous c^+c. The man has normal vision. He has one X and one Y chromosome so the only possibility for his genotype is c^+Y; the fact that his father was color blind is not a factor in this case. For the cross $c^+c \times c^+$Y, the expected progeny are 1/4 c^+c^+ (normal vision female) : 1/4 c^+c (normal vision female) : 1/4 c^+Y (normal vision male) : 1/4 *c*Y (color blind male). In sum, all female offspring will have normal vision and 1/2 of the males will have normal vision and 1/2 will be color blind.

IMR

You learned about amino acids and protein structure in Section 3.4. Figure 3.17 shows the 20 amino acids. Glycine is a nonpolar amino acid, whereas arginine is positively charged polar amino acid. Since the varied properties and functions of proteins depend on the types and locations of the different amino acid side groups in their structures, it would be expected that a change from a nonpolar amino acid to a charged polar amino acid in a functionally important part of a protein would adversely affect the protein's function. Both glycine and alanine are nonpolar amino acids; the two amino acids are therefore similar, chemically speaking. You would speculate that substituting alanine for glycine would have a minimal effect on protein function.

UQ

No, the genes in the del(5q) region still have some activity. In these patients, only one copy of the gene is lacking due to a chromosomal deletion. The other copy, or allele, is present and appears to have its normal function. Since these patients now only have one copy of the genes in this deleted region, there is a lowered gene dosage, referred to as "haploinsufficiency." A lowered gene dosage (and resultant production of lower protein levels) can have profound effects on cell growth. Scientists currently believe that the lowered gene expression of *multiple* genes in the del(5q) region work together to play a role in disease development.

Test Your Knowledge

1. All sons will be color blind, but none of the daughters will be. However, all daughters will be heterozygous carriers of the trait.
2. The chance that her son will be color blind is 1/2, regardless of whether she marries a normal or color-blind male.
3. All these questions can be answered from the pedigree. Polydactyly is caused by a dominant allele, and the trait is not sex-linked. The genotypes of each person are:

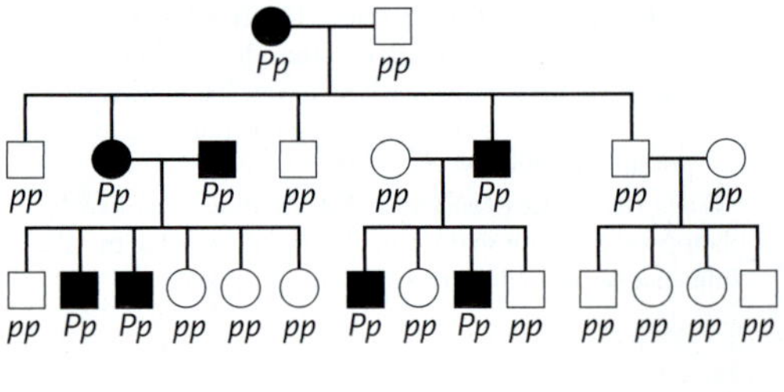

4. The sequence of the genes is ADBC.
5. Let the allele for wild-type gray body color = b^+, and the allele for black body = b. Let the allele for wild-type red eye color = p^+, and the allele for purple eyes = p. Then the parents are:

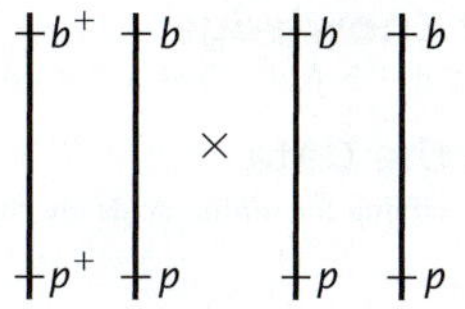

The F_1 flies with black bodies and red eyes are:

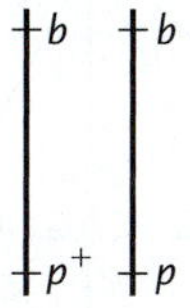

and the flies with gray bodies and purple eyes are:

6. The genes are linked by their presence on the same chromosome (an autosome), but they are not sex-linked. Because the F_1 females must have produced 600 gametes to give these 600 progeny, and because 42 + 30 of these were recombinant, the percentage of recombinant gametes is 72/600, or 12%, which implies that 12 map units separate the two genes.
7. The initial cross is X^wX^w white female 3 $X^{w^+}Y$ red male. The F_1 females produced are $X^{w^+}X^w$, with red eyes, and the F1 males are X^wY, with white eyes.

   The cross of an F_1 female with a parental male, therefore, is $X^{w^+}X^w$ (red) 3 $X^{w^+}Y$ (red). The offspring are $X^{w^+}X^{w^+}$ and $X^{w^+}X^w$ females, both of which have red eyes, and $X^{w^+}Y$ (red) and X^wY (white) males in equal proportions. Thus the phenotypic ratio for females is 2 red : 0 white, and that for males is 1 red : 1 white.

   The cross of an F_1 male with a parental female is X^wY (white) × X^wX^w (white). All progeny of both sexes will be white-eyed.
8. You might suspect that a recessive allele is sex-linked and is carried on one of the two X chromosomes of the female parent in the cross. When present on the single X of the male (or if present on both X chromosomes of a female), the gene is lethal.

Interpret the Data

1. Data are expressed as μg of DNA to control for the amount of sample analyzed from each individual.
2. Smoking promotes the formation of DNA adducts. The more the individual smoked, the more adducts were observed.
3. Stopping smoking lowered the level of adduct formation.
4. The highest point in the data from the nonsmoker most probably reflects an exposure to high levels of passive or second-hand smoke in the nearby environment.

Chapter 14

Study Break

STUDY BREAK 14.1

^{35}S-labeled phages in this scenario will have labeled protein coats and labeled DNA. When these phages infect bacteria, radioactivity enters the cell and is found in the progeny phages. In addition, radioactivity is found in the phage material removed by the blender. ^{32}P-labeled phages in this scenario are like the phages in Hershey and Chase's experiment; they have labeled DNA but unlabeled protein. When these phages infect bacteria, radioactivity enters the cell and is found in the progeny phages. No radioactivity is found in the phage material removed by the blender.

STUDY BREAK 14.2

1. Adenine and guanine are purines. Thymine and cytosine are pyrimidines.
2. Complementary base pairs are held together by hydrogen bonds. Each base is attached to the deoxyribose sugar by a covalent bond.
3. Watson and Crick described the right-handed double helix that consists of two sugar–phosphate backbones on the outside and complementary base pairs between the two backbones. A complementary base pair is a purine paired with a pyrimidine, more specifically, an A with a T, and a G with a C. The two strands of DNA are antiparallel. The key dimensions of the molecule are: diameter = 2 nm; 1 base pair = 0.34 nm; 1 turn of the helix = 10 base pairs = 3.4 nm.
4. The question focuses on the complementary base-pairing rules: A = T and G = C. If A = 20%, then T = 20%, giving 40% of the DNA as A–T base pairs. Therefore, 60% of the base pairs in this DNA molecule are G–C, and the percentage of C is 30%.

STUDY BREAK 14.3

1. Complementary base pairing ensures that the new DNA double helix is a faithful copy of the parental DNA double helix. For whatever base is exposed on the template strand, the DNA polymerase inserts the nucleotide with the complementary base.
2. DNA polymerases cannot initiate a DNA strand; they can add DNA nucleotides only to the 3′ end of an existing strand. The primer serves to provide a short stretch of nucleic acid that can be extended by DNA polymerase. The primer consists of RNA, rather than DNA, and is made by primase.
3. DNA polymerase III is the main DNA polymerase for replication in *E. coli*. This enzyme extends each primer that is synthesized on the lagging strand template, and synthesizes the leading strand. DNA polymerase I is used in lagging strand DNA synthesis. This enzyme replaces DNA polymerase when a new DNA fragment reaches the 5′ end of the Okazaki fragment that was made previously. With its 5′→3′ exonuclease activity, DNA polymerase I removes the RNA primer of that Okazaki fragment, and with its 5′→3′ polymerizing activity it replaces that primer with DNA nucleotides. In this way the Okazaki fragment is converted from an RNA–DNA hybrid into a DNA fragment.
4. Telomeres are buffers against the progressive loss of the ends of chromosomes by repeated rounds of replication. Only when the hundreds to thousands of copies of the telomere repeats have been lost are genes exposed. When those genes are lost by continued chromosome shortening and/or when chromosomes break down in the absence of telomeres, the cell is severely damaged.

STUDY BREAK 14.4

Proofreading prevents errors from being introduced into the DNA sequence. The DNA sequence in an organism's genome specifies everything about that organism—most notably, its function and reproduction. If significant errors occur during replication, gene sequences could be changed and the function of the organism could be adversely affected. Particularly if there is a high rate of errors, as there would be in the absence of proofreading, these errors would have potentially lethal consequences.

As part of the proofreading process, DNA polymerase reverses and removes the mispaired nucleotide. The enzyme then continues to move forward, inserting the correct nucleotide.

DNA repair mechanisms then look for and correct any errors that were not detected by proofreading. For example, repair enzymes remove a section of the newly synthesized DNA with the mismatch, and DNA polymerase synthesizes a replacement section with the correct base pairing.

STUDY BREAK 14.5

1. The nucleosome consists of two molecules each of histones H2A, H2B, H3, and H4 assembled into a nucleosome core particle wrapped with almost two turns of DNA. The diameter of the nucleosome is 10 nm.
2. Histone H1 is responsible for the next level of chromosome packing above the nucleosome. H1 binds to the exit/entry point of DNA on the nucleosome and to the linker DNA and brings about a coiling of the chromatin into the 30-nm chromatin fiber. The coiled structure is called the solenoid.

Think Like a Scientist

FIGURE 14.2

Yes, theoretically this could have been done; that is, the transforming principle is the genetic material. Given that *R* and *S* cells are genetically different, the transforming principle released from heat-killed *R* cells could convert living *S* cells to *R* cells. However, only if all *S* cells in a mouse were converted to *R* cells would the mouse live. This is unlikely so, practically speaking, this proposed experimental design is not a good one.

FIGURE 14.3

Carbon is found in both protein and DNA. Therefore, the radioisotope would be found in phage coats on phage-infected cells, within phage-infected cells, and in progeny phages. The success of the Hershey and Chase experiment relied on using radioisotopes that were specific for DNA and protein which is why the isotopes of phosphorus and sulfur were used. The carbon radioisotope would have not been useful in the experiment to answer the question asked.

FIGURE 14.9

First consider the data in the Figure. After one generation in ^{14}N, all the DNA is ^{15}N–^{14}N. After two generations, 1/2 the DNA is ^{15}N–^{14}N and 1/2 is ^{14}N–^{14}N. Therefore, by understanding how the semiconservative replication mechanism works, after four replications you would expect 1/8 of the DNA to be ^{15}N–^{14}N, and 7/8 to be ^{14}N–^{14}N. After five replications you would expect 1/16 of the DNA to be ^{15}N–^{14}N and 15/16 to be ^{14}N–^{14}N.

FIGURE 14.15

Without normal DNA ligase activity, Okazaki fragments could not be joined and synthesis of a continuous lagging strand could not be completed. In actuality, a mutant DNA ligase of this kind would be lethal to the cell. Some temperature-sensitive mutants have been isolated for bacteria, though, in which DNA ligase activity is normal at normal growth temperatures but nonfunctional or only partially functional at high temperatures. At the high temperatures, such mutants accumulate Okazaki fragments because they cannot be joined into longer chains.

IMR

Males have one X chromosome whereas females have two. For a female to express an X-linked recessive trait, both of her X chromosomes must carry the mutant gene. However, for a male to express such a trait, he just needs one copy of the mutant chromosome. On probability grounds, therefore, it is more likely for a male to have an X-linked recessive trait because he only has to inherit one mutant chromosome in contrast to the situation for a female who must inherit two mutant copies.

UQ

Telomerase alone is not the sole determinant of telomere length. The complex of proteins at the telomere regulates access of telomerase to the telomere. Thus, even if the amount of telomerase in the cell is unchanged, certain conditions could alter the function of one or more proteins at the telomere to increase the ability of telomerase to elongate the chromosome ends.

Test Your Knowledge

1. b 2. d 3. a 4. a 5. d 6. c 7. b 8. a 9. c 10. b

Interpret the Data

1. Schedule B, CDDP (8h) followed by 5-FU, was the most effective treatment because it resulted in the lowest percentage of cell proliferation.
2. B was the only schedule in which cells were first treated with CDDP alone.

Chapter 15

Study Break

STUDY BREAK 15.1

1. Although most enzymes are proteins, not all proteins are enzymes. And, some proteins consist of more than one polypeptide subunit. Each different polypeptide is encoded by a different gene, hence the one gene–one polypeptide hypothesis.
2. There are four different letters in the code (A, U, G, C), so a five-letter code would have 4^5 possible combinations = 1,024 codons.

STUDY BREAK 15.2

1. 5′-GUUUAACCGAAUAAUGGCCUAC-3′
2. The promoter determines where transcription of a gene will begin. In prokaryotes, RNA polymerase binds to the nucleotide sequence of the promoter and orients in the correct way to transcribe the associated gene. In eukaryotes, transcription factors bind to the promoter and then recruit RNA polymerase, which then orients properly for transcription from the transcription start point.

STUDY BREAK 15.3

1. Both pre-mRNAs and mRNAs have a 5′ cap, exons, and a 3′ poly(A) tail. Only pre-mRNAs have introns, which are removed from pre-mRNAs to produce mRNAs.
2. Particular snRNPs bind to the ends of an intron using their contained RNAs to recognize the boundary sequences of the intron. Other snRNPs then bind, causing the intron to loop out, and completing the active spliceosome. Cleavage at each intron-exon junction, looping back of the intron on itself, and joining the two exons together completes the splicing event; the intron and snRNPs are then released.

STUDY BREAK 15.4

1. In eukaryotes, a complex of the small ribosomal subunit, initiator tRNA, initiation factors, and GTP binds to the 5′ cap of the mRNA and scans along the mRNA until it reaches the first AUG codon, which is the start codon. The anticodon of the initiator tRNA binds to the start codon, the large ribosomal subunit binds, and the initiation factors are released when GTP is hydrolyzed.

   In prokaryotes, a complex of the small ribosomal subunit, initiator tRNA, initiator factors, and GTP binds to the region of the mRNA where the AUG start codon is located, directed by a specific RNA sequence upstream of the start codon. The other steps are the same as those in eukaryotes.
2. The P site is where the tRNA with the growing polypeptide is located. Downstream of the P site is the A site. An incoming aminoacyl–tRNA enters the A site and its anticodon base pairs with the codon of the mRNA in that site. When the polypeptide is transferred to the amino acid on the tRNA in the A site, the ribosome translocates one codon along the mRNA. As translocation takes place, the empty tRNA that was in the P site is moved to the E site. It remains there, blocking a new aminoacyl–tRNA entering the A site until translocation is finished. Then the empty tRNA is released from the ribosome.
3. Proteins found in the cytosol are made on free ribosomes.

   Proteins are sorted to the endomembrane system by cotranslational import. The proteins begin their synthesis on free ribosomes. These proteins have signal sequences at their N-terminal ends that direct them and the ribosome to dock with a receptor on the rough ER membrane. Continued translation inserts the growing polypeptide into the lumen of the ER, and the signal sequence is removed by signal peptidase. The proteins are then tagged to target them for sorting to their final destinations. Some proteins remain in the ER, whereas others are transported via the Golgi to vesicles for sorting to lysosomes, secreting them from the cell, or depositing them in the plasma membrane.

   Proteins are sorted to mitochondria, chloroplasts, microbodies, and the nucleus by posttranslational import. Proteins destined for the mitochondria, chloroplasts, and microbodies have short N-terminal transit sequences that target them to the organelle. A transit sequence interacts with a specific transport complexes on the organelle, and the polypeptide is then taken into the organelle. The transit sequence is then removed. Proteins destined for the nucleus have a nuclear localization signal which is bound by a cytosolic protein. The complex then interacts with the nuclear pore complex and the polypeptide then is transported into the nucleus through the pore.

STUDY BREAK 15.5

1. A missense mutation involves a change from a sense codon to another sense codon that specifies a different amino acid. A silent mutation involves a change from one sense codon to another sense codon, but where both codons specify the same amino acid.
2. Genetic recombination occurs by crossing over between two homologous sequences. TE transposition occurs by integration into a new location with which the TE has no sequence homology.
3. Both: (1) have inverted repeats at their ends; (2) contain a transposase gene; and (3) integrate into target sites and cause a duplication of the target site.
4. A transposon is a mobile genetic element that moves from one location to another in the genome as a DNA molecule.

   A retrotransposon is a mobile genetic element that moves from one location to another in the genome using an RNA intermediate; that is, the integrated DNA element is transcribed to produce an RNA copy. The RNA copy is reverse-transcribed into DNA, which then integrates at a new location in that genome.

Think Like a Scientist

FIGURE 15.2

Skin color depends on the synthesis of a pigment. It is reasonable to assume that synthesis of a pigment involves a biosynthesis pathway. A hypothesis to explain the results is that mutations in genes that encode enzymes that control different steps in that pathway can result in albinism. Therefore, the first couple consists of two individuals each of whom are homozygous for the albinism mutation in one of those genes. That pairing can produce offspring who have the same genotypes as the parents with respect to the albinism genes, and, thus, they have albinism.

The individuals of the second couple, however, must be homozygous mutant for different genes involved in albinism. However, their other albinism gene is homozygous normal. Therefore, each offspring of this couple inherits a normal allele and a mutant allele of one albinism gene and a normal allele and a mutant allele of the other albinism gene. Because albinism is a recessive trait, the presence of the two normal genes of the genes results in the children having normal skin color.

FIGURE 15.6

No. Although transcription occurs in highly similar ways in eukaryotes and prokaryotes, the initiation processes in particular are specific; that is, different promoter sequences are used, and eukaryotic RNA polymerases typically recognize only eukaryotic promoter sequences, and prokaryotic RNA polymerases typically recognize only prokaryotic promoter sequences. Therefore, *E. coli* RNA polymerase is unlikely to be able to recognize the promoter of the eukaryotic protein-coding gene in the figure.

IMR

The result provides support for the hypothesis that the conserved bases likely play an important functional role in peptide bond formation and, therefore, supports the conclusions of the study.

UQ

1. The first ribosomes were made only of RNA.
2. Not to make proteins, but to make simple protein fragments (peptides) that bind to RNA to help RNA carry out its biological functions before the evolution of proteins.

Test Your Knowledge

1. b 2. a 3. e 4. d 5. b 6. d 7. b 8. a 9. e 10. d

Interpret the Data

The codons specifying the amino acids are shown in the following figure:

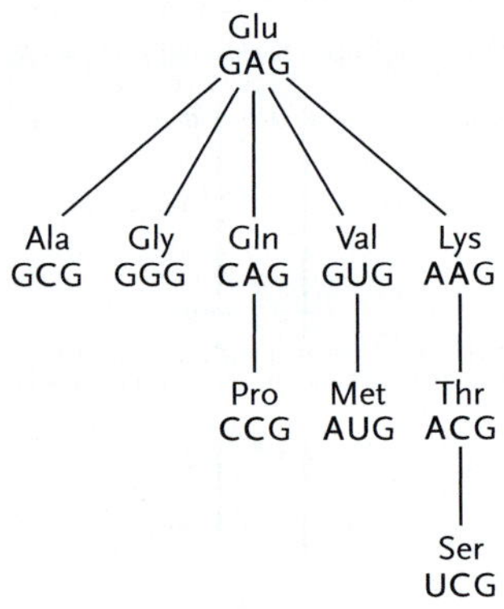

Chapter 16

Study Break

STUDY BREAK 16.1

1. The Lac repressor is active when it is made. In normal cells, in the absence of lactose, the Lac repressor binds to the operator, blocking transcription. In the mutant, the Lac repressor is not made, so transcription can never be blocked because no repressor is available to bind to the operator. As a consequence, the lactose metabolizing enzymes will be made both in the presence and absence of lactose in the medium. The mutation involved is an example of a regulatory mutant. Mutations such as this were valuable to Jacob and Monod in developing the operon model for gene regulation.
2. The Trp repressor is inactive when it is made. In normal cells, in the presence of tryptophan, the Trp repressor is activated, binding to the operator and blocking transcription. In the mutant, the Trp repressor is not made, so the operon cannot be turned off when tryptophan is present. This means that the tryptophan biosynthesis enzymes will be produced both in the presence and absence of tryptophan.

STUDY BREAK 16.2

1. General transcription factors bind to the promoter and recruit RNA polymerase II, orienting the enzyme so that it will begin transcription at the beginning of the gene. Activators bind to regulatory sequences associated with genes and increase the rate of transcription. Activators that bind to regulatory sequences in the proximal promoter region interact directly with the general transcription factors at the promoter to exert their action. Activators that bind to regulatory sequences in the enhancer stimulate transcription indirectly. These latter activators bind to a coactivator that also binds to the complex of proteins at the promoter, and transcription then occurs at the maximal possible rate.
2. Histones are general negative regulators of gene expression. When DNA is complexed with histones in normal chromatin, gene promoters typically are not very accessible to the transcription machinery. By acetylating histones, the chromatin is remodeled, making the promoter now accessible to the transcription machinery. Acetylation of histones occurs in response to the binding of an activator to a regulatory sequence associated with the gene.

STUDY BREAK 16.3

1. A microRNA (miRNA), in a complex with particular proteins, binds to an mRNA by complementary base pairing. Either the proteins cut the mRNA in the region of pairing, thereby destroying that molecule, or the double-stranded RNA region blocks translation.
2. Removal of the poly(A) tail would result in the mRNA not being translated.

STUDY BREAK 16.4

1. Determination is the process by which the developmental fate of a cell is set. Differentiation is the establishment of a cell-specific developmental program in cells. Determination and differentiation are under molecular control. Regulatory genes encode regulatory proteins that bind to promoters of the genes they control, switching the genes on or off depending on the interaction.
2. The segmentation genes subdivide the embryo progressively into regions, thereby determining the segments of the embryo and the adult. In essence, they organize the embryo into segments. The homeotic genes specify the identity of each segment with respect to the body part it will become.

STUDY BREAK 16.5

1. A tumor suppressor gene encodes a product that has an inhibitory role in the cell division cycle. If a tumor suppressor gene is mutated so that its product is nonfunctional or its function is significantly diminished, then the gene's inhibitory control of cell division is lost or reduced. As a result, the cell may progress toward division.
2. A proto-oncogene encodes a product that has a stimulatory role in cell division. Mutations that lead to increased levels of that product have converted a proto-oncogene to an oncogene. The increased amount of product is stimulatory to cell division.
3. Some miRNAs have been shown to regulate the expression of mRNA transcripts of particular tumor suppressor genes. Overexpression of those miRNAs can abnormally inhibit the activity of those tumor suppressor genes, thereby removing or decreasing inhibitory signals for cell proliferation. Other miRNAs have been shown to regulate the expression of mRNA transcripts of particular proto-oncogenes. Inactivation of those miRNAs means that the proto-oncogenes are expressed at higher than normal levels, which can stimulate cell proliferation.

Think Like a Scientist

FIGURE 16.3

If the Lac repressor cannot bind to the operator, then the *lac* operon structural genes will be transcribed even in the absence of lactose because there is no block to the binding of RNA polymerase to the promoter. By contrast, in a normal cell, the Lac repressor binds to the operator and blocks RNA polymerase from transcribing the *lac* structural genes as long as lactose is absent from the medium.

FIGURE 16.5

If the Trp repressor cannot bind tryptophan, then the repressor could not be activated when tryptophan is present in the medium. Without an active repressor in this condition, RNA polymerase cannot be blocked from binding to the promoter. Therefore, the Trp structural genes will be transcribed. By contrast, in a normal cell, the presence of tryptophan in the medium activates the Trp repressor which then binds to the operator and blocks transcription of the *trp* structural genes by RNA polymerase.

IMR

First, the CREB monomer must have a binding site for itself so that it can form a dimer. Second, the dimer must have a binding site for CRE (cAMP response element) sequences in the enhancers for regulatory genes that it controls. Third, CREB must have a binding site for the Tax protein.

UQ

Dr. Kay's lab uses mice to test RNAi therapies against hepatitis C virus. You could test various RNAi agents via an intravenous infusion to find a dose where the viral load (or level) is reduced substantially without causing any dangerous toxicities that have been determined by standard laboratory testing of blood samples. Kay's lab aims to reduce viral load by at least 100-fold.

Test Your Knowledge

1. d 2. a 3. b 4. c 5. e 6. b 7. a 8. d 9. a 10. b

Interpret the Data

1. The woman would have an 8.6% chance of dying of cancer.
2. Her risk of dying of cancer is 18.0% if she carries a mutated *BRCA1* gene.
3. This study suggests a *BRCA1* mutation is more dangerous than a *BRCA2* mutation (in effect, *BRCA1* mutations are associated with a higher percentage of deaths). Sixteen out of 89 women with the *BRCA1* mutation died whereas only 1 out of 35 women with the *BRCA2* mutation died.
4. You would need to know the number of deaths of women who had a preventive mastectomy and the number of deaths of women who had a preventive oophorectomy.

Chapter 17

Study Break

STUDY BREAK 17.1

1. An F^+ cell contains the F factor plasmid in addition to the chromosomal DNA. The F factor enables an F^+ cell to conjugate with an F^- cell, which lacks the F factor. By a special replication mechanism, a copy of the F factor is transferred to the recipient F^- cell in an $F^+ \times F^-$ conjugation, so the recipient is converted to an F^+ cell.

   An Hfr cell has the F factor integrated into the chromosomal DNA. When an Hfr cell conjugates with an F^- cell, the F factor begins its replicative transfer into the recipient as in an $F^+ \times F^-$ conjugation and, by that transfer mechanism brings in chromosomal genes from the Hfr donor. Those chromosomal genes can recombine with the genes in the recipient. Because replication of the F factor begins in the middle of the plasmid, the entire F factor cannot be transferred to the recipient unless the entire chromosome is transferred to the recipient, which occurs only rarely. Hence, the recipient remains F^- in this conjugation experiment.
2. In gene segregation in sexually reproducing organisms, genetic material moves from one generation to the next, basically by descent. In horizontal gene transfer, the movement of genetic material between organisms is other than by descent. For instance, in transformation, a DNA molecule in the environment is taken up by a recipient cell—the DNA has moved horizontally from a donor cell (from which the DNA was released) to the recipient cell. Genetic recombination in the recipient can alter the phenotype of that cell.

STUDY BREAK 17.2

1. A virulent phage always enters the lytic cycle when it infects a bacterial cell. The end result is the assembly of many progeny phages and their release into the surroundings when the cell breaks open.

   A temperate phage enters either the lytic cycle or the lysogenic cycle when it infects a cell. The lytic cycle is the same as that for a virulent phage. The lysogenic cycle involves integration of the phage's chromosome into the bacterial chromosome. In the integrated state the phage—now called the prophage—is inactive and replicates only when the bacterial chromosome replicates. In response to an adverse environmental signal to the cell, the phage chromosome can excise itself from the bacterial chromosome and enter the lytic cycle.
2. Animal cells: Viruses without an envelope bind by their recognition proteins to receptor proteins in the host cell's plasma membrane and are then taken into the cell by receptor-mediated endocytosis. For some enveloped viruses, the genome-containing capsid enters the cell when the envelope fuses with the host cell's plasma membrane. For other enveloped viruses, the complete virus, with the envelope, enters the cell by endocytosis.

   Plant cells: All plant viruses lack envelopes. They enter cells either through mechanical injuries to leaves and stems or by the action of biting and feeding insects.
3. Non-retrovirus RNA viruses have RNA genomes that are replicated in an RNA-to-RNA manner. Retroviruses have RNA genomes that are replicated via a DNA intermediate: that is, the RNA genome is copied to double-stranded DNA by reverse transcriptase and the DNA molecule integrates into the host cell's nuclear chromosomes, from which location new RNA viral genomes are transcribed.

STUDY BREAK 17.3

Although all are infectious agents, viruses have protein coats, whereas viroids and prions do not. A virus consists of a nucleic acid genome surrounded by a protein coat and, in some cases an envelope. Viroids are naked, single-stranded RNA circular molecules. Prions are proteins.

Think Like a Scientist

FIGURE 17.1

The control plates were used to check that the mutant alleles in the two strains did not mutate back to normal alleles; that is, mutation back to the nonmutant state would allow the cells to grow on the minimal medium plates. By having more than one mutant allele in each strain (two in strain 1 and three in strain 2), Lederberg and Tatum were making it extremely unlikely to see colonies on the minimal plate; that is, for strain 1, two mutant alleles would have to mutate to normal alleles, and for strain 2, three mutant alleles would have to mutate to normal alleles. If we made the simple assumption that the rate of mutation from mutant to normal is the same for all genes, then the probability of having two genes mutate from mutant to normal is the square of the probability for one to mutate. Moreover, the probability of having three genes mutate from mutant to normal is the cube of the probability for one to mutate. Therefore, having just one mutant allele in each strain would mean that the probability of generating a nonmutant cell by mutation would be much higher than for the two- or three-mutant allele strains.

FIGURE 17.4

The transfer of chromosomal genes by conjugation is dependent on the F factor. Genes on the F factor direct the one-way transfer of a copy of the F factor to a recipient in $F^+ \times F^-$ conjugations. When the F factor is integrated into the chromosome to produce an Hfr, again the F factor genes direct a one-way transfer of DNA from donor to recipient, in this case involving the F factor attached to the bacterial chromosome. There is no mechanism, therefore, for transfer of chromosome genes in the opposite direction—that is, from recipient to donor.

FIGURE 17.8

As you learned in the text, the prophage state is maintained by a phage-encoded repressor molecule that prevents expression of phage genes for the lytic cycle. When the prophage is transferred into a recipient cell, that cell will not contain repressor proteins and, because it takes time for new repressor proteins to be synthesized, the result will be that the λ lytic cycle genes will be expressed and the recipient will be killed by phage release. If the recipient already contained a λ prophage, then repressor proteins would already be present and the transferred prophage would remain repressed with respect to the lytic cycle genes.

IMR

That the DNase converted the radioactive macromolecules from acid-insoluble to acid-soluble, whereas RNase did not, affirms that the macromolecules synthesized must be DNA and not RNA.

UQ

Researchers can further study the interactions between normal prion protein and prion aggregates and the effect of these interactions on neurons. One possible function for the normal prion protein is to act as a signaling molecule to regulate apoptosis. Interactions with prion particles may induce aberrant signaling through the normal prion protein. Since the normal prion protein also binds to many other molecules (for example, extracellular matrix components), one could also investigate the effects of the conversion process on the interactions of normal prion protein with other molecules.

Test Your Knowledge

1. e 2. a 3. b 4. c 5. d 6. a 7. c 8. c 9. a 10. b

Interpret the Data

The answer is shown in the figure. The segments of donor chromosomes transferred by each Hfr strain are shown, with the arrowheads showing the beginning.

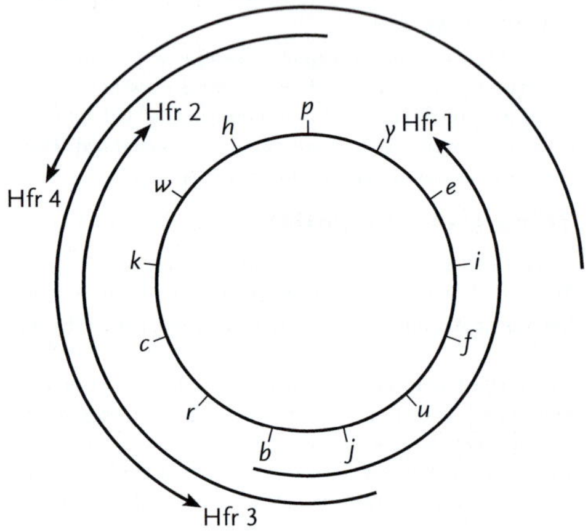

Chapter 18

Study Break

STUDY BREAK 18.1

1. Each restriction enzyme recognizes a specific sequence in DNA, typically in the range of 4 to 6 bp, and cuts both strands of the DNA within the sequence. Restriction enzymes differ with respect to the DNA sequences—the restriction sites—they recognize and cut. The enzymes most useful for cloning produce sticky ends.
2. Replication origin so that the plasmid will replicate in *E. coli*, an antibiotic resistance gene to allow selection of bacteria containing the plasmid, a cluster of restriction sites (a multiple cloning site) to provide choices for inserting fragments of DNA, and the $lacZ^+$ gene to use in blue–white screening, in which colonies containing recombinant plasmids (white) are distinguished from colonies with vectors lacking inserted DNA (blue).
3. A cDNA library is a collection of clones in which DNAs that are complementary to the mRNAs in the cell have been inserted into a cloning vector. The array of particular clones in a cDNA library is directly related to the genes expressed in the cells from which the mRNAs are isolated. Therefore, not all genes are represented in any given cDNA library. By contrast, a genomic library is a collection of clones containing all the sequences of an organism inserted into a cloning vector. A genomic library contains all of the genes of an organism, as well as noncoding sequences found between and within genes.
4. PCR is a method to amplify a specific segment of DNA. The amplification process depends on DNA replication and, therefore, requires primers. A limitation of PCR, then, is that DNA sequence information must be available for the segment of DNA to be amplified. Otherwise, the primers cannot be synthesized. The ingredients of a PCR are the template DNA containing the sequence to be amplified, the two primers, a buffer, and a DNA polymerase that is tolerant to the high temperature used to denature the DNA repeatedly during the cycles of the reaction.

STUDY BREAK 18.2

1. Each human (or other organism) has unique combinations and variations of DNA sequences. DNA fingerprinting exploits these combinations and variations to distinguish between different individuals (with the exception of identical twins). Using DNA technologies to analyze the particular regions of the genome showing sequence variation, it is possible to compare two DNA samples to see if they are from the same person or not. DNA fingerprinting is used in several ways, including forensics, paternity testing, and basic research.
2. A transgenic organism is one into which a gene or genes from an external source have been introduced as a means to modify the organism genetically.
3. Germ-line cells develop into reproductive cells, so modifying this cell type genetically will lead to the genetic modification being passed to offspring. In somatic cell gene therapy, germ-line cells and their products are not involved, so the inserted genes remain with the individual and are not passed to offspring.

Think Like a Scientist

FIGURE 18.14

To demonstrate that the experiment was a success, it was necessary to show that a clone matched that of the donor genetically and, importantly, did not match that of the surrogate mother. Only by starting with genetically distinguishable kinds of sheep could that be demonstrated.

IMR

As discussed in Chapter 7, plasma membrane receptors, when activated by a signal molecule, transduce the signal into the cell and through a signal transduction pathway to bring about a cellular response. By knowing that signal transduction pathway is involved, researchers could set out to elucidate the pathway thereby providing information that could be useful for the design of disease control strategies.

UQ

Tissue-specific stem cell niches serve to both protect stem cells against environmental insults and control stem cell responses. Cellular and neuronal signals that control stem cell dormancy and proliferation are often uniquely adapted to these microenvironments. This ensures that tissue-specific stem cells receive signals in a tightly controlled manner.

Test Your Knowledge

1. e 2. d 3. a 4. a 5. d 6. b 7. e 8. b 9. a 10. d

Interpret the Data

An offspring must have one allele of each gene inherited from the male parent, and the other allele of each gene inherited from the female parent. The same goes for STR locus alleles.

Juvenile 1: Not the offspring of the two adults. The juvenile has one allele in common with the male parent (8 repeats = 72 bp), but its other allele came from neither adult (9 repeats = 81 bp). In other words, it is heterozygous for 8- and 9-repeat alleles.

Juvenile 2: Could be the offspring of the two adults. There was only one DNA fragment size amplified by PCR. The interpretation is that this juvenile is homozygous for a 6-repeat (54-bp) allele, which is why there is more DNA amplified (a thicker band). Each adult has a 6-repeat allele that could have passed to the juvenile.

Juvenile 3: Not the offspring of the two adults. This juvenile is heterozygous for 5-repeat (45-bp) and 11-repeat (99-bp) alleles. The 11-repeat allele could have been inherited from the female adult, but neither adult has the 5-repeat allele.

Juvenile 4: Could be the offspring. This juvenile is heterozygous for 6-repeat (54-bp) and 8-repeat (72-bp) alleles. It could have inherited the 6-repeat allele from either parent, but could have inherited the 8-repeat allele only from the male parent. Therefore, the 6-repeat allele would have had to come from the female parent.

Chapter 19

Study Break

STUDY BREAK 19.1

Having the complete sequence of a genome enables a researcher to ask questions about the organization of genes and of other sequences of importance in the genome. The most important question is how the functions of all of the genes in a genome direct that organism's life. It also allows researchers to compare gene content and gene order between organisms, which provides insight into how genomes evolve.

STUDY BREAK 19.2

1. In whole-genome shotgun genome sequencing, the entire genome is broken into thousands to millions of random, overlapping fragments, each fragment is cloned and sequenced, and then the genome sequence is assembled by computer on the basis of sequence overlaps between fragments.
2. Protein-coding genes, noncoding RNA genes, regulatory sequences associated with genes, origins of replication, transposable elements and related sequences, pseudogenes, and short repetitive sequences.
3. To find protein-coding genes in a bacterium, computer algorithms search a genome sequence for open reading frames (ORFs), which are defined as the DNA equivalent of a start codon (ATG) in frame (a multiple of three away) from a stop codon (TAG, TAA, or TGA). Such a segment of DNA could potentially produce an mRNA that could be translated into a protein. Other analyses would be required to show if any given ORF identified in this way is an actual protein-coding gene.

   To find protein-coding genes in a mammal, one also needs to find ORFs. However, the presence of introns in eukaryotic protein-coding genes complicates matters. Therefore, more sophisticated computer algorithms are needed—in this case ones that attempt to locate exon-intron boundaries while they search for ORFs.
4. Prokaryotic genomes vary widely in size. Nonetheless, their genes are densely packed in their genomes (a fairly consistent density of 900 genes per Mb), with little noncoding space between them. Genomes of eukaryotes also show a great range of sizes; overall, their genomes are much larger than those of prokaryotes. Gene density also varies considerably in eukaryotes, ranging from 500 per Mb for the single-celled yeast to fewer than 10 per Mb in mammals. There is no uniformity to the packing of genes, however.

   Prokaryotic genes are arranged either singly or in operons, depending on the genes. All eukaryotic genes are arranged singly; no operons are found in eukaryotes. The number of regulatory sequences typically is much higher for eukaryotic protein-coding genes than for prokaryotic protein-coding genes. There are no introns in prokaryotic protein-coding genes, whereas most eukaryotic protein-coding genes contain introns. Introns can make up significant proportions of eukaryotic genomes—for example, about 20% in humans. In contrast to prokaryotes, eukaryotes contain many pseudogenes, many transposable element sequences, and a variety of short repeated sequences. A prokaryotic chromosome has one origin of replication, whereas a eukaryotic chromosome may have hundreds to thousands of origins of replication.

STUDY BREAK 19.3

1. The function of a gene identified in a genome sequence may be assigned by a sequence similarity search of sequence databases to find a match with a gene of known function, by comparing the structure of the protein encoded by the gene with the structures of proteins with known function, and by experiments that alter the expression of the gene to see how the phenotype of a cell or organism is affected.
2. A genome-wide analysis of gene expression here could involve DNA microarray assays. The mRNAs could be isolated from untreated tissue culture cells and from cells treated with a steroid hormone. Convert each mRNA preparation to cDNAs using reverse transcriptase, using nucleotide precursors with different fluorescent labels for the two batches; for example, green for the untreated (reference) sample and red for the treated (experimental) sample. Mix the two cDNAs and pump them through a DNA chip prepared to have spots with DNA representing every gene in the genome. Allow hybridization to occur, and analyze the hybridization by laser detection. The colors of the spots indicate which genes are affected by the steroid hormone. Purely red spots represent genes that are active only in hormone-treated cells. Purely green spots represent genes that are

active only in untreated cells. Spots that are a mixture of green and red represent genes that are active in both types of cells. Based on controls, it would be possible to see if any of the genes have higher or lower expression levels in the treated cells.
3. The proteome is complete set of proteins that can be expressed by an organism's genome. Proteomics has the three major goals of determining the number and structures and functions of proteins of the proteome, determining the location of each protein within or outside the cell, and determining the physical interactions between proteins.

STUDY BREAK 19.4

1. Tandem duplication happens when homologous pairs of chromosomes do not line up properly in meiosis. As a result of unequal crossing over, one gamete inherits a chromosome with two copies of the stretch of DNA between the two points of crossing over, and therefore duplicate copies of any genes on that stretch of DNA. Dispersed duplication happens when transposable elements copy the DNA for a gene and insert the new copy somewhere else in the genome.
2. Gene duplication produces two identical copies of a gene that only evolve distinct functions as different mutations occur in each copy. Exon shuffling, by contrast, combines parts of two or more genes in an evolutionarily novel way. The protein coded for by the new gene contains a combination of domains that may never before have existed in one protein. Because a protein's domains usually each perform a different molecular function, the newly evolved protein may have a combination of functionalities unlike any previously existing protein.

Think Like a Scientist

UQ

Knowledge that an individual is genetically predisposed to developing a disease or is at greater risk for the aftereffects of disease based on genotype may cause emotional distress, and fear of discrimination by employers or insurance companies. Similar concerns may exist if genetic information suggests that disease treatment options for an individual are limited. The Genetic Information Nondiscrimination Act (GINA) was recently enacted to help address these concerns.

Test Your Knowledge

1. b 2. a 3. a 4. a 5. d 6. d 7. d 8. a 9. d 10. d

Interpret the Data

To find an open reading frame, you first find a start codon (see Figure 15.5), and then count three nucleotides at a time until you find a stop codon in the same reading frame (see Figure 19.4). This sequence contains 6 start codons and therefore 6 open reading frames. The first open reading frame begins with the start codon at the very beginning of the second line and has 145 codons between the start and stop codon, which makes it large enough to code for a functional protein. The other open reading frames have only 5, 15, 8, 28, and 11 codons between the start and stop codon, so they are very unlikely to code for functional proteins.

Chapter 20

Study Break

STUDY BREAK 20.1

1. Buffon did not understand how "anatomically perfect" animals could have useless structures.
2. Lamarck proposed: all species change through time, the changes are inherited by the next generation, the changes arise in response to environmental conditions, and specific mechanisms caused the changes.
3. The concepts of gradualism and uniformitarianism suggest that Earth's major geological features were produced by the very slow action of geological processes observed today. Thus, it must have taken more than 6,000 years for these features to assume their present forms.

STUDY BREAK 20.2

1. Darwin observed that living organisms often resemble fossils found in the same area; that organisms found on South America resembled one another, even if they occupied different environments; and that many species found on the Galápagos Islands resembled species from the South American mainland.
2. Darwin realized that the effects of competition for resources in nature were similar to the action of a plant or animal breeder who used only certain individuals as parents to produce the next generation.
3. Darwin's theory relied on physical explanations for the origin of biodiversity: he recognized that evolutionary change takes place within a population rather than in individuals; he recognized that natural selection is a multistage process; and he emphasized the importance of environmental conditions to the process of natural selection.

STUDY BREAK 20.3

1. Two problems that slowed the acceptance of Darwin's theory are that Mendel's genetic studies focused on simple traits and that these traits often changed in just a few generations.
2. Microevolution refers to small, genetically based changes within populations. Macroevolution refers to larger-scale evolutionary changes observed in species and more inclusive groups. Current research suggests that microevolutionary change and macroevolutionary change result from the same evolutionary processes.
3. Evidence for evolution comes from studies of adaptation, the fossil record, historical biogeography, comparative morphology, and comparative molecular biology.

Think Like a Scientist

FIGURE 20.12

The continued use of DDT to control mosquito populations would probably result in nearly 100% resistance to DDT in the populations of mosquitoes exposed to that insecticide.

FIGURE 20.14

Most vertebrate biologists agree that the loss of legs in snakes was probably favored by natural selection as a mechanism to facilitate locomotion through environments with dense vegetation or to facilitate burrowing underground.

IMR

The conclusion that the extinct woolly mammoth is similar genetically to living Asian elephants parallels Darwin's conclusion that extinct glyptodonts and living armadillos are closely related. In both cases, living organisms occupy the landmass where fossils of their extinct relatives are found.

UQ

If climate change rapidly disrupts physical environments everywhere on Earth, Arctic species are unlikely to adapt to those changes. Given the frequency of extinction in the fossil record, one might assume that most of the species living today will become extinct eventually, and rapid climate change is likely to speed that process.

Test Your Knowledge

1. c 2. e 3. d 4. b 5. c 6. b 7. a 8. b 9. d 10. d

Interpret the Data

The graph suggests that selective breeding of race horses did improve the speed of horses that won the Kentucky Derby between 1896 and 1960. However, the speed of winners has been stable since that time, suggesting that additional selection would not lead to future improvements in speed.

Chapter 21

Study Break

STUDY BREAK 21.1

1. The variation in skunks is qualitative.
2. The researchers used artificial selection to change the activity levels of the mice.
3. Genetic variation, differing environmental effects on individuals, and interactions between genes and the environment affect phenotypic variation in a population.

STUDY BREAK 21.2

1. Genotype frequencies specify how alleles are combined in individuals, and allele frequencies specify how common the alleles are.
2. The Hardy–Weinberg principle is a null model because it identifies the conditions under which evolution will *not* occur.
3. If genotype frequencies are not already in equilibrium, they will stop changing after one generation of random mating.

STUDY BREAK 21.3

1. Mutation and gene flow tend to increase genetic variation within populations, and natural selection and genetic drift tend to decrease it.
2. Stabilizing selection increases the representation of the average phenotype in a population.
3. Sexual selection, like directional selection, favors extreme phenotypes.

STUDY BREAK 21.4

1. Diploidy protects harmful recessive alleles because dominant alleles mask their effects in heterozygotes.
2. A balanced polymorphism is one in which two or more phenotypes are maintained in fairly stable proportions over many generations.
3. The sickle-cell allele is rare in Northern Europe because, in the absence of the malarial parasite, it confers no advantage on individuals that carry it.

STUDY BREAK 21.5

1. The adaptive value of a trait can be evaluated by comparing closely related species that live in different environments.
2. Natural selection preserves traits that were useful when the organisms subject to selection were alive and reproducing.

Think Like a Scientist

FIGURE 21.6

After nine or 10 generations of artificial selection, the distance run and average speed appear to have leveled off for three of the four experimental lines of mice. The data suggest that continued artificial selection might not have a very strong ongoing effect on these traits.

FIGURE 21.9

Dachshunds are the product of directional selection on two traits: body length was increased and leg length was decreased.

FIGURE 21.10

Natural selection may select against a genetic predisposition for multiple births in humans because newborn triplets or quadruplets are each much smaller than the offspring of a single-child pregnancy and would therefore have a higher mortality rate.

FIGURE 21.11

A decline in bird populations might decrease natural selection against gall-making flies that produced large galls. As a result, the average size of goldenrod galls would likely increase.

FIGURE 21.13

Lengthening the male widowbirds' tails had a much larger effect on the number of mates they attracted than shortening their tails did.

FIGURE 21.15

No, the association between shell color and pattern and the habitat occupied is not absolute. The data in the graph indicate that virtually all populations of snails include individuals with different shell colors and striping patters. The data show that the majority of snails in populations in wooded habitats have brown, unstriped shells and that the majority of snails in populations in hedges and fields have yellow, striped shells.

Abbreviations: IMR = Insights from the Molecular Revolution; UQ = Unanswered Questions

FIGURE 21.16
Yes, the data indicate the fish's preference for the dark brown form of the water boatmen over the medium brown and pale brown forms: fish ate a higher percentage of dark brown water boatmen at every percentage availability. Nevertheless, this preference did not override the frequency-dependence of their food choice.

IMR
If the ban on whaling had taken effect 34 years later than it actually did, the amount of genetic variability in humpback whale populations today would undoubtedly be lower than the level revealed in the 1993 study. Of course, with a worldwide population of only 5,000 individuals in 1966, the species may well have become extinct if the ban had not taken effect at that time.

UQ
If recombination induces mutations, then regions of high recombination should have many differences between species. This suggests a test for whether recombination rate and genetic variation are connected mechanistically vs. connected by the action of natural selection. Both mechanistic and selective explanations predict that regions of high recombination have lots of genetic variation *within* species—more than regions of low recombination—but only mechanistic explanations predict that genetic differences *between* species should be greater in regions of high recombination than in regions of low recombination.

Test Your Knowledge
1. c 2. b 3. c 4. d 5. b 6. e 7. a 8. b 9. c 10. d

Interpret the Data
The graphs illustrate the action of directional selection favoring birds with deeper bills because the distribution of bill depths in the survivors (lower graph) is shifted slightly to the right of the distribution of bill depths in the population before the drought (upper graph).

Chapter 22

Study Break

STUDY BREAK 22.1
1. The morphological species concept defines species based on morphological differences among them. The biological species concept defines species as populations that can successfully interbreed under natural conditions. The phylogenetic species concept defines a species as a cluster of populations with a recent shared evolutionary history.
2. Clinal variation is a pattern of smooth variation along a geographical gradient.

STUDY BREAK 22.2
1. Prezygotic isolating mechanisms either prevent individuals of different species from mating or prevent sperm of one species from fertilizing the eggs of another. Postzygotic isolating mechanisms limit the survivorship or reproductive capability of hybrid individuals.
2. The scenario illustrates a behavioral isolating mechanism.

STUDY BREAK 22.3
1. In the first stage of allopatric speciation, populations become geographically separated. In the second stage, they become reproductively isolated.
2. Some populations of bent grass survive better on unpolluted soil, whereas others survive better on polluted soil.
3. Insects from different host races spend most of their time on different host plant species. Thus, they rarely encounter each other and would be unlikely to mate under natural conditions.

STUDY BREAK 22.4
1. Natural selection cannot promote reproduction isolation in allopatric populations directly, but it can lead to genetic divergence which results in reproductive isolation.
2. Polyploidy changes the number of chromosome sets in the cells of an organism. Polyploid individuals are often reproductively isolated from their parent species because their gametes contain different numbers of chromosomes. When a gamete from a polyploid individual fuses with a gamete from a nonpolyploid individual, the resulting offspring is usually sterile because the odd number of chromosomes cannot segregate properly during meiosis.

Think Like a Scientist

FIGURE 22.13
Yes, the pollination agent should have little impact on temporal reproductive isolation. Copper-tolerant plants, which flower earlier than pasture plants, would simply attract insect pollinators before pasture plants did.

FIGURE 22.16
After a spontaneous doubling of the chromosomes in the hybrid offspring, the allopolyploid would have $2n = 20$.

FIGURE 22.18
The difference in chromosome structure between humans and their closest relatives among the apes represents a postzygotic reproductive isolating mechanism, because, at a minimum, these structural differences would prevent human and ape chromosomes from pairing up during meiosis.

IMR
The eight floral traits that differ between the two monkey-flower species attract different animal pollinators. Genetic differences in anthocyanin and carotenoid pigment concentrations influences petal color; bees are attracted to purple flowers with yellow nectar guides, and hummingbirds are attracted to red flowers. Differences in petal width and flower width allow bees to land on *M. lewisii* flowers, but not on *M. cardinalis* flowers. The small volume of concentrated nectar in *M. lewisii* flowers is suitable for bees, and the large volume of dilute nectar in *M. cardinalis* flowers is suitable for hummingbirds. Differences in the positions of male and female reproductive parts allow bees to collect pollen from *M. lewisii* on their legs and body and hummingbirds to collect pollen from *M. cardinalis* on their foreheads.

UQ
Most of the asexual organisms for which the biological species definition is not a good match are bacteria. Because these organisms do not exhibit as much morphological distinctness as many animals do, biologists rely on comparisons of their DNA or RNA sequences to analyze their evolutionary relationships. These data allow researchers to construct phylogenetic trees for the organisms they study, indicating that any new species definition that applies to them would probably resemble the phylogenetic species definition more than the morphological species definition.

Test Your Knowledge
1. a 2. e 3. c 4. d 5. b 6. e 7. b 8. c 9. a 10. b

Interpret the Data
The graphs suggest that sympatric populations of the three frog species experience temporal prezygotic reproductive isolation. Populations that are allopatric exhibit substantial overlap in breeding times, especially in March, April, and May. By contrast, sympatric populations exhibit almost no overlap in breeding times.

Chapter 23

Study Break

STUDY BREAK 23.1
1. Hard parts, such as the shells or bones of animals, are the materials most likely to fossilize.
2. The fossil record provides an incomplete portrait of life in the past because not all organisms are equally likely to form fossils; fossils do not form in all types of habitats; and fossils are often destroyed by geological processes and erosion.
3. The fossil record provides information about the morphology of ancient organisms; how structures changed over time; and the proliferation and extinction of evolutionary lineages. It also offers indirect evidence about the behavior, ecology, and physiology of organisms that lived in the past.

STUDY BREAK 23.2
1. Continental drift caused large-scale geographical separation of populations and lineages that subsequently evolved in isolation.
2. Sea levels fell whenever a large proportion of Earth's water was incorporated into glaciers.

STUDY BREAK 23.3
1. Continuous distribution requires no special explanations. Biologists infer that lineages with a continuous distribution simply occupy all or one part of their historical range.
2. Distantly related species that live in widely separated parts of the world may resemble each other because convergent evolution fosters similar adaptations to the environments they occupy.

STUDY BREAK 23.4
1. A population of organisms may occupy a new adaptive zone after the evolution of a key morphological innovation that allows it to use the environment in a unique way or after a once-successful group of organisms declines.
2. Huge volcanic eruptions triggered a chain of events that led to the mass extinction at the end of the Permian period. These eruptions warmed the atmosphere and oceans enough to melt frozen undersea methane reserves. The methane entering the atmosphere warmed the climate even further, creating a "runaway greenhouse" effect.
3. The first major adaptive radiation of animals took place in the Cambrian era.

STUDY BREAK 23.5
1. The horse lineage was highly branched, and it included some species that were larger and some that were smaller than their ancestors.
2. Phyletic gradualism predicts that morphological change is slow and steady, producing fossils with morphologies that are intermediate between the structures found in lower and higher strata. The punctuated equilibrium hypothesis predicts that morphological changes occur rapidly as new species form and that most species experience little morphological change for long periods of time.

STUDY BREAK 23.6
1. Allometry refers to the differential growth of body parts. Heterochrony refers to changes in the timing of developmental events.
2. Similar developmental control genes are present in a wide variety of animals, plants, fungi, and prokaryotes. Their widespread distribution suggests that they were present in the ancestor of all these organisms and have been conserved through countless generations.
3. The *Pitx1* gene is expressed in fin buds that later produce spines, and it is not expressed in those that fail to produce spines.

Think Like a Scientist

FIGURE 23.18
No, artificial selection has not reversed the many evolutionary trends observed in the fossil record for horses. A reduction in body size appeared multiple times in horse evolution.

FIGURE 23.19
The changes in rib number over time in *Bergamia* and *Ogyginus* appear to be the products of directional selection that alternated between favoring increased rib number and decreased rib number. However, the observed changes in rib number may have been induced by genetic drift.

FIGURE 23.20
Once they became distinct as different species, *M. colligatum*, new species #7, and new species #9 sometimes exhibited more gradual change in overall morphological difference than did most of the other species.

FIGURE 23.23
The structural difference in the flowers of *D. decorum* and *D. nudicaule* might help maintain reproductive isolation between the two species by restricting bee and hummingbird pollinators to one species or the other.

IMR

The prolonged expression of *Hoxd* genes in the developing limbs of tetrapods is an example of heterochrony, a change in the timing of developmental events.

UQ

For genes that are very pleiotropic—that is, have many diverse functions—regulatory evolution may be more likely. Shifts in gene expression domains can produce rapid morphological change while leaving other functions intact. However, we can never rule out coding changes, especially if gene duplication has reduced pleiotropy.

Test Your Knowledge

1. a 2. e 3. b 4. d 5. c 6. c 7. a 8. d 9. e 10. b

Interpret the Data

The data illustrate allometric growth in this evolutionary lineage. Brain size increased more than 200% over three million years, but body size increased only about 50%.

Chapter 24

Study Break

STUDY BREAK 24.1

1. The system of binomial nomenclature avoids ambiguity in the naming of species because it assigns a unique two-part name to each species.
2. The taxonomic hierarchy helps biologists organize information about different species because it categorizes them into increasingly inclusive groups. Species that are included in a lower taxonomic category share many characteristics, whereas those included only in the same higher taxonomic category share fewer characteristics.

STUDY BREAK 24.2

1. A phylogenetic tree is a formal hypothesis about the evolutionary relationships among species. A classification is an arrangement of organisms into hierarchical groups that reflect their relatedness.
2. A monophyletic taxon contains an ancestor and all of its descendants. A polyphyletic taxon includes species from different clades. A paraphyletic taxon includes an ancestor and some, but not all, of its descendants.

STUDY BREAK 24.3

1. Systematists use homologous characters in their analyses because similarities in homologous characters indicate genetic relatedness and shared ancestry.
2. Morphological traits are often useful for tracing the evolutionary relationships within a group of organisms because they can be observed and measured in fossils as well as in living organisms.
3. Molecular characters provide several advantages in systematic analyses: (1) they provide abundant data; (2) molecular sequences can be compared between distantly related organisms that share no organismal characteristics or between closely related species with only minor morphological differences; and (3) proteins and nucleic acids are not directly affected by developmental or environmental factors that cause nongenetic morphological variation.

STUDY BREAK 24.4

Traditional systematics emphasizes the divergence of groups as well as branching evolution. Groups that have evolved very different morphology are sometimes placed in a taxon that is different from the one to which the rest of their clade is assigned.

STUDY BREAK 24.5

1. Outgroup comparison is a technique that compares the group under study—the ingroup—to more distantly related organisms—the outgroup—to identify the ancestral and derived versions of characters. Character states observed in the outgroup are considered ancestral.
2. In a cladistic analysis, organisms with synapomorphies (shared derived characters) are grouped together within a clade.
3. The principle of parsimony suggests that the cladogram or phylogenetic tree with the fewest number of hypothesized evolutionary changes is the best current hypothesis about the phylogenetic history of a group.

STUDY BREAK 24.6

1. An assumption that underlies the use of genetic sequence differences as molecular clocks is that mutations arise at a relatively constant rate.
2. Birds are more closely related to non-avian dinosaurs than they are to crocodilians.

STUDY BREAK 24.7

1. Different strains of HIV were acquired from chimpanzees and from sooty mangabey monkeys.
2. Phylogenetic analyses of prokaryotes based on morphological data were not very successful because prokaryotes do not have many morphological features. Analyses based on molecular sequence data have been more successful because researchers can identify many molecular differences among the lineages of prokaryotes.

Think Like a Scientist

FIGURE 24.12

No, the data in the table do not contain any information that would allow the development of an hypothesis about the evolutionary relationship of cycads and ginkgophytes. For the five characters listed, these groups exhibit identical character states. The phylogenetic trees illustrated in the figure are actually based on data about many additional characters in all of the plant groups listed. The analysis in the figure is simplified to illustrate the parsimony method for evaluating phylogenetic trees.

FIGURE 24.14

The *Yp1* sequence in *Drosophila* appears to have accumulated mutations slightly faster than the cytochrome *b* sequence in honeycreepers, as revealed by the average numbers of units of genetic distance that evolved per million years.

IMR

The ancestors of whales must have undergone many morphological changes as they made the transition from land to the sea. Three obvious changes include modifications of their limbs into flippers; modification of their overall shape to facilitate more efficient swimming; and modifications of their sensory systems to function under water instead of in the open air.

UQ

The Linnaean system of classification cannot easily incorporate the phenomenon of horizontal gene transfer because it pigeonholes species into a nested hierarchy of rigidly defined taxonomic categories. Although a phylogenetic view of biodiversity also depends on the vertical transmission of derived traits from one generation to the next, horizontal gene transfer can at least be illustrated on phylogenetic trees.

Test Your Knowledge

1. c 2. b 3. a 4. b 5. e 6. a 7. a 8. e 9. d 10. c

Interpret the Data

The European wildcat is the domestic cat's closest relative. The clade that includes jaguars, lions, and leopards is the sister clade to tigers. Bobcats are more closely related to cougars than to ocelots because they share a more recent common ancestor with cougars.

Chapter 25

Study Break

STUDY BREAK 25.1

1. Only in a reducing atmosphere can amino acids be produced from simpler chemicals and energy. Without amino acids there can be no life. Thus, any theory for the origin of life must consider the necessity of reducing conditions.
2. Current thinking is that early Earth's atmosphere *per se* was not reducing; scientists have looked for localized regions where conditions were of a reducing nature, such as ocean floor hydrothermal vents. Another hypothesis is that some amino acids had an extraterrestrial origin.

STUDY BREAK 25.2

Lipid bilayers in the form of membranes are present in all present-day cellular organisms. Therefore, it is more intuitive to conceive of a path from a simple lipid bilayer system to a membrane-bound cell than it is to conceive of a transition of a clay-based system to a cellular system.

STUDY BREAK 25.3

The basic tenet of the theory of endosymbiont origins for mitochondria and chloroplasts is that organelles such as mitochondria and chloroplasts originated from symbiotic relationships between two prokaryotic organisms. An anaerobic prokaryotic is proposed to have ingested an aerobic prokaryote, which persisted in the cytoplasm, continuing to respire aerobically. A gradual process of mutual adaptation transformed the cytoplasmic aerobes into mitochondria.

The same basic mechanism is believed to have led to the appearance of membrane-bound plastids (including chloroplasts) at a later time. In this case, nonphotosynthetic aerobic cells with mitochondria are proposed to have ingested photosynthetic prokaryotes that were perhaps similar to present-day cyanobacteria. Again, through mutual adaptation, the photosynthetic prokaryotes changed into plastids.

Think Like a Scientist

FIGURE 25.2

The experiment used boiling water to create water vapor. Bacteria would have been killed by the boiling water.

IMR

The RNA world model states that the first genes and enzymes were RNA molecules—that is, ribozymes. Ribozymes, which are RNA molecules capable of catalyzing biochemical reactions, may have functioned both as informational molecules and catalysts in protocells, without requiring protein enzymes. Thus, the observation that a ribozyme could become activated within the protocell-like vesicles supports the RNA world model and the possibility that living cells may have developed from protocells.

UQ

Two possible related explanations might be summarized as "division of labor" and "efficiency." In an "RNA world," RNA would have a dual function, both as the transcriptional unit (RNA copy of DNA) and the molecule that carries out the activity of catalyzing biochemical reactions. With a division of labor between proteins, the work unit, and RNA, the informational unit, RNA becomes more specialized and efficient and, hence, the whole process becomes more efficient. Proteins then evolve specialization to catalyze biochemical reactions.

Test Your Knowledge

1. d 2. e 3. d 4. a 5. b 6. c 7. e 8. e 9. d 10. e

Interpret the Data

1. Asteroid impacts declined before levels of atmospheric oxygen began to rise.
2. Oxygen was much less abundant while carbon dioxide was much more abundant in the atmosphere when the first cells arose compared to levels today.
3. Oxygen (21%) is now more abundant in the atmosphere than carbon dioxide (currently 0.038%).

Chapter 26

Study Break

STUDY BREAK 26.1

1. Prokaryotes have no major cytoplasmic organelles that are equivalent to the endoplasmic reticulum, the Golgi complex, mitochondria, or chloroplasts of eukaryotes, nor does it have lysosomes.

   The genetic material of a prokaryote generally is a single, circular DNA molecule localized in a nonmembrane-bound central region of the cell called a nucleoid. By contrast, the genetic material of a eukaryote is distributed among a number of linear chromosomes, which consist of DNA complexed with basic proteins known as histones.

Most prokaryotes are surrounded by a cell wall located outside the plasma membrane. Animal cells do not have a cell wall, but plants, fungi, and some other eukaryotes do. The compositions of the eukaryotic cell walls are chemically different from bacterial cell walls.

2. A chemoheterotroph oxidizes organic molecules as its energy source and obtains carbon in organic form. A photoautotroph uses light as its energy source and carbon dioxide as its carbon source.
3. Obligate anaerobes are poisoned by oxygen. They survive either by fermentation, or by a form of respiration in which inorganic molecules are used as final electron acceptors. Facultative anaerobes use oxygen when it is present but live by fermentation when conditions are anaerobic.
4. Nitrogen fixation is the reduction of atmospheric nitrogen (N_2) to ammonia (NH_3). Nitrification is the conversion of ammonium (NH_4^+) to nitrate (NO_3^-).

   Nitrogen fixation, an exclusively prokaryotic process, is the only means of replenishing the nitrogen sources used by most microorganisms, and by all animals and plants.
5. A biofilm is a complex aggregation of microorganisms (either one or multiple microorganisms, many or all of them prokaryotes) attached, in most cases, to a surface. Biofilms are used in a variety of beneficial applications, including bioremediation of toxic organic chemicals contaminating groundwater. On the other hand, biofilms can have adverse effects on human health. For instance, biofilms can result in antibiotic-resistant infections if they adhere to surgical materials, such as catheters and implants.

STUDY BREAK 26.2

1. An exotoxin is a toxic protein that is secreted from the bacterium that makes it, or is released when that bacterium lyses. An endotoxin is a normal lipopolysaccharide component of the outer membrane of Gram-negative bacteria; it is released when the bacteria die and lyse. An exoenzyme is an enzymatic protein that is released from cells.

   An exotoxin interferes with biochemical processes of body cells. An endotoxin overstimulates the immune system, often causing inflammation. Depending on the bacterium, the endotoxin release has different effects, which may include organ failure and death. An exoenzyme digests plasma membranes, causing cells of the infected host to rupture and die. Exoenzymes may also digest extracellular materials and red and white blood cells.
2. Many pathogenic bacteria develop resistance to antibiotics as a result of mutations. The mutations allow the bacteria to break down the antibiotics or otherwise counteract their effects. Pathogenic bacteria may also develop resistance to antibiotics as a result of horizontal gene transfer when antibiotic resistance genes on plasmids are transferred between bacteria by conjugation, or when DNA containing an antibiotic resistance gene is introduced into the pathogen in other ways, such as by transformation or transduction.

STUDY BREAK 26.3

1. See Table 25.1 for a comparison of the properties of organisms in each of the three domains. The classification of Archaea as a distinct domain was based on comparisons of DNA and rRNA sequences.
2. A methanogen lives in reducing environments, generating energy by converting substrates such as carbon dioxide, hydrogen gas, methanol or acetate into methane gas. All known methanogen belong to the Euryarchaeota.
3. Extreme halophilic archaeans live in high-salt environments, requiring at least 1.5 M NaCl in order to live. Most of these organisms are aerobic chemoheterotrophs, obtaining energy from sugars, alcohols, and amino acids using pathways similar to those of bacteria. All known extreme halophilic archaeans belong to the Euryarchaeota.
4. Extreme thermophiles live in extremely hot environments such as thermal hot springs and hydrothermal ocean floor vents. Psychrophiles grow optimally at temperatures in the range −10°C to −20°C, such as in the Antarctic and Arctic oceans.

Think Like a Scientist

FIGURE 26.13

A likely scenario is that the individuals harbored populations of the bacterium, which, at some point, changed from being harmless to being infectious, resulting in ulcers. Indeed, there is evidence that young children may acquire *Helicobacter* causing a chronic infection which confers on them a lifelong susceptibility to developing ulcers.

IMR

Regulation of gene expression in bacteria was discussed in Section 16.1. In some way the bacteria must "sense" their density and then trigger the transition from planktonic to biofilm growth. Changes in gene expression must involve molecular signals, probably involving operons (see Section 16.1). Thus, down-regulation of a gene or of genes in an operon would require repression of those genes or the removal of a gene activator, whereas up-regulation would require removal of repression or the production of a gene activator.

UQ

One possibility is that the last universal common ancestor (LUCA) of all life on the planet had a genome composed not of DNA but of RNA. If the switch to DNA-based genomes occurred after LUCA, once in the lineage that led to Bacteria and once entirely independently in the lineage that led to Archaea and Eukarya, then the two machineries for replicating DNA need not be related to one another. A second possibility is that LUCA had a DNA-based genome but that the cellular replication machinery was displaced by that of a virus in either the bacterial or archaeal/eukaryal lineages.

Test Your Knowledge

1. d 2. a 3. c 4. b 5. e 6. a 7. d 8. b 9. c 10. d

Interpret the Data

1. Freezing occurs at 0°C, and *Pseudomonas* will not grow at or near freezing temperatures.
2. Yes, the growth rate at 20°C (293 K) is 0.44 and at 4°C (277 K) it is 0.012. The growth rate is more than 36 times slower at refrigeration than room temperature, indicating that refrigeration is sufficient.

Chapter 27

Study Break

STUDY BREAK 27.1

A protist is distinguished from a prokaryote by having typical eukaryotic cell features like a nuclear envelope surrounding its genetic material, and cell organelles such as mitochondria (in most protists), chloroplasts (in some protists), endoplasmic reticulum, Golgi complex, and so on.

Distinguishing protists from fungi, animals, and plants is more blurry. Fungi are nonmotile at all stages of their life cycles, whereas most protists are motile or have motile stages in their life cycles. Cell wall structure is also different from fungi, and from plants. Protists differ from both animals and plants by lacking highly differentiated structures and by not having complex developmental stages. Collagen, the extracellular support protein of animals, is absent in protists.

STUDY BREAK 27.2

1. Excavates' nuclear genomes contain genes that are of mitochondrial origin, arguing that they once had mitochondria.
2. The chloroplast will have two membranes: one derived from the plasma membrane of the engulfing eukaryote and the other from the plasma membrane of the cyanobacterium.

Think Like a Scientist

FIGURE 27.11

Sequence more diatom genomes so as to obtain more data about similarities and differences.

IMR

Binding of a signal molecule—in this case cAMP—to its G-protein–coupled receptor activates the receptor. The activated receptor activates the G protein that was associated with the cytoplasmic end of the receptor. The activated G protein then activates the effector molecule for the pathway, which, in this case, is adenylyl cyclase. The cAMP activates protein kinases in the signal transduction pathway specific for the response, including the kinase discussed in the box.

UQ

1. Evolution is a dynamic field of study utilizing many tools, from fossils to gene sequence data to complicated mathematical models. New ideas about the first eukaryotes and the significance of endosymbioses will continue to emerge by using the scientific method.
2. Anything is possible, but no current evidence suggests prokaryotes evolved from a single-celled eukaryote. Fossil evidence and evidence for the endosymbiotic theory support a prokaryote-first world. Unlike *Giardia* that has mitochondrial genes, for example, prokaryotes do not have "eukaryotic" genes that appear to have no (or lost) function.

Test Your Knowledge

1. e 2. d 3. d 4. b 5. b 6. c 7. b 8. a 9. b 10. c

Interpret the Data

1. The 50 μmol/m^2/s irradiance treatment has a maximum at ~450 μmol/m^2/s. The 1,200 μmol/m^2/s irradiance treatment has a maximum above 1,200 μmol/m^2/s. No, the maximum was different for the low irradiance treatment, because 50 μmol/m^2/s was too low a level of light to maximize photosynthesis, so the diatoms increased their rates of photosynthesis in increased light levels. The 1,200 μmol/m^2/s irradiance treatment reached its photosynthesis maximum at the treatment light level, indicating that this was an optimal light level.
2. The low irradiance treatment has a higher photosynthesis efficiency with light.

Chapter 28

Study Break

STUDY BREAK 28.1

1. Evolution of a root system gave land plants access to minerals and water in soil and provided physical support for aerial parts. The evolving shoot system of land plants, including lignified tissues in stems, allowed vascular plants to grow taller and stay erect, thereby gaining better access to sunlight for photosynthesis. Reproductive structures borne on aerial stems (such as flowers) might serve as platforms for more efficient dispersal of spores from the parent plant. Vascular tissues were innovations for distributing water (xylem) and sugars (phloem) through the plant body.
2. Homosporous plants produce a single type of sexual spore and are in effect bisexual, with each gametophyte capable of producing both sperm and eggs. Heterosporous species, including angiosperms and gymnosperms, produce two types of spores, which develop into sexually different gametophytes that produce either sperm or eggs. Plant scientists associate the evolution of heterospory with several key reproductive innovations in land plant evolution, including the protection of male gametes inside pollen grains and the protection of plant embryos inside seeds.

STUDY BREAK 28.2

1. Like aquatic plants, bryophytes produce flagellated sperm that must swim through water to reach eggs, and they lack a complex vascular system (although some have a primitive type of conducting tissue). Bryophytes do have parts that are rootlike, stemlike, and leaflike, although the

"roots" are rhizoids, and bryophyte "stems" and "leaves" did not evolve from the same structures that vascular plant stems and leaves did. Sporophytes of some species have a water-conserving cuticle and stomata. Like most plants, bryophytes also have both sexual and asexual reproductive modes.
2. In general, mosses are the bryophytes that most closely resemble vascular plants. Some species produce structurally complex gametophytes that have a central strand of primitive water-conducting tissue that resembles the xylem of vascular plants, and in a few species the water-conducting cells are surrounded by sugar-conducting tissue resembling the phloem of vascular plants.

STUDY BREAK 28.3

1. In bryophytes, the gametophyte is much larger than the sporophyte and obtains its nutrition independently. The comparatively tiny sporophyte remains attached to the gametophyte and depends on the gametophyte for much of its nutrition. In modern lycophytes (club mosses and their close relatives), the gametophyte is free-living—though it is nourished by mycorrhizae instead of carrying out photosynthesis—and it is smaller than the sporophyte, which is a photosynthetic autotroph.
2. Fern leaves often take the form of feathery fronds, and roots extend from underground stems called rhizomes. Whisk ferns lack true leaves and roots; instead, small leaf-like scales dot an upright, green, branching stem, which arises from a horizontal rhizome system anchored by rhizoids. Horsetail sporophytes typically have underground rhizomes and roots that anchor the rhizome to the soil. The scalelike leaves are arranged in whorls about a photosynthetic stem.
3. In horsetails, the sporangia that produce spores are borne in strobili, and spores are carried away from the plant by air currents. In ferns, sporangia are produced on the lower surface or margin of leaves, and spores are forcefully dispersed from the parent plant when contraction of a belt-like annulus rips open the sporangium and ejects the spores.

Think Like a Scientist

IMR

Perhaps, because extant lycophytes and known lycophyte fossils all have relatively rudimentary xylem, the water-transporting vascular tissue. Among other impacts, the discovery of an ancient lycophyte species having well-developed xylem would open up numerous questions about genetic variations and evolutionary shifts within the lycophyte group, including the possibility that vascular tissues began to arise earlier than previously thought.

UQ

If the number and density of genes remain largely unchanged, increases in chromosome number should generally increase independent assortment because genes will be distributed on more independent linkage groups. The result is a higher effective recombination rate. Such changes in chromosome number may happen because of fission of existing chromosomes into more than one chromosome. In most eukaryotes, one obstacle that chromosomes produced by fission must overcome is the development of a centromere in which spindle fibers link sister chromatids. Chromosomes produced by duplication do not generally need a new centromere, but must also overcome being the odd chromosome in the population. The likelihood that a chromosome will be duplicated is at least partly based on the dosage sensitivity of the genes located on it. For example, the human genetic disorder Down syndrome is caused by the presence of an extra copy of chromosome 21 (trisomy 21). Thus, the consequences of gene duplication to phenotype and fitness play a significant role in shaping which regions of the genome may duplicate or not—in plants, animals, and all other eukaryotes.

Test Your Knowledge

1. b 2. d 3. e 4. a 5. b 6. b 7. a 8. e 9. b 10. a, c, e

Interpret the Data

1. The data summarized in Figure A support the hypothesis that a single endosymbiosis event correlated with the evolution of all photosynthesizing organisms. This event corresponds with the node at which the clade including green plants, red algae, and glaucophytes diverged from the clade including alveolates and heterokonts (diatoms and their relatives).
2. Yes. The cladogram shows less evolutionary distance between green plants and glaucophytes than between green plants and red algae.

Chapter 29

Study Break

STUDY BREAK 29.1

1. The four major reproductive adaptations that evolved in gymnosperms include pollen and the ovule, both of which shelter spores; pollination rather than dispersal of swimming sperm; and the seed as a "package" that protects and often nourishes the embryo.
2. The three basic parts of a seed are the embryo sporophyte, endosperm, and outer seed coat. The endosperm nourishes the embryo sporophyte, and the seed coat protects it.
3. Features that make conifer sporophytes structurally more complex than other gymnosperms include anatomically complex needlelike or scalelike leaves that are adapted to aridity and the production of resins as metabolic by-products.

STUDY BREAK 29.2

1. The lack of fossil early angiosperms, including obvious transitional forms, has made it difficult to trace a clear evolutionary path for flowering plants. As a result plant scientists have proposed several, often conflicting, classification schemes for angiosperms.
2. Seeds leaves (cotyledons) and pollen morphology are two major features used to distinguish monocots and eudicots. Monocots have a single seed leaf and pollen grain with a single groove. Eudicots have two seed leaves and pollen grains with three grooves. Monocots (such as grasses, lilies, and palms) also generally have fibrous root systems, leaves with parallel veins, flower parts in multiples of three, and scattered vascular bundles in stems. Eudicots (most flowering trees and shrubs, roses, sunflowers, beans) usually have netlike leaf venation, a primary taproot, flower parts in fours or fives, and vascular tissues arranged in a ring.
3. Adaptations that have contributed to the evolutionary success of angiosperms include vascular tissue modifications that make transport of water and nutrients more efficient; double fertilization, which results in enhanced nutrition (endosperm) for embryos; physical protection of embryos within ovaries and seeds; and coevolution with animal pollinators, which increases the likelihood that pollination will occur.

STUDY BREAK 29.3

1. All land plants carry the *LFY* gene, which encodes the regulatory protein LEAFY. Researchers studying the protein's effects in groups as different as mosses and angiosperms have found that while the phenotypic effects—and resulting adaptations—vary markedly in different plant groups, only a few changes in the base sequence of the *LFY* gene have been associated with the shifts.
2. Among other benefits, having the entire sequences of a variety of plant genomes allows researchers to use the resulting to elucidate the relationship between large-scale shifts in genomes and the evolution of fundamental characteristics of major lineages.

Think Like a Scientist

FIGURE 29.14

In fact, several studies have compared angiosperm photoreceptor genes with those of mosses and ferns. Among other outcomes, such comparative studies can help shed light on whether and how phytochromes came to have new or refined functions during the course of plant evolution.

IMR

The study authors proposed that other factors worthy of study include the roles of regulatory elements acting in a genome, of microRNAs, posttranslational modification of gene products, and epigenetic factors (changes in gene expression unrelated to a change in the DNA sequence of a gene or genome).

UQ

One explanation hinges on the theory that all apical meristems (except root meristems), whether vegetative or reproductive, produce lateral organs as an innate function. Vegetative meristems produce leaves that arise laterally on the stem, and flower meristems produce sepals that arise laterally on the floral axis. According to this theory, once a flower meristem is formed, it will automatically produce sepals with no new gene functions needed. Then, if B-gene function is added to the baseline gene activity in the meristem (which includes E function), petals will be produced—so instead of requiring A+B function to produce petals, only B function is required. Stamen and carpel formation is exactly the same as in the ABC model: B+C= stamens, and C = carpels. Thus, only two gene functions, B and C, in addition to E, are required for the formation of all four floral organs types.

Test Your Knowledge

1. d, e 2. b 3. a 4. d 5. a, c, d, e 6. a, b, d 7. e 8. a 9. b 10. b, d

Interpret the Data

1. Wild AN (annual) genotypes, which grow in inland habitats that experience summer drought, are adapted to flower earlier, when moisture is still available. The later flowering times of PE (perennial) plants reflect such plants' adaptation to coastal habitats having year-round moisture.
2. Populations of *M. guttatus* plants that differ with respect to a chromosomal inversion show divergent life history adaptations to contrasting environments. Populations adapted to survive in inland habitats affected by summer drought show a pattern of annual (AN) reproduction, whereas populations adapted to moister coastal habitats show a pattern of perennial (PE) reproduction. Interpreting the results graphed in Figure C, both AN and PE plants grown in an AN setting flourished in the moist early spring, but PE plants died back several months sooner than would be expected if they had been in their wild, coastal habitat.

Chapter 30

Study Break

STUDY BREAK 30.1

1. Some fungi are multicellular while others, the yeasts, are single cells. (Some species alternate between these two forms at different life cycle stages.) The cells of all fungi are surrounded by a hardened wall; in most cases the hardener is the polysaccharide chitin. The body of a multicellular fungus consists of a dense mesh of filaments called hyphae, which in some groups are separated into cell-like compartments by cross walls (septa). Aggregations of hyphae are the structural foundation for all other parts that develop as part of a multicellular fungus. For example, in some species modified hyphae form rhizoids that anchor the fungus to its substrate.
2. Fungal spores are microscopic, usually nonmotile reproductive cells in which haploid nuclei are surrounded by a tough outer wall. They are produced sexually or asexually. Sexual spores are produced by genetically different parent fungi and may unite in a sexual process that gives rise to a diploid life stage. Asexual spores are genetically identical to the parent fungus and may give rise to a new, haploid individual.
3. Many fungal species have a life cycle stage called a dikaryon, which contains two haploid nuclei (a condition expressed as $n + n$). A dikaryon forms as the result of

plasmogamy, a sexual stage in which the cytoplasms of two genetically different partners fuse. This fusion ensures genetic diversity in new individuals. At some point after a dikaryon forms, the nuclei fuse (karyogamy) to form a short-lived zygote. Meiosis in the zygote produces haploid nuclei that become packaged into sexual spores.

STUDY BREAK 30.2

1. The main phyla of fungi are the Chytridiomycota, Zygomycota, Glomeromycota, Ascomycota, and Basidiomycota. Chytrids are the only fungi that produce motile, flagellated spores. Zygomycetes often reproduce asexually, but sometimes reproduce sexually by way of hyphae that occur in + and − mating strains; haploid nuclei in the hyphae function as gametes. Following plasmogamy, further development produces zygospores in which karyogamy gives rise to diploid zygotes (2*n* nuclei), which then undergo meiosis as sexual spores form. Glomeromycetes reproduce asexually, by way of spores that form at the tips of hyphae. In ascomycetes, chains of asexual spores called conidia, each containing a haploid nucleus, develop during asexual reproduction. Ascomycetes produce haploid sexual spores in pouchlike cells called asci. Most basidiomycetes reproduce only sexually: club-shaped basidia develop on a basidiocarp (for example, a "mushroom") and bear sexual spores on their outer surface. When dispersed, the spores may germinate and give rise to a haploid mycelium. Cytoplasmic fusion may occur between hyphae of two compatible mating strains, producing a dikaryotic mycelium from which basidiocarps may grow. Microsporidia, single-celled parasites that may be related to zygomycetes, resemble spores but lack mitochondria. Fungi for which no sexual life stage has been identified are placed in a convenience grouping called "conidial fungi."
2. Anatomically, the simplest fungi are chytrids, which are microscopic, and zygomycetes, which have aseptate hyphae. Ascomycetes, basidiomycetes, and glomeromycetes all form septate (walled) hyphae.
3. Most chytrids are aquatic, but some are parasites of insects, plants, and some animals. Others are symbiotic partners in the gut of cattle and some other herbivores.

   Many zygomycetes are saprobes in soil, feeding on plant detritus. Their metabolic activities release mineral nutrients that plant roots can take up. Some zygomycetes are parasites of insects or spoil stored grains, bread, fruits, and vegetables such as sweet potatoes. Others are used in manufacturing products such as industrial pigments and pharmaceuticals. Many more are highly destructive plant pathogens and several can be human pathogens, causing athlete's foot, ringworm infections, and more serious illnesses. The pink bread mold *Neurospora crassa* has been crucial in genetic research.

   Some basidiomycetes participate in vital mutualistic associations (mycorrhizae) with the roots of forest trees. Others produce prized edible mushrooms.

   Rusts and smuts are parasites that cause serious diseases in wheat, rice, and other plants.

   Glomeromycetes are all specialized to form mycorrhizae with plant roots.

STUDY BREAK 30.3

1. A lichen is a communal life form representing a symbiosis between a photosynthetic green alga or species of cyanobacteria (the photobiont) and a nonphotosynthetic fungus (the mycobiont). The algal cells supply the lichen's carbohydrates, most of which are absorbed by the fungus. In some cases, the alga is protected from desiccation or some other environmental threat.
2. A mycorrhiza is a symbiotic association between a fungus and plant roots. The fungal hyphae make mineral ions and sometimes water available to the plant's roots, and in exchange the fungus absorbs carbohydrates, amino acids, and possibly other growth-enhancing substances provided by the plant. Mycorrhizae greatly enhance the plant's ability to extract various nutrients, especially phosphorus and nitrogen, from soil, and they are crucial to the survival of many plant species.
3. In endomycorrhizae the hyphae of a fungus (typically a glomeromycete) enter plant roots, where their tips branch into clusters called arbuscules where exchanges of nutrients take place. In ectomycorrhizae the fungal partners are basidiomycetes. Their hyphae surround plant roots but do not penetrate them.

Think Like a Scientist

IMR

As the text describes, chytrids may have arisen very early in fungal evolution. Accordingly, a comparison of 18S rRNA sequences among different living chytrids might shed valuable light on the early origins of fungi—and perhaps on the assumed divergence of fungi and animal lineages from a common protist ancestor.

UQ

1. Ecological theory predicts that species with complete overlap in the way they use resources and interact with other species would not coexist in a community. Therefore, it is likely that ectomycorrhizal fungi that coexist on a single tree differ in factors such as how they gain nutrients from the tree, how they gain nutrients from the surrounding soil, how they interact with the tree (for example, what parts of the root system they colonize, what types and amounts of nutrients they provide, or what additional benefits they confer to the tree), where they occur in space (for example, at what soil depth they occur), or how they respond to different environmental conditions (such as rainfall and temperature).
2. Because mycorrhizal symbiosis may be controlled by the plant symbiont, the fungal symbiont, or both, the dominance of fungal symbionts within a community may result from interactions between different fungal symbionts (for example, direct competition between mycelia or indirect competition for nutrients) or from differences in the way they interact with the tree (such as the "selection" by the tree of one symbiont over another).
3. Understanding the nature of interactions between plants, fungi, and the environment can be beneficial for protecting rare fungal or plant species, maintaining forest diversity, predicting the effects of climate or land use change, and improving the outcome of restoration efforts.

Test Your Knowledge

1. a, c 2. b, e 3. a 4. d 5. c, d 6. e 7. c 8. a 9. b, e 10. d

Interpret the Data

1. In the dark, no. The number of fusion events in wild-type pairs showed the same pattern as matings between wild-type and mutant hyphae.
2. In white and blue light, the most fusion events occurred in crosses in which one or both hyphae were mutants, regardless of mating type. The fewest fusion events under these light conditions occurred in crosses in which one or both hyphae were the **a** mating type. In green and red light, the pattern shifted. The most fusion events occurred when the cross included one mutant and one wild-type hypha. The researchers hypothesized that, in general, light inhibits fusion of wild-type (+) strains of *C. neoformans*. These results support that hypothesis. Red light wavelengths predominate in shade, so it is not surprising that overall the most fusion events occurred in darkness and in red light conditions.

Chapter 31

Study Break

STUDY BREAK 31.1

1. Several characteristics distinguish animals from plants: plant cells have cell walls, but animal cells do not; almost all plants are autotrophic, whereas all animals are heterotrophic; and plants are usually sessile, but animals are motile at some stage of their life cycle. Animals differ from fungi in that the cells of fungi have cell walls, and most fungi are sessile.
2. The ability of animals to move through the environment allows them to search for and pursue the food items that supply them with nutrients and energy.

STUDY BREAK 31.2

1. A tissue is a group of cells that share a common structure and function. The three primary tissue layers that contribute to the bodies of most animals are endoderm, mesoderm, and ectoderm.
2. Humans are bilaterally symmetrical.
3. The coelom is a space within which internal organs can move independently of the body wall muscles. The fluid within it provides protection for internal organs. In some animals the coelom functions as a hydrostatic skeleton.
4. Having a segmented body may allow an animal to survive damage to some parts of its body segments and may allow improved control over body movements.

STUDY BREAK 31.3

1. Molecular sequence studies have confirmed the distinctions between the Parazoa and the Eumetazoa, between the Radiata and the Bilateria, and between the Protostomia and the Deuterostomia.
2. A schizocoelom appears to be the ancestral body cavity among the protostomes.

STUDY BREAK 31.4

1. Sponges do not exhibit any kind of body symmetry.
2. A sponge gathers food from its environment by drawing water into its body through numerous small pores and harvesting particulate matter from the water with its choanocytes, or collar cells.

STUDY BREAK 31.5

1. Cnidarians capture animal prey by stinging it with their nematocysts and using their tentacles to pull it into their mouths.
2. The anthozoans, including sea anemones and corals, have only a polyp stage in their life cycle.
3. Ctenophores capture microscopic plankton in sticky filaments on their two tentacles, which are then drawn across the mouth.

STUDY BREAK 31.6

1. Free-living flatworms have digestive, excretory, nervous, and reproductive systems. Tapeworms lack a digestive system.
2. Ectoprocts, brachiopods, and phoronid worms all have a circular or U-shaped feeding structure called a lophophore, a characteristic that reveals their close evolutionary relationship.
3. The anatomical and physiological systems that allow squids and other cephalopods to be more active than other types of mollusks include a closed circulatory system with accessory hearts and a complex nervous system with giant nerve fibers. Many cephalopods use their excurrent siphon to expel jets of water, allowing them to move rapidly through the environment.
4. The organ systems that exhibit segmentation in most annelid worms include: respiratory surfaces; parts of the nervous, circulatory, and excretory systems; and the body wall and coelom.

STUDY BREAK 31.7

1. The cuticle protects a nematode from the digestive enzymes of its host.
2. Although the rigid exoskeletons of arthropods do not expand, these animals grow a new, soft exoskeleton inside the existing one. After shedding the old exoskeleton, they grow to a larger size by expanding the new exoskeleton with either water or air before it hardens.
3. The body regions of the four living subphyla of arthropods differ in how they have become fused. Chelicerates have a fused cephalothorax and an abdomen. Crustaceans show variable patterns, but many have a fused cephalothorax and an abdomen. Myriapods have a head and a trunk. Hexapods have a separate head, thorax, and abdomen.
4. Insects with incomplete metamorphosis hatch from their eggs as wingless nymphs, which vary in how closely they resemble adults; nymphs then undergo metamorphosis

into the adult form. Insects with complete metamorphosis hatch from eggs as larvae, which are always very different from adults. After becoming a pupa, their cells and tissues are reorganized into the adult form.

Think Like a Scientist

IMR

The location of the two genes encoding leucine-carrying tRNAs in trilobites is expected to match that found in chelicerates. That location, which is shared with both mollusks and annelids, appears to be the ancestral state for this character. Because the trilobite clade originated before the other arthropod clades, we would expect it to exhibit the ancestral state for this character.

UQ

1. Recent genetic comparisons of invertebrate taxa confirm the distinction between Parazoa and Eumetazoa and the distinction between Radiata and Bilateria.
2. The nucleotide sequences most likely to shed light on the diversity of living invertebrates are those in small subunit ribosomal RNA, mitochondrial DNA, and *Hox* genes of species within the protostome and deuterostome lineages.

Test Your Knowledge

1. b 2. d 3. c 4. a 5. c 6. e 7. e 8. a 9. d 10. b

Interpret the Data

Leech therapy worked better than drug therapy to relieve the pain of arthritis in the thumbs of the patients tested. The effect of leech therapy persisted for at least two months after just one treatment.

Chapter 32

Study Break

STUDY BREAK 32.1

1. Echinoderms have a water vascular system, which operates the tube feet that are used for locomotion and/or feeding.
2. Water enters the pharynx of a hemichordate through its mouth, and it exits the pharynx through the branchial slits. The animal extracts oxygen and particulate food from the water as it passes through the pharynx.

STUDY BREAK 32.2

1. The animal is not a chordate, because chordates have a dorsal nerve cord.
2. Vertebrates have an internal bony skeleton, including a cranium and vertebral column in most groups, as well as structures derived from neural crest cells.

STUDY BREAK 32.3

1. Vertebrates have multiple *Hox* gene complexes, which provide them with several copies of each *Hox* gene. Cephalochordates have just one *Hox* gene complex.
2. Of the three groups listed, Gnathostomata has the most species, and Amniota has the fewest.

STUDY BREAK 32.4

1. Hagfishes lack bone, paired fins, and scales in their skin. Hagfishes have neither a cranium nor a vertebral column, and lampreys have only rudimentary traces of vertebrae. These observations suggest that their lineages arose before these structures appeared in vertebrates.
2. The derived traits possessed by conodonts and ostracoderms include structures made of bone or a bonelike material and, in some ostracoderms, a brain divided into three regions.

STUDY BREAK 32.5

1. Sharks are more efficient predators than acanthodians or placoderms were because they have well-developed sensory systems to detect prey; their lightweight skeletons and absence of heavy body armor allow them to pursue prey rapidly; and they have numerous teeth that are replaced when damaged or worn, as well as a loosely attached upper jaw that permits them to suck in large chunks of food.
2. The air bladders of ray-finned bony fishes increase their locomotor abilities by allowing them to rise or sink easily in the water. Their fin rays allow them to engage in precise movements during locomotion.
3. The lungs of lungfish allow them to survive in environments with low oxygen content because they can acquire oxygen from the air.

STUDY BREAK 32.6

1. For the first tetrapods, the advantages of moving onto land included abundant food resources, the rarity of predators, and readily available oxygen. The disadvantages included the need for more skeletal support against gravity, mechanisms to prevent dehydration in air, and modifications of sensory systems so that they would function in air.
2. The parts of the amphibian life cycle that are most dependent on water are the egg and larval stages.

STUDY BREAK 32.7

1. The amniote egg freed amniotes from a dependence on standing water because it can survive on land. The shells of amniote eggs mediate gas exchange and water exchange with the environment.
2. Among the three amniote lineages, the Synapsida includes the mammals, the Anapsida includes turtles, and the Diapsida includes lizards, snakes, crocodilians, and birds.
3. Because lizards are lepidosaurs and both birds and crocodilians are archosaurs, crocodilians are more closely related to birds than they are to lizards.

STUDY BREAK 32.8

The overall structure of turtles differs from other amniotes in that their bodies are enclosed within a bony, keratin-covered shell.

STUDY BREAK 32.9

1. Besides their loss of legs, snakes differ from their lizard ancestors in having smaller skull bones, and the connections between them are more elastic.
2. Various species of lizards feed on vegetation, insects, or larger animals. Virtually all snakes are carnivores, and they swallow their prey whole.

STUDY BREAK 32.10

1. Several characteristics reveal the close evolutionary relationship of crocodilians and birds, including a four-chambered heart and maternal care of offspring.
2. The specific adaptations that allow birds to fly either reduce their weight or increase their muscle power. Weight-reducing adaptations include a lightweight skeleton, the absence of teeth and a urinary bladder, and the habit of laying an egg as soon as it has a shell. Power-promoting adaptations include large wing muscles, efficient digestive, respiratory, and circulatory systems, and a high metabolic rate.
3. The structure of a bird's bill reflects its diet. For example, hummingbirds that drink nectar have long thin bills, and parrots that eat hard nuts have stout sharp bills. Wings and feet are adapted to birds' flying habits and habitats. For example, ducks have webbed feet that allow them to paddle in water, and albatrosses have long thin wings that work efficiently for long distance flight.

STUDY BREAK 32.11

1. Most mammals were probably active at night during the Mesozoic era to avoid competition with and predation by dinosaurs, most of which were active during the day.
2. The key adaptations that allow mammals to be active under many types of environmental conditions include insulating fur and fat and a high metabolic rate that generates lots of body heat.
3. The three major groups of living mammals are distinguished on the basis of their reproductive habits. Monotremes lay eggs. Marsupials give birth to relatively undeveloped young after a short period of gestation. Placentals give birth to more developed young after a long period of gestation.

STUDY BREAK 32.12

1. The characteristics that allow many species of primates to spend a lot of time in trees include flexible shoulder and hip joints, grasping hands, and excellent depth perception.
2. The lowest taxonomic group that includes monkeys, apes, and humans is the Anthropoidea. The lowest taxonomic group that includes only apes and humans is the Hominoidea.
3. Gorillas, chimpanzees, and bonobos are the apes that spend the most time on the ground.

STUDY BREAK 32.13

1. Researchers usually use the criterion of bipedal locomotion to distinguish between humans and apes. Humans (that is, hominins) are bipedal, and apes are not.
2. The strongest evidence suggesting that Neanderthals and modern humans belong to different species comes from mtDNA sequence data: the differences between gene sequences of Neanderthals and humans are much greater than the differences between any two modern humans.
3. Independent genetic studies of mitochondrial DNA and the Y chromosome indicate that all human populations are descended from a common ancestor that originated in Africa and then migrated to various regions on Earth.

Think Like a Scientist

FIGURE 32.8

Based on information in the figure, a change from radial to bilateral symmetry accompanied the evolution of an increased number of *Hox* genes in the *Hox* gene complex.

IMR

If early primates were leaping or climbing through trees, increased visual acuity and better depth perception would allow them to estimate the distances between branches, decreasing their likelihood of falling. Increased olfactory function would not contribute to better arboreal locomotion.

UQ

Viviparity could provide several benefits. By carrying her developing young, the mother can protect them from some detrimental environmental factors; she can regulate the temperature at which they develop by regulating her own body temperature; she can provide them with nutrients and energy throughout their development; and she can select a favorable environment into which she releases them when they are born.

Test Your Knowledge

1. a 2. c 3. d 4. b 5. e 6. c 7. e 8. b 9. d 10. a

Interpret the Data

According to this analysis, turtles fall within the diapsid lineage, and they are more closely related to archosaurs (alligator and chicken) than to lepidosaurs (tuatara and iguana). The results of this analysis, therefore, contradict the traditional view.

Chapter 33

Study Break

STUDY BREAK 33.1

1. A land plant's shoot system consists of its photosynthetic tissues and organs—stems, leaves, and buds. Stems are frameworks for upright growth and favorably position leaves for light exposure and flowers for pollination. Leaves increase a plant's surface area and thus its exposure to sunlight. Buds eventually extend the shoot or give rise to a new, branching shoot. The shoot system of a flowering plant also includes flowers and fruits. Parts of the shoot system store carbohydrates manufactured during photosynthesis.

   The root system usually grows below ground. It anchors the plant, and sometimes structurally supports its upright parts. It also absorbs water and dissolved minerals from soil and stores carbohydrates.
2. Meristem tissue is self-perpetuating embryonic tissue. Apical meristems, at the tips of shoots and roots, gives rise to a young plant's stems, buds, roots, and other primary

tissues. In plants that show secondary growth, cylinders of lateral meristem tissue give rise to (often woody) secondary tissues that increase the diameter of older stems and roots.

STUDY BREAK 33.2

1. The ground tissue system makes up most of the plant body. It includes three types of structurally simple tissues—parenchyma, collenchyma, and sclerenchyma—each of which is composed mainly of one type of cell. Parenchyma makes up most of a plant's primary tissue and typically has air spaces between its cells, which are alive at maturity and can continue to divide. Subgroups of parenchyma cells are specialized for photosynthesis, secretion, and storage (of starch). Collenchyma is flexible ground tissue that contains cellulose. Its cells remain alive and metabolically active at maturity. They provide mechanical support for parenchyma and often collectively form strands or a sheathlike cylinder under the dermal tissue of growing shoot regions and leaf stalks. Cells of sclerenchyma are dead at maturity, but they develop thick secondary walls while alive that typically are lignified and provide additional support and protection in mature plant parts.
2. Xylem and phloem are the tissues of the vascular tissue system. The two types of xylem cells, called tracheids and vessel members, both develop thick, lignified secondary cell walls and die at maturity. The empty cell walls of abutting cells serve as pipelines for water and minerals. The conducting cells of phloem, called sieve tube members, form sieve tubes that conduct solutes, mainly sugars made during photosynthesis, throughout a plant.
3. The dermal tissue system serves as a skinlike protective covering for the plant body. Cells of the epidermis are tightly packed and cover the primary plant body. They secrete a cuticle that coats all plant parts except the very tips of the shoot and most absorptive parts of roots. Some epidermal cells become modified for specialized functions. Examples include guard cells, which form stomata; root hairs, which absorb water and minerals; and hairlike trichomes, which function in defense against herbivory or secrete sugars that attract pollinators.

STUDY BREAK 33.3

1. Stems have four main functions: (1) they provide mechanical support for body parts involved in growth, photosynthesis, and reproduction; (2) they house the vascular tissues (xylem and phloem), which transport products of photosynthesis, water and dissolved minerals, hormones, and other substances throughout the plant; (3) they often are modified to store water and food; and (4) they have specific stem regions that contain meristematic tissue, which gives rise to new cells of the shoot.

   A plant stem is divided into modules, each consisting of a node, where leaves are attached, and an internode, the space between nodes. New primary growth occurs in buds—a terminal bud at the apex of the main shoot, and lateral buds, which produce branches (lateral shoots), in the leaf axils. Meristem tissue in buds gives rise to leaves, flowers, or both.

   In eudicots, most primary growth in a stem's length occurs directly below the shoot apical meristem. When a meristematic cell divides, one of its daughter cells becomes an initial, a cell that remains as part of the meristem. The other daughter cell becomes a derivative, which typically divides once or twice and then enters on the path to differentiation. As derivatives differentiate, they give rise to three primary meristems: protoderm, procambium, and ground meristem. These primary meristems produce cells that differentiate into specialized cells and tissues. In eudicots, the primary meristems are also responsible for elongation of the plant body. Each primary meristem occupies a different position in the shoot tip. Outermost is protoderm, which gives rise to the stem's epidermis. Inward from the protoderm the ground meristem gives rise to ground tissue (mostly parenchyma). Procambium, which produces the primary vascular tissues, is sandwiched between ground meristem layers. In most plants, inner procambial cells give rise to xylem and outer procambial cells to phloem. The developing vascular tissues become organized into vascular bundles that are wrapped in sclerenchyma and thread lengthwise through the parenchyma. In the stems and roots of most eudicots and some conifers, the vascular bundles form a stele (vascular cylinder) that vertically divides the column of ground tissue into an outer cortex and an inner pith.
2. Leaves are organs specialized for photosynthesis. In both eudicots and monocots, the leaf blade provides a large surface area for absorbing sunlight and carbon dioxide. Many eudicot leaves have a broad, flat blade attached to the stem by a petiole. Unless a petiole is very short, it holds a leaf away from the stem and helps prevent individual leaves from shading one another. In most monocot leaves, such as those of rye grass or corn, the blade is longer and narrower and its base simply forms a sheath around the stem.
3. Leaves develop on the sides of the shoot apical meristem. Initially, meristem cells near the apex divide and their derivatives elongate. The resulting bulge enlarges into a thin, rudimentary leaf, or leaf primordium. As the plant grows and internodes elongate, the leaves become spaced at intervals along the length of the stem or its branches. Leaf tissues typically form several layers. Uppermost is epidermis, with cuticle covering its outer surface. Just beneath the epidermis is mesophyll, which is composed of loosely packed parenchyma cells that contain chloroplasts. Leaves of many plants, especially eudicots, contain two layers of mesophyll. Palisade mesophyll cells contain more chloroplasts and are arranged in compact columns with smaller air spaces between them, typically toward the upper leaf surface. Spongy mesophyll, which tends to be located toward the underside of a leaf, consists of irregularly arranged cells with a network of air spaces that enhance the uptake of carbon dioxide and release of oxygen during photosynthesis and account for 15% to 50% of a leaf's volume. Below the mesophyll is another cuticle-covered epidermal layer. Except in grasses and a few other plants, this layer contains most of the stomata through which water vapor exits the leaf and gas exchange occurs. Vascular bundles form a network of veins throughout the leaf.
4. Plants that live for many years may spend part of their lives in a juvenile phase, and then shift to a mature, or adult phase. The differences between juveniles and adults often are reflected in leaf size and shape, in the arrangement of leaves on the stem, or in a change from vegetative growth to a reproductive stage. Most woody plants must attain a certain size before their meristem tissue can respond to the hormonal signals that govern flower development.

STUDY BREAK 33.4

1. Most eudicots have a taproot system—a single main root, or taproot, that is adapted for storage and smaller branching lateral roots. As the main root grows downward, its diameter increases, and the lateral roots emerge along the length of its older, differentiated regions. Grasses and many other monocots develop a fibrous root system in which several main roots branch to form a dense mass of smaller roots. Fibrous root systems are adapted to absorb water and nutrients from the upper layers of soil, and tend to spread out laterally from the base of the stem.
2. The root apical meristem and the actively dividing cells behind it form the zone of cell division. Cells in the center of the root tip become the procambium; those just outside the procambium become ground meristem; and those on the periphery of the apical meristem become protoderm. The zone of cell division merges into the zone of elongation, where most of the increase in a root's length occurs. Above the zone of elongation, cells may differentiate further and take on specialized roles in the zone of maturation.
3. Primary root growth produces a system of vascular pipelines extending from root tip to shoot tip. The root procambium produces cells that mature into the root's xylem and phloem. Ground meristem gives rise to the root's cortex, its ground tissue of starch-storing parenchyma cells that surround the stele. In many flowering plants, the outer root cortex cells give rise to an exodermis, a thin band of cells that may limit water losses from roots and help regulate the absorption of ions. The innermost layer of the root cortex is the thin endodermis, which helps control the movement of water and dissolved minerals into the stele. Between the stele and the endodermis is the pericycle, which gives rise to lateral roots. In some cells in the developing root epidermis, the outer surface extends into root hairs.

STUDY BREAK 33.5

1. Secondary growth processes add girth to roots and stems over two or more growing seasons. In plant species that have secondary growth, older stems and roots become more massive and woody through the activity of two types of lateral meristems. One of these meristems, the vascular cambium, produces secondary xylem and phloem. The other, the cork cambium, produces cork, a secondary epidermis that is one element of bark.
2. Vascular cambium consists of two types of cells—fusiform initials and ray initials. Fusiform initials are derived from cambium inside the vascular bundles and give rise to secondary xylem and phloem cells. Secondary xylem forms on the inner face of the vascular cambium, and secondary phloem forms on the outer face. Ray initials are derived from the parenchyma cells between vascular bundles. Their descendants form spokelike rays of parenchyma cells—horizontal channels that carry water sideways through the stem. As the mass of secondary xylem inside the ring of vascular cambium increases, it forms hard tissue known as wood. Bark encompasses all the living and nonliving tissues between the vascular cambium and the stem surface. It includes the secondary phloem and the periderm, the outermost portion of bark that consists of cork, cork cambium, and secondary cortex.
3. Roots of some plant species undergo secondary growth, but the ring of vascular cambium develops differently than it does in stems. When their primary growth is complete, these roots have a layer of residual procambium between the xylem and phloem of the stele. The vascular cambium arises in part from this residual cambium, and in part from the pericycle. Eventually, the cambial tissues arising from the procambium and those arising from the pericycle merge into a complete cylinder of vascular cambium. As in stems, the vascular cambium gives rise to secondary xylem to the inside and secondary phloem to the outside. As secondary xylem accumulates, older roots can become thick and woody.

Think Like a Scientist

FIGURE 33.3

If part of an apical meristem were to be removed, hormonal signals would prompt the remaining meristematic cells to give rise to replacement cells.

IMR

Obvious subjects for subsequent research include identifying specific genes *MIXTA* regulates, and characterizing the protein products that directly influence the shape of epidermal cells in *A. majus* petals.

UQ

Despite producing progeny cells that adopt different fates, both shoot and root stem cells are pluripotent cells that share the properties of continuous division and self-renewal, and thus their molecular signatures are likely to be more similar than different.

Test Your Knowledge

1. d 2. c 3. b 4. a 5. c 6. a 7. d 8. e 9. c 10. b

Interpret the Data

1. The data show wider than usual annual growth rings for the twentieth century. In their research report, Dr. Jacoby and his coworkers concluded that "recent warming is unusual relative to temperatures of the past 450 years."

2. It was important for the study authors to begin by acknowledging that the width of rings formed by individual trees shows natural variations related to each tree's growth. Including a graph of the ring-width measurements from a single 500-year-old tree allows for a clear comparison of that pattern with the graph that charts aggregated data from many trees—which you will notice follows a quite similar pattern.
3. The lower line represents sample size in the study. It shows that relatively few very old trees (with rings formed before 1550) were available to be sampled, with a steady rise in the number of available trees that began forming rings in subsequent centuries. The line also reveals consistency in the number of core samples dating from about 1850, when the Industrial Revolution and more intensive burning of fossil fuels began. This consistency lends credence to the reliability of the study's findings.
4. Airborne smoke and ash reduce the amount of solar energy that reaches Earth's surface. Less available sunlight is likely to translate into reduced photosynthesis and plant growth, which in turn will be reflected in the width of tree rings that form during that period.

Chapter 34

Study Break

STUDY BREAK 34.1

1. After an H^+ gradient is established (by proton pumping out of the cell), the resulting inward flow of H^+ down its concentration gradient provides the energy to actively transport other substances into the cell.
2. Water potential is potential energy stored in water. It is the driving force for osmosis, which in turn is responsible for the movement of water into and out of plant cells, including root cells.

STUDY BREAK 34.2

1. In the apoplastic pathway, water and dissolved substances do not pass through living root cells but instead move through the continuous network of adjoining cell walls and air spaces. When apoplastic water and solutes reach the endodermis, however, they must detour around the impermeable Casparian strip and pass through cells to move into the stele. The symplastic pathway passes through living cells. Water that diffuses into root cells moves in this pathway from cell to cell through plasmodesmata.
2. Epidermal cells of root hairs actively transport most mineral ions into root epidermal cells. These ions travel inward via the transmembrane pathway. Other ions may be dissolved in apoplastic water. They ultimately travel to the xylem in the symplast after crossing into and through endodermal cells of the Casparian strip. Once an ion reaches the stele it enters the xylem.

STUDY BREAK 34.3

1. In the cohesion–tension mechanism, water transport begins as water evaporates from the walls of mesophyll cells inside leaves and into the intercellular spaces. This water vapor escapes by transpiration through open stomata. As water molecules exit the leaf, they are replaced by others from the mesophyll cell cytoplasm. The water loss gradually reduces the water potential in a transpiring cell below the water potential in the leaf xylem. Water from the xylem in the leaf veins then follows the gradient into cells, replacing the water lost in transpiration.
2. Stomata open and close in response to changing environmental cues, such as light levels (detected via blue-light receptors), CO_2 concentration in the air spaces inside leaves, and the amount of water available to the plant. Stomata open when hydrogen ions are pumped out of guard cells, setting up the symport of H^+ and K^+ into the guard cells through ion channels. Water then follows by osmosis. Stomata close when H^+ pumping in guard cells ceases and K^+ moves out of guard cells, with water again following by osmosis. Through their ability to open and close, stomata help regulate water loss by plants and the uptake of carbon dioxide for photosynthesis.

STUDY BREAK 34.4

1. Translocation is the long-distance transport of substances in plants. The term generally applies to the transport of organic compounds, mainly sucrose, in phloem. Transpiration is the evaporation of water from a plant's aerial parts, mainly leaves. This water moves from roots upward to aerial parts in the xylem.
2. The mechanism of pressure flow moves sucrose from a source (such as a leaf or stem) into sieve tubes. Pressure builds up at the source end of a sieve tube system as sucrose enters sieve tubes at sources and water follows by osmosis. Under high pressure, sucrose moves by bulk flow toward a sink (plant parts that take up sucrose as metabolic fuel), where the sugar is unloaded.

Think Like a Scientist

FIGURE 34.7

In general, analysis shows that efforts to directly measure xylem pressure—most notably, by inserting a needlelike pressure probe into xylem conduits of living plants—produce unreliable findings because puncturing the conduit with the probe skews the measurements.

FIGURE 34.12

Recent research has shown that, at least in the case of aphids, the insect's saliva contains substances that inhibit normal mechanisms that might otherwise quickly seal small punctures in plant tissues.

FIGURE 34.13

The main known sucrose channels are sucrose/H^+ antiporters and symporters. Other phloem nutrients enter sink cells by other means, including nonselective cation channels, aquaporins, H^+/ATPases, and amino acid/H^+ symporters.

IMR

Experiments show that environmental factors such as drought and high salinity, both of which affect water balance, help determine which specific PIP genes are expressed in leaf cells.

UQ

The results support a hypothesis that active transport may sometimes move substances through plasmodesmata. Active transport is unidirectional—a transported solute moves either into or out of a cell—and the tracer moved unidirectionally in the experiments. The halt in tracer movement when cell metabolism was inhibited by sodium azide also is consistent with an active transport mechanism, because active transport requires ATP produced by the cell's metabolism.

Test Your Knowledge

1. b 2. d 3. e 4. a 5. a 6. b 7. e 8. c 9. c 10. c

Interpret the Data

1. In both graphs, the darker blue blocks represent the movement of deionized water through girdled branch segments over time. The lighter blue blocks represent the movement of a potassium chloride solution through girdled branch segments over time.
2. The results support the hypothesis because the values for the flow of pure water are consistently lower than the flow values of the KCl solution.
3. Results were substantially the same for both species, in that in both there was a marked increase in flow rates when the researchers switched from deionized water to the KCl solution. There was, however, a species-specific difference in the magnitude of the increased flow.

Chapter 35

Study Break

STUDY BREAK 35.1

1. Plants require relatively large amounts of macronutrients, such as nitrogen, sulfur, potassium, and calcium, and trace amounts of micronutrients, such as iron, chlorine, zinc, nickel, and copper.
2. Plants vary in their nutritional requirements. For example, leafy plants require more nitrogen and magnesium than other plant types do, and alfalfa, a grass, requires significantly more potassium than lawn grasses do. An adequate amount of an essential element for one plant also may be toxic for another. For these reasons, the nutrient content of soils is an important factor determining which plants grow well in a given location.

STUDY BREAK 35.2

1. Humus is important in soil because it generally contains nutrient-rich organic material and because it absorbs water, which contributes to the water-holding capacity of soil.
2. The amount of water that is available in soil to be taken up by plant roots depends primarily on the relative proportions of different soil components. Water moves quickly through sandy soils, whereas soils rich in clay and humus tend to hold the most water.
3. A plant's ability to absorb soil minerals depends partly on cation exchange, in which one cation, usually H^+, replaces a soil cation. As H^+ enters the soil solution, it displaces adsorbed mineral cations attached to clay and humus, freeing them to move into roots. Anions in the soil solution, such as nitrate (NO_3^-), sulfate (SO_4^{2-}), and phosphate ($PO_4—$), generally move more readily into root hairs. Soil pH also affects the availability of some mineral ions because chemical reactions in very acid (pH : soils can trigger chemical reactions that bind various mineral cations in compounds that are insoluble in soil water.

STUDY BREAK 35.3

1. As described in this chapter and in Section 28.3, a mycorrhiza is a symbiotic association between a fungus and plant roots. Most plants form mycorrhizal associations, which facilitate the plant's ability to extract soil nutrients such as nitrogen and phosphorus. As with plant roots, mineral ions enter fungal hyphae by way of transport proteins. Some of the plant's sugars and nitrogenous compounds nourish the fungus, and as the root grows, it takes up a portion of the minerals that the fungus has secured. In some types of mycorrhizae the fungus actually lives inside cells of the root cortex.
2. Nitrogen fixation refers to the incorporation of atmospheric nitrogen into compounds, especially nitrate (NO_3^-), which plants can readily take up. Ammonification is a process in which soil bacteria known as ammonifying bacteria break down decaying organic matter and convert it to ammonium (NH_4^+). In nitrification, nitrifying bacteria oxidize NH_4^+ to NO_3^-. Inside root cells, absorbed NO_3^- is converted by a multistep process back to NH_4^+. In this form, it is rapidly used to synthesize organic molecules, mainly amino acids.
3. Associations with bacteria supply nitrogen to certain types of plants, such as legumes. The host plant provides organic molecules that the bacteria use for cellular respiration, and the bacteria supply NH_4^+ that the plant uses to produce nitrogenous molecules. In legumes the nitrogen-fixing bacteria reside in root nodules. Usually, a single species of nitrogen-fixing bacteria colonizes a single legume species, drawn to the plant's roots by chemical attractants (mainly flavonoids) that the roots secrete. By way of exchanged molecular signals, bacteria then are able to penetrate a root hair and form a colony inside the root cortex. Each cell in a root nodule may contain several thousand bacteria (now called bacteroids). The plant takes up some of the nitrogen fixed by the bacteroids, and the bacteroids use some compounds produced by the plant.

Think Like a Scientist

FIGURE 35.9

Apparently not. Recent studies have identified a few species of nodule-forming bacteria (though not in the genera *Rhizobium* and *Bradyrhizobium*) that lack *nod* genes. The bacteria may enter root cortex cells by infiltrating cracks in the epidermis where lateral roots arise.

IMR

These lines of research may lead to methods for reducing the amount of phosphate fertilizers added to crop plants by identifying mycorrhizal fungi providing the most efficient phosphate uptake—or by engineering crop plants with an improved capacity to take in this essential nutrient.

UQ

Even after microorganisms satisfy their N requirements, they continue to consume nitrogen-containing organic compounds because they need the carbon in these compounds to fuel metabolism and growth.

Test Your Knowledge

1. e 2. c 3. d 4. c 5. b 6. c 7. a 8. d 9. a 10. e

Interpret the Data

1. Overall, aspen leaf litter contained about twice as much nitrogen but the same amount of carbon as leaf litter of pines and spruces, so with aspens the ratio of nitrogen to carbon in leaf litter was about half the value as in the other species. Lodgepole pine litter (the pine needles) was significantly higher in both cellulose and lignin than were trembling aspen or Engelmann spruce litter. The study data showed no statistically different values among the species in either fine or coarse root biomass.
2. Although we see clear species-specific differences in the values, overall the species with the lowest litter nitrogen—the pines and spruces—have significantly higher ratios of lignin to nitrogen.
3. There was no significant difference in root biomass among the three tree species in the study plots.

Chapter 36

Study Break

STUDY BREAK 36.1

1. The two alternating generations of plants are the sporophyte (spore-producing) and gametophyte (gamete-producing) generations.
2. Sporophytes produce spores that give rise to gametophytes. Gametophytes then may produce gametes; male gametophytes produce sperm, and female gametophytes produce eggs. In all seed plants the sporophyte is much larger and longer-lived than the gametophyte, and the gametophyte is protected within sporophyte tissues for all or part of its life. Gametophytes also are dependent upon the sporophyte for their nutrition.

STUDY BREAK 36.2

1. Flowers are specialized for reproduction. Before an angiosperm can produce a flower, biochemical signals (triggered in part by environmental cues such as day length and temperature) travel to the apical meristem of a shoot. In response, cells there change their activity: instead of continuing vegetative growth, the shoot is modified into a floral shoot that will give rise to floral organs.
2. Pollen grains are the mature male gametophytes. They arise by the following steps:
   a. Spores that give rise to male gametophytes are produced in a flower bud's anthers.
   b. Diploid microsporocytes inside an anther's pollen sacs undergo meiosis; eventually each one produces four small haploid microspores.
   c. Microspores then divide by mitosis.
   d. One of the two resulting nuclei divides again by mitosis, yielding a three-celled immature gametophyte: two haploid sperm cells and a third cell that will control the development of a pollen tube after pollen lands on a receptive stigma.
   e. A mature male gametophyte consists of the pollen tube and sperm cells.
3. Female gametophytes develop inside ovules in a flower's carpels. They arise by the following steps:
   a. In an ovule, a diploid megasporocyte divides by meiosis, forming four haploid megaspores.
   b. Three (usually) of these megaspores disintegrate.
   c. The remaining megaspore undergoes three rounds of mitosis without cytokinesis. The result is a single large cell with eight nuclei arranged in two groups of four.
   d. One nucleus in each group migrates to the center of the cell.
   e. After the cell undergoes cytokinesis a cell wall forms around these two polar nuclei, forming a large "central cell."
   f. A wall also forms around each of the remaining nuclei, and three of them, including an egg cell, cluster near the micropyle.
   g. The result is an embryo sac containing seven cells and eight nuclei. This sac is the mature female gametophyte.

STUDY BREAK 36.3

1. A pollen grain that lands on a compatible stigma absorbs moisture and germinates a pollen tube, which burrows through the stigma and style toward an ovule. Chemical cues from the two synergid cells help guide the pollen tube toward the egg. Before or during these events, the pollen grain's haploid sperm-producing cell divides by mitosis, forming two haploid sperm. When the pollen tube reaches the ovule, it enters through the micropyle and an opening forms in its tip. The two sperm are released into the cytoplasm of a disintegrating synergid. Next double fertilization occurs: typically, one sperm nucleus fuses with the egg to form a diploid ($2n$) zygote. The other sperm nucleus fuses with the central cell, forming a cell with a triploid ($3n$) nucleus. Tissues derived from the $3n$ cell are called endosperm.
2. As a seed matures, the embryo inside it develops a root–shoot axis with a root apical meristem at one end and a shoot apical meristem at the other end. Depending on the plant group, one or two cotyledons also develop. In monocots a single large cotyledon develops and stores endosperm; protective tissues arise around the root and shoot apical meristems. In eudicots, two endosperm-storing cotyledons form. Near the micropyle the radicle (embryonic root) attaches to the cotyledon at a region called the hypocotyl. Beyond the hypocotyl is the epicotyl, which has the shoot apical meristem at its tip and which often bears a cluster of tiny foliage leaves, the plumule. At germination, when the root and shoot first elongate and emerge from the seed, the cotyledons are positioned at the first stem node with the epicotyl above them and the hypocotyl below them.
3. Imbibition causes the seed coat to split, and water and oxygen move more easily into the seed. Metabolism switches into high gear as cells divide and elongate to produce the seedling. Enzymes that were synthesized before dormancy become active; other enzymes are produced as the genes encoding them begin to be expressed. The increased gene activity and enzyme production mobilize the seed's food reserves in cotyledons or endosperm. Nutrients released by the enzymes sustain the developing seedling sporophyte until its root and shoot systems are established.

STUDY BREAK 36.4

1. Flowering plants may reproduce asexually (vegetatively) by fragmentation, in which cells in a piece of the parent plant dedifferentiate and then regenerate a whole plant; by apomixis, in which a diploid embryo develops from an unfertilized egg or from diploid cells in ovule tissue; or by the production of structures such as rhizomes or suckers from a nonreproductive plant part, typically meristem tissues in a bud on a root or stem.
2. Totipotency is the capacity of fully differentiated cells to dedifferentiate, return to an unspecialized embryonic state, and then develop into a fully functional mature plant. Plant tissue culture procedures trigger the development of a mass of dedifferentiated cells (a callus), some of which regain totipotency and develop into plantlets.

STUDY BREAK 36.5

1. A homeotic gene is a regulatory gene in the genome of an organism that encodes a transcription factor. In *A. thaliana,* homeotic genes govern the development of the root and shoot tissue systems, as well as of floral organs.
2. The two basic mechanisms of plant morphogenesis are oriented cell division and cell expansion. Oriented cell division establishes the general shape of a plant organ, and cell expansion enlarges the cells in a developing organ in particular directions. They increase in circumference (girth) when new cell walls form parallel to the nearest plant surface (such as the surface of a stem or tree trunk) or when cell walls form at right angles both to the nearest surface and to the transverse plane.
3. Leaves arise through a developmental program that begins with gene-regulated activity in meristematic tissue. Hormones or other signals may arrive at target cells via the stem's vascular tissue, activating genes that regulate development. Small phloem vessels penetrate a young leaf primordium almost immediately after it begins to bulge out from the underlying meristematic tissue, followed by xylem. A growing primordium becomes cone-shaped (wider at its base than at its tip). Rapid mitosis in a particular plane in cells along the flanks of the cone (perpendicular to the surface in eudicots and parallel to the surface in monocots) produce the leaf blade that is characteristic of the particular species. Leaf tip cells typically are the first to stop dividing. By the time a leaf has expanded to its mature size, mitosis has ended and the leaf is a fully functional photosynthetic organ.

Think Like a Scientist

FIGURE 36.9

Researchers may use several strategies for identifying the signaling events that guide pollination. One approach is to identify mutants that show particular defects in the guidance steps, and then analyze the defects to shed light on what kind of signal(s) operate in normal plants. Some investigators also employ microscopy-based real-time assays to directly observe pollen-tube behavior in various mutants.

FIGURE 36.24

It seems they do. Versions of the ABCE system have been discovered in eudicots including petunia *(Petunia)* and snapdragon *(Antirrhinum),* among others. A modified ABC-like system appears to operate in monocots that have been studied, including members of the lily family, corn, and rice, although much less is known about the particular gene classes that may be involved.

IMR

Because trichomes are accessible and transparent, different mutants can (and do) serve as cellular "laboratories" for studying a range of subcellular processes, including interactions among organelles and the behavior of the actin and microtubule cytoskeletons under varying conditions.

UQ

1. Pollen tube guidance signals are likely to be species specific. Since they are involved in mating, it is likely that these signals ensure that successful pollination, and consequently fertilization, occur only within compatible species. It is also likely that the later the guidance event, the more specific is the guidance signal. For example, the guidance signal from an ovule is likely more selective and specific than those that mediate early interactions between a pollen tube and stigma.
2. If pollen tube growth and guidance are defective, seeds are not produced. Around 80% of the world's staple food is derived from seeds of crop plants, so studying this process is very important agriculturally. By understanding pollen tube guidance signals, we are equipping ourselves with knowledge using which we could: (a) improve seed yield; (b) regulate interspecies hybridizations and, thereby, generate novel plant hybrids; and (c) contain pollen spreading from genetically modified crops—a possibility that will reassure concerned public and regulatory agencies.

Test Your Knowledge

1. c 2. a 3. b 4. e 5. d 6. c 7. b 8. e 9. c 10. c

Interpret the Data

1. The variables were the concentration of gibberellin (10 μM GA_3 vs. 1 μM GA_3) and the effect of adding ABA to the test plants' growth medium.
2. Slightly more of the mutant seeds exposed to supplementary ABA germinated, but (as evidenced by the confidence intervals) the difference was not significant.
3. First, more seeds germinated at the higher concentration of GA_3 (10 μM). Second, regardless of the GA_3 concentra-

tion, supplemental ABA made no significant difference in the seed germination rate.

Chapter 37

Study Break

STUDY BREAK 37.1

1. Auxins, gibberellins, cytokinins, and brassinosteroids all promote the growth of plant parts, and ethylene stimulates cell division in seedlings. Abscisic acid is the major growth-inhibiting plant hormone.
2. Ethylene is a good example of a hormone that can stimulate or inhibit growth at various stages of the plant life cycle. In seedlings, it simultaneously slows elongation of the stem and stimulates cell divisions that increase stem girth. In mature plants of deciduous species, it governs senescence (including fruit ripening) and the abscission of flowers, fruits, and leaves. Studies of brassinosteroids have revealed that this family of steroid hormones has different effects in different tissues—for example, promoting the elongation of vascular tissue and pollen tubes, but inhibiting elongation in roots.
3. Two general types of signal response pathways convert chemical signals into altered plant cell activity. Some plant hormones and growth factors bind to receptors at the target cell's plasma membrane. Others cross the plasma membrane and bind to receptors inside the cell. Second messengers enhance the cellular response.

STUDY BREAK 37.2

1. General plant responses to attack include mobilization of jasmonates and salicylic acid, systemin (in tomato), the hypersensitive response, PR proteins, and secondary metabolites (phytoalexins). Gene-for-gene recognition is a pathogen-specific response.
2. Salicylic acid (SA) is considered to be a general systemic response to damage because experiments show that when a plant is wounded, soon thereafter SA can be detected in a variety of its tissues.
3. While the hypersensitive response is underway, SA also is synthesized and operates in other defensive chemical pathways in a plant. This effect includes the synthesis of PR (pathogenesis-related) proteins that attack pathogenic cells.

STUDY BREAK 37.3

1. Directional light of blue wavelengths is the direct stimulus for phototropism. The most widely accepted scientific explanation for gravitropism is the sinking of amyloplasts in cells surrounding vascular bundles in response to gravity. Sinking amyloplasts may provide a mechanical stimulus that triggers a gene-guided redistribution of IAA. The changing auxin gradient in turn adjusts a plant's growth pattern.
2. Unlike tropisms, nastic movements occur in response to nondirectional stimuli, such as mechanical pressure resulting from an insect brushing against hairlike sensory structures in the leaves of a Venus flytrap plant.

STUDY BREAK 37.4

1. Plant responses to changes in photoperiod rely on different chemical forms of the blue-green pigment phytochrome. Daylight converts the inactive phytochrome (P_r) to an active form, (P_{fr}). When light levels fall, P_{fr} reverts to P_r. This switching mechanism helps regulate light-related processes such as photosynthesis.
2. Photoperiod length is a factor in the seasonal flowering of many angiosperms. Spinach and irises are examples of long-day plants, which flower in spring when the period of daylight extends at least 9 to 16 hours. Chrysanthemums and potatoes are examples of short-day plants, which flower as day-length becomes shorter than some critical period, a condition that occurs naturally in the fall.
3. Dormancy is an adaptive response because it attunes a plant's growth to the most favorable environmental conditions for survival.

STUDY BREAK 37.5

1. Some response pathways may directly stimulate gene transcription whereas others might inhibit gene expression. Some pathways have an intermediary step, in which the original signal triggers the synthesis of regulatory proteins that in turn promote or inhibit the expression of still other genes.
2. Second messengers boost cellular responses to a hormonal signal by setting in motion a cascade of activated protein kinases, which in turn activate cell proteins that carry out the cell's response.
3. Although Chapter 7 focuses on signal transduction pathways in animals, in fact the basics of some transduction pathways are similar in animals and plants. In both groups we see examples of one-step signal transduction (in which a signal exerts its effect by binding a cell receptor) and of signaling cascades such as second messenger systems.

Think Like a Scientist

FIGURE 37.3

Quite possibly. When eudicot seedlings are used in phototropism research, it is the lower stem rather than the seedling shoot tip that appears to bend, so the Darwins might have concluded that the "factor" that caused bending was present in the lower stem. Only recently did researchers show convincingly that in eudicots the auxin signal for phototropism also is transmitted downward from the shoot tip. The research, reported in 2011 by a team at Purdue University, relied on sophisticated genetic techniques to establish that, in *A. thaliana* seedlings, auxin accumulates in the shoot tip and then is redistributed laterally and downward. Intervening steps have yet to be resolved and are sure to be the subject of future experiments.

IMR

As described in Section 37.1, the second messenger inositol triphosphate (IP_3), which is involved in ABA responses that close stomata, was first discovered in animals (and fungi). So was the plant second messenger cGMP (cyclic guanosine monophosphate), which serves as a second messenger in animals, fungi, and even some bacteria.

UQ

1. The genome reorganization occurs in a specific subset of the genome and reproducibly occurs under the same stress conditions. Therefore, it is not random, but the mechanism by which these regions of the genome are recognized is currently unknown. Most mutation systems are assumed to be random within the genome although there is evidence that transposons do have preferential sites.
2. The data indicate that the environment can direct higher rates of mutations to particular regions of the genome and, therefore, mutating and selecting plants simultaneously might provide a better frequency of adaptive mutations. Because plants are sessile, do they need mechanisms that enable the generation of diversity in the face of chronic stress growth conditions for evolution? If so, partitioning of the plant genome into regions that are protected from change (because they are essential) and variable regions (in which changes have phenotypic effects but are not lethal) is important to identify. Using the information might permit the more rapid breeding of stress tolerant crops.

Test Your Knowledge

1. e 2. a 3. c 4. b 5. b 6. b, c 7. a 8. e 9. c 10. e

Interpret the Data

1. Overall, many more *Yucca* genes are up-regulated (expression enhanced) in response to BR and auxin than are down-regulated (expression suppressed).
2. The brassinosteroid (BR) down-regulated many more *Yucca* genes than did auxin. Also, a small number of genes that are up-regulated by auxin are down-regulated by BR. In an even smaller number of genes the reverse was true—that is, the genes were up-regulated by BR but down-regulated by auxin.
3. There is no overlap between the "up" and "down" bubbles of each hormone because a given hormone does not have contrasting effects on the same gene. Thus, with respect to each hormone there are two distinct groups of *Yucca* genes—those the hormone up-regulates and those it down-regulates.

Chapter 38

Study Break

STUDY BREAK 38.1

1. Multicellularity made it possible for animals to create an internal fluid environment for fulfilling the nutrient supply, waste removal, and osmotic balance needs of individual cells. As a result, multicellular organisms could evolve to occupy a variety of habitats, including dry terrestrial environments, in which single cells cannot survive. Multicellularity also allowed major life functions to be distributed among specialized groups of cells, with each group having a single activity. The specialized groups of cells are typically organized into tissues, the tissues into organs, and the organs into organ systems.
2. A tissue is a group of cells with the same structure and function. Cells in the tissue work together to perform one or more activities. An organ integrates two or more different tissues into a structure that carries out a specific function. An organ system coordinates the activities of two or more organs to carry out a major body function.

STUDY BREAK 38.2

1. Exocrine and endocrine glands are formed by epithelia. An exocrine gland remains connected to the epithelium by a duct, whereas an endocrine gland is suspended in connective tissue underlying the epithelium, with no ducts leading to the epithelial surface.
2. Loose connective tissue, fibrous connective tissue, cartilage, bone, adipose tissue, and blood.
3. Skeletal, cardiac, and smooth.

STUDY BREAK 38.3

See Figure 36.10.

STUDY BREAK 38.4

A stimulus—a change in the external or internal environment—starts the homeostatic control system. The stimulus is detected by a *sensor*, an *integrator* compares the environmental change with a set point, and the *effector* or effectors become activated by the integrator and function to return the environmental parameter to the set point.

Think Like a Scientist

IMR

When the cells are growing singly, the cells are separate and they remain undifferentiated. When they cover the culture dish and begin piling up on one another, the cells are in contact with one another. In some way, this contact triggers a change in the genetic programs of the cells so that they make a transition from being undifferentiated to becoming differentiated.

UQ

Some organisms needed to anticipate light in order to harvest its energy, whereas other organisms needed to avoid its damaging ultraviolet radiation. Thus, anticipation, via circadian rhythm, was and is crucial for survival.

Test Your Knowledge

1. b 2. a 3. e 4. c 5. b 6. a 7. c 8. e 9. e 10. d

Interpret the Data

Yes, these data are consistent with the hypothesis. Taurine is known to inhibit vasopressin. The figure shows a method by which taurine could act as part of a negative feedback loop. Taurine release is directly stimulated by vasopressin; therefore, its levels are directly increased by vasopressin, which could lead to negative feedback of vasopressin release.

Chapter 39

Study Break

STUDY BREAK 39.1

1. A dendrite receives signals and conducts them toward the cell body. An axon conducts signals away from the cell body toward another neuron or effector.
2. An afferent neuron conducts information from its sensory receptors to interneurons. Efferent neurons conduct signals from interneuron networks to effectors, the muscles and glands that carry out the response. The afferent and efferent neurons constitute the peripheral nervous system (PNS). Interneurons process information from afferent neurons and send a response to the efferent neurons. Interneurons form the brain and spinal cord, the central nervous system (CNS).
3. In an electrical synapse, the plasma membrane of the axon terminal of the presynaptic cell is in direct contact with the postsynaptic cell, allowing ions to pass directly between the cells when an electrical impulse arrives. In a chemical synapse, the presynaptic and postsynaptic cells are separated by a small gap. Neurotransmitters released from the presynaptic cell diffuse across the gap and bind to receptors in the plasma membrane of the postsynaptic cell. If enough neurotransmitter molecules bind to those receptors, the postsynaptic cell generates a new electrical impulse, which travels along its axon to the next neuron or effector in the circuit.

STUDY BREAK 39.2

1. At the peak of an action potential, the plasma membrane of the neuron enters a short refractory period, in which the threshold for generation of an action potential is much higher than normal. Only the region in front of the action potential can fire, meaning that the impulse can only move in one direction—that is, toward the axon tip. The refractory period remains in effect until the membrane again reaches the resting potential. By that time, the action potential has moved too far away to cause a second action potential in the same region.
2. Neurons insulated with myelin sheaths have gaps in the sheaths called nodes of Ranvier, where the axon membrane is exposed to extracellular fluids. The inward movement of sodium ions at a node produces depolarization and an action potential, but the adjacent myelin sheath prevents the excess positive ions from exiting through the membrane. Instead, they diffuse rapidly to the next node where they cause depolarization, inducing an action potential there. Continuation of this process allows the action potential to jump rapidly along the axon from node to node, at a faster rate than a nonmyelinated neuron of the same diameter.

STUDY BREAK 39.3

1. A neurotransmitter is synthesized in a neuron, is released into the synaptic cleft from a presynaptic axon terminal, and binds to receptors in the plasma membrane of the postsynaptic cell. Depending on the type of receptor, a neurotransmitter either stimulates or inhibits the generation of action potentials.
2. A neurotransmitter that binds to an ionotropic neurotransmitter receptor, like all neurotransmitters, is stored in synaptic vesicles in the cytoplasm at an axon terminal. When an action potential arrives at the terminal, the change in membrane potential opens voltage-gated Ca^{2+} channels in the axon terminal, allowing Ca^{2+} to flow back into the cytoplasm. The rise in Ca^{2+} concentration triggers the release of the neurotransmitters into the synaptic cleft by exocytosis. Neurotransmitters that bind to an ionotropic neurotransmitter receptor diffuse across the synaptic cleft and open or close ligand-gated ion channels in the postsynaptic neuron's membrane. Most of the channels regulate Na^+ or K^+ movement through the membrane, although some regulate Cl^-. Depending on the ion flow, action potentials are either stimulated or inhibited in the postsynaptic neuron.

STUDY BREAK 39.4

One way a postsynaptic neuron integrates signals is through summation of EPSPs and IPSPs that alter the neuron's membrane potential. EPSPs move the membrane potential toward the threshold for an action potential, whereas IPSPs move the membrane potential away from the threshold for an action potential. The final change in membrane potential depends on the particular array of EPSPs and IPSPs received. The patterns of synaptic connections made by a neuron also contribute to integration.

Think Like a Scientist

FIGURE 39.7

You could use an experimental setup similar to that of Loewi but with just one heart. Add acetylcholine to the solution surrounding that heart and monitor the strength and rate of its beating. The results should mimic that of Loewi's experiments; that is, the heart should respond to the acetylcholine by exhibiting a rapid decrease in the strength and rate of its beating.

FIGURE 39.11

The activity of the K^+ channel is responsible for causing the membrane potential to fall. Inhibiting the K^+ channel would decrease significantly the outflow of K^+ and, therefore, slow the fall in membrane potential. In other words, it would take much longer than normal for the membrane potential to reach its resting value.

IMR

The chimeric proteins must fold in an appropriate way in order to function. The hydrophilic domain must have a three-dimensional structure that can recognize the neurotransmitter, and the membrane-embedded hydrophobic domain must have a three-dimensional structure that can carry out the ion conduction function. The simplest interpretation of the results is that only two of the five artificial receptors had hydrophobic domains that were folded properly for conducting ions.

UQ

If we understood the biological nature of conscious thought, then perhaps we could design a more rational judicial system, for example.

Test Your Knowledge

1. d 2. b 3. a 4. c 5. e 6. a 7. c 8. b 9. e 10. a

Interpret the Data

1. Cyanide inhibits the transport of Na^+ out of the squid axon, as evidenced by the decrease in the rate of ^{22}Na lost from the axon.
2. The addition of broken-down ATP has no effect on the cyanide-caused inhibition of Na^+ transport out of the axon.
3. The addition of ATP increases the rate of ^{22}Na transport out of the axon, somewhat reversing the effects of the cyanide. These results mean that cyanide likely is poisoning the Na^+/K^+ active transport pump, which depends upon ATP hydrolysis for its action. An injection of ATP overcomes the cyanide inhibition until that ATP is used up, and then the curve shows a decrease again in ^{22}Na transport out of the axon (at around the three hour point).

Chapter 40

Study Break

STUDY BREAK 40.1

1. A nerve net is a loose meshwork of neurons organized in a radial pattern that reflects the radial symmetry of the animal in which it is found. Nerves are bundles of axons surrounded by connective tissue. Nerve cords are bundles of nerves.
2. Cephalization is the formation of a distinct head region containing ganglia, which form a major central control center or brain, and major sensory structures. Cephalization is found in more complex invertebrates and in all vertebrates.

STUDY BREAK 40.2

(a) Sympathetic nervous system; this is the classic "fight or flight" scenario for this autonomic nervous system division. (b) Parasympathetic nervous system.

STUDY BREAK 40.3

1. The blood–brain barrier is a distinct separation between the blood and the brain resulting from tight junctions between the epithelial cells forming the walls of the capillaries in the brain. The blood–brain barrier regulates exchanges between blood and brain, in particular preventing most dissolved substances from entering the cerebrospinal fluid, thereby protecting the brain and spinal cord from viruses, bacteria, and toxic substances that may be in the blood.
2. Gray matter consists of nerve cell bodies and dendrites; white matter consists of axons, many of them surrounded by myelin sheaths.
3. The brain stem regulates a number of the body's vital housekeeping functions without conscious involvement or control by the cerebrum, including heart and respiration rates, blood pressure, constriction and dilation of blood vessels, coughing, and reflex activities of the digestive system.
4. The cerebellum is an outgrowth of the pons, but it is structurally and functionally separate from the brain stem. The cerebellum has extensive connections with other parts of the brain, through which it receives sensory inputs from receptors in muscles and joints, from balance receptors in the inner ear, and from the receptors of touch, vision, and hearing. The cerebellum integrates the various sensory signals and compares them with signals from the cerebrum that control voluntary body movements. Information flow from the cerebellum to the cerebrum, brain stem, and spinal cord modifies and fine-tunes movements of the body. In humans, the cerebellum also is involved in the learning and memory of motor skills.

   The cerebral cortex is a thin, folded layer of gray matter that forms the surface of the cerebrum. The cerebral cortex contains sensory areas that receive and integrate sensory information of many kinds, including touch, pain, temperature, pressure, hearing, vision, smell, and taste. It also contains motor areas that are involved in controlling body movements and position, and association areas, which integrate information from the sensory areas and send responses to the motor area. Most higher functions of the human brain, including critical thinking, abstract thought, musical ability, and aspects of personality, involve activities of many regions of the cerebral cortex.

STUDY BREAK 40.4

Short-term memory involves transient changes in neurons. Short-term memory loss resulting from aging may be caused by a loss of control of the mechanisms involved, or, more likely, by a loss or degeneration of the neurons that constitute the short-term memory system.

Learning first involves storing memories. If the short-term memory system is faulty, perhaps due to loss or degeneration of neurons, then the transfer of short-term memories to the long-term memory system will be impaired. Another possibility is a loss or degeneration of the neurons responsible for long-term memory.

Think Like a Scientist

FIGURE 40.7

A slow response to the knee-jerk reflex test might indicate a defect in nerve conduction.

FIGURE 40.13

Yes. Information from the right visual field crosses over to the left side of the brain, and vice versa, via the optic chiasm. So, when a split-brain patient is shown a flower vase in the right visual field, this information crosses over to the left side of the brain (the optic chiasm is still intact even though the corpus callosum is severed). It is important to realize that each hemisphere of the brain specializes in different things. The right side of the brain is concerned with things such as artistic abilities and intuitive thinking, while the left side of the brain is concerned with language, reasoning, and mathematics. Because the two hemispheres can no longer communicate with each other, information about the flower vase is constrained to the left hemisphere once it crosses over through the optic chiasm. Since this hemisphere is involved in language, the patient will be able to describe the object without the help of the other hemisphere.

IMR

The key aspect of the experimental design was that the knockout mice lacked NMDA receptor function just in the hippocampus.

UQ

A good guess is that the generalist will show better learning. The reason is that it normally must find food across a wide variety of environmental conditions that might vary in unpredictability. Therefore, learning ability would be highly adaptive. The specialist could adapt quite well by "hard wiring" an attraction to all the specific chemical cues that are normally associated with its single food source. Of course, only an experimental test would verify this hypothesis.

Test Your Knowledge

1. d 2. b 3. b 4. b 5. e 6. c 7. a 8. c 9. c 10. a

Interpret the Data

1. In the first 5-minute interval, the control rats were more active than the rats exposed prenatally to MDMA.
2. 76 photobeam breaks
3. The rats exposed to MDMA moved most during the last 5-minute interval.
4. Yes, because the two groups of rats had different patterns of movement. The control rats began to move around the cage almost immediately. The MDMA rats were less active to begin with but continued moving long after the control rats had settled down.

Chapter 41

Study Break

STUDY BREAK 41.1

1. Sensory transduction is the conversion of a stimulus into a change in membrane potential.
2. The signals from particular sensory receptors are routed by afferent neurons to specific regions of the CNS. Processing of the incoming signals by those regions gives the "sense" of the stimulus—a smell, pain, and so forth.

STUDY BREAK 41.2

1. Proprioceptors detect stimuli that are processed to provide the animal with information about movements and position of the body.
2. All proprioceptors are mechanoreceptors because they are stimulated by a mechanical force.

STUDY BREAK 41.3

1. Octopuses have mechanoreceptors on their head and tentacles, similar to the lateral line of fishes. These mechanoreceptors detect vibrations in the water. Many insects have sensory receptors in the form of hairs or bristles that vibrate in response to sound waves, whereas some insects (for example, moths, grasshoppers, and crickets) have auditory organs on either side of the abdomen or on the first pair of walking legs.
2. In humans, vibrations representing sound frequencies are transmitted into the fluid-filled inner ear. They travel through the inner ear and cause the basilar membrane to vibrate in response, bending the sensory hair cells and stimulating them to release a neurotransmitter that triggers action potentials in afferent neurons leading from the inner ear. The vibrations from a particular sound frequency cause the basilar membrane to vibrate maximally at one particular location, stimulating the hair cells in that region to initiate action potentials. That information is sent to the brain, which integrates it into a perception of the sound stimulus.

STUDY BREAK 41.4

(a) A photopigment is an association of retinal with one of several different opsin proteins. (b) A photoreceptor is a receptor specialized for detection of colors (different wavelengths of light). (c) A ganglion cell's receptive field is the specific set of photoreceptors that sends signals to that cell. Receptive fields are usually circular and vary in size; the smaller the receptive field, the more precise the information sent to the brain and the sharper the image.

STUDY BREAK 41.5

1. Most odor perceptions arise from a combination of different olfactory receptors, which are located in the nasal cavities in humans. We have about 1,000 different olfactory receptor types, each of which responds to a different class of chemicals. The stimulated olfactory receptors send signals via the olfactory bulbs to the olfactory centers of the cerebral cortex, where the signals are interpreted as particular smells.
2. Chemicals bind to taste receptors and generate signals. Signals from taste receptors are relayed to the thalamus. From there, some signals go to gustatory centers in the cerebral cortex where they are integrated to produce taste perception. Other signals go to the brain stem and limbic system, which links tastes to involuntary visceral and emotional responses, such as sensations of pleasure or revulsion.

STUDY BREAK 41.6

The other sensory receptors involve specialized receptor structures. Afferent neurons synapse with the receptors. When a receptor is stimulated, the change in membrane potential (sensory transduction) is transmitted to the afferent neuron, which transmits the signal to the interneuron networks of the CNS. By contrast, both thermoreceptors and nociceptors consist of free nerve endings formed by the dendrites of afferent neurons, with no specialized receptor structures involved.

STUDY BREAK 41.7

Electroreceptors are used in aquatic vertebrates for electrolocation (locating other animals such as prey), electrocommunication (communicating with other members of the same species), and killing prey (involving high-voltage discharge).

Think Like a Scientist

IMR

The researchers showed clearly that a particular cDNA expressed in nonneuronal cells resulted in a receptor that responded specifically to capsaicin. Since a cDNA is derived from an mRNA, which encodes a protein, the receptor must be proteinaceous. The researchers also demonstrated that, in the transformed nonneuronal cells, capsaicin stimulation produced an influx of Ca^{2+}, thereby supporting the conclusion that the protein encoded by the cDNA is an ion channel.

UQ

1. Photoreceptors (they will clearly "detect" that the color of the eggs is odd); chemoreceptors (although the taste of the eggs is not altered, the person eating the green eggs may make a facial expression that leads you to believe that he or she has detected an altered taste). Combining an MRI and a PET scan might be a good method for testing synesthesia.
2. Detection and response to light are logically a key sensory system for any living organism—from prokaryotes to plants to animals—but sensing light with photoreceptors does not to have to involve "vision." Yet, the genes for photoreceptor proteins would logically be of early evolutionary origin.

Test Your Knowledge

1. c 2. a 3. c 4. b 5. e 6. b 7. d 8. a 9. e 10. d

Interpret the Data

1. Most of the bird pairs support the hypothesis, with k, l, and m providing the strongest data.
2. Pairs d and h show little difference between species, but pair g does not fit the hypothesis.
3. Overall the data appear to support the hypothesis.

Chapter 42

Study Break

STUDY BREAK 42.1

1. A *hormone* is a signaling molecule secreted by one group of cells and transported through the circulatory system to other, target cells, whose activities they change. Specific target cells react to the hormone because they carry receptors for the hormone. The best-known hormones are secreted by cells of the endocrine system and elicit a response in target cells that have receptors for the hormone. An *exocrine secretion* is a molecule released from an exocrine gland. Such molecules are secreted into ducts that lead outside the body or into the cavities of the digestive tract.
2. In *endocrine signaling,* a hormone secreted by an endocrine gland into the ECF is transported in the blood and causes a response in a target cell that has a receptor for the hormone. In *autocrine signaling,* a local regulatory molecule is released from a cell and binds to cell surface receptors on the same cells that produce the molecule, triggering a response.

STUDY BREAK 42.2

1. Glucagon, a peptide hormone, triggers a response by binding to a surface receptor. Glucagon binding activates the receptor, which triggers a signal transduction pathway inside the cell, leading to phosphorylation of target proteins. The altered activities of the phosphorylated proteins produce responses in the target cells, in this case the breakdown of glycogen in liver cells to glucose.

   Aldosterone, a steroid hormone, passes though the plasma membrane and binds to an internal receptor in the cytoplasm or nucleus, activating it. The hormone-activated receptor complex binds to control sequences in the DNA that the receptor recognizes and either activates or inhibits transcription of the associated target genes. Through binding to a receptor, the hormone affects transcription of specific genes in target cells.
2. A target cell could respond to different hormones if it carries receptors for those different hormones. Turning this around, a target cell will respond only to one hormone if it just has the receptor for that hormone. The same hormone could produce different effects in different cells if there are different receptors for that hormone, each triggering a distinct response pathway.

STUDY BREAK 42.3

1. The hypothalamus and anterior pituitary gland are connected by nerve and vascular tissues. The hypothalamus releases peptide neurosecretory hormones into the linking blood vessels. These hormones regulate the secretion of peptide hormones by the anterior pituitary gland. The pituitary hormones regulate several key body systems.
2. Secretion of hormones from the anterior pituitary is controlled by releasing hormones (RHs) and inhibiting hormones (IHs), which are released by the hypothalamus. For the posterior pituitary, neurosecretory neurons directly secrete the hormones into the circulatory system.

STUDY BREAK 42.4

1. Parathyroid hormone stimulates the dissolution of calcium and phosphate ions from bone and their release into the bloodstream.
2. The adrenal medulla secretes epinephrine and norepinephrine. Epinephrine prepares the body for handling stress or physical activity by, among other actions: (1) increasing heart rate; (2) breaking down glycogen and fats, thereby releasing glucose and fatty acids into the blood for fuel; (3) dilating blood vessels in the heart, skeletal muscles, and lungs to increase blood flow; (4) constricting blood vessels elsewhere, thereby raising blood pressure, reducing blood flow to the intestine and kidneys, and inhibiting smooth muscle contraction, which reduces water loss and slows down the digestive system; and (5) dilating airways in the lungs, thereby increasing air flow.

Norepinephrine has similar effects to epinephrine on heart rate, blood pressure, and blood flow to the heart muscle. In contrast to epinephrine, norepinephrine causes blood vessels in skeletal muscles to contract.

3. Glucocorticoids and mineralocorticoids are the two major classes of steroid hormones secreted by the adrenal cortex. Glucocorticoids help maintain the concentration of glucose and other fuel molecules in the blood, and mineralocorticoids regulate the levels of Na^+ and K^+ in blood and the ECF.
4. Estradiol is an estrogen and progesterone is a progestogen; both are steroid hormones. Estradiol produced by the ovaries stimulates maturation of sex organs at puberty and development of secondary sexual characteristics. Progesterone, also produced by the ovaries, prepares and maintains the uterus for implantation of a fertilized egg and the growth and development of an embryo.

STUDY BREAK 42.5

In general, invertebrates have fewer hormones regulating fewer body processes and responses than vertebrates do. Peptide and steroid hormones are produced in both invertebrates and vertebrates. However, most of those hormones are different in structure and molecular function in the two groups, even though the reaction pathways stimulated by the hormones are the same.

Think Like a Scientist

FIGURE 42.5

Mutant forms of the receptor could be produced *in vitro*, placed in liposomes, and tested for their ability to respond to epinephrine by measuring an increase in cAMP levels within fused cells. For example, specific mutations affecting the ability to bind epinephrine or activate the associated G protein could be found. For the latter, some specific mutations might not be able to activate the G protein, whereas others might activate the G protein constitutively.

IMR

It is possible that tamoxifen caused the breast cancer cells to make more of the receptor, which then resulted in a higher luciferase level.

UQ

Although no animal model perfectly mimics human breast cancer, animals are useful in understanding the effect of specific exposures on breast cancer risk and the risk of recurrence, without having to worry that some unintended factors explain the findings. Further, it is far easier to study the effects of timing of exposure to estrogenic compounds, and multigenerational effects, on breast cancer in animal models than it is in humans. The benefits of cell culture studies include the ability to perform highly mechanistic studies to understand causal effects of estrogens and their biological targets in affecting breast cancer growth. Main weakness of cell culture studies is that breast cancer cells are studied in isolation from other cells in the breast which are known to interact with the cancer cells to impact their growth.

Test Your Knowledge

1. b 2. d 3. a 4. c 5. c 6. e 7. a 8. b 9. d 10. b

Interpret the Data

1. The level of growth hormone was not constant over the 24-hour period. The level of this hormone increased markedly immediately after the onset of sleep and then declined in the middle of the sleep period. Another increase in growth hormone level occurred upon waking and then the level dropped to a level similar to that of presleep.
2. The level of prolactin also is not constant over the 24-hour period. The fluctuations in prolactin levels in the period before sleep are similar to those seen in that period for growth hormone. The level of prolactin increased in the sleep period, showing a rise immediately after sleep onset and continuing to a maximum level at the time of waking. The level of prolactin then decreased to a level similar to that of presleep.
3. Mostly there is no consistent relationship between the levels of the two hormones over the 24-hour period. The only exception is the initial rise in prolactin and in growth hormone that occurred with sleep onset.
4. Yes. The correlation between high level of prolactin with sleep and low level of prolactin with wakefulness supports the hypothesis. However, it cannot be determined from the data if the sleep-related rise in prolactin is directly dependent on sleep, or whether other factors triggered by sleep are involved.

Chapter 43

Study Break

STUDY BREAK 43.1

1. An axon terminal of an efferent neuron makes a synapse called a neuromuscular junction with a muscle fiber. When an action potential arrives at that junction, the axon terminal releases acetylcholine, which triggers an action potential in the muscle fiber that moves in all directions over its surface and penetrates to the interior of the fiber through T tubules. When the action potential reaches the end of the T tubules, ion channels are opened in the sarcoplasmic reticulum that allow Ca^{2+} ions to flow from the sarcoplasmic reticulum into the cytosol. Troponin then binds the ion and undergoes a conformational change that allows tropomyosin to enter the grooves in the actin helix of the thin filaments. As a result, the myosin-binding sites are uncovered, and the crossbridge cycle is turned on, leading to muscle contraction.
2. In the sliding filament mechanism of muscle contraction, the thin filaments on each side of a sarcomere slide over the thick filaments toward the center of the A band, thereby bringing the Z lines closer together. A crossbridge cycle is responsible for the contraction. At the beginning of the cycle, the myosin crossbridge has an ATP bound to it and is not in contact with the actin of the thin filament. The ATP is hydrolyzed, causing the myosin crossbridge to bend away from the tail and bind to a myosin-binding site on the thin filament that was uncovered in response to the release of Ca^{2+} ions into the cytosol from the sarcoplasmic reticulum. When the myosin crossbridge binds to the actin, the crossbridge snaps back toward the myosin tail to produce the power stroke that pulls the thin filament over the thick filament. ADP and phosphate are released from the crossbridge in this step. A new molecule of ATP now binds to the crossbridge, causing the myosin to detach from the actin. The cycle then repeats.
3. Contraction intensity: low in slow aerobic fibers, intermediate in fast aerobic fibers, and high in fast anaerobic fibers.

   Fatigue resistance: high in slow aerobic fibers, intermediate in fast aerobic fibers, and low in fast anaerobic fibers.

   Oxidative phosphorylation capacity: high in slow aerobic fibers, high in fast aerobic fibers, and low in fast anaerobic fibers.

   Number of mitochondria: many in slow aerobic fibers, many in fast aerobic fibers, and few in fast anaerobic fibers.

   Myoglobin content: high in slow aerobic, high in fast aerobic, and low in fast anaerobic.

STUDY BREAK 43.2

1. A hydrostatic skeleton provides support to the body or body part through muscles acting on compartments filled with fluid under pressure. In vertebrates, the penis is a hydrostatic skeletal structure.

   An exoskeleton is a rigid external body covering. The force of muscle contraction against the covering provides support to the body or part of the body. In vertebrates, the shell of a turtle or tortoise and the bony plates in the skin, the abdominal ribs, the collarbones, and most of the bony skull of the American alligator are exoskeletal structures.

   An endoskeleton consists of internal body structures such as bones. The force of muscle contraction against the internal body structures provides support. In vertebrates, the endoskeleton is the primary skeletal system.
2. The bones of the vertebrate endoskeleton provide support for the body and body parts, protect key internal organs, store calcium and phosphate ions, and are the sites where new blood cells form.

STUDY BREAK 43.3

1. Synovial joints consist of the ends of two bones that are enclosed by a fluid-filled capsule of connective tissue. Fluid within the capsule and a smooth layer of cartilage over the ends of the bones enable the bones to slide easily as the joint moves. In these joints, ligaments extend across the joints across the capsule. The ligaments confine the motion of the joint and protect it to an extent from deleterious effects of heavy loads.

   Cartilaginous joints have no fluid-filled capsule surrounding them, and the bones involved are covered with cartilage. The bones of the joint are covered by and connected with fibrous connective tissue. Cartilaginous joints are less movable than synovial joints.
2. Antagonistic muscle pairs are muscles arranged so that bones can be extended, flexed, or rotated in opposite directions around a joint. For example, in humans, the biceps and triceps muscles are an antagonistic muscle pair.

Think Like a Scientist

FIGURE 43.4

ATPase hydrolyzes ATP to ADP and inorganic phosphate. The existence of ATPase activity on myosin suggests that the sliding of thick (myosin) and thin (actin) filaments over each other requires, at least in part, energy released from ATP hydrolysis.

FIGURE 43.5

ATPase hydrolyzes ATP to ADP and inorganic phosphate. The existence of ATPase activity on myosin suggests that the sliding of thick (myosin) and thin (actin) filaments over each other requires, at least in part, energy released from ATP hydrolysis.

IMR

You would want to repeat the same kind of study with other mammals to see which genes change expression in response to training in the same way as was found in the racehorse study. One could use humans and dogs in such a study, and perhaps some other mammals that can be trained in the way needed.

UQ

Skeletal muscle consumes metabolites, including fatty acids and carbohydrates, both for movement and for building new muscle. Thus, maintaining and using new muscle could increase fat metabolism and its breakdown and oxidation for energy, while also diverting carbohydrates, which are often converted to fat, away from fat cells and to muscle cells.

Test Your Knowledge

1. c 2. e 3. a 4. d 5. b 6. b 7. c 8. a 9. d 10. b

Interpret the Data

1. Seven out of the nine treated children showed an increase in vertebral area during the study.
2. None of the six untreated control children showed an increase in vertebral area.
3. The treated children experienced a mean of 2.6 fractures per year, less than half of the mean of 6.3 fractures per year in the untreated children.
4. Yes.

Chapter 44

Study Break

STUDY BREAK 44.1

1. The three basic features of animal circulatory systems are: (1) a specialized fluid medium for transporting molecules, exemplified by the blood of vertebrates; (2) a muscular heart for pumping the fluid; and (3) tubular vessels for distributing the fluid pumped by the heart.
2. In an open circulatory system, there is no distinction between blood and interstitial fluid. Vessels from the heart release hemolymph directly into body spaces and the fluid

is subsequently collected and reenters the heart. In a closed circulatory system, blood is channeled in blood vessels leading to and from the heart and is distinct from the interstitial fluid.

In an open circulatory system, most of the fluid pressure generated by the heart dissipates when the blood is released into the body spaces. Consequently, blood flows relatively slowly. Humans could not function as we do with such a system, because we would not be able to distribute oxygen efficiently throughout the body, nor would we be able to eliminate the wastes we produce.

3. Atria are chambers of the heart that receive blood returning to the heart. Ventricles are chambers of the heart that pump blood from the heart. Arteries are vessels of circulatory systems that conduct blood away from the heart at relatively high pressure. Veins are vessels of circulatory systems that carry blood back to the heart.
4. The systemic circuit of the circulatory system is the circuit from the heart to most of the tissues and cells of the body and back to the heart. The pulmocutaneous circuit goes from the heart to the skin and lungs or gills in amphibians and back to the heart. The pulmonary circuit goes from the heart to the lungs and back to the heart.

STUDY BREAK 44.2

1. Erythrocytes originate as pluripotent stem cells in the red bone marrow. In humans and other mammals, they lose their nucleus, cytoplasmic organelles, and ribosomes as they mature, becoming essentially a membrane-bound hemoglobin reservoir that is not capable of protein synthesis. At the end of their life span—about 4 months—erythrocytes are engulfed by macrophages, a type of leukocyte, in the spleen, liver, and bone marrow.
2. Low oxygen content triggers a negative feedback mechanism to increase erythrocytes in the blood. The kidneys are stimulated to synthesize the hormone erythropoietin (EPO), which stimulates stem cells in the bone marrow to increase erythrocyte production. EPO synthesis is stopped when the oxygen content of the blood rises above normal levels.
3. Leukocytes act as the first line of defense against invading organisms, eliminate dead and dying cells from the body, and remove cellular debris. Platelets assist in blood clotting. When blood vessels are damaged, platelets stick to the collagen fibers exposed to leaking blood and recruit other platelets to the site. Eventually a plug forms at the site, sealing off the damaged area.

STUDY BREAK 44.3

1. The right atrium receives blood in the systemic circuit returning to the heart in vessels coming from the entire body, with the exception of the lungs. The right atrium pumps blood into the right ventricle.

   The right ventricle receives blood from the right atrium, and pumps it into the pulmonary arteries going to the lungs, beginning the pulmonary circuit.

   The left atrium receives blood returning from the lungs in the pulmonary veins, completing the pulmonary circuit. The left atrium pumps this blood into the left ventricle.

   The left ventricle pumps the blood received from the left atrium into the aorta where the blood begins its path in the systemic circuit.
2. Systole is the period of contraction and emptying of the heart. Diastole is the period of relaxation and filling of the heart between contractions.
3. Neurogenic hearts beat under the control of signals from the nervous system. If the signals cease, this type of heart stops beating. Myogenic hearts maintain a contraction rhythm without signals from the nervous system. In the event of a serious trauma to the nervous system, this type of heart keeps beating.
4. Pacemaker cells of the SA node undergo regularly timed depolarizations, which initiate waves of contraction that travel over the heart. The waves stop before they reach the bottom part of the heart as they encounter an insulating layer of connective tissue. The contraction signals at this point excite cells of the AV node, and then pass to the bottom of the heart along Purkinje fibers. There they induce waves of contraction that begin at the bottom of the heart and move upwards, expelling blood from the ventricles into the aorta and pulmonary arteries.

STUDY BREAK 44.4

1. Blood flow through capillary networks is controlled by contraction of smooth muscles in arterioles and by contraction of precapillary sphincters at the junctions of capillaries and arterioles.
2. In most body tissues, there are small spaces between capillary endothelial cells, but in the brain the capillary endothelial cells are tightly sealed together, forming a blood–brain barrier.
3. The two major mechanisms are diffusion along concentration gradients and bulk flow. Diffusion along concentration gradients occurs both through the spaces between the capillary endothelial cells and through the plasma membranes. The direction of movement of the molecule or ion depends on the concentration gradient. Thus, oxygen and glucose, which are at higher concentrations in the capillaries, diffuse into the interstitial fluid and then into the cells of the tissues. Carbon dioxide, which is at a higher concentration in the interstitial fluid, diffuses into the capillaries.

   Bulk flow carries water, ions, and molecules out of the capillaries. Driven by the pressure of the blood, which is higher than the pressure of the interstitial fluid, bulk flow occurs through the spaces between capillary endothelial cells.

STUDY BREAK 44.5

1. Arterial blood pressure must be regulated within limits to provide sufficient blood flow for the brain and other tissues and to prevent damage to blood vessels, tissues, and organs that would occur at high blood pressures.
2. The three main mechanisms for regulating blood pressure are: (1) controlling cardiac output (the pressure and amount of blood pumped by the ventricles); (2) controlling the degree of constriction of the blood vessels (mostly the arterioles); and (3) controlling the total blood volume.
3. The hormone epinephrine raises blood pressure by increasing the strength and rate of the heart rate, and by stimulating vasoconstriction of arterioles in certain parts of the body.

STUDY BREAK 44.6

Lymph is interstitial fluid, an aqueous solution containing molecules, ions, infecting bacteria, damaged cells, cellular debris, and lymphocytes. Because the lymph capillaries consist of a single layer of endothelial cells, the lymphatic fluid can enter them at sites where the endothelial cells overlap when the pressure of the interstitial fluid forces the flaps open. The openings produced are large enough for cells to enter.

Think Like a Scientist

FIGURE 44.12

An SA node dysfunction would involve nonnormal electrical signals. Because the SA node is the pacemaker for cardiac contraction, SA node dysfunction would result in an abnormal heart rhythm. The abnormality could be a rate of heart contraction that is too fast or too slow, depending on the dysfunction.

FIGURE 44.18

You would use the same norepinephrine-treated rabbit aorta preparation without endothelium as in the experiment described. The effect of NO in solution can then be determined. In an experiment like this, Furchgott and his colleagues demonstrated that the NO caused a relaxation of the muscle. Similar experiments by other researchers confirmed the result.

IMR

The key element of the experimental design was that the specific knockout of both α_{1D}-adrenergic receptor genes in the mice enabled the researchers to study just one variable—the loss of the receptor. Studies with drugs could not lead to conclusions specifically about this receptor because no drug selectively affects just one subtype of adrenergic receptors.

UQ

1. Probably not. The most common form of hemophilia is a result of a mutation in clotting factor VIII (see Section 42.2), and lampreys lack that clotting factor.
2. When a blood vessel is injured during surgery or extended bed rest, the body uses platelets and fibrin to form a blood clot. The coagulation process is similar in lampreys and mammals, so the molecular basis of drugs used to prevent DVT could be elucidated by using DVT preventive drugs on lampreys.

Test Your Knowledge

1. d 2. e 3. b 4. b 5. c 6. a 7. e 8. e 9. d 10. a

Interpret the Data

1. No. Nifedipine reduces blood pressure and heart rate prior to noticeable increases in plasma nifedipine (below 10 ng/mL).
2. Both capsules reduce blood pressure, but the GITS capsule seems to also reduce heart rate. Since reducing blood pressure or heart rate will decrease the workload on the heart, both will achieve this goal; however, GITS might be more effective.
3. Contracten XL causes greater elevation of nifedipine, which doesn't appear to be necessary for the beneficial effects. Based on these data alone, GITS capsules would be better at reducing workload with fewer side-effects.

Chapter 45

Study Break

STUDY BREAK 45.1

1. Mucous layers are physical barriers to pathogens. Mucous layers may contain toxins and other chemicals that kill pathogens. The aqueous environment in contact with an epithelial surface may inhibit or kill pathogens by being acidic, or by containing enzymes or bile juices.
2. Innate immunity provides a nonspecific, immediate response to a pathogen. It is the first response system that comes into play when a pathogen is encountered, but it retains no memory of the encounter. The innate immune responses include inflammation and specialized cells that attack pathogens.

   Adaptive immunity, by contrast, provides a specific response to a pathogen, and it retains a memory of exposure to the pathogen so that it can respond more quickly to future attacks. The adaptive immune response takes several days to become protective, in contrast to minutes for an innate immune response.

STUDY BREAK 45.2

1. The inflammatory response is characterized by heat, pain, redness, and swelling at the infection site.
2. Inflammation-mediating molecules released at the infection site cause the dilation of local blood vessels and increase their permeability. As a result, blood flow increases and fluid leaks from the blood vessels. Heat, redness, and swelling are direct consequences of these effects. Pain is caused by the migration of macrophages and neutrophils to the infection site, and their activities at the site.
3. The complement system is a nonspecific defense mechanism in which more than 30 interacting plasma proteins are activated by molecules on the surface of pathogens. Activated complement system proteins produce membrane attack complexes that insert into the plasma membrane of many types of bacterial cells and coat other types of cells with fragments of the complement proteins. Membrane-attack complexes create pores that lead to loss of osmotic balance, which causes the pathogens to lyse. Pathogens coated with complement protein fragments are recognized by phagocytes that then engulf and destroy the pathogens.
4. Viral pathogens do not have surface molecules that are distinct from those of the host. The main strategies used by the host in innate immunity are interferon and natural killer cells.

STUDY BREAK 45.3

1. The antibody-mediated immune response uses antibodies secreted by plasma cells (differentiated from activated B cells) to target antigens in various body fluids. The cell-mediated immune response uses cytotoxic T cells to target and kill cells infected with a pathogen.
2. An antibody is a member of the immunoglobulin (Ig) family of proteins. It consists of four polypeptides, two of which are identical heavy chains and two of which are identical light chains. Structurally, an antibody looks like a Y. Each of the two arms of the Y involves pairing of a light chain with part of the heavy chain. The tail of the Y consists of the remaining parts of the heavy chains paired together. Each heavy chain and each light chain has a variable region and a constant region. The variable regions, which occur at the top half of the arms of the Y, form two identical binding sites for the antigen with which the antibody can react.
3. DNA rearrangements of genes during differentiation are responsible for generating diverse light-chain genes and heavy-chain genes for the B-cell receptor and for the T-cell receptor. In brief, there are many different V DNA segments, one C DNA segment, and a few different J (joining) DNA segments. During differentiation, a light-chain gene is assembled from one V segment, one J segment, and the C segment, creating a gene that can be transcribed. Transcription of the gene produces a pre-mRNA that is spliced to remove introns and generate the translatable mRNA. The various combinations of segments plus imprecision in the joining at the DNA level of the V and J segments produces many types of light-chain genes. A similar DNA rearrangement process occurs for heavy-chain genes. Together, this results in tremendous variability in the antibody molecules that can be made.
4. Clonal selection is the process by which an antigen stimulates the production of a clone of B-cell-derived plasma cells that secrete antibodies against that antigen. Clonal selection accounts for the rapid response, specific action, and diversity of antibodies seen for the adaptive immune system. It also accounts for immunological memory through the production of memory B cells and memory T cells.
5. Immunological memory is the aspect of adaptive immunity that allows for a rapid, intense immune reaction to develop if the body encounters an antigen it has seen before.

STUDY BREAK 45.4

1. Immunological tolerance, a feature of the adaptive immune system, protects the body's own molecules from attack by its immune system. B cells and T cells are involved in the development of immunological tolerance, but the exact process is not understood.
2. Basically, an allergy results from an overreaction of the immune system to a particular antigen. The substances responsible for allergic reactions are a class of antigens known as allergens. Allergens stimulate B cells to secrete IgE antibodies in high amounts. IgE binds to mast cells, which are then induced to oversecrete histamine. Histamine contributes to inflammation, and at high amounts, the histamine-induced inflammation can be severe and even life threatening.

STUDY BREAK 45.5

Invertebrates lack the adaptive immune system seen in mammals; they have no B or T cells, for instance. Invertebrate immune defenses, like the innate immune system of mammals, are nonspecific. For example, all invertebrates have phagocytic cells that engulf pathogens, and many invertebrates produce antimicrobial proteins such as lysozyme. Also, many invertebrates produce proteins of the immunoglobulin family. Although these are not antibody molecules, in some invertebrates they provide a protective function against some pathogens.

Think Like a Scientist

FIGURE 45.5

The key cellular event is the interaction of an antigen-presenting B cell with a helper T cell that differentiated as a result of encountering and recognizing the same antigen as did the B cell.

FIGURE 45.10

Memory cytotoxic T cells in the cell-mediated response are cells that have been programmed to recognize a particular antigen of a pathogen. Similar to the memory cells for the antibody-mediated immune response, memory cytotoxic T cells are long-lived and serve to mount a rapid response should the same antigen be detected on the surface of an infected host cell.

IMR

A logical target would be the expression of FasL in the melanoma cells. A reduction in or absence of FasL receptors on the melanoma cell surface would reduce or abolish the induction of apoptosis of cytotoxic T cells.

UQ

MHC-I genes are highly polymorphic. The large number of MHC-I variants expands the number of antigens that can potentially be targeted by the immune system. A new MHC-I molecule like HLA-C with unique characteristics might be selected for in a population if it allowed the host immune system to better control a pathogen that would otherwise evade the immune system. In fact, some investigators have hypothesized that pathogens put pressure on the evolution of MHC-I sequences.

Test Your Knowledge

1. a 2. c 3. b 4. e 5. e 6. d 7. b 8. c 9. a 10. d

Interpret the Data

1. At 110 months, about 1% of the women who were not infected with any HPV type and about 17% of women who were infected with HPV16 had cervical cancer.
2. Women infected with both HPV16 and HPV18 must be in the HPV16 group.
3. Yes, by adding the percentages of individual groups.
4. Yes.

Chapter 46

Study Break

STUDY BREAK 46.1

1. The respiratory medium is the environmental source of oxygen and the repository for released CO_2. For aquatic animals, water is the respiratory medium; for terrestrial animals, it is air. The respiratory surface is the layer of epithelial cells between the body and the respiratory medium. Gas exchange occurs across the respiratory surface—O_2 in and CO_2 out of the body. In certain small animals, the body surface itself is the respiratory surface. Among larger animals, aquatic animals use gills, insects use tracheal systems, and terrestrial animals use lungs as the respiratory surface.
2. The advantage of water over air as a respiratory medium is that it readily enables the respiratory surface to remain wet at all times. Two key advantages of air over water as a respiratory medium are: (1) there is much more oxygen in air than in water; and (2) air is much less dense and less viscous than water, so significantly less energy is needed to move air over the respiratory surface than to move water.

STUDY BREAK 46.2

1. The advantages of gills to water-breathing animals over skin breathing are greater efficiency of gas exchange, the ability to live in more diverse habitats, and the potential to achieve a greater body mass.
2. In countercurrent exchange, the respiratory medium flows in the opposite direction of the blood flow under the respiratory surface. Examples are the flow of water over the gills in sharks, fishes, and some crabs and the one-way flow of air through the lungs of birds; all these are opposite to the flow of blood. The advantage of this mechanism is that it maximizes the amounts of O_2 and CO_2 exchanged with the respiratory medium.
3. In the tracheal system, fine branches called tracheoles end in tips that are in contact with body cells. The tracheole tips are filled with fluid, and gas exchange occurs through the fluid and the plasma membrane of the body cells in contact with the tips. Air enters the tracheal system through spiracles on the body surface. Movement of air through the tracheal system occurs by muscle-driven contractions and expansions of air sacs within the system.
4. In positive pressure, breathing, gulping, swallowing, or pumping action forces air into the lungs. In negative pressure, breathing, muscular activity expands the lungs, lowering the pressure of air in the lungs and thereby causing air to be pulled inward.

STUDY BREAK 46.3

1. Contraction of the diaphragm and the external intercostal muscles pulls the ribs upward and outward, expanding the chest and lungs. By this negative pressure mechanism, the air pressure within the lungs is lower than outside of the body. The higher outside pressure drives air into the lungs, expanding and filling them. Relaxation of the diaphragm and the intercostal muscles reverses the pressure condition, and the elastic recoil of lungs expels the air.
2. For conscious exhalation, you contract the internal intercostal muscles between the ribs to pull the diaphragm down while simultaneously contracting abdominal muscles to compress the abdominal organs and force them upward against the diaphragm to expel air from the lungs.
3. The most important feedback stimulus for breathing is the level of CO_2 in the blood.
4. The chemoreceptors in the medulla respond to the CO_2 level in the blood. If it rises, the medulla triggers an increase in the rate and depth of breathing. If it falls, the medulla triggers a decrease in the rate and depth of breathing.

STUDY BREAK 46.4

1. Hemoglobin is present in red blood cells. O_2 diffuses from alveoli into the plasma solution in the capillaries and then into the red blood cells. In the red blood cells, the O_2 binds to hemoglobin, thereby lowering the partial pressure of O_2 in the plasma. This leads to more O_2 molecules diffusing down the oxygen concentration gradient from alveolar air to blood.

   The binding of O_2 to hemoglobin is reversible. Further, the affinity of hemoglobin for O_2 increases as the partial pressure of O_2 increases and vice versa. This property is important for determining the release of O_2 from hemoglobin keyed to tissue requirements.
2. CO has much greater affinity for hemoglobin than does O_2, and so it displaces O_2 from hemoglobin, reducing the amount of O_2 carried in the blood. Because the brain does not monitor O_2 levels, and other receptors do not respond until blood O_2 levels are critically low, victims can easily lapse into unconsciousness and death.

STUDY BREAK 46.5

- More blood per unit body weight than land animals
- Additional red blood cells, many stored in the spleen and released during a dive
- More myoglobin in muscles than land animals
- Slowing of the heart by as much as 90%
- Reduction of blood circulation to internal organs and muscles by up to 95%
- Retention of lactic acid in the muscles with no release into the blood until the animal returns to the surface

Think Like a Scientist

FIGURE 46.14

The specific mutations presumably involved in the evolutionary adaptation of deer mice to high altitude had to have occurred in a relatively short period of time, given that the mice could have only inhabited the alpine areas no earlier than 10,000 years ago.

IMR

Expression vectors were described in Chapter 18. They contain a selectable marker for selecting *E. coli* transformants that contain them, an origin of replication sequence for replication of the plasmid independently of the bacterial chromosome, and restriction sites for cloning genes of interest into the vector. Most important here is the expression of myoglobin in *E. coli.* The source of the myoglobin gene is the sperm whale, which is a eukaryote. Eukaryotes use different molecular signals for the transcription of protein-coding genes and for the translation of the resulting mRNAs. Therefore, to express a myoglobin gene in *E. coli,* it must be inserted into an expression vector that has, for transcription, a transcription terminator sequence in addition to the bacterial promoter and, for translation, a ribosome binding site so that protein synthesis will be initiated at the correct start codon. Using an expression vector to synthesize a eukaryotic protein in *E. coli* is diagrammed in Figure 18.11.

UQ

1. Carbon dioxide level.
2. Contraction of smooth muscle in the heart is under the control of the autonomic nervous system. Nicotine stimulates the postganglionic neurons of both the sympathetic (increases heart rate) and parasympathetic (decreases heart rate) divisions and disrupts heart rate rhythm. Therefore, research is focused on both divisions, as nicotine affects the opposing actions of these divisions.

Test Your Knowledge

1. e 2. d 3. a 4. c 5. a 6. d 7. b 8. d 9. e 10. a

Interpret the Data

1. Tobacco smoke significantly increased the lung tissue damage in these mice.
2. The tobacco-exposed mice given tomato juice showed significantly less tissue damage than tobacco-exposed mice without tomato juice. This suggests that something in the tomato juice is protecting again lung tissue damage, which supports the hypothesis that lycopene has a protective effect.

Chapter 47

Study Break

STUDY BREAK 47.1

1. Carnivores primarily eat other animals as their source of organic materials. Herbivores primarily eat plants as their source of organic materials. Omnivores obtain their organic materials from any digestible organic source.
2. Essential nutrients are the amino acids, fatty acids, vitamins, and minerals that the animal cannot make itself and must obtain from its diet. The list of essential nutrients varies among animal types.
3. Deposit feeders pick up or scrape particles of organic matter from solid material they live in or on. Suspension feeders ingest small organisms that are suspended in water. Depending on the animal, the ingested organisms may be bacteria, protozoa, algae, or small crustaceans, or fragments of those organisms.

STUDY BREAK 47.2

1. Intracellular digestion takes place within body cells, whereas extracellular digestion takes place outside the cells, either in a pouch or a tube enclosed by the body.
2. (1) Mechanical processing, to break up the food; (2) secretion of enzymes and other substances that aid digestion; (3) enzymatic hydrolysis of food molecules into simpler molecular subunits; (4) absorption of the molecular subunits into body fluids and cells; and (5) elimination of undigested materials.

STUDY BREAK 47.3

1. The two classes of vitamins are the water-soluble vitamins and the fat-soluble vitamins. Fat-soluble vitamins ingested in the diet in excess of bodily needs can be stored in adipose tissues. In contrast, excess water-soluble vitamins in the diet are excreted in the urine. Hence, it is critical that humans meet their daily requirements for the water-soluble vitamins.
2. The four layers of the gut are the mucosa, the submucosa, the muscularis, and the serosa. The muscularis layer, which is formed from two smooth muscle layers, is responsible for peristalsis.
3. Pepsin is secreted from chief cells into the stomach lumen as an inactive precursor molecule, pepsinogen. The pepsinogen is converted to pepsin by the highly acid conditions of the stomach. Pepsin is a digestive enzyme that begins the digestion of proteins by making breaks in polypeptide chains.
4. In the small intestine, the digestion of macromolecules into their molecular subunits occurs, and those subunits are absorbed. In the large intestine, water and mineral ions are absorbed from the remaining digestive contents, leaving the undigested remnants, the feces, which are expelled from the body.

STUDY BREAK 47.4

1. Gastrin stimulates the secretion of pepsinogen and HCl in the stomach. Pepsin generated by cleavage of pepsinogen digests protein in the swallowed food. Gastrin also stimulates contraction of the stomach and the intestines. Cholecystokinin (CCK), released in the duodenum in response largely to fat content in chyme, inhibits gastric activity; that is, CCK has the opposite effect of gastrin.
2. The hypothalamus has two interneuron centers that play a central role in regulating the digestive processes. One center stimulates appetite and reduces oxidative metabolism. The other center works in opposition to the first center by stimulating the release of a peptide hormone that inhibits appetite.

STUDY BREAK 47.5

1. Incisors nip or cut food. Canines bite and pierce food. Premolars and molars crush, grind, and shear food.
2. Symbiotic microorganisms aid digestion in many herbivores by assisting in the breakdown of plant material. The microorganisms synthesize cellulase, an enzyme vertebrates cannot make, which hydrolyzes the cellulose of plant cell walls into glucose subunits.

Think Like a Scientist

FIGURE 47.21

One could colonize the intestines of the germ-free normal mice with bacteria harvested from the intestines of obese (*ob*/*ob*) and (as a control) wild-type (ob^+/ob^+) mice and determine the effect on body fat. If the intestinal bacteria directly contribute to production of body fat, the prediction is that the percentage of body fat in obese mice would increase significantly whereas it would not in wild-type mice. Gordon's research group did this experiment and observed the result of the prediction.

IMR

Although rats are a highly used model mammal for experiments, rats are different enough from humans that results from experiments with rats cannot necessarily be assumed to parallel results from similar experiments in humans. To test whether a drug found effective in rats is effective in humans would require clinical trials in which administration of the drug to human subjects is done in a carefully designed experiment.

UQ

1. Tortoises are ectotherms and allocate energy to activity associated with maintaining body temperature; hares are endotherms and allocate energy to physiological processes to maintain body temperature.
2. They are used widely as stored energy. In fact, fats are an excellent source of energy in the diet.

Test Your Knowledge

1. b 2. b 3. a 4. d 5. c 6. e 7. c 8. d 9. e 10. a

Interpret the Data

1. About 30% (0.30) of the Hazda had fewer than five copies of the *AMY-1* gene.
2. Yes, about 10% of the European-Americans had more than 10 copies of the *AMY-1* gene.
3. Yes, the data support the hypothesis that a starchy diet favors duplications of the *AMY-1* gene. Cultural groups with a high-starch diet have, proportionally, more copies of the *AMY-1* gene than cultural groups with low-starch diets.

Chapter 48

Study Break

STUDY BREAK 48.1

Osmosis is a process in which water molecules move across a selectively permeable membrane from a region where they are more highly concentrated to one where they are less highly concentrated.

Osmolarity is the total solute concentration of a solution. Osmolarity is measured in osmoles per liter of solution, where an osmole is the number of solute molecules and ions (in moles).

A solution that is hypoosmotic has lower osmolarity than the solution on the other side of a selectively permeable membrane. The solution with the higher osmolarity is said to be hyperosmotic.

An osmoregulator is an animal that uses an active control mechanism to keep the osmolarity of cellular and extracellular fluids the same. This osmolarity value may differ from the osmolarity of the surroundings.

Tubules that carry out osmoregulation and excretion are formed from transport epithelium, a layer of cells with specialized transport proteins in their plasma membranes that move specific molecules and ions into and out of the tubule.

STUDY BREAK 48.2

Protonephridia are found in flatworms and larval mollusks. They are the simplest invertebrate excretory tubules. Body fluids enter the end of protonephridia and are moved through the tubule by movement of cilia on the flame cells. As the fluids move through the protonephridia, some molecules and ions are reabsorbed, and others, including nitrogenous wastes, are secreted into the tubules. Excess fluid is released through pores connecting the network of protonephridia to the body surface.

Metanephridia are found in annelids and most adult mollusks. Body fluids enter the funnel-like proximal end, driven by cilia surrounding that end. Some molecules and ions are reabsorbed as the fluids move through the tubule, and other ions and nitrogenous wastes are secreted into the tubule and excreted from the body surface.

Malpighian tubules are found in insects and other arthropods. Body fluids enter the tubules through spaces between the tubule cells. The distal ends of the tubules empty into the gut. Pressure is not used in the filtration step in Malpighian tubules. The secretion of K^+ from the hemolymph into the lumen draws in Cl^-. The accumulation of KCl causes water to enter the tubule from the hemolymph by osmosis. Organic wastes now are secreted into the tubule. When the fluids reach the hindgut and rectum, K^+ and Cl^- are reabsorbed, followed by water. As water leaves, uric acid precipitates as crystals that are excreted from the rectum in the feces.

STUDY BREAK 48.3

1. At the proximal end, a human nephron forms a cuplike region called Bowman's capsule. Bowman's capsule surrounds the glomerulus, a complex of blood capillaries. The capsule and the glomerulus are located in the renal cortex. The proximal convoluted tubule descends in a U-shaped bend called the loop of Henle into the renal medulla, and ascends again to form the distal convoluted tubule. Up to eight distal convoluted tubules drain into one collecting duct.
2. The collecting ducts are permeable to water but not to salt ions. The ducts, which begin in the cortex and descend into the medulla of the kidney, become surrounded by an ever-increasing solute concentration. As the urine passes

down the collecting ducts, water moves osmotically out of the ducts, causing the concentration of the urine to increase. At the bottom of the collecting ducts, the urine is about four times more concentrated than body fluids.

STUDY BREAK 48.4

The RAAS compensates for excessive loss of salt and body fluids. The ADH system compensates for excessive water intake or loss. Combined, the two systems play an important role in regulating the interactions between the kidneys and the rest of the body.

The RAAS is activated when the Na^+ concentration of body fluids falls, causing the volume of extracellular fluids and blood pressure also to drop. Cells in the juxtaglomerular apparatus secrete renin, which activates angiotensinogen to produce angiotensin I. Angiotensin-converting enzyme converts angiotensin I to angiotensin II, which has three effects: (1) It constricts arterioles around the body, thereby quickly increasing blood pressure; (2) It stimulates the secretion of aldosterone from the adrenal cortex, which increases Na^+ reabsorption in the kidneys and raises the osmolarity of body fluids; as a result, water moves from the tubules into the interstitial fluid, conserving water; and (3) It stimulates thirst so that more water is brought into the body. The RAAS is suppressed if NaCl concentration in body fluids is higher than normal.

In the ADH system, ADH is released from the posterior pituitary when osmoreceptors in the hypothalamus detect an increase in the osmolarity of body fluids. ADH increases water reabsorption from the distal convoluted tubules and collecting ducts of the kidney; as a result, urinary output is reduced and water is conserved. In the case of a decrease in osmolarity, ADH release from the pituitary is inhibited, thereby decreasing water reabsorption in the kidney.

STUDY BREAK 48.5

1. Marine teleosts live in seawater, which is hyperosmotic to their body fluids. They drink seawater continuously to replace water lost to the environment by osmosis. Na^+, K^+, and Cl^- ions from the seawater they drink are excreted by the gills. Nitrogenous wastes are released from the gills, primarily as ammonia, by simple diffusion.

   Freshwater fishes live in fresh water, which is hypoosmotic to their body fluids. They take in water by osmosis and excrete excess water. Salts needed for bodily functions are obtained from food and by active transport through the gills from the water. Nitrogenous wastes are excreted from the gills as ammonia.
2. The excretion of urea in mammals involves expelling the urea in solution—urine. Thus, there is loss of water with excretion of nitrogenous waste as urea. Excretion of nitrogenous waste in the form of uric acid conserves water because uric acid crystals are almost water-free.

STUDY BREAK 48.6

Ectothermy applies to animals that obtain heat energy primarily from the environment; endothermy applies to animals that obtain heat energy primarily from internal reactions. Generally speaking, ectotherms are highly successful in warm environments, but their bodily functions slow down as the temperature drops. Endotherms remain active over a broader range of environmental temperatures than ectotherms do, but they require an almost constant supply of energy to maintain their body temperature.

STUDY BREAK 48.7

1. The thermoregulatory responses shown by ectotherms can be physiological, such as regulating blood flow to internal organs or to the skin, or behavioral, such as physically moving to a location in the environment suitable for their heat energy needs at the time.
2. Thermal acclimatization refers to the physiological changes that many ectotherms make to compensate for seasonal shifts in environmental temperature.

STUDY BREAK 48.8

Temperature regulation in endothermic animals involves mechanisms that balance internal heat production against heat loss from the body. Internal heat production is controlled by negative feedback pathways triggered by thermoreceptors in the skin, the hypothalamus, and the spinal cord. When a deviation from the set point occurs, this system operates to return the core temperature to that set point through changes in metabolic processes, behavioral activities, and control of heat loss at the skin surface.

Think Like a Scientist

FIGURE 48.12

Heritable central diabetes insipidus must be caused by some defect in the amount of ADH. That could result from mutations affecting the hypothalamus or pituitary, which are involved in the synthesis and release of ADH. Heritable nephrogenic diabetes insipidus must be caused by some defect in the action of ADH. Mutations in this case could be in genes encoding the ADH receptor or the aquaporins involved in the response.

IMR

The ORE is at a significant distance from the site where transcription begins, indicating that is likely is in the enhancer for the gene. It is located too far from the transcription site to be a regulatory sequence in the promoter proximal region. (Refer to Section 16.2 for information about regulatory sequences in eukaryotic protein-coding genes.)

UQ

Recurrence of kidney disease in a transplanted kidney suggests that the disease is not intrinsic to the kidney, but reflects factors made elsewhere that act on the kidney in some way. Recurrent disease is in general much less common in inherited forms of kidney disease which, for the most part, are defects in kidney-expressed proteins. One interesting clinical observation occurs in patients with Alport syndrome, a disease of the basement membrane protein type 4 collagen. Alport syndrome patients develop kidney disease, and can also have deafness and eye problems. Patients with kidney failure from Alport syndrome who receive kidney transplants sometimes develop new antibodies directed against normal type 4 collagen (which was absent previously). This can lead to a different form of kidney disease characterized by antibody-mediated inflammation of the glomeruli.

Test Your Knowledge

1. d 2. c 3. a 4. b 5. d 6. e 7. c 8. b 9. d 10. e

Interpret the Data

1. Phase 2
2. Phase 1
3. Yes.

Chapter 49

Study Break

STUDY BREAK 49.1

Recall that asexual reproduction produces offspring with genes from only one parent. In most animals that undergo asexual reproduction, the offspring are produced by mitosis and, therefore, are genetically identical to one another and to the parent. Such genetic homogeneity can be advantageous in stable and uniform environments. Another advantage is that individuals do not need to use energy to produce gametes or to find and select a mate. Further, for individuals in sparse populations, or for sessile animals, asexual reproduction can be an advantage. A disadvantage is that a genetically homogeneous population may not adapt readily to new environments.

By contrast, sexual reproduction always generates genetic diversity among offspring. This provides a population the opportunity to adapt to changing environments, and perhaps to move to and colonize new environments. Disadvantages of sexual reproduction include the expenditure of energy and raw materials to produce gametes and to find and select mates. Another disadvantage of sexual reproduction is that it produces fewer offspring per parent than does asexual reproduction; that is, asexual reproduction of a female produces all-female progeny, each of which can produce offspring for the next generation. However, in sexual reproduction, only one half of the progeny of a mating are females and only they can produce the offspring of the next generation.

STUDY BREAK 49.2

1. Egg coats are surface coats around the egg. They are added during oocyte development or fertilization. Egg coats provide protection against mechanical injury, infection by microorganisms, and, in some species, loss of water.

   Mammalian eggs have an egg coat called the zona pellucida immediately surrounding the egg. This gel-like coating is called the vitelline coat in other organisms. In birds, a thick solution of proteins—the egg white—surrounds a vitelline coat. Bounding the egg white is a hard shell.
2. The slow block to polyspermy occurs in many organisms, including mammals. The fusion of egg and sperm triggers an increase in calcium ions in the cytosol. The calcium ions activate proteins that initiate a high level of metabolic activity in the egg. The released calcium ions also cause the cortical granules to fuse with the plasma membrane of the egg and release their contents to the outside. Enzymes released from those granules alter the egg coats in a matter of minutes after fertilization, and this blocks further sperm from attaching and penetrating the egg.

STUDY BREAK 49.3

- FSH stimulates a number of oocytes to begin meiosis. A follicle forms around each oocyte.
- FSH and LH interact to stimulate follicle cells to secrete estrogens.
- Later in the cycle, an increased level of estrogen leads to a new burst of release of FSH and LH. This new LH stimulates follicle cells to release enzymes that digest away the follicle wall, leading to egg release.
- The new LH also causes the remaining follicle cells to grow into the corpus luteum.
- FSH and LH levels decrease. This removes the stimulatory signal for follicular growth, and no new follicles grow in the ovary.

STUDY BREAK 49.4

The oral contraceptive pill inhibits the secretion of FSH and LH by the pituitary. FSH and LH stimulate oocytes in the ovaries to begin meiosis. They also stimulate follicle cells that surround the oocytes to secrete estrogens. Later, LH causes ovulation—the release of an egg. Therefore, without FSH and LH, ovulation cannot occur.

Think Like a Scientist

FIGURE 49.12

The menstrual cycle would not proceed normally; that is, menstruation would not occur.

FIGURE 49.13

Look for a correlation between vocalization and stage of the menstrual cycle in nonhuman mammals as measured by hormonal levels. One study with giant pandas, for example, showed that the acoustic structure of female chirps differs depending on the stage of the menstrual cycle. Potentially, a male giant panda could use such vocal cues, although that has yet to be determined. (Reference: B. D. Charlton et al. 2010. Female giant panda *(Ailuropoda melanoleuca)* chirps advertise the caller's fertile phase. *Proceedings of the Royal Society B* 277:1101–1106.)

IMR

The study showed clearly a human male superiority in olfactory sensitivity for bourgeonal. (The authors stated that it was the first study to show a male superiority in olfactory sensitivity.) Potentially, this might be due to differences in its behavioral relevance for males and females, although that remains to be shown. (Reference: P. Olsson and M. Laska. 2010. Human male superiority in olfactory sensitivity to the sperm attractant odorant bourgeonal. *Chemical Senses* 35:427–432.)

UQ

It could be used in cases of a low sperm count (few sperm produced), poorly motile or immotile sperm, incomplete sperm maturation, or inability to produce sperm by ejaculation (sperm could be collected surgically from the epididymis).

Test Your Knowledge

1. b 2. a 3. d 4. a 5. e 6. c 7. b 8. e 9. c 10. d

Interpret the Data

1. The lowest sperm count, 58.7 million/mL, was in Columbia, Missouri. The highest, 102.9 million/mL, was in New York, New York.
2. The lowest percent motile sperm was 48.2% in Columbia, Missouri, and the highest was 56.4% in New York, New York.
3. Based on these data, the adverse effects of age and STDs were unlikely explanations for the high sperm count in New York, New York, relative to the other locations, since New York had the highest average age and highest percentage of men with a history of STDs. The data do not show a clear correlation between smoking and sperm count, but it could have been a factor explaining the results.
4. The data support the hypothesis that living near farmland may adversely affect male reproductive function, but further study would be needed to eliminate other variables that could have contributed to the differences.

Chapter 50

Study Break

STUDY BREAK 50.1

1. In cleavage divisions, cycles of DNA replication and division occur without the production of new cytoplasm. Therefore, the cytoplasm of the egg partitions into many cells without increasing in overall size or mass. In cell division in an adult organism, the mitotic cell cycle involves cycles of DNA replication and division interspersed with cell growth; that is, the cell grows, then divides, then the two progeny cells grow, then divide, and so on. New cytoplasm is produced during these cell cycles so that, overall, there is an increase in mass of the cells.
2. The three primary cell layers of the embryo are produced by gastrulation. They are ectoderm, mesoderm, and endoderm.

STUDY BREAK 50.2

1. The gray crescent establishes the dorsal–ventral axis, with the gray crescent marking the future dorsal side.
2. If cells of the dorsal lip of the blastopore are transplanted to another location in the egg, they cause a second blastopore, and subsequently a second embryo, to form in that region.
3. The extraembryonic membranes in birds and their functions are as follows:
    - Yolk sac: Surrounds the yolk, which provides nutrients to the embryo.
    - Chorion: Exchanges oxygen and carbon dioxide with the environment through the egg shell.
    - Amnion: Encloses the embryo and secretes amniotic fluid into the space that provides an aquatic environment in which the embryo can develop.
    - Allantois: Stores nitrogenous wastes, primarily in the form of uric acid, that are derived from the embryo. Part of the allantoic membrane forms a bed of capillaries that is connected to the embryo and that delivers carbon dioxide from the embryo to the chorion and picks up oxygen absorbed through the shell.

STUDY BREAK 50.3

1. The three primary tissues, ectoderm, mesoderm, and endoderm, give rise to the tissues and organs of the embryo.
2. The central nervous system, notably the brain and spinal cord, develops from the neural tube. Neural crest cells give rise to parts of the nervous system (including cranial nerves, ganglia of the autonomic nervous system, peripheral nerves from the spinal cord to body structures, and nerves of the developing gut) and contribute to a variety of other body structures (for example, bones of the inner ear and skull, cartilage of facial structures, teeth, pigment cells in the skin, and the adrenal medulla).

STUDY BREAK 50.4

1. Cells of the trophoblast are responsible for implantation of the blastocyst in the endometrium (uterine lining). When the blastocyst is ready for implantation, trophoblast cells secrete proteases that digest pathways between endometrial cells. Dividing trophoblast cells fill the spaces and, through continued digestion and division, the blastocyst burrows into the endometrium and eventually becomes covered by a layer of endometrial cells. From then on, cells originating in the trophoblast support the development of the embryo/fetus in the uterus through the production of the chorion.

    The inner cell mass becomes the embryo/fetus itself. During implantation, the inner cell mass separates into the epiblast and the hypoblast. The epiblast produces the ectoderm, mesoderm, and endoderm of the developing embryo. The hypoblast gives rise to part of the extraembryonic membranes.
2. Oxytocin. This hormone is responsible for stimulating contractions of the uterus.

STUDY BREAK 50.5

1. Cell division, cell movement, and cell adhesion.
2. Induction is the process whereby a region of the embryo acts on other cells to alter the course of development. Induction is brought about by proteins; hence, induction is under genetic control.

Think Like a Scientist

FIGURE 50.13

The embryo/fetus and mother are different genetically. Therefore, the cells of the embryo/fetus will have surface antigens recognized as foreign by the mother. This will induce an immune response by the mother (see Chapter 45) in which antibodies will be produced that will attack the cells of the embryo/fetus and likely kill it.

FIGURE 50.18

You could put that cloned cDNA into an expression vector and transform it into separated cells that do not demonstrate selective cell adhesion. If the cadherin produced in the transformed cells is responsible for selective cell adhesion, then that property should not be detectable in those cells. To test that, you would test the transformed cells with cells known to adhere to cells that naturally have that cadherin on their surfaces.

FIGURE 50.20

In induction, the presence of one tissue triggers a change in the developmental fate of another tissue. If donor tissue (duck cells) induce the formation of dorsal structures in recipient tissue, the donor tissues will cause the recipient tissue to form dorsal structures and will not themselves form those structures. Therefore, both sets of dorsal structures in the recipient are formed from chick cells.

IMR

The normal (or close-to-normal) HMG box amino acid sequence is necessary for the SRY protein to bind to regulatory regions of genes that it controls. The missense mutations in the experiments described altered the structure of the HMG box region sufficiently to prevent or significantly inhibit binding. (a) A nonsense mutation in the *SRY* gene upstream of the HMG box-encoding gene would result in a truncated protein that does not include the HMG box. This protein could not activate the composite *AMH*–luciferase gene—no luciferase would be produced. (b) A nonsense mutation downstream of the HMG box-encoding region would result in a truncated SRY protein that still includes the HMG box. In this case a luciferase signal would be seen but only if the shortened SRY protein still folds to produce an HMG box that can bind to the *AMH* gene regulatory sequence. (c) A frameshift mutation changes the reading frame of the resulting mRNA by $+1$ or -1. This means that the encoded amino acid sequence downstream of the mutation is completely different from that encoded by the normal gene. Therefore, in this case no HMG box will be present in the protein and luciferase will not be produced.

UQ

No (and, in fact, there are not any). This can be inferred from the description of sex determination in mammals and from the list of gynandromorph animals. The significance of the research is unchanged.

Much can be learned about sex determination in mammals by studying sex determination in other organisms, and one could argue that understanding why it is different in other animals only provides greater insight into sex determination in mammals.

Test Your Knowledge

1. d 2. e 3. b 4. a 5. e 6. c 7. d 8. d 9. b 10. c

Interpret the Data

1. Heart defects were the most common birth defect in the single-birth group.
2. Heart defects were more common in the multiple-birth group than in the single-birth group.
3. Multiples had more than twice the risk of central nervous system defects than did single-birth infants.
4. No, the 0.93 relative risk of chromosomal defects indicates that multiple pregnancy decreases the relative risk of chromosomal defects in offspring.

Chapter 51

Study Break

STUDY BREAK 51.1

1. Studies of ecosystems are more "inclusive" than studies of populations because ecosystems include the populations of many different species.
2. Mathematical models are useful in ecological research because they help scientists formalize hypotheses about the relationships between variables and because they allow researchers to simulate the effects of changing variables before investing time and resources in experiments or observational studies.

STUDY BREAK 51.2

1. Because of Earth's spherical shape, sunlight striking the planet's surface is more concentrated near the equator than at the poles. As a result, temperatures are higher at low latitudes. The concentrated sunlight near the equator heats the atmosphere, causing air masses near the equator to rise, establishing three circulation cells in the Northern Hemisphere and three in the Southern Hemisphere.
2. Earth's fixed tilt on its axis causes seasonal variation in the amount of sunlight striking the temperate zone as the planet orbits the sun.
3. Dry conditions prevail at 30° N and S latitudes because sinking air masses warm as they descend, causing them to absorb water from the land.
4. Mountains affect local precipitation because rising air masses on the windward side of a mountain cool adiabatically and release moisture. When the air masses descend on the leeward side of a mountain, they warm and absorb moisture, causing a rain shadow.

STUDY BREAK 51.3

1. *Anolis* lizards in the Dominican Republic bask more frequently at high elevation than they do at low elevation.
2. Global warming will likely cause the geographical distributions of species to shift or expand to higher latitudes and to higher elevations.

STUDY BREAK 51.4

1. Tropical rainforest and temperate rainforest are the terrestrial biomes that receive the most rainfall.
2. Savannas, chaparral, and temperate grasslands are renewed by periodic fires.
3. Tropical rainforests and temperate rainforests have the tallest vegetation. Arctic and alpine tundra have the shortest vegetation.
4. Trees are usually evergreen in tropical rainforest, temperate rainforest, and taiga.

STUDY BREAK 51.5

1. Dissolved oxygen concentration is usually high in the headwaters of a stream, gradually diminishing as water flows into a river.
2. The factors that cause seasonal overturns in lakes include seasonal changes in environmental temperature, variations in wind velocity, and the fact that water is densest at 4°C.
3. Oligotrophic lakes are better than eutrophic lakes for recreational purposes because the water in oligotrophic lakes is clear, whereas the water in eutrophic lakes is often clogged with strands of algae and cyanobacteria.

STUDY BREAK 51.6

1. The benthic province of the ocean includes all of the bottom sediments. The pelagic province includes all of the water.
2. Estuaries experience the largest fluctuations in salinity over time.
3. Estuaries, the intertidal zone, and the upper layer of the oceanic pelagic zone receive substantial energy inputs from sunlight.
4. The benthos of the oceanic zone receives nutrients and energy from the detritus sinking from the upper layers of water.
5. Chemoautotrophic bacteria are the primary producers of hydrothermal vent communities and cold seep communities. Unlike photosynthetic organisms of the photic zone, they use hydrogen sulfide and other molecules, instead of sunlight, as an energy source for their chemosynthetic activity.

Think Like a Scientist

FIGURE 51.9

If global climate change increased the air temperatures in the habitats occupied by these lizards, they would be likely to bask in patches of sun less frequently.

FIGURE 51.11

Climate warming will not change the distribution of biomes on the climograph, because the biomes will still occur at particular combinations of temperature and precipitation. However, global warming will probably change the distribution of biomes on the Earth's surface, as reflected in the map in Figure 51.12. Biomes may shift toward the north and south as the climate warms at higher latitudes.

FIGURE 51.25

Housing developments on hilly ground pose a danger to nearby lakes because household wastes, such as human sewage and waste water, flow downhill into lake basins. The chemicals in these wastes promote eutrophication of the lakes.

IMR

We would not expect to find antifreeze proteins in fishes that live in the tropics because those fishes probably never experience freezing conditions in their habitats. Biologists interpret the antifreeze as evolutionary adaptations to the freezing conditions that polar-dwelling fishes routinely encounter.

UQ

We should expect to see the loss of species in mountaintop habitats because the species have no cooler habitats into which they can migrate. We would also expect to see the loss of polar species, especially those that depend on sea ice, the extent of which is shrinking rapidly.

Test Your Knowledge

1. c 2. e 3. b 4. a 5. d 6. a 7. d 8. a 9. a 10. d

Interpret the Data

1. SSTs have predicted 19 El Niño events and 22 La Niña events from 1950 through 2008.
2. These events differ greatly in magnitude and duration.
3. There is no absolute pattern in the occurrence of the two types of events, although it appears that very pronounced events of one type are followed by an event of the other type.

Chapter 52

Study Break

STUDY BREAK 52.1

1. A population's size is simply the number of individuals it contains. Its density is the number of individuals per area or volume of habitat occupied.
2. A clumped pattern of dispersion implies that individuals in the population help each other or that some vital resource in the environment also has a clumped distribution. A uniform pattern of dispersion implies that individuals in the population repel each other. A random pattern of dispersion does not imply either positive or negative interactions among individuals in the population.

STUDY BREAK 52.2

1. A life table usually summarizes statistics about the age-specific survival rates, age-specific mortality rates, and age-specific fecundity of a population.
2. Humans in the industrialized countries exhibit Type I survivorship curves because they provide lots of care to their offspring, thus reducing infant and childhood mortality to low levels.

STUDY BREAK 52.3

1. Children spend most of their energy on growth and maintenance; they devote energy to reproduction later in life.
2. Fecundity and the amount of parental care devoted to each offspring exhibit an inverse relationship because organisms that produce few offspring can devote substantial time and energy to each, whereas those that produce many offspring can devote only minimal time and energy to each.

STUDY BREAK 52.4

1. The model of exponential population growth predicts unlimited population growth through time, generating a J-shaped curve of population size versus time. The logistic model predicts that population growth slows down as the population approaches its carrying capacity, generating an S-shaped curve of population size versus time.
2. Carrying capacity is the maximum number of individuals in a population that an environment can support. The carrying capacity is thus a property of the environment with reference to a particular population.
3. A time lag is a delay in a population's response to a changing environment. It may cause a population's size to oscillate around its carrying capacity.

STUDY BREAK 52.5

1. The effects of density-dependent factors get stronger (that is, they affect a larger percentage of the individuals in the population) as the population's density increases. The effects of density-independent factors do not change (that is, they affect the same percentage of the individuals in a population) as the population's density changes.
2. The effects of infectious diseases are usually density-dependent because disease-causing pathogens spread more quickly through dense populations of the organisms they infect.

STUDY BREAK 52.6

1. Humans have sidestepped the controls that regulate the populations of other organisms by expanding their geographical range to include a wide variety of habitats, by increasing the carrying capacity through agricultural production, and by decreasing their death rate through the introduction of medical care and improved sanitation.
2. The age structure of a population influences its future population growth by determining how many individuals will reach reproductive age in the future. Populations with a bottom-heavy age structure (that is, with many young children) will experience a growth spurt when children alive today reach sexual maturity. Populations with a more even age structure will not experience a dramatic future increase in population size.

Think Like a Scientist

FIGURE 52.17

The increases and decreases in spider numbers in the two treatments appear to be unrelated to the presence or absence of lizards. Changes in weather or the availability of prey may be responsible for these overarching changes in spider numbers.

FIGURE 52.19

If unregulated development destroyed suitable habitat patches that are home to source populations or close to source populations, the development would prevent butterflies in source populations from colonizing sink populations. As a result, more sink populations would become extinct.

FIGURE 52.20

The lynx and hare appear to be on a 10-year population cycle. From the data presented in the graph, the lynx population increase followed the hare population increase almost instantaneously in 1865, 1885, and 1905. The lynx population increased more slowly after increases in the hare population in 1855, 1875, 1895, 1915, and 1925.

IMR

The assumption that the adult armadillos living closest to a juvenile were its likely parents may have caused the researchers to underestimate how far individual armadillos move in their lifetimes. If the researchers had identified parents that lived a greater distance from the juvenile, their estimate of how far a young armadillo moved after weaning would have been higher.

UQ

1. The "ideal" conditions under which the equations predict growth simply never exist, as is clear, for example, from the growth of populations illustrated in the chapter (see growth of *Thrips imaginis*). Therefore, unlike a law such as the law of gravity, population growth cannot simply be predicted by applying the logistic or exponential equation for population growth. The mathematical growth models are constructed after basic observations define the variables of the models.
2. The time scales for most population studies are too short to consider selection, genetic drift, or mutation. Molecular studies allow population biologists to measure changes in gene frequencies at the time scales of single generations and before those changes are manifested in noticeable changes in fecundity or survival (or morphology and behavior).

Test Your Knowledge

1. b 2. c 3. b 4. d 5. a 6. e 7. d 8. e 9. d 10. b

Interpret the Data

The data in the *Albertosaurus* life table suggest that it had a survivorship pattern falling between Types I and II, though perhaps leaning more toward Type I. This species was most vulnerable to mortality between hatching and age two.

Chapter 53

Study Break

STUDY BREAK 53.1

1. Some carnivores spend more time and energy capturing large prey than small prey because large prey provides a larger return on their investment of time and energy in the hunt.
2. Cryptic coloration makes an organism inconspicuous, allowing it to blend in with its surroundings. Aposematic coloration makes an organism highly conspicuous, advertising its unpalatability. Mimicry allows one organism, the mimic, to resemble another species, the model; models are usually unpalatable or poisonous. A mimic will have aposematic coloration if it resembles an aposematic model.
3. Field experiments can demonstrate that two species are competing for limiting resources if the removal of one

species increases population size or density in the other or if the addition of a potential competitor decreases the population size or density of the other.

STUDY BREAK 53.2

1. Gleason's individualistic view of communities suggests that they are just chance assemblages of species that happen to be adapted to similar abiotic environmental conditions.
2. Ecologists find more species living in an ecotone than in the communities on either side of it because ecotones contain species from both neighboring communities as well as species that are adapted to transitional environmental conditions.

STUDY BREAK 53.3

1. The plant growth forms found in tropical forests include a canopy of tall trees, an understory of shorter trees and shrubs, an herb layer, vinelike lianas, and epiphytes.
2. The species richness of a community is the number of species it contains. Relative abundance refers to the commonness or rarity of species in the community.
3. Pigeons, which eat grain and other vegetable matter, are included in the second trophic level, primary consumers. Peregrine falcons, which feed on pigeons and other birds, are in the third trophic level, secondary consumers.

STUDY BREAK 53.4

1. On the one hand, the ecological literature may overestimate the importance of competition because ecologists are more likely to study and publish papers on interactions in which competition is important than on interactions in which it is not. On the other hand, the literature may underestimate the importance of competition because, if strong competition between species cannot persist for long periods of time, we are unlikely to find populations competing strongly in nature.
2. Keystone species are those that have a substantial effect on community structure even if their populations are not very dense. Keystone species may either increase or decrease species richness in the communities they occupy.

STUDY BREAK 53.5

1. Strong storms allow coral communities to be rejuvenated through the recruitment of new individuals because they scour the seafloor, removing existing coral colonies from the community. These openings provide spaces where coral larvae may settle and initiate the growth of new colonies.
2. Moderately severe and moderately frequent disturbances increase a community's species richness by creating opportunities for *r*-selected species to colonize the habitat while allowing populations of *K*-selected species to persist.

STUDY BREAK 53.6

1. Primary succession occurs in places without soil; secondary succession occurs after a disturbance has destroyed vegetation.
2. A climax community differs from earlier successional stages in having taller, longer-lived vegetation, generally higher species richness, and a buffered physical environment under the vegetation.
3. Three hypotheses about the underlying causes of succession differ in how they view the role of population interactions. The facilitation hypothesis specifies no particular role for population interactions. The inhibition hypothesis suggests that the species already present prevent other species from joining a community. The tolerance hypothesis suggests that as environmental conditions within the community change during succession, only species that can compete strongly under the changing conditions will persist.

STUDY BREAK 53.7

1. Some explanations of the high species richness in the tropics suggest that the benign climate and historically low levels of severe disturbance have fostered more rapid rates of speciation. Other explanations suggest that the year-round availability of food resources and complex food webs allow more species to coexist in tropical regions.
2. According to the equilibrium theory of island biogeography, large islands will harbor more species than small islands, and islands that are close to the mainland will harbor more species than those that are farther away.

Think Like a Scientist

FIGURE 53.8

Yes, the presence of *P. caudatum* does appear to have an effect on *P. aurelia* in the mixed cultures. In the presence of *P. caudatum,* the population size of *P. aurelia* is lower than it is when the latter species is grown alone.

FIGURE 53.12

Given the results of Connell's experiment, *Balanus* appears to be a stronger competitor than *Chthamalus,* at least in deep water. The former species is able to displace the latter species from that habitat. *Balanus* does not compete with *Chthamalus* at all in shallow water, because the former species is unable to occupy shallow water, presumably because it is not physiologically adapted to do so.

FIGURE 53.23

The removal of *Pisaster* eventually resulted in a drastic reduction in species richness and an increase in the density of *Mytilus* in experimental plots. The full effects of the experiment would not be evident until other species of organisms had been eliminated by old age, predation, or other factors and a new cohort of *Mytilus* became established in the plot. This sequence of changes may have begun began shortly after the experiment was established, but its full effects were apparently not evident until three years had elapsed.

FIGURE 53.24

Lubchenco's result for snails and algae in tidepools is consistent with the predictions of the intermediate disturbance hypothesis. An intermediate density of snails, which establishes an intermediate-level disturbance, results in the highest algal species richness in tidepools.

IMR

The communities of organisms living in bromeliad cups are analogous to those living in British treeholes. Thus, we would expect the species richness of microorganisms and invertebrates living in bromeliad cup communities to exhibit species-area functions similar to those observed in communities living on islands.

UQ

1. Many, but certainly not all, mutualisms involve partners "taking" resources from each other. In pollination, for example, bees take pollen, and plants take the bees' time and direct their movements.
2. Yes, because of shared traits among close relatives. This is one reason cannibalism may be rare.

Test Your Knowledge

1. c 2. c 3. a 4. b 5. b 6. e 7. b 8. d 9. d 10. b

Interpret the Data

The protected shrubs responded to the exclusion of herbivores by steadily reducing the density of thorns on their flowering stems over two years. The unprotected shrubs continued to produce a high density of thorns on their flowering stems. The unprotected shrubs' production of thorns probably protected their flowers and fruits feeding by herbivores, allowing them to produce seeds.

Chapter 54

Study Break

STUDY BREAK 54.1

1. In the generalized compartment model of biogeochemical cycling, nutrient pools are classified as available or unavailable and as organic or inorganic.
2. The advantage of using conceptual models of ecosystem function is that they are a simplification of the processes that determine ecosystem function in nature. The disadvantage of these models is that they do not include processes that carry nutrients and energy out of one ecosystem and into another; neither do they include the details of exactly how specific ecosystems function. Thus, conceptual models do not provide precise predictions about potential changes in ecosystem function.
3. Before constructing a simulation model of an ecosystem, ecologists must collect data about the population sizes of important species, the average energy and nutrient content of each, the food webs in which they participate, the quantity of food each species consumes, and the productivity of each population; the ecosystem's energy and nutrient gains and losses caused by erosion, weathering, precipitation, and runoff; and seasonal and annual variations in these factors.

STUDY BREAK 54.2

1. Gross primary productivity is a measure of the total amount of solar energy converted into chemical energy by the producers in an ecosystem. Net primary productivity is the amount of chemical energy that remains after deducting the producers' maintenance costs from the gross primary productivity.
2. In terrestrial ecosystems, primary productivity may be influenced by the availability of light, water, and nutrients and by the environmental temperature. In aquatic ecosystems, primary productivity is often limited by the joint availability of light and nutrients in the same place.
3. Energy is lost from an ecosystem at every transfer between trophic levels because some of the energy is not assimilated, because organisms use some of the energy they assimilate for maintenance costs, and because biological processes are never 100% efficient.
4. The presence of predators can influence an ecosystem's productivity by consuming herbivores and thereby changing the population dynamics of the herbivores and the plants they eat. These effects can reverberate through an ecosystem in a trophic cascade.

STUDY BREAK 54.3

1. The global hydrologic cycle maintains its balance because the amount of water returned to the atmosphere by evaporation and transpiration is equal to the amount that falls as precipitation. Runoff from the land maintains the balance between terrestrial and marine components of the cycle.
2. Respiration, excretion, leaching, and the burning of fossil fuels move large quantities of carbon from an organic compartment to an inorganic compartment of an ecosystem.
3. Bacteria, cyanobacteria, and fungi drive the global nitrogen cycle through their activities in nitrogen fixation, ammonification, nitrification, and denitrification.
4. Marine sediments are Earth's main reservoir for phosphorus, which is recycled slowly after geological uplifting and erosion make it available to producers.

STUDY BREAK 54.4

1. The greenhouse effect refers to the tendency of certain gases, called greenhouse gases, to foster the accumulation of heat in the lower atmosphere by hindering the escape of infrared heat into space. CO_2 enhances the greenhouse effect because it is one of the greenhouse gases.
2. The combustion of fossil fuels and wood is the human activity that releases the most CO_2 into the atmosphere.
3. Irrigation, the use of synthetic nitrogen-containing fertilizers, the harvesting of crops, and the rearing of livestock contribute to the disruption of the nitrogen cycle.

Think Like a Scientist

FIGURE 54.7

An increased number of trees growing along the banks of Silver Springs would limit the amount of sunlight reaching the primary producers in the water. As a result, productivity and the numbers of organisms at all trophic levels would probably decrease.

Abbreviations: IMR = Insights from the Molecular Revolution; UQ = Unanswered Questions

FIGURE 54.11

The removal of crabs and turtles would allow snail populations to increase in size. The increasing snail populations would consume the cordgrass, thereby destroying the ecosystem.

IMR

Moderate acidification of ocean water appears to affect the transcription of genes that influence biomineralization. Thus, the production of tests (that is, shells) by sea urchins might be adversely affected.

UQ

Large cities release substantial quantities of airborne pollutants that can be carried by prevailing winds even to distant forests. These materials can have a profound influence on the health and biological activity of primary producers in the forest.

Test Your Knowledge

1. c 2. d 3. c 4. e 5. e 6. c 7. a 8. d 9. c 10. a

Interpret the Data

The productivity of the algal community in Walker Branch is controlled by both nutrients and herbivores. The graph illustrates that for grazed algae, the addition of nitrogen, phosphorus, or both nitrogen and phosphorus increased productivity over the level seen in the control treatment. Similarly, for ungrazed algae, the addition of nitrogen or the combination of nitrogen and phosphorus increased productivity over the level measured in the controls. The graph also illustrates the significant impact of grazing by herbivorous snails: in the control treatment and two of the three experimental treatments, ungrazed algae had much higher productivity than grazed algae.

Chapter 55

Study Break

STUDY BREAK 55.1

1. The growing human population in south Florida has increased the likelihood of desertification there by withdrawing groundwater for agricultural, recreational, and residential uses faster than it is replenished.
2. Overexploitation of fish populations typically causes fishes to reach reproductive maturity at a smaller size and younger age, decreases population sizes, and sometimes leads to the extinction of populations.
3. The construction of dams changes the water flow in rivers, leading to changes in the physical structure of the river and the species that inhabit it; disrupting the environmental cues that trigger successful reproduction of organisms; inhibiting the free movement of migrating animals through all parts of the river system; and opening the habitat to invasion by species that originated elsewhere.

STUDY BREAK 55.2

1. Habitat fragmentation in eastern North America has affected breeding songbirds by reducing the variety of habitats available to them, increasing the frequency of brood parasitism of their nests, and increasing the rate of predation on their eggs and young.
2. The use of the anti-inflammatory drug diclofenac on livestock has decimated vulture populations in South Asia. Diclofenac is highly toxic to birds, and the vultures ingest it when they feed on the carcasses of livestock.
3. By one estimate, extinction rates today may be 1,000 times greater than the background extinction rate evident in the fossil record.

STUDY BREAK 55.3

1. Living systems are a storehouse of potentially useful genetic information because naturally occurring compounds may prove to be useful in the treatment of disease, in the manufacture of new products, or in agriculture.
2. Naturally occurring organisms provide many ecosystem services, such as the sequestration of carbon dioxide, fixation of nitrogen into forms that plants can absorb, recycling of nutrients with ecosystems, and the retention of water in ecosystems.

STUDY BREAK 55.4

1. Conservation biologists identify a site as one where extinction is imminent if it houses 95% or more of the individuals in an endangered species and it has definable boundaries that encompass distinctive habitats.
2. Conservation biologists are especially alarmed about deforestation in the New World tropics because these forests harbor many species in imminent danger of becoming extinct.

STUDY BREAK 55.5

1. Population bottlenecks—large, temporary reductions in a population's size—inevitably foster genetic drift, thereby reducing a population's genetic variability, which increases its likelihood of becoming extinct.
2. A population viability analysis allows a conservation biologist to identify the minimum population size that is likely to survive both predictable and unpredictable environmental change. It therefore specifies how many individuals must be conserved for the continued survival of the population and species.
3. Several small preserves would collectively experience more edge effects than one large preserve of the same total size.

STUDY BREAK 55.6

1. The Pine Bush habitat in the state of New York is an example of conservation through preservation.
2. The Royal Chitwan National Park has been judged a success because local residents benefit from the park's existence and therefore support it, and because populations of many animals, including tigers and rhinoceroses, have increased within its borders.
3. Economists can determine the economic values of specific ecosystem services and convince local governments that it is economically beneficial to preserve ecosystems and the services they provide.

Think Like a Scientist

FIGURE 55.6

Wilcove placed nests in different positions because predators look for prey in different parts of the habitat. If most of the predators sought food on or close to the ground, any nests placed high in trees might have experienced lower rates of predation.

FIGURE 55.18

The results of this study suggest that the species richness of small animals might also be higher in tracts of forest connected by landscape corridors than in unconnected patches of similar size. Small animals should be able to use the corridors to move from one patch to another. It is difficult to predict the effects of corridors on exotic small animal species.

IMR

DNA barcoding might be useful in the effort to track the spread of Asian carp through North American rivers and lakes because biological samples can be subjected to DNA barcoding analysis to detect the presence of these species in habitats in which they had not previously been found.

UQ

Student answers may vary. Because students are asked to come up with a list of unanswered questions on their own, we can't really provide an answer to this question.

Test Your Knowledge

1. e 2. b 3. b 4. b 5. a 6. d 7. d 8. a 9. c 10. b

Interpret the Data

Extinctions in mountain habitats are accelerating the most, exhibiting a potential for more than a 20-fold (366/17) increase. Among groups, extinctions are accelerating the most among amphibians, which exhibit a potential 12-fold (408/34) increase. Conservation biologists should therefore focus their efforts on amphibians that live in mountainous regions.

Chapter 56

Study Break

STUDY BREAK 56.1

1. An instinctive behavior is a genetically or developmentally programmed response that appears in complete and functional form the first time it is used. A learned behavior is one that is dependent on having a particular kind of experience during development.
2. Arnold demonstrated that the receptiveness of garter snakes to a meal of banana slugs had a genetic basis by breeding snakes that almost always eat banana slugs with snakes that rarely eat them. The behavior of the hybrid offspring was intermediate between the behaviors of the two parent populations.
3. Tail wagging by a dog when it sees its owner pick up a leash is an example of classical conditioning.
4. Sleeping through an alarm clock is an example of habituation.

STUDY BREAK 56.2

1. The conclusion that certain neurons in a young male bird's brain are influenced only by acoustical signals from its own species and only during a critical period is supported by two observations: (1) young birds that did not hear taped songs during the critical period never produced the full song of their species; and (2) young birds that heard recordings of *other* bird species' songs during the critical period never generated replicas of those songs as they matured.
2. The role of the ZENK enzyme in song learning is to program the nerve cells of the bird's brain to anticipate key acoustical events of potential biological importance.
3. A high estrogen concentration in the brains of young male zebra finches stimulates the production of more neurons in the higher vocal center.
4. The loss of a territory by a male African cichlid fish causes its brain to produce less GnRH.

STUDY BREAK 56.3

1. When piloting, animals use familiar landmarks to guide their journey. In compass orientation, animals use external environmental cues such as the position of the sun or stars as a compass to move in a particular direction, often over a specific distance or for a prescribed length of time. When navigating, animals use a compass as well as a mental map of their position in relation to their destination.
2. Seasonal changes in temperature and food availability are the most likely selection pressures responsible for the evolution of migratory behavior in birds.

STUDY BREAK 56.4

1. Wood lice accumulate in moist habitats because they move much less in moist habitats than they do in dry habitats.
2. The costs of maintaining a territory include the time and energy needed to defend territory borders, the possibility of being injured or killed during a territorial encounter, and the increased likelihood of being noticed and captured by a predator. The benefits of holding a territory include access to all of the resources within the territory.

STUDY BREAK 56.5

1. Humans consciously use acoustical, visual, and tactile channels to communicate with each other.
2. A honey bee uses the waggle dance to communicate the location of a distant food source to its hive mates.
3. A young wandering raven that discovers a food source in the territory of other ravens will call vigorously to attract other ravens from outside the territory. Collectively, the nonterritory holders can overwhelm the defenses of the resident birds and then consume the food.

STUDY BREAK 56.6

1. For monogamous species, the males that are most attractive to females are those that can demonstrate their good genes with large showy morphological characteristics and

elaborate behavioral displays and those that hold territories rich in resources.

2. Male sage grouse perform displays at a lek, trying to attract the attention of females. Female sage grouse go to a lek to evaluate the qualities of the males that are displaying there.
3. A female red-winged blackbird might settle in the territory of a male even in the presence of other resident females if the male's territory is very rich in resources.

STUDY BREAK 56.7

1. The social behavior of musk oxen and sawfly larvae includes cooperative defense of the group against predators.
2. Among the animals in a dominance hierarchy, the most dominant individuals are the most likely to reproduce.
3. The genetic system in eusocial insects promotes altruistic behavior because the workers in a colony are more closely related to each other than are siblings in most other animal species.

Think Like a Scientist

FIGURE 56.2

The sign stimulus/fixed action pattern interaction between parent herring gulls and their offspring ensures that the parent regurgitates food to a chick that is hungry. Herring gulls may have multiple chicks in the nest, and the hungriest chick is likely to be the one that taps most vigorously on the parent's bill.

FIGURE 56.10

The GnRH production and behavior of cichlid fishes each influence the other in a feedback loop. When a male loses status, his GnRH production decreases, leading to a reduction in his aggressive behavior. However, if his aggressive behavior increases because of changing social circumstances, his GnRH production will increase, reinforcing his aggressive behavior.

FIGURE 56.13

The results of Emlen's experiment indicate that the position of the stars only provides cues about the direction in which they should migrate. Birds attempted to take off from the cages to begin a migration even under cloudy conditions when they could not see any stars.

FIGURE 56.29

If the queen in a bee colony mated with two drones instead of just one, the average degree of relatedness among the workers would decrease. Although their relatedness through their shared mother would not change, some workers would be unrelated to each other through their different fathers.

IMR

As described in Section 21.3, a recessive mutation is not easily eliminated from a population because recessive mutations have no impact on heterozygotes. Some form of genetic screening would be necessary to identify people who carry the mutation and counsel them about the possible consequences of having children together. Unidentified heterozygotes could produce offspring who are homozygous for the mutation and would therefore exhibit the condition that it causes.

UQ

If the partner of a female is able to discriminate between offspring that are his own and those that are not, he should bias any paternal care in favor of his true son in preference to his nephew. The production of chimeras could also give rise to conflicts between a male and a female because the offspring would be more closely related to the female parent than to the female's partner (i.e., the offspring's uncle).

Test Your Knowledge

1. b 2. d 3. b 4. e 5. c 6. d 7. b 8. e 9. c 10. a

Interpret the Data

The results suggest that the more male neighbors a Seychelles warbler has, the more time he spends guarding his mate; and the more time he spends guarding his mate, the less time he can devote to collecting food. When the number of neighboring males was experimentally reduced, the focal males spent more time feeding and less time guarding their mates.

Appendix B: Classification System

The classification system presented here is based on a combination of organismal and molecular characters and is a composite of several systems developed by microbiologists, botanists, and zoologists. This classification reflects current trends toward a phylogenetic approach to taxonomy, one that incorporates the ever more detailed information about the relationships of monophyletic lineages provided by new molecular sequence data. In keeping with these trends, we have omitted reference to the traditional taxonomic categories, such as "class" and "order." Instead, we present the major monophyletic lineages in each of the three domains, and we indicate their relationships within a nested hierarchy that parallels that of traditional Linnaean classification.

Although researchers generally agree on the identity of the major monophyletic lineages, the biologists who study different groups have not established universal criteria for identifying the somewhat arbitrary taxonomic categories included in the traditional Linnaean hierarchy. As a result, a "class" or an "order" of flowering plants may not be the equivalent of a "class" or an "order" of animals. In fact, as described in *Unanswered Questions* at the end of Chapter 24, systematic biologists are shifting toward a more phylogenetic approach to taxonomy and classification, such as the one represented here.

Bear in mind that we include this appendix to introduce the diversity of life and illustrate many of the evolutionary relationships that link monophyletic groups. Like all phylogenetic hypotheses, this classification is open to revision as new information becomes available. Moreover, the classification is incomplete because it includes only those lineages that are described in Unit Four.

Prokaryotes and Eukaryotes

Organisms fall into two groups, prokaryotes and eukaryotes, based on the organization of their cells. Prokaryotes consist of the Domains Bacteria and Archaea and are characterized by a central region, the nucleoid, which has no boundary membrane separating it from the cytoplasm, and by membranes typically limited to the plasma membrane. Most prokaryotes are single-celled, although some are found in simple associations. All other organisms are eukaryotes, which make up the Domain Eukarya. Eukaryotes are characterized by cells with a central, membrane-bound nucleus, and an extensive membrane system. Some eukaryotes are single-celled, while others are multicellular.

Domain Bacteria

The largest and most diverse group of prokaryotes. Includes photoautotrophs, chemoautotrophs, and heterotrophs.

Proteobacteria Gram-negative bacteria that include chemoautotraphs, chemoheterotrophs, and photoautotrophs. Some are aerobic and some are anaerobic. Five subgroups of proteobacteria are recognized.

Spirochetes Helically spiraled bacteria that move by twisting in a corkscrew pattern

Chlamydias Gram-negative intracellular parasites of animals, with cell walls that lack peptidoglycans

Gram-Positive Bacteria Chemoheterotrophic bacteria with thick cell walls

Cyanobacteria Photoautotrophic Gram-negative bacteria that carry out photosynthesis using the same chlorophyll as in plants and release oxygen as a by-product

---

Domain Archaea

Prokaryotes that are evolutionarily between eukaryotic cells and the bacteria. Most are chemoautotrophs. None is photosynthetic. Originally discovered in extreme habitats, they are now known to be widely dispersed. Compared with bacteria, the Archaea have a distinctive cell wall structure and unique membrane lipids, ribosomes, and RNA sequences. Some are symbiotic with animals, but none is known to be pathogenic.

Euryarchaeota Includes methanogens, extreme halophiles, and some extreme thermophiles

Nanoarchaeota Extreme thermophile (only one member known). Lives as a symbiont of a thermophilic member of the Crenarchaeota.

Crenarchaeota Includes most of the archaean extreme thermophiles, as well as psychrophiles; mesophilic species comprise a large part of plankton in cool, marine waters

Thaumoarchaeota All living species (four) are mesophilic chemoautotrophs. Other members are represented by DNA sequences in environmental samples.

Korarchaeota Only living member is a thermophile. Other members are represented by DNA sequences in environmental samples.

Domain Eukarya

Protists A collection of single-celled and multicelled lineages, which are almost certainly not a monophyletic group.

Excavata (excavates) Single-celled animal parasites that have greatly reduced mitochondria, or organelles derived from mitochondria, and move by means of flagella; most have a scooped out (excavated) feeding apparatus on the ventral surface of the cell

Metamonada (metamonads)—consist of the Diplomonadida and the Parabasala

Diplomonadida (diplomonads)—cells have two nuclei; move by multiple freely beating flagella

Parabasala (parasabala)—move by freely beating flagella and an undulating membrane formed by a flagellum buried in a fold of cytoplasm

Euglenozoa (euglenozoans)—mostly single-celled, highly motile cells that swim using flagella and have disc-shaped inner mitochondrial membranes

Euglenids—free-living with anterior flagella; most are photosynthetic autotrophs

Kinetoplastids—nonphotosynthetic heterotrophs that live as animal parasites. Their single mitochondrion contains a large DNA–protein deposit called a kinetoplast.

Chromalveolata (chromalveolates) Heterogeneous group with a range of forms and life styles

Alveolata (alveolates)—characterized by small membrane-bound vesicles called alveoli in a layer under the plasma membrane

Ciliophora (ciliates)—motile, primarily single-celled highly complex heterotrophs; swim by means of cilia

Dinoflagellata (dinoflagellates)—nonmotile single-celled marine heterotrophs or autotrophs; shell formed from cellulose plates

Apicomplexa (apicomplexans)—nonmotile parasites of animals with apical complex for attachment and invasion of host cells

Stramenopila (stramenopiles)—characterized by two different flagella, one with hollow tripartite projections, and the other plain

Oomycota (oomycetes)—water molds, white rusts, and mildews; all are funguslike heterotrophs lacking chloroplasts

Bacillariophyta (diatoms)—single-celled autotrophs that carry out photosynthesis by pathways similar to those of plants; covered by a glassy silica shell; most are free-living, whereas some are symbionts

Chrysophyta (golden algae)—colonial; each cell of the colony has a pair of flagella and is covered by a glassy shell consisting of plates or scales; most are autotrophs that carry out photosynthesis by pathways similar to those of plants

Phaeophyta (brown algae)—photosynthetic autotrophs

Rhizaria (rhizarians) Amoebas with stiff, filamentous pseudopodia; some with outer shells

Radiolaria (radiolarians)—marine heterotrophs; glassy internal skeleton with projecting axopods—raylike strands of cytoplasm supported internally by long bundles of microtubules

Foraminifera (forams)—marine heterotrophs; shells, most of which are chambered, spiral structures, consist of organic matter reinforced by calcium carbonate

Chlorarachniophyta (chlorarachniophytes)—green, photosynthetic amoebas

Archaeplastida (archaeplastids) Red algae, green algae, and land plants *(viridaeplantae),* photosynthesizers with a common evolutionary origin

Rhodophyta (red algae)—marine seaweeds, typically multicellular with plantlike bodies; most are reddish in color

Chlorophyta (green algae)—green single-celled, colonial, and multicellular autotrophs that carry out photosynthesis using the same pigments as plants; likely ancestor of land plants

Amoebozoa (amoebozoans) Includes most of the amoebas as well as the slime molds. All use pseudopods for locomotion and for feeding.

Amoebas—single-celled; use lobose (non-stiffened) pseudopods for locomotion and feeding

Cellular slime molds—heterotrophs; primarily individual cells; move by amoeboid motion, or as a multicellular mass

Plasmodial slime molds—heterotrophs; live as plasmodium, a large composite mass with nuclei in a common cytoplasm, that moves and feeds like a giant amoeba

Opisthokonta (opisthokonts) A single posterior flagellum at some stage in the life cycle; consist of the fungi, animals, and two protist groups, the choanoflagellates and the nucleariids.

Choanoflagellata (choanoflagellates)—motile protists with a single flagellum surrounded by collar of closely packed microvilli; likely ancestor of animals and fungi

Nucleariidae (nucleariids)—heterotrophic, mainly spherical amoebae with radiating, fine pseudopods that are not supported by microtubules; more closely related to fungi than to animals

PLANTAE Multicellular autotrophs, mostly terrestrial, and most of which gain energy via photosynthesis; life cycle characterized by alternation of a gametophyte (gamete-producing) generation and sporophyte (spore-producing) generation

Bryophytes (seedless plants)—nonvascular plants and seedless vascular plants

Hepatophyta (liverworts)—leafy or simple flattened thallus with rhizoids; no true leaves, stems, roots, or stomata (porelike openings for gas exchange); spores in capsules[1]

Anthocerophyta (hornworts)—simple flattened thallus, hornlike sporangia[1]

Bryophyta (mosses)—feathery or cushiony thallus; some with hydroids; spores in capsules[1]

Lycophyta (club mosses)—simple leaves, cuticle, stomata, true roots; most species have sporangia on sporophylls; fertilization by swimming sperm[2]

Pterophyta (ferns, whisk ferns, horsetails)—*Ferns:* Finely divided leaves; sporangia in sori. *Whisk ferns:* Branching stem from rhizomes; sporangia on stem scales. *Horsetails:* hollow stem, scalelike leaves, sporangia in strobili.[2]

Spermatophyta (seed plants)—vascular plants in which embryos develop within seeds

Gymnosperms—seeds born on stems, on leaves, or under scales

Cycadophyta (cycads)—shrubby or treelike with palmlike leaves; male and female strobili on separate plants

Ginkgophyta (ginkgoes)—lineage with a single living species *(Ginkgo biloba);* tree with deciduous, fan-shaped leaves; male, female reproductive structures on separate plants

Gnetophyta (gnetophytes)—shrubs or woody vinelike plants; male and female strobili on separate plants

Coniferophyta (conifers)—predominant extant gymnosperm group; mostly evergreen trees and shrubs with needlelike or scalelike leaves; male and female cones usually on the same plant

Anthophyta (angiosperms/flowering plants)—reproductive structures in flowers

Monocotyledones (monocots)—grasses, palms, lilies, orchids and their relatives; a single cotyledon (seed leaf); pollen grains have one groove

Eudicotyledones (eudicots)—roses, melons, beans, potatoes, most fruit trees, others; two cotyledons; pollen grains have three grooves

Other major angiosperm lineages: magnoliids (magnolias and relatives); star anise (Family Illicium); water lilies (Family Nymphaeaceae); *Amborella* (Family Amborellaceae)

FUNGI Heterotrophic, mostly multicellular organisms with cell wall containing chitin and cell nuclei occurring in threadlike hyphae; life cycle typically includes both asexual and sexual phases, with sexual structures used as the basis for phylum-level classification. Single-celled species are known as yeasts.

Chytridiomycota (chytrids)—mostly aquatic; asexual reproduction by way of motile zoospores; sexual reproduction via gametes produced in gametangia; hyphae mostly aseptate

Zygomycota (zygomycetes)—terrestrial; asexual reproduction via nonmotile haploid spores formed in sporangia; sexual spores (zygospores) form in zygosporangia; aseptate hyphae

Glomeromycota (glomeromycetes)—terrestrial; asexual reproduction via spores at the tips of hyphae; form mycorrhizal associations with plant roots

Ascomycota (ascomycetes/sac fungi)—terrestrial and aquatic; sexual spores form in asci; asexual reproduction occurs via conidia (nonmotile spores); septate hyphae

Basidiomycota (basidiomycetes)—terrestrial; reproduction usually via sexual basidiospores produced by basidia; septate hyphae

Basidiomycetes: mushroom-forming fungi and relatives

Teliomycetes: rusts

Ustomycetes: smuts

Conidial fungi—not a true phylum but a convenience grouping of species for which no sexual phase is known

Microsporidia—single-celled sporelike parasites of animals, other groups; phylogeny uncertain

[1]Nonvascular plants (bryophytes)—A polyphyletic group of plants with no specialized structures for transporting water and nutrients; swimming sperm require liquid water for sexual reproduction

[2]Seedless vascular plants—A polyphyletic group of plants in which embryos are not housed inside seeds

Animalia Multicellular heterotrophs; nearly all with tissues, organs, and organ systems; motile during at least part of the life cycle; sexual reproduction in most; embryos develop through a series of stages; many with larval and adult stages in life cycle

Parazoa Animals lacking tissues and body symmetry

Porifera (sponges)—multicellular; extract oxygen and particulate food from water drawn into a central cavity

Eumetazoa Animals with tissues and either radial or bilateral symmetry

Radiata—acoelomate animals with radial symmetry and two tissue layers

Cnidaria (cnidarians)—two tissue layers; single opening into gastrovascular cavity; nerve net; nematocysts for defense and predation; some sessile, some motile; most are predatory, some with photosynthetic endosymbionts; freshwater and marine

Hydrozoa: hydrozoans

Scyphozoa: jellyfishes

Cubozoa: box jellyfishes

Anthozoa: sea anemones, corals

Ctenophora (comb jellies)—two (possibly three) tissue layers; feeding tentacles capture particulate food; beating cilia provide weak locomotion; marine

Bilateria—animals with bilateral symmetry and three tissue layers

Protostomia—acoelomate, pseudocoelomate, or schizocoelomate; many with spiral, indeterminate cleavage; blastopore forms mouth; nervous system on ventral side

Lophotrochozoa—many with either a lophophore for feeding and gas exchange or a trochophore larva

Ectoprocta (bryozoans)—coelomate; colonial; secrete hard covering over soft tissues; lophophore; sessile; particulate feeders; marine

Brachiopoda (lamp shells)—coelomate; dorsal and ventral shells; lophophore; sessile; particulate feeders; marine

Phoronida (phoronid worms)—coelomate; secrete tubes around soft tissues; lophophore; sessile; particulate feeders; marine

Platyhelminthes (flatworms)—acoelomate; dorsoventrally flattened; complex reproductive, excretory, and nervous systems; gastrovascular cavity in many; free-living or parasitic, often with multiple hosts; terrestrial, freshwater, and marine

Turbellaria: free-living flatworms

Trematoda: flukes

Monogenoidea: flukes

Cestoda: tapeworms

Rotifera (wheel animals)—pseudocoelomate; microscopic; complete digestive system; well-developed reproductive, excretory, and nervous systems; particulate feeders; major components of marine and freshwater plankton

Nemertea (ribbon worms)—schizocoelomate; proboscis housed within rhynchocoel; complete digestive tract; circulatory system; predatory; mostly marine

Mollusca (mollusks)—schizocoelomate; many with trochophore larva; many with shell secreted by mantle; body divided into head–foot, visceral mass, and mantle; well-developed organ systems; variable locomotion; herbivorous or predatory; terrestrial, freshwater, and marine

Polyplacophora: chitons

Gastropoda: snails, sea slugs, land slugs

Bivalvia: clams, mussels, scallops, oysters

Cephalopoda: squids, octopuses, cuttlefish, nautiluses

Annelida (segmented worms)—schizocoelomate; many with trochophore larva; segmented body and organ systems; well-developed organ systems; many use hydrostatic skeleton for locomotion; some predatory, some particulate feeders, some detritivores; terrestrial, freshwater, and marine

Polychaeta: marine worms

Oligochaeta: freshwater and terrestrial worms

Hirudinea: leeches

Ecdysozoa—cuticle or exoskeleton is shed periodically

Nematoda (roundworms)—pseudocoelomate; body covered with tough cuticle that is shed periodically; well-developed organ systems; thrashing locomotion; many are parasitic on plants or animals; mostly terrestrial

Onychophora (velvet worms)—schizocoelomate; segmented body covered with cuticle; locomotion by many unjointed legs; complex organ systems; predatory; terrestrial

Arthropoda (arthropods)—schizocoelomate; jointed exoskeleton made of chitin; segmented body, some with fusion of segments in head, thorax, or abdomen; complex organ systems; variable modes of locomotion, including flight; specialization of numerous appendages; herbivorous, predatory, or parasitic; terrestrial, freshwater, and marine

Trilobita: trilobites (extinct)

Chelicerata: horseshoe crabs, spiders, scorpions, ticks, mites

Crustacea: shrimps, crayfishes, lobsters, crabs, barnacles, copepods, isopods

Myriapoda: centipedes, millipedes

Hexapoda: springtails and insects

Deuterostomia—enterocoelomate; many with radial, determinate cleavage; blastopore forms anus; nervous system on dorsal side in many

Echinodermata (echinoderms)—secondary radial symmetry, often organized around five radii; hard internal skeleton; unique water vascular system with tube feet; complete digestive system; simple nervous system; no circulatory or respiratory system; generally slow locomotion using tube feet; predatory, herbivorous, particulate feeders, detritivores; exclusively marine

Asteroidea: sea stars

Ophiuroidea: brittle stars

Echinoidea: sea urchins, sand dollars

Holothuroidea: sea cucumbers

Crinoidea: feather stars, sea lilies

Concentricycloidea: sea daisies

Hemichordata (acorn worms)—pharynx perforated with branchial slits; proboscis; complex organ systems; tube-dwelling in soft sediments; particulate or deposit feeders; exclusively marine

Chordata (chordates)—notochord; segmental body wall and tail muscles; dorsal hollow nerve chord; perforated pharynx; complex organ systems; variable modes of locomotion; extremely varied diets; terrestrial, freshwater, and marine

Cephalochordata: lancelets

Urochordata: tunicates, sea squirts

Vertebrata: vertebrates

Myxinoidea: hagfishes

Petromyzontoidea: lampreys

Placodermi: placoderms (extinct)

Chondrichthyes: sharks, skates, and rays

Acanthodii: acanthodians

Actinopterygii: ray-finned fishes

Sarcopterygii: fleshy-finned fishes

Amphibia: salamanders, frogs, caecelians

Synapsida: mammals

Anapsida: turtles

Diapsida: sphenodontids, lizards, snakes, crocodilians, birds

3′ end The end of a polynucleotide chain at which a hydroxyl group is bonded to the 3′ carbon of a deoxyribose sugar.

3′ untranslated region (3′ UTR) The part of an mRNA between the stop codon and the 3′ end of the molecule; this region does not code for amino acids.

5′ cap In eukaryotes, a guanine-containing nucleotide attached in a reverse orientation to the 5′ end of pre-mRNA and retained in the mRNA produced from it. The 5′ cap on an mRNA is the site where ribosomes attach to initiate translation.

5′ end The end of a polynucleotide chain at which a phosphate group is bound to the 5′ carbon of a deoxyribose sugar.

5′ untranslated region (5′ UTR) The part of an mRNA between the 5′ end of the molecule and the start codon; this region does not code for amino acids.

10-nm chromatin fiber The most fundamental level of chromatin packing of a eukaryotic chromosome in which DNA winds for almost two turns around an eight-protein nucleosome core particle to form a nucleosome and linker DNA extends between adjacent nucleosomes. The result is a beads-on-a-string type of structure with a 10-nm diameter.

30-nm chromatin fiber Level of chromatin packing of a eukaryotic chromosome in which histone H1 binds to the 10-nm chromatin fiber causing it to package into a coiled structure about 30 nm in diameter and with about six nucleosomes per turn. Also referred to as a *solenoid.*

A site The site where the incoming aminoacyl–tRNA carrying the next amino acid to be added to the polypeptide chain binds to the mRNA.

abdomen The region of the body that contains much of the digestive tract and sometimes part of the reproductive system; in insects, the region behind the thorax.

abiotic Nonbiological, often in reference to physical factors in the environment.

abscisic acid (ABA) A plant hormone involved in the abscission of leaves, flowers, and fruits, dormancy of buds and seeds, and closing of stomata.

abscission In plants, the dropping of flowers, fruits, and leaves in response to environmental signals.

absorption Within the digestive tube, the process in which molecular subunits are absorbed from the digestive contents into body fluids and cells.

absorption spectrum Curve representing the amount of light absorbed at each wavelength.

abyssal zone The bottom sediments that lie permanently below deep ocean water.

acclimitization A process by which an animal adjusts to live and function at higher altitudes to permit adequate oxygen delivery to the tissues.

accommodation A process by which the lens changes to enable the eye to focus on objects at different distances.

acid Proton donor that releases H^+ (and anions) when dissolved in water.

acid precipitation Rainfall with low pH, primarily created when gaseous sulfur dioxide (SO_2) dissolves in water vapor in the atmosphere, forming sulfuric acid.

acid-growth hypothesis A hypothesis to explain how the hormone auxin promotes growth of plant cells; it suggests that auxin stimulates H^+ pumps in the plasma membrane to move H^+ from the cell interior into the cell wall, which increases wall acidity, making the wall expandable.

acidity The concentration of H^+ in a water solution, as compared with the concentration of OH^-.

acoelomate A body plan of bilaterally symmetrical animals that lack a body cavity between the gut and the body wall.

acoustical signaling A means of animal communication in which a signaler produces a sound that is heard by a signal receiver.

acquired immunity *See* **adaptive (acquired) immunity.**

acquired immunodeficiency syndrome (AIDS) A constellation of disorders that follows infection by the human immunodeficiency virus (HIV).

acrosome A specialized secretory vesicle on the head of an animal sperm, which helps the sperm penetrate the egg.

acrosome reaction The process in which enzymes contained in the acrosome are released from an animal sperm and digest a path through the egg coats.

actin A protein that, in an interaction with the protein myosin, causes muscle contraction.

action potential The abrupt and transient change in membrane potential that occurs when a neuron conducts an electrical impulse.

action spectrum Graph produced by plotting the effectiveness of light at each wavelength in driving photosynthesis.

activation energy The initial input of energy required to start a reaction.

activator A regulatory protein that controls the expression of one or more genes.

active immunity The production of antibodies in the body in response to exposure to a foreign antigen.

active parental care Parents' investment of time and energy in caring for offspring after they are born or hatched.

active site The region of an enzyme to which substrate(s) bind and where catalysis occurs.

active transport The mechanism by which ions and molecules move against the concentration gradient across a membrane, from the side with the lower concentration to the side with the higher concentration.

adaptation Characteristic that helps an organism survive longer or reproduce more under a particular set of environmental conditions.

adaptation, evolutionary The accumulation of adaptive traits over time.

adaptation, sensory *See* **sensory adaptation.**

adaptive (acquired) immunity The response by the body to eliminate cellular pathogens that involves mechanisms of the adaptive immune system.

adaptive immune system The inherited mechanisms leading to the synthesis of molecules that target pathogens in a specific way. The adaptive immune system is one of two components of the immune system in vertebrates. It is not found in invertebrates.

adaptive radiation A cluster of closely related species that are each adaptively specialized to a specific habitat or food source.

adaptive trait A genetically based characteristic, preserved by natural selection, that increases an organism's likelihood of survival or its reproductive output.

adaptive zone An environment or part of an environment that may be occupied by a group of species exploiting resources in a similar manner.

adductor muscle A muscle that pulls inward toward the median line of the body; in bivalve mollusks, it pulls the shell closed.

adenine A purine that base-pairs with either thymine in DNA or uracil in RNA.

adherens junction Animal cell junction in which intermediate filaments are the anchoring cytoskeletal component.

adhesion The adherence of molecules to the walls of conducting tubes, as in plants.

adiabatic cooling A decrease in temperature without the actual loss of heat energy, occurring in air masses that expand as they rise in the atmosphere.

adipocytes Densely clustered cells within adipose tissue that are specialized for fat storage.

adipose tissue Connective tissue containing large, densely clustered cells called *adipocytes* that are specialized for fat storage.

adrenal cortex The outer region of the adrenal glands, which contains endocrine cells that secrete two major types of steroid hormones, the glucocorticoids and the mineralocorticoids.

adrenal medulla The central region of the adrenal glands, which contains neurosecretory neurons that secrete the catecholamine hormones epinephrine and norepinephrine.

adrenergic receptors G-protein–coupled receptors to which hormones originating in the adrenal gland can bind.

adrenocorticotropic hormone (ACTH) A hormone that triggers hormone secretion by cells in the adrenal cortex.

adventitious root A root that develops from the stem or leaves of a plant.

aerobe An organism that requires oxygen for cellular respiration.

aerobic respiration The form of cellular respiration found in eukaryotes and many prokaryotes in which oxygen is a reactant in the ATP-producing process.

afferent arteriole The vessel that delivers blood to the glomerulus of the kidney.

afferent neuron A neuron that transmits stimuli collected by a sensory receptor to an interneuron.

agar A gelatinous product extracted from certain red algae or seaweed used as a culture medium in the laboratory and as a gelling or stabilizing agent in foods.

agarose gel electrophoresis Technique by which DNA, RNA, or protein molecules are separated in an agarose gel subjected to an electric field.

age structure A statistical description or graph of the relative numbers of individuals in each age class in a population.

age-specific fecundity The average number of offspring produced by surviving females of a particular age.

age-specific mortality The proportion of individuals alive at the start of an age interval that died during that age interval.

age-specific survivorship The proportion of individuals alive at the start of an age interval that survived until the start of the next age interval.

agglutination One of two important mechanisms to clear foreign antigens from the body, the immobilization of pathogens by antibodies.

aggregate fruit A fruit that develops from multiple separate carpels of a single flower, such as a raspberry or strawberry.

agonist A muscle that causes movement in a joint when it contracts.

albumin The most abundant protein in blood plasma, important for osmotic balance and pH buffering; also, the portion of an egg that serves as the main source of nutrients and water for the embryo.

alcohol A molecule of the form R—OH in which R is a chain of one or more carbon atoms, each of which is linked to hydrogen atoms.

alcoholic fermentation Reaction in which pyruvate is converted into ethyl alcohol and CO_2 in a two-step series that also converts NADH into NAD^+.

aldehyde Molecule in which the carbonyl group is linked to a carbon atom at the end of a carbon chain, along with a hydrogen atom.

aldosterone A mineralocorticoid hormone released from the adrenal cortex that increases the amount of Na^+ reabsorbed from the urine in the kidneys and absorbed from foods in the intestine, reduces the amount of Na^+ secreted by salivary and sweat glands, and increases the rate of K^+ excretion by the kidneys, keeping Na^+ and K^+ balanced at the levels required for normal cellular function.

aleurone The thin layer of cells that separates the endosperm of a seed from the pericarp.

algin Alginic acid, found in the cell walls of brown algae.

allantois In an amniote egg, an extraembryonic membrane sac that fills much of the space between the chorion and the yolk sac and store's the embryo's nitrogenous wastes.

allele One of two or more versions of a gene.

allele frequency The abundance of one allele relative to others at the same gene locus in individuals of a population.

allergen A type of antigen responsible for allergic reactions, which induces B cells to secrete an overabundance of IgE antibodies.

allometric growth A pattern of postembryonic development in which parts of the same organism grow at different rates.

allopatric speciation The evolution of reproductive isolating mechanisms between two populations that are geographically separated.

allopolyploidy The genetic condition of having two or more complete sets of chromosomes from different parent species.

all-or-nothing principle The principle that an action potential is produced only if the stimulus is strong enough to cause depolarization to reach the threshold.

allosteric activator Molecule that converts an enzyme with an allosteric site, a regulatory site outside the active site, from the inactive form to the active form.

allosteric inhibitor Molecule that converts an enzyme with an allosteric site, a regulatory site outside the active site, from the active form to the inactive form.

allosteric regulation Specialized control mechanism for enzymes with an allosteric site, a regulatory site outside the active site, that may either slow or accelerate activity depending on the enzyme.

allosteric site A regulatory site outside the active site.

alpha (α) helix A type of secondary structure of a polypeptide in which the amino acid chain is twisted into a regular, right-hand spiral.

α-melanocyte-stimulating hormone (alpha-MSH) A peptide hormone released from the hypothalamus that inhibits appetite.

alpine tundra A biome that occurs on high mountaintops throughout the world, in which dominant plants form cushions and mats.

alternation of generations The regular alternation of mode of reproduction in the life cycle of an organism, such as the alternation between diploid (sporophyte) and haploid (gametophyte) phases in plants.

alternative hypothesis An explanation of an observed phenomenon that is different from the explanation being tested.

alternative splicing Mechanism by which a pre-mRNA in a eukaryotic cell is processed by reactions that join exons in different combinations to produce different mRNAs from a single gene.

altruism A behavioral phenomenon in which individuals appear to sacrifice their own reproductive success to help other individuals.

Alveolata (alveolates) A subgroup of the Chromalveolata protist evolutionary group; characterized by small, flattened, membrane-bound vesicles called alveoli in a layer just under the plasma membrane.

alveolus (plural, *alveoli*) One of the millions of tiny air pockets in mammalian lungs, each surrounded by dense capillary networks.

amacrine cell A type of neuron that forms lateral connections in the retina of the eye, connecting bipolar cells and ganglion cells.

amino acid A molecule that contains both an amino and a carboxyl group.

amino group Group that acts as an organic base, consisting of a nitrogen atom bonded on one side to two hydrogen atoms and on the other side to a carbon chain.

aminoacylation The process of adding an amino acid to a tRNA. Also referred to as *charging*.

aminoacyl–tRNA A tRNA linked to its "correct" amino acid, which is the finished product of aminoacylation.

aminoacyl–tRNA synthetase An enzyme that catalyzes aminoacylation.

aminopeptidase An enzyme that hydrolyzes small peptides to produce individual amino acids.

ammonification A metabolic process in which bacteria convert organic nitrogen compounds into ammonia and ammonium ions; part of the nitrogen cycle.

amniocentesis Technique of prenatal diagnosis in which cells are obtained from the amniotic fluid for DNA testing, for biochemical analysis, or to test for the presence of chromosomal mutations.

amnion In an amniote egg, an extraembryonic membrane that encloses the embryo, forming the amniotic cavity and secreting amniotic fluid, which provides an aquatic environment in which the embryo develops.

Amniota The monophyletic group of vertebrates that have an amnion during embryonic development.

amniote egg A shelled egg that can survive and develop on land.

amoeba A descriptive term for a single-celled protist that moves by means of temporary cellular projections called *pseudopods*.

Amoebozoa (amoebozoans) A protist evolutionary group that includes most of the amoebas as well as the cellular and plasmodial slime molds; characterized by the use of pseudopods for locomotion and feeding for all or part of their life cycles.

amplification An increase in the magnitude of each step as a signal transduction pathway proceeds.

amygdala A gray-matter center of the brain that works as a switchboard, routing information about experiences that have an emotional component through the limbic system.

amyloplast Colorless plastid that stores starch in plants.

anabolic-androgenic steroid (AAS) A synthetic derivative of androgens that mimics their effects. Also called an *anabolic steroid.*

anabolic pathway A metabolic pathway in which energy is used to build complicated molecules from simpler ones; also called a *biosynthetic pathway.* An individual reaction in an anabolic pathway is an anabolic reaction, also called a *biosynthetic reaction.*

anabolic reaction Metabolic reaction that requires energy to assemble simple substances into more complex molecules.

anabolic steroid *See* **anabolic-androgenic steroid (AAS).**

anaerobe An organism that does not require oxygen to live.

anaerobic respiration The form of cellular respiration found in some prokaryotes in which a molecule other than oxygen is used in the ATP-producing process.

anagenesis The slow accumulation of evolutionary changes in a lineage over time.

anal sphincter A muscular ring that controls the opening and closing of the anus.

anaphase The phase of mitosis during which the spindle separates sister chromatids and pulls them to opposite spindle poles.

anaphase-promoting complex (APC) An enzyme complex activated by M phase–promoting factor that controls the separation of sister chromatids and the onset of daughter chromosome separation in anaphase of mitosis.

anaphylactic shock A severe inflammation stimulated by an allergen, involving extreme swelling of air passages in the lungs that interferes with breathing, and massive leakage of fluid from capillaries that causes blood pressure to drop precipitously.

anapsida A lineage of amniotes that lack a bony temporal arch in the skull.

anatomy The study of the structures of organisms.

ancestral character state A trait that was present in a distant common ancestor.

anchoring junction Cell junction that forms belts that run entirely around cells, "welding" adjacent cells together.

androgen One of a family of hormones that promote the development and maintenance of sex characteristics.

aneuploid An individual with extra or missing chromosomes.

angiosperm A flowering plant. Its egg-containing ovules mature into seeds within protected chambers called *ovaries.*

angiotensin A peptide hormone that raises blood pressure quickly by constricting arterioles in most parts of the body; it also stimulates release of the steroid hormone aldosterone.

angiotensin I In the renin-angiotensin-aldosterone system (RAAS), the molecule produced by cleavage of the plasma protein angiotensinogen.

angiotensin II In the renin-angiotensin-aldosterone system (RAAS), the molecule converted from angiotensin I by angiotensin-converting enzyme (ACE); angiotensin II is a hormone that constricts arterioles to raise blood pressure, stimulates synthesis of aldosterone and its secretion from the adrenal cortex, and stimulates thirst.

angiotensin-converting enzyme (ACE) In the renin-angiotensin-aldosterone system (RAAS), the enzyme that converts angiotensin I to angiotensin II.

animal behavior The responses of animals to specific internal and external stimuli.

animal pole The end of the egg where the egg nucleus is located, which typically gives rise to surface structures and the anterior end of the embryo.

Animalia The taxonomic kingdom that includes all living and extinct animals.

anion A negatively charged ion.

annual An herbaceous plant that completes its life cycle in one growing season and then dies.

annulus In ferns, a ring of thick-walled cells that nearly encircles the sporangium and functions in spore release.

antagonistic pair Two skeletal muscles, one of which flexes as the other extends to move joints.

antenna A chemosensory appendage attached to the head of some adult arthropods.

antenna complex (light-harvesting complex) In photosystems, the sites at which light is absorbed and converted into chemical energy during photosynthesis, an aggregate of many chlorophyll pigments and a number of carotenoid pigments that serves as the primary site of absorbing light energy in the form of photons.

anterior Indicating the head end of an animal.

anterior pituitary The glandular part of the pituitary, composed of endocrine cells that synthesize and secrete several tropic and nontropic hormones.

anther The pollen-bearing part of a stamen.

antheridium (plural, *antheridia*) In plants, a structure in which sperm are produced.

Anthocerophyta The phylum comprising hornworts.

Anthophyta The phylum comprising flowering plants.

Anthropoidea The monophyletic lineage of primates comprising monkeys, apes, and humans.

antibody A highly specific soluble protein molecule that circulates in the blood and lymph, recognizing and binding to antigens and clearing them from the body.

antibody class The molecular type of an antibody determined by the constant regions of the heavy chains of the molecule.

antibody-mediated immunity Adaptive immune response in which plasma cells secrete antibodies. Also called *humoral immunity.*

anticodon The three-nucleotide segment in a tRNA that pairs with a codon in an mRNA.

antidiuretic hormone (ADH) A hormone secreted by the posterior pituitary that increases water absorption in the kidneys, thereby increasing the volume of the blood.

antigen A foreign molecule that triggers an adaptive immunity response.

antigen-presenting cell (APC) A cell that presents an antigen to T cells in antibody-mediated immunity and cell-mediated immunity.

antihistamines Substances that block histamine receptors thereby reducing symptoms of an allergic reaction.

antiparallel Strands of double-stranded DNA that run in opposite directions with the 3′ end of one strand opposite the 5′ end of the other strand.

antiport A secondary active transport mechanism in which a molecule moves through a membrane channel into a cell and powers the active transport of a second molecule out of the cell. Also referred to as *exchange diffusion.*

aorta A large artery from the heart that branches into arteries leading to all body regions except the lungs.

aortic body One of several small clusters of chemoreceptors, baroreceptors, and supporting cells located along the aortic arch that measures changes in blood pressure and the composition of arterial blood flowing past it.

aphotic zone Deeper water of a lake or ocean where sunlight does not penetrate.

apical dominance Inhibition of the growth of lateral buds in plants due to auxin diffusing down a shoot tip from the terminal bud.

apical meristem A region of unspecialized dividing cells at shoot tips and root tips of a plant.

apical surface The outer surface of epithelial cells.

Apicomplexa (apicomplexans) A lineage within the Alveolata subgroup of the Chromalveolata protist evolutionary group; nonmotile parasites of animals characterized by the apical complex, a special grouping of fibrils, microtubules, and organelles at one end of the cell that functions in attachment and invasion of host cells.

apomixis In plants, the production of offspring without meiosis or formation of gametes.

apomorphy A derived character state.

apoplastic pathway The route followed by water moving through plant cell walls and intercellular spaces (the apoplast). *Compare* **symplastic pathway.**

apoptosis A type of programmed cell death.

aposematic coloration Bright, contrasting patterns that advertise the unpalatability of poisonous or repellant species.

appendicular skeleton The bones comprising the pectoral (shoulder) and pelvic (hip) girdles and limbs of a vertebrate.

appendix A fingerlike sac that extends from the cecum of the large intestine.

applied research Research conducted with the goal of solving specific practical problems.

aquaporin A specialized protein channel that facilitates diffusion of water through cell membranes.

aquatic succession A process in which debris from rivers and runoff accumulates in a body of fresh water, causing it to fill in at the margins.

aqueous humor A clear fluid that fills the space between the cornea and lens of the eye.

Archaea One of two domains of prokaryotes; archaeans have some unique molecular and biochemical traits, but they also share some traits with Bacteria and other traits with Eukarya.

archaeal chromosome DNA molecule in archaeans in which hereditary information is encoded.

Archaeplastida (archaeplastids) An evolutionary group consisting of the Rhodophyta and Chlorophyta protists, and the land plants; all are photosynthesizers.

archegonium The flask-shaped structure in which bryophyte eggs form.

archenteron The central endoderm-lined cavity of an embryo at the gastrula stage, which forms the primitive gut.

Archosauromorpha A diverse group of diapsids that comprises crocodilians, pterosaurs, and dinosaurs (including birds).

arctic tundra A treeless biome that stretches from the boreal forests to the polar ice cap in Europe, Asia, and North America.

arteriole A branch from a small artery at the point where it reaches the organ it supplies.

artery A vessel that conducts blood away from the heart at relatively high pressure.

artificial selection Selective breeding of organisms to ensure that certain desirable traits appear at higher frequency in successive generations.

ascending reticular formation Part of the reticular formation containing neurons that convey stimulatory signals via the thalamus to arouse and activate the cerebral cortex. It is responsible for the sleep–wake cycle. Also called the *reticular activating system.*

ascocarp In ascomycete (sac) fungi, a reproductive body that bears or contains asci.

ascus (plural, *asci*) A saclike cell in ascomycetes (sac fungi) in which meiosis gives rise to haploid sexual spores (meiospores).

asexual reproduction Any mode of reproduction in which a single individual gives rise to offspring without fusion of gametes; that is, without genetic input from another individual. *See also* **vegetative reproduction.**

association area One of several areas surrounding the sensory and motor areas of the cerebral cortex that integrate information from the sensory areas, formulate responses, and pass them on to the primary motor area.

aster Radiating array produced as microtubules extending from the centrosomes of cells grow in length and extent.

asthma In an allergic reaction, a severe response to allergens that involves constriction of airways in the lungs.

astrocyte A star-shaped glial cell that provides support to neurons in the vertebrate central nervous system.

asymmetrical Characterized by a lack of proportion in the spatial arrangement or placement of parts.

atmosphere The component of the biosphere that includes the gases and airborne particles enveloping the planet.

atom The smallest unit that retains the chemical and physical properties of an element.

atomic nucleus The nucleus of an atom, containing protons and neutrons.

atomic number The number of protons in the nucleus of an atom.

atomic weight The weight of an element in grams, equal to the mass number.

ATP (adenosine triphosphate) The primary agent that couples exergonic and endergonic reactions.

ATP/ADP cycle The continual hydrolysis and resynthesis of ATP in living cells.

ATP synthase A membrane-spanning protein complex that couples the energetically favorable transport of protons across a membrane to the synthesis of ATP.

atrial natriuretic factor (ANF) A peptide hormone that inhibits renin release and increases the filtration rate by dilating the arterioles that deliver blood to glomeruli and by inhibiting aldosterone release.

atrial siphon A tube through which invertebrate chordates expel digestive and metabolic wastes.

atriopore The hole in the body wall of a cephalochordate through which water is expelled from the body.

atrioventricular node (AV node) A region of the heart wall that receives signals from the sinoatrial node and conducts them to the ventricle.

atrium (plural, *atria*) A body cavity or chamber surrounding the perforated pharynx of invertebrate chordates; also, one of the chambers that receive blood returning to the heart.

autoimmune disease A malfunction of the immune system in which the body reacts against its own proteins or cells.

autoimmune reaction The production of antibodies against molecules of the body.

autonomic nervous system A subdivision of the peripheral nervous system that controls largely involuntary processes including digestion, secretion by sweat glands, circulation of the blood, many functions of the reproductive and excretory systems, and contraction of smooth muscles in all parts of the body.

autopolyploidy The genetic condition of having more than two sets of chromosomes from the same parent species.

autosomal dominant inheritance Pattern in which the allele that causes a trait is dominant, and only homozygous recessives are unaffected.

autosomal recessive inheritance Pattern in which individuals with a trait are homozygous for a recessive allele.

autosome Chromosome other than a sex chromosome.

autotroph An organism that produces its own food using CO_2 and other simple inorganic compounds from its environment and energy from the sun or from oxidation of inorganic substances.

autumn overturn A process in which winds mix the water in a lake vertically, equalizing the concentrations of dissolved gases and nutrients at all depths.

auxin Any of a family of plant hormones that stimulate growth by promoting cell elongation in stems and coleoptiles, inhibit abscission, govern responses to light and gravity, and have other developmental effects.

auxotroph A mutant strain that requires for its growth a nutrient supplement that is not needed by the wild-type strain.

Avogadro's number The number 6.022×10^{23}, derived by dividing the atomic weight of any element by the weight of an atom of that element.

***Avr* gene** A gene in certain plant pathogens that encodes a product triggering a defensive response in the plant.

axial skeleton The bones comprising the head and trunk of a vertebrate: the cranium, vertebral column, ribs, and sternum (breastbone).

axil The upper angle between the stem and an attached leaf.

axon The single elongated extension of a neuron that conducts signals away from the cell body to another neuron or an effector.

axon hillock A junction with the cell body of a neuron from which the axon arises.

axon terminal A branch at the tip of an axon that ends as a small, buttonlike swelling.

B cell A lymphocyte that recognizes antigens in the body.

Bacillariophyta (diatoms) A lineage of the Stramenopila subgroup of the Chromalveolata protist evolutionary group; single-celled organisms covered by an intricate, glassy silica shell.

bacillus (plural, *bacilli*) A cylindrical or rod-shaped prokaryote.

background extinction rate The average rate of extinction of taxa through time.

Bacteria One of the two domains of prokaryotes; collectively, bacteria are the most metabolically diverse organisms.

bacterial chromosome DNA molecule in bacteria in which hereditary information is encoded.

bacterial flagellum *See* **flagellum.**

bacteriophage A virus that infects bacteria. Also referred to as a *phage.*

bacteroid A rod-shaped or branched bacterium in the root nodules of nitrogen-fixing plants.

balanced polymorphism The maintenance of two or more phenotypes in fairly stable proportions over many generations.

bark The tough outer covering of woody stems and roots, composed of all the living and nonliving tissues between the vascular cambium and the stem surface.

Barr body The inactive, condensed X chromosome seen in the nucleus of female placental mammals.

basal angiosperm Any of the earliest branches of the flowering plant lineage; includes the star anise group and water lilies.

basal body Structure that anchors cilia and flagella to the surface of a cell.

basal lamina The membrane that fixes the epithelium to underlying tissues (also called the *basement membrane*).

basal nucleus One of several gray-matter centers that surround the thalamus on both sides of the brain and moderate voluntary movements directed by motor centers in the cerebrum.

basal surface The inner surface of epithelial cells.

base Proton acceptor that reduces the H^+ concentration of a solution.

base-pair mismatch An error in the assembly of a new nucleotide chain in which bases other than the correct ones pair together.

basic research Research conducted to search for explanations about natural phenomena in order to satisfy curiosity and to advance collective knowledge of living systems.

basidiocarp A fruiting body of a basidiomycete; mushrooms are examples.

basidiospore A haploid sexual spore produced by basidiomycete fungi.

basidium (plural, *basidia*) A small, club-shaped structure in which sexual spores of basidiomycetes arise.

basilar membrane A part of the floor of the cochlear duct, which anchors sensory hair cells in the organ of Conti and which vibrates in response to vibrations moving through the inner ear.

basophil A type of leukocyte that is induced to secrete histamine by allergens.

Batesian mimicry The form of defense in which a palatable or harmless species resembles an unpalatable or poisonous one.

Bayesian methods A statistical technique that evaluates a given phylogenetic tree by determining the probability that it is correct given the distribution of character states included in the analysis.

B-cell receptor (BCR) The receptor on B cells that is specific for a particular antigen.

behavioral isolation A prezygotic reproductive isolating mechanism in which two species do not mate because of differences in courtship behavior; also known as ethological isolation.

behavioral repertoire The set of actions that an animal can perform in response to stimuli in its environment.

benthic province The bottom sediments in the ocean.

benthos Species living in and on the bottom sediments of the ocean.

beta (β) sheet A type of primary structure in a polypeptide in which the amino acid chain zigzags in a flat plane to form a beta strand, and beta strands then align side by side in the same or opposite direction.

biennial A plant that completes its life cycle in two growing seasons and then dies; limited secondary growth occurs in some biennials.

Bikonta (bikonts) Eukaryotes with two flagella.

bilateral symmetry The body plan of animals in which the body can be divided into mirror image right and left halves by a plane passing through the midline of the body.

bilayer A membrane with two molecular layers.

bile A mixture of substances including bile salts, cholesterol, and bilirubin that is made in the liver, stored in the gallbladder, and used in the digestion of fats.

bile salts A component of bile, consisting of derivatives of cholesterol and amino acids, that aids fat digestion.

bilirubin A component of bile consisting of a waste product derived from worn-out red blood cells.

binomial Relating to or consisting of two names or terms.

binomial nomenclature The naming of species with a two-part scientific name, the first indicating the genus and the second indicating the species.

biodiversity The richness of living systems as reflected in genetic variability within and among species, the number of species living on Earth, and the variety of communities and ecosystems.

biodiversity hot spot An area where biodiversity is both highly concentrated and endangered.

biofilm A prokaryotic multicellular association consisting of a complex aggregation of microorganisms (either one or multiple species) attached, in most cases, to a surface.

biogeochemical cycle Any of several global processes in which a nutrient circulates between the abiotic environment and living organisms.

biogeographical realm A major region of Earth that is occupied by distinct evolutionary lineages of plants and animals.

biogeography The study of the geographical distributions of plants and animals.

bioinformatics The application of mathematics and computer science to extract information from biological data, including those related to genome structure and function.

biological clock An internal time-measuring mechanism that adapts an organism to recurring environmental changes.

biological evolution The process by which some individuals in a population experience changes in their DNA and pass those modified instructions to their offspring.

biological lineage An evolutionary sequence of ancestral organisms and their descendants.

biological magnification The increasing concentration of nondegradable poisons in the tissues of animals at higher trophic levels.

biological research The collective effort of individuals who have worked to understand how living systems function.

biological species concept The definition of species based on the ability of populations to interbreed and produce fertile offspring.

bioluminescent An organism that glows or releases a flash of light, particularly when disturbed.

biomass The dry weight of biological material per unit area or volume of habitat.

biome A large scale vegetation type and its associated microorganisms, fungi, and animals.

bioremediation Applications of chemical and biological knowledge to decontaminate polluted environments.

biosignature Particular organic molecules in sedimentary rocks that could only have been formed by cellular activity.

biosphere All regions of Earth's crust, waters, and atmosphere that sustain life.

biosynthetic reaction An individual reaction in an anabolic pathway (biosynthetic pathway).

biota The total collection of organisms in a geographical region.

biotechnology The manipulation of living organisms to produce useful products.

biotic Biological, often in reference to living components of the environment.

bipedalism The habit in animals of walking upright on two legs.

bipolar cell A type of neuron in the retina of the eye that connects the rods and cones with the ganglion cells.

blade The expanded part of a leaf that provides a large surface area for absorbing sunlight and carbon dioxide.

blastocoel A fluid-filled cavity in the blastula embryo.

blastocyst An embryonic stage in mammals; a single-cell-layered hollow ball of about 120 cells with a fluid-filled blastocoel in which a dense mass of cells is localized to one side.

blastodisc A disclike layer of cells at the surface of the yolk produced by early cleavage divisions.

blastomere A small cell formed during cleavage of the embryo.

blastopore The opening at one end of the archenteron in the gastrula that gives rise to the mouth in protostomes and the anus in deuterostomes.

blastula The hollow ball of cells that is the result of cleavage divisions in an early embryo.

blending theory of inheritance Theory suggesting that hereditary traits blend evenly in offspring through mixing of the blood of the two parents.

blood A fluid connective tissue composed of blood cells suspended in a fluid extracellular matrix called **plasma.**

blood–brain barrier A specialized arrangement of capillaries in the brain that prevents most substances dissolved in the blood from entering the cerebrospinal fluid and thus protects the brain and spinal cord from viruses, bacteria, and toxic substances that may circulate in the blood.

blood pressure The measurement of the hydrostatic pressure on the walls of the arteries as the heart pumps blood through the body.

body system *See* **organ system.**

Bohr effect The reduction in affinity of hemoglobin for O_2 that results from a conformational change caused by lowered pH.

bolting Rapid formation of a floral shoot in plant species that form rosettes, such as lettuce.

bolus The food mass after chewing.

bone The densest form of connective tissue, in which living cells secrete the mineralized matrix of collagen and calcium salts that surrounds them; forms the skeleton.

bony temporal arches Bones that border the temporal fenestrae in the skulls of some amniotes.

book lungs Pocketlike respiratory organs found in some arachnids consisting of several parallel membrane folds arranged like the pages of a book.

boreal forest A biome that is a circumpolar expanse of evergreen coniferous trees in Europe, Asia, and North America.

Bowman's capsule An infolded region at the proximal end of a nephron that cups around the glomerulus and collects the water and solutes filtered out of the blood.

brachiation A pattern of locomotion among primates in which an individual swings below branches from one handhold to another.

brain A single, organized collection of nervous tissue in an organism's head that forms the control center of the nervous system and major sensory structures.

brain stem A stalklike structure formed by the pons and medulla, along with the midbrain, which connects the forebrain with the spinal cord.

branchial slits Openings in the walls of the pharynx that allow water to exit the pharynx.

brassinosteroid Any of a family of plant hormones that stimulate cell division and elongation and differentiation of vascular tissue.

breathing The exchange of gases with the respiratory medium by animals.

bronchiole One of the small, branching airways in the lungs that lead into the alveoli.

bronchus (plural, *bronchi*) An airway that leads from the trachea to the lungs.

brown adipose tissue A specialized tissue in which the most intense heat generation by nonshivering thermogenesis takes place.

brown algae *See* **Phaeophyta.**

brown fat *See* **brown adipose tissue.**

brush border The fingerlike projections of the plasma membrane of the epithelial cells covering the intestinal villi.

Bryophyta The phylum of nonvascular plants to which mosses are assigned.

bryophyte A general term for plants (such as mosses) that lack internal transport vessels.

budding A mode of asexual reproduction in which a new individual grows and develops while attached to the parent.

buffer Substance that compensates for pH changes by absorbing or releasing H^+.

bulbourethral gland One of two pea-sized glands on either side of the prostate gland, which secrete a mucous fluid that is added to semen.

bulk feeder An animal that consumes sizeable food items whole or in large chunks.

bulk flow The group movement of molecules in response to a difference in pressure between two locations.

bulk-phase endocytosis Mechanism by which extracellular water is taken into a cell together with any molecules that happen to be in solution in the water. Also referred to as *pinocytosis.*

C_3 pathway *See* **light-independent reaction;** also referred to as the *Calvin cycle.*

C_4 pathway In C_4 plants the pathway to fix CO_2 into oxaloacetate in mesophyll cells and then produce CO_2 for the Calvin cycle in bundle sheath cells.

Ca^{2+} pump (calcium pump) Pump that pushes Ca^{2+} from the cytoplasm to the cell exterior, and also from the cytosol into the vesicles of the endoplasmic reticulum.

cadherin A cell surface protein responsible for selective cell adhesions that require calcium ions to set up adhesions.

calcitonin A nontropic peptide hormone that lowers the level of Ca^{2+} in the blood by inhibiting the ongoing dissolution of calcium from bone.

callus An undifferentiated tissue that develops on or around a cut plant surface or in tissue culture.

calorie (cal) The amount of heat required to raise 1 g of water by 1°C, known as a "small" calorie; when capitalized, a unit equal to 1,000 small calories.

Calvin cycle *See* **light-independent reaction.**

calyx The outermost whorl of a flower, made up of sepals; early in the development of a flower, it encloses all the other parts, as in an unopened bud.

CAM pathway In CAM plants the pathway to fix CO_2 into oxaloacetate and then produce CO_2 for the Calvin cycle, both occurring in mesophyll cells, but separated by time of day. CAM stands for "crassulacean acid metabolism."

cancer When a tumor becomes malignant and its cells invade and disrupt surrounding tissues.

canines Pointed, conical teeth of a mammal, located between the incisors and the first premolars, that are specialized for biting and piercing.

CAP (catabolite activator protein) Key regulatory molecule involved in positive gene regulation of the *lac* operon.

CAP site Region in the promoter of the *lac* operon and in the promoters of a large number of other operons that control the catabolism of many sugars to which activated catabolite activator protein (CAP) binds, thereby enabling RNA polymerase to bind and transcribe the operon's structural genes.

capillary The smallest diameter blood vessel, with a wall that is one cell thick, which forms highly branched networks well adapted for diffusion of substances.

capsid The protective layer of protein that surrounds the nucleic acid core of a virus in free form; also known as a *coat.*

capsule An external layer of sticky or slimy polysaccharides coating the cell wall in many prokaryotes.

carapace A protective outer covering that extends backward behind the head on the dorsal side of an animal, such as the shell of a turtle or lobster.

carbon cycle The global circulation of carbon atoms, especially via the processes of photosynthesis and respiration.

carbonyl group The reactive part of aldehydes and ketones, consisting of an oxygen atom linked to a carbon atom by a double bond.

carboxyl group The characteristic functional group of organic acids, formed by the combination of carbonyl and hydroxyl groups.

carboxypeptidase An enzyme that digests peptides to produce amino acids.

cardiac cycle The systole-diastole sequence of the heart.

cardiac muscle The contractile tissue of the heart.

carnivore An animal that primarily eats other animals.

carotenoid Molecule of yellow-orange pigment by which light is absorbed in photosynthesis.

carotid body A small cluster of chemoreceptors and supporting cells located near the bifurcation of the carotid artery that measures changes in the composition of arterial blood flowing through it.

carpel The reproductive organ of a flower that houses an ovule and its associated structures.

carrageenan A chemical extracted from the red alga *Eucheuma* that is used to thicken and stabilize paints, dairy products such as pudding and ice cream, and many other creams and emulsions.

carrier A heterozygote—an individual who carries a recessive mutant allele and could pass it on to offspring, but does not display its symptoms.

carrier protein Transport protein that binds a specific single solute and transports it across the lipid bilayer.

carrying capacity The maximum size of a population that an environment can support indefinitely.

Cartagena Protocol on Biosafety An international agreement that promotes biosafety as it relates to the handling and use of genetically modified organisms.

cartilage A tissue composed of sparsely distributed chondrocytes surrounded by networks of collagen fibers embedded in a tough but elastic matrix of the glycoprotein.

cartilaginous joint A joint between bones that is not very movable and in which the ends of the bones are covered with layers of cartilage but with no fluid-filled capsule surrounding them.

Casparian strip A thin, waxy impermeable band that seals abutting cell walls in roots; the strip helps control the type and amount of solutes that enter the stele by blocking the apoplastic pathway at the endodermis and forcing substances to pass through cells (the symplast).

catabolic pathway A metabolic pathway in which energy is released by the breakdown of complex molecules to simpler compounds. An individual reaction in a catabolic pathway is a catabolic reaction.

catabolic reaction Cellular reaction that breaks down complex molecules such as sugar to make their energy available for cellular work.

catalysis The process of accelerating a chemical reaction with a catalyst.

catalyst Substance with the ability to accelerate a spontaneous reaction without being changed by the reaction.

catastrophism The theory that Earth has been affected by sudden, violent events that were sometimes worldwide in scope.

catecholamine Any of a class of compounds derived from the amino acid tyrosine that circulates in the bloodstream, including epinephrine and norepinephrine.

cation A positively charged ion.

cation exchange Replacement of one cation with another, as on a soil particle.

$CD4^{+}$ T cell A type of T cell in the lymphatic system that has CD4 receptors on its surface. This type of T cell binds to an antigen-presenting cell in antibody-mediated immunity.

$CD8^{+}$ T cell A type of T cell in the lymphatic system that has CD8 receptors on its surface. This type of T cell binds to an antigen-presenting cell in cell-mediated immunity.

cDNA library The entire collection of cloned cDNAs made from the mRNAs isolated from a cell.

cecum A blind pouch formed at the junction of the large and small intestine.

cell Smallest unit with the capacity to live and reproduce.

cell adhesion molecule A cell surface protein responsible for selectively binding cells together.

cell adhesion protein Protein that binds cells together by recognizing and binding receptors or chemical groups on other cells or on the extracellular matrix.

cell body The portion of the neuron containing genetic material and cellular organelles.

cell center *See* **centrosome.**

cell culture Living cells growing in a growth medium in a laboratory vessel.

cell cycle The sequence of events during which a cell experiences a period of growth followed by nuclear division and cytokinesis.

cell differentiation A process in which changes in gene expression establish cells with specialized structure and function.

cell expansion A mechanism that enlarges the cells in specific directions in a developing organ.

cell fractionation Technique that divides cells into fractions containing a single cell component.

cell junction Junction that seals the spaces between cells and provides direct communication between cells.

cell lineage Cell derivation from the undifferentiated tissues of the embryo.

cell plate In cytokinesis in plants, a new cell wall that forms between the daughter nuclei and grows laterally until it divides the cytoplasm.

cell signaling The system of communication between cells through signaling pathways.

cell theory Three generalizations yielded by microscopic observations: all organisms are composed of one or more cells; the cell is the smallest unit that has the properties of life; and cells arise only from the growth and division of preexisting cells.

cell wall A rigid external layer of material surrounding the plasma membrane of cells in plants, fungi, bacteria, and some protists, providing cell protection and support.

cell-mediated immunity An adaptive immune response in which a subclass of T cells—cytotoxic T cells—becomes activated and, with other cells of the immune system, attacks host cells infected by pathogens, particularly those infected by a virus.

cellular respiration The process by which energy-rich molecules are broken down to produce energy in the form of ATP.

cellular slime mold Any of a variety of primitive organisms of the phylum Acrasiomycota, especially of the genus *Dictyostelium;* the life cycle is characterized by a slimelike amoeboid stage and a multicellular reproductive stage.

cellulose One of the primary constituents of plant cell walls, formed by chains of carbohydrate subunits.

centimorgan *See* **map unit.**

central canal The central portion of the vertebral column in which the spinal cord is found.

central dogma The name given by Francis Crick to the flow of information from DNA to RNA to protein.

central nervous system (CNS) One of the two major divisions of the nervous system containing the brain and spinal cord.

central vacuole A large, water-filled organelle in plant cells that maintains the turgor of the cell and controls movement of molecules between the cytosol and sap.

centriole A cylindrical structure consisting of nine triplets of microtubules in the centrosomes of most animal cells.

centromere A specialized chromosomal region that connects sister chromatids and attaches them to the mitotic spindle.

centrosome (cell center) The main microtubule organizing center of a cell, which organizes the microtubule cytoskeleton during interphase and positions many of the cytoplasmic organelles.

cephalothorax The anterior section of an arachnid, consisting of a fused head and thorax.

cerebellum The portion of the brain that receives sensory input from receptors in muscles and joints, from balance receptors in the inner ear, and from the receptors of touch, vision, and hearing.

cerebral cortex A thin outer shell of gray matter covering a thick core of white matter within each hemisphere of the brain; the part of the forebrain responsible for information processing and learning.

cerebrospinal fluid Fluid that circulates through the central canal of the spinal cord and the ventricles of the brain, cushioning the brain and spinal cord from jarring movements and impacts, as well as nourishing the CNS and protecting it from toxic substances.

cervix The lower end of the uterus.

channel protein Transport protein that forms a hydrophilic channel in a cell membrane through which water, ions, or other molecules can pass, depending on the protein.

chaparral A biome comprising a scrubby mix of short trees and shrubs that dominates coastal land between 30° and 40° latitude, where winters are cool and wet and summers hot and dry.

chaperone protein (chaperonin) "Guide" protein that binds temporarily with newly synthesized proteins, directing their conformation toward the correct tertiary structure and inhibiting incorrect arrangements as the new proteins fold.

character A specific heritable attribute or property of an organism.

character differences Alternative forms of characters. Also called a *trait.*

character displacement The phenomenon in which allopatric populations are morphologically similar and use similar resources, but sympatric populations are morphologically different and use different resources; may also apply to characters influencing mate choice.

character states One or more forms of a character used in a phylogenetic analysis.

charging *See* **aminoacylation.**

charophyte A member of the group of green algae most similar to the algal ancestors of land plants.

checkpoint Internal control of the cell cycle that prevents a critical phase from beginning until the previous phase is complete.

chelicerae The first pair of fanglike appendages near the mouth of an arachnid, used for biting prey and often modified for grasping and piercing.

chemical bond Link formed when atoms of reactive elements combine into molecules.

chemical equation A chemical reaction written in balanced form.

chemical reaction A reaction that occurs when atoms or molecules interact to form new chemical bonds or break old ones.

chemical signal Any secretion from one cell type that can alter the behavior of a different cell that bears a receptor for it; a means of cell communication.

chemical synapse In the nervous system where the plasma membranes of the presynaptic and postsynaptic cells are separated by a narrow gap. Communication across such a synapse occurs by means of a neurotransmitter; that is, an electrical impulse arriving at an axon terminal of the presynaptic cell triggers release of a neurotransmitter that crosses the gap and binds to a receptor on the postsynaptic cell, triggering an electrical impulse in that cell.

chemiosmotic hypothesis Model proposing that mitochondrial electron transfer produces an H^+ gradient and that the gradient powers ATP synthesis by ATP synthase.

chemoautotroph An organism that obtains energy by oxidizing inorganic substances such as hydrogen, iron, sulfur, ammonia, nitrites, and nitrates and uses carbon dioxide as a carbon source.

chemoheterotroph An organism that oxidizes organic molecules as an energy source and obtains carbon in organic form.

chemokine A protein secreted by activated macrophages that attracts other cells, such as neutrophils.

chemoreceptor A sensory receptor that detects specific molecules, or chemical conditions such as acidity.

chemotroph An organism that obtains energy by oxidizing inorganic or organic substances.

chiasmata *See* **crossover.**

chief cells Cells in the gastric glands of the stomach lining that secrete pepsinogen.

chitin A polysaccharide that contains nitrogen and is present in the cell walls of fungi and the exoskeletons of arthropods.

Chlorarachniophyta (chlorarachniophytes) A lineage within the Rhizaria protist evolutionary group; green, photosynthetic amoebas that also engulf food.

chlorophyll Molecule of green pigment that absorbs photons of light in photosynthesis.

Chlorophyta (green algae) A lineage within the Archaeplastida protist evolutionary group; autotrophs that carry out photosynthesis using the same pigments as plants.

chloroplast The site of photosynthesis in plant cells.

chlorosis An abnormal yellowing of plant tissues due to lack of chlorophyll; a sign of nutrient deficiency or infection by a pathogen.

choanocyte One of the inner layer of flagellated cells lining the body cavity of a sponge.

Choanoflagellata (choanoflagellates) A group of minute, single-celled protists found in water; the flask-shaped body has a collar of closely packed microvilli that surrounds the single flagellum by which it moves and takes in food. Choanoflagellates are a subgroup of the opisthokonts.

cholecystokinin (CCK) Hormone released in response to fat, and to a lesser extent protein, in the chyme that enters the duodenum and inhibits gastric activity.

cholesterol The predominant sterol of animal cell membranes.

chondrocyte A cartilage-producing cell.

chorion In an amniote egg, an extraembryonic membrane that surrounds the embryo and yolk sac completely and exchanges oxygen and carbon dioxide with the environment; becomes part of the placenta in mammals.

chorionic villus (plural, *villi*) One of many treelike extensions from the chorion, which greatly increase the surface area of the chorion.

chorionic villus sampling Technique of prenatal diagnosis in which cells are obtained from portions of the placenta that develop from tissues of the embryo for DNA testing, for biochemical analysis, or to test for the presence of chromosomal mutations.

Chromalveolata (chromalveolates) A protist evolutionary group consisting of the Alveolata and the Stramenopila.

chromatin Any assemblage of eukaryotic nuclear DNA molecules and their associated proteins.

chromoplast Plastid containing red and yellow pigments.

chromosomal mutation A variation from the normal condition in chromosome structure or chromosome number.

chromosomal protein A histone and nonhistone protein associated with DNA in a eukaryotic nuclear chromosome.

chromosome In eukaryotic cells, a linear structure composed of a single DNA molecule complexed with protein. Each eukaryotic species has a characteristic number of chromosomes in the nucleus. Most prokaryotes have a single, usually circular chromosome with few or no associated proteins.

chromosome segregation The equal distribution of daughter chromosomes to each of the two cells that result from cell division.

chromosome theory of inheritance The principle that genes and their alleles are carried on the chromosomes.

Chrysophyta (golden algae) A lineage of the Stramenopila subgroup of the Chromalveolata protist evolutionary group; mostly colonial protists in which each cell of a colony has a pair of flagella and is surrounded by a glassy shell.

chylomicron A small triglyceride droplet coated with a layer of lipoprotein from the endoplasmic reticulum.

chyme Digested content of the stomach released for further digestion in the small intestine.

chymotrypsin An enzyme that digests peptides to produce small peptides.

ciliary body A fine ligament in the eye that anchors the lens to a surrounding layer of connective tissue and muscle.

Ciliophora (ciliates) A lineage of the Alveolata subgroup of the Chromalveolata protist evolutionary group.

cilium Motile structure, extending from a cell surface, that moves a cell through fluid or fluid over a cell.

circadian rhythm Any biological activity that is repeated in cycles, each about 24 hours long, independently of any shifts in environmental conditions.

circulatory system An organ system consisting of a fluid, a heart, and vessels for moving important molecules, and often cells, from one tissue to another.

circulatory vessel An element of the circulatory system through which fluid flows and carries nutrients and oxygen to tissues and remove wastes.

circumcision Removal of the prepuce for religious, cultural, or hygienic reasons.

cisternae Membranous channels and vesicles that make up the endoplasmic reticulum.

citric acid cycle Series of reactions in which acetyl groups are oxidized completely to carbon dioxide and some ATP molecules are synthesized. Also referred to as *Krebs cycle* and *tricarboxylic acid cycle.*

clade A monophyletic group of organisms that share homologous features derived from a common ancestor.

cladistics An approach to systematics that uses shared derived characters to infer the phylogenetic relationships and evolutionary history of groups of organisms.

cladogenesis The evolution of two or more descendant species from a common ancestor.

cladogram A branching diagram in which the endpoints of the branches represent different species of organisms, used to illustrate phylogenetic relationships.

claspers A pair of organs on the pelvic fins of male crustaceans and sharks, which help transfer sperm into the reproductive tract of the female.

class A Linnaean taxonomic category that ranks below a phylum and above an order.

class II major histocompatibility complex (MHC) A collection of proteins that present antigens on the cell surface of an antigen-presenting cell in an antibody-mediated immune response.

classical conditioning A type of learning in which an animal develops a mental association between two phenomena that are usually unrelated.

classification An arrangement of organisms into hierarchical groups that reflect their relatedness.

clathrin The network of proteins that coat and reinforce the cytoplasmic surface of cell membranes.

cleavage Mitotic cell divisions of the zygote that produce a blastula from a fertilized ovum.

climate The weather conditions prevailing over an extended period of time.

climax community A relatively stable, late successional stage in which the dominant vegetation replaces itself and persists until an environmental disturbance eliminates it, allowing other species to invade.

climograph A graph that portrays the particular combination of temperature and rainfall conditions where each terrestrial biome occurs.

cline A pattern of smooth variation in a characteristic along a geographical gradient.

clitoris The structure at the junction of the labia minora in front of the vulva, homologous to the penis in the male.

cloaca The cavity in reptiles, birds, amphibians, and many fishes into which both the intestinal and genital tracts empty.

clonal analysis A method of culturing meristematic tissue that contains a mutated embryonic cell having a readily observable trait, such as the absence of normal pigment.

clonal expansion In antibody-mediated immunity, the proliferation of an activated $CD4^+$ T cell by cell division to produce a clone of cells.

clonal reproduction The type of asexual reproduction in which the parent and offspring are genetically identical to one another.

clonal selection The process by which a lymphocyte is specifically selected for cloning when it encounters a foreign antigen from among a randomly generated, enormous diversity of lymphocytes with receptors that specifically recognize the antigen.

clone An individual genetically identical to an original cell from which it descended.

cloning vector DNA molecule into which a DNA fragment can be inserted to form a recombinant DNA molecule for the purpose of cloning.

closed circulatory system A circulatory system in which the fluid, blood, is confined in blood vessels and is distinct from the interstitial fluid.

clumped dispersion A pattern of distribution in which individuals in a population are grouped together.

cnidocyte A prey-capturing and defensive cell in the epidermis of cnidarians.

CO_2 fixation Process in which electrons are used as a source of energy to convert inorganic CO_2 to an organic form.

coactivator (mediator) In eukaryotes, a large multiprotein complex that bridges between activators at an enhancer and proteins at the promoter and promoter proximal region to stimulate transcription.

coat *See* **capsid.**

coated pit A depression in the plasma membrane that contains receptors for macromolecules to be taken up by endocytosis.

coccus (plural, *cocci*) A spherical prokaryote.

cochlea A snail-shaped structure in the inner ear containing the organ of hearing.

codominance Condition in which alleles have approximately equal effects in individuals, making the alleles equally detectable in heterozygotes.

codon Each three-letter word (triplet) of the genetic code.

coelom A fluid-filled body cavity in bilaterally symmetrical animals that is completely lined with derivatives of mesoderm.

coelomate A body plan of bilaterally symmetrical animals that have a coelom.

coenocytic Condition in which a single cell has many nuclei.

coenzymes Organic cofactors that include complex chemical groups of various kinds.

coevolution The evolution of genetically based, reciprocal adaptations in two or more species that interact closely in the same ecological setting.

cofactor An inorganic or organic nonprotein group that is necessary for catalysis to take place.

cognition A form of learning in which animals use insight to solve a novel problem; sometimes called trial-and-error learning.

cohesion The high resistance of water molecules to separation.

cohesion–tension mechanism of water transport A model of how water is transported from roots to leaves in vascular plants; the evaporation of water from leaves pulls water up in xylem by creating a continuous negative pressure (tension) that extends to roots.

cohort A group of individuals of similar age.

coleoptile A protective sheath that covers the shoot apical meristem and plumule of the embryo in monocots, such as grasses, as it pushes up through soil.

coleorhiza A sheath that encloses the radicle of an embryo until it breaks out of the seed coat and enters the soil as the primary root.

collagen Fibrous glycoprotein—very rich in carbohydrates—embedded in a network of proteoglycans.

collecting duct A location where urine leaving individual nephrons is processed further.

collenchyma A ground tissue that flexibly supports rapidly growing plant parts. Its elongated cells are alive at maturity and collectively often form strands or a sheathlike cylinder under the dermal tissue of growing shoot regions and leaf stalks.

colon The main part of the large intestine.

colony Multiple individual organisms of the same species living in a group.

combinatorial gene regulation The combining of a few regulatory proteins in particular ways so that the transcription of a wide array of genes can be controlled and a large number of cell types can be specified.

commaless The sequential nature of the words of the nucleic acid code, with no indicators such as commas or spaces to mark the end of one codon and the beginning of the next.

commensalism A symbiotic interaction in which one species benefits and the other is unaffected.

community Populations of all species that occupy the same area.

community ecology The ecological discipline that examines groups of populations occurring together in one area.

companion cell A specialized parenchyma cell that is connected to a mature sieve tube member by plasmodesmata and assists sieve tube members both with the uptake of sugars and with the unloading of sugars in tissues.

comparative genomics Comparison of the sequences of entire genomes (or extensive portions of them) to understand evolutionary relationships and the basic biological similarities and differences among species.

comparative morphology Analysis of the structure of living and extinct organisms.

compartment model A graphical depiction of the pathways through which nutrients and energy move between the living and nonliving components of an ecosystem.

compass orientation A wayfinding mechanism that allows animals to move in a particular direction, often over a specific distance or for a prescribed length of time.

competitive exclusion principle The ecological principle stating that populations of two or more species cannot coexist indefinitely if they rely on the same limiting resources and exploit them in the same way.

competitive inhibition Inhibition of an enzyme reaction by an inhibitor molecule that resembles the normal substrate closely enough so that it fits into the active site of the enzyme.

complement system A nonspecific defense mechanism activated by invading pathogens, made up of more than 30 interacting soluble plasma proteins circulating in the blood and interstitial fluid.

complementary base pairing Feature of DNA in which the specific purine–pyrimidine base pairs A–T (adenine–thymine) and G–C (guanine–cytosine) occur to bridge the two sugar–phosphate backbones.

complementary DNA (cDNA) A DNA molecule that is complementary to an mRNA molecule, synthesized by reverse transcriptase.

complete digestive system A digestive system having a mouth at one end, through which food enters, and an anus at the other end, through which undigested waste is voided.

complete flower A flower in which all four whorls (sepals, petals, stamens, carpels) are present.

complete medium A growth medium containing a full complement of nutrient substances that a wild-type microorganism can make for itself.

complete metamorphosis The form of metamorphosis in which an insect passes through four separate stages of growth: egg, larva, pupa, and adult.

complex virus A bacteriophage with a DNA genome that has a tail attached at one side of a polyhedral head.

compound A molecule whose component atoms are different.

compound eye The eye of most insects and some crustaceans, composed of many faceted, light-sensitive units called ommatidia fitted closely together, each having its own refractive system and each forming a portion of an image.

concentration The number of molecules or ions of a substance in a unit volume of space.

concentration gradient A difference in concentration of molecules or ions between two areas.

condensation reaction Reaction during which the components of a water molecule are removed, usually as part of the assembly of a larger molecule from smaller subunits. Also referred to as *dehydration synthesis reaction.*

conduction The flow of heat between atoms or molecules in direct contact.

cone In the vertebrate eye, a photoreceptor in the retina that is specialized for detection of different wavelengths (colors). In cone-bearing plants, a cluster of sporophylls.

conformation The overall three-dimensional shape of a protein.

conformational change Alteration in the three-dimensional shape of a protein.

conformers Animals having internal environments that change as the external environment changes.

conidiophore In ascomycete fungi, a modified hyphal branch that gives rise to conidia.

conidium (plural, *conidia*) An asexual spore produced by many species of ascomycetes.

Coniferophyta The major phylum of cone-bearing gymnosperms, most of which are substantial trees; includes pines, firs, and other conifers.

conjugation In bacteria, the process by which DNA of the donor cell moves through the cytoplasmic bridge into the recipient cell. With some types of donor cells, this can lead to genetic recombination in the recipient cell. In ciliate protists, a process of sexual reproduction in which individuals of the same species temporarily couple and exchange genetic material.

connective tissue Tissue having cells scattered through an extracellular matrix; forms layers in and around body structures that support other body tissues, transmit mechanical and other forces, and in some cases act as filters.

conodont An abundant, bonelike fossil dating from the early Paleozoic era through the early Mesozoic era, now described as a feeding structure of some of the earliest vertebrates.

consciousness Awareness of oneself, identity, and surroundings, with understanding of the significance and likely consequences of events.

conservation biology An interdisciplinary science that focuses on the maintenance and preservation of biodiversity.

constant (C) region For the light and heavy polypeptides of a particular type of antibody molecule, the regions that have the same amino acid sequences for all molecules.

consumer An organism that consumes other organisms in a community or ecosystem.

contact inhibition The inhibition of movement or proliferation of normal cells that results from cell–cell contact.

continental climate Climate not moderated by the distant ocean.

continental drift The long-term movement of continents as a result of plate tectonics.

continuous distribution A geographical distribution in which a species lives in suitable habitats throughout a geographical area.

contraception The prevention of pregnancy.

contractile vacuole A specialized cytoplasmic organelle that pumps fluid in a cyclical manner from within the cell to the outside by alternately filling and then contracting to release its contents at various points on the surface of the cell.

control Treatment that tells what would be seen in the absence of the experimental manipulation.

convection The transfer of heat from a body to a fluid, such as air or water, that passes over its surface.

convergent evolution The evolution of similar adaptations in distantly related organisms that occupy similar environments.

copulation The physical act involving the introduction of the accessory sex organ of a male into the accessory sex organ of a female to accomplish internal fertilization.

coral reef A structure made from the hard skeletons of coral animals or polyps; found largely in tropical and subtropical marine environments.

corepressor In the regulation of gene expression in bacteria, a regulatory molecule that combines with a repressor to activate it and shut off an operon.

cork A nonliving, impermeable secondary tissue that is one element of bark.

cork cambium A lateral meristem in plants that forms periderm, which in turn produces cork.

cornea The transparent layer that forms the front wall of the eye, covering the iris.

corolla The structure formed collectively by the petals of a flower.

coronary arteries The arteries from the aorta that branch extensively over the heart, supplying blood to the cardiac muscle cells.

corpus callosum A structure formed of thick axon bundles that connect the two cerebral hemispheres and coordinate their functions.

corpus luteum Cells remaining at the surface of the ovary during the luteal phase; the structure acts as an endocrine gland, secreting several hormones: estrogens, large quantities of progesterone, and inhibin.

cortex Generally, an outer, rindlike layer. In mammals, the outer layer of the brain, the kidneys, or the adrenal glands. In plants, the outer region of tissue in a root or stem lying between the epidermis and the vascular tissue, composed mainly of parenchyma.

cortical granule A secretory vesicle just under the plasma membrane of an egg cell.

cortical reaction The reaction in which cortical granules fuse with the plasma membrane of the egg and release their contents to the outside.

cortisol The major glucocorticoid steroid hormone secreted by the adrenal cortex, which increases blood glucose by promoting breakdown of proteins and fats.

cotranslational import A mechanism by which a polypeptide being sorted via the endomembrane system in a eukaryotic cell begins its import into the endoplasmic reticulum simultaneously with translation of the mRNA encoding the polypeptide.

cotransport *See* **symport.**

countercurrent exchange A mechanism in which the water flowing over the gills moves in a direction opposite to the flow of blood under the respiratory surface.

coupled reaction Reaction that occurs when an exergonic reaction is joined to an endergonic reaction, producing an overall reaction that is exergonic.

courtship display A behavior performed by males to attract potential mates or to reinforce the bond between a male and a female.

covalent bond Bond formed by electron sharing between atoms.

cranial nerve A nerve that connects the brain directly to the head, neck, and body trunk.

cranium The part of the skull that encloses the brain.

crassulacean acid metabolism (CAM) A biochemical variation of photosynthesis that was discovered in a member of the plant family Crassulaceae. Carbon dioxide is taken up and stored during the night to allow the stomata to remain closed during the daytime, decreasing water loss.

Crenarchaeota A major group of the domain Archaea that contains most of the extreme thermophiles. The group also includes psychrophiles and mesophiles. Members include obligate anaerobes, facultative anaerobes, and aerobes.

criss-cross inheritance The transmission pattern characteristic of an X-linked allele from a parent of one sex to a "child" of the opposite sex to a "grandchild" of the first sex.

crista Fold that expands the surface area of the inner mitochondrial membrane.

critical period A restricted stage of development early in life during which an animal has the capacity to respond to specific environmental stimuli.

crop Of birds, an enlargement of the digestive tube where the digestive contents are stored and mixed with lubricating mucus.

cross-fertilization Fertilization of one plant by a different plant.

crossing-over The recombination process in meiosis, in which chromatids exchange segments.

crossover Site of recombination during meiosis. Also referred to as a *chiasmata.*

cross-pollination *See* **cross-fertilization.**

cross-talk Interaction by which cell signaling pathways communicate with one another to integrate their responses to cellular signals.

cryptic coloration Coloration that allows an organism to match its background and hence become less vulnerable to predation or recognition by prey.

cryptochrome A light-absorbing protein that is sensitive to blue light and that may also be an important early step in various light-based growth responses.

C-terminal end The end of an amino acid chain with a $—COO^-$ group.

cupula In certain mechanoceptors, a gelatinous structure with stereocilia extending into it that moves with pressure changes in the surrounding water; movement of the cupula bends the stereocilia, which triggers release of neurotransmitters.

cuticle The outer layer of plants and some animals, which helps prevent desiccation by slowing water loss.

Cycadophyta A phylum of palmlike gymnosperms known as cycads; the pollen-bearing and seed-bearing cones (strobili) occur on separate plants.

cyclic AMP (cAMP) In particular signal transduction pathways, a second messenger that activates protein kinases, which elicit the cellular response by adding phosphate groups to specific target proteins. cAMP functions in one of two major G-protein–coupled receptor–response pathways.

cyclic electron flow An electron transport pathway associated with photosystem I in photosynthesis that produces ATP without the synthesis of NADPH.

cyclin In eukaryotes, protein that regulates the activity of CDK (cyclin-dependent kinase) and controls progression through the cell cycle.

cyclin-dependent kinase (CDK) A protein kinase that controls the cell cycle in eukaryotes.

cytochrome Protein with a heme prosthetic group that contains an iron atom.

cytokine A molecule secreted by one cell type that binds to receptors on other cells and, through signal transduction pathways, triggers a response. In innate immunity, cytokines are secreted by activated macrophages.

cytokinesis Division of the cytoplasm into two daughter cells following nuclear division in mitosis or meiosis.

cytokinin A hormone that promotes and controls growth responses of plants.

cytoplasm All the parts of the cell that surround the central nucleus (eukaryotes) or nucleoid region (prokaryotes).

cytoplasmic determinants The mRNA and proteins stored in the egg cytoplasm that direct the early stages of animal development in the period before genes of the zygote become active.

cytoplasmic inheritance Pattern in which inheritance follows that of genes in the cytoplasmic organelles, mitochondria, or chloroplasts.

cytoplasmic streaming Intracellular movement of cytoplasm.

cytosine A pyrimidine that base-pairs with guanine in nucleic acids.

cytoskeleton The interconnected system of protein fibers and tubes that extends throughout the cytoplasm of a eukaryotic cell.

cytosol Aqueous solution in the cytoplasm containing ions and various organic molecules.

cytotoxic T cell A T lymphocyte that functions in cell-mediated immunity to kill body cells infected by viruses or transformed by cancer.

daily torpor A period of inactivity and lowered metabolic rate that allows an endotherm to conserve energy when environmental temperatures are low.

dalton A standard unit of mass, about 1.66×10^{-24} grams.

day-neutral plant A plant that flowers without regard to photoperiod.

decomposer A small organism, such as a bacterium or fungus, that feeds on the remains of dead organisms, breaking down complex biological molecules or structures into simpler raw materials.

dedifferentiation A process in which cells lose their normal regulatory controls and revert partially or completely to an embryonic developmental state.

deep vein thrombosis A medical condition that results when sitting for a long period of time causes blood to pool in the veins of the body below the heart and then clot, particularly in the legs.

defecation reflex The opening of the anal sphincter and expulsion of feces in response to feces entering the rectum and stretching the rectal wall.

degeneracy (redundancy) The feature of the genetic code in which, with two exceptions, more than one codon represents each amino acid.

dehydration synthesis reaction *See* **condensation reaction.**

deletion Chromosomal alteration that occurs if a broken segment is lost from a chromosome.

demographic transition model A graphical depiction of the historical relationship between a country's economic development and its birth and death rates.

demography The statistical study of the processes that change a population's size and density through time.

denaturation A loss of both the structure and function of a protein due to extreme conditions that unfold it from its normal conformation.

dendrite The branched extension of the nerve cell body that receives signals from other nerve cells.

dendritic cell A type of phagocyte, so called because it has many surface projections that resemble dendrites of neurons, which engulfs a bacterium in infected tissue by phagocytosis.

denitrification A metabolic process in which certain bacteria convert nitrites or nitrates into nitrous oxide and then into molecular nitrogen, which enters the atmosphere.

density-dependent Description of environmental factors for which the strength of their effect on a population varies with the population's density.

density-independent Description of environmental factors for which the strength of their effect on a population does not vary with the population's density.

deoxyribonucleic acid (DNA) The large, double-stranded, helical molecule that contains the genetic material of all living organisms.

deoxyribonucleotide Nucleotide containing deoxyribose as the sugar; deoxyribonucleotides are components of DNA.

deoxyribose A 5-carbon sugar to which a nitrogenous base and a phosphate group link covalently in a nucleotide of DNA.

depolarized State of the membrane (which was polarized at rest) as the membrane potential becomes less negative.

deposit feeder An animal that consumes particles of organic matter from the solid substrate on which it lives.

derivative One of the daughter cells produced when a plant cell divides; it typically divides once or twice and then enters on the path to differentiation.

derived character A new version of a trait found in the most recent common ancestor of a group.

dermal tissue system The plant tissue system that comprises the outer tissues of the plant body, including the epidermis and periderm; it serves as a protective covering for the plant body.

dermis The skin layer below the epidermis; it is packed with connective tissue fibers such as collagen, which resist compression, tearing, or puncture of the skin.

descending reticular formation Part of the reticular formation that receives information from the hypothalamus and connects with interneurons in the spinal cord that control skeletal muscle contraction, thereby controlling muscle movement and posture.

descent with modification Biological evolution.

desert A sparsely vegetated biome that forms where rainfall averages less than 25 cm per year.

desertification A process in which large tracts of subtropical forest are cleared and overused, the groundwater table recedes to deeper levels, less surface water is available for plants, soil accumulates high concentrations of salts, and topsoil is eroded by wind and water.

desmosome Anchoring junction for which microfilaments anchor the junction in the underlying cytoplasm.

determinate cleavage A type of cleavage in protosomes in which each cell's developmental path is determined as the cell is produced.

determinate growth The pattern of growth in most animals in which individuals grow to a certain size and then their growth slows dramatically or stops.

determination Mechanism in which the developmental fate of a cell is set.

detritivore An organism that extracts energy from the organic detritus (refuse) produced at other trophic levels.

development A series of programmed changes encoded in DNA, through which a fertilized egg divides into many cells that ultimately are transformed into an adult, which is itself capable of reproduction.

diabetes mellitus A disease that results from problems with insulin production or action.

diacylglycerol (DAG) In particular signal transduction pathways, a second messenger that activates protein kinases, which elicit the cellular response by adding phosphate groups to specific target proteins. DAG is involved in one of two major G-protein–coupled receptor–response pathways.

diapsid A member of a group within the amniote vertebrates having a skull with two temporal arches.

Their living descendants include lizards and snakes, crocodilians, and birds.

diastole The period of relaxation and filling of the heart between contractions.

diastolic blood pressure The low point of the arterial blood pressure in the cardiac cycle that occurs between ventricular contractions.

diatoms *See* **Bacillariophyta.**

differentiation Process by which cells that have been committed to a particular developmental fate by the determination process now develop into specialized cell types with distinct structures and functions.

diffusion The net movement of ions or molecules from a region of higher concentration to a region of lower concentration.

digestion The splitting of carbohydrates, proteins, lipids, and nucleic acids in foods into chemical subunits small enough to be absorbed into the body fluids and cells of an animal.

digestive tract Part of an extracellular digestive system; it is tubelike with two openings that form a separate mouth and anus. Digestive contents move in one direction through specialized regions of the tube, from the mouth to the anus.

dihybrid A zygote produced from a cross that involves two characters.

dihybrid cross A cross between two individuals that are heterozygous for two pairs of alleles.

dikaryon The life stage in certain fungi in which a cell contains two genetically distinct haploid nuclei.

Dinoflagellata (dinoflagellates) A lineage of the Alveolata subgroup of the Chromalveolata protist evolutionary group; mostly single-celled marine phytoplankton that live as heterotrophs or autotrophs or sometimes using both modes of nutrition.

dioecious Having male flowers and female flowers on different plants of the same species.

dipeptidase An enzyme that splits dipeptides (two amino acids joined together) into individual amino acids.

diploblastic An animal body plan in which adult structures arise from only two cell layers, the ectoderm and the endoderm.

diploid An organism or cell with two copies of each type of chromosome in its nucleus.

directional selection A type of selection in which individuals near one end of the phenotypic spectrum have the highest relative fitness.

disaccharidase An enzyme that splits disaccharides (two sugar molecules joined together) into individual monosaccharides (single sugar molecules).

disclimax community An ecological community in which regular disturbance inhibits successional change.

discontinuous replication Replication in which a DNA strand is formed in short lengths that are synthesized in the direction opposite of DNA unwinding.

disjunct distribution A geographical distribution in which populations of the same species or closely related species live in widely separated locations.

dispersal The movement of organisms away from their place of origin.

dispersed duplication Gene copies that are found in different places in the genome, often on two different chromosomes.

dispersion The spatial distribution of individuals within a population's geographical range.

disruptive selection A type of natural selection in which extreme phenotypes have higher relative fitness than intermediate phenotypes.

dissociation The separation of water to produce hydrogen ions and hydroxide ions.

distal convoluted tubule The tubule in the human nephron that drains urine into a collecting duct that leads to the renal pelvis.

disulfide linkage Linkage that occurs when two sulfhydryl groups interact during a linking reaction.

DNA *See* **deoxyribonucleic acid.**

DNA chip *See* **DNA microarray.**

DNA fingerprinting Technique in which DNA samples are used to distinguish between individuals of the same species.

DNA helicase An enzyme that catalyzes the unwinding of DNA template strands.

DNA hybridization Technique in which a gene or sequence of interest is identified in a set of clones when it base pairs with a single-stranded DNA or RNA molecule called a nucleic acid probe.

DNA ligase In DNA replication, an enzyme that seals the nicks left after RNA primers are replaced with DNA.

DNA methylation Process in which a methyl group is added enzymatically to cytosine bases in the DNA.

DNA microarray A solid surface divided into a microscopic grid of thousands of spaces each containing thousands of copies of a DNA probe. DNA chips are used commonly for analysis of gene activity and for detecting differences between cell types. Also referred to as a *DNA chip.*

DNA polymerase I In *E. coli,* the replication enzyme that replaces the RNA primer at the start of a new DNA segment with DNA.

DNA polymerase III The principal replication polymerase in *E. coli* that synthesizes the majority of the new DNA.

DNA repair mechanism Mechanism to correct base-pair mismatches that escape proofreading.

DNA technologies Techniques to isolate, purify, analyze, and manipulate DNA sequences.

domain In protein structure, a distinct, large structural subdivision produced in many proteins by the folding of the amino acid chain. In systematics, the highest taxonomic category; a group of cellular organisms with characteristics that set it apart as a major branch of the evolutionary tree.

dominance The masking effect of one allele over another.

dominance hierarchy A social system in which the behavior of each individual is constrained by that individual's status in a highly structured social ranking.

dominant The allele expressed when paired with a recessive allele.

dominant species The species that is represented by a large proportion of the individuals present in an ecological community.

dormancy A period in the life cycle in which biological activity is suspended.

dorsal Indicating the back side of an animal.

dorsal lip of the blastopore A crescent-shaped depression rotated clockwise 90° on the embryo surface that marks the region derived from the gray crescent, to which cells from the animal pole move as gastrulation begins.

dosage compensation mechanism Mechanism in placental mammals by which the effects of most genes carried on the X chromosome in females are equalized in females (who have two X chromosomes) and males (who have one X chromosome).

double fertilization A characteristic feature of sexual reproduction in flowering plants. In the embryo sac, one sperm nucleus unites with the egg to form a diploid zygote from which the embryo develops, and another unites with two polar nuclei to form the primary endosperm nucleus.

double helix Two nucleotide chains wrapped around each other in a spiral.

double-helix model Model of DNA consisting of two polynucleotide strands twisted around each other.

downy mildews One of the subgroups of the Oomycota; parasites of plants.

duodenum A short region of the small intestine where secretions from the pancreas and liver enter a common duct.

duplication Chromosomal alteration that occurs if a segment is broken from one chromosome and inserted into its homolog.

E site The site where an exiting tRNA binds before its release from the ribosome in translation.

ecdysis Shedding of the cuticle, exoskeleton, or skin; molting.

ecdysone A steroid hormone secreted by the prothoracic glands of insects.

echolocation A technique for locating prey by making squeaking or clicking noises, and then listening for the echoes that bounce back from objects in their environment.

ecological community An assemblage of species living in the same place.

ecological efficiency The ratio of net productivity at one trophic level to net productivity at the trophic level below it.

ecological isolation A prezygotic reproductive isolating mechanism in which species that live in the same geographical region occupy different habitats.

ecological niche The resources a population uses and the environmental conditions it requires over its lifetime.

ecological pyramid A diagram illustrating the effects of energy transfer from one trophic level to the next.

ecological succession A somewhat predictable series of changes in the species composition of a community over time.

ecology The study of the interactions between organisms and their environments.

ecosystem Group of biological communities interacting with their shared physical environment.

ecosystem ecology An ecological discipline that explores the cycling of nutrients and the flow of energy between the biotic components of an ecological community and the abiotic environment.

ecosystem services The ecological processes on which all life depends, which include decomposition of wastes, nutrient recycling, oxygen production, maintenance of fertile topsoil, and air and water purification.

ecosystem valuation A process in which ecosystem services are assigned an economic value.

ecotone A wide transition zone between adjacent communities.

ecotourism An activity in which visitors, often from wealthy countries, pay a fee to visit a nature preserve.

ectoderm The outermost of the three primary germ layers of an embryo, which develops into epidermis and nervous tissue.

ectomycorrhiza A mycorrhiza that grows between and around the young roots of trees and shrubs but does not enter root cells.

ectoparasite A parasite that lives on the exterior of its host organism.

ectotherm An animal that obtains its body heat primarily from the external environment.

edge effect A phenomenon in which the removal of natural vegetation disrupts the local physical environment, exposing the borders of the remaining habitat to additional sunlight, wind, and rainfall.

effector In signal transduction, a plasma membrane–associated enzyme, activated by a G protein, that generates one or more second messengers. In homeostatic feedback, the system that returns the condition to the set point if it has strayed away.

effector T cell A cell involved in effecting—bringing about—the specific immune response to an antigen.

efferent arteriole The arteriole that receives blood from the glomerulus.

efferent neuron A neuron that carries the signals indicating a response away from the interneuron networks to the effectors.

egg cell The female reproductive cell.

elastin A rubbery protein in some connective tissues that adds elasticity to the extracellular matrix—it is able to return to its original shape after being stretched, bent, or compressed.

electrical signaling A means of animal communication in which a signaler emits an electric discharge that can be received by another individual.

electrical synapse In the nervous system, where plasma membranes of the presynaptic and postsynaptic cells are in direct contact and communication across such a synapse occurs by the direct flow of an electrical signal.

electrocardiogram (ECG) Graphic representation of the electrical activity within the heart, detected by electrodes placed on the body.

electrochemical gradient A difference in chemical concentration and electric potential across a membrane.

electromagnetic spectrum The range of wavelengths or frequencies of electromagnetic radiation extending from gamma rays to the longest radio waves and including visible light.

electron Negatively charged particle outside the nucleus of an atom.

electron microscope Microscope that uses electrons to illuminate the specimen.

electron transfer system Stage of cellular respiration in which high-energy electrons produced from glycolysis, pyruvate oxidation, and the citric acid cycle are delivered to oxygen by a sequence of electron carriers.

electronegativity The measure of an atom's attraction for the electrons it shares in a chemical bond with another atom.

electroreceptor A specialized sensory receptor that detects electrical fields.

element A pure substance that cannot be broken down into simpler substances by ordinary chemical or physical techniques.

elimination Referring to the digestive tube, the process in which undigested materials are expelled through the anus.

elongation In transcription, the step in which RNA polymerase (RNA polymerase II in eukaryotes) moves along the gene extending the RNA chain, with the DNA continuing to unwind ahead of the enzyme. In translation, the step in which the assembled translation complex reads the string of codons in the mRNA one at a time while joining the specified amino acids into the polypeptide.

elongation factor (EF) A protein that aids in an elongation step of translation.

embryo An organism in its early stage of reproductive development, beginning in the first moments after fertilization.

embryo sac The female gametophyte of angiosperms, within which the embryo develops; it usually consists of seven cells: an egg cell, an endosperm mother cell, and five other cells with fleeting reproductive roles.

embryogenesis Stages of development from a fertilized egg to an embryo.

embryophyte Any plant in which the embryo is retained within maternal tissue.

emergent property Characteristic that depends on the level of organization of matter, but does not exist at lower levels of organization.

emigration The movement of individuals out of a population.

enantiomers Isomers that are mirror images of each other. Also referred to as *optical isomers.*

endangered species A species in danger of extinction throughout all or a significant portion of its range.

endemic species A species that occurs in only one place on Earth.

endergonic reaction Reaction that can proceed only if free energy is supplied.

endocrine gland Any of several ductless secretory organs that secrete hormones into the blood or extracellular fluid.

endocrine system The system of glands that release their secretions (hormones) directly into the circulatory system.

endocytic vesicle Vesicle that carries proteins and other molecules from the plasma membrane to destinations within the cell.

endocytosis In eukaryotes, the process by which molecules are brought into the cell from the exterior involving a bulging in of the plasma membrane that pinches off to form an endocytic vesicle.

endoderm The innermost of the three primary germ layers of an embryo, which develops into the gastrointestinal tract and, in some animals, the respiratory organs.

endodermis The innermost layer of the root cortex; a selectively permeable barrier that helps control the movement of water and dissolved minerals into the stele.

endomembrane system In eukaryotes, a collection of interrelated internal membranous sacs that divide a cell into functional and structural compartments.

endomycorrhiza A mycorrhiza in which the fungal hyphae penetrate into cells of the root.

endoparasite A parasite that lives in the internal organs of its host organism.

endoplasmic reticulum (ER) In eukaryotes, an extensive interconnected network of cisternae that is responsible for the synthesis, transport, and initial modification of proteins and lipids.

endorphin One of a group of small proteins occurring naturally in the brain and around nerve endings that bind to opiate receptors and thus can raise the pain threshold.

endoskeleton An internal body structure, such as bones, that provides support.

endosperm Nutritive tissue inside the seeds of flowering plants.

endospore A small, metabolically inactive, asexual spore that develops within some bacterial cells when environmental conditions become unfavorable.

endosymbiont Organism that lives symbiotically within a host cell.

endosymbiont theory The proposal that the membranous organelles of eukaryotic cells (mitochondria and chloroplasts) may have originated from symbiotic relationships between two prokaryotic cells.

endotherm An animal that obtains most of its body heat from internal physiological sources.

endothermic Referring to a reaction that absorbs energy, that is, a reaction in which the products have more potential energy than the reactants.

endotoxin A lipopolysaccharide released from the outer membrane of the cell wall when a bacterium dies and lyses.

end-product inhibition *See* **feedback inhibition.**

energy The capacity to do work.

energy budget The total amount of energy that an organism can accumulate and use to fuel its activities.

energy coupling The process in living cells by which the hydrolysis of ATP is coupled to an endergonic reaction so that energy is not wasted as heat.

energy levels Regions of space within an atom where electrons are found. Also referred to as *shells.*

enhancer In eukaryotes, a region at a significant distance from the beginning of some protein-coding genes that contain regulatory sequences to which regulatory proteins bind to stimulate or inhibit the rate of transcription initiation over and above that seen for regulatory events at the promoter proximal elements.

enterocoelom In deuterostomes, the body cavity pinched off by outpocketings of the archenteron.

enthalpy The potential energy in a system.

entropy Disorder, in thermodynamics.

enveloped virus A virus that has a surface membrane derived from its host cell.

enzymatic hydrolysis A process in which chemical bonds are broken by the addition of H^+ and OH^-, the components of a molecule of water.

enzyme Protein that accelerates the rate of a cellular reaction.

enzyme specificity The ability of an enzyme to catalyze the reaction of only a single type of molecule or group of closely related molecules.

eosinophil A type of leukocyte that targets extracellular parasites too large for phagocytosis in the inflammatory response.

epiblast The top layer of the blastodisc.

epicotyl The upper part of the axis of an early plant embryo, located between the cotyledons and the first true leaves.

epidermis A complex tissue that covers an organism's body in a single, continuous layer or sometimes in multiple layers of tightly packed cells.

epididymis A coiled storage tubule attached to the surface of each testis.

epigenetics A phenomenon in which a change in gene expression does not involve a change in the DNA sequence of the gene or of the genome.

epiglottis A flaplike valve at the top of the trachea.

epilimnion The top layer of the limnetic zone in a lake.

epinephrine A nontropic amine hormone secreted by the adrenal medulla.

epiphyte A plant that grows independently on other plants and obtains nutrients and water from the air.

epistasis Interaction of genes, with one or more alleles of a gene at one locus inhibiting or masking the effects of one or more alleles of a gene at a different locus.

epithelial tissue Tissue formed of sheetlike layers of cells that are usually joined tightly together, with little extracellular matrix material between them. They protect body surfaces from invasion by bacteria and viruses, and secrete or absorb substances.

epitope The small region of an antigen molecule to which BCRs or TCRs bind.

equilibrium potential The electrical potential necessary to balance the diffusional potential of an ion at the plasma membrane of an axon.

equilibrium theory of island biogeography A hypothesis suggesting that the number of species on an island is governed by a give and take between the immigration of new species to the island and the extinction of species already there.

ER (endoplasmic reticulum) lumen The enclosed space surrounded by a cisterna.

erythrocyte A red blood cell, which contains hemoglobin, a protein that transports O_2 in blood.

erythropoietin (EPO) A hormone that stimulates stem cells in bone marrow to increase erythrocyte production.

esophagus A connecting passage of the digestive tube.

essential amino acid Any amino acid that is not made by the human body but must be taken in as part of the diet.

essential element Any of a number of elements required by living organisms to ensure normal reproduction, growth, development, and maintenance.

essential fatty acid Any fatty acid that the body cannot synthesize but needs for normal metabolism.

essential mineral Any inorganic element such as calcium, iron, or magnesium that is required in the diet of an animal.

essential nutrient Any of the essential amino acids, fatty acids, vitamins, and minerals required in the diet of an animal.

estivation Seasonal torpor in an animal that occurs in summer.

estradiol A form of estrogen.

estrogen Any of the group of female sex hormones.

estuary A coastal habitat where tidal seawater mixes with fresh water from rivers, streams, and runoff.

ethology A discipline that focuses on how animals behave in their natural environments.

ethylene A plant hormone that helps regulate seedling growth, stem elongation, the ripening of fruit, and the abscission of fruits, leaves, and flowers.

eudicot A plant belonging to the Eudicotyledones, one of the two major classes of angiosperms; their embryos generally have two seed leaves (cotyledons), and their pollen grains have three grooves.

Euglenids A lineage of the Euglenozoa subgroup of the Excavata protist evolutionary group; free-living with anterior flagella.

Euglenozoa (euglenozoans) A subgroup of the Excavata protist evolutionary group; single-celled, highly motile cells that swim by means of flagella and that contain mitochondria characterized by disc-shaped cristae.

Eukarya The domain that includes all eukaryotes, organisms that contain a membrane-bound nucleus within each of their cells; all protists, plants, fungi, and animals.

eukaryote Organism in which the DNA is enclosed in a nucleus.

eukaryotic chromosome A DNA molecule, with its associated proteins, in the nucleus of a eukaryotic cell.

euploid An individual with a normal set of chromosomes.

Euryarchaeota A major group of the domain Archaea, members of which are found in different extreme environments. They include methanogens, extreme halophiles, and some extreme thermophiles.

eusocial A form of social organization, observed in some insect species, in which numerous related individuals—a large percentage of them sterile female workers—live and work together in a colony for the reproductive benefit of a single queen and her mate(s).

eutrophic lake A lake that is rich in nutrients and organic matter.

evaporation Heat transfer through the energy required to change a liquid to a gas.

evolutionary developmental biology A field of biology that compares the genes controlling the developmental processes of different animals to determine the evolutionary origin of morphological novelties and developmental processes.

evolutionary divergence A process whereby natural selection or genetic drift causes populations to become more different over time.

exaptation A trait that is adaptive in a context different from the context in which it originally evolved.

Excavata A protist evolutionary group; single-celled animal parasites with greatly reduced mitochondria, or organelles derived from mitochondria, that move by means of flagella, and most of which have a scooped-out feeding apparatus on the ventral surface of the cell.

exchange diffusion *See* **antiport.**

excitatory postsynaptic potential (EPSP) The change in membrane potential caused when a neurotransmitter opens a ligand-gated Na^+ channel and Na^+ enters the cell, making it more likely that the postsynaptic neuron will generate an action potential.

excretion The process that helps maintain the body's water and ion balance while ridding the body of metabolic wastes.

exergonic reaction Reaction that has a negative ΔG because it releases free energy.

exocrine gland A gland that is connected to the epithelium by a duct and that empties its secretion at the epithelial surface.

exocytosis In eukaryotes, the process by which a secretory vesicle fuses with the plasma membrane and releases the vesicle contents to the exterior.

exodermis In the roots of some plants, an outer layer of root cortex that may limit water losses from roots and help regulate the absorption of ions.

exoenzyme An enzymatic protein released by a bacterium that digests plasma membranes and causes cells of the infected host to rupture and die.

exon An amino acid-coding sequence present in pre-mRNA that is retained in a spliced mRNA that is translated to produce a polypeptide.

exon shuffling Molecular evolutionary process that combines exons of two or more existing genes to produce a gene that encodes a protein with an unprecedented function.

exoskeleton A hard external covering of an animal's body that blocks the passage of water and provides support and protection.

exothermic Referring to a reaction that releases energy, that is, a reaction in which the products have less potential energy than the reactants.

exotic species A nonnative organism.

exotoxin A toxic protein that leaks from or is secreted from a bacterium and interferes with the biochemical processes of body cells in various ways.

experimental data Information that describes the result of a careful manipulation of the system under study.

experimental variable The variable in a scientific study that is manipulated by the experimenter.

exploitative competition Form of competition in which two or more individuals or populations use the same limiting resources.

exponential model of population growth Model that describes unlimited population growth.

expression vector Cloning vector that, in addition to normal features, contains regulatory sequences for transcription and translation of a gene.

extensor muscle A muscle that extends a joint, thereby increasing the angle between the two bones; with a flexor muscle, constitutes the antagonistic pair of muscles that control a joint's movement.

external environment The environment outside of the bodies of multicellular organisms.

external fertilization The process in which sperm and eggs are shed into the surrounding water, occurring in most aquatic invertebrates, bony fishes, and amphibians.

external gill A gill that extends out from the body and lacks a protective covering.

extinction The death of the last individual in a species or the last species in a lineage.

extracellular digestion Digestion that takes place outside body cells, in a pouch or tube enclosed within the body.

extracellular fluid The fluid occupying the spaces between cells in multicellular animals.

extracellular matrix (ECM) A molecular system that supports and protects cells and provides mechanical linkages.

extraembryonic membrane A primary tissue layer extended outside the embryo that conducts nutrients from the yolk to the embryo, exchanges gases with the environment outside the egg, or stores metabolic wastes removed from the embryo.

F factor Plasmid in a donor bacterial cell that confers on that cell the ability to conjugate with a recipient bacterial cell.

F pilus Structure on the cell surface that allows an F^+ donor bacterial cell (a cell containing an F factor) to attach to an F^- recipient bacterial cell (a cell lacking an F factor). Also referred to as a *sex pilus*.

F^- cell Recipient cell in conjugation between bacteria; it lacks an F factor.

F^+ cell Donor cell in conjugation between bacteria; it contains an F factor.

F_1 generation The first generation of offspring from a genetic cross.

F_2 generation The second generation of offspring from a genetic cross produced by interbreeding F_1 individuals.

facilitated diffusion Mechanism by which polar and charged molecules diffuse across membranes with the help of transport proteins.

facilitation hypothesis A hypothesis that explains ecological succession, suggesting that species modify the local environment in ways that make it less suitable for themselves but more suitable for colonization by species typical of the next successional stage.

facultative anaerobe An organism that can live in the presence or absence of oxygen, using oxygen when it is present and living by fermentation under anaerobic conditions.

familial (hereditary) cancer Cancer that runs in a family.

family A Linnaean taxonomic category that ranks below an order and above a genus.

family planning program A program that educates people about ways to produce an optimal family size on an economically feasible schedule.

fast block to polyspermy The barrier set up by the wave of depolarization triggered when sperm and egg fuse, making it impossible for other sperm to enter the egg.

fast muscle fiber A muscle fiber that contracts relatively quickly and powerfully.

fat Neutral lipid that is semisolid at biological temperatures.

fate map Mapping of adult or larval structures onto the region of the embryo from which each structure developed.

fat-soluble vitamin A vitamin that dissolves in liquid fat or fatty oils, in addition to water.

fatty acid One of two components of a neutral lipid, containing a single hydrocarbon chain with a carboxyl group linked at one end.

feather A sturdy, lightweight structure of birds, derived from scales in the skin of their ancestors.

feces Condensed and compacted digestive contents in the large intestine.

feedback inhibition In enzyme reactions, regulation in which the product of a reaction acts as a regulator of the reaction. Also referred to as *end-product inhibition.*

fermentation Process in which electrons carried by NADH are transferred to an organic acceptor molecule rather than to the electron transfer system.

fertilization The fusion of the nuclei of an egg and sperm cell, which initiates development of a new individual.

fetus A developing human from the eighth week of gestation onward, at which point the major organs and organ systems have formed.

fever A condition characterized by a rise in body temperature above the normal range.

fiber In sclerenchyma, an elongated, tapered, thick-walled cell that gives plant tissue its flexible strength.

fibrin A protein necessary for blood clotting; fibrin forms a weblike mesh that traps platelets and red blood cells and holds a clot together.

fibrinogen A plasma protein that plays a central role in the blood-clotting mechanism.

fibroblast The type of cell that secretes most of the collagen and other proteins in the loose connective tissue.

fibronectin A class of glycoproteins that aids in the attachment of cells to the extracellular matrix and helps hold the cells in position.

fibrous connective tissue Tissue in which fibroblasts are sparsely distributed among dense masses of collagen and elastin fibers that are lined up in highly ordered, parallel bundles, producing maximum tensile strength and elasticity.

fibrous root system A root system that consists of branching roots rather than a main taproot; roots tend to spread laterally from the base of the stem.

filament In flowers, the stalk of a stamen, which supports the anther.

filtration The nonselective movement of some water and a number of solutes—ions and small molecules, but not large molecules such as proteins—into the proximal end of the renal tubules through spaces between cells.

first law of thermodynamics The principle that energy can be transferred and transformed but it cannot be created or destroyed.

first messenger The extracellular signal molecule in signal transduction pathways controlled by G-protein–coupled receptors.

fission The mode of asexual reproduction in which the parent separates into two or more offspring of approximately equal size.

fixed action pattern A highly stereotyped instinctive behavior; when triggered by a specific cue, it is performed over and over in almost exactly the same way.

flagellum (plural, *flagella*) A long, threadlike cellular appendage responsible for movement; found in both prokaryotes and eukaryotes, but with different structures and modes of locomotion.

flatulence Gas expelled through the anus.

flexor muscle A muscle that decreases the angle between the two bones at a joint, thereby performing the opposite action to an extensor muscle;

with an extensor muscle, constitutes the antagonistic pair of muscles that control a joint's movement.

flower The reproductive structure of angiosperms, consisting of floral parts grouped on a stem; the structure in which seeds develop.

fluid feeder An animal that obtains nourishment by ingesting liquids that contain organic molecules in solution.

fluid mosaic model Model proposing that the membrane consists of a fluid phospholipid bilayer in which proteins are embedded and float freely.

fMRI *See* **functional magnetic resonance imaging.**

follicle cell A cell that grows from ovarian tissue and nourishes the developing egg.

follicle-stimulating hormone (FSH) The pituitary hormone that stimulates oocytes in the ovaries to continue meiosis and become follicles. During follicle enlargement, FSH interacts with luteinizing hormone to stimulate follicular cells to secrete estrogens.

food chain A depiction of the trophic structure of a community; a portrait of who eats whom.

food vacuole A membrane-bound sac used for digestion.

food web A set of interconnected food chains with multiple links.

Foraminifera (forams) A lineage of the Rhizaria protist evolutionary group; heterotrophic marine organisms with shells consisting of organic matter reinforced by calcium carbonate.

forebrain The largest division of the brain, which includes the cerebral cortex and basal ganglia. It is credited with the highest intellectual functions.

foreskin A loose fold of skin that covers the glans of the penis.

formula The name of a molecule written in chemical shorthand.

fossil The remains or traces of an organism of a past geologic age embedded and preserved in Earth's crust.

foundation species A species that defines the nature of a community by creating locally stable environmental conditions.

founder effect An evolutionary phenomenon in which a population that was established by just a few colonizing individuals has only a fraction of the genetic diversity seen in the population from which it was derived.

fovea The small region of the retina around which cones are concentrated in mammals and birds with eyes specialized for daytime vision.

fragmentation A type of vegetative reproduction in plants in which cells or a piece of the parent break off, then develop into new individuals.

frameshift mutation Mutation in a protein-coding gene that causes the reading frame of an mRNA transcribed from the gene to be altered, resulting in the production of a different, and nonfunctional, amino acid sequence in the polypeptide.

free energy The energy in a system that is available to do work.

freeze-fracture technique Technique in which experimenters freeze a block of cells rapidly, then fracture the block to split the lipid bilayer and expose the hydrophobic membrane interior.

frequency-dependent selection A form of natural selection in which rare phenotypes have a selective advantage simply because they are rare.

fruit A mature ovary, often with accessory parts, from a flower.

fruiting body In some fungi, a stalked, spore-producing structure such as a mushroom.

functional genomics The study of the functions of genes and of other parts of the genome.

functional groups The atoms in reactive groups.

functional magnetic resonance imaging (fMRI) A scanning technique for studying activities of the cerebral cortex.

fundamental niche The range of conditions and resources that a population can possibly tolerate and use.

fungal spore In fungi, a haploid reproductive cell that can germinate and produce a mycelium.

furrow In cytokinesis, a groove that girdles the cell and gradually deepens until it cuts the cytoplasm into two parts.

G_0 phase The phase of the cell cycle in eukaryotes in which many cell types stop dividing.

G_1 phase The initial growth stage of the cell cycle in eukaryotes, during which the cell makes proteins and other types of cellular molecules, but not nuclear DNA.

G_2 phase The phase of the cell cycle in eukaryotes during which the cell continues to synthesize proteins and to grow, completing interphase.

gallbladder The organ that stores bile between meals, when no digestion is occurring.

gametangium In certain plants and fungi, a cell or organ in which gametes are produced.

gamete A haploid cell, an egg or sperm. Haploid cells fuse during sexual reproduction to form a diploid zygote.

gametic isolation A prezygotic reproductive isolating mechanism caused by incompatibility between the sperm of one species and the eggs of another; may prevent fertilization.

gametogenesis The formation of male and female gametes.

gametophyte In organisms in which alternation of generations occurs, notably plants and green algae, the multicellular haploid generation that produces gametes. *Compare* **sporophyte.**

ganglion A functional concentration of nervous system tissue composed principally of nerve-cell bodies, usually lying outside the central nervous system.

ganglion cell A type of neuron in the retina of the eye that receives visual information from photoreceptors via various intermediate cells such as bipolar cells, amacrine cells, and horizontal cells.

gap gene In *Drosophila* embryonic development, one of the first activated set of segmentation genes that progressively subdivide the embryo into regions, determining the segments of the embryo and the adult. Gap genes subdivide the embryo along the anterior–posterior axis into broad regions that later develop into several distinct segments.

gap junction Junction that opens direct channels allowing ions and small molecules to pass directly from one cell to another.

gastric glands Glands in the stomach wall that contain cells which secrete some of the products needed to digest food.

gastric juice A substance secreted by the stomach that contains the digestive enzyme pepsin.

gastric pits Entrances to gastric glands.

gastrin A hormone secreted in response to stimulation of chemoreceptors in the stomach by the presence of food molecules, particularly proteins.

gastroesophageal sphincter Ring of muscle at the junction of the esophagus and the stomach that controls the movement of a bolus of food into the stomach.

gastrovascular cavity A saclike body cavity with a single opening, a mouth, which serves both digestive and circulatory functions.

gastrula The developmental stage resulting when the cells of the blastula migrate and divide once cleavage is complete.

gastrulation The second major process of early development in most animals, which produces an embryo with three distinct primary tissue layers.

gated channel Ion transporter in a membrane that switches between open, closed, or intermediate states.

gemma (plural, *gemmae*) Small cell mass that forms in cuplike growths on a thallus.

gene A unit containing the code for a protein molecule or one of its parts, or for functioning RNA molecules such as tRNA and rRNA.

gene cloning DNA cloning when it involves a gene.

gene duplication Any process that produces two identical copies of a gene in an organism's genome. It is the simplest and most common mechanism for the evolution of new genes.

gene flow The transfer of genes from one population to another through the movement of individuals or their gametes.

gene pool The sum of all alleles at all gene loci in all individuals in a population.

gene therapy Correction of genetic disorders using genetic engineering techniques.

gene-for-gene recognition A mechanism in which plants can detect an attack by a specific pathogen; the product of a specific plant gene interacts with the product of a specific pathogen gene, triggering the plant's defensive response.

general transcription factor (basal transcription factor) In eukaryotes, a protein that binds to the promoter of a gene in the area of the TATA box and recruits and orients RNA polymerase II to initiate transcription at the correct place.

generalized compartment model A model used to describe nutrient cycling in which two criteria—organic versus inorganic nutrients and available versus unavailable nutrients—define four compartments where nutrients accumulate.

generalized transduction Transfer of bacterial genes between bacteria using virulent phages that have incorporated random DNA fragments of the bacterial genome.

generation time The average time between the birth of an organism and the birth of its offspring.

genetic code The nucleotide information that specifies the amino acid sequence of a polypeptide.

genetic counseling Counseling that allows prospective parents to assess the possibility that they might have a child affected by a genetic disorder.

genetic drift Random fluctuations in allele frequencies as a result of chance events; usually reduces genetic variation in a population.

genetic engineering The use of DNA technologies to alter genes for practical purposes.

genetic equilibrium The point at which neither the allele frequencies nor the genotype frequencies in a population change in succeeding generations.

genetic recombinants Nonparental combinations of alleles. In eukaryotes they result from crossing-over in meiosis.

genetic recombination The process by which the combinations of alleles for different genes in two parental individuals become shuffled into new combinations in offspring individuals.

genetic screening Biochemical or molecular tests for identifying inherited disorders after a child is born.

genetically modified organism (GMO) A transgenic organism.

genomic imprinting Pattern of inheritance in which the expression of a nuclear gene is based on whether an individual organism inherits the gene from the male or female parent.

genomic library A set of clones that contains a copy of every DNA sequence in a genome.

genotype The genetic constitution of an organism in terms of its genes and alleles.

genotype frequency The percentage of individuals in a population possessing a particular genotype.

genus A Linnaean taxonomic category ranking below a family and above a species.

geographical range The overall spatial boundaries within which a population lives.

germ cell An animal cell that is set aside early in embryonic development and gives rise to the gametes.

germ layers The three primary cell layers of the embryo: the outer ectoderm, the inner endoderm, and the mesoderm between the ectoderm and the endoderm.

germ-line cells Cells that develop into sperm or eggs.

germ-line gene therapy Experiment in which a gene is introduced into germ-line cells of an animal to correct a genetic disorder.

gestation The period of mammalian development in which the embryo develops in the uterus of the mother.

gibberellin Any of a large family of plant hormones that regulate aspects of growth, including cell elongation.

gill A respiratory organ formed as evagination of the body that extends outward into the respiratory medium.

gill arch One of the series of curved supporting structures between the slits in the pharynx of a chordate.

gill slit One of the openings in the pharynx of a chordate through which water passes out of the pharynx.

Ginkgophyta A plant phylum with a single living species, the ginkgo (or maiden-hair) tree.

gizzard The part of the digestive tube that grinds ingested material into fine particles by muscular contractions of the wall.

gland A cell or group of cells that produces and releases substances nearby, in another part of the body, or to the outside.

glans A soft, caplike structure at the end of the penis, containing most of the nerve endings producing erotic sensations.

glial cell A nonneuronal cell contained in the nervous tissue that physically supports and provides nutrients to neurons, provides electrical insulation between them, and scavenges cellular debris and foreign matter.

globulin A plasma protein that transports lipids (including cholesterol) and fat-soluble vitamins; a specialized subgroup of globulins, the immunoglobulins, constitute antibodies and other molecules contributing to the immune response.

glomerular filtration The process by which plasma filters through the glomerular capillaries into Bowman's capsule as blood flows through the glomerulus; it is the first step of urine formation.

glomerulus A ball of blood capillaries surrounded by Bowman's capsule in the human nephron.

glucagon A pancreatic hormone with effects opposite to those of insulin: it stimulates glycogen, fat, and protein degradation.

glucocorticoid A steroid hormone secreted by the adrenal cortex that helps maintain the blood concentration of glucose and other fuel molecules.

glucose-dependent insulinotropic peptide (GIP) Hormone released in response to a meal entering the digestive tract that triggers insulin release from the pancreas.

glycocalyx A carbohydrate coat covering the cell surface.

glycogen Energy-providing carbohydrates stored in animal cells.

glycolipid A lipid molecule with carbohydrate groups attached.

glycolysis Stage of cellular respiration in which sugars such as glucose are partially oxidized and broken down into smaller molecules.

glycoprotein A protein with carbohydrate groups attached.

glycosidic bond Bond formed by the linkage of two α-glucose molecules with oxygen as a bridge between a carbon of the first glucose unit and a carbon of the second glucose unit.

Gnathostomata The group of vertebrates with moveable jaws.

golden algae *See* **Chrysophyta.**

Golgi complex In eukaryotes, the organelle responsible for the final modification, sorting, and distribution of proteins and lipids.

Golgi tendon organ A proprioceptor of tendons.

gonad A specialized gamete-producing organ in which the germ cells collect. Gonads are the primary source of sex hormones in vertebrates: ovaries in the female and testes in the male.

gonadotropin A hormone that regulates the activity of the gonads (ovaries and testes).

gonadotropin-releasing hormone (GnRH) A tropic hormone secreted by the hypothalamus that causes the pituitary to make luteinizing hormone (LH) and follicle-stimulating hormone (FSH).

G-protein–coupled receptor In signal transduction, a surface receptor that responds to a signal by activating a G protein.

graded potential A change in membrane potential that does not necessarily trigger an action potential.

gradualism The view that Earth and its living systems changed slowly over its history.

gradualist hypothesis The hypothesis that large changes in either geological features or biological lineages result from the slow, continuous accumulation of small changes over time.

Gram stain technique A technique of staining bacteria to distinguish between types of bacteria with different cell wall compositions.

Gram-negative Describing bacteria that do not retain the stain used in the Gram stain technique.

Gram-positive Describing bacteria that appear purple when stained using the Gram stain technique.

granum Structure in the chloroplasts of higher plants formed by thylakoids stacked one on top of another.

gravitropism A directional growth response to Earth's gravitational pull that is induced by mechanical and hormonal influences.

gray crescent A crescent-shaped region of the underlying cytoplasm at the side opposite the point of sperm entry exposed after fertilization when the pigmented layer of cytoplasm rotates toward the site of sperm entry.

gray matter Areas of densely packed nerve cell bodies and dendrites in the brain and spinal cord.

greater vestibular gland One of two glands located slightly below and to the left and right of the opening of the vagina in women. They secrete mucus that provides lubrication, especially when the woman is sexually aroused.

green algae *See* **Chlorophyta.**

greenhouse effect A phenomenon in which certain gases foster the accumulation of heat in the lower atmosphere, maintaining warm temperatures on Earth.

gross primary productivity The rate at which producers convert solar energy into chemical energy.

ground meristem The primary meristematic tissue in plants that gives rise to ground tissues, mostly parenchyma.

ground tissue system One of the three basic tissue systems in plants; includes all tissues other than dermal and vascular tissues.

growth factor A molecule (typically a peptide hormone) that stimulates cell division of a target cell.

growth hormone (GH) A hormone that stimulates cell division, protein synthesis, and bone growth in children and adolescents, thereby causing body growth.

growth-inhibiting factor A molecule that inhibits cell division of a target cell.

guanine A purine that base-pairs with cytosine in nucleic acids.

guard cell Either of a pair of specialized crescent-shaped cells that control the opening and closing of stomata in plant tissue.

guttation The exudation of water from leaves as a result of strong root pressure.

gymnosperm A seed plant that produces "naked" seeds not enclosed in an ovary.

H^+ pump *See* **proton pump.**

habitat The specific environment in which a population lives, as characterized by its biotic and abiotic features.

habitat fragmentation A process in which remaining areas of intact habitat are reduced to small, isolated patches.

habituation The learned loss of responsiveness to stimuli.

half-life The time it takes for half of a given amount of a radioisotope to decay.

haplodiploidy A pattern of sex determination in insects in which females are diploid and males are haploid.

haploid An organism or cell with only one copy of each type of chromosome in its nuclei.

Hardy–Weinberg principle An evolutionary rule of thumb that specifies the conditions under which a population of diploid organisms achieves genetic equilibrium.

haustorium (plural, *haustoria*) The hyphal tip of a parasitic fungus that penetrates a host plant and absorbs nutrients from it; likewise in parasitic flowering plants, a root that can penetrate a host's tissues and absorb nutrients.

head The anteriormost part of an organism's body, containing the brain, sensory structures, and feeding apparatus. For a bacteriophage, the usually polyhedral part of the virus containing the genetic material.

head–foot In mollusks, the region of the body that provides the major means of locomotion and contains concentrations of nervous system tissues and sense organs.

heart The muscular organ that pumps fluid through the circulatory system.

heartwood The inner core of a woody stem; composed of dry tissue and nonliving cells that no longer transport water and solutes and may store resins, tannins, and other defensive compounds.

heat of vaporization The heat required to give water molecules enough energy of motion to break loose from liquid water and form a gas.

heat-shock protein (HSP) Any of a group of chaperone proteins that are present in all cells in all life forms. They are induced when a cell undergoes various types of environmental stresses such as heat, cold, and oxygen deprivation.

heavy chain The heavier of the two types of polypeptide chains that are found in immunoglobulin and antibody molecules.

helical virus A virus in which the protein subunits of the coat assemble in a rodlike spiral around the genome.

helper T cell A clonal cell that assists with the activation of B cells.

hemolymph The circulatory fluid of invertebrates with open circulatory systems, including mollusks and arthropods.

hepatic portal vein The blood vessel that leads to capillary networks in the liver.

Hepatophyta The phylum that includes liverworts and their bryophyte relatives.

herbicide A compound that, at proper concentration, kills plants.

herbivore An animal that obtains energy and nutrients primarily by eating plants.

herbivory The process in which herbivores consume plants.

hermaphroditism The mechanism in which both mature egg-producing and mature sperm-producing tissue are present in the same individual.

heterochrony Changes in the relative rate of development of morphological characters.

heterogametic sex The sex that produces two types of gametes with respect to the sex chromosomes.

heterosporous Producing two types of spores, "male" microspores and "female" megaspores.

heterotroph An organism that acquires energy and nutrients by eating other organisms or their remains.

heterozygote An individual with two different alleles of a gene.

heterozygote advantage An evolutionary circumstance in which individuals that are heterozygous at a particular locus have higher relative fitness than either homozygote.

heterozygous State of possessing two different alleles of a gene.

Hfr cell A special donor cell that can transfer genes on a bacterial chromosome to a recipient bacterium.

hibernation Extended torpor during winter.

high-throughput technique A method that facilitates the handling of many samples simultaneously, such as DNA molecules for sequencing or genes for analysis.

hindbrain The lower area of the brain that includes the brain stem, medulla oblongata, and pons.

hippocampus A gray-matter center that is involved in sending information.

histamine An inflammatory signaling molecule.

histone A small, positively charged (basic) protein that is complexed with DNA in the chromosomes of eukaryotes.

histone code A regulatory mechanism for altering chromatin structure and, therefore, gene activity, based on signals in histone tails represented by chemical modification patterns.

historical biogeography The study of the geographical distributions of plants and animals in relation to their evolutionary history.

homeobox A region of a homeotic gene that corresponds to an amino acid section of the homeodomain.

homeodomain An encoded transcription factor of each protein that binds to a region in the promoters of the genes whose transcription it regulates.

homeostasis A steady internal condition maintained by responses that compensate for changes in the external environment.

homeostatic control system The processes and activities responsible for homeostasis.

homeotic gene Any of the family of genes that determines overall body plan (the structure of body parts) during embryonic development.

hominin A member of a monophyletic group of primates, characterized by an erect bipedal stance, that includes modern humans and their recent ancestors.

Hominoidea The monophyletic group of primates that includes apes and humans.

homogametic sex The sex that produces only one type of gamete with respect to the sex chromosomes.

homologies Characteristics shared by a set of species because they inherited them from their common ancestor.

homologous chromosomes The two chromosomes of each pair in a diploid cell—one of the pair derives from the maternal parent and the other derives from the paternal parent. Homologous chromosomes have the same genes, in the same order, in their DNA.

homologous traits Characteristics that are similar in two species because they inherited the genetic basis of the trait from their common ancestor.

homoplasies Characteristics shared by a set of species, often because they live in similar environments, but not present in their common ancestor; often the product of convergent evolution.

homosporous Producing only one type of spore.

homozygote An individual with two copies of the same allele.

homozygous State of possessing two copies of the same allele.

horizon A noticeable layer of soil, such as topsoil, having a distinct texture and composition that varies with soil type.

horizontal cell A type of neuron that forms lateral connections among photoreceptor cells in the retina of the eye.

horizontal gene transfer Movement of genetic material between organisms other than by descent.

hormone A signaling molecule secreted by a cell that can alter the activities of any cell with receptors for it; in animals, typically a molecule produced by one tissue and transported via the bloodstream to another specific tissue to alter its physiological activity.

host A species that is fed on by a parasite.

host race A population of insects that may be reproductively isolated from other populations of the same species as a consequence of their adaptation to feed on a specific host plant species.

***Hox* gene** A type of homeotic gene that controls an organism's overall body plan.

human chorionic gonadotropin (hCG) A hormone that keeps the corpus luteum in the ovary from breaking down.

human immunodeficiency virus (HIV) A retrovirus that causes acquired immunodeficiency syndrome (AIDS).

humoral immunity *See* **antibody-mediated immunity.**

humus The organic component of soil remaining after decomposition of plants and animals, animal droppings, and other organic matter.

hybrid breakdown A postzygotic reproductive isolating mechanism in which hybrids are capable of reproducing, but their offspring have either reduced fertility or reduced viability.

hybrid inviability A postzygotic reproductive isolating mechanism in which a hybrid individual has a low probability of survival to reproductive age.

hybrid sterility A postzygotic reproductive isolating mechanism in which hybrid offspring cannot form functional gametes.

hybrid zone A geographical area where the hybrid offspring of two divergent populations or species are common.

hybridoma A B cell that has been induced to fuse with a cancerous lymphocyte called a *myeloma cell,* forming a single, composite cell.

hydration layer A surface coat of water molecules that covers other polar and charged molecules and ions.

hydrocarbon Molecule consisting of carbon linked only to hydrogen atoms.

hydrogen bond Noncovalent bond formed by unequal electron sharing between hydrogen atoms and oxygen, nitrogen, or sulfur atoms.

hydrologic alterations Human-induced changes in the pathways through which water moves in the hydrologic cycle.

hydrologic cycle The global cycling of water between the ocean, the atmosphere, land, freshwater ecosystems, and living organisms.

hydrolysis Reaction in which the components of a water molecule are added to functional groups as molecules are broken into smaller subunits.

hydrophilic In chemistry and biology, referring to polar molecules that associate readily with water.

hydrophobic In chemistry and biology, referring to nonpolar substances that are excluded by water and other polar molecules.

hydroponic culture A method of growing plants not in soil but with the roots bathed in a solution that contains water and mineral nutrients.

hydrosphere The component of the biosphere that encompasses all the waters on Earth, including oceans, rivers, and polar ice caps.

hydrostatic skeleton A structure consisting of muscles and fluid that, by themselves, provide support for the animal or part of the animal; no rigid support, such as a bone, is involved.

hydroxyl group Group consisting of an oxygen atom linked to a hydrogen atom on one side and to a carbon chain on the other side.

hymen A thin flap of tissue that partially covers the opening of the vagina.

hyperpolarized The condition of a neuron when its membrane potential is more negative than the resting value.

hypersensitive response A plant defense that physically cordons off an infection site by surrounding it with dead cells.

hypertension Commonly called *high blood pressure,* a medical condition in which blood pressure is chronically elevated above normal values.

hyperthermia The condition resulting when the heat gain of the body is too great to be counteracted.

hypertonic Solution containing dissolved substances at higher concentrations than the cells it surrounds.

hypha (plural, *hyphae*) Any of the threadlike filaments that form the mycelium of a fungus.

hypoblast The bottom layer of a blastodisc.

hypocotyl The region of a plant embryo's vertical axis between the cotyledons and the radicle.

hypodermis The innermost layer of the skin that contains larger blood vessels and additional reinforcing connective tissue.

hypolimnion The deep water of the profundal zone of a lake.

hypothalamus The portion of the brain that contains centers regulating basic homeostatic functions of the body and contributing to the release of hormones.

hypothermia A condition in which the core temperature falls below normal for a prolonged period.

hypothesis A tentative explanation for an observation, phenomenon, or scientific problem that can be tested by further investigation.

hypotonic Solution containing dissolved substances at lower concentrations than the cells it surrounds.

ice lattice A rigid, crystalline structure formed when a water molecule in ice forms four hydrogen bonds with neighboring molecules.

IgA The class of antibodies found mainly in secretions at particular locations in the body; IgA functions to bind to surface groups on pathogens and block their attachment to body surfaces.

IgD The class of antibodies that occurs with IgM as a receptor on the surfaces of B cells; the function of IgD is uncertain.

IgE The class of antibodies secreted by plasma cells of the skin and tissues lining the gastrointestinal tract and respiratory tract that binds to basophils and mast cells to trigger the release of histamine, thereby causing an inflammatory response.

IGF *See* **insulin-like growth factor.**

IgG The most abundant antibody type circulating in the blood and lymphatic system that is involved in primary and secondary immune responses.

IgM The first antibody type secreted by B cells in a primary immune response.

imbibition The movement of water into a seed as the water molecules are attracted to hydrophilic groups of stored proteins; the first step in germination.

immigration Movement of organisms into a population.

immune response The defensive reactions of the immune system.

immune system The body's system of defenses against disease, composed of certain white blood cells and antibodies.

immunoglobulin A specific protein substance produced by plasma cells to aid in fighting infection.

immunological memory The capacity of the immune system to respond more rapidly and vigorously to the second contact with a specific antigen than to the primary contact.

immunological tolerance The process that protects the body's own molecules from attack by the immune system.

imperfect flower A type of incomplete flower that has stamens or carpels, but not both.

imprinting The process of learning the identity of a caretaker and potential future mate during a critical period early in life.

inbreeding A special form of nonrandom mating in which genetically related individuals mate with each other.

incisors Flattened, chisel-shaped teeth of mammals, located at the front of the mouth, that are used to nip or cut food.

incomplete dominance Condition in which the effects of recessive alleles can be detected to some extent in heterozygotes.

incomplete flower A flower lacking one or more of the four floral whorls.

incomplete metamorphosis In certain insects, a life cycle characterized by the absence of a pupal stage between the immature and adult stages.

incurrent siphon A muscular tube that brings water containing oxygen and food into the body of an invertebrate.

incus The second of the three sound-conducting middle ear bones in vertebrates, located between the malleus and the stapes.

independent assortment Mendel's principle that the alleles of the genes that govern two characters assort independently during formation of gametes. Mechanistically this is the case because any combination of chromosomes may be segregated to the spindle poles during meiosis I.

indeterminate cleavage A type of cleavage, observed in many deuterostomes, in which the developmental fates of the first few cells produced by mitosis are not determined as soon as cells are produced.

indeterminate growth Growth that is not limited by an organism's genetic program, so that the organism grows for as long as it lives; typical of many plants. *Compare* **determinate growth.**

inducer Concerning regulation of gene expression in bacteria, a molecule that turns on the transcription of the genes in an operon.

inducible operon Operon whose expression is increased by an inducer molecule.

induction A mechanism in which one group of cells (the inducer cells) causes or influences another nearby group of cells (the responder cells) to follow a particular developmental pathway.

infection thread In the formation of root nodules on nitrogen-fixing plants, the tube formed by the plasma membrane of root hair cells as bacteria enter the cell.

infertility The inability for the female of a couple to get pregnant after 12 months of frequent, contraceptive-free intercourse.

inflammation The heat, pain, redness, and swelling that occur at the site of an infection.

ingestion The feeding methods used to take food into the digestive cavity.

inheritance The transmission of DNA (that is, genetic information) from one generation to the next.

inhibin A peptide that, in females, is an inhibitor of FSH secretion from the pituitary thereby diminishing the signal for follicular growth. In males, inhibin inhibits FSH secretion from the pituitary, thereby decreasing spermatogenesis.

inhibiting hormone (IH) A hormone released by the hypothalamus that inhibits the secretion of a particular anterior pituitary hormone.

inhibition hypothesis A hypothesis suggesting that new species are prevented from occupying a community by whatever species are already present.

inhibitory postsynaptic potential (IPSP) A change in membrane potential caused when hyperpolarization occurs, pushing the neuron farther from threshold.

initial A plant cell that remains permanently as part of a meristem and gives rise to daughter cells that differentiate into specialized cell types.

initiation In transcription, the step in which the molecular machinery that carries out transcription assembles at the promoter and begins synthesizing an RNA copy of the gene. In translation, the step in which the translation components assemble on the start codon of the mRNA.

initiation factor (IF) A protein that aids an initiation step of translation.

initiator codon *See* **start codon.**

innate immune system A nonspecific line of defense against pathogens that includes inflammation, which creates internal conditions that inhibit or kill many pathogens, and specialized cells that engulf or kill pathogens or infected body cells. The innate system is found in all animals and is one of two components of the immune system in vertebrates.

innate immunity The initial response by the body to eliminate cellular pathogens that involves mechanisms of the innate immune system.

inner boundary membrane Membrane lying just inside the outer boundary membrane of a chloroplast, enclosing the stroma.

inner cell mass The dense mass of cells within the blastocyst that will become the embryo.

inner ear That part of the ear, particularly the cochlea, that converts mechanical vibrations (sound) into neural messages that are sent to the brain.

inner mitochondrial membrane Membrane surrounding the mitochondrial matrix.

inorganic molecule Molecule without carbon atoms in its structure.

inositol triphosphate (IP_3) In particular signal transduction pathways, a second messenger that activates transport proteins in the endoplasmic reticulum to release Ca^{2+} into the cytoplasm. IP_3 is involved in one of two major G-protein–coupled receptor–response pathways.

insertion sequence A transposable element that contains only genes for its transposition.

instinctive behavior A genetically "programmed" response that appears in complete and functional form the first time it is used.

insulin A hormone secreted by beta cells in the islets, acting mainly on cells of nonworking skeletal muscles, liver cells, and adipose tissue (fat) to lower blood glucose, fatty acid, and amino acids levels, and promote the storage of those molecules.

insulin-like growth factor (IGF) A peptide released by target tissues in response to binding by growth hormone that directly stimulates growth processes.

integral protein Protein embedded in a phospholipid bilayer.

integration The third component of neural signaling involving the sorting and interpretation of neural messages and the determination of the appropriate response(s).

integrator In homeostatic feedback, the control center that compares a detected environmental change with a set point.

integument Skin.

intercalated disk The joining point between two cells in cardiac muscle.

interference competition Form of competition in which individuals fight over resources or otherwise harm each other directly.

interferon A cytokine produced by infected host cells affected by viral dsRNA, which acts both on the infected cell that produces it, an autocrine effect, and on neighboring uninfected cells, a paracrine effect.

interkinesis A brief interphase separating the two meiotic divisions.

intermediate disturbance hypothesis Hypothesis proposing that species richness is greatest in communities that experience fairly frequent disturbances of moderate intensity.

intermediate filament A cytoskeletal filament about 10 nm in diameter that provides mechanical strength to cells in tissues.

intermediate-day plant A plant that flowers only when day length falls between the values for long-day and short-day plants.

internal environment The fluid within an organism that supplies all the needs of individual cells.

internal fertilization The process in which sperm are released by the male close to or inside the entrance of the reproductive tract of the female.

internal gill A gill located within the body that has a cover providing physical protection for the gills. Water must be brought to internal gills.

interneuron A neuron that integrates information to formulate an appropriate response.

internode The region between two nodes on a plant stem.

interphase The first stage of the mitotic cell cycle, during which the cell grows and replicates its DNA before undergoing mitosis and cytokinesis.

interspecific competition The competition for resources between species.

interstitial fluid The fluid occupying the spaces between cells in multicellular animals.

intertidal zone The shoreline that is alternately submerged and exposed by tides.

intestinal villus A microscopic, fingerlike extension in the lining of the small intestine.

intestine The portion of digestive system where organic matter is hydrolyzed by enzymes secreted into the digestive tube. As muscular contractions of the intestinal wall move the mixture along, cells lining the intestine absorb the molecular subunits produced by digestion.

intracellular digestion The process in which cells take in food particles by endocytosis.

intraspecific competition The dependence of two or more individuals in a population on the same limiting resource.

intrinsic rate of increase The maximum possible per capita population growth rate in a population living under ideal conditions.

intron A non-protein-coding sequence that interrupts the protein-coding sequence in a eukaryotic gene. Introns are removed by splicing in the processing of pre-mRNA to mRNA.

invagination The process in which cells changing shape and pushing inward from the surface produce an indentation, such as the dorsal lip of the blastopore.

inversion Chromosomal alteration that occurs if a broken segment reattaches to the same chromosome from which it was lost, but in reversed orientation, so that the order of genes in the segment is reversed with respect to the other genes of the chromosome.

invertebrate An animal without a vertebral column.

involution The process by which cells migrate into the blastopore.

ion A positively or negatively charged atom.

ionic bond Bond that results from electrical attractions between atoms that have lost or gained electrons.

ionotropic neurotransmitter receptor A receptor in the postsynaptic membrane to which a neurotransmitter binds directly, causing an associated ion channel gate to open or close thereby altering the flow of a specific ion or ions in the postsynaptic cell.

iris Of the eye, the colored muscular membrane that lies behind the cornea and in front of the lens, which by opening or closing determines the size of the pupil and hence the amount of light entering the eye.

islets of Langerhans Endocrine cells that secrete the peptide hormones insulin and glucagon into the bloodstream.

isomers Two or more molecules with the same chemical formula but different molecular structures.

isotonic Equal concentration of water inside and outside cells.

isotope A distinct form of the atoms of an element, with the same number of protons but different number of neutrons.

jasmonate Any of a group of plant hormones that help regulate aspects of growth and responses to stress, including attacks by predators and pathogens.

juvenile hormone (JH) A peptide hormone secreted by the corpora allata, a pair of glands just behind the brain in insects.

juxtaglomerular apparatus A structure in the kidney near a point where the distal convoluted tubule contacts the afferent arteriole carrying blood to the glomerulus; specialized tubule cells in the juxtaglomerular apparatus monitor the salt level of the fluid flowing past them in the tubule.

karyogamy In plants, the fusion of two sexually compatible haploid nuclei after cell fusion (plasmogamy).

karyotype A characteristic of a species consisting of the shapes and sizes of all the chromosomes at metaphase.

keeled sternum The ventrally extended breastbone of a bird to which the flight muscles attach.

ketone Molecule in which the carbonyl group is linked to a carbon atom in the interior of a carbon chain.

keystone species A species that has a greater effect on community structure than its numbers might suggest.

kilocalorie (kcal) The scientific unit equivalent to a Calorie and equal to 1,000 small calories.

kin selection An explanation for altruistic behavior to close relatives, allowing them to produce proportionately more surviving copies of the altruist's genes than the altruist might otherwise have produced on its own.

kinesis A change in the rate of movement or the frequency of turning movements in response to environmental stimuli.

kinetic energy The energy of motion.

kinetochore A specialized structure consisting of proteins attached to a centromere that mediates the attachment and movement of chromosomes along the mitotic spindle.

Kinetoplastids A lineage of the Euglenozoa subgroup of the Excavata protist evolutionary group; nonphotosynthetic, heterotrophic cells that live as animal parasites and that contain a single mitochondrion with a large DNA–protein deposit called a kinetoplast.

kingdom A Linnaean taxonomic category that ranks below a domain and above a phylum.

Kingdom Animalia The taxonomic kingdom that includes all living and extinct animals.

Kingdom Fungi The taxonomic kingdom that includes all living or extinct fungi.

Kingdom Plantae The taxonomic kingdom encompassing all living or extinct plants.

Korarchaeota A group of the domain Archaea consisting of one living species, with other members represented in sequences in environmental samples. This group diverged very early in the evolution of the Archaea.

Krebs cycle *See* **citric acid cycle.**

***K*-selected species** Long-lived species that thrive in more stable environments.

labia majora A pair of fleshy, fat-padded folds that partially cover the labia minora.

labia minora Two folds of tissue that run from front to rear on either side of the opening to the vagina.

lactate fermentation Reaction in which pyruvate is converted into lactate.

lactose intolerance The unpleasant intestinal symptoms that result when an individual loses the capacity to synthesize lactase, the enzyme that breaks down the milk sugar.

lagging strand The new DNA strand synthesized discontinuously during replication in the direction opposite to that of DNA unwinding.

lagging strand template The DNA template strand for the lagging strand.

landscape ecology The field that examines how large-scale ecological factors—such as the distribution of plants, topography, and human activity—influence local populations and communities.

larva A sexually immature stage in the life cycle of many animals that is morphologically distinct from the adult.

larynx The voice box.

latent state The time during which a pathogen such as a virus remains in an infected organism in an inactive form and cannot be isolated and identified.

lateral bud A bud on the side of a plant stem from which a branch may grow.

lateral geniculate nuclei Clusters of neurons located in the thalamus that receive visual information from the optic nerves and send it on to the visual cortex.

lateral inhibition Visual processing in which lateral movement of signals from a rod or cone proceeds to a horizontal cell and continues to bipolar cells with which the horizontal cell makes inhibitory connections, serving both to sharpen the edges of objects and enhance contrast in an image.

lateral line system The complex of mechanoreceptors along the sides of some fishes and aquatic amphibians that detect vibrations in the water.

lateral meristem A plant meristem that gives rise to secondary tissue growth. *Compare* **primary meristem.**

lateral root A root that extends away from the main root (or taproot).

lateralization A phenomenon in which some brain functions are more localized in one of the two hemispheres.

lateral-line system The complex of organs and sensory receptors along the sides of many fishes and amphibians that detects vibrations in water.

leaching The process by which soluble materials in soil are washed into a lower layer of soil or are dissolved and carried away by water.

leading strand The new DNA strand synthesized during replication in the direction of DNA unwinding.

leading strand template The DNA template strand for the leading strand.

leaf primordium A lateral outgrowth from the apical meristem that develops into a young leaf.

learned behavior A response of an animal that is dependent on having a particular kind of experience during development.

learning A process in which experiences stored in memory change the behavioral responses of an animal.

leghemoglobin An iron-containing, red-pigmented protein produced in root nodules during the symbiotic association between *Bradyrhizobium* or *Rhizobium* and legumes.

lek A display ground where males each possess a small territory from which they court attentive females.

lens The transparent, biconvex intraocular tissue that helps bring rays of light to a focus on the retina.

Lepidosauromorpha A monophyletic lineage of diapsids that includes both marine and terrestrial animals, represented today by sphenodontids, lizards, and snakes.

leukocyte A white blood cell, which eliminates dead and dying cells from the body, removes cellular debris, and participates in defending the body against invading organisms.

Leydig cell A cell that produces the male sex hormones.

lichen A single vegetative body that is the result of an association between a fungus and a photosynthetic partner, often an alga.

life cycle The sequential stages through which individuals develop, grow, maintain themselves, and reproduce.

life history The lifetime pattern of growth, maturation, and reproduction that is characteristic of a population or species.

life table A chart that summarizes the demographic characteristics of a population.

ligament A fibrous connective tissue that connects bones to each other at a joint.

ligation The process of sealing two DNA molecules together with DNA ligase to produce a recombinant DNA molecule.

light chain The lighter of the two types of polypeptide chains found in immunoglobulin and antibody molecules.

light microscope Microscope that uses light to illuminate the specimen.

light-dependent reaction The first stage of photosynthesis, in which the energy of sunlight is absorbed and converted into chemical energy in the form of ATP and NADPH.

light-independent reaction The second stage of photosynthesis, in which electrons are used as a source of energy to convert inorganic CO_2 to an organic form. Also referred to as the *Calvin cycle.*

lignification The deposition of lignin in plant cell walls; it anchors the cellulose fibers in the walls, making them stronger and more rigid, and protects the other wall components from physical or chemical damage.

lignin A tough, rather inert polymer that strengthens the secondary walls of various plant cells and thus helps vascular plants to grow taller and stay erect on land.

limbic system A functional network formed by parts of the thalamus, hypothalamus, and basal nuclei, along with other nearby gray-matter centers—the amygdala, hippocampus, and olfactory bulbs—sometimes called the "emotional brain."

limiting nutrient An element in short supply within an ecosystem, the shortage of which limits productivity.

limnetic zone The sunlit, open water in a lake, beyond the zone where plants rooted in the bottom can grow.

linkage The phenomenon of genes being located on the same chromosome.

linkage map Map of a chromosome showing the relative locations of genes based on recombination frequencies.

linked genes Genes on the same chromosome.

linker A short segment of DNA extending between one nucleosome and the next in a eukaryotic chromosome.

lipase A pancreatic enzyme that hydrolyzes fats.

lipopolysaccharide A large molecule that consists of a lipid and a carbohydrate joined by a covalent bond.

lithosphere The component of the biosphere that includes the rocks, sediments, and soils of the crust.

littoral zone The shallow, sunlit water near the shore of a lake or pond.

liver A large organ whose many functions include aiding in digestion, removing toxins from the body, and regulating the chemicals in the blood.

loam Any well-aerated soil composed of a mixture of sand, clay, silt, and organic matter.

local homeostatic controls Controls that operate only within an organ where a change in the internal environment needs to be addressed.

local inflammation Inflammation that occurs at the site of an infection.

locus The particular site on a chromosome at which a gene is located.

logistic model of population growth Model of population growth that assumes that a population's per capita growth rate decreases as the population gets larger.

long-day plant A plant that flowers in spring when dark periods become shorter and day length becomes longer.

long-term memory Memory that stores information from days to years or even for life.

long-term potentiation A long-lasting increase in the strength of synaptic connections in activated neural pathways following brief periods of repeated stimulation.

loop of Henle A U-shaped bend of the proximal convoluted tubule.

loose connective tissue A tissue formed of sparsely distributed cells surrounded by a more or less open network of collagen and other glycoprotein fibers.

lophophore The circular or U-shaped fold with one or two rows of hollow, ciliated tentacles that surrounds the mouth of brachiopods, bryozoans, and phoronids and is used to gather food.

lumen The inside of the digestive tube.

lung One of a pair of invaginated respiratory surfaces, buried in the body interior where they are less susceptible to drying out; the organs of respiration in mammals, birds, reptiles, and most amphibians.

luteinizing hormone (LH) A hormone secreted by the pituitary that stimulates the growth and maturation of eggs in females and the secretion of testosterone in males.

Lycophyta The plant phylum that includes club mosses and their close relatives.

lymph The interstitial fluid picked up by the lymphatic system.

lymph node One of many small, bean-shaped organs spaced along the lymph vessels that contain macrophages and other leukocytes that attack invading disease organisms.

lymphatic system An accessory system of vessels and organs that helps balance the fluid content of the blood and surrounding tissues and participates in the body's defenses against invading disease organisms.

lymphocyte A leukocyte that carries out most of its activities in tissues and organs of the lymphatic system. Lymphocytes play major roles in immune responses.

lysogenic cycle Cycle in which the DNA of the bacteriophage is integrated into the DNA of the host bacterial cell and may remain for many generations.

lysosome Membrane-bound vesicle containing hydrolytic enzymes for the digestion of many complex molecules.

lytic cycle The series of events from infection of one bacterial cell by a phage through the release of progeny phages from lysed cells.

M phase–promoting factor (MPF) A complex of M cyclin and cyclin-dependent kinase 1 (Cdk1). The complex initiates mitosis and orchestrates some of its key events.

macroevolution Large-scale evolutionary patterns in the history of life, producing major changes in species and higher taxonomic groups.

macronucleus In ciliophorans, a single large nucleus that develops from a micronucleus but loses all genes except those required for basic "housekeeping" functions of the cell and for ribosomal RNAs.

macronutrient In humans, a mineral required in amounts ranging from 50 mg to more than 1 gram per day. In plants, a nutrient needed in large amounts for normal growth and development.

macrophage A phagocyte that takes part in nonspecific defenses and adaptive immunity.

magnetoreceptor A receptor found in some animals that navigate long distances which allows them to detect and use Earth's magnetic field as a source of directional information.

magnification The ratio of an object as viewed to its real size.

magnoliids An angiosperm group that includes magnolias, laurels, and avocados; they are more closely related to monocots than to eudicots.

major histocompatibility complex A large cluster of genes encoding the MHC proteins.

malleus The outermost of the sound-conducting bones of the middle ear in vertebrates.

malnutrition A condition resulting from a diet in which intake of organic fuels is inadequate or whose assimilation of such fuels is abnormal

(= **undernutrition**), or in which there is excessive intake of specific nutrients (= **overnutrition**).

Malpighian tubule The main organ of excretion and osmoregulation in insects, helping them to maintain water and electrolyte balance.

mammary glands Specialized organs of female mammals that produce energy-rich milk, a watery mixture of fats, sugars, proteins, vitamins, and minerals.

mantle One or two folds of the body wall that lines the shell and secretes the substance that forms the shell in mollusks.

map unit The unit of a linkage map, equivalent to a recombination frequency of 1%. Also referred to as a *centimorgan*.

maritime climate Climate tempered by ocean winds.

marsupium An external pouch on the abdomen of many female marsupials, containing the mammary glands, and within which the young continue to develop after birth.

mass The amount of matter in an object.

mass extinctions The disappearance of a large number of species in a relatively short period of geological time.

mass number The total number of protons and neutrons in the atomic nucleus.

mast cell A type of cell dispersed through connective tissue that releases histamine when activated by the death of cells, caused by a pathogen at an infection site.

maternal chromosome The chromosome derived from the female parent of an organism.

maternal inheritance A type of uniparental inheritance in which all progeny (both males and females) have the phenotype of the female parent.

maternal-effect gene One of a class of genes that regulate the expression of other genes expressed by the mother during oogenesis and that control the polarity of the egg and, therefore, of the embryo.

mating The pairing of a male and a female for the purpose of sexual reproduction.

mating systems The social systems describing how males and females pair up.

mating type A genetically defined strain of an organism (such as a fungus) that can only mate with an organism of the opposite mating type; mating types are often designated + and −.

matter Anything that occupies space and has mass.

maximum likelihood methods A statistical technique that compares alternative phylogenetic trees with specific models of evolutionary change.

mechanical isolation A prezygotic reproductive isolating mechanism caused by differences in the structure of reproductive organs or other body parts.

mechanical processing In the digestive tube, the process of chewing, grinding, and tearing food chunks into smaller pieces.

mechanoreceptor A sensory receptor that detects mechanical energy, such as changes in pressure, body position, or acceleration. The auditory receptors in the ears are examples of mechanoreceptors.

medusa The tentacled, usually bell-shaped, free-swimming sexual stage in the life cycle of a coelenterate.

megapascal A unit of pressure used to measure water potential.

megaspore A plant spore that develops into a female gametophyte; usually larger than a microspore.

meiosis The division of diploid cells to haploid progeny, consisting of two sequential rounds of nuclear and cellular division.

meiosis I The first division of the meiotic cell cycle in which homologous chromosomes pair and undergo an exchange of chromosome segments, and then the homologous chromosomes separate, resulting in two cells, each with the haploid number of chromosomes and with each chromosome still consisting of two chromatids.

meiosis II The second division of the meiotic cell cycle in which the sister chromatids in each of the two cells produced by meiosis I separate and segregate into different cells, resulting in four cells each with the haploid number of chromosomes.

melanocyte-stimulating hormone (MSH) A hormone secreted by the anterior pituitary that controls the degree of pigmentation in melanocytes.

melatonin A peptide hormone secreted by the pineal gland that helps maintain daily biorhythms.

membrane attack complexes (MAC) An abnormal activation of the complement (protein) portion of the blood, forming a cascade reaction that brings blood proteins together, binds them to the cell wall, and then inserts them through the cell membrane.

membrane potential An electrical voltage that measures the potential inside a cell membrane relative to the fluid just outside; it is negative under resting conditions and becomes positive during an action potential.

memory The storage and retrieval of a sensory or motor experience, or a thought.

memory B cell In antibody-mediated immunity, a long-lived cell expressing an antibody on its surface that can bind to a specific antigen. A memory B cell is activated the next time the antigen is encountered, producing a rapid secondary immune response.

memory cell An activated lymphocyte that circulates in the blood and lymph, ready to initiate a rapid immune response on subsequent exposure to the same antigen.

memory cytotoxic T cells Cells differentiated from cytotoxic T cells that are stored for subsequent cell-mediated immunity involving the same antigens.

memory helper T cell In cell-mediated immunity, a long-lived cell differentiated from a helper T cell, which remains in an inactive state in the lymphatic system after an immune reaction has run its course and ready to be activated on subsequent exposure to the same antigen.

meninges Three layers of connective tissue that surround and protect the spinal cord and brain.

menstrual cycle A cycle of approximately 1 month in the human female during which an egg is released from an ovary and the uterus is prepared to receive the fertilized egg; if fertilization does not occur, the endometrium breaks down, which releases blood and tissue breakdown products from the uterus to the outside through the vagina.

meristem An undifferentiated, permanently embryonic plant tissue that gives rise to new cells forming tissues and organs.

mesenteries Sheets of loose connective tissue, covered on both surfaces with epithelial cells, which suspend the abdominal organs in the coelom and provide lubricated, smooth surfaces that prevent chafing or abrasion between adjacent structures as the body moves.

mesoderm The middle layer of the three primary germ layers of an animal embryo, from which the muscular, skeletal, vascular, and connective tissues develop.

mesohyl The gelatinous middle layer of cells lining the body cavity of a sponge.

mesophyll The ground tissue located between the two outer leaf tissues, composed of loosely packed parenchyma cells that contain chloroplasts.

messenger RNA (mRNA) An RNA molecule that serves as a template for protein synthesis.

metabolism The biochemical reactions that allow a cell or organism to extract energy from its surroundings and use that energy to maintain itself, grow, and reproduce.

metabotropic neurotransmitter receptor A receptor in the postsynaptic membrane, typically of the G-protein–coupled receptor type, to which a neurotransmitter binds as a first messenger, activating the receptor and triggering generation of a second messenger that brings about changes in ion-conducting channels in the membrane.

metagenomics Analysis of DNA sequences of the genomes in entire communities of microbes that have been isolated from the environment.

Metamonada (metamonads) A subgroup of the Excavata protist evolutionary group containing the Diplomonadida and the Parabasala.

metamorphosis A reorganization of the form of certain animals during postembryonic development.

metanephridium The excretory tubule of most annelids and mollusks.

metaphase The phase of mitosis during which the spindle reaches its final form and the spindle microtubules move the chromosomes into alignment at the spindle midpoint.

metapopulation A group of neighboring populations that exchange individuals.

metastasis The spreading of a malignant tumor through the blood system or lymphatic system, forming new tumors at other locations in the body.

micelles The tiny droplets produced when fats are emulsified.

microbody Small, membrane-bound organelle that carries out vital reactions linking metabolic pathways.

microclimate The abiotic conditions immediately surrounding an organism.

microevolution Small-scale genetic changes within populations, often in response to shifting environmental circumstances or chance events.

microfilament A cytoskeletal filament composed of actin.

microfossil The remains of a cell that has decayed and been filled in by calcium carbonate or silica.

micronucleus In ciliophorans, one or more diploid nuclei that contains a complete complement of genes, functioning primarily in cellular reproduction.

micronutrient Any mineral required by an organism only in trace amounts.

micropyle A small opening at one end of an ovule through which the pollen tube passes before fertilization.

microRNA (miRNA) One of the major types of small regulatory RNAs in eukaryotes involved in RNA interference (RNAi).

microsatellite *See* **short tandem repeat (STR).**

microscope Instrument of microscopy with different magnifications and resolutions of specimens.

microscopy Technique for producing visible images of objects that are too small to be seen by the human eye.

microspore A plant spore from which a male gametophyte develops; usually smaller than a megaspore.

microsporidium (plural, *microsporidia*) A fungal parasite of animals; many mycologists believe they make up a possible sixth phylum within the Kingdom Fungi.

microtubule A cytoskeletal component formed by the polymerization of tubulin into rigid, hollow rods about 25 nm in diameter.

microtubule organizing center (MTOC) An anchoring point near the center of a eukaryotic cell from which most microtubules extend outward.

microvilli Fingerlike projections forming a brush border in epithelial cells that cover the villi.

midbrain The uppermost of the three segments of the brainstem, serving primarily as an intermediary between the rest of the brain and the spinal cord.

middle ear The air-filled cavity containing three small, interconnected bones: the malleus, incus, and stapes.

middle lamella Layer of gel-like polysaccharides that holds together walls of adjacent plant cells.

migration The predictable seasonal movement of animals from the area where they are born to a distant and initially unfamiliar destination, returning to their birth site later.

mimic The species in Batesian mimicry that resembles the model.

mimicry A form of defense in which one species evolves an appearance resembling that of another.

mineralocorticoid A steroid hormone secreted by the adrenal cortex that regulates the levels of Na^+ and K^+ in the blood and extracellular fluid.

minimal medium A growth medium containing the minimal ingredients that enable a nonmutant organism, such as *E. coli,* to grow.

minimum viable population size The smallest population size that is likely to survive both predictable and unpredictable environmental variation.

miRNA-induced silencing complex (miRISC) Protein complex containing an miRNA that binds to sequences in the 3′ UTRs of target mRNAs, resulting in either inhibition of translation of the mRNAs or their degradation.

mismatch repair Repair system that removes mismatched bases from newly synthesized DNA strands.

missense mutation A base-pair substitution mutation in a protein-coding gene that results in a different amino acid in the encoded polypeptide than the normal one.

mitochondrial electron transfer system Series of electron carriers that alternately pick up and release electrons, ultimately transferring them to their final acceptor, oxygen.

mitochondrial matrix The innermost compartment of the mitochondrion.

mitochondrion Membrane-bound organelle responsible for synthesis of most of the ATP in eukaryotic cells.

mitosis Nuclear division that produces daughter nuclei that are exact genetic copies of the parental nucleus.

model The species in Batesian mimicry that is resembled by the mimic.

model organism An organism with characteristics that make it a particularly useful subject of research because it is likely to produce results widely applicable to other organisms.

modern synthesis A unified theory of evolution developed in the middle of the twentieth century.

molarity (*M*) The number of moles of a substance dissolved in 1 L of solution.

molars Posteriormost teeth of mammals, with a broad chewing surface for grinding food.

mold Asexual, spore-producing stage of many multicellular fungi.

mole (mol) Amount of substance that contains as many atoms or molecules as there are atoms in exactly 12 g of carbon-12, which is 6.022×10^{23}.

molecular clock A technique for dating the time of divergence of two species or lineages, based on the number of molecular sequence differences between them.

molecular phylogenetics Approach of using DNA or amino acid sequence comparisons to determine evolutionary relationships among organisms.

molecular weight The weight of a molecule in grams, equal to the total mass number of its atoms.

molecule A unit composed of atoms combined chemically in fixed numbers and ratios.

molt-inhibiting hormone (MIH) A peptide neurohormone secreted by a gland in the eye stalks of crustaceans that inhibits ecdysone secretion.

monoclonal antibody An antibody that reacts only against the same segment (epitope) of a single antigen.

monocot A plant belonging to the Monocotyledones, one of the two major classes of angiosperms; monocot embryos have a single seed leaf (cotyledon) and pollen grains with a single groove.

monocyte A type of leukocyte that enters damaged tissue from the bloodstream through the endothelial wall of the blood vessel.

monoecious Having both "male" flowers (which possess only stamens) and "female" flowers (which possess only carpels).

monogamy A mating system in which one male and one female form a long-term association.

monohybrid An F_1 heterozygote produced from a genetic cross that involves a single character.

monohybrid cross A genetic cross between two individuals that are each heterozygous for the same pair of alleles.

monophyletic taxon A group of organisms that includes a single ancestral species and all of its descendants.

monoploid An individual with one set of chromosomes instead of the usual two.

monosaccharides The smallest carbohydrates, containing three to seven carbon atoms.

monotreme A lineage of mammals that lay eggs instead of bearing live young.

monounsaturated Fatty acids with one double bond.

monsoon cycle A wind pattern that brings seasonally heavy rain to a region by blowing moisture-laden air from the sea to the land.

morphogenesis Orderly, genetically programmed changes in the size, shape, and proportion of body parts of an organism; the process by which specialized tissues and organs form.

morphological species concept The concept that all individuals of a species share measurable traits that distinguish them from individuals of other species.

morphology The form or shape of an organism, or of a part of an organism.

morula The first stage of animal development, a solid ball or layer of blastomeres.

motif A highly specialized region in a protein produced by the three-dimensional arrangement of amino acid chains within and between domains.

motile Capable of self-propelled movement.

motor neuron An efferent neuron that carries signals to skeletal muscle.

motor unit A block of muscle fibers that is controlled by branches of the axon of a single efferent neuron.

mRNA splicing Process that removes introns from pre-mRNAs and joins exons together.

mucosa The lining of the gut that contains epithelial and glandular cells.

mucous cells Cells of gastric pits that secrete alkaline mucus, which protects the mucosal layer of the stomach.

Müllerian duct The bipotential primitive duct associated with the gonads that leads to a cloaca.

Müllerian mimicry A form of defense in which two or more unpalatable species share a similar appearance.

multicellular organism Individual consisting of interdependent cells.

multigene family A family of homologous genes in a genome. The members of a multigene family have all evolved from one ancestral gene, and therefore have similar DNA sequences and produce proteins with similar structures and functions.

multiple alleles More than two different alleles of a gene.

multiple sclerosis An autoimmune disease resulting from an attack against a protein of the myelin sheaths insulating the surfaces of neurons.

multipotent In development, referring to cells derived from pluripotent cells; multipotent cells give rise to cells with particular functions.

muscle fiber A bundle of elongated, cylindrical cells that make up skeletal muscle.

muscle tissue Cells that have the ability to contract (shorten) forcibly.

muscle twitch A single, weak contraction of a muscle fiber.

muscularis The muscular coat of a hollow organ or tubular structure.

mutation A spontaneous and heritable change in DNA.

mutualism A symbiotic interaction between species in which both partners benefit.

mycelium A network of branching hyphae that constitutes the body of a multicellular fungus.

mycobiont The fungal component of a lichen.

mycorrhiza A mutualistic symbiosis in which fungal hyphae associate intimately with plant roots.

myofibril A cylindrical contractile element about 1 mm in diameter that runs lengthwise inside the muscle fiber cell.

myogenic heart A heart that maintains its contraction rhythm with no requirement for signals from the nervous system.

myoglobin An oxygen-storing protein closely related to hemoglobin.

myosin A protein that interacts with the protein actin to cause muscle contraction.

Na^+/K^+-ATPase *See* **Na^+/K^+ pump.**

Na^+/K^+ pump Pump that pushes 3 Na^+ out of the cell and 2 K^+ into the cell in the same pumping cycle. Also referred to as the *sodium–potassium pump* or as *Na^+/K^+-ATPase*.

Nanoarchaeota A group of the domain Archaea consisting of one living species that lives as a symbiont and species represented by DNA samples.

nastic movement In plants, a reversible response to nondirectional stimuli, such as mechanical pressure or humidity.

natural history The branch of biology that examines the form and variety of organisms in their natural environments.

natural killer (NK) cell A type of lymphocyte that destroys virus-infected cells.

natural selection The evolutionary process by which alleles that increase the likelihood of survival and the reproductive output of the individuals that carry them become more common in subsequent generations.

natural theology A belief that knowledge of God may be acquired through the study of natural phenomena.

navigation A wayfinding mechanism in which an animal moves toward a specific destination, using both a compass and a "mental map" of where it is in relation to the destination.

negative feedback The primary mechanism of homeostasis, in which a stimulus—a change in the external or internal environment—triggers a response that compensates for the environmental change.

negative pressure breathing Muscular contractions that expand the lungs, lowering the pressure of the air in the lungs and causing air to be pulled inward.

nekton Animals that can actively swim against water currents.

nematocyst A coiled thread, encapsulated in a cnidocyte, that cnidarians fire at prey or predators, sometimes releasing a toxin through its tip.

nephron A specialized excretory tubule that contributes to osmoregulation and carries out excretion, found in all vertebrates.

neritic zone The shallow water of the oceans above the continental shelves.

nerve A bundle of axons enclosed in connective tissue and all following the same pathway.

nerve cord A bundle of nerves that extends from the central ganglia to the rest of the body, connected to smaller nerves.

nerve net A simple nervous system that coordinates responses to stimuli but has no central control organ, or brain.

nervous tissue Tissue that contains neurons, which serve as lines of communication and control between body parts.

net primary productivity The chemical energy remaining in an ecosystem after a producer's cellular respiration is deducted from the gross primary productivity.

neural circuit A connection between axon terminals of one neuron and the dendrites or cell body of a second neuron; typically a neural circuit consists of an afferent (sensory) neuron connected to one or more interneurons, and an efferent neuron.

neural crest A band of cells that arises early in the embryonic development of vertebrates near the region where the neural tube pinches off from the ectoderm; later, the cells migrate and develop into unique structures.

neural plate Ectoderm thickened and flattened into a longitudinal band, induced by notochord cells.

neural signaling The process by which an animal responds appropriately to a stimulus.

neural tube A hollow tube in vertebrate embryos that develops into the brain, spinal cord, spinal nerves, and spinal column.

neurogenic heart A heart that beats under the control of signals from the nervous system.

neuromuscular junction The junction between a nerve fiber and the muscle it supplies.

neuron An electrically active cell of the nervous system responsible for controlling behavior and body functions.

neuroscience The integrated study of the structure, function, and development of the nervous system.

neurosecretory neuron A neuron that releases a neurohormone into the circulatory system when appropriately stimulated.

neurotransmitter A chemical released by an axon terminal at a chemical synapse.

neurulation The process in vertebrates by which organogenesis begins with development of the nervous system from ectoderm.

neutral lipid Energy-storing molecule consisting of a glycerol backbone and three fatty acid chains.

neutral variation hypothesis An evolutionary hypothesis that some variation at gene loci coding for enzymes and other soluble proteins is neither favored nor eliminated by natural selection.

neutralization One of two important mechanisms that clear foreign antigens from the body; toxins produced by an invading pathogen are inactivated (neutralized) by antibodies.

neutron Uncharged particle in the nucleus of an atom.

neutrophil A type of phagocytic leukocyte that attaches to blood vessel walls in massive numbers when attracted to the infection site by chemokines.

nicotinamide adenine dinucleotide (NAD^+) A coenzyme that serves as an electron carrier.

nitrification A metabolic process in which certain soil bacteria convert ammonia or ammonium ions into nitrites that are then converted by other bacteria to nitrates, a form usable by plants.

nitrogen cycle A biogeochemical cycle that moves nitrogen between the huge atmospheric pool of gaseous molecular nitrogen and several much smaller pools of nitrogen-containing compounds in soils, marine and freshwater ecosystems, and living organisms.

nitrogen fixation A metabolic process in which certain bacteria and cyanobacteria convert molecular nitrogen into ammonia and ammonium ions, forms usable by plants.

nitrogenous base A nitrogen-containing molecule with the properties of a base.

nociceptor A sensory receptor that detects tissue damage or noxious chemicals; their activity registers as pain. In systematics, the point of intersection of two branches of a phylogenetic tree.

node In plants, the point on a stem where one or more leaves are attached.

node of Ranvier The gap between two Schwann cells, which exposes the axon membrane directly to extracellular fluids.

noncoding RNA gene A gene encoding an RNA that is not translated; that is, a gene other than a protein-coding gene.

noncompetitive inhibition Inhibition of an enzyme reaction by an inhibitor molecule that binds to the enzyme at a site other than the active site and, therefore, does not compete directly with the substrate for binding to the active site.

noncyclic electron flow Pathway in photosynthesis in which electrons travel in a one-way direction from H_2O to $NADP^+$.

nondisjunction The failure of homologous pairs to separate during the first meiotic division or of chromatids to separate during the second meiotic division.

nonhistone protein All the proteins associated with DNA in a eukaryotic chromosome that are not histones.

nonpolar association Association that occurs when nonpolar molecules clump together.

nonpolar covalent bond Bond in which electrons are shared equally.

nonsense codon *See* **stop codon.**

nonsense mutation A base-pair substitution mutation in a gene in which the base-pair change results in a change from a sense codon to a nonsense codon in the mRNA. The polypeptide translated from the mRNA is shorter than the normal polypeptide because of the mutation.

nonshivering thermogenesis The generation of heat by oxidative mechanisms in nonmuscle tissue throughout the body.

nonvascular plant *See* **bryophyte.**

norepinephrine A nontropic amine hormone secreted by the adrenal medulla.

northern blot analysis Method of analysis in which RNA molecules extracted from cells or a tissue are separated by size using electrophoresis, transferred to a filter paper, and then probed to reveal the presence of particular RNAs.

notochord A flexible rodlike structure, constructed of fluid-filled cells surrounded by tough connective tissue, which supports a chordate embryo from head to tail.

N-terminal end The end of a polypeptide chain with an $—NH_3^+$ group.

nuclear envelope In eukaryotes, membranes separating the nucleus from the cytoplasm.

Nucleariidae (nucleariids) A protist group consistsing of heterotrophic, predominantly spherical amoebae with radiating, fine pseudopods that are not supported by microtubules. Nucleariids are a subgroup of the opisthokonts and are more closely related to fungi than to animals.

nuclear localization signal A short amino acid sequence in a protein that directs the protein to the nucleus.

nuclear pore complex A large, octagonally symmetrical, cylindrical structure that functions to exchange molecules between the nucleus and cytoplasm and prevents the transport of material not meant to cross the nuclear membrane. A nuclear pore—a channel through the complex—is the path for the exchange of molecules.

nucleoid The central region of a prokaryotic cell with no boundary membrane separating it from the cytoplasm, where DNA replication and RNA transcription occur.

nucleolus The nuclear site of rRNA transcription, processing, and ribosome assembly in eukaryotes.

nucleoplasm The liquid or semiliquid substance within the nucleus.

nucleoside Chemical structure containing only a nitrogenous base and a five-carbon sugar.

nucleosome The basic structural unit of chromatin in eukaryotes, consisting of DNA wrapped around a histone core.

nucleosome core particle An eight-protein particle formed by the combination of two molecules each of H2A, H2B, H3, and H4, around which DNA winds for almost two turns.

nucleosome remodeling complex Multiprotein complex that uses the energy of ATP hydrolysis to slide a nucleosome along the DNA of a eukaryotic chromosome to expose the promoter of a gene, or to restructure the nucleosome without moving it to allow transcription factors to bind.

nucleotide The monomer of nucleic acids consisting of a five-carbon sugar, a nitrogenous base, and a phosphate.

nucleus The central region of eukaryotic cells, separated by membranes from the surrounding cytoplasm, where DNA replication and messenger RNA transcription occur. In the nervous system, a nucleus is a concentration of nerve cells within the central nervous system that have related functions.

null hypothesis A statement of what would be seen if the hypothesis being tested were wrong.

null model A conceptual model that predicts what one would see if a particular factor had no effect.

nutrition The processes by which an organism takes in, digests, absorbs, and converts food into organic compounds.

obligate aerobe A microorganism that uses oxygen for cellular respiration and requires oxygen in its surroundings to support growth.

obligate anaerobe A microorganism that cannot use oxygen and can grow only in the absence of oxygen.

observational data Basic information on biological structures or the details of biological processes.

oceanic zone The deep ocean water beyond the continental shelves.

ocellus (plural, *ocelli*) The simplest eye, which detects light but does not form an image.

oil Neutral lipid that is liquid at biological temperatures.

Okazaki fragments The short lengths of lagging strand DNA produced by discontinuous replication.

olfactory bulb A gray-matter center that relay inputs from odor receptors to both the cerebral cortex and the limbic system.

oligodendrocyte A type of glial cell that populates the CNS and is responsible for producing myelin.

oligosaccharin A complex carbohydrate that in plants serves as a signaling molecule and as a defense against pathogens.

oligotrophic lake A lake that is poor in nutrients and organic matter, but rich in oxygen.

omasum One of the four stomach chambers of a ruminant; in this chamber, water is absorbed from the mass of digesting material.

ommatidium (plural, *ommatidia*) A faceted visual unit of a compound eye.

omnivore An animal that feeds at several trophic levels, consuming plants, animals, and other sources of organic matter.

oncogene A gene capable of inducing one or more characteristics of cancer cells.

one gene–one enzyme hypothesis Hypothesis showing the direct relationship between genes and enzymes.

one gene–one polypeptide hypothesis Restatement of the one gene–one enzyme hypothesis, taking into account that some proteins consist of more than one polypeptide and not all proteins are enzymes.

oocyte A developing gamete that becomes an ootid at the end of meiosis.

oogenesis The process of producing eggs.

oogonium A cell that enters meiosis and gives rise to gametes, produced by mitotic divisions of the germ cells in females.

Oomycota (water molds, white rusts, and downy mildews) A lineage of the Stramenopila subgroup of the Chromalveolata protist evolutionary group; funguslike organisms that lack chloroplasts and live as heterotrophs.

open circulatory system An arrangement of internal transport in some invertebrates in which the vascular fluid, hemolymph, is released into sinuses, bathing organs directly, and is not always retained within vessels.

open reading frame (ORF) Segment of a protein-coding gene or an mRNA transcribed from such a gene that involves a start codon separated by a multiple of three nucleotides from one of the stop codons. The ORF is a potential polypeptide-coding sequence.

operant conditioning A form of associative learning in which animals learn to link a voluntary activity, an operant, with its favorable consequences, the reinforcement.

operator A DNA regulatory sequence that controls transcription of an operon.

operculum A lid or flap of the bone serving as the gill cover in some fishes.

operon A cluster of prokaryotic genes organized into a single transcription unit and their associated regulatory sequences.

Opisthokonta (opisthokonts) A protist evolutionary group; a broad group of eukaryotes in which a single posterior flagellum is found at some stage in the life cycle.

opsin One of several different proteins that bond covalently with the light-absorbing pigment of rods and cones (retinal).

optic chiasm Location just behind the eyes where the optic nerves converge before entering the base of the brain, a portion of each optic nerve crossing over to the opposite side.

optical isomers *See* **enantiomers.**

optimal foraging theory A set of mathematical models that predict the diet choices of animals as they encounter a range of potential food items.

oral hood Soft fleshy structure at the anterior end of a cephalochordate that frames the opening of the mouth.

orbital The region of space where the electron "lives" most of the time.

order A Linnaean taxonomic category of organisms that ranks above a family and below a class.

organ Two or more different tissues integrated into a structure that carries out a specific function.

organ of Corti An organ within the cochlear duct that contains the sensory hair cells detecting sound vibrations transmitted to the inner ear.

organ system The coordinated activities of two or more organs to carry out a major body function such as movement, digestion, or reproduction.

organelles The nucleus and other specialized internal structures and compartments of eukaryotic cells.

organic acid (carboxylic acid) Acid for which the characteristic functional group is a carboxyl group (—COOH).

organic molecule Molecule based on carbon.

organismal ecology An ecological discipline in which researchers study the genetic, biochemical, physiological, morphological, and behavioral adaptations of organisms to their abiotic environments.

organogenesis The development of the major organ systems, giving rise to a free-living individual with the body organization characteristic of its species.

oriented cell division Cell division in different planes; establishes the overall shape of a plant organ.

origin of replication A specific region at which DNA replication commences. Bacterial chromosomes have single origins of replication (*ori*) whereas eukaryotic chromosomes have multiple origins.

orthogenesis An obsolete theory that evolution is goal oriented, striving to perfect organisms.

oscula One or more openings in a sponge through which water is expelled.

osmoconformer An animal in which the osmolarity of the cellular and extracellular solutions matches the osmolarity of the environment.

osmolarity The total solute concentration of a solution, measured in osmoles—the number of solute molecules and ions (in moles)—per liter of solution.

osmoreceptor A sensory receptor in the hypothalamus that responds to changes in the osmolarity of the fluid surrounding it, which reflects the osmolarity generally of the body fluids.

osmoregulation The regulation of water and ion balance.

osmoregulator An animal that uses control mechanisms to keep the osmolarity of cellular and extracellular fluids the same, but at levels that may differ from the osmolarity of the surroundings.

osmosis The passive transport of water across a selectively permeable membrane in response to solute concentration gradients, a pressure gradient, or both.

osmotic pressure A state of dynamic equilibrium in which the pressure of the solution on one side of a selectively permeable membrane exactly balances the tendency of water molecules to diffuse passively from the other side of the membrane due to a concentration gradient.

osteoblast A cell that produces the collagen and mineral of bone.

osteoclast A cell that removes bone minerals and recycles them through the bloodstream.

osteocyte A mature bone cell.

osteon The structural unit of bone, consisting of a minute central canal surrounded by osteocytes embedded in concentric layers of mineral matter.

ostracoderm One of an assortment of extinct, jawless fishes that were covered with bony armor.

otolith One of many small crystals of calcium carbonate embedded in the otolithic membrane of the hair cells.

outer boundary membrane A smooth membrane that surrounds a chloroplast, enclosing the stroma.

outer ear The external structure of the ear, consisting of the pinna and meatus.

outer membrane In Gram-negative bacteria, an additional boundary membrane that covers the peptidoglycan layer of the cell wall.

outer mitochondrial membrane The smooth membrane covering the outside of a mitochondrion.

outgroup comparison A technique used to identify ancestral and derived characters by comparing the group under study to more distantly related species that are not otherwise included in the analysis.

ova *See* **ovum.**

oval window An opening in the bony wall that separates the middle ear from the inner ear.

ovarian cycle The cyclic events in the ovary leading to ovulation.

ovary In animals, the female gonad, which produces female gametes and reproductive hormones. In flowering plants, the enlarged base of a carpel in which one or more ovules develop into seeds.

overexploitation The excessive harvesting of an animal or plant species, potentially leading to its extinction.

overnutrition A form of malnutrition. The condition caused by excessive intake of specific nutrients.

oviduct The tube through which the egg moves from the ovary to the outside of the body.

oviparous Referring to animals that lay eggs containing the nutrients needed for development of the embryo outside the mother's body.

ovoviviparous Referring to animals in which fertilized eggs are retained within the body and the embryo develops using nutrients provided by the egg; eggs hatch inside the mother.

ovulation The process in which oocytes are released into the oviducts as immature eggs.

ovule In plants, the structure in a carpel in which a female gametophyte develops and fertilization takes place.

ovum (plural, *ova*) A female sex cell, or egg.

oxidation The partial or full loss of electrons from a substance.

oxidative phosphorylation Synthesis of ATP in which ATP synthase uses an H^+ gradient built by the electron transfer system as the energy source to make the ATP.

oxidized Substance—the electron donor—from which the electrons are lost during oxidation.

oxytocin A hormone that stimulates the ejection of milk from the mammary glands of a nursing mother.

P generation The parental individuals used in an initial genetic cross.

P site The site in the ribosome where the tRNA carrying the growing polypeptide chain is bound during translation.

pacemaker cell A specialized cardiac muscle cell in the upper wall of the right atrium that sets the rate of contraction in the heart.

paedomorphosis A common form of heterochrony in which juvenile characteristics are retained in a reproductive adult.

pairing Process in meiosis in which homologous chromosomes come together and pair. Also referred to as *synapsis.*

pair-rule gene In *Drosophila* embryonic development, one of the second activated set of segmentation genes that progressively subdivide the embryo into regions, determining the segments of the embryo and the adult. Pair-rule genes control the division of the embryo into units of two segments each.

paleobiology The study of ancient organisms.

pancreas A mixed gland composed of an exocrine portion that secretes digestive enzymes into the small intestine and an endocrine portion, the islets of Langerhans, that secretes insulin and glucagon.

pancreatic enzymes Digestive enzymes secreted by exocrine cells in the pancreas into ducts that empty into the lumen of the duodenum.

parapatric speciation Speciation between populations with adjacent geographical distributions.

paraphyletic taxon A group of organisms that includes an ancestral species and some, but not all, of its descendants.

parapodia Fleshy lateral extensions of the body wall of aquatic annelids, used for locomotion and gas exchange.

parasite An organism that feeds on the tissues of or otherwise exploits its host.

parasitism A symbiotic interaction in which one species, the parasite, uses another, the host, in a way that is harmful to the host.

parasitoid An insect species in which a female lays eggs in the larva or pupa of another insect species, and her young consume the tissues of the living host.

parasympathetic division The division of the autonomic nervous system that predominates during quiet, low-stress situations, such as while relaxing.

parathyroid gland One of a pair of glands that produce parathyroid hormone (PTH) (found only in tetrapod vertebrates).

parathyroid hormone (PTH) The hormone secreted by the parathyroid glands in response to a fall in blood Ca^{2+} levels.

parenchyma A ground tissue with cells having a thin primary wall, which is pliable and permeable to water. Parenchyma cells may be specialized for photosynthesis, storage, secretion, or other tasks.

parental Phenotypes identical to the original parental individuals.

parental investment The time and energy devoted to the production and rearing of offspring.

parietal cells Cells of gastric glands that secrete H^+ and Cl^-, which combine to form HCl in the lumen of the stomach.

parthenogenesis A mode of asexual reproduction in which animals produce offspring by the growth and development of an egg without fertilization.

partial diploid A condition in which part of the genome of a haploid organism is diploid. Recipients in bacterial conjugation between an Hfr and an F^- cell become partial diploids for part of the Hfr bacterial chromosome.

partial pressure In a mixture of gases, the pressure of each individual gas.

parturition The process of giving birth.

passive immunity The acquisition of antibodies as a result of direct transfer from another person.

passive parental care The amount of energy invested in offspring—in the form of the energy stored in eggs or seeds or energy transferred to developing young through a placenta—before they are born.

passive transport The transport of substances across cell membranes without expenditure of energy, as in diffusion.

paternal chromosome The chromosome derived from the male parent of an organism.

pathogenesis-related (PR) protein A hydrolytic enzyme that breaks down components of a pathogen's cell wall.

pattern formation The arrangement of organs and body structures in their proper three-dimensional relationships.

pectoral girdle A bony or cartilaginous structure in vertebrates that supports and is attached to the forelimbs.

pedicellariae Small pincers at the base of short spines in starfishes and sea urchins.

pedigree Chart that shows all parents and offspring for as many generations as possible, the sex of individuals in the different generations, and the presence or absence of a trait of interest.

pelagic province The open water in the oceans.

pellicle A layer of supportive protein fibers located inside the cell, just under the plasma membrane, providing strength and flexibility instead of a cell wall.

pelvic girdle A bony or cartilaginous structure in vertebrates that supports and is attached to the hind limbs.

pepsin An enzyme made in the stomach that breaks down proteins.

pepsinogen The inactive precursor molecule for pepsin.

peptic ulcer Lesion in the stomach wall resulting from attack by HCl and pepsin.

peptide bond A link formed by a dehydration synthesis reaction between the $—NH_2$ group of one amino acid and the —COOH group of a second.

peptidoglycan A polymeric substance formed from a polysaccharide backbone tied together by short polypeptides, which is the primary structural molecule of bacterial cell walls.

peptidyl transferase An enzyme that catalyzes the reaction in which an amino acid is cleaved from the tRNA in the P site of the ribosome and forms a peptide bond with the amino acid on the tRNA in the A site of the ribosome.

peptidyl–tRNA A tRNA linked to a growing polypeptide chain containing two or more amino acids.

per capita growth rate The difference between the per capita birth rate and the per capita death rate of a population.

perception The conscious awareness of our external and internal environments derived from the processing of sensory input.

perennial A plant in which vegetative growth and reproduction continue year after year.

perfect flower A flower that has both male (stamen) and female (carpel) sexual organs.

perfusion The flow of blood or other body fluids on the internal side of the respiratory surface.

pericarp The fruit wall.

pericycle A tissue of plant roots, located between the endodermis and the phloem, which gives rise to lateral roots.

periderm The outermost portion of bark; consists of cork, cork cambium, and secondary cortex.

peripheral nervous system (PNS) All nerve roots and nerves (motor and sensory) that supply the muscles of the body and transmit information about sensation (including pain) to the central nervous system.

peripheral protein Protein held to membrane surfaces by noncovalent bonds formed with the polar parts of integral membrane proteins or membrane lipids.

peristalsis The rippling motion of muscles in the intestine or other tubular organs characterized by the alternate contraction and relaxation of the muscles that propel the contents onward.

peritoneum The thin tissue derived from mesoderm that lines the abdominal wall and covers most of the organs in the abdomen.

peritubular capillary A capillary of the network surrounding the glomerulus.

permafrost Perpetually frozen ground below the topsoil.

peroxisome Microbody that produces hydrogen peroxide as a by-product.

PET *See* **positron emission tomography.**

petal Part of the corolla of a flower, often brightly colored.

petiole The stalk by which a leaf is attached to a stem.

pH scale The numerical scale used by scientists to measure acidity.

Phaeophyta (brown algae) A lineage of the Stramenopila subgroup of the Chromalveolata evolutionary subgroup; photosynthetic autotrophs that range from microscopic forms to giant kelps.

phage *See* **bacteriophage.**

phagocytosis Process in which some types of cells engulf bacteria or other cellular debris to break them down.

pharyngeal arches Embryonic features of all vertebrates; they contribute to the formation of the face, neck, mouth, larynx, and pharynx.

pharynx The throat. In some invertebrates, a protrusible tube used to bring food into the mouth for passage to the gastrovascular cavity; in mammals, the common pathway for air entering the larynx and food entering the esophagus.

phenotype The observable or measurable (biochemical, molecular) characteristics of an organism that are produced by an interaction between the genotype and the environment.

phenotypic variation Differences in appearance or function between individual organisms.

pheromone A distinctive volatile chemical released in minute amounts to influence the behavior of members of the same species.

phloem The food-conducting tissue of a vascular plant.

phloem sap The solution of water and organic compounds that flows rapidly through the sieve tubes of plants.

phosphate group Group consisting of a central phosphorus atom held in four linkages: two that bind —OH groups to the central phosphorus atom,

a third that binds an oxygen atom to the central phosphorus atom, and a fourth that links the phosphate group to an oxygen atom.

phosphodiester bond The linkage of nucleotides in polynucleotide chains by a bridging phosphate group between the 5′ carbon of one sugar and the 3′ carbon of the next sugar in line.

phospholipid A phosphate-containing lipid.

phosphorus cycle A biogeochemical cycle in which weathering and erosion carry phosphate ions from rocks to soil and into streams and rivers, which eventually transport them to the ocean, where they are slowly incorporated into rocks.

phosphorylation The addition of a phosphate group to a molecule.

photic zone Surface water of a lake or ocean that sunlight penetrates.

photoautotroph Photosynthetic organism that uses light as its energy source and carbon dioxide as its carbon source.

photobiont The photosynthetic component of a lichen.

photoheterotroph An organism that uses light as the ultimate energy source but obtains carbon in organic form rather than as carbon dioxide.

photoperiodism The response of plants to changes in the relative lengths of light and dark periods in their environment during each 24-hour period.

photophosphorylation The synthesis of ATP coupled to the transfer of electrons energized by photons of light.

photopigment Light-absorbing pigment.

photopsin One of three photopigments in which retinal is combined with different opsins.

photoreceptor A sensory receptor that detects the energy of light.

photorespiration A process that metabolizes a by-product of photosynthesis.

photosynthesis The conversion of light energy to chemical energy in the form of sugar and other organic molecules.

photosystem A large complex into which the light-absorbing pigments for photosynthesis are organized with proteins and other molecules.

photosystem I In photosynthesis, a protein complex in the thylakoid membrane that uses energy absorbed from sunlight to synthesize NADPH.

photosystem II In photosynthesis, a protein complex in the thylakoid membrane that uses energy absorbed from sunlight to synthesize ATP.

phototroph An organism that obtains energy from light.

phototropism The tendency of a plant shoot to bend toward a source of light.

phyletic gradualism hypothesis The hypothesis that most morphological change occurs gradually over long periods of time.

PhyloCode A formal set of rules governing phylogenetic nomenclature.

phylogenetic species concept A concept that seeks to delineate species as the smallest aggregate population that can be united by shared derived characters.

phylogenetic tree A branching diagram depicting the evolutionary relationships of groups of organisms.

phylogeny The evolutionary history of a group of organisms.

phylum (plural, *phyla*) A major Linnaean division of a kingdom, ranking above a class.

physical barriers The first of three lines of defense that humans and other mammals have against threats of pathogens; parts of the organism that physically block access to a pathogen. Examples are skin and epithelial surfaces covering internal body cavities and ducts. Physical barriers are not part of the immune system.

physiological respiration The process by which animals exchange gases with their surroundings—how they take in oxygen from the outside environment and deliver it to body cells, and remove carbon dioxide from body cells and deliver it to the environment.

physiology The study of the functions of organisms—the physico-chemical processes of organisms.

phytoalexin A biochemical that functions as an antibiotic in plants.

phytochrome A blue-green pigmented plant chromoprotein involved in the regulation of light-dependent growth processes.

phytoplankton Microscopic, free-flowing aquatic plants and protists.

phytosterol A sterol that occurs in plant cell membranes.

piloting A wayfinding mechanism in which animals use familiar landmarks to guide their journey.

pilus (plural, *pili*) A hair or hairlike appendage on the surface of a prokaryote.

pinacoderm In sponges, an unstratified outer layer of cells.

pineal gland A light-sensitive, melatonin-secreting gland that regulates some biological rhythms.

pinna The external structure of the outer ear, which concentrates and focuses sound waves.

pinocytosis *See* **bulk-phase endocytosis.**

pith The soft, spongelike, central cylinder of plant stems, composed mainly of parenchyma.

pituitary A gland consisting mostly of two fused lobes suspended just below the hypothalamus by a slender stalk of tissue that contains both neurons and blood vessels; it interacts with the hypothalamus to control many physiological functions, including the activity of some other glands.

placenta A specialized temporary organ that connects the embryo and fetus with the uterus in mammals, mediating the delivery of oxygen and nutrients.

plasma The clear, yellowish fluid portion of the blood in which cells are suspended. Plasma consists of water, glucose and other sugars, amino acids, plasma proteins, dissolved gases, ions, lipids, vitamins, hormones and other signal molecules, and metabolic wastes.

plasma cell A large antibody-producing cell that develops from B cells.

plasma membrane The outer limit of the cytoplasm responsible for the regulation of substances moving into and out of cells.

plasmid A DNA molecule in the cytoplasm of certain prokaryotes, which often contains genes with functions that supplement those in the nucleoid and which can replicate independently of the nucleoid DNA and be passed along during cell division.

plasmodesma (plural, *plasmodesmata*) A minute channel that perforates a cell wall and contains extensions of the cytoplasm that directly connect adjacent plant cells.

plasmodial slime mold A slime mold of the class Myxomycetes.

plasmodium The composite mass of plasmodial slime molds consisting of individual nuclei suspended in a common cytoplasm surrounded by a single plasma membrane.

plasmogamy The sexual stage of fungi during which the cytoplasms of two genetically different partners fuse.

plasmolysis Condition due to outward osmotic movement of water, in which plant cells shrink so much that they retract from their walls.

plastids A family of plant organelles that includes chloroplasts, amyloplasts, and chromoplasts.

plastron The ventral part of the shell of a turtle.

plate tectonics The geological theory describing how Earth's crust is broken into irregularly shaped plates of rock that float on its semisolid mantle.

platelet An oval or rounded cell fragment enclosed in its own plasma membrane, which is found in the blood; they are produced in red bone marrow by the division of stem cells and contain enzymes and other factors that take part in blood clotting.

pleiotropy Condition in which single genes affect more than one character of an organism.

pleura The double layer of epithelial tissue covering the lungs.

ploidy The number of chromosome sets of a cell or species.

plumule The rudimentary terminal bud of a plant embryo located at the end of the hypocotyl, consisting of the epicotyl and a cluster of tiny foliage leaves.

pluripotent In development, referring to cells derived from totipotent cells; pluripotent cells can give rise to most adult cell types.

polar association Association that occurs when polar molecules attract and align themselves with other polar molecules and with charged ions and molecules.

polar body A nonfunctional cell produced in oogenesis.

polar covalent bond Bond in which electrons are shared unequally.

polar nucleus In the embryo sac of a flowering plant, one of two nuclei that migrate into the center of the sac, become housed in a central cell, and eventually give rise to endosperm.

polar transport Unidirectional movement of a substance from one end of a cell (or other structure) to the other.

polarity The unequal distribution of yolk and other components in a mature egg.

pollen grain The male gametophyte of a seed plant.

pollen sac The microsporangium of a seed plant, in which pollen develops.

pollen tube A tube that grows from a germinating pollen grain and carries sperm cells to the ovary.

pollination The transfer of pollen to a seed plant's female reproductive parts by air currents or on the bodies of animal pollinators.

pollutant Materials or energy in a form or quantity that organisms do not usually encounter.

poly(A) polymerase The enzyme in eukaryotes that adds a chain of adenine nucleotides—the poly(A) tail—to the cleaved 3′ end of a pre-mRNA.

poly(A) tail The string of A nucleotides added posttranscriptionally to the 3′ end of a cleaved pre-mRNA molecule and retained in the mRNA produced from it that enables the mRNA to be translated efficiently and protects it from attack by RNA-digesting enzymes in the cytoplasm.

polyadenylation signal Sequence near the 3′ end of a eukaryotic gene which, in the pre-mRNA transcript of the gene, specifies where the transcript should be cleaved. Once cleaved, a poly(A) tail is added to the 3′ end of the RNA.

polyandry A polygamous mating system in which one female mates with multiple males.

polygamy A mating system in which either males or females may have many mating partners.

polygenic inheritance Inheritance in which several to many different genes contribute to the same character.

polygyny A polygamous mating system in which one male mates with many females.

polyhedral virus A virus in which the coat proteins form triangular units that fit together like the parts of a geodesic sphere.

polymerase chain reaction (PCR) Process that amplifies a specific DNA sequence from a DNA mixture to an extremely large number of copies.

polymorphism The existence of discrete variants of a character among individuals in a population.

polyp The tentacled, usually sessile stage in the life cycle of a coelenterate.

polypeptide The chain of amino acids formed by sequential peptide bonds.

polyphyletic taxon A group of organisms that belong to different evolutionary lineages and do not share a recent common ancestor.

polyploid An individual with one or more extra copies of the entire haploid complement of chromosomes.

polyploidy The condition of having one or more extra copies of the entire haploid complement of chromosomes.

polysaccharide Chain with more than 10 linked monosaccharide subunits.

polysome The entire structure of an mRNA molecule and the multiple associated ribosomes that are translating it simultaneously.

polyunsaturated Fatty acid with more than one double bond.

population All the individuals of a single species that live together in the same place and time.

population bottleneck An evolutionary event that occurs when a stressful factor reduces population size greatly and eliminates some alleles from a population.

population density The number of individuals per unit area or per unit volume of habitat.

population dynamics Changes in the characteristics of populations through time or over space.

population ecology The ecological discipline that focuses on how a population's size and other characteristics change in space and time.

population genetics The branch of science that studies the prevalence and variation in genes among populations of individuals.

population size The number of individuals in a population at a specified time.

population viability analysis A mathematical analysis used by conservation biologists to determine the minimum viable population size for threatened or endangered species.

positive feedback A mechanism that intensifies or adds to a change in internal or external environmental condition.

positive pressure breathing A gulping or swallowing motion that forces air into the lungs.

positron emission tomography (PET) A scanning technique for studying activities of the cerebral cortex.

posterior Indicating the tail end of an animal.

posterior pituitary The neural portion of the pituitary, which stores and releases two hormones made by the hypothalamus, antidiuretic hormone and oxytocin.

postsynaptic cell The neuron or the surface of an effector after a synapse that receives the signal from the presynaptic cell.

posttranslational import A mechanism by which proteins are sorted to their final cellular locations in a eukaryotic cell after they have been made on free ribosomes in the cytosol.

postzygotic isolating mechanism A reproductive isolating mechanism that acts after zygote formation.

potential energy Stored energy.

predation The interaction between animals and the animal prey they consume.

prediction A statement about what the researcher expects to happen to one variable if another variable changes.

pregnancy The period of mammalian development in which the embryo develops in the uterus of the mother.

premolars Teeth located in pairs on each side of the upper and lower jaws of mammals, positioned behind the canines and in front of the molars.

pre-mRNA (precursor-mRNA) The primary transcript of a eukaryotic protein-coding gene, which is processed to form messenger RNA.

prenatal diagnosis Techniques in which cells derived from a developing embryo or its surrounding tissues or fluids are tested for the presence of mutant alleles or chromosomal alterations.

prepuce Foreskin; a loose fold of skin that covers the glans of the penis.

pressure flow mechanism In vascular plants, pressure that builds up at the source end of a sieve tube system and pushes solutes by bulk flow toward a sink, where they are removed.

presynaptic cell The neuron with an axon terminal on one side of the synapse that transmits the signal across the synapse to the dendrite or cell body of the postsynaptic cell.

prezygotic isolating mechanism A reproductive isolating mechanism that acts before the production of a zygote, or fertilized egg.

primary active transport Transport in which the same protein that transports a substance also hydrolyzes ATP to power the transport directly.

primary cell layers The ectoderm, mesoderm, and endoderm layers that form the embryonic tissues.

primary cell wall The initial cell wall laid down by a plant cell.

primary consumer An herbivore, a member of the second trophic level.

primary endosymbiosis In the model for the origin of plastids in eukaryotes, the first event in which a eukaryotic cell engulfed a photosynthetic cyanobacterium.

primary growth The growth of plant tissues derived from apical meristems. *Compare* **secondary growth.**

primary immune response The response of the immune system to the first challenge by an antigen.

primary meristem Root and shoot apical meristems, from which a plant's primary tissues develop. *Compare* **lateral meristem.**

primary motor area The area of the cerebral cortex that runs in a band just in front of the primary somatosensory area and is responsible for voluntary movement.

primary plant body The portion of a plant that is made up of primary tissues.

primary producer An autotroph, usually a photosynthetic organism, a member of the first trophic level.

primary somatosensory area The area of the cerebral cortex that runs in a band across the parietal lobes of the brain and registers information on touch, pain, temperature, and pressure.

primary structure The sequence of amino acids in a protein.

primary succession Predictable change in species composition of an ecological community that develops on bare ground.

primary tissue A plant tissue that develops from an apical meristem.

primase An enzyme that assembles the primer for a new DNA strand during DNA replication.

primer A short nucleotide chain made of RNA that is laid down as the first series of nucleotides in a new DNA strand, or made of DNA for use in the polymerase chain reaction (PCR).

primitive groove In development of birds, the sunken midline of the primitive streak that acts as a conduit for migrating cells to move into the blastocoel.

primitive streak In development of birds, the thickened region of the embryo produced by cells of the epiblast streaming toward the midline of the blastodisc.

Principle of Independent Assortment Mendel's principle that the alleles of the genes that govern two characters assort independently during formation of gametes.

principle of parsimony A principle of systematic biology that states that a particular trait is unlikely to evolve independently in separate evolutionary lineages.

Principle of Segregation Mendel's principle that the pairs of alleles that control a character segregate as gametes are formed, with half the gametes carrying one allele, and the other half carrying the other allele.

prion An infectious agent that contains only protein and does not include a nucleic acid molecule.

probability The possibility that an outcome will occur if it is a matter of chance.

procambium The primary meristem of a plant that develops into primary vascular tissue.

product An atom or molecule leaving a chemical reaction.

product rule Mathematical rule in which the final probability is found by multiplying individual probabilities.

profundal zone The perpetually dark layer below the limnetic zone in a lake.

progesterone A female sex hormone that stimulates growth of the uterine lining and inhibits contractions of the uterus.

progestin A class of sex hormones synthesized by the gonads of vertebrates and active predominantly in females.

prokaryote Organism in which the DNA is suspended in the cell interior without separation from other cellular components by a discrete membrane.

prokaryotic chromosome The genetic material of prokaryotes, in most cases a single, circular DNA molecule.

prolactin (PRL) A peptide hormone secreted by the anterior pituitary that stimulates breast development and milk secretion in mammals.

prometaphase A transition period between prophase and metaphase during which the microtubules of the mitotic spindle attach to the kinetochores and the chromosomes shuffle until they align in the center of the cell.

promiscuity A mating system in which individuals do not form close pair bonds, and both males and females mate with multiple partners.

promoter The site to which RNA polymerase binds (prokaryotes) or to which general transcription factors bind and recruit RNA polymerase (eukaryotes) for initiating transcription of a gene.

promoter proximal element Regulatory sequence within the promoter proximal region, a region upstream of the promoter of a eukaryotic protein-coding gene. Regulatory proteins bind to promoter proximal elements and stimulate or inhibit the rate of transcription initiation.

promoter proximal region Upstream of a eukaryotic gene, a region containing regulatory sequences—promoter proximal elements—for transcription called promoter proximal elements.

proofreading mechanism Mechanism during DNA replication in which DNA polymerase backs up and removes a mispaired nucleotide from a newly synthesized DNA strand and then adds the correct nucleotide to the growing chain.

prophage A viral genome inserted in the host cell DNA.

prophase The beginning phase of mitosis during which the duplicated chromosomes within the nucleus condense from a greatly extended state into compact, rodlike structures.

proprioceptor A mechanoreceptor that detects stimuli used in the CNS to maintain body balance and equilibrium and to monitor the position of the head and limbs.

prostaglandin One of a group of local regulators derived from fatty acids that are involved in paracrine and autocrine regulation.

prostate gland An accessory sex gland in males that adds a thin, milky fluid to the semen and adjusts the pH of the semen to the level of acidity best tolerated by sperm.

proteasome Large cytoplasmic protein complex in eukaryotic cells that degrades ubiquitinylated proteins.

protein Molecules that carry out most of the activities of life, including the synthesis of all other biological molecules. A protein consists of one or more polypeptides depending on the protein.

protein chip *See* **protein microarray.**

protein kinase Enzyme that transfers a phosphate group from ATP to one or more sites on particular proteins.

protein microarray Similar in concept to a DNA microarray, a solid surface with a microscopic grid with thousands of spaces containing probes for analyzing the proteome, the complete set of proteins encoded by the genome of an organism. Also referred to as a *protein chip.*

protein phosphatase Enzyme that removes phosphate groups from target proteins.

protein-coding gene A gene encoding a protein.

proteome The complete set of proteins that can be expressed by the genome of an organism.

proteomics The study of the proteome.

prothoracicotropic hormone (PTTH) A peptide hormone secreted by neurosecretory neurons in the brains of insects that is one of the hormones regulating molting and metamorphosis.

protists A diverse and polyphyletic group of single-celled and multicellular eukaryotic species.

protocell A primitive cell-like structure that has some of the properties of life and that might have been the precursor of cells.

protoderm The primary meristem that will produce stem epidermis.

proton Positively charged particle in the nucleus of an atom.

proton pump Pump that moves hydrogen ions across membranes and pushes hydrogen ions across the plasma membrane from the cytoplasm to the cell exterior. Also referred to as *H^+ pump.*

protonema The structure that arises when a liverwort or moss spore germinates and eventually gives rise to a mature gametophyte.

protonephridium The simplest form of invertebrate excretory tubule.

proton-motive force Stored energy that contributes to ATP synthesis, as well as to the cotransport of substances to and from mitochondria.

proto-oncogene A gene that encodes various kinds of proteins that stimulate cell division. Mutated proto-oncogenes—oncogenes—contribute to the development of cancer.

protoplast The cytoplasm, organelles, and plasma membrane of a plant cell.

protoplast fusion A plant breeding process in which protoplasts are fused into a single cell.

provirus DNA copy of a retrovirus RNA genome that becomes inserted into host DNA.

proximal convoluted tubule The tubule between the Bowman's capsule and the loop of Henle in the nephron of the kidney, which carries and processes the filtrate.

pseudocoelom A fluid- or organ-filled body cavity between the gut (a derivative of endoderm) and the muscles of the body wall (a derivative of mesoderm).

pseudocoelomate A body plan of bilaterally symmetrical animals with a body cavity that lacks a complete lining derived from mesoderm.

pseudopod (plural, *pseudopodia*) A temporary cytoplasmic extension of a cell.

pseudogene A gene that is very similar to a functional gene at the DNA sequence level but that has one or more inactivating mutations that prevent it from producing a functional gene product.

psychrophile An archaean or bacterium that grows optimally at temperatures in the range $-10°C$ to $-20°C$.

Pterophyta The plant phylum of ferns and their close relatives.

pulmocutaneous circuit In amphibians, the branch of a double blood circuit that receives deoxygenated blood and moves it to the skin and lungs or gills.

pulmonary arteries The arteries from the right ventricle of the heart that lead to the lungs.

pulmonary circuit The circuit of the cardiovascular system that supplies the lungs.

pulmonary veins The veins that carry oxygenated blood from the lungs to the right atrium of the heart.

pulse The measurable rate of ventricular contraction of the heart detected at the peak of high pressure in an artery.

pulvinus (plural, *pulvini*) A jointlike, thickened pad of tissue at the base of a leaf or petiole; flexes when the leaf makes nastic movements.

punctuated equilibrium hypothesis The evolutionary hypothesis that most morphological variation arises rapidly during speciation events in isolated populations at the edge of a species' geographical distribution.

Punnett square Method for determining the genotypes and phenotypes of offspring and their expected proportions by combining gametes and their probabilities of occurrence in a matrix table.

pupa The nonfeeding stage between the larva and adult in the complete metamorphosis of some insects, during which the larval tissues are completely reorganized within a protective cocoon or hardened case.

pupil The dark center in the middle of the iris through which light passes to the back of the eye.

purine A type of nitrogenous base with two carbon–nitrogen rings.

pyloric sphincter The muscular ring at the junction between the stomach and small intestine that controls the flow of materials from the stomach.

pyramid of biomass A diagram that illustrates differences in standing crop biomass in a series of trophic levels.

pyramid of energy A diagram that illustrates the amount of energy that flows through a series of trophic levels.

pyramid of numbers A diagram that illustrates the number of individual organisms present in a series of trophic levels.

pyrimidine A type of nitrogenous base with one carbon–nitrogen ring.

pyrogens Chemicals released by macrophages in response to infection that stimulate prostaglandin release from the hypothalamus thereby leading to an increase in body temperature (fever).

pyruvate oxidation (pyruvic acid oxidation) Stage of cellular respiration in which the three-carbon molecule pyruvate is converted into a two-carbon acetyl group that is completely oxidized to carbon dioxide.

qualitative variation Variation that exists in two or more discrete states, with intermediate forms often being absent.

quantitative trait A character that displays a continuous distribution of the phenotype involved, typically resulting from several to many contributing genes.

quantitative trait loci (QTLs) The individual genes that contribute to a quantitative trait.

quantitative variation Variation that is measured on a continuum (such as height in human beings) rather than in discrete units or categories.

quaternary structure The arrangement of polypeptide chains in a protein that contains more than one chain.

quiescent center A region in a root apical meristem where there is no cell division.

***R* gene** A resistance gene in a plant; dominant *R* alleles confer enhanced resistance to plant pathogens.

R plasmid A bacterial plasmid containing genes that provide resistance to unfavorable conditions.

radial cleavage A cleavage pattern in deuterostomes in which newly formed cells lie directly above and below other cells of the embryo.

radial symmetry A body plan of organisms in which structures are arranged regularly around a central axis, like spokes radiating out from the center of a wheel.

radiation The transfer of heat energy as electromagnetic radiation.

radicle The rudimentary root of a plant embryo.

radioactivity The giving off of particles of matter and energy by decaying nuclei.

radioisotope An unstable, radioactive isotope.

Radiolaria (radiolarians) A lineage of the Rhizaria protist evolutionary group; characterized by axopods, slender, raylike strands of cytoplasm supported internally by long bundles of microtubules.

radiometric dating A dating method that uses measurements of certain radioactive isotopes to calculate the absolute ages in years of rocks and minerals.

radula The tooth-lined "tongue" of mollusks that scrapes food into small particles or drills through the shells of prey.

rain shadow An area of reduced precipitation on the leeward side of a mountain.

random coil An arrangement of the amino acid chain providing flexible regions that allow sections of the chain to bend.

random dispersion A pattern of distribution in which the individuals in a population are distributed unpredictably in their habitat.

rapid eye movement (REM) sleep The period during deep sleep when the delta wave pattern is replaced by rapid, irregular beta waves characteristic of the waking state. The person's heartbeat and breathing rate increase, the limbs twitch, and the eyes move rapidly behind the closed eyelids.

reabsorption The process in which some molecules (for example, glucose and amino acids) and ions are transported by the transport epithelium back into the body fluid (animals with open circulatory systems) or into the blood in capillaries surrounding the tubules (animals with closed circulatory systems) as the filtered solution moves through the excretory tubule.

reactants The atoms or molecules entering a chemical reaction.

reaction center Part of photosystems I and II in chloroplasts of plants. In the light-dependent reactions of photosynthesis, the reaction center receives light energy absorbed by the antenna complex in the same photosystem.

reading frame The series of codons for a polypeptide encoded by the mRNA.

realized niche The range of resources and environmental conditions actually used by a population in nature.

receptacle The expanded tip of a flower stalk that bears floral organs.

reception In signal transduction, the binding of a signal molecule with a specific receptor in a target cell. In neural signaling, the first of four components in which a stimulus is detected by specialized sensory receptors.

receptive field The region surrounding a receptor within which the receptor responds to a stimulus.

receptor potential The change in membrane potential exhibited by a sensory receptor in response to a stimulus in its receptive field.

receptor protein Protein that recognizes and binds molecules from other cells that act as chemical signals.

receptor tyrosine kinase In signal transduction, a surface receptor with built-in protein kinase activity.

receptor-mediated endocytosis The selective uptake of macromolecules that bind to cell surface receptors concentrated in clathrin-coated pits.

recessive An allele that is masked by a dominant allele.

reciprocal altruism Form of altruistic behavior in which individuals help nonrelatives if they are likely to return the favor in the future.

reciprocal cross A genetic cross in which the two parents are switched with respect to which trait is associated with each sex.

recognition protein Protein in the plasma membrane that identifies a cell as part of the same individual or as foreign.

recombinant Phenotype with a different combination of traits from those of the original parents.

recombinant chromosomes Chromosomes that contain nonparental combinations of alleles. In eukaryotes they are generated by crossing-over in meiosis.

recombinant DNA DNA from two or more different sources joined together.

recombination *See* **genetic recombination.**

recombination frequency In the construction of linkage maps of diploid eukaryotic organisms, the percentage of testcross progeny that are recombinants.

rectum The final segment of the large intestine.

red tide A growth in dinoflagellate populations that causes red, orange, or brown discoloration of coastal ocean waters.

redox reaction Coupled oxidation–reduction reaction in which electrons are removed from a donor molecule and simultaneously added to an acceptor molecule.

reduced Substance—the electron acceptor—that gains electrons during reduction.

reduction The partial or full gain of electrons to a substance.

reflex A programmed movement that takes place without conscious effort, such as the sudden withdrawal of a hand from a hot surface.

refractory period A period that begins at the peak of an action potential and lasts a few milliseconds, during which the threshold required for generation of an action potential is much higher than normal.

regulators Animals that maintain factors of the internal environment in a relatively constant state.

regulatory gene Gene that encodes a protein that regulates the expression of a structural gene or genes.

regulatory protein DNA-binding protein that binds to a regulatory sequence and affects the expression of an associated gene or genes.

regulatory sequence DNA sequence involved in the regulation of a gene or genes to which a regulatory protein binds to control the transcription of the gene or genes.

reinforcement The enhancement of reproductive isolation that had begun to develop while populations were geographically separated.

relative abundance The relative commonness of populations within a community.

relative fitness The number of surviving offspring that an individual produces compared with the number left by others in the population.

release The process in which urine is released into the environment from the distal end of the excretory tubule.

release factor A protein that recognizes stop codons in the A site of a ribosome translating an mRNA and terminates translation. Also referred to as the *termination factor.*

releasing hormone (RH) A peptide neurohormone that control the secretion of hormones from the anterior pituitary.

renal artery An artery that carries bodily fluids into the kidney.

renal cortex The outer region of the mammalian kidney that surrounds the renal medulla.

renal medulla The inner region of the mammalian kidney.

renal pelvis The central cavity in the kidney where urine drains from collecting ducts.

renal vein The vein that routes filtered blood away from the kidney.

renaturation The reformation of a denatured protein into its folded, functional state.

renin An enzyme secreted by cells in the juxtaglomerular apparatus into the bloodstream that converts a blood protein into the peptide hormone angiotensin.

renin-angiotensin-aldosterone system (RAAS) The most important hormonal system involved in regulation of Na^+ in mammals.

replica plating Technique for identifying and counting genetic recombinants in conjugation, transformation, or transduction experiments in which the colony pattern on a plate containing solid growth medium is pressed onto sterile velveteen and transferred to other plates containing different combinations of nutrients.

replicates Multiple subjects that receive either the same experimental treatment or the same control treatment.

replication bubble The two Y-shaped replication forks joined together at the tops of the Ys after DNA is unwound at an origin of replication.

replication fork The region of DNA synthesis where the parental strands separate and two new daughter strands elongate.

repressible operon Operon whose expression is prevented by a repressor molecule.

repressor A regulatory protein that prevents the operon genes from being expressed.

reproduction The process in which parents produce offspring.

reproductive isolating mechanism A biological characteristic that prevents the gene pools of two species from mixing.

reproductive strategy One of a set of behaviors that lead to reproductive success.

residual volume The air that remains in lungs after exhalation.

resilin An elastic protein related to elastin, found only in insects and some crustaceans.

resolution The minimum distance two points in a specimen can be separated and still be seen as two points.

resource partitioning The use of different resources or the use of resources in different ways by species living in the same place.

respiratory medium The environmental source of O_2 and the "sink" for released CO_2. For aquatic animals, the respiratory medium is water; for terrestrial animals, it is air.

respiratory surface A layer of epithelial cells that provides the interface between the body and the respiratory medium.

respiratory system All the parts of the body involved in exchanging air between the external environment and the blood.

response In signal transduction, the last stage in which the transduced signal causes the cell to change according to the signal and to the receptors on the cell. In neural signaling, the fourth and last component involving the action resulting from the integration of neural messages.

resting potential A steady negative membrane potential exhibited by the membrane of a neuron that is not stimulated—that is, not conducting an impulse.

restriction endonuclease (restriction enzyme) An enzyme that cuts DNA at a specific sequence.

restriction fragment A DNA fragment produced by cutting a long DNA molecule with a restriction enzyme.

restriction fragment length polymorphisms When comparing different individuals, restriction enzyme–generated DNA fragments of different lengths from the same region of the genome.

reticular activating system *See* **ascending reticular formation.**

reticular formation A complex network of interconnected neurons that runs through the length of the brain stem, connecting to the thalamus at the anterior end and to the spinal cord at the posterior end.

reticulum One of the four chambers of the stomach of a ruminant; in this chamber, fermentation reactions by symbiotic microorganisms begin digesting boluses of swallowed plant matter.

retina A light-sensitive membrane lining the posterior part of the inside of the eye.

retrotransposon A transposable element that transposes via an intermediate RNA copy of the transposable element.

retrovirus A virus with an RNA genome that replicates via a DNA intermediate.

reverse transcriptase An enzyme that uses RNA as a template to make a DNA copy of the retrotransposon. Reverse transcriptase is used to make DNA copies of RNA in test tube reactions.

reversible The term indicating that a reaction may go from left to right or from right to left, depending on conditions.

rheumatoid arthritis An autoimmune disease that results by a self-attack on connective tissues, particularly in the joints, causing pain and inflammation.

Rhizaria (rhizarians) A protist evolutionary group; amoebas with filamentous pseudopods.

rhizome A horizontal, modified stem that can penetrate a substrate and anchor the plant.

Rhodophyta (red algae) A lineage of the Archaeplastida protist evolutionary group; mostly free-living autotrophic marine seaweeds.

rhodopsin A photopigment consisting of retinal bound covalently to an opsin protein.

ribonucleic acid (RNA) A polymer assembled from repeating nucleotide monomers in which the five-carbon sugar is ribose. Cellular RNAs include mRNA (which is translated to produce a polypeptide), tRNA (which brings an amino acid to the ribosome for assembly into a polypeptide during translation), and rRNA (which is a structural component of ribosomes). The genetic material of some viruses is RNA.

ribonucleotide Nucleotide containing ribose as the sugar; ribonucleotides are components of RNA.

ribose A five-carbon sugar to which the nitrogenous bases in nucleotides link covalently.

ribosomal RNA (rRNA) The RNA component of ribosomes.

ribosome A ribonucleoprotein particle that carries out protein synthesis by translating mRNA into chains of amino acids.

ribosome binding site In translation initiation in prokaryotes, a sequence just upstream of the start codon that directs the small ribosomal subunit to bind and orient correctly for the complete ribosome to assemble and start translating in the correct spot.

ribozyme An RNA-based catalyst that is part of the biochemical machinery of all cells.

ring species A species with a geographical distribution that forms a ring around uninhabitable terrain.

RNA *See* **ribonucleic acid.**

RNA interference (RNAi) The phenomenon of silencing a gene posttranscriptionally by a small, single-stranded RNA that is complementary to part of an mRNA.

RNA polymerase An enzyme that catalyzes the assembly of ribonucleotides into an RNA strand.

RNA world model A model which states that the first genes and enzymes were RNA molecules.

rod In the vertebrate eye, a type of photoreceptor in the retina that is specialized for detection of light at low intensities.

root An anchoring structure in land plants that also absorbs water and nutrients and (in some plant species) stores food.

root cap A dome-shaped cell mass that forms a protective covering over the apical meristem in the tip of a plant root.

root hair A tubular outgrowth of the outer wall of a root epidermal cell; root hairs absorb much of a plant's water and minerals from the soil.

root nodule A localized swelling on a root in which symbiotic nitrogen-fixing bacteria reside.

root pressure The pressure that develops in plant roots as the result of osmosis, forcing xylem sap upward and out through leaves. *See also* **guttation.**

root primordium A rudimentary root.

root system An underground (or submerged) network of roots with a large surface area that favors the rapid uptake of soil water and dissolved mineral ions.

rough ER Endoplasmic reticulum with many ribosomes studding its outer surface.

round window A thin membrane that faces the middle ear.

***r*-selected species** A short-lived species adapted to function well in a rapidly changing environment.

RuBP carboxylase/oxygenase (rubisco) An enzyme that catalyzes the key reaction of the Calvin cycle, carbon fixation, in which CO_2 combines with RuBP (ribulose 1,5-bisphosphate) to form 3-phosphoglycerate.

rumen One of the four chambers of the stomach of a ruminant; in this chamber, fermentation reactions by symbiotic microorganisms begin digesting boluses of swallowed plant matter.

ruminant An animal that has a complex, four-chambered stomach.

S phase The phase of the eukaryotic cell cycle during which DNA replication occurs.

saccule Saclike, fluid-filled structure that is part of the vestibular apparatus. With the utricle, provides information about the up–down position of the head, as well as changes in the rate of linear movement of the head.

salicylic acid (SA) In plants, a chemical synthesized following a wound that has multiple roles in plant defenses, including interaction with jasmonates in signaling cascades.

salivary amylase A substance that hydrolyzes starches to the disaccharide maltose.

salivary gland A gland that secretes saliva through a duct on the inside of the cheek or under the tongue; the saliva lubricates food and begins digestion.

salt marsh A tidal wetland dominated by emergent grasses and reeds.

saltatory conduction A mechanism that allows small-diameter axons to conduct impulses rapidly.

saprobe An organism nourished by dead or decaying organic matter.

sapwood The newly formed outer wood located between heartwood and the vascular cambium. Compared with heartwood, it is wet, lighter in color, and not as strong.

sarcomere The basic unit of contraction in a myofibril.

sarcoplasmic reticulum In vertebrate muscle fibers, a complex system of vesicles modified from the smooth endoplasmic reticulum that encircles the sarcomeres. The sarcoplasmic reticulum is part of the pathway for the stimulation of muscle contraction by neural signals.

saturated enzymes Enzymes for which increases in substrate concentration have no effect on the reaction rate.

saturated fatty acid Fatty acid with only single bonds linking the carbon atoms.

savanna A biome comprising grasslands with few trees, which grows in areas adjacent to tropical deciduous forests.

schizocoelom In protostomes, the body cavity that develops as inner and outer layers of mesoderm separate.

Schwann cell A type of glial cell in the PNS that wraps nerve fibers with myelin and also secretes regulatory factors.

scientific method An investigative approach in which scientists make observations about the natural world, develop working explanations about what they observe, and then test those explanations by collecting more information.

scientific name A two-part name identifying the genus to which a species belongs and designating a particular species within that genus.

scientific theory A broadly applicable idea or hypothesis that has been confirmed by every conceivable test.

sclereid A type of sclerenchyma cell; sclereids typically are short and have thick, lignified walls.

sclerenchyma A ground tissue in which cells develop thick secondary walls, which commonly are lignified and perforated by pits through which water can pass.

scrotum The baglike sac in which the testes are suspended in many mammals.

scutellum The shield-shaped cotyledon of a grass.

second law of thermodynamics Principle that for any process in which a system changes from an initial to a final state, the total disorder of the system and its surroundings always increases.

second messenger In particular signal transduction pathways, an internal, nonprotein signal molecule that directly or indirectly activates protein kinases, which elicit the cellular response.

secondary active transport Transport indirectly driven by ATP hydrolysis.

secondary cell wall A layer added to the cell wall of plants that is more rigid and may become many times thicker than the primary cell wall.

secondary consumer A carnivore that feeds on herbivores, a member of the third trophic level.

secondary endosymbiosis In the model for the origin of plastids in eukaryotes, the second event, in which a nonphotosynthetic eukaryote engulfed a photosynthetic eukaryote.

secondary growth Plant growth that originates at lateral meristems and increases the diameter of older roots and stems. *Compare* **primary growth.**

secondary immune response The rapid immune response that occurs during the second (and subsequent) encounters of the immune system of a mammal with a specific antigen.

secondary plant body The part of a plant made up of tissues that develop from lateral meristems.

secondary productivity Energy stored in new consumer biomass as energy is transferred from producers to consumers.

secondary structure Regions of alpha helix, beta strand, or random coil in a polypeptide chain.

secondary succession Predictable changes in species composition in an ecological community that develops after existing vegetation is destroyed or disrupted by an environmental disturbance.

secondary tissue In plants, the tissue that develops from lateral meristems.

secretin Hormone released into the bloodstream from glandular cells in the small intestine in response to stimulation by acidic chyme.

secretion A selective process in which specific small molecules and ions are transported from the body fluids (in animals with open circulatory systems) or blood (in animals with closed circulatory systems) into the excretory tubules.

secretory vesicle Vesicle that transports proteins to the plasma membrane.

seed The structure that forms when an ovule matures after a pollen grain reaches it and a sperm fertilizes the egg.

seed coat The outer protective covering of a seed.

segment polarity gene In *Drosophila* embryonic development, one of the third activated set of segmentation genes that progressively subdivide the embryo into regions, determining the segments of the embryo and the adult. Segment polarity genes set the boundaries and anterior–posterior axis of each segment in the embryo, thereby determining the regions that become segments of larvae and adults.

segmentation The production of body parts and some organ systems in repeating units.

segmentation genes Genes that work sequentially, progressively subdividing the embryo into regions, determining the segments of the embryo and the adult.

segregation The separation of the pairs of alleles that control a character as gametes are formed.

selective cell adhesion A mechanism in which cells make and break specific connections to other cells or to the extracellular matrix.

selectively neutral *See* **neutral variation hypothesis.**

selectively permeable Membranes that selectively allow, impede, or block the passage of atoms and molecules.

self-fertilization (self-pollination) Fertilization in which sperm nuclei in pollen fertilize egg cells of the same plant.

self-incompatibility In plants, the inability of a plant's pollen to fertilize ovules of the same plant.

semen The secretions of several accessory glands in which sperm are mixed before ejaculation.

semicircular canal Fluid-filled structures that are part of the vestibular apparatus. The three semicircular canals are positioned at angles corresponding to the three planes of space and detect rotational (spinning) motions.

semiconservative replication The process of DNA replication in which the two parental strands separate and each serves as a template for the synthesis of new progeny double-stranded DNA molecules.

seminal fluid Fluid secreted by the seminal vesicles that contains prostaglandins, which when ejaculated into the female trigger contractions of the female reproductive tract that help move the sperm into and through the uterus.

seminal vesicle A vesicle that secretes seminal fluid.

seminiferous tubule One of the tiny tubes in the testes where sperm cells are produced, grow, and mature.

senescence The biologically complex process of aging in mature organisms that leads to the death of cells and eventually the whole organism.

sense codon A codon that specifies an amino acid.

sensitization Increased responsiveness to mild stimuli after experiencing a strong stimulus; one of the simplest forms of memory.

sensor A tissue or organ that detects a change in an external or internal factor such as pH, temperature, or the concentration of a molecule such as glucose.

sensory adaptation A condition in which the effect of a stimulus is reduced if it continues at a constant level.

sensory neuron A neuron that transmits stimuli collected by the sensory receptors to interneurons.

sensory receptor A receptor formed by the dendrites of afferent neurons, or by specialized receptor cells making synapses with afferent neurons that pick up information about the external and internal environments of the animal.

sensory transduction The conversion of a stimulus into a change in membrane potential.

sepal One of the separate, usually green parts forming the calyx of a flower.

septum (plural, *septa*) A thin partition or cross wall that separates body segments.

sequential hermaphroditism The form of hermaphroditism in which individuals change from one sex to the other.

serosa The serous membrane: a thin membrane lining the closed cavities of the body; has two layers with a space between that is filled with serous fluid.

Sertoli cell One of the supportive cells that completely surrounds developing spermatocytes in the seminiferous tubules. Follicle-stimulating hormone stimulates Sertoli cells to secrete a protein and other molecules that are required for spermatogenesis.

sessile Unable to move from one place to another.

set point The level at which the condition controlled by a homeostatic pathway is to be maintained.

seta (plural, *setae*) A chitin-reinforced bristle that protrudes outward from the body wall in some annelid worms.

sex chromosomes Chromosomes that are different in male and female individuals of the same species.

sex pilus *See* **F pilus.**

sex ratio The relative proportions of males and females in a population.

sex-linked gene Gene located on a sex chromosome.

sexual dimorphism Differences in the size or appearance of males and females.

sexual reproduction The mode of reproduction in which male and female parents produce offspring through the union of egg and sperm generated by meiosis.

sexual selection A form of natural selection established by male competition for access to females and by the females' choice of mates.

shells *See* **energy levels.**

shoot system The stems and leaves of a plant.

short tandem repeat (STR) Short 2- to 6-bp DNA sequence repeated in series.

short-day plant A plant that flowers in late summer or early autumn when dark periods become longer and light periods become shorter.

short-term memory Memory that stores information for seconds.

sieve tube A series of phloem cells joined end to end, forming a long tube through which nutrients are transported; a feature mainly of flowering plants.

sieve tube member Any of the main conducting cells of phloem that connect end to end, forming a sieve tube.

sign stimulus A simple cue that triggers a fixed action pattern.

signal peptide A short segment of amino acids to which the signal recognition particle binds, temporarily blocking further translation. A signal peptide is found on polypeptides that are sorted to the endoplasmic reticulum. Also referred to as *signal sequence.*

signal recognition particle (SRP) Protein-RNA complex that binds to signal sequences and targets polypeptide chains to the endoplasmic reticulum.

signal sequence *See* **signal peptide.**

signal transduction The series of events by which a signal molecule released from a controlling cell causes a response (affects the function) of target cells with receptors for the signal. Target cells process the signal in the three sequential steps of reception, transduction, and response.

silencing Phenomenon in which methylation of cytosines in eukaryotic promoters inhibits transcription and turns the genes off.

silent mutation A base-pair substitution mutation in a protein-coding gene that does not alter the amino acid specified by the gene.

simple diffusion Mechanism by which certain small substances diffuse through the lipid part of a biological membrane.

simple fruit A fruit that develops from a single ovary; in many of them at least one layer of the pericarp is fleshy.

simulation modeling An analytical method in which researchers gather detailed information about a system and then create a series of mathematical equations that predict how the components of the system interact and respond to change.

simultaneous hermaphroditism A form of hermaphroditism in which individuals develop functional ovaries and testes at the same time.

single-lens eye An eye type that works by changing the amount of light allowed to enter into the eye and by focusing this incoming light with a lens.

single-stranded binding protein (SSB) Protein that coats single-stranded segments of DNA, stabilizing the DNA for the replication process.

sink Any region of a plant where organic substances are being unloaded from the phloem and used or stored.

sink population In metapopulation analysis, a population that routinely declines in size after being replenished by immigrants from a source population.

sinoatrial node (SA node) The region of the heart that controls the rate and timing of cardiac muscle cell contraction.

sinus A body space that surrounds an organ.

siRNA-induced silencing complex (siRISC) Protein complex containing an siRNA that binds to a sequence in a target RNA resulting in cleavage of that RNA.

sister chromatid One of two exact copies of a chromosome duplicated during replication.

sister clades Two evolutionary lineages (that is, clades) that emerge from the same node in a phylogenetic tree.

sister species Two species that are descended from the same recent ancestral species.

skeletal muscle A muscle that connects to bones of the skeleton, typically made up of long and cylindrical cells that contain many nuclei.

sliding DNA clamp A protein that encircles the DNA and binds to the DNA polymerase to tether the enzyme to the template, thereby making replication more efficient.

slime layer A coat typically composed of polysaccharides that is loosely associated with bacterial cells.

slime molds Members of the Amoebozoa; heterotrophic protists that, at some stage of their life cycle, exist as individuals that move by amoeboid motion but that exist in more complex forms the remainder of the time.

slow block to polyspermy The process in which enzymes released from cortical granules alter the egg coats within minutes after fertilization, so that no other sperm can attach and penetrate to the egg.

slow muscle fiber A muscle fiber that contracts relatively slowly and with low intensity.

small interfering RNA (siRNA) One of the major types of small, single-stranded regulatory RNAs in eukaryotes involved in RNA interference (RNAi).

small ribonucleoprotein particle A complex of RNA and proteins.

smooth ER Endoplasmic reticulum with no ribosomes attached to its membrane surfaces. Smooth ER has various functions, including synthesis of lipids that become part of cell membranes.

smooth muscle A relatively small and spindle-shaped muscle cell in which actin and myosin molecules are arranged in a loose network rather than in bundles.

social behavior The interactions that animals have with other members of their species.

sodium–potassium pump *See* **Na^+/K^+ pump.**

soil solution A combination of water and dissolved substances that coats soil particles and partially fills pore spaces.

solenoid *See* **30-nm chromatin fiber.**

solute The molecules of a substance dissolved in water.

solution Substance formed when molecules and ions separate and are suspended individually, surrounded by water molecules.

solvent The water in a solution in which the hydration layer prevents polar molecules or ions from reassociating.

somaclonal selection A procedure in which somatic embryos derived from tissue culture are screened to identify those having desired characteristics, such as disease resistance.

somatic cell Any of the cells of an organism's body other than reproductive cells.

somatic embryo A plant embryo that is genetically identical to the parent because it arose through asexual means.

somatic gene therapy Gene therapy in which genes are introduced into somatic cells.

somatic nervous system A subdivision of the peripheral nervous system controlling body movements that are primarily conscious and voluntary.

somites Paired blocks of mesoderm cells along the vertebrate body axis that form during early vertebrate development and differentiate into dermal skin, bone, and muscle.

soredium (plural, *soredia*) A specialized cell cluster produced by lichens, consisting of a mass of algal cells surrounded by fungal hyphae; soredia function like reproductive spores and can give rise to a new lichen.

sorus (plural, *sori*) A cluster of sporangia on the underside of a fern frond; reproductive spores arise by meiosis inside each sporangium.

source In plants, any region (such as a leaf) where organic substances are being loaded into the sieve tube system of phloem.

source population In metapopulation analyses, a population that is either stable or increasing in size.

Southern blot analysis Technique in which labeled probes are used to detect specific DNA fragments that have been separated by gel electrophoresis.

spatial summation The summation of EPSPs produced by firing of different presynaptic neurons.

specialized transduction Transfer of bacterial genes between bacteria using temperate phages that have incorporated fragments of the bacterial genome as they make the transition from the lysogenic cycle to the lytic cycle.

speciation The process of species formation.

species A group of populations in which the individuals are so closely related in structure, biochemistry, and behavior that they can successfully interbreed.

species cluster A group of closely related species recently descended from a common ancestor.

species composition The particular combination of species that occupy a site.

species diversity A community characteristic defined by species richness and the relative abundance of species.

species richness The number of species that live within an ecological community.

specific epithet The species name in a binomial.

specific heat The amount of heat required to increase the temperature of a given quantity of water.

spermatocyte A developing gamete that becomes a spermatid at the end of meiosis.

spermatogenesis The process of producing sperm.

spermatogonium (plural, *spermatogonia*) A cell that enters meiosis and gives rise to gametes, produced by mitotic divisions of the germ cells in males.

spermatozoan Also called *sperm;* a haploid cell that develops into a mature sperm cell when meiosis is complete.

sphincter A powerful ring of smooth muscle that forms a valve between major regions of the digestive tract.

spinal cord A column of nervous tissue located within the vertebral column and directly connected to the brain.

spinal nerve A nerve that carries signals between the spinal cord and the body trunk and limbs.

spindle The structure that separates sister chromatids and moves them to opposite spindle poles.

spindle pole One of the pair of centrosomes in a cell undergoing mitosis from which bundles of microtubules radiate to form the part of the spindle from that pole.

spinneret A modified abdominal appendage from which spiders secrete silk threads.

spiracle An opening in the chitinous exoskeleton of an insect through which air enters and leaves the tracheal system.

spiral cleavage The cleavage pattern in many protostomes in which newly produced cells lie in the space between the two cells immediately below them.

spiral organ *See* **organ of Corti.**

spiral valve A corkscrew-shaped fold of mucous membrane in the digestive system of elasmobranchs, which slows the passage of material and increases the surface area available for digestion and absorption.

spirillum (plural, *spirilla*) Any flagellated aerobic bacterium twisted helically like a corkscrew.

spliceosome A complex formed between the pre-mRNA and small ribonucleoprotein particles, in which mRNA splicing takes place.

spongocoel The central cavity in a sponge.

spontaneous reaction Chemical or physical reaction that occurs without outside help.

sporadic (nonhereditary) cancer Cancer that is not inherited.

sporangium (plural, *sporangia*) A single-celled or multicellular structure in fungi and plants in which spores are produced.

spore A haploid reproductive structure, usually a single cell, that can develop into a new individual without fusing with another cell; found in plants, fungi, and certain protists.

sporophyll A specialized leaf that bears sporangia (spore-producing structures).

sporophyte An individual of the diploid generation produced through fertilization in organisms that undergo alternation of generations; it produces haploid spores.

sporopollenin A tough polymer in the walls of spores and pollen grains, the presence of which helps such structures resist decay.

spring overturn The mixing of surface water with deep water in a lake or pond, causing oxygen at the surface to move to the bottom, and nutrients from the bottom to move to the surface.

SRP (signal recognition particle) receptor A protein on the membrane of the endoplasmic reticulum that binds the signal recognition particle.

stability The ability of a community to maintain its species composition and relative abundances when environmental disturbances eliminate some species from the community.

stabilizing selection A type of natural selection in which individuals expressing intermediate phenotypes have the highest relative fitness.

stamen A "male" reproductive organ in flowers, consisting of an anther (pollen producer) and a slender filament.

standing crop biomass The total dry weight of plants present in an ecosystem at a given time.

stapes The smallest of three sound-conducting bones in the middle ear of tetrapod vertebrates.

starch A storage polysaccharide in plants consisting of branched or unbranched chains of glucose subunits.

start codon The first codon read in an mRNA in translation—AUG. Also referred to as the *initiator codon.*

statocyst A mechanoreceptor in invertebrates that senses gravity and motion using statoliths.

statolith A movable starch- or carbonate-containing stonelike body involved in sensing gravitational pull.

stele The central core of vascular tissue in roots and shoots of vascular plants; it consists of the xylem and phloem together with supporting tissues.

stem cells Cells capable of undergoing many divisions in an unspecialized, undifferentiated state and that also have the ability to differentiate into specialized cell types.

stereocilia Microvilli covering the surface of hair cells clustered in the base of neuromasts.

stereoisomers Molecules that are mirror images of one another are an example of stereoisomers.

steroid A type of lipid derived from cholesterol.

steroid hormone receptor Internal receptor that turns on specific genes when it is activated by binding a signal molecule.

steroid hormone response element The DNA sequence to which the hormone-receptor complex binds.

sterol Steroid with a single polar —OH group linked to one end of the ring framework and a complex, nonpolar hydrocarbon chain at the other end.

sticky end End of a DNA fragment generated by digestion with a restriction enzyme, with a single-stranded structure that can form hydrogen bonds with a complementary sticky end on any other DNA molecule cut with the same enzyme.

stigma The receptive end of a carpel where deposited pollen germinates.

stimulus A component of a negative feedback control system maintaining homeostasis, specifically an environmental change that triggers a response.

stoma (plural, *stomata*) The opening between a pair of guard cells in the epidermis of a plant leaf or stem, through which gases and water vapor pass.

stomach The portion of the digestive system in which food is stored and digestion begins.

stomach ulcer *See* **peptic ulcer.**

stop codon A codon that does not specify amino acids. The three nonsense codons are UAG, UAA, and UGA. Also referred to as the *nonsense codon* and *termination codon.*

Stramenopila (stramenopiles) A subgroup of the Chromalveolata protist evolutionary group; protists with two different flagella, one with hollow tripartite projections, and one that is plain.

stratification Horizontal layering of sedimentary rocks beneath the soil surface.

strict aerobe Cell with an absolute requirement for oxygen to survive, unable to live solely by fermentations.

strict anaerobe Organism in which fermentation is the only source of ATP.

strobilus *See* **cone.**

structural gene Gene that encodes a protein that has a function other than gene regulation.

structural isomers Two molecules with the same chemical formula but atoms that are arranged in different ways.

style The slender stalk of a carpel situated between the ovary and the stigma in plants.

submucosa A thick layer of elastic connective tissue that contains neuron networks and blood and lymph vessels.

subsoil The region of soil beneath topsoil, which contains relatively little organic matter.

subspecies A taxonomic subdivision of a species.

substrate The particular reacting molecule or molecular group that an enzyme catalyzes.

substrate-level phosphorylation An enzyme-catalyzed reaction that transfers a phosphate group from a substrate to ADP.

sugar–phosphate backbone Structure in a polynucleotide chain that is formed when deoxyribose sugars (in DNA) or ribose sugars (in RNA) are linked by phosphate groups in an alternating sugar–phosphate–sugar–phosphate pattern.

sulfhydryl group Group that works as a molecular fastener, consisting of a sulfur atom linked on one side to a hydrogen atom and on the other side to a carbon chain.

sum rule Mathematical rule in which final probability is found by summing individual probabilities.

surface tension The force that places surface water molecules under tension, making them more resistant to separation than the underlying water molecules.

survivorship curve Graphic display of the rate of survival of individuals over a species' life span.

suspension feeder An animal that ingests small food items suspended in water.

suspensor In seed plants, a stalklike row of cells that develops from a zygote and helps position the embryo close to the nourishing endosperm.

swallowing reflex The involuntary action produced by contractions of muscles in the walls of the pharynx that direct food into the esophagus.

swim bladder A gas-filled internal organ that helps bony fishes maintain buoyancy.

symbiosis An interspecific interaction in which the ecological relations of two or more species are intimately tied together.

symmetry Exact correspondence of form and constituent configuration on opposite sides of a dividing line or plane.

sympathetic division Division of the autonomic nervous system that predominates in situations involving stress, danger, excitement, or strenuous physical activity.

sympatric speciation Speciation that occurs without the geographical isolation of populations.

symplastic pathway The route taken by water that moves through the cytoplasm of plant cells (the symplast). *Compare* **apoplastic pathway.**

symport The transport of two molecules in the same direction across a membrane. Also referred to as *cotransport.*

synapomorphy A derived character state found in two or more species.

synapse A site where a neuron makes a communicating connection with another neuron or an effector such as a muscle fiber or gland.

synapsida A lineage of amniotes that have one bony temporal arch near the back of the skull; includes mammals.

synapsis *See* **pairing.**

synaptic cleft A narrow gap that separates the plasma membranes of the presynaptic and postsynaptic cells.

synaptonemal complex A protein framework that tightly holds together homologous chromosomes as they pair.

systematics The branch of biology that studies the diversity of life and its evolutionary relationships.

systemic acquired resistance A plant defense response to microbial invasion; defensive chemicals including salicylic acid may spread throughout a plant, rendering healthy tissues less vulnerable to infection.

systemic circuit In amphibians, the branch of a double blood circuit that receives oxygenated blood and provides the blood supply for most of the tissues and cells of a body.

systemic homeostatic controls Controls that are initiated outside of an organ or organ system to control that organ's or organ system's activity.

systemic inflammation Inflammation that occurs throughout the body.

systemic lupus erythematosus (lupus) An autoimmune disease caused by production of a wide variety of anti-self antibodies against blood cells, blood platelets, and internal cell structures and molecules such as mitochondria and proteins associated with the DNA in the cell nucleus; characterized by anemia and problems with blood circulation and kidney function.

systemin A plant peptide hormone that functions in defense responses to wounds.

systems biology An area of biology that studies the organism as a whole to unravel the integrated and interacting network of genes, proteins, and biochemical reactions responsible for life.

systole The period of contraction and emptying of the heart.

systolic blood pressure The peak of high blood pressure in the cardiac cycle that occurs as the heart ventricles contract and move blood outward through the arteries.

T (transverse) tubule The tubule that passes in a transverse manner from the sarcolemma across a myofibril of striated muscle.

T cell A lymphocyte produced by the division of stem cells in the bone marrow and then released into the blood and carried to the thymus. T cells participate in adaptive immunity.

tactile signal A means of animal communication in which the signaler uses touch to convey a message to the signal receiver.

taiga *See* **boreal forest.**

tandem duplication Duplicate genes clustered together in the same region of the same chromosome. Unequal crossing-over produces such an arrangement.

taproot system A root system consisting of a single main root from which lateral roots can extend; often stores starch.

target site The location in a genome to which a transposable element moves when it transposes.

taxis A behavioral response that is directed either toward or away from a specific stimulus.

taxon (plural, *taxa*) A name designating a group of organisms included within a category in the Linnaean taxonomic hierarchy.

taxonomic hierarchy A system of classification based on arranging organisms into ever more inclusive categories.

taxonomy The science of the classification of organisms into an ordered system that indicates natural relationships.

T-cell receptor (TCR) A receptor that covers the plasma membrane of a T cell, specific for a particular antigen.

tectorial membrane A membrane that extends the length of the cochlear canal of the inner ear in which the stereocilia of the sensory hair cells of the organ of Corti are embedded.

telomerase An enzyme that adds telomere repeats to chromosome ends.

telomeres Repeats of simple-sequence DNA that maintain the ends of linear chromosomes.

telophase The final phase of mitosis, during which the spindle disassembles, the chromosomes decondense, and the nuclei reform.

temperate bacteriophage Bacteriophage that may enter an inactive phase (lysogenic cycle) in which the host cell replicates and passes on the bacteriophage DNA for generations before the phage becomes active and kills the host (lytic cycle).

temperate deciduous forest A forested biome found at low to middle altitudes at temperate latitudes, with warm summers, cold winters, and annual precipitation between 75 and 250 cm.

temperate grassland A nonforested biome that stretches across the interiors of most continents, where winters are cold and snowy and summers are warm and fairly dry.

temperate rainforest A coniferous forest biome supported by heavy rain and fog, which grows where winters are mild and wet and the summers are cool.

template A nucleotide chain used in DNA replication for the assembly of a complementary chain.

template strand The DNA strand that is copied into an RNA molecule during gene transcription.

temporal fenestrae Openings near the back of the skull in some amniote vertebrates.

temporal isolation A prezygotic reproductive isolating mechanism in which species live in the same habitat but breed at different times of day or different times of year.

temporal summation The summation of several EPSPs produced by successive firing of a single presynaptic neuron over a short period of time.

tendon A type of fibrous connective tissue that attaches muscles to bones.

terminal bud A bud that develops at the apex of a shoot.

termination In transcription, the step in which transcription ends and the RNA transcript and RNA polymerase (RNA polymerase II in the case of eukaryotes) are released from the DNA template. In translation, the step in which the translation complex disassembles after the last amino acid of the polypeptide specified by the mRNA has been added to the polypeptide.

termination codon *See* **stop codon.**

termination factor *See* **release factor.**

terminator Specific DNA sequence for a gene that signals the end of transcription of a gene. Terminators are common for prokaryotic genes.

territory A plot of habitat, defended by an individual male or a breeding pair of animals, within which the territory holders have exclusive access to food and other necessary resources.

tertiary consumer A carnivore that feeds on other carnivores, a member of the fourth trophic level.

tertiary structure The overall three-dimensional folding of a polypeptide chain.

testcross A genetic cross between an individual with the dominant phenotype and a homozygous recessive individual.

testis (plural, *testes*) The male gonad. In male vertebrates, they secrete androgens and steroid hormones that stimulate and control the development and maintenance of male reproductive systems.

testosterone A hormone produced by the testes, responsible for the development of male secondary sex characteristics and the functioning of the male reproductive organs.

tetanus A situation in which a muscle fiber cannot relax at all between stimuli, and twitch summation produces a peak level of continuous contraction.

tetrad Homologous pair of eukaryotic chromosomes during the first meiotic division consisting of four chromatids.

Tetrapoda A monophyletic lineage of vertebrates that includes animals having four feet, legs, or leglike appendages.

T-even bacteriophage Virulent bacteriophages, T2, T4, and T6, that have been valuable for genetic studies of bacteriophage structure and function.

thalamus A major switchboard of the brain that receives sensory information and relays it to the regions of the cerebral cortex concerned with motor responses to sensory information of that type.

thallus A plant body not differentiated into stems, roots, or leaves.

Thaumarchaeota A group of the domain Archaea consisting of four living species, all of which are mesophilic, chemoautotraphic ammonia oxidizers, and other members represented by DNA sequences in environmental samples.

thermal acclimatization A set of physiological changes in ectotherms in response to seasonal shifts in environmental temperature, allowing the animals to attain good physiological performance at both winter and summer temperatures.

thermocline The narrow depth range in a lake where water temperature changes abruptly at the boundary between the epilimnion and the hypolimnion.

thermodynamics The study of the energy flow during chemical and physical reactions.

thermoreceptor A sensory receptor that detects the flow of heat energy.

thermoregulation The control of body temperature.

thick filament A type of filament in striated muscle composed of myosin molecules; they interact with thin filaments to shorten muscle fibers during contraction.

thigmomorphogenesis A plant response to a mechanical disturbance, such as frequent strong winds; includes inhibition of cellular elongation and production of thick-walled supportive tissue.

thigmotropism Growth in response to contact with a solid object.

thin filament A type of filament in striated muscle composed of actin, tropomyosin, and troponin molecules; they interact with thick filaments to shorten muscle fibers during contraction.

thorax The central part of an animal's body, between the head and the abdomen.

thorn forest A forested biome that grows at the arid borders of true savanna, where large mammals are less abundant.

threshold potential In signal conduction by neurons, the membrane potential at which the action potential fires.

thymine A pyrimidine that base-pairs with adenine.

thymus An organ of the lymphatic system that plays a role in filtering viruses, bacteria, damaged cells, and cellular debris from the lymph and bloodstream, and in defending the body against infection and cancer.

thyroid gland A gland located beneath the voice box (larynx) that secretes hormones regulating growth and metabolism.

thyroid-stimulating hormone (TSH) A hormone that stimulates the thyroid gland to grow in size and secrete thyroid hormones.

thyroxine (T_4) The main hormone of the thyroid gland, responsible for controlling the rate of metabolism in the body.

Ti (tumor-inducing) plasmid A plasmid used to make transgenic plants.

tidal volume The volume of air entering and leaving the lungs during inhalation and exhalation.

tight junction Region of tight connection between membranes of adjacent cells.

time lag The delayed response of organisms to changes in environmental conditions.

tissue A group of cells and intercellular substances with the same structure that function as a unit to carry out one or more specialized tasks.

tolerance hypothesis Hypothesis asserting that ecological succession proceeds because competitively superior species replace competitively inferior ones.

tonicity The effect a solution has on a cell when the solution surrounds it.

tonoplast The membrane that surrounds the central vacuole in a plant cell.

topoisomerase An enzyme that relieves the overtwisting and strain of DNA ahead of the replication fork.

topsoil The rich upper layer of soil where most plant roots are located; it generally consists of sand, clay particles, and humus.

torpor A sleeplike state produced when a lowered set point greatly reduces the energy required to maintain body temperature, accompanied by reductions in metabolic, nervous, and physical activity.

torsion The realignment of body parts in gastropod mollusks that is independent of shell coiling.

totipotent Having the capacity to produce cells that can develop into or generate a new organism or body part.

trace element An element that occurs in organisms in very small quantities (less than 0.01%); in nutrition, a mineral required by organisms only in small amounts.

tracer Isotope used to label molecules so that they can be tracked as they pass through biochemical reactions.

trachea In insects, an extensively branched, air-conducting tube formed by invagination of the outer epidermis of the animal, and reinforced by rings of chitin. In vertebrates, the windpipe, which branches into the bronchi.

tracheal system A branching network of tubes that carries air from small openings in the exoskeleton of an insect to tissues throughout its body.

tracheid A conducting cell of xylem, usually elongated and tapered.

tracheophyte A plant with xylem, phloem, and usually well-developed roots, stems, and leaves.

traditional systematics An approach to systematics that uses phenotypic similarities and differences to infer evolutionary relationships, grouping together species that share both ancestral and derived characters.

trait One of the forms of a genetic character.

transcription The mechanism by which the information encoded in DNA is made into a complementary RNA copy.

transcription factor (TF) In eukaryotes, the proteins required for RNA polymerase to initiate transcription or that regulate that process. One class of transcription factors recognizes and binds to the promoter in the area of the TATA box and then recruit RNA polymerase.

transcription initiation complex Combination of general transcription factors with RNA polymerase II.

transcription unit A region of DNA that transcribes a single primary transcript.

transcriptional regulation The fundamental level of control of gene expression that determines which genes are transcribed into mRNA.

transduction In cell signaling, the process of changing a signal into the form necessary to cause the cellular response. In prokaryotes, the process in which DNA is transferred from donor to recipient bacterial cells by an infecting bacteriophage.

transfer cell Any of the specialized cells that form when large amounts of solutes must be loaded or unloaded into the phloem; they facilitate the short-distance transport of organic solutes from the apoplast into the symplast.

transfer RNA (tRNA) The RNA that brings amino acids to the ribosome for addition to the polypeptide chain.

transformation The conversion of the hereditary type of a cell by the uptake of DNA released by the breakdown of another cell.

transgene A gene introduced into an organism by genetic manipulation to alter its genotype.

transgenic An organism that has been modified to contain genetic information from an external source.

transition state An intermediate arrangement of atoms and bonds that both the reactants and the products of a reaction can assume.

transit sequence Short amino acid sequence on a protein that serves to direct the protein to the appropriate organelle (other than the nucleus or ER) in a eukaryotic cell.

translation The use of the information encoded in mRNA to assemble amino acids into a polypeptide.

translocation In genetics, a chromosomal alteration that occurs if a broken segment is attached to a different, nonhomologous chromosome. In vascular plants, the long-distance transport of substances by xylem and phloem.

transmembrane pathway The path followed by water when it enters root cells by crossing the cells' plasma membranes.

transmission The second component of neural signaling involving the sending of a message along a neuron, and then to another neuron or to a muscle or gland.

transpiration The evaporation of water from a plant, principally from the leaves.

transport The controlled movement of ions and molecules from one side of a membrane to the other.

transport epithelium A layer of cells with specialized transport proteins in their plasma membranes.

transport protein A protein embedded in the cell membrane that forms a channel allowing selected polar molecules and ions to pass across the membrane.

transposable element (TE) A sequence of DNA that can move from one place to another within the genome of a cell.

transposase An enzyme that catalyzes some of the reactions inserting or removing the transposable element from the DNA.

transposition The movement of a transposable element from one site to another in a genome.

transposon A bacterial transposable element with an inverted repeat sequence at each end enclosing a central region with one or more genes.

tricarboxylic acid cycle *See* **citric acid cycle.**

trichocyst A dartlike protein thread that can be discharged from a surface organelle for defense or to capture prey.

trichome A single-celled or multicellular outgrowth of the plant epidermis; examples include root hairs and leaf trichomes that resemble hairs.

triglyceride A nonpolar compound produced when a fatty acid binds by a dehydration synthesis reaction at each of glycerol's three —OH-bearing sites.

triiodothyronine (T_3) A hormone secreted by the thyroid gland that regulates metabolism.

trimester A division of human gestation, three months in length.

triploblastic An animal body plan in which adult structures arise from three primary germ layers, endoderm, mesoderm, and ectoderm.

trochophore The small, free-swimming, ciliated aquatic larva of various invertebrates, including certain mollusks and annelids.

trophic cascade The effects of predator–prey interactions that reverberate through other population interactions at two or more trophic levels in an ecosystem.

trophic level A position in a food chain or web that defines the feeding habits of organisms.

trophoblast The outer single layer of cells of the blastocyst.

tropical deciduous forest A tropical forest biome that occurs where winter drought reduces photosynthesis, and most trees drop their leaves seasonally.

tropical forest Any forest that grows between the Tropics of Capricorn and Cancer, a region characterized by high temperature and rainfall and thin, nutrient-poor topsoil.

tropical montane forest A tropical forest biome of short trees, which are frequently enveloped in mist; also known as a "cloud forest."

tropical rainforest A dense tropical forest biome that grows where some rain falls every month, mean annual rainfall exceeds 250 cm, mean annual temperature is at least 25°C, and humidity is above 80%.

tropics The latitudes between 23.5° N and 23.5° S, the Tropics of Cancer and Capricorn.

tropism The turning or bending of an organism or one of its parts toward or away from an external stimulus, such as light, heat, or gravity.

true-breeding Individual that passes traits without change from one generation to the next.

trypsin A digestive enzyme that digests peptides to produce small peptides.

tubular reabsorption The selective process in which some molecules and ions are transported by the transport epithelium from the lumen of the tubule back into the extracellular fluid and eventually into the blood as the solution produced by filtration moves through an excretory tubule.

tubular secretion The selective process in which specific small molecules and ions are transported from the extracellular fluid and blood into an excretory tubule.

tumor Tissue mass that results with differentiated cells of complex multicellular organisms deviate from their normal genetic program and being to grow and divide.

tumor-suppressor gene A gene that encodes proteins that inhibit cell division.

turgor pressure The internal hydrostatic pressure within plant cells.

turnover rate The rate at which one generation of producers in an ecosystem is replaced by the next.

tympanic membrane Also called the *eardrum,* a sheet of tissue forming the boundary between the outer ear and the middle ear that vibrates in response to sound waves that move through the auditory canal.

tympanum A thin membrane in the auditory canal that vibrates back and forth when struck by sound waves.

umbilical cord A long tissue with blood vessels linking the embryo and the placenta.

umbilicus Navel; the scar left when the short length of umbilical cord still attached to the infant after birth dries and shrivels within a few days.

undernutrition A form of malnutrition. A condition in animals in which intake of organic fuels is inadequate, or whose assimilation of such fuels is abnormal.

undulating membrane In parabasalid protists, a finlike structure formed by a flagellum buried in a fold of the cytoplasm that facilitates movement through thick and viscous fluids; an expansion of the plasma membrane in some flagellates that is usually associated with a flagellum.

unequal crossing-over The rare phenomenon in meiosis in which, instead of crossing-over occurring at the exact same point on each homolog of a homologous pair of chromosomes, crossing-over occurs at different points.

unicellular organism Individual consisting of a single cell.

uniform dispersion A pattern of distribution in which the individuals in a population are evenly spaced in their habitat.

uniformitarianism The concept that the geological processes that sculpted Earth's surface over long periods of time—such as volcanic eruptions, earthquakes, erosion, and the formation and movement of glaciers—are exactly the same as the processes observed today.

Unikonta (unikonts) Eukaryotes with a single flagellum.

uniparental inheritance A pattern of inheritance in which all progeny (both males and females) have the phenotype of only one of the parents.

universal A feature of the nucleic acid code, with the same codons specifying the same amino acids in all living organisms.

unreduced gamete A gamete that contains the same number of chromosomes as a somatic cell.

unsaturated fatty acid Fatty acid with one or more double bonds linking the carbons.

ureter The tube through which urine flows from the renal pelvis to the urinary bladder.

urethra The tube through which urine leaves the bladder. In most animals, the urethra opens to the outside.

urinary bladder A storage sac located outside the kidneys.

uterine cycle The menstrual cycle.

uterus A specialized saclike organ, in which the embryo develops in viviparous animals.

utricle Saclike, fluid-filled structure that is part of the vestibular apparatus. With the saccule, provides information about the up–down position of the head, as well as changes in the rate of linear movement of the head.

vaccination The process of administering a weakened form of a disease to patients as a means of giving them immunity to a more serious form of the disease.

vagina The muscular canal that leads from the cervix to the exterior.

valence electron An electron in the outermost energy level of an atom.

Van der Waals forces Weak molecular attractions over short distances.

variable An environmental factor that may differ among places or an organismal characteristic that may differ among individuals.

variable (V) region For the light and heavy polypeptides of a particular type of antibody, the regions that have different amino acids sequences from molecule to molecule.

vas deferens The tube through which sperm travel from the epididymis to the urethra in the male reproductive system.

vascular bundle A cord of plant vascular tissue; often multistranded with both xylem and phloem.

vascular cambium A lateral meristem that produces secondary vascular tissues in plants.

vascular plant *See* **tracheophyte.**

vascular tissue system One of the three tissue systems in plants that provide the foundation for plant organs; it consists of transport tubes for water and nutrients.

vegetal pole The end of the egg opposite the animal pole, which typically gives rise to internal structures such as the gut and the posterior end of the embryo.

vegetative reproduction Asexual reproduction in plants by which new individuals arise (or are created) without seeds or spores; examples include fragmentation from the parent plant or the use of cuttings by gardeners.

vein In a plant, a vascular bundle that forms part of the branching network of conducting and supporting tissues in a leaf or other expanded plant organ. In an animal, a vessel that carries the blood back to the heart.

ventilation The flow of the respiratory medium (air or water, depending on the animal) over the respiratory surface.

ventral Indicating the lower or "belly" side of an animal.

ventricle In the brain, an irregularly shaped cavity containing cerebrospinal fluid. In the heart, a chamber that pumps blood out of the heart.

venule A capillary that merges into the small veins leaving an organ.

vernalization The stimulation of flowering by a period of low temperature.

vertebrae The series of bones that form the vertebral column of vertebrate animals.

vertebral column The series of vertebrae that surrounds and protects the dorsal nerve cord and forms the supporting axis of the body.

vertebrate A member of the monophyletic group of tetrapod animals that possess a vertebral column.

vesicle A small, membrane-bound compartment that transfers substances between parts of the endomembrane system.

vessel In plants, one of the tubular conducting structures of xylem, typically several centimeters long; most angiosperms and some other vascular plants have xylem vessels.

vessel member Any of the short cells joined end to end in tubelike columns in xylem.

vestibular apparatus The specialized sensory structure of the inner ear of most terrestrial vertebrates that is responsible for detecting rotational movement of the head and provides information about the up–down positioning of the head as well as changes in the rate of linear movement of the head. The vestibular apparatus consists of three semicircular canals, the utricle, and the saccule.

vestigial structure An anatomical feature of living organisms that no longer retains its function.

vibrio Any of various short, motile, S-shaped or comma-shaped bacteria of the genus *Vibrio.*

vicariance The fragmentation of a continuous geographical distribution by nonbiological factors.

virion A complete virus particle.

viroid A plant pathogen that consists of strands or circles of RNA, smaller than any viral DNA or RNA molecule, that have no protein coat.

virulent bacteriophage Bacteriophage that kills its host bacterial cells during each cycle of infection.

virus An infectious agent that contains either DNA or RNA surrounded by a protein coat.

visceral mass In mollusks, the region of the body containing the internal organs.

visual signal A means of communication in which animals use facial expressions or body language to send messages to other individuals.

vital capacity The maximum tidal volume of air that an individual can inhale and exhale.

vitamin An organic molecule required in small quantities that the animal cannot synthesize for itself.

vitamin D A steroidlike molecule that increases the absorption of Ca^{2+} and phosphates from ingested food by promoting the synthesis of a calcium-binding protein in the intestine; it also increases the release of Ca^{2+} from bone in response to PTH.

vitelline coat A gel-like matrix of proteins, glycoproteins, or polysaccharides immediately outside the plasma membrane of an egg cell.

vitreous humor The jellylike substance that fills the main chamber of the eye, between the lens and the retina.

viviparous Referring to animals that retain the embryo within the mother's body and nourish it during at least early embryo development.

voltage-gated ion channel A membrane-embedded protein that opens and closes as the membrane potential changes.

vulva The external female sex organs.

water lattice An arrangement formed when a water molecule in liquid water establishes an average of 3.4 hydrogen bonds with its neighbors.

water molds *See* **Oomycota.**

water potential The potential energy of water, representing the difference in free energy between pure water and water in cells and solutions; it is the driving force for osmosis.

watershed An area of land from which precipitation drains into a single stream or river.

water-soluble vitamin A vitamin with a high proportion of oxygen and nitrogen able to form hydrogen bonds with water.

wax A substance insoluble in water that is formed when fatty acids combine with long-chain alcohols or hydrocarbon structures.

weight A measure of the pull of gravity on an object.

wetland A highly productive ecotone often at the border between a freshwater biome and a terrestrial biome.

white blood cell *See* **leukocyte.**

white matter The myelinated axons that surround the gray matter of the central nervous system.

white rusts *See* **Oomycota.**

wilting The drooping of leaves and stems caused by a loss of turgor.

wobble hypothesis Hypothesis stating that the complete set of 61 sense codons can be read by fewer than 61 distinct tRNAs because of particular pairing properties of the bases in the anticodons.

Wolffian duct A bipotential primitive duct associated with the gonads that leads to a cloaca.

wood The secondary xylem of trees and shrubs, lying under the bark and consisting largely of cellulose and lignin.

X chromosome Sex chromosome that occurs paired in female cells and single in male cells.

X-linked gene A gene on the X chromosome.

X-linked inheritance The pattern of inheritance of an X-linked gene.

X-linked recessive inheritance Pattern in which displayed traits are due to inheritance of recessive alleles carried on the X chromosome.

X-ray diffraction Method for deducing the position of atoms in a molecule.

xylem The plant vascular tissue that distributes water and nutrients.

xylem sap The dilute solution of water and solutes that flows in the xylem.

Y chromosome Sex chromosome that is paired with an X chromosome in male cells.

yeast A single-celled fungus that reproduces by budding or fission.

yolk The portion of an egg that serves as the main energy source for the embryo.

yolk sac In an amniote egg, an extraembryonic membrane that encloses the yolk.

zero population growth A circumstance in which the birth rate of a population equals the death rate.

zona pellucida A gel-like matrix of proteins, glycoproteins, or polysaccharides immediately outside the plasma membrane of the egg cell.

zone of cell division The region in a growing root that consists of the root apical meristem and the actively dividing cells behind it.

zone of elongation The region in a root where newly formed cells grow and elongate.

zone of maturation The region in a root above the zone of elongation where cells do not increase in length but may differentiate further and take on specialized roles.

zooplankton Small, usually microscopic, animals that float in aquatic habitats.

zygospore A multinucleate, thick-walled sexual spore formed in zygomycete fungi (Zygomycota).

zygote A fertilized egg.

Index

The letter i designates illustrations; t designates tables; **bold** designates defined or introduced term.

A band, 967i, 968, 969i
A horizon, 799, 799i
A site (aminoacyl site), **324**, 324i–326i, 325, 327, 328i–329i
AAS. *See* Anabolic-androgenic steroid
ABA. *See* Abscisic acid
Abdomen, 695, 697, 697i, 698i
Abiotic, **1146**
ABO blood group, genetics of, 254–255, 254t, 255i
Abomasum, 1063i, 1065
Aboral end, 679
Abrahamson, Warren G., 468, 469i
Abscisic acid (ABA), 785, 821, 840t, **848**–849, 850i–851i
Abscission, **846**–847
Abscission zone, 847, 847i
Absorption
by fungi, 652
of nutrients, **1047**, 1055
Absorption spectrum, **187**, 188i
Abyssal zone, **1165**, 1165i, 1167
Acacia *(Acacia cornigera),* 1207, 1208i
Acanthamoeba, 603
Acanthodii, 717
Accessory fruit, 820, 821i
Accessory glands, human male, 1109
Accessory pigment, 188
Acclimatization, **1037**, 1088
Accommodation, **930**, 931, 931i
ACE. *See* Angiotensin converting enzyme
Acetabularia, 601i
Acetaldehyde, 44t
Acetic acid, 44t
Acetone, 44t
Acetyl-coenzyme A (acetyl-CoA), 168, 168i–171i, 171
Acetylation of histones, 351
Acetylcholine, 887i, 896, 898t, 970i
Achene, 820
Achondroplasia, 251, 251i, 277, 278
Acid, **36**
Acid-growth hypothesis, **844**
Acid precipitation, **1259**, 1260i
Acid rain, 37, 793, 801
Acid reflux, 1053
Acidity, **36**–37
Acoelomate animal, **673**, 673i, 684
Aconitase, 170i
Acorn worm, 709, 710i
Acoustical signal, **1293**
Acquired characteristics, inheritance of, 442
Acquired immune deficiency syndrome (AIDS), 384, 1002, 1019, 1020
Acquired immunity. *See* Adaptive (acquired) immunity
Acrosome, **1099**, 1099i, 1102, 1102i
Acrosome reaction, **1102**, 1102i, 1110, 1114
ACTH. *See* Adrenocorticotropic hormone
Actin, **874**, 967–968, 967i, 969i, 970i
Actin, in cytoskeleton, 107, 108i, 109
Actinopterygii, 719, 719i, 720i, 721
Action potential
all-or-nothing principle, 889
defined, **857, 888**
frequency, 893
magnitude, 893
muscle fiber, 970i
nastic sleep movement, 857
production, 890–892, 890i–891i
propagation, 892–893, 892i, 894i
saltatory conduction, 893, 894i
sensory system and, 922, 922i, 923
Action spectrum, 182i, 183, **187**, 188i
Activation energy
defined, **79**
physical example of, 78i, 79
reduction by enzymes, 79, 79i, 80–81
Activation gate, 890, 890i, 891i
Activator, **341**, 343i, **344**, 347–349, 348i, 349i
Active immunity, **1014**
Active parental care, **1179**
Active site, **80**, 80i, 82, 82i
Active transport
antiport, 776, 776i
defined, **126, 131**
in phloem, 787
in plants, 776–777, 776i, 780–781, 784, 787, 788i
primary, **131**–132, 132i, 133i
into proximal convoluted tubule, 1077
in roots, 780–781
secondary, **131**, 133–134, 134i, 776
symport, 776, 776i, 784, 787
Acyclovir, 383
ADA. *See* Adenosine deaminase deficiency
Adaptation
angiosperm, 639, 641
constraints on adaptive evolution, 475
convergent evolution, 510–511, 511i
cryptic coloration example, 7–8, 8i
defined, **7, 475**
by natural selection, 448–449, 454
Adaptive (acquired) immunity, 1007–1017
antibody-mediated immunity, 1007, 1008–1015, 1012i–1014i
antibody use in research, 1015
cell-mediated immunity, 1007, 1015–1017, 1017i
components of, 1015–1016
defined, **1003**
HIV evasion of, 1021
steps in response, 1008
Adaptive radiation, 488, **512**–514, 512i
Adaptive trait
defined, **445, 475**
evolution of, 475
natural selection favoring of, 445–446
Adaptive zone, **512**
Adductor muscle, bivalve, **689,** 689i
Adelgids, 1261, 1262i
Adenine, 63, 64i, 66i, **290**, 290i
Adenosine deaminase deficiency (ADA), 406–407
Adenosine diphosphate (ADP), 64
Adenosine monophosphate (AMP), 64
Adenosine triphosphate (ATP)
in active transport, 131, 132i, 133i, 776
in coupled reactions, 76–77, 77i, 78i
defined, **76**
formation/synthesis
chemiosmosis and, 172i, 175
in citric acid cycle, 4–165, 164i, 168–171, 168i, 170i, 171i, 175i
from electron transfer system, 164i, 165, 171, 172i, 173, 173i, 175–176
in glycolysis, 164, 164i, 165–166, 167i, 168, 175–176, 175i
by oxidative phosphorylation, 164i, 165, 172i, 175i, 176
by photophosphorylation, 191
in photosynthesis, 183i, 184, 185, 190i, 191–192, 192i
by substrate-level phosphorylation, 164, 164i, 165i, 166, 167i, 169, 170i, 175i, 176
generation of, 78
hydrolysis of, 76–78, 77i, 131, 132i, 133i
in muscle contraction process, 970i, 971–972
structure of, 64, 76, 77i
use in Calvin cycle, 193–194, 195i, 196–197
Adenovirus, 378i, 382
Adenylyl cyclase, 151, 151i
ADH. *See* Antidiuretic hormone
Adherens junction, **114**, 114i
Adhesion, **34**
Adhesive property of water, 781
Adiabatic cooling, **1150**
Adipocyte, **872**
Adipose tissue, **872**–873, 973i, 1065, 1089
ADP. *See* Adenosine diphosphate
Adrenal cortex, 950i, 951t, **956**, 957–958, 958i
Adrenal medulla, 950i, 951t, **956**, 957
Adrenergic receptor, **948**, 949i, 995
Adrenocorticotropic hormone (ACTH), 951t, **952**, 953i, 958
Adsorption, 800
Adventitious root, **765**, 765i
Aequorin, 153
Aerobe, **178, 573**
Aerobic respiration, **162**
Afferent arteriole, 1075i, **1076**, 1080
Afferent neuron, **884**, 885
Aflatoxin, 659
African buffalo *(Syncerus caffer),* 1206i
African cichlid fish *(Haplochromis burtoni),* 1286–1287, 1288i
African lungfish *(Neoceratodus forsteri),* 721i
African pancake tortoises *(Malacochersus tornieri),* 1202
African sleeping sickness, 592, 592i
Agama stellio, 1086, 1086i
Agar, 371, **601**
Agaricus arvensis, 662i
Agarose gel electrophoresis, **399**, 400i
Age-specific fecundity, **1177**, 1177t
Age-specific mortality, **1176**–1177, 1177t
Age-specific survivorship, **1177**, 1177t
Age structure, **1175**, 1193–1194, 1194i
Agglutination, **1011**, 1014i
Aggregate fruit, **820**, 821i
Agnathan, 713, 714i, 715–716, 716i
conodonts and ostracoderms, 716, 716i
hagfishes and lampreys, 715–716, 715i
Agonist, **976**
Agrawal, Anurag, 1225
Agre, Peter, 127, 138
Agriculture, benefits of biodiversity to, 1263–1264
Agrobacterium, 409, 409i, 577–578
AIDS. *See* Acquired immune deficiency syndrome
Air, as respiratory medium, 1026–1027
Air circulation, 1148–1149, 1149i
Air control system, 141
Air pollution, lichens as monitors of, 664
Air sacs, 1029, 1030i
Alanine, 44t
Albatross, 732i
Albinism, 251, 251i
Albumin, 60, **724**, **986**
Alcohol, 44t
Alcoholic fermentation, **177**, 178i
Aldehyde, 44t
Aldolase, 167i
Aldose reductase, 1078–1079
Aldosterone, 945, 947–948, 948i, 951t, **958**, **1081**, 1081i, i
Aleurone, **822**, 822i
Alfalfa *(Medicago),* 760i
Algae, 92i
brown, 588i, 598, 598i, 599i
golden, 597–598, 598i
green, 182–183, 182i, 601, 601i, 602i
meiosis, time and place of, 234i, 235
photosynthesis, 182–183, 182i, 184
red, 600–601, 600i
Algin, **598**
Alimentary canal, 1047
Alkaline solution, 37
Alkaloid, 853
Alkaptonuria, 311
All-or-nothing principle, **889**
Allantois, **1125**, 1126i, 1129, 1131
Allard, Henry, 858
Allele, **226, 243**, 249–251, 250i, 251i
imprinted, 351
shuffling of existing alleles, 460
Allele frequency, **460**–461, 461t, 464t, 464–467
Allergen, **1019**–1020, 1020i
Allergy, 1019–1020
Allium, 423
Allolactose, **341**, 342i, 343i
Allometric growth, 518, 519i, 522
Allopatric speciation, **486**–489, 486i–489i, 491–492, 496
Allopolyploidy, 493i, **494**
Allosteric activator, **82**, 83i
Allosteric inhibitor, **82**, 83i
Allosteric regulation, 81, **82**–83, 83i
Allosteric site, **82**–83, 83i
Alpha cell, 952t, 959
Alpha helix, **58**, 58i, 59i
α-ketoglutarate, 170i
α-ketoglutarate dehydrogenase, 170i
α-melanocyte-stimulating hormone (MSH), 1060
α-tropomyosin, 320, 321i
Alpha wave, 916, 917i
Alphaproteobacteria, 577–578
Alpine chipmunk *(Tamius alpinus),* 1154i
Alpine tundra, **1161**, 1161i
Alternation of generations, 234, **615**, 615i, 812i, **813**
Alternative hypothesis, **15**
Alternative splicing, **320**–321, 320i, 352
Altman, Sidney, 85
Altruism
defined, **1300**
genetic basis of, 1302, 1303i
kin selection, 1300–1302, 1301i
reciprocal, 1302
Alveolata, **592**–595, 606i
apicomplexa, 594–595, 595i
ciliophora, 591i, 592–594
dinoflagellates, 594, 594i
Alveoli, **1029**, 1030, 1031i
Alzheimer disease, 179
Amacrine cell, **932**, 933i
Amanita muscaria, 661–662, 661i
Amanita phalloides, 662
Amanitin, 662
Amantadine, 383
Amazon basin, deforestation in, 1255i
Amber, 501, 501i
Amborella trichopoda, 639, 639i, 755
Ambystoma talpoideum, 519i
American alligator *(Alligator mississippiensis),* 730, 730i
American black bear, 9, 9i
AMH. *See* Anti-Müllerian hormone
Amine, 44t
biogenic, 898t
hormone, 945
as neurotransmitter, 898t

Amino acid
  chain formation, 57, 57i
  defined, **55**
  essential, 1044, 1048, 1049i
  formation in Miller-Urey apparatus, 555, 556i
  genetic code and, 314–315
  in meteorites, 556
  as neurotransmitter, 898t
  stereoisomers, 44–45, 45i
  structure of, 55, 56–57, 56i
Amino group, **43,** 44t, 57i
Aminoacyl-tRNA, **322, 323,** 323i
Aminoacyl-tRNA synthetase, **323,** 323i
Aminoacyl site (A site), **324,** 324i–326i, 325, 327, 328i–329i
Aminoacylation, **323,** 323i
Aminopeptidase, 1053i, 1056, **1057,** 1057i
Amish, Old Order, 465, 470
Ammonia, 804–805, 1072, 1072i, 1084
Ammonification, **804,** 804i, **1243,** 1244i, 1245t
Amniocentesis, **279,** 279i, 1131
Amnion, **1125,** 1126i, 1129, 1130i
Amniota, **715**
Amniote
  adaptations to terrestrial life, 724, 724i
  gastrulation in, 1124
  lineages, 724
  phylogeny, 723, 724, 725i, 726
Amniote egg, **724,** 724i
Amniotic cavity, 1125, 1126i, 1129, 1130i
Amniotic fluid, 1125, 1131
Amoeba proteus, 137, 603, 603i
Amoebas, **599,** 602–603, 603i, 655
Amoebic dysentery, 603
Amoeboid motion, 109
Amoebozoa, **602**–603
  amoebas, 599, 602–603, 603i
  cellular slime molds, 603, 604i, 605
  plasmodial slime molds, 603
AMP. *See* Adenosine monophosphate
Amp^R gene, 394, 395i
Amphibian, 722–723, 723i
  chytrid infection, 1262, 1262i, 1274
  circulatory system, 984, 984i
  development, 1120i, 1123, 1124i
  ectothermy, 1086, 1088
  lateral line system, 926
  lungs, 1029, 1029i
  nervous system development, 1126i
  reproduction, 1100–1101, 1101i
  water and salt conservation, 1084–1085
Amplexus, 1100–1101, 1101i
Amplification, **946,** 949
  of DNA by PCR, 398, 398i, 399
  in signal transduction pathway, **147,** 147i
Ampulla
  echinoderm, 707–708, 709i
  of semicircular canals, 925, 925i
Amygdala, 911i, **912,** 916, 1062
Amylase, 1046, 1053, 1053i, 1056
Amylopectin, 49, 249
Amyloplast, **111,** 855–856
Amylose, 49, 49i
Anabaena, 574, 1243
Anabolic-androgenic steroid (AAS), **958,** 959
Anabolic pathway, **76**
Anabolic reaction, **76**
Anabolic steroid, **958,** 959
Anaerobe
  defined, **573**
  facultative, **177**
  strict, **177**
Anaerobic respiration, 162
Anagenesis, 531, 536
Anal sphincter, **1059,** 1059i
Anaphase, of mitosis, **209,** 209i, **210,** 210i, 215i
Anaphase I, of meiosis, 229i, 230i
Anaphase II, of meiosis, 229i
Anaphase-promoting complex (APC), 218
Anaphylactic shock, **1019**–1020
Anapsida, **724**
Anatomy, **868**
Ancestral character state, **538**
Anchiornis huxleyi, 500, 500i
Anchoring junction, **113,** 114i, 868
Anderson, Richard G. W., 136
Anderson, W. French, 406
Andersson, Malte, 470, 471i
Androgen, 951t, **958,** 1108
Anemia, 987
Anesthetic, 893
Aneuploid, **273,** 275
ANF. *See* Atrial natriuretic factor
Anfinsen, Christian, 60, 61i
Angiosperm, 637–643
  adaptive success of, 639, 641
  basal, 639, 639i
  characteristics of, 631t
  clades, 638–639
  coevolution with pollinators, 639, 641, 643i
  grouping by growth type and lifespan, 751–752
  homeosis and origin of, 525
  leaf structure, 763i
  life cycle, 642i, 812–813, 812i, 815i
  origin of, 637–638
  overview, 637
  phylogenetic tree, 638, 638i
  pressure flow mechanism in, 787–789, 788i
  reproduction
    asexual reproduction, 824–826, 824i, 825i
    overview, 812–813, 812i
    sexual reproduction, 812–822, 815i
Angiotensin, 152–153
Angiotensin converting enzyme (ACE), 1081
Angiotensin I, **1081,** 1081i
Angiotensin II, **1081,** 1081i
Anglerfish, 1043, 1043i, 1167i
Angraceum sesquipedale, 641, 643i
Anguillicola crassus, 693i
Anhidrotic ectodermal dysplasia, 271
Animal(s)
  ancestral forms, 672, 672i
  characteristics of, 671
  development in, 355–360, 356i–360i
  diversification of, 671
  diversity, history of, 514i
  ecdysozoan protosomes, 693–702
  eumetazoa with radial symmetry, 679–682
  evolution, innovations in, 672–675
    body cavities, 673
    chordates, 710, 710i
    developmental patterns, 674–675, 674i
    segmentation, 675
    symmetry in body plan, 672–673, 673i
    tissues and tissue layers, 672
  genetic engineering of, 405–408, 406i–408i
  genomes sequenced, 421t
  lophotrochozoan protosomes, 683–692
  meiosis, time and place of, 234, 234i
  nutrition, 1043–1068
    defined, 1043
    requirements, 1044
  organization of body, 868, 869i
  parazoa, 672, 676i, 677–678, 677i
  phylogeny and classification, 675–677, 676i
  tissues, 868–875
    connective tissue, 871–874, 872i
    epithelial tissues, 869–871, 869i, 870i
    junctions, 868
    muscle, 874, 874i
    nervous, 875, 875i
    organs and organ systems, 868, 869i, 875, 876i
    stem cells, 869–870, 871
Animal behavior, 1278–1295
  communication, 1293–1295
    acoustical signals, 1293
    chemical signals, 1293–1294, 1294i
    dance, 1294–1295, 1294i
    electrical signals, 1294
    evolutionary analysis, 1295
    tactile signals, 1294, 1294i
    visual signals, 1293, 1293i
  defined, **1279**
  fixed action patterns, 1280
  genetic basis, 1280, 1282i, 1292
  habitat selection, 1292, 1292i
  hormones and, 1285–1287, 1286i, 1287i, 1288i
  instinctive, 1279–1282, 1280t, 1281i, 1282i
  learned, 1280, 1280t, 1282–1284, 1283i, 1284i
  migration, 1288–1291, 1289i–1291i
  neurophysiological control, 1285–1286
  proximate causes, 1279
  reproduction
    mating systems, 1297
    nesting, 1297–1298, 1298i
    sexual selection, 1296–1297, 1296i, 1297i
  social, 1299–1302
    altruistic behavior, 1300–1302, 1301i, 1308–1310, 1309i
    benefits and costs of group living, 1299, 1299i
    defined, 1289
    dominance hierarchies, 1300
    human, 1304
  stereotyped, 1280–1281, 1281i
  territoriality, 1285, 1293, 1297
  ultimate causes, 1279
Animal cells, 92i, 98i, 113–115, 114i, 115i
Animal clones, 356, 407, 408i
Animal development, 1118–1144
  after birth, 1132–1133
  cellular events, 1133–1141
    cell adhesions, selective, 1121, 1136, 1137i
    cell division orientation, 1133
    cell division rate, 1133
    cell migration, 1135
    cell shape changes, 1135, 1135i
    differentiation, 1138–1139
    induction, 1136, 1138, 1139i
  cleavage, 1119–1121, 1120i, 1122–1124, 1122i
  cytoplasmic determinants, 1119
  fate mapping cell lineages, 1136, 1138i
  gastrulation, 1120i, 1122–1125, 1122i, 1124i
  human, 1118–1119, 1128–1133, 1129i–1133i
  information storage in egg, 1119
  mechanisms, 1119–1121
  organogenesis, 1126–1128, 1126i–1128i
  patterns of cleavage and gastrulation, 1122–1125
    amphibian, 1120i, 1123, 1124i
    bird, 1124–1125, 1125i, 1126i
    sea urchin, 1122–1123, 1122i
  primary tissue layers, 1121, 1121t
Animal pole, **1119,** 1122i, 1123, 1124i, 1133
Animal reproduction, 1096–1117
  asexual, 1097–1098, 1097i
  human, 1104–1112
    contraception, 1112–1114, 1112t
    copulation, 1110
    female sexual organs, 1104, 1104i
    fertilization, 1110–1112, 1110i
    hormonal regulation in males, 1109, 1109i
    male sexual organs, 1108–1109, 1108i
    ovarian cycle, 1105i, 1106–1107, 1106i
    uterine cycle, 1105, 1106i, 1107–1108
  sexual
    cellular mechanisms, 1099–1104
    fertilization, 1100–1103, 1100i, 1102i, 1114
    gametogenesis, 1098i, 1099–1100
    hermaphroditism, 1103–1104, 1103i
    oviparous animals, 1103
    ovoviviparous animals, 1103
    viviparous animals, 1103
Animal viruses, 379t, 382, 382i
Animalia (kingdom), 12, 12i, **670**
Anion, **29,** 800
Ankylosaurus, 726
Annelid, 690–692
  anatomy, 690–691, 691i
  digestive system, 1047, 1047i
  hydrostatic skeleton, 974, 974i
  lineages
    Hirudinea, 691–692, 692i
    Oligochaeta, 691, 692i
    Polychaeta, 691, 692i
  metanephridium, 1073, 1073i
Annotation, 421
Annual, **752**
Annulus, **624,** 625
Anolis, 15–17, 17i, 727, 729, 1153, 1153i
Anopheles mosquito, 528–529, 529i, 594, 595, 702
Ant, 670i, 700i, 748i, 1207, 1208i, 1294, 1294i
Antagonistic pair, **976**–977, 977i
Anteater, 734
Antenna complex, 188–**189,** 188i
Antennae
  defined, **695**
  trilobite, 695
Antennapedia gene, 359, 359i
Anterior end, **673**
Anterior pituitary, **950,** 950i, 951t, 952–953, 953i
Anther, 241i, 242, 813i, **814**
Antheridium
  description, **618,** 618i
  fern, 624i, 625
  liverwort, 619
  moss, 620i, 621
Anthocerophyta, **619**
Anthoceros, 620i
Anthocyanin, 758
Anthophyta, **638**
Anthozoa, 680, 682i
Anthropoidea, 531i, 532i, **737**
Antibiotic resistance, 456–457, 581
Antibiotic resistance genes, 332–333
Antibody
  antigen clearance and, 1011–1014, 1014i
  classes, 1009–1010
  defined, **1007**
  diversity, 1010, 1010i
  monoclonal, 1015, 1016i
  structure, 1009, 1009i
  use in research, 1015
Antibody-mediated immunity
  active immunity, 1014
  antibodies, 1009–1012, 1009i, 1015i
  antigen clearance, 1011–1014, 1014i
  antigen recognition, 1008–1009, 1009i
  defined, **1007**
  drug effects on, 1015
  immunological memory, 1012–1014, 1014i, 1015i
  lymphocyte activation, 1010–1011, 1012i–1013i
  passive immunity, 1014
Anticodon, **322,** 322i, 323, 323i, 325i
Antidiuretic hormone (ADH), 152–153, 951t, 953i, **954, 1082**–1084, 1082i–1083i
Antiestrogen, 954–955
Antifreeze molecules, 851t, 854, 1088, 1152
Antigen, **1007**
Antigen-binding sites, 1008, 1009i

Antigen-presenting cell (APC), **1011,** 1012i, 1016
Antigenic determinant, 1009
Antihistamine, **1019**
Antimicrobial peptide, 1005
Anti-Müllerian hormone (AMH), 1132, 1134
Antiparallel strands, **291**
Antipodal cells, 815i, 816, 816i
Antiport, **133,** 134i, **776,** 776i
Anura, 722–723, 723i
Anus, 1047, 1047i, 1048, 1049i, 1059, 1059i
Aorta, **988,** 988i
Aortic body, **1032,** 1033i
Apatosaurus, 726
APC. *See* Anaphase-promoting complex; Antigen-presenting cell
Apes, 738–739, 738i
APETALA2 gene, 832, 832i–833i
Aphid, 787, 787i
Aphotic zone, **1162,** 1163i
Apical cell, plant, 818, 826, 828i
Apical dominance, **758**–759
Apical meristem
  description, **613, 750,** 751i
  embryo, 819i, 820i
  root, 765–766, 766i, 819i, 820i
  shoot, 757i, 759, 759i, 819, 819i, 820i
Apical surface, **869,** 870i
Apicomplexa, 594–595, 595i, 607
APL gene, 272–273, 273i
Apomixis, **824**
Apomorphy, **539**
Apoplast, 780, 787, 788
Apoplastic pathway, **780,** 780i, 781
Apoptosis, 1002i, **1007, 1121,** 1126, 1139–1140, 1140i
Aposematic coloration, **1202,** 1202i
Appendicular skeleton, **712**
Appendix, **1058,** 1059i
Appetite, 1060–1061, 1065
Apple *(Malus domestica),* 748i
Apple maggot *(Rhagoletis pomonella),* 490, 490i
Applied research, **13**
Aquadro, Chip, 476
Aquaporin, **127,** 128i, 138, **779,** 1077, 1082i–1083i
Aquatic succession, **1219**
Aqueous humor, **930,** 931i
Arabidopsis Information Resource, 827
Arabidopsis thaliana
  bri1 gene, 848, 848i
  development, 826–828, 828i
  floral organ development, 832, 832i–833i
  flower formation, 645
  FT gene, 862
  genome, 643, 644
  genome sequence, 826, 827
  ion transporters in, 802
  miRISC mutants, 360
  as model research organism, 17, 827
  OPT/YSL multigene family, 433–434, 434i
  phytoremediation, 801
  positional information for embryo sac development, 816, 816i
  trichomes, 829
  YODA enzyme, 756–757
Arachnid, 697, 697i
Arbuscule, 658, 665, 665i
Archaea (domain), 582–584
  Bacteria and Eukarya compared, 548t
  characteristics of, 11, 12i, 571, 582, 582t, 584
  Crenarchaeota, 583–584
  defined, **567**
  Euryarchaeota, 582, 583
  extreme habitats of, 582, 582i
  genome size, 423–424, 423i, 423t
  genomes sequenced, 421t
  Korarchaeota, 583, 584
  Methanococcus jannaschii, genome sequence of, 96
  Nanoarchaeota, 583
  origin of, 563
  Thaumarchaeota, 583, 584
  in Tree of Life, 10i, 548, 548i
Archaefructus, 637, 638i
Archaeal chromosome, **220**
Archaeopteris, 632, 632i
Archaeopteryx, 449, 450i, 519–520, 730
Archaeplastida, **600**–601, 607
Archegonium
  description, **618,** 618i
  fern, 624i, 625
  liverwort, 619
  moss, 620i, 621
Archenteron, **674,** 674i, **1122,** 1122i, 1123, 1124i
Archosauromorpha, **724**
Archosaurs, 546, 546i, 724, 726, 728–732
Arctic tern *(Sterna paradisaea),* 1289, 1289i
Arctic tundra, 808, **1161,** 1161i
Ardipithecus, 739–740
Arginine, auxotrophic mutants, 311, 312i–313i
Aristotle, 440
Armadillo, 443, 443i
Armillaria ostoyae, 651, 651i
Arnold, Stevan, 1282
Arousal, sexual, 1109, 1110
Arteriole, **991,** 992, 992i, 993i, 995–996
Artery, **983,** 991–992, 991i, 994, 994i
Arthington, Angela H., 1257
Arthritis, rheumatoid, 1018
Arthrobotrys dactyloides, 659i
Arthropod, 694–702
  anatomy, 694–695
  closed circulatory system of, 983
  exoskeleton, 974–975, 975i
  hydrostatic skeletal elements, 974
  migrating, 1289, 1289i
  nervous systems, 904i, 905
  phylogenetics, 696
  subphyla, 695–702
    Chelicerata, 695–697, 697i
    Crustacea, 697–699, 699i
    Hexapoda, 699–702, 700i, 701i
    Myriapoda, 699, 699i
    Trilobita, 695, 695i
Artificial selection, **6,** 7i, **445,** 459, 459i
Artiodactyla, 734
Asci, 658, 659i, 660i, 661
Ascocarp, 659i, 660i, **661**
Ascomycota (ascomycetes), 655t, 658–661, 659i–660i
Asexual reproduction, **212**
  angiosperm, 824–826, 824i, 825i
  animal, 1097–1098, 1097i
  cloned reproduction, 1097
  defined, **1097**
  fungi, 653, 653i, 657i, 658, 660–661, 660i, 662
  hornworts, 619
  liverworts, 619, 619i
  mosses, 620i, 621
  parthenogenesis, 686
  in prokaryotes, 574
Asexual "species," 496
Asian carp, 1258
Aspartame, 277
Aspergillus, 92i, 659
Assimilation efficiency, 1235
Association areas, brain, 912i, **914**
Asteroid impact, 509, 513
Asteroidea, 708, 708i
Asters, **213,** 213i
Asthma, **1019**
Astrocyte, **885,** 885i
Asymmetric carbon, 44
Asymmetrical body plan, **672**
Atherosclerosis, 52, 54, 136, 994, 994i
Athletic menstrual irregularities, 1112
Atlantic cod *(Gadus morhua),* 1256, 1257i
Atmosphere
  defined, **1146**
  primordial Earth, 555, 556i
  reducing, 555, 556i
Atmospheric cycles, 1241
Atmospheric pressure, 1037–1038
Atom
  chemical bonds, 28–32
  chemical symbols, 24, 24t
  defined, **24**
  electron distribution, 26–28, 27i
  electronegativity of, 30
  structure, 24–28, 24i
Atomic nucleus, **24**–25, 25i
Atomic number, **24**–25, 24t
Atomic symbol, 24, 24t
Atomic weight, **36**
ATP. *See* Adenosine triphosphate
ATP/ADP cycle, **78,** 78i
ATP synthase, **165,** 172i, 174i, 175, 189i, 190i, 191, 201
Atria, **983,** 984i, 988–989, 988i–990i
Atrial natriuretic factor (ANF), **1081**
Atrial siphon, 711, 711i
Atriopore, **711,** 711i
Atrioventricular node (AV node), **989,** 990i
Atrioventricular valves (AV valves), **988**
Atrium, urochordate, **711,** 711i
Auditory tube, 929
AUG codon, 314–315, 315i
Australian saltwater crocodile *(Crocodylus porosus),* 730
Australopithecus, 740–741, 741i
Austrobaileyales, **639**
Autocrine regulation, 142, 143i, 944i, 945
Autoimmune disease, **1017**–1019
Autoimmune reaction, **1017**
Autonomic nervous system, **907,** 907i, 908i
Autophagy, 104
Autophosphorylation, 147, 148, 148i, 153, 154i
Autopolyploidy, **493**–494, 493i
Autoregulation system, 1080
Autosomal dominant inheritance, **277,** 278
Autosomal recessive inheritance, **276**–277
Autosome, **266**
Autotroph, **183, 573,** 589, **1211**
Autumn overturn, 1163i, **1164**
Auxin, 151, 360
  cytokinin-auxin ratio, 846, 846i
  defined, **841**
  discovery, 841–842
  effects, 840t, 842–843, 842i, 855–856, 855i
  mechanism of action, 844, 845i
  positional information for embryo sac development, 816, 816i
  signal transduction, 843–844, 844i
  transport, 843, 843i
Auxotroph, **311,** 312i–313i, 371, 375, 376i
AV node. *See* Atrioventricular node
AV valves. *See* Atrioventricular valves
Avery, Oswald, 287, 375
Avian sarcoma virus, 384
Avogadro's number, **36**
Avr gene, **853,** 853i
Axel, Richard, 936
Axial skeleton, **712**
Axil, **757,** 757i
Axon, neuron, 875, 875i, **884**–885, 885i, 889i–892i, 892–893, 894i, 895i
Axon hillock, **884,** 885i
Axon terminal, **884**–885, 885i
Ayres, Edward, 798
Azotobacter, 573, 1243

B cell
  activation, 1011, 1013i
  defined, **1007**
  differentiation, 1007, 1013i
  immunological tolerance, 1017
  memory, 1011, 1013–1014, 1013i
  plasma cells, 1007, 1011, 1013i, 1014i
B-cell receptor, **1008**–1009, 1009i, 1010, 1011, 1013i
B horizon, 799, 799i
B vitamins, 1050t, 1051, 1059
Baboon, 738i, 1295, 1295i, 1300
Baciliariophyta (diatoms), **596**–597, 596i, 597i
Bacilli (prokaryote cell shape), **569,** 569i
Bacillus anthracis, 580
Bacillus subtilis, 580
Bacillus thuringiensis, 409
Background extinction rate, 513
Background selection, 476
Bacon, Sir Francis, 440
Bacteria
  flagella, 92i, 95i, **96**
  genetics, 369–377
    conjugation, 371, 372, 373i, 374, 375, 376
    recombination, 370–377, 371i, 374
    transduction, 375, 376
    transformation, 375, 376
  genome sizes, 423–424, 423i, 423t
  genomes sequenced, 421t
  intestinal, 1064i
  nitrogen cycling and, 804–806
  transposable elements, 332–333, 333i
  viral infection of, 379–382, 380i, 381i
Bacteria (domain), 577–581
  antibiotic resistance, 581
  Archaea and Eukarya compared, 548t
  characteristics of, 11, 12i, 582t
  chlamydias, 578, 580i
  cyanobacteria, 580, 580i
  defined, **567**
  evolutionary groups, 577–580
  Gram-positive bacteria, 580, 580i
  pathogenicity, mechanism of, 581
  Proteobacteria, 577–578, 578i
  shapes, 569, 569i
  size, 567, 568i
  spirochetes, 578
  in Tree of Life, 10i, 548, 548i
Bacterial chromosome
  defined, **220, 306**
  replication, 220–221, 221i, 306, 306i
Bacterial population growth, 1180–1181
Bacteriochlorophyll, 187
Bacteriophage (phage), **369**
  defined, **288**
  of *Escherichia coli,* 369, 379–382, 380i, 381i
  Hershey-Chase experiment, 288–289, 289i
  structure, 378, 378i
  temperate, 380–382, 381i
  transduction and, 375, 377, 380, 382
  virulent, 379, 380i
Bacteriorhodopsin, 122i, 174i
Bacteroid, **805,** 805i
Baculum, 1109
Balanced polymorphism
  defined, **472**
  maintenance of, 472–474
    frequency-dependent selection, 474, 474i
    heterozygote advantage, 472–473, 472i
    selection in varying environments, 473–474, 473i
Balantidium coli, 594
Bald eagle *(Haliaeetus leucocephalus),* 732i, 1043i
Baltimore, David, 385
Baltimore oriole *(Cterus galbula),* 489i
Bamboo, 758, 760
Banana, 825
Banana slug *(Ariolimax columbianus),* 1282, 1282i
Banding pattern, chromosome, 494–495, 495i
Barbie pagoda fungus *(Podoserpula miranda),* 528
Barcode system, DNA, 1267
Bardet–Biedl syndrome, 258

Bark, **768,** 769i
Barker, Michael S., 627
Barley *(Hordeum vulgare),* 802, 822, 822i
Barnacle, 447, 698–699, 699i, 1205–1206, 1206i
Baron, Gerald, 388
Baroreceptor, 994–995
Barr body, **271,** 275, 306, 350
Barton, Nick H., 496
Basal angiosperm, 639, 639i
Basal body, **110,** 110i
Basal cell, plant, 818, 826, 828i
Basal lamina, **869,** 870i
Basal nuclei, 909, 909i, 911i, **912**
Basal surface, **869,** 870i
Basal transcription factor, 347
Base, **36**
Base, DNA, 290, 290i, 313
Base, RNA, 313
Base-pair mismatch, **303,** 303i, 304i
Base-pair substitution, 331–332, 331i, 332i
Basic Local Alignment Search Tool (BLAST), 422
Basic research, **13**
Basidia, 661–662, 662i
Basidiocarp, **662,** 662i, 663
Basidiomycota (basidiomycetes), 655t, 661–663, 661i, 662i, 663i
Basidiospore, **662,** 662i
Basilar membrane, **928**–929, 928i
Basket star, 708
Basking, 1153, 1153i
Basophil, 1004t, **1019**
Bat, 641, 643i, 734, 736i, 921
Bates, Henry W., 1202
Batesian mimicry, **1202,** 1203i
Bateson, William, 311
Bath sponge *(Spongia),* 678
Batrachochytrium dendrobatis, 655, 656i
Baum, David, 525
Bay checkerspot butterfly *(Euphydryas editha bayensis),* 464, 1190, 1190i
BCR gene, 273, 273i
Beadle, George, 18, 311, 312i–313i
Beagle, H.M.S., 439i, 443–445, 447
Bean *(Phaseolus),* 859i
Beaver *(Castor fiber),* 443
Beckles, Diane, 790
Bee hummingbird *(Mellisuga helenae),* 731
Bees, pollination by, 641, 643i
Behavioral isolation, **485,** 485i
Behavioral repertoire, **1278**–1279
Bell, Stephen D., 585
Bell, Thomas, 1224
Bell pepper *(Capsicum annuum),* 814
Bengal tiger *(Panthera tigris),* 1273
Benign tumor, 361
Benson, Andrew A., 193
Bent grass *(Agrostis tenuis),* 489, 490i
Benthic province, **1164,** 1165i, 1167–1168, 1168i
Benthos, **1167**
Benton, Michael, 513
Benzene, 43i
Berg, Paul, 394
β-carotene, 410, 410i
β-catenin, 430, 430i
Beta cell, 952t, 959
Beta diversity, 1270
β-galactosidase, 394, 395i
β-interferon gene, 349, 349i
β-lactoglobulin promoter, 407
β-sheet, **59,** 59i
β-strand, 58–59, 59i, 60i
Beta wave, 917, 918i
Betaproteobacteria, 578
Bicarbonate, 37–38, 800, 1053, 1055, 1056, 1057, 1060
Biceps brachii, 977, 977i
Bicoid gene, 358, 358i
Biennial, **752**
Big Bang Theory, 553
Big laughing fungus *(Gymnopilus),* 650i
Bikonta, **588,** 589i
Bilateral symmetry
description, **672**–673, 673i
ecdysozoan protosomes, 693–702
lophotrochozoan protosomes, 683–692
Bilayer, membrane, **34**
defined, **120**
formation of, 34, 35i
phospholipid, 53, 120, 121i, 122i, 124i, 125i
Bile, **1057**
Bile salts, **1057,** 1058i
Bilirubin, **1057**
Binary fission, **220,** 221, 221i, 574
Binomial, **529**
Binomial nomenclature, **529**
Biodiversity
beta diversity, 1270
classification systems, 8–12, 9i
crisis, causes of
deforestation, 1255–1256, 1255i
desertification, 1256, 1256i
disease introduction, 1261–1262, 1262i
exotic species, 1260–1261, 1261i, 1262i
extinction, 1262–1263
habitat fragmentation, 1258–1259, 1259i
overexploitation, 1256–1257, 1257i
pollution, 1259–1260, 1260i
defined, **9, 512, 1255**
historical, 513–514, 514i
hotspots, 1264
value of
ecosystem services, 1264
to humans, 1263–1264
intrinsic worth, 1264
Biodiversity hotspots, **1264**
Biofilm, **575,** 575i
Biogenic amines, 898t
Biogeochemical cycle
atmospheric, 1241
carbon, 1242, 1244i–1245i, 1246–1248, 1250–1251
compartment model, 1231–1232, 1231i
defined, **1231**
hydrologic, 1241–1242, 1241i
nitrogen, 1243–1245, 1244i, 1245t, 1248–1250
phosphorus, 1245–1246, 1246i
sedimentary, 1241
Biogeographical realms, **510,** 510i
Biogeography
continuous distribution, 509
defined, **440**
in development of evolutionary thought, 440
disjunct distribution, 509–510
dispersal, 509
equilibrium theory of island biogeography, 1221–1225, 1222i, 1223i
historical, 449, 509–511
realms, 510, 510i
vicariance, 509, 510
Bioinformatics, 62, **421**
Biological clock
circadian rhythm, 880
defined, **858**
dormancy, 861–863
flowering responses, 859–861
hypothalamus and, 912
melatonin and, 960
photoperiodism, 858–859, 859i
stomata opening, 785
Biological evolution. *See* Evolution
Biological insurance hypothesis, 667
Biological lineage, **449**
Biological magnification, **1239**
Biological molecules, 42–67
carbohydrates, 47–50, 47i–49i
carbon backbones, 43
formation and modification, 43–47
functional groups, 43–44, 44t–45t, 46
isomers, 44–45, 45i, 46i
lipids, 50–55, 51i–54i
macromolecules, 46–47
nucleotides and nucleic acids, 63–67, 64i–66i
proteins, 55–63, 55t, 56i–60i, 62i, 63i
Biological reaction. *See also* Chemical reaction
enzymes and, 71–72, 78–81
minimum energy content of products, 73
Biological research. *See* Research, biological
Biological species concept, **481**–482
Bioluminescent, **594**
Biomass
description, 1233
by ecosystem, 1234t
net primary productivity and, 1235i
of prokaryotes, 567
pyramids of, 1237–1238, 1237i
standing crop, 1233, 1234t
Biome
change with anthropogenic global warming, 1168–1169
defined, **1154**
freshwater, 1161–1164
lakes, 1162–1164, 1162i, 1163i, 1164i
streams and rivers, 1161–1162, 1162i
marine, 1164–1168
abyssal zone, 1165, 1167
benthic province, 1167–1168, 1168i
estuaries, 1165, 1165i
intertidal zone, 1165–1166, 1166i
neritic zone, 1166–1167, 1166i
oceanic zone, 1167, 1167i
zonation, ocean, 1164–1165, 1165i
terrestrial, 1154–1161
boreal forest/taiga, 1160, 1160i
chaparral, 1159, 1159i
desert, 1158, 1158i
savanna, 1157–1158, 1158i
temperate deciduous forest, 1160, 1160i
temperate grassland, 1159, 1159i
temperate rain forest, 1160–1161
tropical forest, 1155–1157, 1156i
tundra, 1161, 1161i
Bioremediation, **23,** 38–39
Biosignatures, **560**
Biosphere, 1145–1171
defined, 2i, **3, 1146**
environmental diversity of, 1147–1152
air circulation, 1148–1149, 1149i
ocean currents, 1150, 1150i
organismal responses to, 1152–1154
precipitation, 1149–1150, 1149i
regional and local effects, 1150–1151, 1151i
seasonality, 1148, 1148i
solar radiation, 1147–1150, 1148i
terrestrial biome distribution, 1155, 1155i
freshwater biomes, 1161–1164
lakes, 1162–1164, 1162i, 1163i, 1164i
streams and rivers, 1161–1162, 1162i
marine biomes, 1164–1168
abyssal zone, 1165, 1167
benthic province, 1167–1168, 1168i
estuaries, 1165, 1165i
intertidal zone, 1165–1166, 1166i
neritic zone, 1166–1167, 1166i
oceanic zone, 1167, 1167i
zonation, ocean, 1164–1165, 1165i
terrestrial biomes, 1154–1161
boreal forest/taiga, 1160, 1160i
chaparral, 1159, 1159i
desert, 1158, 1158i
savanna, 1157–1158, 1158i
temperate deciduous forest, 1160, 1160i
temperate grassland, 1159, 1159i
temperate rain forest, 1160–1161
tropical forest, 1155–1157, 1156i
tundra, 1161, 1161i
Biosphere, 2, 568
Biosynthetic pathway, 76
Biota, **510,** 510i
Biotechnology, **18, 392**
Biotic, **1146**
Bipolar cell, **932,** 933, 933i
Bipolar disorder, 153
Bird
circulatory system, 984i, 985
classification of, 537, 538i, 546
colonial living, 1299, 1299i
development, 1124–1125, 1125i, 1126i
digestive system, 1047i, 1048
diversity, 731–732, 732i
DNA barcoding analysis, 1267
evolution, 449, 450i, 519–521, 521i
excretion, 1085
flightless, 441i
habitat selection, 1292, 1292i
high-altitude adaptations, 1038
key adaptations, 730, 731i
lungs, 1029, 1030i
mating systems, 1297
migration, 1289–1291, 1289i, 1290i
nest-building behavior, 1297–1298, 1298i
origin of, 730
pollination by, 641, 643i
retina, 931
sex chromosomes of, 267
sex determination, 1141
sexual selection, 1296–1297, 1296i, 1297i
songbird decline, 1259, 1259i
songs, 1278–1280, 1279i, 1285–1286, 1285i, 1286i, 1293
stereotyped behaviors, 1280–1281, 1281i
thermoregulation, 1086, 1088–1090, 1088–1091
Bird of paradise *(Paradisaea raggiana),* 480, 480i
Birth, 1118, 1131–1132, 1132i, 1141–1142
Birth control methods, 1112–1114, 1112t
Birth rate, per capita, 1182
Bison *(Bison bison),* 1159i
Bisphenol A (BPA), 236
Bivalve, 688i, 689, 689i
Bizily, Scott, 801
Black-capped chickadee, 732i
Black cottonwood *(Populus trichocarpa),* 643, 644
Black-handed gibbon *(Hylobates agilis),* 738i
Black saw-wing *(Psalidoprocne holomelas),* 1298
Black widow *(Latrodectus mactans),* 697
Blackburn, Elizabeth H., 303
Bladder, 1074i, 1075
Blade, leaf, **762,** 762i
BLAST (Basic Local Alignment Search Tool), 422
Blastema, 205
Blastocoel, **1121,** 1122, 1122i, 1124, 1124ii–1125i, 1129, 1129i, 1130i
Blastocyst, **1129,** 1129i, 1130i
Blastodisc, **1124,** 1125i
Blastomere, **1119,** 1133
Blastopore, **674,** 674i, **1122,** 1122i, 1123, 1124i
Blastula, **1121,** 1122i, 1124i
Blaustein, Jeffrey, 157–158
Blending theory of inheritance, **240**
Blind spot, 931i, 932
Blobel, Günter, 330
Blood, 985–987
clotting, 987, 987i, 997–998
as connective tissue, 873–874, 873i
defined, **873**
erythrocytes, 985i, 986–987, 986i, 987i
flow control, 992, 992i
gas exchange, 1034–1036, 1036i
hemoglobin, 1034–1037, 1035i, 1036i, 1038, 1038i
leukocytes, 985i, 986i, 987
plasma, 985, 985i, 986

platelets, 986i, 987
temperature regulation, 878–879, 879i
Blood–brain barrier, **908, 993**
Blood groups, genetics of, 254–255, 254t, 255i
Blood pH, 37–38
Blood pressure, 878, **991,** 995, 1080, 1081
Blood vessel, 991–994
arteries, 983, 991–992, 992i, 994, 994i
capillaries, 983, 991, 992–993, 992i, 993i
disorders, 994, 994i
veins, 983, 992i, 993–994, 994i
Blubber, 1090
Blue crab *(Callinectes sapidus),* 1238, 1240i
Blue jay *(Cyabicutta cristata),* 464i
Blue-light receptors, 784–785
Blue tit *(Parus caeruleus),* 1292, 1292i
Blue-white screening, 394, 395i
Bluegill sunfish *(Lepomis macrochirus),* 1201, 1201i
Bodmer, Sir Walter, 468i
Body axes, 1119, 1119i
Body cavities, 673, 673i
Body plan
asymmetrical, 672
bilateral symmetry, 672–673, 673i
cnidarian, 679, 679i
diploblastic, 672
insect, 699, 700i
radial symmetry, 672, 673i
sac-within-a-sac, 672, 672i
segmented, 675
simple, 678, 678i
sponge, 678, 678i
triploblastic, 672
tube within a tube, 673
Body size, 518–519
Body system, **868**
Body temperature
homeostasis of, 4–5
mammalian, 733
regulation of, 867–868, 878–879, 879i, 911
regulation of by lizards, 15–17, 17i
Bohr effect, **1035**
Boiling point, of water, 33
Bolting, **845,** 845i
Bolus, **1053,** 1054i, 1063
Bond. *See* Chemical bond
Bone (tissue), **712, 872,** 873i
Bone marrow, 975
Bone morphogenic proteins (BMPs), 1141
Bones
functions, 975–976
human body, 975–976, 975i
as levers, 976, 976i
movement, 976–977, 976i
Bonner, James, 859
Bonobo *(Pan paniscus),* 739
Bony fishes, 719–721, 719i–721i
Bony temporal arches, **724**
Book lung, **697,** 697i
Boore, Jeffry L., 696
Bordetella pertussis, 578
Boreal forest, **1160,** 1160i
Bormann, Herbert, 1242, 1243
Boron, 796t, 797
Botfly, 1199
Botox, 581
Bottleneck, population, **465,** 465i, 466
Botulism, 177, 581
Botulism toxin, 969
Bourgeonal, 1111
Bovine spongiform encephalopathy (BSE), 387i
Bowman's capsule, **1075,** 1075i, 1076, 1076i, 1077t
Box jellyfish, 680
Boyer, Herbert, 394
Brachiation, **738**
Brachiopoda, 683i, 684
Bradshaw, H. D., 492
Bradyrhizobium, 805, 806i
Brain
blood–brain barrier, 908, 993
cephalization, 904i, 905
cephalopod, 690, 690i
defined, **905**
embryonic development, 905, 906i
evolution, 900
functions
association areas, 912i, 914
consciousness, 916–918, 917i
coordination of body movement, 911
distribution of, 912i, 914, 915i
housekeeping, 909–910
lateralization, 914
learning, 914, 916, 918
memory, 914, 916, 917, 918
motor, 913i, 914
sensory, 913, 913i
higher vocal center, 1286, 1286i
mammalian, 733
PET scans, 903i, 912, 913i, 914
steroid hormone action on, 157–158
structure, 909, 909i
basal nuclei, 909, 909i, 911i, 912
brain stem, 909–910
cerebellum, 911, 911i
cerebrum, 912–914, 912i, 913i
hypothalamus, 911, 911i, 912
limbic system, 911i, 912
reticular formation, 910, 911i
thalamus, 909i, 911, 911i
wiring diagram, 900
Brain stem, **909**–910, 1032–1033, 1033i
Branchial slits, **709,** 710i
Branching evolution concept, 530, 530i
Branchiostoma, 711, 712
Brassinosteroid, 840t, **847**–848, 848i
Brazilian Atlantic Forest, 1272
BRCA1 gene, 363, 363i
Breast cancer, 962–963
Breathing
control of, 1032–1033, 1033i
defined, **1025**
negative pressure, 1029, 1030
positive pressure, 1029, 1029i
prenatal nicotine exposure, effect of, 1039
Breen, Gail, 179
Bri1 gene, 848, 848i
Briggs, Robert, 1121, 1138
Briggs, Winslow, 843
Bright field microscopy, 93i
Brinster, Ralph, 406
Bristle worms, 691
Bristlecone pine *(Pinus aristata),* 633
Brittle star, 708, 708i
Broca's area, 912i, 914
Brodribb, Timothy, 615, 785
Broeckling, Corey D., 802
Bronchiole, **1030,** 1031i
Bronchus, **1030,** 1031i
Brood parasite, 1199–1200, 1200i, 1208, 1259, 1281, 1281i
Brown, Michael S., 67, 136
Brown, Robert, 90
Brown adipose tissue (brown fat), 1089
Brown algae, 588i, 598, 598i, 599i
Brown fat, 176
Brown-headed cowbird *(Molothrus ater),* 1208, 1259
Brown recluse spider *(Loxosceles reclusa),* 697
Brugia, 686
Bruno, John F., 1210
Brush border, 1053i, **1055,** 1056i, 1057, 1057i, 1058i
Brutlag, Douglas L., 303
Bryant, Gregory, 1107i
Bryophyta, **620**
Bryophyte, 618–621
defined, **613**
hornworts, 619–620, 620i
liverworts, 619, 619i
moss, 618i, 620–621, 620i
overview, 618–619, 618i, 619i, 622t
Bryozoa. *See* Ectoprocta
BSE. *See* Bovine spongiform encephalopathy
Bt toxin, 409–410, 415
Buck, Linda, 936
Bud, 749, 750i, 757, 757i
Bud scale, 848, 848i
Bud suppression by abscisic acid, 848
Budding, **1097,** 1097i
Buffer, **37**–38, 38i, 1036
Buffering capacity, 38
Buffon, le Comte de, 441
Bulb, 760, 761i
Bulbourethral gland, 1108i, 1109
Bulk feeder, 1045i, **1046**
Bulk flow, **777,** 786, 787, 788i
Bulk-phase endocytosis, **134,** 135i
Bullhead catfish *(Ameiurus),* 1088
Bullock's oriole *(Cterus bullockii),* 489i
Bull's horn acacia *(Acacia cornigera),* 1207, 1208i
Bult, Carol J., 96
Bundle sheath cells, 198, 198i, 199i
Bunn, Stuart E., 1257
Burgess Shale formation, 670
Burnet, F. Macfarlane, 1011
Bursa of Fabricus, 1007
Bush lily *(Clivia miniata),* 756i
Buttercup *(Ranunculus),* 753i, 766i
Butterfly
DNA barcoding analysis, 1267
mouthparts, 701i
Byssus, 310

C horizon, 799, 799i
C_3 pathway, **194**
C_4 pathway, 197–198, 199i
C_3 plants, **194,** 198, 198i
C_4 plants, 197–198, 199i
C-terminal end, amino acid, **57**
Cabbage *(Brassica oleracea),* 845
Cacao, 811–812, 811i, 817, 820
Cactus, 510, 511i, 762
Cactus finch *(Geospiza conirostris),* 470, 470i
Caddis fly, 539i
Cadherins (calcium-dependent adhesion molecules), **1136**
Caecelian, 723, 723i
Caenorhabditis elegans
apoptosis, 1139–1141, 1140i
cell lineages, 1136, 1138i
as model organism, 693, 694
Caetano-Anollés, Gustavo, 62
Cain, A. J., 473i
Calamites, 622, 623i
Calcitonin, 951t, **956,** 957i
Calcium
cortical granule reaction, 1102, 1110
deficiency in plants, 797i
in gravitropism, 856
hormonal control of blood level, 956, 957i
loss in watershed, 1243
muscle contraction and, 968, 969, 970i
plant macronutrient, 795, 796t
soil, availability in, 800
Calcium carbonate
otoliths, 926
shell, 974
statoliths, 855
Calcium channel, 895–896, 895i
Calcium ions (Ca^{2+})
detecting release in cells, 153
release from endoplasmic reticulum, 151, 152i, 153
as second messenger, 157
Calcium pump (Ca^{2+} pump), **131**
Calico cat, 271, 271i
California poppy *(Eschscholzia californica),* 765i
Calliphlox evelynae, 643i
Callus, 408, 825i, **826,** 846i
Calmodulin, 856, 857
Calorie, **33, 34,** 1044
Calvin, Melvin, 193
Calvin cycle, 183, **193,** 193–196, 195i, 197–199, 199i
Calyx, flower, 813i, **814**
CAM plants, **198**–199, 199i. *See also* Crassulacean acid metabolism
Cambia, 767
Camel *(Camelus dromedarius),* 736i, 1090
Camouflage, 7–8, 8i
cAMP. *See* Cyclic adenosine monophosphate; cyclic AMP
cAMP-response element (CRE), 362
Campbell, Keith H. S., 407, 408i
Campylobacter, 578
CAMs. *See* Cell adhesion molecules
Canada lynx *(Lynx candensis),* 1191, 1192i
Canary grass *(Phalaris canariensis),* 841
Cancer, **361**
breast, 962–963
cell cycle controls, loss of, 219
chromosomal mutations and, 272–273, 273i, 282
colorectal, 364, 364i
cytotoxic T cells and, 1002i, 1016, 1018
diacylglycerol levels and, 153
genetics of, 360–364
microRNAs, 363–364
multistep model for development of cancer, 364, 364i
proto-oncogenes, 362
tumor suppressor genes, 362–363, 363i
metastasis, 219
oncogenes, 219
radiation therapy, 26
RNA interference therapies, 365
skin, 304
telomeres and, 302, 307
transposable elements and, 335
tumors, 219
virus-caused, 220, 362
Candida albicans, 659, 659i
Canine, **733,** 1061, 1061i
Cann, Rebecca, 743
Canopy, 1157
CAP. *See* Catabolite activator protein
CAP site, **344**
Capillary, **983,** 991, 992–993, 992i, 993i
Capping enzyme, 318
Capsaicin, 937, 938
Capsicum annuum, 938
Capsid, **377**
Capsule, **95,** 95i, **571,** 571i
Capybara *(Hydrochoerus hydrochaeris),* 736i
Carapace, **698, 727,** 727i
Carbohydrates, 47–50
disaccharides, 47, 48, 48i
enzymatic digestion of, 1053i
monosaccharides, 47, 47i, 48i
oxidative pathways, 170, 171i, 173
polysaccharides, 48–50, 49i
Carbon
asymmetric, 44
bonding, 43
chains and rings, 43, 43i
covalent bonding of, 29, 29i
isotopes of, 25, 25i
as plant macronutrient, 795, 796t
recycling of, 649, 652
Carbon cycle, **1242,** 1244i–1245i, 1246–1248, 1250–1251
Carbon dioxide
atmospheric concentration of, 42
in carbon cycle, 1242
fixation, **183,** 184, 193, 194, 197–199, 199i
gas exchange, 1035–1036, 1036i
as greenhouse gas, 1247–1248, 1247i, 1255
homeostatic control of concentration, 877
ocean acidification, 1248–1249
partial pressure, 1034, 1034i, 1035, 1036
plant uptake of, 784–785
receptors, 1032–1033
stomata and, 756

Carbon fixation, **183,** 184, 193, 194, 197–199, 199i
Carbon monoxide, 986–987, 1036
Carbon sequestration, 1264, 1274
Carbonic acid, 800, 1036
Carbonic acid–bicarbonate buffer system, 37–38, 38i
Carbonic anhydrase, 1036
Carboniferous period, 622, 623, 631
Carbonyl group, **43,** 44t
Carboxyl group, **43,** 44t, 56, 57i
Carboxylase, 194, 197–198
Carboxylic acids, 44t
Carboxypeptidase, 1053i, **1056**
Cardiac cycle, **989**–990, 989i, 990i
Cardiac muscle, **874,** 874i, 966
Cardiac output, 994–995
Carnivora, 734
Carnivore
defined, **1044**
intestinal length, 1061, 1063i
teeth, 1061, 1061i
Carnivorous plant, 806, 807i
Carotenoids, 55, **186,** 187, 187i, 188
Carotid body, **1032,** 1033i
Carpel, 241i, **637,** 813i, **814**
Carrageenan, **601**
Carrier, **270,** 276
Carrier protein, 125–126, **127,** 128i
Carroll, Sean, 522
Carruth, Laura, 1142
Carrying capacity, **1183**–1184, 1183i, 1258
Carson, Hampton, 488
Cartagena Protocol on Biosafety, **410**
Carter, Patrick A., 459
Cartilage, **872,** 873i
Cartilaginous joint, 976
Cascade frog *(Rana cascadae),* 1292
Casein, 352
Casparian strip, **780**–781, 780i, 783
Castor bean *(Ricinus communis),* 819, 820i
Cat
calico, 271, 271i
Siamese, 84i
Catabolic pathway, **76**
Catabolic reactions, **76**
Catabolite activator protein (CAP), 343i, **344**
Catalase, 107
Catalysis
defined, **71, 79**
by enzymes, 78–81, 80i
by ribozymes, 79, 85–86
Catalyst, 71, **79,** 85
Catalytic RNA, 326, 335
Catastrophism, theory of, **441**
Catecholamine, **957**
Caterpillar, 1202, 1202i
Cation, **28,** 776, 776i, 777
Cation exchange, **800,** 800i
Catshark *(Scyliorhinus canicula),* 523
Cattalya rojo, 630i
Cattle egret *(Bubulcus ibis),* 1206–1207, 1206i
Caudata, 723
Cavallari, Larisa H., 436
Cavalli-Sforza, Luigi, 468i
Cavitation, 786
CCK. *See* Cholecystokinin
$CD4^+$ T cell, **1011,** 1012i, 1019
$CD8^+$ T cell, **1015,** 1016, 1017i, 1021
Cdc genes, 214–215
CDK. *See* Cyclin-dependent kinase
cDNA. *See* Complementary DNA
cDNA library, **396,** 396i, 422, 429i
Cech, Thomas R., 85
Cecum, **1058,** 1059i, 1061, 1063i
CED protein, 1140–1141, 1140i
Celera Genomics, 417
Cell(s)
defined, **2**–3, 2i
early observations of, 90–91, 91i
examples, 92i
origin of
archaean, 563
eukaryotic, 561–564, 561i, 562i
prokaryotic, 559–560, 560i
protocells, 554, 557–559, 557i
timeline, 554i
size, 91–92, 92i
structure and function, 91–94
Cell adhesion molecules (CAMs), **113, 122, 1136**
Cell adhesions, selective, **1121,** 1136, 1137i
Cell body, neuron, 875, 875i, 884, 885i
Cell–cell recognition, 142
Cell center, **107**
Cell communication, 141–158
air traffic control analogy, 141
components of systems, 145–147
cross-talk, 156–157, 157i
evolution, 144–145
gap junctions, 142, 157
in immune system, 141i, 142
internal receptors, 154–156, 156i
methods, 142
overview, 142–145
pathway integration, 156–157, 157i
plasmodesmata, 142
signal processing steps, 143, 145i
steroid hormone receptors, 154, 156i, 157, 158
surface receptors, 145–154
characteristics of, 145–147, 146i
G-protein-coupled, 149–154, 149i, 152i
tyrosine kinase, 147–149, 148i, 153, 154i
Cell culture, **207**
Cell cycle
defined, **206**
mitotic, 207–214
overview, 206–207, 207i
regulation, 214–219, 216i, 217i, 222
Cell differentiation, defined, **1119**
Cell division
in archaeans, 584
binary fission, 220, 221, 221i
cancer and, 360, 361i
cell cycle, 207, 207i, 214–219
meiosis, 225–236
mitosis, 207–219, 207i–210i, 212i–217i
mitotic spindle, 212–214, 213i, 214i
orientation of, 830, 830i, 1133
in prokaryotes, 220–221, 221i
rate of, 207, 1133
telomeres, role of, 302, 307
Cell expansion, **830**–831, 831i
Cell fractionation, 97i, 173
Cell junctions, **113,** 114i
Cell lineage, **1136,** 1138i
Cell-mediated immunity, **1007, 1015**–1017, 1017i
Cell plate, **211,** 212i
Cell shape, changes in, 1135, 1135i
Cell signaling, **142,** 944–945, 944i. *See also* Cell communication
Cell theory, **91**
Cell wall
algal, 112
fungal, 112
plant, 99i, 112–113, 113i
cell expansion and, 830–831, 831i
lignification, 749
primary, 749, 753, 830
secondary, 749, 754
primary, **112,** 113i
prokaryote, **95,** 95i, 570–571, 571i
secondary, **112,** 113i
Cellular proteome, 428
Cellular respiration, 1025, 1025i
ATP yield, 175–176, 175i
cellular location of, 106
citric acid cycle, 164, 164i, 168–171, 168i, 170i, 171i, 175i, 176
defined, **4, 162**
efficiency of, 176
electron flow, 162i
electron transfer system, 164i, 165, 171, 172i, 173, 173i, 175–176
evolution of, 200–201
glycolysis, 164, 164i, 165–166, 166i, 167i, 168, 175–176, 175i
overall reaction, 164
overview, 4i
oxidative phosphorylation, 164i, 165, 172i, 175i, 176
photosynthesis compared, 199, 200i
pyruvate oxidation, 164, 164i, 168–169, 168i, 169i, 171i
stages, 164–165, 164i
Cellular slime molds, 603, 604i, 605
Cellulase, 1063
Cellulose, **47,** 49, 49i, 50, 112
in plant cell wall, 749, 754, 830–831, 831i
Cenarchaeum symbiosum, 584
Centimorgan, **265**
Centipede, 699, 699i
Central canal, **905,** 906i
Central cell, 814
Central dogma, **312**
Central nervous system (CNS)
brain, 908–918, 909i–913i, 915i
defined, **885, 905**
flatworm, 905
glial cells, 885
spinal cord, 909, 909i
Central vacuole, 99i, **111, 777,** 778i
Centriole, 98i
in cell division, 213, 213i
defined, **107, 213**
flagella and cilia development, 109–110
structure, 109, 111i
Centromere, **210**
Centrosome, **107,** 213, **213,** 214i
Centruroides sculpturatus, 697i
Cepaea nemoralis, 457i, 458, 473i, 481
Cephalization, **673,** 904i, 905
Cephalochordata, 711, 711i
Cephalopod
description, 688i, 690, 690i
eyes, 930, 930i
vibration-detection system, 927
Cephalothorax, **695,** 697i
Cerebellum, 905, 906i, 911, 911i
Cerebral cortex
association areas, 912i, 914
description, **733, 909,** 909i
functional studies, 912–913, 913i
motor regions, 913i, 914
sensory regions, 913, 913i
structure, 912, 912i
Cerebral hemisphere, 909, 909i, 912, 914, 915i
Cerebrospinal fluid, **907**
Cerebrum, 905, 909, 912–914, 912i, 913i, 915i
Cerione, Richard, 949
Cervical cap, 1112t, 1113
Cervix, **1104,** 1104i, 1111, 1113, 1131, 1132i
Cestoda, 685, 685i
Cetaceans, 537, 734, 736i
CFTR. *See* Cystic fibrosis transmembrane conductance regulator
CGG repeats, 301
Chagas disease, 592
Chain fern *(Woodwardia),* 624i
Chambered nautilus, 690, 690i
Chance, genetic drift and, 464, 464t
Chang, Sherwood, 557
Channel protein, 125, **127,** 128i, 776–777, 776i, 779
Chaos carolinense, 137i
Chaparral, **1159,** 1159i
Chaperone (chaperonin), **60,** 61i, **328**
Character, **240,** 242, 242i, **538**
Character differences, **240**
Character displacement, **1204**–1205, 1205i
Character state, **538**
Chargaff, Erwin, 290
Chargaff's rules, 291
Charging, of tRNA, **323**–324, 323i
Charlesworth, Brian, 476
Charophyte, **601,** 602i, 612, 612i
Chase, Martha, 288–289, 289i
Checkpoints, **216,** 217–218, 217i
Cheetah *(Acinonyx jubatus),* 465
Cheetham, Alan, 517
Chelicerae, **695,** 697, 697i
Chelicerta, 695, 697, 697i
Cheliped, 698, 698i
Chemical bond, 28–32
covalent, 29–30, 29i
defined, **28**
hydrogen bonds, 30–31, 31i, 32–34, 33i
ionic, 28–29, 28i
van der Waals forces, 31, 32i
Chemical energy, 72, 73, 73i
Chemical equation, **32**
Chemical equilibrium, 75
Chemical modification, of enzymes, 83
Chemical reaction, 31–32
activation energy, 78–81, 78i, 79i
anabolic, 76
biosynthetic, 76
catabolic, 76
catalysis, 79–81
coupled reactions, 76–78, 78i
endergonic, 75, 76i, 78i
endothermic, 74
equilibrium point, 75, 75i
exergonic, 75, 76i, 78i
exothermic, 74
oxidation-reduction (redox), 162–163, 163i
reversible, 75
spontaneous, 74–75
Chemical signal, **1293**–1294, 1294i
Chemical synapse
conduction across, 885i, 895–897
description, **886,** 886i, 887
integration at, 889, 897, 899i
structure, 895i
Chemiosmosis, **165,** 172i, 175, 190i, 191, 192
Chemiosmotic hypothesis, 174, **175**
Chemoautotroph, **573,** 573i, 582
Chemoheterotroph, **573,** 573i
Chemokines, 1005i, **1006**
Chemoreceptor
defined, **923**
osmoreceptors, 1082–1083
oxygen concentration, 995
respiration control and, 1032
stomach, 1060
for taste and olfaction, 934–936, 935i, 936i
Chemotherapy, 361
Chemotroph, **573**
Cherry *(Prunus),* 641i, 812i, 813i, 815i
Chestnut-headed oropendola *(Zarhynchus wagleri),* 1199–1200, 1200i
Chickweed *(Stellaria),* 816i
Chicxulub crater, 513
Chief cell, **1054,** 1055i
Childbirth, positive feedback in, 879–880
Chilopoda, 699
Chimpanzee *(Pan troglodytes),* 518, 519i, 738–739, 738i
Chinese liver fluke *(Opisthorchis sinensis),* 685i
Chiroptera, 734, 736i
Chitin, 49i, 50, 112, **650,** 694, 808
Chiton, 688, 688i
Chlamydia trachomatis, 578, 580i
Chlamydias, 578, 580i
Chloranthales, **639**
Chlorarachniophyta, **600,** 606i
Chlorella, 193
Chloride cell, 1084
Chlorine, as plant micronutrient, 795, 796t
Chlorophyll
absorption spectrum, 187, 188i
antenna complex, 188–189, 188i
of bacteria, 580

in chlorarachniophyta, 600
in chloroplasts, 111
chlorosis and, 797
defined, **186**
photosystems, 188–192, 188i–190i, 192i
reaction center, 188–189, 188i
structure, 55, 187, 187i
Chlorophyll *a*, 187–188, 187i, 188i
P680 chlorophyll *a*, 189, 189i–190i, 191
P700 chlorophyll *a*, 189, 189i–190i, 191, 192i
Chlorophyll *b*, 187, 187i, 188i
Chlorophyta (green algae), 601, 601i, 602i
Chloroplast, 99i
ATP synthesis, 189i–190i, 191–192, 192i
defined, **111**
endosymbiont hypothesis and, 561i, 562, 563i, 605–607, 606i
genome, 426
horizontal gene transfer and, 549, 549i
maternal inheritance, 280
origin of, 111
protein sorting to, 330
structure, 111, 112i, 182i, 184, 185i
Chloroplast DNA (cpDNA), 545, 626
Chlorosis, **797**
Choanocyte, **678,** 678i, 1046
Choanoflagellate, 435, **604,** 604i, 672
Chocolate, 811
Cholecystokinin (CCK), **1060,** 1060i
Cholera toxin, 150
Cholesterol, 1088
in animal membranes, 121, 121i, 123
animal nutrition and, 1057, 1058, 1065
coronary artery disease, role in, 52, 54
defined, **121**
in membranes, 54
receptor-mediated endocytosis, 136, 137
structure, 54i
synthesis, regulation of, 67
Chondrichthyes, 718–719, 718i
Chondrocyte, **872**
Chondroitin sulfate, 872
Chondromyces crocatus, 578i
Chondrus, 1215–1215, 1215i
Chordate
invertebrate, 710, 710i
morphological innovations, 710, 710i
phylogeny, 714i
vertebrate tissues, unique, 711–712
Chorion, **1125,** 1126i, 1129, 1130i
Chorionic villi, **1129,** 1130i
Chorionic villus sampling, **279,** 1131
Choroid, 930, 931i
Chromalveolata, **592**–598
Chromatin
appearance, 98i, 99i
defined, **100, 304**
gene regulation and, 351, 351i
Chromatin fibers, **305,** 305i
Chromatin remodeling, 339i, 347i, **351,** 351i
Chromatography, two-dimensional paper, 193
Chromoplast, **111**
Chromosomal mutations, **271**–276, 272i–275i, 275t
Chromosomal proteins of eukaryotes, **304**
Chromosome(s)
bacterial, 220–221, 221i
banding patterns, 494–495, 495i
condensation of, 209–210, 228i
defined, **206**
DNA methylation of, 350
eukaryotic, **100**
homologous, 226–227, 227i–230i, 230–232, 232i, 235
human and great apes, differences among, 495–496, 495i
human genome, 425, 425t
independent assortment, 232–233, 232i
karyotype, 210–211, 211i
linkage map, 264–265, 266i
linked genes, 262–266, 263i, 265i, 266i
movement by mitotic spindles, 213–214, 215i
mutations that affect inheritance, 271–276, 272i–275i, 275t
nondisjunction of, 227, 273–275, 274i, 275i
nucleoproteins, 63
number of, 626–627
partitioning of, 206–207, 206i
prokaryotic, **95,** 569
replication, 207
in archaeans, 584
staining of, 211i
Chromosome recombination, 262
Chromosome segregation, **207**
Chromosome theory of inheritance, 249–251, **250,** 250i
Chronic myelogenous leukemia (CML), 272–273, 273i
Chronic wasting disease (CWD), 388
Chrysophyta (golden algae), **597**–598, 598i
Chua, Nam-Hai, 850–851
Chylomicron, **1058,** 1058i, 1065
Chyme, 1052i, **1055,** 1056, 1060
Chymotrypsin, 1053i, **1056**
Chytridiomycota (chytrids), 655–657, 655t, 656i, 1262, 1262i, 1274
Chytriomyces hyalinus, 656i
Cicada *(Graptopsalatsia nigrofusca),* 695i
Ciechanover, Aaron, 354
Cilia, 109–110, 110i
of ctenophores, 681
epithelial tissue, 869
of protists, 590
respiratory system, 1030
of rotifers, 685
Ciliary body, **931,** 931i
Ciliary muscle, 931, 931i
Ciliophora, **592**–594, 593i, 606i
Circadian rhythm
in plants, **858**
stomata opening, 785
vascular, 880
Circulatory system, 981–1001
annelid, 691, 691i
blood, 985–987
clotting, 987, 987i, 997–998
erythrocytes, 985i, 986–987, 986i, 987i
leukocytes, 985i, 986i, 987
plasma, 985, 985i, 986
platelets, 986i, 987
blood pressure, 991
blood vessels, 991–994, 992i–994i
cephalopod, 690
closed, **690,** 983, 983i
defined, **981**–982
disorders, 994, 994i
double circuit, 984–985, 984i
elements of, 982–983
embryonic, 1130i
heart, 987–991, 988i–990i
invertebrate, 983
lymphatic system, 982, 996–997, 997i
mollusk, 687
nemertean, 687
open, 687, 982–983, 983i
single circuit, 983–984, 984i
vertebrate, 983–985, 984i
Circulatory vessels, of nemerteans, **687**
Circumcision, **1109**
Cisternae, **102,** 102i, 103
Citrate, 169, 170i
Citrate synthase, 169, 170i
Citric acid cycle, **164**–165, 164i, **168,** 168–171, 168i, 170i, 171i, 175i, 176
CJD. *See* Creutzfeldt-Jakob disease
Clade, **531,** 532
Cladistics, 538–543
defined, **538**
focus on recently evolved character states, 538–539
outgroup comparison, 539, 539i
phylogenetic tree construction, 539–540, 540i–541i
synapomorphy use in, 539–541
Cladogenesis, 531, 536
Cladogram, chordate, 714i
Cladonia rangiferina, 665i
Clam, 689, 689i, 1027i, 1028
Claret cup cactus *(Echinocereus triglochidratus),* 641i
Claspers, **719**
Class, **9**
Class I major histocompatibility complex (MHC), 1015, 1016, 1017i, 1021
Class II major histocompatibility complex (MHC), **1011,** 1012i
Classical conditioning, **1284**
Classification. *See also* Phylogenetic tree; Phylogenetics
by Aristotle, 440
cladistics, 538–543, 539i–541i
defined, **530**
Linnaean system, 529–530, 549–550
overview of animal, 675–677, 676i
Classification systems, **9**–12, 9i, 529–530, 549–550
Clathrin, 135i, **137**
Claviceps purpurea, 659
Clay soil, 798, 799, 799i, 800, 800t
Clays, molecular aggregation on, 557
Cleaner wrasse *(Labroides dimidiatus),* 1207i
Cleavage
amphibian, 1120i, 1123, 1124i
bird, 1124, 1125i
blastula, 1121
cell division orientation, 1133
cell division rate, 1133
defined, **674, 1119**
determinate, 674
human, 1128
indeterminate, 674
morula, 1121, 1122i
patterns of, 1122–1125
radial, 674, 674i
sea urchin, 1122, 1122i
spiral, 674, 674i
Cleavage furrow, 1133
Clements, Frederic, 1154, 1208
Cleome droserifolia, 1177i, 1178
Cliff swallows *(Petrochelidon pyrrhonota),* 1299
Climate
continental, 1150
defined, **1147**
Earth's history, 508–509, 508i
El Niño, 1145–1146, 1146i
global warming, 1153–1154, 1168–1169
maritime, 1150
microclimate, 1151
ocean proximity, 1150–1151, 1151i
topography, 1151, 1151i
Climate change, 1153–1154, 1168–1169. *See also* Global warming
Climatius, 718i
Climax community, **1219**
Climograph, **1155,** 1155i
Cline, **483,** 483i
Clitoris, **1104,** 1104i, 1110
Cloaca, **1101**
Clonal analysis, **833**
Clonal expansion, **1011**
Clonal reproduction, **1097**
Clonal selection, 1011, 1014i
Clone
animal, 356, 407, 408i
bacterial, **371**
defined, **207, 392**
Cloning, DNA. *See* DNA cloning
Cloning animals, 356
Cloning vector, 394, 395i
Closed circulatory system, **690, 983,** 983i
Closed system, 72, 73i
Clostridium botulinum, 177, 581, 969
Clostridium tetani, 574i
Clotting, 987, 987i, 997–998
Clotting factor, 407
Cloud forest, 1157
Club moss, 623, 623i
Clumped dispersion, **1174,** 1174i
CML. *See* Chronic myelogenous leukemia
Cnidarian
chemoreceptor, 934
description, 679–681, 679i, 680i, 681i, 682i
gastrovascular cavity, 982, 982i, 1046
hydrostatic skeleton, 974
intracellular digestion, 1046
nerve net, 904, 904i
Cnidocyte, **680,** 680i
CNS. *See* Central nervous system
CO_2 fixation. *See* Carbon dioxide, fixation
CoA. *See* Coenzyme A
Coactivator, **348,** 349i
Coal Age, 622, 623i
Coal tit *(Parus ater),* 1292, 1292i
Coast redwood *(Sequoia sempervirens),* 12i, 632, 774, 774i
Coated pit, 135i, **137**
Cobra lily *(Darlingtonia californica),* 806, 807i
Cocci (prokaryote cell shape), **569,** 569i
Cochlea, **927,** 928i
Cochlear duct, 927–928, 928i
Cochlear nerve, 929
Cocklebur *(Xanthium strumarium),* 859
Cocoanut Grove nightclub, 456
Coconut palm *(Cocus nucifera),* 1189i
Codominance, **254**
Codon, **314**–315, 315i
Coelacanth *(Latimeria chalumnae),* 721, 721i
Coelom
description, **673**
echinoderm, 707
enterocoelom, 674i, 675
schizocoelom, 674i, 675
Coelomate animal, **673,** 673i
Coencytic fungi, **650,** 651i
Coenzyme A (CoA), 168, 168i, 169i
Coenzyme Q (CoQ), 172i, 173
Coenzymes, **80,** 1044
Coevolution, 639, **641,** 643i, **1200,** 1202–1203, 1203i
Cofactor, **80**
Cognition, **1284,** 1284i
Cohen, Stanley, 394
Cohesins, 207
Cohesion, 32, **34**
Cohesion-tension mechanism of water transport, **781–783,** 782i
Cohesive property of water, 781
Cohn, Martin, 450, 523
Coho salmon *(Oncorhynchus kisutch),* 1178
Cohort, **1176**
Colaiácovo, Monica, 236
Colchicine, 211i
Coleochaete, 601
Coleoptile, **819,** 820i, 822i, 823i
Coleorhiza, **819,** 820i, 822i
Coleus, 834i
Collagen, 115, 115i, **872**
Collar cells, 672
Collecting duct, **1075,** 1075i, 1076i, 1077t, 1078–1079, 1082i, 1083
Collenchyma, 753, 753i, 753t, **754**
Collins, Kathleen, 1021
Colon, **1058**–1059, 1059i
Colonial living, 1299, 1299i
Colony, **589**–590, 1097
Color blindness, 268, 277
Coloration
aposematic, 1202, 1202i
cryptic, 1202
Colorectal cancer, 364, 364i
Columnar epithelium, 869, 870i
Comb jellies, 681, 683i
Combinatorial gene regulation, **348**–349, 349i
Commaless, feature of genetic code, **315**
Commensalism, **1206**–1207, 1206i

Communication
acoustical signals, 1293
chemical signals, 1293–1294, 1294i
dance, 1294–1295, 1294i
eavesdroppers, 1303
electrical signals, 1294
evolution of, 1293–1295
evolutionary analysis, 1295
tactile signals, 1294, 1294i
visual signals, 1293, 1293i
Community
climax, 1219
defined, 2i, **3**
disclimax, 1220
Community ecology
defined, **1147**
population interactions, 1200–1208, 1200t
Companion cell, **755,** 756i, 760i, 787–788, 788i
Comparative genomics, **417,** 431–436
Comparative morphology, 440–441, 449–450
Compartment model, **1231**–1232, 1231i
Compass orientation, **1289**–1290, 1290i
Competence, 1138
Competitive exclusion principle, **1203,** 1204i
Competitive inhibition, **82,** 82i
Complement system, **1006**
Complementarity, in DNA molecule, 66
Complementary base pairing, **291,** 292i
Complementary DNA (cDNA), **396,** 396i, 405i
Complete digestive tract, **685**
Complete flower, **813,** 813i
Complete medium, **375,** 376i
Complete metamorphosis, **701,** 701i
Complex virus, **378,** 378i
Compound eye, **695, 930,** 930i
Compounds, **24**
Concentration, defined, **35**
Concentration gradient, **126**
Concentricycloidea, 708
Condensation, of chromosomes, 209–210, 228i
Condensation reaction, **46,** 46i, 557
Conditioned stimulus, 1284
Conditioning
classical, 1284
operant, 1284
Condom, 1112t, 1113
Conduction, **1085,** 1086i
Cone
conifer, 634–635, 636i
Ephedra, 634i
ginkgo, 634i
photoreceptor, **931**–934, 932i, 933i
Confocal laser scanning microscopy, 93i
Conformation, protein, **59**–60
Conformational changes, **60**
Conformers, **877**
Conidia, 654, 660–661, 660i
Conidial fungi, 654, 663
Conidiophore, **660**–661, 660i
Conifer, 42, 43i, 633, 634–635, 636i
Coniferophyta, **634**
Conjugate, **593**
Conjugation, 371, 373i, 374, 375, 376
in bacteria, 574
defined, **372**
F factor, 371, 373i, 374
Hfr cells, 372, 373i
mapping genes by, 374
micrographs of, 372i
in *Paramecium,* 593, 593i
Connective tissue, 871–874
adipose, 872–873, 873i
blood, 873–874, 874i
bone, 872, 873i
cartilage, 872, 873i
defined, **871**
extracellular matrix, 872–873
fibrous, 872, 873i
loose, 872, 873i
Connell, Joseph, 1205–1206, 1206i, 1214, 1216–1217
Conodont, **716,** 716i
Consciousness, 900, **916**–918, 917i
Conservation biology
conservation approaches
mixed-use conservation, 1272, 1272i
preservation, 1271–1272, 1272i
restoration, 1272
defined, **1266**
economic factors
ecosystem valuation, 1273–1274
ecotourism, 1273
local involvement, 1273
genetic drift, implications of, 464–465
landscape ecology, 1269–1270
population ecology and behavior, 1267–1268
population genetics, 1266
population viability analysis, 1268, 1269
species/area relationships, 1268–1269, 1268i
systematics, 1266
Conservative replication model, 293, 293i
CONSTANS gene, 860, 861i, 862
Constant region, of antibody molecule, **1009,** 1009i, 1010i
Consumer, **4,** 183, FIXME
Contact inhibition, 219
Continental climate, **1150**
Continental drift, **506**–508, 507i, 511i
Continental shelf, 1166
Continuous distribution, **509**
Contraception, 1112–1114, 1112t
Contractile vacuole, **590,** 590i
Control, **15,** 16i
Controlling cell, 141, 143, 143i
Convection, **1085,** 1086i
Convergent evolution, **510**–511, 511i
Cook, Lyn, 509
Cooksonia, 613–614, 614i, 622
Cooling, adiabatic, 1150
Copepod, 698, 698i
Copernicus, Nicolaus, 440
Copper, as plant micronutrient, 795, 796t
Copulation
defined, **1101**
earthworms, 1103i
frogs, 1101i
human, 1110
CoQ. *See* Coenzyme Q
Coral fungus *(Ramaria),* 661i
Coral reef
biome, **1167,** 1167i
disturbance of, 1216–1217, 1216i
Corals, 680–681, 682i
Corepressor, **344,** 345i, **348**
Corey, Robert, 58
Coriolis effect, 1149
Cork, **767,** 768, 769i
Cork cambium, **767,** 768, 769i, 770
Cork cells, 90, 91i
Cork oak *(Quercus suber),* 768
Corm, 761, 761i
Corn *(Zea mays)*
gravitropism in, 856, 856i
perennial, 1263
prop roots, 765
root structure, 766i, 767
seed germination, 822, 823i
seed structure, 819, 820i
shoot apical meristem, 761
stem organization, 760i
teosinte compared, 1263
transposable elements in, 333, 334i
zeatin, 846
Cornea, **930,** 930i, 931i, 1127–1128, 1128i
Cornflower *(Centaurea montana),* 848i
Corolla, flower, 813i, **814**
Corona, rotifer, 685, 687i
Coronary artery, 52, **988**
Coronary circulation, **989,** 994, 994i
Corpora allata, 961
Corpus callosum, 909i, **912,** 914, 915i
Corpus luteum, **1105,** 1105i, 1106i, 1111
Correns, Carl, 249, 280
Cortex, plant
defined, **759**
root, 766i, 767, 767i, 769i, 780, 780i
stem, 759i, 760, 760i
Cortical granule, **1102,** 1102i, 1110, 1110i
Cortical reaction, **1102**
Cortisol, 945, 951t, **958,** 958i
Cotranslational import, **330**
Cotransport, **133**
Cotton plant *(Gossypium),* 630, 630i
Cotyledon
eudicot, 638, 752t, 818–819, 819i, 820i
monocot, 638, 752t, 819, 820i, 822i
seed germination and, 822, 822i
Cough reflex, 1053
Countercurrent exchange, **1028**
gas exchange, 1028, 1028i, 1029, 1030i
Countercurrent heat exchanger, 1091–1092, 1091i
Coupled reaction, ATP and, 76–78, 77i, 78i
Courtship displays, 484, **1296**–1297, 1297i
Covalent bond
carbon, 29, 29i
defined, **29**
diagram designation of, 29
formation of, 29
nonpolar, 30
polar, 30
Cowbird, 1199–1200, 1208, 1259
Cowpox, 1002, 1014
Coyne, Jerry, 496
cpDNA. *See* Chloroplast DNA
Crab, 698, 698i, 962, 962i, 1045–1046, 1045i
Crane fly, 700i
Cranial nerves, **906,** 908i
Cranium, **712**
Crassulacean acid metabolism (CAM), **198**–199, 199i, **786**
Crayfish, 698
CRE. *See* cAMP-response element
CRE-binding protein (CREB), 362
CREB. *See* CRE-binding protein
Crenarchaeota, **583**–584
Creosote *(Larrea tridentata),* 838, 838i, 1174, 1174i
Cretaceous mass extinction, 513
Creutzfeldt-Jakob disease (CJD), 387
Crick, Francis H. C., 18, 65, 286–287, 287i, 290–292, 312, 323
Cri-du-chat disorder, 272
Crinoidea, 708i, 709
Crisp, Michael, 509
Criss-cross inheritance, **268**
Cristae, **106,** 106i
Critical period, 1282
Crocodilians, 724, 730, 730i
Crocodilians, classification of, 537, 538i, 546
Crocus *(Crocus),* 761, 761i
Crop
bird, 1047i, 1048
earthworm, **1047,** 1047i
insect, 1047, 1047i
Cross
dihybrid, 246, 248i, 249
monohybrid, 243
procedure in pea plants, 241i, 242
reciprocal, 242
testcross, 246, 247i
Cross-fertilization, 241i, **242**
Cross-pollination, 241i, **242**
Cross-talk, **156**–157, 157i
Crossbridges, 967, 968–969, 970i, 972
Crossing-over, 227, 227i, 228i, 230–232, 230i, 232i, 235, 262, 265, 265i
unequal, 433, 433i
Crowding, 1185–1186, 1185i
Crown gall disease, 409
Crustacean
description, 697–699, 698i, 699i
molting, 962, 962i
neurogenic hearts, 989
Cryptic coloration, 7–8, 8i, **1202**
Cryptochrome, **855,** 859
Cryptococcus neoformans, 662
Crystallin, 1127–1128
Crystalline cone, 930, 930i
Ctenophore, 681, 683i
Cuboidal epithelium, 869, 870i
Cubozoa, 680, 682i
Cuckoo, 1208, 1281i
Cud, 1065
Cullis, Christopher, 863
Cupula, **925,** 925i, 926i
Currents, ocean, 1145–1146, 1146i, 1150, 1150i
Cusps, 1061
Cuticle
exoskeleton, 974
plant, 197, **613,** 613i, **756,** 756i, 762, 763i, 784
roundworm, **693**
Cuttings, 825
Cuttlefish, 690, 1027i, 1028
Cuvier, Georges, 441
CWD. *See* Chronic wasting disease
Cyanide, 82, 173
Cyanobacteria, 580
biosignature molecules, 560
in eutrophic lake, 1164, 1164i
fossil, 560, 560i
heterocyst, 574, 580i
in lichens, 664
nitrogen fixation, 580
in *Paulinella,* 607
photosynthesis in, 184, 187, 188, 192, 560, 573, 580
plastid evolution from, 606
Cyanophora paradoxa, 563i
Cycad, 633, 633i
Cycadophyta, **633**
Cyclic adenosine monophosphate (cAMP), 605, 840
Cyclic ADP-ribose, 850–851
Cyclic AMP (cAMP), 949i
CAP and, 344
as second messenger, **151**–152, 151i, 156, 157, 362
Cyclic AMP signaling, memory and, 918
Cyclic electron flow, **191**–192, 192i
Cyclin, 217–219, 217i, **218,** 222
Cyclin-dependent kinase (CDK), 217i, **218,** 219, 222
Cyclooxygenase-1, 1142
Cyclosporin A, 1015
Cyp26b1 gene, 231
Cyst, 594
Cystic fibrosis, 119–120, 127, 251, 276–277, 276i, 412
Cystic fibrosis transmembrane conductance regulator (CFTR), 119–120, 120i, 277
Cytochrome *c,* 173
Cytochrome oxidase, 82
Cytochrome oxidase 1 gene, 1267
Cytochromes, **173,** 189, 189i, 190i, 192i
Cytogenetics, 282
Cytokine, **1006,** 1016
Cytokinesis, **206,** 207, 207i, 211, 212i
Cytokinin, 153, 840t, **846,** 846i
Cytoplasm, **94,** 570
Cytoplasmic determinants, **355,** 1119
Cytoplasmic inheritance, **280,** 280i
Cytoplasmic streaming, 109, 603, **650**
Cytosine, 63, 64i, 66i, **290,** 290i
DNA methylation of, 350
Cytoskeleton
defined, **107**
membrane protein attachment to, 123–124
motor proteins, 107–108, 109i
prokaryotic, 96, 569
structural elements of, 107–109, 108i

Cytosol, **94,** 97, 98i, 99i
protein sorting to, 329
Cytotoxic T cell, **1016,** 1017i, 1018, 1021

DADP. *See* Deoxyadenosine diphosphate
DAG. *See* Diacylglycerol
Daily torpor, **1090**–1091, 1091i
Dall mountain sheep *(Ovis dalli),* 1177i, 1178
Dalton, **25**
Dalton, John, 25
Daly, R. C., 959
dAMP. *See* Deoxyadenosine monophosphate
Dams, 1257–1258, 1257i
Damschen, Ellen I., 1270, 1271i
Dance communication, 1294–1295, 1294I
Danio rerio, 205, 205i
Daphnia, 1201, 1201i
Dark field microscopy, 93i
Darwin, Charles, 6–7
on artificial selection, 445
barnacle study, 447
Beagle voyage, 442–445, 443i, 447
branching evolution concept, 530, 530i
on descent with modification, 446
on earthworms, 691
evolutionary classification and, 549–550
on flowering plant evolution, 637
Galápagos finches and, 444–445, 445i, 446
illustration of, 439i
life as a scientist, 447
Linnaean Society presentation, 439
on natural selection, 446, 448–449
observations and inferences, 445–446, 446i
On the Origin of Species by Means of Natural Selection, 439–440, 446, 447, 481
phototropism studies by, 841–842, 841i
Data
experimental, 13
observational, 13, 15, 17i
Dating
radiometric, **504**–505, 504i
relative, 501, 504
DATP. *See* Deoxyadenosine triphosphate
Daughter chromosomes, 211
Dayflower *(Commelina communis),* 784i
Day-neutral plant, **860**
De Vries, Hugo, 249
Deamer, David W., 558
Death cap mushroom *(Amanita phalloides),* 662
Death rate, per capita, 1182
Death signal, 1140, 1140i
Decapoda, 698, 698i
Decay, radioactive, 25
Deciduous forest
temperate, 1160, 1160i
tropical, 1156
Deciduous plant, 764
Decomposer, **4,** 183, **1212,** 1231
Dedifferentiation, 361, 824, 825, 825i
Deep vein thrombosis, **993**–994
Deer fly *(Chrysops),* 930i
Deer mouse *(Peromyscus maniculatus),* 1038, 1038i, 1091i
Defecation reflex, **1059**
Defense behavior, social, 1299, 1299I
Defensin, 1005
Deforestation, 1255–1256, 1255i, 1274
Degeneracy, in genetic code, **315**
Dehydration, 1069, 1084
Dehydration synthesis, **46,** 46i, 557
Dehydrogenases, 164
Delay, Rona, 940
Deletion, chromosomal, **272,** 272i, 282
Delphinium decorum, 519, 520i
Delphinium nudicaule, 519, 520i
Delta wave, 916, 917i
Deltaproteobacteria, 578
Demographic transition model, **1194**–1195, 1195i
Demography, 1175–1178
defined, **1175**
life tables, 1176–1177, 1177t
survivorship curve, 1177–1178, 1177i
Denaturation, protein, **60,** 61i, 84
Dendrite, neuron, 875, 875i, **884,** 885i, 922i
Dendritic cell, **1010,** 1012i
Denitrification, 1244i, 1245, 1245t
Density-dependent factors, **1185**–1186, 1185i, 1186i, 1188i
Density-independent factors, **1187,** 1187i
Deoxyadenosine diphosphate (dADP), 64
Deoxyadenosine monophosphate (dAMP), 64
Deoxyadenosine triphosphate (dATP), 64
Deoxyribonuclease, 305
Deoxyribonucleic acid (DNA). *See* DNA
Deoxyribonucleoside triphosphates, 295, 295i
Deoxyribonucleotides, **64**
Deoxyribose, **64,** 290, 295
Dephosphorylation, 83
Depolarization, **889,** 889i, 890i, 892, 892i, 893, 894i, 897
Deposit feeder, 1045i, **1046**–1047
Derivative cell, plant meristem, **759,** 768i
Derived character state, **538,** 539
Dermal tissue system
components and functions, 753t
defined, **752**
epidermis, 756–757
location of, 753i
Dermatophagoides pteronyssinus, 697i
Dermis, **1090,** 1090i
Descartes, René, 440
Descending reticular formation, 910
Descent with modification, **446**
Descriptive science, 13
Desert, **1158,** 1158i
Desertification, **1256,** 1256i
Deshaies, Raymond, 222
Desmosome, **114,** 114i
Determinate cleavage, **674**
Determinate growth, **750,** 1065
Determination, **355,** 356, 356i, **1121,** 1138
Detrital food web, 1230i, 1231
Detritivore, **1212,** 1231
Deuterium, 25, 25i
Deuteromycetes, 663
Deuterostome, 706–747
blastopore development in, 1122–1123
developmental patterns, 674–675, 674i
invertebrate, 707–710, 708i–710i
nervous system, 675
Development
defined, **5**
embryonic development pathways, evolution of, 454
evolutionary developmental biology, 521–524
patterns
in animals, 674–675, 674i
in insects, 701, 701i
regulation of, 355–360, 356i–360i
determination, 355, 356, 356i
in *Dictyostelium,* 605
differentiation, 355, 356, 356i
induction, 355–356
location of developmental information, 355
microRNAs, 359–360
pattern formation, 357–359, 357i–360i
Developmental biology, evolutionary, 521–524
Developmental plasticity, 639
Developmental switch, 522
Diabetes mellitus, 149, **959**–960, 977–978, 1018
Diacylglycerol (DAG), 151–153, 152i, 157i
Diagnostic characters, 481, 481i
Diaphragm, 1031, 1031i, 1032i, 1112t, 1113
Diapsid
crocodilians, 728, 730, 730i
defined, **724**
diversification during mesozoic era, 724, 726
sphenodontids, 727, 728i
squamates, 727–728, 728i
turtles as, 726–727, 727i
Diapsida, **724**
Diarrhea, 1059
Diastole, **989,** 989i
Diastolic blood pressure, 991
Diatom, **596**–597, 596i, 597i
Diatomaceous earth, 597
Diatoms, 221
Dicer, 353, 353i, 360
Dichloro-diphenyl-trichloroethane (DDT), 1239
2,4-D (2,4-dichlorophenoxyacetic acid), 843
Dickerson, Richard E., 560
Dickinsonia, 501i
Dictyostelium discoideum, 603, 604i, 605
Dideoxy (Sanger) method for DNA sequencing, 418, 419i
Didinium, 588i
Diencephalon, 906i
Diethylstilbestrol (DES), 962
Differential interference contrast (Nomarski) microscopy, 93i
Differentiation, 339–340, **355, 356,** 356i, **1121**
animal, 1128, 1138–1139
plant cell, 749, 828
Diffusion
across respiratory surface, 1025
defined, **126**
exchange, 133
facilitated, 126t, 127, 128i, 129
passive, 1070, 1077
simple, 126t, 127
Digestion
defined, **1043**
enzymatic hydrolysis, 1046, 1047, 1053i
extracellular, 1046
intracellular, 1046
steps, 1046–1047
Digestive system
annelid, 691, 691i, 1047, 1047i
bird, 1047i, 1048
echinoderm, 707, 709i
flatworm, 684, 684i
gastrovascular cavity, 1046, 1046i
insect, 1047, 1047i
mammals, 1048–1059, 1049i, 1052i–1059i
regulation, 1059–1061, 1060i
rotifers, 685
ruminant, 1063, 1063i, 1065
saclike, 1046
specializations in vertebrates, 1061–1065
intestine length, 1061, 1063, 1063i
symbiotic microorganisms, 1063, 1063i, 1065
teeth, 1061, 1061i
wall layers, 1052, 1052i
Digestive tract, **685,** 1046, **1047**
Digestive tube, 1046
Digger wasp *(Philanthus triangulum),* 1289
Dihybrid, **249**
Dihybrid cross, 246, 248i, **249**
Dikaryon, **653,** 653i, 654, 657i, 658, 660i, 661, 662i, 663
Dimer, 147, 148i
Dinerstein, Eric, 1274
Dinoflagellates, 221, **594,** 594i, 606i
Dinosaur, 724, 726
Anchiornis huxleyi, 500, 500i
Hadrosaurus foulkii, 499
lambeosaurines, 505–506, 505i
parental care in, 546, 546i
Dioecious species, **814**
Dipeptidase, 1053i
Diploblastic body plan, **672**
Diploda, 699
Diploid, **206,** 226, 234–235, 235i, 471–472, 472t
Diplomonadida, 591, 591i
Directional selection, **467**–468, 467i, 469i
Disaccharidase, 1053i, **1057,** 1057i
Disaccharides, **47,** 48, 48i
Disclimax community, **1220**
Discontinuous replication, **298**
Disease
bacterial, 581
defense against, 1002–1023
adaptive immunity, 1003, 1007–1017
innate immunity, 1003, 1004–1007, 1005i
invertebrates, 1020
lines of defense, 1003–1004
in sharks and rays, 1020
exotic species introduction and, 1261–1262, 1262i
disheveled gene, 1283
Disjunct distribution, **509**–510
Disorder, 73–74
Dispersal, **509**
Dispersed duplication of genes, **433**
Dispersion, **1173**–1174
Dispersive replication model, 293, 293i
Disruptive selection, 467i, **468,** 470, 470i
Dissociation
defined, **36**
of water, 36
Distal convoluted tubule, **1075,** 1075i, 1076i, 1077t, 1078
Distal-less gene, 699
Disturbance
ecological succession, 1217–1220, 1218i, 1219i
effect on community characteristics, 1216–1217, 1216i, 1217i
intermediate disturbance hypothesis, 1217, 1217i
Disturbance climax, 1220
Disulfide linkage, 57, 57i, 60i, 61i
Diversity, evolution of, 744
Diving mammal, 1038–1039
Dixson, Henry, 781
DNA
base-pairing, 66, 66i
defined, **3, 63, 286**
development from RNA, 558
discovery of, 286–287
formation flow, role in, 3, 3i
fragments, separation by agarose gel electrophoresis, 399, 400i
as the hereditary molecule, 287–289, 288i, 289i
Avery, MacLeod, and McCarty experiments, 287
Griffith experiment, 287, 288i
Hershey and Chase experiment, 288–289, 289i
inheritance of, 5
library, 394, 396–397
methylation, **350**–351
mutations in, 7
organization
in eukaryotes, 304–306, 305i
in prokaryotes, 306, 306i
packing, in eukaryotes, 304–306
recombinant, 393
repair
mismatch repair, 303–304, 304i
mutations, 304
proofreading by DNA polymerase, 303, 303i
replication, 66, 66i, 293–304, 293i–304i
in archaeans, 584
bacterial, 220–221, 306, 306i
continuous and discontinuous synthesis, 297–298, 297i
enzymes of
DNA polymerase, 295–299, 295i–299i
helicase, 296, 297i
ligase, 299i

DNA *(Continued)*
replication *(Continued)*
enzymes of *(Continued)*
primase, 297, 297i
table of, 300t
telomerase, 302, 302i
topoisomerases, 297, 297i
errors, 303–304, 303i
Meselson and Stahl experiment, 293, 294i, 295
origin, 296, 300, 300i, 306, 306i
plasmid, 306
primers for, 297
S phase of cell cycle, 207, 207i
semiconservative, 292i, 293, 293i, 294i, 295
speed of, 299
at telomeres, 300, 302, 302i
transcription compared, 318
Watson and Crick model for, 292i, 293, 293i
sequence, molecular phylogenetic use of, 536, 543, 544i
structure, 3, 3i, 18, 65–66, 66i, 290–292, 290i, 291i, 292i
antiparallel strands, 291
complementary base pairing, 291, 292i
double helix, 291–292, 292i
Watson and Crick model, 286, 287i, 290–292, 293i
Wilkins and Franklin, 290, 291i
T DNA, 409, 409i
DNA barcoding, 1267
DNA-binding motifs, 347–348, 348i
DNA chip, 396, **428**
DNA cloning, **392,** 392–400, 392i, 395i
DNA library, 394, 396–397
identifying clones by DNA hybridization, 397, 397i
overview of process, 392–393, 392i
polymerase chain reaction and, 398, 398i, 399,
restriction enzyme use in, 393–394, 393i, 395i
DNA fingerprinting
defined, **401**
forensics, use in, 391–392, 403–404
paternity testing, use in, 404
principles of, 401, 403
DNA helicase, **296,** 297i, 298i, 300t, 584
DNA hybridization
defined, **397**
for identification of DNA sequence of interest, 397, 397i
Southern blot analysis, **401,** 402i
DNA library, 394, 396–397
DNA ligase
in DNA cloning, 394, 395i
DNA repair and, 304, 304i
in replication, 299i
DNA microarray, **428,** 429i, 432, 862
DNA polymerase, 295–299, 295i–299i, 300t
DNA polymerase I, 298i, **299**
DNA polymerase III, 298i–299i, **299,** 303
error rate, 303
in PCR, 398i
proofreading mechanism, 303, 303i
sliding clamp, 296, 296i, 300t
structure, 295–296, 296i
thermostable, 583
DNA repair mechanisms, **303**–304, 304i
DNA sequencing
massively parallel, 421
methods
dideoxy (Sanger) method, 418, 419i
Illumina/Solexa method, 420i, 421
whole-genome shotgun sequencing, 418, 418i
DNA technology, 391–413
agarose gel electrophoresis, 399, 400i
applications of, 393, 400–412
DNA fingerprinting, 401, 403–404
genetic engineering, 404–412
molecular testing for genetic disease, 401
defined, **392**
DNA cloning, 392–400
DNA fingerprinting, 401, 403–404
DNA hybridization, 397, 397i
PCR (polymerase chain reaction), 398, 398i, 399, 400i, 401
RFLP (restriction fragment length polymorphism) analysis, 401, 402i
Southern blot analysis, 402i
terminology, 401
Dobberstein, B., 330
Dobzhansky, Theodosius, 1225
Dodder, 806, 807
Dog rose *(Rosa canina),* 813i
Dogs, classical conditioning in, 1284
Dogwood *(Cornus),* 757i
Dolly, the cloned sheep, 356, 407, 408i
Dolomedes sp., 12i
Dolphin, 734
Domain
Archaea, 10i, 11, 12i
Bacteria, 10i, 11, 12i
defined, **9**
differences among, 548t
Eukarya, 10i, 11–12, 12i
in Tree of Life, 548, 548i
Domains, protein, **61**–62, 63i
Dominance
codominance, 254
defined, **243**
hierarchies, **1300**
incomplete, 251–254, 253i, 460
Dominant allele, **243,** 277, 278, 279
Dominant species, **1210**–1211
Doolittle, Russell, 998
Dopamine, 157, 898t
Dormancy
plant, **861**–863, 862i
seed, **821**–822
Dorsal hollow nerve cord, 710, 710i
Dorsal lip of the blastopore, **1123,** 1124i, 1139i
Dorsal surface, **673**
Dosage compensation mechanism, **270**–271
Dosidicus gigas, 690i
Double bond, 43
Double fertilization, 639, 642i, 815i, **817**–818, 818i
Double helix, DNA, **65**–66, 65i
Double helix model, **291,** 292i
Double muscling, 977
Douglas fir *(Pseudotsuga menziesii),* 634, 861i, 862
Down syndrome, 227, 273, 274i
Downy mildew, **596,** 596i
Downy woodpecker *(Dendrocopus pubescens),* 469i
Dragonfly, 499i, 700i, 701
Dray, Nicolas, 677
Dromaeosaurus, 449, 450i
Drone fly *(Eristalis tenax),* 1203i
Drosophila melanogaster
development in, 354, 357–359, 357i–360i
homeotic genes, 359, 359i, 360i
maternal-effect genes, 358, 358i
microRNAs, 359
segmentation genes, 358, 358i, 359
eye color phenotypes in, 268, 268i, 269i
gametic isolation, 485
genome sequencing, 264
genome size, 423
hybrids, 486, 491, 496
life cycle, 357, 357i
linkage studies in, 262–266, 263i, 265i, 266i
as model research organism, 17, 264
mutations, 262, 264, 268i
reproductive system, 1101i
sex-linked genes in, 268, 268i, 269i
speciation in Hawaiian, 488
taste receptors, 935i
temporal isolation, 484
wingless/Wnt pathway, 1283
Yp1 sequence in, 545i
dT primer, 396, 396i
Dubnau, Josh, 918
Duboule, Denis, 523
Duchenne muscular dystrophy, 277, 277i, 401, 968, 977
Duck-billed platypus *(Ornithorhynchus anatinus),* 706–707, 707i, 733i, 733, 734
Ducks, sexual selection in, 491i
Dugesia, 1046, 1046i, 1073, 1073i
Dunkleosteus, 718, 718i
Duodenum, **1056,** 1056i, 1057i, 1060
Duplication, chromosomal, **272,** 272i
Dust mite, 697, 697i
Dutch elm disease, 652, 659
Dwarfing, achondroplastic, 277, 278
Dwarfism, 251
Dynamic equilibrium, 127
Dyneins, 107, 109, 110i
Dystrophin, 277

E site, **324,** 324i–326i, 325, 328i–329i
Ear, human, 927–929, 928i
Earth
age of, 553
calendar of 505, 505i, 506–508, 506i, 507i
environmental changes, 508–509, 508i
geological time scale, 502t–503t
as closed system, 72
plate tectonics, 506–508, 506i, 507i
primordial, 554–555, 556i
viewed from space, 1i
Earthworm
deposit feeders, 1045
description, 691, 691i, 692i
digestive system, 1047, 1047i
hermaphroditism, 1103, 1103i
hydrostatic skeleton, 974i
metanephridium, 1073, 1073i
taste/smell receptors, 934
vibration-detection system, 927
Eastern gray kangaroo *(Macropus giganteus),* 734i
Eastern hemlock *(Tsuga canadensis),* 1261, 1262i
Eastern painted turtle *(Chrysemys picta),* 727i
Eastern prairie fringed orchid *(Platanthera leucophaea),* 641i
Ecdysis, **677,** 694, 695i, 701
Ecdysone, **961,** 961i
Ecdysozoa
Arthropoda, 694–702
description, 677
Nematoda, 693, 694
Onychophora, 693, 694i
ECF. *See* Extracellular fluid (ECF)
ECG. *See* Electrocardiogram
Echidna, 733i, 734
Echinocereus, 511i
Echinoderm
anatomy, 707–708, 709i
endoskeleton, 975
groups
Asteroidea, 708, 708i
Concentricycloidea, 708
Crinoidea, 708i, 709
Echinoidea, 708i, 709
Holothuroidea, 708i, 709
Ophiuroidea, 708, 708i
hydrostatic skeletal elements, 974
nerve ring, 904, 904i
overview, 707–709
Echinoidea, 708i, 709
Echinolittorina peruviana, 1087
Echolocation, **921**
ECM. *See* Extracellular matrix
Ecological community
boundaries, 1209, 1209i
characteristics
change during succession, 1219, 1219i
disturbance, effects of, 1216–1217, 1216i–1217i
foundation species, 1210, 1211i
plant growth forms, 1210, 1210i
population interactions, effects on, 1213–1216
relative abundance, 1210–1211
species diversity, 1211, 1211i
species richness, 1210–1211, 1213–1216, 1214i, 1215i, 1221–1225, 1221i–1223i
trophic structure, 1211–1213, 1212i
defined, **1200**
ecological succession, 1217–1220, 1218i, 1219i
ecotones, 1209
individualistic view of, 1208–1209, 1208i
interactive view of, 1208, 1209i
nature of, 1208–1209
positive *versus* negative interactions, 1225
species composition, 1208
stability, 1213
Ecological efficiency, 1235–1237
assimilation efficiency, 1235
defined, **1235**
harvesting efficiency, 1235
production efficiency, 1235
Ecological isolation, **484**
Ecological niche
defined, **1204**
fundamental niche, **1204,** 1205, 1205i, 1206i
realized niche, **1204,** 1205, 1205i, 1206i
Ecological pyramid, **1237**–1238, 1237i, 1238i
Ecological succession
in animals, 1219, 1219i
aquatic, 1219
community characteristics, changes in, 1219, 1219i
defined, **1217**
disclimax community, 1220
facilitation hypothesis, 1220
inhibition hypothesis, 1220
local, 1220
primary, 1217–1218, 1218i
secondary, 1219
tolerance hypothesis, 1220
Ecologists, 1147
Ecology
applied, 1147
basic, 1147
community, 1208–1225
defined, **1146**
levels of organization, 1147
population, 1172–1198
scale, importance of, 1196
science of, 1146–1147
Ecomorph, 729
*Eco*RI restriction enzyme, 393–394, 393i
Ecosystem, 1229–1253
defined, 2i, **3, 1229**
energy flow, 1232–1240
consumers, 1238, 1240i
ecological efficiency, 1235–1237
ecological pyramids, 1237–1238, 1237i, 1238i
food webs, 1230i, 1231
primary productivity, 1233–1235, 1233i, 1234i, 1235i
secondary productivity, 1235
Silver Springs ecosystem, 1229i, 1235, 1236i, 1237
sunlight, 1233
human disruption of, 1246–1250
hydrologic alterations of, 1257–1258, 1257i
Lake Erie, 1229–1230, 1229i

  modeling, 1230–1232, 1231i
  nutrient cycling, 1241–1250
    carbon cycle, 1242, 1244i–1245i, 1246–1248, 1250–1251
    compartment model, 1231–1232, 1231i
    hydrologic cycle, 1241–1242, 1241i
    nitrogen cycle, 1243–1245, 1244i, 1245t, 1248–1250
    phosphorus cycle, 1245–1246, 1246i
Ecosystem ecology, **1147**
Ecosystem services, **1264**
Ecosystem valuation, **1273**–1274
Ecotherm, 1152–1153
Ecotone, **1209**
Ecotourism, **1273**
Ectoderm, **672,** 673i, **1121,** 1121t, 1122–1129, 1122i, 1124i–1128i, 1137i
Ectomycorrhizae, **666,** 666i
Ectoparasite, **685, 1208**
Ectoprocta, 683, 683i
Ectotherm
  amphibians, 1086, 1088
  defined, **1086**
  endotherms compared, 1086–1087, 1087i
  fishes, 1086, 1088
  invertebrates, 1086, 1087
  reptiles, 1086, 1087i, 1088
  thermal acclimatization, 1088
Edge effects, **1259,** 1270, 1270i
Edidin, Michael A., 124, 124i
Edwards syndrome, 275
EEG. *See* Electroencephalogram
EF. *See* Elongation factor
Effector, **149,** 150i, 152i, 878i, **879,** 879i, **884**
Effector T cell, **1011,** 1016
Efferent arteriole, 1075i, **1076**
Efferent neuron, **884,** 885
Egg
  amniote, 724, 724i
  conifer, 635, 636i
  fertilization, 1100–1103, 1100i, 1102i
    in humans, 1110–1111, 1110i
  information storage in, 1119
  oogenesis, 1098i, 1099–1100
  yolk, 1119
Egg activation, 1102
Egg cell
  human, 339, 339i
  plant, 815i, **816,** 816i, 817
Egg coat, 1100, 1100i, 1102, 1114
Egg jelly, 1100, 1101i, 1102i
Ehlers-Danlos syndrome, 462
Ehrlich, Paul R., 1190i
Eiselé, Jean-Luc, 896
Eisenberg, Robert, 1186
Ejaculation, 1108, 1110
El Niño, 1145–1146, 1146i
El Niño Southern Oscillation (ENSO), 1146
Elasmobranchii, 718–719
Elastin, **872**
Eldredge, Niles, 515
Electric eel *(Electrophorus electricus),* 1294
Electrical signal, **1294**
Electrical synapse, **885,** 886i
Electrocardiogram (ECG), 981, 982I, **990,** 990i
Electrochemical gradient, **132,** 776
Electroencephalogram (EEG), 916–917, 917i, 918i
Electromagnetic spectrum, **185,** 186i
Electron
  defined, **24**
  energy levels (shells), 186, 186i
  excited state, 186–187, 186i
  ground state, 186–187, 186i
  orbitals, 26–28, 27i
  photosynthesis, role in, 183–187, 189, 189i, 190i, 191–192, 192i
  valence, 27–28
Electron acceptor, 162, 163–164
Electron carrier
  cytochrome, 189, 189i, 190i, 192i
  ferredoxin, 190i, 191, 192i
  flavin adenine dinucleotide, 168–169, 170i–173i, 171, 173, 175–176, 175i
  NADPH, 183i, 184–185, 189, 189i, 190i, 191–193, 195i, 196, 198
  nicotinamide adenine dinucleotide, 164i, 166, 166i, 167i–173i, 168–169, 173, 175i, 176–178
  in photosynthesis, 189, 190i, 191, 192i
  plastocyanin, 190i, 191, 192i
  plastoquinone, 190i, 191, 192i
Electron donor, 162
Electron flow
  cyclic, 191–192, 192i
  electron transfer system, 164i, 165, 172i, 173, 173i, 175–176
  to final electron acceptors, 163–164
  linear, 189i, 191
  oxidation-reduction (redox) reactions, 162–163, 163i
Electron microscope, **92**
Electron microscopy, 93i
Electron transfer system, 164i, 165, **165,** 171, 172i, 173i, 175–176
  in photosynthesis, 189, 189i, 190i, 191
Electronegativity, **30**
Electrophoresis, agarose gel, 399, 400i
Electrophysiology, 981
Electroporation, 375
Electroreceptor, **719,** 939
Element
  chemical symbols, 24, 24t
  defined, **23**
  proportions in living organisms, 23, 23i
  structure of, 24
  trace, 23
Elephant, 445, 452–453
Elephantiasis, 686, 686i
Elimination, **1047**
Elk *(Cervus elaphus),* 943
Ellis-van Crevald syndrome, 465, 470
Elm *(Ulmus),* 769i
Elongation, in translation, **324,** 325–326, 326i
Elongation factor (EF), **325,** 326i
Embden, Gustav, 165
Embden-Meyerhof pathway, 165. *See also* Glycolysis
Embryo
  animal (*See* Animal development)
  defined, **671**
  plant
    development, 818–819, 819i, 826–832
    seed germination, 821–822, 822i, 823i
    somatic, 824, 825i
    structure, 818–819, 819i, 820i
Embryo sac, 815i, **816,** 834
Embryogenesis, **357**
Embryology, 1122
Embryonic development pathways, evolution of, 454
Embryonic disc, 1129, 1130
Embryonic lethal mutations, 357
Embryonic stem cells, 871
Embryophyte, **613**
Emergency contraception pill, 1113
Emergent properties, **3**
Emigration, **1175**
Emlen, Stephen, 1289, 1290i
Emotions, 912
Emu *(Dromaius novaehollandiae),* 441i
Emulsification, 1057
Encephalartos, 633i
Endangered species, **1265**
Endemic species, **510**
Endergonic reaction, **75,** 76i, 78i
Endler, John, 1180–1181
Endocrine glands, 870i, **871, 944,** 950i, 951t–952t
Endocrine signaling, classical, 944–945, 944i
Endocrine system, 943–966
  defined, **944**
  human glands and humans, 951t–952t
  invertebrates, 961–962, 961i
  signaling, types of, 944–945, 944i
Endocytic vesicle, **104,** 104i, 105i, 135i, 137, 137i
Endocytosis, **104,** 104i, 134–137, 135i, 562, 562i
  bulk-phase (pinocytosis), **134,** 135i
  in cancer cells, 119i
  receptor-mediated, **134,** 135i, 136–137, 382
  of receptors and signal molecules, 152
Endoderm, **672,** 673i, **1121,** 1121t, 1122i, 1123–1126, 1124i, 1125i, 1129, 1137i
Endodermis, 766i, **767,** 780–781, 780i, 783
Endoduplication, 829
Endolymph, 925, 925i
Endomembrane system, **102,** 102–106, 105i, 330
Endometrium, **1104,** 1129, 1129i, 1130i
Endomycorrhizae, **665,** 665i, 666
Endoparasite, **685, 1208**
Endoplasmic reticulum
  appearance, 98i, 99i, 102i
  Ca^{2+} release by, 151, 152i, 153
  defined, **102**
  lumen, **102,** 102i
  origin of, 562, 562i
  protein sorting and, 330, 330i
  rough ER, **102**–103, 102i
  smooth ER, 102i, **103**
Endorphin, 898t, 937, 951t, 952, 953i
Endoskeleton, **975**–976, 975i
Endosperm, 639, **817,** 818–819, 819i, 820i, 822i
Endospore, **574,** 574i
Endosymbiont hypothesis, **561**–563, 561i, 606i, 650–607
Endosymbiosis
  plastid evolution by, 561i, 562, 563i, 605–607, 606i
  primary, 606–607, 606i
  secondary, 606i, 607
Endosymbiotic theory, 106, **561**–563, 561i–563i
Endothelium, 991, 992i
Endothelium derived relaxing factor, 996i
Endotherm, 1153
  countercurrent heat exchanges, 1091–1092, 1091i
  defined, **1086**
  ectotherms compared, 1086–1087, 1087i
  estivation, 1091
  hibernation, 1091
  responses to temperature changes, 1089–1090
  set point variation, 1090–1091
  skin, 1090
  thermoreceptors, 1085, 1088–1089, 1089i
  torpor, 1090–1091, 1091i
Endothermic, **74**
Endotoxins, 581
End-product inhibition, **82,** 83i
Energy
  activation, 78–81, 78i, 79i
  conservation of, 72–73
  conversions, 72
  defined, **72**
  forms of, 72
  free, 74–76
  harvesting chemical, 161–179
  heat, 73, 73i, 176, 178
  kinetic, 72, 73
  potential, 72
  reversible reactions and, 75
  solar/sunlight, 73, 73i, 162i, 183, 186, 201
  thermodynamics, laws of, 72–74
Energy budget, **1178**
Energy coupling, **76**–78, 77i
Energy flow
  ecosystem, 1232–1240
  from sunlight to ATP, 162i
  through living systems, 4, 5i
Energy levels (shells), electron, **26**–28, 27i, 186, 186i
Energy partitioning, in animals, 1065
Energy transfer, in photosynthesis, 187
Engelhardt, John F., 413
Engelmann, Theodor, 182–183, 182i, 186
English walnut *(Juglans regia),* 824i
Enhancer, **346,** 346i, 348, 348i, 349i
Enkephalin, 898t
Enolase, 167i
Ensatina eschscholtzii, 482–483, 483i
ENSO. *See* El Niño Southern Oscillation
Entamoeba histolytica, 603
Enterocoelom, 674i, **675**
Enteromorpha, 1215–1216, 1215i
Enthalpy, **74**
Entropy, **74**
Enveloped viruses, **378,** 378i
Environment, phenotypic variation due to, 458–459, 458i
Enzymatic hydrolysis, **1046,** 1047
Enzyme(s), 31
  activation energy reduction by, 79, 79i, 80–81
  active site, 80, 80i, 82, 82i
  activity, influences on, 81–84
    concentration of enzyme and substrate, 81, 81i
    inhibitors, 81–83, 82i, 83i
    pH, 83, 84, 84i
    temperature, 83, 84, 84i
  catalysis by, 71, 79–81, 80i
  cofactors/coenzymes, 80
  conformation, 79, 80
  defined, **55, 79**
  denaturation, 84
  exoenzymes, 581
  lysosomal, 104–106
  membrane-associated, 125
  naming of, 79
  one gene-one enzyme hypothesis, 311
  prosthetic groups, 80
  regulation
    activators, 82, 83i
    allosteric, 82–83, 83i
    chemical modification, 83
    feedback inhibition, 82, 83i
    inhibitors, 81–83, 82i, 83i
  relationship to genes, 311, 312i–313i
  role in biological reactions, 71–72, 78–81
  saturation of, 81
  specificity of, 80
  transition state induction by, 79, 80–81
  unfolding, 84
Enzyme specificity, **80**
Enzyme-substrate complex, 80, 80i
Eosinophil, 1004t, **1006**
Ephedra, 629, 629i, 633–634, 634i, 638
Epiblast, **1124,** 1125, 1125i, 1129, 1130i
Epicotyl, **819,** 820i
Epidermal growth factor, 149
Epidermis
  animal skin, **1090,** 1090i
  cnidarian, 679–680, 679i
  plant
    components and functions, 753t
    cork, 767, 769i
    cuticle, 756, 756i, 762, 763i
    defined, **756**
    flower petal, 757
    guard cells, 756, 756i
    leaf, 756, 756i, 762, 763i, 764
    root, 766i, 767, 767i
    root hairs, 756i, 757
    stomata, 756, 756i
    structure, 756–757
    trichomes, 757
Epididymis, **1108,** 1108i
Epigenetics, **350**
Epiglottis, **1053,** 1053i

Epilimnion, **1163,** 1163i
Epinephrine, 142–143, 144i, 946, 948, 949i, 951t, **957,** 958i, 995, 1020, 1089
Epiphyte, **618,** 793i, **806,** 807i
Epistasis, **255**–256, 255i
Epithelial tissue, **869**–871, 869i, 870i, 1003
Epitope, **1009**
Epsilonproteobacteria, 578
EPSP. *See* Excitatory postsynaptic potential
Equilibrium point, 75, 75i
Equilibrium potential, **889**
Equilibrium theory of island biogeography, **1221**–1225, 1222i, 1223i
Equisetum, 625, 625i
Equus, 516i
Erectile tissue, 974
Erection, 1109, 1110
Ergotism, 659
Erwin, Terry, 1266
Erythrocyte, **873,** 985I, **986**–987, 986i, 987i
  in sickle-cell disease, 239–240, 239i, 254, 257, 257i, 276
Erythromycin, 324
Erythropoietin, **987,** 1037
Eschenmoser, Albert, 564
Escherich, Theodor, 369
Escherichia coli, 578
  bacteriophages, 369, 379, 380, 380i, 381i
  cell structure, 95, 95i
  culturing, 369, 371, 379–380
  discovery of, 369
  DNA replication, 220
  as facultative anaerobes, 177
  food contamination by, 567
  genetic engineering of, 393, 394, 395i, 404–405
  genome, 370–377, 423i, 424–425, 424i, 425t
  geometric population growth, 1180–1181
  horizontal gene transfer, 375–377
  lac operon, 341–342, 341i–343i, 344
  in mammalian intestine, 1059
  as model research organism, 17–18, 369–370, 369i
  recombination in, 371–377, 371i, 374
  strain O157:H7, 376
  structure, 11i
  trp operon, 344, 345i
ESCRT system, 584
Esophagus
  bird, 1047i, 1048
  earthworm, **1047,** 1047i
  food passage through, 1053
  human, 1049i
  insect, 1047, 1047i
Essay on the Principles of Population (Malthus), 445
Essential amino acid, **1044,** 1048, 1049i
Essential element, **794**–796, 796t
Essential fatty acid, **1044,** 1048
Essential mineral, **1044,** 1051, 1051t
Essential nutrient, **1044**
Ester linkage, **50**
Esterman, Michael, 939
Estivation, **1091**
Estradiol, 54, 54i, 157, 945, 952t, **958, 1105**
Estrogen receptor, 954–955, 962
Estrogens, 952t, **958,** 958–959, 962–963, **1105,** 1106i, 1107, 1111, 1113, 1132
Estrous cycle, 1107
Estuary, **1165,** 1165i
Ethane, 43i
Ethene, 43i
Ethology, **1279**
Ethyl alcohol, 44t
Ethylene, 840t, **846**–847
Ethyne, 43i
EU. *See* European Union
Eucalyptus, 764, 764i
Eucheuma, 601
Euchromatin, 305
Eudicot
  defined, **638**
  examples of, 639, 641i, 752
  leaves, 762, 762i, 764
  life cycle, 815i
  monocots compared, 752t
  pollen grains, 638, 638i
  primary growth, 759i
  secondary growth, 751
  stem growth, 758
  stem organization, 759–760, 760i
Euglena gracilis, 591, 592i
Euglenid, **591**–592, 606i, 607
Euglenozoans, **591**–592
  euglenid, 591–592, 592i
  kinetoplastid, 592, 592i
Eukarya (domain), **567**
  Bacteria and Archaea compared, 548t
  characteristics of, 11–12, 12i, 582t
  genome sizes, 423i, 423t, 424
  in Tree of Life, 10i, 548, 548i
Eukaryote(s)
  cellular components, 97–115, 98i–101i, 103i–111i, 112i–115i
  cellular respiration, location of, 165
  defined, **11, 94**
  DNA organization in, 304–306, 305i
  first, 607
  gene organization, 346, 346i
  genomes sequenced, 421t
  multicellularity, origin of, 563
  photosynthesis location in, 184, 185i
  protein sorting, 329–330, 330i
  regulation of transcription in, 344–351, 346i–351i
  ribosomes, 324
  transcription in, 313, 314i, 316i, 317–319
  translation in, 314i, 321, 324–328, 325i
  transposable elements, 333–335, 334i
Eukaryotic cells, origin of, 561–564, 561i, 562i
Eumetazoan
  with bilateral symmetry, 683–702
    ecdysozoan protosomes, 693–702
    lophotrochozoan protosomes, 683–692
  with radial symmetry, 679–682
    cnidarians, 679–681, 679i, 680i, 681i, 682i
    ctenophores, 681, 683i
Eupenicillium, 659i
Euphorbia, 511i
Euploid, **273**
European cuckoo *(Cuculus canorus),* 1281i
European hare *(Lepus europaeus),* 736i
European mouflon sheep *(Ovis musimon),* 1185i
European rabbit *(Oryctolagus cuniculus),* 1172–1173, 1173i
European starling *(Sturnus vulgaris),* 1172, 1260–1261, 1260i
European Union (EU), GMO regulation in, 410
Euryarchaeota, **582,** 583
Eusocial insect, **1302,** 1302i, 1303i
Eustachian tube, 929
Eutherian, 734, 736i, 1103
Eutrophic lake, **1164,** 1164i
Evans, David, 505
Evaporation
  in animals, **1086,** 1086i
  macromolecule formation by, 557
  transpiration, 781, 782i, 783–785
Everglades, 1256
Evergreen, 634–635
Evergreen coniferous forest, 1160, 1160i
Evernia lichens, 664
Evolution
  adaptations and, 7–8
  of adaptive traits, 475
  angiosperms coevolution with animal pollinators, 639, 641, 643i
  animal, key innovations in, 672–675
  of antibiotic resistance, 456–457
  of birds, 448, 449, 450i
  blood clotting, 997–998
  brain, 900
  branching, 530, 530i
  of cellular life, timeline for, 554i
  coevolution, 639, 641, 643i, 1200, 1202–1203, 1203i
  of common themes, 744
  of communication, 1293–1295
  convergent, 510–511, 511i
  defined, **6, 440**
  development of evolutionary thought, 439–455
  of diversity, 744
  ecological interactions and, 1196
  of fins, 717
  of flowering plants, 637–638, 640
  of flowers, 641, 645
  fossil record, 500–506
  genome, 376–377, 431–436
  geological time scale, 502t–503t
  of gynosperms, 631–632
  of horses, 515, 516i
  of human social behavior, 1304
  of humans, 739–743
  of insecticide resistance, 449i
  invertebrate, 702
  of jaws, 717–718, 717i
  Lamarck's theory, 442
  of leaves, 614–615, 614i
  life history traits in guppies, 1180–1181
  of limbs, 722, 722i
  macroevolution, 499–527
  of memory and learning, 918
  microevolution, 456–479
  misinterpretation of theory, 453
  of mitosis, 221
  modern synthesis, 448
  multicellularity, 868
  mutationism theory, 448
  mutations and, 7
  natural selection, 6–8
  of nest-building behavior, 1297–1298, 1298i
  orthogenesis, 453
  phylogenetic trees and, 10, 10i, 675–677, 676i
  of plants, 612–618, 616t, 626, 644, 645
  of plastids, 605–607, 606i
  punctuated equilibrium hypothesis, 515, 517, 518i
  of reproductive behavior and mating systems, 1296–1298
  response to climate change, 1169
  senses, 939–940
  of snakes, 450–451, 450i
  of social behavior, 1299–1302
  tetrapod, 721–722, 722i
  as unifying theme of biology, 6
  vertebrate, 712–715, 713i, 714i
  viral, 386–387
Evolutionary change, evidence for
  adaptation by natural selection, 448–449
  comparative morphology, 449–450
  fossil record, 449
  historical biogeography, 449
  molecular techniques and, 450–453, 450i
Evolutionary developmental biology, **521**–524
Evolutionary divergence, **446**
Exaptation, **519**
Excavata, **591**–592
Exchange diffusion, **133**
Excitatory postsynaptic potential (EPSP), **897,** 899, 899i
Excited state, electron, 186–187, 186i
Excretion
  defined, **1070**
  invertebrates, 1072–1074, 1073i, 1074i
  in mammals, 1074–1080, 1074i–1076i, 1077t, 1080i
  nitrogenous wastes, 1071–1072, 1072i
  osmoregulation and, 1071, 1071i
  tubule structure and function, 1071, 1071i
Excretory system
  annelid, 691, 691i
  flatworm, 684, 684i
Exergonic, **75,** 76i, 78i
Exocrine gland, **870**–871, 870i, 944
Exocytosis, **104,** 104i, 134, 135i, 137, 895–897
Exodermis, **767,** 781
Exoenzyme, 581
Exon(s), **319,** 319i, 352
  human genome, 426
Exon shuffling, **320,** 432, **434**–436
Exoskeleton
  arthropod, 694
  barnacle, 699
  carapace, 698
  crustacean, 697–698
  description, **694, 974**–975, 975i
  ecdysis, 694, 695i
Exothermic, **74**
Exotic species, **1260**–1261, 1261i, 1262i
Exotoxin, 581
Expansin, 844, 845i
Experimental data, **13,** 1147
Experimental variable, **15**
Exploitative competition, **1203**
Exponential model of population growth, 1181i, **1182**
Expression vector, **404**–406, 405i
Extensor muscle, **977**
External environment, **868,** 877
External fertilization, **1100**–1101, 1101i, 1102
External gills, 1026i, **1027**–1028, 1027i
External intercostal muscle, 1031, 1032i
Extinction
  background rate, 513, 1262–1263
  current rate, 1263
  defined, **513**
  determinants of, 453
  imminent, 1264–1265, 1265i, 1274
  mass, 513, 513i, 1263
  Miss Waldron's red colobus monkey, 1254–1255, 1254i
  natural theology and, 440, 441
Extracellular digestion, 652, **1046**
Extracellular fluid (ECF), 134, 142, **867,** 876i, 877, 1070, 1071
Extracellular matrix (ECM), **113,** 115, 115i, 868, 869, 872–873, 1135
Extraembryonic membranes, **1125,** 1126i, 1129, 1130i, 1131
Extreme thermophile, 583
Extremophile, 582
Eye
  compound, 695, 930, 930i
  development, 1127–1128, 1128i
  invertebrate
    compound eye, 930, 930i
    ocellus, 929–930, 929i
    single-lens eye, 930, 930i
  vertebrate, 930–934, 931i–934i
Eyecup, 929–930
Eyespot, 591–592, 592i, 929–930

F^- cell, 372, 374, 373i,
F^+ cell, 372, 373i
F factor, 371, 372, 373i 374
F_1 generation, **243**
F_2 generation, **243**
F pilus, **372,** 373i
Facilitated diffusion, 126t, **127,** 128i, 129
Facilitation hypothesis, **1220**
Facultative anaerobe, **177, 573**
FAD. *See* Flavin adenine dinucleotide
Fallén, C. F., 264
Faloona, F., 398
Familial (hereditary) cancers, **361**
Familial hypercholesteremia, 136
Family, **9**
Family planning programs, **1195**
Fan worm *(Sabella melanostigma),* 692i
Fanworm *(Myxicola),* 893

Fas-FasL system, 1018
Fast block to polyspermy, **1102,** 1102i
Fast muscle fiber, 971t, **972**
Fat, **50,** 51, 52, 1053i, 1057, 1058
Fat-soluble vitamins, **1049,** 1050t
Fate map, **1136**
Fats, 171, 171i
Fatty acid
  defined, **50**
  essential, 1044, 1048
  from fat digestion, 1058, 1058i
  prostaglandins from, 945
  saturated, 50–52, 50i
  trans, 52
  unsaturated, 50–52, 50i
Fe/S protein, 172i, 173i
Feather, 519–520, 521i, **730,** 730i
Feather star, 708i, 709
Feces, **1059**
Fecundity
  age-specific, **1177,** 1177t
  versus parental care, 1179
Feedback inhibition, **82,** 83i
Feeding
  adaptations, 1201
  carnivores, 1044, 1061, 1061i, 1063i
  defenses against herbivory and predation, 1201–1203, 1202i, 1203i
  defined, 1044
  food choice, 1201, 1201i
  herbivores, 1044, 1061, 1061i, 1063, 1063i
  methods, 1044–1046
    bulk feeders, 1045i, 1046
    deposit feeders, 1045–1046, 1045i
    fluid feeders, 1045, 1045i
    suspension feeders, 1045, 1045i
  omnivores, 1044
  optimal foraging theory, 1201
  regulation of, 1065
Feild, Taylor, 615
Female condom, 1112t, 1113
Female sex organs, development of, 1132, 1133i
Feng, Da-Fei, 998
Fermentation, 176–179
  alcoholic, 177, 177i
  by anaerobes, 177
  defined, **176**
  lactate, 176–177, 177i
Fern, 623–625, 624i, 626–627
Fernald, Russell, 1287, 1288i
Ferraris, Joan D., 1078–1079
Ferredoxin, 190i, 191, 192i
Fertilization, **225,** 226i, 233, 234–235, 234i
  in animals, 1100–1103, 1101i, 1102i
  defined, **1099**
  double, 639, 642i, 815i, 817–818, 818i
  external, 1100–1101, 1101i, 1102
  in humans, 1110–1112, 1110i
  internal, 1101, 1102
  plant, 817–818
  prevention of, 1112–1114, 1112t
  steps in, 1102, 1102i
Fertilizer, 798, 801
Fetus, **1128,** 1131–1132, 1131i
Fever, **1006**
FGFRs. *See* Fibroblast growth factor receptors
Fibers, plant sclerenchymal, **754,** 754i
Fibrin, **987,** 987i
Fibrinogen, **986**
Fibroblast, **872**
Fibroblast growth factor receptors (FGFRs), 278
Fibronectin, 115, 115i, **872,** 1135
Fibrous connective tissue, **872,** 873i
Fibrous root system, **765,** 765i
Fiddler crab *(Uca),* 1045–1046, 1045i
Fight-or-flight, 956–957
Filament, flower, 813i, **814**
Filarial worm, 686, 686i, 693
Filtration, **1071,** 1071i, 1076
Fin, 717, 717i
Final acceptor for electrons, 163, 190i
Fire, Andrew, 352
Fire ant, 700i
Firefly, behavioral isolation in, 484, 485i
First law of thermodynamics, **72**–73
First messenger, **142, 149,** 150i, 895
Fischer, Edmond, 147
Fish
  anatomy, 720i
  antifreeze proteins, 1152
  circulatory system, 983–984, 984i
  ectothermy, 1086, 1088
  evolution, 717–718
  gills, 1027i, 1028, 1028i
  hermaphroditism in, 1103–1104
  jawed, 717–721
    actinopterygii, 719, 719i, 720i, 721
    chondrichthyes, 718–719, 718i
    origin of, 717–718, 718i
    sarcopterygii, 719, 721, 721i
  jawless, 715–716
  kidney function
    freshwater fishes, 1084i
    marine fishes, 1084, 1084i
  lateral line system, 926, 926i
  lungfish, 1029
  osteolepiforms, 721, 722i
  overexploitation, 1256–1257, 1257i
Fisher, Donald, 787i
Fishing spider (*Dolomedes* species), 12i
Fission, **1097**
Fitzroy, Robert, 444
5′ cap, **318,** 319i
5′ end, **290,** 290i, 317
5′ untranslated region (5′ UTR), **318,** 319i
Five-lined skink *(Eumeces fasiatus),* 1177i, 1178
Fixed action pattern, **1280**
Flabellina iodinea, 689i
Flagella
  beating patterns, 110i
  eukaryotic, 109–110, 110i
  prokaryotic, 92i, 95i, **96, 572,** 572i
  protist, 590
Flame cell, 684, 684i, 1073, 1073i
Flamingo *(Phaenicopterus ruber),* 1172i
Flatulence, **1059**
Flatus, 1059
Flatworm. *See also* Platyhelminthes
  gastrovascular cavity, 1046, 1046i
  hydrostatic skeleton, 974
  nervous system, 904i, 905
  ocellus, 929–930, 929i
  protonephridium, 1073, 1073i
  respiratory surface, 1026i
Flavin adenine dinucleotide (FAD), 168–169, 168i, 170i–173i, 171, 173, 175–176, 175i
Flavin mononucleotide (FMN), 172i, 173i
Flavonoid, 805, 806i
Flax *(Linum),* 863
Flea, 700i
Fleming, Alexander, 456
Fletcher, Jennifer, 771
Flexor muscle, **977**
Fliesher, Robert C., 545i
Flight
  adaptations for, 730, 730i
  wing movements during, 731i
Floral organ identity program, 525
Floral shoot, 812, 813, 820
Florida panther *(Puma concolor coryi),* 1254i, 1270
Florigen, 860, 862
Flounder *(Platichthys flesus),* 720i
Flower
  chemical signals for flowering, 860
  color, 757, 758
  complete flower, 813, 813i
  development, 813, 832–833, 832i–833i
  dioecious species, 814
  epidermal cells, 757
  evolution of, 641, 645
  imperfect, 804
  incomplete flower, 813i, 814
  monocots and eudicots compared, 752t
  monoecious species, 814
  perfect, 814
  photoperiodism, 859–860, 860i
  pollination, 811, 817, 834
  structure, 813–814, 813i, 816
  vernalization, 860–861
Flowering plants, 637–643. *See also* Angiosperm
  evolution, 637–638, 640
  phylogenetic tree, 638, 638i
  pressure flow mechanism in, 787–789, 787i, 788i
Fluid feeder, **1045,** 1045i
Fluid mosaic model, 122i, **123**–124
Flukes, 685, 685i, 686
Fluorescence, **186,** 186i, 595
Fluorescence microscopy, 93i
Fly agaric mushroom, 661, 661i
FMN. *See* Flavin mononucleotide
FMR-1 gene, 301
FMRI. *See* Functional magnetic resonance imaging
Folding, protein, 328
Follicle, **1105,** 1105i, 1106i, 1107
Follicle cell, **1100**
Follicle-stimulating hormone (FSH), 951t, **952,** 953i, 958, **1105,** 1106i, 1109, 1109i, 1113
Food chain, **1213**
Food vacuole, **590,** 590i
Food web, 1239
  description, 1212i, **1213**
  detrital, 1230i, 1231
  grazing, 1230i, 1231
Foot, mollusk, **687,** 688i
Foraminifera (forams), **599**–600, 599i
Forebrain, **905,** 906i, 909
Forelimbs, mammalian, 441i, 450
Forensics, DNA fingerprinting in, 391–392, 403–404
Foreskin, 1108i, **1109**
Forest
  biomes
    boreal forest/taiga, 1160, 1160i
    temperate deciduous, 1160, 1160i
    temperate rain forest, 1160–1161
    thorn forest, 1158
    tropical, 1155–1157, 1156i
  carbon sequestration, 1264
  deforestation, 1255–1256, 1255i
  habitat fragmentation, 1258–1259, 1259i
Formula, **24**
Fossil fuel, 622, 1246, 1247, 1264
Fossil record, 500–506
  amniotes, 724, 726
  biological lineages, 449
  as evidence of evolutionary change, 449
  flowering plants, 637–638
  geological time scale, 502t–503t
  for horses, 515
  incompleteness of, 501
  information provided by, 505–506
  phyletic gradualist hypothesis, 516, 517i
  punctuated equilibrium hypothesis, 515, 517, 518i
  seedless vascular plants, 621
Fossils
  Burgess Shale formation, 670
  dating, 501, 504–505, 504i
  defined, **441**
  endomycorrhizae, 666
  examples, 501i
  formation of, 500–501
  microfossils, 560
  plants, 612, 613–614, 614i, 616, 619, 623i, 631, 632, 632i, 637–638, 638i
  role in development of evolutionary thought, 441
  stomatolites, 559–560, 560i
  trilobites, 695, 695i
Foundation species, **1210,** 1211i
Founder cell, 1136
Founder effect, **465,** 487
Four-o' clock *(Mirabilis)* plant, 280, 280i, 858
Fovea, **931,** 931i
Fragile X syndrome, 301
Fragment reaction, 327
Fragmentation, habitat, 1258–1259, 1259i, 1268–1270, 1270i
Fragmentation (reproduction), **824, 1097**
Frameshift mutation, 331i, **332**
Franklin, Rosalind, 65, 286, 291, 291i, 292
Free energy
  changes, 75–76
  in coupled reactions, 77–78, 77i
  defined, **74**
  exergonic and endergonic reactions, 75, 76i
  release with ATP hydrolysis, 77–78, 77i
  in reversible reactions, 75
Free water, 129
Freeze-fracture technique, **124**–125, 125i
Fregosi, Ralph, 1039
Freitas, Renata, 523
Frequency-dependent selection, **474,** 474i
Freshwater biomes, 1161–1164
  lakes, 1162–1164, 1162i, 1163i, 1164i
  streams and rivers, 1161–1162, 1162i
Freshwater ecosystem, hydrologic alterations of, 1257–1258, 1257i
Friedman, Jeffrey, 1060
Frisbee, 883, 883i
Frog, 966, 966i
  chytrid infection, 1262, 1262i, 1274
  crowding effects, 1185i
  description, 722–723, 723i
  development, 1120i, 1123, 1124i
  habitat preferences, 1292
  lungs, 1029, 1029i
  metamorphosis, 956, 956i
  phenotypic variation in, 456i
  reproduction, 1100–1101, 1101i
Frontal lobe, brain, 912, 912i
Fructose, 46i
Fruit, 639
  defined, **819**
  ripening, 847
  types, 820, 821i
Fruit fly. *See also Drosophila*
  Hox genes, 522, 524
  temporal isolation, 484
Fruiting body, 578, 578i, **603**
Frye, L. David, 124, 124i
FSH. *See* Follicle-stimulating hormone
FT gene, 862, 862i
Fulcrum, 976, 976i
Fumarase, 170i
Fumarate, 169, 170i
Functional groups, 43–44, 44t–45t, 46
  amino, 43, 44t
  carbonyl, 43, 44t
  carboxyl, 43, 44t
  defined, **43**
  hydroxyl, 43, 44t
  phosphate, 43, 45t
  sulfhydryl, 43, 45t
Functional magnetic resonance imaging (fMRI), 903i, 912–913, 913i, 914
Fundamental niche, **1204,** 1205, 1205i, 1206i
Fungal cells, 92i
Fungal mycorrhizae, 803–804, 806
Fungal spore, **652**–654, 653i
Fungi, 649–669
  characteristics of, 650–654
  as decomposers, 649, 652
  defined, **649**
  diversity, 649, 650i
  genomes sequenced, 421t
  groups, 654–663, 655t
    Ascomycota (ascomycetes), 655t, 658–661, 659i–660i
    Basidiomycota (basidiomycetes), 655t, 661–663, 661i, 662i
    Chytridiomycota (chytrids), 655–657, 655t, 656i
    conidial fungi (imperfect fungi/deuteromycetes), 654, 663
    Glomeromycota (glomeromycetes), 655t, 658

- Fungi *(Continued)*
  - groups *(Continued)*
    - microsporidia, 654, 663, 663i
    - Zygomycota (zygomycetes), 655t, 657–658, 657i, 658i
  - Kingdom Fungi, 11, 12i
  - life cycle
    - basidiomycete, 662–663, 662i
    - generalized, 652, 653i
  - meiosis, time and place of, 234–235, 234i
  - nutrient acquisition, 651–652
  - phylogeny of, 654–655, 654i
  - reproduction, 652–654, 653i
  - structure, 650–651, 651i
  - symbiotic associations, 663–667
    - lichen, 664, 665i
    - mycorrhizae, 665–666, 665i, 666i
- Furchgott, Robert, 996, 996i
- Furrow, **211,** 212i
- *Fused Toes* (FTO) gene, 258
- Futuyma, Douglas J., 454

- G_0 phase, **207,** 207i, 219
- G_1 phase, **207,** 207i, 208i, 209i, 214, 216–220, 216i, 217i, 222
- G_1/S checkpoint, 216–217, 217i
- G_2 phase, **207,** 207i, 208i, 214, 215, 217, 217i, 218
- G_2/M checkpoint, 217, 217i
- G-protein–coupled receptor, **149**–154, 149i–152i, 932, 946–947, 995
- GABA. *See* Gamma aminobutyric acid
- Galagos, 736
- Galápagos finches
  - adaptive radiation, 512, 512i
  - bill shape and food habits, 444–445, 445i, 446
  - character displacement, 1205, 1205i
  - disruptive selection in cactus finches, 470, 470i
- Galápagos Islands
  - Darwin and, 443–445
  - finches, 1205, 1205i
  - map of, 443i
- Galilei, Galileo, 440
- Gallbladder, 1049i, 1056i, **1057**
- Gallmaking fly *(Eurosta solidaginis),* 468, 469i
- *Gambusia,* 529
- Gametangium, **618,** 618i, 621, 657i, **658**
- Gametes, **225,** 226, 231, 233–235, 234i, 494, **671, 1097**
- Gametic isolation, **485**
- Gametogenesis, **226**
  - defined, **1099**
  - oogenesis, 1098i, 1099–1100
  - spermatogenesis, 1098i, 1099
- Gametophyte, **234**–235, 234i, **598**
  - alternation of generations, 615, 615i
  - angiosperm, 639, 642i
  - angiosperms, 812i, 813, 815i, 816
  - bryophyte, 618, 618i
  - conifer, 635, 636i
  - defined, **615, 813**
  - fern, 624i, 625
  - gymnosperm, 632
  - from heterosporous plants, 616
  - from homosporous plants, 616
  - hornworts, 619
  - kelp, 598
  - liverwort, 619, 619i
  - moss, 618i, 620i, 621
  - seedless vascular plants, 621
- Gamma aminobutyric acid (GABA), 898t
- Gammaproteobacteria, 578
- Ganglion, **684,** 684i, 710, **885, 905**
- Ganglion cell, **932,** 933i
- Gap genes, 358i, **359**
- Gap junction, **114,** 114i, 142, 157, 868, 874, 886, 886i
- Garden pea *(Pisum sativum),* 240, 241i
- Garland, Theodore, Jr., 459
- Garner, Wightman, 858
- Garrod, Archibald, 311
- Gars, 719, 719i
- Garter snake *(Thamnophis elegans),* 1282, 1282i
- Gas exchange, 1024–1042
  - adaptation to maximize, 1026, 1026i
  - in air, 1026–1027
  - countercurrent flow, 1028, 1028i, 1029, 1030i
  - function of, 1025–1027
  - gills, 1025, 1026, 1026i, 1027–1028, 1027i, 1028i
  - lungs, 1025, 1026i, 1029–1033, 1030i–1032i
  - in mammals, 1030–1033, 1031i–1033i
  - mechanisms of, 1034–1037, 1034i, 1035i, 1036i
  - tracheal system, 1025, 1028–1029, 1029i
  - in water, 1026–1027
- Gassner, Gustav, 860–861
- Gastric ceca, 1047, 1047i
- Gastric gland, **1054,** 1055i
- Gastric juice, **1054**
- Gastric pit, **1054,** 1055i
- Gastrin, 1059i, **1060**
- Gastrodermis, 679, 679i, 680
- Gastroesophageal sphincter, 1052i, **1053**
- Gastrointestinal tract, 1047
- Gastropod, 688–689, 688i, 689i
- Gastrovascular cavity, **679, 982,** 982i, **1046,** 1046i
- Gastrula, **1121,** 1122i
- Gastrulation
  - amphibian, 1120i, 1123, 1124i
  - bird, 1124–1125, 1125i
  - defined, **1121**
  - mammalian, 1129
  - sea urchin, 1122–1123, 1122i
- Gated channel, **127,** 128i
- Gause, G. F., 1203, 1204i
- Gazzaniga, Michael S., 914, 915i
- Gecko, 31, 32i, 728i
- Gemmae, **619,** 619i
- Gemmule, 678
- GenBank, 417
- Gene(s)
  - alleles, 243, 249–251, 250i, 251i
  - cloning, 392–393, 392i, 394, 395i
  - defined, **7**
  - homologous, 422
  - linked, 262–266, 263i, 265i, 266i
  - locus, 251, 251i
  - as Mendel's factors, 243, 249
  - non-protein coding, 318
  - organization, 346, 346i
  - protein-coding, 311, 313
  - relationship to enzymes, 311, 312i–313i
  - sex-linked, 266–271, 268i–271i
  - trapped, 334
- Gene activation, steroid hormones and, 154, 155–156, 156i
- Gene cloning, **392,** 392i, 393, 394, 395i. *See also* DNA cloning
- Gene duplication, 432, **433,** 645, 712, 713i, 998
- Gene expression
  - nonhistone proteins, role of, 306
  - regulation of, 359–365
    - chromatin remodeling, 351, 351i
    - in eukaryotes, 344–351, 346i–351i
    - posttranscriptional regulation, 351–354, 353i
    - posttranslational regulation, 354, 354i
    - in prokaryotes, 340–344, 341i–345i
    - transcriptional regulation, 340–351
    - translational regulation, 353–354
  - training effects on racehorses, 973
- Gene flow, **463**–464, 464t, 464i
- Gene-for-gene recognition, 851t, **853**–854
- Gene function determination, 426–431
  - by gene knockout/gene knockdown, 427–428
  - phenomics, 428
  - from protein structure, 427
  - proteomics, 428, 430–431
  - by sequence similarity search, 427
  - transcriptomics, 428
- Gene knockdown, 427–428
- Gene knockout, 427–428
- Gene of interest, 392–393, 392i, 394, 395i, 397, 397i
- Gene pool, **460**
- Gene regulation
  - combinatorial, 348–349, 349i
  - negative, 344
  - positive, 343, 343i, 344
- Gene therapy, **406**–407, 412
- Gene transfer, 370–377
- General transcription factors, **347,** 347i, 349i
- Generalized transduction, **380**
- Generation time, **1175,** 1175i, 1183i
- Genetic code, **313**–315, 315i, 317
- Genetic counseling, **279**
- Genetic disorders
  - cancer as, 361–362
  - correcting with gene therapy, 406–407
  - molecular testing for, 401
- Genetic divergence, 491–492
- Genetic drift
  - as agent of microevolution, 464t, 464–465, 466
  - conservation implications, 464–465
  - defined, **464**
  - founder effect, 465
  - population bottlenecks, 465, 465i, 466
- Genetic engineering, **392,** 404–412
  - of animals, 405–408, 406i–408i
  - of bacteria to produce proteins, 404–405, 405i
  - of plants, 408–412, 409i, 410i
  - public concern over, 410–411
- Genetic equilibrium, **461**
- Genetic recombinants, **232,** 232i
- Genetic recombination, **232,** 232i, **264,** 282. *See also* Recombination
  - in *Escherichia coli,* 371–377, 371i, 374
  - genetic diversity from, 1099
- Genetic screening, **279**
- Genetic tool kit, 521, 522
- Genetic variability
  - generation by independent assortment of chromosomes, 232–233, 232i
  - generation by random joining of gametes in fertilization, 233
  - generation by recombination, 231–232, 232i
- Genetic variation
  - agents of microevolutionary change, 462–471, 464t
  - balanced polymorphisms, 472–474
    - frequency-dependent selection, 474, 474i
    - heterozygote advantage, 472–473, 472i
    - selection in varying environments, 473–474, 473i
  - biodiversity, 1255
  - gene flow effects, 463–464, 464i
  - generation of, 460
  - level in population, 460
  - maintenance of, 471–474
  - mutations as source of, 462–463
  - natural selection effects, 465–470, 467i–470i
  - neutral, 474
  - recombination rate and, 476
  - reduction with genetic drift, 465,465i
- Genetically modified organisms (GMOs), **410**
- Genetics. *See also* Inheritance
  - bacterial, 369–377
  - of cancer, 360–364
  - human, 276–279, 276i, 277i, 279i
  - of human disease, 258
  - population, 460–461, 461t
  - viral, 377–387
- Genome(s)
  - *Escherichia coli,* 423i, 424–425, 424i, 425t
  - human, 423i, 424, 425–426, 425i, 425t
  - landscapes, 423–426
  - number sequenced, 421t
  - size, 423–424, 423i, 423t
- Genome analysis, of diatoms, 597i
- Genome annotation
  - identifying open reading frames by computer search of genome sequences, 421–422
  - identifying protein-coding genes from sequences of gene transcripts, 422–423
- Genome evolution, 375–377
- Genome sequence determination and annotation, 417, 418–426
- Genomic imprinting, 280–**281,** 281i, **351**
- Genomic library, **394**
- Genomics
  - comparative, 417, 431–436
  - defined, **416**
  - evolution of genome, 417, 431–436
  - gene function determination, 426–431
  - genome sequence determination and annotation, 417, 418–426
  - overview, 417
- Genotype, **243**
- Genotype frequencies, **460**–461, 461t, 464–465
- Genus, **9, 529**
- Geoduck *(Panope generosa),* 689i
- Geographical range, **1173**
- Geographical variation, in species, 482–483, 482i, 483i
- Geography of speciation, 486–491, 486i–490i
- Geological strata, 500, 500i, 501
- Geological time scale, 502t–503t
- Geology
  - catastrophism, 441
  - gradualism, 442
  - uniformitarianism, 442
- Geometric population growth, 1180–1181, 1181i
- *Geospiza,* 1205i
- Germ cell, **1098, 1099,** 1125
- Germ layers, **1121**
- Germ-line cells, genetic engineering of, **405**–406, 406i
- Germ-line gene therapy, **406**
- Germination, seed, 821–822, 822i, 823i
- Gestation, **1128**
- Giant saguaro *(Carnegia gigantea),* 643i
- Giant sequioa *(Sequoiadendron giganteum),* 632
- Giant swallowtail butterfly *(Papilio cresphontes),* 1202i
- *Giardia lamblia,* 591, 591i, 607
- Gibberellin, 252, 822, 840t, **844**–846
- Gibbon, 738, 738i
- Gibbs, Josiah Willard, 74
- Gilbert, Walter, 418
- Gill arches, **717,** 717i
- Gill slits, 709
- Gills, **1025,** 1026, 1026i, 1027–1028, 1027i, 1028i
  - bivalves, 689, 689i
  - crustacean, 698
  - external, 1026i, 1027–1028, 1027i
  - internal, 1027i, 1028, 1028i
  - mollusk, 687, 688i
  - salt excretion by marine fishes, 1084
- Gilman, Alfred G., 149
- Ginger *(Zingiber officinale),* 761, 761i
- Gingivitis, 576
- *Ginkgo biloba,* 633, 634i
- Ginkgophyta, **633**
- GIP. *See* Glucose-dependent insulinotrophic peptide (GIP)
- Girdling a tree, 768
- Givol, David, 278
- Gizzard
  - bird, 1047i, 1048
  - earthworm, **1047,** 1047i
  - insect, 1047, 1047i
- *GL2* gene, 828i, 829

Glaciations, 508
Glacier retreat, 1217, 1218i
Gland
defined, **870**
endocrine, 870i, 871
exocrine, 870–871, 870i
Gland cells, 142
Glans, penis, **1109**
Gleason, Henry A., 1208–1209
Glial cell, **875, 885,** 885i
Global climate change, 1153–1154, 1168–1169, 1187, 1247–1248
Global warming, 1255, 1260, 1264
Globulin, **986**
Glomeromycota (glomeromycetes), 655t, 658
Glomerular filtration, **1075,** 1076
Glomerular filtration rate, 1080
Glomerulus, 1069i, **1075,** 1075i, 1076, 1076i, 1080
Glomus versiforme, 665i, 803
Glor, Rich, 550
Glottis, 1053, 1054i
Glover, Beverly J., 758
Glucagon, 152, 946, 947, 948, 952t, **959,** 1056, 1056i
Glucocorticoid, 350, 946, 951t, 957–958, 958i
Glucose
glucagon and, 959, 960i
glucocorticoids and, 957–958, 958i
insulin and, 959, 960i
phosphorylation of, 80, 80i
from photosynthesis, 184, 194
ring formation, 47, 48i
Glucose-1-phosphate, 75, 75i
Glucose-6-phosphate, 75, 75i, 80, 80i
Glucose-dependent insulinotrophic peptide (GIP), 1060, 1060i
Glutamate, 898t, 932, 935, 937
Glutamine, 77, 77i
Glutathione, 934
Glyceraldehyde, 47i
Glyceraldehyde-3-phosphate (G3P), 45t, 166, 166i, 167i, 193–194, 195i, 197, 198, 199
Glycerol, 50, 51, 51i, 53
Glyceryl palmitate, 51i
Glycocalyx, **95, 122, 571**
Glycogen, **47,** 49, 49i, 170, 948
Glycogen phosphorylase, 142, 143, 144i
Glycogen synthase, 152
Glycolate, 197
Glycolipid, 55, **122,** 122i
Glycolysis, 164i, 165–166, 168, 175–176, 175i
defined, **164, 166**
energy-requiring and energy-releasing steps, 166
in fast muscle fibers, 972
reactions, 166i, 167i
regulation of, 166, 168
Glycoprotein
description, 63
extracellular matrix, 115
membrane, **122,** 122i
surface receptors, 145, 149
Glycosidic bond, **48,** 48i
Glyoxysome (glycosome), 107
Glyphosate, 409–410
Glyptodont, 443, 443i
GMOs. *See* Genetically modified organisms
Gnathostomata, **713,** 717–721
Gnetophyta, 633–634, 634i
Gnetum, 633–634
Goiter, 23–24, 956
Goldacre, R. J., 557
Golden algae, **597**–598, 598i
Golden plover *(Pluvialis dominica),* 732
Goldenrod plant *(Solidago altissima),* 468, 469i
Goldingay, Russ, 1269
Goldschmidt, Richard, 524
Goldstein, Joseph L., 67, 136
Golgi, Camillo, 103
Golgi complex, 98i, 99i, **103**–104, 103i, 562
Golgi tendon organ, **924,** 925i
Gonadotropin, **952,** 953i
Gonadotropin releasing hormone (GnRH), **958, 1105,** 1106i, 1107, 1109, 1109i, 1287, 1288i
Gonads, **226, 958, 1098**
bipotential, 1132, 1133i
development, 1132, 1133i
Gondwana, 507, 507i, 509
Gooseneck barnacles *(Lepas anatifera),* 699i
Gorbsky, G.S., 215i
Gordon, Jeffrey, 1064i
Gorilla *(Gorilla gorilla),* 738
Gossypium, 630, 630i
Gould, Stephen Jay, 517, 524
G3P. *See* Glyceraldehyde-3-phosphate
Graded potential, **897**
Gradualism, **442**
Grafting, 824i, 825
Graham dog cactus *(Opuntia grahamii),* 785i
Grain, **819**
Grain borer beetle *(Rhizopertha dominica),* 1185i
Gram, Hans Christian, 570
Gram-negative bacteria, **570,** 571, 571i, 572, 572i, 577
Gram-positive bacteria, **570,** 571, 571i, 572, 580, 580i
Gram stain technique, **570**
Grana, **111,** 112i, 184, 185i
Grand Banks, 1256, 1257i
Grant, Peter and Rosemary, 470
Grass
asexual reproduction, 824
pollen grain, 816i
rhizomes, 760
root system, 765
stem growth, 758
Grasshopper, 700i, 701, 701i
digestive system, 1047, 1047i
Malpighian tubules, 1074i
Grasshopper mouse *(Onychiomys),* 1203, 1203i
Grassland
savanna, 1157–1158, 1158i
temperate, 1159, 1159i
Graves' disease, 956
Gravitropism, **855**–856, 856i
Gray crescent, **1123,** 1124i
Gray matter, **909,** 909i, 911–912
Gray whale *(Eschrichtius robustus),* 1045i, 1289
Grazing food web, 1230i, 1231
Great blue heron *(Ardea herodias),* 442i, 1045i
Great tit *(Parus major),* 1186i, 1299
Great white shark *(Carcharodon carcharias),* 718
Greater vestibular glands, **1104,** 1104i, 1110
Green, Michael R., 362
Green algae, 588i, 601, 601i, 602i, 664
alternation of generations, 615
photosynthesis in, 182–183, 182i
plant evolution, 612
Green sulfur bacteria, 192
Greenhouse effect, **1247,** 1255, 1260, 1264
Greenhouse gases, 1247–1248
Greider, Carol W., 303
Grey reef shark *(Carcharhinus amblyrhynchos),* 718i
Greylag geese *(Anser anser),* 1284, 1284i
Griffin, Kevin, 1251
Griffith, Frederick, 287, 288i, 374
Grinnell, Joseph, 1154
Gross primary productivity, **1233,** 1250
Ground meristem, **759,** 759i, 766
Ground state, electron, 186–187, 186i
Ground tissue
collenchyma, 753, 753i, 753t, 754
components and functions, 753t
defined, **752**
location of, 753i
meristem, 759, 759i, 766
parenchyma, 753–754, 753i, 753t
sclerenchyma, 753, 753i, 754, 754i
Group living, 1299, 1299i
Growth
determinant, 1065
indeterminant, 1065
Growth factor, 149, 219, **360,** 361i, **945,** 951t
Growth hormone, 946, 951t, **952,** 953i
Growth hormone gene, 406, 407i
Growth-inhibiting factor, **360,** 361i
GTP. *See* Guanosine triphosphate
Guanine, 63, 64i, 66i, **290,** 290i
Guano, 1246
Guanosine triphosphate (GTP), 64, 149–150, 150i–152i, 153, 154i, 157i
Guard cell, **756,** 756i, 784–785, 784i
Guerrant, Edward O., 520i
Guppy *(Poecilia reticulata),* 1180–1181
Gurdon, John B., 1121, 1139
Gut, 1047
Guttation, **783,** 783i
Gymnophilus, 12i, 650i
Gymnophiona, 723, 723i
Gymnosperm, 631–636
ancestral forms, 631–632, 632i
characteristics of, 631t
conifers, 634–635, 636i
cycads, 633, 633i
defined, **631**
ginkgo, 633, 634i
gnetophytes, 633–634, 634i
overview, 631
reproductive adaptations, 632
secondary growth in, 751

H zone, 967i, 968, 969i
Haber, Edgar, 60, 61i
Habitat
defined, **1173**
patch, 1268–1270i, 1270i
of protists, 590
Habitat fragmentation, **1258**–1259, 1259i, 1268–1270
Habitat selection, 1292, 1292i
Habituation, **1284**
Haddock *(Melanogrammus aegelfinus),* 1256
Hadrosaurus, 499, 726
Haeckel, Ernst, 672, 1147
Haemophilus influenzae, 578
Hagfish, 713, 715, 715i
Hahn, Beatrice, 547
Haikouichthys, 712
Hainsworth, F. Reed, 1153
Hake, Sarah, 761
Haldane, J. B. S., 555
Haleakala silversword, 845i
Half-life, **504,** 504i
Haliclona, 678i
Hallucigenia, 670, 671i
Halobacterium, 583
Halophile, 583
Hamadryas baboon *(Papio hamadryas),* 738i
Hamilton, William D., 1300, 1302
Hamner, Karl, 859
Hanson, Jean, 969i
Haplinsufficiency, 282
Haplodiploidy, **1302,** 1303i
Haploid, **206, 226,** 234–235, 235i
Haplorhini, 736–739, 737i, 738i
Hardy, G. H., 461
Hardy-Weinberg principle, **461,** 462, 463, 464–465
Harlequin frog/toad *(Atelopus),* 655, 656i, 1202i
Harrison, Maria J., 803
Harrison, Susan, 1190i
Hartwell, Leland, 214
Harvesting efficiency, 1235
Hasegawa, Mike, 821
Haselton, Martie, 1107i
Hatena, 608
Haustoria, **651**
Haustorial roots, **806,** 807i
Hawaiian honeycreepers, cytochrome *b* sequence in, 545i
Hawaiian Islands, fruit fly speciation in, 488
Hawkmoth, 639, 643i
Hawthorn (*Crataegus* species), 490, 490i
HCG. *See* Human chorionic gonadotropin
HDL. *See* High-density lipoprotein
Head
arthropod segment, 695
insect, 699, 700i
mollusk, **687,** 688i
virus, 378, 378i, 380
Hearing, 927–929, 928i
Heart, 987–991
amphibian, 984, 984i
bird, 985
blood pressure, 990–991
cardiac cycle, 989–990, 989i, 990i
circulation, 988, 988i
defined, **982**
electrical control, 989–990, 990i
fish, 983–984, 984i
internal structure, 988, 988i
mammals, 987–991, 988i–990i
myogenic, 989
neurogenic, 989
reptile, 985
sounds, 989
Heart attack, 880, 994
Heart disease, cholesterol and, 52
Heart murmur, 989
Heartbeat, 989, 994
Heartburn, 1053
Heartwood, **768,** 769i
Heat
specific, 33
units of measurement, 33–34
of vaporization, 34
Heat energy, 73, 73i, 176, 178
Heat exchange mechanisms
conduction, 1086, 1086i
convection, 1086, 1086i
evaporation, 1086, 1086i
radiation, 1086, 1086i
Heat-shock proteins, 851t, **854**
Heavy chain, antibody, **1009,** 1009i
Hebert, Paul, 1267
Hedden, Peter, 252
Heinrich, Bernd, 1284, 1295
HeLa cells, 214, 216i
Helianthus annuus, 92i
Helical viruses, **377,** 378i
Helicase, **296,** 297i, 298i, 300t
Helicobacter pylori, 12i, 578, 579i, 1054–1055
Heliconius erato, 1203i
Heliconius melpone, 1203i
Helix pomatia, 689i
Helix-turn-helix, 348, 348i
Helper T cells, **1011,** 1012i–1013i, 1016, 1019
Heme group, 986
Hemichordata, 709, 710i
Hemlock woolly adelgid *(Adelges tsugae),* 1261, 1262i
Hemocoel, 695
Hemoglobin, 986–987, 1034–1037, 1035i, 1036i, 1038, 1038i
gene duplication, 272
genes, 339, 350
heterozygote advantage and, 472–473, 472i
sickle-cell, 239–240, 254, 257, 257i, 276, 332, 332i, 401 402i
structure of, 57, 58i, 61
Hemoglobin-O_2 dissociation curve, 1035, 1035i
Hemolin, 1020
Hemolymph, **687,** 690, 695, **982**–983, 1073, 1073i, 1074i
Hemophilia, 270, 270i, 987
Hennig, Willi, 538

Henslow, John, 443
Hepatic portal vein, **1058**
Hepatophyta, 619
Herbaceous plant, 752
Herbicide, **843**
Herbivore
defined, **1044**
intestinal length, 1061, 1063, 1063i
symbiotic microorganisms, 1063, 1063i, 1065
teeth, 1061, 1061i
Herbivory
adaptation for feeding, 1201
defenses against, 1201–1202
defined, **1201**
interspecific competition among plants, effect on, 1215–1216, 1215i
in trophic cascade, 1238, 1240i
Herbs, 622
Hereditary enamel hypoplasia, 279, 279i
Herhsko, Avram, 354
Hermaphroditic
annelids, 691
barnacle, 699
ctenophores, 681
defined, **678**
sponges, 678
trematodes, 685i
turbellaria, 685
Hermaphroditism, **1103**–1104, 1103i
Herpesvirus, 220, 378i, 382, 383
Herrick, James, 239–240
Herring gull *(Larus argentatus),* 509, 1280–1281, 1281i
Hershey, Alfred D., 288–289, 289i
Hertz, Paul E., 15–16, 1153, 1153i
Heterochromatin, 305–306
Heterochrony, 518–519, 522
Heterocyst, 574, 580i
Heterogametic sex, **267**
Heterosporous, **616,** 623, 632
Heterotroph, **183, 573,** 589, 651, **671, 1213,** 1242
Heterozygote, **243,** 246
Heterozygote advantage, **472**–473, 472i
Heterozygous, **243**
Hexaploids, 273
Hexapoda, 699–702, 700i, 701i
Hexokinase, 80, 80i, 167i
Hexose, 47
Hfr cell, **372,** 373i, 374
HGP. *See* Human Genome Project
Hibernation, **1091**
Hibiscus, 643i
High-density lipoprotein (HDL), 52, 1066
High-throughput techniques, **417**
Higher vocal center, 1286, 1286i
Hilakivi-Clarke, Leena, 963
Hileman, Lena, 525
Hill, Emmeline, 973
Hindbrain, **905,** 906i, 909
Hinkle, Gregory, 656
Hippocampus, 911i, **912,** 916, 917
Hippocrates, 528
Hirudin, 692
Hirudinea, 691–692, 692i
Hirudo medicinalis, 692, 692i
Histamine, 1005i, **1006,** 1019, 1020i
Histone code, **351**
Histones, **304**–305, 305i, 351, 351i
Historical biogeography, **449**
HIV. *See* Human immunodeficiency virus
HMG box, 1134
Hoatzin *(Opisthocomus hoazin),* 1065
Hoffmann, Gretchen, 1248
Holothuroidea, 708i, 709
Holtfreter, Johannes, 1136, 1137i
Homeobox, **359,** 523
Homeodomain, **359**
Homeosis, 524, 525
Homeostasis, **5, 142,** 876–880
calcium, 957i
defined, **867, 877**
environmental variation, responses to, 1152–1153, 1153i
factors regulated by, 877–878
glucose, 959, 960i
hormone regulation, 946
local controls, 878
negative feedback mechanisms, 878–879, 878i, 879i
positive feedback mechanisms, 879–880
systemic controls, 878
Homeostatic control system, **867,** 877–879
Homeotic genes, **359,** 359i, 360i, 522, 712, 713i
defined, **521, 826**
floral organ development, 832–833, 832i–833i
in plants, 761, 826, 832–833, 832i–833i
shoot apical meristems, 761
Homing pigeon *(Columba livia),* 1290
Hominins
Ardipithecus, 739–740
Australopithecus, 739–740, 741i
bipedal locomotion, 739–740, 739i
defined, **739**
Homo erectus, 742, 742i
Homo habilis, 741–742
Homo neaderthalensis, 742, 743
Homo sapiens, 742–743
Orrorin tugenensis, 739
Paranthropus, 740, 741
time line, 740i
tool use, 742, 742i
Hominoidea, **738**
Homo erectus, 742, 742i
Homo floresiensis, 742, 743
Homo habilis, 741–742
Homo neanderthalensis, 742, 743
Homogametic sex, **267**
Homologies (homologous characters), **534,** 534i, 535i
Homologous chromosomes, 226–227, 227i–230i, 230–232, 232i, 235
crossing-over between, 262, 265, 265i
defined, **206**
Homologous gene, **422**
Homologous structures, 441i
Homologous traits, **449**–450
Homoplasies (homoplasious characters), **534,** 535i
Homosporous, **616,** 623, 626–627
Homozygote, **243,** 246
Homozygous, **243**
Honey mushroom *(Armillaria ostoyae),* 651, 651i
Honeybee *(Apis mellifera),* 52, 1203i, 1286, 1287i, 1294–1295, 1302, 1302i, 1303i
Honeydew, aphid, 787, 787i
Hooke, Robert, 90, 91i
Hooker, Joseph, 439
Horizon, soil, **799,** 799i
Horizontal cells, **932,** 933, 933i
Horizontal gene transfer, **375**–377, **549,** 549i, **574,** 577, 581
Hormone
behavior and, 1285–1287, 1286i, 1287i, 1288i
classes, 945
amines, 945
fatty acid derived molecules, 945
peptides, 945
steroids, 945
coordination of action, 946
defined, **142, 350, 839, 943**
endocrine glands, 870–871, 870i
feedback regulation, 945, 946i, 957i
gene expression control by, 350, 350i
invertebrate, 961–962, 962i, 964i
male reproductive functions, regulation of, 1109, 1109i
mechanism of action, 946–949
amplification, 946
hydrophilic hormones, 946–947, 947i
hydrophobic hormones, 947, 947i, 948, 948i
receptors, 944, 946–948, 947i, 948i, 949i, 954–955
peptide, 145, 152
plant, 152, 153
abscisic acid, 840t, 848–849, 850i–851i
auxins, 840t, 841–844, 842i–844i, 845i
brassinosteroids, 840t, 847–848, 848i
cytokines, 840t, 846, 846i
ethylene, 840t, 846–847
gibberellin, 840t, 844–846
jasmonates, 840t, 849–852, 851t, 852i
oligosaccharins, 840t, 849, 851t
salicylic acid, 840t, 850–851, 851t, 852, 854, 854i
second messengers, 840–841
systemin, 840t
steroid, 54, 54i, 154–156, 156i, 157–158
table of human, 951t–952t
target cells, 944, 948
Hormone-binding domain, 156
Hornwort, 619–620, 620i
Horses, evolutionary lineage of, 515, 516i
Horseshoe crab (Merostomata), 697, 697i
Horsetail, 612i, 625–626, 625i
Host, **1207**
Host race, **490**
House sparrow *(Passer domesticus),* 483i, 1172
Housefly *(Musca domestica),* 701i, 903–904
Housekeeping genes, 339
Hox genes, 359, 360i, 450–451, 523, 676, 699, 712, 713i
HTLV. *See* Human T-cell leukemia virus
Hubbard Brook watershed, 1242–1243
Huey, Raymond B., 1153, 1153i
Human(s)
antagonistic pairs of muscles, 976–977, 977i
biodiversity benefits to, 1263–1264
blood pressure, 990–991
body temperature variation, 1089
bones, 975i
chromosomal differences among great apes, 495–496, 495i
development, 1118–1119, 1128–1133, 1129i–1133i
digestive system, 1048–1059, 1049i, 1052i–1059
ear, 927–929, 928i
endocrine glands and hormones, 950–962, 950i, 951t–952t, 957i, 958i, 960i
evidence for stabilizing selection in, 468i
evolution, 739–743
eye, 931–934, 931i–934i
genome, 423i, 424, 425–426, 425i, 425t
heart, 988–991, 988i–990i
kidneys and urinary system, 1074i, 1075, 1076, 1092
lymphatic system, 996–997, 997i
minerals, 1051t
organ systems, 875, 876i
population growth, 1191–1195, 1193i–1195i
as primates, 739
respiratory system, 1030–1033, 1031i–1033i
skin structure, 1090, 1090i
skull, changes in, 518, 519i
social behavior, 1304
taste receptors, 935–936, 936i
vitamins, 1050t
Human chorionic gonadotropin (hCG), **1111**
Human genetics, 276–279, 276i, 277i, 279i
Human Genome Project (HGP), 417, 421
Human immunodeficiency virus (HIV), **378,** 384, 386, 386i, 547, 547i, 1002, 1019, 1021
Human reproduction, 1104–1112
contraception, 1112–1114, 1112t
copulation, 1110
female sexual organs, 1104, 1104i
fertilization, 1110–1112, 1110i
hormonal regulation in males, 1109, 1109i
male sexual organs, 1108–1109, 1108i
ovarian cycle, 1105–1107, 1105i, 1106i
uterine cycle, 1105, 1106i, 1107–1108
Human T-cell leukemia virus (HTLV), 362
Hummingbird, 643i, 732i, 748i, 1045, 1045i, 1091, 1153
Humoral immunity, **1007**
Humpback whale *(Megaptera novaeangliae),* 466, 1167i
Humulin, 404
Humus, **798,** 799, 800
Hunt, R. Timothy, 218
Huntington disease, 467
Hutchinson, G. Evelyn, 1195
Hutton, James, 442
Huxley, Andrew, 969i
Huxley, Hugh, 969i
Huxley, Thomas, 447, 519
Hyacinth macaw *(Anodorhynchus hyacinthinus),* 1294i
Hybrid, polyploid, 493i, 494
Hybrid breakdown, **486**
Hybrid inviability, **485**
Hybrid sterility, **485**–486, 491
Hybrid zone, **489,** 489i
Hybridization, DNA, 397, 397i
Hybridoma, **1015**
Hydra, 961
chemoreceptor, 934
description, 680, 680i
gastrovascular cavity, 982, 982i
reproduction, 1097, 1097i
Hydrangea macrophylla, 458i
Hydration layer, **35,** 35i
Hydrocarbons, **43,** 43i
Hydrochloric acid, 1054
Hydrogen
H^+ pumps, **131**
ions, 36–38
cation exchange, 800
isotopes of, 25, 25i
as plant macronutrient, 795, 796t
Hydrogen bond, 31i
defined, **30**
diagram designation of, 30
intermolecular, 31
intramolecular, 31
strength of, 31
in water, 32–34, 33i
Hydrogen ion gradient, 190i, 191, 192
Hydrogen ion pump, 776, 776i, 843–844, 843i, 845i
Hydrogen peroxide, 107
Hydrogenation, of plant oils, 51
Hydrologic alternations, **1257**–1258, 1257i
Hydrologic cycle, **1241**–1242, 1241i
Hydrolysis
of ATP, 76–78, 77i, 131, 132i, 133i
defined, **46,** 46i
of GTP, 150, 150i, 154i
Hydrolytic enzyme, 104–106, 652
Hydrophilic hormone, 946–947, 947i
Hydrophilic substances, **30,** 120, 121, 121i, 122i, 127
Hydrophobic hormone, 947, 947i, 948, 948i
Hydrophobic substances, **30,** 120, 121, 121i, 122i, 127
Hydroponic culture, **794,** 795i
Hydrosphere, **1146**
Hydrostatic pressure, 990
Hydrostatic skeleton, **673,** 690, **974,** 974i
Hydrothermal vent community, 553i, 556, 1167–1168, 1168i
Hydroxide ion, 36–37
Hydroxyapatite, 872
Hydroxyl group, **43,** 44t
Hydrozoa, 680, 680i
Hyla species, 535–536, 535i
Hymen, **1104**
Hyomandibula, 534, 534i
Hyperosmotic, 1070, 1072, 1073, 1076, 1078–1079, 1084
Hyperpolarization, **889,** 889i, 891i, 932
Hypersensitive response, **852,** 852i

Hypertension, **991**
Hyperthermia, 867–868, **1090**
Hyperthyroidism, 956
Hypertonic solution/condition, **130**–131, 130i
Hyphae
  fungus, **650**–653, 653i
  in Oomycota, 596, 596i
Hypoblast, **1124,** 1125, 1125i, 1129, 1130i
Hypocotyl, **819,** 820i, 823i
Hypodermis, **1090,** 1090i
Hypolimnion, **1163,** 1163i
Hypoosmotic, 1070, 1077
Hypothalamus
  description, **911,** 911i, 912
  digestive system control, 1060–1061
  endocrine functions, 950–952, 950i, 951t, 953i
  homeostasis and, 878, 879, 905, 906i
  temperature regulation, 1088–1090, 1089i
Hypothermia, 867, 869, **1090**
Hypothesis, 13–17
  alternative, 15
  defined, **14**
  falsifiable, 14–15
  null, 14, 15, 16, 17i
  testing, 14–17, 14i, 1147
Hypothyroidism, 956
Hypotonic solution/condition, **130,** 130i
Hyracotherium, 515, 516i

I band, 967i, 968, 969i
IAA. *See* Indoleacetic acid
Ice, melting of, 74
Ice-binding motif, 1152
Ice lattice, **33,** 33i
Iceburg lettuce *(Lactuca sativa),* 845
Ichthyosaurs, 726
Ichthyostega, 722, 722i
IDEAL-labeling, 388
IgA, **1009**–1010, 1014
IgD, **1010**
IgE, **1010,** 1019, 1020i
IGF (insulin-like growth factor), **952**
IGF-2 (insulin-like growth factor-2), 281, 281i
IgG, **1009**
IgM, **1009**
Ignarro, Louis, 996
Iguana, 2i, 444i, 1088
Illumina/Solexa method for DNA sequencing, 420i, 421
Imbibition, **822**
Imino group, 56
Immigration, **1175**
Immune response, **1003,** 1013–1014, 1015i
Immune system
  adaptive, 1003
  defined, **1002**
  innate, 1003
  invertebrates, 1020
  malfunctions and failures
    in AIDS, 1019, 1021
    allergies, 1019–1020, 1020i
    autoimmune disease, 1017–1019
  in sharks and rays, 1020
Immunity
  active, 1014
  adaptive, 1003, 1007–1017
  cell-mediated, 1015–1017
  innate, 1003, 1004–1007, 1005i
  passive, 1014
Immunoglobulin, 986, **1009**–1010, 1009i, 1012i–1016i
Immunological memory, 1012–1014, **1013,** 1014i, 1015i
Immunological tolerance, **1017**
Imperfect flower, **814**
Imperfect fungi, 663
Implantation, embryo, 1129, 1129i
Imprinted allele, 351
Imprinting
  behavioral, **1282,** 1284, 1284i
  genomic, 280–**281,** 281i
Inactivation gate, 891, 890i, 891i
Inborn error of metabolism, 311
Inbreeding, **470**
Incisor, **733,** 1061, 1061i
Inclusive fitness, **1301**
Incomplete dominance, **251**–254, 253i, 460
Incomplete flower, 813i, **814**
Incomplete metamorphosis, **701,** 701i
Incurrent siphon, **711,** 711i
Incus, **927,** 928i
Independent assortment, 248i, **249,** 250i, 266, 1099
Independent assortment of chromosomes, **232**–233, 232i
Indeterminate cleavage, **674**
Indeterminate growth, **750,** 1065
Indian pipe *(Monotropa uniflora),* 637i
Indigo bunting *(Passerina cyanea),* 1289–1290, 1290i
Indoleacetic acid (IAA), 841–844, 842i–844i, 855–856
Induced fit, 80
Inducer, **341,** 342i, 343i
Inducible operon, **341,** 342i, 344
Induction, **355**–356, **1121,** 1136, 1138, 1139i
Infection thread, **805,** 806i
Infectious disease, phylogenetic analysis for tracking origin and spread of, 547, 547i
Inferior vena cava, 988, 988i
Infertility, human, **1111**–1112
Inflammation, **1005**–1006, 1005i
Inflorescence, 812
Influenza virus, 155, 377, 382, 382i, 383–384
Ingestion, **1043**
Ingram, Vernon, 998
Inheritance
  of acquired characteristics, 442
  autosomal dominant, 277, 278
  autosomal recessive, 276–277
  blending theory of, 240
  chromosomal mutations affecting, 271–276, 272i–275i, 275t
  chromosome theory of, 249–251, 250i
  criss-cross, 268
  cytoplasmic, 280, 280i
  defined, **5**
  genetic counseling, 279
  genomic imprinting, 280–281, 281i
  human genetics, 276–279, 276i, 277i, 279i
  linked genes, 262–266, 263i, 265i, 266i
  maternal, 280
  Mendelian, 239–258
    codominance, 254
    dominance, 243
    epistasis, 255–256, 255i
    incomplete dominance, 251–254, 253i
    multiple alleles, 254–255, 254i, 254t, 255i
    pleiotropy, 257, 257i
    polygenic inheritance, 256–257, 256i
    Principle of Independent Assortment, 248i, 249, 250i
    Principle of Segregation, 243, 244i, 250i
    probability use with, 245–246, 245i, 246i
  non-Mendelian patterns, 280–281
  sex-linked genes, 266–271, 268i–271i
  uniparental, 280
  X-linked, 268, 277, 279
Inhibin, **1107,** 1109, 1109i
Inhibiting hormones, **950,** 951t, 952
Inhibition hypothesis, **1220**
Inhibitor, enzyme, 81–83, 82i, 83i
Inhibitory postsynaptic potential (IPSP), **897,** 899, 899i
Initial cell, plant meristem, **759,** 768i
Initiation, of translation, **324**–325, 325i
Initiation factors, **324**
Initiator codon, **315**
Innate immunity, 1004–1007
  antimicrobial peptides, 1005
  complement system, 1006
  defensins, 1005
  defined, **1003**
  inflammation, 1005–1006, 1005i
  interferon, 1006
  natural killer cells, 1004t, 1006–1007
  recognition of pathogen, 1004–1005
  viral pathogens, response to, 1006–1007
Inner boundary membrane, chloroplast, **111**
Inner cell mass, **1129,** 1129i, 1130i
Inner ear, **927**–929, 928i
Inner mitochondrial membrane, **106,** 106i
Inorganic molecule, **43**
Inositol triphosphate (IP_3), **151**–153, 152i, 156, 157, 157i
Insect
  anatomy, 699, 700–701, 700i, 701i
  circulatory system, 983, 983i
  compound eye, 930, 930i
  development patterns, 701, 701i
  digestive system, 1047, 1047i
  diversity, 699, 700i
  ecdysis, 695i, 701
  eusocial, 1302, 1302i, 1303i
  exoskeleton, 974–975, 975i
  exotic species, 1261, 1262i
  hydrostatic skeletal elements, 974
  Malpighian tubule, 1073–1074, 1074i
  mouthparts, 699, 701, 701i
  parasitoids, 1208
  pollination by, 641, 643i
  taste and smell receptors, 934–935, 935i
  tracheal system, 1025, 1028–1029, 1029i
  vibration-detecting systems, 927, 927i
Insecticide resistance, evolution of, 448–449, 449i
Insertion sequence (IS), **332,** 333i
Instar, 701
Instinctive behavior, **1279**–1282, 1280t, 1281i, 1282i
Insulin, 58, 58i, 149, 354, 404, 946, 948, 952t, **959,** 960, 960i, 977, 1056, 1056i, 1060, 1060i
Insulin-like growth factor (IGF), **952**
Insulin-like growth factor-2 (IGF2), 281, 281i
Insulin receptor, 149
Insulin resistance, 977–978
Integral membrane proteins, 122i, **123,** 145, 146i
Integrase, 386
Integration, neural signaling, **884,** 884i
Integrator, **878,** 878i, 879i
Integrin, 115, 115i, 1136
Integument, 632, **1088**–1089, 1090, 1090i
Intensity, sound, 927
Intercalated disk, **874,** 874i
Intercostal muscle, 1031, 1031i, 1032i
Intercourse, 1110
Interference competition, **1203**
Interferon, **1006**
Interkinesis, **227**
Interleukin, 1011
Intermediate-day plant, **860**
Intermediate disturbance hypothesis, **1217,** 1217i
Intermediate filaments, 107, 107i, **108,** 114, 114i
Internal environment, **868,** 877
Internal fertilization, **1101,** 1102
Internal gills, 1027i, **1028,** 1028i
Internal intercostal muscle, 1032, 1032i
Interneuron, **884,** 885, 899, 909, 910i, 922, 1032–1033
Internode, plant, **757,** 757i
Interphase
  defined, **207**
  premitotic, 207, 208i
Intersexual selection, 470, 471i
Interspecific competition, 1203–1206, 1204i, 1206i
  character displacement, 1204–1205, 1205i
  competitive exclusion, 1203, 1204i
  defined, **1203**
  exploitative competition, 1203
  interference competition, 1203
  niche concept, 1204, 1205–1206, 1205i, 1206i
  resource partitioning, 1204, 1205i
  species richness, reduction of, 1213–1214, 1214i
Interstitial fluid, **876,** 876i, 877, 982–983, 992–993, 1078
Intertidal zone, **1164,** 1164–1165, 1165–1166, 1165i, 1166i
Intestinal villi, **1055,** 1056i
Intestine
  bird, 1047i, 1048
  earthworm, **1047,** 1047i
  insect, 1047, 1047i
  length, 1055, 1061, 1063, 1063i
Intracellular digestion, **1046**
Intrasexual selection, 470
Intraspecific competition, **1184**
Intrauterine device (IUD), 1112t, 1113
Intrinsic rate of increase, **1182,** 1183i
Introduced species, 1172–1173, 1173i, 1260–1261, 1261i, 1262i
Intron(s), 352
  absence in cDNA, 396, 404
  in archaeans, 563
  defined, **319**
  discovery of, 319
  human genome, 426
  splicing, 319–321, 319i, 320i
Invagination, **1123,** 1124i
Inversion, chromosomal, **272,** 272–273, 272i
Invertebrate(s), 973
  chordates, 710–711, 710i, 711i
  circulatory system
    closed, 983
    open, 983, 983i
  defined, **671**
  deuterostomes, 707–710, 708i–710i
  diversity, genetic basis of, 702
  ectothermy, 1086, 1087
  endocrine systems, 961–962, 961i, 962i
  evolutionary relationships, 702
  eyes, 929–930, 929i, 930i
  immune defenses, 1020
  nervous system, 905, 905i
  osmoregulation and excretion in, 1072–1074, 1073i, 1074i
  statocysts in, 926, 927i
  taste and smell, 934–935, 935i
  vibration-detecting systems, 927, 927i
Involution, **1123,** 1124i
Iodine, 23–24, 26, 956
Ion channel
  gated, **127,** 128i
  ligand-gated, 894
  voltage-gated, 888, 889i–892i, **890**–893, 894i, 895, 896
Ionic bond, **28**–29, 28i
Ionotropic neurotransmitter, **894**
Ions, **28**
  root uptake by active transport, 780–781
  transport, 776–777, 776i
IP_3. *See* Inositol triphosphate
IPSP. *See* Inhibitory postsynaptic potential
Irby, William S., 702
Iridium, 513
Iris, **930,** 930i, 931i
Iris *(Iris),* 860, 860i
Iron
  deficiency in plants, 797, 797i
  plant micronutrient, 796t
Irons, Ernest, 239–240
Irreversible inhibition, enzyme, 82
IS. *See* Insertion sequence
Islands
  allopatric speciation on, 487, 487i
  equilibrium theory of island biogeography, 1221–1225, 1222i, 1223i
Islet of Langerhans, 950i, 952t, 959

Isocitrate, 169
Isocitrate dehydrogenase, 170i
Isolated system, 72, 73i
Isoleucine, 82–83, 83i, 1048, 1049i
Isomers
defined, **44**
of monosaccharides, 45, 46i, 47
stereoisomers, 44–45, 45i, 47
structural, 45, 46i
Isosmotic, 1070, 1078
Isotonic solution/condition, 130i, **131**
Isotope
defined, **25**
examples, 25i
oxygen, 184
radioisotope, 25, 26
Isotope ratios, 559
IUD. *See* Intrauterine device
Ivy *(Hedera),* 765, 765i

Jacob, François, 221, 340, 341, 370, 374
Jade plant *(Crassula),* 824
Jagendorf, Andre T., 192
Jansen, Robert K., 638
Japanese blood fluke *(Schistosoma japonicum),* 686, 686i
Jarrow's spiny lizard *(Sceloporus jarrovi),* 1293
Jasmonate, 840t, **849**–852, 851t, 852i
Jaws
evolution of, 717–718, 717i, 718i
mammalian specializations, 733
Jefferson, Thomas, 441
Jeffreys, Sir Alec, 401
Jellyfish, 680, 682i, 904
Jenner, Edward, 1002, 1014
Jerne, Niels, 1011
Johnson, Robert T., 214, 216i
Joint, 976–977, 976i
Julius, David, 938
Junction
anchoring, 868
gap, 868, 874, 886, 886i
tight, 868, 908, 1003
Juvenile hormone, **961,** 961i, 1286, 1287i
Juxtaglomerular apparatus, **1080,** 1081i

K-cyclin, 220
K-selected species, **1188**–1189, 1189i, 1189t, 1214, 1217–1220
Kamen, Martin, 184
Kandel, Eric, 916
Kangaroo rat *(Dipodymus),* 1079, 1080i
Kaposi's sarcoma, 220
Karenia brevis, 594i
Karner Blue butterfly *(Lycaeides melissa samuelis),* 1272i
Karyogamy, 653i, **654,** 657, 657i, 658, 660i, 661, 662i, 663
Karyotype, **210**–211, 211i
Katsanis, Nicholas, 258
Katydid *(Mimetica),* 1202i
Katz, Paul S., 900
Kay, Mark A., 365
Kcal. *See* Kilocalorie
Keeled sternum, **730,** 730i
Kelp, 598, 1166, 1166i
Keratin, 109, 869, 1079, 1090, 1128
Kesterson Wildlife Refuge, 22
Ketone, 44t
Keystone species, **1215,** 1267
Khorana, H. Ghobind, 314
Kidney
diseases of, 1092
function, 1075–1080
in nonmammalian vertebrates, 1083–1085, 1084–1085, 1084i, 1085i
regulation, 1080–1084, 1081i–1083i
structure, 1074–1075, 1074i, 1075i
Kidney bean *(Phaseolus),* 763i, 819, 820i, 823i
Killifish *(Rivulus hartii),* 1180–1181
Kilocalorie (kcal), **34**
Kin selection, **1300**–1302, 1301i
Kinase
action of, 83
CDK, 217i, 218, 219, 222
defined, **147**
MAP, 153, 154i, 155
signal transduction pathways and, 146, 146i, 147–148, 148i, 149, 150i, 152i, 153, 154, 155, 156, 157i
tyrosine, 147–149, 148i, 153, 154i
Kinesin, 107, 109i
Kinesis, **1292**
Kinetic energy, **72,** 73
Kinetochore, 208i, **210,** 210i, 213, 214, 214i, 215i
Kinetochore microtubule, 208i, **210,** 210i, 213–214, 215i
Kinetoplastid, **592,** 592i
King, Thomas, 1121, 1138
Kingdom
Animalia, 12, 12i
defined, **9**
Fungi, 11, 12i
Plantae, 11, 12i
Kissing Gourami *(Helostoma temmincki),* 720i
Klinefelter syndrome, 275, 275t
Knee-jerk reflex, 909, 910i
Knifefish, 205
Knockout, gene, 406, 1008, 1283
knotted-1 gene, 761
Koch, George W., 783
Koch's postulates, 579
Kohler, Georges, 1015
Komodo dragon *(Varanus komodoensis),* 727
Koopman, Peter, 231
Korarchaeota, **583,** 584
Kornberg, Arthur, 299, 303
Kornberg, Roger, 317
Kramer, David, 201
Kramer, Elena M., 525
Krebs, Charles, 1191
Krebs, Edwin, 147
Krebs, Hans, 169
Krebs cycle, **169,** 170i. *See also* Citric acid cycle
Kring, Matthias, 742
Kudzu *(Pueraria lobata),* 1261, 1261i
Kuru, 387
Kushner, Peter, 954

La Niña, 1146
Labia majora, **1104,** 1104i
Labia minora, **1104,** 1104i, 1110
Labium, 701, 701i
Labrador retrievers, epistasis in coloration, 255–256, 255i
Labrum, 701, 701i
Lac operon, 341–342, 341i, 342i, 343i, 344
Lac repressor, **341,** 341i, 342i, 343i
Lactase, 1057
Lactate fermentation, **176–177,** 177i
Lactic acid, 1039
Lactobacillus, 177, 580
Lactose, 48, 48i
Lactose intolerance, 1057
LacZ$^+$ gene, 394, 395i
Lady Be Good (aircraft), 1069
Lady-of-the-night orchid *(Brassavola nodosa),* 807i
Ladybird beetle *(Coccinella septempunctata),* 1199i
Lagging strand, of DNA replication, 297i–300i, **298**
Lagging strand template, **298**
Lagomorpha, 734, 736i
Lahav, Noam, 557
Lake
seasonal changes, 1162–1164, 1163i
trophic nature of, 1164, 1164i
zonation, 1162–1163, 1163i
Lake Erie, 1229–1230, 1229i
Laloi, Maryse, 178
Lamarck, Jean Baptiste de, 442
Lambda (bacteriophage λ), 380–382, 381i
Lambeosaurines, 505–506, 505i
Lamin, 99
Laminaria, 599i
Lamprey, 715–716, 715i, 998
Lancelet, 711, 711i
Landscape corridor, 1270, 1271i
Landscape ecology, conservation and, 1269–1270
Landsteiner, Karl, 254–255
Language, brain areas associated with, 912i, 914
Large intestine, 1049i, 1058–1059, 1059i
Larva, **672,** 701, 701i, 961, 961i
Larynx, **1030,** 1031i, 1053, 1054i
Latent stage, **382**
Latent state, **1019**
Lateral bud, **757,** 757i, 758
Lateral geniculate nuclei, **934,** 934i
Lateral inhibition, **933**
Lateral line system, **719, 926,** 926i
Lateral meristem, **750,** 751i
Lateral root, **765,** 767, 767i
Lateralization, **914**
Laurasia, 507, 507i
LCAT. *See* Lecithin cholesterol acyltransferase
LDL. *See* Low-density lipoprotein
Le gene, 252
Le Quéré, Corinne, 1247
Leaching, **793**
Leading strand, of DNA replication, 297i–300i, **298**
Leading strand template, **298**
Leaf
anatomy, 782i, 783
compound, 762i
deciduous plants, 764
development, 833–834, 834i
epidermis, 76i, 756
evolution of, 614–615, 614i
functions, 762
growth, 762, 764
guttation, 783, 783i
internal structure, 762, 763i, 764
modifications, 762, 763i
monocots and eudicots compared, 752t
morphology and anatomy, 762, 762i
phase changes, 764, 764i
simple, 762i
transpiration, 781, 782i, 783–785
veins, 752t
Leaf anatomy, 185i, 197–198, 198i
Leaf primordium, 759i, **762,** 833–834, 834i
LEAFY protein, 643
Leakey, Louis, 742
Learned behavior, **1280,** 1280t, 1282–1284, 1284i
Learning
classical conditioning, 1284
cognition, 1284, 1284i
defined, **914, 1282**
habituation, 1284
imprinting, 1282, 1282i, 1284, 1284i
memory and, 916, 918
operant-conditioning, 1284
Le Beau, Michelle, 282
Leber's hereditary optic neuropathy (LHON), 280
Lecanora conizaeoides, 664
Lecithin cholesterol acyltransferase (LCAT), 1066
Leclerc, George-Louis, 441
Leder, Philip, 314
Lederberg, Esther, 375
Lederberg, Joshua, 370, 371, 371i, 375
Leech, 691–692, 692i
Leeuwenhoek, Anton van, 90, 91i
Leghemoglobin, **805**–806
Legionnaire's disease, 575, 578
Legumes, 805, 805i, 1243
Leishmaniasis, 592
Leks, **1296,** 1297i
Lemaître, George, 553
Lemming *(Lemmus lemmus),* 1191
Lemur, 736, 737i
Lens, **930,** 930i, 931, 1127–1128, 1128i
Lens placode, 1127–1128, 1128i
Lens vesicle, 1127, 1128i
Lentic system, 1161
Lenticel, 768
Leopard, 1300i
Leopard frog *(Rana pipiens),* 723i, 1101i
Lepidodendron, 622, 623i
Lepidosauromorpha, **724**
Lepidosaurs, 724, 726, 727–728
Leptin, 1060–1061
Lester, Diane, 252
Leucine, 1048, 1049i
Leucine zipper, 348, 348i
Leukocyte, **873,** 985i, 986i, **987, 1003**–1004, 1004t
Lewis, Edward B., 357
Leydig cell, **1108,** 1109, 1109i
LFY gene, 643
LH. *See* Luteinizing hormone
Li, Li, 39
Library, DNA, 394, 396–397
Lichen, 601, **664,** 665i, 1217
Lidocaine, 893
Life
characteristics of, 2–6
defined, **1**
inevitability of, 563
organization of, 2–3, 2i
origin of, 553–566
energy-harnessing reactions, 558
extraterrestrial, 554, 556–557
hydrothermal vents and, 556
inevitability of life, 563
Oparin-Haldane hypothesis, 555
organic molecule formation, 555–556, 556i
protein-first hypothesis, 564
protocells, 554, 557–559, 557i
RNA world, 558, 559
timeline, 554i
tree of, 446, 446i, 547–549, 548i, 549i
Life cycle
algae
brown, 598, 599i
green, 601, 602i
animal
hydrozoan, 681i
parasitic worms, 686, 686i
defined, **6**
Dictyostelium discoideum, 604, 604i
fungi
ascomycete, 660–661, 660i
basidiomycete, 662–663, 662i
generalized, 652, 653i
zygomycete, 657i
meiosis, time and place of, 233–235, 234i
moth, 6i
plant
angiosperm, 642i
bryophyte, 618
cherry *(Prunus),* 815i
conifer, 635, 636i
fern, 624i
general, 615, 615i
moss, 620i
Plasmodium, 595i
protist, 589
Life history
defined, **1178**
early *versus* late reproduction, 1179
energy budget, 1178
evolution of traits in guppies, 1180–1181
fecundity *versus* parental care, 1179
population size, effect on, 1188–1189, 1189i, 1189t
reproductive episodes, number of, 1179
variation among species, 1178–1179
Life table, **1176**–1177, 1177t
Ligaments, **872,** 976, 976i
Ligand-gated ion channel, 894
Ligation, **394**

Light
  absorption in photosynthesis, 185–189, 185i, 186i, 187i, 188i
  fluorescence, 186
  induction of stomata opening, 784–785
  visible spectrum, 185, 186i
Light chain, antibody, **1009,** 1009i
Light-dependent reactions of photosynthesis, **183,** 183i, 184, 185–192, 188i, 189i, 190i, 192i
Light-harvesting complex, 188
Light-independent reactions, **183,** 183i, 184, 193–196, 195i
Light microscope, **91**–92, 93i
Lignification, **749,** 754
Lignin, 112, **613,** 614
Lijam, Nardos, 1283
Likens, Gene, 1243
Lilac *(Syringa vulgaris),* 862i
Limbic system, 911i, **912**
Limbs, evolution of, 722, 722i
Limiting nutrient, **1234**
Limnetic zone, **1162,** 1163i
Limulus polyphemus, 697i
Linear electron flow, **191,** 198i
LINEs. *See* Long Interspersed Elements
Linkage, **262**–266, 263i, 265i, 266i
Linkage map, **264**–265, 266i
Linked genes, **262**–266, 263i, 265i, 266i
Linker, DNA, **305,** 305i
Linnaeus, Carolus, 440, 529–530
Linoleic acid, 1048
Linolenic acid, 1048
Linum insertion sequence-1 (LIS-1), 863
Lion–tiger hybrids, 484
Lipase, 1046, 1053i, **1057,** 1058i
Lipid bilayer assembly, protocell, 557–558, 557i
Lipids, 50–55
  defined, **50**
  fatty acids, 50–52, 51i
  membrane, 120–121, 121i, 122i, 123–124
  neutral, 50–53
  phospholipids, 53, 53i
  steroids, 54–55, 54i
  triglyceride, 50, 51, 51i, 52i
  waxes, 52–53
Lipopolysaccharide, **570,** 581
Lipoprotein, 52, 54, 62, 136, 1057
Lipoprotein lipase, 1065–1066
Liposomes, 949
Lipotyphla, 734, 736i
LIS-1 (*Linum* insertion sequence-1), 863
Listeria monocytogenes, 115
Lithium, 153
Lithosphere, **1146**
Litt, Amy, 645
Littoral zone, **1162,** 1163i
Liver, 1049i, 1056i, **1057,** 1058
Liverwort, 619, 619i
Living systems, characteristics of, 2–6
  chemical instructions, 3–4, 3i
  energy and matter cycling, 4, 5i
  evolution of populations, 5–6
  metabolic activity, 4, 4i
  organization, 2–3, 2i
  reproduction and development, 5, 6i
  response to environmental change, 4–5
Lizard
  description, 727–728, 728i
  ectothermy in, 1086, 1087, 1087i, 1088
  iguana, 2i
  as model research organism, 728, 729
  predation on spiders, 1187–1188, 1188i
  temperature regulation, 15–17, 17i
  thermoregulation, 1152–1153, 1153i
Loam, **798,** 800t
Lobaria verrucosa, 665i
Lobe-finned fish, 721, 721i
Lobster, 698, 698i
Local homeostatic controls, **878**
Local inflammation, **1005**
Local regulation, 142, 945
Locus, **251,** 251i
Locust *(Locusta migratoria),* 1186, 1186i
Locust *(Schistocerca gregaria),* 423
Lodgepole pine *(Pinus contorta),* 666i
Loewi, Otto, 887i
Loggerhead sea turtles *(Caretta caretta),* 939, 939i
Logistic model of population growth, 1182–1184, **1183,** 1183i, 1184t, 1185i
Lohmann, Kenneth, 939, 939i
Long-day plant, **859**–860, 860i
Long Interspersed Elements (LINEs), 537
Long-tailed widowbird *(Euplectes progne),* 470, 471i
Long-term memory, **916,** 918
Long-term potentiation, **916,** 917
Loop of Henle, **1075,** 1075i, 1076i, 1077–1078, 1077t
Loose connective tissue, **872,** 873i
Lophophore, 677, **683,** 683i
Lophotrochozoa, 677, 683–692
  feeding structure, 683, 683i
  phyla
    Annelida, 690–692, 691i, 692i
    Brachiopoda, 683i, 684
    Ectoprocta, 683, 683i
    Mollusca, 687–690, 688i–690i
    Nemertea, 686–687, 687i
    Phoronida, 683i, 684
    Platyhelminthes, 684–685, 684i, 685i, 686
    Rotifera, 685–686, 687i
Loreau, Michel, 667
Lorenz, Konrad, 1279, 1284, 1284i
Lorises, 736
Losos, Jon, 729
Lotic system, 1161
Lotka, Alfred J., 1203
Lotus *(Nelumbo lutea),* 822
Low-density lipoprotein (LDL), 52, 136, 137, 1065–1066
Lubchenco, Jane, 1215, 1215i
Luciferase, 594, 1078–1079, 1134
Luciferin, 594
Lucy (hominid fossil), 740, 741i
Ludwig, Stephen, 155
Luft, Rolf, 161
Luft syndrome, 161
Lumbricus, 692i
Lumen, digestive tube, **1047**
Luna moth, 700i
Lungfish, 721i, 1029
Lungs, **1025,** 1026i, 1029–1033, 1030i, 1031i, 1032i
Lupine *(Lupinus arcticus),* 822
Luteal phase, 1105, 1106i
Luteinizing hormone (LH), 951t, **952,** 953i, 958, **1105,** 1106i, 1107, 1109–1110, 1109i, 1113
Lycophyta, **622,** 622t
Lycophytes, 622, 622t, 623, 623i, 626–627
Lycopodium, 623, 623i
Lyell, Charles, 439, 442
Lyme disease *(Borrelia burgdorferi),* 569, 578
Lymph, **996**–997
Lymph capillaries, 996–997
Lymph duct, 996
Lymph node, **997,** 997i
Lymph vessel, 996–997, 997i
Lymphatic system, **982, 996**–997, 997i, 1004
Lymphocyte
  activation, 1010–1011, 1012i–1013i
  B cells, 1007
  defined, **1006**
  function, 1004t
  natural killer cells, 1006
  T cells, 1007
Lymphoid stem cell, 986, 986i
Lynch, Tom, 413
Lynx, 1191, 1192i
Lysine, 1048, 1049i
Lysogenic cycle, **380**–382, 381i
Lysosomal storage diseases, 106
Lysosome, 98i, **104**–106, 104i, 105i, 1011, 1046
Lysozyme, 60, 60i, 1053
Lytic cycle, **380,** 380i, 380, 381i, 382
M cyclin, 218
M line, 967i, 968, 969i
M phase, 207, 207i, 216i, 217, 217i, 218. *See also* Mitosis
M phase-promoting factor (MPF), **218**
MacArthur, Robert, 1195, 1213, 1221–1225
MacLeod, Colin, 287
Macrocystis, 598i
Macroevolution
  biodiversity trends, 512–514, 514i
    adaptive radiations, 512–514, 512i
    extinction, 513
  defined, **448**
  interpretation of lineages, 514–518, 516i
  in modern synthesis view, 448
  morphological trends
    allometric growth, 518, 519i, 522
    exaptation, 519
    heterochrony, 518–519, 522
    novelties, 518–524
    paedomorphosis, 519, 519i, 520i
  tempo of morphological change, 515, 517, 518i
Macromineral, 1051, 1051t
Macromolecules, **46**–47, 557
Macronucleus, 590i, **593**
Macronutrient, **795,** 796, 796t
Macrophage, 987, 1004t, 1005i, **1006**
Macular degeneration, 365
Mad cow disease, 387
Madreporite, 707, 709i
Magnesium
  deficiency in plants, 797, 797i
  plant macronutrient, 795, 796t
  soil, availability in, 800
Magnetite, 939
Magnetoreceptor, **938**–939
Magnification, **91**
Magnolia *(Magnolia grandiflora),* 764
Magnoliids, **639,** 639i
Maize. *See* Corn *(Zea mays)*
Maizel, Alexis, 643
Major histocompatibility complex (MHC), 1007, **1011,** 1012i, 1015, 1016, 1017i
Malaria, 472–473, 472i, 528–529, 594, 595, 1239
Malate, 169, 170i, 198–199, 199i
Malate dehydrogenase, 170i
Male reproductive structures, human
  anatomy, 1108–1109, 1108i
  development of, 1132, 1133i, 1134
Malignant tumor, 361
Malleus, **927,** 928i
Malnutrition, **1044**
Malpighian tubule, **700,** 700i, **1073**–1074, 1074i
Malthus, Thomas, 445
Maltose, 48, 48i
Mammal(s), 732–743
  adaptive radiation, 512
  auditory structures, 927–929, 928i
  convergent evolution, 511, 511i
  digestive system, 1048–1059, 1049i, 1052i–1059i
  diving, 1038–1039
  embryonic development, 1128–1133, 1129i–1133i
  fertilization in, 1110i
  heart, 987–991, 988i–990i
  high-altitude adaptations, 1038, 1038i
  hominins, 739–743, 739i–742i
  key adaptations, 733
  kidney regulation, 1080–1084, 1081i–1083i
  lungs, 1030–1032, 1031i, 1032i
  marsupials, 734, 734i
  monotremes, 733i, 734
  nonhuman primates, 734–739, 736i–738i
  osmoregulation and excretion, 1074–1080, 1074i–1076i, 1077t, 1080i
  placentals, 734, 736i
  reproductive system, 1101i
  respiratory system, 1030–1033, 1031i–1033i
  retina, 931–933, 931i, 933i
  sense of smell in, 735
  teeth specialization, 1061, 1061i
  thermoregulation, 1088–1092
Mammary gland, **733,** 1132
Mammoth, 501i
Mandible, crustacean, 698
Manganese, as plant micronutrient, 796t
Mangold, Hilde, 1138, 1139i
Mangrove, 765, 765i, 1165, 1165i
Mani, Shaila, 157
Manic depression, 153
Mannose, 47i
Manta ray *(Manta birostris),* 718, 718i
Mantle, **687,** 688i
Mantle cavity, 687, 688, 688i
MAP kinases. *See* Mitogen-activated protein kinases
Map unit, **265**
Maple *(Acer),* 820, 821i, 847i
Mapping genes, by conjugation, 374
Marchantia, 619, 619i
Margulis, Lynn, 561
Marine animals, diversity of, 514, 514i
Marine biomes, 1164–1168
  abyssal zone, 1165, 1167
  benthic province, 1167–1168, 1168i
  estuaries, 1165, 1165i
  intertidal zone, 1165–1166, 1166i
  neritic zone, 1166–1167, 1166i
  oceanic zone, 1167, 1167i
  zonation, ocean, 1164–1165, 1165i
Marine mammal, salt excretion in, 1079
Maritime climate, **1150**
Mark-release-recapture technique, 1174i
Marler, Catherine, 1293
Marler, Peter, 1280, 1285
Marsh, Othniel C., 515, 516i
Marsh marigold, 643i
Marshall, Barry, 579i, 1054
Marsupials (Metatheria), 511, 511i, 734, 734i, 1103
Marsupium, **724, 1103**
Martin, David, 252
Martins, 1297–1298, 1298i
Mass, **25**
Mass extinction, **513,** 513i, 1263
Mass number, 24t, **25,** 35
Mast cell, 1005i, **1006,** 1020i
Mastax, rotifer, 685, 687i
Maternal chromosome, **226,** 232–233, 233i
Maternal-effect genes, **358,** 358i
Maternal inheritance, 280
Mating, nonrandom, 470–471
Mating behavior, 943, **1099**
Mating system, 488, **1297**
Mating type, 267, **652,** 653i, 657–658, 657i, 661
Matter, **23**–24
Maurel, Christophe, 779
Maxam, Allan M., 418
Maxillae, crustacean, 698
Maxilliped, 698, 698i
Maximum likelihood methods, **543**
Mayfly *(Hexagenia),* 1230
Mayr, Ernst, 480, 481, 486
McAdam, Scott A. M., 785
McCarty, Maclyn, 287
McClintock, Barbara, 333–334, 334i
McFadden, Geoff, 608
McGraw, William S., 1254
Mclr gene, 8
MCM helicase, 584
Meagher, Richard, 801
Mean, 458, 458i
Measure, units of, 92i
Mechanical isolation, **484**–485, 485i
Mechanical processing, of food, **1047**
Mechanoreceptor
  baroreceptors, 994–995
  defined, **923**

Mechanoreceptor *(Continued)*
  hearing, 927–929, 928i
  lateral line system, 926, 926i
  proprioreceptors, 924–926
  sensory adaptation, 923
  statocysts, 926, 927i
  tension receptors, 924–925, 925i
  touch and pressure, 924, 924i
  vestibular apparatus, 925–926, 925i
  vibration detection, 927, 927i
Mediator, 348
Medicago trunculata, 802
Medulla oblongata, 905, 906i, 909, 911i
Medullosa, 622, 623i
Medusae, **679**–680, 679i, 681i, 682i
Megapascal, **777**
Megaphylls, 614, 614i
Megaspore, **635,** 636i, **674, 814,** 815i
Meiosis, **206,** 225–236
  in angiosperms, 812–813, 812i, 814, 815i
  chromosome theory of inheritance and, 249–251, 250i
  defined, **225**
  environmental chemicals and, 235–236
  gametogenesis, 1098–1099, 1098i
  genetic diversity from, 1098–1099
  mechanisms of, 226–233, 227i–229i
  meiosis-inhibiting factor gene, 231
  mitosis compared, 230i
  nondisjunction during, 273, 274i, 275
  regulation of, 235
  time and place in organismal life cycles, 233–235, 234i
Meiosis I, **226,** 227i–230i, 232i
Meiosis II, **226,** 227i–230i, 232i
Meissner's corpuscle, 924i
Melanocyte, 952, 1018
Melanocyte-stimulating hormone (MSH), 951t, **952,** 953i, 1060
Melanoma, 1018
Melanophore, 962
Melatonin, 952t, **961**
Mello, Craig, 352
Membrane. *See also* Plasma membrane
  origin of internal eukaryotic, 561i, 562, 562i
  outer of Gram-negative bacteria, 570, 571
  water polarity, effect of, 34, 35i
Membrane attack complex, **1006,** 1012
Membrane potential, **131,** 776
  action potential and, 888–893, 889i–892i, 894i
  defined, **888**
  depolarization, 889, 889i, 890i, 892, 892i, 893, 894i, 897
  excitatory postsynaptic potential, 897, 899, 899i, 900i
  graded potential, 897
  hyperpolarization, 889, 889i, 891i, 932
  inhibitory postsynaptic potential, 897, 899, 899i
  measuring, 888, 888i
  resting potential, 888–889, 889i
  threshold potential, 889, 889i
Memory, **914,** 916, 917, 918
Memory B cell, **1011,** 1013–1014, 1013i
Memory cell, **1008,** 1013i
Memory helper T cells, **1013,** 1014
Mendel, Gregor, 240–252, 240i, 448
  dihybrid crosses, 246, 248i, 249
  dominance, 243
  predictions of offspring proportions, 245, 245i, 246i
  Principle of Independent Assortment, 248i, 249, 250i
  Principle of Segregation, 243, 244i, 250i
  rediscovery of Mendel's research, 249
  single-character crosses, 242–245, 242i, 244i
  testcross use, 246, 247i
Mendelian inheritance. *See* Inheritance, Mendelian
Meninges, **907**
Menotti-Raymond, Marilyn, 391
Menstrual cycle, **1105,** 1106i, 1107
Menstruation, 1107
Mental retardation, 301
Mercaptoethanol, 45t
Mercury, soil contamination with, 801
Meristem
  apical, 750, 751i, 757i, 759, 759i, 765–766, 766i
  buds, 757
  cork cambium, 767, 768, 769i, 770
  defined, **750**
  ground, 759, 759i, 766
  lateral, 750, 751i
  primary, 759
  procambium, 759, 759i, 766i
  protoderm, 759, 759i, 766i
  root, 765–766, 766i
  stem cell reservoir, 771
  vascular cambium, 767–769, 768i, 769i
MERRF. *See* Myoclonic epilepsy and ragged-red fiber
Meselson, Matthew, 293, 294i, 295
Mesencephalon, 906i
Mesentery, **673,** 673i, **872,** 1052
Mesoderm, **672,** 673i, **1121,** 1121t, 1122–1129, 1122i, 1124i, 1125i, 1127i, 1137i
Mesoglea, 679, 679i, 680
Mesohyl, **678**
Mesonephric duct, 1132, 1133i
Mesophile, 582
Mesophyll, **762,** 763i, 764
Mesophyll cells
  in C_4 pathway, 197–198, 198i, 199i
  in CAM pathway, 198–199, 198i, 199i
Mesquite *(Prosopis),* 786
Messenger RNA (mRNA)
  artificial, 314
  breakdown, 352
  defined, **313**
  modification
    5′ cap, 318, 319i
    poly(A) tail, 318, 319i
  posttranscriptional regulation and, 351–354, 353i
  prokaryotic, 95
  protein-coding gene identification from, 422–423
  splicing, **319**–321, 319i, 320i, 352
Metabolic pathway
  anabolic, 76
  biosynthetic, 76
  catabolic, 76
  defined, **75**
  feedback inhibition, 82, 83i
Metabolic rate, 733, 1087, 1087i
Metabolic water, 1071
Metabolism
  as characteristic of living systems, 4
  defined, **4, 72**
  inborn error, 311
  in protists, 589
Metabolomics, in plant science, 789–790
Metabotropic neurotransmitter, **895**
Metamonada, **591**
  diplomonadida, 591, 591i
  parabasala, 591, 591i
Metamorphosis, **956,** 956i, 961, 961i
  amphibian, 722
  complete, 701, 701i
  defined, **678**
  incomplete, 701, 701i
  sponge, 678
Metanephridia, **691,** 691i
Metanephridium, **1073,** 1073i
Metaphase, in mitosis, **209,** 209i, **210,** 210i **210**–211
Metaphase I, in meiosis, 229i, 230i, 233i
Metaphase II, in meiosis, 229i
Metaphase plate, 210
Metapopulation, **1189**–1190, 1190i
Metastasis, 219, 361
Metatheria, 733
Metatherian, 1103
Metencephalon, 906i
Meteorite, 554, 556–557, 564
Methane, 29, 29i, 43, 74, 163, 163i, 1065
Methanobacterium, 583
Methanococcus, 96, 583
Methanogen, 583, 583i
Methanosarcina, 423, 583i
Methionine, 1048, 1049i
Methylation
  of DNA, 281, 350–351
  of histones, 351
Methylmercury, 801
Methyltestosterone, 959
Metrarabdotos, 517, 518i
Mexican leaf frog *(Pachymedusa dacnicolor),* 966, 966i
Meyerhof, Otto, 165
Meyerowitz, Elliot, 832
MHC. *See* Major histocompatibility complex
Michel, Andrew P., 490
Micrasterias, 92i, 588i
Microbody, 98i, **106–107,** 107i, 330
Microclimate, **1151**
Microevolution, 456–479
  adaptation, 475
  agents of, 462–471, 464t
    gene flow, 463–464, 464t, 464i
    genetic drift, 464t, 464–465, 465i, 466
    mutation, 462–463, 464t
    natural selection, 465–470, 467i–470i
    nonrandom mating, 470–471
  antibiotic resistance, 456–457
  defined, **448, 457**
  maintenance of variation, 471–474
    balanced polymorphisms, 472–474, 472i–474i
    diploidy, 471–472, 472t
    neutral variation hypothesis, 474
  population genetics, 460–461, 461t
  variation in natural populations, 456i, 457–461, 457i, 458i
    genetic, 460
    phenotypic, 456i, 457–460, 457i, 458i
    qualitative, 458, 458i
    quantitative, 457–458, 458i
Microfilaments, 98i, 107, 108i, **109,** 114, 115i, 1135, 1135i
Microfossils, **560**
Micronucleus, 590i, **593**
Micronutrient, **795,** 796, 796t
Microphylls, 614, 614i
Micropyle, **814,** 815i, 817, 817i
Microraptor gui, 520, 521i
microRNAs (miRNAs), **352**–353, 353i
  cancer, role in, 363–364
  in development, 359–360
  development, role in, 359–360
Microsatellite, **401**
Microscope, **91**
Microscopy
  bright field, 93i
  confocal laser scanning, 93i
  dark field, 93i
  defined, **91**
  electron, **91**–92, 93i
  fluorescence, 93i
  light, **91**–92, 93i
  Nomarski, 93i
  phase-contrast, 93i
Microsporangia, 635, 636i
Microspore, **635,** 636i, **814,** 815i
Microsporidia, **654, 663,** 663i
Microtubule organizing center (MTOC), **213**
Microtubules
  appearance, 98i, 108i, 109i
  cell shape and, 1135, 1135i
  defined, **107**
  flagella, 109–110, 110i
  kinetochore, 208i, 210, 210i, 213–214, 215i, 227, 228i–229i
  motor proteins and, 107, 109i
  in oriented cell division, 830, 831i
  spindle, 205i, 208i–210i, 210–211, 212–214, 213i–214i
Microvilli, 869, **1055,** 1056i
Midbrain, **905,** 906i, 909, 911i
Middle ear, **927,** 928i
Middle lamella, **112,** 113i
Miescher, Johann Friedrich, 286
Mifepristone (RU-486), 1113–1114
Migration, 1288–1291
  bird, 732
  defined, **1288**
  food supply variation and, 1290–1291, 1291i
  seasonal cycle, 1288–1289, 1289i1281i
  wayfinding mechanisms, 1289–1290
    compass orientation, 1289–1290, 1290i
    navigation, 1290
    piloting, 1289
MIH. *See* Molt-inhibiting hormone
Mildew, downy, 596, 596i
Milkweed, 1201
Miller, Stanley L., 555
Miller, UIrich, 86
Miller–Urey apparatus, 555, 556i
Millipede, 699, 699i
Milstein, Cesar, 1015
Mimic, **1202**
Mimicry, **1202,** 1202i, 1203i
Mimosa pudica, 857, 858i
Mineralocorticoid, 958
Minerals
  availability in soil, 800–801, 800i
  cation exchange, 800, 800i
  deficiencies in plants, 796t, 797–798, 797i
  essential, 1044, 1051, 1051t
  plant requirements, 794–796, 795i, 796t
  transport in xylem, 755, 755i
Minimal medium, **371,** 371i, 375–376, 376i
Minimum viable population size, **1268**
miRISC. *See* miRNA-induced silencing complex
miRNA-induced silencing complex (miRISC), **352,** 353i
miRNAs. *See* MicroRNAs
Mismatch repair, **303,** 304, 304i
Miss Waldron's red colobus monkey *(Procolobus badius waldroni),* 1254–1255, 1254i
Missense mutations, **331**–332, 331i, 332i
Mistletoe *(Phoradendron),* 1208
Mitchell, Peter, 175
Mite, 695, 697, 697i
Mitochondria
  cell respiration and, 164i, 165, 168–176, 168i–173i
  defined, **106**
  disorders of, 161–162
  endosymbiont hypothesis and, 561–562, 561i
  genome, 426
    mammoth, 452
  horizontal gene transfer and, 549, 549i
  Luft syndrome, 161
  protein sorting to, 330
  structure, 98i, 106, 106i, 165i
Mitochondrial DNA studies
  arthropods, 696
  evolution rate of, 544–545
  human evolution, 742, 743
  humpback whales, 466
  Neanderthal, 742
Mitochondrial electron transfer system, 165, **171,** 172i, 173, 173i, 175–176
Mitochondrial Eve, 743
Mitochondrial genes, inheritance of, 280
Mitochondrial matrix, **106,** 106i
Mitogen, **153**
Mitogen-activated protein kinases (MAP kinases), 153, 154i, 155
Mitosis
  asexual reproduction by, 1097
  in cell cycle, 206, 207i
  clones, 207
  cytokinesis, 206, 207, 207i, 211, 212i

defined, **206**
evolution from binary fission, 221
meiosis compared, 230i
oriented cell division, 830, 831i
spindles, 208i–210i, 210–211, 212–214, 213i–214i
stages of, 208–211, 208i–210i
vegetative (asexual) reproduction, 212
Mitotic spindle. *See* Spindle
Mitotic spindle checkpoint, 217, 217i
Miura, Kenji, 821
Mixed-use conservation, 1272, 1272i
mixta (snapdragon mutant), 758
MN blood group gene, 254
Mnium, 618i
Model, **1202**
Model organisms, **17**–18
Modeling, ecosystem, 1230–1232, 1231i
Modern synthesis, **448**
Molar, **733,** 1061, 1061i
Molarity, **36**
Mold, **653**
Molecular clock, **544**–546, 545i
Molecular fossils, 86
Molecular phylogeny, 676–677, 676i
Molecular switch, 149
Molecular weight, **36**
Molecules. *See also* Biological molecules
compounds, 24
defined, **24**
formula, 24
inorganic, 43
nonpolar, 30
organic, 43
polar, 30
Moles, 734, 736i
Mollusk, 687–690, 688i–690i
closed circulatory system, 983
exoskeleton, 974
metanephridium, 1073
nervous system, 904i, 905
thermoregulation, 1087
Molting, 961–962, 962i
Molt-inhibiting hormone (MIH), **962,** 962i
Molybdenum, as plant micronutrient, 796t
Monarch butterfly *(Danaus plexippus),* 539i, 1291, 1291i
Monera, 547
Monkey, 737–738, 738i
Monkey-flower *(Mimulus),* 484–485, 485i, 492
Monoclonal antibody, **1015,** 1016i
Monocot
defined, **638**
eudicots compared, 752t
examples of, 639, 641i, 751
leaves, 762, 762i
pollen grains, 638
stem growth, 75, 759
stem organization, 760, 760i
Monocyte, 1004t, 1005i, **1006**
Monod, Jacques, 340, 341, 370
Monoecious species, **814**
Monogamous species, **1297**
Monogenoidea, 685
Monoglycerides, 1058, 1058i
Monohybrid, **243**
Monohybrid cross, **243**
Monomers, **46**
Monophyletic taxa, **533,** 533i
Monoploids, **273,** 275
Monosaccharide
defined, **47**
isomers, 45, 46i, 47
linear and ring forms of, 47, 47i, 48i
Monosiga brevicollis, 435, 435i
Monosodium glumatate (MSG), 935
Monotremes (Prototheria), 734, 734i, **1100,** 1103
Monounsaturated fat/fatty acid, **50**
Monsoon cycles, **1151**
Montane forest, tropical, 1157
Moore, Michael, 1293
Moray eel *(Gymnothorax moringa),* 720i
Morel *(Morchella esculenta),* 659, 659i
Morgan, Thomas Hunt, 262, 263i, 264–265, 268, 269i, 448
Moritz, Craig, 1154
Morning-after pill, 1113
Morphogenesis, **829**–832, **1119,** 1133
Morphological species concept, **481**
Morphology
comparative, 440–441, **449,** 450
defined, **440, 1118**–1119
evolution of novelties, 518–524
evolutionary relationship clues in, 534–535
Mortality, age-specific, **1176**–1177, 1177t
Morula, **1121,** 1122i, 1129, 1129i
Mosquitoes, 528–529, 529i, 594, 595, 701i
Moss, 612i, 618i, 620–621, 620i
Moth, development of, 5, 6i
Mother of thousands plant *(Kalanchoe daigremontiana),* 824, 824i
Motifs, **62, 347**–348, 348i
Motile, **671**
Motor neuron, **884,** 885i
Motor protein, 107–108, 109i, 214
Motor regions of the cerebral cortex, 913i, 914
Motor unit, **972,** 972i
Mouse *(Mus musculus)*
activity level, genetic basis of, 459, 459i
genome size, 423i, 424
giant, 406, 407i
knockout, 406, 1008, 1283
as research organism, 1008, 1015
Mouth, 1049i, 1053
Movement, in vertebrates, 976–977, 976i, 977i
Movement, plant, 855–858
MPF. *See* M phase-promoting factor
mRNA. *See* Messenger RNA
MSG. *See* Monosodium glumatate
MSH. *See* Melanocyte-stimulating hormone
mtDNA. *See* Mitochondrial DNA studies
MTOC. *See* Microtubule organizing center
Mucosa, **1052,** 1052i
Mucous cell, **1054,** 1055i
Mucous membrane, 1003
Mud crab *(Eurytium limosum),* 1238
Muglia, Louis J., 1142
Mule, 486
Müller, Fritz, 1202
Müllerian duct, **1132,** 1133i
Müllerian mimicry, **1202,** 1203i
Mullis, Kary B., 398
Multicellular associations, 574–576
Multicellular organisms, 2i, **3**
Multicellularity, evolution of, 563, 868
Multigene family, 432, **433**–434, 434i
Multiple alleles, **254**–255, 254i, 254t, 255i
Multiple fruit, 820, 821i
Multiple sclerosis, 893, **1018**–1019
Multipotent cells, **1121**
Munnich, Arnold, 278
Murad, Ferid, 996
Murphy, Dennis D., 1190i
Muscle(s)
agonist, 976
antagonist, 976–977, 977i
bone interactions, 976–977
Muscle fibers
contraction, 968–972, 968i–971i
defined, **874, 967**
structure, 967–968, 967i, 968i
tetanus, 969, 971, 971i
twitch, 969, 971i
types
fast aerobic, 971t, 972
fast anaerobic, 971t, 972
motor units, 972, 972i
slow, 971–972, 971t
Muscle spindles, 924, 925i
Muscle tissue
cardiac, 874, 874i
defined, **874**
skeletal, 874, 874i, 967–973
smooth, 874, 874i
Muscle twitch, **969,** 971i
Muscular dystrophy, 968, 977
Duchenne, 277, 277i
Muscularis, **1052,** 1052i
Mushroom, 12i, 649, 650i, 651, 651i, 661–663, 661i, 662i
Musk oxen *(Ovibos moschatus),* 1278i, 1299
Mussel, 310, 310i, 689, 1214, 1214i
Mutagen, **311**
Mutation
adaptations, production of favorable, 7–8
advantageous, 463
as agent of microevolution, 462–463, 464t
auxotrophs, 371, 375, 376i
base-pair substitution, 331–332, 331i–332i
cancer and, 362–363
defined, **7, 304, 331, 462**
deleterious, 462
detrimental, 476
frameshift, 331i, 332
human genetic disorders, 258
lethal, 462
missense, 331–332, 331i, 332i
neutral, 463
nonsense, 331i, 332
null, 427
recessive embryonic lethal, 357
replication errors, 302–304
reproductive isolation and, 491, 492
sickle-cell, 401, 402i
silent, 331i, 332
somaclonal, 826
as source of variability, 7
Mutationism, 448
Mutualism, 1063, **1207,** 1207i, 1208i
Mycelium, **596, 650,** 651, 651i
Mycobacterium tuberculosis, 1019
Mycobiont, **664,** 665i
Mycology, 650
Mycoplasma, 580
Mycoplasma genitalium, 423, 569
Mycorrhizae, 655, 658, 665–666, 665i, 666i, **803**–804, 806
Myelencephalon, 906i
Myelin sheath, 885, 886i, 893, 894i
Myeloid stem cell, 986, 986i
Myeloma cells, 1015
Myers, Norman, 1257, 1264
Myllokunmingia, 712, 713i
Myoblasts, 356i
Myoclonic epilepsy and ragged-red fiber (MERRF), 280
myoD gene, 356, 356i
Myofibril, **967,** 967i
Myogenic heart, **989**
Myoglobin, **972,** 1037, 1039
Myosin, 107, 109, **874,** 967–968, 970i
Myostatin, 977
Myriapoda, 699, 699i
Mytilus, 310, 310i
Myxinoidea, 713, 715
Myxobacteria, 578
Myxoma virus, 1172

N-terminal end, amino acid, **57**
Na^+/K^+ pump (sodium-potassium pump), **132,** 133i
Nachman, Michael W., 7–8
NAD^+. *See* Nicotinamide adenine dinucleotide
NADP. *See* Nicotinamide adenine dinucleotide phosphate
Nanoarchaeota, **583**
Nanoplankton, 598
Nanos gene, 358
Narwhal *(Monodon monoceros),* 736i
Nasal cavities, 936, 936i
Nastic movements, **857**–858, 857i, 858i
National Institutes of Health (NIH), 406
Natrosobacterium, 583
Natural history, **440**
Natural killer cell, 1004t, **1006**–1007
Natural selection
adaptation and, 8, 448–449, 454
adaptive traits favored by, 445–446
balanced polymorphisms, maintenance of, 472–474, 472i–474i
Darwin and, 6–7, 446, 448–449
defined, **6, 445, 465**
genetic variability shaped by, 465–470
human behavior, 1304
kin selection, 1300–1302, 1301i
modes
directional, 467–468, 467i, 469i
disruptive, 467i, 468, 470, 470i
stabilizing, 467i, 468, 468i, 469i
recessive alleles masked by diploidy, 471–472, 472t
relative fitness and, 466
Natural theology, **440,** 441
Nautilus macromphalus, 690i
Nautiluses, 690, 690i
Navarro, Arcadi, 496
Navigation, **1290**
Necrotizing fasciitis, 581
Nectar, 641, 788
Nef protein, 1021
Negative feedback, **878**–879, 878i, 879i, 945–946, 946i, 950, 957i, 959, 960i, 1085, 1105
Negative gene regulation, 344
Negative pressure breathing, **1029,** 1030
Neisseria gonorrhoeae, 572, 578
Neisseria meningitidis, 578
Nekton, **1167**
Nematocyst, 680, 680i
Nematode, 686, 686i, 693, 693i, 694
Nematode-trapping fungi, 659, 659i
Nemertea, 686–687, 687i
Neotiella, 228i
Nephron, **1074**–1080, 1074i–1076i, 1077t
Nereis, 692i
Neritic zone, **1164,** 1165i, 1166–1167, 1166i
Nerve, **904**
Nerve cord, 710, 710i, 904i, **905**
Nerve growth factor, 149
Nerve net, **680, 904,** 904i
Nerve ring, 904, 904i
Nervous system, 903–920
annelid, 691, 691i
autonomic, 907, 907i, 908i
bivalves, 689
central, 905, 908–918, 909i–913i
cephalized, 904i, 905
cephalopod, 690
crustacean, 698
development, 1126–1127, 1126i, 1127i
development, hormone effects on, 1286, 1286i
flatworm, 684, 684i
nerve nets, 904, 904i
peripheral, 905, 906–907, 907i, 908i
vertebrate, 904i, 905
Nervous tissue, **875,** 875i
Nest-building behavior, evolution of, 1297–1298, 1298i
Nest predation, 1259, 1259i
Net primary productivity, **1233,** 1233i, 1234, 1234i, 1234t, 1235i, 1250
Neural circuit, 918
Neural crest, **712,** 1126i, **1127,** 1135, 1136
Neural plate, **1126,** 1126i, 1135, 1136
Neural signaling, 157–158, **884,** 884i
Neural tube, **905,** 906i, 1126i, **1127,** 1127i, 1135, 1136, 1139i
Neuroendocrine signaling, 944i, **945**
Neurogenic heart, **989**
Neurohormone, 945, 962
Neuromasts, 926
Neuromuscular junction, **968,** 968i, 970i, 972i

Neuron, 883–901
afferent, 884, 885, 922, 922i, 923
defined, **875**
efferent, 884, 885
glial cell support, 885, 885i
integration of signals, 897–899, 899i
interneurons, 884, 885, 899, 909, 910i, 922, 1032–1033
neurosecretory, 945, 952, 961
number in human brain, 900
organization, 884–885
patterns of synaptic connections, 899
refractory period, 889, 889i, 890, 893
respiratory, 1039
signaling by, 888–893
steroid hormone receptors in, 157
structure, 875, 875i, 884–885, 885i
synapses, 885–887, 886i, 894–899, 895i, 897i
Neuronal circuit, **885,** 900
Neuropeptide, 898t, 1062
Neuropeptide Y, 1062, 1065
Neuropeptide Y receptor, 1062
Neurophysiology, of behavior, 1285–1286
Neuroscience, **1279**
Neurosecretory neuron, **945,** 952, 961
Neurospora crassa, 18, 206, 311, 312i–313i, 649, 651, 659, 660i, 661
Neurotransmitter, 145, 149, 153, 158
action, 886–887, 897
description, **886**
ionotropic, 894
metabotropic, 895
receptor, 894–895, 895i, 896, 897
release by exocytosis, 895–897
types, 897, 898t
Neurulation, **1126,** 1129, 1136
Neutral lipid, **50**–53
Neutral variation hypothesis, **474**
Neutralization, by antibodies, **1011,** 1014i
Neutron, **25**
Neutrophil, 1004t, **1006**
Newt, 723, 723i
Newton, Sir Isaac, 440
Next-generation DNA sequencing, 421
Ngorongoro Conservation Area, 1272
Niche cells, 771
Nicholas II (Czar), 404
Nickel, as plant micronutrient, 795–796, 796t
Nicolson, Garth L., 123
Nicotinamide adenine dinucleotide (NAD^+), **164,** 164i, 166, 166i, 167i–173i, 168–169, 173, 175i, 176–178
Nicotinamide adenine dinucleotide phosphate (NADP), 183–185, 183i, 189, 189i, 190i, 191–193, 195i, 196, 198
Nicotine, 1039
Niedergerke, Ralph, 969i
Nielsen, Peter, 564
NIH. *See* National Institutes of Health
Niimura, Yoshihito, 735
Nine-banded armadillo *(Dasypus novemcinctus),* 443i, 1176
Nirenberg, Marshall, 314
Nitella, 601
Nitrate, 800, 804–805
Nitric oxide (NO), 433, 996, 996i
Nitric oxide synthase, 433
Nitrification, **573,** 574, 804i, **805, 1243,** 1244i, 1245t
Nitrifying bacteria, 573
Nitrobacter, 573
Nitrogen
deficiency in plants, 797, 797i
metabolic wastes, 1071–1072, 1072i
plant acquisition of, 804–806, 804i–806i, 807–808
plant macronutrient, 795, 796t
Nitrogen cycle, 804, 804i, 807–808, 1243–1245, 1244i, 1245t
ammonification, 1243, 1244i, 1245t
defined, **1243**
denitrification, 1244i, 1245, 1245t
human disruption, 1248–1250
nitrification, 1243, 1244i, 1245t
nitrogen fixation, **573,** 580, 1243, 1244i, 1245t
in terrestrial ecosystem, 1244i
Nitrogen fixation, **573,** 580, **804,** 805–806, 806i, **1243,** 1244i, 1245t
Nitrogen metabolism, 573–574
Nitrogen radioisotopes, 25
Nitrogenase, 805–806
Nitrogenous base, **63,** 64, 64i, 290, 290i
Nitrosomonas, 573, 578
Nitrous oxide, 898t
NK cell. *See* Natural killer cell
NMDA receptor, 917
Nociceptors, **923,** 936, 937, 938
nod genes, 805, 806i
Noda, Ken-ichi, 758
Node, phylogenetic tree, **531,** 532
Node, plant, **624, 757,** 757i
Node of Ranvier, **885,** 886i, 893, 894i
Noller, Harry F., 85, 327, 335
Nomarski (differential interference contrast) microscopy, 93i
Nomenclature, 529–530
Noncompetitive inhibition, **82,** 82i
Nondisjunction, **227, 273**–275, 274i, 275i
Nonhistone proteins, 306
Nonkinetochore microtubules, 208i, **210,** 213, 214i
Nonpolar associations, **30**
Nonpolar covalent bond, **30**
Nonrandom mating, 470–471
Nonsense codon, **315**
Nonsense mutations, 331i, **332**
Nonshivering thermogenesis, **1089**
Nonvascular plant, **613,** 618–621
Noor, Mohamed, 476
Norell, Mark A., 546
Norepinephrine, 152–153, 898t, 946, 951t, **957,** 995
Northern blot analysis, **401**
Northern elephant seal *(Mirounga angustirostris),* 465, 1175, 1296
Northern spotted owl *(Strix occidentalis caurina),* 1175, 1273
Nostoc, 1243
Notch1 protein, 435–436, 435i
Notochord, **710,** 710i, **1126**–1127, 1126i, 1127i, 1139i
Notopterus notopterus, 205
Nucelariids, **655**
Nuclear envelope, 98i, **99,** 99i, 100i
Nuclear localization signal, **100,** 101i
Nuclear localization signals, **330**
Nuclear pore complex, 98i, **99**–100, 100i
Nucleariids (Nucleariidae), **604,** 655
Nuclease, 1046, 1057
Nuclei, **885**
Nuclei, brain, 1285
Nucleic acid, 63–67, 65i. *See also* DNA; Ribonucleic acid
Nucleic acid probe, 397, 397i
Nuclein, 286
Nucleoid, **94,** 95, 95i, **306,** 569
Nucleolus, 98i, 99i, **100,** 100i
Nucleoplasm, **100,** 100i
Nucleoporins, 99
Nucleoproteins, 63
Nucleosidase, **1057**
Nucleoside, **64**
Nucleoside analogs, 383
Nucleosome, **305,** 305i, 351, 351i
Nucleosome core particle, **305**
Nucleotidase, **1057**
Nucleotide, **63**–65, 64i, 290, 290i, 313
Nucleus, **94,** 97, 98i, 99, 99i, 330, 562, 562i
Nudibranch (sea slug), 689, 689i, 1027i
Null hypothesis, **14,** 15, 16, 17i
Null model, **461**
Null mutation, 427
Nurse, Paul, 214–215
Nüsslein-Volhard, Christiane, 357
Nutria *(Myocastor coypus),* 443, 444i
Nutrient
concentration, homeostatic control of, 877
cycling, 1241–1250
compartment model, 1231–1232, 1232i
hydrologic cycle, 1241–1242, 1241i
nitrogen cycle, 1243–1245, 1244i, 1245t, 1248–1250
phosphorus cycle, 1245–1246, 1246i
deficiency, plant, 796t, 797–798, 797i
essential, 1044
limiting, 1234
Nutrient recycling, 4, 5i
Nutrition
animal, 1043–1068
defined, **1043**
essential amino acids, 1044, 1048, 1049i
essential fatty acids, 1044, 1048
malnutrition, 1044
minerals, 1044, 1051, 1051t
overnutrition, 1044
undernutrition, 1044
vitamins, 1044, 1048–1051, 1050t
Nutritional modes, 573–574, 573i
Nutritional mutant, 311, 312i–313i
Nymph, 701, 701i
Nymphaeales, **639**

O horizon, 799, 799i
Oak *(Quercus),* 464i, 666, 1178
Oates, John, 1254
Oats *(Avena sativa),* 841
Obelia, 681i
Obesity, 960, 1062, 1064i
Obligate aerobe, **573**
Obligate anaerobe, **573**
O'Brien, Stephen J., 391
Observational data, **13,** 15, 17i, 1147
Occipital lobe, brain, 912, 912i
Ocean acidification, 1248–1249
Oceanic zonation, 1164–1165, 1165i
Oceanic zone, **1164,** 1167, 1167i
Ocelli (eye spots), **680,** 684, 684i
Ocellus, **929**–930, 929i
Octopamine, 1286, 1287i
Octopus, 225–226, 690, 690i, 904i, 905, 930i
Odorant, 1111
Odum, Howard T., 1236i
Oil spill, 1260
Oils, **50,** 51
Okada, Norihiro, 537
Okazaki, Reiji, **298**
Okazaki fragments, **298,** 298i
Old man's beard *(Usnea trichodea),* 664, 664i
Oleander *(Nerium oleander),* 785, 785i
Oleic acid, 51i
Olfaction (smell), 735, 934–935, 935i, 936, 936i
second-messenger pathways in, 156–157
Olfactory bulb, 911i, **912,** 936, 936i
Oligochaeta, 691, 692i
Oligodendrocyte, **885**
Oligosaccharin, 840t, **849,** 851t
Oligotrophic lake, **1164**
Olive baboon *(Papio anubis),* 1300i
Olson, John S., 1037
Omasum, 1063i, **1065**
Ommatidia, **930,** 930i
Omnivore, **1044, 1212**
On the Origin of Species by Means of Natural Selection (Darwin), 439–440, 446, 447, 481
Oncogenes, **219, 362**
One gene–one enzyme hypothesis, **311**
One gene–one polypeptide hypothesis, **311,** 321
One-horned rhinoceros *(Rhinocerus unicornis),* 1273
1000 Genomes Project, 416
Onion *(Allium),* 760, 761i, 831i
Onychophora, 670, 693, 694i
Oocyte, **1100**
Oogenesis, 1098i, **1099**–1100
Oogonia, **1099**
Oomycota, **596,** 596i
Opabinia, 670, 671i
Oparin, Aleksandr I., 555
Oparin–Haldane hypothesis, 555
Open circulatory system, **687, 982**–983, 983i
Open reading frame (ORF), **421**–422, 422i
Open system, 72, 73i, 74
Operant conditioning, **1284**
Operator, **341**
Operculum, **719,** 1028
Operon, 425
components of, 340–341
defined, **340**
lac, 341–342, 341i, 342i, 343i, 344, 345i
in *Methanococcus,* 96
trp, 344, 345i
Ophiostoma ulmi, 658
Ophiuroidea, 708, 708i
Opisthokonta, 603–604, 604i, 655
Opossum *(Didelphis virginiana),* 734
Opsin, **930,** 933
OPT/YSL multigene family, 433–434, 434i
Optic chiasm, **934,** 934i
Optic cup, 1127, 1128i
Optic nerve, 893, 932, 934, 934i
Optic vesicle, 1127–1128, 1128i
Optimal foraging theory, **1201**
Oral contraceptive pill, 1112t, 1113
Oral hood, **711,** 711i
Orange palm dart butterfly *(Cephrenes auglades),* 539i
Orangutan *(Pongo pygmaeus),* 738
Orbital, **26**–28, 27i
Orca *(Orcinus orca),* 1212i, 1213
Orchid, 630i, 639, 641i, 643i, 666, 804, 806, 807i
Order, **9**
Ordovician extinction, 513
ORE. *See* Osmotic response element
ORF. *See* Open reading frame
Organ, **749, 868,** 869i
Organ of Corti, **928,** 928i
Organ system, **868,** 869i, 875, 876i
Organelles, **11, 94**
Organic acid, 44t
Organic molecules
defined, **43**
formation of early, 555–556, 556i
movement in plants, 786–789, 787i, 788i
Organismal ecology, **1147**
Organismal performance, 1085–1086, 1086i
Organogenesis
defined, **1120**
eyes, 1127–1128, 1128i
nervous system, 1126–1127, 1126i, 1127i
Orgasm, 1110
Oriented cell division, **830,** 830i
Origin of replication, 221i, **296,** 300, 300i, 306, 306i, 394, 395i, 584
Ornithischians, 726
Orrorin tugenensis, 739
Orthogenesis, **453**
Oscula, **678,** 678i
Osmoconformer, **1070,** 1072
Osmolarity, **1070,** 1070i, 1078, 1082, 1083
Osmolyte, 1079
Osmoreceptor, **1082**–1083
Osmoregulation
defined, **1070**
excretion and, 1071
invertebrates, 1072–1074
in mammals, 1074–1080, 1074i–1076i, 1077t, 1080i
tubule structure and function, 1071, 1071i
water conservation, 1084, 1084i, 1085

Osmoregulator, **1070,** 1072–1074
Osmosis
defined, **129,** 130, **777**
osmoregulation, 1070, 1073, 1077
physical demonstration of, 129–130, 129i
in plants, 775, 777, 778i
plasmolysis, 130i, 131
tonicity and, 130–131, 130i
Osmotic pressure, **130**
Osmotic response element (ORE), 1079
Osmotic stress, 1078–1079
Osmundson, Todd, 667
Ossicles, 707, 709i, 975
Osteoblast, **872**
Osteoclast, **872**
Osteocyte, **872**
Osteolepiform, 721, 722i
Osteon, **872**
Ostracoderm, **716,** 716i
Ostrich *(Struthio camelus),* 441i, 731
Ostrom, John, 519
Otolith, 925i, **926**
Otolithic membrane, 925i, 926
Outer boundary membrane, chloroplast, **111**
Outer ear, **927,** 928i
Outer membrane, **570,** 571
Outer mitochondrial membrane, **106,** 106i
Outgroup comparison, **539,** 539i
Ova (eggs), **226,** 226i, 231, **1099**
Oval window, **927**–928, 928i
Ovarian cycle, **1105**–1107, 1105i, 1106i
Ovary
animal, 950i, 952t, **958**–**959, 1099,** 1104–1105, 1104i, 1107
flower, **639,** 642i, 813i, **814,** 815i, 819–820
meiosis, **226,** 226i
Overexploitation, 1256–1257, 1257i
Overnutrition, **1044**
Oviduct, **1100,** 1104, 1104i, 1110–1111, 1113
Oviparous animal, **1103**
Oviraptor, 546
Ovoviviparous animal, **1103**
Ovulation, **1104**–1107, 1105i, 1106i, 1107i, 1113
Ovule
angiosperm, 639, 642i
conifer, 635, 636i
diploid, 494
flower, 813i, **814,** 815i, 819, 834
gymnosperm, **632**
Owen, Sir Richard, 499
Oxaloacetate, 169, 170i, 197, 199i
Oxidation
defined, **162**
redox reactions, 162–163, 163i
Oxidation-reduction reaction. *See* Redox reaction
Oxidative phosphorylation, 164i, **165,** 172i, 175i, 176
Oxygen
aircraft cabin pressure and, 1024–1025
blood transport, 986–987, 1034–1035, 1035i
capillary exchange, 993
gas exchange, 1024–1042
homeostatic control of concentration, 877, 878
partial pressure, 1034–1035, 1034i, 1035i, 1037
as plant micronutrient, 795, 796t
prokaryotic metabolism of, 573
receptors for, 1032–1033
release by water-splitting, 560
release in photosynthesis, 184
Oxygenase, activity of rubisco, 197, 197i
Oxytocin, 880, 945, 951t, 951t, 953i, **954,** 1110, 1132, 1132i, 1141
Oyster, 689
Oyster mushrooms *(Pleurotus ostreatus),* 661
p53, 363
P680 chlorophyll *a,* 189, 189i–190i, 191
P700 chlorophyll *a,* 189, 189i–190i, 191, 192i
P generation, **243**
P site (peptidyl site), **324,** 324i–326i, 325, 327, 328i–329i
Pääbo, Svante, 742, 743
Pacemaker cell, **989**
Pachycephalosaurus, 726
Pacific sea hare *(Aplysia californica),* 916, 918
Pacific yew tree *(Taxus brevifolia),* 1263, 1263i
Pacinian corpuscle, 924, 924i
Paedomorphosis, **519,** 519i, 520i
Pagnussat, Gabriella C., 816, 816i
Pain receptor, 923, 936, 937, 938
Paine, Robert, 1214, 1214i
Pair-rule genes, 358i, **359**
Palanivelu, Ravi, 835
Paleobiology
Cuvier and, 441
defined, **441**
fossil record, 500–506
historical biogeography, 509–511
Palisade mesophyll, 763i, 764
Palmiter, Richard, 406
Palolo worm *(Eunice viridis),* 1096–1097, 1096i
Palp, bivalve, 689, 689i
Pampas, 1159
Pancreas, 950i, 952t, 959–960, 1049i, **1056,** 1056i
Pancreatic amylase, 1053i
Pancreatic enzymes, **1056**
Pangaea, 507, 511
Panting, 1090
Paper chromatography, 193
Papillae, 935, 936i
Parabasalids, 591, 591i
Paracrine regulation, 142, 143i, 944i, 945
Paramecium, 11i, 93i, 588i, 590i, 593i, 594, 940, 1203, 1204i
Paramesonephric duct, 1132, 1133i
Paranthropus, 740, 741
Parapatric speciation, **489,** 496
Paraphyletic taxa, **533,** 533i, 537
Parapodia, **691,** 692i
Parasite, **1207**–1208
Parasitism, 1186, **1207**–1208
Parasitoid, **1208**
Parasympathetic division, **907,** 907i, 908i
Parathyroid gland, 950i, 951t, **956,** 957i
Parathyroid hormone (PTH), 951t, **956,** 957i
Parazoa, 672, 676, 676i, 677–678, 677i
Parenchyma
companion cells, 755, 756i
cortex, 759
description, **753**–754, 753i
function, 753t
pith, 759
root, 767
stem, 759–760
in xylem, 755
Parental care, 730, 733, 1179, 1297
Parental chromosomes, **232**–233, 233i
Parental investment, **1296**
Parental phenotypes, **262,** 263i, 264, 265i
Parietal cell, **1054,** 1055i
Parietal lobe, brain, 912, 912i
Parkinson disease, 912
Parmesan, Camille, 1154, 1169
Parsimony, principle of, 536, **541,** 542i, 543
Parthenogenesis, **686, 1097**–1098
Partial diploid, **372,** 373i, 374
Partial pressure, **1034,** 1034i
Partridge, Linda, 1292
Parturition, **1131**
Passionflower *(Passiflora),* 857, 857i
Passive diffusion, 1070, 1077
Passive immunity, **1014**
Passive parental care, **1179**
Passive transport
defined, **126**
diffusion, 126–127, 126t
facilitated diffusion, 126t, 127, 128i
osmosis, 129–131, 129i, 130i
in plants, 776–777
Patau syndrome, 274
Patch, habitat, 1268–1270, 1270i
Patch-clamp technique, 891–892
Paternal chromosome, **226,** 232–233, 233i
Paternity testing, DNA fingerprinting use in, 404
Paterson, Eric, 802
Pathogenesis-related proteins (PR proteins), 851t, **852**
Pathogens, 98, 115, 1003
Pattern formation, **357**–359, 357i–360i, **828,** 828i
Pattern recognition receptors, 1005
Paulinella, 607
Pauling, Linus, 58
Pavlov, Ivan, 1284
Pax-6 gene, 522
PCBs. *See* Polychlorinated biphenyls
PCR. *See* Polymerase chain reaction
Pea *(Pisum sativa),* 820, 845
Peach *(Prunus),* 820, 821i
Peafowl *(Pavo cristatus),* 1296, 1296i, 1297
Peat moss, 621
Pectin, 749, 754, 766
Pectoral girdle, **712**
Pedal locomotion, 739–740, 739i
Pedicellariae, **708,** 709i
Pedigree, **268,** 270i
Pedipalp, 695, 697, 697i
Pelagic province, **1164,** 1165i
Pellicle, **590,** 593
Pelvic girdle, **712**
Penguins, triglyceride secretion in, 52, 52i
Penicillin, 456, 457i, 571
Penicillium, 659, 659i
Penis, 1108–1109, 1108i, 1110
Pentose, 47
PEP. *See* Phosphoenolpyruvate
PEP carboxylase, 198, 199i
Pepsin, 84, 329, 354, **1054,** 1055i
Pepsinogen, 329, 354, **1054,** 1055i
Peptic or stomach ulcer, 1054–1055
Peptic ulcer, 579i
Peptide bond, **57,** 57i, 325, 327, 335
Peptide hormone, 145, 152, 219, 945
Peptidoglycan, **570,** 571, 571i
Peptidyl transferase, **325,** 326, 326i, 327
Peptidyl-tRNA, **325,** 326i
Per capita growth rate, 1181i, **1182,** 1183i
Perception, of stimuli, **923**
Perdeck, A. C., 1281i
Perennial, **752**
Perfect flower, **814**
Perforin, 1006, 1016
Perfusion, **1026**
Pericarp, **819,** 820
Pericycle, 766i, **767,** 767i, 769, 770
Periderm, 753t, **768,** 769i, 770
Periodontitis, 576
Peripheral nervous system (PNS)
autonomic nervous system, 907, 907i, 908i
defined, **885, 905**
flatworm, 905
glial cells, 885
somatic system, 906, 907i
Peripheral proteins, 122i, **123**
Peripheral vision, 931
Perissodactyla, 734
Peristalsis, **1052,** 1052i, 1053, 1054i, 1055
Peritoneum, **673,** 673i
Peritubular capillary, **1075,** 1075i
Periwinkle snail, 1215–1216, 1215i, 1238, 1240i
Permafrost, **1161**
Permian extinction, 513, 514, 695
Peroxisomes, **107**
Perry, Donald, 1157
PET. *See* Positron emission tomography
Petal, flower, 813i, **814**
Petiole, 754, **762,** 762i
Petrified wood, 501i
Petromyzontoidea, 713, 715
pH
blood, 37–38
enzyme activity, effect on, 83, 84, 84i
homeostatic control of, 878
lysosomal, 104
soil, 801
pH scale, **36**–37, 37i
Phaeodactylum tricornutum, 597i
Phaeophyta (brown algae), 598, 598i, 599i
Phage, **288.** *See also* Bacteriophage
Phagocyte, 106, 137, **1004,** 1016
Phagocytosis, **104,** 106, **137,** 137i, **1004,** 1012
Pharmacogenomics, 436
Pharming, 407–408, 410
Pharyngeal arches, **1131**
Pharyngeal slits, 709
Pharynx
bird, 1048
description, **685, 1030,** 1031i, **1046,** 1046i
earthworm, 1047, 1047i
human, 1049i
insect, 1047, 1047i
perforated chordate, 710, 710i
swallowing reflex, 1053, 1054i
Phase change, plant, 764, 764i
Phase-contrast microscopy, 93i
Phenotype, **243**
Phenotypic variation
causes, 458–460, 458i
defined, **457**
maintenance of, 471–474
qualitative, 458, 458i
quantitative, 457–458, 458i
in snails, 457, 457i
Phenylalanine, 277, 1048, 1049i
Phenylketonuria (PKU), 277, 279, 420
Pheromone, 935, **1294,** 1294i
Philadelphia chromosome, 272–273, 273i
Phillips, George, 1037
Phloem
angiosperm, 639
components and functions, 753t
defined, **613, 755**
in eudicots, 760i
in monocots, 760i
primary, 759i
from procambial cells, 759
root, 766i, 767
secondary, 767–770, 768i, 769i
sieve tube members, 755, 756i
transport in, 786–789, 787i, 788i
whisk fern, 625
Phloem sap, **786**–789, 787i, 788i
Phorbol ester, 153
Phoronida, 683i, 684
Phoronopsis californica, 683i
Phosphatase, 71, 83, 147, 152
Phosphate
organic, 45t
soil, 800
Phosphate group, **43,** 45t
Phosphate transport protein, 803
Phosphatidyl ethanolamine, 53i
Phosphatidylcholine, 121i
Phosphodiester bond, **65,** 65i, 290, 290i
Phosphodiesterase, 151, 151i
Phosphoenolpyruvate (PEP), 198, 199i
Phosphofructokinase, 166, 167i, 168
Phosphoglucomutase, 167i
3-phosphoglycerate (3PGA), 193, 195i, 197
Phosphoglycerate kinase, 167i
Phosphoglyceromutase, 167i
Phosphoglycolate, 197, 197i
Phospholipase C, 151, 152i
Phospholipid, protocell, 557–558, 557i
Phospholipids, **53,** 53i, 94i, 120–121, 121i, 123, 124i

Phosphorus
deficiency in plants, 797, 797i
lake eutrophication and, 1164, 1164i
plant macronutrient, 795, 796t
Phosphorus cycle, 1245–**1246**, 1246i
Phosphorus radioisotopes, 25
Phosphorylases, 143, 144i
Phosphorylation
autophosphorylation, 147, 148, 148i, 153, 154i, 155
defined, **76,** 164
of enzymes, 83
oxidative, 164i, 165, 172i, 175i, 176
photophosphorylation, 191
in signal transduction, 146, 146i, 147, 148i, 150i, 152i
substrate-level, 164, 164i, 165i, 166, 167i, 169, 170i, 175i, 176
Phosphorylation cascade, 146
Photic zone, **1162,** 1163i
Photoautotroph, **183, 573,** 573i
Photobiont, **664,** 665i
Photoheterotroph, **573,** 573i
Photoinhibition, 201
Photon, **186,** 189
Photoperiodism, **858**–859, 859i
Photophosphorylation, **191**
Photopigment, **930,** 931–932, 932i, 933
Photoprotection, 187, 201
Photoreceptor, **923,** 929–934, 929i–934i
Photorespiration, 196–197, **197,** 197i, 198
Photosynthesis
action spectrum, 182i, 183, 187, 188i
in bacteria, 580
in brown algae, 598
C_4 pathway, 197–198, 199i
in CAM plants, 786
cellular respiration compared, 199, 200i
in chlorarachniophyta, 600
in chloroplasts, 111
defined, **4, 162, 182**
in dinoflagellates, 594
efficiency of, 198, 201
electron flow, 162i, 163
in euglenids, 591–592
evolution of, 200–201, 560
freshwater biome, 1162–1163, 1164
in golden algae, 598
in green algae, 601
light-dependent reactions, 183–184, 183i, 185–192, 188i, 189i, 190i, 192i
light-independent reactions, 183–184, 183i, 193–196, 195i
location of, 184, 185i
open ocean, 1167
overall reaction of, 32, 184
overview, 4i, 183–185
in parenchyma cells, 754
photoinhibition, 201
photorespiration, 196–197, 197i, 198
primary productivity, 1233–1234, 1233i
in prokaryotes, 95–96, 580
in red algae, 600
water-splitting reaction, 184, 189
Photosystems
components of, 188–189, 188i
defined, **188**
photosystem I, **188,** 189, 189i–190i, 191–192, 192i
photosystem II, **188,** 189, 189i–190i, 191–192
Phototroph, **573**
Phototropism
Darwin's experiments, 841–842, 841i
defined, **841**
description, 855, 855i
Phycobilin, 600
Phyletic gradualist hypothesis, **516,** 517i
PhyloCode, **541,** 549
Phylogenetic species concept, **482**
Phylogenetic tree
amniotes, 725i
animal, 675–677, 676i
for arthropods, 696
chordates, 714i
cladistic analysis and, 539–541, 540i–541i
comparative method, 546, 546i
converting into a classification, 533, 533i
defined, **10**–11, 10i, **530**
evolutionary history depicted in, 530–532
for flowering plants, 638, 638i
genetic distance, 543, 544i
land plants, 617i
maximum likelihood method, 543
optimal, identification of, 541–543
parsimony approach, 541–543
statistical approach, 543
primates, 737i
principle of parsimony, 541, 542i, 543
of prokaryotes, 577, 577i
public health and, 547, 547i
as research tool, 544–546
seed plants, 635i
structure of, 530–532, 531i, 532i
of swallows and martins, 1298i
of tetrapod vertebrates, 536–537, 538i
Phylogenetics
data sources for analysis of, 533–536
behavioral characters, 535–536, 535i
homologous characters, 534, 534i, 535i
molecular sequences, 536
morphological characters, 534–535
molecular, 547–549
principle of parsimony, **541,** 542i, 543
protists, 588, 589i
Phylogeny
animals, 675–677, 676i
defined, **530**
molecular, 676–677, 676i
Phylum, **9, 671**
Physarum, 588i, 603
Physical barriers, **1003**
Physiological respiration, **1025,** 1025i
Physiology, **868**
Phytoalexin, 851t, **853**
Phytochrome, 640i, **858**–859, 859i
Phytohormone, 839i. *See also* Plant(s), hormones
Phytophthora infestans, 587
Phytophthora ramorum, 587
Phytoplankton
carbon sequestration, 1264
defined, **590**
diatoms, 596
dinoflagellates, 594
estuary, 1165
golden algae, 598
Lake Erie, 1229–1230
open ocean, 1167
as primary producers, 1238
Phytoremediation, 801
Phytosterol, **54,** 121
Pigeon, 6, 7i, 1047i, 1048
Pigment
accessory, 188
photosynthetic, 186i–188i, 187–189 (*see also* Chlorophyll)
Pike-cichlid *(Crenicichla alta),* 1180–1181
Pili, 95i, **96, 572,** 572i
Pillbug, 697
Pilobolus, 658, 658i
Piloting, **1289**
Pilus, sex, 372, 373i
Pinacate beetle *(Eloides longicallis),* 1202–1203, 1203i
Pinacoderm, **678,** 678i
Pine *(Pinus),* 484, 633, 635, 636i, 755i, 765, 769, 769i
Pine Bush, 1271–1272, 1272i
Pineal gland, 950i, 952t, **960**
Pineapple *(Ananus comosus),* 820, 821i
Pinna, **927,** 928i
Pinocytosis, **134,** 135i
Pinworm *(Enterobius),* 693
Pistil, flower, 814, 834
Pit, 749, 754, 755
Pit organ, 937, 937i
Pitch, 767, 927, 929
Pitchfork, Colin, 404
Pith
defined, **759**
root, 766i, 767
stem, 759, 759i, 760, 760i
Pituitary giants and dwarfs, 952
Pituitary gland, 911
anterior, 950i, 951t, 950–953, 953i
defined, **950**
posterior, 950, 950i, 951t, 952, 953i, 954
Pitx1 gene, 524, 524i
PKU. *See* Phenylketonuria
Placenta, **733,** 734, **1103,** 1129, 1130i, 1131, 1132i
Placental mammals (Eutheria), 511, 511i, 734, 736i, 1103
Placoderm, 717–718, 718i
Plan B, contraceptive pill, 1113
Planaria, 684i, 904i, 929–930, 929i, 1073, 1073i
Plankton, 681
Planktonic crustacean, 698
Plant(s), 611–648
adaptations to life on land, 612–618
angiosperms, 637–643
adaptive success of, 639, 641
clades, 638–639
coevolution with pollinators, 639, 641, 643i
origin of, 637–638
overview, 637
phylogenetic tree, 638, 638i
bryophytes, 618–621
hornworts, 619–620, 620i
liverworts, 619, 619i
moss, 618i, 620–621, 620i
overview, 618–619, 618i
carnivorous, 806, 807i
cell structures, 99i, 111–113, 112i, 113i
chemical defenses, 849–854, 851t
gene-for-gene recognition, 851t, 853–854
heat-shock protein, 851t, 853, 854
hypersensitive response, 852, 852i
jasmonates, 849–852, 851t, 852i
oligosaccharins, 849, 851t
pathogenesis-related proteins, 851t, 852
salicylic acid, 850–851, 851t, 852, 854, 854i
secondary metabolites, 853
systemic acquired resistance, 851t, 854, 854i
systemin, 851–852, 851t, 852i
chlorophylls of, 186, 187
convergent evolution in, 510, 511i
diploid dominance, 615, 616i
diversity, history of, 514i
environmental response
abscisic acid, 848–849
biological clocks, 858–863
movements, 855–858
epiphytes, 793i, 806, 807i
evolution of, 612–618, 616t, 626, 644, 645
genetic engineering of, 408–412, 409i, 410i
genome sequencing, 421t, 643–644
gymnosperms, 631–636
ancestral forms, 631–632, 632i
conifers, 634–635, 636i
cycads, 633, 633i
ginkgo, 633, 634i
gnetophytes, 633–634, 634i
overview, 631
reproductive adaptations, 632
herbaceous, 751
herbivory, 1201–1202
hormones
abscisic acid, 840t, 848–849, 850i–851i
auxins, 840t, 841–844, 842i–844i, 845i
brassinosteroids, 840t, 847–848, 848i
cytokines, 840t, 846, 846i
ethylene, 840t, 846–847
G-protein–coupled receptors for, 153
gibberellin, 840t, 844–846
jasmonates, 840t, 849–852
oligosaccharins, 840t, 849, 851t
second messengers, 840–841
introduced, 1261, 1261i
life cycle, 615, 615i
angiosperm, 642i
conifer, 635, 636i
fern, 624i
general, 615, 615i
moss, 620i
life span, 751–752
meiosis, time and place of, 234–235, 234i
metabolomics, 789–790
movement, 855–858
gravitropism, 855–856, 856i
nastic, 857–858, 857i, 858i
phototropism, 841–842, 841i, 854, 854i, 855, 855i
thigmotropism, 856–857, 857i
mutualism, 1207
mycorrhiza, 655, 658, 665–666, 665i, 666i
nutrition, 793–810
deficiencies, 796t, 797–798, 797i
hydroponic culture, 794, 795i
obtaining and absorbing nutrients, 801–807, 803i–807i
requirements, 794–796, 795i, 796t
soil, 798–801, 799i–801i
phyla and major characteristics, 622t
phylogenetic trees, 617i
polyploid, 276
polyploidy, 491, 493–494, 493i
reproduction in angiosperms, 812–826
asexual reproduction, 824–826, 824i, 825i
overview, 812–813, 812i
sexual reproduction, 812–822, 815i
seedless vascular plants, 621–626
ancestral forms, 622, 623i
ferns, 623–625, 624i
horsetails, 625, 625i
lycophytes, 623, 623i
overview, 621–622, 622t
whisk ferns, 625, 625i
structure
cell wall, 749, 752, 753, 754
eudicots, 751, 752t
monocots, 751, 752t
organs, 749
overview, 749–752
root system, 750, 750i, 753t, 764–767
secondary growth, 767–770, 768i–770i
shoot system, 749, 750i, 757–764
transport in, 774–792
active transport, 776–777, 776i, 780–781, 787, 788i
apoplastic pathway, 780, 780i, 781
auxin transport, 843, 843i
osmosis, 775, 777, 778i
passive transport, 776–777
in phloem, 755, 756i, 786–789, 787i, 788i
root pressure, 783, 783i
in roots, 779–781, 780i
short-distance and long-distance, 775, 775i
symplastic pathway, 780, 780i, 781
translocation, 786–789, 787i, 788i
transmembrane pathway, 780, 780i
transpiration, 781, 782i, 783–785
in xylem, 755, 755i, 780–786, 782i
woody, 751–752, 768, 768i, 769i

Plant cells, 92i, 99i, 111–113, 112i, 113i
Plant development, 826–834
body plan, 826–828, 828i
radial organization of tissue layers, 827–828, 828i
root-shoot axis, 826–827, 828i
floral organs, 832–833, 832i–833i
leaf, 833–834, 834i
morphogenesis, 829–832
cell expansion, 830–831, 831i
oriented cell division, 830, 830i
patterns of cell division, 831
pattern formation, 828, 828i
Plant growth
determinate, 750
hormone influences
jasmonates, 840t, 849
oligosaccharins, 840t, 849
indeterminate, 750
leaf, 762, 764
meristems, 750, 751i
primary, 750, 753, 758–767, 759i, 766i, 770i
secondary, 750, 751–752, 767–770, 768i–770i
Plant tissue
complex, 752
cut sections, 752i
general features, 749
location, 753i
radial organization of layers, 827–828, 828i
simple, 752
systems, 752–757
dermal tissue, 752, 753t, 756–757, 756i
ground tissue, 752, 753–754, 753i, 753t, 754i
vascular tissue, 752, 753t, 754–755, 755i, 756i
terminology of cut sections, 752i
Plant viruses, 379t, 383
Plantae (kingdom), 11, 12i, **611**
Plasma, 874, **876**, 876i, 877, 878, **985**, 985i, 986
Plasma cell, **1007, 1011**, 1013i
Plasma membrane, 119–138
asymmetry, 123
defined, **92, 119**
eukaryotic, 97–98, 98i
fluidity, 123–124, 124i
freeze fracture of, 124–125, 125i
functions, 124–125
glycolipids and glycoproteins, 122, 122i
lipids, 120–121, 121i, 122i, 123–124
prokaryotic, 95, 95i
proteins, 121–126, 122i, 124i, 127, 128i, 129, 131–133, 132i–134i, 137
selective permeability, 127
structure, 94i, 120–125, 121i, 122i, 124i–125i
transport mechanisms
active, 126, 126t, 131–134, 132i–134i
characteristics, 126t
diffusion, 126–127, 126t
endocytosis, 134–137, 135i
exocytosis, 134, 135i, 137
facilitated diffusion, 126t, 127, 128i, 129
osmosis, 129–131, 129i, 130i
passive, 126–131, 126t, 128i–130i
phagocytosis, 137, 137i
Plasma membrane intrinsic protein, 779
Plasmids
antibiotic resistance, 581
cloning DNA fragment in, 392i, 393, 394, 395i
cloning vector, 394, 395i
defined, **306, 570**
expression vector, 404–406, 405i
F factor, 371, 372, 373i, 374
micrograph, 372i
R, **374**
Ti, 409, 409i
Plasmodesmata, 99i, **112**, 113i, 142, 383, 749, 755, 767, 780, 780i, 788
Plasmodial slime molds, 588i, 603
Plasmodium, **603**
Plasmodium species, 472i, 528, 594, 595, 1239
Plasmogamy, **653**, 653i, 657, 657i, 658, 660i, 661, 662–663, 662i
Plasmolysis, 130i, **131**, 778i
Plasmopara viticola, 596i
Plasticity, in plant indeterminate growth, 750
Plastid
defined, **111**
endosymbiont hypothesis, 561i, 562, 563i, 605–607, 606i
Plastocyanin, 190i, 191, 192i
Plastoquinone, 190i, 191, 192i
Plastron, **727**, 727i
Plate tectonics, **506**–508, 506i, 507i
Platelet, **874**, 986i, **987**
Platelet-derived growth factor, 149
Platy *(Xiphophorus maculatus)*, 491
Platyhelminthes
lineages
Cestoda, 685, 685i
Monogenoidea, 685
Trematoda, 685, 685i, 686
Turbellaria, 684–685, 684i
organ systems, 684, 684i
parasitic, 684, 685, 685i, 686i
Platypus, 706–707, 707i, 733i, 734
Pleiotropy, **257**, 257i
Pleschka, Stephen, 155
Plesiosaurs, 726
Plethysmograph, 1039
Pleura, **1031**, 1031i
Pleurotus ostreatus, 661
Ploidy, **207**
Plumatella repens, 683i
Plumule, **819**, 820i, 822i
Pluripotent cells, 986, 986i, **1121**
PNS. *See* Peripheral nervous system
Pohorille, Andrew, 564
Poinar, Hendrick, 452
Polar associations, **30**
Polar body, 1098i, **1100**, 1100i
Polar covalent bonds, **30**
Polar nuclei, **814**, 815i, 816i
Polar transport, **843**, 843i
Polarity, egg, **1119**
Polarity of water, 30, 30i, 34–35
Poliovirus, 382
Pollak, Martin, 1092
Pollen
diploid, 494
gymnosperm, 632
Pollen grain, 752t, **814**, 816i, 817i, 818i
conifer, 635, 636i
defined, **632**
eudicot, 638, 638i
as male gametophyte, 616
monocot, 638
Pollen sac, 813i, **814**, 815i
Pollen tube, 635, 636i, 642i, **814**, 815i, 817, 817i, 818i, 834
Pollination, 811, 817, 818i, 834
angiosperms coevolution with animal pollinators, 639, 641, 642i
conifer, 635, 636i
defined, **632**
as gymnosperm adaptation, 632
Pollutants, **1259**
Pollution, 38, 1259–1260, 1260i
Polstelsia palmaeformis, 598
Poly(A) polymerase, **318**, 319i
Poly(A) tail, **318**, 319i, 354
Polyadenylation signal, **318**, 319i
Polyandry, **1297**
Polychaeta, 691, 692i
Polychlorinated biphenyls (PCBs), 1239
Polydactyly, 465
Polygamous species, **1297**
Polygenic inheritance, **256**–257, 256i
Polygyny, **1297**
Polyhedral viruses, **377**, 378i
Polymer(s)
defined, **46**
first, 564
Polymerase chain reaction (PCR), **398**, 398i, 399, 401
use in DNA fingerprinting, 391, 401, 403
Polymerization, **46**
Polymorphism
balanced, 472–474, 472i–474i
defined, **458**
Polynucleotide chain, 65, 290, 290i
Polyp, **679**–681, 679i, 681i, 682i
Polypeptide
defined, **57**, 311
one gene–one polypeptide hypothesis, 311, 321
processing and folding, 328–329
Polyphyletic taxa, **533**, 533i
Polyplacophora, 688, 688i
Polyploid, **273**, 276, 626–627
allopolyploidy, 493i, 494
autopolyploidy, 493–494, 493i
defined, **491**
Polysaccharide, **47**, 48–50, 49i
Polysome, 327–**328**, 328i
Polyspermy, blocks to, 1102, 1102i
Polytrichum, 620i
Polyunsaturated fat/fatty acid, **50**
Ponderosa pine *(Pinus ponderosa)*, 630i
Pons, 905, 907i, 909, 911i
Pools, 1162
Poppy *(Papaver rhoeas)*, 811i
Population
age structure, 1175, 1193–1194, 1194i
carrying capacity, 1183–1184, 1183i
defined, 2i, **3, 457, 1147**
density, 1173, 1173i
dispersion, 1173–1174, 1174i
generation time, 1175, 1175i, 1183i
geographical range, 1173
habitat, 1173
recruitment, 1216
sex ratio, 1175
size, 1173, 1173i, 1174i
source/sink, 1190
variation, 457–461
genetic, 460
phenotypic, 456i, 457–460, 457i, 458i, 459i
qualitative, 458, 458i
quantitative, 457–458, 458i
Population bottleneck, **465**, 465i, 466
Population density, **1173**, 1173i
Population dynamics, **1173**, 1184–1191
density-dependent factors, 1185–1186, 1185i, 1186i, 1188i
density-independent factors, 1187, 1187i
interacting environmental factors, 1187–1188
life history characteristics, 1188–1189, 1189i, 1189t
metapopulations, 1189–1190, 1190i
Population ecology
age structure, 1175, 1193–1194, 1194i
conservation and, 1267–1268
defined, **1147**
demography, 1175–1178
life tables, 1176–1177, 1177t
survivorship curve, 1177–1178, 1177i
dispersal patterns, 1173–1174, 1174i
generation time, 1175, 1175i, 1183i
human population growth, 1191–1195, 1193i–1195i
life history, 1178–1179, 1180–1181
population dynamics, 1184–1191
population growth models, 1180–1184
exponential growth, 1180–1183, 1182i
geometric, 1180–1181, 1181i
logistic model, 1182–1184, 1183
population size and density, 1173, 1173i, 1174i
sex ratio, 1175
Population genetics, **448**, 460–461, 461t, 1266
Population growth
carrying capacity, 1183–1184, 1183i
generation time, 1175
human, 1191–1195, 1193i–1195i
age structure, 1193–1194, 1194i
economic development and, 1194–1195
family planning programs, 1195
zero population growth, 1193
intraspecific competition and, 1184
intrinsic rate of increase, 1182, 1183i
models, 1180–1184
exponential growth, 1181i, 1182
geometric, 1180–1181, 1181i
logistic model, 1182–1184, 1183i, 1184t, 1185i
per capita rates, 1181i, 1182, 1183i
Population interactions, 1200–1208, 1200t
coevolution, 1200, 1202–1203, 1203i
community characteristics, effects on, 1219, 1219i
herbivory, 1201–1202
interspecific competition, 1203–1206, 1204i, 1206i
character displacement, 1204–1205, 1205i
competitive exclusion, 1203, 1204i
exploitative competition, 1203
interference competition, 1203
niche concept, 1204, 1205–1206, 1205i, 1206i
resource partitioning, 1204, 1205i
predation, 1201–1203
symbiosis, 1206–1208
commensalism, 1206–1207, 1206i
mutualism, 1207, 1208i
parasitism, 1207–1208
Population size
carrying capacity, 1258
cycles in, 1191, 1192i
description, **1173**, 1173i, 1174i
extrinsic control, 1191
intrinsic control, 1191
K-selected species, 1188–1189, 1189i, 1189t
life history characteristics and, 1187, 1188–1189, 1189i, 1189t
minimum viable, 1268
predator-prey interactions, 1191, 1192i
r-selected species, 1188–1189, 1189i, 1189t
regulation
crowding, 1185–1186, 1185i, 1186i
density-dependent factors, 1185–1186, 1185i, 1186i, 1188i
density-independent factors, 1187, 1187i
interacting environmental factors, 1187–1188
Population viability analysis (PVA), **1268**, 1269
Porcupine *(Erethizon)*, 1202
Porifera, 677–678
Porocyte, 678, 678i
Porphyra, 601
Porphyromonas gingivalis, 576
Portal vein, 950, 953i
Positional information, 816, 828–829, 828i
Positive feedback, **879**–880, 1105
Positive gene regulation, 343, 343i, 344
Positive pressure breathing, **1029**, 1029i
Positron emission tomography (PET), 903i, 912, 913i, 914
Possingham, Hugh, 1269
Postelsia palmaeformis, 588i
Posterior end, **673**
Posterior pituitary gland, 945, 950, 950i, 951t, 952, 953i
Postsynaptic cell, **885**, 886, 887, 895i, 897
Postsynaptic membrane, 895i

Posttranscriptional regulation, 340, 351–354, 353i
Posttranslational import, **330**
Posttranslational regulation, 340, 354, 354i
Postzygotic isolating mechanisms, **484,** 484t, 485–486, 486i, 491
Potassium
aldosterone effects, 958
deficiency in plants, 797, 797i
guard cell expansion and, 784–785, 784i
plant macronutrient, 795, 796t
soil, availability in, 800
transport channels, 802
Potassium channel, 888–893, 889i–892i, 897
Potato *(Solanum tuberosum),* 178, 760, 761i, 803
Potato cod *(Epinephelus tukula),* 1207i
Potato famines, 587
Potential energy, **72**
Poxvirus, 382
PR proteins. *See* Pathogenesis-related proteins
Prado, Marco, 388
Prairie, 1159, 1159i
Prairie gentian *(Gentiana),* 816i
Prakash, Om, 495i
Praying mantid, 700i
Preadaptation, 519. *See* also Exaptation
Precapillary sphincter, 992, 992i
Precipitation
monsoon cycles, 1151
rain shadow, 1151, 1151i
variations in, 1149–1150, 1149i
Precursor-mRNA (pre-mRNA), 313, **318**–320, 319i, 320i, 321i, 352
Predation
adaptations for feeding, 1201
by cnidarians, 680, 680i
community stability and, 1213
defenses against, 1201–1203, 1202i, 1203i
defined, **1201**
density-dependent population regulation and, 1186
keystone species, 1215
on songbirds, 1259, 1259i
species richness, effect on, 1214–1215, 1214i
trophic cascade, 1238, 1240i
Predator-prey model, 1191, 1192i
Predictions, **15**
Pregnancy
contraception, 1112–1114, 1112t
defined, **1128**
Premolar, **733,** 1061, 1061i
Prenatal diagnosis, **279**
Preprophase band, 830, 831i
Prepuce, 1108i, **1109**
Preservation, 1271–1272, 1272i
Pressure flow mechanism, in phloem, **787**–789, 787i, 788i
Pressure potential, 777
Pressure receptor, 924, 924i
Presynaptic cell, **885,** 886, 895, 895i
Presynaptic membrane, 895, 895i, 896
Prezygotic isolating mechanisms
behavioral isolation, 484, 485i
defined, **484**
ecological isolation, 484
gametic isolation, 485
genetic divergence and, 491
mechanical isolation, 484–485, 485i
sexual selection and, 491, 491i
temporal isolation, 484
Primary acceptor, in photosynthesis, 187, 188i, 189, 190i
Primary active transport, **131**–132, 132i, 133i
Primary cell layers, **672**
Primary cell wall, **112,** 113i
Primary consumer, **1212,** 1212i, 1238
Primary endosymbiosis, **606**–607, 606i
Primary growth
defined, **750**
roots, 765–766, 766i
summary of, 770i
Primary immune response, **1013**–1014, 1015i
Primary meristem, **759**
Primary mesenchyme cell, 1122, 1122i
Primary motor area, 913i, **914**
Primary plant body, **750**
Primary producer, **4,** 183, **1211,** 1212i, 1238
Primary productivity
consumer regulation of, 1238, 1240i
by ecosystem, 1234t
gross, 1233, 1250
net, 1233, 1233i, 1234, 1234i, 1234t, 1235i, 1250
variations in, 1233–1235, 1233i, 1234i, 1235i
Primary somatosensory area, **913,** 913i
Primary structure, protein, **57,** 58i
Primary succession, **1217**–1218, 1218i
Primary thickening meristem, 759
Primary tissue, plant, **750**
Primase, **297,** 297i, 298i, 300t
Primates
Haplorhini, 736–739, 738i
key derived traits, 734
phylogeny, 737i
Strepsirhini, 736, 737i
Primer
defined, **297**
in DNA replication, 297, 297i, 300, 302
use in PCR, 398i, 399
Primitive groove, **1124,** 1125i
Primitive streak, **1124,** 1125i, 1129
Principle of Independent Assortment, 248i, **249,** 250i
Principle of parsimony, **541,** 542i, 543
Principle of Segregation, **243,** 244i, 250i
Principles of Geology (Lyell), 442, 443
Prion, **387,** 388
Probability
defined, **245**
in Mendel's crosses, 245–246, 246i
product rule in, 245, 245i, 246i
sum rule in, 245, 246i
Proboscis worm, 686–687, 687i
Procaine, 893
Procambium
root, 766, 766i, 769
stem, **759,** 759i
Prodöhl, Paulo A., 1176
Product, **31**–32
Product rule, **245,** 245i, 246i
Production efficiency, 1235
Profundal zone, **1162**–1163, 1163i
Progeria, 261, 262i, 277
Progesterone, 157, 952t, **959,** 1106i, **1107,** 1111, 1113, 1132
Progestin, **958,** 1113
Proglottid, 685, 685i
Programmed cell death, 141, 1007
Progymnosperm, 631–632, 632i
Proinsulin, 354
Prokaryote(s), 567–584
biofilms, 575, 575i
cell division in, 220–221, 221i
cellular components, 95–97, 95i
defined, **11, 94**
diversity, 568–569
DNA organization in, 306, 306i
Domain Archaea, 582–584
characteristics of, 582, 582t, 584
Crenarchaeota, 583–584
Euryatchaeota, 582, 583
Korarchaeota, 584
Nanoarchaeota, 583
Thaumarchaeota, 583, 584
Domain Bacteria, 577–581
antibiotic resistance, 581
characteristics of, 582t
chlamydias, 578, 580i
cyanobacteria, 580, 580i
Gram-positive bacteria, 580, 580i
pathogenicity, mechanism of, 581
Proteobacteria, 577–578, 578i
spirochetes, 578
genomes sequenced, 421t
metabolic activity, 568, 572–574
nitrogen metabolism, 573–574
nutrition modes, 572–573, 573i
oxygen use, 573
phylogenetic tree, 577, 577i
protein sorting, 330–331
regulation of gene expression, 340–344, 341i–345i
reproduction, 574
ribosomes, 324, 328
size, 567, 568i
structure, 568–572, 570i
capsule, 571, 571i
cell walls, 570–571, 571i
endospore, 574, 574i
flagella, 572, 572i
internal, 569–570
nucleoid, 569
pili, 572, 572i
ribosomes, 570
shape of cells, 569, 569i
slime layer, 571
transcription in, 313, 314i, 318
translation in, 314i, 321, 324–328
Prokaryotic cells, origin of, 559–560, 560i
Prokaryotic chromosome, **95, 220,** 569
Prolactin, 352, **950,** 951t, 953i, 1132
Proliferative phase, 1106i, 1107
Prometaphase, in mitosis, 208i, **209,** 210, **210,** 210i
Prometaphase I, in meiosis, 228i
Prometaphase II, in meiosis, 228i
Promiscuous species, **1297**
Promoter, **340**
β-lactoglobulin, 407
defined, **317**
DNA methylation of, 350
eukaryotic gene regulation and, 346–348
operon, 340–341, 341i–344, 345i
TATA box, 346, 346i, 347i
Promoter proximal elements, **346,** 346i, 347
Promoter proximal region, **346,** 346i, 348, 349i
Pronghorn *(Antilocapra americanus),* 748i, 1063i
Prop root, 765
Propagation
action potential, 892–893, 892i, 894i
vegetative, 825–826, 825i
Prophage, **380,** 381i
Prophase, in mitosis, **208,** 208i, 209–210, 210i, 213i
Prophase I, in meiosis, 228i, 230, 230i, 231
Prophase II, in meiosis, 228i
Propranolol, 146
Proprioreceptor
defined, **924**
lateral line system, 926, 926i
statocysts, 926, 927i
tension receptors, 924–925, 925i
vestibular apparatus, 925–926, 925i
Prostaglandin, **945,** 952t, 1006, 1107, 1110, 1132i, 1141
Prostate gland, 1108i, **1109**
Prosthetic groups, 80
Protease, 1006–1007, 1016, 1046, 1056
Proteasome, **354,** 354i
Protein, 55–63
amino acids, 55, 56–57, 56i
defined, **3**
degradation, 354, 354i
denaturation, 60, 61i
enzymatic digestion of, 1053i
folding, 328
functions, 55, 55t, 57
information flow, role in, 3, 3i
integral, 122i, 123
membrane, 121–126, 122i, 124i, 127, 128i, 129, 131–133, 132i–134i, 137
one gene–one polypeptide hypothesis, 311, 321
oxidative pathways, 171, 171i
peptide bonds, 57, 57i
peripheral, 122i, 123
processing, 328–329
sorting, 329–331, 330i
structure
domains and motifs, 61–62, 63i
folding, 59–60, 61, 62i
primary, 57, 58i
quaternary, 57, 58i, 61
secondary, 58–59, 58i, 59i
tertiary, 57, 58i, 59–60, 60i
Protein-coding gene, 311, **313**
Protein-first hypothesis, 564
Protein kinase, 83, 145, **146,** 146i, 147–149, 148i, 150i, 152i, 154i, 156, 849, 917, 946, 949i. *See also* Kinase
Protein phosphatase, 83, **147,** 152
Protein synthesis. *See* Translation
Proteinases (proteases), 79
Proteobacteria, 577–578, 578i
Proteoglycan, 115, 115i
Proteome, 416, **428,** 430
Proteomics
defined, 416, **430**
protein function, 431, 431i
protein interactions, 430, 430i
protein location within cells, 430
protein structure determination, 430
Prothoracic gland, 961, 961i
Prothoracicotropic hormone (PTTH), **961,** 961i
Protist, 11, 12i, 587–610
classification, 588–589
diversity of, 588, 588i
genomes sequenced, 421t
groups, 591–604
Alveolata, 592–595, 593i, 594i, 595i
Amoebozoa, 602–603, 603i
Archaeplastida, 600–601
Chlorarachniophyta, 600
Chlorophyta (green algae), 601, 601i, 602i
Chromalveolata, 592–598
Euglenozoans, 591–592, 592i
Excavata, 591–592
Foraminifera (forams), 599–600, 599i
Metamonada, 591, 591i
Opisthokonta, 603–604, 604i
Radiolaria, 598, 599i
Rhodophyta (red algae), 600–601, 600i
Stramenopila, 595–598, 596i–598i
habitat, 590
metabolism, 589
pathogenic, 587
plastid evolution from endosymbionts, 605–607, 606i
reproduction, 589
structure, 589–590
"Protista," 547, 548
Protocell, 53, **554,** 557–559, 557i
Protoctists, 587
Protoderm, **759,** 759i, 766i
Proton
defined, **24**
gradients in photosynthesis, 190i, 191–192
number of, 24–25
Proton-motive force, **175,** 191, 201
Proton pump, **131,** 174i, 776, 776i
Protonema, 620i, **621**
Protonephridium, **1073,** 1073i
Proto-oncogene, **362**
Protoplast, **749**
Protoplast fusion, **826**
Protostome
blastopore development in, 1123
developmental patterns, 674–675, 674i
Ecdysozoa, 677, 693–702
Lophotrochozoan, 677, 683–692
nervous system, 675
Prototheria, 733
Protozoa, 92i, 587i, 588
Proventriculus, 1047i, 1048

Provirus, **384,** 384i, 386, 386i
Proximal convoluted tubule, **1075,** 1075i, 1076i, 1077, 1077t
Prusiner, Stanley, 387
Pseudoceros dimidiatus, 684i
Pseudocoelom, **673,** 673i
Pseudocoelomate animal, **673,** 673i, 702
Pseudogene, **421,** 422
Pseudohermaphrodite, 1103
Pseudomonas fluorescens, 92i
Pseudomyrmex ferruginea, 1207, 1208i
Pseudonitzschia, 598
Pseudopod, **590,** 603, 603i
Psilotum, 625, 625i
Psychrophile, **583**
Pterophyta, 622t, **623,** 623–626, 624–626
Pterosaurs, 724, 726
PTH. *See* Parathyroid hormone
PTTH. *See* Prothoracicotropic hormone
Pufferfish *(Takifugu rubripes),* 423, 893, 998
Pulmocutaneous circuit, **984,** 984i
Pulmonary artery, 988, 988i
Pulmonary circuit, 984i, **985,** 988, 988i
Pulmonary vein, **988,** 988i
Pulvini, **857**
Punctuated equilibrium hypothesis, **515,** 517, 518i
Punnett square, **245,** 246i, 248i
Pupa, **701,** 701i, 961, 961i
Pupil, **930,** 930i, 931i
Pure-breeding, 242
Purine, **63,** 64i, **290,** 290i
Purkinje fiber, **989,** 990i
Puromycin, 327
Purple martin *(Progne subis),* 480i
Purple sulfur bacteria, 192
PVA. *See* Population viability analysis
Pyloric sphincter, 1052i, **1055**
Pyramid of biomass, **1237**–1238, 1237i
Pyramid of energy, **1237,** 1237i
Pyramid of numbers, **1238,** 1238i
Pyrimidine, **63,** 64i, **290,** 290i
Pyrobolus, 583
Pyrococcus, 12i, 583
Pyrogen, **1006**
Pyrophilus, 583
Pyrophosphate, 295i
Pyruvate
in C_4 cycle, 198, 199i
fermentation reactions and, 177, 177i
in glycolysis, 165–166, 166i, 167i
Pyruvate kinase, 167i
Pyruvate oxidation, **164,** 164i, **168**–169, 168i, 171i, 175, 175i

Qiu, Yin-Long, 626
QTLs. *See* Quantitative trait loci
Qualitative variation, **458,** 458i
Quantitative trait loci (QTLs), **256**
Quantitative traits, **256**–257
Quantitative variation, **457**–458, 458i
Quaternary structure, protein, **57,** 58i, 61
Quiescent center, **765,** 766i
Quillwort, 623
Quinoa *(Chenopodium quinoa),* 1189i
Quorum sensing, 144, **575**

R gene, **853,** 853i
R group, of amino acids, 56
R-MuLV. *See* Rauscher murine leukemia virus
R plasmid, **374**
R-selected species, **1188**–1189, 1189i, 1189t, 1214, 1217–1220
R symbol, in organic molecules, 44
RAAS. *See* Renin-angiotensin-aldosterone system
Rabbit, 1172–1173, 1173i
Racehorses, training effects on gene expression in, 973
Racker, Efraim, 174i
Radial canal, 707, 709i
Radial cleavage, **674,** 674i
Radial nerve, 904, 904i
Radial symmetry
description, **672,** 673i
echinoderms, 707
eumetazoans with, 679–682
secondary, 707
Radiation, 185, 186i, **1086,** 1086i
Radiation therapy, 26
Radicle, 765, **819,** 820i, 822, 822i
Radioactivity, **25,** 26
Radioisotope, **25,** 26
Radiolaria, **598,** 599i
Radiometric dating, **504**–505, 504i, 559
Radula, **687,** 688i
Ragweed *(Ambrosia),* 816i
Rain forest
temperate, 1160–1161
tropical, 1155–1157, 1156i
Rain shadow, **1151,** 1151i
Ramalina, 661i, 664
Random coil, **58,** 62
Random dispersion, **1173**–1174, 1174i
Rao, Potu N., 214, 216i
Raphanobrassica, 494
Rapid-eye movement (REM) sleep, **917**
Ras G protein, 153, 154i
Raspberry *(Rubus),* 820, 821i
Rat snake *(Elaphe obsoleta),* 482i
Rattlesnake, 728i, 937, 937i, 1201
Raup, David, 513
Rauscher murine leukemia virus (R-MuLV), 385
Raven *(Corvus corax),* 1284, 1284i, 1295
Ray, plant, 767
Rays, 718, 718i, 1020, 1084
Reabsorption, **1071,** 1071i, 1076, 1077, 1080–1081
Reactant, **31**–32
Reaction. *See* Chemical reaction
Reaction center, **188**–189, 188i
Reactive oxygen species, 192, 201
Read, Betsy A., 196
Reading frame, 315, **325,** 331i, 332
Realized niche, **1204,** 1205, 1205i, 1206i
Realms, biogeograhical, 510, 510i
Receptacle, flower, **813,** 813i
Reception, **143,** 145i, 146i, 147i, 150i–152i, 156i, **884,** 884i
Receptive field, **922,** 932–933
Receptor
defined, **4**
hormone, 944, 946–948, 947i, 948i, 949i, 954–955
internal, 154–156, 156i
peptide hormone, 146
sensory, 922–939
steroid hormone, 154, 156i, 157–158
surface, 145–154
G-protein-coupled, 149–154, 149i–152i
tyrosine kinase, 147–149, 148i, 153, 154i
Receptor-mediated endocytosis, **134,** 135i, 136–137, 382
Receptor potential, **922**
Receptor protein, **122,** 126
Receptor tyrosine kinase, **147**–149, 148i, 153, 154i
Recessive allele, **243,** 276–277, 471–472, 472t
Reciprocal altruism, **1302**
Reciprocal cross, **242, 268**
Reciprocal translocation, 272, 272i
Recognition proteins, **122**
Recombinant chromosomes, **232,** 232i
Recombinant DNA, **393**
Recombinant phenotypes, **262,** 263i, 264, 265i
Recombinants, **232,** 232i
Recombination
in bacteria, 370–377, 371i, 374
conjugation, 371, 372, 373i, 374, 375, 376
crossing-over, 227, 227i, 228i, 230–232, 230i, 232i
linked genes and, 262–266, 263i, 265i, 266i
Recombination frequency, **264**–266, 265i, 266i
Recombination rate, 476
Recruitment, population, 1216
Rectal salt gland, 1084
Rectum, 1049i, **1059,** 1059i
Red algae, 600–601, 600i
Red blood cells. *See* Erythrocyte
Red deer *(Cervus elaphus),* 1179
Red fir *(Abies magnifica),* 1187i
Red-legged frog *(Rana aurora),* 1292
Red mangrove *(Rhizophora),* 765i
Red marrow, 975
Red oak *(Quercus rubra),* 755i, 802
Red-spotted newt *(Notophthalmus viridescens),* 723i
Red tide, **594**
Redox reaction, **163,** 163i
Reduced Emissions from Deforestation and Degradation (REDD), 1274
Reduction
in Calvin cycle, 194, 195i
defined, **162**
redox reactions, 162–163, 163i
Redundancy, in genetic code, 315
Redwood, 774, 774i, 783
Reflex, **909,** 910i
Refractory period, 889i, **889,** 893, 1110
Regulators, **877**
Regulatory gene, **341,** 341i, 345i
Regulatory proteins, **341**
Regulatory sequences, **340**
Reinforcement, **489,** 1284
Relatedness, calculating degrees of, 1300, 1301i
Relative abundance, **1210**–1211
Relative fitness, **466**
Relative humidity, effect on transpiration, 783
Release factor, **327,** 328i–329i
Releasing hormones, **950,** 951t, 950
REM. *See* Rapid-eye movement (REM) sleep
Renal artery, 1074i, **1075**
Renal cortex, 1074i, **1075**
Renal medulla, 1074i, **1075**
Renal pelvis, 1074i, **1075**
Renal vein, 1074i, **1075**
Renaturation, **60**
Renin, **1081,** 1081i
Renin-angiotensin-aldosterone system (RAAS), **1081,** 1081i
Replica plating, **375**–376, 376i
Replicates, **15**
Replication, RNA world and, 558
Replication bubble, **300,** 300i
Replication fork, **296,** 297i, 298i, 300, 300i, 302i
Replication origin, 221i, 296, 300, 300i, 306, 306i, 394, 395i
Repressible operon, 344, **345i**
Repressor, **341**–342, 344, 341i–345i, 348
Reproduction
animal, 1096–1117
asexual, 1097–1098, 1097i
human, 1104–1112
sexual, 1098–1117
asexual, 212, 574, 1097–1098, 1097i
cephalopod, 690
defined, **5**
early *versus* late, 1179
flatworm, 684i
fungi, 652–654, 653i
asexual, 653, 653i
sexual, 653–654, 653i
number of episodes, 1179
protists, 589
sexual, 225–236
Reproductive behavior
mating systems, 1297
nesting, 1297–1298, 1298i
sexual selection, 1286–1287, 1296i, 1297i
Reproductive isolation
genetic divergence and, 491–492
isolating mechanisms
defined, **484**
postzygotic, 484, 485–486, 485i, 486i, 491
prezygotic, 484–485, 484i, 485i, 491, 491i
table of, 484t
Reproductive strategies, **1296**
Reptile
circulatory system, 984i, 985
classification of, 537, 538i
ectothermy, 1086, 1087i, 1088
excretion, 1085
Research, biological, 13–19
applied, 13
basic, 13
controlled experiments, 15–17, 16i
defined, **13**
experimental data, 13
hypotheses, 13–17, 14i, 17i
molecular techniques, 18
motivations for, 18–19
observational data, 13, 15, 17i
scientific method, 13, 14i
theories, 18
Residual volume, **1032**
Resilin, **872**
Resin, 633, 635
Resolution, **92**
Resource partitioning, **1204,** 1205i
Respiration
adaptations for, 1027–1029
cellular (*see* Cellular respiration)
countercurrent exchange, 1028, 1028i, 1029, 1030i
in diving mammals, 1038–1039
at high altitudes, 1037–1038
perfusion, 1026
physiological, 1025, 1025i
ventilation, 1026
Respiratory medium, **1025,** 1026–1027
Respiratory surface, **1025,** 1026, 1026i
Respiratory system, 1024–1042
defined, **1025**
gills, 1025, 1026, 1026i, 1027–1028, 1027i, 1028i
lungs, 1025, 1026i, 1028–1033, 1030i, 1031i, 1032i
mammalian, 1030–1033, 1031i–1033i
tracheal system, 1025, 1028–1029, 1029i
Respiratory tree, sea cucumbers, 709
Response, **143,** 145i, 146i, 147i, 150i–152i, 156i, **884,** 884i
Resting potential, 888–889, 889i
Restoration, 1272
Restriction endonuclease (restriction enzyme)
defined, **393**
in DNA cloning, 393–394, 393i, 395i
use in Southern blot analysis, 401, 402i
Restriction fragment length polymorphisms (RFLPs), **401,** 402i
Restriction fragments, **393**
Restriction sites, 393, 393i, 395i
Reticular activating system, 910
Reticular formation, 910, 911i
Reticulum, **1063,** 1063i, 1065
Retina, **930,** 930i, 931–933, 931i, 933i
Retinal, 929, 931, 932i, 933i
Retrotransposons, **334,** 334i, 384
Retroviruses, **384**–386, 384i, 386i
Reverse transcriptase, 1019
in cDNA library construction, 396, 396i, 404
of retroviruses, 384, 384i, 385, 386, 386i
Reversible inhibition, enzyme, 82
Reversible reaction, **36, 75**
Reynolds, Eric, 576
Reznick, David, 1180–1181, 1196
RFLPs. *See* Restriction fragment length polymorphisms
Rhea *(Rhea americana),* 441i
Rheumatoid arthritis, **1018**

Rhinoceros, 734
Rhinoceros beetle, 700i
Rhizaria, **598**–600
Rhizobium, 571i, 573, 577, 801, 805, 806i, 1207, 1243
Rhizoid
  as adaptation for anchoring, 614
  bryophyte, 618, 618i
  liverwort, 619
  lycophyte, 623
  moss, 621
  whisk fern, 625
Rhizome
  defined, **614**
  description, 760–761, 761i
  fern, 624, 624i
  horsetail, 625
  whisk fern, 625
Rhizomorph, 651, 651i
Rhizopus stolonifer, 657i, 658, 658i
Rhodophyta (red algae), **600**–601, 600i
Rhodopsin, **930,** 932, 932i, 933, 933i
Rhynchoceol, 687, 687i
Rhyniophytes, 614, 614i
Rhythm method, 1112, 1112t
Ribbon worm, 686–687, 687i
Ribonuclease, 60, 61i
Ribonucleic acid (RNA)
  defined, **3, 63**
  information flow, role in, 3, 3i
  ribozymes, 79, 85–86, 326, 335
  RNA world, 558, 559
  sequence, molecular phylogenetics use of, 536
  structure, 65i, 66–67
Ribonucleoprotein particle, 319
Ribonucleoside triphosphate, 295
Ribonucleotides, **64**
Ribose, 47i, **64,** 295
Ribosomal RNA (rRNA), 95, 100, 318, **324,** 326, 548, 548i, 656
Ribosome
  antibacterial agents targeting, 324
  defined, **95, 311, 324**
  eukaryotic, 98i, 99i, 100–102, 102i
  free, 330
  mechanism of action, 335
  polysome, 327–328, 328i
  prokaryotic, 95, 95i, 570
  structure, 324, 324i
  translation mechanism and, 321, 322i, 324–328, 324i–326i, 328i–329i
Ribosome binding site, **324**
Ribozyme, 79, **85**–86, **326,** 335, 558, 559
Ribulose 1,5-bisphosphate (RuBP), 193, 194, 195i, 196
Rice, genetically engineered, 410–411, 410i
Rice blight, 411
Ricketts, Taylor, 1274
Rickettsia, 578
  R. parkeri, 115
  R. rickettsii, 1239
Riffles, 1162
Rigor mortis, 969
Ring canal, 707, 709i
Ring species, **482**–483, 483i
Ring-tailed lemur *(Lemur catta),* 737i
Ripening, fruit, 847
Ritualized signals, 1293
River biome, 1161–1162, 1162i
Rivers, damming of, 1257–1258, 1257i
RNA. *See* Ribonucleic acid; Ribonucleic acid (RNA)
RNA interference (RNAi), **352,** 353, 353i, 365, 427
RNA polymerase, 313, **317,** 317i, 318
RNA-seq, **428,** 432
RNA tumor viruses, 385
RNA world, ribozymes in, 86
RNA world model, **558**
RNAi. *See* RNA interference
Roberts, Richard, 319
Robinson, Scott K., 1259
Rock dove, 6, 7i
Rock needlebush *(Hakea gibbosa),* 784i
Rock pocket mouse, 7–8, 8i, 458
Rocky shore habitat, 1166, 1166i
Rod, **931**–933, 932i, 933i
Rodbell, Martin, 149
Rodent (Rodentia), 734, 736i
Rodgers, Buel (Dan), 978
Romwer, Alfred Sherwood, 717
Ronald, Pamela, 411
Root, phylogenetic tree, **530**
Root, plant
  apical meristem, 819i, 820i
  defined, **615**
  embryonic (radicle), 765, 819, 820i, 822, 822i
  haustorial roots, 806, 807i
  lateral, 765, 767, 767i
  secondary growth, 769–770, 769I
  transport in, 779–781, 780i
Root cap, **765,** 766i
Root exudates, 802
Root hair, 756i, **757,** 767, 800, 802, 828–829, 828i
Root nodule, **804**–805, 805i, 806i
Root pressure, **783,** 783i
Root primordium, **767**
Root-soil interactions, 799–801
Root system, 764–767
  adventitious system, 765, 765i
  defined, **615**
  eudicot and monocot compared, 752t, 766i
  functions, 764–765
  mycorrhizae, 803–804, 806
  nutrient absorption, 802
  overview, **749**–750, 750i
  primary growth, zones of, 765–766, 765i
  taproot system, 764–765, 764i
  tissues, 766–767, 766i
Rose, Irwin, 354
Rose *(Rosa),* 641i
Rotifera, 685–686, 687i
Rough ER, 98i, **102**–103, 102i
Round window, **928,** 928i
Roundup, 410
Roundworm, 693, 694
  filarial worms, 686, 686i
  hydrostatic skeleton of, 974
Rous sarcoma virus, 385
Royal Chitwan National Park, 1273, 1273i
Royal penguin *(Eudyptes schlegeli),* 1299, 1299i
RRNA. *See* Ribosomal RNA (rRNA)
RU-486. *See* Mifepristone
Ruben, Samuel, 184
Rubisco. *See* RuBP carboxylase/oxygenase (rubisco)
RuBP. *See* Ribulose 1,5-bisphosphate
RuBP carboxylase/oxygenase (rubisco), **194,** 195i, 196–198, 197i, 199i
Rudic, R. Daniel, 880
Ruffini ending, 924i
Rumen, **1063,** 1063i, 1065
Ruminant, **1063,** 1063i, 1065
Runs, 1162
Russian royal family, 270, 270i
Rust, white, 596
Ryan, Michael J., 1304
Rye *(Secale cereale),* 764

S cyclin, 218
S phase, **207,** 207i, 216–217, 216i, 217i, 218, 222, 300
SA node. *See* Sinoatrial node
Sabre squirrelfish *(Sargocentrum spiniferum),* 1174i
Sac-within-a-sac body plan, 672, 672i
Saccharomyces cerevisiae, 96, 116, 177, 206, 214, 218, 222, 267, 423, 649, 650i, 659–660
Saccule, 925i, **926**
Sachs, Julius von, 794
Saddle-billed stork, 732i
Sage grouse *(Centrocercus urophasianus),* 1296–1297, 1297i
Sahara Desert, 1256, 1256i
Sahel, 1256, 1256i
Sahelanthropus tchadensis, 739
Salamander, 482–483, 483i, 519, 519i, 722, 723, 1026i, 1029
Salicylic acid, 840t, **850**–851, 851t, 852, 854, 854i
Salinization, 1256
Saliva, 1053, 1060
Salivary amylase, **1053,** 1053i
Salivary gland, 1049i, **1053**
Salk, Jonas, 1014
Salmon, 1179, 1230, 1232, 1258
Salmonella, 578, 581
  S. typhimurium, 375
Salt
  conservation by freshwater fishes and amphibians, 1084–1085
  excretion by marine fishes, 1084
  excretion by marine mammals, 1079
Salt gland, 1085, 1085i
Salt marsh, **1165,** 1165i
Salt marsh cordgrass *(Spartina alterniflora),* 1210, 1211i, 1238, 1240i
Saltatory conduction, **893,** 894i
Saltbush *(Atriplex spongiosa),* 762, 763i
Samson, Rachel Y., 585
Sand dollar, 709
Sand tiger shark *(Eugomphodus taurus),* 718i
Sandy shore habitat, 1166
Sandy soil, 799, 799i, 800, 800t
Sanger, Frederick, 58, 418
Saprobe, **651**
Saprolegnia parasitica, 596i
Sapwood, **768,** 769i
Sarcomere, **968,** 969i, 973
Sarcoplasmic reticulum, **968,** 968i
Sarcopterygii, 719, 721, 721i
Sargassum, 598
Saturated fat/fatty acid, **50**–52, 50i
Saturation, enzyme, **81**
Saurischians, 726
Savanna, 1155–1157, 1158i
Sawfly *(Perga dorsalis),* 1299, 1299i
SBE-1 (starch-branching enzyme 1), 249
Scala Naturae, 440, 453
Scallop, 689
Scanlan, Thomas, 954
Scanning electron microscopy (SEM), 93i
Scarlet cup fungus *(Sarcoscypha),* 659i
Scarlet hood *(Hygrophorus),* 661i
Scenedesmus, 93i
Scent marking, 1293–1294
Schindler, David, 1164, 1164i
Schistosomiasis, 686
Schizocoelom, 674i, **675**
Schizosaccharomyces pombe, 215
Schleiden, Matthias, 90
Schoener, Thomas, 1187, 1188i, 1214
Schwann, Theodor, 90
Schwann cell, **885,** 886i
Scientific method, **13,** 14i
Scientific name, **9**
Scientific theory, **18**
Scion, 825
Sclera, 930, 931i
Sclerenchyma, 753i, 753t, **754,** 754i
Sclerid, **754,** 754i
Scolex, 685, 685i
Scorpion, 695, 697i
SCR gene, 828
Scrapie, 387, 388
Scrotum, **1108,** 1108i
Scutellum, **819,** 820i, 822i
Scyphozoa, 680, 682i
Sea anemone, 681, 682i
Sea cucumber, 708i, 709
Sea daisies (Concentricycloidea), 708
Sea gooseberry *(Pleurobrachia pileus),* 683i
Sea hare *(Aplysia),* 1284
Sea horse *(Hippocampushistrix),* 720i
Sea level, changes in, 508–509, 508i
Sea lilies, 709
Sea nettle (*Chrysaora* quinquecirrha), 682i
Sea otter *(Enhydra lutris),* 1267–1268, 1268i
Sea squirt, 711
Sea star, 708, 708i, 709i, 1214–1215, 1214i
Sea urchin, 708i, 709, 1248–1249, 1267
  development, 1122–1123, 1122i
  fertilization, 1102, 1102i
Sea wasp *(Chironex fleckeri),* 682i
Seal, 734, 1038–1039, 1175
Seasonality, 1148, 1148i
Second law of thermodynamics, 73–**74**
Second messenger, 840–841, 895, 946–947
  cyclic AMP, 151–152, 151i, 156, 362
  defined, **149**
  diacylglycerol, 151–153, 152i
  inositol triphosphate, 151–153, 152i, 156
  Sutherland's discovery of, 142–143, 144i
Secondary active transport, **131,** 133–134, 134i
Secondary cell wall, **112,** 113i
Secondary consumer, **1212,** 1212i
Secondary contact, 486i, 489, 491
Secondary endosymbiosis, 606i, 607
Secondary growth
  as adaptive response, 770
  defined, **750**
  gymnosperm, 632
  plant life span and, 751–752
  roots, 769–770, 769i
  stems, 767–768, 768i, 769i
  summary of, 770i
  tree rings, 768, 769i
Secondary immune response, **1013**–1014, 1015i
Secondary mesenchyme cell, 1122, 1122i, 1123
Secondary metabolite, 853
Secondary plant body, **750**
Secondary productivity, **1235**
Secondary structure, protein
  alpha helix, 58, 58i, 59i
  beta strand, 58–59, 59i
  defined, **57**
  random coil, 58, 62
Secondary succession, **1219**
Secondary tissue, plant, **750**
Secretin, **1060,** 1060i
Secretion, **1071,** 1071i, 1076, 1078
Secretory phase, 1106i, 1107
Secretory vesicle, **104,** 104i, 105i
Sedimentary cycles, 1241
Sedimentary rock, 441, 500, 500i
Sedimentation, 500, 500i
Sedum, 785i
Seed
  angiosperm, 633i
  conifer, 635, 636i
  defined, **632, 819**
  dispersal, 819–820
  germination, 821–822, 822i, 823i
  gymnosperm, 632, 633i
  sporophyte embryo development, 818–819, 819i
  suppression by abscisic acid, 848
Seed coat, 635, 815i, **819,** 819i, 820i, 821, 822
Seed fern, 632
Seed plants, 630–648. *See also* Angiosperm; Gymnosperm
  phyla and characteristics, 631t
  phylogenetic relationships, 635i
Seedless plants, 611–629
Seedless vascular plants, 621–626
  ancestral forms, 622, 623i
  ferns, 623–625, 624i
  horsetails, 625–626, 625i
  lycophytes, 623, 623i
  overview, 621–622, 622t
  whisk ferns, 625, 625i
Segment polarity genes, 358i, **359**
Segmentation
  in annelids, 690–691, 691i
  in chordates, 710, 710i
  description, **675**
  in onychophorans, 693, 694i

Segmentation genes, 358, 358i, **359**
Segregation, principle of, **243,** 244i, 250i
Selection
artificial, 445, 459, 459i
background, 476
frequency-dependent, 474, 474i
sexual, **470,** 471i, 484, 488, 491–492, 491i, 1296–1297, 1296i, 1297i
Selective breeding, 445, 446
Selective permeability, **127**
Selective sweep, 476
Selectively neutral, **474**
Selenium, 22, 23i, 796
Self-fertilization, **240,** 242
Self-incompatibility, **817,** 817i
Self-pollination, **240,** 242
SEM. *See* Scanning electron microscopy
Semen, **1109**
Semicircular canal, **925,** 925i, 926
Semiconservative replication, **293,** 292i, 293i, 294i, 295
Semilunar valve, **988**
Seminal fluid, **1109**
Seminal vesicle, 1108i, **1109**
Seminiferous tubule, **1108,** 1108i, 1109i
Senescence, **846,** 847i
Sense codon, **314**
Sensilla, 934, 935i
Sensitization, **916**
Sensor, **878,** 878i, 879i
Sensory adaptation, **923**
Sensory drive, 1304
Sensory hair cell, 925, 926, 926i, 928–929, 928i
Sensory neuron, **884,** 885
Sensory receptor
adaptation, 923
afferent neuron link to CNS, 923
chemoreceptors, 923, 934–936, 935i, 936i
defined, **922**
electroreceptors, 939
magnetoreceptors, 938–939
mechanoreceptors, 922–929
hearing, 927–929, 928i
lateral line system, 926, 926i
proprioreceptors, 924–926
statocysts, 926, 927i
tension receptors, 924–925, 925i
touch and pressure, 924, 924i
vestibular apparatus, 925–926, 925i
vibration detection, 927, 927i
nociceptors, 923, 936, 937, 938
overview, 922–923
photoreceptors, 923, 929–934, 929i–934i
thermoreceptors, 923, 936–937, 937i
Sensory regions of the cerebral cortex, 913, 913i
Sensory system, 921–940
hearing, 927–929, 928i
vision, 929–934, 929i–934i
Sensory transduction, **922,** 931–932, 933i
Sepal, flower, 813i, **814**
Separase, 211
Septa, 650, **690,** 691i
Sequence similarity search, 422, 427
Sequential hermaphroditism, **1103**
Sequoia sempervirens, 12i
Serengeti National Park, 1272
Serosa, **1052,** 1052i
Serotonin, 896, 898t
Sertoli cell, **1108,** 1108i, 1109i
Sessile, **671**
Set point, **878,** 878i, 879, 1085, 1089–1091
Setae, **690,** 691i
Sex, genetic determinants of, 227
Sex chromosomes. *See also* X chromosome; Y chromosome
aneuploidy of, 275, 275i, 275t
defined, **227**
gonad development, 1132, 1134
in meiosis, 227
sex determination and, 267–268, 267i
sex-linked genes, 266–271, 268i–271i
Sex determination in humans, 267–268, 267i
Sex-determining region of the Y (SRY), 1132, 1134
Sex hormones, 54, 54i, 945, 958–959
Sex-linked genes, 266–271
defined, **266**
in *Drosophila,* 268, 268i, 269i
in humans, 268, 270, 270i
X-chromosome inactivation, 271, 271i
Sex pilus, 96, **372,** 373i, 572
Sex ratio, **1175**
Sexual dimorphism, **470**
Sexual reproduction
angiosperm, 639, 642i, 812–822, 815i
animal, 1098–1114
cellular mechanisms, 1099–1104
fertilization, 1100–1103, 1100i, 1102i, 1110i, 1114
gametogenesis, 1098i, 1099–1100
hermaphroditism, 1103–1104, 1103i
oviparous animals, 1103
ovoviviparous animals, 1103
viviparous animals, 1103
conifers, 635, 636i
conjugation in ciliates, 593, 593i
defined, **225, 1097**
ferns, 624–625, 624i
fungi, 653–654, 653i, 657, 657i, 658, 660i, 661, 662–663, 662i
genetic diversity from, 1098–1099
gnetophytes, 634
hornworts, 619
liverworts, 619
meiosis, 225–236
mosses, 620i, 621
octopus, 225–226
plants, 616
variability generated by, 225, 230–233
Sexual selection, **470,** 471i, 484, 488, 491–492, 491i, **1296**–1297, 1296i, 1297i
Shampay, Janis, 307
Shapiro, Michael, 522
Shark
antibody production, 1020
description, 718–719, 718i
spiny, 717, 718i
urea concentration, 1084
Sharp, Phillip, 319
Sheldon, Frederick H., 1297–1298
Sheldon, Peter R., 517i
Shelf fungus, 650i, 661, 661i, 662
Shelford, Victor, 1154
Shell, electron, **26**
Shell, turtle, 727, 727i
Shepard, P. M., 473i
Shepherd's purse *(Capsella burse-pastoris),* 818, 819i, 1186i
Sheridan, Mark, 1066
Shigella dysenteriae, 578
Shivering, 178, 879
Shoot
apical meristem, 819, 819i, 820i
floral, 812, 813, 820
Shoot apical meristem, 759, 761
Shoot stem cells, 771
Shoot system, 613, **614**
overview, **749,** 750i
primary, 757–764
leaves, 762–764, 762i, 763i, 764i
phase changes, 764, 764i
stems, 757–761, 757i, 759i–761i
secondary growth, 767–768, 768i, 769i
Short-day plant, **860,** 860i
Short-eared owl *(Asio flammeus),* 1202i
Short Interspersed Elements (SINEs), 537
Short-nosed echidna *(Tachyglossus aculeatus),* 733i
Short repetitive sequences, 421
Short tandem repeat (STR) sequence, **401,** 403i, 421, 1176
Short-term memory, **916,** 918
Shrimp, 698
Siamang, 738
Sicheri, Frank, 1152
Sickle-cell anemia, 239–240, 239i, 254, 257, 257i, 276, 332, 332i, 401, 402i, 472–473, 472i
Sickle-cell trait, 254, 276, 401, 402i
SIDS. *See* Sudden infant death syndrome
Sieve plate, 755, 756i
Sieve tube, **755,** 787–789, 788i
Sieve tube member, **755,** 756i, 760i, 787, 788
Sign stimuli, **1280,** 1281i
Signal molecule, 142–143, 145–147, 145i–146i, 148i, 149i
Signal peptidase, 330i
Signal peptide, **330,** 330i
Signal recognition particle (SRP), 330i
Signal sequence, **330,** 330i
Signal transduction
amplification of, 147, 147i
conduction across chemical synapses, 895–897, 895i
defined, **143**
hormones and, 946–947, 947i, 949i
phosphorylation and, 145, 146, 146i, 147, 148i, 150i, 152i
in plants, 839–840, 839i
auxin, 843–844, 844i
steps, 146–147
Signaling
by neurons, 888–893
in pollen-tube guidance, 834
Signaling cascade, 143
Signaling pathway, 143
cross-talk, 156–157, 157i
evolution of, 145
reproductive behavior and, 157–158
steps, 143, 145i
viral infections and, 155
Silencing, **350,** 352–353, 353i
Silent mutation, 331i, **332**
Silicon, 796
Silk, 59
Silkworm, 935, 935i, 961i
Silkworm moth *(Bombyx mori),* 1294
Sillén-Tullberg, Birgitta, 1299
Silliman, Brian R., 1238
Silt, 798
Silverfish, 700i, 701i
Silversword *(Argyroxiphium sandwicense),* 845, 845i
Simberloff, Daniel, 1223, 1223i, 1224
Simian immunodeficiency virus (SIV), 547, 547i
Simple diffusion, 126t, **127**
Simple epithelium, 869, 870i
Simple fruit, **820,** 821i
Simulation modeling, **1232**
Simultaneous hermaphroditism, **1103,** 1103i
SINEs. *See* Short Interspersed Elements
Singer, S. Jonathan, 123
Single-lens eye, 930, 930i
Single-stranded binding proteins, **296,** 297i, 300t
Sink (plant region), **787**–789, 788i
Sink population, **1190**
Sinoatrial node (SA node), **989,** 990i
Sinuses, 687, **982**–983
Siphon, 689, 689i, 690, 690i
siRISC. *See* siRNA-induced silencing complex
siRNA. *See* Small interfering RNA
siRNA-induced silencing complex (siRISC), 353
Sister chromatids, **207,** 208i, 210, 210i, 211, 227, 227i, 228i, 230i, 231
Sister clades, **532**
Sister species, **532**
SIV. *See* Simian immunodeficiency virus
Skate, 718
Skeletal muscle, 320, 320i, 356, 356i
contraction, 968–972, 969i–971i
description, **874,** 874i, 966, **967**
double muscling, 977
invertebrates, 973
motor units, 972, 972i
structure, internal, 967–968, 967i, 968i
tetanus, 969, 971, 971i
twitch, 969, 971i
types
fast aerobic, 971t, 972
fast anaerobic, 971t, 972
motor units, 972, 972i
slow, 971–972, 971t
Skeletal system
endoskeleton, 975–976, 975i
exoskeleton, 974–975, 975i
hydrostatic skeleton, 974, 974i
Skeleton
bird, 730, 731i
echinoderm, 707
human, 975i
hydrostatic, 673, 690
Skin
receptors in human, 924, 924i
structure, 1090, 1090i
Skin cancer, 304
Sleep
consciousness, 916–918, 917i
rapid-eye movement, 917
Sleep movements, 857, 857i
Sliding DNA clamp, **296,** 296i, 300t
Sliding filament model, muscle contraction, **968**–969, 969i
Slime layer, **95, 571**
Slime mold, **603**
cellular, 603, 604i, 605
plasmodial, 588i, 603
Slow block to polyspermy, **1102,** 1102i
Slow muscle fiber, **971**–972, 971t
Small interfering RNA (siRNA), **352, 353,** 365
Small intestine, 1055–1058, 1056i, 1057i, 1058i
Small nuclear RNA (snRNA), 319
Small ribonucleoprotein (snRNP), **319,** 320i
Smallpox, 1002, 1014
Smell (olfaction), 735, 934–935, 935i, 936, 936i
Smith, Alan, 101
Smith, Neal, 1199
Smooth ER, 98i, 102i, **103**
Smooth muscle, 320, 321i, **874,** 874i, 966
Snail
balanced polymorphisms in, 473, 473i
description, 688–689, 689i
Elysia, 562
herbivory by, 1215, 1215i
phenotypic variation, 457, 457i
Snake
description, 727–728, 728i
limb loss in, 450–451, 451i
pit organs of, 937, 937i
subspecies of rat snake, 482i
Snapdragon *(Antirrhinum),* 460–461, 461t, 462, 758
incomplete dominance in, 251–254, 253i
SNAREs, 1114
Sneeze reflex, 1053
Snow geese *(Chen caerulescens),* 458, 458i
Snow monkey *(Macaca fuscata),* 706i
Snow plant *(Sarcodes sanguinea),* 806, 807i
Snowshoe hare *(Lepus americanus),* 1191, 1192i
SnRNA. *See* Small nuclear RNA
SnRNP. *See* Small ribonucleoprotein
Social behavior
altruistic behavior, 1300–1302, 1301i
defined, **1299**
dominance hierarchies, 1300
group living, 1304
human, 1312–1313
Sodium
aldosterone effects, 947–948, 948i, 958, 1081
atrial natriuretic factor and, 1081
reabsorption, 1080–1082
Sodium channels, 888–893, 889i–892i, 897, 899

Sodium chloride
homeostatic control of, 877
hydration layers in water, 35, 35i
ionic bonds in, 28–29, 28i
molecular weight, 36
Sodium-potassium pump (Na^+/K^+ pump), **132,** 133i, 888, 891
Sogin, Mitchell L., 656
Soil, 798–801
components, 798
horizons, 799, 799i
influence of organisms on, 798–799
mineral availability, 800–801, 800i
nitrogen in, 804, 804i
particle size, 798
pH, 801
phytoremediation, 801
salinization of, 1256
water availability, 799–800, 799i, 800t
Soil solution, **799**–800, 799i
Solar energy, 73, 73i, 1233
Solar radiation, 1147–1150, 1148i
Solar/sunlight energy, 183, 186, 201
Solar tracking, 858
Solenoid model, **305,** 305i
Solenostemon, 759i
Solute, **35**
Solute potential, 777
Solution, **35**
Solvent, 32, **35**
Somaclonal mutation, 826
Somaclonal selection, **826**
Somatic cell, **225, 405, 1099**
Somatic embryo, **824,** 825i
Somatic embryogenesis, 826
Somatic gene therapy, **406**–407
Somatic system, **906,** 907i
Somite, **1127,** 1127i
Song sparrow *(Melospiza melodia),* 1278, 1279i
Songbirds
population declines in, 1259, 1259i
songs of, 1278, 1279i
Sorbitol, 1078, 1079
Sordino, Paolo, 523
Soredia, 664, 665i
Sorting of proteins, 329–331, 330i
Sorus, **624**
Sound spectrogram, 1279i
Source (plant region), **787**–789, 788i
Source population, **1190**
Southern, Edward, 401
Southern beech tree *(Nothofagus),* 509, 510, 510i
Southern blot analysis, **401,** 402i
Southern magnolia *(Magnolia grandiflora),* 634i
Sowbug, 697
Soybean *(Glycine max),* 805, 805i, 847i, 854, 856
Spanish moss *(Tillandsia usneoides),* 793i
Spatial summation, 899i
Specialized transduction, **382**
Speciation, 480–498
allopatric, 486–489, 486i–489i, 491–492, 496
defined, **481**
genetic mechanisms of, 491–496
chromosome alterations, 494–496, 495i
genetic divergence, 491–492
polyploidy, 491, 493–494, 493i
geography of, 486–491
in Hawaiian fruit flies, 488
parapatric, 489, 496
reproductive isolation, 484–486, 484t, 485i, 486i
sympatric, 489–491, 490i, 493–494, 494i, 496
Speciation genes, 496, 550
Species
asexual, 496
biological species concept, 481–482
defined, **9**
endemic, 510
geographical variation, 482–483, 482i, 483i
morphological species concept, 481
phylogenetic species concept, 482
sister, 532
subspecies, 482, 482i
Species/area relationships, 1224, 1268–1269, 1268i
Species cluster, **487,** 487i
Species composition, **1208**
Species diversity, **1211,** 1211i
Species pool, 1222
Species richness
biodiversity, 1255
defined, **1210**
equilibrium theory of island biogeography, 1221–1225, 1222i, 1223i
herbiory, effect of, 1215
intermediate disturbance hypothesis, 1217, 1217i
interspecific competition, effect of, 1213–1214, 1214i
landscape corridors, effect of, 1270, 1271i
latitudinal gradients, 1221, 1221i
predation, effect of, 1214–1215, 1214i
species diversity, 1211, 1211i
variations in, 1219–1225
Specific epithet, **529**
Specific heat, **33**
Spehr, Marc, 1111
Spemann, Hans, 1138, 1139i
Spencer, Chris, 476
Sperm
angiosperm, 814, 815i, 817
bryophyte, 618
chemical attractants, 1110, 1111
conifer, 635, 636i
fern, 624i, 625
fertilization of egg, 1100–1102, 1100i, 1102i, 1110–1112, 1110i, 1114
gymnosperm, 632
liverwort, 619
recognition of egg coat, 1114
in semen, 1109
structure, 1099, 1099i
Sperm adhesion molecule 1, 62
Spermatide, 1098i, 1099, 1108i
Spermatocyte, 1098i, 1108, 1108i
Spermatogenesis, 1098i, **1099,** 1108, 1108i
Spermatogonia, **1099**
Spermatozoa, **226,** 226i, 231, **1099,** 1099i. *See also* Sperm
Spermicide, 1113
Sperry, Roger, 914, 915i
Sphagnum, 612i, 621
Sphenodontid, 727, 728i
Sphincter, **1052**
Spicule, sponge, 678, 678i
Spider
description, 695–697, 697i
lizard predation on, 1187–1188, 1188i
Spider monkey *(Ateles geoffroyi),* 738i
Spike moss, 623
Spiller, David, 1187, 1188i
Spinach *(Spinacia oleracea),* 860i
Spinal cord, **905,** 909, 909i
Spinal nerve, **906,** 908i
Spindle
action, 210–211, 210i, 213–214
defined, **210**
formation, 208i, 210, 213, 213i
mitotic, 205i, 208i–210i, 210–211, 212–214, 213i–214i
Spindle poles, 208i, **210,** 210i
Spinneret, **697,** 697i
Spiny anteater, 734
Spiny lobster *(Panulirus),* 1289, 1289i
Spiny shark, 717, 718i
Spiracle, **1028**–1029, 1029i
Spiral cleavage, **674,** 674i
Spiral organ, **928**
Spiral valve, **719**
Spirilla (prokaryote cell shape), **569,** 569i
Spirochetes, 578
Spirogyra, 601, 602i, 612, 612i
Spliceosome, **319,** 320i
Splicing, mRNA, **319**–321, 319i, 320i, 352
Spo11 protein, 235
Sponge, 672, 677–678, 678i, 982, 982i, 1046
Spongocoel, **678,** 678i
Spongy mesophyll, 763i, 764
Spontaneous reaction, **74**–75
Sporadic (nonhereditary) cancers, **361**
Sporangia
club moss, 623
conifer, 635
defined, **612**
fern, 624, 625
fungal, 657i, 658
hornworts, 619
horsetail, 625, 625i
liverwort, 619
moss, 620, 620i
whisk fern, 625, 625i
Spore, **234,** 234i, 235
angiosperm, 812i, **813**
fungal, **652**–654, 653i
horsetail, 625–626, 625i
lycophyte, 623, 623i
plant, **615,** 616
whisk fern, 625
Sporocarp, 651, 651i
Sporophyll, **623,** 635
Sporophyte, **234**–235, 234i
alternation of generations, 615, 615i
angiosperms, 639, 642i, 812–813, 813i, 818–819, 819i
bryophyte, 618, 618i
defined, **598, 615, 812**
fern, 624–625, 624i
gymnosperm, 632, 633
hornworts, 619
horsetail, 625, 625i
kelp, 598
liverwort, 619
lycophytes, 623, 623i
moss, 620i, 621
seedless vascular plants, 621
whisk fern, 625, 625i
Sporopollenin, **612,** 814
Sporozoite, 594
Spring overturn, **1163,** 1163i
Spurge, 510, 511i
Squamate, 727–728, 728i
Squamous epithelium, 869, 870i
Squash *(Cucurbita),* 813i
Squid, 690, 690i
SREBPs. *See* Sterol regulatory element-binding proteins
SRP. *See* Signal recognition particle
SRY gene (sex-determining region of the *Y*), 267–268, 275
Stability, community, **1213**
Stabilizing selection, **468,** 468i, 469i
Stahl, Franklin, 293, 294i, 295
Stamen, 241i
Stamen, flower, 813i, **814**
Standing crop biomass, **1233,** 1234t
Stapes, **722, 927,** 928i
Staphylococcus, 456, 580, 581
Star anise *(Illicium floridanum),* 639, 639i
Starch, **47,** 48, 49, 49i, 760
Starch-branching enzyme 1 (SBE1), 249
Start codon, **315,** 324, 325i
Startle defenses, 1202, 1202i
Statocyst, **680, 926,** 927i
Statolith, **855,** 856i, **926,** 927i
Stearic acid, 51i
Stegosaurus, 726
Stele
defined, **759**
root, 766i, 767, 780, 780i, 781, 783
stem, 759, 760i
Stem
eudicot *versus* monocot, 759–760, 760i
functions, 757
homeobox genes and, 761
modifications, 760–761, 761i
organization, modular, 757, 757i
primary growth, 759, 759i
primary shoot system, 757–761
secondary growth, 767–768, 768i, 769i
structure, 759–760, 759i–760i
Stem cells, **405,** 771, 869–870, 871, 986, 986i
adult, 405
embryonic, 405
genetic engineering of, 405
therapy, 405–406
Steppe, 1159
Stereocilia, **925,** 925i, 926i, 928, 928i
Stereoisomers, **44**–45, 45i, 47
Stereotyped behaviors, 1280–1281, 1281i
Steroid
anabolic-androgenic, 54–55
defined, **54, 350**
hormones, 54, 54i, 945, 947, 948, 948i
gene expression control by, 350, 350i
structure, 54, 54i
Steroid hormone receptors, **154,** 156i, 157–158, **350,** 350i
Steroid hormone response element, **350,** 350i
Sterol regulatory element-binding proteins (SREBPs), 67
Sterols, **54.** *See also* Cholesterol
membrane, 120–121, 121i
Steward, Frederick C., 825–826
STICHEL gene, 829
Stickel, Shawn K., 656
Stickleback *(Gasterosteus aculeatus),* 522, 524, 524i
Sticky ends, **394**
Stigma, 241i, 813i, **814,** 815i, 817, 817i, 818i, 834
Stimulus, **878,** 878i, 879i, **922,** 922i
Stoddart, Angela, 282
Stoeckenius, Walther, 174i
Stolon, 761, 761i
Stoma (stomata), 184, 185i, 197, 198, 198i, 199i
adaptation, 613, 613i, 615
defined, **613, 756**
genetics of, 75–757
in hornworts, 619
location, 764
opening and closing, 784–785, 784i
structure, 756i
transpiration and, 781, 782i
water loss, regulation of, 783–785, 784i, 785i
Stomach
bird, 1047i, 1048
defined, **1048**
human, 1049i
insect, 1047, 1047i
mammalian, 1054–1055, 1055i
ruminant, 1063, 1063i, 1065
structure and function, 1054–1055, 1055i
wall layers, 1052, 1052i
Stomata. *See* Stoma (stomata)
Stonecrops, 198
Stoneking, Mark, 742, 743
Stop codon, **315,** 327, 328i, 332
Stork, Nigel, 1266
STR. *See* Short tandem repeat
Stramenopila, **595**–598, 606i
Baciliariophyta (diatoms), 596–597, 596i
Chrysophyta (golden algae), 597–598, 598i
Oomycota, 596, 596i
Phaeophyta (brown algae), 598, 598i, 599i
Strata, 500, 500i, 501
Stratification, **441**
Stratified epithelium, 869, 870i
Strawberry *(Fragaria ananassa),* 761, 761i, 783i, 820, 821i, 824, 852i
Stream biome, 1161–1162, 1162i
Strepsirhini, 736, 737i
Streptococcus, 571, 580, 580i, 581
S. pneumoniae, 287, 288i, 374

Streptomycin, 324
Stress, 957, 958i
Stress response, plant, 848–849, 850–851, 863
Stretch receptor, 924, 925i
Striated muscle, 973
Strict aerobe, **178**
Strict anaerobe, **177**
Strobilus
  conifer, 635
  cycad, 633
  description, **623**
  gnetophyte, 634
  horsetail, 625, 625i
Stroke, 880, 994
Stroma, chloroplast, 111, 112i, 184, 185i
Stromal lamellae, 184, 185i
Stromatolites, **559**–560, 560i
Structural gene, **341,** 341i, 344, 345i
Structural isomers, **45,** 46i
Strychnine, 853
Sturgeon, 719, 719i
Sturtevant, Alfred, 264–265
Style, flower, 813i, **814**
Styracosaurus, 726
Suberin, 780
Submucosa, **1052,** 1052i
Subsoil, **799**
Subspecies
  defined, **482**
  rat snake, 482i
  ring species, 482–483, 483i
Substance P, 898t, 937
Substrate, **80**
Substrate-level phosphorylation, **164,** 164i, 165i, 166, 167i, 169, 170i, 175i
Succinate, 169, 170i
Succinate dehydrogenase, 169, 170i
Succinyl CoA, 170i
Succinyl CoA synthetase, 170i
Sucrose
  breakdown of, 78
  as disaccharide, 48, 48i
  production in photosynthesis, 194
  transport in plants, 787–784, 788i
Sudden infant death syndrome (SIDS), 1039
Sudden oak death, 587
Sugar, transport in phloem, 755, 756i
Sugar-phosphate backbone, **290,** 290i
Sulfate, soil, 800
Sulfhydryl group, **43,** 45t, 57
Sulfolobus, 92i, 584
Sulfur
  deficiency in plants, 797i
  plant macronutrient, 795, 796t
  radioisotopes, 25
Sum rule, **245,** 246i
Summation, 897, 899, 899i, 969, 971i
Sundew, 71i
Sunflower *(Helianthus annuus),* 753i, 818, 838i
Superior vena cava, 988, 988i
Suprachiasmatic nucleus, 912, 960
Surface area-to-volume ratio, 92, 94i
Surface tension, 32, **34,** 34i
Surgeonfish *(Acanthurus lineatus),* 1293
Surroundings, 72, 73i, 74
Survivorship, age-specific, **1177,** 1177t
Survivorship curve, **1177**–1178, 1177i
Suspension feeder, 710–711, **1045,** 1045i
Suspensor, **818,** 819i
Sutherland, Earl, 142–143, 144i
Sutherland, Grant R., 301
Sutton, Walter, 249–250
Swallow, John G., 459
Swallowing reflex, 1053, 1054i
Swallows, 1297–1298, 1298i, 1299
Swamp sparrow *(Melospiza georgiana),* 1278, 1279i
Sweat gland, 271, 1090, 1090i
Sweet pea *(Lathyrus odoratus),* 763i
Swell shark *(Cephaloscylium ventricosum),* 718i
Swim bladder, **719,** 720i
Swimmeret, 698, 698i
Switch, developmental, 522
Swordtail *(Xiphophorus helleri),* 491
Symbiosis, 1206–1208. *See also* Endosymbiosis
  commensalism, 1206–1207, 1206i
  defined, **663, 1206**
  herbivore digestive system microorganisms, 1063, 1063i, 1065
  lichen, 664, 665i
  mutualism, 1207, 1207i, 1208i
  mycorrhizae, 665–666, 665i, 666i
  parasitism, 1207–1208
Symmetrical body plan, **672**
Sympathetic division, **907,** 907i, 908i
Sympatric speciation, **489**–491, 490i, 493–494, 494i, 496
Symplast, 780, 787, 788
Symplastic pathway, **780,** 780i, 781
Symport, **133,** 134i, **776,** 776i, 784, 787
Synapomorphy, **539**–541
Synapse
  chemical, 886–887, 886i, 887i, 894–899, 897i
  defined, **885**
  electrical, 886, 886i
Synapsida (synapsids), **724,** 733
Synapsis, 228i
Synaptic cleft, **886,** 886i, 895–897, 895i
Synaptic vesicle, 895–896, 895i
Synaptonemal complex, 228i, 235
Synergid, 815i, 816, 816i, 817, 834
Synesthesia, 939
Synovial joint, 976, 976i
Synura, 598i
Syphilis, 578
System, defined, **72,** 73i, 74
Systematics, 528–552. *See also* Phylogenetic tree
  cladistics, 538–543, 539i–541i
  conservation biology and, 1266
  defined, **529**
  Linnaean system of taxonomy, 529–530, 549–550
  molecular phylogenetics, 536, 547–549
  traditional, 537–538
Systematics and the Origin of Species (Mayr), 480
Systemic acquired resistance, 851t, **854,** 854i
Systemic circuit, **984,** 984i, 988, 988i
Systemic homeostatic controls, **878**
Systemic inflammation, **1006**
Systemic lupus erythematosus, **1018**
Systemin, 840t, **851,** 851t, 852, 852i
Systole, **989,** 989i
Systolic blood pressure, **991**
Szostak, Jack, 303, 559

T cell
  activation, 1011, 1012i, 1015–1016, 1017i
  CD4$^+$, 1011, 1012i, 1019
  CD8$^+$, 1015, 1016, 1017i, 1021
  cytotoxic, 1016, 1017i, 1018, 1021
  defined, **1007**
  effector, 1011, 1016
  helper, 1011, 1012i–1013i, 1016, 1019
  immunological tolerance, 1017
  memory, **1013,** 1014
T-cell receptor, **1008**–1009, 1009i, 1012i–1013i, 1015, 1017i
T DNA, 409, 409i
T-even bacteriophage, 378i, **379,** 380i
T2 phage, 288, 289i
T tubule. *See* Transverse tubule
Tabita, Robert, 196
Tactile signal, **1294,** 1294i
Tadpole, 956, 956i
Taiga, **1160**
Tail
  chordate, 710, 710i
  virus, **378,** 378i, 379, 380, 380i
Takifugu rubripes (pufferfish), 423
Talmage, David, 1011
Tamoxifen, 954–955
Tandem duplication, **433**
Tannin, 853
Tapeworm, 685, 685i
Taproot system, **765,** 765i
Target cell, 141, 142, 143, 143i, 944, 948
Target protein, 146, 149
Tarsiers *(Tarsius bancanus),* 736, 737i
Taste, 934–936, 935i, 936i
Taste bud, 935, 935i, 936i
TATA box, **317,** 346, 346i, 347i
Tatum, Edward, 18, 311, 312i–313i, 370, 371, 371i
Taubenberger, Jeffrey, 383
Tax protein, 362
Taxis, **1292**
Taxol, 1263, 1263i
Taxon, **530**
Taxonomic hierarchy, **530**
Taxonomy, **440, 529**
Tawny frogmouth, 732i
Tay-Sachs disease, 106
Tectorial membrane, **928,** 928i
Teeth, 576, 733, 1061, 1061i
Telencephalon, 905, 906i, 909
Teleosts, 719, 720i, 721
Telomerase, **302,** 302i, 307, 871
Telomere, **302,** 302i, 307
Telomere repeat, 302
Telophase, in mitosis, 209i, 210i, **211**
Telophase I, in meiosis, 229i, 230i
Telophase II, in meiosis, 229i
Telson, 698, 698i
TEM. *See* Transmission electron microscopy
Temin, Howard, 385
Temperate bacteriophages, **379**–382, 381i
Temperate deciduous forest, **1160,** 1160i
Temperate grassland, **1159,** 1159i
Temperate rain forest, **1160**–1161
Temperature
  enzyme activity, effect on, 83, 84, 84i
  membrane fluidity, effect on, 120, 123
  regulation in lizards, 15–17, 17i
Template, in DNA replication, **66**
Template strand, **312**
Temporal fenestrae, **724**
Temporal isolation, **484,** 489
Temporal lobe, brain, 912, 912i, 929
Temporal summation, 899i
Tendon, **872,** 967
Tendril, 763i, 856–857, 857i
Tension, on water in xylem sap, 781, 782i, 783
Tension receptor, 924–925, 925i
Teosinte *(Zea diploperennis),* 1263
Terebraulina septentrionalis, 683i
Terminal bud, **757,** 757i, 758
Termination, of translation, **324,** 327, 328i–329i
Termination codons, **315**
Termination factor, **327**
Terminators, **318**
Terpene, 853
Terrapin *(Malaclemys terrapin),* 1238, 1240i
Terrestrial animal
  olfactory receptors of, 936
  taste and smell receptors of, 935
Terrestrial biomes, 1154–1161
  boreal forest/taiga, 1160, 1160i
  chaparral, 1159, 1159i
  desert, 1158, 1158i
  distribution of, 1155, 1155i
  savanna, 1157–1158, 1158i
  temperate deciduous forest, 1160, 1160i
  temperate grassland, 1159, 1159i
  temperate rain forest, 1160–1161
  tropical forest, 1155–1157, 1156i
  tundra, 1161, 1161i
Territory, **1293,** 1297
Terry, Norman, 22, 23i
Tertiary consumer, **1212,** 1212i
Tertiary structure, protein, **57,** 58i, 59–60, 60i
TEs. *See* Transposable elements
Testcross, **246,** 247i
Testes, **226,** 226i, 950i, 951t, 958, **1099,** 1108, 1108i, 1109i
Testosterone, 54, 54i, 945, 951t, **958,** 959, **1108,** 1109, 1109i, 1132
Testudines, 726–727, 727i
Tetanus, **969,** 971, 971i
Tetraploid, 273
Tetrapod, 523, 536, 538i
Tetrapoda, **715,** 721–722, 722i
Tetrodotoxin, 893
Thalamus, 905, 906i, 909i, **911,** 911i
Thalassiosira pseudonana, 597i
Thallium, 26
Thallus
  description, **664,** 665i
  hornworts, 619
  liverwort, 619
Thaumarchaeota, **583,** 584
Theissen, Günter, 525
Theobroma cacao, 811–812, 811i
Theory, scientific, **18**
Therapsid, 732
Therapy-related myeloid neoplasms (t-MN), 282
Thermal acclimatization, **1088**
Thermocline, **1163**–1164, 1163i
Thermodynamics
  defined, **72**
  first law of, 72–73
  second law of, 73–74
Thermoplasma, 423, 583
Thermoreceptors, **923,** 936–937, 937i, 1085, 1088–1089, 1089i
Thermoregulation
  behavioral and physiological responses, 1152–1153, 1153i
  defined, **1070**
  ectothermy, 1086–1088, 1087i
  endothermy, 1086–1087, 1087i, 1088–1092, 1089i–1091i
  heat exchange mechanisms, 1086, 1086i
  optimal organismal performance, 1085–1086, 1086i
Theta wave, 916, 917i
Thick filament, myofibril, **967**–968, 969i, 970i
Thigmomorphogenesis, 857
Thigmotropism, **856**–857, 856i, 857i
Thin filament, myofibril, **967**–968, 970i
Thiols, 45t
Thompson, James A., 871
Thorax, 695, 699, 700i
Thorn forest, **1158**
Threat display, 1295, 1295i
3′ end, **290,** 290i, 317
3′ untranslated region (3′ UTR), **318,** 319i, 352
Three Gorges Dam, 1257i
Threonine, 82–83, 83i, 1048, 1049i
Threonine deaminase, 82–83, 83i
Threshold potential, **889,** 889i
Thrips, 1187, 1187i, 1188
Thrombosis, 993–994
Thrush *(Turdus ericetorum),* 473
Thylakoid membrane, 184, 185i, 187, 188, 190i, 191
Thylakoids, 111, 112i, 184, 185i
Thymine, 63, 64i, 66i, **290,** 290i, 313, 314
Thymus, **1007**
Thyroid gland, 950i, 951t, **955,** 957i
Thyroid gland disease, 23–24, 26
Thyroid hormones, 945, 946i, 947, 951t, 955–956, 956i
Thyroid-releasing hormone (TRH), 945, 946i, 950, 956i
Thyroid-stimulating hormone (TSH), 945, 946i, 951t, **952,** 953i, 956, 956i
Thyroxine, 945, 951t, **955,** 956, 959
Ti (tumor inducing) plasmid, **409,** 409i
Tick, 695, 697
Tickle, Cheryll, 450
Tidal volume, **1032**
Tight junction, **114,** 114i, 868, 908, 1003
Tilman, David, 1217
Time lag, **1184**

Tinbergen, Niko, 1279, 1281i, 1289
TIR1 gene, 843
Tissue
animal, 868–875
defined, **672, 749, 868**
evolution of, 672
Tissue culture, plant, 825–826, 825i
Titanic, 867, 879
Titmouse, 1297–1298
TMAO. *See* Trimethylamine oxide
Tmesipteris, 625
TNTs. *See* Tunneling nanotubes
Toad, 491, 722
Tobacco mosaic virus, 378i, 383
Todgham, Anne, 1248
Tolerance hypothesis, **1220**
Toll-like receptors, 1005
Tomato *(Lycopersicum esculentum),* 750i, 763i, 764, 797i, 851, 855i
Tomato bushy stunt virus, 383
Tonegawa, Susumu, 917
Tongue, 935–936, 936i
Tonicity, 130–131, 130i
Tonoplast, 99i, **112, 777,** 778i
Tonoplast intrinsic protein, 779
Tool-kit gene, 522, 699
Topography, effect on climate, 1151, 1151i
Topoisomerase, **297,** 297i, 300t
Topsoil, **799**
Torpor, **1090**–1091, 1091i, 1153
Torsion, **688**
Tortoises, Galápagos, 443, 444i
Totipotent, **355,** 356
Totipotent cell, **824,** 825i, 826, **1121**
Touch receptor, 924, 924i
Townes, P. L., 1136, 1137i
Townsend, Colin R., 1217
Toxin
biological magnification of, 1239
of centipedes, 699
of cnidarians, 680
endotoxin, 581
exotoxin, 581
G proteins targeted by, 150
neutralization, 1011
of spiders, 697
Toxoplasma, 594–595
TP53 gene, 363
Trace elements, **23,** 23i
Trace mineral, 1051, 1051t
Tracer, **25,** 26
Trachea(e), 983, **1028**–1029, 1029i, 1030, 1031i
Tracheal system, **700, 1025,** 1028–1029, 1029i
Tracheid, 613, 639, **755,** 755i
Tracheole, 1028, 1029i
Tracheophyte, **613**
Traditional systematics, **536**–538
Trait, **240,** 242, 242i, 243
Trans fat/fatty acid, 52
Transcription
defined, **311**
as DNA-directed RNA synthesis, 316i, 317–318
in eukaryotes, 313, 314i, 316i, 317–319
of non-protein-coding genes, 318
in prokaryotes, 313, 314i, 318
stages, 316i, 317
termination, 316i, 317, 318
Transcription factors, 67, **317, 318, 346,** 347i, 348–349, 349i, 351
Transcription initiation complex, **347,** 347i
Transcription unit, 316i, **317,** 340, 341i
Transcriptional regulation
defined, **340**
in eukaryotes, 344–351, 346i–351i
in prokaryotes, 340–344, 341i–345i
Transcriptome, **428**
Transcriptomics, **428**
Transduction, in signaling pathway, **143,** 145i, 146i, 147i, 150i–152i, 156i
Transduction, viral, **375,** 377
generalized, 380
specialized, 382
Transfer cell, plant, **788**
Transfer RNA (tRNA)
charging, 323–324, 323i
defined, **321**
gene mutation, 280
initiator, 324, 325i, 326i
peptidyl, 325, 326i
structure, 322–323, 322i
transcriptio of, 318
Transformation
artificial, 375
Avery, MacLeod, and McCarty experiments, 287
defined, **289, 374**
by electroporation, 375
Griffith's experiment, 287, 288i
natural, 374–376
Transforming principle, 287, 288i
Transgene, **404,** 406i, 409
Transgenic organisms, **404,** 406–410, 406i, 409i
Transit sequences, **330**
Transition state, reaction, 79, 80–81
Translation
defined, **312**
elongation stage, 325–326, 326i
in eukaryotes, 314i, 321, 324–328, 325i
initiation, 324–325, 325i
mRNA-directed polypeptide synthesis, 321–331
polysomes and, 327–328, 328i
in prokaryotes, 314i, 321, 324–328
termination, 324, 327, 328i–329i
Translational regulation, 353–354
Translocation
chromosomal, **272**–273, 272i, 273i, 362
plant transport, **786**–789, 787i, 788i
Transmembrane pathway, **780,** 780i
Transmission, neural signaling, **884,** 884i
Transmission electron microscopy (TEM), 93i
Transpiration, 755, **781,** 782i, 783–785
Transpiration stream, 781
Transport, **126**
Transport epithelium, **1071**
Transport in plants, 774–792
Transport mechanisms
active, 126, 126t, 131–134, 132i–134i
characteristics, 126t
diffusion, 126–127, 126t
endocytosis, 134–137, 135i
exocytosis, 134, 135i, 137
facilitated diffusion, 126t, 127, 128i, 129
osmosis, 129–131, 129i, 130i
passive, 126–131, 126t, 128i–130i
phagocytosis, 137, 137i
Transport proteins
in active transport, 131–134, 132–134i
aquaporins, 127, 128i, 138
carrier proteins, 125–126, 127, 128i
channel proteins, 125, 127, 128i
cystic fibrosis transmembrane conductance regulator, 119–120, 120i
defined, **121**
saturation of, 127
Transposable elements (TEs), **332**–335, 333i, 334i, 426, 433, 537
bacterial, 332–333, 333i
eukaryote(s), 333–335, 334i
Transposase, **332**
Transposition, **332,** 333i
Transposons, **332,** 333, 333i
Transverse tubule, **968,** 968i, 970i
Tree fern, 623, 624
Tree of Life, 10i, 11, 446, 446i, 547–549, 548i, 549i
Tree rings, 768, 769i
Tree swallow *(Tachycineta bicolor),* 480i
Treefrog, 535–536, 535i
Trematoda, 685, 685i, 686
Trembling aspen *(Popuus tremuloides),* 798
Treponema pallidum, 578
TRH. *See* Thyroid-releasing hormone
Tricarboxylic acid cycle, **169.** *See also* Citric acid cycle
Triceps brachii, 977, 977i
Trichocyst, **594**
Trichome, **757,** 762, 763i, 785, 829
Trichomonas vaginalis, 591, 591i, 607
Trichonympha, 12i, 92i
Trigger species, 1265, 1274
Triglyceride, **50,** 51, 51i, 52i
Triiodothyronine, 951t, **955,** 956
Trilobite, 515, 694, 695, 695i
Trimesters, **1128**
Trimethylamine oxide (TMAO), 1084
Triose, 47
Triosephosphate dehydrogenase, 167i
Triosephosphate isomerase, 167i
Triple bond, 43
Triple-X syndrome, 275, 275t
Triploblastic body plan, **672**
Triploid, 273
Trisomy-13, 274
Trisomy-18, 274
Trisomy-21, 273, 274i. *See also* Down syndrome
Triticale, 637i
Tritium, 25, 25i
Trivers, Robert, 1302
tRNA. *See* Transfer RNA
Trochophore larvae, 677, 688, 688i, 691
Trophic level, **1211**–1213, 1212i
Trophoblast, **1129,** 1129i, 1130i
Tropic cascade, **1238,** 1240i
Tropical deciduous forest, **1156**
Tropical forest, **1156,** 1156i
Tropical montane forest, **1157**
Tropical rain forest, 793–794, 794i, **1156,** 1156i, 1157
Tropics, **1148**
Tropism
defined, **841**
gravitropism, 855–856, 856i
phototropism, 841–842, 841i, 855, 855i
thigmotropism, **856**–857, 857i
Tropomyosin, 967i, 968
Troponin, 967i, 968
Trp operon, 344, 345i
Trp repressor, 344, 345i
True-breeding, **242**
Truffle *(Tuber melanosporum),* 659, 666
TRY gene, 829
Trypanosoma brucei, 592, 592i
Trypsin, 84, 1053i, **1056**
Tryptophan, 1048, 1049i
biosynthesis, 344, 345i
TSH. *See* Thyroid-stimulating hormone
Tsujimoto, Gozoh, 995
Tuatara *(Sphenodon punctatus),* 727, 728i
Tubal ligation, 1113
Tube feet, 707–708, 709i
Tube-within-a-tube body plan, 673
Tube worm, 1167–1168, 1168i
Tuber, 760, 761i
Tuberculosis, 1019
Tubular reabsorption, **1071,** 1071i
Tubular secretion, **1071,** 1071i, 1076
Tubulin, 107, 108i, 214
Tulip *(Tulipa),* 641i
Tumor, 219, **361**
Tumor-suppressor gene, **362**–363, 363i
Tumor-suppressor protein, 220
Tundra, **1161,** 1161i
Tunicate, 711, 711i, 712
Tunneling nanotubes (TNTs), 388
Turbellaria, 684–685, 684i
Turgor pressure, **130,** 130i, **777,** 778i, 779, 781, 788i
Turnbull, Andrew, 1062
Turner syndrome, 275, 275t
Turnover rates, **1238**
Turtle, 726–727, 727i, 975
Twitch, muscle, **969,** 971i
Twitch summation, 969, 971i
Twitchett, Richard, 513
Two-dimensional paper chromatography, 193
Tympanic canal, 927, 928, 928i
Tympanic membrane, **927,** 928i, 929
Tympanum, **722,** 927, 927i
Typhoid fever, 581
Typhus, 1239
Tyrannosaurus, 726
Tyrosine, 945
Tyrosine kinase, **147**–149, 148i, 153, 154i, 273, 946

Ubiquinone, 172i, 173
Ubiquitin, 222, 354, 354i
UCPs. *See* Uncoupling proteins
Ulcer, 579i, 1054–1055
Ulothrix, 602i
Ultrabithorax (Ubx) gene, 699–700
Ulva, 601, 601i
Umami receptor, 935
Umbilical cord, 1130i, **1131**
Umbilicus, **1132**
Unconditioned stimulus, 1283
Uncoupling proteins (UCPs), 176, 178
Undecanal, 1111
Undernutrition, **1044**
Undulating membrane, **591,** 592, 592i
Unequal crossing-over, **433,** 433i
Ungulate, 734
Unicellular organisms, **3**
Uniform dispersion, **1174,** 1174i
Uniformitarianism, **442**
Unikonta, **588,** 589i, 601–605, **602**
Uniparental inheritance, 280
Uniport transport, 127
Universal, nature of genetic code, **315,** 317
Universe, origin of, 553
Unreduced gamete, **494**
Unsaturated fat/fatty acid, **50**–52, 50i
Uracil, 63, 64i, 313, 314
Uranium, bioremediation and, 38–39
Urea, 1072, 1072i, 1078, 1084
Ureter, 1074i, **1075**
Urethra, 1074i, **1075**
Urey, Harold, 555
Uribe, Ernest, 192
Uric acid, 1072, 1072i, 1085
Urinary bladder, 1074i, **1075**
Urinary system, 1074i, 1075
Urine, 1071, 1075–1076, 1079
Urochordata, 711, 711i
Uropods, 698, 698i
Ursus americanus, 9, 9i
Urtica (stinging nettle), 757
Use and disuse, principle of, 442
Uterine cycle, **1105,** 1106i, 1107
Uterus, **1103,** 1104i, 1105, 1106i, 1107, 1111
Utricle, 925i, **926**

Vaccination, **1002,** 1014
Vagina, **1104,** 1104i, 1110
Valence electron, **27**–28
Valine, 1048, 1049i
Valve, heart, 988, 988i
Valve, vein, 993, 994i
Van Buuren, Marianne L., 803
Van der Hoeven, Frank, 523
Van der Waals forces, **31,** 32i
Van Mantgem, Phillip J., 1187
Vaporization, heat of, **34**
Variable region, of antibody molecule, **1009,** 1009i, 1011i
Variables, **15**
Variant Creutzfeldt-Jakob disease, 387
Variation
maintenance of, 471–474
balanced polymorphisms, 472–474, 472i–474i
diploidy, 471–472, 471t
neutral variations, 474
population, 457–461
genetic, 460
phenotypic, 456i, 457–460, 457i, 458i, 459i
qualitative, 458, 458i
quantitative, 457–458, 458i
Vas deferens, **1108,** 1108i, 1110

Vascular bundle
  defined, **759**
  development, 759–760
  eudicot, 752t, 759, 760i
  monocot, 752t, 760, 760i
  veins, 764
Vascular cambium, **767**–769, 768i–769i
Vascular cylinder, 759
Vascular damage, 880
Vascular endothelial growth factor (VEGF), 365
Vascular plants, **613**
Vascular tissue system, plant, 613
  components and functions, 753t
  defined, **752**
  location of, 753i
  phloem, 755, 756i
  from procambial cells, 759
  xylem, 755, 755i
Vasectomy, 1113
Vasodilation, 995–996, 996i
Vasopressin, 153
Vegetal pole, **1119,** 1122, 1122i, 1123, 1124i, 1133
Vegetarian diet, 1048–1051, 1049i
Vegetative reproduction, **212, 824**–826
VEGF. *See* Vascular endothelial growth factor
Veins
  animal, **983,** 992i, 993–994, 994i
  plant, **614**–615, 752t, 763i, **764,** 781, 782i, 783, 783i
Veldt, 1159
Veliger, 688
Velvet worm, 693, 694i
Venter, J. Craig, 96, 417
Ventilation, **1026**
Ventral surface, **673**
Ventricle
  brain, **905,** 909i
  heart, **983,** 984i, 988–990, 988i–990i
Venturia inaequalis, 658, 659i
Venule, **991,** 992i, 993, 993i
Venus' flower basket (*Euplectella* species), 678i
Venus fly trap *(Dianaea muscipula),* 762, 763i, 1245
Vernalization, **860**–861
Vertebrae, **712**
Vertebral column, **712**
Vertebrate(s)
  agnathans, 715–716, 715i, 716i
    conodonts and ostracoderms, 716, 716i
    hagfishes and lampreys, 715–716, 715i
  amniote origin and diversification, 723–726, 725i
  amphibians, 722–723, 723i
  birds, 730–732, 731i, 732i
  blood clotting evolution, 997–998
  circulatory system, 983–985, 984i
  crocodilians, 730, 730i
  defined, **670**
  digestive specializations, 1061–1065, 1061i, 1063i
  diversification of, 712–715
  endoskeleton, 975, 975i
  eyes, 930–934, 931i–934i
  hydrostatic skeletal elements, 975
  jawed fishes, 717–721
    actinopterygii, 719, 719i, 720i, 721
    chondrichthyes, 718–719, 718i
    origin of, 717–718, 718i
    sarcopterygii, 719, 721, 721i
  kidney function in nonmammalian, 1084–1085, 1084i, 1085i
  mammals, 732–743
    hominins, 739–743, 739i–742i
    key adaptations, 733
    marsupials, 734, 734i
    monotremes, 733i, 734
    nonhuman primates, 734–739, 736i–738i
    placentals, 734, 736i
  movement, 976–977, 976i
  nervous system, 904i, 905
  origin and diversification, 712–715, 713i, 714i
  phylogeny, 714i
  skeletal muscle, 967–973
  sphenodontids, 727, 728i
  squamates, 727–728, 728i
  taste receptors, 935, 935i
  tension receptors, 924–925, 925i
  tetrapods, evolution of, 721–722, 722i
  turtles, 726–727, 727i
  unique tissues of, 711–712
  vestibular apparatus in, 925–926, 925i
Very-low-density lipoprotein (VLDL), 1065
Vesicle
  appearance, 98i, 102i
  bilayer, 120, 121i
  in cell plate formation, 212i
  defined, **102**
  endocytic, 135i, 137, 137i
  exocytosis and endocytosis, **104,** 104i
  secretory, **104,** 104i, 105i
  traffic in the cytoplasm, 102–104, 103i, 104i, 105i
Vessel elements, in angiosperms, 639
Vessel member, **755,** 755i
Vessels, **755,** 755i, **982**
Vestibular apparatus, **925**–926, 925i
Vestibular canal, 927–928, 928i
Vestigial structure, **441**
Vibration-detecting system, 927, 927i
Vibrio cholerae, 150, 578
Vibrios (prokaryote cell shape), **569**
Vicariance, **509,** 510
Victoria (Queen of England), 270, 270i
Villi, intestinal, 1055, 1056i
Virchow, Rudolf, 90
Viridiplantae, 611
Virion, **377**
Viroid, **387**
Virulent bacteriophage, **379,** 380i
Virus
  agglutination by antibody, 1011
  animal, 379t, 382, 382i
  bacteriophages, 369, 378, 378i, 379–382, 380i, 381i
  cancer causing, 362
  classification, 377, 378, 379t
  defined, **288, 377**
  effects of infection on host, 378–379
  evolution of, 386–387
  infection pathways, 379–383, 380i, 381i, 382i
  innate immune response to pathogens, 1006–1007
  living or nonliving question, 377
  as nonliving systems, 91
  plant, 379t, 383
  retroviruses, 384–386, 384i, 386i
  RNA interference therapies, 365
  RNA tumor viruses, 385
  structure, 377, 378, 378i
  transduction and, 375, 377, 380, 382
  treatment of infection, 383–384
Visceral mass, **687,** 688, 688i
Visible light, 185, 186i
Vision, 929–934, 929i–934i
  accommodation, 930, 931, 931i
  color, 933–934
  lateral inhibition, 933
  neural pathway, 934, 934i
  peripheral, 931
Visual cortex, 934, 934i
Visual signal, **1293,** 1293i
Vital capacity, **1032**
Vitamin A deficiency, 410
Vitamin C, 1050, 1050t
Vitamin D, **956,** 1050, 1050t
Vitamin K, 1050, 1050t, 1059
Vitamins, **1044,** 1048–1051, 1050t
Vitelline coat, **1100,** 1102i
Vitreous humor, **930,** 931i
Viviparous animal, **1103**
VLDL. *See* Very-low-density lipoprotein
Volcanic eruptions, 509
Voltage, 888
Voltage-gated ion channel, 888, 889i–892i, **890**–893, 894i, 895, 896
Voltage-gated ion transport proteins, 127, 128i
Volterra, Vito, 1203
Volvox, 601i
Von Frisch, Karl, 1279, 1294–1295
Von Tischermak, Erich, 249
Vulpia fasciculata, 1185i
Vulture *(Gyps),* 1260
Vulva, **1104**

Waggle dance, 1294–1295, 1294i
Wagner, Susanne, 362
Wainwright, Patricia O., 656
Wake, Marvalee H., 744
Wallace, Alfred Russel, 6–7, 439, 439i, 447
Waller, Augustus, 981, 982i, 989–990, 990
Walrus *(Obodenus rosmarus),* 734, 736i
Wang, Minglei, 62
Ward, Jamie, 939
Warfarin, 436
Warren, J. Robin, 579i, 1054–1055
Wasp *(Eurytoma gigantea),* 468, 469i
Waste chemical concentration, homeostatic control of, 877
Water
  adhesive property, 781
  biomes
    freshwater, 1161–1164
    marine, 1164–1168
  in chemical reactions, 46
  cohensive property, 781
  evaporation of, 557
  flow in xylem, 755, 755i
  homeostatic control of concentration, 877
  hydrogen bonds of, 32–34, 33i
  hydrologic cycle, 1241–1242, 1241i
  ionization of, 36–37
  metabolic, 1071
  net primary productivity and, 1234, 1234i
  passive transport (osmosis), 129–131, 129i, 130i
  polarity of, 30, 30i, 34–35
  of primordial Earth, 554–555
  properties of, 32–36
  as respiratory medium, 1026
  soil solution, 799–800, 799i
  splitting in photosynthesis, 184, 189
  transpiration, 781, 782i, 783–785
  transport in plants, 775i, 777, 779, 780, 780i, 781–786, 782i, 783i
  water-splitting reaction in photosynthesis, 560
Water balance, 1069, 1082–1083
Water boatmen *(Sigara distincta),* 474, 474i
Water conservation
  marine fishes, 1084, 1084i
  marine mammals, 1079
  reptiles and birds, 1085, 1085i
  terrestrial mammals, 1079, 1080i
Water flea *(Daphnia),* 1185i
Water lattice, 32–34, **33,** 33i
Water lily, 639, 639i
Water mold, 587, 596, 596i
Water pollution, 1260
Water potential, **777,** 778i, 779, 782i
Water-soluble vitamin, **1049,** 1050t
Water stress, in plants, 785, 785i
Water vascular system, 707, 709i
Watershed, **1242**–1243
Watson, James D., 18, 65, 286–287, 287i, 290–292
Wavelength, 185–186, 186i
Wax, **52**–53
Wayfinding, 1289–1290
Weaver ants *(Oecophylla longinoda),* 670i
Webbed fingers, 251, 251i
Weeping willow *(Salix babylonica),* 787i
Weight, **25**
Weintraub, Michael, 808
Weis, Arthur E., 468
Weismann, Jonathan, 116
Weiss, Michael, 1134
Welch, Matthew, 116
Welwitschia, 633–634, 634i
Went, Fritz, 842, 842i
Wernicke's area, 912i, 914
Western diamondback rattlesnake *(Crotalus atrox),* 937i
Wetlands, **1161**
Whale, 537, 734, 1038–1039, 1045i
Whale shark *(Rhincodon typus),* 718
Wheat *(Triticum),* 494, 494i, 641i
Whiptail lizard *(Cnemidophorus),* 1098
Whisk fern *(Psilotum),* 625, 625i
White, Tim, 739–740
White-crowned sparrow *(Zonotrichia leucophrys),* 1278, 1279i, 1280, 1285
White-egg bird's next fungus *(Crucibulum laeve),* 661i
White matter, **909,** 909i
White pelican *(Pelecanus erythrorhynchos),* 1299
White rust, **596**
Whitfield, L. Simon, 743
Whittaker, Robert, 1208
Whole-genome shotgun sequencing, **418,** 418i
Whooping crane *(Grus americana),* 1266, 1266i
Whorl, flower, 813–814, 813i
Widowbird, 470, 471i
Wienberg, Wilhelm, 461
Wieschaus, Eric, 356
Wigge, Philip A., 862
Wilcove, David S., 1259, 1259i
Wild-type allele, 262
Wilkins, Maurice H. F., 65, 291, 292
Williams, Ernest E., 729
Willow *(Salix),* 767i
Wilmut, Ian, 407, 408i, 1121, 1139
Wilson, Allan, 743
Wilson, Edward O., 1221–1225, 1263, 1274
Wilting, **779,** 781
Wind
  air circulation, 1148–1149, 1149i
  ocean proximity and, 1150–1151, 1151i
  topography and, 1151
Wingless/Wnt pathway, 1283
Wings, 534, 535i, 730, 731, 731i
Winkler, David W., 1297–1298
Winter flounder *(Pseudopleuronectes americanus),* 1152
Winter rye *(Secale cereale),* 861
Wishbone flower (*Torenia* fournieri), 834
Withdrawal, birth control method, 1112–1113, 1112t
Wobble hypothesis, **323**
Woese, Carl R., 96, 548, 548i
Wolcott, Charles, 670
Wolf, Larry, 1153
Wolf spider, 697i
Wolfenden, Richard, 71
Wolffian duct, **1132,** 1133i
Wollman, Elie L., 374
Wolly mammoth, 452–453, 452i, 453i
Wood
  defined, **768**
  heartwood, 768, 769i
  sapwood, 768, 769i
Wood duck, 54i
Wood lice, 1292
Woodward, F. Ian, 757
Woody plant, 751–752, 768, 768i, 769i
Work, 72
Worm, Boris, 1257
Wrasse *(Thalassoma),* 487i
Wright, John, 787i
Wuchereria, 686, 686i

X chromosome, **267**
fragile X syndrome, 301
inactivation, 270–271, 271i, 275, 306, 350–351
in meiosis, 227
X-gal, 394, 395i
X-linked dominant trait, 279, 279i
X-linked gene, **268**
X-linked inheritance, **268**
X-linked recessive inheritance, **277**
X-linked trait, 301
X-ray diffraction, **290,** 291i
Xa21 gene, 411
Xanthomonas oryzae, 411
Xanthopan morgani praedicta, 643i
Xenopus laevis, 896
Xeroderma pigmentosum, 304
Xerophyte, 786
Xylem, 781–786
angiosperm, 639
club moss, 623
cohesion-tension mechanism of water transport, 781–783, 782i
components and functions, 753t
defined, **613, 755**
in eudicots, 760i
ions loaded into, 780–781
primary, 759i
from procambial cells, 759
root, 766i, 767
secondary, 767–770, 768i, 769i
structure, 755
tracheids, 755, 755i
vessel members, 755, 755i
water travel to root, 780, 780i
whisk fern, 625
Xylem sap, **777,** 781, 782i, 783
XYY syndrome, 275t

Y chromosome, **267,** 743
in meiosis, 227
role in development, 1132, 1134
sex determination in humans, 267–268, 267i
Yachi, Shigeo, 667
Yang, D. S. C., 1152
YDA gene, 756
Yeast, **650,** 650i, 659–660, 659i
alcoholic fermentation, 177
as facultative anaerobes, 177
mitosis, 221
model research organism, 218
Yellow-bellied glider *(Petaurus australis),* 1269
Yellow bush lupine *(Lupinus arboreus),* 641i
Yellow marrow, 975
Yellow pond lily *(Nuphar polysepala),* 639i
Yellow-throated warbler *(Dendroica dominica),* 481i
Yellowfin tuna *(Thunnus albacares),* 720i
Yellowtail flounder *(Limanda ferruginea),* 1256
Yi, Keke, 802
YODA enzyme, 756–757
Yohe, Gary, 1154
Yolk, **724, 1119**
Yolk plug, 1123, 1124i
Yolk sac, **1125,** 1126i, 1129, 1130i
Yucca *(Yucca),* 1207i
Yucca moth *(Tegeticula),* 1207i
Yunis, Jorge J., 495i

Z line, 967i, 968, 968i
Z scheme, 191
Zeatin, 846
Zebra finch *(Taeniopygia guttata),* 1285, 1285i, 1286, 1286i
Zebra mussel *(Dreissina polymorpha),* 1230
Zebrafish *(Danio rerio),* 205, 205i, 523
Zebroid, 486, 486i
Zenk gene, 1285, 1285i
Zero population growth, **1182,** 1193
Zhang, Billiang, 85
Zhang, GuangJun, 523
Zinc, as plant micronutrient, 796t
Zinc finger, 348, 348i
Zinder, Norton, 375
Zona pellucida, **1100,** 1100i, 1102i, 1110, 1114, 1128, 1129i, 1130i
Zonation
intertidal, 1166, 1166i
lake, 1162–1163, 1163i
ocean, 1164–1165, 1165i
Zone of cell division, **766,** 766i
Zone of elongation, **766,** 766i
Zone of maturation, **766,** 766i
Zooplankton
defined, **590**
open ocean, 1167
as primary consumers, 1238
ZW sex-determination system, 1141
Zygomycota (zygomycetes), 655t, 657–658, 657i, 658i
Zygospore, 657, 657i, **658**
Zygote, **225,** 227, 235, 242, 355
angiosperm, 812, 818, 819i
defined, **671**
fungal, 653–654, 653i

Experimental Research

Figure	Description	Page
1.15	Hypothetical Experiment Illustrating the Use of Control Treatment and Replicates	16
3.24	Anfinsen's Experiment Demonstrating That the Amino Acid Sequence of a Protein Specifies Its Tertiary Structure	61
5.11	Discovery of the Nuclear Localization Signal	101
6.6	The Frye–Edidin Experiment Demonstrating That the Phospholipid Bilayer Is Fluid	124
7.2	Sutherland's Experiments Discovering a Second Messenger Molecule	144
8.15	The Racker and Stoeckenius Experiment Demonstrating That an H^+ Gradient Powers ATP Synthesis by ATP Synthase	174
10.12	Movement of Chromosomes during Anaphase of Mitosis	215
10.14	Demonstrating the Existence of Molecules Controlling the Cell Cycle by Cell Fusion	216
12.5	The Principle of Segregation: Inheritance of Flower Color in Garden Peas	244
12.8	Testing the Predicted Outcomes of Genetic Crosses	247
12.9	The Principle of Independent Assortment	248
12.13	Experiment Showing Incomplete Dominance of a Trait	253
13.2	Evidence for Gene Linkage	263
13.8	Evidence for Sex-Linked Genes	269
14.2	Griffith's Experiment with Virulent and Nonvirulent Strains of *Streptococcus pneumoniae*	288
14.3	The Hershey and Chase Experiment Demonstrating That DNA Is the Hereditary Molecule	289
14.9	The Meselson and Stahl Experiment Demonstrating the Semiconservative Model for DNA Replication to Be Correct	294
15.2	Relationship between Genes and Enzymes	312
17.1	Genetic Recombination in Bacteria	371
18.14	The First Cloning of a Mammal	408
20.12	How Exposure to Insecticide Fosters the Evolution of Insecticide Resistance	449
21.6	Using Artificial Selection to Demonstrate That Activity Level in Mice Has a Genetic Basis	459
21.13	Sexual Selection in Action	471
21.16	Demonstration of Frequency-Dependent Selection	474

Experimental Research

25.2	The Miller–Urey Apparatus Demonstrating That Organic Molecules Can Be Synthesized Spontaneously under Conditions Simulating Primordial Earth	556
26.13	Demonstration of the Bacterial Cause of Peptic Ulcers	579
27.11	The Evolutionary History of Diatoms as Revealed by Genome Analysis	597
29.14	Analyzing Phytochromes to Explore Angiosperm Evolution	640
34.12	Translocation Pressure	787
36.24	Probing the Roles of Floral Organ Identity Genes	832
37.3	The Darwins' Experiments on Phototropism	841
39.7	Demonstration of Chemical Transmission of Nerve Impulses at Synapses	887
40.13	Investigating the Functions of the Cerebral Hemispheres	915
42.5	Demonstration That Binding of Epinephrine to β-Adrenergic Receptors Triggers a Signal Transduction Pathway within Cells	949
43.4	The Sliding Filament Model of Muscle Contraction	969
44.18	Demonstration of a Vasodilatory Signal Molecule	996
46.14	Demonstration of a Molecular Basis for High-Altitude Adaptation in Deer Mice	1038
47.21	Association of Intestinal Bacteria Populations with Obesity in Humans	1064
48.12	ADH-Stimulated Water Reabsorption in the Kidney Collecting Duct	1082
49.13	Vocal Cues of Ovulation in Human Females	1107
50.18	Demonstrating the Selective Adhesion Properties of Cells	1137
50.20	Spemann and Mangold's Experiment Demonstrating Induction in Embryos	1139
51.25	What Causes Lake Eutrophication?	1164
52.17	Evaluating Density-Dependent Interactions between Species	1188
53.8	Gause's Experiments on Interspecific Competition in *Paramecium*	1204
53.12	Demonstration of Competition between Two Species of Barnacles	1206
53.23	Effect of a Predator on the Species Richness of Its Prey	1214
53.24	The Complex Effects of an Herbivorous Snail on Algal Species Richness	1215